2nd International Conference on New Material and Chemical Industry (NMCI2017)

IOP Conference Series: Materials Science and Engineering
Volume 292

Sanya, China
18 - 20 November 2017

Editor:

Shixuan Xin

ISBN: 978-1-5108-5749-0
ISSN: 1757-8981

Printed from e-media with permission by:

Curran Associates, Inc.
57 Morehouse Lane
Red Hook, NY 12571

proceedings.com

Some format issues inherent in the e-media version may also appear in this print version.

Printed by Curran Associates, Inc. (2018)

Institute of Physics
Dirac House, Temple Back
Bristol BS1 6BE UK

Phone: 44 1 17 929 7481
Fax: 44 1 17 920 0979

techtracking@iop.org

Additional copies of this publication are available from:

Curran Associates, Inc.
57 Morehouse Lane
Red Hook, NY 12571 USA
Phone: 845-758-0400
Fax: 845-758-2633
Email: curran@proceedings.com
Web: www.proceedings.com

Table of contents

Volume 292

2nd International Conference on New Material and Chemical Industry (NMCI2017)

18–20 November 2017, Sanya, China

Accepted papers received: 21 December 2017
Published online: 5 February 2018

Preface

2nd International Conference on New Material and Chemical Industry (NMCI2017)

Peer review statement

Papers

Environment Protection

The application of waste fly ash and construction-waste in cement filling material in goaf

W X Chen, F K Xiao, X H Guan, Y Cheng, X P Shi, S M Liu and W W Wang.....1

Heat-resistant agent used for control sand of steam huff and puff heavy oil well

F S Zhang, G L Liu, Y J Lu, X C Xiong, J H Ma and H M Su.....7

Antibacterial activity of anthraquinone from cassia seed on spiced pig head

L Y Xu, X Li, Y Q Cui, M X Pang, F Wang and J H Qi.....13

Corrosion behaviors of pure copper and Cu-Ni-Zn alloy in NaCl solution and artificial salt water

J M Shi, C L He, G P Li, H Z Chen, X Y Fu, R Li, G F Ma and J M Wang.....19

Calculation and affection of pH value of different desulfurization and dehydration rates in the filling station based on Aspen Plus

J X Lv, B F Wang, L H Nie, R R Xu, J Y Zhou and Y J Hao.....24

Plasma electrolytic oxidation treatment mode influence on corrosion properties of coatings obtained on Zr-1Nb alloy in silicate-phosphate electrolyte

R G Farrakhov, V R Mukaeva, A R Fatkullin, M V Gorbatkov, P V Tarasov, D M Lazarev, N Ramesh Babu and E V Parfenov.....28

Research on breakdown characteristics of oil-paper insulation in compound field at different temperatures

L Li, M Y Chen, X C Zhu, Z W Gao, H D Zhang, G X Li, J Zhang, C L Yu and Y M Feng.....35

Thermal degradation and plasticizing mechanism of poly(vinyl chloride) plasticized with a novel cardanol derived plasticizer

J Chen, X A Nie, J C Jiang and Y H Zhou.....42

Dependance of naphthenic acid corrosion of SA106B on temperature and turbulence

Y F Shi, Q K Zheng, J Liu, L B Zhang, Z W Pan and S X Rao.....50

Research on breakdown characteristics of converter transformer oil-paper insulation under compound electric field in alpine region

C Xu, Z W Gao, S Lan, H X Guo and M C Gong.....61

Experimental research on sealing performance of sulfur hexafluoride electrical equipment under low-temperature environment

G X Li, L Li, R Zhang, Z Q Jiang, X C Zhu, Z W Gao, H D Zhang and J Zhang.....69

Anti-bacteria effect of active ingredients of siraitia grosvenorii on the spoilage bacteria isolated from sauced pork head meat

X Li, L Y Xu, Y Q Cui, M X Pang, F Wang and J H Qi.....77

The use of pyro-polymeric catalysts and a new cylindrical cell design in oxygen-aluminum generators

E A Kiseleva, A Z Zhuk, B V Kleymenov and V G Oudaltsov.....83

Investigation on the Room-temperature preparation of Cobalt hybrid/Graphene Nanocomposite and application in wastewater purification: Highly Efficient Removal of Congo Red

L X Wang, Y F Zhao and Q M Meng.....90

Materials Synthesis

One-step preparing magnesium hydroxide particles from mother liquor of salt production

H Guo, C S Peng, Z W Ding, H T Yuan and K Yang.....96

Preparation and electrochemical properties of perovskite $Ce_xLi_{0.5-x}Ca_{0.5}TiO_3$ composites

J H Cheng and C A Tian.....103

Green synthesis of silver nanoparticles using wolfberry fruits extract and their photocatalytic performance

L Sun, P C Lv, Y C Yin, H N Li and F Wang.....109

MnO_2/multiwall carbon nanotube/Ni-foam hybrid electrode for electrochemical capacitor

L H Chen, L Li, W J Qian and C K Dong.....116

Synthesis of Fe_3O_4 core/alumina shell nanospheres for partial hydrogenation of benzene

T T He, S L Mu, Q T Fu and C G Liu.....122

Effects of B site doping on electronic structures of $InNbO_4$ based on hybrid density functional calculations

M F Lu, C P Zhou, Q Q Li, C L Zhang and H F Shi.....129

Zeolite-encapsulated Co(II), Mn(II), Cu(II) and Cr(III) salen complexes as catalysts for efficient selective oxidation of benzyl alcohol

F H Li, H Bi, D X Huang, M Zhang and Y B Song.....135

Optimization of preparation of NDV F Gene encapsulated in N-2-HACC-CMC nanoparticles

S S Li, Y Zhang, K Zhao and X H Wang.....143

Direct atomic layer deposited Gd_2O_3 on grapheme and effects of rapid thermal annealing on its property

L Zheng, D L Zhang, X H Cheng, L Y Shen, Q Wang, W Zhou, D P Wu and Y H Yu.....148

Surfactant-assisted growth and optical properties of ZnO hexagonal bilayer disk-like microstructures

Q P Zhu, X Y Shen, L L Wang, L P Zhu, L J Wang and G H Liao.....155

Comparative dynamics analysis on xonotlite spherical particles synthesized via hydrothermal synthesis

F Liu, S Chen, Q Lin, X D Wang and J X Cao.....160

Room-temperature synthesis and photoluminescence of hexagonal $CePO_4$ nanorods

J Zhu, K Zhang and H Y Zhao.....168

Green synthesis of Pd nanoparticles supported on LDHs and its application in the Suzuki coupling reaction

G H Liu and X F Bai.....173

SRT research of the crystal defects on the TGT synthesized diamond doped with boron and phosphorus

W L Yu, X P Jia, Q X Yuan and W X Huang.....178

Synthesis of nanoparticle emulsion collector HNP and its application in microfine chalcopyrite flotation

G C He, J Ding, C H Huang and Q Kang.....183

Preparation of Ag superhydrophobic surface on metal substrates

J Y Li, S X Lu, W G Xu, Y Q Duan, X C Yang, Y Y Cheng, G He and S Cui.....191

Hydrogen production by sodium borohydride in NaOH aqueous solution

Q Wang, L F Zhang and Z G Zhao.....198

Explosive performance of HMX/NTO co-crystal

J C Li, Q J Jiao, Y G Gong, Y Y Wang, T Liang and J Sun.....202

Manufacture of a ceramic paper for art applications

K Dölle, A Honig, J Piatkowski and C Kuempel.....207

An electrochemical sensor based on nitrogen doped carbon material prepared from nitrogen-containing precursors

G Y Cui, C Y Wang, G Q Xiang and B Zhou.....213

Effect of synthesizing temperature on the electrochemical properties of nitrogen-doped Co-B alloys

Y L Liao, W Zhao, S J Qiu, H L Chu, Y J Zou, C L Xiang, H Z Zhang, F Xu and L X Sun.....219

The first solid state glycosylation of quercetin

D. V. Orlov, O. I. Lomovsky and I. O. Lomovskiy.....226

Novel $Co(OH)_2$ with cotton-like structure as anode material for alkaline secondary batteries

W Zhao, Y L Liao, S J Qiu, H L Chu, Y J Zou, C L Xiang, H Z Zhang, F Xu and L X Sun.....231

Recent progress in the synthesis of thiazolo[3,2-a]pyrimidine compounds

F Y Wu, Y Luo and C B Hu.....237

Materials Property & Applicaton

Analysis of material parameter effects on fluidlastic isolators performance

Q Y Cheng, J H Deng, Z Z Feng and F Qian.....243

Effect of addition of Proline, ionic liquid [Choline][Pro] on CO_2 separation properties of poly(amidoamine) dendrimer / poly(ethylene glycol) hybrid membranes

S H Duan, T Kai, F A Chowdhury, I Taniguchi and S Kazama.....249

Influence of heat treatment and hot extrusion on the microstructure and tensile properties of rare earth modified Mg-Zn based alloy

L Y Sheng, B J Wang, B N Du, C Lai and T F Xi.....256

Simulation of the temperature field distribution in medium-voltage vacuum interrupter and experimental verification

W Sun, H L Liu, Y G Cai, Y L Li, H Y Zhou and Y Zhou.....263

Influence of stacking sequence and number of pin on load distribution of composite joints

F Liu, L Qi and S F Feng.....273

Breakdown characteristics of SF_6/N_2 in severely non-uniform electric fields at low temperatures

Y Wang, Z W Gao, G X Li, X C Zhu, C L Yu, J Q Liang and L Li.....279

The effect of material matching on the stress-induced power degradation for light-redirecting-ribbon-based silicon photovoltaic modules

X F Gou, H Zhuang, J Zhu and X Y Li.....285

Influence of B_2O_3 content on sintering behaviour and dielectric properties of La_2O_3-B_2O_3-CaO/Al_2O_3 glass-ceramic composites for LTCC applications

F L Wang, Y W Zhang, X Y Chen, H J Mao and W J Zhang.....292

Interband optical transitions in Gd_2O_3: Er nanoparticles – prospective system for energy convertors

A F Zatsepin, Yu A Kuznetsova, M A Mashkovtsev and V N Rychkov.....299

Measurement and correlation of the solubility of gossypol acetic acid and gossypol acetic acid of optical activity in different solvents

B Zhang, H Tang, X Y Liu, X Zhai and X C Yao.....305

Effect of GBCD on IGC susceptibility of solution treated 2024 Al alloy

G P Li, C L He, R Li, J M Shi, G F Ma, X Y Fu, H Z Chen and J M Wang.....313

Study on hydrothermal synthesis dynamics of xonotlite spherical particles

F Liu, S Chen, Q Lin, X D Wang and J X Cao.....317

The CRG-PVA hydrogels study of properties with various nanoparticles and their application for cultivation of phototrophic microorganisms

K V Gorin, G U Badranova, P M Gotovtsev, A Yu Shatalova, T E Grigoriev, S V Krasheninnikov, S A Tihomirov, O A Kondratev, M V Vishnevskaya and R G Vasilov.....327

Life prediction for high temperature low cycle fatigue of two kinds of titanium alloys based on exponential function

G Y Mu, X Z Mi and F Wang.....333

Improvement in hydrogen storage properties of MgH_2 catalyzed with $BaTiO_3$ additive

J S Wang, W Zhang, S M Han and F Qin.....340

Thermodynamic calculation study on phase structures of Pt-Ir-M(Y, Zr, Mo)

S B Wang, Y Sun, J L Chen, S Wang, M J Peng, S Chen, MM Liu, J Q Hu, Y T Chen, J M Zhang, Y C Yang and M Xie.....347

Properties investigation and microstructures characterization of SiCp/6061Al composites produced by PM route

A Q Wang, H W Tian and J P Xie.....352

Research on connection structure of aluminumbody bus using multi-objective topology optimization

Q Peng, X Ni, F Han, K Rhaman, C Ulianov and X Fang.....362

Infiltration sintering properties of Ni-4B-4Si(wt.%) alloy powders

Q Yang, X C Zhang, F L Wang and J T Zou.....369

Enhancement of red emission intensity of $Ca_2Al_2SiO_7:Eu^{3+}$ phosphor by MoO_3 doping or excess SiO_2 addition for application to white LEDs

H Y Jiao, C R LiMao, Q Chen, P Y Wang and R C Cai.....374

Effect of gate bias sweep rate on the threshold voltage of in-plane gate nanowire transistor

H X Liu, J Li and R R Tan.....380

Effect of different fibers on mechanical properties and ductility of alkali-activated slag cementitious material

J Zhu, W Z Zheng, C Z Qin, Z Z Xu and Y Q Wu.....385

The design of wideband metamaterial absorber at E band based on defect

L S Wang, D Y Xia, X Y Ding and Y Wang.....392

Synthesis and microwave absorbing properties of Cobalt ferrite

H Y Liu and Y S Li.....398

An experimental investigation of a thermoelectric power generation system with different cold-side heat dissipation

Y H Li, Z H Wu, H Q Xie, J J Xing, J H Mao, Y Y Wang and Z Li.....403

Microstructure analysis of snail trails in photovoltaic modules

J Tao, Z X Shi, S Y Chen, J X Kong, Q S Huang and X Fg Gou.....412

Fe-doped ZnO synthesized by parallel flow precipitation process for improving photocatalytic activity

Q M Meng, Q L Lu, L X Wang and J Wang.....416

Development of material formula and structure property indicators for low cold-resistant characterization of Cables' Material

W Sun, Y G Cai, Y M Feng, Y L Li, H Y Zhou and Y Zhou.....420

Diagnostics of the thickness of a plasma electrolytic oxidation coating on a nanostructured Mg-Sr alloy

V R Mukaeva, O B Kulyasova, R G Farrakhov, E V Parfenov and Y F Zheng.....431

A study of low cycle fatigue life prediction method for 1Cr11Ni2W2MoV at 360°C

G D Gao, W X Zhang and D Chen.....437

Effect of physicochemical action on the aggregative properties of detonation-synthesized nanodiamonds

Z W Fan, H Ilnitska, V Lysakovskyi, S Ivakhnenko and T Kovalenko.....444

Application of P.O and R-SAC mortar for 3D printing in construction

J C Lin, X Wu, W Yang, R X Zhao and L G Qiao.....451

Use B-spline interpolation fitting baseline for low concentration 2, 6-di-tertbutyl p-cresol determination in jet fuels by differential pulse voltammetry

D S Wen, H Wen, Y G Shi, B Su, Z C Li and G Z Fan.....458

Analysis and selection of materials for the design of lightweight railway vehicles

C Ulianov, A Önder and Q Peng.....465

Performance characteristics of rubber seed oil biodiesel

P Liu, M Qin, J Wu and B S Chen.....472

Load distributions analysis of muti-pin composite joints: single-lap

F Liu, L Qi and S F Feng.....480

Influences of the current density on the performances of the chrome-plated layer in deterministic electroplating repair

H Xia, X M Shen, X C Yang, Y Xiong and G L Jiang.....485

Preparation and mechanical properties of carbon nanotube-silicon nitride nano-ceramic matrix composites

C Y Tian and H Jiang.....492

Study the effect of $Ni_{0.5}Zn_{0.5}Fe_2O_4$ doped electromagnetic and microwave absorbing properties on RGO based composites

B Mei, X J Su, S Bi, G L Hou and J Li.....500

Numerical analysis of temperature field in the high speed rotary dry-milling process

N X Wu, L J Deng and D H Liao.....505

The influence of fine aggregates on the 3D printing performance

J C Lin, X Wu, W Yang, R X Zhao and L G Qiao.....510

The research of penetration testing in the FRP pipe applications

M L Xin, Y Chen, M D Li, Z G Wang and G H Gu.....516

Research progress of self-healing intelligent composite materials

P Li, G Z Liu and J Y Huang.....522

Influences of glass content on the microstructure and properties of $BaO-CaO-Al_2O_3-SiO_2$ glass/alumina composite for LTCC applications

Z X Li, H J Mao, Y W Zhang and W J Zhang.....529

Adsorption equilibrium and dynamics of toluene vapors onto three kinds of silica gels

K L Yan and Q Wang.....536

Comparison of sound absorbing performances of copper foam and iron foam with the same parameters

X C Yang, X M Shen, P J Xu, X N Zhang, P F Bai, K Peng, Q Yin and D Wang.....543

Study on bamboo gluing performance numerical simulation

Z R Zhao, W H Sun, X M Sui and X F Zhang.....550

Fabrication of graphene/polydopamine/copper foam composite material and its application as supercapacitor electrode

Y Zheng, S X Lu, W G Xu, G He, Y Y Cheng, F Y Xiao and Y Zhang.....555

Research on influence of water-cement ratio on workability and mechanical properties of geopolymer grouting material

X Q Wang, P H Wen, Z W Gao and C H Wang.....561

Influence of temperature and moisture joint action on oil-paper insulation compound electric field distribution

H D Zhang, L Li, M J Wu, Y Zhou, H Zhang and Y M Feng.....566

Raman spectroscopic study on structure and phase transition of zirconium tungstate synthesized with a CO_2 laser

J P wang and Q D chen.....575

Magnetic effect on oxide-scale growth of Fe-5Cr alloy

C H Zhou, X W Li, S H Wang and H T Ma.....581

Long term mechanical properties of alkali activated slag

J Zhu, W Z Zheng, Z Z Xu, Y F Leng and C Z Qin.....587

Numerical simulation of canned extrusion process of the W-25wt. %Cu composites under low temperature and liquid sintering

D Q Ma, J P Xie and Y H Song.....594

Preparation and characterization of flexibility and lightweight CF-RGO composites

J Li, S Bi, X J Su, G L Hou, B Mei and W Q Ma.....601

The study of size and stability of n-butylcyanoacrylate nanocapsule suspensions encapsulating green grass fragrance

G Y Zhu, C T Lin, J M Chen, D M Lei and G X Zhu.....606

Study on the quality of FRP fishing vessel based on improved Fishbone Chart

J H Sui, Y F Yu, Q F Du and D W Jiang.....612

Ultra-thin carbon-fiber paper fabrication and carbon-fiber distribution homogeneity evaluation method

L F Zhang, D Y Chen, Q Wang, H Li and Z G Zhao.....619

Investigation on photoluminescence emission of (reduced) graphene oxide paper

J J Ding, H X Chen, D Q Feng and H W Fu.....626

High-temperature creep properties and life predictions for T91 and T92 steels

J P Pan, S H Tu, G L Sun, X W Zhu, L J Tan and B Hu.....631

Study on the temperature field of large-sized sapphire single crystal furnace

J P Zhai, J W Jiang, K G Liu, X B Peng, D L Jian and I L Li.....635

Experimental exploration of the hydrodynamic effect polishing machinability for different types of material

W Q Peng, Y Li, Z Wang and S Y Li.....642

Galvanealed coating evolution for hot forming steel

F Fang, X F Du and Y M Chen.....649

Structure and Raman scattering of chained carbon films on copper substrate: ab initio approach

E A Buntov, A F Zatsepin, M B Guseva, D A Boqizoda and B L Oksengendler.....653

Enhanced photoluminescence properties of Al doped ZnO films

H X Chen and J J Ding.....657

Design of transmissive lens with nearly 100% efficiency by using Huygens metasurface

B C Lin, G M Wang, T Cai, H S Hou, W L Guo, P Xie, S Y Shuai, Z H Liao, K Li, L Zhu and X J Gao.....662

Effect of Zn doping on the microwave absorption of BFO multiferroic materials

S Bi, J Li, B Mei, X J Su, C Z Ying and P H Li.....668

Polymer Chemistry & Physics

Study on the effect of polymeric rheology modifier on the rheological properties of oil-based drilling fluids

C Ma, L Li, Y P Yang, W W Hao, Q Zhang and J Lv.....673

Cobalt-cadmium bimetallic porphyrin coordination polymers for electrochemistry application

C Y Wang, G Y Cui, D Ding and B Zhou.....679

Preparation of low viscosity epoxy acrylic acid photopolymer prepolymer in light curing system

P Li, J Y Huang and G Z Liu.....684

Synthesis and properties of amphiphilic hyperbranched polyethers as pigment dispersant

Q Xu, Y J Zhou, S J Long, Y G Liu and J H Li.....691

The application of epoxy resin coating in grounding grid

Q Hu, Z R Chen, L J Xi, X Y Wang and H F Wang.....696

The synthesis and characterization of p(nBA-DEAEMA) polymer

X M Jiang, J Zhu and Y Y Duan.....700

Experimental and numerical analysis of the constitutive equation of rubber composites reinforced with random ceramic particle

D M Luo, Y Xie, X R Su and Y L Zhou.....707

Experimental study on emission of VOCs from tanker using hollow fiber membrane absorption method with different absorbents

J Y Zhou, B F Wang, L H Nie, J X Lu, Y J Hao and R R Xu.....714

The relationship between compressive strength and flexural strength of pavement geopolymer grouting material

L Zhang, X X Han, J Ge and C H Wang.....720

Kinetics of waterborne fluoropolymers prepared by one-step semi-continuous emulsion polymerization of chlorotrifluoroethylene, vinyl acetate, butyl acrylate and Veova 10

H Z Liu, M H Wang, Z F Wang and J M Bian.....725

Quantum chemical calculation of imidization mechanism of polyamic acid

H Y Gu, L X Song, J L Zhang, L Wang, Z Y Qi and T Zhang.....734

Hydrogenolysis of α-methylbenzyl alcohol to ethylbenzene over Pd/C catalyst

J Feng, Y H Zhong and S H Dai.....740

H_2O-Polyaluminium chloride-TBAB as synergistic catalysts for the synthesis of cyclic carbonate

X J Liu, P Yan and Y Han.....746

Hydrothermal synthesis and photocatalytic degradation ability of nickel phosphide micro/nano materials

J S Yan, H B Liu, Y S E and F W He.....753

Research status and development of application fields in enzyme technology

Y B Ji, S W Wang, M Yu, X Ru, C Wei, H J Zhu, Z Y Li, H Zhao, A N Qiao, S Z Guo and L Lu.....758

The preparation and activity of Cu-Fe-Zr-Ce based catalysts for water gas shift

H D Wu, T S Liu and H Z Liu.....763

A new sandwich type transition metal complex based on tungstobismuthates and 1-methylimidazole ligands exhibiting photocatalytic properties

Q Wu, D Q Liang, H D Ju, K Liu, Y Cui, W L Li, B L Wang, F P Ye, Y Zheng, Y M Wu, J Y Li, Y Lei, Y Xu, T Liu and Y Wang.....770

Preparation and characterization of antimony barium composite oxide photocatalysts

X P Han, B H Yao, Q H Pan, C Pen and C L Zhang.....777

IOP Conf. Series: Materials Science and Engineering **292** (2017) 011001 doi:10.1088/1757-899X/292/1/011001

Preface

Shixuan Xin

Editor, Proceedings of the Second International Conference on New Material and Chemical Industry

Petro China Company Limited, China
East China University of Science and Technology, China
Dalian University of Technology, China

E-mail: xinshixuan@petrochina.com.cn

The Sencond International Conference on New Material and Chemical Industry (NMCI2017) was successfully held in Sanya, China during November 18-20, 2017. The conference was an annual forum for researchers and application developers in the area of material and chemical industry.

This conference proceeding included 123 accepted articles selected from 193 submissions.

We would like to express our gratitude to the reviewers of these manuscripts, who provided constructive criticism and stimulated comments and suggestions to the authors. We are extremely grateful as organizers, technical program committee and editors and extend our most sincere thanks to all the authors for their excellent contribution and work. Our sincere gratitude also goes to the IOP Publishing editors and managers for their helpful cooperation during the preparation of the proceeding.

On behalf of The Organizing Committees of NMCI2017

NMCI2017-photos from the conference

Peer review statement

All papers published in this volume of *IOP Conference Series: Materials Science and Engineering* have been peer reviewed through processes administered by the proceedings Editors. Reviews were conducted by expert referees to the professional and scientific standards expected of a proceedings journal published by IOP Publishing.

IOP Conf. Series: Materials Science and Engineering **292** (2017) 012001 doi:10.1088/1757-899X/292/1/012001

The application of waste fly ash and construction-waste in cement filling material in goaf

W X Chen[1], F K Xiao[1], X H Guan[1,4], Y Cheng[1], X P Shi[2], S M Liu[1] and W W Wang[3]

[1] College of mining engineering, Heilongjiang University of Science and Technology,harbin 150022,China
[2] College of materials engineering, Heilongjiang University of Science and Technology,harbin 150022,China
[3] College of Safety Engineering, Heilongjiang University of Science and Technology,harbin 150022,China

[4] E-mail: 450655485@qq.com

Abstract. As the process of urbanization accelerated, resulting in a large number of abandoned fly ash and construction waste, which have occupied the farmland and polluted the environment. In this paper, a large number of construction waste and abandoned fly ash are mixed into the filling material in goaf, the best formula of the filling material which containing a large amount of abandoned fly ash and construction waste is obtained, and the performance of the filling material is analyzed. The experimental results show that the cost of filling material is very low while the performance is very good, which have a good prospect in goaf.

1. Introduction

Coal mining which have brought huge pressure to the environment and the prominent environmental problem is subsidence and underground water loss. The filling mining is an effective approach for coal mine marching toward "harmless" and "safe" production. However, the main reason why it cannot be popularized in a large area is that the cost of filling material is too high. There is about 650 million m^2 newly built building every year in China, and the construction waste arising from the construction of newly built building in a year is about 40 million tons. The construction waste arising from the dismantling of old building is about 40 million tons in a year. The combination of them reaches a huge quantity [1-3]. The survey showed that for every 10,000 construction waste, about 0.067 km^2 of land is required. However, the supply of fly ash produced by the plant in the remote area is larger than the demand, which resulted in a large quantity of fly ashes wasted in the open, all about of which brought huge pressure to the environment. The survey showed that fly ash contained a variety of heavy metals, compounds and other toxic substances, which will pollute air, water and soil, endanger human health, and aggravate secondary geological disasters. If the construction waste and fly ash is mixed with filling material after recovery processing and reaches the performance requirement of filling material, which will solve not only the environmental pollution from construction waste,but also the problem that the cost of filling material is too high. Thus, this paper obtains the best formula of construction waste-fly ash based cementing filling material through orthogonal experiment, and analyzes the mechanical property of filling material.

2. Raw material and its composition relationship

The filling aggregate is the crushed construction waste–recycled aggregate (brick, glass, tile and concrete are get from 1 m^3 construction waste as filling aggregate), which is shown in figure 1. The cement is Swan brand 32.5 complex Portland cement manufactured by Tatai Group Harbin Cement Co., Ltd, which properties are that the content of sulfur trioxide was 2.45, the content of magnesium oxide was 2.11%, the burning loss was 4.27, the initial setting time was 172 hours, the final setting time was 231 hours.The fly ash is abandoned fly ash from Harbin Power Plant, the main chemical composition of waste fly ash and physical properties, which is shown in table 1. The mixing water is tap water, and the additives are sodium chloride and anhydrous sodium carbonate (the mass ratio is 1:3). The raw material composition relationship is shown in figure 2.

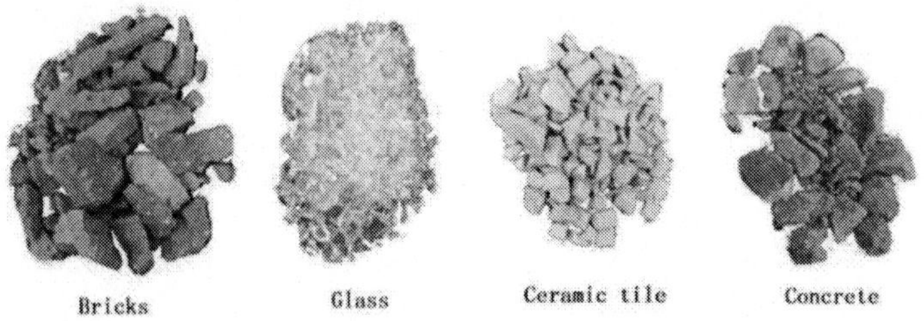

Figure 1. Construction waste after crushed.

Table 1. Chemical composition and physical properties of cribble fly ash.

	SiO_2	Al_2O_3	Fe_2O_3	CaO	MgO	K_2O	Na_2O	SO_3	Loss	Total	Fineness (m2/kg)	Density (kg/m^3)
Coarse fly ash	37.2	24.5	8.4	8.3	1.6	1.2	1.6	12.2	2.7	97.7	119.0	2.2

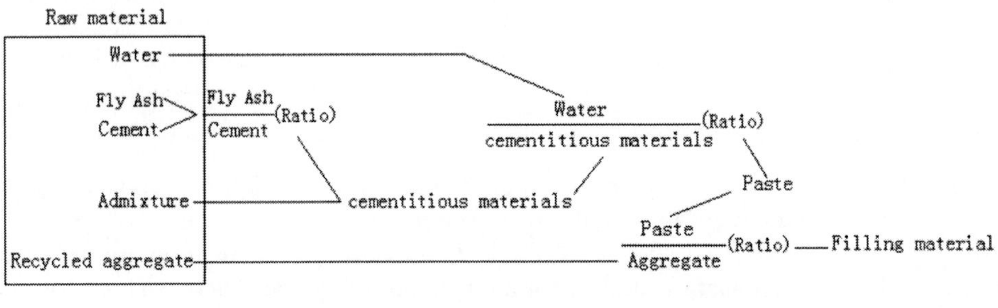

Figure 2. The composition of filling raw material.

Table 2. Levels and factors of the orthogonal test.

Levels	Factors			
	A/g	B	C	D
1	1600	0.40	0.70	7/3
2	1300	0.45	0.75	6/3
3	1000	0.50	0.80	5/3

3. Experimental scheme and result analysis

3.1. Orthogonal experiment factor level

This paper take construction waste (A), mortar-aggregate ratio (B), water-binder ratio (C) and Fly ash-Cement ratio (D) as the experimental factor. Each factor has three levels, the orthogonal experiment factor levels are shown in table 2.

3.2. Preparation of filling material

According to experiment scheme, which is shown in table 3, nine groups of different proportions of raw materials for experiment are weighed using electronic scales. The bottom of 7.07 cm×7.07 cm×7.07 cm trigeminy trial mould is plugged with a wooden plug to prevent the slurry discharging. Then, pouring the weighed water, fly ash, cement and additives to the forced stirrer and stir them for 5 min, pour the weighed construction waste to the stirrer and stir them for 10 min and finally make them into filling slurry. Next, we pour the prepared filling slurry to the compression-resistant trial mould, and vibrate them using handheld slab vibrator so as to shape up. After that, we float the top surface using the scraper, cover the surface with a block of plastic cloth, and put the trial mould in a standard curing box for curing.

Table 3. Programs and results of orthogonal test.

Experiment number	Factors				σ_c /MPa		
	A	B	C	D	3 days	7 days	28 days
1	1(1600 g)	1(0.40)	1(0.70)	1(7/3)	1.90	3.66	5.34
2	1(1600 g)	2(0.45)	2(0.75)	2(6/3)	1.49	3.32	4.46
3	1(1600 g)	3(0.50)	3(0.80)	3(5/3)	1.03	2.95	3.75
4	2(1300 g)	1(0.40)	2(0.75)	3(5/3)	2.05	3.90	6.73
5	2(1300 g)	2(0.45)	3(0.80)	1(7/3)	1.19	2.60	3.47
6	2(1300 g)	3(0.50)	1(0.70)	2(6/3)	1.49	3.94	5.82
7	3(1000 g)	1(0.40)	3(0.80)	2(6/3)	1.70	4.01	4.82
8	3(1000 g)	2(0.45)	1(0.70)	3(5/3)	1.87	5.03	6.46
9	3(1000 g)	3(0.50)	2(0.75)	1(7/3)	1.44	3.27	4.39

3.3. Experimental scheme and result

The experimental scheme will be obtained according to the orthogonal experiment factor level,the filling slurry was poured into test mode and then put it in the curing box. After curing for 3days, the test block of filing material is measured the uniaxial compressive strength with the full automatic cement press machine, and record the experimental data. After curing for 7days and 28days, measuring the uniaxial compressive strength one more time. The experimental scheme and result is shown in table 3.

3.4. Range analysis

The range analysis of compressive strength is shown in table 4. That $\sigma_{c,3days}$, $\sigma_{c,7days}$ and $\sigma_{c,28days}$ in the table 4, which mean the compressive strength of the test block after curing for 3days, 7days and 28 days. A means construction waste, B means mortar-aggregate ratio, C means water-binder ratio, D means Fly ash-Cement ratio, I means three levels of influencing factors, K_i means the sum of compressive strength of three test blocks when the influencing factor selects ith level, k_i is the mean of K_i, and R is the range of influencing factor [4-6].

Table 4. Range analysis of compressive strength.

item	3 days $\sigma_{c,3days}$/MPa				7 days $\sigma_{c,7days}$/MPa				28 days $\sigma_{c,28days}$/MPa			
	A	B	C	D	A	B	C	D	A	B	C	D
K_1	4.42	5.65	5.26	4.53	9.93	11.57	12.63	9.53	13.55	16.89	17.62	13.20
K_2	4.73	4.55	4.98	4.68	10.44	10.95	10.49	11.27	16.02	14.39	15.58	15.10
K_3	5.01	3.96	3.92	4.95	12.31	10.16	9.56	11.88	15.67	13.96	12.04	16.94
k_1	1.47	1.88	1.75	1.51	3.31	3.86	4.21	3.18	4.52	5.63	5.87	4.40
k_2	1.58	1.52	1.66	1.56	3.48	3.65	3.50	3.76	5.34	4.80	5.19	5.03
k_3	1.67	1.32	1.31	1.65	4.10	3.39	3.19	3.99	5.22	4.65	4.01	5.65
R	0.2	0.56	0.44	0.14	0.79	0.47	1.02	0.81	0.82	0.98	1.86	1.25

According to the range analysis result of compressive strength, the primary and secondary sequence of influencing factors will be obtained. According to table 3, the range of compressive strength with the age of 3days is B>C>A>D, and the primary and secondary sequence of influencing factors of compressive strength is mortar-aggregate ratio, water-binder ratio, construction waste and fly ash-cement ratio. B approximates to C, about double A and D.so, the compressive strength of test block at the age of 3days mainly depends on the value of mortar-aggregate ratio and water-binder ratio. The compressive strength range of test block with the age of 7 days is C>D>A>B and the primary and secondary sequence of influencing factors of compressive strength is water-binder ratio, fly ash-cement ratio, construction waste and mortar-aggregate ratio. The compressive strength range of test block with the age of 28days is C>D>B>A and the primary and secondary sequence of influencing factors of compressive strength is water-binder ratio, fly ash-Cement ratio, mortar-aggregate ratio and construction waste.

According to analyzing table 3, it is known that the corresponding result of factor A at different levels indicates that the less the construction waste is, the compressive strength of test block is bigger; the mortar-aggregate ratio and water-binder ratio increase gradually, and the compressive strength of test block decreases gradually; the fly ash-cement ratio decreases gradually and the compressive strength of test block increased gradually. Thus, the optimization level combination of various factors is $A_3B_1C_1D_3$, hat is construction waste is 1000g, mortar-aggregate ratio is 0.4, water-binder ratio is 0.7and fly ash-cement ratio is 5/3, the construction waste and fly ash account for 71.4% and 5.4% of total mass respectively.

4. Material setting time and mechanical characteristic analysis

4.1. Relationship between setting time and water-binder ratio

Since construction waste has no gelation characteristic, the setting time of material mainly directs at the binding slurry composed cementing material mixed with water. Hydration reaction happens after mixing cementing material and water. The binding slurry is solidified gradually with the progress of hydration reaction, and becomes the solid with certain strength. The time used by the slurry setting process is called setting time. The setting time includes initial setting time and final setting time. The initial setting time affects the conveyance performance of filling slurry directly. When the slurry is at the initial setting stage, it is not transported. To prevent blockage, we must ensure the initial setting time of material greater than the slurry conveyance time, so the initial setting time of binding slurry need to determine. The final setting time decides the demoulding time, so the final setting time need to determine [7-8].

The influencing factor of setting time of binding slurry is water-binder ratio. Three levels of water-binder ratio 0.7, 0.75 and 0.8 are selected to prepare three groups of binding slurry and determine the setting time, the result is shown in figure 3. The initial setting time is 141, 180 and 232 minutes, and the final setting time is 205, 231 and 276 minutes. According to figure 3, it is known when the water-binder ratio is 0.7, the setting time is a minimum, and the setting time increases with the increasing of water-binder ratio. Water-binder ratio increases, the concentration of cementing material decreases, and the possibility for producing greater gelling structure in the hydration reaction of slurry becomes small, so the initial setting time and final setting time increase. According to figure 3, it is also known that the final setting time of material is within 8hours, while the working system of coal mine is usually "three eight-hour" system. It indicates that filling and demoulding can be completed in a work shift. Therefore, the material has quick setting characteristic.

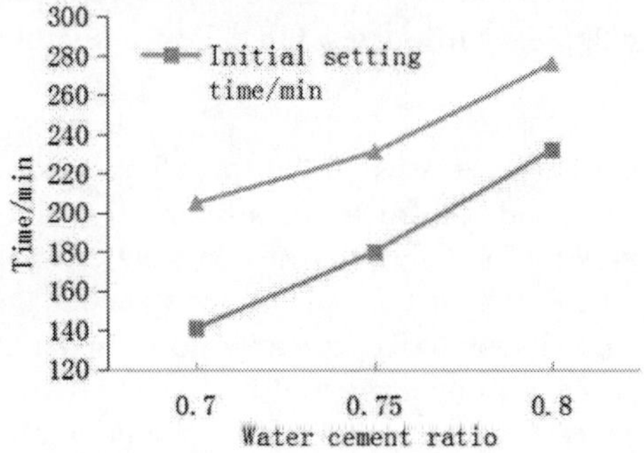

Figure 3. Relationship between water cement ratio and setting time.

4.2. Mechanical property of material

After the test block is obtained according to formula, 3 days stress-time curve of materials can be obtained according to press machine. As is shown in figure 4, that the curve can be divided into four stages: the first stage is compression stage, curve curvature increases gradually. The second stage is elastic stage, the slope of curve approximates to a constant and the load and time appear basic linear relationship. The third stage is plastic deformation stage. The curvature of curve decreases gradually and reaches the load peak. It indicates that test block behaves strong plasticity in the compressive failure process. The failure process is slow and progressive, not sudden. It is extremely beneficial to the safety of mine filling. The fourth stage is yield failure stage. The curvature of curve increases gradually, test block produces obvious crack until it is destroyed totally and loses strength, as shown in figure 5. However, the residual strength after the yield of material is high, about 70% to 90% of the ultimate strength. Thus, the filling body has good carrying ability [9-10].

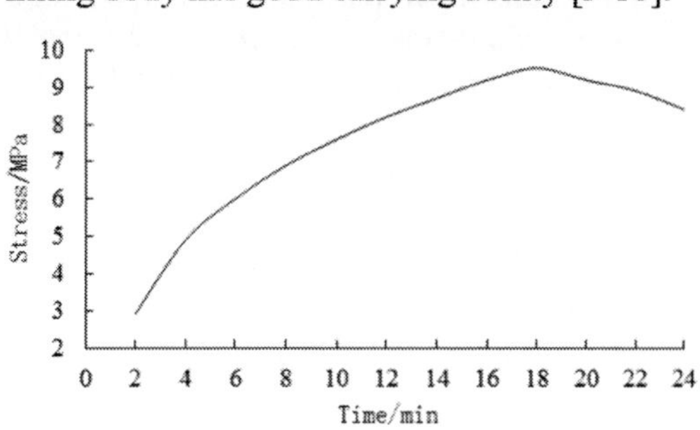

Figure 4. 3 days stress-time curve of materials.

(a) (b) (c)

Figure 5. 3 days filling test block destructive shape.

5. Conclusion

1)The results show that construction waste was 1000g, pulp bone ratio was 0.4, water-cement ratio was 0.7, powder mud ratio was 5/3, and construction waste and fly ash accounted for 1.4% and 5.4% weight of the material, which was the best formula of construction-waste and fly-ash-based cement filling material.

2) This construction-waste and fly-ash-based cement filling material is provided to have strong plastic, high residual strength and fast-setting properties.

3) This formula reduces the cost of cemented filling material effectively and digests a large quantity of construction waste and abandoned fly ash, which is conducive to protecting the environment and has a good application prospect.

Funding

Research on energy transfer and guiding law of rock burst (51574115).

References

[1] Fengyi L, Tan L and Weiwei W 2012 Filling experiment of high-water material in Taoshan coal mine *Journal of Heilongjiang Institute of Science and Technology* **(5)** 221-224

[2] Guangming Feng, Chengzhen Wang. Surface subsidence law of filling with super-high-water material and skip mining *Shandong University of Science and Technology* **(4)** 1-8

[3] Cui Zengdi, Sun Henghu. The preparation and properties of coal gangue based sialite paste-like backfill material *Journal of China Coal Society* **(6)** 896-899

[4] Guangming Feng, Chengzhen Wang, Fengkai Li, Kaijun Jia. Research on Bag-Type Filling Mining with Super-High-Water Material *Mining and Safety Engineering* **(12)** 602-607

[5] Fengyi Li, Xianrui Wang, Changliang Yao. Experimental study on filling mines under buildings with "three wastes" cemented material *Journal of Heilongjiang Institute of Science and Technolog y* **(11)** 527-530

[6] Fengyi Li, Congfang Han, Guohua Zhang 2003 Rock excavation and maintenance *Xuzhou: China University of Mining and Technology Press*189–224

[7] Wang C, Tannant D D, Padrutt A, etal. Influence of admixtures on cemented backfill strength *Mineral Resources Engineering* **(3)** 261-270

[8] Yuzi Lei 2008 Chongqing low activity fly-ash application research at the grassroots level *Chang'an university*

[9] Zifang Xu 2010 Fabrication experimental study on new insulation building materials made from fly ash and Styrofoam *Anhui university of science and technology*

[10] Shuxia Ren. Study on stimulating methods and mechanism of activity of fly ash *Fly ash comprehensive utilization* **(4)** 34-36

Heat-resistant agent used for control sand of steam huff and puff heavy oil well

F S Zhang[1,3], G L Liu[1], Y J Lu[1], X C Xiong[2], J H Ma[1] and H M Su[1]

[1] Oilfield Chemistry Key Laboratory, CNPC (RIPED, PetroChina), Beijing, China
[2] Office of Chiefs, RIPED, PetroChina, Beijing, China

[3] E-mail:zhfsh@petrochina.com.cn

Abstract. Heat-resistant agent containing hydroxymethyl group was synthesized from coal tar, which has similar structure with phenolic resin and could improve the heat resistance of phenolic resin sand control agent. The results showed that the heat resistance of the sand control agent was improved by adding 10% to 30% heat-resistant agent, after 280℃ high temperature treatment for 7d, the compressive strength of consolidated core was increased to more than 5MPa. The compressive strength of consolidation core was not decreased after immersion in formation water, crude oil, acid or alkaline medium, which showed good resistance to medium immersion. The sand control agent had small core damage and the core permeability damage ratio of sand control agent consolidation was only 18.7%.

1. Introduction

With the increase of oil demand, people are growing to strengthen the development of heavy oil reservoir. Steam stimulation is a kind of effective method to improve heavy oil reservoir in the heat. In the process of steam stimulation, oil reservoir bears reservoir rock and repeated incentive, resulting in easy sand of the oil well [1-2]. With the increase of stimulation rounds, increasing the degree of the sand, it brings a lot of harm to the production. In order to alleviate the contradiction between development and easy sand of oil well, oil field increases the experiment and application of sand control technology.

Current chemical sand control agent can be divided into inorganic kind and organic class [3-6]. (phenolic resin[7-8], epoxy resin, urea-formaldehyde resin, furan resin and so on).Although sand control technology obtains a certain result in oil field, there is a series of problem, for example, the inorganic sand control agent costs low and heat resistance performance is good, but the consolidation of strength is low. The consolidation strength of organic sand control agent is high, but the heat resistance performance is poor. The temperature of strata is high in the process of steam stimulation, and high strength steam has large scouring force to strata, so oil filed needs big cementing strength and high temperature resistant sand control agent [9-11]. In order to improve the performance of the phenolic resin heat-resistant [12], combining the coal tar with phenolic resin, but coal tar is different from phenolic resin in the structure, so compatibility performance is not ideal. Naphthalane, anthracene and rich in polynuclear aromatic hydrocarbons asphalt as raw material to synthesis of condensed polynuclear aromatic resin has excellent heat resistance [13-14], heat resistance up to 400℃ above.

The author uses the coal tar which riches in phenolic compounds as raw materials [15], to synthetize heat-resistant agent which is similar to the structure of phenolic resin. This kind of heat resistance agent and phenolic resin has good compatibility performance. Combine heat resistance agent and phenolic resin to get high temperature resistant control-sand agent. Not only it improves the heat resistance performance of phenolic resin, but also it can reduce the cost of control-sand agent.

2. Experimental

2.1. The synthesis of heat-resistant agent

The first step, put right amount of coal tar into three mouth bottle which installs blender. Then join the trioxymethylene whose quantity score is 5% to 10%, using N_2 to protect it, heating up to 110-120℃, adding 5% to 8%(of coal tar quality)of sulphuric acid, to react 6 to 8 hours. The second step, cooling to 80 to 90℃, to join KH550 whose quality score is 1% to 3%,continuing to 3-5 hours, to join right amount of ethanol when the reaction ends.

2.2. The evaluation of the compressive strength

(1)The equipment of the core: Take appropriate amount of find sand, to add to a certain amount heat resistance control-sand agent (phenolic resin + heat-resistant agent), quality score for 10% of curing agent and 1% of coupling agent, stirring well, and then put it in the glass tube(the lower end of pared with copper wire).Compact it well ,then put it into airtight containers(equipped with right amount water) .Keep the temperature of 40℃ to 80℃ for a certain time, waiting until after consolidation, then break the glass tube, using cutting machine to cut concretion body into the core of 25 mm long, then test the strength.

(2)The temperature resistance of the core: Put the core made into high pressure reactor (filled with proper amount of water)and keep the temperature of 280℃ for 7d,then take it out to test the strength.

(3)Take the core made to different medium, immersing 30d under 45℃, then take it out to test the compressive strength.

(4)Compressive strength test: According to the enterprise standard oil SY/T 5276-2000"chemical sand control artificial flexural strength, compressive strength and the test of the gas permeability", using the electronic pressure test machine to test the compressive strength of the core.

2.3. The retention test of core permeability

The use of petroleum enterprise standard SY/6672"sand resin performance evaluation method “tests the retention of core permeability, using kerosene and 2% solution as medium, to test the permeability of the oil and water. The permeability retention ratio is the average value of oil phase and water phase permeability retention ratio. The permeability retention ratio = (the core permeability after consolidation / the core permeability before consolidation)×100%.

3. The synthesis of heat-resistance agent

Coal tar riches in classification of compound, phenol, cresol, two cresol and alpha-naphthol, so including its derivatives, obtaining the heat-resistant agent with trioxane under the action of strong acid.

Chemical reaction mechanism and the use of the mechanism of phenol to phenolic resin are similar [16]. The trioxane happens the open loop reaction under the action of heat and strong acid, forming carbocation ($^+CH_2$-O-CH_2-O-CH_2-OH or $^+CH_2$-OH), carbocation ($^+CH_2$-O-CH_2-O-CH_2-OH) occur electrophilic substitution reactions with the aromatic ring of phenolic compounds (Ar-OH), then generate products (HO-Ar-CH_2-O-CH_2-O-CH_2-OH). The product forms new carbocation (HO-Ar-CH_2-O-CH_2-O-CH_2^+) in the action of acid, reacting with another aromatic ring (Ar-OH), then it forms (HO-Ar-CH_2-O-CH_2-O-CH_2-Ar-OH). Its structure is easy to break because of C-O-C, forming relatively stable carbocation (HO-Ar-CH_2^+), then it continues reacting with other aromatic electrophilic substitution (Ar-OH) to form products (HO-Ar-CH_2-Ar-OH). It repeats the electrophilic substitution reaction, to make product have the structure of hydroxymethyl (HO-Ar-CH_2-OH).The structure of hydroxymethyl is as followed formula (1):

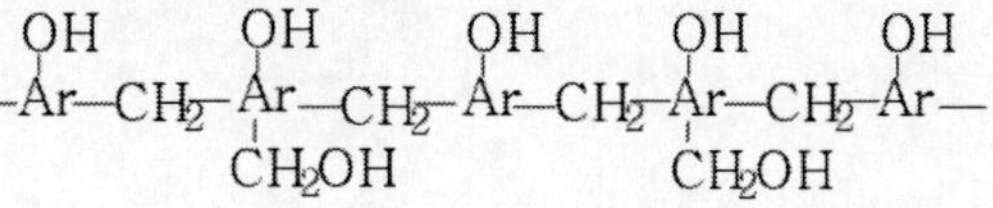

(1)

4. Experimental results

4.1. The influence of heat-resistant to the temperature performance of phenolic resin

In the process of steam stimulation oil well steam injection, the formation temperature is around 280℃ and the injection cycle is generally 5 to 7 d. Adding to the heat-resistant agent whose structure is similar to can improve its heat resistant performance in the resin, extending the validity of the sand operations. Make sand control processing of sand well in the production process. Reservoir temperature is lower (low curing temperature), usually 40 to 50℃. We should pay attention to make sand control processing of sand well in the production process. Reservoir temperature is higher (high curing temperature), about 180℃. In order to adapt to the reservoir of control agent temperature requirements in different construction conditions. Examine the 40℃ low temperature curing and 180℃ heat curing conditions, the consolidation strength of the control-sand agent and core experiences high-temperature processing of 180℃ for 7d.The results are shown in figure 1 and figure 2.

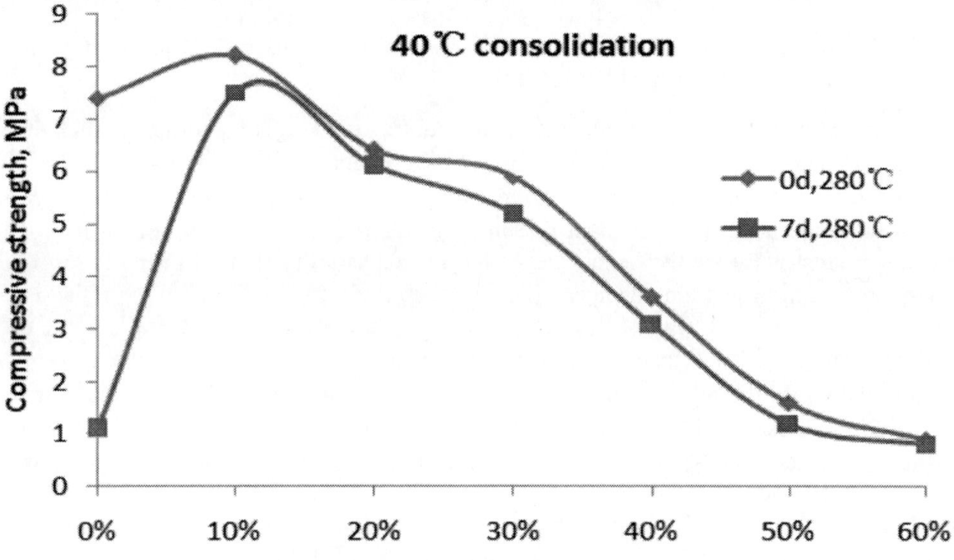

Figure 1. Influence of heat-resistance agent to the temperature performance of phenolic resin at 40℃ consolidation

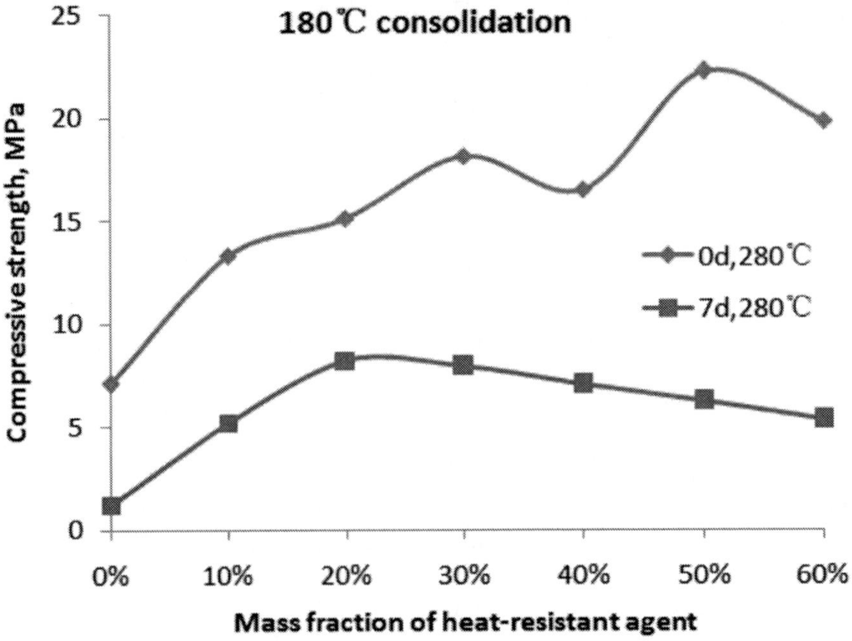

Figure2. Influence of heat-resistance agent to the temperature performance of phenolic resin at 180℃ consolidation

[a]**Note:** Heat resistance agent content = [The quality of heat-resistance agent / (the quality of phenolic resin + the quality of heat-resistance agent)]×100%.

From figure 1 and figure 2, we can see the control-sand agent solidify in the 40℃. The consolidation strength decreases with the increase of the amount of heat-resistant agent. It solidifies in 180℃ high temperature, then the consolidation strength tends to increase with the increase of the amount of heat-resistant agent. The main reason is that kerosene for the synthesis of heat-resistant agent has a similar structure with phenolic resin, but the activity than that of phenolic resin is low. In the low temperature high-resistant agent can participate in the activity of quality score that is not high, so it can't participate in the curing of the cores after the consolidation, decreasing the strength of the core. In the case of 40℃ low temperature curing, heat-resistant whose quality scores is too much will lead to strength loss. In the high temperature of 180℃, its quality score of activity which participates in curing with phenolic resin is relatively high. In a certain extent, it reflects the heat resistance performance of the heat-resistant agent. With the increase of the heat-resistant agent content, the compressive strength increases.

From figure 1 and figure 2, the strength of the consolidation core of phenolic resin decreases greatly to 1.0MPa approximately after a 280℃ processing for 7 d. In the heat-resistance agent quality score of 10% to 30% of the case, the core which was obtained by the curing of 40℃ reacts after 280℃ for 7d. The strength decreases greatly, but it can keep above 5.0MPa.It can improve the heat resistance performance of phenolic resin greatly by joining in the heat-resistant agent and it also can reduce the cost of control-sand agent.

4.2. The stability of consolidation core medium resistance

The resistance to media immersion stability of the core of consolidation has a certain impact on the cycle of the control-sand agent. From table 1, we can see the compressive strength of the core which was soaked 30d in different media of 45℃.

Table1. The influence of different media to the compressive strength of the core.

Core curing temperature /℃	The strength before soaking /MPa	The strength of different media soaking /MPa			
		Formation Water	Crude oil	HCl solution (5%)	Lye solution pH=11.2
45	9.5	9.4	9.5	9.3	9.2
160	13.6	13.5	13.4	13.2	13.3

From table 1, crude oil, formation water and hydrochloric acid have no effect on the compressive strength of the core of consolidation. Alkaline medium of the core encountered in normal circumstances is the dissolution of the alkaline steam in the steam stimulation process. The PH of alkaline steam won't more than 11. The lye of PH=11.2 has no effect on the compressive strength of consolidation. It explains that the control-sand agent has good resistance to media immersion stability.

4.3. The permeability and retention of consolidation of core

In the condition of control-sand agent of 6%, curing agent mass fraction of 10%and coupling agent mass fraction of 1%, to get the core by curing of 180℃, testing the compressive strength and the retention of permeability.

Table 2. The permeability retention rate of the consolidation core.

The length of the core, /cm	The diameter of the core, /cm	The compressive strength, /MPa	Permeability retention ratio, /%	Permeability loss rate, /%
5.3	2.5	9.5	81.3	18.7

From table 2, we can see that the permeability retention rate and permeability loss rate are 81.3% and 18.7% respectively after consolidation. The sand control agent has less damage to the core permeability.

5. Conclusion

The heat-resistance agent which is using coal tar as raw material and similar to the structure of phenolic resin can happen condensation reaction with phenolic resin.

The strength of consolidation core of high temperature control-sand agent which consists of heat-resistant agent and phenolic resin is still larger than 5.0MPa after 280℃ of aging 7d in the condition of the quality score 10% to 30% of the heat-resistant agent. It explains that phenolic resin blends with heat-resistant properly, but also can reduce the cost of control-sand agent.

The core by consolidation of heat resistance control-sand agent was soaked by the formation water, crude, acidic medium and alkaline soak. The strength doesn't decrease. It explains that it has good resistance to media immersion performance.

After the consolidation of heat-resistant, sand control agent has less damage to core permeability.

References

[1] Kotlar H K, Haavind F, Springer M 2005 *SPE 95723*
[2] Zhou Q H, Zhao B, Xu J F 2005 *Oil Gas Field Surf. Eng.* **24** pp 15-15
[3] Zheng Y C, Zhao X T, Li K H 2000 *J. Petro.Nat gas Chem.* **29** pp 144-146
[4] Abdel-Alim H, Musaed N 2001 *SPE 68225*
[5] Heng W L, Huang X, Wang Wenkai, et al. 2002 *Oil. Chem.* **19** pp 287-292

IOP Conf. Series: Materials Science and Engineering **292** (2017) 012002 doi:10.1088/1757-899X/292/1/012002

[6] Nguyen, Rickman, Dusterhoft, et al. 2008 *US 7448451*
[7] Zhang F S, Zhang H B, Xie H Z, et al. 2006 *Fine Oil Chem.Prog.* **7** pp 20-22
[8] Li G R 1997 *Synthetic resin and fiberglass* (Beijing: Chemical Industry Press) pp 76-79
[9] Gao R 2015 *Appl.Chem.Ind.* **44** pp 1775-8
[10] Zhao J H, Liu W 2014 *Appl.Chem.Ind.* **43** pp 1675-8
[11] Zhao Y Z, Sun S D, Gao A H, et al. 2014 *Pet.Drilling Tech* **42** pp 90-94
[12] Zheng Y C, Zhao X T, Li K H, et al. 2000 *Fine oil Chem. Prog* **1** pp 6-8
[13] Chou R C, Ma J H, Kong R R, et al. 2004 *J. Tianjin Univ.* **37** pp 647-50
[14] Otani S, Raskovic V, Oya A 1986 *J. Mater Sci.* **21** pp 2027-32
[15] Li B L, Song H H, Liu L 1996 *Coal Convers.***19** pp 70-74
[16] Huang F R, 2004 *Phenolic resin and its application*(Beijing:Chemical Industry Press) pp 64-66

IOP Conf. Series: Materials Science and Engineering **292** (2017) 012003 doi:10.1088/1757-899X/292/1/012003

Antibacterial activity of anthraquinone from cassia seed on spiced pig head

L Y Xu[1,2,3,4], X Li[1,2,3,4], Y Q Cui[1,2,3,4], M X Pang[1,2,3,4], F Wang[1,2,3,4] and J H Qi[1,2,3,4,5]

[1] College of Food Science and Engineering, Beijing Key Laboratory of Detection and Control of Spoilage Organisms and Pesticide Residues in Agricultural Products, Beijing 102206, China

[2] Food Chemistry teaching team, Beijing University of Agriculture, Beijing 102206, China

[3] Pig innovation team, Modern agricultural industry technology system Beijing innovation team, Beijing 100102, China

[4] High Level Projects of Beijing Universities Cross Training, Beijing 102206, China

[5] E-mail:abc960718@sina.com

Abstract. [Objective] To optimize the extraction of anthraquinone from cassia seed by ultrasonic extraction and its antibacterial activity. [Method] The influences of different extraction time and ethanol concentration, on anthraquinone content were evaluated by a single factor experiment. And anthraquinone content was determined by ultraviolet spectrophotometry. The bacteriostasis of anthraquinone on spiced pig head's common putrefying bacteria: Staphylococcus, Serratieae, Bacillus, Proteus and the minimal inhibitory concentration (MIC) were studied by oxford plate assay system. [Result] The best extraction time was 30 minutes and the best ethanol concentration was 80%. The antibacterial activity of the cassia seed anthraquinone on Staphylococcus Aureus, Bacillus Proteus is obviously, the minimum inhibitory concentrations were 0.125 g/mL, 0.125 g/mL, 1 g/mL respectively and no inhibitory effect on Serratieae.[Conclusions] The anthraquinones from Cassia seed can inhibit a part of spoilage bacteria in spiced pig heads.

1. Introduction

Low-temperature meat products are popular in public, because of its rich nutrition, strong flavor and rich sense of mastication. However, in order to maintain its flavor, the processing of sterilization condition needs to be controlled at 68-70 degrees for 30 minutes. This leads to some spores and some heat-resistant bacteria are not all killed, and their higher water activity was also beneficial for putrefying bacteria to grow and reproduce, which causes that the low-temperature meat products are more perishable than other meat products and affects their shelf life [1-5].

To prolong the shelf life of low-temperature meat products, people generally choose to add some synthetic antibacterial agents, but with the improvement of living standards, people's health awareness gradually increased, more and more pollution-free natural products are demanded by public [6-7].

Cassia seed, also known as cassia obtusifolia, is the dried ripe seeds of legume of cassia seed. It is cultivated in most parts of the country, such as Anhui, Guangxi, Zhejiang and Sichuan. The main active ingredient in cassia seed is anthraquinone. It has many physiological functions such as antioxidation, bacteriostasis and anti mutation [8-10].

In the food industry, the traditional method of anthraquinone from Cassia seed is used ethanol or alkaline aqueous solution to extract, but it has many disadvantages, such as many processes, large losses, long cycle, and the low extraction rate. Ultrasonic extraction has many advantages, such as high extraction rate, easy operation, fast extraction speed and less loss of active components. It has been widely used in the extraction of natural ingredients [11-12]. This experiment was used ethanol as extraction agent, and extracted the anthraquinones form Cassia seed by ultrasonic extraction method. The extraction of anthraquinones in Cassia seed was determined by using 1,8-dihydroxyanthraquinone as the reference substance. The extraction process was optimized. It can provide scientific basis for further research and production. The oxford plate assay system was used to investigate the inhibition of anthraquinone compounds form Cassia seed to the main spoilage bacteria, Staphylococcus, Serratia, Bacillus and Proteus in spiced pig head, for the future, added the Cassia seed as a bacteriostatic agent into the low-temperature meat products to extend its shelf life.

2. Materials and methods

2.1. Materials

Cassia seed blocks are purchased at Mengzheng Pharmacy. All organic reagents came from Beijing land bridge Company and all Inorganic reagents came from Beijing Chemical Works. The indicator bacteria were isolated and purified from spiced pig head by Beijing University of Agriculture.

2.2. Alcohol extract

Took out the dried Cassia seed block and grinded into powder by pulverizer, after 40 mesh sieve, it was placed in a sealed place at normal temperature.

2.2.1 Ethanol concentration. The 2.5g samples were divided into five groups on average. Each group according to the solid-liquid ratio of 1:25, were added for 50%, 60%, 70%, 80%, 90% ethanol, soaking at room temperature, overnight. The samples were extracted by ultrasonic power 480W at 30 ℃ for 40 min. The extraction with ethanol was repeated thrice. The ethanol extracts were filtered and evaporated by using a rotary evaporator at 50℃, kept at 4 ℃.

2.2.2 Ultrasonic time. The 2.5g samples were divided into five groups on average. Each group according to the solid-liquid ratio of 1:25, were added for 60% ethanol soaking at room temperature, overnight. Each group was extracted by ultrasonic power 480W at 30 ℃ for 20min, 30min, 40min, 50min, 60min. The extraction with ethanol was repeated thrice. The ethanol extracts were filtered and evaporated by using a rotary evaporator at 50℃, kept at 4 ℃.

2.3. Ultraviolet spectrophotometry

1, 8-dihydroxyanthraquinone standard was accurately weighed 10mg and dissolved in 100mL methanol, which was the standard solution of 1,8-dihydroxyanthraquinone. Added 0.5 magnesium acetate-methanol solution as chromogenic agent, with 0.5 Magnesium Acetate-methanol solution as blank sample, the standard solution was spectrally scanned in 300-700nm [13], to determine the optimum absorption wavelength.

The standard solution was removed 0.5mL, 1.0mL, 1.5mL, 2.0mL, 2.5mL, 3.0mL, 3.5mL, 4.0mL and added 0.5 magnesium acetate-methanol solution as chromogenic agent, then constant-volumed in I0mL volumetric flask. With 0.5 Magnesium Acetate-methanol solution as a blank solution, it was measured the absorbance. Absorbance (A) and concentration (C) was used as coordinates and fitted equation of linear regression.

2.4. Actication of culture
Strains (Staphylococcus、 Serratieae、Bacillus, Proteus) were recovered to normal temperature and poured into 50mL sterilized nutrient broth, cultured 12 h at 30 ° C. And then the strain added to new nutrient broth medium with 2% inoculation amount, activated again.

2.5. Bacteriostat of anthraquinone by Oxford cup method
Sterile nutrient agar was inoculated with microbial cells (1 mL of microbial cell suspension in 100 mL agar medium) at 46 ° C and then poured into sterile petri dishes each about 15-20mL.

The sterilized Oxford cup was placed on the carrier plategently.100 μL of 1 g / mL of Cassia seed active ingredients and distilled water were added to the Oxford cups (three times). After incubating at 37℃ for 12h,the antibacterial activity was evaluated by the diameter of inhibition zones.

2.6. Determination of minimum inhibitory concentration
1g / mL of Cassia seed active ingredients were diluted 2 times into different mass concentration gradient. 100mL Cassia seed extraction of different concentrations were added in different Oxford cups on the same plate, and the purified distilled water was used as blank control (three times). After culturing at 30℃for 12h, the antibacterial activity was evaluated by the diameter of inhibition zones [14].

3. Results and discussion
As seen in figure 1, with the increase of ethanol concentration, the concentration of anthraquinone in the extract is also slowly increasing, and when the increase to 60%, the concentration of anthraquinone drops rapidly. When the concentration of ethanol was 60%, the concentration of the extract was the highest. The reason is that the form of anthraquinone in Cassia seed is conjugated anthraquinone glycoside and free anthraquinone, and the content of conjugated anthraquinone glycoside is higher. The conjugated anthraquinone glycosides are dissolved in water easily, while the free anthraquinone is soluble in organic solvent easily [15]. So when the ethanol concentration is less than 80%, with the increase of the concentration of ethanol, the free anthraquinone is soluble more. Total concentration of anthraquinone is gradually increased. But when the ethanol concentration is higher than 60%, anthraquinone glycosides which dissolved in water easily are dissolved decreased, making the concentration of the anthraquinone decreased.

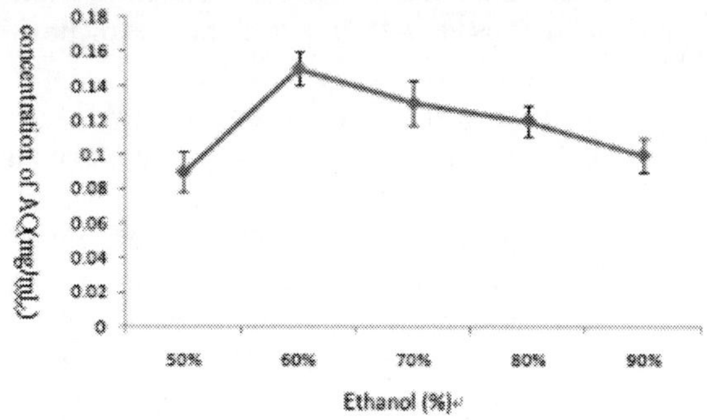

Figure 1. The effect of ethanol concentration on extraction of Anthraquinone.

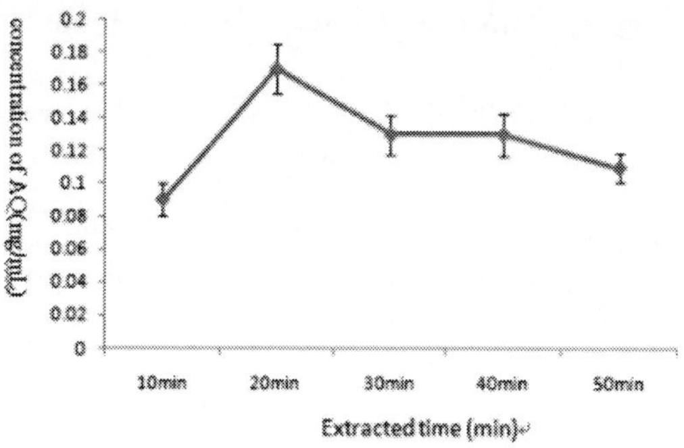

Figure 2. The effect of extracted time time on extraction of Anthraquinone.

As seen in figure 2, before 20min, the concentration of anthraquinone increases with the increase of extracted time. After 30min, the concentration of anthraquinone decreases slowly with the increase of extraction time. The reason may be that some anthraquinones are unstable in heat. With the increase of extraction time, it will lead to decomposition [16].

Table 1. Inhibitory zone diameter of anthraquinone in Cassia seed.

strain	Inhibitory zone diameter /mm
Proteus	12±0.1
Staphylococcus	12±0.2
Bacillus	18±0.5
Serratieae	0

[a]Note: 0 means no inhibition zone

As can be seen from table 1, 1g/mL of anthraquinone in Cassia seed has obvious inhibitory effect on Proteus, Staphylococcus and Bacillus, but it has no bacteriostasis effect on Serratieae.

Table 2. MIC of anthraquinone in Cassia seed.

anthraquinone in Cassia seed (g/mL)		1	0.05	0.25	0.125	0.0625	0.03125
Inhibitory zone diameter /mm	Proteus	12±0.1	0	0	0	0	0
	Bacillus	12±0.2	10±0.5	8±0.3	6±0.1	0	0
	Serratieae	18±0.5	11±0.2	10±0.2	6±0.1	0	0

[a]Note: 0 means no inhibition zone

As can be seen from table 2, the anthraquinones in Cassia seed has well bacteriostatic efficacy on Staphylococcus which minimum inhibitory concentration is 0.125g/mL and the maximum inhibition zone is 18mm; the effect of Bacillus is second which minimum inhibitory concentration is 0.125g/mL and the maximum inhibition zone is 12mm; the bacteriostatic efficacy on Proteus is ordinary which minimum inhibitory concentration is 1g/mL and the maximum inhibition zone is 12mm. The reason may be that Staphylococcus and Bacillus are gram positive bacteria, while Proteus and Serratieae belong to gram negative bacteria and the anthraquinone in Cassia seed is better for gram positive bacteria.

The components of anthraquinones in Cassia seed is complex. There are more than 20 kinds of substances [17] at least, such as aloin, Cassia seed sin, emodin and so on. The antibacterial mechanism of anthraquinone in Cassia seed might be two. One is to change the permeability of cell membrane leading to the cell damage: bacteria are through the selective permeability of cell to absorb nutrition from the outside world for its growth and reproduction. When the permeability of the cell membrane changing, the leakage of nucleic acids causes the cell deformation and nutrient flowing will result irreversible damage to bacteria themselves, and make bacterial die ultimately [18-19]. The other one is to inhibit the respiratory metabolism of bacteria: the growth and reproduction of bacteria cannot be separated from respiratory metabolism. Respiratory metabolism is the carbohydrates being oxidized and discomposed to release energy. Anthraquinone in Cassia seed can inhibit the Krebs cycle of oxidation of sugar and inhibit the synthesis of ATP and NADH in cell, which obstructs the oxidation of sugar. The respiratory metabolism of bacteria is affected, and then the bacteria can even lead to death [20].

4. Conclusion

In the single factor analysis of ultrasonic extraction, it can be concluded that 60% ethanol concentration is the best concentration of extraction and the extraction time of 20min is the best time. The main component of the extraction liquid is anthraquinone.

The anthraquinone in Cassia seed has better bacteriostatic efficacy for gram positive bacteria. And with the decrease of the concentration of active ingredients, the inhibitory effect gradually decreased, which indicated that high concentration extracts have certain bacteriostatic effect.

References

[1] Jos H J and Huisin't Veld 1996 Microbial and biochemical spoilage of foods: an overview *International Journal of Food Microbiology* **33** 1-18
[2] Gram L and Dalgaard P 2002 Fish spoilage bacteria problems and solutions *Current Opinion in Biotechnology* **13** 262-6
[3] Cao Yin 2017 Extraction process of anthraquinone from Aloe *Textile Auxiliaries* **34** 46-49
[4] Zhang G Q 2011 Study on the extraction of the active substance from aloe barbadensis and its antimicrobial effect *Shan Xi: Northwest A & F University* 1-2
[5] Liu Z X, Wang L J, Yang Y Z, Bai X and Li R J 2007 Research on bacteriostasis effect of Aloe *Journal of Shanghai University of Engineering Science* **21**(3) 219-221
[6] Ganzler K, Szonal I and Salgo A 1990 Effetive sample preparation method for extracting biologically active compounds from different matrices by a microwave technique *Journal of Chromatography* **520** 257-262
[7] Zhang L et al 2012 Study on the Antibacterial Activity of Aloe Microwave Extracts *Journal of Anhui Agricultural Sciences* **40** (21) 11032-34
[8] Wu C H, Hsileh C H, Song T Y, et al 2001 Inhibitory effect of Cassia tora L on benzo(a) pyrene-nediated DNA danage tow and H epG 2 cells *J Agric Food Chen* **49**(5) 2579-86
[9] Weissinger W R, McWatters K H, Beuchat and L R 2001 Evaluation of volatile chemical treatments for lethality to Salmonella on alfalfa seeds and sprouts *Journal of Food Protection* **64** (4) 442-50
[10] Zhou M L, Shang G, Xin C, Zhan Y Q and Yang W Y 2013 Optimization of Ultrasonic — assisted Extraction for Total Anthraquinones from Aloe Leaves by Response Surface MethodologyF *Acta Agriculturae Universitatis Jiangxiensis* **35** (5) 1055-62
[11] Wang Y, Wang Q, Wang X J, Zhang W D, Wang P Y and Gu Y 2012 Optimization of extraction of anthraquinones in Cassiae Semen by multi-index orthogonal test *Chinese Traditional Patent Medicine* **34**(9) 1684-16
[12] Liu W M, Jiang B, Zeng Y, Chen R D and Lin P J 2009 Study on Effect of Water and Alcohol Extractiig Process on Anthragahone Content in Semen cassiae *Lishizhen Medicine & Materia Medica Research* **20**(9) 2269-71
[13] Jia Z B, Chen W W, Jiang J X, Ding and Xiao L 2009 Study on Anthraquinone Constituents in the Seed of Cassia tora L *Chemistry and Industry of Forest Products* **29**(3) 100-102
[14] Melvyn G and Peter M M 2001 New fungal anthraduinonesl *Adcivoc* p145-156
[15] Fu Jmoto H, Nakamura E Okuyana E, et al 2004 Six inmunosuppressive features from an asarn yczte, Zopfiella loragicaadata, found in a screening shady monitored by imnunanodulaty activity *Chan Pham Bull* **52**(8) 1005-8
[16] Panwnyk L, Bearfoy E, Lornmer J P and Nason T J 2001 The extraction of rutin from flower buds of Sophora japonica *Ultrasomcs Sonochemistry* **8** 299-301
[17] Kieth P 2001 Multidrug resistance in gram-negative bacteria *Current Opinion in Microbiology* **4** 500-508
[18] Vivek K, Bajpai, Ajay S and Bokyung M 2014 Chemical Composition analysis and antibacterial mode of action of taxes cuspidata leaf essential oil against foodbrone pathogens *Journal of food safety* **34** p9-20
[19] Wei J, YingFeng W, RiLi G, et al 2007 Simultaneous analysis of multiple bioactiveconstituents

in Rheum tanguticum Maxim. ex Balf. By high-performance liquid chromatography coupled to tandem mass spectrometry *Rapid Communications in mass Spectrometry* **21** 2351-2360

[20] Ultee A 2000 Adaptation of the food-borne bacillus cereus to carvacrol *Archives of Microbiology* **174** (4) 233-238

Corrosion behaviorsof pure copper and Cu-Ni-Zn alloy in NaCl solution and artificialsaltwater

J M Shi, C L He[1], G P Li, H Z Chen, X Y Fu, R Li, G F Ma and J M Wang

Liaoning Provincial Key Laboratory of Advanced Materials, Shenyang University, Shenyang 110044, China

[1] E-mail: ccllhhe@126.com

Abstract. Corrosion behaviors of pure Cu and Cu-Ni-Zn alloy were investigated in 3.5% NaCl solution and artificial seawater by electrochemical impedance spectroscopy and potentiodynamic polarization technologies, and the corrosion morphologies were observed by field emission scanning electron microscopy. The results revealed that Cu-Ni-Zn alloy possessed better corrosion resistance than that of pure Cu in both 3.5% NaCl solution and artificial seawater. The corrosion morphology displayed the corrosion product films on Cu-Ni-Zn alloy were more compact and uniform than that on the pure Cu in both 3.5% NaCl and artificial seawater media.

1. Introduction

Copper alloys have been widely used in deep sea pipelines, bearings and electronics in marine engineering, due to their high strength, high thermal conductivity,good mechanicalworkability,and excellent corrosion resistance[1,2]. Copper alloys are also used in marine environments to defend against biofouling by inhibiting microbial induced corrosion [3, 4]. In the ocean, microbial communities rapidly colonize and strongly adhere to metal surfaces, forming a layer of biofilm. The biofilms can damage expensive equipment, cause loss of production and increase maintenance costs [5, 6]. Therefore, it is necessary to study the corrosion behaviors of copper and copper alloys in marine environment.

In this work, the electrochemical corrosion behaviors of pure Cu and Cu-Ni-Zn alloy were investigated in both 3.5% NaCl solution and artificial seawater by measuring electrochemical impedance spectra (EIS) and potentiodynamic polarization curves. Field emission scanning electron microscopy (FESEM, S-4800) was used to examine the corrosion morphologyof the samples after polarization measurement.

2. Materials and methods

The experimental material wasCu-Ni-Zn alloy plate, and its composition was listed in table 1. Before the electrochemical tests, the pure Cu and Cu-Ni-Zn alloy samples were machined into dimensions of 50 mm × 40 mm × 3 mm. The working surface was mechanically grinded with sandpaper from 400# to 1500# and then polishedby using2.5μm diamond paste. The corrosion solutions were 3.5% NaCl solution and artificial seawater.

Table 1. Composition of Cu-Ni-Zn alloy (mass fraction, %).

Element	Ni	Cu	Zn	Total impurities
Composition	16.03	56.81	26.35	0.82

The electrochemical experiments were performed on a PARSTAT2273 electrochemical workstation in both 3.5% NaCl solution and artificial seawater at room temperature. The area of working electrode was 1 cm^2. A conventional three-electrode system was used with pure Cu and Cu-Ni-Zn alloy as the working electrode, saturated calomel electrode as the reference electrode, and graphite rod as the auxiliary electrode. EIS was acquired over a frequency range of 100 kHz−10 mHz with an applied sinusoidal perturbation of 10 mV root-mean-square potential. Potentiodynamic polarization measurement was carried out at 0.332 mV/s. The corrosionmorphology of the sample after polarization test was examined by an FESEM.

3. Results and discussion

3.1. Potentiodynamic polarization measurements

Figure 1 shows the representative potentiodynamic polarization curves of pure Cu and Cu-Ni-Zn alloy in 3.5% NaCl and artificial seawater. It is clearly seen from figure 1a and 1b that the curve shapes of pure Cu and Cu-Ni-Zn alloy in 3.5% NaCl solution are similar to those in artificial seawater. The pure Cu sample is quickly dissolved after the scanning potential over the corrosion potential in both 3.5% NaCl and artificial seawater media, showing no protective product film appears on the surface. On the contrary, the polarization curves of the Cu-Ni-Zn alloy have an obvious passive behavior, showing protective product films are formed on the Cu-Ni-Zn alloy in the two solutions.

The relative electrochemical parameters obtained from the polarization curves are listed in table 2. The corrosion potentials (E_{corr}) of pure Cu and Cu-Ni-Zn alloy are -0.242V and -0.226 V in 3.5% NaCl, and -0.230 V and -0.214 Vin artificial seawater, respectively. A more positiveE_{corr} means that the sample is nobler in the solution. As shown in table 2, the corrosion current density (I_{corr}) values of pure Cu and Cu-Ni-Zn alloy are 7.84×10^{-6} A/$cm^2$and 1.73×10^{-6} A/cm^2 respectively in 3.5% NaCl, 4.79×10^{-6}A/cm^2 and 4.66×10^{-6}A/cm^2 respectively in artificial seawater. I_{corr} is an important parameter commonly used to assess the kinetics of corrosion reactions, and it is almost directly used to characterize the corrosion rate at E_{corr}. Compared with the pure Cu, the Cu-Ni-Zn alloy exhibits higher $E_{corr}$and lower I_{corr}in both 3.5% NaCl and artificial seawater,meaning it has higher corrosionresistance in the solutions. The results show that improved corrosion resistance in Cu-Ni-Zn alloy is brought about possibly by Zn and Ni additions [7]. Ni enhances the corrosion resistance of the alloy by its segrega-tion into the defective lattice of Cu_2O, The addition of zinc to copper enhances both its mechanical characteristics and its corrosion resistivity to artificial seawater [8].

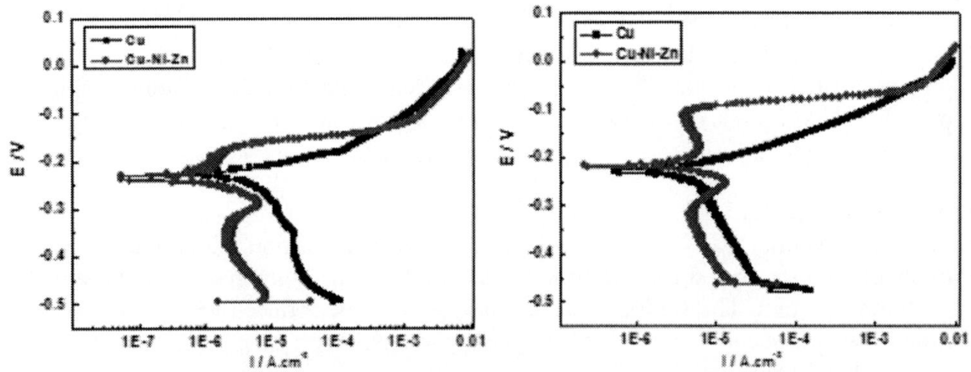

(a) 3.5 % NaCl (b) artificial seawater

Figure 1. P olarization curves for pure Cu and Cu-Ni-Zn alloy in different solutions.

It also clearly shows that E_{pit} of the Cu-Ni-Zn alloy is much higher than that of pure Cu in both 3.5% NaCl and artificial seawater electrolytes, especially in the artificial seawater. Moreover, the E_{pit} values of the pure Cu and Cu-Ni-Zn alloy are a little lower in 3.5% NaCl solution than that in artificial seawater.

Table 2. Tafel fitting results of pure Cuand Cu-Ni-Zn alloy.

Solution	Material	I_{corr} /(A.cm^{-2})	E_{corr}/V	E_{pit}/V
3.5%NaCl	Cu	7.84×10^{-6}	-0.242	-0.163
	Cu-Ni-Zn	1.73×10^{-6}	-0.226	-0.147
artificial seawater	Cu	4.79×10^{-6}	-0.230	-0.139
	Cu-Ni-Zn	4.66×10^{-6}	-0.214	-0.082

3.2. Electrochemical impedance spectroscopy (EIS) measurements

Figure 2 presents the EIS spectra of pure Cu and Cu-Ni-Zn alloy in 3.5% NaCl and artificial seawater. The EIS plots exhibit two capacitive loops, presented in the high frequency area and low frequency area, respectively. The EIS plots of the pure Cu sample also show the appearance of Warburg diffusion in the two solutions. Ma et al [9] also reported the EIS of 90 Cu-10 Ni alloy presented Warburg diffusion impedance in 3.5% NaCl solution. Whereas for the Cu-Ni-Zn alloy, there seems not to appear Warburg diffusion behavior, probably associated with the compact product film on the alloy. It is remarkable that the semicircle diameters of the Cu-Ni-Zn alloy are much bigger than those of pure Cu in both 3.5% NaCl and artificial seawater. Capacitive semicircle at high frequency is related to the formation of corrosion products whereas at low frequency is related to the double electric layer.It is obvious that Cu-Ni-Zn alloy displays much larger capacitive loops than pure Cu. Larger capacitive loop diameters mean higher impedance, indicating the effect of thecorrosion product film on inhibiting corrosion. The result of EIS is good agreement with the polarization curves, they both display that the Cu-Ni-Zn alloy possesses better corrosion resistance in both 3.5% NaCl solution and artificial seawater compared with the pure Cu sample.

3.3. Corrosion morphologies

Figure 3 shows the corrosion morphologies of pure Cu and Cu-Ni-Zn alloy after polarization test in 3.5% NaCl solution and artificial seawater. It can be seen from the figure 3 that the surface of corroded Cu sample is covered with loose and non-uniform corrosion products; however, the corroded surface of Cu-Ni-Zn alloy is covered with dense and uniform products. Obviously, the dense and uniform corrosion product layer can effectively protect the Cu-Ni-Zn alloy from further corrosion in the two solutions.

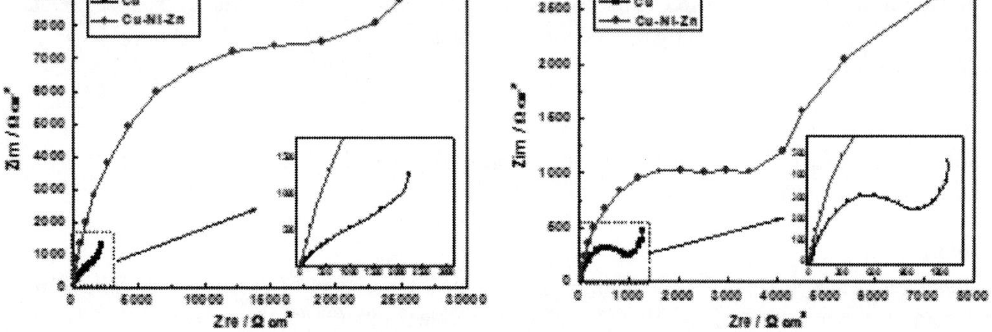

(a) 3.5 % NaCl (b) artificial seawater

Figure 2. Nyquist plots of pure Cu and Cu-Ni-Zn alloy.

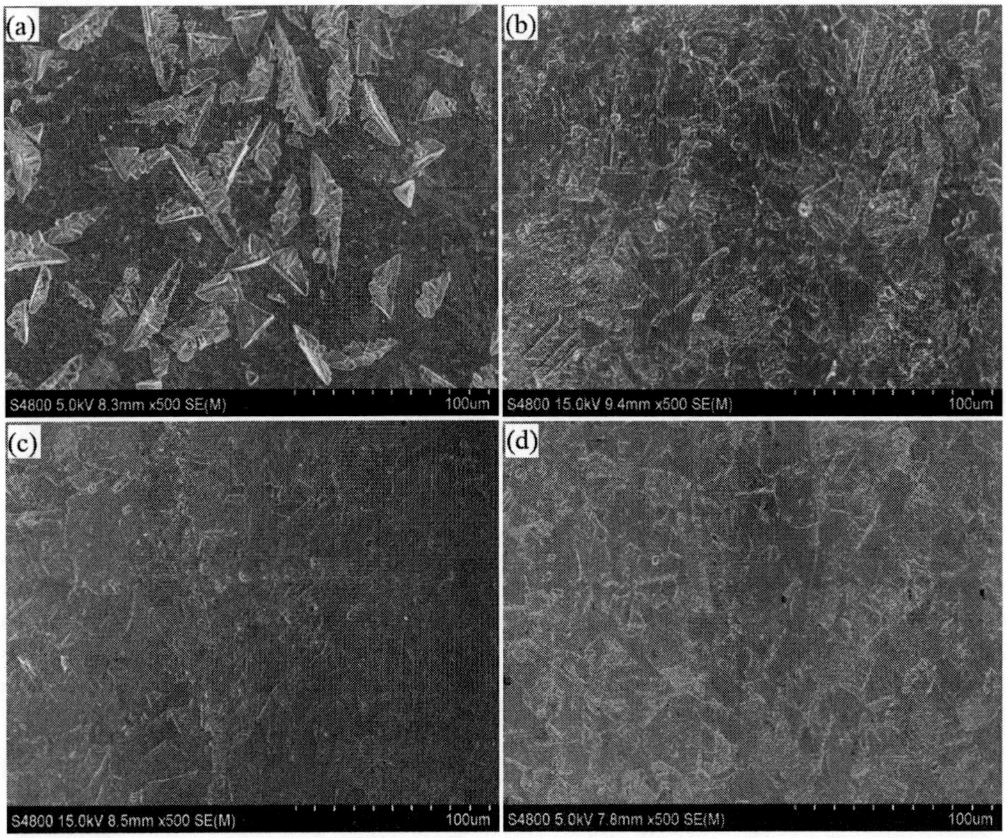

Figure3. Corrosion morphologies of pure Cu (a, c) and Cu-Ni-Zn alloy(b,d)in3.5%NaCl solution(a,b)and in artificial seawater (c, d).

4. Conclusions

The electrochemical corrosion behaviors of pure Cu and Cu-Ni-Zn alloy were studied in 3.5% NaCl solution and artificial seawater by electrochemical impedance spectroscopy and potentiodynamic polarization tests. The polarization curves of the Cu-Ni-Zn alloy presented an obvious passive behavior in the two solutions, showing protective product films were formed. Both polarization and EIS results revealed that Cu-Ni-Zn alloy possessed higher resistance than pure Cu in 3.5% NaClsolution and artificial seawater. The corrosion morphology displayed the corrosion product films on Cu-Ni-Zn alloy were more compact and uniform than that on the pure Cu in both 3.5% NaCl and artificial seawater media.

Acknowledgments

This work was financially supported by the Project-sponsored by SRF for ROCS, SEM and the National Natural Science Foundation (No. 51171118) of China.

References

[1] Y Zhang Z Xiao Y Zhao Z Li Y Xing and K Zhou 2017 Effect of thermo-mechanical treatments on corrosion behavior of Cu-15Ni-8Sn alloy in 3.5 wt% NaCl solution *Materials Chemistry and Physics* **199** 54-56

[2] NK Awad EA Ashour and NK Allam 2015 Unravelling the composition of the surface layers formed on Cu Cu-Ni Cu-Zn and Cu-Ni-Zn in clean and polluted environments *Applied Surface Science* **346** 158-164

[3] NO San G Dönmez and H Nazir 2011 Microbial corrosion of Ni-Cu alloys by aeromonas eucrenophila bacterium *Corrosion Science* **53** 2216-2221

[4] HY Chen 2007 Pitting behavior of Cu - Ni alloy in NaCl solution *Materials Protection* **40 (6)** 17-19

[5] P Nath DK Sahu and A Mallik 2016 Physicochemical and corrosion properties of sono-electrodeposited Cu-Ni thin films *Surface & Coatings Technology* **307** 772-780

[6] M Heshmati D Seifzadeh P Shoghi and M G Gheshlagi 2017 Duplex Ni-Zn-Cu-P/Ni-P electroless coating on magnesium alloy via maleic acid pretreatment *Surface & Coatings Technology* **328** 20-29

[7] D J Horton H Ha LL Foster HJ Bindig and JR Scully 2015 Tarnishing and Cu Ion release in Selected Copper-Base Alloys Implications towards Antimicrobial Functionality *Electrochimica Acta* **169** 351-366

[8] NK Awad EA Ashour AS Fouda and NK Allam 2014 Effect of alloying elements on the electrochemical behavior of Cu-Ni-Zn ternary system in sulfide-polluted saltwater *Applied Surface Science* **307** 621-630

[9] AL Ma SL Jiang YG Zheng and W Ke 2015 Corrosion product film formed on the 90/10 copper-nickel tube innatural seawater Composition/structure and formation mechanism *Corrosion Science* **91** 245-261

IOP Conf. Series: Materials Science and Engineering **292** (2017) 012005 doi:10.1088/1757-899X/292/1/012005

Calculation and affection of pH value of different desulfurization and dehydration rates in the filling station based on Aspen Plus

J X Lv, B F Wang[1], L H Nie, R R Xu, J Y Zhou and Y J Hao

School of Petrochemical and Energetic Engineering, Zhejiang Ocean University, Zhoushan316000, China

[1] E-mail: wangbeifu@zjou.edu.cn

Abstract. The simulation process of the whole CNG filling station are established using Aspen Plus V7.2. The separator (Sep) was used to simulate the desulfurization and dehydration equipment in the gas station, and the flash module separator Flash 2 was used to simulate the gas storage well with proper temperature and environmental pressure. Furthermore, the sensitivity module was used to analyse the behaviour of the dehydration and desulfurization rate, and the residual pH value of the gas storage wells was between 2.2 and 3.3. The results indicated that the effect of water content on pH value is higher than that of hydrogen sulphide in the environment of gas storage wells, and the calculation process of the pH value is feasible. Additionally, the simulation process provides basic data for the subsequent anticorrosive mechanism and work of gas storage well and has great potential for practical applications.

1. Introduction

Trace amounts of hydrogen sulfide, carbon dioxide and water components in compressed natural gas can dissolve in each other in the process of storing in gas storage well which will form bottom hole liquid loading and causing corrosion of gas storage well, directly related to the safe operation of CNG (compressed natural gas) filling station [1]. As the domestic gas storage wells have are not been used for a long time, there are few studies on the corrosion status of practical use of gas storage wells. In 2010, Cen et al. introduced the corrosion research of abandoned gas storage wells by macrography, scanning electron microscopy (SEM) analysis and energy spectrum (EDX) analysis. But the author only pays attention to the damage and results of internal and external surface which caused by corrosion, and there are few research on the production and composition analysis of etching solution due to the special situation of the gas storage well [2].

Next years, with the increasingly extensive application of chemical simulation technology and the problem caused by hydrogen sulfide stress corrosion cracking (SCC) in the gas storage wells [3, 4], more researches should be done [5]. Thus, in this work, Aspen Plus software was used to establish a simulation process and the sensitivity analysis was utilized to obtain the solubility data of hydrogen sulfide and carbon dioxide with different desulfurization and dehydration conditions in water, and the pH of the mixture was also obtained. This study provides basic data for the corrosion mechanism and anticorrosion work of hydrogen sulfide stress corrosion cracking in gas storage well [3, 4].

2. Establishment of simulation process and material calculation

2.1. Establishment of simulation process

The simulation process of the whole CNG filling station established by Aspen Plus V7.2 is shown in figure 1. Separator (Sep) was used to simulate the desulfurization and dehydration equipment in the gas station, flash module separator Flash 2 was used to simulate the gas storage well with proper temperature of 283K and environmental pressure of 25MPa. As shown in figure 1, firstly, natural gas entered into the separator (B1) for desulfurization and exported to the compressor (B2), then entered the heat exchanger (B3) cooling and exported to the separator (B4) for dewatering, and finally entered the flash module (B5).

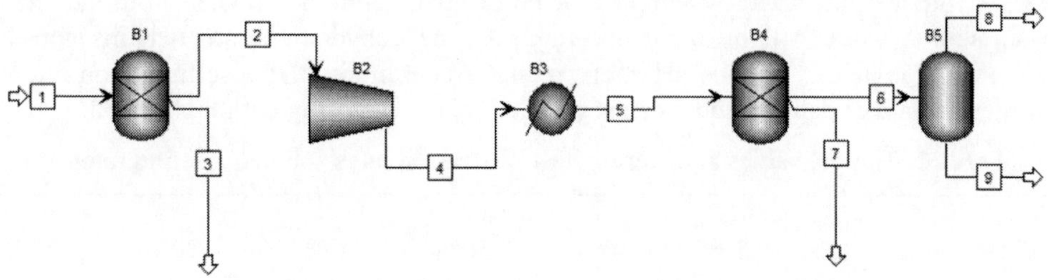

Figure 1. Simulation process of CNG filling station.

2.2. Operating conditions

Aspen Plus V7.2 software operating conditions are as follows: methane as the main component, with a small amount of hydrogen sulfide, carbon dioxide and water [6], feed temperature of 293K, pressure of 0.8MPa, volumetric flow rate of 1000m^3/hr, the compressor (B2) output pressure of 25MPa, the heat exchanger output temperature of 293K, flash module operating conditions for 25MPa, 283K.

According to the mandatory national standard GB17820-2012 [6], in the situation of 0.1MPa, 293K, the hydrogen sulfide content in natural gas should be less than or equal to 20mg / m^3, the carbon dioxide content should less than or equal to 3% in Mole, water dew point should lower 5K than the lowest atmosphere temperature. All of these are shown in table 1.

Because the gas state equation:

$$V = nRT \tag{1}$$

$$\rho = \frac{PM}{RT} \tag{2}$$

In the situation of 0.1MPa, 293K, Solution ρ (H_2S) = 1.1165kg / m^3.

It is found that the hydrogen sulfide in the natural gas of 1000 m^3 is 20g, that is 0.0143369 m^3.

The water content of the natural gas is according to the water dew point 288K and GB/T 22634-2008 [7] .

In the gas storage wells, the chemical reaction of H_2O-H_2S-CO_2 solution system is:

$$H_2O + CO_2 \rightarrow H^+ + HCO_3^- \tag{3}$$

$$H_2S \rightarrow H^+ + HS^- \tag{4}$$

$$HCO_3^- \rightarrow H^+ + HS^- \tag{5}$$

$$HS^- \rightarrow H^+ + S^{2-} \tag{6}$$

$$CO_2 + OH^- \rightarrow HCO_3^- \tag{7}$$

3. Results and discussion

The desulfurization rate and dewatering rate of the two separators are important parameters in this simulation, and they have a decisive correlation in the pH value of the residual solution.

At present, post-dewatering is usually used in the filling station, but due to the powdery caused by high pressure and the entrance of natural gas after the gasification of the lubricating oil in the compressor, the adhesion of adsorbent to the oil will lead to the decrease of the dehydration capacity. So, there is a replacement for about half a year and the performance of dehydration will gradually deteriorate over time. As for more solid desulfurizing agent was used in the desulfurization equipment, the desulfurization agent will lose activity in the process of application. Therefore, different desulfurization rate and dehydration rate was used in this process to simulate the actual application process, which demonstrated that efficiency of desulfurization and dehydration equipment was reduced, so as to examine its impact on the gas storage equipment. In order to obtain different effects of desulfurization and dewatering rates for the pH value of the residual solution, the sensitivity module of Aspen Plus was used to analyze the sensitivity of these two parameters.

The pH values at different desulfurization rates and dehydration rates calculated from Aspen Plus are shown in Table 1. Corresponding to table 1, we can see that although the pH values at different dehydration and desulfurization rates are different, the overall pH value of the solution is in the range of 2.2 to 3.3. Moreover, analyzed by samples of the residual solution extracted from the gas storage well, the simulation value will be more accurate. As the dehydration rate remains constant, the desulfurization rate increases and the pH keeps rising. Accordingly, the desulfurization rate remains constant while, the rate of dehydration increases and the pH keeps rising with high speed.

Table 1. The pH values at different desulfurization rates and dehydration rates.

Dehydration / Desulfurization	20%	30%	40%	50%	60%	70%	80%
20%	2.257	2.416	2.451	2.661	2.843	3.082	3.296
30%	2.264	2.423	2.463	2.676	2.854	3.089	3.289
40%	2.274	2.431	2.469	2.685	2.866	3.071	3.281
50%	2.285	2.440	2.478	2.698	2.873	3.064	3.277
60%	2.290	2.451	2.484	2.706	2.886	3.075	3.265
70%	2.298	2.456	2.491	2.711	2.895	3.079	3.269
80%	2.306	2.461	2.496	2.724	2.902	3.086	3.279

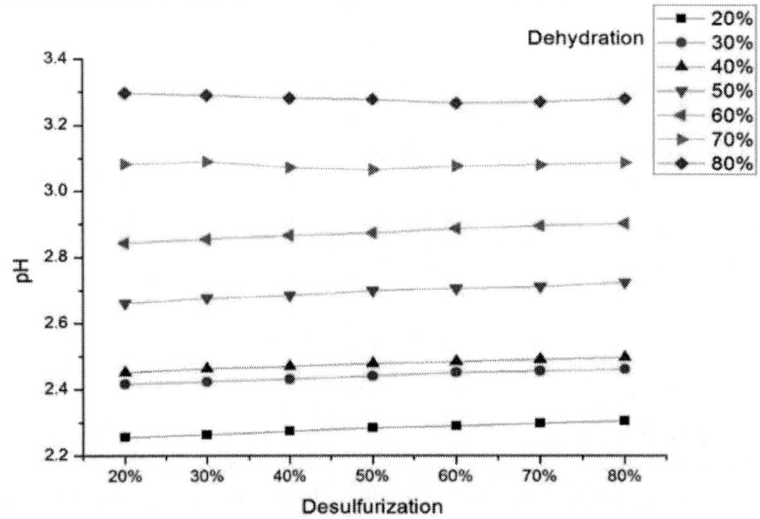

Figure 2. The PH values at different desulfurization rates.

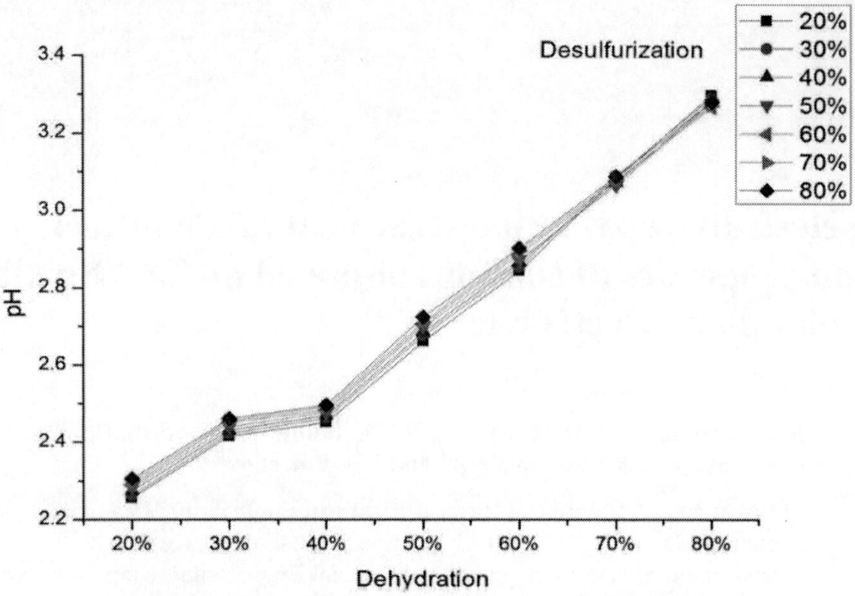

Figure 3. The pH values at different dehydration rates.

The pH values at different desulfurization rates and dehydration rates were shown in figure 2 and figure 3. As shown in figure 2, in the same desulfurization rate, with the increase of the dehydration rate, the pH value increases rapidly. According to figure 3, in the condition of different dehydration rates, the pH values of residual liquid with different desulfurization rates are similar and have minimal distinction. Comparing figure 2 with figure 3, it can be seen that the dehydration rate has a much greater effect on the pH value of the residual liquid than the desulfurization rate.

4. Conclusion

In this study, the simulation process of CNG filling station was established by Aspen Plus software, sensitivity function was used to analyze the sensitivity of pH value of residual liquid under different dehydration rate and desulfurization rate conditions. It concluded that the pH value of residual liquid in the gas storage well is as small as 2.2. Furthermore, by comparing the pH value of the residual solution in different desulfurization rate and different dehydration rate, it can be seen that the effect of the dehydration rate on the pH value in the gas storage as well as environment is much greater than the desulfurization rate. Therefore, in the actual production process of the gas station, the solid desulfurizer and dehydrating adsorbent in the desulfurization and dewatering equipment should be frequently regenerated and the interval of replacement should be reduced.

References

[1] Cen K, Liao K X, Li Q and Li W 2010 *J. Corrosion & Protection* **31** 552
[2] Nakayama T, Nagashima Y, Shimoda K and Bannai H 2002 *P. Conditioning metal surfaces before phosphating them* US US6478860
[3] Liu W, Pu X L, Bai X D, Zhao H W 2008 *J. Petroleum Drilling Techniques* **36** 83
[4] Chen Z Z, Shi Q, Li B X, Han H W, Cui G Y, Chen Y H and Li W B 2013 *J. Process Equipment & Piping* **50** 20
[5] Liu Q Y, He X and Meng S H 2005 *J. Natural Gas Industry* **25** 138
[6] GB17820-2012 2012 S. *Natural gas*
[7] GB/T 22634-2008 2008 S. Conversion between water content and water dew point of natural gas

IOP Conf. Series: Materials Science and Engineering **292** (2017) 012006 doi:10.1088/1757-899X/292/1/012006

Plasma electrolytic oxidation treatment mode influence on corrosion properties of coatings obtained on Zr-1Nb alloy in silicate-phosphate electrolyte

R G Farrakhov[1], V R Mukaeva[1], A R Fatkullin[2], M V Gorbatkov[1], P V Tarasov[3], D M Lazarev[1], N Ramesh Babu[4] and E V Parfenov[1, 5, 6]

[1] Department of Theoretical Basis of Electrical Engineering, Ufa State Aviation Technical University, 12 Karl Marx Street, Ufa, 450008, Russia
[2] Department of Natural Sciences and General Engineering, Kumertau Campus of Ufa State Aviation Technical University, 22A Maxim Gorky Street, Kumertau, 453300, Russia
[3] Department of Engineering Technology, Ufa State Aviation Technical University, 12 Karl Marx Street, Ufa, 450008, Russia
[4] Department of Metallurgical and Materials Engineering, National Institute of Technology, Tiruchirappalli, NIT Quarters, Tiruchirappalli, 620015, India
[5] Academy of Sciences of Republic of Bashkortostan, 15 Kirov Street, Ufa, Russia

[6] E-mail: evparfenov@mail.ru

Abstract. This research is aimed at improvement of corrosion properties for Zr-1Nb alloy via plasma electrolytic oxidation (PEO). The coatings obtained in DC, pulsed unipolar and pulsed bipolar modes were assessed using SEM, XRD, PDP and EIS techniques. It was shown that pulsed unipolar mode provides the PEO coatings having promising combination of the coating thickness, surface roughness, porosity, corrosion potential and current density, and charge transfer resistance, all contributing to corrosion protection of the zirconium alloy for advanced fuel cladding applications.

1. Introduction

Zirconium and its alloys are widely used as nuclear energy materials owing to their excellent neutron transparency, high mechanical strength, creep resistance, corrosion resistance in boiling water and steam environment. The major application of zirconium alloys both in PWR and BWR type reactors is fuel cladding tubes which enclose nuclear fuel and emit neutrons into coolant water which heats up and transfers energy from the reactor [1].

Nuclear reactor safety is the major concern; therefore, the fuel cladding failures should be minimized by application of advanced energy materials and appropriate corrosion resistant coatings. One of the failure mechanisms associated with Zr alloy cladding tube is hydrogen embrittlement which follows from ZrH_2 formation within the tube due to pickup of the hydrogen resulting from the Zr water corrosion [2]. During the life cycle, the Zr alloy components gain ZrO_2 oxide layer on its surface as a result of the water corrosion reaction; this layer protects from further corrosion; however, simultaneously with its growth, the hydrogen pickup also takes place. Therefore, alternative oxidation techniques are developed in order to produce protective zirconia coatings [3, 4].

Plasma electrolytic oxidation (PEO) is one of the promising methods to develop oxide ceramic coatings on aluminum, titanium, magnesium and zirconium and their alloys. PEO of Zr alloys is currently gaining more attention since the methodology of the process modelling, optimization and mechanism understanding have been rigorously developed for Al and Ti alloys [5, 6].

This paper is devoted to investigation into the influence of the PEO treatment mode: DC, pulsed unipolar, pulsed bipolar, on the corrosion properties of the coatings obtained on Zr-1Nb alloy in silicate and phosphate containing electrolyte widely used for plasma electrolytic oxidation of various alloys, in order to choose the best mode for further coating optimization.

2. Experimental

2.1. Plasma electrolytic oxidation

In this study, reactor grade Zr-1wt%Nb zirconium alloy plate (GOST E110) was used. The material is supplied after a vacuum annealing; no texture differences appear in the plate; the grain size is 5-7 μm [7]. The samples were cut out of the plate to the size 20x20x0.8 mm; the sample roughness was 0.1 μm. The PEO process was performed in a 6 liter tank at the electrolyte temperature of 20±1 °C kept constant by heating and cooling system under microcontroller regulation (TRM202, Owen, Russia). The treatment time was 10 min. The electrolyte was composed of 1 g l^{-1} KOH, 2 g l^{-1} Na_2SiO_3 and 2 g l^{-1} $Na_4P_2O_7$, all of puris grade.

The plasma electrolytic oxidation was realized using 50 kW automated equipment providing PC process control and data acquisition for voltage, current and electrolyte temperature. The following PEO treatment modes were used: DC at constant current density j = 10 A·dm^{-2}; pulsed unipolar (PUP) at the same average current density and duty cycle d_p = 50 %; and pulsed bipolar (PBP) at positive / negative pulse voltages U_p / U_n = 550 / 100 V, and duty cycles d_p / d_n = 50 / 30 %. For the pulsed modes the frequency was f = 500 Hz. For the constant current density modes the maximal voltage achieved was 600 V.

2.2. Surface characterization

The surface morphology was assessed using scanning electron microscope (SEM) JEOL JSM-6490LV. The coating porosity was calculated from the SEM images using ImageJ software. The surface roughness *Ra* was measured with TR 220 profilometer. The coating thickness *h* was measured using Defelsko Positector 6000 eddy current gauge with accuracy +/– 0.1 μm. X-ray diffractometry was performed using Shimadzu XRD-6100 diffractometer in Cu-Kα radiation from 10 to 80 degrees 2θ, step size 0.02 degrees 2θ and step scan time 1 s.

To evaluate the corrosion properties, 1M LiOH solution was used, because lithium hydroxide is added to the coolant water in nuclear reactors. Elins P-5X potentiostate-impedancemeter was used (Elins, Russia). The open circuit potential (OCP) was measured for 2 hours to achieve a steady state value. The potentiodynamic polarization (PDP) tests were performed from –1.5 to +0.5 V with respect to OCP at the scan rate 0.25 mV/s. The corrosion potential E_{corr} and corrosion current density I_{corr} were obtained from Tafel parts of the PDP curves. The electrochemical impedance spectroscopy (EIS) was performed from $2\cdot10^{-2}$ to $1\cdot10^{5}$ Hz at 10 mV perturbation with respect to OCP. The EIS spectra were fitted with EIS Spectrum Analyzer 1.0 using the equivalent circuit shown in Figure 1. As a reference, silver chloride electrode Ag/AgCl in saturated KCl with E_0=200 mV vs. standard hydride electrode was used; as a counter electrode, a Pt plate was used.

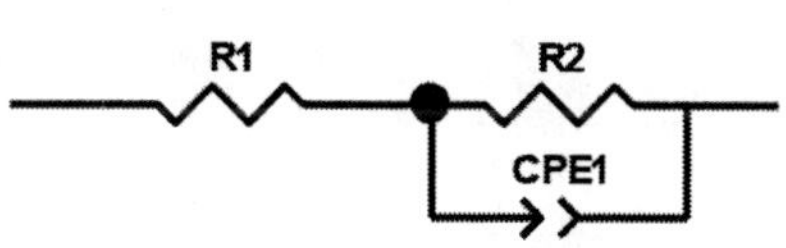

Figure 1. Equivalent circuit used for the EIS analysis.

3. Results and discussion

3.1. Influence of the PEO treatment mode on the coating morphology

Figure 2 shows the PEO coating morphology in top view and cross-sections for different treatment modes. The coating thickness and roughness are shown in Figure 3. The highest coating thickness comes out of the DC mode; this mode also provides the highest roughness. Numerous cracks can be seen in the cross-section for this mode. The opposite results appear for the PBP mode: it provides high thickness and low roughness. This suggests that the PBP mode appears to be promising; however, as follows from Figure 2, the coating has uneven morphology and cracks in the cross-section; moreover, in macro scale this mode yields burned edges of the samples. Therefore, the research into the PBP mode requires additional optimization. For the PUP mode, the coating has the middle value of the roughness and lowest thickness; no burned edges appear on the samples. In the cross-section the PUP coating appears solid, without cracks.

The coatings have porous structure (Figure 2) which is typical for the PEO process since the surface layer becomes intensively modified with numerous microdischarges contributing to zirconium oxidation, coating melting and resoldifying.

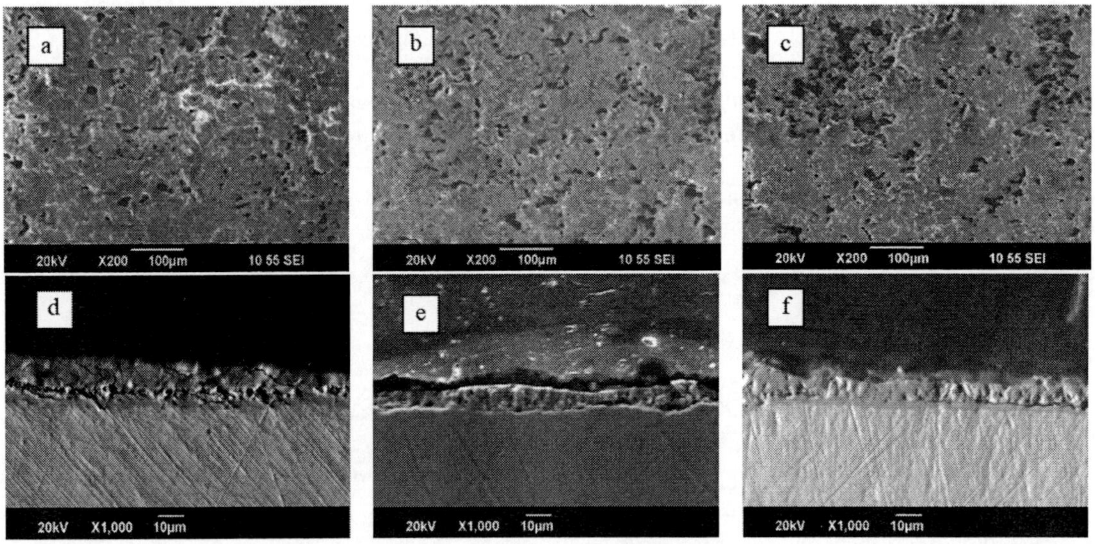

Figure 2. PEO coating morphology: top view (a-c) and cross-sections (d-f) for different PEO treatment modes: DC (a, d), PUP (b, e), PBP (c, f).

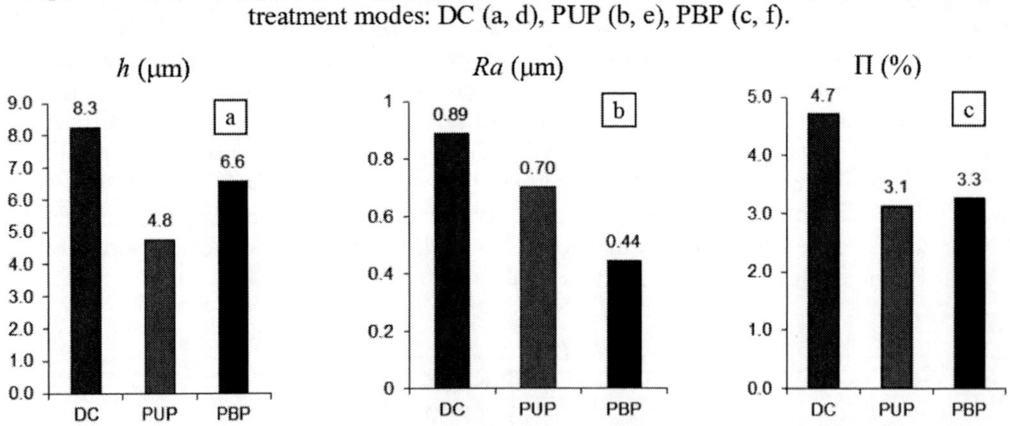

Figure 3. PEO coating thickness *h* (a), roughness *Ra* (b) and porosity Π (c) for different PEO treatment modes.

The largest pore diameter reaches 10 μm, and the average pore size is 2 μm. SEM image analysis with ImageJ software shows quantitative difference in the coating porosity П (Figure 3). DC PEO yields the highest porosity correlated to the highest roughness and coating thickness. Pulsed PEO of both types results in 1.5-fold lower porosity and more compact coatings [8]; the lowest value belongs to PUP mode due to the lowest coating thickness.

The results of X-ray diffractometry (Figure 4) show that the PEO coating contains crystalline ZrO_2 phase baddeleyite which is a monoclinic form of the oxide. The content of tetragonal ZrO_2 phase is quite small and does not exceed 2-3% as follows from its strongest peak at 30.2 degrees 2θ; therefore, in these experiments, the influence of this high-temperature oxide phase on the coating properties appears to be minor.

3.2. Analysis of corrosion properties for the PEO coatings obtained in different modes

Table 1 and Figure 5 show the electrochemical test results for the uncoated and PEO coated samples for different treatment modes. The OCP stabilizes quite fast – within 1 hour – and stays constant for all the treatment modes. According to the OCP values (Table 1), all the coated samples have nobler surface compared to the substrate due to passive oxide layer presence. The PUP mode provides the most difference with the uncoated sample.

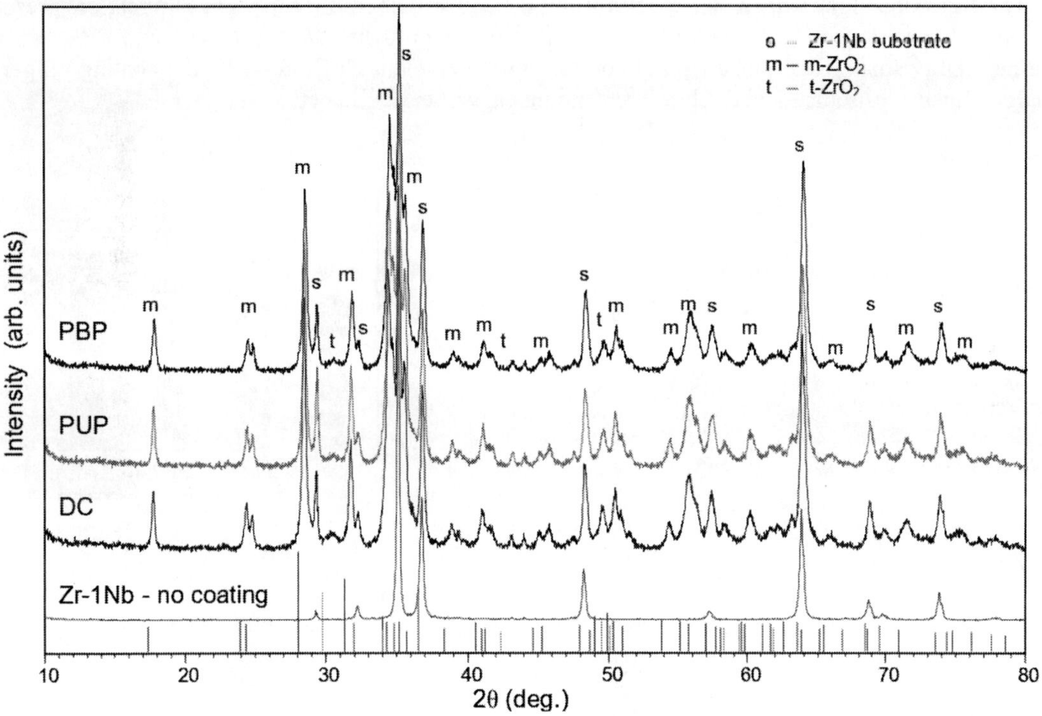

Figure 4. XRD patterns for the substrate and coatings obtained in different PEO treatment modes.

Table 1. Open circuit potential for Zr-1Nb alloy samples, uncoated and coated by PEO in different modes.

PEO treatment mode	$-E_{OCP}$ (V) vs. Ag/AgCl electrode
DC	0.355
PUP	0.189
PBP	0.379
no coating	0.482

The PDP test results are presented in Figure 5a. The uncoated sample has much higher values of the corrosion current. For the coated samples, at high polarizations, the PDP curves overlap, showing the similarity of the anodic and cathodic corrosion processes. The curve tips have different positions and depths, so E_{corr} and I_{corr} vary among the PEO treatment modes (Figure 5b, 5c). The PDP analysis shows that the corrosion current for all the coated samples is much lower than that for the substrate; this indicates an increase in the corrosion protection. As follows from E_{corr} values (Figure 5b), both DC and PUP modes provide surface passivation compared to the substrate, while PBP mode results in more active surface, probably, due to incorporation of the electrolyte cations K^+ and Na^+ into the coating during the negative pulse. Corrosion current I_{corr} appears to be in a good correlation with the coating porosity Π and coating thickness h: DC mode yields the highest values, while PUP results in the lowest values. Excessive porosity and cracks in thicker coating welcome corrosive media into the coating impurities, thus, resulting in less surface protection.

Figure 6 shows the results of the EIS tests for uncoated and coated samples. The impedance modulus at the lowest frequencies, uncovering corrosion processes, shows the same tendency as the corrosion current. The coated samples exhibit two orders of magnitude higher impedance modulus; this indicates corrosion resistance improvement. The PUP mode provides the highest impedance modulus at the lowest frequencies; this can be explained by thinner coating having denser structure.

The fit results of the impedance spectra using the equivalent circuit shown in Figure 1 are presented in Table 2. Resistance R_1 stands for the electrolyte of the corrosive media; compared to R_2, its values are negligibly small. Resistance R_2 and constant phase element CPE_1 describe the coating properties: charge transfer resistance and double layer impure capacitance, respectively.

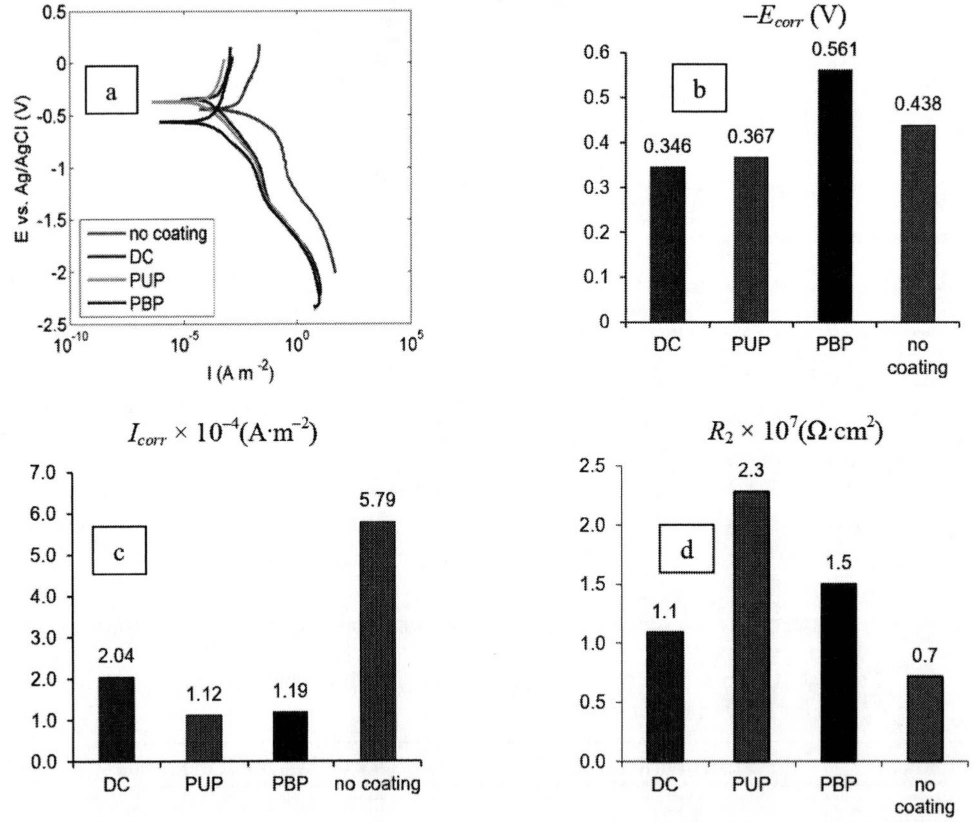

Figure 5. Electrochemical test results for Zr-1Nb alloy samples, uncoated and coated by PEO in different modes: PDP curves (a); corrosion potential E_{corr} (b); corrosion current density I_{corr} (c); charge transfer resistance R_2 (d).

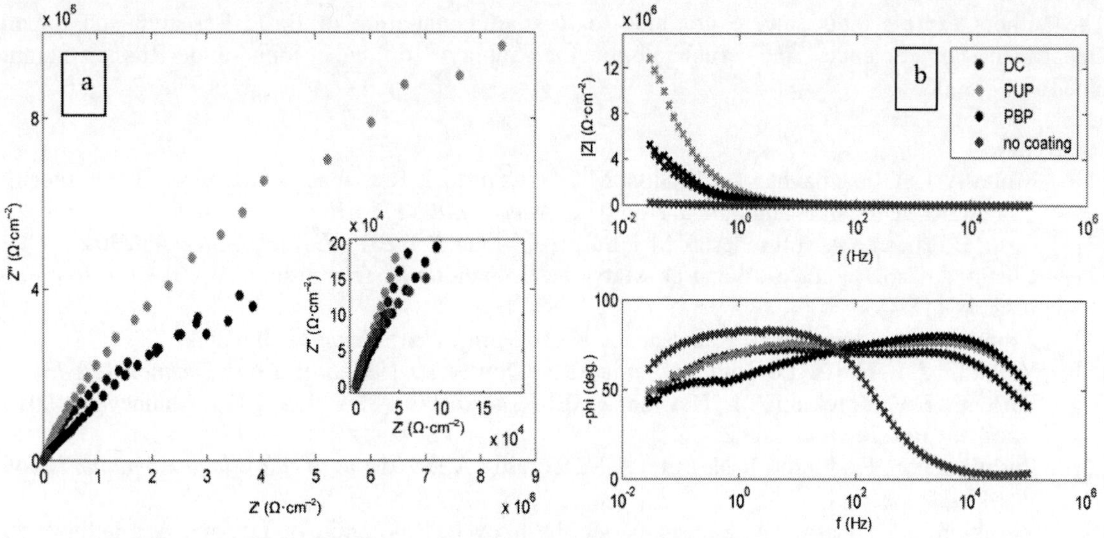

Figure 6. Complex (a) and Bode (b) plots of impedance spectra for Zr-1Nb alloy samples, uncoated and coated by PEO in different modes.

Table 2. Equivalent circuit elements describing EIS results for Zr-1Nb alloy samples, uncoated and coated by PEO in different modes.

PEO treatment mode	R_1 ($\Omega\cdot cm^2$)	R_2 ($\Omega\cdot cm^2$)	CPE$_1$	
			Q ($\mu F^{n-1}\cdot cm^{-2}$)	n
DC	120	$1.09\cdot10^7$	$6.00\cdot10^{-7}$	0.80
PUP	155	$2.29\cdot10^7$	$1.30\cdot10^{-7}$	0.86
PBP	160	$1.51\cdot10^7$	$4.39\cdot10^{-7}$	0.75
no coating	37	$7.15\cdot10^5$	$2.64\cdot10^{-5}$	0.92

As seen from Figure 5d and Table 2, the value of R_2 generally increases for less porous and thinner coating; it is the highest for PUP mode. Moreover, the difference between PUP and PBP in R_2 is stronger than in porosity Π. This suggests that the coating is not only less porous for PUP, but also more uniform as it can be seen in Figure 2. Also, this statement is supported by the CPE power n. Its value is closest to one – pure capacitor – for the uncoated substrate having ideal double layer with the highest "capacitance" Q among the tested samples. The next higher value of n belongs to the coating developed in PUP mode. Moreover, this coating has the lowest "capacitance" Q among the coated samples suggesting that the specific surface area is lower; therefore, the coating is more compact, and it can provide better corrosion protection.

4. Conclusions

In this study, three treatment modes of plasma electrolytic oxidation (DC, PUP, PBP) were compared regarding development of corrosion resistant coatings on Zr-1Nb reactor grade alloy in a silicate-phosphate electrolyte. Based on the SEM morphology, cross-section and porosity assessment, coating thickness and roughness measurements, X-ray diffractometry, potentiodynamic polarization, and electrochemical impedance spectroscopy, it was shown that pulsed unipolar PEO treatment promises the best advances in developing corrosion resistant coatings for this alloy.

Acknowledgements
The authors express their sincere gratitude to Russian Foundation of Basic Research and Indian Department of Science and Technology for support through joint Indo-Russian grant No. 16-53-48008.

References

[1] Motta A T, Yilmazbayhan A, da Silva M J G, Comstock R J, Was G S, Busby J T, Gartner E, Peng Q, Jeong Y H and Park J Y 2007 *J. Nucl. Mater.* **371** 61
[2] Ensor B, Lucente A M, Frederick M J, Sutliff J, Motta A T 2017 *J. Nucl. Mater.* **496** 301
[3] Cheng Y, Cao J, Peng Z, Wang Q, Matykina E, Skeldon P, Thompson G E 2014 *Electrochim. Acta* **116** 453
[4] Arun S, Arunnellaiappan T, Rameshbabu N 2016 *Surf. Coat. Technol.* **305** 264
[5] Yerokhin A L, Nie X, Leyland A, Matthews A, Dowey S J 1999 *Surf. Coat. Technol.* **122** 73
[6] Parfenov E V, Yerokhin A L, Nevyantseva R R, Gorbatkov M V, Liang C J, Matthews A 2015 *Surf. Coat. Technol.* **269** 2
[7] Pushilina N S, Kudiiarov V N, Lider A M, Teresov A D 2015 *J. of Alloys and Compounds* **645** s476
[8] Yerokhin A L, Shatrov A, Samsonov V, Shashkov P, Pilkington A, Leyland A, Matthews A 2005 *Surf. Coat. Technol.* **199** 150

IOP Conf. Series: Materials Science and Engineering **292** (2017) 012007 doi:10.1088/1757-899X/292/1/012007

Research on breakdown characteristics of oil-paper insulation in compound field at different temperatures

L Li[1,3], M Y Chen[2], X C Zhu[1], Z W Gao[1], H D Zhang[1], G X Li[1], J Zhang[1], C L Yu[1] and Y M Feng[2]

[1] State Grid Heilongjiang Electric Power Company Limited Electric Power Research Institute, Harbin, Heilongjiang, 150030, China
[2] State Grid Heilongjiang Electric Power Company Limited, Harbin, Heilongjiang, 150090, China

[3] E-mail: lilin77@yeah.net

Abstract. The breakdown characteristics of oil-paper insulation in AC, DC and compound field at different temperatures were studied. The breakdown mechanism of oil-paper insulation at different temperatures and in AC and DC electric fields was analyzed. The breakdown characteristic mechanisms of the oil-paper insulation in the compound field at different temperatures were obtained: the dielectric strength of oil-paper compound insulation is changed gradually from dependence on oil dielectric strength to dependence on paperboard dielectric strength at low temperature. The dielectric strength of oil-paper compound insulation is always related to the oil dielectric strength closely at high temperature with decrease of AC content.

1. Introduction

Oil-gap breakdown characteristic plays a decisive role in the oil-paper insulation structure in the AC electric field. Therefore, oil gap breakdown characteristics are mainly focused in most studies [1-5]. Since the oil-paper insulation withstanding electric field at compound voltage is different from that at pure AC voltage, transformer oil gap breakdown characteristics, oil-paper surface flashover characteristics and insulation paperboard breakdown characteristics should be taken into account comprehensively when oil-paper insulation breakdown and flash characteristics are studied [6-8]. The breakdown and surface flashover characteristics of oil-paper insulation under the action of compound electric field are studied mostly at room temperature. Temperature is an important factor affecting oil-paper insulation of transformer according to previous research on relationship between ordinary AC transformer oil-paper breakdown characteristics and temperature as well as the investigation and analysis on transformer operation accidents [9-14]. In recent years, several insulation faults appeared in China when transformers were put into operation at low temperature. The breakdown data about transformer oil-paper insulation at low temperature or extremely low temperature is also urgent for transformer operation departments. It is necessary to study the aspect since there is no test data of oil - paper insulation in compound electric field about breakdown and flashover characteristics at alpine region temperature at present. In the paper, the breakdown characteristics of oil-paper insulation AC, DC and compound field at different temperatures were studied. The breakdown mechanism of oil-paper insulation at different temperatures in AC and DC electric fields was analyzed. The breakdown characteristic mechanisms of the oil-paper insulation in the compound field at different temperatures were obtained.

2. Sample preparation and testing platform

2.1. Treatment of transformer oil

KIX45 transformer oil produced by Kunlun Company was selected in the experiment according to the temperature requirements of the project. Transformer oil underwent standardized treatment before testing; therefore it can meet various requirements regulated in national standards.

Oil filter was selected for treating transformer oil. The solubility of moisture and air in oil was reduced under vacuum and high temperature conditions, and the principle was utilized for dehydrating and degassing oil in the treatment equipment. The oil filter was used for filtering impurity particles with tiny particle size in oil. The transformer oil treated by the oil filter met the national standard GB 2535-90 Transformer Oil. The treatment process was shown in figure 1.

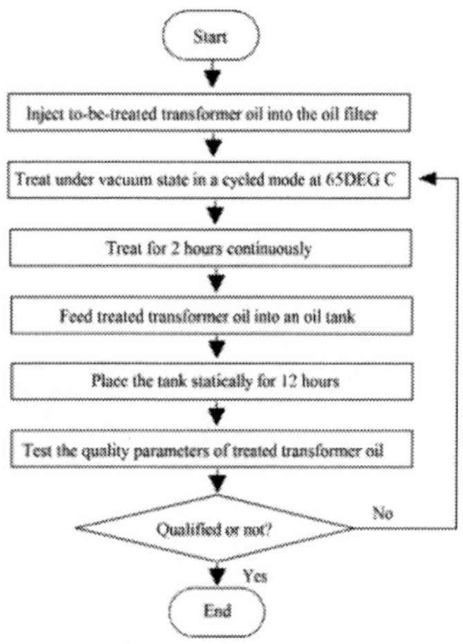

Figure 1. Treatment flow of transformer oil.

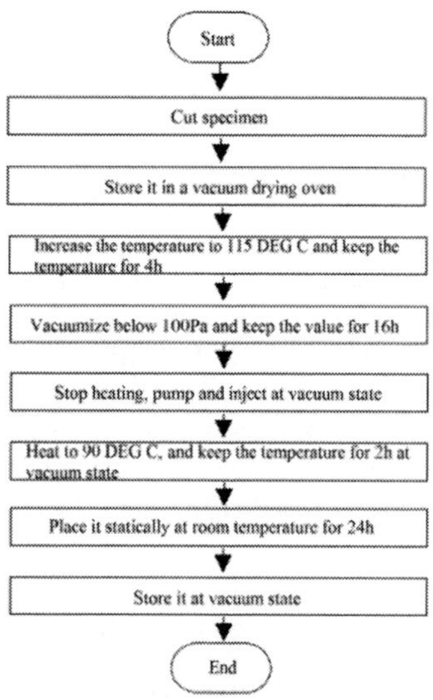

Figure 2. Treatment flow of oil impregnated paperboard.

The moisture content was measured firstly for the treated transformer oil, which was measured again at certain interval, thereby ensuring no change in the moisture content. The transformer oil was finally stored in a sealed oil tank.

2.2. Treatment of oil impregnated paperboard

The vacuum drying oven transformed in the laboratory was utilized for treating paperboard according to the national standard GB10580-89. The treatment process included vacuum drying and oil immersion. Related parameters of treated oil impregnated paperboard were tested. The operation flow was shown in figure 2.

The moisture content was measured firstly after treatment, which was measured again at certain interval, thereby ensuring no change in the moisture content. The treated paperboard was stored in transformer oil which was qualified in the test.

2.3. Testing platform

The main insulation structure of large exchange transformer is more complicated. In order to simplify the experimental model, the typical oil paper composite insulation structure is adopted. The simplified structure of the electrode structure is shown in figure 3. For the sake of preventing the surface discharge during the breakdown test, the sample size is larger than the discharge distance of the surface discharge.

According to the temperature requirements of this paper, the temperature of sample temperature is divided into two parts: cooling and heating. The heating test section is implemented by temperature sensor, thermostat, contactor and heater. The low temperature test box is used for the cooling test. The testing platform is shown in figure 4.

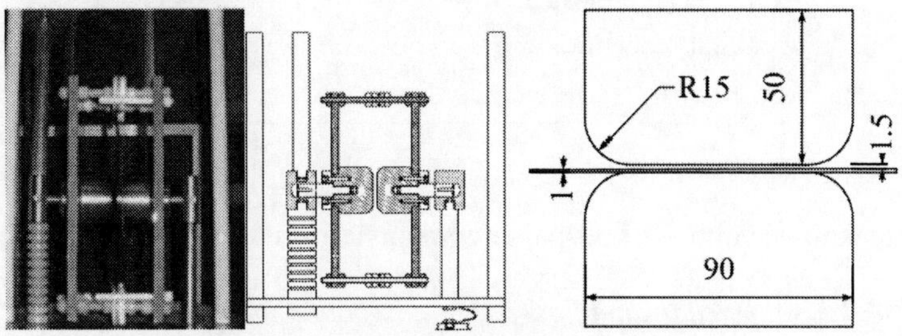

Figure 3. The electrode structure (mm).

Figure 4. The testing platform.

3. Breakdown characteristics of oil-paper compound insulation

3.1. Breakdown characteristics of oil-paper compound insulation at different temperatures

In the paper, AC/DC superposition equipment was used for studying the influence of temperature (-40 °C to 100 °C) on oil-paper compound insulation breakdown characteristics under AC content η=0% (pure DC), η=50% (DC superposed with AC) and η=100% (pure AC). Typical compound insulation structures composed of 1 mm oil-impregnated paperboard and 1.5mm transformer oil were adopted as specimens. The paperboard moisture content was between 0.45% and 0.6%. The moisture content was between 4.32ppm and 5.16ppm in transformer oil.

The average value of breakdown voltage data of oil-paper compound insulation under three voltage wave-forms was shown in figure 5: the breakdown voltage of oil-paper compound insulation showed a trend of decrease-increase-decrease with the temperature increase under the action of pure AC voltage and AC-DC superimposition voltage. Their trends were the same. Two extreme value points appeared at 0 °C and 60 °C or so. However, their change amplitude was not the same. The breakdown voltage of the oil-paper compound insulation showed a trend of increase-decrease under the action of pure DC voltage, and the maximum value appeared at 50 °C or so.

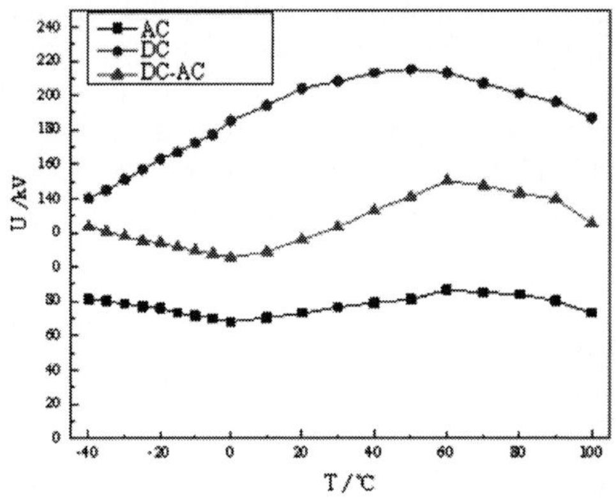

Figure 5. Breakdown voltage of oil-paper compound insulation at different temperatures.

3.2. Analysis on breakdown mechanism of oil-paper compound insulation under AC electric field

In the paper, corresponding theoretical values of field intensity in transformer oil and oil-impregnated paperboard was calculated according to the electric field distribution content in paper [15], wherein oil-paper compound insulation was broken down in pure AC electric field. The result was compared with the breakdown field intensity actual value of transformer oil and oil-impregnated paperboard at different temperatures to analyze the breakdown process of oil-paper compound insulation in the AC electric field quantitatively. The comparison was shown in figure 6 and figure 7.

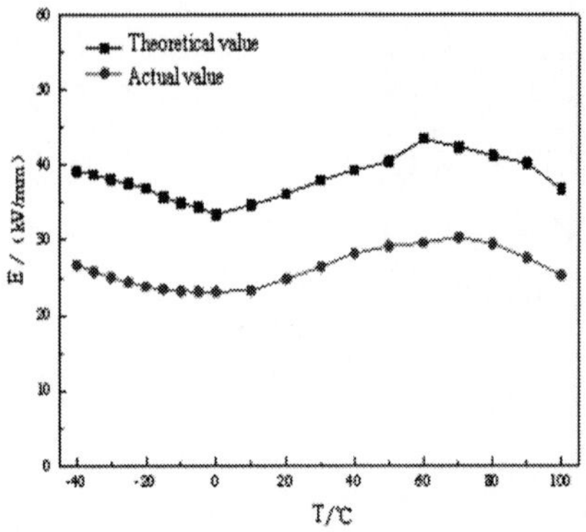

Figure 6. Field intensity in oil under different conditions.

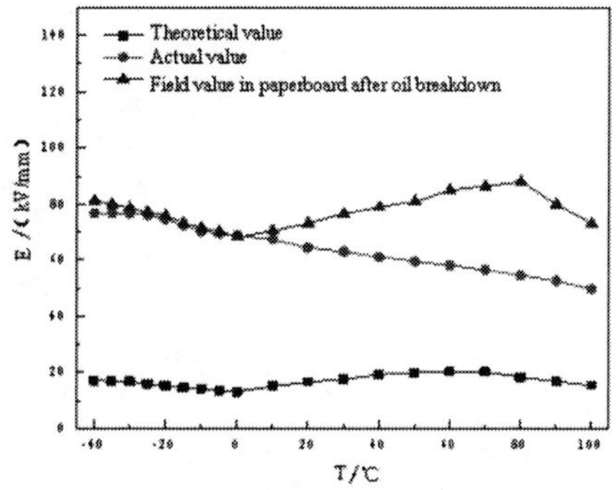

Figure 7. Field strength in oil and paperboard under different conditions.

Figure 6 showed the breakdown field intensity of pure oil gaps at different temperatures under the action of AC voltage and theoretical values of corresponding electric field distribution in oil gaps during breakdown of oil-paper compound insulation. Figure 7 showed the breakdown field intensity of oil-impregnated paperboard at different temperatures under AC voltage, theoretical values of corresponding electric field distribution in the oil-impregnated paperboard during breakdown of oil-paper compound insulation and the field intensity in oil-impregnated paperboard after breakdown of transformer oil.

Data comparison showed that transformer oil withstood most field intensity in oil-paper compound insulation under the action of pure AC voltage. The field intensity withstood by the paperboard was far lower than the breakdown field intensity thereof. Therefore, the breakdown process in AC can be concluded as follows: discharge breakdown occurred in the transformer oil firstly, and the field strength falling on the paperboard was higher than or close to the actual value of breakdown field strength in AC electric field after single medium breakdown in the oil, thereby leading to overall breakdown of compound insulation.

Figure 7 showed that the pure oil gap breakdown field intensity in AC voltage at different temperatures was much lower than the electric field intensity in oil-paper compound insulation oil gap possibly because AC breakdown field intensity was the result, which was obtained from breakdown voltage conversion of 2.5 mm standard oil gap. However, the oil gap thickness was 1.5mm in compound insulation. The field strength in the oil gap was higher due to influence of volume effect, wherein the strength was converted from oil-paper compound insulation breakdown voltage.

3.3. Analysis on breakdown mechanism of oil-paper compound insulation under AC electric field

In the paper, corresponding theoretical values of field intensity in transformer oil and oil-impregnated paperboard was calculated according to the electric field distribution content in paper [15], wherein the oil-paper compound insulation was broken down under pure DC electric field. The result was compared with the breakdown field intensity actual value of transformer oil and oil-impregnated paperboard at different temperatures to analyze the breakdown process of oil-paper compound insulation in DC electric field quantitatively. The comparison was shown in figure 8 and figure 9.

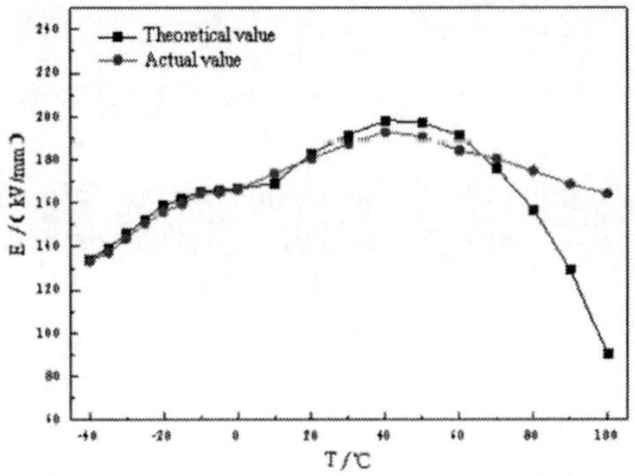

Figure 8. Field intensity in paperboard under different conditions.

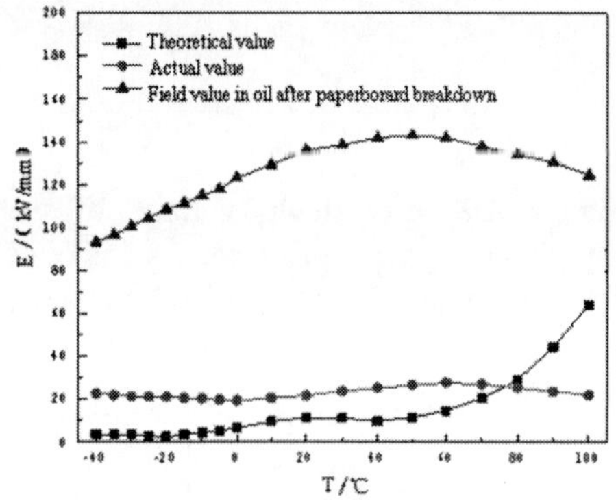

Figure 9. Field strength in oil and paperboard under different conditions.

Figure 8 showed breakdown field intensity of oil-impregnated paperboard at different temperatures under DC voltage and theoretical values of corresponding electric field distribution in oil-impregnated paperboard during breakdown of oil-paper compound insulation. Figure 9 shows the breakdown field intensity of pure oil gap at different temperatures under the action of DC voltage, theoretical values of corresponding electric field distribution in oil gap during breakdown of oil-paper compound insulation and the field intensity in oil gap after breakdown of oil-impregnated paperboard.

Data comparison showed that: paperboard withstood most field intensity between -40℃ and 60℃ under the action of pure DC voltage. The field intensity in the transformer oil did not reach the actual value of the breakdown field intensity. Therefore, it can be concluded that the paperboard suffered from breakdown firstly, and all voltage was applied on the oil after breakdown, thereby leading to overall breakdown of compound insulation. The field intensity in oil between 60℃ and 100℃ was higher than the actual breakdown voltage, and oil suffered from breakdown firstly, thereby leading to overall breakdown of compound insulation.

Figure 9 showed that the theoretical calculation field strength was greatly different from actual breakdown field intensity since the resistivity change law of the paperboard and oil was changed greatly in high field intensity. The difference magnitude order was significant between paperboard resistivity and oil resistivity at the low temperature interval, which had low influence on the calculation results. The resistivity difference was insignificant at high temperature, which had high influence on the calculation results, thereby leading to significant deviation in the calculation results.

3.4. Analysis on breakdown mechanism of oil-paper compound insulation under DC-AC superposition electric field

The AC content of voltage withstood by oil and paperboard was also always changed with temperature change, and it was difficult to analyze the AC content quantitatively in the oil-paper compound insulation structure under the action of DC-AC superposition voltage (η=50%). The change trend of breakdown voltage of oil, paperboard and oil-paper compound insulation structure with temperature was shown in figure 10, figure 11 and figure 12 respectively under different AC contents.

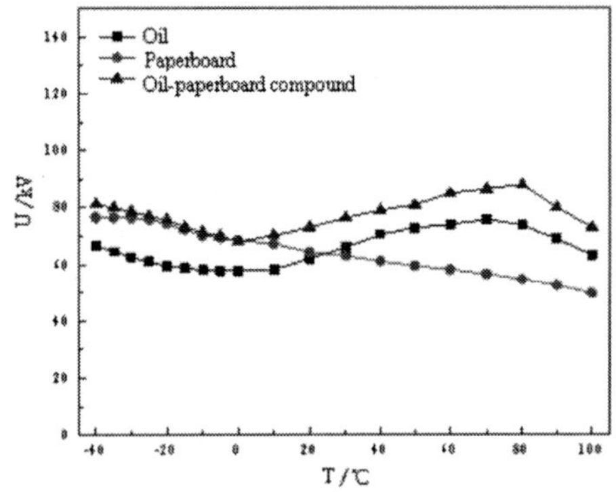

Figure 10. Breakdown voltage of oil, paperboard and oil-paper compound insulation under the action of AC voltage.

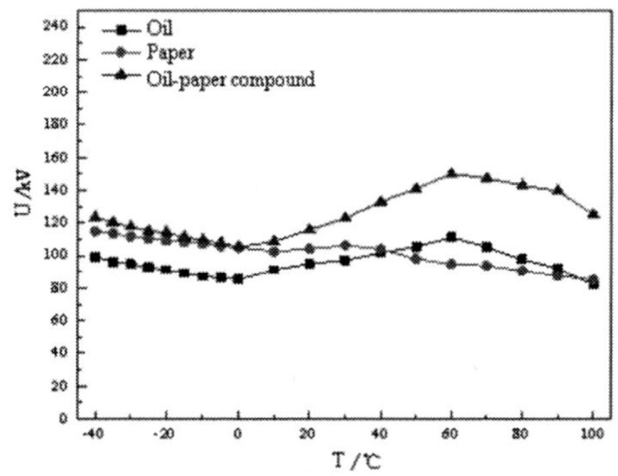

Figure 11. Breakdown voltage of oil, paperboard and oil-paper compound insulation under the action of DC-AC superposition voltage.

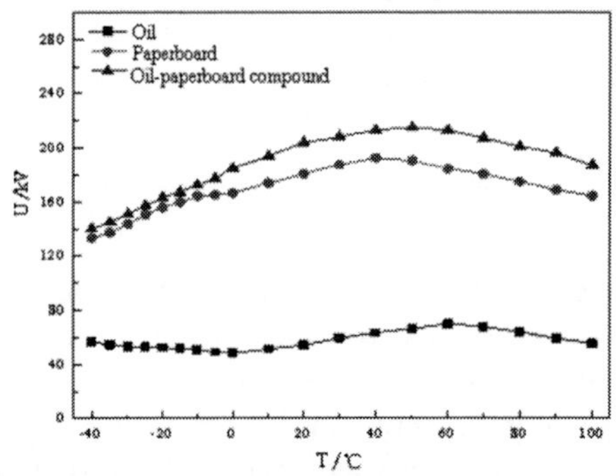

Figure 12. Breakdown voltage of oil, paperboard and oil-paper compound insulation under the action of DC voltage.

It was concluded in the above context that oil gaps were broken down firstly in oil-paper compound insulation under the action of AC voltage, and the compound insulation breakdown was consistent with oil change trend as a result. However, the paperboard was broken down firstly at low temperature, and oil was firstly broken down at high temperature in oil-paper compound insulation under the action of DC voltage. The breakdown law of the compound insulation was consistent with the paperboard at low temperature as a result, and the law was tended to oil at high-temperature. Figure 11 showed that the breakdown law of oil-paper compound insulation was similar with that under the action of pure AC voltage when η =50%, and they were consistent with the breakdown trend of oil. It can be concluded according to the law that the breakdown of oil-paper compound insulation gradually tended to breakdown law of paperboard at low temperature with the decrease of AC content. However, the breakdown of oil-paper compound insulation was always consistent with the breakdown law of oil at high temperature.

4. Conclusion
The dielectric strength of oil-paper compound insulation is changed gradually from dependence on oil dielectric strength to dependence on paperboard dielectric strength at low temperature. The dielectric strength of oil-paper compound insulation is always related to the oil dielectric strength closely at high temperature with decrease of AC content.

References
[1] Watanbe T, Ishii T, Yamada N, Nitta T, Tamura R and Miura Y 1980 *IEEE Trans. On Power Apparatus and Systems* **99** 335
[2] Takahashi E, Tsutsumi Y, Okuyama K and Ogata F 1976 *IEEE Transactions on Power Apparatus and Systems* **95** 411
[3] Zouaghi A, Beroual A 1998 *IEEE International Symposium on Electrical Insulation* p 640
[4] Zhou Y X, Sha Y C, Jiang X X, Sun Q H, Wang Y S and Tian J H 2011 *IEEE 2011 Annual Report Conference on Electrical Insulation and Dielectric Phenomena* p 718
[5] Ohshima I, Motegi S, Honda M, Yanari T and Ebisawa Y 1983 *IEEE Transactions on Power Apparatus and System* **3** 44
[6] Pahlavanpour, Martins M and Eklund 2003 *Proceedings of the 7th International Conference on Properties and Applications of Dielectric Materials* vol 3 p 1124
[7] Gafvert U, Jaksts A, Tornkvist C and Walfridsson L 1992 *IEEE Trans. on Electrical Insulation* **27** 646
[8] Hasegawa T, Yamaji K 1997 *IEEE Transactions on Power Delivery* **12** 1526
[9] Wang Y H, Wei X L and Chen Q G 2011 *High Voltage Engineering* **37** 2205
[10] Li B, Cheng H C, Sun Q, Li J Z, Li G F, Deng J Y, Fan W and Xu H Y 2011 *Transactions of China Electrotechnical Society* **26** 36
[11] Guan L, Li B, Cheng H C, Li J Z and Li G F 2012 *IEEE Annual Report Conference on Electrical Insulation and Dielectric Phenomena* p 403-406
[12] Zhou Y X, Sun Q H, Li G F, Wang Y S, Jiang X X, Chou Y Z, Li J Z and Wang N H 2011 *Transactions of China Electrotechnical Society* **26** 27
[13] Zhou Y X, Sha Y C, Jiang X X, Sun Q H, Wang Y S and Tian J H 2011 *IEEE 2011 Annual Report Conference on Electrical Insulation and Dielectric Phenomena* p 718
[14] Raghuveer M R, Kolaczkowski Z, Jin Weifang 1990 *IEEE Transactions on Electrical Insulation* **25** 341
[15] Chen Q G, Chi M H, Gao Y, Wei X L, Wang Y H, Zhang S 2013 *Proceedings of the Chinese society for electrical engineering* **33** 170

IOP Conf. Series: Materials Science and Engineering **292** (2017) 012008 doi:10.1088/1757-899X/292/1/012008

Thermal degradation and plasticizing mechanism of poly(vinyl chloride) plasticized with a novel cardanol derived plasticizer

J Chen[1], X A Nie[1, 2, 3], J C Jiang[1, 2, 3] and Y H Zhou[1, 2]

[1] Institute of Chemical Industry of Forestry Products, Chinese Academy of Forestry; National Engineering Laboratory for Biomass Chemical Utilization;Key Laboratory of Biomass Energy and Material, Nanjing, Jiangsu 210042, China
[2] Institute of New Technology of Forestry, Chinese Academy of Forestry, Beijing 100091, China

[3] E-mail: niexiaoan@126.com

Abstract. A natural plasticizer cardanol derivatives glycidyl ether (CGE) was synthesized and employed as a plasticizer for the poly(vinyl chloride).The effect of CGE on thermal degradation of PVC films and its plasticizing mechanism were firstlyreported.The molecular structure of CGE was characterized with Fourier transform infrared spectroscopy (FTIR). Thermal properties, degradation properties and compatibility of the PVC films were investigated byDifferential scanning calorimeter analysis (DSC), Thermogravimetric analysis (TGA) and FTIR, respectively. Compared with the commercial plasticizers dioctylphthalate (DOP), CGE can endow PVC film with a decrease of 4.31 °C in glass transition temperature (Tg), an increase of 24.01 °C and 25.53 °C in10% weight loss(T_{10}) and 50% weight loss (T_{50}) respectively, and a higher activetion energy of thermal degradation (E_a).

1. Introduction

Plasticizers have become the most consumable additive in PVC which can improve the mechanical properties, processability, and thermal stability of the poly(vinyl chloride) (PVC). Since their potentially toxic effects on human health and environmental impact [1-5], the use of phthalates is incontrovertibly becoming more limited. Hence, there is a trend to replace phthalates by alternative plasticizers [6-8]. Currently, there is growing interest in the research and use of natural-based plasticizers which are characterized by low toxicity [8-11].They mainly include modified and epoxidized triglyceride oils from soybean oil, linseed oil, sunflower oil, glycerol monoester, and fatty acid esters [12-14]. As the plastics industry and environmental awareness continues to grow, there is an urgent and unmet need to develop new natural plasticizers with improved properties and cost competitiveness.

Cardanolis an extracted product from cashew nut shell liquid, which is one of the most commonly used renewable agricultural resource materials [15]. Due in part to its versatile chemical structure and low cost, cardanol can be considered to be a very promising raw-material to develop new derived materials to be used in environmentally friendly processes [16,17]. Cardanol and its derivatives usually have major applications insynthetic resins [18, 19] bio-composites [20], epoxy curing agents [21] and coatings [22-24]. Recently, cardanol with or without modification has been used as a plasticizer in the polymer and rubber industries, showing significant plasticizing effects [25, 26].

Furthermore, cardanol-based plasticizers have been reported to be a kind of efficient plasticizer for PVC [17, 27-29].

The present work provided a novel method to prepare cardanolglycidyl ether (CGE), and firstly reportedthe effect of CGE on thermal degradation of PVC films and its plasticizing mechanism.The thermal stability, degradation properties and compatibility of the plasticized PVC films were evaluated by the DSC, TGA and FTIR. The plasticizing effect of CGE was comparable to that of commercial phthalate ester plasticizer DOP. The objective of this work is to demonstrate that the heat resistant plasticizer derived from renewable resource cardanol can offer excellent thermal stability and compatibility for PVC and provide new idea for development of special bio-plasticizer.

2. Experimental

2.1. Materials

Cardanol (stabilized, 88.5%) was purchased from Shanghai Meidong Biological Material Co., Ltd., China. The refractive index (n_D^{25}) of Cardanol is 1.509, viscosity is 46 mPa·s (25°C), density is 0.928 g/ cm^3 (25°C).Cardanol was used after distillation. Epichlorohydrin, sodium hydroxide (98.7%), calcium oxide (98%), benzyltriethylammonium chloride (97%) were purchased from Sinopharm Chemical Reagent Co., Ltd., China and used as received. PVC (DG-1000K) was purchased from the Tianjin Dagu Chemical Co., Ltd, China. Calcium stearate and zinc stearate were supplied by Changzhou JiaRenWo Chemical Co., Ltd, China. Di (2-ethylhexyl) phthalate (DOTP) (97%) and Diisononyl phthalate (DINP) (99%) were obtained from Aladdin Chemical Reagent Co., Ltd., China, and were used as received.

2.2. Preparation of CGE

300 g Cardanol was charged into a flask equipped with a fractional distillation device. A pale yellow distillate was obtained within 240.0–260.0 °C under a high vacuum of 8 Torr. 100.0 g (0.332 mol) of cardanol, 245.0 g (2.64 mol) of epichlorohydrin and 1.8 g (0.0097 mol) benzyltriethyl ammonium chloride were introduced into a 250 ml flask equipped with mechanical stirrer, reflux condenser and thermometer. Then the mixture was slowly heated to 117 °C and kept reacting for 2.0 h. After the mixture was cooled to 60 °C, 13.3g (0.332 mol) sodium hydroxide was charged, and the reaction was continued at 60 °C for 3h. The excess epichlorophydrin was distilled to be recycled and the solids were filtered. 108.6g of a yellowish low-viscosity compound was obtained (yield: 91.01 % relative to Cardanol).

2.3. Preparation of plasticized PVC test specimens

A series of plasticized PVC specimens with different plasticizers at different concentrations were prepared. The compositions are shown in table 1. First, PVC powder (100 g), plasticizers (25g), and thermal stabilizers(2.0g, Ca soap/Zn soap = 3/1) were mixed using a mechanical mixer at room temperature (RT) for 5 min. Second, the mixture was compounded into a homogeneous mixture at 160 °C for 5 min by aPlasti-Corder Lab-Station (South Hackensack, NJ). Finally, the films were made using a Carver 3969 mini hot press (Wabash, IN) at 170 °C for 5 min.

Table 1. The composition of PVC blends.

Component	CGE/ PVC	DOP/ PVC
PVC (g)	100.00	100.00
CGE (g)	25.00	0.00
DOP (g)	0.00	25.00
Thermal stabilizer (g)	2.00	2.00

2.4. Characterizations

Fourier transform infrared (FTIR) analysis was conducted using a Nicolet IS10 spectrometer (Thermo Fisher Scientific Inc., USA) by an attenuated total reflectance method. The samples were scanned from 4000 to 500 cm^{-1}.

Differential scanning calorimetry (DSC) analysis was performed on a Diamond instrument (Perkin-Elmer Co., USA). Each sample was scanned from -60 °C to 80 °C under a nitrogen atmosphere at a heating rate of 10 °C/min.

Tensile properties were measured using a SANS CMT-4303 universal testing machine (Shenzhen XinsansiJiliang Instrument Co., China) according to ISO 527-2: 1993. The cross-head speed was set at 10 mm/min. All samples were conditioned at 23 °C for 1 day prior to tensile testing. Five sample pieces were prepared for each group to obtain an average value.

Thermogravimetric analysis (TGA) was carried out in a thermogravimetric analyzer (NetzschCo., Germany). Each sample was scanned from ambient temperature to 400 or 600°C at a heating rate of 5, 10, 15, 20, and 25 °C/min, respectively. The TGA experiment was performed under a 60 mL/min nitrogen flow. The kinetic parameters such as activation energy (E_a, KJ/mol) and pre-exponential factor (A) were evaluated by the Kissinger equation:

$$\ln\left(\frac{\beta}{T_{\max}^{2}}\right)=\ln\frac{AR}{E_a}-\frac{E_a}{R}\frac{1}{T_{\max}} \quad (1)$$

where R = Gas constant (8.314 J/mol•K). Equation (1) indicates that ln (β/T_{max}^2) and $1/T_{max}$ are in a linear correlation. E and A can be calculated from the slope (a) and intercept (d) on the curves of ln (β/T_{max}^2) change versus $1/T_{max}$.

$$E_a=-a\bullet R \quad (2)$$

$$\ln(\mathrm{A})=d+\ln(\frac{E_a}{R}) \quad (3)$$

3. Results and discussion

3.1. Charactization

Figure 1 shows the FTIR spectra of cardanol and CGE. Compared with CGE, the FT-IR spectra of cardanol showed abroad band at around 3332cm^{-1}correspoding to phenolic hydroxyl group.The unsaturated alkyl chains at 3009 cm^{-1}appeared in the FT-IR of cardanol. The broad band phenolic hydroxyl group disappeared in the spectra of CGE, and ether group were found at 772, 849 and 909cm^{-1}, which indicated that the epoxidation reaction occurred. Furthermore, new peaks appear at 1264 and 1030 m^{-1} due to the presence of cyclic C—O—C asymmetric and symmetric stretching vibrations, respectively. All of theses data may suggest that the cardanol had reacted and formed ether group.

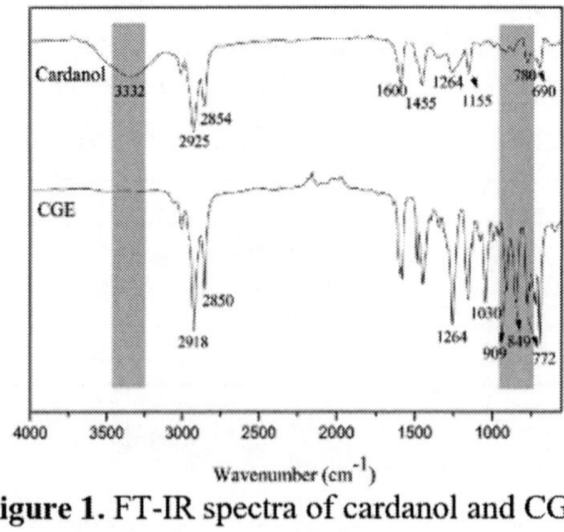

Figure 1. FT-IR spectra of cardanol and CGE.

3.2. DSC

Figure 2 compares the DSC thermograms of PVC films with and without plasticizers, the traces demonstrate a characteristic change in themalpropertie assigned to a glass transition (T_g). The pure PVC film had a T_g temperature of approximately 85.63°C (table 2) whilst the DOP/ PVC and CGE/ PVC films have significantly lower T_g of 30.87 °C and 26.56 °C respectively. It can be seen that CGE/ PVC had a maximum decrease of 4.31 °C in T_g compared with that of DOP/ PVC, indicating a more efficient plastization system[17].

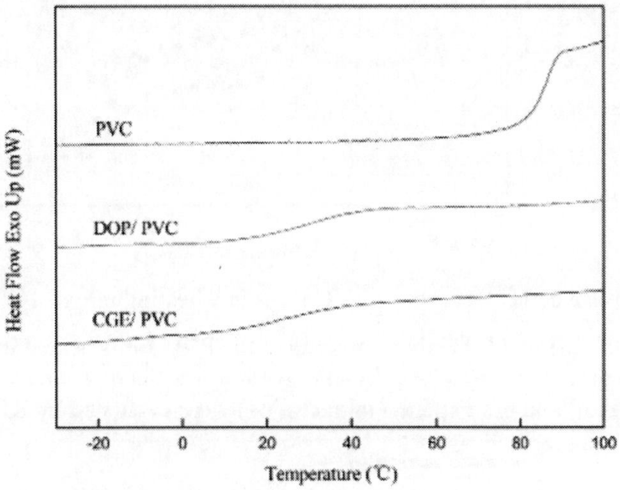

Figure 2. DSC curves of the PVC films.

Table 2. Glass transition temperature (T_g), 10% and 50% weight loss temperatures (T_{10} and T_{50}) for different PVC samples.

Sample	T_g(℃)	T_{10} ^a (℃)	T_{50} ^a (℃)	Carbon residue (%)
CGE/ PVC	26.56	280.39	326.69	15.54
DOP/ PVC	30.87	256.38	301.16	11.69

^aValues of 10 ℃/min. T_i: onset temperature of TGA.

3.3. Thermal stability and degradation kinetics

Figure 3 shows the TGA curves of the films with different plasticizers heated in nitrogen at the rate of 10 °C/min. It can be observed that both of the samples are thermally stable in nitrogen gas below 200 °C and mainly displayed three-stage thermal degradation above this temperature. The first stage degradation (about 230-350 °C) is the fastest and corresponded to dechlorination of PVC, with formation and stoichiometric elimination of HCl and a few chlorinated hydrocarbons. [30] The formation of aromatic compounds by the cyclization of conjugated polyene[31, 32] mainly occurred in the second stage (about 350-420 °C). In the last stage (> 420 °C), the second degradation was shown, which is mainly contributed to the degradation of the complex structures resulting from aromatization.[33] Furthermore, the TGA characteristic temperatures of PVC films are illustrated in Table 2, including the 10% and 50% weight loss temperatures (T_{10} and T_{50}), and carbon residue for the degradation. It clearly can be seen that CGE improves the thermal stability of PVC films. Compared with DOP/ PVC, the CGE/ PVC film had an increase of 24.01 °C and 25.53 °C in T_{10} and T_{50} respectively. As well as the carbon residue increased from 11.69% to 15.54%. This may be mainly due to the epoxy groups of CGE which can scavenge for HCl and delay the degradation events[34-36].

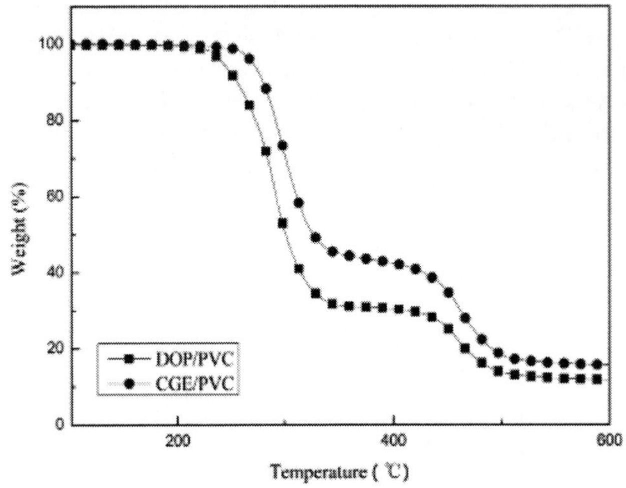

Figure 3. TGA curves of PVC films at a heating rate of 10 °C/min.

Figure 4 show the TGA and DTG thermograms of the PVC films heated in nitrogen at heating rates at 5, 10, 15, 20 and 25 °C/min. The kinetic parameters such as activation energy of thermal degradation (E_a, KJ/mol) and pre-exponential factor (A) were evaluated by Kissinger equation. [37]

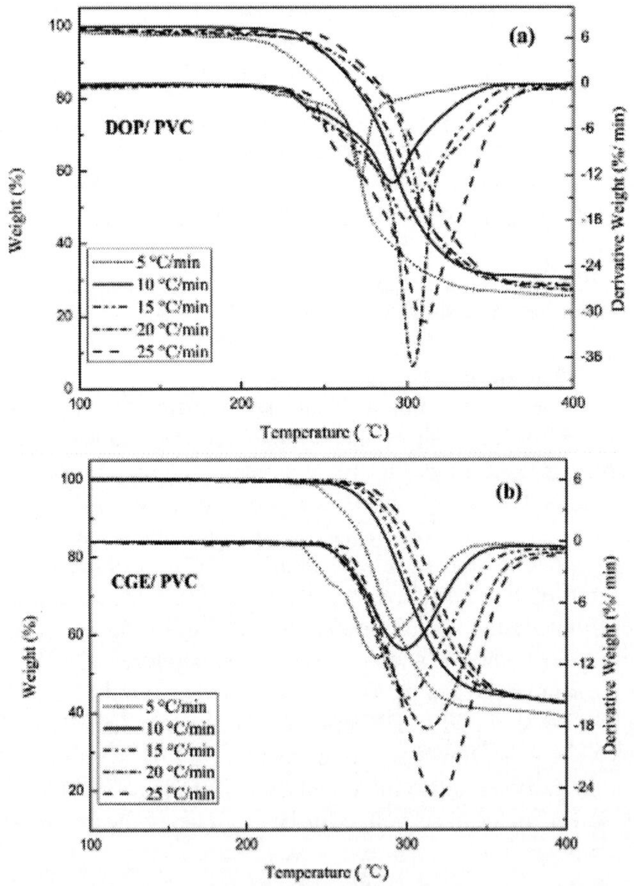

Figure 4. TGA and DTG curves at different heating rates of the (a) DOP/ PVC and (b) CGE/ PVC films.

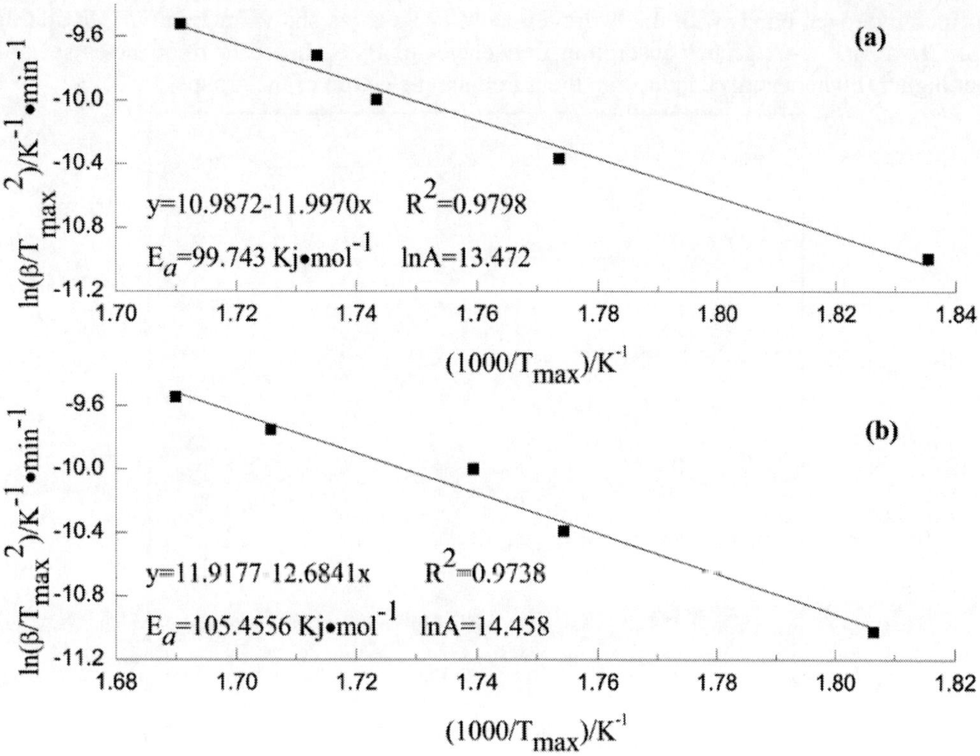

Figure 5. Plots of $\ln(\beta/T_{max}^2)$ versus $1000/T_{max}$and kinetic parameters for the PVC films(a) DOP/ PVC and (b) CGE/ PVC.

The kinetic parameters for the two plasticizers are summarized in figure 5. Compared with the DOP/ PVC, the CGE/ PVC had higher thermal degradation temperature. It was found that the E_a of GCE/ PVC(105.4556 $Kj{\cdot}mol^{-1}$) was higher than that of DOP/ PVC(99.7430 $Kj{\cdot}mol^{-1}$). Generally, a higher E_a suggests a higher thermal degradation resistance. The increase in the E_a could be explained by the enhancement of the intermolecular distance and chain mobility [38]. Hence, CGE could increase the thermal stability of the PVC matrix effectively.

3.4. FT-IR

Figure 6 shows the FTIR spectra of pure PVC film, pure plasticizers, DOP/ PVC andCGE/ PVC films in the wavenumber range 2000 to 600 cm^{-1}. FTIR would be sensitive in situation where complexation has occurred in amorphous phase.[39] In the PVC films there are some characteristic peaks: CH_2-Cl angular deformation at 1426 cm^{-1}, CH-Cl out of plane angular deformation at 1254 cm^{-1}, C-H out of plane trans deformation at 960 cm^{-1}, and C-Cl bond stretching at 831, 691, and 615 cm^{-1}. These characteristic bonds are in accordance with data found in the literature for PVC films.[36]After the incorporation of plasticizers into the formulations, the 1724 cm^{-1} band of C=O groups in DOP shifted about 4-5 cm^{-1} toward the lower frequency and its width increases. Furthermore, the glycidylether group in CGE shifted from 910, 859, 847 cm^{-1} to 908, 858, 834 cm^{-1}. This phenomenon can be mainly explained by equation (4) derived from Hooke's Law:

$$\nu = \frac{1}{2\pi}\sqrt{\frac{k}{m}} \tag{4}$$

Where ν is the vibration frequency, k is the force constant, and m is the mass. Eq. (4) shows the relationship of force constant to the wavenumber (vibration frequency). As the wavenumber decreases, the force constant decreased, indicating the electronic was given by carbonyl group or epoxide groups, and formed hydrogen bonds with the hydrogen atom in PVC (as shown in figure 5). In addition, the shift of CH-Cl and C-Cl groups absorption frequencies of PVC shifted to frequencies several wave number higher (higher energy), indicating the introducing electron of the groups.

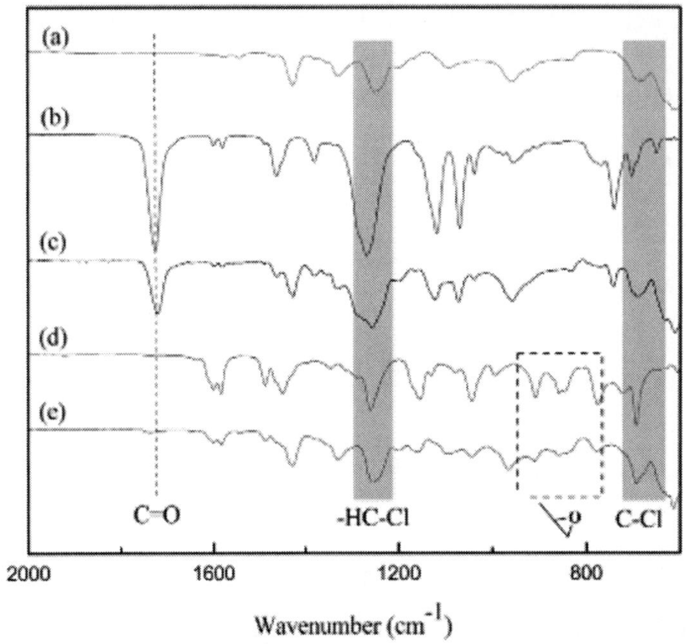

Figure 6. FTIR spectra of PVC films with different plasticizers: (a) pure PVC, (b) DOP plasticizer, (c) DOP/ PVC, (d) CGE plasticizer, and (e) CGE/ PVC.

This result is consistent with the proposed plasticization mechanism shown in figure 7. Besides that, it can be observed that the bands shift of CGE/ PVC was clearly than that of DOP/ PVC, showing stronger interaction between the polar parts of CGE (glycidylether group) and the PVC (carbon-chloride bond), the results has showed that the PVC plastization is the process of formatting the hydrogen-bonds between the carbonyl of the plasticizer and the hydrogen of PVC and weaker intermolecular bonding force of PVC-PVC.

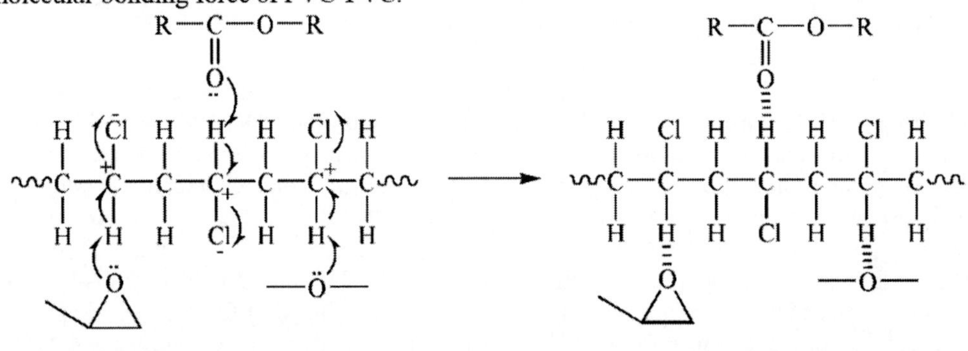

Figure 7. Possible interaction between plasticizers and PVC molecular in the plasticized system.

4. Conclusions

In summary, the CGE derived from cardanol has been evidenced as a high- performance plasticizer for improving thermal stability and degradation properties of PVC resin. Furthermore, DSC and FTIR analysis suggested that CGE has higher plasticizing effect and good compatibility on PVC compared with the commmercial plasticizer DOP. The results indicate that CGE could be a new bio-based plasticizer which can endow the PVC with higher thermal properties, lower cost and more eco-friendly.

Acknowledgments
This research was supported bythe Key laboratory of biomass energy and materials of Jiangsu province (grant number: JSBEM-S-201604).

References

[1] Sampson J and Korte D 2011 *Transfusion. Med.* **21** 73
[2] Lithner D, Larsson Å and Dave G 2011 *Sci. Total. Environ.* **409** 3309
[3] Hines CJ, Hopf NB, Deddens JA, Silva MJ and Calafat AM 2012*Int. arch. Occ. Env. Hea.* **85** 317
[4] Chiellini F, Ferri M, Morelli A, Dipaola L and Latini G 2013 *Prog. Polym. Sci.* **38** 1067
[5] Chiellini F, Ferri M and Latini G 2011 *Int. J. Pharm.* **409** 7
[6] Wu D, Chang PR and Ma X 2011 *Carbohydr. Polym.* **86** 877
[7] Li M, Yuan J, Pan M, Song H and Tang L 2012 *Polym. Mat. Sci. Eng.* **1** 4
[8] Li Z, Liu N, Yao Y, Chen S, Wang H and Wang H 2015 *RSC Adv.***5** 901
[9] Rahman M, Brazel C S 2004 *Prog. Polym. Sci.***29** 1223
[10] Panda B, Parihar A S, Mallick S 2014 *Int. J. Biol. Macromol.***67** 295
[11] Niazi M B K, Zijlstra M, Broekhuis A A 2013 *Carbohyd. Polym.***97** 571
[12] Garg B, Bisht T and Ling YC 2014 *RSC Adv.* **4** 57297
[13] Yang D, Peng X, Zhong L, Cao X, Chen W and Zhang X 2014 *Carbohydr. Polym* **103** 198
[14] Fenollar O, García D, Sánchez L, López J and Balart R 2009 *Eur. Polym.J.* **45** 2674
[15] Rao B and Palanisamy A 2013 *Eur. Polym. J.* **49** 2365
[16] Voirin C, Caillol S, Sadavarte NV, Tawade BV, BoutevinB and Wadgaonkar PP 2014 *Polym. Chem.* **5** 3142
[17] Greco A, Brunetti D, Renna G, Mele G and Maffezzoli A 2010 *Polym. Degrad. Stab.***95** 2169
[18] Jaillet F, Darroman E, Ratsimihety A and Auvergne R 2014 *Eur. J. Lipid. Sci. Technol.***116** 63
[19] Yadav R and Srivastava D 2009 *Eur. Polym. J.* **45** 946
[20] Barreto A, Rosa D, Fechine P and Mazzetto S 2011 *Compos. Part A: Appl. Sci. Manuf.* **42** 492
[21] Sultania M, RaiJ and Srivastava D 2012 *Mater. Chem. Phys.* **132** 180
[22] Kathalewar M and Sabnis A J 2014 *Coat. Technol. Res.* **1** 601
[23] Kanehashi S, Yokoyama K, Masuda R, Kidesaki T, Nagai K and Miyakoshi T 2013 J.*Appl. Polym. Sci.***130** 2468
[24] Kathalewar M, Sabnis A and DMello D 2014 *Eur. Polym. J.* **57** 99
[25] Mohapatra S and Nando GB 2013 *Ind. Eng. Chem. Res.* **52** 5951
[26] Alexander M and Thachil ET 2006 *J. Appl. Polym. Sci.* **102** 4835
[27] Calò E, Greco A and Maffezzoli A 2011 *Polym. Degrad. Stab.* **96** 784
[28] Semsarzadeh M, Mehrabzadeh M and Arabshahi S 2002 *Eur. Polym. J.* **38** 351
[29] Yao Q and Wilkie CA 2001 *J. Vinyl Addit. Techn.* **7** 26
[30] Soudais Y, Moga L, Blazek J and Lemort F J 2007 *Anal. Appl. Pyrolysis.* **78** 46
[31] Pitchaimari G, Sarma K, Varshney L and VijayakumarC 2014 *Thermochim. Acta.* **597** 8
[32] Grossman RF 1993 *J. Vinyl Techn.* **15** 25
[33] Lerke I and Szymański W 1977 *J. Appl. Polym. Sci.* **21** 2067
[34] Ayrey G, Head B and Poller R 1974 *J. Polym. Sci. Macro. Rev.***8** 1
[35] Li M, Zhang J, Huang K, Li S, Jiang J and Xia J 2014 *RSC Adv.* **4** 63576
[36] Dai L, Wang LY, YuanTQ and He J 2014 *Polym. Degrad. Stab.* **99** 233
[37] Ramesh S, Leen KH, Kumutha K and ArofAK 2007 *Spectrochim. Acta A.* **66** 1237
[38] Coltro L, Pitta JB and Madaleno E 2013 *Polym. Test.* **32** 72
[39] Silva M, Vieira M, Maumoto A and Beppu M 2011 *Polym. Test.* **30** 478

IOP Conf. Series: Materials Science and Engineering **292** (2017) 012009 doi:10.1088/1757-899X/292/1/012009

Dependance of naphthenic acid corrosion of SA106B on temperature and turbulence

Y F Shi[1], Q K Zheng[1], J Liu[1], L B Zhang[2], Z W Pan[1] and S X Rao[1, 3]

[1] School of mechanical engineering,Anhui university of technology, Ma'an'shan, 243002, P R China
[2] HuaiHai Industry group Co. ltd, Changzhi, 046012, P R China

[3] E-mail: snowdrio@126.com

Abstract. Analysis about the kinetic mechanism of Naphthenic Acid Corrosion (NAC) indicated that the natural logarithm of corrosion rate (ln*r*) was proportional to negative reciprocal of temperature(-1/T,K), linear regression on the corrosion data in API581 and experiments both proved the linear relationship between ln*r* -(-1/T,K) which could be applied to predicate the average corrosion rate under different temperature. Corrosion morphology observations proved the influence of turbulence on NAC, measurements on the maximum corrosion depth in different turbulence area indicated that in 7.5% turbulence area the corrosion depth could be 2-3 times of the corrosion depth in 2% turbulence area, existence of turbulence could lead to serious local NAC.

1. Introductions

Naphthenic Acid Corrosion(NAC) is a special corrosion mode mainly existed in high acid crude oil processing refineries, although NAC has been discovered for nearly one century, but up to now it keeps to be a reliability issue in the oil refineries.

Existing researches have confirmed the main corrosion mechanism and influencing factors of NAC, the main influencing factors are listed in figure 1 [1-3]. Temperature is the most important factor that could significantly influence NAC, NAC would not occur below 473.15K, as NAC is an endothermal reaction and high activation energy barrier could inhibit obvious NAC under low temperature. When temperature exceeds 673.15K, Most organic Naphthenic Acids (NA) would decompose and NAC vanishes [4-5].

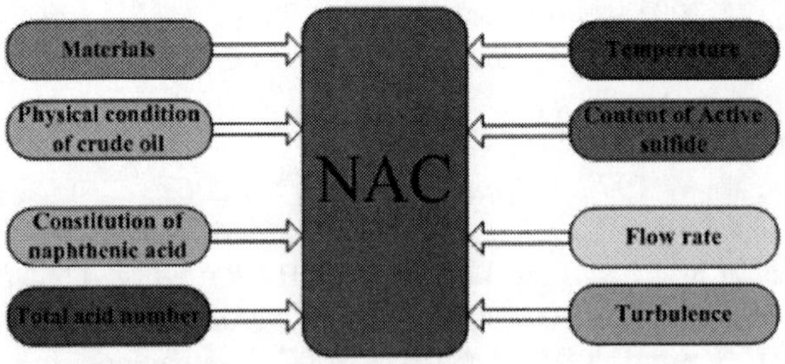

Figure 1. The main influencing factors of NAC.

Total acid number(TAN) is related with the content of NA in high acid crude oil. In fact NAis a complex mixture of hundreds of carboxylic acid, the content of NA could determine the NAC rate [6-7]. Available researches indicated that obvious NAC only occurred above TAN of 0.5mgKOH/g.

NA constitution of high acid crude oil from different origins is not the same. As mentioned above, NA is a complex mixture of various carboxylic acid, the corrosivity of each carboxylic acid is quite different which means the corrosivity of crude oil from different origins may differ obviously even under the same TAN. Researches by Omar indicated that the corrosivity of carboxylic acid depended on the solubility of iron naphthenate in crude oil[8].

Sulfides also exist in high acid crude oil, the existence of active sulfides could lead to sulfide corrosion, meanwhile active sulfides could influence NAC. Appropriate content of active sulfides may suppress NAC through forming an integrate sulfide film on the metal, but the content must be controlled accurately. If the content of active sulfide is too low, the sulfide film formed on the metal is not thick enough and imperfect, the incomplete sulfide film could not play enough protection role. Meanwhile too high content of active sulfides would cause serious sulfide corrosion [9-11].

Although existing researches had confirmed the main factors of NAC, but the mechanism of NAC still could not be determined [12], the dependence of corrosion rate on each influencing factor were not clear[13]. With the drying up of high quality crude oil, in the future high acid crude oil processing would be inevitable. Thus NAC should be further researched. Here the mechanism of NAC was discussed firstly, then the dependence of NAC rate on temperature and turbulence was researched.

2. Experiment preparations

2.1. Experimental materials and specimens

As ASME SA106B low carbon steel was widely used as pipeline material in oil refineries and the corrosion resistance of SA106B to NAC was poor, SA106B was selected as experimental material. SA106B was quenched and tempered, the metallurgical structure was shown in figure2. The nominal chemical compositions of SA106B were listed in table1. The specimens were machined to thin ring structure (As the model was shown in figure 3, the external diameter 18mm, internal diameter 3mm, thickness 3mm.).

Table 1. Chemical composition of test specimens (mass%).

Alloy	C	Si	Mn	P	S	Cu	Ni	Fe
SA106B	0.18	0.20	0.36	≤0.03	≤0.030	≤0.020	≤0.025	Balance

After experiments the specimens were taken out, first the specimens were rinsed by acetone to remove the oil remained on the surface, subsequently the specimens were cleaned by ultrasonic wave cleaner to remove the corrosion products, then the specimens were rinsed again by distilled water and dried for 24hours in glass dryer, finally the weight of the specimens were re-scaled and recorded. Then the weight loss of the specimens could be obtained. The average corrosion rate could be calculated by equation (1).

$$Corrosion\ rate(mm/a)=\frac{3650\times(Weight\ loss/g)}{Density\ of\ metal(g/cm^3)\times Area(cm^2)\times time(day)} \tag{1}$$

2.2. Experimental parameters and procedures

As NAC mainly occurred in the temperature range from 473.15K to 673.15K, the experimental temperature selected were 513.15K, 533.15K, 553.15K, 573.15K, 593.15K, 613.15K and 633.15K, the corrosion time was controlled at 8 hours. In order to investigate the influence of turbulence on NAC, the specimens were installed at two flushing angles(0° and 90°) in pipelines(as shown in figure3).

Figure 2. The metallurgical structure of SA106B.

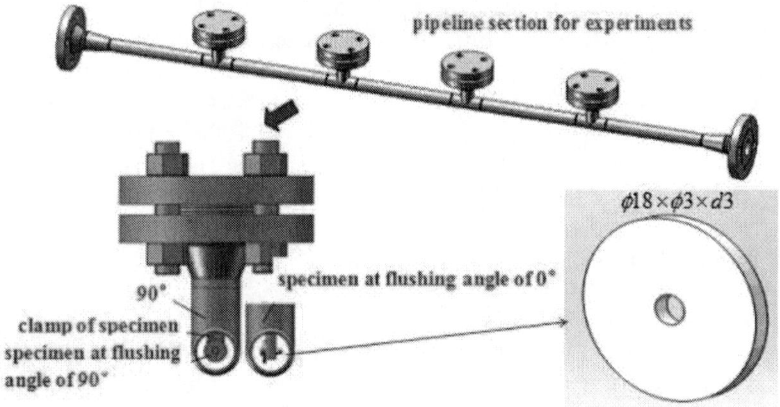

Figure 3. Installation diagram of specimens in the pipelines.

2.3. *Analysis of turbulence and topography*

Turbulence intensity was defined as the ratio of turbulent fluctuation velocity and average velocity, it could be calculated by equation (2). Here I is the turbulence intensity of fluid, Re is the Reynolds number and Re could be calculated by equation (3).

$$I = 0.16 * \mathrm{Re}^{-1/8} \tag{2}$$

$$\mathrm{Re} = \rho v d / \mu \tag{3}$$

ρ is the density of fluid, *v* is the flow velocity of fluid, d is the internal diameter of pipeline and μ is the dynamic viscosity of fluid(the values of the parameters above were listed in Table 2).

Table 2. The parameters needed in flow velocity and turbulence distribution simulation.

ρ	*v*	d	μ
0.95g/cm^3	20m/s	36mm	1×10-2Pa·s

3. Discussion on the mechanisms of NAC

3.1. *Corrosion mechanism of NAC*

The accepted corrosion mechanism of NAC is listed in the chemical reactions below.

$$\mathrm{Fe} + 2\mathrm{RCOOH} \rightarrow \mathrm{Fe(RCOO)_2} + \mathrm{H_2} \tag{4}$$

$$\mathrm{Fe} + \mathrm{H_2S} \rightarrow \mathrm{FeS} + \mathrm{H_2} \tag{5}$$

$$\mathrm{Fe(RCOO)_2 + H_2S \rightarrow FeS + 2RCOOH} \quad (6)$$

Reaction(4) produces iron naphthenate which is oil-soluble and could be taken away by high flow velocity fluid, under most circumstances that is the reason that no obvious corrosion products remain on the metal. On the spots that highflow velocityand strong turbulence exists, the metal keeps corroding under flushing of high flow velocity fluid, finally streamline groove would be formed on the surface.

Reaction(5) indicates that existence of active sulfides could influence NAC, the active sulfides under high temperature could be transformed to H_2S, H_2S could react with iron and produces FeS, as FeS does not dissolve in the oil and could combine with the substrate tightly, if the content of active sulfides is appropriate, integrated FeS protection film could be formed on the substrate, the FeS protection film could insulate NA from the metal and suppress NAC to some extent, that is why NAC rate would decrease under some circumstances when active sulfides exist. However when the content of active sulfides is too low, the FeS protection film formed is not integrated, then NAC still exists, H_2S would react with iron naphthenate and release NA again as shown in reaction(6).

The mechanism above integrates the influence of active sulfide on NAC, but when no active sulfides exist, NAC still occurs. To uncover the corrosion law of NAC, the influence of active sulfide must be removed. In the following experiments no active sulfides were added, then only reaction(4) existed.

Some researchers thought that NAC belonged to electrochemical reaction, the fact was not true. As NAC occurs in crude oil instead of in water, the electro-conductibility of crude oil is less that 0.01μs/cm, obviously lower than the electro-conductibility of pure water(0.05μs/cm), meanwhile most of the iron naphthenate would ionize in crude oil. Generally NAC is a kind of pure chemical reaction instead of electrochemical reaction.

3.2. Kinetic mechanism of NAC

Up to now the kinetics mechanism of NAC is still indecisive, what is the control step of NAC needs to be confirmed.

In static laboratory experiments, the mass transfer, adsorption and desorption of reactants and resultants could be the control step.

On the spots that serious NAC exists, the flow velocity of fluid is quite high and strong turbulence exists, mass transfer, adsorption and desorption of reactants and resultants would not be the corrosion control step, the corrosion is controlled by the corrosion reaction itself. The total chemical reaction of pure NAC is listed in reaction (4), but in fact reaction(4) could be divided into several elementary reactions, here it is supposed that two elementary reactions existed (listed in equation (7) and in equation (8)).

$$\mathrm{Fe + RCOOH \rightarrow FeRCOO + H} \quad (7)$$

$$\mathrm{FeRCOO + RCOOH \rightarrow Fe(RCOO)_2 + H} \quad (8)$$

Equation (7) and equation (8) indicated that only the first elementary reaction (equation (7)) was directly related with corrosion of metal, the second elementary reaction(equation (8)) only transformed the intermediates FeRCOO into final resultant $Fe(RCOO)_2$. In pipelines crude oil flowed at high rate and the intermediates of equation (7) could be taken downstream and then transformed to final resultant as equation (8), so it sounded that corrosion was mainly controlled by the first elementary reaction, then the corrosion rate could be expressed as $r= k_1[\mathrm{RCOOH}][\mathrm{Fe}]$.

In the same refining pipelines the constitutions and TAN of crude oil keep steady, so the concentration of RCOOH could be supposed to be constant, the reactant of Fe was solid, the first elementary reaction could be considered as zero order reaction. Thus total NAC rate was only controlled by the rate constant k_1 of the first elementary reaction. According to Arrhenius equation k_1=Aexp($-E_a$/RT), corrosion rate r was proportional to k_1 and exp($-E_a$/RT). So it was deduced that the natural logarithm of corrosion rate (ln$r$) was proportional to negative reciprocal of temperature(-1/T,K) and the slope of ln$r$ -(-1/T) line was related to activation energy E_a.

4. Experiment results and discussion

4.1. Dependance of NAC rate on temperature

To verify the linear relationship between lnr and (-1/T) deduced above, the naphthenic acid corrosion data at five different TAN levels(TAN≤0.3,0.3≤TAN≤1.0,1.1≤TAN≤2.0, 2.1≤TAN≤4.0, TAN≥4.0) of low carbon steel and 1.25Cr in API581 were analyzed. The analysis results of corrosion data were shown in figure 4 and figure 5.

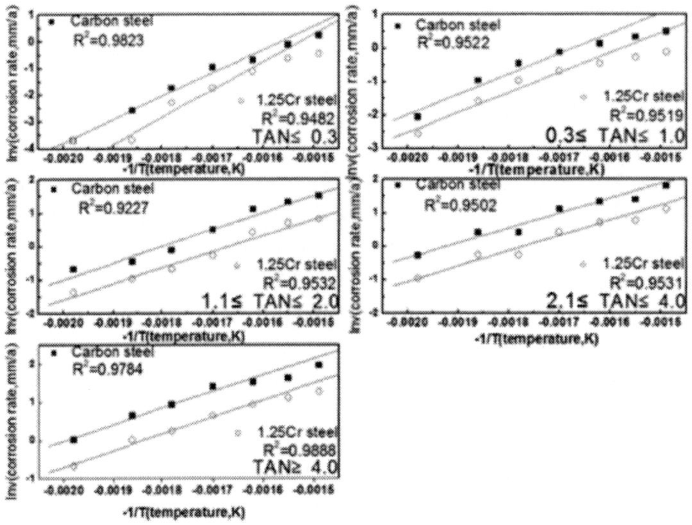

Figure 4. ln*r* and (-1/T) of low carbon steeland 1.25Crlow alloy steel atdifferent TAN (content of S≤0.2%).

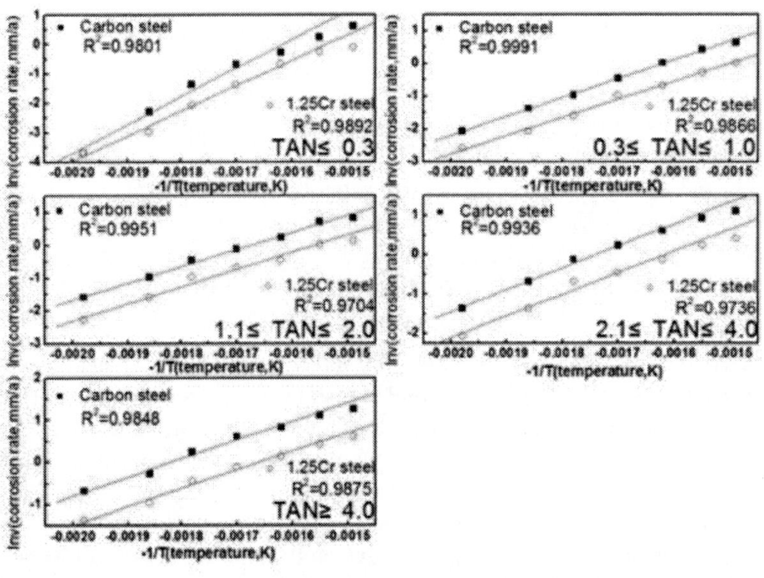

Figure 5. ln*r* and (-1/T) of low carbon steel and 1.25Cr low alloy steel at different TAN (0.2%≤ content of S ≤0.6%).

Figure 4 and figure 5 indicated that the relationship between ln*r* and (-1/T) of low carbon steel and 1.25Cr low alloy steel accorded with linear relation quite well, it was confirmed that linear relationship between ln*r* and (-1/T) did exist.

As mentioned above the slope of ln*r* -(-1/T) line was related with the activation energy E_a. After linear regression, the slope of all the fitting lines could be obtained. It was found that the slope of the fitting lines decreased with the increase of temperature and TAN, this phenomenon could be attributed to the fact that NA is a complex mixture of hundreds of carboxylic acid, with increasing temperature, more carboxylic acid participated in NAC.

To further prove the linear relationship between ln*r* and (-1/T), naphthenic acid corrosion experiments of SA106B were carried out. The relationship between ln*r* and(-1/T) was listed in figure 6, the experiment results also proved the linear relationship of ln*r* and (-1/T). The linear relationship between ln*r* and (-1/T) had been occasionally reported by some researchers, but those researchers did not find the reason why the linear relationship existed.

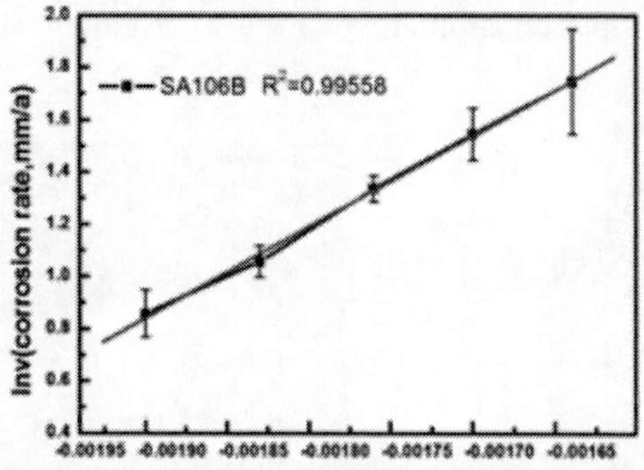

Figure 6. ln*r* and (-1/T) of SA106B at TAN of 4.6mgKOH/g.

4.2. Dependance of NAC on flow rate and turbulence

4.2.1. Analysis of flow rate and turbulence around the specimens. As severe NAC occurred onthe spots that high flow velocity and remarkable turbulence existed. The corrosion rate must be related with flow velocity and turbulence. In experiments two flushing angles were applied(0° and 90°) to create different turbulence distribution. To correlate NAC rate with flow velocity and turbulence, the flow velocity distribution and turbulence distribution must be simulated.

The inlet velocity of fluid was controlled at 20m/s, then the flow velocity distributions under two flushing angles were simulated and shown in figure 7 and figure 8. On the surface of 90° specimen in figure 7, due to the blockage of specimen, the flow velocity on the specimen was decreased to lower than 10m/s; while on the surface of 0° specimen in figure 8, the block effect was not so obvious, the flow velocity on the specimen was still close to 20m/s.

The specimens installed in the pipeline varied the direction of fluid flow, highly increased the flow velocity around the specimens and formed strong turbulence. In figure 8 the max flow velocity could exceed 45m/s which was much higher than the inlet velocity of 20m/s.

The turbulence distributions on the specimens under two flushing angles were simulated and shown in figure 9. From figure 9a it was confirmed that under flushing angle of 90° the highest turbulence existed in the central ring of the specimen, while under flushing angle of 0° the highest turbulence existed in the edge facing the flow direction (shown in figure 9b). On both specimens turbulence intensity could reach 7.5%.

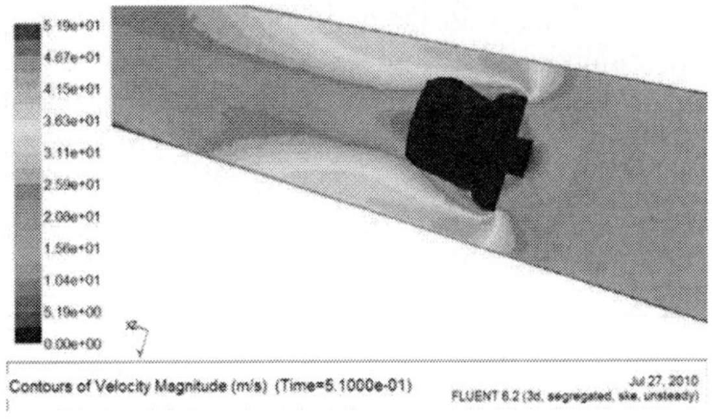

Figure 7. Flow velocity distribution aroundthe specimens at 90°.

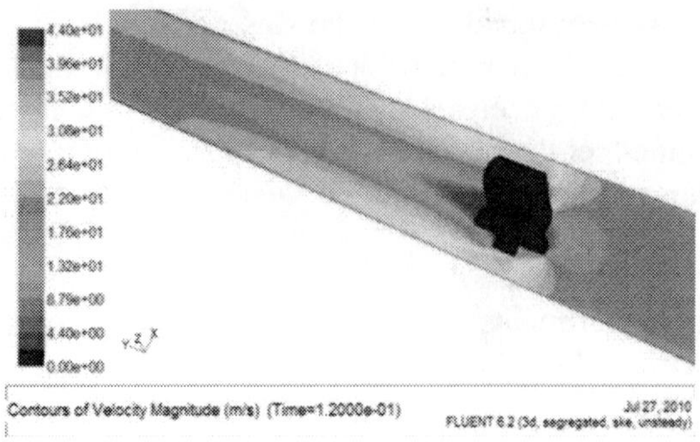

Figure 8. Flow velocity distribution around thespecimens at 0°.

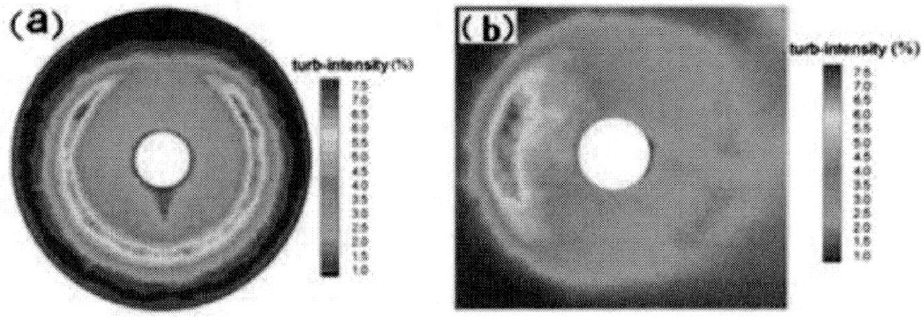

Figure 9. Turbulence distribution on the work area of the specimens(a- 90°,b- 0°)

4.2.2. Dependance of NAC rate on flow velocity and turbulence. As the turbulence intensity on the specimen was directly related with the inletvelocity of fluid, it was difficult to separate the influence of flow velocity and turbulence. As in straight pipelines no obvious turbulence existed, NAC was not serious even under high flow velocity, the corrosion morphology on the specimen was quite smooth and no obvious local NAC existed, it was deduced that the influence of flow velocity on NAC was mainly expressed in the increment of average corrosion rate. As the flow velocity distribution on the specimens under two flushing angle was quite different, the average corrosion rate under two flushing angle was compared with each other. Under most circumstances the corrosion rate under 90°was slightly higher than the corrosion rate under 0°, the max difference was limited in 20%. Even when the flow velocity of fluid was increased to 80m/s, the corrosion rate difference under two flushing angle was limited in 50%. That is to say, the increase of flow velocity could aggravate average NAC rate, but the influence was limited.

When no turbulence existed, NAC is almost an uniform corrosion, but when strong turbulence existed, the corrosion morphology was quite different. The corrosion morphology of SA106B at 513.15K under flushing angle of 90° was recorded and listed in figure 10. In figure 10a the ferrite grains could be distinguished easily, but the pearlites could not be obviously observed, this fact indicated that the pearlites dissolved preferentially in NA, as the turbulence intensity in inner ring was quite low(less than 2%), NAC was quite slight. In central ring the turbulence intensity was increased to 7%-8%, the corrosion was more obvious relative to the corrosion in inner ring, almost all the pearlites disappeared, most ferrite grains could still be distinguished, but some grain boundaries became fuzzy and some ferrite grains had be corroded(shown in figure 10b). In outer ring the turbulence intensity was about 5%, the corrosion morphology was similar to the morphology in central ring(shown in figure 10c).

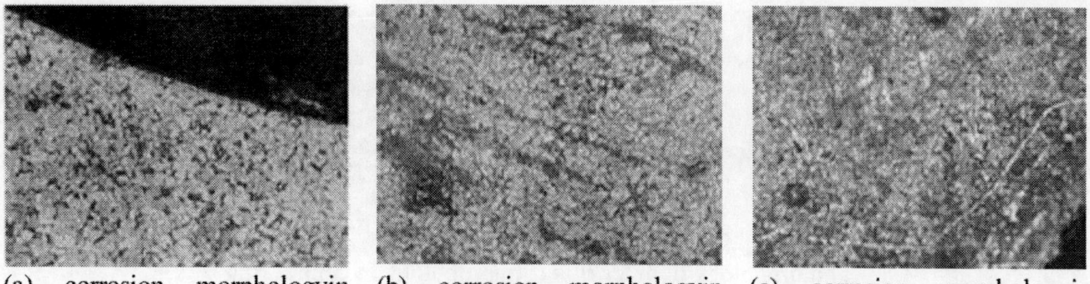

(a) corrosion morphologyin inner ring (b) corrosion morphologyin central ring (c) corrosion morphologyin outer ring

Figure 10. Corrosion morphology of SA106B at 513.15K under flushing angle of 90°.

The preferential dissolution of pearlites indicated pearlites were more sensitive to NAC, it is known that the main chemical constitution of pearlites is Fe_3C which is soluble in acids. In hydrochloric acid the pearlites would react with acid as equation (9). But NAC occurs in crude oil in which the content of water was quite low, NA would not ionize and could not react with pearlites as equation(9). Since active sulfides are not added in the experiments, the possible reaction between NA and pearlites could be expressed by equation (10).

$$2Fe_3C + 6HCl + 12H_2O = 6CO_2\uparrow + 2FeCl_3 + 9H_2\uparrow \tag{9}$$

$$Fe_3C + 6RCOOH \rightarrow 3Fe(RCOO)_2 + CH_4\uparrow + H_2\uparrow \tag{10}$$

The ferrites could also react with NA, but the corrosion rate was much lower than the corrosion rate of pearlites in NA. From the corrosion difference in different turbulence distribution area, it was confirmed that turbulence could aggravate NAC.

When the temperature was increased to 553.15K, NAC became more serious relative to 513.15K. In inner ring as the turbulence intensity was quite low, although the pearlites dissolved completely, but most ferrites still kept integrate, grain boundaries could still be distinguished. In central ring the pearlites and ferrites corroded seriously, no grain boundaries could be seen, a clear boundary (dot dash line shown in figure 11a) could be observed between the inner ring and central ring due to the different corrosion severity induced by turbulence intensity distribution. The same phenomenon existed between central ring and outer ring, as the corrosion rate at 553.15K increased obviously and the corrosion rate in central ring and outer ring differed obviously, clear boundary (dot dash line shown in figure 11b) also formed between central ring and outer ring.

The corrosion morphology on the ring specimen could be related with the turbulence distribution in figure 8 directly. The boundaries in different area on the specimen further proved the significant influence of turbulence on corrosion rate.

Even when the temperature was increased to 593.15K, some ferrite grains in inner ring kept quite integrate and some grain boundaries could be observed (shown in figure 12a). But in central ring most ferrite grains had disappeared (shown in figure12b), corrosion in central ring became quite serious.

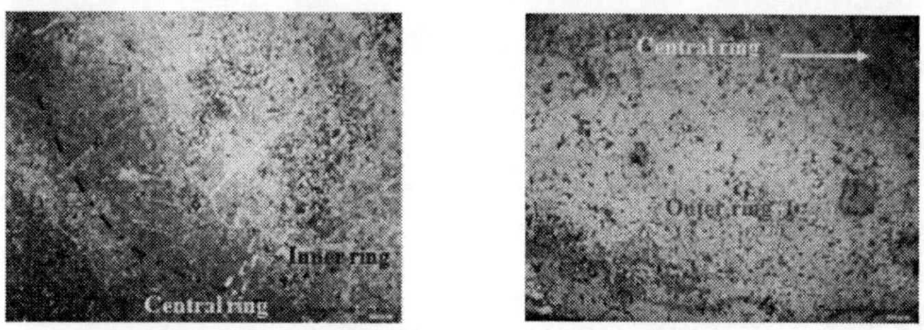

(a) corrosion morphology ininner ring and central ring (b) corrosion morphology inouter ring

Figure 11. Corrosion morphology of SA106B at 553.15K under flushing angle of 90°.

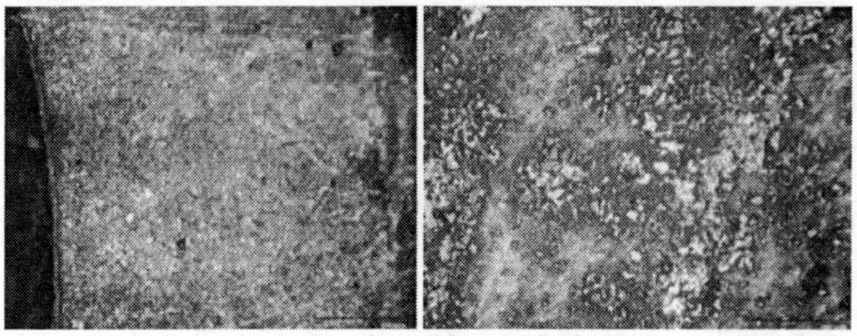

(a) morphology in inner ring (b)morphology in central ring

Figure 12. Corrosion morphology of SA106B at 593.15K under flushing angle of 90°.

When temperature was increased to 633.15K, the corrosion became more serious, in inner ring most ferrite grains had been corroded (shown in figure 13a), in central ring almost all the ferrite grains had disappeared and no grain boundaries could be observed (shown in figure 13b).

(a) morphology in inner ring and central ring (b) morphology in central ring

Figure 13. Corrosion morphology of SA106B at 633.15K under flushing angle of 90°.

Although significant influence of turbulence on local NAC had been confirmed throughcorrosion morphology analysis, the dependence of local corrosion rate on turbulence had not been determined quantitatively. In local corrosion, the maximum corrosion depth was the most important indicator. To uncover the relationship between maximum corrosion depth and turbulence intensity, the maximum corrosion depth in different turbulence area must be analyzed.

As the experimental corrosion time was controlled at 8 hours, the average corrosion rate could be transformed to average corrosion depth DA in 8 hours Maximum corrosion depth in different turbulence intensity area were measured and listed in Table3. Table3 indicated that the ratio between maximum corrosion depth in 7.5% turbulence area(DM,7.5%) and 2% turbulence area(DM,2%) increased with temperature, this phenomenon indicated that under action of turbulence, local NAC became more serious with increasing temperature, the maximum corrosion depth in 7.5% turbulence area could be 2-3 times of the corrosion depth in 2% turbulence area.

Table 3. Influence of turbulence on maximum corrosion depth.

Temperature	513.15K	553.15K	593.15K	633.15K
Average corrosion rate（mm/a）	2.52	3.94	6.89	5.81
Average corrosion depth(μm), DA	2.30	3.6	6.29	5.30
Maximum corrosion depth in 2% turbulence area（mm/a）, DM,2%	3.71	4.02	4.67	4.90
Maximum corrosion depth in 7.5% turbulence area（mm/a）, DM,7.5%	5.73	8.69	12.03	13.78
Ratio between DM,7.5% and DA	2.49	2.41	1.91	2.6
Ratio between DM,7.5% and DM,2%	1.54	2.16	2.58	2.81

Restricted by the safety of experimental devices, the experimental time had to be controlled at 8 hours, if the corrosion time was long enough, the corrosion rate difference on the specimen would lead to the formation of groove, this fact could explain why streamline grooves often appears in NAC.

5. Conclusion

(1)Analysis about kinetic mechanism of NAC indicated natural logarithm of corrosion rate (ln*r*) was proportional to negative reciprocal of temperature(-1/T,K) . Linear regression on the corrosion data in API581 and experiments proved the existence of the linear relationship between ln*r* and -1/T, this linear relationship could be used to predicate the average corrosion rate under different temperature.

(2)The corrosion morphology indicated that preferential corrosion of pearlites existed, when NAC occurred, corrosion rate of pearlites was much higher than the corrosion rate of ferrites. Meanwhile the corrosion morphology proved the influence of turbulence on local NAC, in high turbulence area local NAC became serious.

(3)Maximum corrosion depth in different turbulence area indicated that in 7.5% turbulence area the maximum corrosion depth could be 2-3 times of the corrosion depth in 2% turbulence area. The ratio between DM,7.5% and DM,2% increased with increasing temperature. The relationship between turbulence and local corrosion depth could be applied to predicate the maximum corrosion depth in local NAC.

Acknowledgements

Thanks for the support by National High Technology Research and Development Program of China(No.2012AA040103) and Anhui province natural scienceresearch project(KJ2016SD09).

Reference

[1] Qu D R, Zheng Y G, Jing H M, Yao Z M and Ke W 2006 High temperature naphthenic acid corrosion and sulphidic corrosion of Q235 and 5Cr1/2Mo steels in synthetic refining media *Corrosion Science* **48** 1960-1985

[2] Wu X Q, Jing H M, Zheng Y G, Yao Z M and Ke W 2004 Erosion Corrosion of Various Oil refining Materials in Naphthenic acid *Wear* **256** 133-144

[3] Slavcheva E, Shone B and Turbull A 1999 Review of naphthenic acid corrosion in oil refining British *Corrosion Journal* **34** 125-131

[4] Rao S X, Zhou Y, AiZhibin, Pan Z W, Cen Y W and Chen X D 2014 High temperature Naphthenic Acid Corrosion of SA210C and A335-P5 *Materials and Corrosion* **65** 619-625

[5] Wu X Q, Jing H M, Zheng Y G, Yao Z M and Ke W 2004 Study on high-temperature naphthenic acid corrosion and erosion-corrosion of aluminized carbon steel *Journal of Materials Science* **39** 975-985

[6] Nugent M J, Dobis J D 1998 Experience with Naphthenic Acid Corrosion in low TAN Crudes *Corrosion 98 Houston NACE International* **577**

[7] OmarYepez 2005 Influence of Different Sulfur Compounds on Corrosion Due toNaphthenic Acid *Fuel* **84** 97-104

[8] OmarYepez 2007 On the Chemical Reaction between Carboxylic Acids and Iron, Including Special case of Naphthenic Acid *Fuel* **86** 1162-1168

[9] Craig H L 1996 Temperature and Velocity Effects in Naphthenic Acid Corrosion *Corrosion 96 Houston NACE International* **603**

[10] Liu G Q, ZhengY G, Jiang S L, Jing J H, Dong W J, Zeng H and Si P X 2015 Stability and Erosion Corrosion Behavior of Corrosion Product Film of Q235 Carbon Steel and Cr5Mo Low Alloy Steel in Simulated Oil Refinery Media *Journal of Chinese Society for Corrosion and Protection* **35** 122-128

[11] Peng J, Nesic S and Wolf H A 2015 Analysis of corrosion scales formed on steel at high temperatures in hydrocarbons containing model naphthenic acids and sulfur compounds *Surface and Interface Analysis* **47** 454-465

IOP Conf. Series: Materials Science and Engineering **292** (2017) 012009 doi:10.1088/1757-899X/292/1/012009

[12] Rebak R B 2011 Sulfidic corrosion in refineries a review *Corros Rev* **29** 123-133
[13] Trasatti S P, Gabetta G 2006 Study of naphthenic acid corrosion by neural network *Science and Technology* **4** 200-211

Research on breakdown characteristics of converter transformer oil-paper insulation under compound electric field in alpine region

C Xu[1], Z W Gao, S Lan, H X Guo and M C Gong

State Grid Heilongjiang Electric Power Company Limited Electric Power Research Institute, 7 Xiangjiang River Road, Harbin, China

[1] E-mail: 704777887@qq.com

Abstract. In the paper, existing research and operating experience was summarized. On the basis, the particularity of oil-paper insulation operation condition for converter transformer was combined for studying the influence of temperature on oil-paper insulation field intensity distribution of converter transformers under different AC contents within wide temperature scope (-40℃~105℃). The law of temperature gradients on space charge accumulation was analyzed. The breakdown or flashover characteristics of typical oil-paper compound insulation structure under the action of DC, AC and AC-DC superposition voltage at different temperatures were explored. The design principles of converter transformer oil-paper insulation structures in alpine region was proposed. The principle was adjusted and optimized properly according to the operation temperature scope and withstood AC-DC proportion. The reliability of transformer operation was improved on the one hand, and the insulating medium can be rationally utilized for reducing the manufacturing cost of the transformer on the other hand.

1. Introduction

According to the relation between the characteristics of oil paper breakdown and temperature in common AC transformer, and the investigation of transformer operation accident, it shows that temperature is an important factor affecting oil paper insulation of transformer [1]. The transformer operation department urgently needs the breakdown data of transformer oil paper insulation at low temperature, because there are several insulation faults when the transformer is put into operation at lower temperature in recent years. At present, there is no oil paper insulation experimental data on breakdown and flashover characteristics under complex electric field at low temperature. It is necessary to carry out research in this field [2-3].

2. Influence of temperature on electric field distribution of oil-paper compound insulation

2.1. Physical model of oil-paper insulation structure

In the paper, a typical double-layer oil-paper insulation structure model was adopted for facilitating research on electric field distribution of oil-paper insulation. Relative dielectric constant, conductivity and thickness of paperboard and oil were ε_p, γ_p, d_p and ε_o, γ_o, d_o respectively. Its equivalent circuit was shown in figure 1. The voltage was U at both ends of the oil-paper compound insulation structure. The equivalent voltage, capacitance and resistance at two ends of the paperboard and oil were U_p, C_p, R_p and U_o, C_o, R_o respectively.

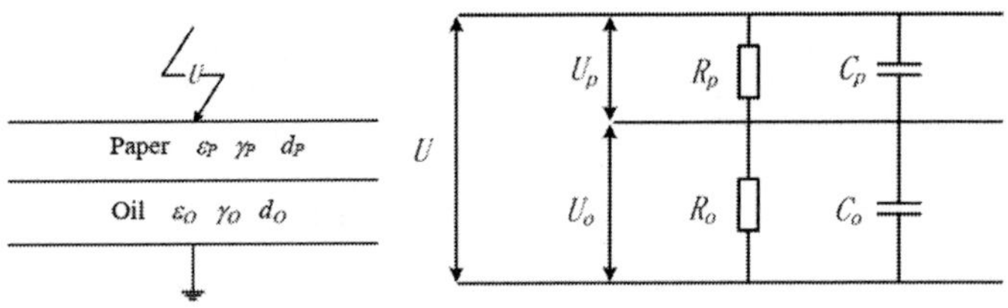

Figure 1. Oil-paper insulation model and equivalent circuit.

The space charge was set as zero on the oil-paper interface before voltage application. Namely, the equivalent circuit was in the zero state as shown in figure 1. The follows was concluded according to Ohm's law and first-order circuit solution:

Voltage on the paperboard was shown as follows at any time:

$$u_p = \frac{R_p}{R_p + R_0} U_{dc} + \left(\frac{C_0}{C_p + C_0} - \frac{R_p}{R_p + R_0} \right) U_{dc} e^{-\frac{t}{\tau}} + u_{ac} \frac{C_0}{C_p + C_0} \tag{1}$$

Voltage on the oil was shown as follows:

$$u_0 = \frac{R_0}{R_p + R_0} U_{dc} + \left(\frac{C_p}{C_p + C_0} - \frac{R_0}{R_p + R_0} \right) U_{dc} e^{-\frac{t}{\tau}} + u_{ac} \frac{C_p}{C_p + C_0} \tag{2}$$

In the formula, U_{dc} and U_{ac} were DC voltage and AC voltage at both ends of the oil-paper compound insulation structure, $\tau = \left(C_p + C_0\right)\left(\frac{1}{R_p} + \frac{1}{R_0}\right)^{-1}$, was a time constant.

2.2. *Mathematical expression of electric field distribution of oil-paper compound insulation*

The mathematical expression of voltage on the paperboard and oil was shown as follows in the steady state:

$$\begin{cases} u_p = \dfrac{\rho_p d_p}{\rho_p d_p + \rho_0 d_0} U + \left[\dfrac{\varepsilon_0 d_p}{\varepsilon_0 d_p + \varepsilon_p d_0} \sin(\omega t) - \dfrac{\rho_p d_p}{\rho_p d_p + \rho_0 d_0} \right] \eta U \\ u_0 = \dfrac{\rho_0 d_0}{\rho_p d_p + \rho_0 d_0} U + \left[\dfrac{\varepsilon_p d_0}{\varepsilon_0 d_p + \varepsilon_p d_0} \sin(\omega t) - \dfrac{\rho_0 d_0}{\rho_p d_p + \rho_0 d_0} \right] \eta U \end{cases} \tag{3}$$

Maximum value was obtained for u_p and u_o when $\sin(\omega t) = 1$. Then,

$$\begin{cases} u_p = \dfrac{\rho_p d_p}{\rho_p d_p + \rho_0 d_0} (1-\eta) U + \dfrac{\varepsilon_0 d_p}{\varepsilon_0 d_p + \varepsilon_p d_0} \eta U \\ u_0 = \dfrac{\rho_0 d_0}{\rho_p d_p + \rho_0 d_0} (1-\eta) U + \dfrac{\varepsilon_p d_0}{\varepsilon_0 d_p + \varepsilon_p d_0} \eta U \end{cases} \tag{4}$$

It was believed that AC and DC electric fields can be superposed linearly in the project since compound electric fields were distributed between oil and paperboard, and equivalent parallel impedance value was much higher than capacitive reactance value. It was assumed that the electric fields were distributed evenly in the media, the electric field intensities Ep and Eo in oil-impregnated paperboard and transformer oil were respectively concluded as follows according to Ep and Eo [4].

$$\begin{cases} E_p = \dfrac{\rho_p}{\rho_p d_p + \rho_0 d_0}(1-\eta)U + \dfrac{\varepsilon_0}{\varepsilon_0 d_p + \varepsilon_p d_0}\eta U \\ E_0 = \dfrac{\rho_0}{\rho_p d_p + \rho_0 d_0}(1-\eta)U + \dfrac{\varepsilon_p}{\varepsilon_0 d_p + \varepsilon_p d_0}\eta U \end{cases} \quad (5)$$

In the formula, ρ_0, ρ_p, ε_0 and ε_p were functions of temperature T. Therefore, the field intensities Ep and Eo in the paperboard and oil were functions of temperature T and AC content η.

2.3. Numerical calculation results of electric field distribution of oil-paper compound insulation

The expression of resistivity and relative dielectric constant change with temperature was obtained through fitting. The expression was brought into formula (5). U=150 kV, dp=1mm and do=1.5mm were taken, and the change trend of Ep and Eo with η and T was shown in figure 2.

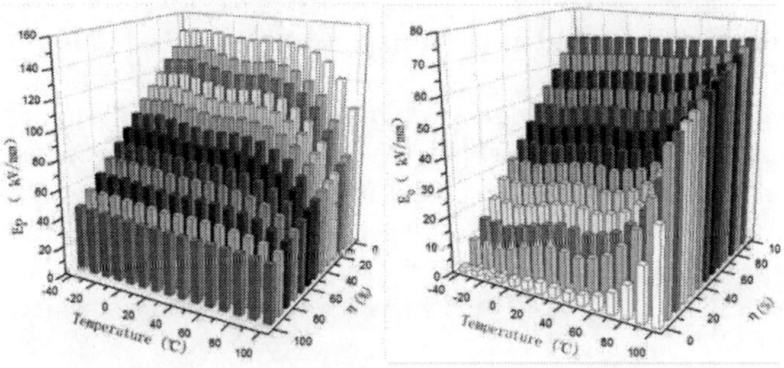

Figure 2. Field intensity in transformer oil and oil-impregnated paperboard at different AC contents and temperatures.

When AC content was a fixed value, the field intensity in the paperboard was decreased with temperature increase as a whole. When temperature was a fixed value, the field intensity in the paperboard was increased monotonously with decrease of AC content. However, the field intensity change in oil was just opposite because electric field distribution mainly depended on oil-paper relative dielectric constant ratio mainly during high AC contents, and the electric field distribution mainly depended on oil-paper resistivity ratio during low AC content. The maximum value of field intensity in paperboard appeared at low temperature and low AC content (η= 0,T= -20 ℃), and the maximum value of field intensity in oil appeared at high temperature and high AC content (η = 100%,T= 100 ℃). Therefore, the insulation margin at high temperature and high AC content should be focused aiming at oil gaps when the converter transformer was designed. The insulation margin at low temperature and DC should be focused aiming at insulation paperboard.

3. Space charge characteristics of oil-paper insulation under different temperature gradients

Voltage of -20kV was applied on double-layer insulation paperboard at different temperature gradients (Δ=0℃, Δ=20℃ and Δ=40℃). The space charge accumulation condition was studied. The voltage was applied for one hour. The space charge distribution was shown in figure 3.

The measurement data showed that prominent space charge was produced in double-layer paperboard interface at different temperature gradients. Different temperature gradients had different influences on space charge accumulation. The interface space charge was the least when the temperature difference was 20 ℃. The space charge of lower insulation paperboard near the low electrode was the maximum when Δ=0℃, and they belonged to like charges. The electrical polarity was gradually changed from like charges to opposite charge with temperature difference increase in the lower paperboard. The space charge accumulation near the upper electrode was decreased with temperature difference increase.

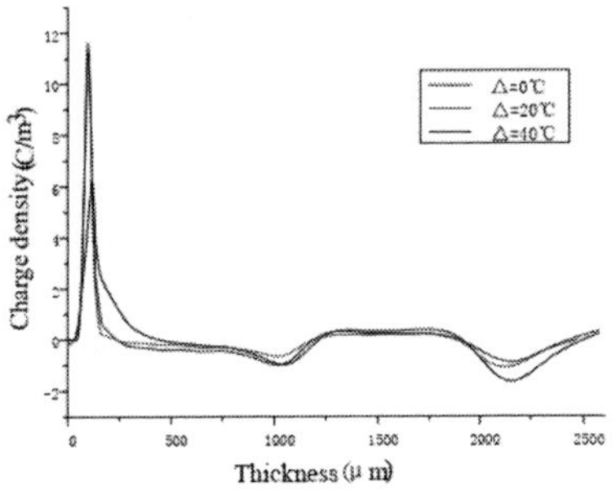

Figure 3. Space charge distribution of double-layer insulation paperboard at different temperature gradients.

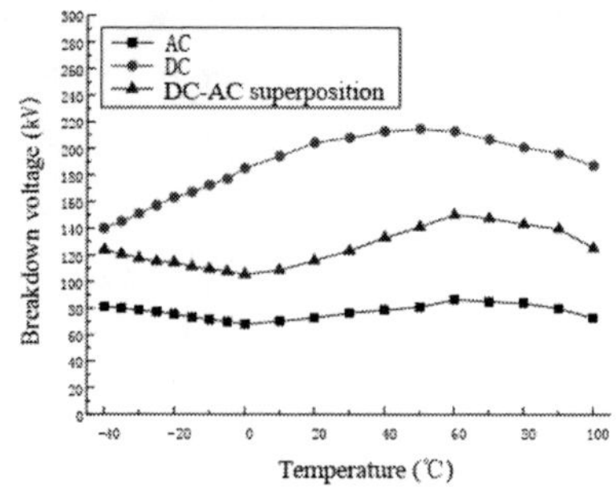

Figure 4. Breakdown voltage of oil-paper compound insulation at different temperatures.

4. Influence of temperature on breakdown characteristics of oil-paper compound insulation

4.1. Breakdown characteristics of oil-paper compound insulation at different temperatures

In the paper, AC/DC superposition equipment was utilized for studying the influence of temperature (-40℃ to 100℃) on breakdown characteristics of oil-paper compound insulation at AC content η=0% (pure DC), η=50% (DC/AC superposition) and η=100% (pure AC). Specimens were typical compound insulation structures composed of 1mm oil-impregnated paperboard and 1.5mm transformer oil. The paperboard moisture content was between 0.45% and 0.6%, and moisture content in transformer oil was between 4.32 PPM and 5.16 PPM.

The average value of breakdown voltage data of oil-paper compound insulation under three voltage wave-forms was shown in figure 4: the breakdown voltage of oil-paper compound insulation showed a trend of decrease-increase-decrease with the temperature increase under the action of pure AC voltage and AC-DC superimposition voltage. Their trends were the same. Two extreme value points appeared at 0 ℃ and 60 ℃ or so. However, their change amplitude was not the same. The breakdown voltage of the oil-paper compound insulation showed a trend of increase-decrease under the action of pure DC voltage, and the maximum value appeared at 50 ℃ or so.

4.2. Analysis on breakdown mechanism of oil-paper compound insulation under AC electric field

Corresponding theoretical values of field intensity in transformer oil and oil-impregnated paperboard was compared with the breakdown field intensity actual value of transformer oil and oil-impregnated paperboard at different temperatures to analyze the breakdown process of oil-paper compound insulation in the AC electric field quantitatively, wherein oil-paper compound insulation was broken down in pure AC electric field. The comparison results were shown in figure 5 and figure 6.

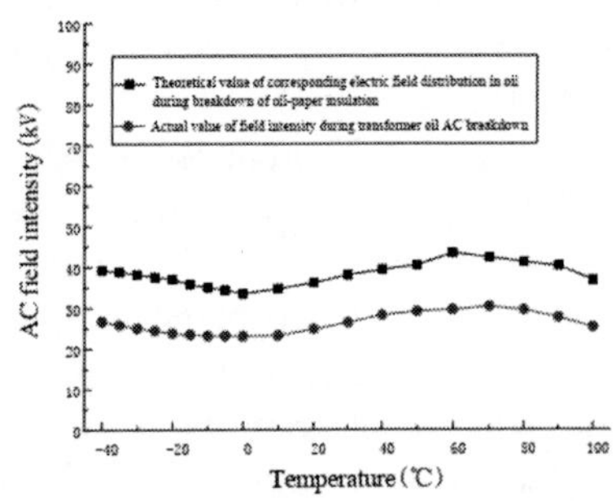

Figure 5. Field intensity in oil under different conditions.

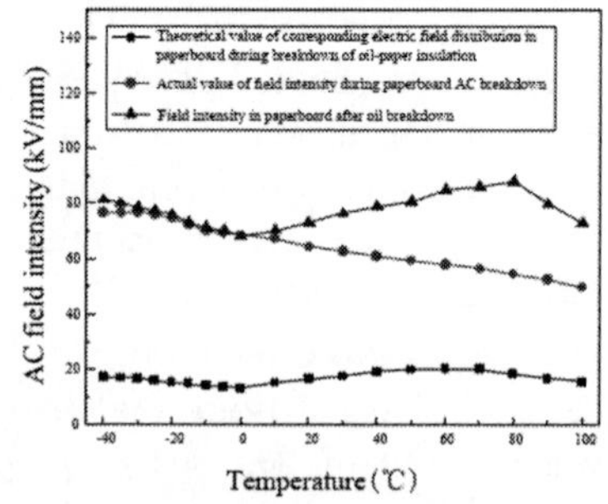

Figure 6. Field strength in oil and paperboard under different conditions.

Figure 5 showed the breakdown field intensity of pure oil gaps at different temperatures under the action of AC voltage and theoretical values of corresponding electric field distribution in oil gaps during breakdown of oil-paper compound insulation. Figure 6 showed the breakdown field intensity of oil-impregnated paperboard at different temperatures under AC voltage, theoretical values of corresponding electric field distribution in the oil-impregnated paperboard during breakdown of oil-paper compound insulation and the field intensity in oil-impregnated paperboard after breakdown of transformer oil.

Data comparison showed that transformer oil withstood most field intensity in oil-paper compound insulation under the action of pure AC voltage. The field intensity withstood by the paperboard was far lower than the breakdown field intensity thereof. Therefore, the breakdown process in AC can be concluded as follows: discharge breakdown occurred in the transformer oil firstly, and the field strength falling on the paperboard was higher than or close to the actual value of breakdown field strength in AC electric field after single medium breakdown in the oil, thereby leading to overall breakdown of compound insulation.

Figure 5 showed that the pure oil gap breakdown field intensity in AC voltage at different temperatures was much lower than the electric field intensity in oil-paper compound insulation oil gap possibly because AC breakdown field intensity was the result, which was obtained from breakdown voltage conversion of 2.5 mm standard oil gap. However, the oil gap thickness was 1.5mm in compound insulation. The field strength in the oil gap was higher due to influence of volume effect, wherein the strength was converted from oil-paper compound insulation breakdown voltage.

4.3. Analysis on breakdown mechanism of oil-paper compound insulation in DC electric field

Corresponding theoretical values of field intensity in transformer oil and oil-impregnated paperboard was compared with the breakdown field intensity actual value of transformer oil and oil-impregnated paperboard at different temperatures to analyze the breakdown process of oil-paper compound insulation in the DC electric field quantitatively, wherein oil-paper compound insulation was broken down in pure DC electric field. The comparison results were shown in figure 7 and figure 8.

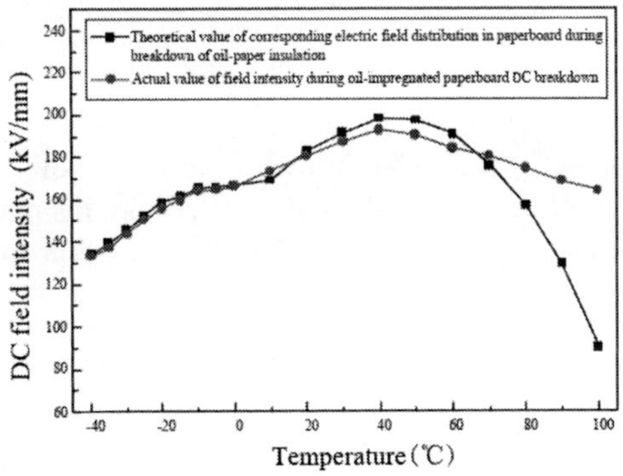

Figure 7. Field intensity in paperboard under different conditions.

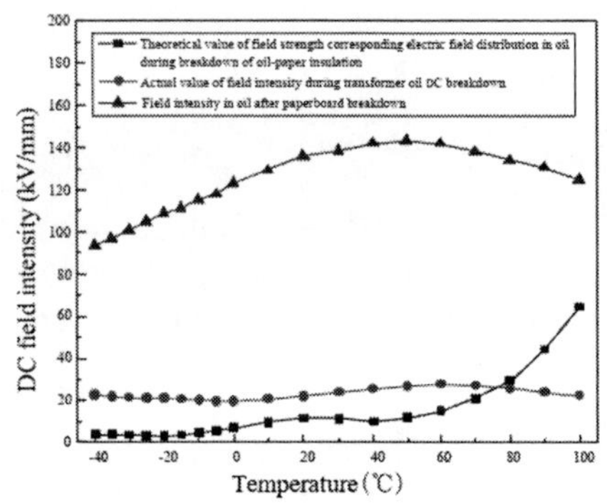

Figure 8. Field strength in oil and paperboard under different conditions.

Figure 7 showed breakdown field intensity of oil-impregnated paperboard at different temperatures under DC voltage and theoretical values of corresponding electric field distribution in oil-impregnated paperboard during breakdown of oil-paper compound insulation. Figure 8 shows the breakdown field intensity of pure oil gap at different temperatures under the action of DC voltage, theoretical values of corresponding electric field distribution in oil gap during breakdown of oil-paper compound insulation and the field intensity in oil gap after breakdown of oil-impregnated paperboard.

Data comparison showed that: paperboard withstood most field intensity between -40℃ and 60℃ under the action of pure DC voltage. The field intensity in the transformer oil did not reach the actual value of the breakdown field intensity thereof. Therefore, it can be concluded that the paperboard suffered from breakdown firstly, and all voltage was applied on the oil after breakdown, thereby leading to overall breakdown of compound insulation. The field intensity in oil between 60℃ and 100℃ was higher than the actual breakdown voltage, and oil suffered from breakdown firstly, thereby leading to overall breakdown of compound insulation.

Figure 8 showed that the theoretical calculation field strength was greatly different from actual breakdown field intensity since the resistivity change law of the paperboard and oil was changed greatly in high field intensity. The difference magnitude order was significant between paperboard resistivity and oil resistivity at the low temperature interval, which had low influence on the calculation results. The resistivity difference was insignificant at high temperature, which had high influence on the calculation results, thereby leading to significant deviation in the calculation results.

4.4. Analysis on breakdown mechanism of oil-paper compound insulation under DC-AC superposition electric field

The AC content of voltage withstood by oil and paperboard was also always changed with temperature change, and it was difficult to analyze the AC content quantitatively in the oil-paper compound insulation structure under the action of DC-AC superposition voltage (η=50%). The change trend of breakdown voltage of oil, paperboard and oil-paper compound insulation structure with temperature was shown in figure 9, figure 10 and figure 11 respectively under different AC contents.

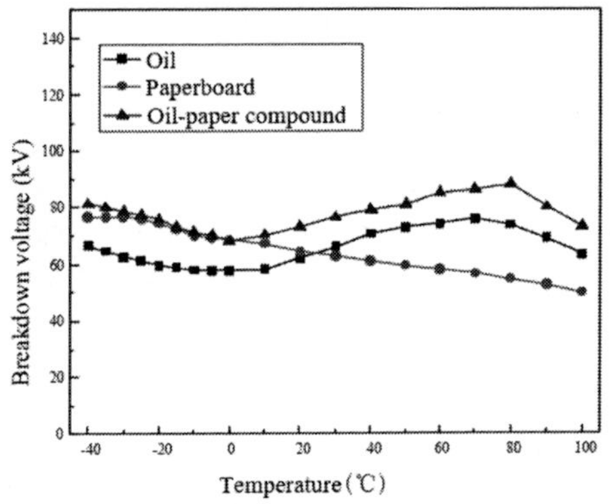

Figure 9. Breakdown voltage of oil, paperboard and oil-paper compound insulation under the action of AC voltage.

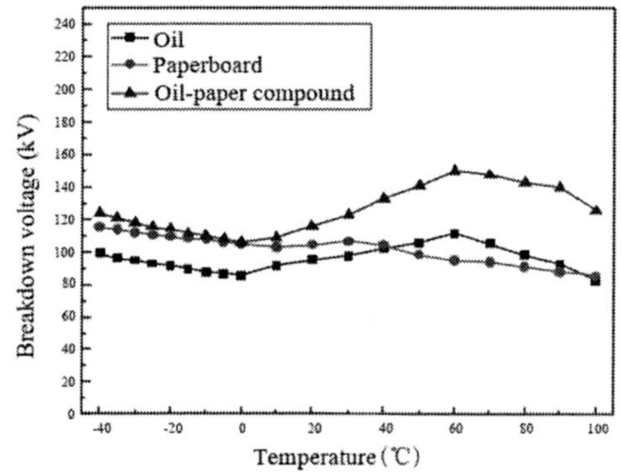

Figure 10. Breakdown voltage of oil, paperboard and oil-paper compound insulation under the action of DC-AC superposition voltage.

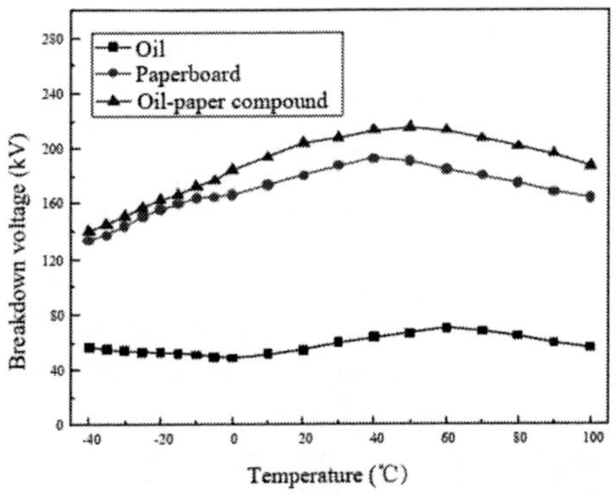

Figure 11. Breakdown voltage of oil, paperboard and oil-paper compound insulation under the action of DC voltage.

It was concluded in the above context that oil gaps were broken down firstly in oil-paper compound insulation under the action of AC voltage, and the compound insulation breakdown was consistent with oil change trend as a result. However, the paperboard was broken down firstly at low temperature, and oil was firstly broken down at high temperature in oil-paper compound insulation under the action of DC voltage. The breakdown law of the compound insulation was consistent with the paperboard at low temperature as a result, and the law was tended to oil at high-temperature. Figure 10 showed that the breakdown law of oil-paper compound insulation was similar with that under the action of pure AC voltage when η=50%, and they were consistent with the breakdown trend of oil. It can be concluded according to the law that the breakdown of oil-paper compound insulation gradually tended to breakdown law of paperboard at low temperature with the decrease of AC content. However, the breakdown of oil-paper compound insulation was always consistent with the breakdown law of oil at high temperature.

In other words, the dielectric strength of oil-paper compound insulation is changed gradually from dependence on oil dielectric strength to dependence on paperboard dielectric strength at the low temperature interval, and the dielectric strength of oil-paper compound insulation is always related to the oil dielectric strength more closely at the high temperature interval with the reduction of AC content.

5. Conclusion

The breakdown of oil-paper compound insulation under AC-DC superimposition voltage is jointly determined by transformer oil and oil-impregnated paperboard. The whole insulation structure can be broken down if any one of the two media is broken down firstly. Since the electric field distribution in the two media is affected by temperature, AC content and oil-paper proportion, and the dielectric strength of the two media is also affected by temperature and AC contents at the same time, the dielectric properties of one medium cannot be fully utilized [5]. The overall breakdown voltage of the oil-paper insulation structure is decreased. The 'bucket effect' reflects the irrationality of the oil-paper insulation structure.

The most ideal insulation structure is shown as follows because of uneven distribution of electric fields and the dielectric strength differences between the two media: oil and paperboard are broken down synchronously, and the oil-paper compound insulation structure is broken down, therefore the utilization rate of oil-paper insulation material can be improved. However, the electric field distribution is greatly changed with AC content and oil-paper proportion between -40℃ and 100℃. For example, 95.1% field intensity is concentrated on the paperboard at most under the action of low temperature and low AC content voltage [6]. In theory, the oil gap thickness should be reduced to nearly one tenths of the original value for synchronous breakdown of the two media. However, the field intensity distribution between two media is also extremely similar at high temperature interval. The insulating medium is wasted if the oil-paper proportion is too low under the circumstance.

The converter transformers in different converter stations are operated under different conditions. Therefore, the main insulation structure on the side of converter transformer valve should be designed according to actual condition of transformer operation in a targeted mode to improve the reliability of the equipment and reduce the manufacturing cost. The design should be adjusted properly according to the operation temperature scope and the withstood AC-DC proportion, thereby ensuring safe, stable and reliable operation of the converter transformer.

References

[1] Jinhua Z 2009 Transactions on Power Systems. Optimization study on voltage level and transmission capacity *IEEE* **24** 193-197

[2] Qingguo C, Minghe C and Yuan G 2013 Breakdown characteristics and its mathematical model of oil-pressboard insulation under compound electric field *Proceedings of the CSEE* **33** 170-776

[3] Tanaka Y, Ono T and Sampel M 1999 Transactions on Power Delivery. Long-term test of the equipment for ±500kV DC converter station *IEEE* **6** 196-202

[4] Hasegawa T and Yamaji K 1997 Transactions on Power Delivery. Dielectric strength of transformer insulation at DC polarity reversal *IEEE* **12** 1525-31

[5] Du Y, Zahn M and Lesieutre B C 1999 Electric Insulation Magazine. Moisture equilibrium in transformer paper-oil systems *IEEE* **15** 11-19

[6] Ebisawa Yoshihito, Yamada Shin and Mori SHigekazu 2009 Transactions on Dielectrics and Electrical Insulation. DC creepage breakdown characteristics of oil-immersed insulation *IEEE* **16** 1685-92

IOP Conf. Series: Materials Science and Engineering **292** (2017) 012011 doi:10.1088/1757-899X/292/1/012011

Experimental research on sealing performance of sulfur hexafluoride electrical equipment under low-temperature environment

G X Li, L Li[1], R Zhang, Z Q Jiang, X C Zhu, Z W Gao, H D Zhang and J Zhang

State Grid Heilongjiang Electric Power Company Limited Electric Power Research Institute, Harbin, Heilongjiang, 150030, China

[1] E-mail: lilin77@yeah.net

Abstract. It is discovered that the sealing performance of the sulfur hexafluoride electrical equipment worsens under low temperature, but research in the aspect is reported rarely when corresponding measures are adopted for improving the sealing performance of electrical equipment. Therefore, sealing and leakage test of sulfur hexafluoride electrical equipment was implemented under low-temperature environment according to the analysis on gas sealing methods of sulfur hexafluoride electrical equipment. Causes of gas leakage of sulfur hexafluoride electrical equipment under low temperature environment were analyzed according to test results. Effective measures were proposed for improving the gas sealing performance of the equipment, thereby providing valuable reference for gas sealing design, installation, operation maintenance and failure analysis of sulfur hexafluoride electrical equipment under low - temperature environment.

1. Introduction

Sulfur hexafluoride gas is used in high-voltage electrical equipment widely, which is applied for insulating and extinguishing arc mainly [1]. The insulation strength of sulfur hexafluoride gas is increased with gas pressure increasing. The insulation performance and arc extinguishing performance of the equipment will be reduced greatly if the gas pressure is decreased due to equipment leakage, which even can threaten safe operation of electrical equipment if the leakage is serious [2]. Moisture penetrates into the equipment due to water vapor pressure difference inside and outside the sulfur hexafluoride electrical equipment when the equipment suffers from poor sealing performance, and the humidity of sulfur hexafluoride gas is increased, thereby directly affecting the insulation performance thereof [3]. Therefore, the gas leak is serious hidden troubles to operation of sulfur hexafluoride electrical equipment.

North Heilongjiang, Tibetan Plateau, Gansu, Xinjiang and Inner Mongolia partially belong to alpine region [4] in China. Winter is cold and long in the above-mentioned areas. These regions also belong to important energy bases. Therefore, it is of great significance to guarantee safe operation of sulfur hexafluoride electrical equipment in winter.

In cold regions, outdoor sulfur hexafluoride electrical equipment suffers from as liquefaction, gas leakage, actuating mechanism abnormal action and other failure [5]· The following circumstances are common during gas leakage detection of sulfur hexafluoride electrical equipment and equipment defect treatment: equipment gas leakage rate is low during daytime and high during nighttime [6], or equipment can be operated normally in summer, but gas leakage is discovered in winter. Causes are

IOP Conf. Series: Materials Science and Engineering **292** (2017) 012011 doi:10.1088/1757-899X/292/1/012011

investigated, namely the circumstances are produced mainly because the sealing performance of the equipment worsens under low temperature operation environment [7]. It is discovered that research in the aspect is reported rarely when corresponding measures are adopted for improving the sealing performance of sulfur hexafluoride electrical equipment at low temperature. Therefore, it has important practical significance to conduct sealing and gas leakage test of sulfur hexafluoride electrical equipment at low temperature, investigate the sealing performance of the equipment at low temperature, analyze gas leakage causes of sulfur hexafluoride electrical equipment at low temperature, and adopt corresponding measures to improve the sealing performance of sulfur hexafluoride electrical equipment. Therefore, it is of an actual significance to improve the sealing performance of sulfur hexafluoride electrical equipment, implement sealing and gas leakage test for sulfur hexafluoride electrical equipment under low temperature environment, and investigate the sealing performance of the equipment under low-temperature environment.

2. Analysis on gas sealing methods of sulfur hexafluoride electrical equipment

Sealing methods of hexafluoride electrical equipment include static sealing and dynamic sealing [8]. Static sealing refers to sealing between two static surfaces, and sealing gaskets are used for sealing mainly, such as sealing between two flange surfaces. Dynamic sealing refers to sealing between relative moving parts, which is divided into reciprocating dynamic sealing, rotary dynamic sealing and compound motion sealing. For example, the sealing between dynamic contact draw bar and the equipment belongs to compound motion sealing.

Equipment sealing structure, sealing method, sealing requirements and interrelation among the whole equipment, sub-package parts and components-sealing reciprocal diagram should be comprehended in details during supervision and inspection when sulfur hexafluoride electrical equipment is installed and after the equipment is put into operation. The equipment state should be consistent with actual operation condition as far as possible during sealing test. The sub-package parts or components can be tested if it is difficult to test the sealing performance of the whole equipment. The relationship between the allowable leakage rate of the sub-package parts or components and gas leakage rate of the equipment should be determined according to the sealing reciprocal diagram. For example, the sealing reciprocal diagram of a three-phase single-pressure porcelain knob circuit breaker produced by one manufacturer is shown in figure 1.

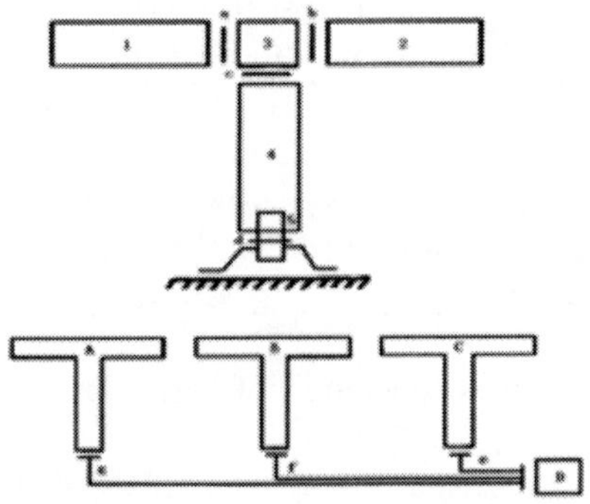

1- arc extinguish chamber; 2- arc extinguish chamber; 3- transmission case; 4- strut porcelain bushing; 5- operation draw bar;
A- phase A breaker; B- phase B breaker; C- phase C breaker; D- control cabinet;
a,b,c,d- O -shaped sealing ring; e,f,g-gas pipeline

Figure 1. Sealing reciprocal diagram of sulfur hexafluoride circuit breaker.

3. Sealing and leakage test of sulfur hexafluoride electrical equipment under low-temperature environment

3.1. Low temperature sealing test of tested device

Figure 2 shows that a tank circuit breaker without casing was adopted as the tested device (effective gas volume is 250L). Flange sealing surfaces 4, 5, 6 and 7, gas charging and discharging connecting part 11, pressure gauge connecting part 10, temperature sensor 3 and breaking-closing connecting part 9 (the sealing structure is similar to that listed in figure 2) were selected for sealing test. The qualified O-shaped sealing ring was installed in the sealing groove according to the requirements of the installation technology indoors in the test (ambient temperature was between 15 and 25 ℃), and the bolt was evenly tightened to meet the tightness requirements after the sealing blind flange was installed. Then, sulfur hexafluoride gas with certain pressure was filled to the tested device. Qualitative leak detection method by quantitative leak detector and binding method were adopted for testing the sealing performance of the tested device after the device was placed statically for 12h. The tested device was pushed outdoors (the temperature was lower than -30 ℃) and maintained outdoors for 24h, and then its sealing performance was tested under low temperature. The gas leakage rate under low temperature was increased compared with that under normal temperature. However, it should not be higher than the allowable leakage rate by three times. The device was switched on and off for 50 times respectively under low temperature after the sealing test was finished. Then, the sealing performance was tested again, and the gas leakage rate of the breaking-closing connecting part should be the same as the original value.

The device was moved back indoors after the low temperature sealing test was finished. The sealing performance was tested again 48h later. Then, the device was pushed outdoors again and maintained for 24h, and the sealing performance was tested again under low temperature. The results of the tested device in low temperature sealing test are shown in table 1.

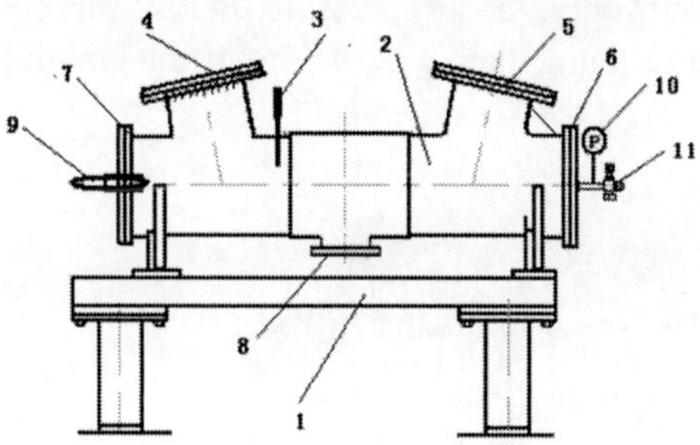

1- braced frame; 2- tested device body; 3- temperature sensor; 4, 5, 6, 7, 8-flanges of sealing junction surfaces;
9-breaking-closing connecting part; 10 - pressure gauge connecting part; 11- gas charging and discharging connecting part

Figure 2. Device for low temperature sealing test.

Table 1. Results of tested results in low temperature sealing test.

Ambient Temperature (℃)	Gas Temperature in the Equipment (℃)	Leakage Test Methods	Sulfur Hexafluoride Gas Concentration (μL/L)									
			Parts for Leakage Test									
			3	4	5	6	7	8	9 (close)	9 (break)	10	11
17.5 to 21.0	17.5 to 20.5	Qualitative leak detection method by quantitative leak detector	0	0	0	0	0	0	0	0	0	0
		Binding method	0	0	0	0	0	0	0	0	0	0
-32.2 to -38.5	-34.5 to -37.5	Qualitative leak detection method by quantitative leak detector	0	0	0	0	0	0	0	0	0	0
		Binding method	0	0	0	0	0	0	0	0	0	0
18.5 to 21.0	17.5 to 20.9	Qualitative leak detection method by quantitative leak detector	0	0	0	0	0	0	0	0	0	0
		Binding method	0	0	0	0	0	0	0	0	0	0
-33.5 to -37.5	-34.8 to -37.0	Qualitative leak detection method by quantitative leak detector	0	0	0	0	0	0	0	0	0	0
		Binding method	0	0	0	0	0	0	0	0	0	0

3.2. Low temperature sealing and leakage test of sulfur hexafluoride electrical equipment in service
The sealing performance of sulfur hexafluoride electrical equipment is always affected by low-temperature environment in winter, wherein the equipment is operated outdoors. The sealing defect of the equipment will be highlighted especially in the environment with temperature lower than -30 ℃. Therefore, 1500 sets of outdoor sulfur hexafluoride electrical equipment underwent sulfur hexafluoride gas leak detection analysis according to the natural condition of low ambient temperature in winter. The tested equipment was classified according to different equipment manufacturers, equipment types, sealing structures and sulfur hexafluoride gas inflation pressures. The distribution of 35 gas leakage points was analyzed, wherein the leakage points were discovered in the test. The test results showed that the gas leakage points were mainly discovered on flange sealing surface, dynamic sealing connecting part, density relay connecting part, tank weld sand holes, porcelain insulator cracks, etc. in the tested sulfur hexafluoride electrical equipment such as porcelain knob circuit breaker, tank circuit breaker, transformer, GIS, etc. Lower leakage was discovered in some leakage points when the ambient temperature was higher, which even cannot be discovered. However, the leakage was increased, and even safe operation of the equipment was affected under low temperature environment. For example, when a 500kV sulfur hexafluoride circuit breaker was operated at the winter environmental temperature of less than -30 ℃ for 5 hours only, the gas pressure thereof was decreased from 0.43Pa to 0.35MPa. The leakage was discovered in flange facing at the top of the circuit breaker according to inspection as shown in figure 3.

Figure 3. Infrared imaging figure of sulfur hexafluoride gas leakage test.

4. Analysis on gas leakage reasons of sulfur hexafluoride electrical equipment under low temperature environment
The sealing performance of sulfur hexafluoride electrical equipment is always affected by low-temperature environment in winter, wherein the equipment is operated outdoors. The results of electrical equipment in sulfur hexafluoride gas leakage test showed that the gas sealing performance stability of electrical equipment under low temperature was determined by the following factors.

4.1. Compression ratio of sealing ring
O-shaped sealing ring is installed in the sealing groove. An initial compression should be reserved in order to ensure its sealing performance. The pre-compression is also different corresponding to sealing surfaces with different cross section diameters in different applications. Generally speaking, the initial compression of the O-shaped sealing ring is approximately 15% ~ 30% under static sealing in a self-tight sealing structure [9], which is about 9% ~ 25% in dynamic sealing. Different requirements are proposed for the compression of O-shaped sealing rings aiming at metal-metal flange sealing and metal flange-porcelain sealing. Rational compression of O-shaped sealing rings can be controlled according to table 2 generally when the sealing ring hardness is within the scope of 65-75 in shore hardness.

Table 2. Compression ratio of O-shaped sealing ring.

Sealing Mode	Material of Sealing Surface	Compression Ratio (%)
Static sealing	Metal - metal flange	25
	Porcelain insulator - metal flange	30
Dynamic sealing and lateral sealing	Metal shaft- metallic sheath	12

The requirements of dynamic sealing assembly and requirements of controlling rotation friction resistance of motion axis are considered. Larger value should not be taken for the compression ratio of dynamic sealing, which can be limited at 12% or so generally. O-shaped sealing ring is applied for dynamic sealing since dynamic sealing has lower compression ratio, and the sealing performance of the sealing ring is poor, which should not be used in key parts.

The sealing ring for sulfur hexafluoride electrical equipment is made of G22B EPDM generally [10]. The compression ratio of the sealing ring should be controlled between 25 and 30% during design of the sealing groove. The compression ratio should be determined according to the material of the sealing surface and working ambient temperature, and the stability of the sealing performance can be affected due to too high or too low compression ratio. Rubber elasticity decreases with temperature drop, which will become vitreous if the temperature is lower than -60 ℃, thereby losing flexibility completely. The reaction of normal force vertical to the sealing surface will be formed by the atmospheric pressure in the sealing ring on the sealing ring, it can be offset by the elastic force, the sealing force will be reduced, which even can be zero or less than zero, and leakage is caused when the ambient temperature is lower if the compression ratio is lower. The compression permanent deformation will be increased when the compression ratio is too large, and the service life of the sealing ring can be reduced as a result. When the depth of the sealing ring along the circumference is not consistent, or the thickness of the sealing ring along the circumference is not uniform, too high or too low compression ratio of the sealing ring can be caused locally, thereby affecting the sealing performance stability of the product.

4.2. Strict control of surface roughness of equipment sealing surface

The sealing effect of the sealing structure may be affected by too high or too low roughness of the sealing surface of the sulfur hexafluoride electrical equipment. The sealing structure of the sulfur hexafluoride electrical equipment was tested. The results showed that the concentric circle formed by tool marks was beneficial for air tightness in suitable scope of surface roughness (Ra 1.6 to 6.3 μm) on the sealing surface processed by turning tool (boring tool). The rubber ring and the sealing ring were combined well in the scope of surface roughness Ra 1.6 to 6.3μm. The distance of SF_6 leakage path was increased by concentric circular knife grains, thereby the leakage was reduced. However, O-shaped sealing ring had poor sealing contact when the surface roughness exceeded Ra 12.5 μm; thereby gas leakage was caused easily. On the contrary, the 'resistance' effect of the concentric circular knife grain cannot be obtained if the rolling extrusion mirror surface was too smooth, and the air tightness cannot meet requirements.

The grinding cracks of the porcelain sealing surface ground by a grinding wheel are not closed concentric circles, but rambling and intermittent short lines. Considerable amount of grinding cracks are directed along SF_6 leakage path (radial direction). Therefore, porcelain insulator sealing is more difficult compared with that of metal sealing. The following effective measures can be adopted in order to achieve ideal sealing effect: (1) the sealing ring compression ratio is appropriately increased. (2) The roughness of the porcelain sealing surface is strictly controlled, and the roughness of the fitting surface is polished to Ra 1.6μm; (3) the dirt and porcelain powder are wiped by non-wool cloth (paper), and the sealing surface should be touched by hand without feeling of porcelain powder rolling, and the compression ratio of the sealing surface is increased properly.

4.3. Guarantee of sealing ring materials

Two indicators are mainly adopted for judging the material performance of the sealing ring aiming at sulfur hexafluoride electrical equipment, namely hardness and compressive permanent deformation [11]. The hardness of the sealing ring should be moderate. It is generally required that the Shaw hardness is 65 ~ 75, it is difficult to achieve excellent combination performance between the sealing ring and sealing surface due to too high hardness, insufficient contact pressure (sealing force) can be caused easily due to too low hardness, and excellent sealing performance cannot be achieved. The compression permanent deformation of the sealing ring should be lower than 20% within the working temperature scope (-55℃ to 80℃), and excellent sealing performance is available.

When materials of O-shaped rings are selected concretely, the compatibility with the working medium should be considered on the one hand, and the pressure, temperature, continuous working hours, operation cycle and other working conditions also should be comprehensively considered on the other hand. High-quality rubber rings should be adopted for sealing in order to prevent gas leakage. The sealing ring must have the features of oil resistance, SF_6 gas decomposition, low permanent deformation, and high compressive strength, tension and extension strength.

EPDM is characterized by good resistance to water, polar solvent resistance, acid resistance, and poor oil resistance. Its working temperature scope is - 55 ~ 125 ℃. Therefore, it can be used for sealing sulfur hexafluoride electrical equipment. 427 neoprene also can be selected in general regions (the lowest temperature of - 30 ℃), 518 nitrile butadiene rubber or G22B EPDM must be used in regions with low temperature (the lowest temperature of - 40 ℃).

If the sealing ring is made of poor materials, the sealing ring itself can be penetrated by SF_6 gas on the one hand, the sealing force can be reduced due to too large compression permanent deformation after being used for certain time even though the compression ratio is controlled within the rational scope on the other hand, and gas leakage can be caused.

4.4. Correct use of lubricating grease

Lubricating grease is required in the sealing ring for dynamic sealing generally due to demand of relative motion among sealing surfaces of dynamic sealing. Vacuum silicone must be used for dynamic sealing in SF_6 environment. Low temperature lubricating grease can be used for the sealing ring in the operating mechanism. However, lubricating grease is not used for static sealing generally. No grease can be used for cold-proof sealing rings.

4.5. Manufacturing quality of control equipment parts

Equipment parts are not accepted and inspected strictly in procurement, processing and delivery, therefore nonconforming or defective products are transferred to next procedure or delivered, thereby gas leakage is caused after installation of whole set equipment. For example, sand holes and cracks are produced on the tank and weld in sulfur hexafluoride gas chamber, performance and materials of sealing gaskets cannot meet sealing requirements sufficiently, etc. The defects can be eliminated in two aspects: firstly, manufacturers should control quality strictly in all links of production, thereby preventing delivery of substandard products. Secondly, the delivery acceptance and test should be controlled well, equipment should be accepted and tested strictly during delivery after the equipment is installed, problems can be discovered immediately, and all related parties should be assisted for perfection.

4.6. Reduction of equipment design defects

Sulfur hexafluoride gas leakage also can be caused due to improper design of the equipment. For example, the ability of bus bars to buffer temperature changes is weakened aiming at GIS buses due to irrational quantity and layout of expansion joints, the flange surfaces are dislocated, and leakage is

IOP Conf. Series: Materials Science and Engineering **292** (2017) 012011 doi:10.1088/1757-899X/292/1/012011

caused. In a 220 kV porcelain knob circuit breaker, the resin materials connected with the brass pipe or the connection part between the resin material and metal material suffer from unreasonable external force, and cracks are caused due to influence of conductor dead weight, wind swinging and other factors, thereby leading to gas leakage. Many gas chambers are connected, and one density relay is shared in some equipment. Gas leakage can be caused easily as a result due to too long connection pipeline and more contacts of the equipment. When leakage accidents occur, it is difficult to discover the leakage points due to the connection mode. Meanwhile, many gas chambers decompress as a result, thereby leading to greater hidden danger to safe operation of the equipment.

4.7. Strict control of equipment installation process and installation quality

Site installation quality of sulfur hexafluoride electrical equipment is a key factor to determine the sealing performance thereof [12]. On-site construction personnel must be familiar with GIS technology data and design drawings, make clear the relevant technical requirements and quality standards, and implement the requirements and standards in the construction strictly in order to ensure the equipment installation quality.

5. Conclusion

The sealing performance of sulfur hexafluoride electrical equipment worsens when it is operated under low temperature environment. The reasons are investigated, namely the low temperature resistance performance of equipment sealing materials is poor, equipment installation defects are exposed under low temperature, or the installation process and quality of equipment are not controlled well under low-temperature environment, etc. The follows are concluded through sealing performance test of sulfur hexafluoride electrical equipment:

(1) Natural environment in the substation installation region must be fully considered for selecting sulfur hexafluoride electrical equipment. Equipment inflation pressure, sealing structure and sealing materials should meet the requirements of equipment gas sealing performance.

(2) Sulfur hexafluoride electrical equipment should not be installed or disassembled for overhaul treatment under low temperature generally. The equipment can be installed or treated under the guidance of equipment manufacturers due to requirements on construction schedule or because the defects must be treated under low temperature.

(3) The construction organization should be scientific and rigorous for installation of sulfur hexafluoride electrical equipment and overhaul engineering. The recovery time of equipment disassembly should be shortened as far as possible.

(4) The sealing ring must be made of 518 nitrile butadiene rubber or G22B EPDM. Its compression ratio is controlled at 30%, and the lubricating grease cannot be used.

(5) Corresponding measures must be equipped for resisting wind, dust and snow. Installation personnel should wear gauze masks to prevent exhalant water vapor from condensation inside the equipment and on the sealing ring, thereby avoiding influence on sealing performance of equipment and gas humidity in the equipment.

(6) The integrity of the sealing surface with waterproof glue injection slots and exposed fastener waterproof glue should be inspected comprehensively before winter is coming. The sealing surface with poor gumming should be filled carefully. Water accumulated in the slots should be dried by blowing, thereby preventing rainwater freezing in winter, expansion force, deformation of sealing ring due to extrusion, invalid bolt pre-tightening force and gas leakage in the sealing surface, or the basin-type insulator is also partially forced by expansion force, and cracks are caused, thereby leading to gas leakage in the sealing surface.

(7) The gas leakage points should be checked and handled completely before winter is coming aiming at equipment with gas leakage, therefore sudden increase of gas leakage rate under low temperature and threat on safe operation of the equipment can be avoided.

References

[1] Liang F J, Wang Y, Wang Z L and Jia X J 2010 *Insulating Materials* **43(3)** 43- 46
[2] Huang T, Wang T J and Kong F H 2013 *Jilin Electric Power* **40 (3)** 35-37
[3] Zhang G T and Liu P 2014 *East China Electric Power* **42(6)** 1257-61
[4] Li Yi, Liu J J, Li G Q, Chen H L and Li X B 2009 *Heilongjiang Electric Power* **31(2)** 90-93
[5] Li Qi 2004 *Qinghai Electric Power* **23(4)** 31-34
[6] Jia M Y, Guo F, Liu S X and Jiang X G 2011 *High Voltage Apparatus* **47(11)** 106 -109.
[7] Ye T, Yin J B and Wang R J 2010 *Inner Mongolia Electric Power* **28 (4)** 31-33
[8] Zhou S H, Li Y H, Lu Z Q and Li H S 2013*Qinghai Electric Power* **32(1)** 57-59
[9] Xiao H, Wu X B, Zeng X J and Tong X J 2006 *High Voltage Engineering* **32(2)** 108-110
[10] Long J Z, Wang K T, Zhao X and Zhou J 2013 *Power System and Clean Energy* **29(6)** 67-70
[11] Zhang S J, Wang J and Song X W 2009 *Power System and Clean Energy* **25(10)** 56-58
[12] Tong Z Y, Zhen L and Zhang Y C 2010 *High Voltage Apparatus* **46(5)** 92-97

IOP Conf. Series: Materials Science and Engineering **292** (2017) 012012 doi:10.1088/1757-899X/292/1/012012

Anti-bacteria effect of active ingredients of siraitia grosvenorii on the spoilage bacteria isolated from sauced pork head meat

X Li[1,2,3,4], L Y Xu[1,2,3,4], Y Q Cui[1,2,3,4], M X Pang[1,2,3,4], F Wang[1,2,3,4] and J H Qi[1,2,3,4,5]

[1] College of Food Science and Engineering, Beijing Key Laboratory of Detection and Control of Spoilage Organisms and Pesticide Residues in Agricultural Products, Beijing 102206, China

[2] Food Chemistry teaching team, Beijing University of Agriculture, Beijing 102206, China

[3] Innovation team, Modern agricultural industry technology system Beijing innovation team, Beijing 100102, China

[4] High Level Talents of Beijing Universities Cross Training "Real Training Plan" Project, Beijing 102206, China

[5] Email: abc960718@sina.com

Abstract. Extraction and anti-bacteria effect of active ingredients of Siraitia grosvenorii were studied in this paper. Extraction combined with ultrasonic was adopted. The optimum extraction condition was determined by single factor test; the anti-bacteria effect of active ingredients and minimum inhibitory concentration (MIC) were valued by Oxford-cup method. The results indicated that optimum extraction condition of active ingredients extracted from Siraitia grosvenorii were described as follows: ethanol concentrations of sixty-five percent and twenty minutes with ultrasonic assisted extraction; the active ingredients of Siraitia grosvenorii had anti-bacteria effect on Staphylococcus epidermidis, Proteus vulgaris, Bacillus sp, Serratia sp and MIC was 0.125g/mL, 0.0625g/mL, 0.125g/mL and 0.125g/mL. The active constituent of Siraitia grosvenorii has obvious anti-bacteria effect on the spoilage bacteria isolated from Sauced pork head meat and can be used as a new natural food preservation to prolong the shelf-life of Low-temperature meat products.

1. Introduction

Low-temperature meat products are processed with the sterilizing condition of 68-72℃, which can retain mostly nutrition and flavour with short shelf life [1, 2]. Therefore, the preservatives have high lights on prolonging the shelf life of Low-temperature meat production recent years [3]. In this paper, Sauced pork head meat is the object of study whose main spoilage microorganisms include Staphylococcus epidermidis, Proteus vulgaris, Bacillus sp and Serratia sp. as is shown in the research [4-6].

Up to now, there are two kinds of food preservations, including natural and chemical additives. Chemical additives were used widely in the past, but with the promotion of life standard, consumers prefer the natural, safe, efficient preservatives to chemical preservatives which is harmful [7]. At present, spices [8, 9] and Chinese herbal medicine [10, 11] were mostly used as natural bacteriostatic agent. In this paper, the anti-bacteria effect of Siraitia grosvenorii was investigated ,which was a kind of medicinal and edible plant in China and obtained widespread application on Natural sweetener [12,

13] and Chinese herbal medicine [14, 15]. The previous studies showed that flavonoid [16, 17], mogroside, triterpene benzoate [18], saponins [19] were the main chemical substances of Siraitia grosvenorii and a new antibacterial compound named siraitiflavandiol was found by Zheng Y et al. [20]. Kaempferol was a kind of flavonoid [21, 22] which would be used to determine the extraction condition. In the paper, antibacterial effects of active ingredients of Siraitia grosvenorii on the spoilage bacteria isolated from Sauced pork head meat was studied aiming to prolong the shelf-life of Sauced pork head meat.

2. Material and methods

2.1. Sample

Siraitia grosvenorii purchased from Mengzheng parity pharmacy in Beijing, China was smashed by Pulverizer(Tianjin Tasite Instruments Co., Ltd, China) and 40 meshes.

2.2. Test strains

All the test strains including Staphylococcus epidermidis(X12), Proteus vulgaris (T4), Bacillus sp. (X10) and Serratia sp.(N11) were isolated from Sauced pork head meat,and identified by Henan Academy of Sciences Institute of Biology Limited Liability Company, China.

2.3. Extraction of active substances

Five grams of sample were soaked in ethanol for 15 hours and extracted with ultrasonic (Kunshan Ultrasonic Instruments Co., Ltd, China) of 480 W and 50°C.Finally the supernatant was taken to evaporate until paste aftercentrifuged with 3000 rpm for 5minutes and then preserved at 4°C until tested.

2.3.1. Standard curve of kaempferol. Standard solution of kaempferol(97%;Aladdin Industrial Corporation, Shanghai, China) was prepared for 0.0500 mg/mL. Ultraviolet absorption wavelength was valued by spectrum scanning standard solution Standard curve was drawn by detecting absorbance values of standard solution of different concentration [23].

2.3.2. Concentration of ethanol. Five grams of sample were extracted with 125mL ethanol [24] of60%, 65%, 70%, 75% and 80% respectively with ultrasonic for 30minutes. Finally, the optimum concentration of ethanol was determined by the yield of kaempferol .

2.3.3. Time of ultrasonic extraction. Five grams of sample were extracted with 125mL ethanol of optimum concentration, extraction time with ultrasonic were 20minutes, 30minutes, 40minutes, 50minutes and 60minutes respectively. Finally, the optimum extraction time was determined by the yield of kaempferol.

2.3.4. Determination of the yield of kaempferol. The paste of 0.00125g/mL was suspended with 70% ethanol whose yield of kaempferol was determined by standard curve.

2.4. Anti-bacteria experiments

All the strains preserved at -80°C were thawed at room temperature and mixed with sterile nutrient broth of 50mL which was prepared by dissolving of 19g/L in water, next the sterile nutrient broth(Beijing land bridge technology CO., Ltd, China)inoculated with strains was incubated for 12 h at 37 °C which was repeated two times.

The sterile nutrient agar(Beijing land bridge technology CO., Ltd, China)which was prepared by dissolving of 33 g/L in water was inoculated with bacterial suspension of 1×10^9 CFU/mL diluted by plate counts method [25] (1mL bacterial suspension in 100mL agar medium) and poured into sterile petri dishes. Two sterile Oxford cups were placed on the surface of agar medium and marked with L and CK (L: antibacterial solution; CK: blank control), then 100μL sterile water and antibacterial

IOP Conf. Series: Materials Science and Engineering **292** (2017) 012012 doi:10.1088/1757-899X/292/1/012012

solution which was prepared by dissolving the paste of 1g/mL in sterile water were added into Oxford cups respectively. Finally all the plates were incubated for 16~18h at 37°C. Thereafter the anti-bacteria effect of active ingredients was determined by measuring the diameter of inhibition zone in the plates.

2.5. Determination of minimum inhibitory concentration

Minimum inhibitory concentration (MIC) was determined by Oxford-cup method [26]. The sterile nutrient agar prepared by dissolving of 33g/L in water was inoculated with bacterial suspension (1mL bacterial suspension in 100mL agar medium) and poured into sterile petri dishes. Eight sterile Oxford cups marked numbers were placed on the surface of agar medium symmetrically and steadily, next 100μL antibacterial solution of different concentrations were added into Oxford cups respectively and sterile water was blank control. All plates were incubated for 16~18h at 37°C. The lowest concentration that had no inhibition zone compared with blank control was regarded as MIC.

2.6. Statistic analysis

Data analysis and charts were made by Microsoft Office Excel 2007, significant difference analysis was valued by IBM SPSS Statistic 21.

3. Results and Discussions

Figure 1 demonstrated that the concentration of ethanol had an obvious impact on the yield of kaempferol ($P<0.01$), with the increase of concentration, the kaempferol yield increased first and then decreased. The highest yield of kaempferol was 1.32mg/g with 70% ethanol.

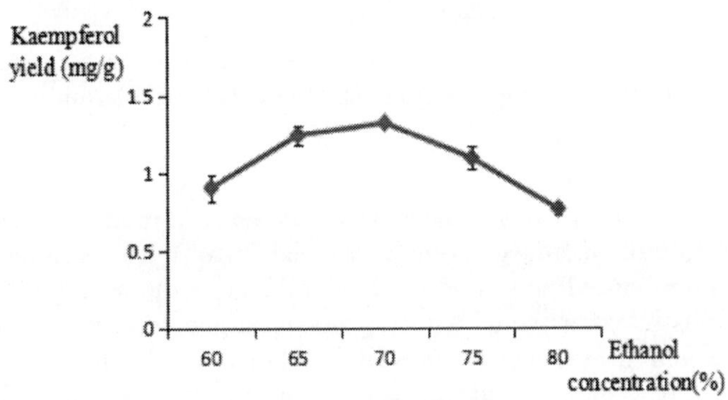

Figure 1. Effect of concentration of ethanol on Kaempferol yield.

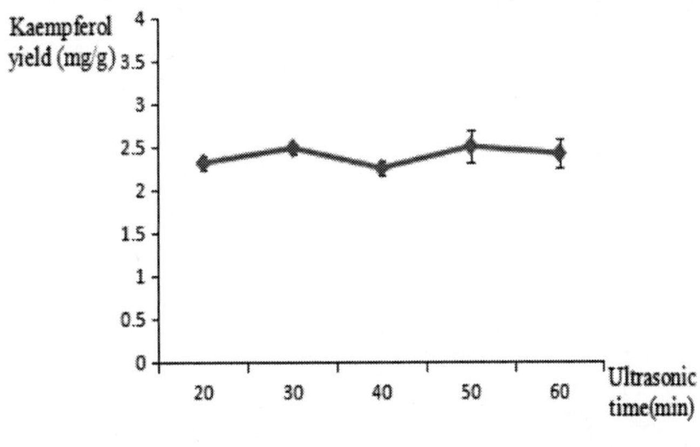

Figure 2. Effect of ultrasonic extraction time on Kaempferol yield.

IOP Conf. Series: Materials Science and Engineering **292** (2017) 012012 doi:10.1088/1757-899X/292/1/012012

Kaempferol is a kind of flavonols and slightly dissolved in water, easily dissolved in hot ethanol, exsisted in Siraitia grosvenorii with free form mainly and glycosides including kaempferol-3-c and kaempferol-3,7-double-glucoside[27].As is showed in figure 1 , the reason that yield of kaempferol decreased when the concentration of ethanol over 70% was that higher ethanol concentration restrained the hydrolysis of glycosides. Finally, the concentration of ethanol used in extraction was determined with 70%.

Figure2 indicated that ultrasonic time had no obvious impact on the yield of kaempferol (P>0.05), so ultrasonic extraction time was 20 minutes with kaempferol of 2.31mg/g. Compared to ethanol extraction, ultrasonic assisted extraction obtained higher yield. As is showed in figure 1 and figure 2,the extraction condition of active constituents of Siraitia grosvenorii was determined as follows: solid-liquid ratio of 1:25, 70% ethanol, temperature of 50℃,ultrasonic time of 20 minutes, power of 480W.

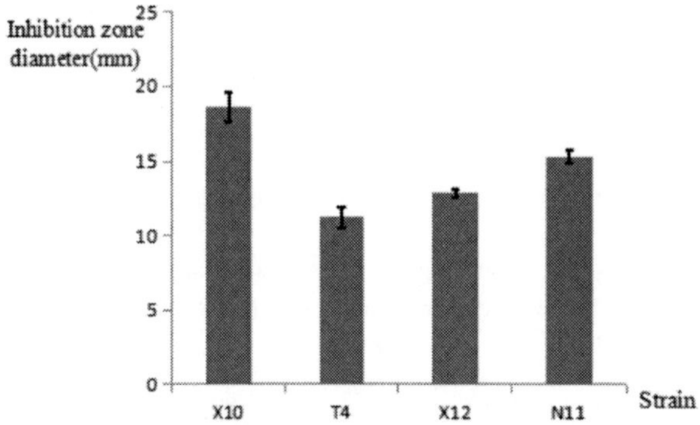

Figure 3. The anti-bacteria effect of active ingredients of Siraitia grosvenorii.
X10=Bacillus sp, T4=Proteus vulgaris, X12=Staphylococcus epidermidis, N11=Serratia sp.

As is showed in figure 3, the active ingredients of Siraitia grosvenorii had different anti-bacteria effects on allstrains, especially the best anti-bacteria effect on X10 with inhibition zone of 18.6 mm, while the anti-bacteria effect on T4 was weakest whose diameter of inhibition zone was only 11.23mm.

The active ingredients of Siraitia grosvenorii showed distinct anti-bacteria effect on the strains which were isolated from Sauced pork head meat. What's more, the results indicated that there were flavonols and flavonoids in extractive of Siraitia grosvenorii according to the chromogenic reactions with sodium hydrate and magnesium acetate. Asif Ahmad et al. studied that the molecular of flavonoids had included benzene ring and hydroxy, which can enter the cells of bacteria to destroy the plasma membrane and restrain the synthesis of nucleic acid [28]. Thereby the effective components of Siraitia grosvenorii can be used as a new natural food preservation and to delay the shelf-life of Sauced pork head meat through inhibiting microorganism in the future.

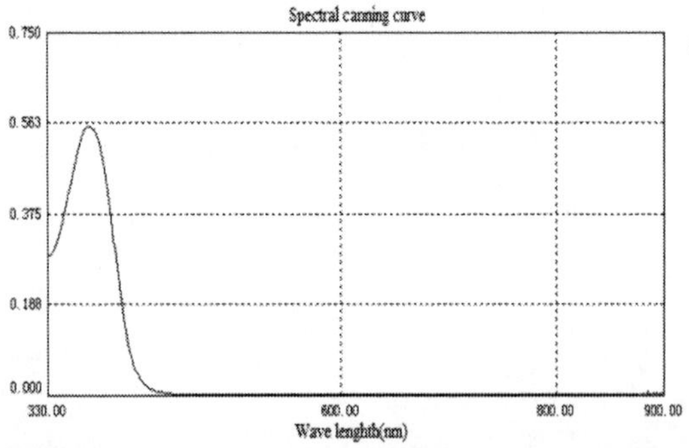

Figure 4. Spectral scanning curve of Kaempferol.

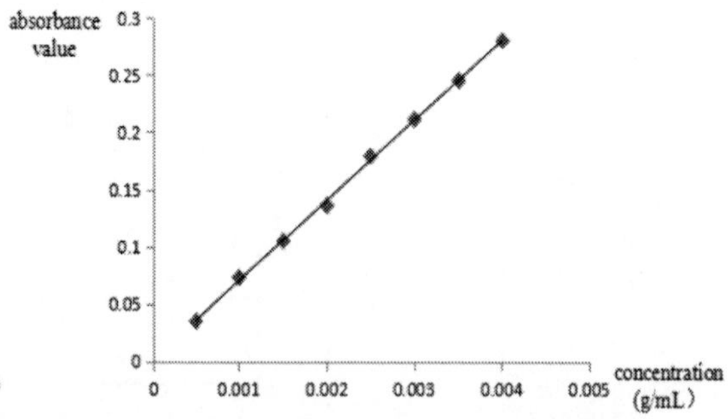

Figure 5. Thestandard curve of Kaempferol.

As was shown in figure 4, ultraviolet absorption wavelength of kaempferol was 369 nm.
Figure 5 demonstrated linear regression equation: y = 69.92872 x + 0.0007, R^2=0.9996.

Table 1. Determination of minimum inhibitory concentration of active substances of S. grosvenorii.

Strains	The concentration of antibacterial solution（g/mL）						
	1.0	0.5	0.25	0.125	0.0625	0.0313	0.0157
X10	++	++	+	+-	-	-	-
T4	+	+	+-	+-	+-	-	-
X12	++	+	+	+-	-	-	-
N11	+	+	+-	+-	-	-	-

[a] “d” means diameter of inhibition zone,d≥20mm =“+++”; 15mm≤d≤20mm=“++”;10mm≤d≤15mm =“+”;d≤10mm =“+-”;“-”means no inhibition zone.

Table 1 showed that MIC of effective component extracted from Siraitia grosvenorii was 0.125g/mL, 0.0625g/mL, 0.125g/mL and 0.125g/mL respectively on X10, T4, X12 and N11. Effective components showed higher anti-bacteria effect on T4 than others. While with the concentration of effective component decreased, the anti-bacteria effects on X10 and X12 changed significantly.

4. Conclusion

The results showed that the extraction condition of Siraitia grosvenorii was determined as follows: solid-liquid ratio 1:25,70% ethanol, ultrasonic temperature of 50℃,ultrasonic time of 20 minutes, ultrasonic power of 480W.The active ingredients of Siraitia grosvenori showed bacteriostatic effects on Staphylococcus epidermidis, Proteus vulgaris, Bacillus sp, Serratia sp. and MIC was 0.125g/mL, 0.0625g/mL, 0.125g/mL and 0.125g/mL respectively.

The study results provide reliable data for the application of natural plant bacteriostatic agent on low-temperature meat products. But active components of Siraitia grosvenorii and anti-bacteria mechanisms need further research.

References

[1] Yuan X Q 2012 Research on quality changes of stewed meat in seasoning Chongqing: Southwest University
[2] SB/T 10481-2008 Requirements of quality and safety on Pasteurized meat products *China*
[3] Zhang D Q, He Z 2008 Research progress of the preservation technology of processed meat products *J.Meat Research* **5** 3-7
[4] Li Y L 2009 Isolation and identification of spoilage bacteria in low-temperature meat products *J.Inner Mongolia Agricultural Science and Technology* **3** 62-63
[5] Xiao X 2013 Bacterial diversity and the inhibitory mechanism of antibacterial plant extract against specific spoilage organism in Yao meat
[6] Lei M, Jiang R R, Ju R H and Xiao Y 2013 Microbiology detection of pasteurized meat products anddominant microflom analysis and identification *China Brewing* **32(3)** 83-86
[7] Fan J S 1993 Functional food additive *Tianjin Science and Technology Press* 318
[8] Tian F, et al. 2017 Study on antibacterial activity of 7 spices on Pseudomonas,Staphylococcus epidermidis and Brochothrix *Journal of Beijing University of Agriculture* **32(2)** 10-14
[9] Liu L, Kong B H, et al. 2008 The inhibition effect of spices extraction on Listeria monocytogenes in culture medium and in chilled pork *J.Science and Technology of Food Industry* **29(9)** 87-90
[10] Liu F, Cao X Z, et al. 2015 Study on bacteriostatic activity of total flavonol from angelica sinensis *J.China Food Additives* **1** 60-63
[11] Zhou J X 2006 Reviews on research progresses, actual problem and prospects on natural food preservatives from plant materials *J.Food Science* **27(1)** 263-268
[12] Chen Y 2005 Chemical and functional studies of Siraitia grosvenorii Shanghai:Shanghai Jiao

Tong University
[13] Qi Y P and Tang M Y 2001 The chemical component in *Siraitia grosvenorii* (Swingle)C. Jeffrey fruit and utility *Fujian Medical Journal* **23(5)** 158-160
[14] Song F, et al. 2006 A natural sweetener, *Momordica grosvenori* attenuates the imbalance of cellular inmune functions in alloxan-induced diabeticmice *J.Phytothory Research* **20** 552-560
[15] Wei R C, et al. 2013 Advances on Medicinal Plant *Siraitia grosvenorii Hubei Agricultural Sciences* **52(23)** 5669-5672
[16] He K X,et al. 2013 Changes of total flavonoids in *Siraitia grosvenorii* and its antibaterial activity *Amino Acids and Biological Resources* **35(4)** 16-20
[17] Chen Q B,et al. 2004 Study on tea value of *Siraitia grosvenorii* leaf *Tea In Fujian* **3** 12-15
[18] Motohiko U.et al. 2002 Inhibitory effects of cucurbitane glycosides and other triterpenoids from the fruit of *Momordica grosvenori* on Epstein-Barr Virus early antigen induced by tumor promoter 12-O-tetradecanoylphorbol-13-acetate *Agric Food Chem* **50** 6 710-6 715
[19] Shi D F 2016 Studies on the extraction, separation, structural identification and biological activities of the chemical compounds from Dictamnus dasycarpus Turcz. and *Siraitia grosvenorii* swingle Changchun:Northeast Normal University
[20] Y Zheng,et al. 2009 A new antibacterial compound from Luo Han Kuo fruit extract (*Siraitia grosvenori*) *J Asian Nat Prod Res* **11(8)** 761-5. doi: 10.1080/10286020903048983.
[21] Huang X S 2007 Chemical constituents of *Siraitia grosvenori*i (Swingle) C. Jeffrey Guangxi:Guangxi Normal University
[22] Zhang H,et al. 2011 Research progress on chemical compositions of fruits Momopdicae *Journal of Anhui Agri* **39(8)** 4555-4556,4559
[23] Chen Z G,et al. 2016 Determination of quercetin content in Raw drug by ultraviolet spectrophotometry *J.Progress in Veterinary Medicine* **37(1)** 122-124
[24] Cui B 2012 Study on the extraction of purification technology of flavone of Mangosteen
[25] GB 4789.2-2010 National food safety standard Food microbiological examination: Aerobic plate count
[26] Li Y, Liu J, et al. 2015 Antioxidation and antibacterial property of flavonoids compounds extracted from Nitraria leaves *J.Food Industry* **36(3)** 94-98
[27] Ma M,Wang L,et al. 2014 Optimization of extraction technology of kaempferol from Sophora japonica by orthogonal test *J.China Pharmacy* **25(43)** 4069-71
[28] Asif Ahmad,Muhammad Kaleem,et al. 2015 Therapeutic potential of flavonoids and their mechanism of action against microbial and viral infections-A review *J.Food Research International* **77** 221-235

IOP Conf. Series: Materials Science and Engineering **292** (2017) 012013 doi:10.1088/1757-899X/292/1/012013

The use of pyro-polymeric catalysts and a new cylindrical cell design in oxygen-aluminum generators

E A Kiseleva[1], A Z Zhuk, B V Kleymenov and V G Oudaltsov

Joint Institute for High Temperatures of the Russian Academy of Sciences, Izhorskaya st.13 Bd. 2, Moscow, Russia 125412

[1] E-mail: kanna787@mail.ru

Abstract. The increase in energy consumption, the economic crisis, the development of certain areas of engineering and energy, as well as the related deterioration of the environmental situation, require the development of new electrochemical current sources with high specific characteristics. In the field of creating air-hydrogen fuel cells, the problems of safety and mobile storage of hydrogen have not been completely solved, stagnation in the development of lithium-ion, lithium-air and lithium-sulfur batteries has been outlined. All this requires searching for new technological solutions, ways to increase the energy and resource characteristics of electrochemical current sources (ECS), reducing their cost [1-2]. The use of metals (aluminum, zinc, magnesium) as an energy carrier is due to their high energy intensity (in combination with the lack of transport, storage and on-board storage problems) of the relatively low cost of metals, their availability, storage safety and the absence of harmful emissions when used. As in the chemical and electrochemical use of metals, safe products (oxides, hydroxides) are formed, which are reduced to metals within the framework of traditional production technologies. Thus, a closed cycle of energy use is organized. The task of this paper is to evaluate the possibility of reducing the cost and increasing the specific power of ECS using oxygen depolarization. The goal is achieved by using non-platinum catalysts and optimizing the design of the current source.

1. Introduction

The task of this paper is to evaluate the possibility of reducing the cost and increasing the specific power of ECS using oxygen depolarization. The goal is achieved by using non-platinum catalysts and optimizing the design of the current source.

One of the most promising electrochemical systems is oxygen (air) -aluminium (AA) [1], which has a high specific energy intensity (in the future up to 800 W-h / kg), which is noticeably higher than for the most common lithium-ion battery batteries (80-150 W • h / kg). The high efficiency of obtaining electric energy air-aluminum systems allows the heat to be released to be fully used, for example, for heating (conditioning) the passenger compartment of an electric vehicle [2-3]. The disadvantages of such systems include low specific power and a high price of platinum oxygen reduction catalysts.

The most probable total current-forming process in an air-aluminum element in an alkaline medium is described by the following equation:

$$4Al+3O_2+6H_2O+4NaOH \rightarrow 4Na[Al(OH)_4] \quad (1)$$

IOP Conf. Series: Materials Science and Engineering **292** (2017) 012013 doi:10.1088/1757-899X/292/1/012013

Electrochemical oxidation (anodic dissolution) of aluminum in an alkaline medium, according to the literature [1-5], proceeds with the formation of the complex hydroxoion $Al(OH)_4^-$ and is described by equation

$$Al + 4OH^- \rightarrow Al(OH)_4^- + 3e^- \quad (2)$$

The cathodic reduction of oxygen on a positive electrode in an alkaline medium is described by the following electrochemical reaction with a standard reaction potential:

$$O_2 + 2H_2O + 4e \rightarrow 4OH^- \quad (3)$$

The reduction of oxygen can be accelerated by applying electrodes with a highly developed surface, for example porous electrodes [6], whose active layer consists usually of a catalyst, a hydrophobizer (fluoroplastic), and a conductive component. As a catalyst in cathodes, platinum is used [7].

Alkaline electrolyte, along with platinum metals, can be used more readily available catalysts - skeleton nickel, activated carbon and soot, modified by various promoters. Pyro-polymers, cobalt porphyrins deposited on dispersed carbon carriers and subjected to pyrolysis, are considered as effective catalysts for gas-diffusion oxygen (air) electrodes for an alkaline electrolyte [8, 9].

In this work, the synthesized catalysts were tested in the composition of the active layer of the gas-diffusion cathode of a semi-element with a liquid alkaline electrolyte. The hydrophobized gas-diffusion electrode consisted of three parts: a shut-off hydrophobized layer, an active layer and a nickel substrate (current collector). Such an electrode structure makes it possible to realize a stably developed three-phase catalyst-electrolyte-oxygen boundary, which is necessary for the oxygen reduction reaction [8]. As a cathode catalyst, a pyro-polymer of cobalt tetra (pmethoxyphenyl) porphyrin (PP TMPPS) was used. The main carbon carrier was acetylene carbon black AD-100 and for comparison – activated carbon UAF.

2. Materials and methods

The laboratory technology for manufacturing cathodes was as follows. A nickel mesh substrate (electrode area 5 cm2) with a pre-welded current collector was placed in a mold and a hydrophobic layer of hydrophobized carbon black (HPCB) with 35 wt% fluoroplastic F-4D was applied thereto. The hydro-clay layer placed in the mold was pressed at low pressure and an active layer was applied over it. Previously, the active layer components, HPCB and catalyst were mixed in a certain ratio. The electrode was then pressed at a pressure of 280 kg / cm2 and held for 3 minutes. Similarly, electrodes based on other carbon materials.

The volt-ampere characteristics of air cathodes with a working surface of 2 cm2 were obtained with the help of the «PR-50-1» potentiostat and the «PR-8» programmer. The measurements were carried out in the potentiostatic regime from the stationary potential in the cathode direction with a step of 25 mV to a polarization of 300 mV with an exposure at each given potential of 3 min.

3. Results and discussion

In figure 1 presents comparative polarization curves for the recovery of air oxygen for cathodes with various catalysts synthesized on carbon UAF: 10% PP TMFPSO, 15% PP TMFPSO and carbon black AD-100: 20% PP TMPPSO. Here, too, is a curve for an electrode with a UAF carbon without a promoter in the active layer.

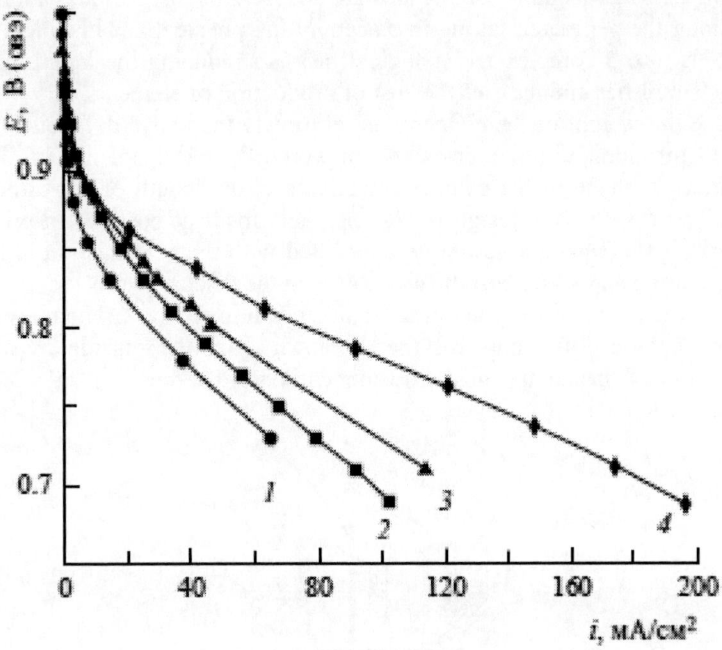

Figure 1. Figure with short caption (caption centred). Comparative volt-ampere curves of cathode gas diffusion electrodes in an air atmosphere with various catalysts based on carbon of UAF and carbon black AD-100, measured at 60 °C in 5M NaOH: 1-UAF; 2–UAF + 15% PP of the TMPPSo; 3–UAF + 10% PP of the TMPPSo; 4–UAF + 20% PP of TMPPSo.

As can be seen from figure 1, the positive effect of catalytic additions begins to appear already at the initial section of the curves. At a potential of 850 mV, the currents obtained at the cathode with the TMPPSo / AD-100 PP are approximately 20 mA/cm^2 higher than the currents obtained at the carbon electrode and at 10 mA / cm^2 higher than on the TMPPSo-promoted angle. The overall dimensions of the currents for large polarizations differ even more: at a potential of 730 mV for the same types of cathodes, the differences are 90 and 50 mA/cm^2, respectively. On the basis of the data, all the catalysts for the activity in the oxygen reaction can be arranged in the following order: 20% PP TMPPSO / AD-100> 10% PP TMPPSo / UAF> 15% PP TMPPSo / UAF. The maximum current on the most active catalyst with a 20% FS of TMPPS with a potential of 730 mV is -155 mA/cm^2.

The optimized gas-diffusion cathode with the catalyst TMPPSo / AD-100 reaches a current density of 300 mA / cm^2 at E = 0.73 V and a temperature of 60 °C.

The cathode with PP catalyst TMPPSo / AD-100 was tested in an air-aluminum electrochemical current source.

Widely known AA generators use flat electrodes in their design when the anode and cathode of the same area are opposite each other [10-13], or the anode is located between two electrically connected cathodes [14-17]. At the same time, the equality of the cathode and the anode over the area of the working surface should be attributed to the shortcomings that do not allow the efficient use of aluminum having better energy yield characteristics than air cathodes. In the elements of a flat construction, the stock of consumable electrolyte is limited by the value of the interelectrode gap and to ensure complete solubility of aluminum in the electrolyte volume it is necessary to introduce additional capacity and an electrolyte circulation system into the generator design. In this case, the organization of a uniform (without stagnant zones) circulation of electrolyte presents a complex technical problem. The circulation speed of the electrolyte is determined not by the speed of consumption of the components, but by the rate of heat removal from the interelectrode gap. The main share in the weight of such, certainly, compact elements is occupied by their case (more than 30% by weight) [18], which provides rigidity of the design and sealing of cathodes along the perimeter. Sealing cathodes along the perimeter, taking into account the physical and chemical properties of the alkaline electrolyte, is also a complex technological task, so reducing the length of the sealed zone increases reliability, simplifies and reduces the cost of production of elements.

In order to reduce the weight of the element and eliminate the above deficiencies, we proposed a cylindrical design of air-aluminum elements [18], in which the casing of the air-aluminum element performed the cylindrical shape of the cathode. In this case, the length of the cathode sealing seam decreased by a factor of 3. Such a design of the elements made it possible in principle to solve a number of engineering problems in a new way, connected with the organization of heat removal and adjusting the concentration and volume of the electrolyte in the discharge process.

The object of the study is an air-aluminum element, shown in figure 2. The body of the element is an air gas diffusion cathode of cylindrical shape, blown from the outside by air. A cylindrical aluminum anode is placed inside the gas diffusion cathode. The space inside the anode and the interelectrode gap are filled with electrolyte.

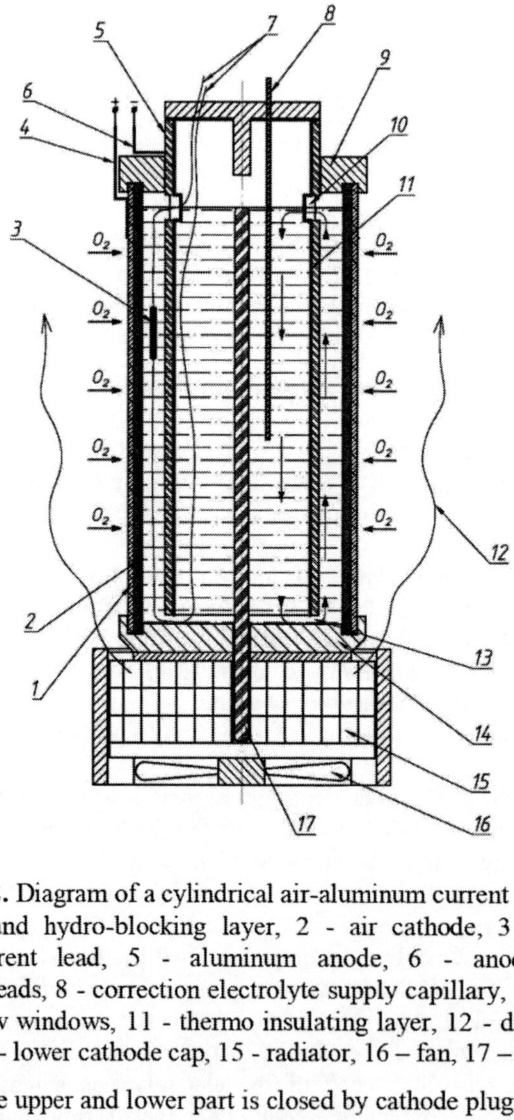

Figure 2. Diagram of a cylindrical air-aluminum current source.
1- Nickel mesh and hydro-blocking layer, 2 - air cathode, 3 - thermocouple, 4 - cathode current lead, 5 - aluminum anode, 6 - anode current lead, 7 - thermocouple leads, 8 - correction electrolyte supply capillary, 9 - upper cathode plug, 10 - overflow windows, 11 - thermo insulating layer, 12 - discharge air flow, 13 - lower gap, 14 - lower cathode cap, 15 - radiator, 16 – fan, 17 – heat pipe.

The air cathode (2) in the upper and lower part is closed by cathode plugs (9, 14), forming a closed volume. End seams on the border with plugs are sealed with a sealant. The lower plug 14 is provided with protrusions for mounting the cylindrical anode 2 with the formation of a gap 12 between its end and the bottom of the bottom plug 13. An electrolyte is enclosed in the cathode and cap-capped area and an aluminum anode 5 is coaxially inserted. In the upper part of the anode, its windows are provided with windows 10 for overflowing Electrolyte from the interelectrode gap to the inner cavity of the anode. The inner surface of the anode is provided with a heat-insulating layer 11.

In the inner cavity of the anode, a heat pipe 17 is provided, provided with a radiator 15 outside. The radiator warms the air pumped by the fan (16), which was directed to the outside of the cathode, thereby increasing its activity. Fan 16 switches on when the set temperature is reached.

The installation scheme for recording the discharge characteristics of a cylindrical air-aluminum current source is shown in figure 3. The discharge of the element was carried out at a constant load of Rn (49.5 mΩ) + Rш (2.5 mOhm). Recorded discharge parameters of a cylindrical air-aluminum element (shunt 30A; 75mV; 2.5 mΩ) and temperature (thermocouple) were controlled by an 8-channel analog input module, signals from which were transmitted via an interface converter for registration to a computer. The time was counted by a computer.

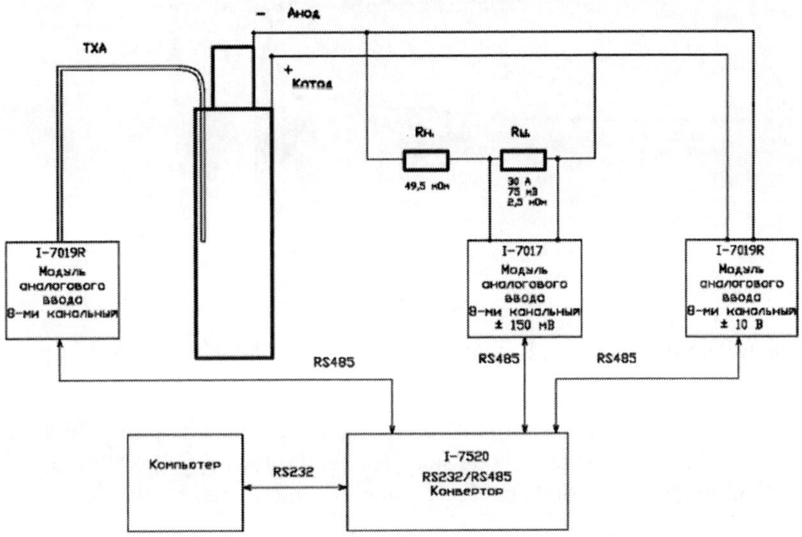

Figure 3. Scheme of installation for registering discharge characteristics.

In the process of discharging an air-aluminum current source, the electrolyte is expended on the formation of sodium aluminates and evaporation. This leads to a gradual decrease in the electrolyte level in the element and a decrease in the concentration of NaOH [19]. When the electrolyte level falls below the edge of the overflow windows, its circulation between the volume inside the anode and the interelectrode gap stops. In order to maintain a constant level and concentration of NaOH in the electrolyte, a corrective electrolyte was supplied during the tests. For an air-aluminum element of 30W power, the correct concentration of the correcting electrolyte (15M NaOH) was set and its feed rate (0.12 ml / min).

The discharge curve of a cylindrical air-aluminum element with simultaneous recording of the potentials of the Al anode and the air electrode (with the TPMPPso / AD-100 PP catalyst) is shown in figure 4. The element was activated by pouring 112 ml of electrolyte (8M NaOH + 0.015 M Na2SnO3 · 3H2O). In the discharge process, another 20 ml of 15M NaOH was added as a correction. According to calculations, 22 g of aluminum is dissolved in 132 ml of electrolyte, which corresponds to an electrolyte capacity of 66 A·h.

After activation, the element was held for 8 minutes without load to stabilize the voltage and potential of the electrodes. The voltage of the open circuit of the element at a temperature of 25 ° C

IOP Conf. Series: Materials Science and Engineering **292** (2017) 012013 doi:10.1088/1757-899X/292/1/012013

was 1.88V. Concerning the Ag / AgCl reference electrode, the anode potential was -1.75V, the cathode 0.13V. When the load was switched on, the anode potential shifted to the cathode side to -0.93V and was set at -1.25 ÷ -1.3V. After 12 minutes after the load was switched on, after the power of the element was reached 30 W and the temperature of the electrolyte was 53 ° C, a fan was switched on. At the same time, the temperature increase slowed down, and by the end of the discharge it reached a value of 69 ° C.

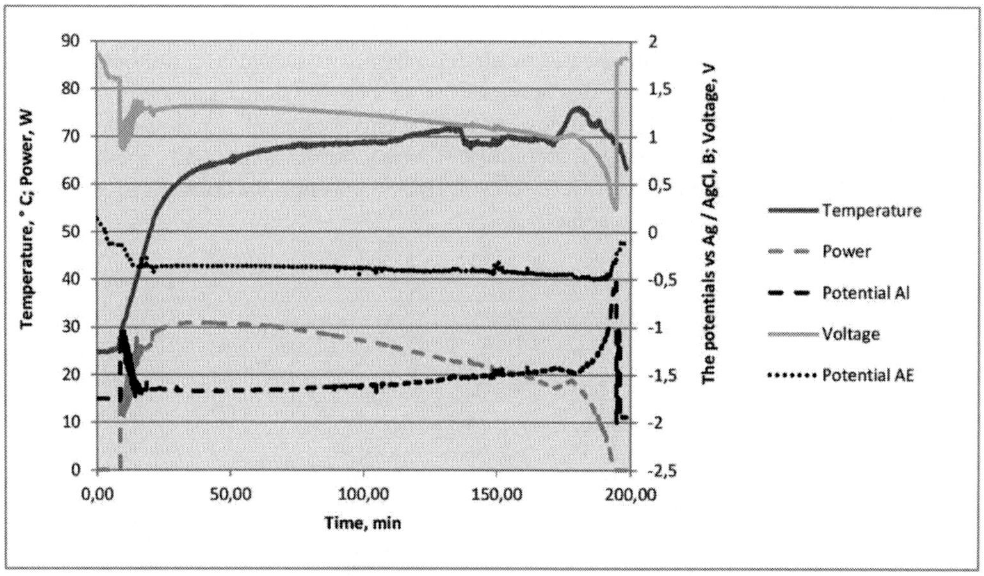

Figure 4. Parameters of the discharge of a cylindrical air-aluminum current source.

The average power of the element was 23 W. All in the process of discharge are realized: capacity - 64 A · ч; electricity - 76 Wh. The discharge was terminated due to a sharp drop in the cell's power due to a shift in the anode potential to the cathode side. The reason for the passivation of the anode was the development of the installed electrolyte capacity, which is determined by the limiting concentration of aluminates in the electrolyte.

Taking into account the total weight of the element 440 g (including the weight of the cathode, anode, electrolyte and heat sink system), the specific power was 52 W / kg. The average specific power of a dry cell without electrolyte is 72 W / kg, and the maximum - 116 W / kg. The specific energy content of the element, taking into account the total weight, is 173 W · h / kg and 285 W · h / kg of element without electrolyte.

4. Conclusion

1. The electrocatalytic activity of the oxygen depolarization catalyst - pyrolypolymer tetra (pmethoxyphenyl) porphyrin cobalt under model conditions and as part of a gas diffusion electrode.

2. The characteristics of gas-diffusion cathodes based on various carbon materials and promoters.

3. The optimized gas-diffusion cathode with the TPMPPso / AD-100 PP catalyst achieves a current density of 300 mA / cm2 at E = 0.73V and a temperature of 60 °C.

4. The design of a cylindrical air-aluminum current source with a specific power of 116 W / kg and a specific energy capacity of 285 W · h / kg.

Acknowledgements
This work has been conducted under financial support from the Ministry of Education and Science of the Russian Federation (Contract No. 14.604.21.0152, identifier RFMEFI6414X0152).

References

[1] Yang S, et al. 2002 Design and analysis of aluminum/air battery system for electric vehicles. *Journal of Power Sources* **112** p 162
[2] Zhuk A, Kleimenov B and Fortov V 2012 *Electric car on aluminum fuel* (Moscow:Nauka) p171
[3] Ilyukhina A, Kleymenov B and Zhuk A 2017 *Journal of Power Sources* **342** pp 741-749
[4] Rahman Md.A, et al 2013 *Journal of the Electrochemical Society* **160** pp 17-59
[5] Zhuk A, Kleimenov B and Shkolnikov E 2007 *Aluminum hydride power engineering* (Moscow: Izd-vo ZAVT RAN*)* p 278
[6] Korovin N 2005 *Fuel Cells and Electrochemical Power Plants.* (Moscow: MEI) pp 126-129
[7] Michel Lef evre, Jean-Pol Dodelet 2003 *Electrochimi Acta* **48** pp 27-49
[8] Tarasevich M, et al. 2004 *Electrochemistry* **40** no.6 p 612
[9] Zhen-qian Fang, et al. 2006 *Electrochim* **51** pp 54-56
[10] Lin W, Metal-air fuel cell module, US Patent *2011/0117455 A1* (publ. 2011)
[11] Dougherty T, et al. 1996 *Proc. 31st Intersociety Energy Conversion Engineering Conf.* (Moscow: NAMI) pp 1176-1180
[12] Pino M, Cuadrado C, Chacon J, Rodriguez P, Fatas E, Ocon P, 2014 *J. Appl. Electroche* **44** pp 1371-1380
[13] Fan L, Lu H, Leng J, 2015 *Electrochim. Acta* **165** pp 22-28
[14] Kharlamov S, Bobylev A, Galkin P, Dobroselsky K, Vlasenko M, Markovich D 2015 *J. Eng. Thermophys* **24** pp 313-321
[15] Ilyukhina A, Zhuk A, Kleymenov B, Ilyukhin A, Nagayama M, 2016 *Fuel Cells* **16** pp 384-394
[16] Doche M, Novel-Cattin F, Durand R, Rameau J, 1997 *J Power Sources* **65** pp 197-205
[17] O'Callaghan W, Metal/air battery with recirculating electrolyte, US Patent 4.908.281 (publ. 1990)
[18] Kleymenov B, et all. Metal-air electrochemical element / Patent for utility model RU No. 169334, pub. 2017
[19] Zhuk A, Kleymenov B, Zakharov V, Chursin A, Kapustin A, 2012 *International scientific journal alternative energy and ecology* **4** pp 108-115

IOP Conf. Series: Materials Science and Engineering **292** (2017) 012014 doi:10.1088/1757-899X/292/1/012014

Investigation on the Room-temperature preparation of Cobalt hybrid/Graphene Nanocomposite and application in wastewater purification: Highly Efficient Removal of Congo Red

L X Wang[1, 3], Y F Zhao[2] and Q M Meng[1]

[1] College of Chemistry and Chemical Engineering, Bohai University, Jinzhou 121013, China
[2] College of Engineering, Bohai University, Jinzhou 121013, China

[3] E-mail: wlxdqpi@163.com

Abstract. Here, we are going to report a simple, low-cost and environmental friendly process to prepare the cobalt hybrid/graphene (Co/G) nanocomposite at room temperature. $NaBH_4$ was used as the reducing agent. Such an approach can be extended to grow some other metal/G nanocomposites, for example, Ni/G, Co/G nanocomposite possesses narrow size-distribution and good dispersion. Because of the special appearance with large surface area, and the special synthesis process of the productions, adsorption experiments for Congo Red were carried out in synthetic wastewater. The CR removal ability of Co/G nanocomposite can reach 263.2 mg/g.

1. Introduction

As an excellent two-dimensional material, graphene (G) has unusual features excellent thermal property and fracture strength, high specific surface area and charming spin transport. In composite G chip also can be used as electrical conduction channel, so as to improve the conductivity of the material and greatly reducing ohmic losses [1-5]. In addition, graphene has the G slice of its unique structure and excellent softness can not only provide enough space volume change, reduce buffer volume change brought about by the pressure, can also prevent Co particles gathered in circulation process, reduce the electrode degradation [6-7]. Rich electronic Co atoms, of course, the electrostatic attraction between the surface of Co can be firmly fixed G, combined with stability of the composites. The most important of all, the performance of the G can also with Co nanoparticles modified further strengthened. Because of the π - π interaction between adjacent slices, G in use process usually serious phenomenon of reunion. Co nanoparticles exists between G, effectively prevent G piece together, to maintain their highly active surface area, to improve G lithium storage capacity of composite materials [8-10]. Recently, Co_3O_4/grapheme [11], $Co(OH)_2$/graphene and Co/G composites have been synthesized and applied to lithium ion batteries [12-14]. Preparation of cobalt/graphene(CoG) composite is rare at room temperature.

Will G and Co nanoparticles composite is a suitable method, will compound used as a lithium-ion battery anode active material with high reversible lithium storage capacity and coulomb efficiency. This kind of Co/G nanocomposite provides a unique performance, to improve the prospect of application and extension provides greater.

In addition to as lithium ion battery anode material, the Co/G nanocomposite ability to remove dyes from aqueous solution are also studied. Co/G nanocomposite material has high magnetic, provides efficient separation after removing Congo red (CR) possible suspension. Nanocomposites are large specific surface area is conducive to the adsorption CR in the solution.

In this article, through the simple a pot of hydrothermal synthesis method of synthesis of Co/G nanocomposite, this method can be extended to other metal/G the synthesis of nanocomposites. In this method, sodium borohydride as reducing agent, the oxidation of graphene (GO) and $CoCl_2$ in the process of hydrothermal reduction at the same time. Co nanoparticles dispersed homogeneously in G nanocomposite structure is beneficial to inhibit gathered, thus provide higher available surface area and enhanced the lithium storage capacity and adsorption.

2. Experimental section

Please follow these instructions as carefully as possible so all articles within a conference have the same style to the title page. This paragraph follows a section title so it should not be indented.

2.1. Sample Preparation

The graphite oxide (GO) was synthesized from natural graphite powder based on a modified Hummers method [15], specific preparation method is as follows: natural graphite powder weighing 1.0 g and 6.0 g $KMnO_4$ into the mixture of concentrated of 9:1 H_2SO_4/H_3PO_4 (120.13.3mL) in the mixed acid, heated to 50 ℃, the mechanical stirring 12 h. Reaction after cooling to room temperature, add about 400 mL ice, drip into the 3 mL30 % H_2O_2. Centrifugal separation after 20 min (10000 RPM), clean with 5% HCl solution, and then use distilled water cleaning and centrifugal separation for many times, make pH neutral. Finally, the synthesis of uniform GO solution, after the vacuum drying for characterization and further chemical reactions.

Preparation Cobalt/graphene nanocomposite (Co/G): use of $NaBH_4$ as a reducing agent by hydrothermal synthesis of Co/G nanocomposites. First of all, 15 mg GO dissolved in 15 ml of distilled water, using ultrasonic stirring 30 min to exfoliate oxidized graphite particles to GO sheets. Then 0.1 g of $CoCl_2{\cdot}6H_2O$ dissolved in 5 ml of distilled water, and add to GO solution, ultrasonic stirring for 1 hour. Then 0.1 g $NaBH_4$ dissolved in 5 ml of distilled water, one drop to join the GO and $CoCl_2$ solution, and continue to ultrasound for 30 min at room temperature. The reaction is responsive quickly. A black material formed in the bottom of the beaker and followed along with the generating gases. After the reaction of hydrogen generation was completed, the black material was magnetically attracted to the bottom of the beaker by a magnet, removed the upper solution and washed twice with deionized water and ethanol. Black Co/G nanocomposites were prepared.

2.2. Materials Characterizations.

The phases were identified by means of X-ray diffraction (XRD) with a Rigaku D/max 2500pc X-ray diffractometer with Cu Kα radiation (λ) 1.54156 (Å) at a scan rate of 0.02°/1(sec), morphologies were characterized by a JEOL JSM-6700F field emission scanning electron microscopy (FESEM) operated at an acceleration voltage of 8.0 kV. The hysteresis loops were measured on a VSM-7300 vibrating sample magnetometer (VSM) (Lakeshore, USA) in room temperature. An Agilent Cary 50 UV–vis spectrophotometer was used for determination of CR concentration in the solutions.

2.3. Batch adsorption experiments

Stock solution preparation: Congo red is the chemical name of diphenyl - 4, 4 '- (azo - 2 -) 2-1 - amino naphthalene - 4 - sodium sulfonate, molecular formula is $C_{32}H_{22}N_6Na_2O_6S_2$ [16], structured as shown in figure 1 [17]. Accurate weighing Congo red dissolved in distilled water, using volumetric flask configuration 1 $g{\cdot}L^{-1}$ mother liquor as a stock solution.

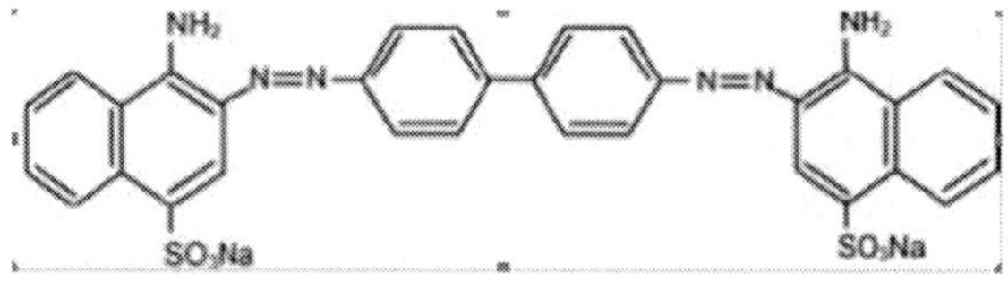

Figure 1. Structure of Congo red molecule.

Using UV–vis spectrophotometer at lambda max = 497 nm (maximum absorbance value of CR) test known absorbance value of different concentration of CR, determine the standard curve. Determination of adsorption quantity: according to take a certain amount of magnetic powder (W), added to the known concentration (C_0) of a certain volume (V) of Congo red solution, using mechanical mixing device, every once in a certain time interval (t) using the magnet to absorb magnetic powder, will pour into the upper Congo red solution than color plate, using UV–vis spectrophotometer to measure the concentration of a solution of CR (C_t). After measuring, recording, the colorimetric dish fell back to the original beaker of solution, continue to stir, measurement, until adsorption reached equilibrium. The adsorption of CR (q_t) used to calculate the following formula:

$$q_t = \frac{(C_0 - C_t)\times V}{W} \tag{1}$$

where q_t is per gram of adsorbent adsorption ability in time t time, $mg{\cdot}g^{-1}$; C_0 is the initial concentration of Congo red solution, $mg{\cdot}dm^{-3}$; C_t is t moment of Congo red solution concentration, $mg{\cdot}dm^{-3}$; V is the volume of a solution, dm^3; W is the quality of magnetic powder, g.
Standard solution with initial concentrations of 150 mg/L was prepared. Then, 10 mg of Co/G nanocomposite was added to 50 mL of above solution under stirring. After a specified time, the solid and liquid were separated by a magnet.

3. Results and discussion

3.1. Structural and Morphological Characterization

Briefly, the Co/G nanocomposite was prepared using a very facile reduced method at room temperature. The crystal structure was further investigated by XRD technique. Figure 2a and 2b is graphite, GO and Co/G nanocomposite of XRD diagrams. Graphite in 26.4° and 54.6° diffraction peak corresponding is graphite (002) and (110) crystal plane (figure 2a). After oxidation treatment, graphite (002) diffraction peak disappear, a GO at 2θ = 11.4° (figure 2a). Figure 2b compared with pure Co nanoparticles XRD figure, was found at 2θ=26.0° added a weak and wide (002) diffraction peak, it is caused by G nano of accumulation [18]. Nevertheless, the broad nature of the reflection indicates poor ordering of the nanosheets along the stacking direction, implying that the sample is composed of mostly few layers of G nanosheets [19, 20]. The diffraction peaks for the as-formed sample at 2θ =41.63, 44.57, 47.51, and 75.97°, which match the reflection planes of (100), (002), (101), and (110), respectively, could be indexed to a pure hcp-Co (JCPDS 05-0727).

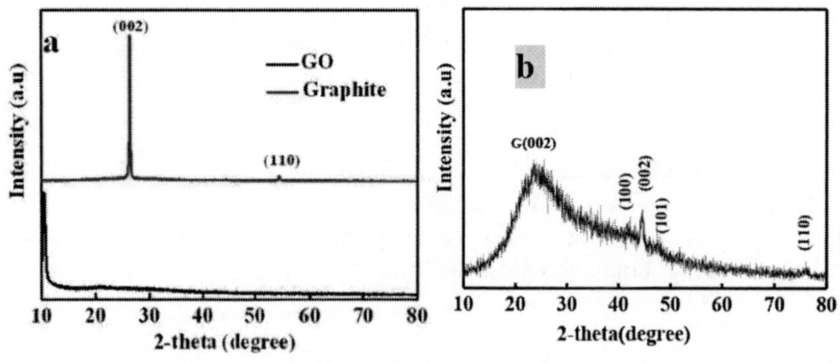

Figure 2. (a) XRD patterns of Graphite and GO; (b) XRD pattern Co/G nanocomposite.

The Co nanoparticles with a mean diameter of about 20-100 nm are closely and homogeneously anchored on the surface of G nanosheets (figure 3). A folding nature of G is clearly visible from figure 3, the relatively light area is due to the stacking of several G nanosheets.

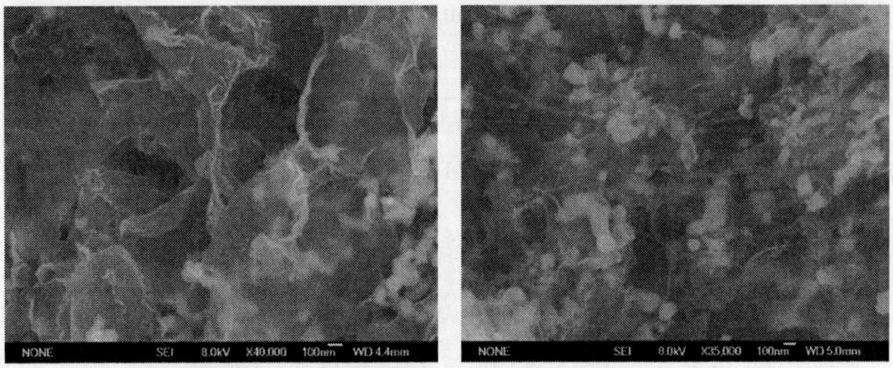

Figure 3. SEM image of Co/G nanocomposite.

3.2. Magnetic Properties

By using VSM magnetic performance of Co/G nanocomposite was measured. From figure 4 room temperature measuring hysteresis loop can be determined, Co/G nanocomposite of saturation magnetization (Ms) is 57.5 emu·g^{-1}, coercive force (Hc) is 667.6 Oe. Ms value of Co/G is lower than Co nanoparticles, and the Hc value is higher than Co nanoparticles. Ms value is lower due to the complex of nonmagnetic material graphene. The Hc value increased significantly, it is because of constraints in G wafer Co nanoparticles of the influence of the rotation of the magnetic vibrator. The result is need to change the alignment of the direction of magnetization of the dipole energy increase the intensity of magnetic field (or higher). Therefore, Co/G nanocomposite exhibited higher Hc [21, 22].

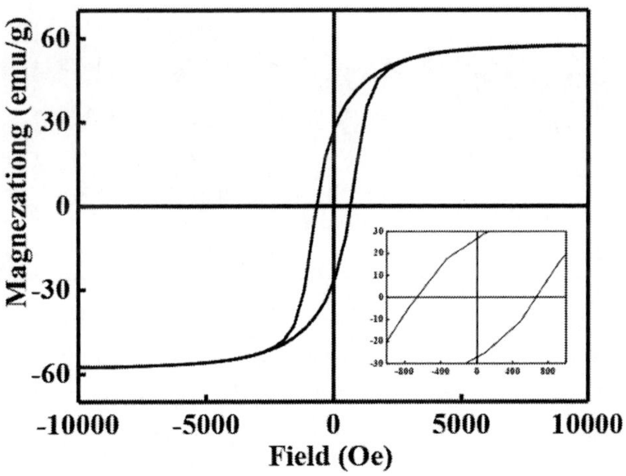

Figure 4. Magnetization curve measured at room temperature of Co/G nanocomposite.

3.3. Evaluation of CR Removal Ability

Because of the Co/G nanocomposite material with high specific surface area and magnetic, would be an ideal adsorbent for sewage treatment. Herein, the Co/G nanocomposite was used to remove CR from aqueous solution. The CR removal abilities of Co/G nanocomposite were shown in figure 5. Aqueous solutions with initial CR concentration 150 mg/L was used for the experiment at neutral pH=7, adsorption different time from 0 to150 minutes, and T =15 ℃. The final CR removal ability of

Co/G nanocomposite is 263.2 mg/g. The maximum adsorption capacity (q_{max}) of Co/G nanocomposite for CR is listed in table 1 with literature values of q_{max} of other magnetic materials for CR adsorption [23-26]. The adsorption efficiency is far higher than the values as reported in the literatures. In addition, Co/G nanocomposite loaded with CR can be quickly separated by a magnet, and then leaves a clear and colorless solution (Insets of figure 5).

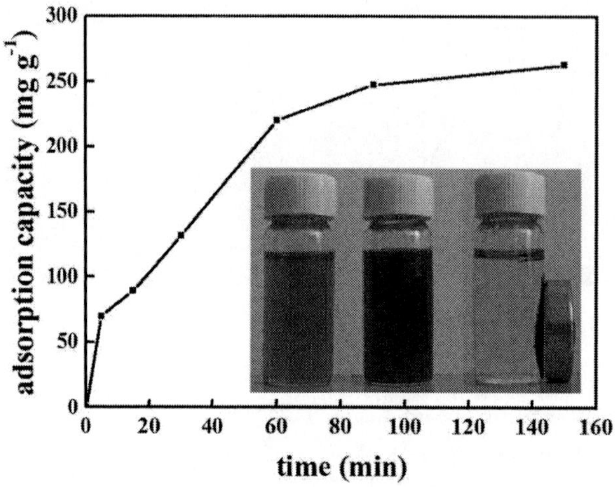

Figure 5. Removal ability for CR on Co/G nanocomposite. Insets: photos of initial CR solution, with Co/G nanocomposite treatment and magnetic separation.

Table 1. Adsorption capacities of cr dye on various adsorbents.

Adsorbent(nanoscale)	q_{max} [$mg \cdot g^{-1}$]	Reference
Co/G nanocomposite	263.2	This study
Fe_2O_3	208.3	[23]
Magnetic core–MnO shell	42.0	[24]
$Fe_{2.95}La_{0.05}O_4$	107.6	[25]
$CoFe_2O_4$	244.5	[26]
Fe_3O_4	149.7	[26]
$NiFe_2O_4$	97.1	[26]
$MnFe_2O_4$	92.4	[26]

4. Conclusions

We have presented a facile one-pot method for the synthesis of Co/G nanocomposite based on a simple reduction process using $NaBH_4$ as reducing agent at room temperature. Uniform Co nanoparticles embedded in G matrix of Co/G nanocomposite are excellent magnetic properties. Furthermore, the regular magnetic properties and high specific surface area of Co/G nanocomposite also make it have a potential application in the field of water treatment.

References

[1] Wu X L, Wang L, Chen, Ch L, Xu A W and Wang X K 2011 *J. Mater. Chem.* **21** 17353

[2] Turchanin A, Weber D, Büenfeld M, Kisielowski C, Fistul M V, Efetov K B, Weimann T, Stosch R, Mayer J and Gölzhäuser A 2011 *ACS nano.* **5** 3896

[3] Kim T Y, Lee H W, Stoller M, Dreyer D R, Bielawski C W, Ruoff R S and Suh K S 2011 *ACS nano.* **1** 436

[4] Yang X, Xu M S, Qiu W M, Chen X Q, Deng M, Zhang J L, Iwai H, Watanabe E and Chen Z H

2011 *J. Mater. Chem* **21** 8096
[5] Wu Z S, Ren W C, Wen L, Gao L B, Zhao J P, Chen Z P, Zhou G M, Li F and Cheng H M 2010 *ACSnano*. **4** 3187
[6] Zhang X R, Gong J L, Zeng G M and Deng J H 2013 *China Environmental Sci.***33** 1379
[7] Wu H J, Wu G L, Ren Y Y, Yang L, Wang LD and Li X H 2015 *J. Mater. Chem. C* **3** 7677
[8] Hu B W, Huang C C, Li X, Sheng G D and Li H 2017 *Chem. Eng. J.* **313** 527
[9] Hu W L, Wang L D, Wu Q F and Wu H J 2014 *Adv. Powder Technol.* **25** 1780
[10] Wen H F, Yang C, Yu D G, Li X Y and Zhang D F 2016 *Chem. Eng. J.* **290** 263
[11] Hu B W, Ye F, Ren X M, Zhao D L and Sheng G D 2016 *Environ. Sci. Nano.* **3** 1460
[12] González J R, Alcántara R, Nacimiento F and Tirado J L 2011 *Electrochimica Acta.* **56** 9808
[13] Zhang J, Yang H, Shen G, Cheng P, Zhang J and Guo S 2010 *Chem. Commun.* **46** 1112
[14] Wang L X, Li J C, Mao C S, Zhang L S, Zhao L J and Jiang Q 2013 *Dalton Transactions.* **42** 8070
[15] Marcano D C, Kosynkin D V, Berlin J M, Sinitskii A, Sun Z Z, Slesarev A, Alemany L B, Lu W J and Tour M 2010 *ACS Nano* **4** 4806
[16] Purkait M K, Maiti A, DasGupta S and De S 2007 *J. Hazard. Mater.* **145** 287
[17] Chatterjee S, Lee M W and Woo S H 2010 *Bioresource Technol.* **101** 1800
[18] Dikin D A, Stankovich S, Zimney E J, Piner R D, Dommett G H B, Evmenenko G, Nguyen S T and Ruoff R S 2007 *Nature* **448** 457
[19] Kundu P, Nethravathi C, Deshpande P A, Rajamathi M, Madras G and Ravishankar N 2011 *Chem. Mater.* **23** 2772
[20] Murugan A V, Muraligant T and Manthiram A 2009 *Chem. Mater.* **21** 5004
[21] Duan L F, Jia S S, Cheng R M and Zhao L J 2011 *Chem. Eur. J.* **173** 233
[22] Duan L F, Jia S S, Cheng R M and Zhao L J 2010 *Eur. J. Inorg. Chem* **156** 1957
[23] Afkhami A and Moosavi R 2010 *J. Hazard. Mater.* 174 398
[24] Zhai Y, Zhai J F, Zhou M and Dong S J 2009 *J. Mater. Chem.* **19** 7030
[25] Wang L X, Li J C, Wang Y Q and Zhao L J 2011 *Hazard. Mater.* **196** 342
[26] Wang L X, Li J C, Wang Y Q, Zhao L J and Jiang Q, 2012 *Chem. Eur. J.* **181** 72

IOP Conf. Series: Materials Science and Engineering **292** (2017) 012015 doi:10.1088/1757-899X/292/1/012015

One-step preparing magnesium hydroxide particles from mother liquor of salt production

H Guo[1, 2, 3], C S Peng[1], Z W Ding[2], H T Yuan[1] and K Yang[1]

[1] Key Laboratory of Marine Environment and Ecology, Ministry of Education, College of Environmental Science and Engineering, Ocean University of China, Qingdao 266100, China
[2] Beijing Institute of High-tech Environmental Science, Beijing 101111, China

[3] E-mail: guohao8432@126.com

Abstract. In this study, MH particles were prepared from mother liquor of salt production in one-step through employing ammonia gas as precipitant and stearic acid as dispersant respectively. Since adopting microporous plate to bubble ammonia gas, the percent conversion of magnesium was boosted obviously. The influence of operating condition of reacting temperature, stirring rate, ammonia flowrate and pore size of plate to magnesium percent conversion were investigated, the maximum is 88.1 % at optimum condition according to experimental results. The MH particle preparing from mother liquor in optimum condition was characterized by XRD, the result indicated the volume of brucite was reach to 99.7% within the composition of the product. In addition, the size distribution and crystal morphology was also detected, the median particle diameter d50 is 883 nm and possessing good dispersibility. From the thermogravimetric analysis of MH particles, the thermostability of product is suitable as flame-retardant composite materials.

1. Introduction

Magnesium hydroxide (MH) particles is wildly used in flame-retardant composite materials due to its ability to endothermically decompose with water release at high temperatures without production of toxic or corrosive substances [1, 2]. Normally, MH particles is produced by hydrothermal synthesis from magnesium chloride with alkaline solution as precipitant [3, 4]. On account of magnesium metal ion ranks only second to sodium among cations in mother liquor of salt production and occupies approximately 43 % of the total amount of cations [5], hence there are a number of processes through using limewater or ammonia hydroxide for precipitation of MH particles from mother liquor of salt production [6-8].

However, the agglomeration and uniformity of MH particles is difficult to control in conventional precipitation process [9]. Adjusting zeta-potential of the particles, adding dispersant in the reaction system and mixing water-soluble solvent in mother liquor was generally employed to solve above problem [10]. In addition, employing liquid precipitant has greatly increased the amount of magnesium hydroxide slurry and aggravated the disposing burden of solid-liquid separation process, washing and a filtration step to concentrate the slurry [11]. Yi used magnesium chloride and ammonia water to produce MH particles with employing reverse-titration precipitation method and the size of product has an excellent conformity [12]. Hu et al. selected sodium dodecyl benzene sulfonate (SDBS) and octaphenyl polyoxyethyiene (OP-10) as composite dispersant to produce MH particles in one-step and obtained the product with commendable dispersibility [13]. Zhai et al. dissolved magnesium chloride in ethanol-water mixed liquor with sodium hydroxide as precipitant to MH particles with the purpose of enhancing the separability of MH precipitate [14]. He et al. also proved the additional ethanol not only enhanced the dispersibility, but also improved the filtrability of the MH particles [15]. Tai et al. utilized spinning disk reactor strengthened the mass transfer to produce MH particles [16].

In this research, a novel method has been proposed regarding of preparing magnesium from mother liquor of salt production in one-step. Microporous plate for bubbling ammonia gas was assembled in reaction tank and employed to precipitation process, the gaseous ammonia and stearic acid was used as precipitant and dispersant respectively. The magnesium percent conversion, uniformity of particle size, shape, agglomeration and thermostability of product are well under controlled due to bubbling and dispersant addition. With this method, the conventional magnesium preparing steps from mother liquor of salt production which including chemical precipitation, hydrothermal treatment and surface modification were developed and simplified to one-step.

2. Experimental

2.1. Materials preparation

The mother liquor of salt production was collected from the saltwork which located in Laizhou bay and filtered with 0.45 μm microfiltration membrane before conducting experiment. The concentration of magnesium ion in mother liquor which has been measured is 43.5 $g{\cdot}L^{-1}$ approximately. The Ammonia gas as precipitant was purchased from Hebei Yongchang Chemical Engineering Co. Ltd., Hebei. Stearic acid as dispersant was purchased from Shanghai Sinopharm Chemical Reagent Co. Ltd., Shanghai. The reaction vessel is a self-made 4-necks stainless-steel tank which is used for preparing magnesium from mother liquor. The microporous plate is also self-made with drilling various pore diameters and being assembled in the reaction vessel. The tail-gas of ammonia was absorbed by diluted hydrochloric acid for security.

2.2. Experimental method

The flow chart of experimental apparatus was shown in figure 1. 100 mL mother liquor and 5 g stearic acid were added in 4-necks stainless-steel tank which was placed in a thermostatic water bath whose temperature was controlled at 40, 50, 60, 70 and 80 °C respectively, then ammonia gas was inlet in mother liquor via microporous plate which has been assembled in the tank and replaced with pore diameters of 0.5, 1.0, 1.5, 2.0, and 2.5 mm. The flowrate of ammonia gas was adjusted to 600, 800, 1000, 1200 and 1400 $mL{\cdot}min^{-1}$ respectively which was measured by flowmeter. During the whole process, the reaction solution was constantly being stirred with the speed of 100, 200, 300, 400 and 500 $r{\cdot}min^{-1}$. At last, the MH particles were separated and washed by deionized water, dried under vacuum at 45 °C. The magnesium concentration of percolate was measured under different molar ratio between ammonia and magnesium (NH_3/Mg^{2+} molar ratio).

2.3. Calculation of magnesium percent conversion

The percent conversion of magnesium under different molar ratio between ammonia and magnesium with the change of ammonia flow, reacting temperature, stirring rate, pore size of the microporous plate was measured. The percent conversion of magnesium [η[%(mass)]] and NH_3/Mg^{2+} molar ratio (α) is defined by equation (1) and (2) respectively as follows.

$$\eta=1-\frac{c_p}{c_0} \tag{1}$$

where c_0($mg{\cdot}L^{-1}$) represents the magnesium concentration of mother liquor, however c_p($mg{\cdot}L^{-1}$) is the magnesium concentration of percolate after solid-liquid separation.

$$\alpha = \frac{n_{NH_3}}{n_{Mg^{2+}}} \quad (2)$$

where n_{NH3} represents the molar mass of ammonia gas which was inlet in mother liquor, and n_{Mg2+} is the molar mass of mother liquor which was added in tank.

2.4. Characterization

The laser particle analyzer (Mastersizer 2000, Malvern, UK) was used to analyze the volume percent and cumulative frequency distribution of the particles. The X-ray diffractometer (XRD, D/max-rB, Rigaku, Japan) equipped with monochromator giving Cu-*ka1*=1.5406 Å radiation was used to quantify the phase composition of MH particles. The crystal morphology of MH particles also has been observed by scanning electron microscope (SEM, S4800, Hitachi, Japan). Moreover, the thermostability of MH particles prepared from mother liquor was tested by thermogravimetric analyzer (TGA, Q5000IR, TA Instruments, US).

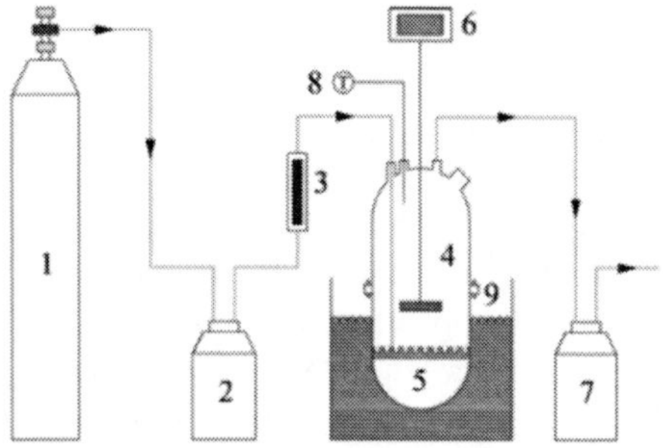

Figure 1. Flow chart of experimental apparatus. 1. Ammonia bottle, 2. Buffer flask, 3. Gas flowmeter, 4. 4-necks stainless-steel tank, 5. Microporous plate, 6. Mechanical stirrer, 7. Tail-gas absorption bottle, 8. Thermometer, 9. Thermostatic water bath.

3. Results and discussion

3.1. pH value change with NH_3/Mg^{2+} molar ratio

The percent conversion of magnesium is closed related with the concentration of OH^- ion, the pH value of mother liquor change with NH_3/Mg^{2+} molar ratio was measured and shown in figure 2. It indicates that the pH value was increasing with NH_3/Mg^{2+} molar ratio, the mother liquor changed from neutral to alkalinity. As well as the pH value was increasing quickly at beginning of inletting gas, then it was turn to change slowly when the NH_3/Mg^{2+} molar ratio was greater than 1.8. The maximum of pH value is 9.78 (<9.8) when NH_3/Mg^{2+} molar ratio is 2.6.

3.2. Solubility of ammonia gas dissolved in water

As above, the concentration of OH^- ion is depended on the NH_3/Mg^{2+} molar ratio. However, the temperature of reaction system is also important, it influenced the solubility of ammonia gas due to the dissolving is an exothermic process from chemical equation (3).

$$NH_3 + H_2O \rightarrow NH_4OH$$
$$\Delta H\text{=-34.5kJ} \cdot \text{mol}^{-1} \quad (3)$$

Theoretically, the solubility of ammonia can be calculated by the following formula (4).

$$S = A\exp\left[-\frac{\Delta H_A}{RT}\right] \quad (4)$$

where A(mol·kg^{-1}·Pa^{-1}) is pre-exponential factor, ΔH_A(J·mol^{-1}) represents the enthalpy of solution, R(8.314 J·mol^{-1}·K^{-1}) is the universal gas constant.

The solubility of ammonia dissolved in water can be calculated by the following formula (5) which was corrected from formula (4).

IOP Conf. Series: Materials Science and Engineering **292** (2017) 012015 doi:10.1088/1757-899X/292/1/012015

$$S = 0.574 \exp\left[\frac{1452}{T} - 28.14\right] \tag{5}$$

where S[%(mass)] is the solubility of ammonia at usual atmospheric pressure, T(K) represents the environment temperature.

At usual atmospheric pressure, the solubility of ammonia dissolved in water was shown in figure 3. As shown in this figure, the solubility of ammonia is influenced by the temperature to a great extent. With temperature increasing, the solubility of ammonia is decreasing quickly. When the temperature is higher than 80, the solubility of ammonia is only 7.0 %(<10.0 %) in mass.

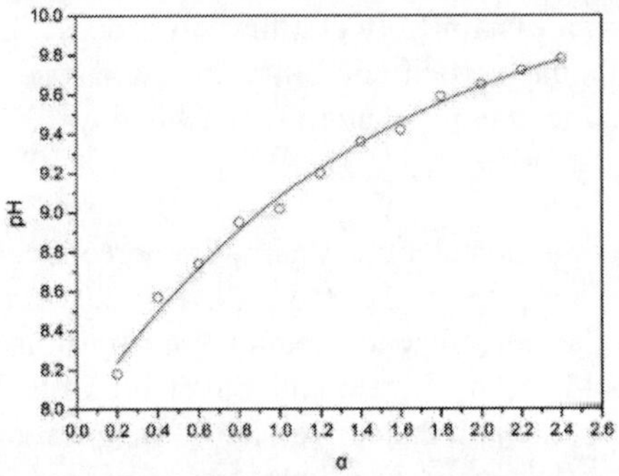

Figure 2. The pH value of mother liquor change with NH_3/Mg^{2+} molar ratio.

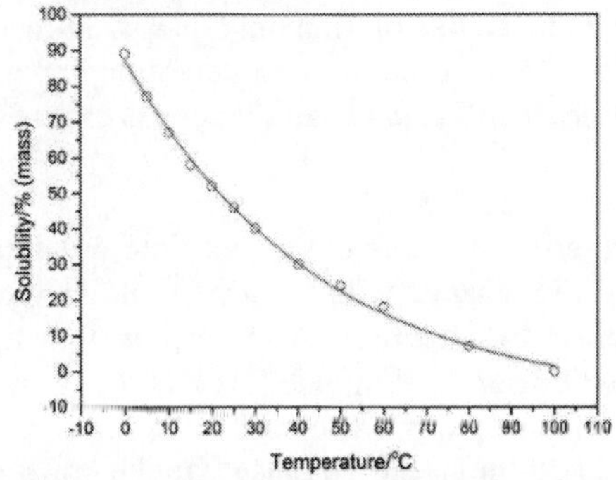

Figure 3. Solubility of ammonia dissolved in water at usual atmospheric pressure.

3.3. Percent conversion of magnesium change with operating conditions

3.3.1. Reacting temperature

The effect of percent conversion of magnesium change with NH_3/Mg^{2+} molar ratio at temperature of 40, 50, 60, 70 and 80 °C was shown in figure 4. From this figure, we can see that the optimum value of percent conversion of magnesium is 88.1 % (<90 %) at temperature of 60 °C. In addition, the percent conversion of magnesium is approached in higher NH_3/Mg^{2+} molar ratio at temperature of 40, 50 and 60 °C, it was because that the solubility of ammonia dissolved in water was influenced by temperature seriously. With the reacting temperature increasing, the solubility of ammonia is decreasing and resulting in the shortage of OH^- ion in the reaction system, hence the percent conversion of magnesium is lower at temperature of 70 and 80 °C. Furthermore, the viscosity of water is decreasing with temperature increasing, and lower viscosity is beneficial to the formation and growth crystal nucleus, therefore the percent conversion of magnesium at temperature of 40 (85.3 %) and 50 °C (87.0 %) is slightly lower than temperature of 60 °C (88.1 %).

3.3.2. Stirring rate

The effect of percent conversion of magnesium change with NH_3/Mg^{2+} molar ratio at stirring rate of 100, 200, 300, 400 and 500 r·min^{-1} was shown in figure 5. As shown in this figure, the optimum value of percent conversion of magnesium appeared at stirring rate of 500 r·min^{-1}, and it was increasing with the stirring rate accelerating. In addition, the percent conversion of magnesium influenced by the stirring rate was even more obviously along with the increasing of NH_3/Mg^{2+} molar ratio. The percent conversion of magnesium is only 60.4 % at stirring rate of 100 r·min^{-1} when the NH_3/Mg^{2+} molar ratio is 2.6, however it was reach up to 88.1 % at stirring rate of 500 r·min^{-1}. This was probably because the rapidly stirring could promote the dissolving of ammonia gas. In the condition of higher NH_3/Mg^{2+} molar ratio, the ammonia gas more possibly escaped from the reaction system due to there is not enough time to dissolve.

3.3.3. Ammonia flowrate

The flowrate of ammonia gas as a significant operating parameter was also investigated in this experiment. The effect of percent conversion of magnesium change with NH_3/Mg^{2+} molar ratio at ammonia flowrate of 600, 800, 1000, 1200 and 1400 mL·min^{-1} was shown in figure 6. In this figure, the percent conversion of magnesium is superior with the ammonia flowrate of 600 mL·min^{-1} towards to all levels of NH_3/Mg^{2+} molar ratio. The percent conversion of magnesium is decreasing gradually along with ammonia flowrate increasing. At NH_3/Mg^{2+} molar ratio of 2.4, the percent conversion of magnesium is 88.1 % and 76.4 % respectively when the ammonia flowrate is 600 and 1400 mL·min^{-1}, that demonstrated the percent conversion of magnesium was influenced by ammonia flowrate evidently. The course of ammonia dissolved in water is a dynamically equilibrium process, hence the faster flowrate will shorten the retention time due to the certain solubility in a constant reacting temperature, it will lead the ammonia gas escaped from the reaction system in the same way.

3.3.4. Pore size of plate

In this experiment, the microporous plate was integrated in the reaction system, the percent conversion of magnesium change with NH_3/Mg^{2+} molar ratio at plate pore size of 0.5, 1.0, 1.5, 2.0 and 2.5 mm was displayed in figure 7. As shown in this figure, the percent conversion of magnesium gravely influenced by pore size of plate. At NH_3/Mg^{2+} molar ratio of 2.4, the percent conversion of magnesium is 88.1 % when the pore size of plate is 0.5 mm, however the percent conversion of magnesium is only 49.7 %. In addition, the NH_3/Mg^{2+} molar ratio which effect on percent conversion of magnesium is more distinct with using larger pore size than smaller pore size. This result was because that the smaller pore size will cut the bubbles more subtly, thus ammonia bubbles were capable of distributing more evenly and dissolved in water quickly. The point of higher or lower ammonia concentration is non-existent in this reaction system, therefore the process of ammonia dissolution is rapidly and evenly.

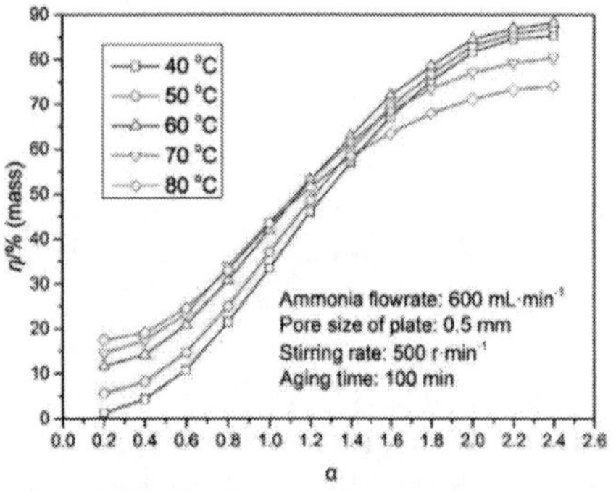

Figure 4. Percent conversion of magnesium in different reacti-ng temperature condition.

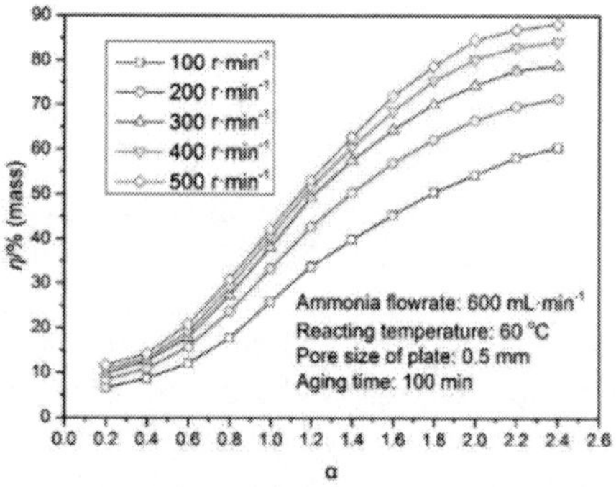

Figure 5. Percent conversion of magnesium with different stirring rate.

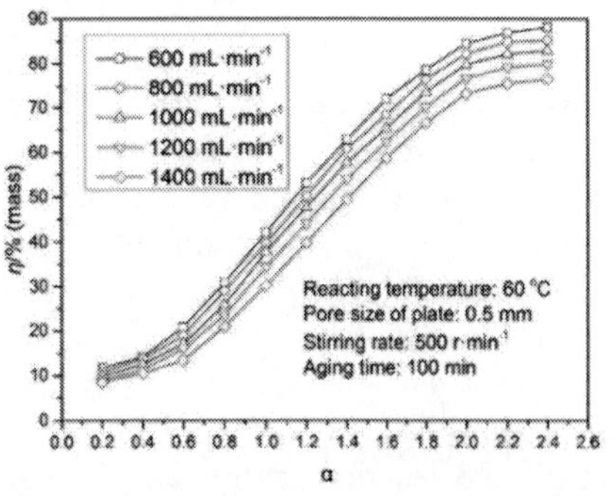

Figure 6. Percent conversion of magnesium with different ammonia flowrate.

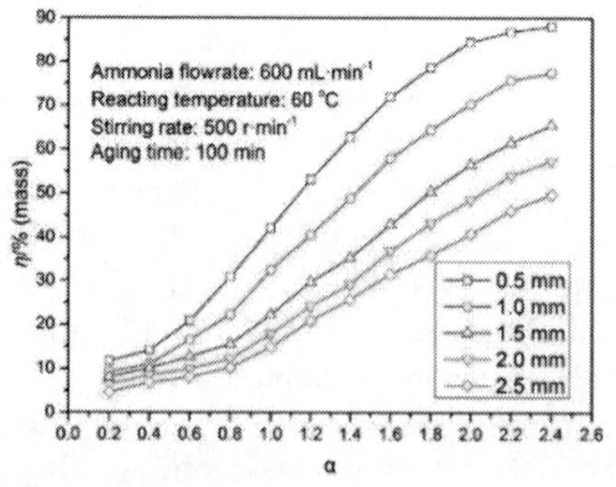

Figure 7. Percent conversion of magnesium with different pore size of plate.

IOP Conf. Series: Materials Science and Engineering **292** (2017) 012015 doi:10.1088/1757-899X/292/1/012015

3.4. Characterization of MH particles

3.4.1. Size distribution

The volume percent and cumulative frequency distribution of MH particles which prepared at NH_3/Mg^{2+} molar ratio of 2.0 (reacting temperature 60 °C, stirring rate 500 $r \cdot min^{-1}$, ammonia flowrate 600 $mL \cdot min^{-1}$ and pore size of plate 0.5 mm) was analyzed, the size distribution was shown as figure 8. The diameter of d10, d50 and d90 is 430, 883 and 1749 nm respectively, from that we can calculated the unevenness of this sample is 1.49. On the whole, the distribution of particle size is from 224 to 3557 nm, and the maximum of volume percent is among at 893 and 1002 nm, approximately 8.2% of total. From this result, we can see that the particle size and dispersibility of MH particle is excellently used as flame-retardant.

3.4.2. XRD characterizing

The MH particles was characterized by XRD and the diffraction pattern was shown in figure 9. By comparing with standard pattern, the volume of brucite is reach to 99.7 %. This result demonstrated the type of the product is high-purity magnesium hydroxide. In addition, the crystal growing at crystal face 001 whose polarity is weak is more obviously than crystal face 101 whose polarity is strong. That indicates the polarity of crystal surface is relatively weak, hence the attractive force of the crystal particle to each other is lower, and the structure of particle is stabilized with better dispersibility.

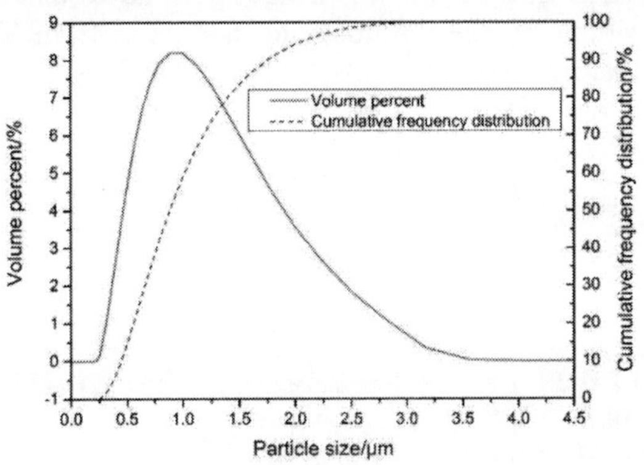

Figure 8. Volume percent and cumulative frequency distribution of MH particles.

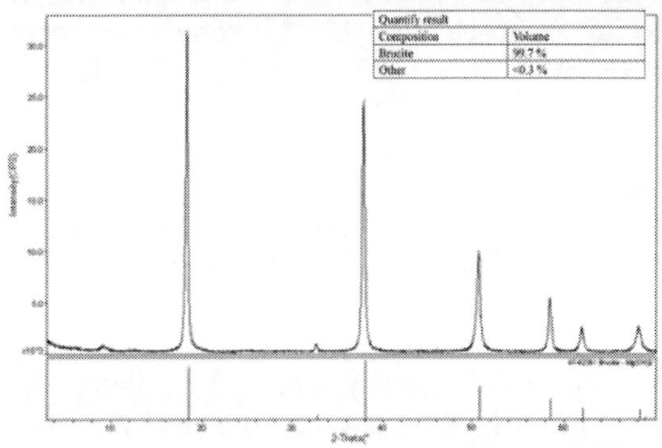

Figure 9. Diffraction pattern of MH particles.

3.4.3. Crystal morphology

The crystal morphology of MH particles also has been observed by SEM and the images were shown as figure 10. From figure 10(a), we can see that the morphology of MH crystal displayed irregularly lamellar and the size of particles is also evenly, and this result is consistent with the analysis of size distribution in section 2.4.1. At high magnification, figure 10 (b) showed that the crystal particles were well independent with each other and the MH particles were dispersing excellently. It indicated that using microporous plate to bubble ammonia gas continually, the concentration of OH^- ion will not appears a larger fluctuation, hence the distributional difference of OH^- ion concentration all round crystal nucleus is smaller, that is to say there is no high energy growing points of crystal nucleus and the degree of crystal anisotropy is relatively weak.

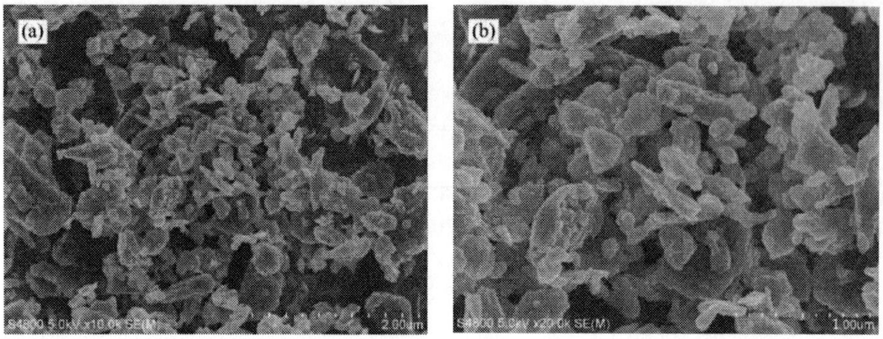

Figure 10. SEM images of MH particles.

4. Conclusion

The present work shows that the maximum of percent conversion of magnesium is 88.1 % at operating condition of reacting temperature 60 ℃, stirring rate 500 $r \cdot min^{-1}$, ammonia flowrate 600 $mL \cdot min^{-1}$ and pore size of plate 0.5 mm. The size distribution and crystal morphology of MH particle preparing from mother liquor in optimum condition was detected, and the results showed the d50 diameter of particles is 883 nm and possessing good dispersibility. The XRD diffraction pattern indicated the volume of brucite is reach to 99.7 % within the composition of the product. In addition, from the thermogravimetric analysis of MH particles, the thermostability of product is suitable as flame-retardant composite materials.

References

[1] Rothon R N & Hornsby P R 1996 *Polym. Degrad. Stabil.* **54(2-3)** 383-385
[2] Durin-France, A, Ferry L, Cuesta J & Crespy A 2000 *Polym. Int.* **49(10)** 1101-1105
[3] Henrist C, Mathieu, J P, Vogels C, Rulmont, A & Cloots, R 2003 *J. Cryst. Growth.* **249(1-2)** 321-330
[4] Song X F, Sun S Y, Zhang D K, Wang J & Yu J G 2011 *Front. Chem. Sci. Eng.* **5(4)** 416-421
[5] Yin X L, Jia Q & Yuan J J 2009 *J. Sal. Chem. Ind.* **05** 14-17
[6] Shahmansouri A, Min J, Jin L & Bellona C 2015 *J. Clean. Prod.* **100** 4-16
[7] Zhao N, Zhang Q, Zhai J & Zong J 2015 *Inorg. Chem. Ind.* **47(5)** 35-37
[8] Yang X Y, Yu P F, Xu H & Shi X C 2014 J. *Centr. Sou. Univer. Sci. Tech.* **01** 25-32
[9] Su M Y & Li S S 2015 *J. Sal. Chem. Ind.* **44(02)** 23-27
[10] Tang L S, Su M Y & Yu K 2010 *J. Sal. Chem. Ind.* **39(04)** 38-42
[11] Song X X, Li L J, Liu Z Q, Shi D, Ji L M, Zhang B & Nie F 2014 *Chem. Eng.* **06** 74-78
[12] Yi Q S. 2001 *Chem. Reagen.* **23(04)** 197-199
[13] Hu Z W, Wang L X, Yang B J & Shan C X 2007 *Chin. J. Proc. Eng.* **7(03)** 561-565
[14] Zhai D W, Chen A M & Ni Z M 2007 *Bull. Sci. Tech.* **23(01)** 141-145
[15] He C H, Zhang M L & Liu J G 2004 *J. Sal. Lak. Res.* **12(02)** 33-38
[16] Tai C Y, Tai C, Chang M & Liu H 2007 *Ind. Eng. Chem. Res.* **46(17)** 5536-5541

IOP Conf. Series: Materials Science and Engineering **292** (2017) 012016 doi:10.1088/1757-899X/292/1/012016

Preparation and electrochemical properties of perovskite $Ce_xLi_{0.5-x}Ca_{0.5}TiO_3$ composites

J H Cheng[1], C A Tian

Department of Chemistry and Materials Engineering, Hefei University, Hefei, 230601 Anhui, China

[1] E-mail: cjh@hfuu.edu.cn

Abstract. Composites with perovskite structure based on lithium and cerium co-doped $CaTiO_3$, $Ce_xLi_{0.5-x}Ca_{0.5}TiO_3$ was prepared by the sol-gel method. Phase structure and morphology of the powders were carried out by X-ray diffraction (XRD) and transmission electron microscopy (TEM). The results show that the $Ce_xLi_{0.5-x}Ca_{0.5}TiO_3$ powders have cubic perovskite structures after sintered above 800 °C and the mean size of the $Ce_xLi_{0.5-x}Ca_{0.5}TiO_3$ powders is about 60 nm. A study of ionic conductivity by electrochemical impedance spectroscopy (EIS) implies that the conductivity of $Ce_xLi_{0.5-x}Ca_{0.5}TiO_3$ increases with the increasing of the contents of Ce^{3+} dopant and reaches a maximum value of 2.1×10^{-4} S·cm^{-1} at x = 0.05, and then decreases for x>0.05.

1. Introduction

Solid electrolyte is a key component of solid-state electrochemical devices, which are widely applied in energy conversion, chemical processing, sensing, and combustion control [1-4]. For most of these applications, high ionic conductivity of the solid electrolyte is required for device performance, and more and more studies have been carried out for the improvement of its ionic conductivity [5].

The perovskite structure is one of the most wonderful to exist in nature. At present, the application of perovskite structure composites is applied largely because of their superior conductivities, and the influence of various additives on the electrical properties of calcium, strontium and barium titanate has been widely investigated [6-9]. $CaTiO_3$ is ABO_3 type perovskite structure, it is cubic crystal system at high temperature. In cubic system, the doping of metal cations with different valances can induce the distortions of oxygen stoichiometric ratio and crystal lattice to form nonstoichiometric compounds easily. High-temperature electrical conductivity of undoped $CaTiO_3$ were studied by W.L. George [10]. According to this research the defect structure of calcium titanate at high temperatures maybe described by majority defect pairs of oxygen vacancies and excess electrons. More recently, several studies on doped calcium titanates were reported [11-14]. Among all doped $CaTiO_3$ materials, lithium doped $CaTiO_3$ composite has a higher conductivity at low temperatures, and has attracted much attention in recent years [15-18].

Lithium doped $CaTiO_3$, $Li_xCa_{1-x}TiO_3$ have been prepared in our previous studies [19]. The object of the present work was to study the preparation and characterization of $Ce_xLi_{0.5-x}Ca_{0.5}TiO_3$ composites. Solid solutions $Ce_xLi_{0.5-x}Ca_{0.5}TiO_3$ were synthesized by a sol-gel method in this study, and the phase constitution and shape of the powders were characterized by X-ray diffraction and transmission electron microscopy, ionic conductivity of the pellets was investigated through AC impedance.

IOP Conf. Series: Materials Science and Engineering **292** (2017) 012016 doi:10.1088/1757-899X/292/1/012016

2. Experimental

$Ce_xLi_{0.5-x}Ca_{0.5}TiO_3$ powders were prepared by a sol-gel method in the present study. All the raw materials were purchased from Sinopharm Chemical Reagent Co., Ltd., China. In this process, $Ti(OC_4H_9)_4$ and acetylacetone were dissolved in absolute alcohol (volume ratio of employed $Ti(OC_4H_9)_4$, absolute alcohol and acetylacetone was 4:4:1) by stirring and heating at 40°C, then the mixture, which formed by $Ca(NO_3)_2{\cdot}4H_2O$, $LiNO_3$ and $Ce(NO_3)_3{\cdot}6H_2O$ dissolved in deionized water, were added into the solution drop by drop under constant stirring. After the precursor solution reacted at 40 °C for 4 h, the $Ce_xLi_{0.5-x}Ca_{0.5}TiO_3$ sol was obtained. Subsequently, the sol was kept in an oven at 50 °C for 48 h to get the dried gel. Then the dried gel was calcined in air at 600 °C, 700 °C and 800 °C for 2 h respectively, the needed powders were obtained. The powders were then pressed into discs of 12 mm in diameter and 1 mm in thickness at a pressure of 30 MPa. At last, the green discs are sintered in a high-temperature furnace at at 1400 °C for 6 h to formed the ceramic discs. To measure the ionic conductivity of the samples, the silver paste conductive adhesive was coated on both surfaces of the calcined pellets.

X-ray diffraction (XRD) analysis was measured with a Japanese D/max-γB rotating diffractometer, and the diffractograms were scanned in 2θ from 10° to 70° at a rate of 6° min^{-1}. Transmission electron microscopy (TEM) photographs were taken on a JEM-100SX transmission electron microscope. The measurement of electrochemical impedance (EIS) was performed with an electrochemical workstation CHI660B at a frequency range of 1 Hz to 100 kHz at room temperature in air.

3. Results and discussion

3.1. Phase of $Ce_xLi_{0.5-x}Ca_{0.5}TiO_3$

The X-ray diffraction patterns of $Ce_{0.05}Li_{0.45}Ca_{0.5}TiO_3$ powders sintered at different temperature are shown in figure 1. The patterns were indexed on the basis of a distorted perovskite structure with cubic symmetry (space group $Fd\bar{3}m$). Figure 1 is the XRD patterns of $Ce_{0.05}Li_{0.45}Ca_{0.5}TiO_3$ powders sintered at different temperatures. Although the typical diffraction peaks of cubic perovskite structure occurred at 600 °C, but the intensity of the diffraction peaks were weak, and some characteristic peaks were not observed, which indicate that the sample was not well crystallized at this temperature. When the sintering temperature increased to 800 °C, the appearance of typical diffraction peaks indicated that the powders have crystallized well and formed cubic perovskite structure with space group $Fd\bar{3}m$. The typical three diffraction lines ((111), (200), and (311) reflections) are assigned, suggesting that the $Ce_{0.05}Li_{0.45}Ca_{0.5}TiO_3$ powders have a cubic perovskite structure and the $Ce_{0.05}Li_{0.45}Ca_{0.5}TiO_3$ powders conduct the transition from amorphous to crystalline at 600 °C-800 °C.

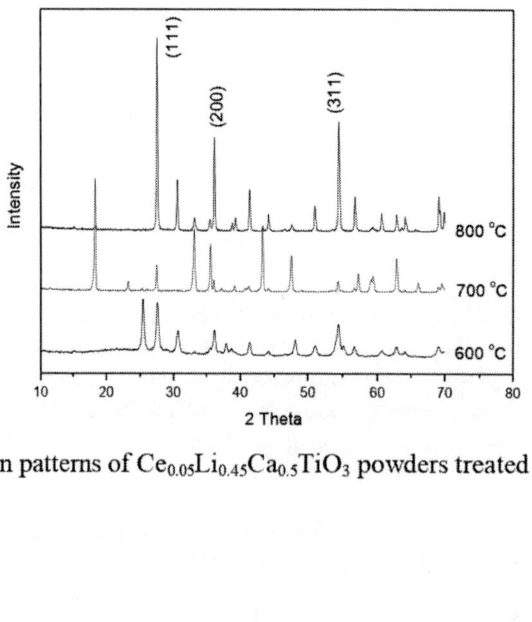

Figure 1. X-ray diffraction patterns of $Ce_{0.05}Li_{0.45}Ca_{0.5}TiO_3$ powders treated at different temperatures.

IOP Conf. Series: Materials Science and Engineering **292** (2017) 012016 doi:10.1088/1757-899X/292/1/012016

The mean size of the prepared powders was calculated to be about 60nm based on Scherrer formula.

$$D=0.89\lambda/B\cos\theta \qquad (1)$$

Where λ is the wavelength of X-ray (Cu Kα radiation, λ=0.154056 nm), θ is the diffraction angle, $B=\sqrt{Bm^2-Bs^2}$ is the corrected halfwidth of the observed halfwidth: Bm is that of the (111) reflection in samples and Bs is that of the (111) reflection in a standard sample.

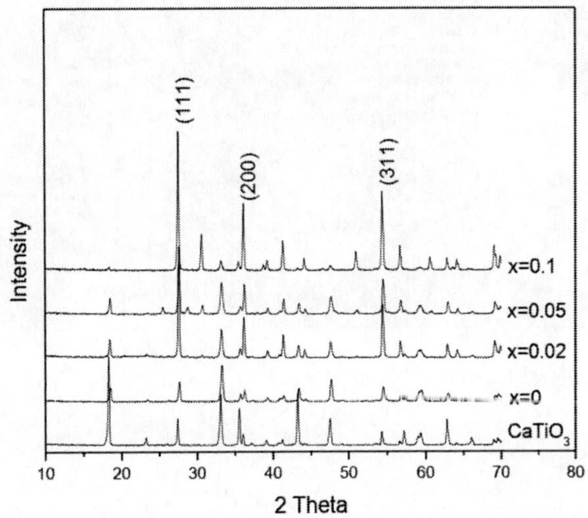

Figure 2. X-ray diffraction patterns of $Ce_xLi_{0.5-x}Ca_{0.5}TiO_3$(x=0~0.1) sintered at 800 °C.

Figure 2 is the X-ray diffraction patterns of $Ce_xLi_{0.5-x}Ca_{0.5}TiO_3$ with different content x prepared by the sol-gel synthesis. As seen from the figure, all these XRD patterns are similar to that of $CaTiO_3$, and there is no evident secondary phases or superstructure could be identified. The result indicates that a relatively low calcinations temperature is satisfiable to prepare pure phase of $Ce_xLi_{0.5-x}Ca_{0.5}TiO_3$ powders with perovskite structure. When $CaTiO_3$ doped with Li an Ce, Ca^{2+} were substituted by Li^+ and Ce^{3+} ions partially formed $Ce_xLi_{0.5-x}Ca_{0.5}TiO_3$ solution. But the amount of Li^+ and Ce^{3+}ions is not more, so this change is not strong enough to distort the $CaTiO_3$ structure, and $Ce_xLi_{0.5-x}Ca_{0.5}TiO_3$ remains cubic perovskite structure alike $CaTiO_3$ [20].

3.2. Morphology of $Ce_{0.05}Li_{0.45}Ca_{0.5}TiO_3$

Figure 3 is the TEM photographs of $Ce_{0.05}Li_{0.45}Ca_{0.5}TiO_3$ powders calcined at different temperatures. The $Ce_{0.05}Li_{0.45}Ca_{0.5}TiO_3$ powders calcined at 600 °C are agglomerated in lump, moreover, the shape of each grain is irregular and the grain boundary is obscure (shown as figure 3a). But when calcining temperature reached at 800 °C, the powders are mono-dispersed and the mean grain size is about 60 nm shown as figure 3b. These results indicate that the $Ce_{0.05}Li_{0.45}Ca_{0.5}TiO_3$ powders are amorphous at 600 °C, but have crystallized well at 800 °C, and the particle size of the powders with a narrow particle size distribution is small and increases with the increase of the calcining temperature .This fact agrees the conclusion drown from XRD analysis.

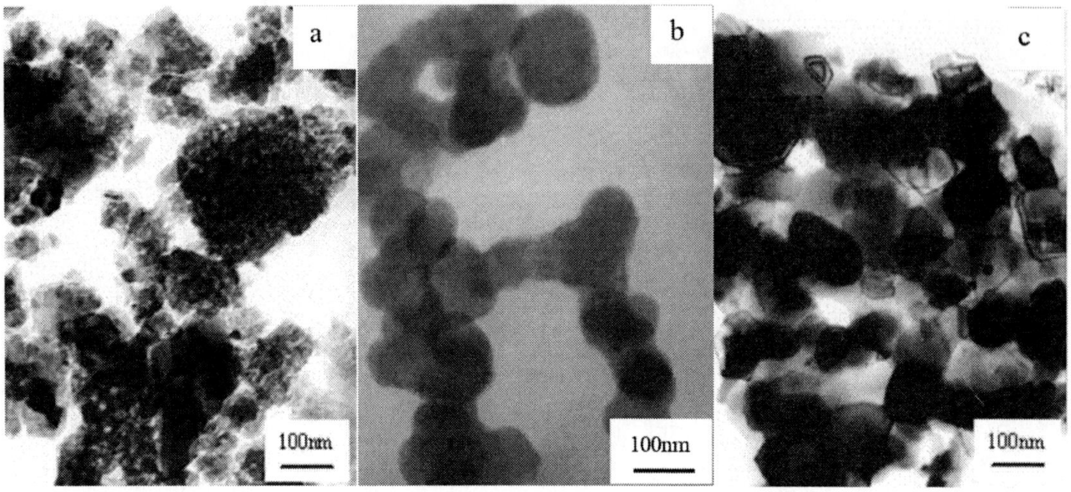

Figure 3. TEM photographs of $Ce_{0.05}Li_{0.45}Ca_{0.5}TiO_3$ powders calcined at 600 °C (a), 700 °C (b) and 800 °C (c).

3.3. Sinterability analysis

The bulk density (ρ) of sintered pellets was calculated using Archimedes' method in deionized water (D.I. water):

$$\rho = \frac{W_{Dry} \times \rho_{Water}}{W_{Wet} - W_{Sus}} \tag{2}$$

where W_{Dry} is the dry weight of sintered discs, ρ_{water} is the density of D.I. water (0.997 g·cm^{-3}), W_{Wet} represents the wet weight (put the pellet in water and boiling for 30 min), and W_{sus} shows the suspended weight of sintered discs in D.I. water. The values of the relative densities reached above 95% when sintering temperature was 1400 °C, which resulted in the formation of dense discs with low porosity.

3.4. Ionic conductivity of $Ce_xLi_{0.5-x}Ca_{0.5}TiO_3$

An interesting approach to the study of the electrical behavior of compounds is the electrochemical impedance spectroscopy (EIS) technique [21]. Figure 4 is the electrochemical impedance spectra of $Ce_xLi_{0.5-x}Ca_{0.5}TiO_3$ pellets sintered at 1400 °C. Each spectrum consists of an incomplete semicircle and an inclined line shown as figure 4. In principle, a semicircle between the origin and the intercept would be observed if much higher frequency were used in the experiment, but the high-frequency limit used in this study (100 kHz is the upper limit for electrochemical workstation CHI660B) is not sufficient and caused the incomplete semicircle [22].

The replacement of A or B-metal in a ABO_3 perovskite by a cation of lower valence, oxygen vacancies are formed, whose concentration is equal to the defects. Conductivity dependence on oxygen partial pressure may be divided into two areas: in the intermediate oxygen partial pressure area where $CaTiO_3$ conductivity is proportional to $p^{1/6}O_2$, and the low partial pressures area, where the obtained experimental dependences increase [23, 24].

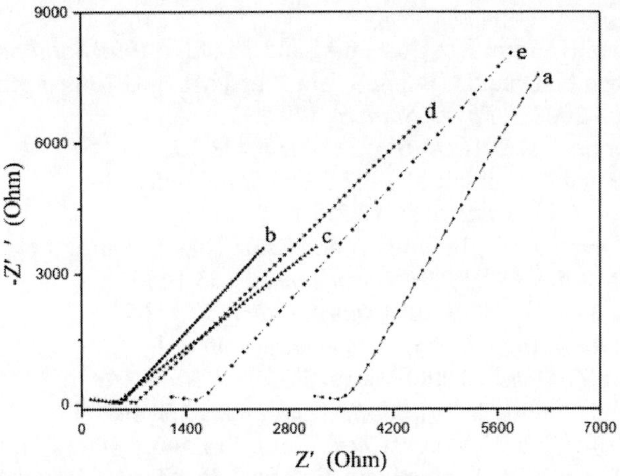

Figure 4. AC impedance curves of pellets, (a) x=0, (b) x =0.02, (c) x =0.05, (d) x =0.1, (e) x =0.2.

The intercepts of the semicircular arcs at "z" axes provide the resistance values corresponding to bulk(R_{gi}) and grain boundaries (R_{gb}). The conductivities can be calculated using the resistance obtained by fitting the impedance spectra using ZSimpWin software. Then the conductivity datums at different temperatures can be obtained using the equation (3). The ionic conductivities of the $Li_{0.5}Ca_{0.5}TiO_3$, $Ce_{0.02}Li_{0.48}Ca_{0.5}TiO_3$, $Ce_{0.05}Li_{0.45}Ca_{0.5}TiO_3$, $Ce_{0.1}Li_{0.4}Ca_{0.5}TiO_3$ and $Ce_{0.2}Li_{0.3}Ca_{0.5}TiO_3$ samples prepared by sol–gel technique at room temperature can be estimated to be 8.66×10^{-5} S·cm^{-1}, 1.63×10^{-4} S·cm^{-1}, 2.1×10^{-4} S·cm^{-1}, 1.9×10^{-4} S·cm^{-1} and 9.81×10^{-5} S·cm^{-1} respectively. These values are near to that of other fluorite type electrolytes such as YSZ, SDC or GDC which have been studied in-depth for SOFCs applications.

$$\sigma = L/R \cdot S \quad (3)$$

Where L and S is the thickness and the area of the pellet, respectively. R is the resistance of the sample (The resistances $R=R_{gi}+R_{gb}$). These datum indicate that the $Ce_xLi_{0.5-x}Ca_{0.5}TiO_3$ ionic conductivity increases with the increasing of the contents of Ce^{3+} dopant and reaches a maximum value at x = 0.05, and then decreases for $x > 0.05$. When the oxygen vacancies were substituted by Li^+ and Ce^{3+} ions, the ionic conductivities of $Ce_xLi_{0.5-x}Ca_{0.5}TiO_3$ solid solutions were significantly improved, but when the amount of the substituted Li^+ and Ce^{3+} ions exceeded a certain value, the interactions among dopant cations (Li^+ and Ce^{3+}) and higher concentrations of oxygen vacancies caused the aggregating of oxygen vacancies, so the effective oxygen vacancies concentration did not increase and the ionic conductivity was lowing for $x > 0.05$.

4. Conclusions

The perovskite structure solution with chemical formula of $Ce_xLi_{0.5-x}Ca_{0.5}TiO_3$ were prepared by the sol-gel method in ethanol and water mixture medium. The products formed pure cubic perovskite structure after calcined at 800 °C, and the mean size of the $Ce_xLi_{0.5-x}Ca_{0.5}TiO_3$ powders calculated to be about 60 nm. The ionic conductivity at room temperature of $Ce_xLi_{0.5-x}Ca_{0.5}TiO_3$ increases with the increase of the contents of Ce^{3+} dopant and reaches a maximum value of 2.1×10^{-4} S·cm^{-1} at x = 0.05, and then decreases for $x > 0.05$.

Acknowledgements

Research supported by the Natural Science Foundation of Education Department of Anhui Province under contract No. KJ2016A591, the Nature Science Foundation of Anhui Province of China under contact No. 1708085ME112 and the Key projects of Foreign visits to key Teachers of Universities in Anhui Province under contract No. gxfxZD2016220.

References

[1] Ishihara T, Matsuda H, Azmi Bin Bustam M and Takita Y 1996 *Solid State Ionics* **86-88** 197
[2] Xie K, Yan R, Chen X, Dong D, Wang S, Liu X and Meng G 2009 *J. Alloys Compd.* **472** 551
[3] Liou Y and Yang S 2008 *J. Power Sources* **179** 553
[4] Li S, Li Z and Bergman B 2010 *J. Alloys Compd.* **492** 392
[5] Huang W, Shuk P and Greenblatt M 1997 *So lid State Ionics* **100** 23
[6] Fang L 2017 *Int. J. Electrochem. Sci.* **12** 218
[7] Dunyushkina L, Gorbunov V, Babkina A and Esina N 2003 *Ionics* **9** 67
[8] Gorelov V and Balakireva V 1997 *Electrochemistry* **33** 1450
[9] Dunyushkina L and Gorbunov V 2001 *Inorg. Mater.* **57** 1165
[10] George W and Grace R 1969 *J. Phys. Chem. Solids* **30** 881
[11] Arita Y, Nagarajan K, Ohashi T and Matsui T 1997 *J. Nucl. Mater.* **247** 94
[12] Dunyushkina L and Gorelov V 2013 *Solid State Ionics* **253** 169
[13] Ahmed M and Bishay S 2009 *Mater. Chem. Phys.* **114** 446
[14] McCammon C, Becerro A, Langenhorst F, Angel R, Marion S and Seifert F 2000 *J. Phys. Condens. Matter* **12** 2969
[15] Cheng J, Bao W, Zhu D and Ding M 2009 *J. Alloys Compd.* **486** 895
[16] Hreniak D, Strek W, Chmielowiec J, Pasciak G, Pazik R, Gierlotka S and Lojkowski W 2006 *J. Alloys Compd.* **408-412** 637
[17] Mei A, Jiang Q, Lin Y and Nan C 2009 *J. Alloys Compd.* **486** 871
[18] Shin K, Ko Y and Shin D 2011 *J. Alloys Compd.* **509S** S478
[19] Wang H, Cheng J, Zhai L and Cheng 2007 *J Solid State Commun.* **142** 710
[20] Liang K, Du Y and Nowick A 1994 *Solid State Ionics* **69** 117
[21] Muccillo R and Carmo J 2012 *Mater. Res. Bull.* **47** 1204
[22] Wu X, Li X, Zhang Y, Xu M and He Z 2004 *Mater. Lett.* **58** 1227
[23] Murashkina A, Demina A, Demin A, Maragou V and Tsiakaras P 2008 *Solid State Ionics* **179** 1615
[24] Vashook V, Vasylechko L, Ullmann H and Guth U 2003 *Solid State Ionics* **158** 317

IOP Conf. Series: Materials Science and Engineering **292** (2017) 012017 doi:10.1088/1757-899X/292/1/012017

Green synthesis of silver nanoparticles using wolfberry fruits extract and their photocatalytic performance

L Sun[1], P C Lv, Y C Yin, H N Li and F Wang

School of Energy and Power Engineering, University of Shanghai for Science and Technology, Jun Gong Road 516, Shanghai 200093, China

[1] E-mail: marybrother@yeah.net

Abstract. Spherical Ag nanoparticles (NPs) was synthesized using the Chinese wolfberry fruits extract (WFE) as surfactant and reductant. The hydrolysis time and the reaction time were systematically investigated. The whole synthesis process must be kept boiling in order to enhance the reducing ability of WFE and decrease the reaction time. Under the optimized hydrolysis time of 10 min, the reaction time just need 5 min. Ag NPs were characterized by means of UV-vis, EDX, TEM, HR-TEM, and SAED. It was found that Ag NPs exhibited the surface plasmon resonance wavelength of 408 nm, a small average size of 12.1 nm, and the face-centered cubic structure. Obtained Ag NPs were used as photocatalyst for the reduction of 4-nitrophenol and revealed a high photocatalytic activity and stability. The rate constant was 0.043 min^{-1}, which suggested the reduction of 4-nitrophenol was fast. Clearly, Ag NPs synthesized in the totally green method have a potential application in the field of the wastewater treatment.

1. Introduction

4-nitrophenol (4-NP), a harmful and toxic pollutant, widely exists in pharmaceutical industry, dyeing industry, food industry, and so on [1, 2]. Traditional wastewater treatment methods are difficult to remove 4-NP with a low concentration [3]. Up to now, photocatalysis is the main method to degrade 4-NP using semiconductor materials such as TiO_2 and WO_3 as photocatalysts [4, 5]. However, most of semiconductor can only absorb ultraviolet light with shorter wavelength due to their relatively wide band gap, indicating a lower utilization of solar energy. In recent years, noble metal nanoparticles (NPs) have attracted wide attention in the field of photocatalysis, which can make full use of solar photons in the visible spectral region [6-8]. Ag NPs can be much more suitable to be chose as the photocatalyst to treat wastewater because of their high photocatalytic properties as well as bactericidal action [9, 10].

At present, several methods, used to synthesize Ag NPs, are divided into physical, chemical and biosynthesis methods. Among these methods, the first two methods rely on complex process, high cost, or strong ligands, which can pose potential environmental and biological risks [11, 12]. Biosynthesis method is regarded as an excellent method due to its simplicity, cost-effectiveness, compatibility for the biomedical and pharmaceutical applications [13]. Especially, the green synthesis methods based on plant extracts provided a promising choice for the preparation of Ag NPs. Nevertheless, only a few kinds of plants, such as Lavandula intermedia, the leaf extract of Cinnamomum tsoi, the leaf and bark extract of Syzygium Jambos Alston and the Chinese wolfberry fruits extract (WFE) have been used to synthesize Ag NPs [14-17]. Among these plants, the Chinese wolfberry, an ancient Chinese herbal medicine, has bactericidal action and a wide distribution in China. Therefore, Ag NPs synthesized by the WFE can possess superior performance in the wastewater treatment. Usually, the reduction of 4-NP is considered as a model to evaluate the photocatalytic properties of biosynthesized Ag NPs in the wastewater treatment [2]. Dong et al. reported the long synthesis time up to 10 h was needed to prepare Ag NPs due to the weak reducing ability of the WFE, which limited their applications in the wastewater treatment [17]. In order to decrease the synthesis time of Ag NPs, the synthesis method based on the WFE must be improved.

In this paper, we tried a fast green synthesis of Ag NPs using the WFE as reductant and surfactant without any chemical agent. The hydrolysis process was adopted to enhance the reducing ability of the WFE and decrease the reaction time. The optimal hydrolysis time was found by the UV-vis spectra. Based on the optimal hydrolysis time, the reaction time was systematically investigated. The composition, morphology and micro-structure of Ag NPs were characterized by UV-vis, EDX, TEM, HR-TEM and SEAD analyses.

2. Experimental methods

2.1. Materials

Silver nitrate ($AgNO_3$, GR), sodium borohydride ($NaBH_4$, AR), 4-NP ($C_6H_5NO_3$, AR) and ethanol (C_2H_5OH, GR) were purchased from Aladdin Reagent Inc. (Shanghai, China) and used directly as the raw materials without further treatment. Deionized water was used throughout this work. In our experiment, the Chinese wolfberry fruits (Lycium barbarum L.) were produced in Ningxia province of China.

2.2. Preparation of the Chinese wolfberry fruits extract

30 g of wolfberry fruits and 250 mL of deionized water were mixed and brought to boiling for 8 min under vigorous stirring. The above boiling solution was cooled in air until to room temperature. Then the cooled solution was centrifuged at 8,000 rpm for 8 min to remove any undesired impurities. The obtained WFE was diluted by deionized water to 10 vol.% for further experiments.

2.3. The synthesis of Ag NPs

Typically, 25 mL of 10 vol.% WFE was taken in a flask and gently heated to boiling with stirring. After the hydrolysis, 5mL of 0.01 mM $AgNO_3$ solution was added into the above boiling solution. Then the synthesis system was kept boiling with continuous stirring till the end of the reaction. The hydrolysis time was 10 min. The reaction time was set at 5, 10, 15, 20 and 25 min. It could be found that the colour of solution transformed from colourless to brown. In the experiment, the amount of WFE and $AgNO_3$ solution was fixed as 25 and 5 mL, respectively.

2.4. Characterization of synthesized Ag NPs

Ultraviolet visible spectrum (UV-vis) analysis was carried out on a spectrophotometer (Cary 50, Varian) using deionized water as the dispersing agent to detect the SPR absorption peak of Ag NPs. Energy-dispersive X-ray spectrum (EDX), Transmission electron microscope (TEM), High resolution transmission electron microscope (HR-TEM), and Selected area electron diffraction (SAED) analyses were performed on a transmission electron microscopy (JEM-2100F, JEOL) operated at 200 kV. The sample for TEM was prepared by placing a drop of dilute solution of NPs on an amorphous carbon-copper grid and subsequent evaporation in air at room temperature.

2.5. The reduction of 4-nitrophenol

To investigate the photocatalytic activity of synthesized Ag NPs, the reduction reaction of 4-nitrophenol was used as a model. The reaction was carried out in a quartz cuvette. 3 mL of 1 M $NaBH_4$ aqueous solution was mixed with 2.5 mL of 10 mM 4-NP. Then 200 μL of 0.1 mM Ag NPs were added into the above mixture. The reduction of 4-NP as a function of time was monitored by recording the absorption spectra at every 5 min interval in the range of 200-600 nm under sunlight irradiation at room temperature.

3. Results and discussion

3.1. The synthesis of Ag NPs

3.1.1. The effect of hydrolysis time. Keeping the reaction time as 10 min, the effect of the hydrolysis time on the synthesis of Ag NPs was investigated. Figure 1 (a) displays the UV-vis absorption spectra of Ag NPs prepared at different hydrolysis time from 4 to 14 min. It can be seen from figure 1 that the absorption peaks of Ag NPs keep a constant as 407 nm with increasing hydrolysis time. Figure 1 (b) shows the influence of hydrolysis time on the absorbance of the SPR peak of Ag NPs. With the enhancement of the hydrolysis time from 4 to 10 min, the absorbance of SPR peak was gradually increased, indicating that the reduction amount of Ag ions was enhanced. After that, the absorbance of absorption peak tends to be stable even increasing the hydrolysis time from 10 to14 min, implying the complete hydrolysis of the WFE. As a result, the optimal hydrolysis time is 10 min, which can guarantee the yield of Ag NPs. The main compositions of WFE are polysaccharides and proteins [18], whose hydrolysis products are monosaccharide and amino acids, respectively. It is well known that monosaccharide and amino acids has strong reducing ability, which can benefit for the decease of the synthesis time.

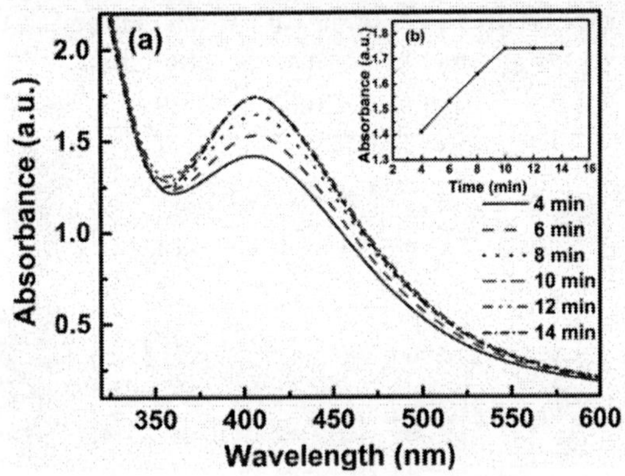

Figure 1. (a) UV-vis spectra of Ag NPs synthesized at 10 min with different hydrolysis time. (b) Relationship of the hydrolysis time and the absorbance of SPR peak.

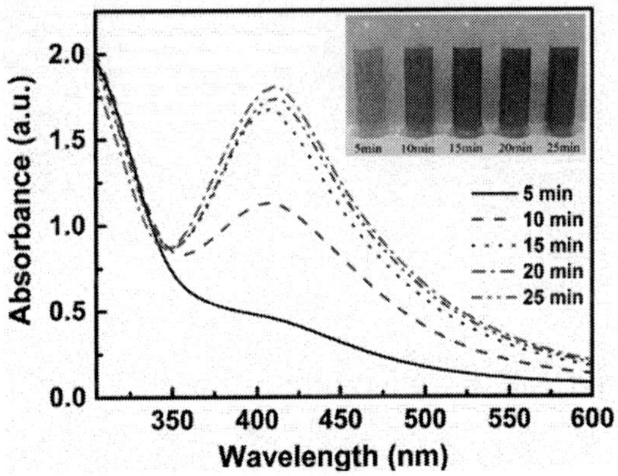

Figure 2. UV-vis spectra of Ag NPs synthesized at different reaction time from 5 to 25 min. Inset is the photographs of corresponding Ag NPs.

3.1.2. The effect of reaction time. The reaction time is an important parameter to evaluate the synthesis system to prepare Ag NPs. Under the optimized hydrolysis time (10 min), the effect of the reaction time on the synthesis of Ag NPs was investigated as shown in figure 2. The reaction time was changed from 5 to 25 min. When the reaction time was 5 min, a weak absorption peak had appeared, implying the formation of Ag NPs. Enhancing the reaction time to 15 to 25 min, the absorbance of the SPR peak exhibit a constant value, which suggests the reduction of Ag ions can be totally completed just at 15 min. With increasing reaction time, the absorption peak of Ag NPs is red-shifted from 407 to 412 nm. According to Mie theory, the SPR absorption peak of spherical Ag NPs was red-shifted, indicating the increase of the size of Ag NPs [19]. However, the difference of the SPR absorption peak was only 5 nm, which meant the growth of Ag NPs was slow. Obviously, the WFE was a strong surfactant. Dong et al reported Ag NPs could also be synthesized using the WFE [17]. However, the reaction time reached up to 10 h because of the weak reducing ability shown by the unhydrolyzed WFE. Clearly, the hydrolysis was a key process for the enhancement of the reducing ability of the WFE and the decrease of the reaction time. Therefore, the synthesis system must be kept boiling in order to ensure the sufficient hydrolysis.

3.2. Characterization

Ag NPs synthesised at the optimized reaction conditions were characterized by means of EDX, UV-vis, TEM, HR-TEM and SEAD analyses. The hydrolysis time and the reaction time was fixed at 10 and 15min, respectively. In order to observe the composition, EDX analysis was carried out. In figure 3a, an intense peak at 22.0 keV and 24.8 keV is shown, which can be ascribe to silver element. Due to the absence of the O element, the product is considered as monometallic Ag NPs. In figure 3b, the SPR absorption peak at 408 nm was further confirm the formation of the spherical Ag NPs [20].

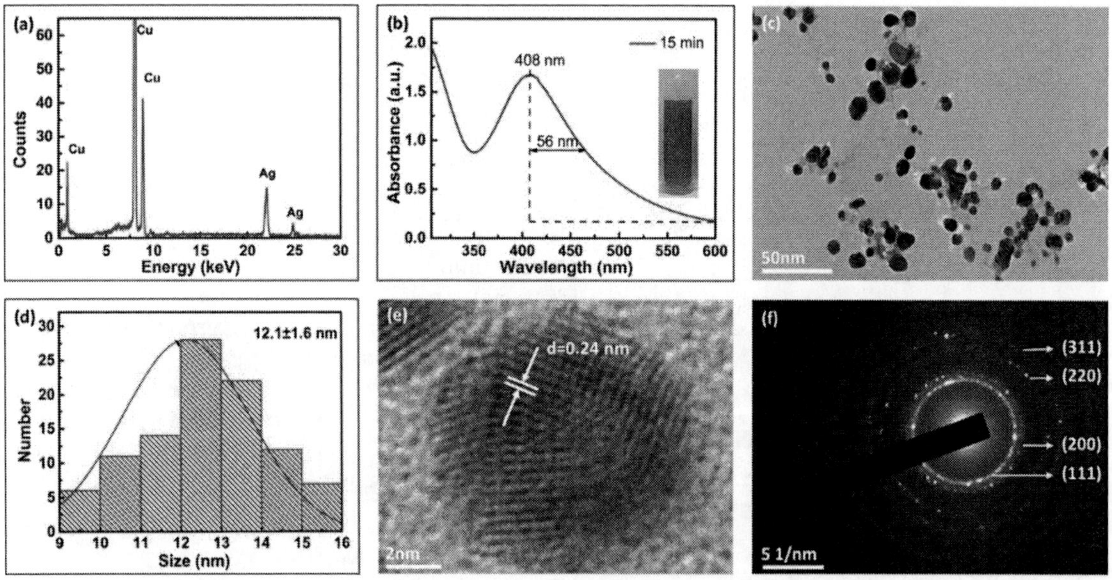

Figure 3. Characterization of Ag NPs synthesized in 10 min hydrolysis time and 15min reaction time: (a) EDX analysis; (b) UV-vis spectrum; (c) TEM image; (d) the particle size distribution; (e) HR-TEM image; (f) SAED pattern.

According to Mie, the variety of the size distribution can also be reflected from the half width at half maximum (HWHM) of the UV-vis absorption spectrum [21, 22]. And the HWHM of Ag NPs is only 56 nm, which indicates that the size distribution of synthesized Ag NPs is narrow [23]. The morphology of Ag NPs was further determined by TEM. As can be seen from TEM image (figure 3c), monodisperse and spherical Ag NPs is obtained in the absence of condensation. It can be found from figure 3d, the average size of Ag NPs is 12.1 nm and the standard deviation is 13%, which further suggests obtained Ag NPs have a narrow size distribution. Obviously, the WFE has a strong surfactant function. Nevertheless, Dong et al reported that Ag NPs synthesized by the WFE without hydrolysis exhibited the average size in the range from 5 to 40 nm. Clearly, Ag NPs synthesized in the boiling WFE showed a smaller size, which could be attributed to the increasing reducing ability of the WFE under the boiling condition. The microstructure of Ag NPs was characterized by HR-TEM and SAED analyses. Figure 3e is the HR-TEM image, which exhibits Ag NPs tend to adopt an icosahedral morphology with a multiple-twinned structure [24]. The interplanar distance is approximately 0.24 nm, corresponding to the (111) lattice planes of the face-centered cubic (fcc) Ag [25]. In fcc metallic NPs with a small size, the multi-twinned icosahedral structure is frequently observed [26]. This multiple-twinned crystalline nature is further indicated via the SAED analysis in figure 3f. The electron diffraction pattern displays four diffuse rings corresponding to (111), (200), (220), and (311) lattice planes of Ag NPs with the fcc structure, confirming the multiple-twinned crystalline nature of Ag NPs.

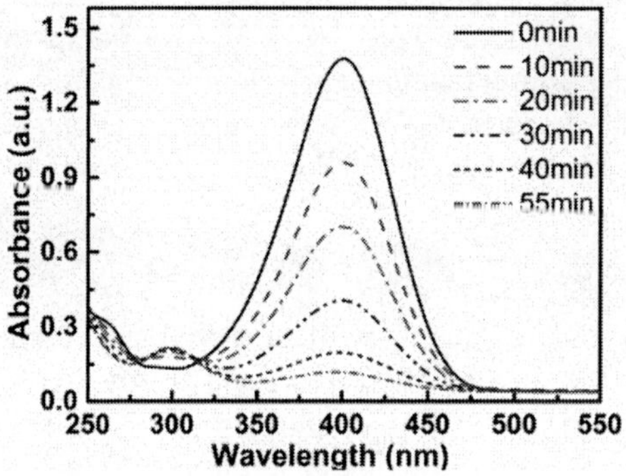

Figure 4. The UV-vis absorption spectra of the reduction of 4-NP solution mixed with Ag NPs.

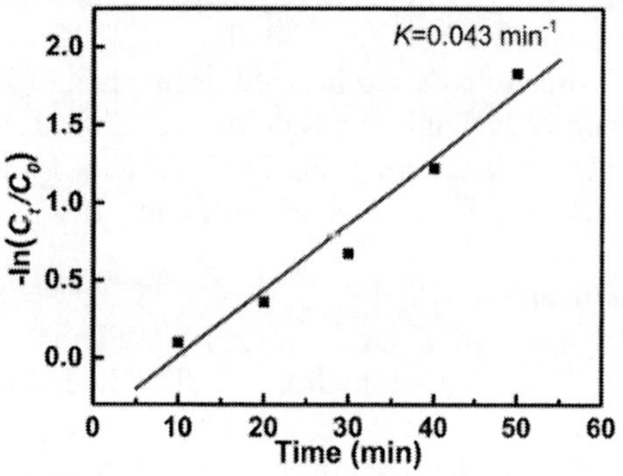

Figure 5. Plot of -ln(C_t/C_0) versus time for the reduction of 4-NP.

3.3. The photocatalytic activity of Ag NPs

As shown in figure 4, the photocatalytic activity of Ag NPs was evaluated in the degradation of the 4-NP under sunlight irradiation.

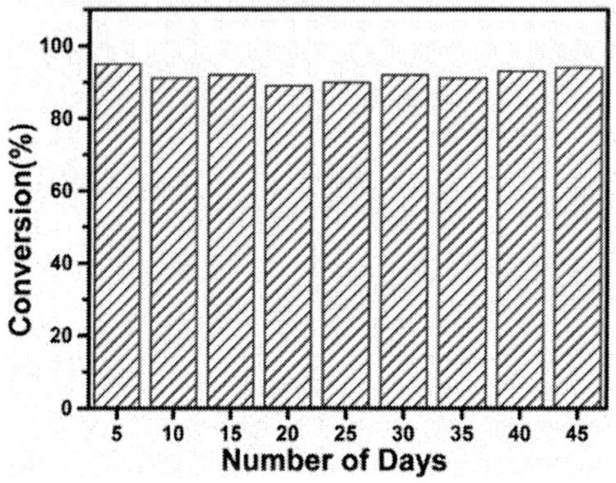

Figure 6. The 4-NP converted to 4-AP with Ag NPs as catalyst in 45 days.

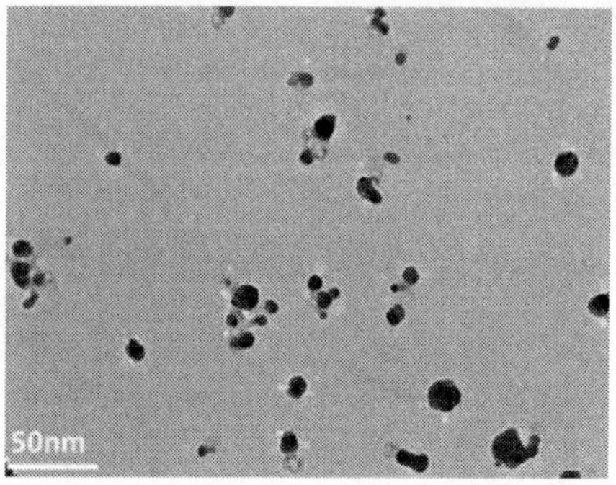

Figure 7. TEM image of Ag NPs used as catalyst after 45 days.

With the addition of $NaBH_4$ as reductant in 4-NP solution, the UV-vis absorption peak of 400 nm was appeared immediately, implying the formation of nitrophenolate ions. In the absence of the catalyst, the peak at 400 nm remained undiminished even for several hours. However, when Ag NPs were added into the reaction system, the absorbance of the peak at 400 nm was decreased quickly. And the new peak at 300 nm appeared, which implied the formation of 4-aminophenol [27]. When

$NaBH_4$ is in excess in the reaction system, the rate constant can be assumed to follow pseudo first order and can be described by

$$\ln(C_t/C_0)=-Kt \quad (1)$$

Here, K is the rate constant, t-reaction time, C_t and C_0-concentration of the 4-NP at time t and 0, respectively. The C_t/C_0 ratio is equal to the ratio of the absorbance A_t/A_0. In figure 5, the rate constant is given as 0.043 min^{-1}, which is corresponding to the slope of fitted linear relationship between $-\ln(C_t/C_0)$ and the reaction time [28]. Generally, Ag NPs are easily oxidized if the surfactant is weak. In order to observe the stability of Ag NPs, the photocatalytic experiment was repeated every 5 days until 45 days. In figure 6, Ag NPs still show a high catalytic activity after 45 days, which indicates that obtained Ag NPs is stable. Figure 7 is the TEM image of Ag NPs used as catalyst after 45 days. Compared with that of the fresh Ag NPs, it can be found that the morphology of Ag NPs has no obvious change and there are in absence of agglomerates, which ensures a high catalytic activity of Ag NPs after 45 days. Guo et al reported the catalytic performance of hollow porous Au NPs (80 nm) had no significant change after 30 days [29]. Clearly, Ag NPs synthesized by WFE has comparable stability with Au NPs synthesized using chemical reagents as reductant and surfactant. Compared with commercially available Ag NPs, the stability of Ag NPs synthesized by WFE was superior [30]. Based on the high activity and stability, Ag NPs synthesized by the above totally green method have a potential application in wastewater treatment.

4. Conclusions

Ag NPs were synthesized using the WFE as reductant and surfactant without any other chemical reagent in the case of hydrolysis. The hydrolysis was a key process for the enhancement of the reducing ability of the WFE and the decreasing of the reaction time. At the optimal hydrolysis time of 10 min, Ag NPs were synthesized just through 5 min, whose obvious SPR absorption peak could be observed in their UV-vis spectrum. When the reaction time was up to15 min, Ag ions could be completely reduced. Obtained Ag NPs showed the SPR absorption peak of 408 nm, the average size of 12.1 nm and the multi-twinned icosahedral microstructure. In the degradation of 4-NP, Ag NPs exhibited excellent photocatalytic activity and stability. The rate constant was calculated as 0.043 min^{-1} according to pseudo first order. The reported green synthesis method is simple and eco-friendly and obtained Ag NPs have the potential applications in the wastewater treatment and the medical field.

Acknowledgement

This work was financially supported by National Natural Science Foundation of China (Grant No. 11404210).

References

[1] Yang Y, Zhang W, Ma X H, Zhao H R and Zhang X 2015 *ChemCatChem* **7** 3454-9
[2] An X, Long Y D and Ni Y H 2017 *Carbohyd. Polym.* **156** 253-8
[3] Ankita G, Ashish K N and Anjali P 2017 *Curr. Pollution Rep.* **3** 17-30
[4] Tian L H, Xu J L, Alnafisah A, Wang R, Tan X Y, Oyler A, Liu L and Chen X B 2017 *Chem. Eur. J.* **23** 5345-51
[5] Hungea Y M, Mahadikb M A, Moholkara A V and Bhosalea C H 2017 *Ultrason. Sonochem.* **35** 233-42
[6] Qin S D, Liu Y C, Zhou Y M, Cha T and Guo J H 2017 *J. Mater. Sci.* **28** 7609-14
[7] Sun H, Zhen S, He Q R, She P, Xu K L and Liu Z N 2017 *Dalton Trans.* **46** 3887-94.
[8] Yin D G, Zhang L, Cao X Z, Chen Z W, Tang J X, Liu Y M, Zhang T T and Wu M H 2016 *Dalton Trans.* **45** 1467-75
[9] Lee T, Chao B K, Kuo Y L and Hsueh C H 2017 *Mater. Chem. Phy.* **192** 78-85
[10] Zhu W Y, Liu J C, Yu S Y, Zhou Y and Yan X L 2016 *J. Hazard. Mater.* **318** 407-16
[11] Pedro Q, Leonor S, Lívia C, Adelaide M, Inês O, Patrícia A. C, Ricardo F and Eulália P 2009 *Green Chem.* **11** 1889-93

[12] Adil S, Assal M, Khan M, Al-Warthan A, Siddiqui M and Liz-Marzán L 2015 *Dalton Trans.* **44** 9709-17
[13] AbdelRahim K, Mahmoud S Y, Ali A M, Almaary K S, Mustafa A E and Husseiny S M 2017 *Saudi J. Biol. Sci.* 24 208-16
[14] Elemike E, Onwudiwe D, Ekennia A and Katata-Seru L 2017 *Res. Chem. Intermed* **43** 1383-94
[15] Maddinedi S, Mandal B K and Maddili S K 2017 *J. Photoch. Photobio. B* **167** 236-41
[16] Dutta P, Bordoloi M, Gogoi K, Roy S, Narzary B, Bhattacharyya D R, Mohapatra P K and Mazumder B 2017 *Biomed. Pharmacother.* **91** 567-80
[17] Dong C F, Cao C L, Zhang X L, Zhan Y L, Wang X J, Yang X Z, Zhou K, Xiao X H and Yuan B 2017 *Optik-Int. J. Light Electron Opt.* **130** 162-70
[18] Ahluwalia V, Kumar J, Sisodia R, Shakil N A and Walia S 2014 *Ind. Crops Prod.* **55** 202-6.
[19] Jia X W, Shen J Q and Yu H T 2017 *J. Quant. Spectrosc. Radiat. Transfer.* **195** 44-54
[20] Li T, Ma G H and Peng T J 2015 *Rare Metal Mat. Eng.* **44** 1071-74
[21] Wang C, Peng S, Chan R and Sun S H 2009 *Small* **5** 567-70
[22] Zempila M M, Koukouli M E, Bais A, Fountoulakis I, Arola A, Kouremeti N and Balis D 2016 *Atmos. Environ.* **140** 283-97
[23] Shulga Y M, Matyushenko D V, Kabachkov E N, Kolesnikova A M, Kurkin E N, Domashnev I A and Brichkin S B 2010 *Tech. Phys.* **55** 141-3
[24] Malathi S, Ezhilarasu T, Abiraman T and Balasubramanian S 2014 *Carbohyd. Polym.* **111** 734-43
[25] Deepak S, Niladri S, Gyanaranjan S, Priyaranjan M and Sarat K S 2017 *Sens. Actuators B: Chem.* **246** 96-107
[26] Raymundo-Pereira P A, Campos A M, Prado T M, Furini L N, Boas N V, Calegaro M L and Machado S A S 2016 *Anal. Chim. Acta* **926** 88-98
[27] Khan F, Asimullah, Khan S, Kamal T, Asiri A, Khan I and Akhtar K 2017 *Int. J. Biol. Macromol.* **102** 868-77
[28] Kumari M M, Jacob J and Philip D 2015 *Spectrochim. Acta A* **137** 185-92
[29] Guo M Z, He J, Li Y, Ma S and Sun X H 2016 *J. Hazard. Mater.* **310** 89-97
[30] Liu W, Zhou Q F, Liu J Y, Fu J J and Jiang G B 2011 *Chinese Sci. Bull.* **56** 27-34

MnO_2/multiwall carbon nanotube/Ni-foam hybrid electrode for electrochemical capacitor

L H Chen, L Li, W J Qian and C K Dong[1]

Institute of Micro-nano Structures & Optoelectronics, Wenzhou University, Wenzhou, Zhejiang 325035, People's Republic of China

[1] E-mail: dck@wzu.edu.cn

Abstract. The ternary composites of manganese dioxide/multiwall carbon nanotube/Ni-foam (MnO_2/MWNT/Ni-foam) for supercapacitors were fabricated via a hydrothermal method after direct growth of MWNTs on the Ni-foam. The structural properties of the electrodes were characterized by SEM and TEM. The electrode exhibited excellent electrochemical properties from the investigation based on the three-electrode setup. Low contact resistance R_s of about 0.291 Ω between MnO_2/MWNT and Ni-foam was reached benefited from the direct growth structure. High capacitance of 355.1 F/g at the current density of 2 A/g was achieved, with good capacitive response at high current density. The MnO_2/MWNT/Ni-foam electrode exhibits good stability performance after 2000 cycles at a current of 40 mA.

1. Introduction

Electrochemical capacitors or supercapacitors (ECs) has been studied widely in recent years as a new type of renewable energy source. Electrochemical capacitors are divided into electrochemical double-layer capacitors (EDLCs) and pseudo-capacitors according to the charge storage mechanism. Carbon-based materials are investigated intensively as the electrode materials for EDLCs with excellent cyclic stability and high specific power due to their superior physical and chemical properties. The transition metal oxide/hydroxide and conducting polymers are commonly employed as pseudo-capacitive active materials for high energy density properties [1, 2]. Comparing with single active materials, the hybrid materials enhance the supercapacitor performance significantly [3, 4]. Carbon nanotubes (CNTs), due to the unique pore structure, superior electrical properties, and good mechanical and thermal stabilities, have attracted a great deal of attentions for supercapacitor electrode applications [5, 6]. MnO_2 is regarded as a promising material for energy storage due to advantages of low cost, low toxicity, environmental safety, and high theoretical capacities ranging from 1100 to 1300 F g^{-1} [7]. The CNT/MnO_2 composite electrodes have showed great electrochemical properties in pseudocapacitors. S. Kong at el have made a MnO_2-CNT/Ni electrode showing a high specific capacitance of 402.5 F/g at 1 A/g and a good cycling performance at a current density of 2 A/g [8]. Zhao et al reported a MnO_x /CNT/Ni-foam electrode with MnO_x showed a high specific capacitance (SC) of 462 F/g with good long cycle stability at a current density of 5 A/g [9]. Normally, CNT based supercapacitor electrode is made either from the synthesis of CNTs on catalyst film [9], or by mix of CNT powder with active materials [10]. The growth of CNTs directly on the current collector without extra catalyst film, which benefits the enhancement of the current collector/carbon contact, has not attracted serious attentions in the development of supercapacitor electrodes.

In this report, we prepared the MnO_2/MWNT/Ni-foam electrodes by hydrothermal synthesis of MnO_2 on MWNTs which were grown directly on the Ni-foam substrate with enhanced MWNT/Ni-foam binding properties. The electrochemical performances of the composite electrode were investigated by cyclic voltammetry (CV), galvanostatic charge discharge (GCD), and electrochemical impedance spectroscopy (EIS) tests.

2. Experimental

In the first step, the MWNTs were grown directly on the Ni-foam of 2.0×2.0 cm2 by CVD technique at 650 ℃ from the source gases of acetylene and argon, then the mass m_1 of the MWNT-Ni foam substrate was weighed. Secondly, the MWNT-Ni foam substrates were put into a reaction kettle, with the addition of liquid mixture of $KMnO_4$ and sodium dodecyl sulfate (SDS). The MnO_2 nanostructures were synthesized by heating the reaction kettle to 130 ℃ for 10 h, followed by the drying process in the oven at 50 ℃. This composite substrate was weighted as m_2, and the difference between m_2 and m_1 was the mass of MnO_2. Going through a foresaid experimental procedures, the Ni-foam was still ductile. Finally, the composite material was pressed under the pressure of 30 MPa to form solid MnO_2/MWNT/Ni-foam electrode. The structural properties of the electrodes were characterized by SEM (JEOL 6700 F) and TEM (JEM-2100 F). The electrochemical properties were studied based on the three-electrode system employing the electrochemical workstation (Zennium, German Zahner Co). In the three-electrode system with 6 M KOH electrolyte, the Hg/HgO electrode was used as the reference electrode, Pt electrode as the counter electrode, and the test sample as the working electrode. CV performances were tested in a potential range of 0 V up to 0.65 V under different scan rates. EIS measurements were scanned over a frequency range of 10^{-2}–10^{-6} Hz at equilibrium conditions with a superimposed sinusoidal voltage signal of 5 mV. The cycling stability was tested in LANHE CT2001A (5 V 50 mA) battery station.

3. Results and discussion

As shown in figure 1a, the curly MWNTs grown densely out off the Ni-foam base. The high-magnification SEM image shows that the diameters of MWNTs are around 100 nm and the length is approximately tens of microns (figure 1b). After the MnO_2 synthesis over the MWNTs, the diameters of the composite structure increase to about 200 nm (figure 1c). TEM analysis shows nanoneedle and nanoflake structures of MnO_2 (figure 1d).

Figure 2a shows the cyclic voltammetric (CV) curves of the MnO_2/MWNT/Ni-foam electrode and MWNT/Ni-foam electrode with the stable potential window form 0 V to 0.65 V at the same scan rate of 50 mV/s. Comparing with the MnO_2/MWNT/Ni-foam, the capacitance of the MWNT/Ni-foam is very small, showing the dramatic capacitance enhancement from MnO_2. The CV curves of MnO_2/MWNT/Ni-foam at different scan rates are shown in figure 2b. The CV curves tend to behave rectangle shapes at high scan rates, representing good capacitive performances of the hybrid electrode. There are significant peaks at slow scan rate of 10 mV/s, marked by arrows. This pair of tiny redox peaks could be attributed to Mn^{+3}/Mn^{+4} conversion reaction. With increasing the scan rate, the electric double layer effect grows and surpasses the faradic pseudocapacitance effect.

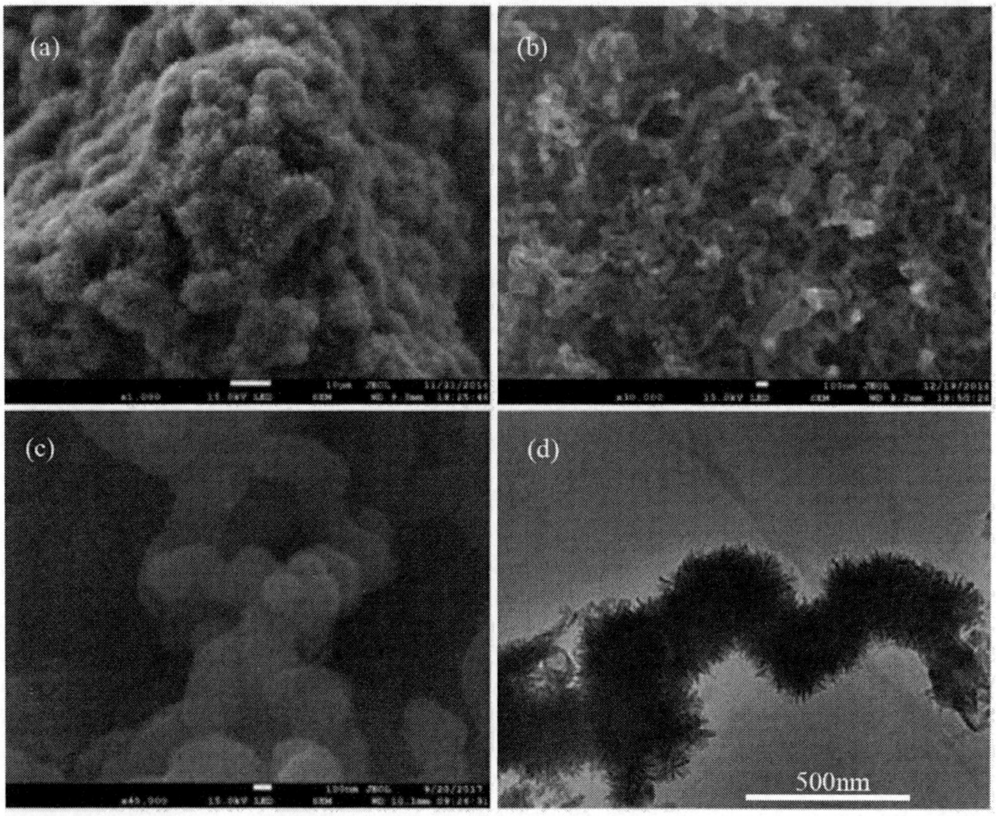

Figure 1. SEM images of MWNT/Ni-foam (a, b) and MnO_2/MWNT/Ni-foam (c). (d) TEM of MnO_2/MWNT nanostructures.

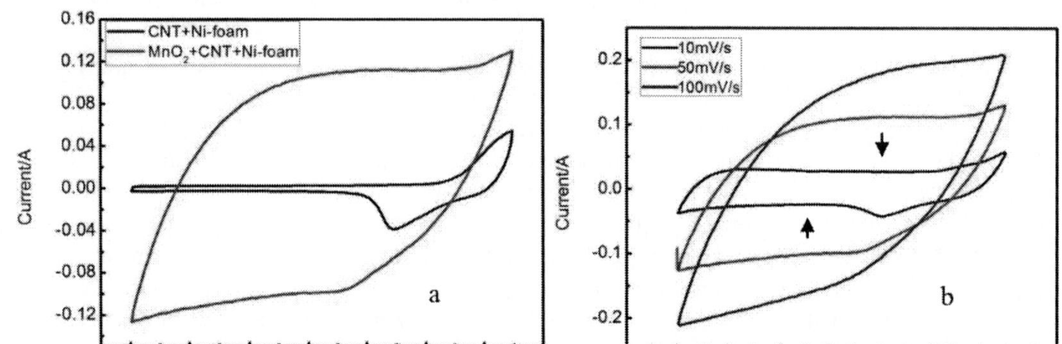

Figure 2. (a) CV plots of two types of electrodes at 50 mV/s. (b) CV plots of MnO_2/MWNT/Ni-foam electrode at different scan rates.

Typical electrochemical impedance spectroscopy (EIS) Nyquist plots for two types of electrodes are shown in figure 3a, which consists of an arc at the high frequency region and a nearly straight line at the low frequency region. The impedances are obtained from the equivalent circuit. At high frequencies, the intercept at real part (Z') is a combinational resistance from the ionic resistance of electrolyte, intrinsic resistance of substrate, and contact resistance at the active material/current collector interface (R_s). The arc at high frequency range corresponds to the charge-transfer resistance (R_{ct}) due to the charge transfer reaction on the grain surface [11]. The higher the R_{ct} is, the lower the specific capacitance of the EC is [12]. The Warburg line at low frequency region corresponds to the frequency dependent ion diffusion resistance in the electrolyte to the electrode interface [13]. The MnO_2-MWNT-Ni foam electrode presented a smaller R_s of 0.291 Ω, comparing with 0.371 Ω for the MnO_2–Ni foam electrode. Meanwhile, R_{ct} of the MnO_2/MWNT/Ni-foam electrode was smaller, which could be attributed to the high conductive property of carbonaceous materials and large surface area for MnO_2 synthesis, benefiting the fast ion transportation [14]. The smaller ion diffusion resistance for the MnO_2/MWNT/Ni-foam electrode at low frequency region demonstrated better capacitive behavior over the MnO_2/Ni-foam electrode.

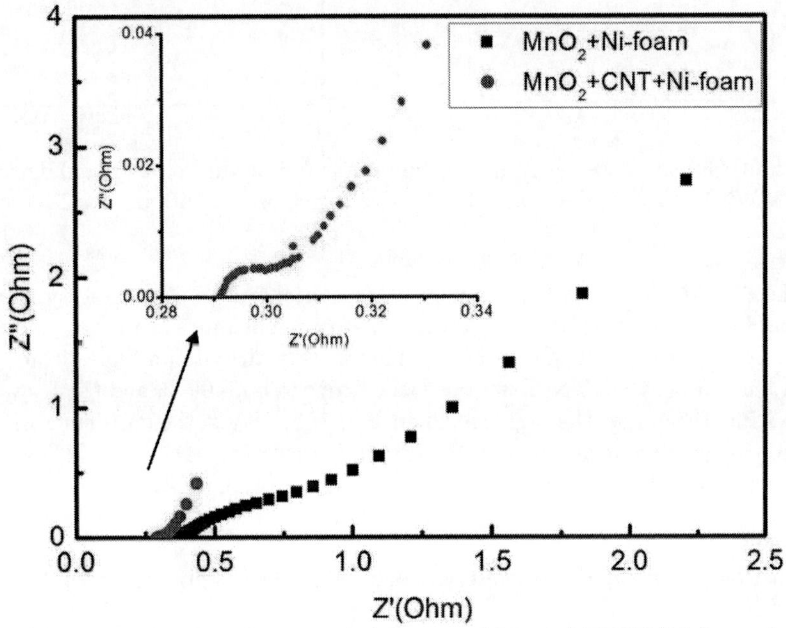

Figure 3. EIS Nyquist plots for MnO_2/Ni-foam and MnO_2/MWNT/Ni-foam electrodes.

In order to evaluate further the electrochemical performances of the MnO_2/MWNT/Ni-foam electrode, the galvanostatic charging/discharging behavior was tested. Based on the galvanostatic charge/discharge curve, we can calculate the specific capacitances (C_s in F/g) of the composite material by follow Equation [15]:

$$C_S = \frac{I}{m\left(\frac{\Delta V}{\Delta t}\right)} \quad (1)$$

where I is the applied current density (A/g), m is the active mass (g), and $\left(\frac{\Delta V}{\Delta t}\right)$ is the slope of the discharge curve. Figure 4a is the galvanostatic charging/discharging curves at different current densities of 2 A/g, 5 A/g, 10 A /g, 15 A/g, and 30 A/g, corresponding to C_s of 355.1 F/g, 263.0 F/g, 254.1 F/g, 214.5 F/g, and 161.0 F/g, respectively. The irregular triangle shapes of the curves under low current densities could be related to the synergistic effect of the faradaic pseudocapacitance and double-layer capacitance [16]. With the increase of the current density, the double-layer capacitance takes the leading role, resulting in the regular triangle shape, which means the good capacitive property of the composite electrode [17]. The cycling stability of the supercapacitor is demonstrated in figure 4b. It is clearly that the MnO_2/MWNT/Ni-foam electrode hold a stable capacity of about 170 F/g after 2000 cycles at the current of 40 mA.

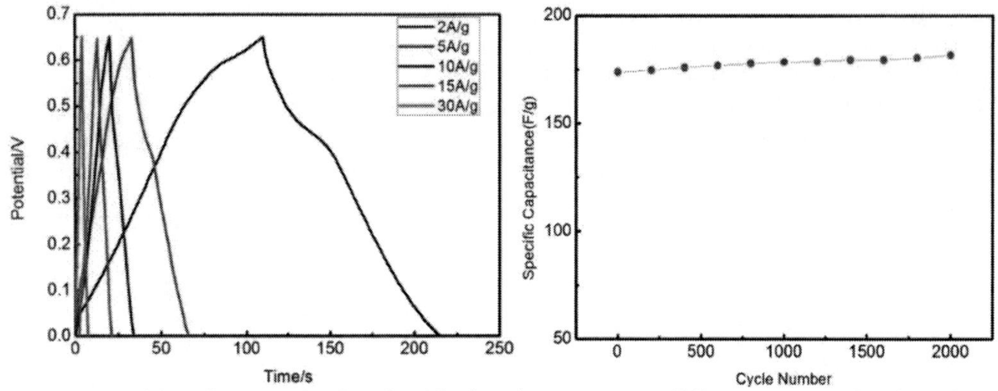

Figure 4. (a) Galvanostatic charging/discharging curves at different current densities for MnO_2/MWNT/Ni-foam electrode. (b) The cycling stability at 40 mA for 2000 cycles.

4. Conclusions

In summary, the MnO_2/MWNT/Ni-foam composite material was synthesized by the hydrothermal method after the direct growth of MWNTs on the Ni-foam substrate. The high conductive MWNTs not only reduce the EIS resistances of pure MnO_2 material, but also offer a high surface area for MnO_2 synthesis. R_s of the MnO_2/MWNT/Ni-foam electrode dropped by 8.09 % and the R_{ct} was also waned, comparing with MnO_2/Ni-foam. The high capacitance of 355.1 F/g at the current density of 2 A/g was reached, with good capacitive response at high current density. The MnO_2/MWNT/Ni-foam electrode exhibited good stability performance after 2000 cycles at the current of 40 mA.

Acknowledgement

This work is partially supported by National Natural Science Foundation of China (Grant No. 61620106006).

References

[1] Simon P and Gogotsi Y 2008 Materials for electrochemical capacitors *Nature materials* **7** 845-53

[2] Ramachandran R, Chen S and Kumar G. G 2015 An Overview of Electrochemical Energy Storage Devices of Various Electrodes and Morphological Studies of Supercapacitors *International Journal of Electrochemical Science* **10** 10355-88

[3] Wang Y G, Song Y F and Xia Y Y 2016 Electrochemical capacitors: mechanism, materials, systems, characterization and applications *Chemical Society Reviews* **45** 5925-50

[4] Zhi M J, Xiang C C and Li J T 2013 Nanostructured carbon–metal oxide composite electrodes for supercapacitors:a review *Nanoscale* **5** 72-88

[5] Zhang L L and Zhao X S 2009 Carbon-based materials as supercapacitor electrodes *Chemical Society Reviews* **38** 2520-31

[6] Bose S, Kuila T and Mishra A K 2017 Carbon-based nanostructured materials and their composites as supercapacitor electrodes *Journal of Materials Chemistry* **22** 767-84

[7] Xiong P, Zhu J W and Wang X 2015 Recent advances on multi-component hybrid nanostructures for electrochemical capacitors *Journal of Power Sources* **294** 31-50

[8] Kong S Y, Cheng K and Ouyang T 2017 Freestanding MnO_2 nanoflakes on carbon nanotube covered nickel foam as a 3D binder-free supercapacitor electrode with high performance *Journal of Electroanalytical Chemistry* **786** 35-42

[9] Zhao D D, Yang Z and Zhang L Y 2011 Electrodeposited Manganese Oxide on Nickel Foam-Supported Carbon Nanotubes for Electrode of Supercapacitors *Electrochemical and Solid-State Letters* **14** A93-A96

[10] Wang K, Gao S and Du Z L 2016 MnO_2-carbon composite for high-areal-density

supercapacitor with high rate performance *Journal of Power Sources* **305** 30-36

[11] Wu M S, Huang C Y and Lin K H 2009 Electrophoretic deposition of nickel oxide electrode for high-rate electrochemical capacitors *Journal of Power Sources* **186** 557-64

[12] Luo J Y and Xia Y Y 2009 Electrochemical profile of an asymmetric supercapacitor using carbon-coated $LiTi_2(PO_4)_3$ and active carbon electrodes *Journal of Power Sources* **186** 224-7

[13] Deng L J, Hao Z P and Wang J F 2013 Preparation and capacitance of graphene/multiwall carbon nanotubes/MnO_2 hybrid material for high-performance asymmetrical electrochemical capacitor *Electrochimica Acta* **89** 191-8

[14] Ramezani M, Fathi M and Mahboubi F 2015 Facile synthesis of ternary MnO_2 /graphene nanosheets/carbon nanotubes composites with high rate capability for supercapacitor applications *Electrochimica Acta* **174** 345-55

[15] Meher S K, Justin P and Rao G R 2010 Pine-cone morphology and pesudocapacitive behavior of nanoporous nickel oxide *Electrochimica Acta* **55** 8388-96

[16] Zhu G Y, He Z and Chen J 2014 Highly Conductive Three-Dimensional MnO_2-CarbonNanotube-Graphene-Ni Hybrid Foams as Binder-Free Supercapacitor Electrodes *Nanoscale* **6** 1079-85

[17] Vinny R T, Chaitra K and Venkatesh K 2016 An excellent cycle performance of asymmetric supercapacitor based on bristles like α–MnO_2 nanoparticles grown on multiwalled carbon nanotubes *Journal of Power Sources* **309** 212-20

IOP Conf. Series: Materials Science and Engineering **292** (2017) 012019 doi:10.1088/1757-899X/292/1/012019

Synthesis of Fe_3O_4 core/alumina shell nanospheres for partial hydrogenation of benzene

T T He[1], S L Mu[1], Q T Fu[1, 3] and C G Liu[2]

[1] School of Chemistry and Resources Environment, LinYi University, Linyi 276000, Shandong, China
[2] School of chemical Engineering, China University of Petroleum (East), Qingdao 266580, Shandong, China

[3] E-mail: fqtmail@aliyun.com

Abstract. We report a novel synthesis of Fe_3O_4 core/alumina shell nanosphere composite for partial hydrogenation of benzene. Fe_3O_4 core/alumina shell (MFeCA) structured nanospheres were obtained by reducing a hematite core/alumina precursor shell (HFeCAP) nanosphere precursor under H_2/N_2 gas flow. The magnetic alumina nanospheres (MFeCAs) possess not only uniform size (180~350nm) but also adjusted saturation magnetization value. A novel Ru-based magnetic catalyst was synthesized for the partial hydrogenation of benzene in magnetically stabilized bed (MSB). The performance of magnetic catalyst in MSB demonstrates that it's an effective and highly selective method for partial hydrogenation of benzene. The chain regime of the MSB reactor's operation conditions is responsible for the high selectivity of cyclohexene.

1. Introduction

The selective hydrogenation of benzene to cyclohexene has been extensively investigated [1–3], which is considered to be ideal reaction process due to their simple, environmental process and great industrial interest. Such liquid-phase selective hydrogenation of benzene to cyclohexene with different catalyst has been demonstrated to be effective method. For its applications, the high cyclohexene yield was achieved. A magnetically stabilized bed (MSB) offer many advantages, including low pressure drop, high mass-transfer efficiency, and the absence of catalyst attrition and catalyst separation step [4-6]. Therefore, the applications of MSB have been investigated widely in the chemical industry. MSB has been demonstrated to be effective hydrogenation method, such as selective acetylene hydrogenation [7], purification of caprolactam [8] and CO methanation [9], etc. The fabrication of microsphere from alumina and magnetic particles has been reported due to their unique magnetic responsivity and excellent physical strength [10-12]. The results showed that magnetic alumina microsphere could be used in a magnetically stabilized bed (MSB) as a catalyst or catalyst support [10, 11].

In this paper, the combination of MSB and magnetic core-shell Ru-based catalyst were employed, and one new method for selective hydrogenation of benzene in an MSB reactor was developed. Firstly, we fabricated uniform nanocomposite sphere with a hematite core/alumina precursor shell (HFeCAP) structure. The alumina shell was formed via a combination of the aluminum isopropoxide sol-gel process and stearic acid self-assembly. After reducing of the nanosphere hematite core in gas flow mixture of H2 and N2, this novel magnetic carrier named as MFeCA (magnetic core/alumina shell) nanosphere was employed to synthesis Ru/Fe_3O_4-Al_2O_3 catalysts, which were investigated in the partial hydrogenation of benzene in MSB.

2. Materials and methods

2.1. Materials

1-propanol, aluminum isopropoxide(Al[OCH(CH3)2]3),ruthenium chloride, zinc sulfate and ethanol were all of analytical grade and purchased from the Shanghai Chmeical Corp.

Hematite particles were hydrothermally synthesized by a surfactant-directing approach previously reported16. In a typical synthesis of hematite core/porous alumina shell composite particles, hematite particles (diameter 300 nm) were dispersed by ultrasonication into 100 mL of 1-propanol containing stearic acid. Aluminum isopropoxide was added under vigorous stirring, and the reaction was stirred at 70°C for 6 h. The resulting stearic acid-incorporated core-shell particles were separated by filtration, and dried at room temperature. This intermediate product was named HFeCAP (hematite core/alumina precursor shell spheres). The thermal treatment of HFeCAP in a mixed H2 (4 vol %) /N2 atmosphere at 673 K for 3 h, yielded the final magnetic core/alumina particles (MFeCA). The preparation of magnetic Ru/Fe3O4-Al2O3 catalyst was by impregnation method with RuCl3 solution, followed by drying at 373 K for 8 h and calcination in air at 573 K for 2 h, and them loading was about 0.3 wt %. Before using catalyst, the reduction of catalyst was performed in H2 fluid (30 ml min-1) at 473 K for 10 h. Subsequently, 2 h of preliminary treatment in the hydrogen 4/5000 hydrogen atmosphere at 423 K with a ZnSO4 aqueous solution at the flow velocity of 1ml min-1 was investigated. The process was employed at 3.0 MPa and 423 K. Samples were analyzed by GC-FID.

2.2. Analysis and characterization

X-ray powder diffraction (XRD) was investigated by a PANalytical X' Pert Pro MPD diffractometer. The magnetic properties of the samples were measured at room temperature by a Lake Shore vibrating sample magnetometer (VSM). Particle morphology and size were examined by a Hitachi S-4800 scanning electron microscope (SEM) and a JEM-2100UHR transmission electron microscope (TEM). Low temperature N2 sorption experiments were investagited by a Micromeritics ASAP-2020 Quantachrome Autosorb-1 system.

3. Results and discussion

3.1. Characterization

TEM images of samples are shown in figure 1. It can be seen that the sizes of samples s are quite uniform. The average sizes of samples vary in a range of ca. 180-250nm as seen from the TEM images. TEM images show that magnetic microspheres are successfully embedded in the MFeCAs. A TEM image of the MFeCA nanospheres (figure 1(a)) showed that most cores shrunk while their shell remained unbroken. The EDX spectrum of the product core–shell nanospheres (figure 1(b)) revealed the containing of Fe, Al, and O, further indicating the existence of the alumina shell on the Fe3O4 nanosphere surface. Elemental analysis indicated an Al/Fe mass fraction of 0.34, which was consistent with that of the synthesis mixture. Nitrogen sorption isotherm of the MFeCAs samples in figure 2 displays typical type II adsorption isotherms with the superposition of type H1 and H3 hysteresis loops [13]. The products have high Brumauer–Emmett–Teller (BET) surface areas (140 m^2g^{-1}) and large pore volumes (0.22 cm^3g^{-1}). The narrow and sharp pore size distribution curves (inset of figure 2) suggest that the mesopores possess very uniform sizes.

IOP Conf. Series: Materials Science and Engineering **292** (2017) 012019 doi:10.1088/1757-899X/292/1/012019

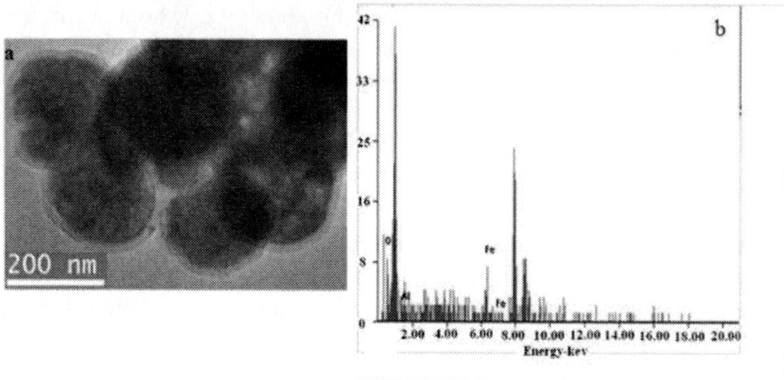

Figure 1. (a) TEM image of MFeCA nanospheres; (b) EDX of MFeCA nanospheres.

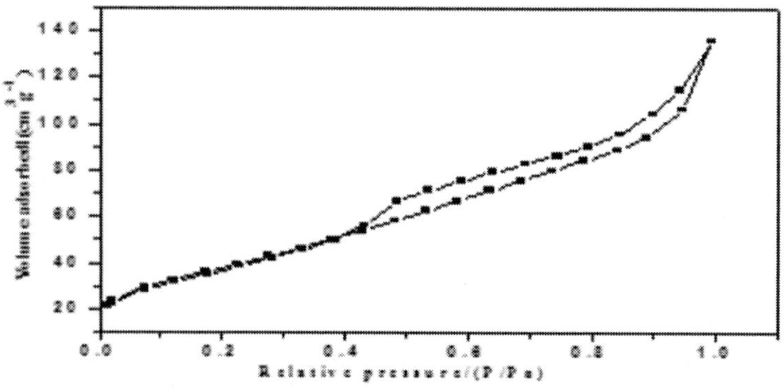

Figure 2. N_2 adsorption isotherms of MFeCA nanospheres.

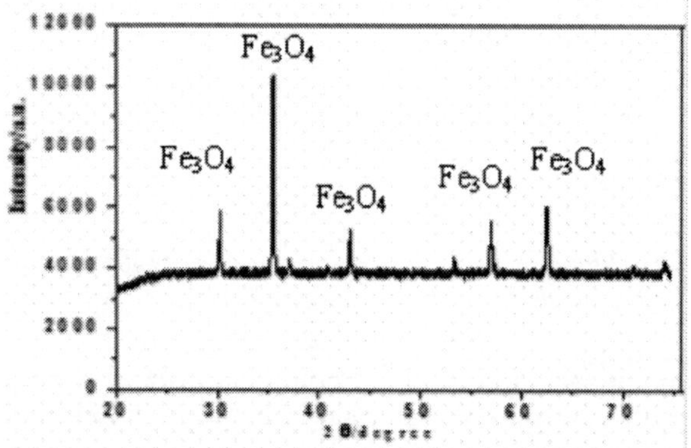

Figure 3. XRD pattern of MFeCA nanospheres.

Figure 3 shows the wide-angle XRD patterns of magnetic nanospheres. The characteristic diffraction peaks of Fe3O4 are clearly observed, which is consistent with TEM observations and further confirms the existence of magnetic microspheres in MFeCAs. The composition of the product could be adjusted by altering the reduction temperature or time, as described previously [14]. The magnetization curve at room temperature (figure 4) shows an obvious magnetic hysteresis loop, which possess a rapid magnetic response to the magnetic field. The saturation magnetization value of the MFeCA nanospheres (~50.2 emu g-1) could also be adjusted by reduction temperature or time, according to the typical reduction processes of hematite. The intrinsic ferromagnetism of MFeCA provided one straightforward way to separate particles from miscible liquids.

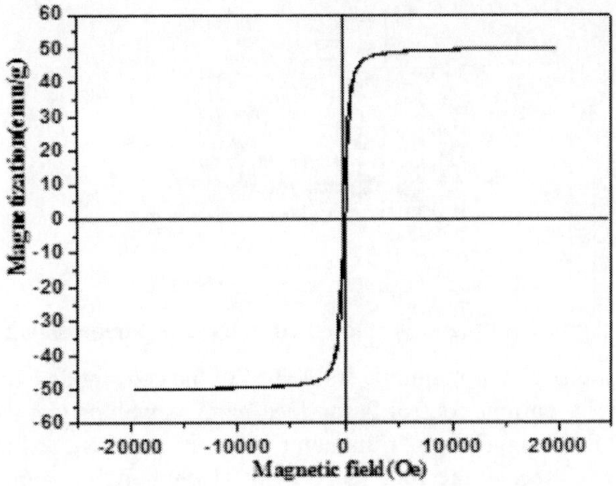

Figure 4. Magnetization hysteresis loops of MFeCA nanospheres.

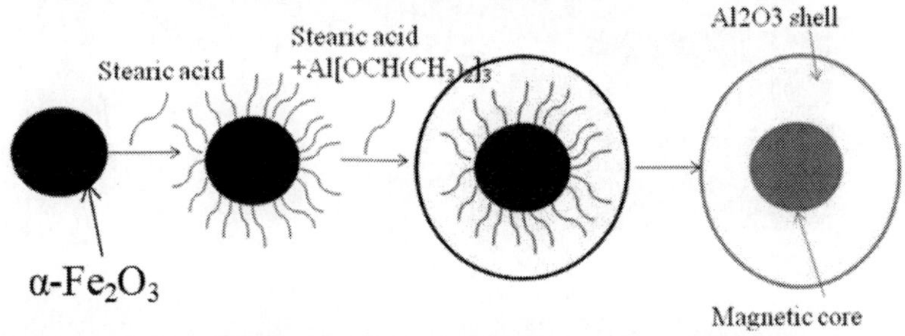

Scheme 1. The formation process of the MFeCA nanospheres.

Scheme 1 illustrates the formation process of the $Fe_2O_3@Al_2O_3$ core-shell nanospherestructure. We propose that HFeCA nanospheres were produced by the self-assembly of mixture α-Fe2O3-surfactant particles and surfactant/alumina compound. The nanosphere/stearic acid capped particles contained a negative charge[15,16], due to the carboxyl groups of the anionic surfactant. Hydrolyzed alumina and template directed self-assembly of alumina/template phases. A specific description is as follows: hydrolyzed aluminum isopropoxide monomers condense to form oligomeric alumina precursors. When the precursors reach a certain size, precursors and surfactants are assembled to produce an alumina shell on the surface of the nanosphere-surfactant composites. The residual alumina/template compound then react favorablly with alumina hydroxyl groups on the surface of the crystal that have been formed, preventing the generation of any additional crystals, thus forming uniform magnetic core-shell alumina nanospheres.

3.2. Performances of the catalyst

In order to assess the activity of the magnetic core–shell Ru/Fe_3O_4-Al_2O_3 catalyst, a number of experiments were employed in MSB under operation conditions ranged within a liquid hourly space velocity (LHSV) of 3h-1 to7h-1, a reaction temperature of 423 K, a system pressure of 3.0 MPa, and a magnetic field intensity of 5 Oe – 20 Oe.

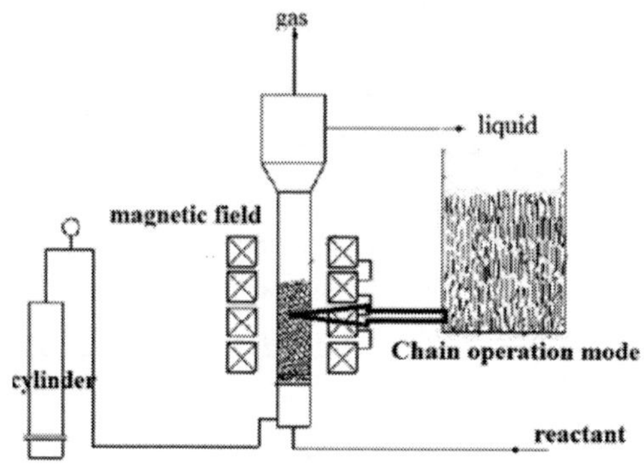

Scheme 2. Diagram of magnetic stable bed in the cold-mode experiment.

Scheme 2 shows the flow chart of of magnetic stable bed in the cold-mode experiment. Its includes the reactor part, magnetic field, current control Parts, feed parts as well as test parts. The reactor is a glass tube with a diameter of 20 mm.The inner diameter is 10mm and Length for 500 mm. The axial uniform magnetic field is generated by the four shafts of the Helmholtz copper wire. The number of turns per coil is 400. The current strength in the coil is controlled by the transformer.

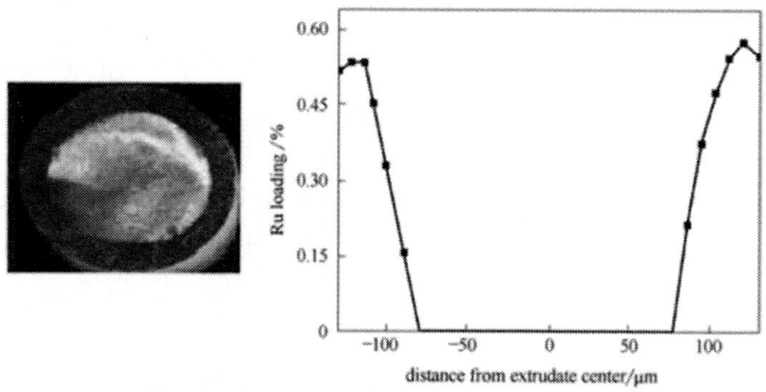

Figure 5. Distribution of Ru/Al_2O_3 by SEM.

According to the figure 5, The dark area in the edge, which is the region of the active constituent Ru. The distribution of the active component is mainly distributed in the outer surface of the catalyst. The EDS element analysis of the scanning electron microscope was analyzed in figure 5, which is further confirmed the surface distribution of the catalyst.

Table 1. Effect of the LHSV on selective hydrogenation of benzene.

(LHSV/h^{-1})	conversion of benzene, %	selectivity of cyclohexene, %	yield of cyclohexene, %
3.0	17.4	35.8	6.2
4.0	13.0	56.0	7.3
5.0	8.8	61.2	5.4
6.0	5.5	63.1	3.5
7.0	3.2	64.3	2.1

[a]Reaction conditions: P = 3.0MPa, H_2O/benzene = 1 (v/v), H_2/benzene = 2 (mol/mol), T =423K, catalyst amount=5ml. magnetic field intensity 10Oe， zinc sulfate solution concentration0.3wt%

During the assessment of catalyst performance, the MSB reactor was under the chain regime station, which has advantage of high mass transfer property by adjusting the magnetic field intensity as necessary. The chain regime state of the MSB reactorwere investigated by the cold model experiments in a liquid–solid MSB reactor, which is favor to provide good heat and mass transfer property according to the reaction parameters [17, 18]. So during the catalytic activity assessment, the MSB reactor was controlled under the chain regime according to adjusting the magnetic field intensity. Furthermore, the assessment was employed with different magnetic field intensity. As listed in table 1, under magnetic field of 10 Oe, the MSB maintains the chain regime state, which is favor to desorption of cyclohexene so as to achieve high cyclohexene selectivity.

3.2.1. Effect of LHSV. The relationship between the catalyst activity and the liquid liquid hourly space velocity space velocity reveals the fact of mass transfer in selective hydrogenation of benzene processes [8]. The yield of cyclohexene decreases with an increase of LHSV, as shown in table 1. An increase of LHSV makes the concentration of cyclohexene in the oil phases, which reduces mass-transfer rate of cyclohexene from the oil phase to the catalyst surface and consequently enhances the cyclohexene selectivity. Under much high space velocity, the residence time is so short that it can't provide much reaction time for benzene hydrogenation, so that the benzene conversion would be decreased. The appropriate liquid hourly space velocity (LHSV) is very important for benzene hydrogenation.

3.2.2. Effect of ruthenium loading. Several catalysts Ru/Fe_3O_4-Al_2O_3 with ruthenium loadings of 0.25–0.7(mass)% were investigated in partial hydrogenation of benzene. As shown in table 2, benzene conversion follows linear relationship with ruthenium loading from 0.25 to 0.55%, while the selectivity of cyclohexene has one breaking point for the highest cyclohexene selectivity at a ruthenium loading of 0.25%. Indeed, the Ru atom possesses highly dispersed state with many active sites leading to high hydrogenation activity along with the increase of Ru loadings. Therefore, the accelerated rate of hydrogenation of benzene and cyclohexene result in increasing benzene conversion and decreasing cyclohexene selectivity. Thus, the proper content of Ru is significant to ensure the best yield of cyclohexene.

Table 2. Effect of the Ru loadings on selective hydrogenation of benzene.

(Ru loadings, %)	conversion of benzene, %	selectivity of cyclohexene, %	yield of cyclohexene, %
0.25	13.0	56.0	7.3
0.4	14.1	68.5	9.7
0.55	21.5	48.6	10.4
0.70	40.3	45.8	18.5

[a]Reaction conditions: P = 3.0MPa, H2O/benzene = 1 (v/v), H2/benzene = 2 (mol/mol), T =423K, catalyst amount=5ml. magnetic field intensity 10Oe， zinc sulfate solution concentration0.3wt%

3.2.3. Effect of magnetic field. The performance of magnetic catalyst was assessed under different magnetic field. As listed in table 3, benzene conversion was only 9.1% when the magnetic field intensity was 5 Oe, which is so low to maintain the catalyst particles in the apparatus. Under high magnetic field intensities, the MSB reactor are stably in the above-mentioned chain regime, which are superior to mass transfer of reactants and products so that it achieve high yield of cyclohexene. No good result in catalyst performance is found by further increasing the magnetic field intensity; moreover, very large magnetic field intensities caused high operating costs.

Table 3. Effect of the magnetic field on selective hydrogenation of benzene.

magnetic field intensity (Oe)	conversion of benzene, %	selectivity of cyclohexene, %	yield of cyclohexene, %
5	9.1	35.5	3.2
10	13.0	56.0	7.3
20	10.5	24.2	2.5

[a]Reaction conditions: P = 3.0MPa,T =423K, LHSV=4h−1, H2/benzene = 2 (mol/mol),H2O/benzene = 1 (v/v), catalyst amount=5ml, zinc sulfate solution concentration 0.3wt%

4. Conclusions

Uniform magnetic core/alumina shell nanospheres were prepared by a combination of the sol-gel process and surfactants self-assembly approach. The magnetic reduced iron core endowed the nanoparticles with magnetic properties. MFeCA possessed a uniform alumina layer of thickness ~30 nm, a large pore volume (~ 0.22 cm^3 g^{-1}) and a large surface area (~140 m2g-1). Through studying selective hydrogenation of benzene in MSB, it was noted that high selectivity of cyclohexene can be obtained due to the chain regime. Magnenetic Ru catalyst with MSB provides a convenient and useful means method for selective hydrogenation by magnetic field. The materials prepared in this work may also have applications prospect in hydrogenation.

References

[1] Xia Y, Fan C, Zhou and Zhou Z L 2013 *J. Mol. Catal. A: Chem* **370** 44
[2] Sun H J, Dong Y Y, Li S H, et al 2013 *J. Mol. Catal. A: Chem* **368** 119
[3] Sun H J, Jiang H B, and Li S H 2013 *Chem. Eng. J* **218** 415
[4] Zong B N, Meng X K and Mu X H 2007 *Chin. Particuology* **5** 116
[5] Li W, Zong B N and Li X F 2006 *Chin. J. Chem. Eng* **14** 734
[6] Zong B N, Meng X K and Mu X H 2013 *Chin. J. Catal* **34** 61
[7] Dong M H, Pan Z Y and Peng Y 2008 *AIChE. Journal* **54** 1358
[8] Meng X K, Mu X H and Zong B N 2003 *Catal. Today* **79** 21
[9] Pan Z Y, Dong M H and Meng X K 2007 *Chem. Eng. Sci* **62** 2712
[10] Feng J T, Lin Y J and Li F 2007 *Appl. Catal .A Gen* **329** 112
[11] Feng J T, Lin Y J and Li F 2009 *Ind. Eng. Chem. Res* **48** 692
[12] Li Y, Liu Y C and Tang J 2007 *J. Chromatogr. A* **1172** 57
[13] Gregg S J, Sing K S W 1982 Adsorption, surface area and porosity, 2rd ed.; New York: 10 Academic Press
[14] Guo X H, Deng Y H, and Gu D 2009 *J. Mater. Chem* **19** 6706
[15] Ohmori M, Matijevi E 1993 *J. Colloid. Interface. Sci* **160** 288
[16] Frederic V, Shervin K, and Mark E D 1996 *Chem. Mater* **8** 1451
[17] Lu X S 1999 *Eng Chem Metall* **20** 129
[18] Zong B N 2007 *Catal Surveys Asia* **11** 87

IOP Conf. Series: Materials Science and Engineering **292** (2017) 012020 doi:10.1088/1757-899X/292/1/012020

Effects of B site doping on electronic structures of $InNbO_4$ based on hybrid density functional calculations

M F Lu[1], C P Zhou[1], Q Q Li[1], C L Zhang[1, 2] and H F Shi[1, 2, 3]

[1] School of science, Jiangnan University, Wuxi, P. R. China, 214122
[2] Department of Physics, Nanjing University, Nanjing, P. R. China, 210093

[3] E-mail: hfshi@jiangnan.edu.cn

Abstract. In order to improve the photocatalytic activity under visible light irradiation, we adopted first principle calculations based on density functional theory (DFT) to calculate the electronic structures of B site transition metal element doped $InNbO_4$. The results indicated that the complete hybridization of Nb 4d states and some Ti 3d states contributed to the new conduction band of Ti doped $InNbO_4$, barely changing the position of band edge. For Cr doping, some localized Cr 3d states were introduced into the band gap. Nonetheless, the potential of localized levels was too positive to cause visible-light reaction. When it came to Cu doping, the band gap was almost same with that of $InNbO_4$ as well as some localized Cu 3d states appeared above the top of VB. The introduction of localized energy levels benefited electrons to migrate from valence band (VB) to conduction band (CB) by absorbing lower energy photons, realizing visible-light response.

1. Introduction

Since the energy shortage is becoming a big headache of threatening the human's sustainable development, semiconductor photocatalysis has brought about widespread attention on account of its potential applications for new energy resources like CO_2 reduction, hydrogen generation and pollutants decomposition [1~10]. Recently, indium niobate ($InNbO_4$) has been studied for its numerous advantages of wide absorption range, stable thermodynamic property and simple catalytic system [11-14]. Unfortunately, $InNbO_4$ was inert in visible-light region owing to its relatively wide band gap (~ 3.83 eV) [15]. Therefore, it is an emergency of founding an appropriate way to modify the band structures of $InNbO_4$ to realize visible-light response. Lee *et al* [16] utilized solid phase reaction method to produce $InTaO_4$ samples as well as $InNbO_4$ synthesized with NiO and Co_3O_4 cocatalysts, aiming at characterizing the activity for photo-reduction of carbon dioxide. Zhou *et al* [17] have studied the electronic structures of N doped $InNbO_4$ with first principle and pointed out that the occurrence of localized N 2p states was beneficial to realizing photocatalytic reactions under visible light.

Actually, B site transition metal elements can be introduced into the crystal lattices of semiconductor photocatalytic materials, bringing localized energy levels into energy gap. Through these levels, electrons in VB can transit to CB by absorbing lower energy photons, consequently expanding spectral response range. Herein, the electronic structures and photocatalytic activities of the first series of transition metals (Ti, Cr, Cu) doped $InNbO_4$ have been investigated by DFT first principles.

2. Calculations and models

In this work, $InNbO_4$ with supercell structure (2×1×2) was adopted as a calculation model (figure 1) and calculated by Cambridge Sequential Total Energy Package (CASTEP). As for (Ti, Cr, Cu) doped $InNbO_4$, one transition metal atom took the place of Nb atom. The GGA-PBE was served to describe the exchange correlation potential. The electronic wave function was expanded with a kinetic energy cut-off (370 eV) and the k point mesh of the Brillouin zone integration was set to 5×4×5. Geometry optimization was implemented to pick the doping structure with the minimum total energy as the model. The maximum stress was smaller than 0.1 GPa, the maximum displacement was 2×10^{-4} nm, and the convergence criteria parameter of the maximum force on atoms was not greater than 0.05 eV Å, respectively.

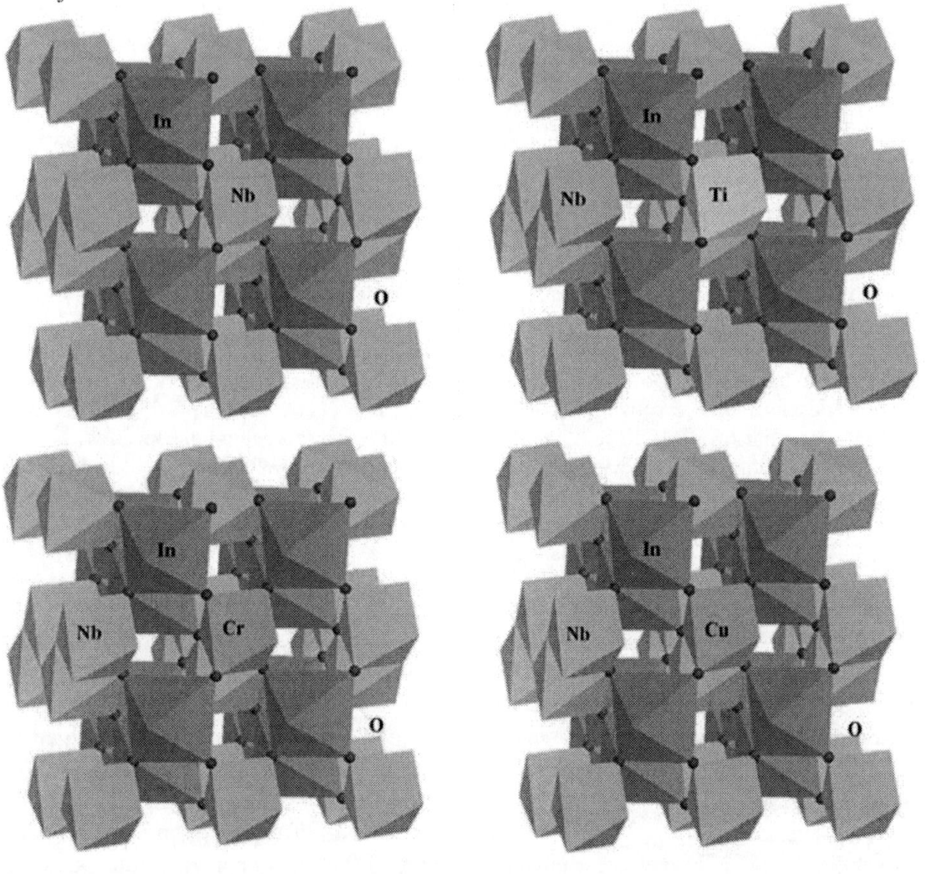

Figure 1. The crystal structure diagrams of $InNbO_4$ and (Ti, Cr, Cu) doped $InNbO_4$.

3. Results and discussions

3.1. Geometry optimization

Table 1 lists the optimized lattice parameters of $InNbO_4$ and (Ti, Cr, Cu) doped $InNbO_4$. The calculated crystal parameters of $InNbO_4$ were a=10.450 Å, b=11.846 Å, c=4.946 Å, in accordance with the experimental values of a=10.288 Å, b=11.542 Å, c=4.834Å. This indicated that the present computing method were reasonable and credible. Additionally, the atom sites and lattice parameters have not significantly changed after introducing the impurity atoms.

Table 1. The optimized lattice parameters of $InNbO_4$ and (Ti, Cr, Cu) doped $InNbO_4$.

	a (Å)	b (Å)	c (Å)	α (Å)	β (Å)	γ (Å)
$InNbO_4$	10.450	11.846	4.946	89.999	91.332	90.000
Ti-$InNbO_4$	10.457	11.861	4.945	90.003	91.215	90.026
Cr-$InNbO_4$	10.477	11.892	4.967	90.014	91.303	90.059
Cu-$InNbO_4$	10.473	11.867	4.955	90.007	91.302	90.041

3.2. Electronic structures

The band structures of B site (Ti, Cr, Cu) doped $InNbO_4$ (figure 2) were discussed combined with their band gaps. As shown in figure 2(a), the calculated band gap of $InNbO_4$ (3.38 eV) was much lower than the experimental value (3.83 eV). The calculated band gap energy is generally underestimated simply because of the inherent error of the DFT algorithm [18]. However, it is still considered to be persuasive to investigate the change trends of energy band structures of materials. It could also be inferred that $InNbO_4$ was a sort of indirect band gap semiconductors because the positions of its conduction band minimum (CBM) and valence band maximum (VBM) were not located in the same G point.

It could be observed from figure 2(b-d) that the localized energy levels were introduced after B site doping. After Ti doping, the impurity levels composed of Ti 3d states and Nb 4d states resulted in the formation of a new CB. It also turned out that there was little change of the bang gap of Ti doped $InNbO_4$ (3.28 eV), which was just 0.10 eV smaller than that of $InNbO_4$ (3.38 eV). As for Cr doping, the band gap (3.33 eV) kept almost unchanged compared with that of $InNbO_4$ as well as the localized levels lay near the bottom of CB. In contrast with $InNbO_4$, the band gap of Cu doped $InNbO_4$ (3.07 eV) minished 0.31 eV along with the appearance of localized Cu 3d states above the top of VB. The introduction of localized energy levels changed the electrons transition method in the band gap so that $InNbO_4$ could absorb photons of longer wavelength to realize visible-light reaction, improving the utilization efficiency of solar energy and photocatalytic activity.

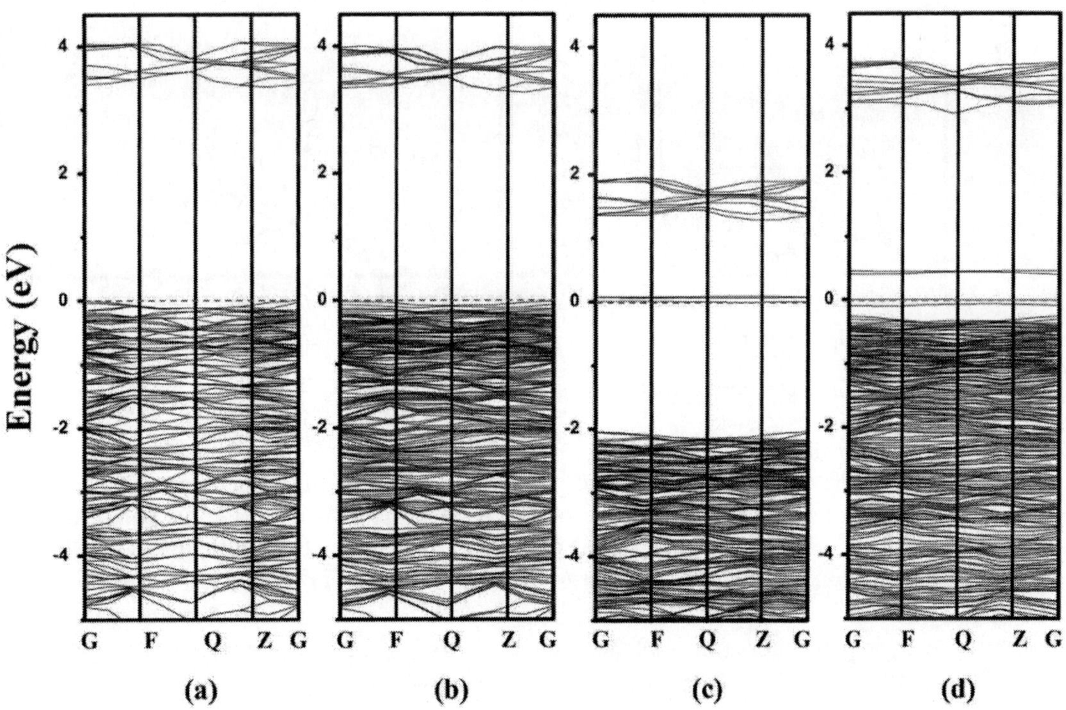

Figure 2. The band structures of $InNbO_4$ and (Ti, Cr, Cu) doped $InNbO_4$.

IOP Conf. Series: Materials Science and Engineering **292** (2017) 012020 doi:10.1088/1757-899X/292/1/012020

Figure 3 exhibits the density of states (DOS) of $InNbO_4$ and (Ti, Cr, Cu) doped $InNbO_4$. Figure 3(a) implied that CB of $InNbO_4$ mainly consisted of Nb 4d states while VB was composed of O 2p states. There was partial overlap of Nb 4d and O 2p states in both CB and VB because of the existence of Nb-O covalent bonds. In the Ti doping case (figure 3(b)), the complete overlap of Ti 3d and Nb 4d peaks barely changed the position of CB and thereby had little influence on the photocatalytic activity of $InNbO_4$, which confirmed the former analysis in figure 2(b). It was perceived from figure 3(c, d) that the Cr 3d peak was formed in the band gap of Cr doping while the impurity peak constituted of Cu 3d states turned up beside the VB edge of Cu doping. This fact was also consistent with the calculated results of the above band structures.

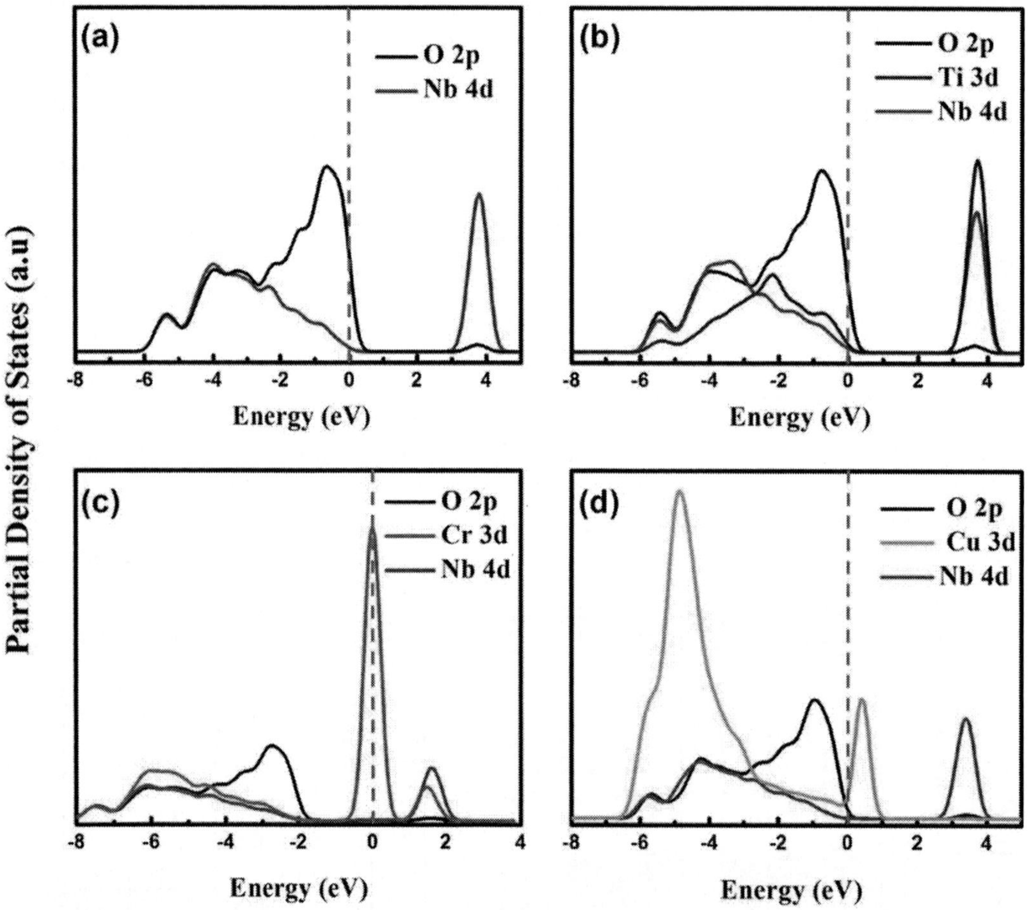

Figure 3. The density of states (DOS) of $InNbO_4$ and (Ti, Cr, Cu) doped $InNbO_4$.

3.3. Schematic band structures

Redox potential and the position of band edge played a leading role in the photocatalytic activity of semiconductor photocatalysts. The band edge positions of semiconductor materials before and after doping were determined by the calculated band gap, and the schematic band structures were shown in figure 4. The calculation formula of the energy band position is as follows: $E_{VB} = \chi + 0.5E_g - E_e$, where E_{VB}, E_g, E_e are valence band maximum, band gap energy, and the free electron energy on the surface of the hydrogen atoms (4.5 eV), respectively; $\chi = \sqrt[n]{\chi_1 \chi_2 \chi_3 \cdots \chi_n}$ represents the geometric mean of the Mulliken electronegativity of the constituent atoms [19, 20].

As we know, the redox ability of water is measured by the normal hydrogen electrode (NHE). From the schematic diagram, we learnt that the top of VB was fairly higher than the electric potential of O_2/H_2O. B site transition metal element doping barely changed the band gap values while the positions of their CB and VB were a bit different from each other. Thereinto, for Ti doping, the mixture of Ti 3d and Nb 4d states slightly narrowed the band gap. As for Cr doping, the positions of CB and VB obviously shifted up and the localized energy levels were introduced to the band gap. Unfortunately, the potential of localized Cr 3d states was below the standard electrode potential of hydrogen. Although the band gap of Cu doped $InNbO_4$ was almost same with that of $InNbO_4$, the electrons in VB could transfer to CB through the localized energy levels to realize visible-light response.

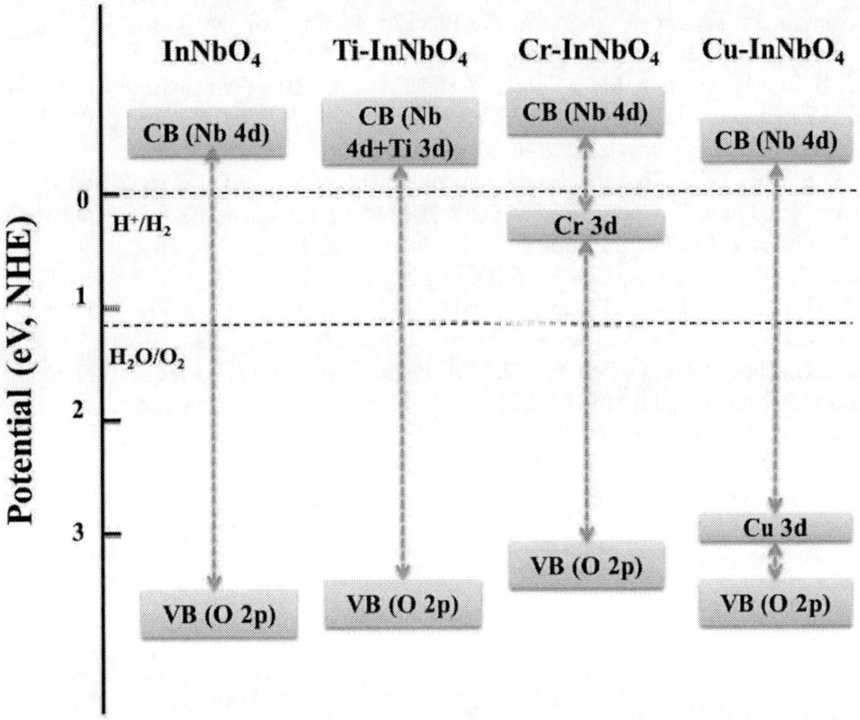

Figure 4. The schematic band structures of $InNbO_4$ and (Ti, Cr, Cu) doped $InNbO_4$.

4. Conclusions

In this paper, we mainly investigated the electronic structures of B site (Ti, Cr, Cu) doped $InNbO_4$ on the basis of DFT.

For Ti doped $InNbO_4$, the Ti 3d states mixed with Nb 4d states changed the position of conduction band edge so little that there was almost no effect on the photocatalytic activity of $InNbO_4$.

For Cr doping, the localized energy levels consisting of Cr 3d states helped to expand the spectral response range. Whereas, the potential of localized Cr 3d states could not achieve the standard electrode potential of hydrogen, which indicated that Cr doped $InNbO_4$ was not an appropriate candidate for photocatalytic reaction.

As for Cu doping, the appearance of localized energy levels changed the electron transition way, making for the red shift of absorption spectrum and the realization of visible-light reaction. Above all, the Cu doped $InNbO_4$ was the most promising photocatalyst for $InNb_{1-x}M_xO_4$ (M= Ti, Cr, Cu).

Acknowledgements

The work was supported by the financial support from National Natural Science Foundation of China (No. 21203077), the Qing Lan Project of Jiangsu Province, and the Fundamental Research Funds for the Central Universities (No. JUSRP51716A).

IOP Conf. Series: Materials Science and Engineering **292** (2017) 012020 doi:10.1088/1757-899X/292/1/012020

References

[1] Peng L P, Xu L and Xia Z C 2014 *Comp. Mater. Sci.* **83** p 309-317
[2] Peng L P, Yue H and Xu L 2016 *Sci. Adv. Mater.* **8** p 2223–2230
[3] Meng F, Li J, Cushing S K, Bright J, Zhi M, Rowley J, Hong Z, Manivannan A, Bristow A D and Wu N Q 2013 *ACS Catal.* **3** p 746-751
[4] Meng A Y, Zhang J, Xu D F, Cheng B and Yu J G 2016 Appl. *Catal. B-Environ.* **198** p 286-294
[5] Garg A B, Errandonea D, Popescu C, Marine-Garcia D, Pellicer-Porres J, Rodriguez-Hernandez P, Munoz A, Botella P, Cuenca V P and Sans J A 2017 *Inorg. Chem.* **56** p 5420-5430
[6] Perronella F, Truppi A, Ingrosso C, Placido T, Striccoli M, Curri M L, Agostiano A and Comparelli R 2017 *Catal. Today* **281** p 85-100
[7] Li Z S, Luo W J, Zhang M L, Feng J Y and Zou Z G 2013 *Energy Environ. Sci.* **6** p 347-370
[8] Shi H F, Zhou C P, Zhang C L and Ye E J 2016 *J. Mod. Phys. B* **30** p 1650060
[9] Osterloh F E 2013 *Chem. Soc. Rev.* **42** p 2294-2320
[10] Shi H F, Chen G Q, Zhang C L and Zou Z G 2014 *ACS Catal.* **4** p 3637-3643
[11] Zhang X F, Quan X, Chen S and Yu H T 2011 *Appl. Catal. B* **105** p 237-242
[12] Fu L, Wu Y N, Lin F and Zhang B 2013 *Mater. Lett.* **109** p 225-228
[13] Zhou M, Zu X T, Sun K, Liu W and Xiang X 2017 *Chem. Eng. J.* **313** p 99-108
[14] Feng H Y, Hou D F and Huang Y H 2014 *J. Alloy. Compd.* **592** p 301-305
[15] Lv J, Kako T, Zou Z G and Ye J H 2010 *J. Mater. Res.* **25** p 159-166
[16] Lee D S, Chen H J and Chen Y E 2012 *J. Phys. Chem. Solids* **73** p 661-669
[17] Zhou C P, Lan B Y and Shi H F 2015 *J. At. Mol. Phys.* **32** p 1055-1058
[18] Cohen A J, Sánchez P M and Yang W 2008 *Science* **321** p 792-794
[19] Nethercot A H 1974 *Phys. Rev. Lett.* **33** p 1088-1091
[20] Butler M A, Ginley D S 1978 *J. Electrochem. Soc.* **125** p 228-232

IOP Conf. Series: Materials Science and Engineering **292** (2017) 012021 doi:10.1088/1757-899X/292/1/012021

Zeolite-encapsulated Co(Ⅱ), Mn(Ⅱ), Cu(Ⅱ) and Cr(Ⅲ) salen complexes as catalysts for efficient selective oxidation of benzyl alcohol

F H Li[1], H Bi[1], D X Huang[1], M zhang[2] and Y B Song[1, 3]

[1] College of Science, Shantou University, Shantou, Guangdong, 515063, China
[2] Entry-exit inspection and quarantine bureau, Shantou, Guangdong, 515041, China

[3] E-mail: ybsong@stu.edu.cn

Abstract. Co(Ⅱ), Mn(Ⅱ), Cu(Ⅱ) and Cr(III) salen type complexes were synthesized in situ in Y zeolite by the reaction of ion-exchanged metal ions with the flexible ligand molecules that had diffused into the cavities. Data of characterization indicates the formation of metal salen complexes in the pores without affecting the zeolite framework structure, the absence of any extraneous species and the geometry of encapsulated complexes. The catalytic activity results show that CosalcyenY exhibited higher catalytic activity in the water phase selective oxidation of benzyl alcohol, which could be attributed to their geometry and the steric environment of the metal actives sites.

1. Introduction

Transition metal complex-catalyzed oxidation of organic compounds has been extensively investigated in the past decades [1-5]. Metal salen complexes have been reported as efficient catalysts for the epoxidation of various olefins, for their well-established ability to bind oxygen [6]. However, metal alkoxides are sensitive to hydrolysis and the presence of adventitious water can result in the formation of dimmer species. The dimmer species formation caused deactivation of the catalysts and as a consequence the re-used catalysts leads to lower substrate conversion [7-9]. Thus, heterogenization of metal salen-type catalysts, which can avoid the formation of dimerization species, has attracted many researcher attentions [10-12].

Heterogenization of homogenous catalysts has been an interesting area of research in the academic and industrial, which display potential for practical application. As this method can provide an ideal way for taking the advantages of homogenous catalysts and simultaneously avoid its disadvantages like handling and reusability. Encapsulation in porous solids such as zeolite is an attractive technique for heterogenization, since no leaching is observed when the complex is confined exclusively in the zeolite pores [13-14]. In addition, zeolite-encapsulated complexes promise better control of selectivity of the reaction and higher stability, due to NaY cavity encapsulated metal complexes, avoids direct contact with the metal complexes, reduced dimerization of complexes in the cavity. [15-19]. This nature for the homogeneous catalyst multiphase provides a good thinking and benzyl alcohol oxidation reaction received extensive attention of the academia in recent years. In this work, synthesis, characterization and catalytic activities of zeolite-encapsulated Co(Ⅱ), Mn(Ⅱ), Cu(Ⅱ) and Cr(III) salen complexes catalysts for selective oxidation of benzyl alcohol to benzaldehyde were systematically investigated.

2. Experimental

2.1. Materials

$Mn(CH_3COO)_2$, $Co(CH_3COO)_2$, $CrCl_3 \cdot 6H_2O$, $Cu(CH_3COO)_2$, ethanol, hydrogen peroxide and benzyl alcohol were from Sinopharm Chemical Reagent Co, Ltd, while ethylenediamine, O-phenylenediamine,1,2-Diaminocyclohexane and salicylic aldehyde were from Aladdin Reagent Co., Ltd. All these chemicals and reagentsused in present work were of analytical grade and used as received without further purication.The sodium form of Y zeolite (NaY) obtained from Nankai University Catalyst Co, Ltdwas dried at 723 K for 4 h prior to being used as the host material for encapsulating complexes.

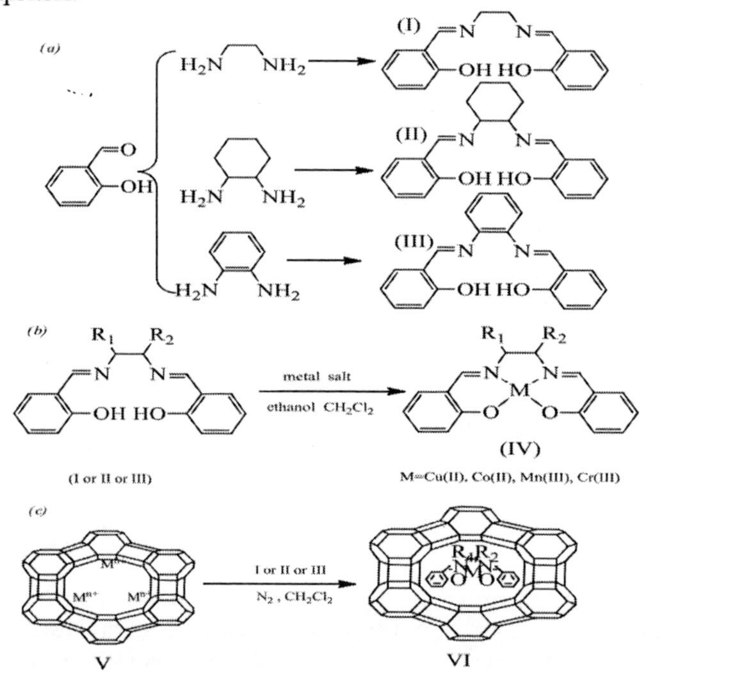

(I) Ligand1,(II) Ligand2, (III) Ligand3
(IV) Metal salen-type complexes:Ligands=I or II or III, Metal=Cu, Co, Mn, Cr
(V) Ion-exchanged YM zeolite, M=Cu(II), Co(II), Mn(II), Cr(III)
(VI) Zeolite encapsulated metal complexes catalyst

Figure 1. Preparation of zeolite encapsulated metal salen complexes.

2.2. Preparation of zeolite encapsulated metal salen catalysts

2.2.1. Synthesis of Salen ligands. The ligands were synthesized as following steps [20]. The 10 mmol salicylaldehyde and 20 mmol ethylenediamine (or O-phenylenediamine, 1,2-Diaminocyclohexane) were dissolved in 20 ml ethanol solution, respectively. Under nitrogen atmosphere, the ethanol solution of corresponding diamine was dropwise to the ethanol solution of salicylaldehyde. The resultant mixture was refluxed for 2.5 h at 333 K and then was concentrated to filtration in ethanol solution and dried in vacuum. The resultant corresponding ligand crystals were collected, marked as Ligand 1, Ligand 2 and Ligand 3, respectively, described in figure 1 (a).

2.2.2. Synthesis of metal Salen complexes. The salen ligand compound in dichloromethane solution was heated to reflux under nitrogen atmosphere at 333 K for insertion of metal center, as indicated in figure 1 (b). A solution of Co(II), Mn(II), Cu(II) metal salt were dissolved in 20 ml ethanol solution, respectively. The mixture was added to the ligand dichloromethane solution to reflux for 2 h. The mixture was concentrated to wash with ethanol to remove the side product and dryness in vacuum, than the corresponding metal-salen complexes were obtained.

2.2.3. Encapsulation of metal salen complexes in Y zeolite. NaY zeolite was ion-exchanged with 0.07 M metal salts solution at 368 K for 24 h and then washed and dehydrated at 673 K for 4 h to obtain CoY, CuY, MnY and CrY zeolites. The anhydrous metal exchanged zeolite was mixed an excess of salen type ligand dissolved in 60 ml CH_2Cl_2 solution, as indicated in figure 1 (c). These solids were heated in sealed glass ampoules for a period of 12 h at 313 K. The resultant mass was soxhlet-extraction with CH_2Cl_2 to remove surface complexes and the free ligand. The extracting in each case was continued for 24 h after the extracting solvent became color less. The purified sample was ion-exchanged with 0.1 M NaCl for 24 h to remove uncoordinated metal ions from the lattice. The zeolite encapsulation complex thus obtained was filtered, washed free of chloride ions and then dried in vacuum.

2.3. Catalytic reaction

The benzyl alcohol oxidation reactions were carried out in a stirred reactor. The vessel was charged with benzyl alcohol (10 mmol), 0.55 mol/L Na_2CO_3 water aqueous solution (30 ml) and catalyst (100 mg). The stirrer was set at 1000 r.p.m and 353K. A certain amount of hydrogen peroxide was dropwise slowly to the reactor, samples from the reactor were taken periodically. The products and the unconverted reactants were analyzed by gas chromatography with a flame ionization detector, using a HP-5 column and decane as the internal standard.

2.4. Characterization

The synthesized materials were characterized by using XRD, FT-IR, DRS UV-vis. XRD experiments were performed using a Bruker-D8 type powder X-ray diffractometer (XRD) operating with Cu Kα radiation (λ=0.154056 nm, 40 kV). The spectra patterns were recorded on a Lambda 950 UV-vis spectrophotometer and a MAGNA-IR 750 Fourier transform infrared spectrometer.

3. Results and discussion

3.1. Characterization of the catalyst

3.1.1. FI-IR. The FT-IR spectra figure 2 (a) shows the C=N characteristic peaks of Schiff base in the range of 1600-1640 cm^{-1}, 1612 cm^{-1} associated with stretching vibrations of C=N appears. Peaks at 1462 and 1443 cm^{-1} could be ascribed to extension vibration of C=C in benzene, while peak at 1381 cm^{-1} was ascribed to flexural vibration of C-H. All these ligands and metal-salophen complexes exhibit the strong peaks at 1612, 1583, 1525, 1462, 1443 and 1382 cm^{-1} corresponding with reported literature [21]. Peaks at 1323, 1186 and 1151 cm^{-1} were ascribed to the C-O bond, and peak at 750.2 cm^{-1} corresponded with benzene plane deformation absorption. Peaks at 3000 and 2900 cm^{-1} were assigned to extension vibration of C-H and C-H stretching vibration of CH = N respectively [22]. The weak peaks exhibited at 470-500 cm^{-1} and 541 cm^{-1} were assigned to the extension vibration of metal-N (metal include Mn, Cu and Co) and metal-O respectively. The FT-IR spectra of Co(Salen) and CoSalenY in 1600-1200 cm^{-1} range are presented in figure 2 (b). Compared with the spectra of Co(salen), the CoSalenY catalyst exhibit the weak peaks at corresponding peaks. The Co(Salen) were successfully encapsulated in zeolite Y in accordance with the reports [23].

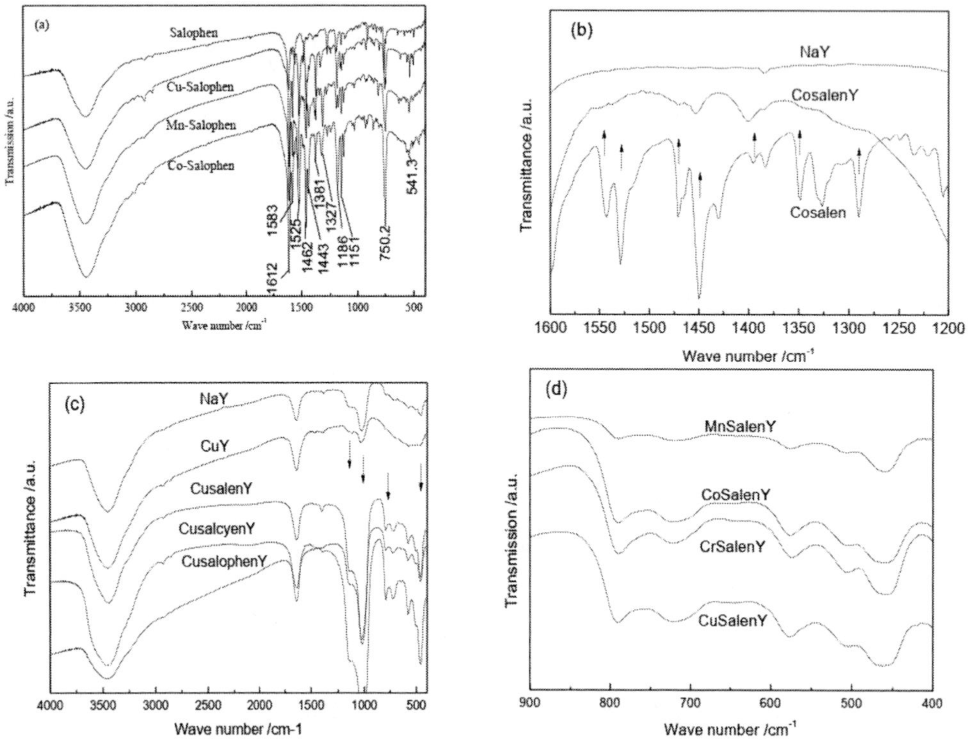

Figure 2. FT-IR spectra of: (a) ligand 3 complexes with different transition. (b) Co(Salen) and CoSalenY; (c) different salen ligand encapsulated in zeolite; (d) different metal complexes encapsulated in zeolite.

The IR spectra of different salen ligand and different metal complexes encapsulated in zeolite exhibit in figure 2 (c) and (d). The sample zeolite encapsulated Cu complexes catalysts exhibit weak peaks in the range of 1650-1520 cm^{-1} corresponding to C=N extension vibrations. The broad bands at the region 1650 and 3400 cm^{-1} are due to lattice water molecules and surface hydroxylic groups. No shift of zeolite lattice bands is observed in the spectra of encapsulated complexes, which implies that the zeolite framework has remained unchanged upon the encapsulated of complexes. The encapsulated metal complexes in the cage could be identified from the bands observed in the region 1200-400cm^{-1} where several peaks at has been observed in 1200, 1000 cm^{-1}, 800-790, 735-715, 580-570, 520, 470-450cm^{-1} as shown in figure 2 (c) and (d). These two results clearly indicate that NaY zeolite structure still remains intact, also further illustrate NaY is suitable for packaging this metal complexes.

3.1.2. DRS UV-vis. The DRS UV-vis electronic spectra are presented in figure 3 (a) and (b). The encapsulated complexes exhibit adsorption bands at 200～310 nm attributable to ligand transition. This indicates the existence of salophen complex in the supercage. The DRS UV-vis spectra of the free and encapsulated Co complexes show intense ligand-metal charge transfer adsorption in the range of 310～420 nm. All the samples exhibit a broad d-d electron transition adsorption at 550～600 nm.

The UV-vis spectra of Co-salophen exhibits a strong peak at ~250 nm, while MnSalophen exhibits a strong peak at ~265 nm and a broad peak at ~420 nm ascribed to the ligand to metal charge transfer transition bonds as shown in figure 3 (b), akin to related metal salen complexes described in the literature. The peaks or broad peak of Co(II), Mn(II), Cu(II) and Cr(III)- salen-type complexes in zeolite Y are observed in the region 250～530 nm, while zeolite has no observed. The FT-IR and DRS

UV-vis spectra results manifest the well synthesis of corresponding salen ligand, metal salen complexes and Metal(Salen)Y catalysts.

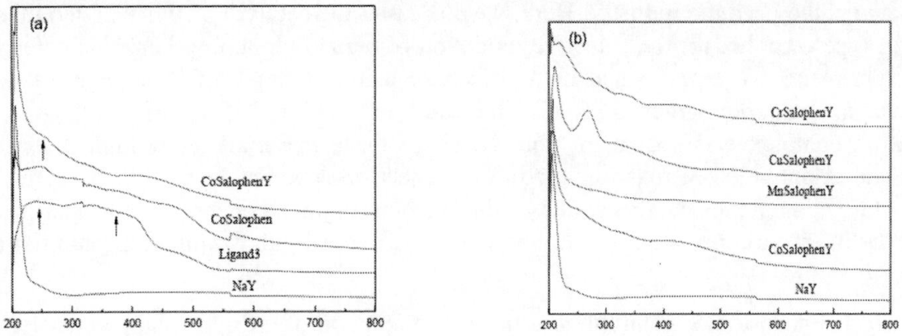

Figure 3. DRS UV-vis spectra of: (a) Ligand 3, CoSalophen and CosalophenY catalyst, (b) Co(II), Mn(II), Cu(II) and Cr(III) salophen complexes catalysts.

3.1.3. XRD. XRD patterns in figure 4 showed that diffraction peak strength corresponding to NaY and Co(Salen)Y in figure 4 (a), in which no change between NaY and Co(salen)Y was observed indicating that NaY zeolite structure did not change. Cu(Salen)Y, Mn(Salen)Y and Cr(Salen)Y in figure 4 (b), (c) and (d) have similar diffraction peaks and peak strength. While the change of difraction peak strength corresponding to Co(salen)Y ($I_{311}>I_{220}$) suggested that some trace of Na^+ ions were replace by Co(salen). XRD measurements showed that the encapsulation of Co(salen) in NaY had no obvious inflence on the structure of the zeolitic matrix.

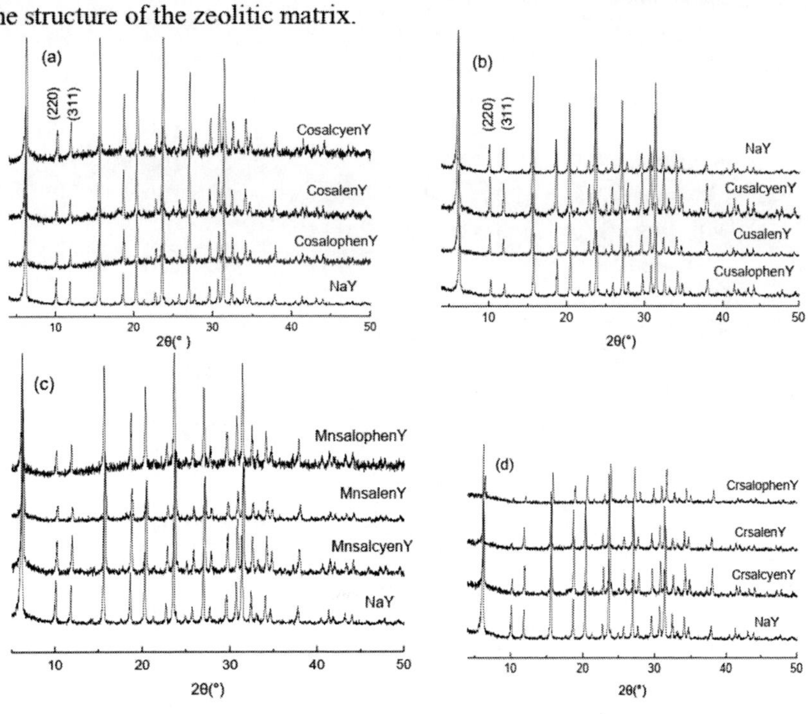

Figure 4. XRD patterns of zeolite encapsulated metal salen type complexes catalysts and NaY. (a) Co(salen) , (b) Cu(salen), (c) Mn(salen), (d) Cr(salen).

3.2. Catalytic activity on oxidation of benzyl alcohol

The oxidation of alcohol is an important reaction in organic chemistry for both laboratory and industrial manufacturing, in particular with the synthesis of aldehydes, which are valuable intermediates in the perfume industry. Here, we discussed the catalytic activity of zeolite encapsulated metal salen-type complex for the selective oxidation of benzyl alcohol to benzaldehyde. The catalytic activity experiments were performed in water phase using 30 ml Na_2CO_3 aqueous as solvents, and H_2O_2 acted as oxidants under 353 K. The catalytic activity of metal complexes and zeolite encapsulation complex was shown in table 1. All of these materials show high selectivity towards benzaldehyde at our selected reaction conditions. The CuSalcyenY shows highest conversion among all encapsulated complexes catalysts, due to the surrounding high electron cloud density of Cu metal atom and the interaction between metal atom and Schiff base ligand, which attribute to the activity of substrate.

Table 1. Encapsulated catalytic in water phase reaction of the conversion and selectivity.

Entry	Catalysts	Conversion (%)	Selectivity(%)	
			Benzaldehyde	Benzoic acid
1	NaY	0	0	0
2	Cusalophen	31.21	100	0
3	Cosalophen	18.76	100	0
4	Mnsalophen	17.15	100	0
5	Crsalcyen	10.22	100	0
6	CusalcyenY	38.14	100	0
7	CusalophenY	37.56	100	0
8	CusalenY	14.89	100	0
9	CosalophenY	15.02	100	0
10	CosalenY	8.84	100	0
11	CosalcyenY	6.62	100	0
12	MnsalophenY	3.21	100	0
13	MnsalenY	0.47	100	0
14	MnsalcyenY	0	0	0
15	CrsalenY	3.83	100	0
16	CrsalophenY	3.23	100	0
17	CrsalcyenY	0	0	0

[a]Reaction condition: 353 K, 10 mmol benzyl alcohol, 30 ml Na_2CO_3 aqueous, 100 mg catalysts, added 10 ml H_2O_2 slowly with stirred at 1000 r.p.m.

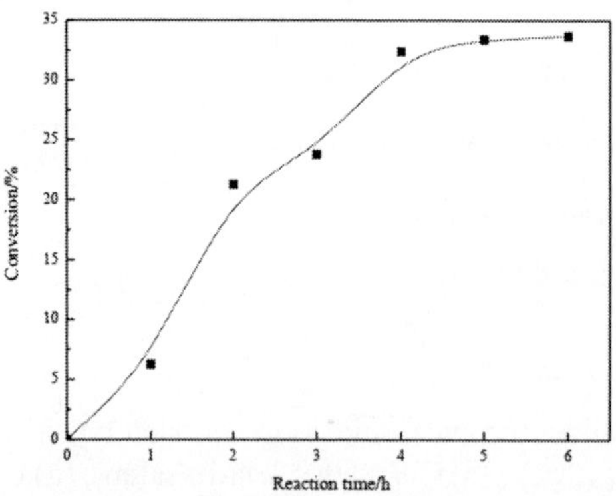

Figure 5. Influence of reaction time on substrate conversion.
[a]Reaction condition: 353 K, 10 mmol benzyl alcohol, 30 ml Na_2CO_3 aqueous, 100 mg catalysts, added 10 ml H_2O_2 slowly with stirred at 1000 r.p.m.

Influence of reaction time was investigated in figure 5. The conversion of benzyl alcohol increase as the reaction time increase in range 1～6 h. Hydrogen peroxide is the preferred oxygen donor in zeolite systems, since they are highly mobile in the pores due to their smaller size. Furthermore, they are cheaper and sufficiently environment-friendly to be used on a commercial scale. The catalytic nature of encapsulated complex catalyst was confirmed by running the reactions in the absence of any catalyst. Hydrogen peroxide alone is unable to oxidize the substrates to a significant extent, indicating that the reactions can perform only when a catalyst promotes them. However, the decomposition of hydrogen peroxide has efficient influence on oxidation of benzyl alcohol.

Table 2. The catalyst of lifetime.

Entry	CoSalenY	CoSalophenY	CoSalcyenY
First run	54.34	32.87	42.32
Second run	47.82	32.67	41.57
Third run	40.65	32.45	41.19
Forth run	22.85	32.22	34.84
Fifth run	19.95	32.14	13.92

[a]Reaction condition: 353 K, 10 mmol benzyl alcohol, 30 ml Na_2CO_3 aqueous, 100 mg catalysts, added 10 ml H_2O_2 slowly in 6 h with stirred at 1000 r.p.m.

In addition, a further study on the catalytic reusability of CoSalenY, CoSalcyenY and CoSalophenY was carried out. After the first catalytic run, the heterogenized catalyst was separated from the reaction solution, washed several times to remove any Physical adsorption of molecular, dried and reused in another four catalytic run in table 2. The CoSalenY and CoSalcyenY begin to loss catalytic activity after three catalytic run, while the CoSalophenY has no much change in the catalytic performance after five catalytic run. This indicated that the zeolite encapsulated metal complex catalysts with high catalytic activity and stability for the selective oxidation benzyl alcohol was developed.

4. Conclusions

Co(Ⅱ), Mn(Ⅱ), Cu(Ⅱ) and Cr(III) salen type complexes were synthesized in situ in Y zeolite, and the characterization and catalytic activity on the selective oxidation of benzyl alcohol were systematically surveyed. It has been found that the formation of metal salen complexes in the pores without affecting the zeolite framework structure, the absence of any extraneous species and the geometry of encapsulated complexes. The catalytic activity results show that CoSalcyenY exhibited higher catalytic activity in the water phase selective oxidation of benzyl alcohol. The catalytic behavior could be mainly attributed to the geometry of encapsulated complexes and steric environment of the metal actives sites. To summarize, zeolite encapsulated complexes have interesting catalytic potential particularly with respect to the activity for selective oxidation and stability, and offer further scope to design efficient and environment-friendly catalyst systems.

Acknowledgements

Financial supports from the National Natural Science Foundation of China (No. 21043006) and the Science and technology planning project of Guangdong Province (No. 2014A020216045, 2012CXZD0024) are gratefully acknowledged.

References

[1] Habibi D and Faraji A R 2013 *Comptes Rendus Chimie.* **16** 888-896
[2] Noshiranzadeh N, Emami M, Bikas R, et al. 2014 *Polyhedron* **72**: 56-65
[3] Narita M, Yoon I, Aoyagi M, Goto M, Shimizu T and Asakawa M 2007 *Eur J. Inorg. Chem.* **26** 4229-37

[4] Zhang X A, Zhou H Q, Su X, Chen X G, Yang C L, Qin J G and Inokuchi M 2007 *J. Alloy Compd.* **432** 247-252
[5] Chen Y, Liu Y, Zhang X, et al. 2015 *Inorg. Chem. Commun.* **53** 1-3
[6] Nandi M, Roy P, Uyama H and Bhaumik A 2011 *Dalton Trans.* **40** 12510-518
[7] Xu Y, Lin L, Kanai M, Matsunaga S and Shibasaki M 2011 *J. Am. Chem. Soc.* **133** 5791-93
[8] Palopoli C, Bruzzo N, Hureau C, Ladeira S, Murgida D and Signorella S 2011 *Inorg. Chem.* **50** 8973-83
[9] Zhang Z, Li X, Wang C, Zhang C, Liu P, Fang T, Xiong Y and Xu W 2012 *Dalton Trans.* **41** 1252-58
[10] Bhunia S, Saha D and Koner S 2011 *Langmuir* **27**: 15322-329
[11] Purse B W, Tran L H, Piera J, Akermark B and Backvall J E 2008 *Chemistry* **14** 7500-03
[12] Jain S, Reiser O 2008 *ChemSuChem.* **1** 534-541
[13] Kowalak, Stanislaw, Ralph C Weiss, and Kenneth J Balkus 1991 *Chem. Commun.* **1** 57-58
[14] Joseph T, Halligudi S, Satyanarayan C, Sawant D P and Gopinathan S 2001 *J. Mol. Catal. A: Chem.* **168** 87-97
[15] Johansson M, Purse B W, Terasaki O and Bäckvall J E 2008 *Adv. Synth. Catal.* **350** 1807-15
[16] Jin C, Fan W, Jia Y, Fan B, Ma J and R L 2006 *J. Mol. Catal. A: Chem.* **249** 23-30
[17] Xavier K 2004 *Appl. Catal. A: Gen.* **258** 251-259
[18] Dominguez I, Fornes V and Sabater M 2004 *J. Catal.* **228** 92-99
[19] Zhan B Z, White M A, Sham T K, Pincock J A, Doucet R J, Rao K R, Robertson K N and Cameron T S 2003 *J. Am. Chem. Soc.* **125** 2195-99
[20] Huang D X, Song Y B , Fang Y W and Sun C Y 2014 *Adv. Mater. Res.* **881** 292-296
[21] Johansson M, Purse B W, Terasaki O, et al. 2008 *Adv. Synth. Cataly.* **350** 1807-15
[22] Ou X, Jiang L, Chen P, et al. 2013 *Adv. Funct. Mater.* **23** 2422-35
[23] Zhang N and Zhou X F 2012 *J. Mol. Catal. A: Chem.* **365** 66-72

Optimization of preparation of NDV F Gene encapsulated in N-2-HACC-CMC nanoparticles

S S Li[1], Y Zhang[1], K Zhao[1, 3] and X H Wang[2, 3, 4]

[1] Key Laboratory of Microbiology, School of Life Science, Heilongjiang University, Harbin 150080, China
[2] Heilongjiang University Library, Harbin 150080, China
[3] School of Biological Science and Technology, University of Jinan, Jinan 250022, China

[4] E-mail: xiaohua020623@foxmail.com

Abstract. In this study, the biodegradable materials *N*-2-hydroxypropyl trimethyl ammonium chloride chitosan (N-2-HACC) and *N, O*-carboxymethyl chitosan (CMC) are used as delivery carrier for the pVAX Ⅰ -F(o)-C3d6. The optimal preparation condition is as follows: concentration of N-2-HACC is 1.0 mg/ml, concentration of CMC is 0.85 mg/ml, concentration of pVAX Ⅰ -F(o)-C3d6 is 100 μg ml. The results show that the prepared N-2-HACC-CMC/pFDNA NPs have regular round shape, smooth surface and good dispersion, the particle size is 310 nm, Zeta potential is 50 mV, the entrapment efficiency is 92 %, the loading capacity is 51 % (n=3).

1. Introduction

Newcastle disease is an important disease of poultry and can cause severe economic losses to the poultry industry around the globe [1]. However, the prevention of this disease requires an effective way, chitosan nanoparticles as an excellent delivery carrier appeared in front of people [2].

Nano vaccines have more long-term significance than the traditional inactivated vaccine and live vaccine, inactivated vaccines may cause disease transmission because of inadequate inactivation, and live vaccines are destroyed by enzymes during immunization. The preparation of chitosan nanoparticles has laid a theoretical and practical basis for the study of nano-delivery system [3], which is of great significance.

Chitosan is a promising carrier material for vaccine mucosal delivery, but its narrow soluble range limits the further application of chitosan. In this study, N-2-HACC-CMC / pFDNA NPs were prepared using N-2-HACC and CMC as carriers. N-2-HACC was a quaternary ammonium salt derivative of chitosan. This derivative, under the premise of maintaining the other biological properties of chitosan, greatly improves the solubility in neutral or alkaline environments [4]. And the modified chitosan nanoparticles have the following advantages: non-toxic, biodegradable [5], bacteriostatic [6], biocompatibility [7], and so on. In conclusion, chitosan based modified nanoparticles have strong advantages, so we have synthesized N-2-HACC-CMC/pFDNA NPs nanoparticles containing Newcastle disease F gene in this laboratory.

IOP Conf. Series: Materials Science and Engineering **292** (2017) 012022 doi:10.1088/1757-899X/292/1/012022

2. Materials and methods

2.1. Materials

The eukaryotic expression plasmid pVAX-optiF with C3d6 molecular adjuvant (pVAX Ⅰ -F(o)-C3d6), *N*-2-hydroxypropyl trimethyl ammonium chloride chitosan (N-2-HACC) and *N, O*-carboxymethyl chitosan (CMC) were provided by our Laboratory in School of Life Science of Heilongjiang University, China.

2.2. Single factor test

There are five single factors influencing the preparation of N-2-HACC-CMC/pFDNA NPs: the concentration of N-2-HACC (0.5, 1.0, 1.5 and 2.0 mg/ml); the concentration of pVAX Ⅰ -F(o)-C3d6 (50, 100, 150, 200 and 250 μg/ml); the concentration of CMC (0.75, 1.0, 1.25, 1.5 and 1.75 mg/ml); stirring speed (900, 1000, 1100, 1200, 1300 r/min) and stirring time (20, 30, 40 and 50 min).

2.3. Orthogonal test

Acording to the single factor results, three factors were selected for the orthogonal test (table 1), the morphology, particle size, encapsulation efficiency (EE) and loading capacity (LC) of N-2-HACC-CMC/pFDNA NPs were taken as evaluation indexes.

2.4. Preparation of NDV F gene encapsulated in N-2-HACC-CMC nanoparticles

N-2-HACC-CMC/pFDNA NPs were prepared by polyelectrolyte complex method, but the specific concentration of each component was determined according to the single factor test and orthogonal test results. The specific flow is as follows: Add 5 ml of a certain concentration of N-2-HACC solution to 50 ml erlenmeyer flask, slowly add 2 ml of a certain concentration of pVAX Ⅰ -F(o)-C3d6 solution with 1 ml syringe, 300 r/min for 3 min, then raise the stirring speed to the target speed for 1min, then 2 ml of a certain concentration of CMC solution was added dropwise slowly with 1 ml syringe, and kept in 5min drops finished, continue magnetic stirring, and maintain a certain mixing time. The solution will be 12000 r/min 4 ℃ for 20 min, discard the supernatant and wash the precipitate with 4 ℃ pre-cooled sterile deionized water, repeat 3 times, and finally resuspend the precipitate with 2 ml of pre-cooled sterile deionized water and lyophilize in vacuo to obtain N-2-HACC-CMC/pFDNA NPs.

3. Results

3.1. Single factor test

(1) Effects of N-2-HACC concentration on the preparation of N-2-HACC-CMC/pFDNA NPs. As shown in figure 1a, with the increasing of N-2-HACC concentration, the EE firstly increased and then decreased. When the concentration of N-2-HACC was 1.0 mg/ml, EE and LC were desirable.

(2) Effects of CMC concentration on the preparation of N-2-HACC-CMC/pFDNA NPs. As shown in figure 1b, with the increasing of CMC concentration, the EE firstly increased and then decreased. When the concentration of CMC was 1.0 mg/ml, the EE and LC were desirable.

(3) Effects of DNA concentration on the preparation of N-2-HACC-CMC/pFDNA NPs. As shown in figure 1c, with the increasing of pVAX Ⅰ -F(o)-C3d6 concentration, the EE firstly increased and then decreased. When the DNA concentration is 100 μg/ml, the EE and LC were desirable.

(4) Effects of stirring rate on the preparation of N-2-HACC-CMC/pFDNA NPs. As shown in figure 1d, with the increasing of stirring rate, the EE firstly increased and then decreased. When the speed was 1200 r/min, the EE and LC were desirable.

(5) Effects of stirring time on the preparation of N-2-HACC-CMC/pFDNA NPs. As shown in figure 1e, with the stirring time prolonging, the EE and LC increased to 40 min, and reached the highest.

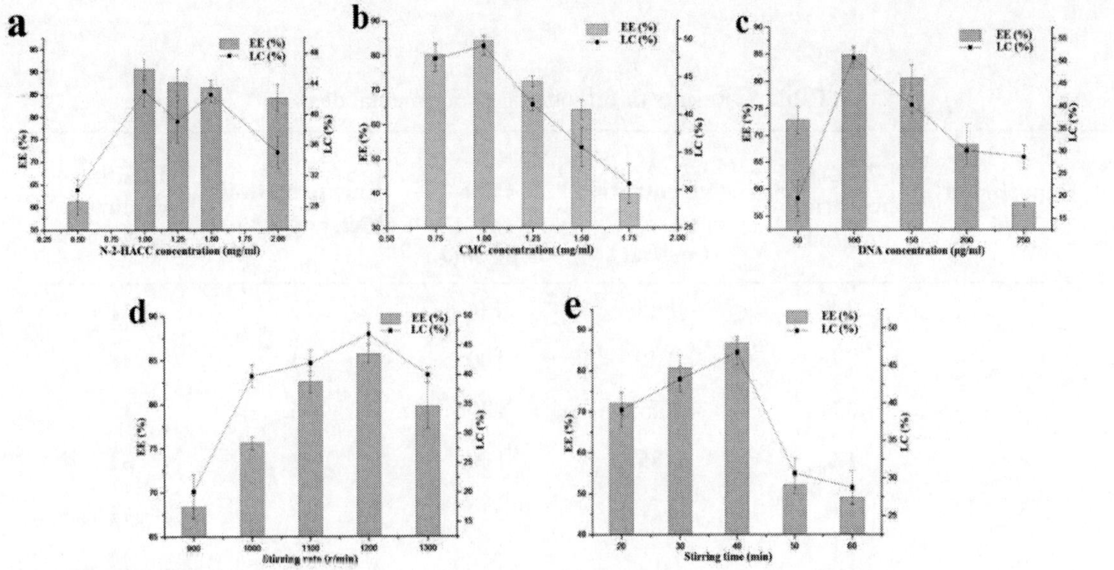

Figure 1. The effect of N-2-HACC concentration (a), CMC concentration (b), pVAX I -F(o)-C3d6 concentration (c), stirring rate (d) and stirring time (e) on N-2-HACC-CMC/pFDNA NPs preparation.

3.2. *Orthogonal experiment*

Three factors that influence the preparation of N-2-HACC-CMC/pFDNA NPs are N-2-HACC concentration, CMC concentration and DNA concentration. The optimal preparation condition of N-2-HACC-CMC/pFDNA NPs were as follows: N-2-HACC concentration was 1.0 mg/ml, CMC concentration was 0.85 mg/ml, DNA concentration was 100 μg/mL (table 1), the entrapment efficiency was 92 %, the loading capacity was 51 % (n=3).

3.3. *Preparation of NDV F Gene Encapsulated in N-2-HACC-CMC Nanoparticles*

The optimal preparation process is as follows: Add 5 ml 1.0 mg/ml N-2-HACC solution to 50 ml erlenmeyer flask, slowly add 2 ml 100 μg/mL pVAX I -F(o)-C3d6 solution with 1 ml syringe, 300 r/min for 3 min, then raise the stirring speed to 1200 r/min for 1min, then 2 ml 0.85 mg/ml CMC solution was added dropwise slowly with 1 ml syringe, and kept in 5min drops finished, continue magnetic stirring, and maintain for 40 min. The solution will be 12000 r/min 4 ℃ for 20 min, discard the supernatant and wash the precipitate with 4 ℃ pre-cooled sterile deionized water, repeat 3 times, and finally resuspend the precipitate with 2 ml of pre-cooled sterile deionized water and lyophilize in vacuo to obtain N-2-HACC-CMC/pFDNA NPs.

It can be seen from figure 2, the N-2-HACC-CMC/pFDNA NPs prepared by the optimal condition had regular round shape, smooth surface and good dispersion. The average particle size was 310 nm (figure 3a) and the Zeta potential was 50 mV (figure 3b).

IOP Conf. Series: Materials Science and Engineering **292** (2017) 012022 doi:10.1088/1757-899X/292/1/012022

Table 1. Results of orthogonal experimental design.

Experiment No.	N-2-HACC concentration (mg/ml)	CMC concentration (mg/ml)	pVAXⅠ-F(o)-C3d6 concentration (μg/ml)	Encapsulation efficiency (%)	Loading capacity (%)
1	0.8	0.85	80	79	31
2	0.8	1.0	100	84	38
3	0.8	1.15	120	66	32
4	**1.0**	**0.85**	**100**	**92**	**51**
5	1.0	1.0	120	88	41
6	1.0	1.15	80	86	47
7	1.2	0.85	120	88	41
8	1.2	1.0	80	75	42
9	1.2	1.15	100	83	39
$K_{EE\,1}$	76%	86%	80%		
$K_{EE\,2}$	89%	82%	86%		
$K_{EE\,3}$	82%	78%	81%		
R_{EE}	13%	8%	6%		
$K_{LC\,1}$	34%	41%	40%		
$K_{LC\,2}$	46%	40%	43%		
$K_{LC\,3}$	41%	39%	38%		
R_{LC}	13%	1%	5%		

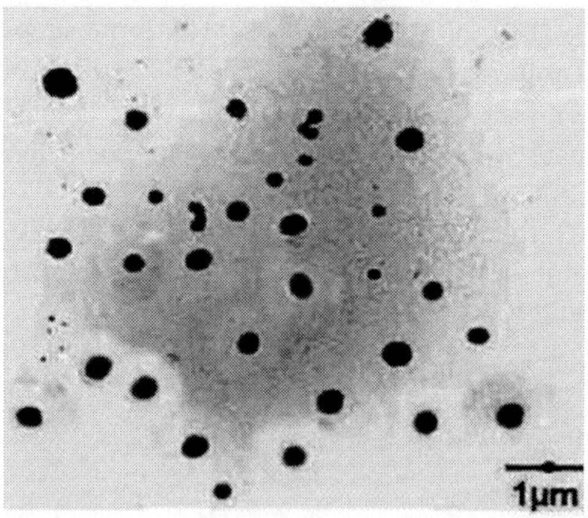

Figure 2. TEM morphology of N-2-HACC-CMC/pFDNA NPs.

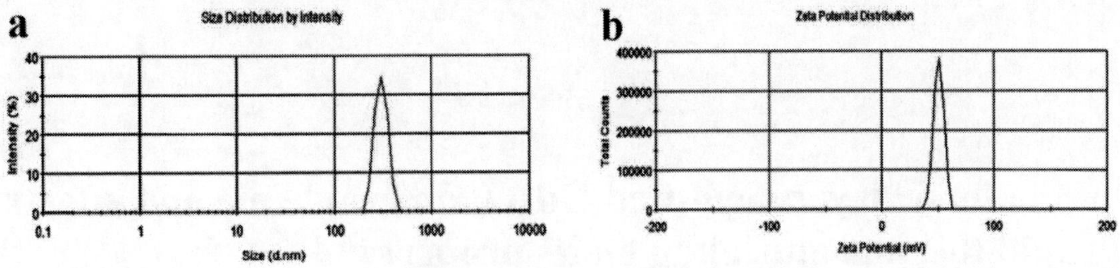

Figure 3. Size distribution (a) and Zeta potential (b) of N-2-HACC-CMC/pFDNA NPs.

4. Conclusion

The optimal preparation condition of N-2-HACC-CMC/pFDNA NPs was achieved, the N-2-HACC-CMC/pFDNA NPs have regular round shape, smooth surface and good dispersion, the particle size is 310 nm, Zeta potential is 50 mV, entrapment efficiency is 92 %, loading capacity 1s 51 %.

Acknowledgments

This work was supported in part by the Key Program of Natural Science Foundation of Heilongjiang Province of China (ZD2015002).

References

[1] Sachin Kumar and Monika Koul 2016 Newcastle disease virus: A constant threat to the poultry industry in India *J. Vaccine* **34** 597–598

[2] Namasivayam Subhapradha and Annaian Shanmugam 2017 Fabrication of β-chitosan nanoparticles and its anticancer potential against human hepatoma cells *J. Int. J. Biol. Macromol.* **94** 194–201

[3] Chandran S P, Natarajan S B, Chandraseharan S et al. 2017 Nano drug delivery strategy of 5-fluorouracil for the treatment of colorectal cancer *J. BMJ Case Rep.* **4** 45–48

[4] Zhang X Z, Yang X Y, Ji J B, Liu A C and Zhai G X 2016 Tumor targeting strategies for chitosan-based nanoparticles *J. Colloids Surf B Biointerfaces.* **148** 460–473

[5] A Landriscina, J Rosen and A J Friedman 2015 Biodegradable chitosan nanoparticles in drug delivery for infectious disease *J. Nnm.* **10** 1609–1619

[6] Feilong Yan, Qifeng Dang, Chengsheng Liu, Jingquan Yan, Teng Wang, Bing Fan, Dongsu Cha, Xiaoli Li, Shengnan Liang and Zhenzhen Zhang 2016 3,6-O-[N-(2-Aminoethyl)-acetamide-yl]-chitosan exerts antibacterial activity by a membrane damage mechanism *J. Carbohydr. Polym.* **149** 102–111

[7] P Pereira, S S Pedrosa, A Correia, C F Lima, M P Olmedo, Á González-Fernández, M Vilanova and F M Gama 2015 Biocompatibility of a self-assembled glycol chitosan nanogel *J. Toxicol. Vitr.* **29** 638–646

Direct atomic layer deposited Gd_2O_3 on grapheme and effects of rapid thermal annealing on its property

L Zheng[1, 2], D L Zhang[1, 2], X H Cheng[1, 3], L Y Shen[1], Q Wang[1, 2], W Zhou[1, 2], D P Wu[1, 2] and Y H Yu[1]

[1] State Key Laboratory of Functional Materials for Informatics, SIMIT, Chinese Academy of Sciences, Changning Road 865, Shanghai 200050, P. R. China
[2] University of Chinese Academy of Sciences, Beijing 100049, P. R. China

[3] E-mail: xh_cheng@mail.sim.ac.cn

Abstract. Atomic layer deposition (ALD) has been accepted as one of the optimal methods for dielectrics growth but it is challenging for ALD to deposit dielectrics on graphene due to its chemically inert property. In this work, Gd_2O_3 films were grown on graphene directly by atomic layer deposition with assistant of pre-H_2O water treatment. In addition, we investigated the effects of rapid thermal annealing (RTA) on microstructure, optical and electrical properties of Gd_2O_3 films on graphene. Raman spectroscopy showed no defects were introduced into graphene during the ALD or RTA processes. The adhesion of Gd_2O_3 films on graphene was enhanced and the crystalline of Gd_2O_3 was improved. In addition, the relative permittivity of Gd_2O_3 on graphene is 14.5 and its EOT can be down to 5.4 nm. This technique may expand the application of graphene in microelectronic devices.

1. Introduction

The International Technology Roadmap for semiconductor (ITRS) reported that the silicon-based semiconductor integrated process will reach its physical limits by 2020s [1, 2, 3]. At the meantime, graphene caused a boom in research since its discovery [4]. Grephene structure is a hexagonal lattice of carbon monolayer with ultrahigh carrier mobility up to 200000 $cm^2V^{-1}s^{-1}$, excellent thermal conductivity and ballistic transport effects [5, 6, 7]. Due to its superior properties, graphene has become the indispensable part of advanced complementary metal oxide semiconductor(CMOS) devices. For suppressing the gate leakage current of conventional SiO_2 CMOS technology, high-k dielectrics are the ideal replacements for SiO_2 as the gate dielectric layer [8]. High k dielectrics, especially the rare earth oxides, have higher dielectric constant (Gd_2O_3=12~17) and wider band gap (ΔE_{Gd2O3}=6.0 eV,) [9, 10, 11]. Therefore, it can be said that the replacement of Si channel by graphene and conventional SiO_2 gate dielectrics by high-k materials maybe a trend of the development of future microelectronics. Atomic layer deposition (ALD) technology has been considered as a very effective and convenient technique to deposit nano-sized and uniform high-k materials on various substrates with its precise control of thickness [12]. However, due to the lack of dangling bonds on graphene surface, it is difficult to deposit high-k films on graphene directly [13]. Several common improved methods include electron-beam physical vapor deposition (EBPVD), surface functionalization of graphene and introduced ultra-thin transition films have been tried to deposit high-k films on graphene [14, 15, 16]. Nevertheless, these methods will introduce impurities into graphene, break graphene lattice structures and markedly reduce its carrier mobility [17]. Our recent work reported that, H_2O can be physical adsorbed on graphene surface working as deposition sites and oxidants without introducing defects or disorder in graphene [18, 19]. With the assistance of pre-H_2O treatment, Al_2O_3 and HfO_2 films have been successfully deposited on graphene directly by ALD. In this work, Gd_2O_3 was deposited on graphene with assistant of pre-H_2O treatment, and rapid thermal annealing (RTA), which is an indispensable part in advanced CMOS integration process, was performed to analyze its effects on Gd_2O_3/graphene heterojunctions. This technique may expand the application of graphene in microelectronic devices.

2. Experimental

Graphene samples on SiO_2 (300 nm)/Si substrates were prepared by the following steps (figure 1). First, the high-quality monolayer graphene was grown on Cu foils by ultra-high vacuum chemical vapor deposition (CVD) system. Using the high concentrations of $FeCl_3$ solution, the Cu foils were etched, and the graphene covered by residual photoresist-polymethyl methacrylate (PMMA) was obtained. Subsequently, the graphene covered with PMMA was transferred onto rinsed SiO_2/Si substrates immediately. And the acetone was utilized to remove the photoresist. After the preparation of graphene samples, the Gd_2O_3 films were deposited on graphene by an ALD system. The gadolinium precursor (Gd[N(SiMe)$_2$]$_3$) and the ALD chamber were heated to 177°C and 200°C, respectively. The chamber temperature was based on sublimation temperature of Gd[N(SiMe)$_2$]$_3$. Then, the graphene samples were pushed into the vacuum chamber. The optimized pre-H_2O treatment on graphene for ALD growth was verified to be 4 cycles in our previous work, which contributed to the formation of a uniform dielectric films morphology on graphene [18, 19]. Therefore, before the deposition of Gd_2O_3 films, 4 cycles of H_2O treatment were performed to work as deposition sites on graphene surface. 1 cycle of pre-H_2O treatment includes 1 s pulse time and 10 s purge time. After 4 cycles of pre-H_2O treatment, Gd_2O_3 films were deposited on graphene surface by the following steps as a complete cycle: 1 s Gd[N(SiMe)$_2$]$_3$ pulse time and 10 s N_2 purge, followed by 1s H_2O pulse and 10 s N_2 purge. And it ran for 100 cycles, totally. After the deposition of Gd_2O_3 films, annealing at 800°C for 30 s was utilized to eliminate internal residual stress of Gd_2O_3 films on graphene. For electrical measurements, Au/Ti alloys were grown on Gd_2O_3 surface to fabricate a metal-oxide-graphene (MOG) structure. In addition, the Gd_2O_3/graphene heterojunctions were also characterized by Raman spectroscopy, transmission electron microscopy (TEM) and spectroscopic ellipsometer (SE) to investigate the microstructure and optical properties.

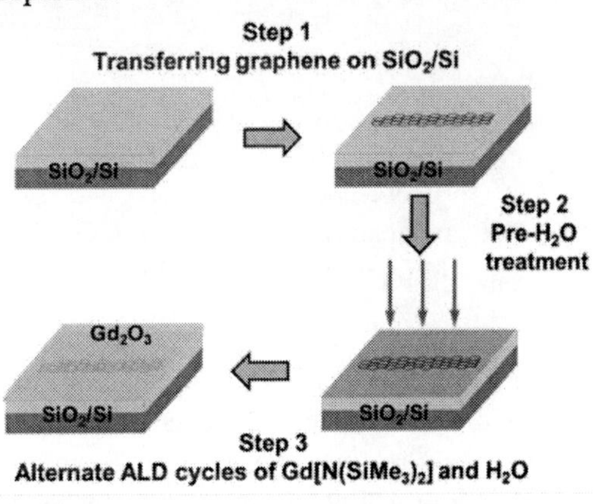

Figure 1. Gd_2O_3/graphene heterojunctions fabrication procedure by H_2O-assised ALD.

3. Results and discussion

3.1. Raman analysis

Raman spectroscopy was an effective measurement to evaluate the quality of carbon-based materials and determine the layers of graphene films [20, 21]. As illustrated in figure 2a, the pristine graphene

IOP Conf. Series: Materials Science and Engineering **292** (2017) 012023 doi:10.1088/1757-899X/292/1/012023

showed a weak D-band peak at approximately 1350cm^{-1}, and two sharp peaks at 1580cm^{-1} and 2670cm^{-1} representing as G-band peak and 2D-band peak, respectively. It was worth saying that the D-band peak was related to the defects of graphene [22, 23]. The 2D-band peak was from double resonance electron-photon scattering process, and it was sensitive to the layers of graphene [24] and for monolayer graphene, 2D-band peak was a single Lorentzion peak. The G peak was from the plane vibration of first-order E_{2g} phonons, and it was sensitive to internal stress of graphene and the change of graphene layers [25]. Therefore, the I_D/I_G ratio could greatly reflect the crystalline of graphene [26] and the I_{2D}/I_G ratio indicated the adhesion of high-k films on graphene [19]. Briefly, the pristine graphene was monolayer with almost no defects or impurities. After Gd_2O_3 films growth on graphene, as shown in figure 2b, compared with pristine graphene, no change of D-band peak was detected implying that no excess defects or impurities were introduced into graphene during the ALD process. The I_{2D}/I_G ratio was decreased due to the Kohn anomaly from the Γ point by variation of stresses [27]. Figure 2c showed that, with RTA at 800°C for 30s, no raise of defect-related D peak was detected, indicating the RTA process would not introduce any defects into graphene. As the same with figure 2b, the ratio of I_{2D}/I_G was further decreased, which may be due to the stronger adhesion of Gd_2O_3 films on graphene after the post RTA process.

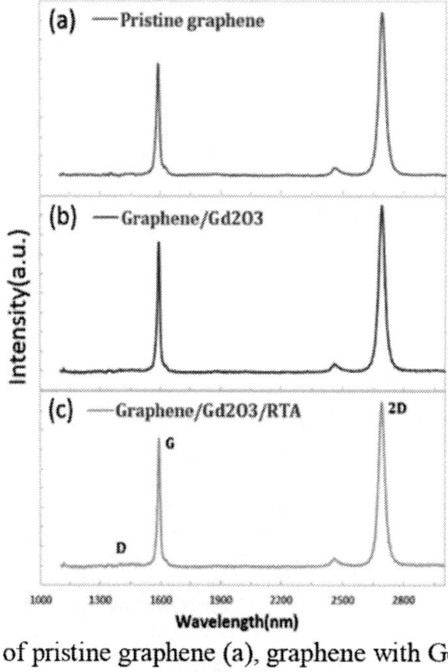

Figure 2. Raman spectra of pristine graphene (a), graphene with Gd_2O_3 films and the graphene with Gd_2O_3 after the post RTA.

3.2. TEM analysis

TEM was performed to verify the thickness of Gd_2O_3 on graphene and analyze the effects of the post RTA on the microstructure of Gd_2O_3 on graphene. The cross-section microstructure of Gd_2O_3 on graphene was displayed in figure 3. The thickness of Gd_2O_3 was ~20 nm with fairly uniform morphology. The growth rate of Gd_2O_3 on graphene could be extracted as 0.2 nm/cycle. After the post RTA, Gd_2O_3 showed partial crystallization (red circles in figure 3), which was due to the low crystallization temperature of Gd_2O_3 film. The presence of Gd_2O_3 on graphene further verified that with pre-H_2O treatment, it was possible to directly deposit high-k dielectric films on graphene by ALD, which was in consistence with our previous work [18, 19].

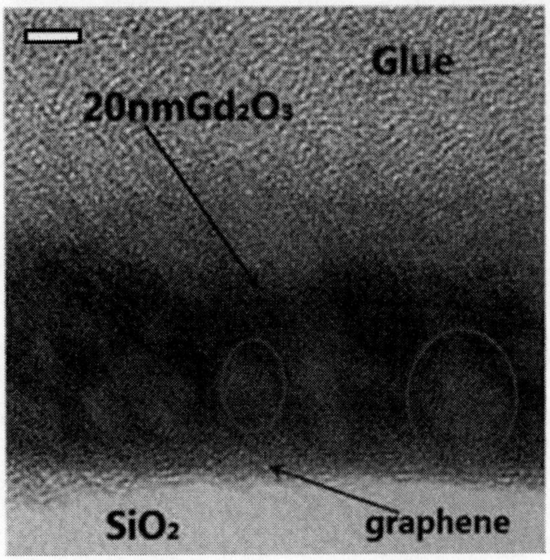

Figure 3. The TEM image of Gd_2O_3 on graphene (the scale bar is 5 nm).

3.3. SE analysis

Spectroscopic ellipsometry (SE) was a convenient implement for the investigation of the thickness and optical constants (including refractive index n, extinction coefficient k, absorption coefficient α, and complex permittivity ε) of thin films. The thickness of Gd_2O_3 (20 nm) obtained from SE results was in great agreement with the consequence of TEM analysis. The refractive index n and extinction coefficient k can be directly measured by SE me[asurement. As displayed in figure 4a, in the near ultraviolet region (about 250-400 nm) the refractive index increased with the increase of wavelength and the refractive index tend to be 3.5 in the visible light region. Compared with the refractive index without RTA, as shown in figure 4b, there was an obvious increase of refractive index after the post RTA. It was reported that the shift of the refractive index is related to the microstructure of the films such as the sample size, doping, and crystalline quality [28]. Thus, figure 5a showed that the crystalline quality was improved after the post RTA, which was in aggrement with TEM results. The absorption coefficient α could be calculated by the following formula [29]:

$$\alpha = \frac{4\pi k}{\lambda} \quad (1)$$

where λ referred to the wavelength. As shown in figure 5b, the Gd_2O_3 film films on graphene were almost transparent in the visible light region while RTA would enhance the ultraviolet absorption of Gd_2O_3. Figure 5c illustrated the complex permittivity $\varepsilon=\varepsilon_r+i\varepsilon_i$, where ε_r was the real part of the complex permittivity and ε_i was the imaginary part. The complex permittivity was calculated from the complex refractive index

$$n_{complex}= \sqrt{\varepsilon}=n+ik \quad (2)$$

where the n and k could be obtained from figure 5a and 5b, respectively. Thus, $\varepsilon_r = n^2-k^2$, and $\varepsilon_i= 2nk$, respectively. It could be calculated that the optical permittivity (real part) ε_r with and without RTA at 800°C was 12.5 and 12, respectively. The difference might be the results of lattice vibration after thermal annealing. The real part of permittivity increased slightly after RTA, indicating the crystallinity improvement of Gd_2O_3 after RTA.

IOP Conf. Series: Materials Science and Engineering **292** (2017) 012023 doi:10.1088/1757-899X/292/1/012023

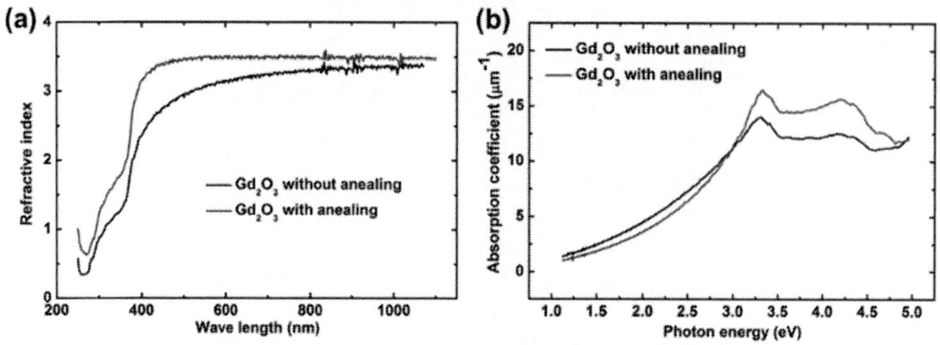

Figure 4. (a) The refractive index (n) of Gd_2O_3 on graphene with and without RTA. (b) The absorption coefficient (α) of Gd_2O_3 on graphene with and without RTA.

3.4. C-V analysis

As shown in figure 5a, the MOG structure was prepared for C-V measurements to evaluate the electrical properties of Gd_2O_3 film on graphene. Figure 5b illustrated the standard circuit model (left) and equivalent series model (right) of C-V measurements, where R_s and R_p referred to the resistance unrelated to the influence of gate dielectric and the resistance related to gate leakage current, interface traps and other factors caused by gate dielectric, respectively; and R_m and C_m referred to the measured values of resistance and capacitance, respectively. Therefore, the impedance of the standard model (left) could be calculated by the following formula:

$$Z = R_s + 1/(\frac{1}{R_p} + jwC) = R_s + \frac{R_p(1 - j2\pi fCR_p)}{1 + (2\pi fCR_p)^2} \tag{3}$$

The impedance could also be from the series model as the following formula:

$$Z = R_m + \frac{1}{j2\pi fC_m} \tag{4}$$

By combing the imaginary of formulas (3) and (4), the relationship between capacitance and frequency could be obtained. Thus, the accurate value of capacitance C could be calculated by the following formula:

$$C = \frac{f_1^2 C_1 - f_2^2 C_2}{f_1^2 - f_2^2} \tag{5}$$

where the C_1 and C_2 referred to the measured values at two different frequency f_1 and f_2, (200 kHz and 500 kHz), respectively. The accurate capacitance C could be drawn as the green curve in figure 5c, called dual-frequency modification curve. It was worth mentioning that the V-shape of the C-V curves was caused by quantum capacitance of graphene [30] and it was 7-10 μF/cm^2 and had negligible effect in the series model. Therefore, the capacitance of Gd_2O_3 film was 0.64 μF/cm^2, and the relative permittivity and the equivalent oxide thickness (EOT) could be extracted as 14.5 and 5.4 nm, respectively.

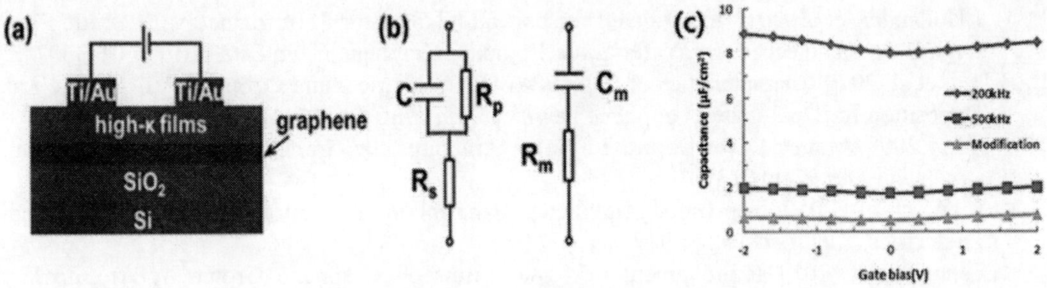

Figure 5. (a) The schematic diagram of a metal-oxide-graphene (MOG). (b) The standard circuit model (left) and series model (right) (c) The C-V curve of Gd_2O_3 film on graphene after the post RTA.

4. Conclusion

In summary, with pre-H_2O treatment, Gd_2O_3 has been successfully grown on graphene directly by ALD. Raman analysis indicates that no defects are introduced into the sp^2 hybridized carbon structure of graphene during both the ALD processs and the post RTA, while the post RTA will enhance the adhesion of Gd_2O_3 films on graphene. TEM and SE measurements both reveal crystalline improvement of Gd_2O_3 films on graphene after the post RTA. In addition, the relative permittivity of Gd_2O_3 on graphene is 14.5 and its EOT can be down to 5.4 nm. This technique may expand the application of graphene in microelectronic devices.

Acknowledgements

This work was supported by the Shanghai Sailing Program (Grant No. 17YF1422700) and the Natural Scientific Foundation of Shanghai (Grant No. 16ZR1442300). We would like to thank Prof. Xiaoming Xie, Prof. Zengfeng Di and Dr. Tianbo Wang for their generous help.

References

[1] C Macilwain 2005 Computer Hardware: Silicon Down to Wire. *Nature* **436**(7047) 22-23
[2] R F Service 2009 Is Silicon's Region Nearing its End *Science* **323**(5917) 1000-1002
[3] P A Packan 1999 Pushing the Limits. *Science* **285**(5436) 2079-2081
[4] K S Novoselov, et al. 2004 Electric Field Effect in Atomically Thin Carbon Films. *Science* **306**(5696) 666-669
[5] A A Balandin 2011 Thermal Properties of Graphene and Nanostructured Carbon Materials. *Nat Mater* **10**(8) 569-581
[6] G Eda, et al. 2008 Large-Area Ultrathin Films of Reduced Graphene Oxide as a Transparent and Flexible Electronic Material *Nat Nanotech* **3**(5) 270-274
[7] A S Mayorow, et al. 2011 Micrometer-Scale Ballistic Transport in Encapsulated Graphene at Room Temperature *Nano Lett* **11**(6) 2396-2399
[8] K Zou, et al. 2010 Deposition of High-Quality HfO_2 on Graphene and the Effect of Remote Oxide Phonon Scattering *Phys Rev Lett* **105**(12) 126601
[9] G D Wilk, et al. 2001 High-Kappa Gate Dielectrics: Current Status and Materials Properties Considerations *J Appl Phys* **89**(10) 5243-5275
[10] A Laha, et al. 2006 Impact of Si Substrate Orientations on Electrical Properties of Crystalline Gd_2O_3 Thin Films for High-K Application *Appl Phys Lett* **89**(14) 143154
[11] J Kido, et al. 2002 Organo Lanthanide Metal Eomplexes for Electroluminescent Materials *Chem Rev* **102**(6) 2357-2368
[12] C Marichy, et al. 2012 Atomic Layer Deposition of Nanostructured Materials for Energy and Environmental Applications *Adv Mater* **24**(8) 1017-1032
[13] L Liao, et al. 2010 Graphene-Dielectric Integration for Graphene Transistor *Mater Sci Eng R Rep* **70**(3) 354-370

[14] M J Hollander, et al. 2011 Enhanced Transport and Transistor Performance with Oxide Seeded High-k Gate Dielectrics on Wafer-Scale Epitaxial Graphene *Nano Lett* **11**(9) 3601-3607
[15] B Lee, et al. 2010 Characteristics of High-k Al_2O_3 Dielectric using Ozone-Based Atomic Layer Deposition for Dual-Gated Graphene Devices *Appl Phys Lett* **97**(4): 043107
[16] Y Xuan 2008 Atomic-Layer-Deposited Nano Structures for Graphene-Based Nanoelectronics *Appl Phys Lett* **92**(1) 013101
[17] L Liao, et al. 2010 Top-Gated Graphene Nanoribbon Transistors with Ultrathin High-k Dielectrics *Nano Lett* **10**(5) 1917
[18] L Zheng, et al. 2014 Improvement of Al_2O_3 Films on Graphene Grown by Atomic Layer Deposition with Pre-H_2O Treatment *ACS Appl Mater Interfaces* **6**(10) 7014-7019
[19] L Zheng, et al. 2014 Property transformation of graphene with Al_2O_3 films deposited directly by atomic layer deposition *Appl Phys Lett* **104**(2) 023112
[20] T Jawhari, et al. 1995 Raman Spectroscopic Characterization of some Commercially Available Carbon Black Materials *Carbon* **33**(11) 1561-1565
[21] A C Ferrari, et al. 2006 Raman Spectrum of Graphene and Graphene Layers *Phys Rev Lett* **97**(18) 187401
[22] Y Wang, et al. 1990 Raman Spectroscopy of Carbon Materials: Structural Basis of Observed Spectra *Chem Mater* **2**(5) 557-563
[23] C Casiraghi, et al. 2009 Raman Spectroscopy of Graphene Edges *Nano Lett* **9**(4) 1433-1441
[24] C Thomsen, et al. 2000 Double Resonant Raman Scattering in Graphite *Phys Rev Lett* **85**(24) 5214-5217
[25] M S Dresselhaus, et al. 2008 Raman Spectroscopy as a Probe of Graphene and Carbon Nanotubes. *Phil Trans R Soc A* **36**(1813) 231-236
[26] P Elitza, et al. 2011 Interference Effects on the I_D/I_G ratio of the Raman Spectra of Diamond-like Carbon Thin Films *Diam Relat Mater* **62** 717-723
[27] S Pisana, et al. 2007 Breakdown of the Adiabatic Born-Oppenheimer Approximation in Graphene *Nat Mater* **6**(3) 198-201
[28] C A Adib, et al. 2014 Tuning Optical Properties of Al_2O_3/ZnO Nanolaminates Synthesized by ALD *J Phys Chem C* **118**(7) 3811-3819
[29] L Miao, et al. 2013 Ellipsometric Studies of Optical Properties of Er-Doped ZnO Thin Films Synthesized by Sol-Gel Method *Thin Solid Films* **543**(10) 125-129
[30] J Xia, et al. 2009 Measurement of the Quantum Capacitance of Graphene *Nat Nanotech* **4**(8) 505-509

IOP Conf. Series: Materials Science and Engineering **292** (2017) 012024 doi:10.1088/1757-899X/292/1/012024

Surfactant-assisted growth and optical properties of ZnO hexagonal bilayer disk-like microstructures

Q P Zhu[1, 2], X Y Shen[1], L L Wang[1], L P Zhu[1, 4], L J Wang[1] and G H Liao[3]

[1] School of Environmental and Materials Engineering, Shanghai Polytechnic University, Shanghai, 201209, China
[2] School of Civil Engineering, Chongqing Institute of Engineering, Chongqing, 400054, China
[3] Technical Institute of Physics and Chemistry, CAS, Beijing 100190, China

[4] E-mail: lpzhu@sspu.edu.cn

Abstract. ZnO hexagonal bilayer disk-like microstructures are successfully fabricated using a simple solvothermal method assisted with surfactant. The structure and morphology were investigated by XRD, SEM, and EDS. XRD result indicated that the as-obtained samples were well-crystallized wurtzite hexagonal ZnO structure. SEM images showed that the ZnO hexagonal bilayer disk-like assembles consist of two uniform and smooth disks with an average edge length of 6 μm and thickness of ~4 μm. UV - vis spectrum reveals that ZnO sampls show an appreciable red shift and the band gap energy of the obtained ZnO samples were about 3.15 eV. A very strong UV emission at the ultraviolet (UV) region was observed in the photoluminescence (PL) spectrum of the as-prepared ZnO samples tested at room-temperature. A possible growth process of the ZnO hexagonal bilayer disk-like microstructures was schematically illustrated.

1. Introduction

Zinc oxide (ZnO), is an important semiconductor metal oxide with a wide direct bandgap of 3.37 eV [1]. It has stimulated great research interest over the past few years due to its unique properties that are useful for the fabrication of nanolasers, optoelectronic devices, and photocatalysts [2]. As is well-known, the physical and chemical properties of materials are strongly depending on their size, size distribution, structures, and dimensions. Controllable preparation of well crystallized and novel morphological ZnO micro/nanostructures is highly desired for many applications. Considerable effort has been devoted to the morphology-controllable synthesis of ZnO [3]. In our previous works, we reported the control synthesis of semiconductor metal oxide using different methods [4-12]. One of the most promising methods is a biomimetic strategy that uses surfactants or polymers as the structure-directing agent to control the nucleation, growth, and alignment of crystals [3]. For example, Ni et al. prepared flowerlike ZnO nanocrystals via a simple hydrothermal route in the presence of sodium dodecyl sulfate (SDS) [13]. Liu et al. synthesized 3D ZnO tubular flowerlike architectures and radial nanorods in the presence of polyethylene glycol [14]. Oh and co-workers prepared micrometer-sized ZnO hexagonal rings using the coordination-polymer template-directed approach [15]. Zhang et al. reported the shape-selective synthesis of ZnO particles with distinctive shapes by a novel polyvinylpyrrolidone-directed crystallization route [8]. In the current paper we propose a facile route for fabricating ZnO hexagonal bilayer disk-like microstructures (HBDMs) via a solvothermal method in the presence of polyvinylpyrrolidone (PVP). The optical properties were systematically measured by UV-vis spectroscopy and fluorometer.

2. Experimental

Firstly, 0.5 g of polyvinylpyrrolidone (PVP) were completely dispersed in 80 mL of water-alcohol (V/V=1) solvent. Then, 20 mmol of ZnAc2 and 20 mL 2.5mol/L NaOH solution was added into the above solvent with stirring. Finally, the solution was transferred into a Teflon-lined stainless-steel autoclave. The autoclave was sealed and maintained at 110 ° C for 10 h. After the reaction was completed, the resulting precipitates were collected and washed thoroughly and dried at 80 ° C in air. The purity of the product was characterized by X-ray diffraction diffractometer (XRD, Rigaku D/max 2500). The morphologies and microstructures of as-prepared samples were examined by scanning electron microscope (SEM, Hitachi S-4800) equipped with an energy dispersive X-ray spectrometer (EDS). Diffuse reflectance UV - vis and photoluminescence (PL) spectrums of the samples were performed by UV-vis spectrophotometer (Shimadzu 2550) and fluorometer (Hitachi F-4500) with a Xe lamp as the excitation light source at room temperature.

3. Results and Discussion

Figure 1(a) shows the typical XRD patterns of the as-prepared ZnO HBDMs. The ZnO powders displayed a good crystallinity and can be assigned to the wurtzite hexagonal ZnO structure (JCPDS card no. 36-1451). EDS result shown in figure 1(b) reveals that the as-prepared sample contains Zn and O elements, and the atomic ratio of Zn/O is about 1:1.

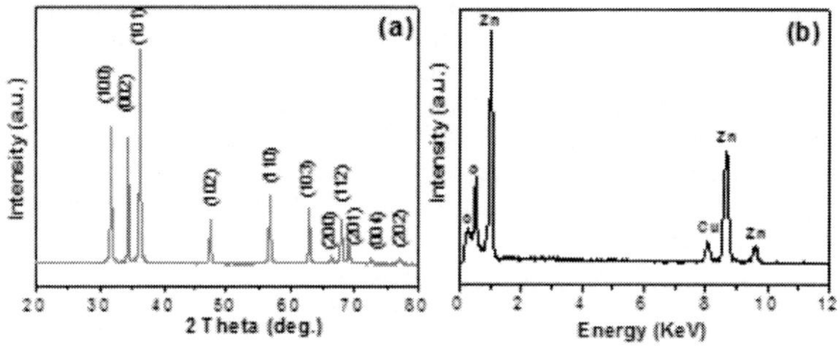

Figure 1. XRD(a) and EDS (b) patterns of the ZnO HBDMs.

The general morphologies of the ZnO microstructurs synthesized by using PVP are shown in figure 2(a). Many HBDMs can be observed. An individual ZnO hexagonal bilayer disk is shown in the figure 2(b)-(c). The hexagonal bilayer disk-like assembles consist of two uniform and smooth disks with an average edge length of 6 μm and thickness of ~4 μm.

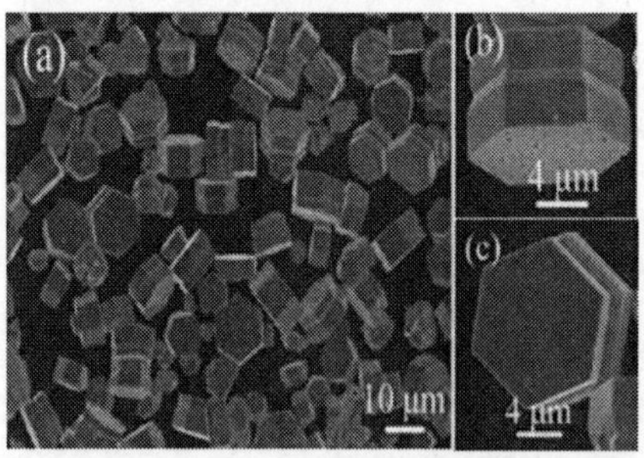

Figure 2. Low-magnification SEM images of the ZnO HBDMs (a), SEM image of a single ZnO HBDM at various angles: (b) declivitous, (c)left-oblique.

On the basis of previous studies [3, 14-23] and the above results, the possible growth processes of the ZnO HBDMs can be schematically illustrated, as shown in figure 3. First, numerous $ZnAc_2$ was hydrolyzed in the highly basic reaction solution, and $Zn(OH)_4^{2-}$ ions were first formed. Then, under solvothermal conditions, a large number of tiny ZnO nanoparticles spontaneously appeared from the dehydration of $Zn(OH)_4^{2-}$ ions by initial homogeneous nucleation. The freshly crystalline nanoparticles are unstable and have a tendency to aggregate together and developed into larger crystals, which may be driven by the minimization of interfacial energy. As we know, PVP is a typical nonionic surfactant. In the presence of certain amount of PVP, an intense coordination occurs between –C=O of PVP and Zn atom of ZnO, due to the O atom in the –C=O functional group of PVP has a negative charge and the Zn in the precursor particles is positively charged. As the reaction proceeded, some active sites on the surface of the initially formed ZnO aggregations would grow along the oriented direction and form a hexagonal disk-like microstructure. Afterward, the formed hexagonal disk-like microstructures further oriented assembly to form ZnO HBDMs. Such a process was also observed in the synthesis of ZnO particles with distinctive shapes [3]. As Zhang and co-workers believed it is the dependence on the amount of surfactant in the reaction that makes the controllable nucleation and growth of ZnO possible, thus affording size control of the final product [3]. Here, PVP played the template-directed crystallization role [8]. The exact mechanism for the formation of the final samples need be further investigated.

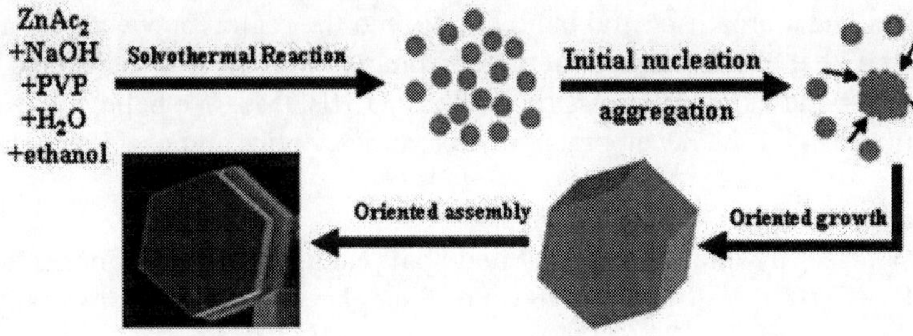

Figure 3. Schematic illustration of the growth process of the ZnO HBDMs.

Figure 4(a) shows the diffuse reflectance spectrum of the ZnO HBDMs. The inset in figure 4 shows the (αhv)2 versus photon energy hv curve for the ZnO samples. The band gap energies (Eg values) of the ZnO samples by extrapolating the straight portion of (αhv)2 against hv plot to the point $\alpha = 0$, which is 3.15 eV for the ZnO HBDMs. These Eg values were a little different from the theoretic value (3.37eV) [1], which could be connected with the distinctive shape and structure. Figure 4(b) shows the room-temperature photoluminescence (PL) spectrum of the as-prepared ZnO HBDMs. It can be observed that the PL spectrum consists of two characteristic parts, that is, a narrow and sharp peak located at the ultraviolet (UV) region and a broad spectral band falling into the region between 480 and 620 nm. According to the literature [24–26], the near-band-edge UV emission typically results from the recombination of free or bound excitons through exciton–exciton collision [24]. The deep-level visible emission should be attributed to the presence of various point defects, such as an oxygen vacancy [27, 28]. Besides, the distinctive contrast between the strong UV band edge emission and weak green emission indicates that the as-prepared ZnO crystals have high crystallinity [29], which is consistent with the XRD result.

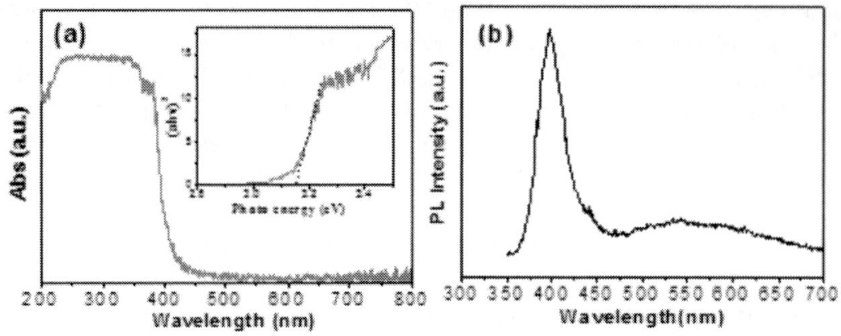

Figure 4. UV-vis diffuse reflectance spectrum (a) and PL spectrum (b) of the ZnO HBDMs. Inset : the $(\alpha h\nu)^2$-$h\nu$ curve for the samples.

4. Conclusions

In summary, ZnO HBDMs with high crystallinity are successfully fabricated using a simple solvothermal method assisted with surfactant. UV–vis spectrum reveals that ZnO sampls show an appreciable red shift and the band gap energy of the obtained ZnO samples were about 3.15 eV. The PL spectrum consists of two characteristic parts, that is, a narrow and sharp peak located at the ultraviolet (UV) region and a broad spectral band falling into the region between 480 and 620 nm.The PL spectrum indicated that the existence of the recombination of free or bound excitons were responsible for the near-band-edge UV emission in the ZnO HBDMs. We believe these ZnO HBDMs will offer new opportunities for the development of electronics, optics and gas sensing applications.

Acknowledgements

This work was supported by the Shanghai Municipal Natural Science Foundation (Grant No. 17ZR1411000, 14ZR1417100), and the key subject of Shanghai Polytechnic University (Grant No. 4: Material Science and Engineering, XXKZD1601, A11NH170310-12).

References

[1] Huang MH, Mao S, Feick H, Yan H, Wu Y, Kind H, Weber E, Russo R and Yang PD 2001 Room-temperature ultraviolet nanowire nanolasers *Science* **292** 1897–9

[2] Jagadish C, Pearton SJ 2006 Zinc Oxide Bulk, Thin Films and Nanostructures *Elsevier New York*

[3] Zhang J, Liu H, Wang Z, Ming N, Li Z and Biris AS 2007 Polyvinylpyrrolidone-directed crystallization of ZnO with tunable morphology and bandgap *Adv Funct Mater* **17** 3897–05

[4] Zhu LP, Bing NC, Wang LL, Jin HY, Liao GH and Wang LJ 2012 Self-assembled 3D porous flowerlike α-Fe_2O_3 hierarchical nanostructures: Synthesis, growth mechanism, and their application in photocatalysis *Dalton Trans* **41** 2959–65

[5] Zhu LP, Liao GH, Bing NC, Zhao X and Gu YY 2011 Synthesis of monodisperse shuttle-like α-Fe_2O_3 nanorods via the EDA-assisted method *Mater Lett* **65** 1287–90

[6] Zhu LP, Liao GH, Bing NC, Wang LL, Yang Y and Xie HY 2010 Self-assembled 3D BiOCl hierarchitectures: tunable synthesis and characterization *CrystEngComm* **12** 3791–06

[7] Zhu LP, Xiao HM, Zhang WD, Yang Y and Fu SY 2008 Synthesis and Characterization of Novel Three-Dimensional Metallic Co Dendritic Superstructures by a Simple Hydrothermal Reduction Route *Cryst Growth Des* **8** 1113–08

[8] Zhu LP, Zhang WD, Xiao HM, Yang Y and Fu SY 2008 Facile Synthesis of Metallic Co Hierarchical Nanostructured Microspheres by a Simple Solvothermal Process *J Phys Chem C* **112** 10073–7

[9] Zhu LP, Zhang WD, Xiao HM, Yang G and Fu SY 2008 One-Pot Template-Free Synthesis of Monodisperse and Single-Crystal Magnetite Hollow Spheres by a Simple Solvothermal Route *Cryst Growth Des* **8** 957–61

[10] Zhu LP, Liao GH, Bing NC, Wang LL and Xie HY 2011 Self-Assembly of Fe_3O_4 Nanocrystal-clusters into Cauliflower-like Architectures: Synthesis and Characterization *J Solid State Chem* **184** 2405–9

[11] Zhu LP, Wang LL, Bing NC, Jin HY, Liao GH and Wang LJ 2012 Preparation and characteristics of core-shell F-doped γ-Fe_2O_3 hollow microspheres *Mater Lett* **88** 164–7

[12] Zhu LP, Wang LL, Bing NC, Huang C, Wang LJ and Liao GH 2013 Porous Fluorine doped γ-Fe_2O_3 Hollow Spheres: Synthesis,Growth Mechanism, and their Application in Photocatalysis *ACS Appl Mater Interfaces* **5** 12478–87

[13] Ni Y, Wu G, Zhang X, Cao X, Hu G, Tao A, Yang Z and Wei X 2008 Hydrothermal preparation, characterization and property research of flowerlike ZnO nanocrystals built up by nanoflakes *Mater Res Bull* **43** 2919–28

[14] Liu J, Huang X, Li Y, Sulieman KM, Sun F and He X 2006 Selective growth and properties of zinc oxide nanostructures *Scripta Mater* **55** 795–8

[15] Jung S, Cho W, Lee HJ and Oh M 2009 Self-Template-Directed Formation of Coordination-Polymer Hexagonal Tubes and Rings, and their Calcination to ZnO Rings *Angew Chem Int Ed* **48** 1459 –62

[16] Ge JC, Tang B, Zhuo LH and Shi ZQ 2006 A rapid hydrothermal route to sisal-like 3D ZnO nanostructures via the assembly of CTA^+ and $Zn(OH)_4^{2-}$: growth mechanism and photoluminescence properties *Nanotechnology* **17** 1316–22

[17] Zhou XF, Chen SY, Zhang DY, Guo XF, Ding WP and Chen Y 2006 Microsphere organization of nanorods directed by PEG linear polymer *Langmuir* **22** 1383–7

[18] Li ZQ, Xiong YJ and Xie Y 2005 Selective growth of ZnO nanostructures with coordination polymers *Nanotechnology* **16** 2303-8

[19] Ko SH, Lee D and Kang HW 2011 Nanoforest of Hydrothermally Grown Hierarchical ZnO Nanowires for a High Efficiency Dye-Sensitized Solar Cell *Nano Lett* **11** 666–71

[20] Liu X 2009 Zinc Oxide Nano- and Microfabrication from Coordination-Polymer Templates *Angew Chem Int Ed* **48** 2–6

[21] Sun Y, Hu J, Wang N, Zou R, Wu J, Song Y, Chen H and Chen Z 2010 Controllable hydrothermal synthesis, growth mechanism, and properties of ZnO three-dimensional structures *New J Chem* **34** 732–7

[22] Zhou Q, Chen W, Peng S and Zeng W 2014 Hydrothermal Synthesis and Acetylene Sensing Properties of Variety Low Dimensional Zinc Oxide Nanostructures *The Sci World J* **489170** 1–8

[23] Yin YX, Jiang LY, Wan LJ, Li CJ and Guo YG 2011 Polyethylene glycol-directed SnO_2 nanowires for enhanced gas-sensing properties *Nanoscale* **3** 1802–6

[24] Fonoberov VA, Alim KA, Balandin AA and Liu JL 2006 Photoluminescence investigation of the carrier recombination processes in ZnO quantum dots and nanocrystals *Phys Rev B* **73** 165317–21

[25] Dingle R 1969 Luminescent Transitions Associated With Divalent Copper Impurities and the Green Emission from Semiconducting Zinc Oxide *Phys Rev Lett* **23** 579–81

[26] Fonoberov VA, Balandin AA 2004 Origin of ultraviolet photoluminescence in ZnO quantum dots: Confined excitons versus surface-bound impurity exciton complexes *Appl Phys Lett* **85** 5971–3

[27] Cao BQ, Cai WP, Li Y, Sun QF and Zhang LD 2005 Ultraviolet-light-emitting ZnO nanosheets prepared by a chemical bath deposition method *Nanotechnology* **16** 1734–8

[28] Chen SJ, Liu YC, Shao CL, Mu R, Lu YM and Zhang JY 2005 Structural and optical properties of uniform ZnO nanosheets *Adv Mater* **17** 586–590

[29] Zhang Y, Mascarenhas A 1999 Scaling of exciton binding energy and virial theorem in semiconductor quantum wells and wires *Phys Rev B* **59** 2040–4

Comparative dynamics analysis on xonotlite spherical particles synthesized via hydrothermal synthesis

F Liu[1, 2, 3], S Chen[1, 3], Q Lin[1, 2], X D Wang[2] and J X Cao[1, 2, 3, 4]

[1] School of Chemistry and Chemical Engineering, Guizhou University, Guiyang, Guizhou 550025, PR China
[2] Key Laboratory of Green Chemical and Clean Energy Technology, Guiyang, Guizhou 550025, PR China
[3] Key Laboratory of Efficient Utilization of Mineral and Green Chemical Technology, Guiyang, Guizhou 550025, PR China

[4] E-mail: jxcao@gzu.edu.cn

Abstract. The xonotlite crystals were synthesized via the hydrothermal synthesis manner from CaO and SiO_2 as the raw materials with their Si/Ca molar ratio of 1.0. Comparative dynamics analysis on xonotlite spherical particles synthesized via hydrothermal synthesis process was explored in this paper. The accuracy of the dynamic equation of xonotlite spherical particles was verified by two methods, one was comparing the production rate of the xonotlite products calculated by the dynamic equation with the experimental values, and the other was comparing the apparent activation energies calculated by the dynamic equation with that calculated by the Kondo model. The results indicated that the production rates of the xonotlite spherical particles calculated by the dynamic equation were in good agreement with the experimental values and the apparent activation energy of the xonotlite spherical particles calculated by dynamic equation (84 $kJ \cdot mol^{-1}$) was close to that calculated by Kondo model (77 $kJ \cdot mol^{-1}$), verifying the high accuracy of the dynamic equation.

1. Introduction

Xonotlite crystal belongs to monoclinic system (a=1.67nm, b=0.73nm, c=0.695nm), and it is a mineral of calcium silicate hydrate with the lowest crystalline water content, the best heat resistance and thermal stability (decomposition temperature of 1050-1100 ℃) [1-3]. In general, the xonotlite products used in industry were fabricated using dynamic process from CaO and SiO_2 as the raw materials with their Si/Ca molar ratio of 1.0 [4-19]. Due to the dissolution characteristics of siliceous material, the xonotlite crystals from H_2SiO_3 and K_2SiO_3 as siliceous materials could grow to form the various morphologies of spherical particles and fibers respectively [20]. In recent years, our group has completed the researches on the morphology and growth mechanism of xonotlite crystals [5, 7, 15 , 19, 20], moreover, by masterly measuring the pH and the calcium concentrations of product slurries, our group achieved the dynamic test of xonotlite spherical particles and pointed out that the dynamic equation of xonotlite spherical particles synthesized via hydrothermal synthesis process was of $-dc_A/dt = kc_A^{0.9}$, but the verification of this conclusion was not enough. Accordingly, in this paper, with H_2SiO_3 solution and $Ca(OH)_2$ suspension as siliceous material and calcareous material respectively, comparative dynamics analysis on xonotlite spherical particles synthesized via hydrothermal synthesis process was explored by taking the Kondo model ($[1-(1-\alpha)^{1/3}]^{-1}=k(t-t_0)$) into account in this paper, aiming to verify the accuracy of the dynamic equation of xonotlite spherical particles.

2. Experiment

2.1. Materials and methods

The xonotlite spherical particles were synthesized via the dynamic hydrothermal synthesis method. H_2SiO_3 solution and $Ca(OH)_2$ suspension were used as siliceous material and calcareous material respectively. $Ca(OH)_2$ suspension was prepared by mixing $CaCl_2$ with KOH $100g{\cdot}L^{-1}$ at the molar ratio of 1:2. The calcareous material, distilled water, and siliceous material were homogeneously mixed, and then the mixture was transferred into a reaction chamber with a stirrer at a heating rate of 1.5 ℃·min^{-1} and a stirring rate of 200 rpm. The hydrothermal treatment proceeded under the following conditions: the Ca/Si molar ratio of 1.0, the reaction temperature of 190 ℃, 210 ℃ and 230 ℃ respectively. After the hydrothermal treatment, the suspensions were filtrated and washed three times by distilled water. The resultant powders were dried at 75 ℃ for 12 h. Technological flow diagram of hydrothermally synthesizing xonotlite products was shown in figure 1.

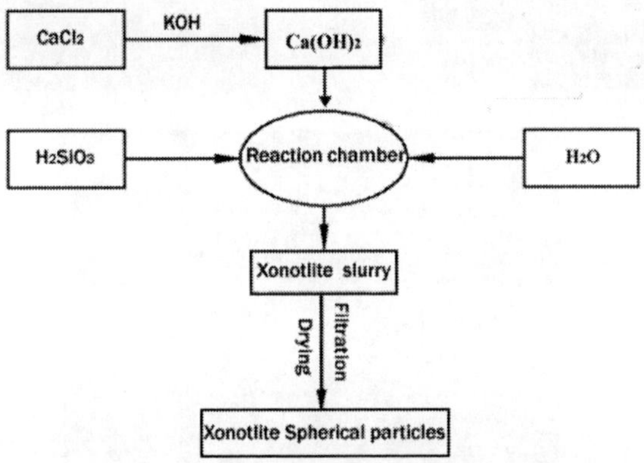

Figure 1. Technological flow diagram of hydrothermally synthesizing xonotlite spherical particles.

In this paper, the accuracy of the dynamic equation of xonotlite spherical particles synthesized via the dynamic hydrothermal synthesis method was verified by two methods, one was comparing the production rate of the xonotlite products calculated by the dynamic equation with the experimental values, and the other was comparing the apparent activation energies calculated by the dynamic equation with that calculated by the Kondo model ($[1-(1-\alpha)^{1/3}]^{-1}=k(t-t_0)$).

2.2. The Determination Method of Samples

The reaction equation for the formation of the xonotlite spherical particles could be simplified as equation (1).

$$6Ca(OH)_2+6H_2SiO_3+H_2O\rightarrow 6CaO\cdot 6SiO_2\cdot H_2O+12H_2O \quad (1)$$

Due to the invisibility of the hydrothermal synthesis reaction, it was impossible to observe the reaction process directly. With H_2SiO_3 solution and $Ca(OH)_2$ suspension as siliceous material and calcareous material respectively, the dynamic test of the hydrothermal synthesis process of the xonotlite spherical particles can be achieved by measuring the change of the pH during the reaction process.

The product slurries synthesized by hydrothermal synthesis process at various reaction temperatures and time were kept at 25 ℃ in thermostat water bath cauldron, and the pH of various slurries were measured using a PHS-3C precise pH meter. The product slurries were accurately weighed by FA2004 electronic balance after drying. After the induction time (t_0), the qualities of the products obtained under various reaction times can be defined as m(Total)=m(HCl)+m($Ca(OH)_2$)+m(C-S-H(I))+m(xonotlite). A scanning electron microscope (SEM, Model JSM-6490LV, operating at 25 KV, Japan) was used to observe the morphologies of xonotlite crystals.

3. Results and discussion

Figure 2 shows the SEM images of the products hydrothermally synthesized from H_2SiO_3 as siliceous material via the dynamic processes at 190 ℃, 210 ℃ and 230 ℃ respectively.

(a) 230 ℃ (b) 210 ℃

(c) 190 ℃

Figure 2. SEM images of the products hydrothermally synthesized from H_2SiO_3 as siliceous material at various temperatures.

In the hydrothermal synthesis process of xonotlite spherical particles, the reaction process can be expressed by the following reaction steps:

$$Ca(OH)_2 + H_2SiO_3 + H_2O \rightarrow C-S-H(gel) \tag{2}$$

$$C-S-H(gel) + Ca(OH)_2 + H_2SiO_3 + H_2O \rightarrow C-S-H(\text{II}) \tag{3}$$

$$C-S-H(\text{II}) + Ca(OH)_2 + H_2SiO_3 + H_2O \rightarrow C-S-H(\text{I})(5CaO\cdot 6SiO_2\cdot 5H_2O) \tag{4}$$

$$Ca(OH)_2 + C-S-H(\text{I})(5CaO\cdot 6SiO_2\cdot 5H_2O) \rightarrow 6CaO\cdot 6SiO_2\cdot H_2O + 5H_2O \tag{5}$$

The pH of the system were respectively measured after various reaction times, shown in figure 3. It can be seen from figure 3, with prolonging reaction time, the pH of hydrothermally synthesized products at various temperatures decreased, and its descending trend can be divided into three stages, that was, the stage forming amorphous C-S-H gels (equation (2, 3 and 4)), the stage forming the most of spherical xonotlite crystals (equation (5)), the stage forming a little xonotlite spherical particles.

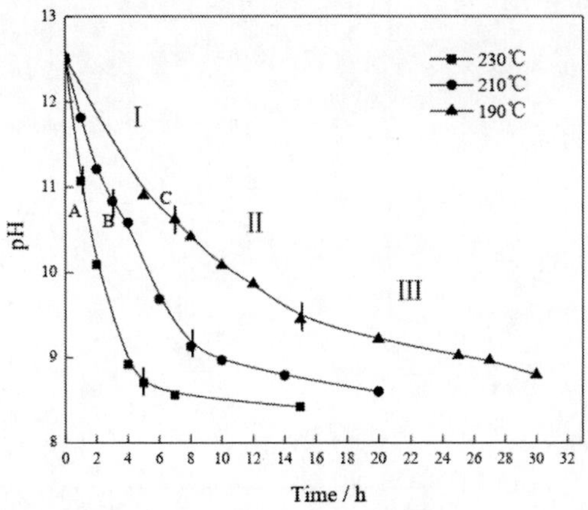

Figure 3. Time dependence of the pH of hydrothermally synthesized products at various temperatures.

Based on chemical reaction dynamic, using fourth-order Runge-Kutta method, spline interpolation and least-squares fitting method, the dynamic equation of xonotlite spherical particles synthesized via hydrothermal synthesis process was:

$$-dc_A / dt = kc_A^{0.9} \tag{6}$$

The dynamic equation (6) was a differential form which was easy to carry out the theoretical analysis. In practical applications, it was necessary to know the relationship between the production rate and the reaction time, thus this required the differential form to be transformed into integral form. The integral form of dynamic equation can be obtained by integrating the differential form of dynamic equation (6).

$$10C_{A,0}^{0.1}[1-(1-\alpha)^{0.1}] = k(t-t_0) \tag{7}$$

In the equation: α —the production rate of the xonotlite spherical particles

t_0—the induction time

k—the rate constant

The values of t_0 and k are related to the reaction temperature.

The production rates of the xonotlite spherical particles calculated by the dynamic equation (7) were compared with the experimental values, shown in figure 4. Obviously, the calculated values were in good agreement with the experimental values, verifying the high accuracy of the dynamic equation (6).

According to the indefinite integral form of Arrhenius (Arrhenius S A) equation (8), if the values of the rate constants (k) at various temperatures was known, the diagram of $\ln k$—$1/T$ could be drawn and it should be a straight line, then the apparent activation energy (Ea) could be obtained from the slope and intercept of the straight line.

$$\ln k = -\frac{Ea}{RT} + \ln A \tag{8}$$

In the equation: k—the rate constant

A—the apparent frequency factor, the unit of it is the same as that of k

Ea—the apparent activation energy, the unit of it is $J \cdot mol^{-1}$

R—the molar gas constant

T—the thermodynamic temperature

The fitted values of dynamic parameters (k) at various temperatures were shown in table 1.

IOP Conf. Series: Materials Science and Engineering **292** (2017) 012025 doi:10.1088/1757-899X/292/1/012025

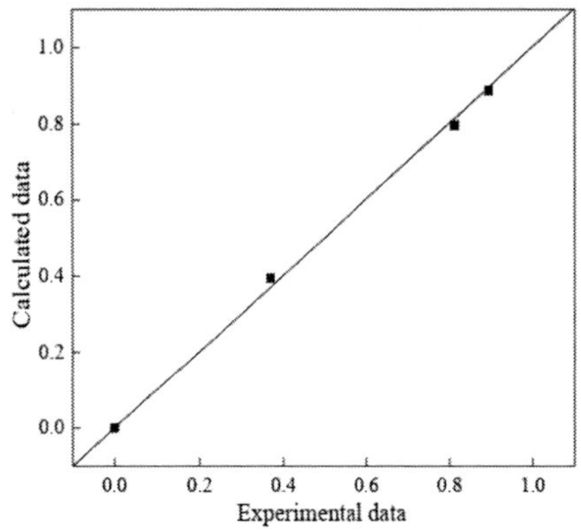

(a) 230 ℃

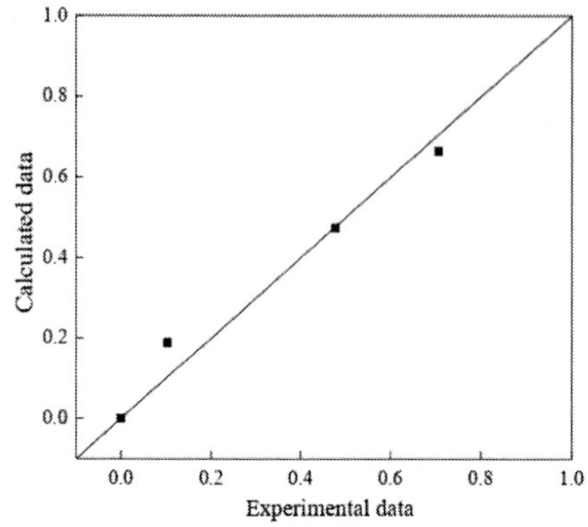

(b) 210 ℃

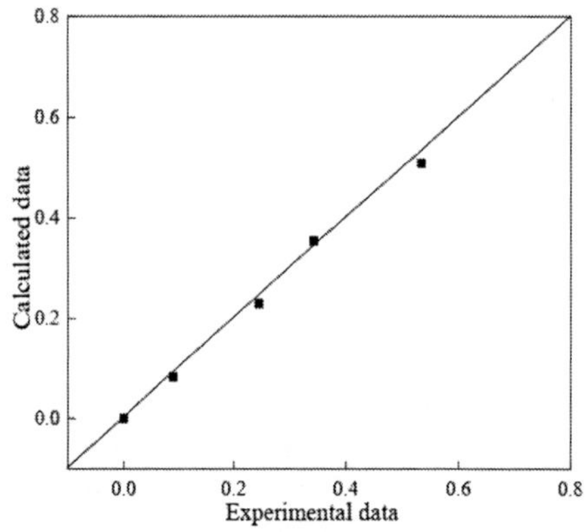

(c) 190 ℃

Figure 4. The experimental data comparison with calculated data of xonotlite spherical particles hydrothermally synthesized at various temperatures.

Table 1. The fitted values of k at various temperatures.

Temperature /℃	k / $(mol \cdot m^{-3})^{1/10} \cdot s^{-1}$
230	0.3147
210	0.1330
190	0.0549

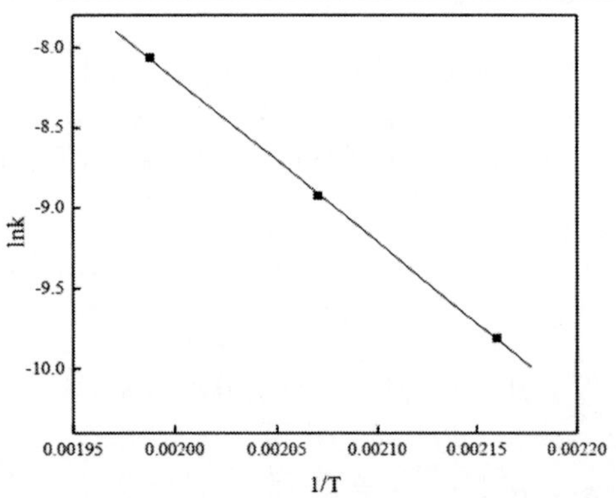

Figure 5. The $1/T$ dependence of $\ln k$ of hydrothermally synthesized xonotlite spherical particles.

According to the data shown in table 1 and equation (8), the diagram of $\ln k$—$1/T$ could be drawn, as shown in figure 5. Finally, the apparent activation energy (Ea) calculated by above-mentioned method was of 84 kJ·mol^{-1}. Through above analysis we can infer that the change of the pH on the curve in the second stage might reflect the production rates of xonotlite crystals. Therefore, the starting points A, B and C of the second stage could be treated as the benchmark of the formation of xonotlite crystal at the hydrothermal synthesis temperature of 230 ℃, 210 ℃ and 190 ℃ respectively. With the process of reaction, the pH in the third stage tended to a fixed value, then the production rates of hydrothermal synthesis products (α) at various holding time can be obtained by setting this fixed value as the benchmark which indicated the complete formation of xonotlite crystals.

$$\alpha = \frac{pH_0 - pH}{pH_0 - pH_1} \tag{9}$$

In the equation: pH —the pH of product slurries synthesized via hydrothermal synthesis at various holding time

pH_0 —the pH of product slurries in the second stage when the xonotlite crystals began to form

pH_1 —the final constant pH of the product slurries synthesized via hydrothermal synthesis in the third stage

As shown in figure 6, the production rates of xonotlite crystals synthesized at various holding time could be calculated by equation (9).

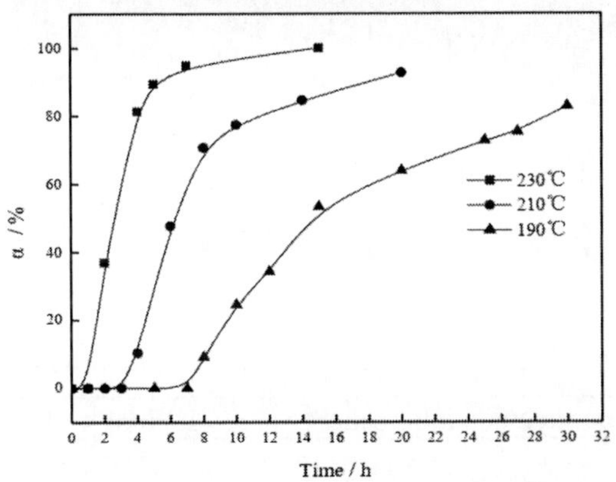

Figure 6. Time dependence of the conversion of hydrothermally synthesized products at various temperatures.

The second stage of the curve in figure 3 was analyzed and discussed by Kondo model commonly used for the hydrothermal reaction of lime-quartz system:

$$[1-(1-\alpha)^{1/3}]^{A} = k(t-t_0) \tag{10}$$

In the equation: A=1 or 2 respectively expressed that the interface reaction of crystal dominated the reaction rate and the diffusion controlled the reaction rate.

If the activation energy of the reaction was extremely small, the reaction rate was quite fast, then the reaction rate was dominated by the diffusion, on the contrary, if the activation energy of the reaction was extremely large, the reaction rate was quite slow, then the reaction rate was controlled by the interface reaction of crystal. In the studies of dynamic hydrothermal synthesis, Hongxun Liang and Hiroaki Noma have calculated that the xonotlite crystal has a large nucleation energy (117-118kJ·mol^{-1}), so the reaction rate was dominated by the interface reaction of crystal. Accordingly, the value of A was 1 in the equation (10) and the equation (10) could be simplified as:

$$[1-(1-\alpha)^{1/3}] = k(t-t_0) \tag{11}$$

As shown in figure 7, the relationship between the various holding time and 1-(1-α)$^{1/3}$could be obtained by applying the equation of Kondo model (11), then the corresponding regression rate parameter k, the induction time t_0 and its correlation coefficient γ could be obtained by regression analysis, shown in table 2. As can be seen from table 2, as the temperature decreased, the induction time increased. The induction time was the period required for the nucleation of the crystal at a certain temperature, and its reciprocal could represent the growth rate of the crystal nucleus. Therefore, the decrease of temperature resulted in the decrease of the rate of crystal growth.

Thus, as shown in figure 8, the diagram of $\ln k$ —$1/T$ could be drawn by using the values of k at various temperature in table 2, then the apparent activation energy (*Ea*) could be calculated to be of 77 kJ·mol^{-1}.

Table 2. The regression rate parameter, induction time and correlation coefficient.

Temperature /℃	k / ×10^{-3}min^{-1}	t_0 / min	γ
230	2.23	57	0.99819
210	1.15	191	0.99589
190	0.46	416	0.99822

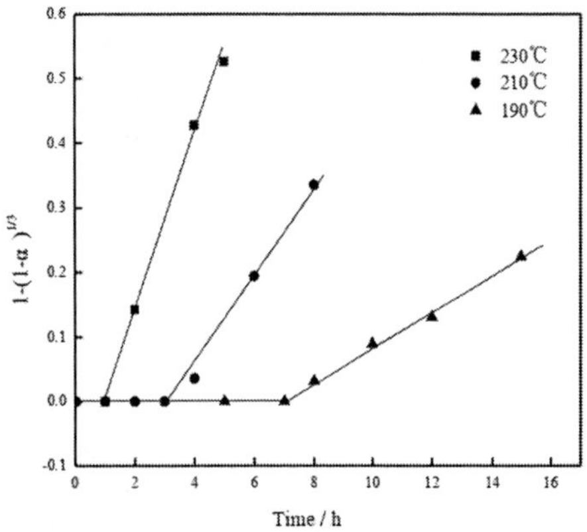

Figure 7. $1-(1-\alpha)^{1/3}$ versus time of Kondo model.

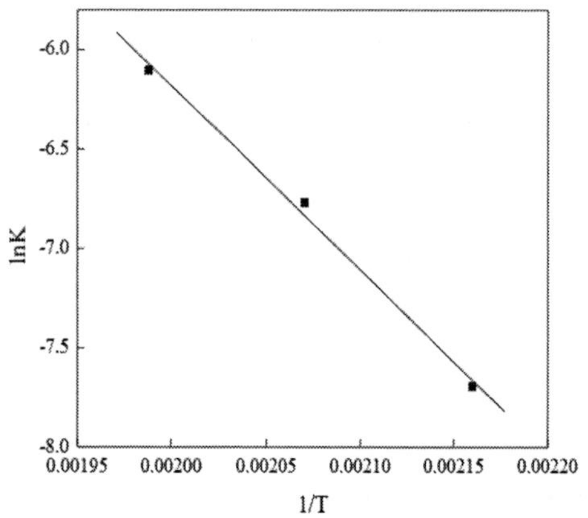

Figure 8. $\ln k$ versus *1/T* of Kondo model.

Table 3. The apparent activation energy of hydrothermally synthesized xonotlite spherical particles.

Models	Apparent activation energy (*Ea*) / kJ·mol^{-1}
$10C_{A_0}^{0.1}[1-(1-\alpha)^{0.1}]=k(t-t_0)$	84
$[1-(1-\alpha)^{1/3}]=k(t-t_0)$	77

The apparent activation energies of the xonotlite spherical particles calculated by dynamic equation (7) and equation of Kondo model (11) respectively were shown in table 3. The data of table 3 indicated that the apparent activation energy of the xonotlite spherical particles (84 kJ·mol^{-1}) calculated by dynamic equation was close to that (77 kJ·mol^{-1}) calculated by Kondo model. This result further verified the high accuracy of the dynamic equation (6).

4. Conclusions

(1) According to the three stages of the reaction process of xonotlite spherical particles, the change of the pH on the curve in the second stage could reflect the production rates of xonotlite crystals. The accuracy of the dynamic equation of xonotlite spherical particles was verified by comparing the activation energies calculated by the Kondo model.

(2) The production rates of the xonotlite spherical particles calculated by the dynamic equation were in good agreement with the experimental values, verifying the high accuracy of the dynamic equation.

(3) The apparent activation energy of the xonotlite spherical particles calculated by dynamic equation (84 kJ·mol^{-1}) was close to that calculated by Kondo model (77 kJ·mol^{-1}), further verifying the high accuracy of the dynamic equation.

Acknowledgments
We gratefully acknowledge the supports of this work by the National Natural Science Foundation of China (Grant No. 21666007), Science Technology Foundation of Guizhou (Grant No. 2014-2007) and Program for "Hundred" High-level Talents in Guizhou (Grant No. 2016-5655).

References
[1] Churakov S V and Mandalie P 2008 *Cement. Concrete. Res* **38** 300-11
[2] Dachowskia R and Stępień A 2011 *Procedia Engineering* **21** 1173-8
[3] Wei G S, Zhang X X and Yu F 2009 J. *Therm. Sci* **18** 142-9
[4] Zou J, Guo C, Jiang Y, Wei C and Li F 2016 *Mater. Chem. Phys* **172** 121-8
[5] Liu F, Zhu B and Cao J X 2011 *Adv. Mater. Res* **148-149** 1755-8
[6] Yang J, Zhang X, Ma H W, Wang M W and Wu H 2014 *Key Eng. Mater* **633** 7-10
[7] Liu F, Wang X D and Cao J X 2013 *Mater. Int. J. Miner. Metall. Mater* **20** 88-93
[8] Hartmann A, Schulenberg D and Buhl J C 2015 *J. Mater. Sci. Chem. Eng* **3** 39-55
[9] Spudulis E, Šavareika V and Špokauskas A 2013 *Mater. Sci-medzg* **19** 190-6
[10] Zou J J, Guo C B, Wei C D and Jiang Y S 2015 *Res Chem Intermed* **18** 1-12
[11] Guo X Y, Ma S H, Lü S Q, Zheng S L and Zou X 2015 *Chin. J. Nonferrous. Metals* **25** 534-44
[12] Yue H Z, Wang X, Yang Z Z and Wei C C 2017 *Key. Eng. Mater* **726** 569-75
[13] Black L, Garbev K and Stumm A 2009 *Adv. Appl. Ceram* **108** 137-44
[14] Liang H X and Li M Q 2002 *J. Chin. Ceram. Soc* **30** 294-9
[15] Liu F, Zeng L K, Cao J X and Li J 2010 *J. Wuhan. Univ. Technol* **25** 295-7
[16] Kurdowski W 2014 *Cement and Concrete Chemistry* (Dordrecht: Springer Netherlands) pp 205-77
[17] Shaw S, Clark S M and Henderson C M B 2000 *Chem. Geol* **167** 129-40
[18] Milestone N B and Ahari G K 2007 *Adv. Appl. Ceram* **106** 302-8
[19] Liu F, Chen S, Lin Q, Wang X D and Cao J X 2017 *Optoelectron. Lett* **13** 81-3
[20] Liu F, Wang X D and Cao J X 2011 *Adv. Mater. Res* **382** 379-83

IOP Conf. Series: Materials Science and Engineering **292** (2017) 012026 doi:10.1088/1757-899X/292/1/012026

Room-temperature synthesis and photoluminescence of hexagonal $CePO_4$ nanorods

J Zhu[1], K Zhang and H Y Zhao

Yunnan Key Laboratory for Micro/nano Materials & Technology, School of Materials Science and engineering, Yunnan University, Kunming 650091, China

[1] E-mail: jzhu@ynu.edu.cn

Abstract. Hexagonal $CePO_4$ nanorods were synthesized via a simple chemical precipitation route at room-temperature without the presence of surfactants and then characterized by powder X-ray diffraction (XRD), energy-dispersive X-ray (EDX) spectrometry, scanning electron microscopy (SEM), transmission electron microscopy (TEM), ultraviolet-visible (UV–vis) absorption and photoluminescence (PL) spectroscopy. Hexagonal $CePO_4$ nanorods exhibit strong ultraviolet absorption and ultraviolet luminescence, which correspond to the electronic transitions between 4*f* and 5*d* state of Ce^{3+} ions.

1. Introduction

Rare earth phosphate nanomaterials have been extensively reported due to their high-performance photoluminescence [1, 2]. Especially, pure and rare-earth ion doped $CePO_4$ nanomaterials have received a lot of research interest [3-6]. $CePO_4$ possesses two common crystallographic structures: monoclinic and hexagonal [7]. Recently, the monoclinic phase was prepared by reflux method [8], hydrothermal synthesis [9] and microwave radiation [10] et al., and the hexagonal phase was synthesized in a sequential, aqueous procedure under continuous flow using a rotating tube processor and a narrow channel reactor by Fang and co-workers [11]. To the best of our knowledge, most of available open literatures focus on the hydrothermal synthesis of pure and rare-earth ion doped $CePO_4$ nanomaterials with different crystallographic structures. However, chemical precipitation, as an important one of chemical synthesis routes, is rarely used for the preparation of rare earth phosphate nanomaterials. There are not reported yet for room-temperature chemical precipitation of $CePO_4$ nanomaterials. So we intend to explore the feasibility of room-temperature precipitation synthesis for rare earth phosphate nanomaterials.

In this work, we synthesized $CePO_4$ nanorods with hexagonal structure by a simple chemical precipitation route at room-temperature without using any surfactant. The product was characterized by XRD, EDX, SEM and TEM. Moreover, luminescent properties of the product were investigated.

2. Experimental section

2.1. Synthesis

All of the chemicals were purchased from Aladdin Chemistry Co. Ltd and used as received without further purification. 5 mmol $Ce(NO_3)_3 \cdot 6H_2O$ and 5 mmol $Na_3PO_4 \cdot 12H_2O$ were dissolved in 10 ml deionized water to produce precipitation. Subsequently, the precipitation disappeared when the pH value of the solution was carefully adjusted to 1 with a given concentrated HNO_3 solution under vigorous stirring. The solution was transferred into a sealed beaker and preserved at room temperature. After 20 days, the as-synthesized precipitate was separated by centrifugation and washed five times with deionized water. Finally, the products were dried at 60 °C in air for further characterization.

2.2. Characterization

The powder X-ray diffraction (XRD) pattern of the as-synthesized sample was determined on a RIGAKU DMAX-3B diffractometer with CuKα radiation (λ = 1.54056 Å). The operation voltage and current were maintained at 40 KV and 40 mA, respectively. The sample was scanned from 10 to 90° (2θ) at a scanning rate of 0.02° s-1. Scanning electron microscopy (SEM) images and element analysis were obtained with a FEI QUANTA200 microscope operated at an accelerating voltage of 30 kV. Transmission electron microscopy (TEM) patterns were taken on a JEOL JEM-2100 microscope under a working voltage of 200 kV. The samples for TEM measurement were supported on carbon-coated copper grids by dropping the ethanol suspension containing uniformly dispersed powder. The ultraviolet-visible (UV–vis) absorption spectrum was recorded on a HITACHI U-4100 UV/VIS/NIR spectrophotometer. The photoluminescence (PL) measurement was carried out on a HITACHI F-4500 fluorescence spectrophotometer using Xe lamp as the excitation source at room temperature.

3. Results and discussion

3.1. Phase identification

The crystalline phase and elemental composition of as-synthesized $CePO_4$ sample have been identified by powder XRD and EDX measurements, as shown in figure 1. All diffraction peaks can be easily indexed with respect to the hexagonal CePO4 (space group: P6222 (180) with cell parameters: a1 = 7.055 Å and c = 6.439 Å (JCPDS 34-1380), suggesting that no secondary phase co-exists. Moreover, the EDX analysis (the inset of figure 1) confirms the presence of Cerium (Ce), phosphorus (P) and oxygen (O) in the product, which can effectively support the XRD result. All above analysis means that the as-synthesized sample have the nature of high purity in our synthesis route.

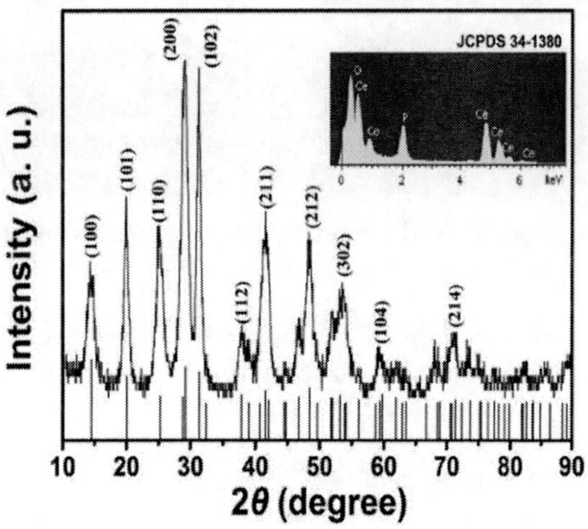

Figure 1. XRD pattern of $CePO_4$ sample (EDX image as inset).

It is with regret that the crystallinity of the sample prepared at room-temperature is poor. The observed broad reflection peaks should be attributed to the small size of the sample and the distribution in particle sizes, as revealed by the below SEM and TEM images (figure 2).

3.2. Morphology

The typical low-magnification SEM image of the as-prepared pure hexagonal $CePO_4$ sample is displayed in figure 2a, from which nanosized-structure morphology with serious aggregation is detected. The morphology and microstructure of the sample was further characterized by TEM and HRTEM. The typical TEM and HRTEM images of hexagonal $CePO_4$ sample are shown in figure 2b and figure 2c, respectively. Nanorods with a diameter of 7–10 nm and length of 60–150 nm are demonstrated, suggesting a high growth rate and single growth direction along the axial direction of $CePO_4$ crystal. The TEM image (figure 2b) clearly unfolds the form of the bundle that is composed of some neighbouring nanorods. The interaction among these nanorods might arise from the synergic effects of chemical binding and physical attachment. The high density of nanorods is representative of the high yield, which may result from this simple room-temperature preparation approach. The HRTEM image (figure 2c) shows that the nanorod is structurally uniform with the interplanar spacing of 0.30 nm, which corresponds to the standard value of d_{200} spacing in hexagonal $CePO_4$ (JCPDS: 34-1380), meaning that the growing [001] direction of the hexagonal $CePO_4$ nanorods is parallel to the (200) lattice plane. This result is in good correlation with the anisotropic character of the (200) peak in the XRD pattern (figure 1).

Based on our experimental results, from a structural point of view, the growth mechanism of these nanorods is analyzed. In the crystal structure of hexagonal $CePO_4$ [12], there are infinite linear chains which are alternating CeO_8 polyhedra and PO_4 tetrahedra along the c axis. Wong et al. [3] claimed that the bonding between these chains is considerably weaker than that within the chains from a thermodynamic perspective, meaning that the activation energy for the c axis direction of growth of hexagonal $CePO_4$ is lower than that for a growth direction perpendicular to the c-axis itself. Hence, this analysis result indicates a higher growth rate along the c axis. In other hexagonal rare earth phosphate nanomaterials [13, 14], the same preferential growth along the c axis is also observed. Therefore, the hexagonal crystal structure of rare earth phosphate should be responsible for the anisotropic growth of $CePO_4$, leading to the observed rod-like morphology with high aspect ratio.

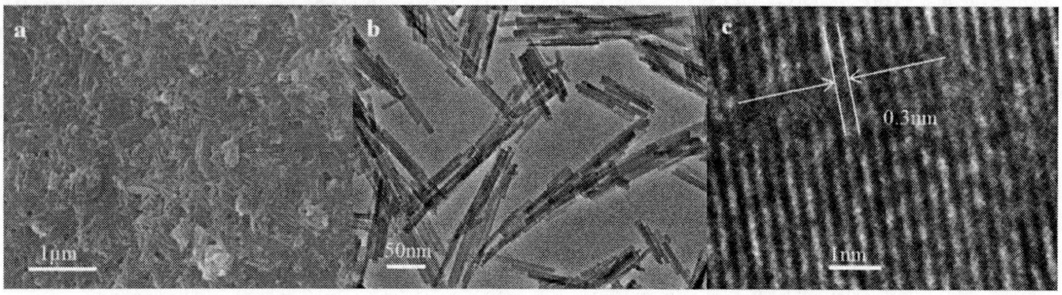

Figure 2. SEM (a), TEM (b) and HRTEM (c) images of $CePO_4$ nanorods.

3.3. Optical properties

Figure 3a shows the UV-vis spectrum of the as-prepared hexagonal $CePO_4$ nanorods from the ultraviolet to visible region, from which a strong ultraviolet absorption peak at around 275 nm is observed. According to the previous related literatures [10, 15], the optical absorption of $CePO_4$ has two electron transition modes (*f–f* and *f–d*) of Ce^{3+} ions in IR region and ultraviolet region. So the ultraviolet absorption peak is attributed to the $4f^1(^2F_{5/2})$–$5d^1(^2D_{5/2})$ electron transition of Ce^{3+} ions. The absorption cutoff wavelength is about 389 nm, indicating the hexagonal $CePO_4$ nanorods have a wider ultraviolet absorption range with the optical band gap of 3.19 eV than hydrothermally synthesized monoclinic $CePO_4$ nanorods (the optical band gap of 4.1 eV) [10] due to the effect of the crystal structure and surface area. This result implies the potential applicability of hexagonal $CePO_4$ nanomaterials as UV absorber, making it a promising material for personal care products [16]. The emission spectrum of hexagonal $CePO_4$ nanorods monitored at 293 nm is illustrated in figure 3b. The photoluminescence curve exhibits a strong ultraviolet broad emission band centered at 356 nm, which may be linked with the electronic transitions from 5*d* excited state to ^2F ground state. Similar broadening optical transitions are observed in the early reports [7].

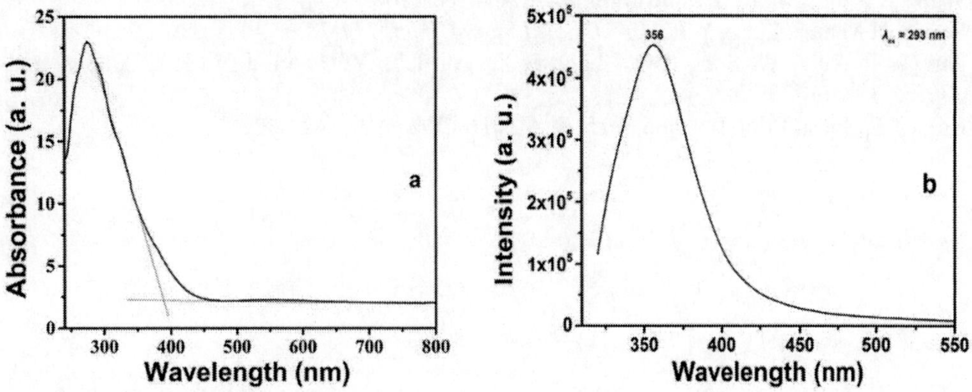

Figure 3. UV-vis absorption (a) and emission (b) spectra of $CePO_4$ nanorods.

4. Conclusions

The present work describes a method for preparing hexagonal $CePO_4$ nanorods by a simple chemical precipitation route at room-temperature without the presence of surfactants. This is the first report on the room-temperature precipitation synthesis of hexagonal $CePO_4$ nanomaterials. Powder XRD and EDX measurement confirm the purity and of crystalline phase of $CePO_4$ nanorods with hexagonal $P6_222$ structure. According to SEM and TEM observations, the hexagonal $CePO_4$ sample is composed of uniform nanorods with a diameter of 7–10 nm and length of 60–150 nm. The formation of rod-like morphology results from the anisotropic growth along the [001] direction. Hexagonal $CePO_4$ nanorods exhibit strong ultraviolet absorption (the optical band gap of 3.19 eV) and ultraviolet luminescence, which correspond to the electronic transitions between 4*f* and 5*d* state of Ce^{3+} ions.

Acknowledgments

This investigation was supported by the Training Program for Young Academic and Technical Leader in Yunnan Province and the Science Foundation of Key Laboratory of Micro/Nano materials and Technology in Yunnan Province (No. ZZ2016002).

References

[1] Sahu N K, Singh N S, Ningthoujam R S and Bahadur D 2014 *ACS Photonics 1*337
[2] Sun C T, Li X X, Wang H and Xue D F 2016 *Inorg. Chem.* **55** 2969
[3] Han J, Wang L and Wong S S 2014 *J. Phys. Chem. C* **118** 5671
[4] Pusztai P, Haspel H, TóthI Y, Tombácz E, László K, Kukovecz Á and Kónya Z 2015 *Appl. Mater. Interfaces* **7** 9947
[5] Chen H Q, Yuan F, Xu J, Zhang Y Y, Wu Y and Wang L 2013 *Spectrochim. Acta, Part A* **107** 151
[6] Chen H Q and Ren J C 2012 *Analyst* **137** 1899
[7] Bao J, Yu R, Zhang J, Yang X, Wang D, Deng J, Chen J and Xing X 2009 *Cryst. Eng. Comm.* **11** 1630
[8] Kirubanithy M, Irudayaraj A A, Raj A D and Manikandan S 2015 *Mater. Today: Proceedings* **2** 4344
[9] Palma-Ramírez D, Domínguez-Crespo M A, Torres-Huerta A M, Dorantes-Rosales H, Ramírez-Meneses E and Rodríguez E 2015 *J. Alloys Compd.* **643** S209
[10] Ekthammathat N, Thongtem T, Phuruangrat A and Thongtem S 2011 *Rare Met.* **30** 572
[11] Fang J, Evans C W, Willis G J, Sherwood D, Guo Y L, Lu G Z, Raston C L and Iyer K S 2010 *Lab Chip* **10** 2579
[12] Achary S N, Bevara S and Tyagi A K 2017 *Coord. Chem. Rev.* **340** 266

[13] Wang X J and Gao M Y 2006 *J. Mater. Chem.* **16** 1360
[14] Guana H M and Zhang Y J 2004 *J. Solid State Chem.* **177** 781
[15] Fang Y P, Xu A W, Song R Q, Zhang H X, You L P, Yu J C and Liu H Q 2003 *J. Am. Chem. Soc.* **125** 16025
[16] Lima J F, Sousa Filho P C and Serra O A 2016 *Ceram. Int.* **42** 7422

IOP Conf. Series: Materials Science and Engineering **292** (2017) 012027 doi:10.1088/1757-899X/292/1/012027

Green synthesis of Pd nanoparticles supported on LDHs and its application in the Suzuki coupling reaction

G H Liu[1] and X F Bai[2, 3]

[1] Key Laboratory of Aerospace Materials and Performance (Ministry of Education), School of Materials Science and Engineering, Beihang University, Beijing, 100191, China

[2] Institute of Petrochemistry, Heilongjiang Academy of Sciences, Harbin 150040, China

[3] E-mail: tommybai@126.com

Abstract. Palladium nanoparticles supported on layered double hydroxides (PdNPs/LDHs) were prepared via a green route employing *Pine* needle extract (PNE) as a reducing and capping agent. The prepared PdNPs/LDHs were characterized by XRD, TEM, EDS, N_2-adsorption analysis. The PdNPs with an average size of 1.75 nm are uniformly dispersed on the LDHs surface. It exhibits high catalytic activity toward the Suzuki coupling reaction with only 0.05 mol% of the catalyst. And the catalyst could be reused for 8 times without an obvious loss of activity.

1. Introduction

Pd catalyzed C-C coupling reactions have been well-known for many years. In these reactions, heterogeneous Pd catalysts are preferred over homogeneous catalysts due to their facile separation from the reaction mixture and the possibility of recycling [1-5].

The catalytic process is a surface phenomenon, a sufficiently high dispersion of metal nanoparticles on the surface of support is crucial for the enhancement of catalytic activity.Layered double hydroxides (LDHs) have attracted interest as a novel support for the dispersal and stabilization of nanoparticles in nanocomposites preparation [6-7], due to its specific interparticle mesoporosity, the ability to exchange anions, and the structure memory effect.

In the recent years, the utilization of biological systems has been became a new method for the synthesis of metal NPs, because it is suitable to develop a low-cost, reliable and eco-friendly route for preparation nano materials [8-10].

In this work, we present a 'green' method for the synthesis of supported Pd NPs using *Pine* needle extract as the reducing agent and the LDHs as the carrier. As expected and shown here, the as-prepared Pd/LDHs complexes could serve as a heterogeneous catalyst, which exhibited excellent catalytic activity for the Suzuki coupling reactions.

2. Experimental

2.1. Preparation of PNE

The sundried *Pine* needles were milled. Then, 2 g of the milled powder was added into 100 ml deionized water and heated at 60 °C for 1 h. The PNE obtained was centrifuged at 4000 rpm and filtered through Whatman No.1 filter paper. The filtrate was collected in an Erlenmeyer flask and stored at 4 °C for further use. The main FT-IR absorption bands of PNE appear at 3415, 2923, 1627, 1515, 1405, 1262 and1074 cm^{-1}. The broad peak at 3415 cm−1 is attributed to the O - H (phenolic) stretch used as a reducing agent and the peaks at1627, 1515, 1405, 1262 and 1074 cm^{-1} are assigned to the stretching vibrations of (NH)-C=O (protein), C=C (aromatic), C-O-C (esters, ethers) and C-OH/C-H (polyols) which can serve as astabilizing agent for protecting the Pd NPs.

2.2. Preparation of Pd/LDHs catalyst
To synthesize PdNPs/LDHs(magnesium-aluminumhydrotalcite) with 3 wt% Pd ($Pd_{0.03}$/LDHs-PNE), LDHs (0.5 g) was dispersed in the Na_2PdCl_4 solution (15 ml, 0.01 M) and impregnated for 12h at room temperature. Subsequently, 22.5 ml of the PNE was added into this precursor mixture with stirring at 60 °C for 3 h.

2.3. Catalytic Suzuki Coupling Reaction
Under air atmosphere, 4-bromotoluene (1 mmol), K_2CO_3(2 mmol) and the catalyst (0.05 mol%) were added into a mixed solvent of 15 ml of EtOH:H_2O (1:1) with constant stirring at room temperature for 10 min to obtain an evenly disperse system. Then, phenylboronic acid (1.5 mmol) was added into the mixture. After completion of the reaction, the obtained products were determined by HPLC analysis.

3. Result and discussion
Figure 1 shows the XRD patterns of the Mg_3Al_2-CO_3-LDHs and $Pd_{0.03}$/LDHs-PNE. The LDHs sample exhibits the characteristic reflections of typical and well-ordered layered double hydroxides with a basal spacing (003) of 0.76 nm, indicating that only CO_3^{2-} ions occupy the gallery region. However, no diffraction peak of Pd species was observed in Figure1b because of the lower Pd loading amount and small Pd particle size.

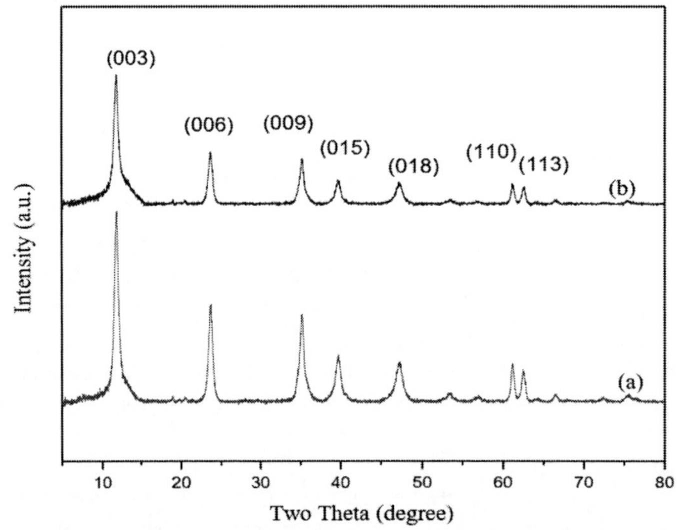

Figure 1. XRD patterns of (a) LDHs, (b) $Pd_{0.03}$/LDHs-PNE

The N_2 adsorption-desorption isotherms for LDHs and $Pd_{0.03}$/LDHs-PNE are shown in figure2. All of the characteristic isotherms correspond to a type IV isotherm with a type H3 hysteresis loop.

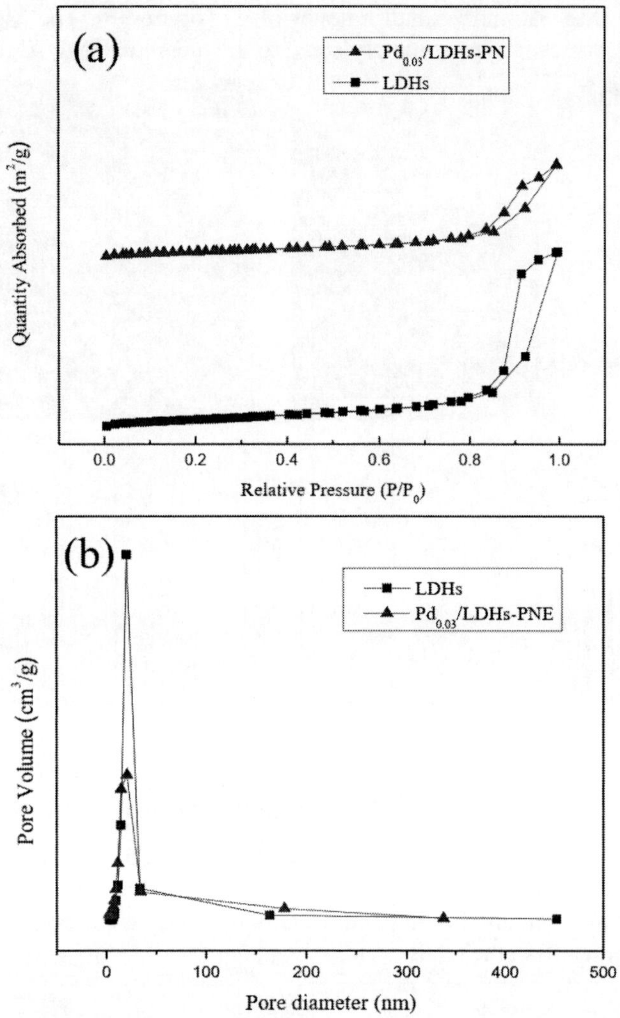

Figure2. N_2 adsorption-desorption isotherms (a) and Pore size distributions curves (b).

As shown in table1, the BET surfaces decreases from175.24 m^2g^{-1}to118.23 m^2g^{-1} and the pore volume also decreases from 20.49 nm to 16.97 nm. The decrease suggests that Pd at least partially located inside the channels of the carrier.

Table1. Textural properties of the LDHs and $Pd_{0.03}$/LDHs-PNE

Sample	BET surface area /m^2g^{-1}	Pore volume /cm^3g^{-1}	Pore size /nm
LDHs	175.24	0.76	20.49
$Pd_{0.03}$/LDHs-PNE	118.23	0.53	16.97

The image offigure 3a shows the nano-laminatedstructure of LDHs. Figure 3c clearly demonstrates that in $Pd_{0.03}$/LDHs-PNE nanocomposite,PdNPs were highlydispersed on the LDHs surface. The size of PdNPs ranges from 1.24 nm-3.11 nm(an average diameter of about 1.75 nm from Figure 3d).EDS was used to identify the elemental composition of Mg_3Al_2-CO_3-LDHs and the EDS spectrumindicated the presence of C, O, Mg, Al and a small amount of Cl (figure 3b). The Mg/Al atomic ratio of the sample was 25.2:15.4, which is approximately equal to the theoretical Mg/Al ratio (3:2).

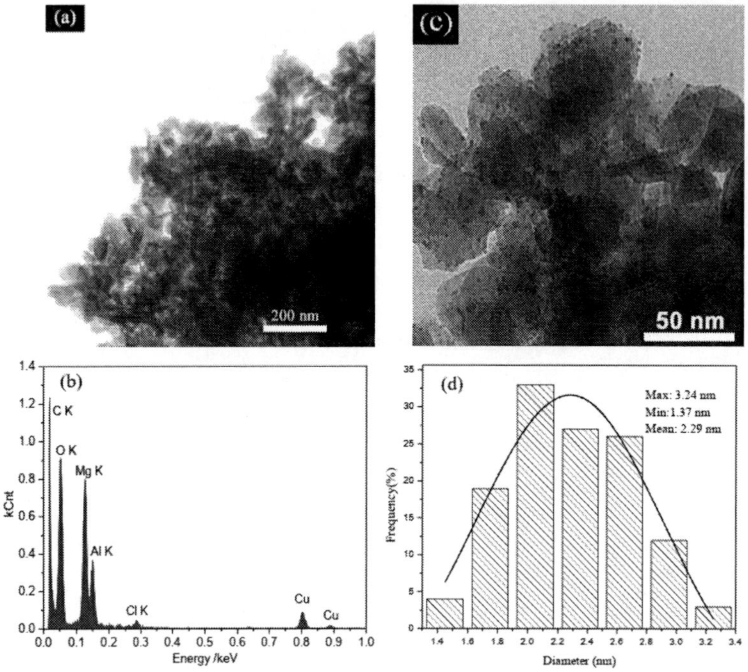

Figure 3. TEM images of LDHs (a),EDS pattern (b),$Pd_{0.03}$/LDHs-PNE (c) and corresponding size distributions (d).

Figure 4a shows the yields at different intervals with the catalyst. The conversion of 4-bromotoluene reached to 99% at 30 min with 0.05 mol% catalyst. This catalyst could be recycled simply by centrifugation. In addition, the yield decreased to 95% (figure 4b) in the eighth cycle, but it decreased to 81% (figure 4b) in the tenth cycle.

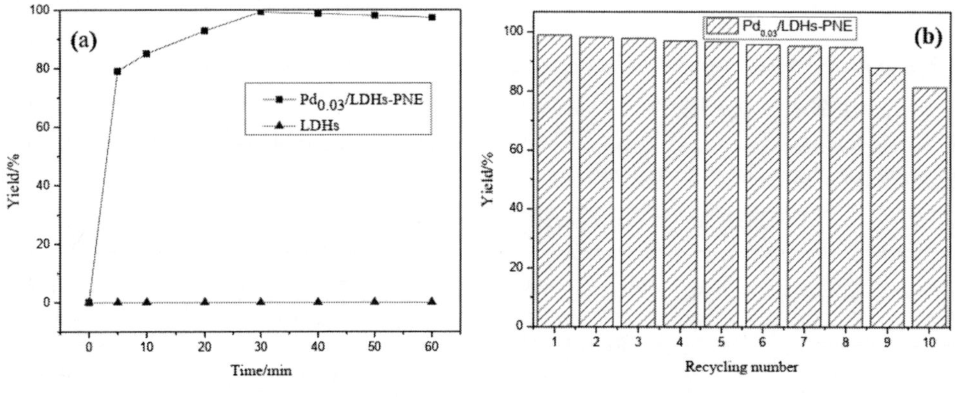

Figure 4. (a) The effect of time in the Suzuki reactionand (b) the reusability of the $Pd_{0.03}$/LDHs-PNE catalyst.

4. Conclusions

In summary, an effective and reusable Pd/LDHs catalyst was designed and successfully prepared by a low-cost and green method using Pine needle extract as the reducing agent and the LDHs as the carrier. The PdNPs with an average particle size of 1.75 nm.The conversion of 4-bromotoluene catalyzed by $Pd_{0.03}$/LDHs-PNE reached 99% with 0.05 mol% catalyst under mild reaction conditions. In addition, the yield of $Pd_{0.03}$/LDHs-PNE nanocomposite was 95% after recycling eight times.This method provides a simple and eco-friendly route for synthesis supported PdNPs, which may find potential application in other nano metal materials.

Acknowledgements
This work was supported by the National Natural Science Foundation of China (No.21676074), Heilongjiang Provincial Science and Technology Projects (ZY17A06) and Science Foundation of Heilongjiang Academy of Sciences (YY2017SH01& ZNBZ2018SH01).

References
[1] Nemamcha A, Rehspringer JL and Khatmi D 2006*J. Phy. Chem. B.* **110**383
[2] Yin L, and Liebscher J 2007 *Chem. Rev.* **107** 133
[3] Molnar A 2011 *Chem. Rev.* **111** 2511
[4] Mastrorilli P, Dell'Anna M M, Rizzuti A,et al. 2015 *Molecules* **20** 18661
[5] Dell'Anna M M, Malia M, Mastrorilli P,et al. 2013 *J. of Mol.Catal.A:Chem.***366** 186
[6] Zhu J F, Tao G H, Liu H Y, et al. 2014 *Green Chem.* **16** 2664
[7] Nalawade P, Aware B, Kadam V J,et al. 2009 *J.Sci.Ind.Res.* **68** 267
[8] Nasrollahzadeh M, Sajadi S M, Rostami-Vartooni A, Alizadeh M and Bagherzadeh M 2016 *J. Colloid Interface Sci.* **466** 360
[9] Rahisuddin, Al-Thabaiti S A, Khan Z, and Manzoor N 2015 *Bioproc. Biosyst. Eng.* **38** 1773
[10] Liu G H, Bai X F and Lv H F 2017 *Appl. Organomet. Chem.* **31** 3587

IOP Conf. Series: Materials Science and Engineering **292** (2017) 012028 doi:10.1088/1757-899X/292/1/012028

SRT research of the crystal defects on the TGT synthesized diamond doped with boron and phosphorus

W L Yu[1, 4], X P Jia[2], Q X Yuan[3] and W X Huang[3]

[1] Yanshan Univ, State Key Lab Metastable Mat Sci & Technol, Qinhuangdao 066004, PR China

[2] National Lab of Superhard Materials, Jilin University, Changchun 130012, PR China

[3] Institute of High Energy Physics, Beijing 100039, PR China

[4] E-mail: wlyu@ysu.edu.cn

Abstract. Crystal defects of two synthesized diamonds doped with boron and phosphorus are compared by synchrotron radiation topography (SRT). The crystal defects in the specimen are mainly dislocations. The dislocations are assembled in bundles in the cone-shape and distributed in the directions approximately towards to <112> and <111>. The Burgers vectors of most of the dislocations are parallel to [202] and oblique to the dislocations, showing that the dislocations are of mixed type. The features of the crystal defects are quite different in several aspects compared with the diamonds without phosphorous doping. The ionic radius of phosphorous is 118.75 percent larger than that of carbon and intensive distortion may occur near the phosphorous ions. The dislocations thus originated in bundles to release the stress caused by the phosphorous doping.

1. Introduction

Diamond is widely used in various areas, such as mechanics, thermotics, electrics, magnetics and optics as a result of the excellent characters it is possessed. Since semiconductor component made from diamond can be used at high temperature, high pressure and strong radiation conditions, its application is wide and fascinating in the semiconductor and microelectronics areas [1-3]. The techniques of the P-type doped diamond are well developed nowadays which are mainly doped with boron but the N-type doped diamond are more difficult so its application is restricted [4, 5]. The atoms of the N-type doped diamond are lithium, sodium, nitrogen, phosphorus, oxygen and sulfur, etc.[6, 7]. Phosphorous is the atom mostly be researched except nitrogen in the N-type diamond. Diamond film doped with phosphorous is widely be considered nowadays but the diamond crystal doped with phosphorous is rarely be concerned, especially the crystals synthesized by HPHT techniques [8, 9]. The diamond crystals synthesized by HPHT techniques doped with phosphorous and boron is studied by SRT in this paper. X-ray topography is a convenient method used to reveal the features of the crystal defects, although some other methods have also been used [10-13]. The features of the defects are revealed with the synchrotron radiation topography and compared with that in the diamonds without phosphorous doping. It is concluded that the unique feature in the specimen tested in the experiments maybe resulted from the doping of phosphorous.

2. Experiments

2.1. Specimens

The specimens D5 and D3, as shown in figure 1 (a) and (b), were synthesized in the State Key Laboratory for Superhard Materials, Jilin University by HPHT temperature gradient growth method. The crystal D5 is a octahedron combined with cube and small faces of rhombododecahedron. Specimen D3 is a cube combined with octahedron and small trapezohedron and rhombododecahedron. The crystal D5 is tawny in color, 3.96mm×3.72mm×1.64mm in dimension and 0.058gram in weight. The specimen D3 is yellowish green, weighted 0.052 gram and 3.84mm×3.25mm×1.72mm in dimension.

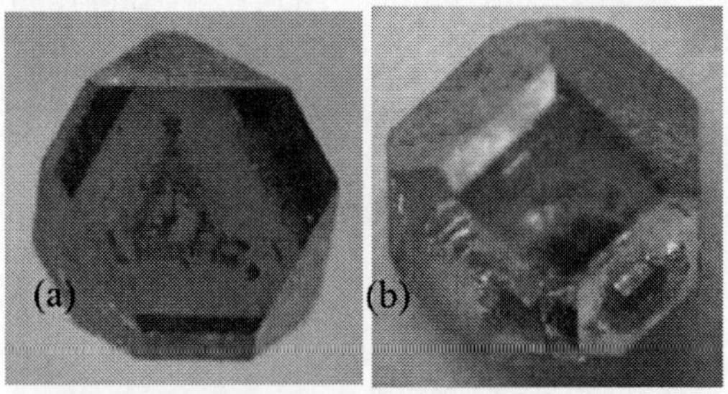

Figure 1. The diamond crystals D5 and D3. D5 is tawny in color, 3.96mm × 3.72mm × 1.64 mm in dimension and 0.058 gram in weight. D3 is yellowish green, 3.84mm × 3.25mm × 1.72mm and weighted 0.052gram.

2.2. Infra-red spectrum experiments

A Tensor 27 infrared spectrometer was used to complete the experiments at The College of Materials Science and Engineering, Yanshan University. The resolution ratio is less than 0.5 cm^{-1}, wave-number ranges from 400 cm^{-1} to 4000 cm^{-1}. Compressed potassium bromide tablet is used as the background. Thirty two times had been scanned for both back ground and the specimen. The equipment was preheated for thirty minutes before the experiments.

2.3. Synchrotron radiation topography experiments

The synchrotron radiation topography experiments were completed at Beijing Synchrotron Radiation Facility (BSRF). The length of beam 4W1 we used was 43m and the maximum receipt angles were 1.0mrad and 0.3mrad in horizon and vertical, respectively. The energy and intensity of the beam were 2.2GeV and 59.6mA. The Laue transmission setting was employed in the experiments. The distance between the specimen and the film was in the range of 25mm and 35mm. The exposure time was 0.5 to 1 seconds. The FUJI50 film was used. In order to avoid ghost images the back surfaces of the films were not developed.

A series of topographs were taken with different incident directions of the synchrotron radiation beam. The specimens were installed on a triaxiality mount. A photograph was taken when [111] of the specimens was parallel to the incident beam and $[11\bar{2}]$ downward. Photographs were then taken after the specimens were turned 15, 30 and 45 degrees clockwise and anticlockwise around the vertical axis.

3. Results and discussions

3.1. The infra-red spectrum experiments

The infrared spectrum diagrams are shown in figure 2. The diamond specimen D3 and D5 belongs to Ib type since absorption bands near 1130 cm^{-1} can be seen on both curves of them [14]. Weak absorptions at 469 cm^{-1} and 893 cm^{-1} in D5 are resulted from the doped boron. The absorption at 661cm^{-1} is induced by the impurities of both boron and phosphorous doped within the diamond. The IR result shows that the diamond D5 is doped with both boron and phosphorous. Compared with D5, the absorption at 661cm^{-1} in curve of D3 is disappeared. It can be concluded that the specimen D3 was doped with boron but without phosphorous doping [15].

IOP Conf. Series: Materials Science and Engineering **292** (2017) 012028 doi:10.1088/1757-899X/292/1/012028

3.2. The synchrotron radiation topography experiments

The synchrotron radiation topographic photos of specimen D5 are shown in figures 3 and 4. The crystal defects in the diamond are mainly dislocations and most of them are distributed in <112> and <111> growing directions. In each area the dislocations are gathered in several bundles and each bundle is composed by dozens of dislocations. The dislocations in a bundle are generated from a same point and extended to cone shaped surface. The vertex of the cone is towards the inner of the crystal and the bottom towards the surface. The angles of the cones are between 20 to 40 degrees. The lengths of the dislocations are between 0.46mm to 0.57mm, average at about 0.5mm. The images of these dislocations are weakened or nearly disappeared in photos (c) and (d) in figure 3, so the product of the diffracted vectors of (c) and (d) in figure 3, $\vec{g}_{\bar{1}11} \times \vec{g}_{\bar{1}\bar{1}1}$= [202], can be a probable estimate of the Burgers vectors of most of the dislocations. The angles between the Burgers vector and the dislocations are bevel so most of them are mixed dislocations.

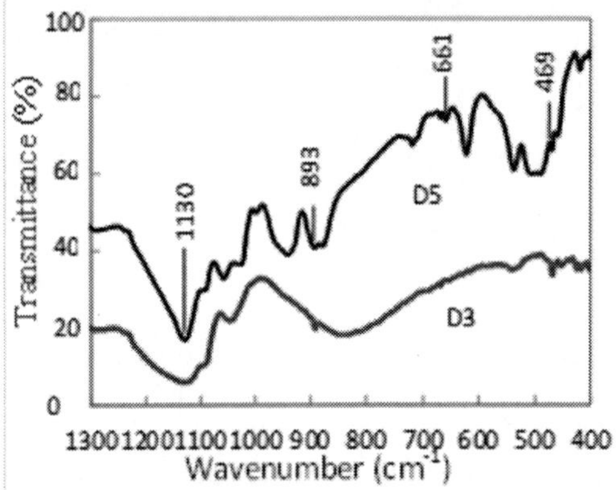

Figure 2. The absorption near 1130 cm-1 indicates that the diamonds D5 and D3 belong to Ib type. Absorptions at 469 cm-1 and 893 cm-1 were resulted from the boron doped in both D3 and D5. The absorption at 661cm-1 was induced by the impurities of both boron and phosphorous in D5

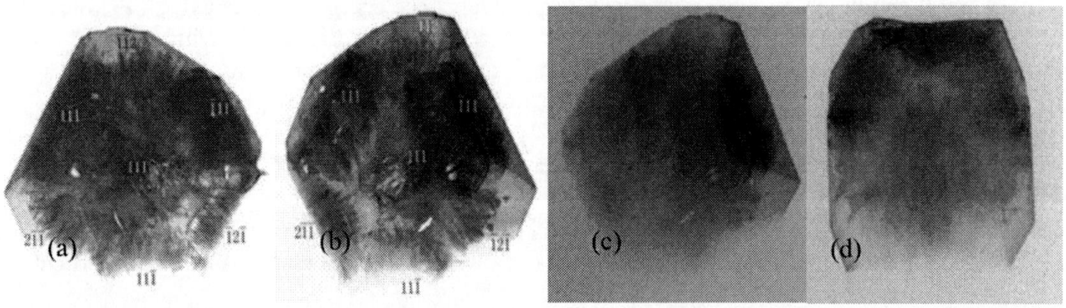

Figure 3. Synchrotron radiation photographs of D5. The diffracted vectors of (a) and (b) are [1$\bar{1}$1] and [$\bar{1}$11]. The dislocations are distributed in the areas of <112> and <111>. The lengths of the dislocations are about 0.5mm. The diffracted vectors of (c) and (d) are [$\bar{1}$11] and [$\bar{1}\bar{1}$1]. In photos (c) and (d), the images of the dislocations are very weak or nearly vanished.

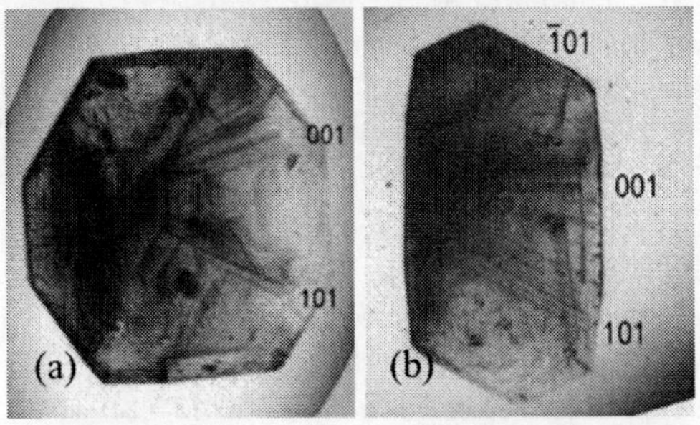

Figure 4. The synchrotron radiation topographic photos of D3 with diffracted vector $\vec{g}_{3\bar{1}1}$ in (a) and $\vec{g}_{202}$ in (b). The dislocations are generated from the seed and extended towards to <100> and <110>. The dislocations are parallel within each bundle.

Compared with D5, the dislocations in D3 are quite different in several aspects. The directions of the dislocations in the diamond D3 are mainly extended to <100> and <110>, while in the specimen D5 are to <112> and <111>. The dislocations in the diamond D3 are generated from the surface of the seed, often, all the dislocations are generated from a unique source, while in D5, numerous sources are observed. The dislocations in D3 are parallel within a bundle, while in D5 the dislocations are scattered in a cone-shaped bundle in the range about 40 degrees.

The differences are resulted probably from the phosphorous doping since the two specimens are synthesized at similar conditions except the phosphorous doping. It could be imagined that the vertex of the cone-shaped dislocation bundle might be the site of the phosphorous ion be doped. The ionic radius of phosphorous and carbon are 0.035nm and 0.016nm, respectively [16]. The ionic radius of phosphorous is about 118.75 percent larger than that of carbon. The crystal lattice near the phosphorous ion doped in the diamond must be strongly distorted. The dislocations bundles are thus generated so that the distortion could be released.

4. Conclusions

Diamond specimens D5 and D3 synthesized by HPHT method with and without phosphorous doping are analyzed with IR and SRT in this paper. The IR experiments show that the specimen D5 is doped with boron and phosphorous while D3 is only boron doped. The crystal defects are revealed by the SRT experiment. In specimen D5, the defects are mainly dislocations which are bundled in cone-shape and situated in the areas of <112> and <111>. The Burgers vectors of most dislocations are parallel to [202], bevel to almost all the dislocations, reveal that the dislocations are of mixed types. Compared with D5, the features of the dislocations in D3 are quite different in several aspects. In D3 almost all dislocations are generated from the seed and grown to <100> and <110>. The phosphorous doping is the most probably reason that resulted all the differences mentioned above. The ionic radius of phosphorous is about 118.75 percent larger than that of carbon, so the crystal lattice is strongly distorted near the phosphorous ions. The dislocations are thus generated around the doped phosphorous ions in order to release the stress brought from the lattice distortion.

Acknowledgment

The SRT experiments were completed at BSRF. The diamond specimens were synthesized by the National Laboratory, Superhard Materials, Jilin University.

References

[1] Govindaraju N and Singh R N *Mater. Sci. Eng. B* 2011 **176** 1058-72
[2] Chicot G, Eon D and Rouger N *Diamond Relat. Mater.* 2016 **69** 68-73
[3] Shakhov F, Abyzov A, Kidalov S, Krasilin A, Lähderantac E, Lebedevd V, Shamshur D and Takaie K *J. Phys. Chem. Solids* 2017 **103** 224-237
[4] Sherehiy A, Dumpala S, Sunkara M, Jasinski J, Cohn R and Sumanasekera G *Diamond Relat. Mater.* 2014 **50** 66-76
[5] Yan B, Jia X, Fang C, Chen N, Li Y, Sun S and Ma H *Int. J. Refractory Metals and Hard Materials* 2016 **54** 309-314

[6] Halliwell S C, May P W, Fox N A and Othman M Z *Diamond Relat. Mater*. 2017 **76** 115-122
[7] Othman M Z, May P W, Fox N A and Heard P J *Diamond Relat. Mater*. 2014 **44** 1–7
[8] Shah J M and Mainwood A *Diamond Relat. Mater*. 2008 **17** 1307-10
[9] Wang J, Li M, Zhang X, Cai X, Yang L, Li J and Jia Y *Tribology International* 2015 **86** 85-90
[10] Khokhryakov A, Nechaev D, Palyanov Y and Kuper K *Diamond Relat. Mater*. 2016 **70** 1-6
[11] Tatsumi N, Tamasaku K, Ito T and Sumiya H *J. Cryst. Growth* 2017 **458** 27–30
[12] Voloshin A E, Baskakova S S and Rudneva E B *J. Cryst. Growth* 2017 **457** 337–342
[13] Masuya S, Hanada K, Oshima T, Sumiya H and Kasu M *Diamond Relat. Mater*. 2017 **75** 155-160
[14] Yu R, Ma H, Liang Z, Liu W, Zheng Y and Jia X *Diamond Relat. Mater*. 2008 **17** 180-184
[15] Hu X, Li R, Shen H, Dai Y and He X *J. Inorg. Mater*. 2004 **19**(4) 895-901
[16] William D. Callister, Jr *Materials science and engineering: an introduction* New York 2000

IOP Conf. Series: Materials Science and Engineering **292** (2017) 012029 doi:10.1088/1757-899X/292/1/012029

Synthesis of nanoparticle emulsion collector HNP and its application in microfine chalcopyrite flotation

G C He[1], J Ding, C H Huang and Q Kang

Faculty of Resource and Environmental Engineering,Jiangxi University of Science and Technology, Ganzhou 341000,China

[1] E-mail: 447671348@qq.com

Abstract. Hydrophobic polystyrene nanoparticles bearing thiazole groups named HNP were used as collectors to improve recovery of microfine chalcopyrite in flotation. HNP adsorbs onto microfine particles selectively, which were modified hydrophobically to induce flotation effectively. Particle size and scanning electron microscope analysis for HNP show that HNP is a spherical nano particles with small size, uniform distribution and good dispersion. Infrared spectrum analysis for HNP proved that functional monomer 2-mercapto styrene acrylic thiazole was bonded chemically onto styrene. Flotation test results indicate that HNP is the right collector of chalcopyrite. Especially, the recovery of chalcopyrite is higher than 95% in neutral and acid media. FTIR results reveal that the flotation selectivity of collector HNP is due to strong chemical absorption onto chalcopyrite surface. Zeta potential analysis shows that the zeta potential of chalcopyrite decreased more quickly after interaction with HNP with the increase of pulp pH value, confirming that collector HNP is an anionic collector. Scanning electron microscope conform that HNP has good selective adsorption on chalcopyrite.

1. Introduction

With the depletion of richer or more easily separation minerals, it is much more necessary than anytime before to utilize refractory ore, especially ultra-fine ones. Much of the undissociated minerals outflow when treating ultra-fine particles, where the efficiency of separation is relatively low. In order to make the fine embedded ores finely liberate, it is necessary to comminute the particles less than 30 μm in diameter. However, most typical mineral processing techniques with traditional agents fail in separating the ultra-fine minerals. It is urgely to find out a new method to recover the ultra-fine ores efficiently [1-4].

A variety of reasons why fine-grained mineral is difficult to choose have been studied by mineral processing workers from different viewpoints [5-7]. The research work mainly includes four aspects: the flotation solution chemistry, the agglomeration flotation technology, flotation equipment, and flotation reagents including collector, regulator, flocculant, and frother were conducted in this paper. However, the reported methods namely carrier flotation, selective flocculation flotation and column flotation cannot be applied to industrial practice at present.

Carrier flotation can increase mineral particle size indirectly by carriers, thereby increasing its collision probability with bubbles. Due to the technological requirements of high strength of dough, strict carrier particle size and high proportion of carrier, it is difficult to implement [8, 9]. Non-selective coagulation among slurry is one of the important obstacles in fine-grained selective flocculation. Pulp dispersion in advance is dispensable before flocculation. The key point to complete separation of minerals and gangue is to find effective flocculant and dispersant. The inevitable ion in the water could produce a great impact on the selective flocculation. Selective flocculation uses reunion between purpose mineral particles to increase mineral particle size, but gangue minerals mixed in the floccules make it difficult to obtain the high grade concentrate [10-13]. Micro-bubble flotation column can produce more small bubbles and then increase collision probability of the mineral and bubble, which has a certain effect on the recycling of micro fine particle minerals. But its large-scale application is limited by some shortcomings such as large elevation difference, complex configuration, easy block of aerator, tumble inside the cylinder of pulp, poor adaptability and stability of the minerals [14, 15].

In this article, starved-feed semi-batch emulsion polymerization was applied to synthesize nanoparticle flotation collector with functional MBT moieties which can adsorb on chalcopyrite selectively. The polymerization of styrene and MBT is shown in figure 1. Addition reaction of C=C double bond is predicted to happen in the polymerization [16-18] effect of the emulsifiers on flotation was studied.

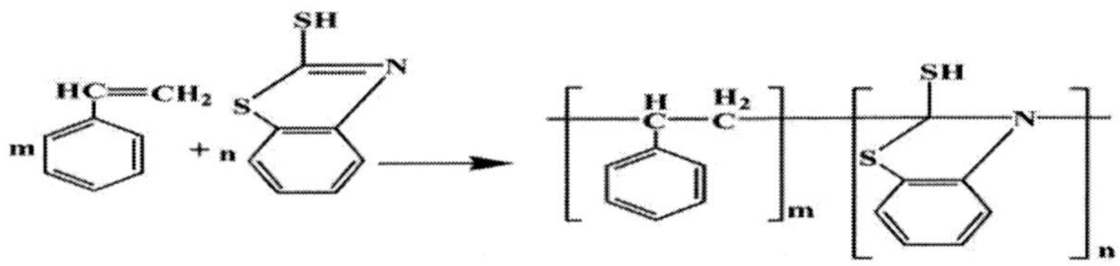

Figure 1. Polymerization reaction equation of styrene and MBT.

2. Materials and methods

2.1. Materials

Chemicals: Styrene (St) with analytic grade were used as monomers. Sodium dodecyl sulfate (SDS), polyethylene glycol and sodium lauryl sulfate with analytic grade were used as emulsifiers. Ammonium persulphate (APS) with analytic grade was used as initiators. 2-mercapto benzothiazole (MBT) with analytic grade were used as functional monomer that can be copolymerized into the composite particle. Styrene was purified further by vacuum distillation. water used throughout the experiment was deionized.

Mineral: Pure mineral purchased from Dexing copper mine had a Cu grade of 33.40%. Note that the Cu grade of 100% pure chalcopyrite is 34.56. The purity of quartz is 99.40%, and the purity is 95.96%. The minerals were ground to less than 23 μm with ceramic ball mill.

2.2. Synthesis of nanoparticle collector

The polystyrene-co-SDS latex was prepared by monomer-starved semi-batch emulsion polymerization. The polymerization was conducted in a three-neck flask equipped with a condenser, two rubber stoppers holding syringe needles (one for nitrogen, the other for monomer addition). The flask was heated in a water bath with magnetic stirring, 100 mL deionized water containing 0.08 g of SDS and 0.5 g of styrene were charged into the flask, followed by nitrogen purging for 30 min at 70 ℃ with 400 r/min stirring. 10 mL of water containing 0.025 g of APS initiator was injected into the reactor to initiate the polymerization 0.125 g of MBT, dissolved in 4.9 g of water and an additional 4.5 g of styrene were added for 5 h. The reaction was stirred at 70 ℃ with 400 r/min for another 19 h. The resulting latex was dialyzed for at least 5 days against deionized water, after which the dialysate conductivity was less than 20 μS/cm [19, 20].

2.3. Scanning electron microscope measurement

The nanoparticles and flotation concentrate were characterized by scanning electron microscope (SEM, MLA650F,FEI) to obtain the surface morphology of nanoparticles and its adsorption form on the mineral surface.

2.4. Particle size analysis
Particle size of nano latexes was measured by Malvern Zetasizer NanoZS laser particle size analyzer. 5 ml emulsion was filled in the tank with 200 ml distilled water to prepare the solution. Measurement angle is 90°, and laser source is He-Ne at a wavelength of 633 nm.

2.5. Infrared spectra measurement
The infrared spectra of the samples were measured using a ALPHA Fourier transform infrared spectrometer. Sample and KBr with the ratio of 1:100 were mixed and ground in an agate mortar. The mixture which formed a transparent sheet by a pressure mold was put on the sample holder and measured. The sample wavenumbers range was from 400 to 4000 cm^{-1}.

2.6. Flotation experiment
In a conventional flotation experiment, 2 g of chalcopyrite washed by ultrasonic for 5 minutes to remove oxides on the mineral surface was added into a 35 mL effective volume microflotation cell in each test. The washing solution of samples was decanted, and the remaining part was floated for at least 3 min. Then, an appropriate amount of the synthesized HNP collector was added to the slurry while agitating at about 1500 rpm followed by a period of 3 min for slurry conditioning. At the end of conditioning, air was fed and the concentrate was collected for 6 minutes. The concentrates and tailings were filtered, dried, weighted and calculated successively for recovery.

2.7. Zeta potential measurement
Zeta potential was measured by means of ZetaProbe (Colloidal Dynamics). The pure minerals were ground to be smaller than 5 μm in the agate mortar. 2 g mineral was added to a beaker with 50 mL distilled water, then the mineral was cleaned ultrasonically for 5 min. Next, 1 ml emulsion was added into the solution, which was stirred for 3 minutes. Finally, pH of the suspension was regulated to a desired value with HCl or NaOH solution which was ready for measurement.

3. Results and discussion

3.1. Nanoparticle emulsion properties

3.1.1. Particle size and scanning electron microscope analysis for HNP. Shape and surface characteristics of the nanoparticles exert a great influence on its properties. Scanning electron microscope (figure 2) illustrates that size of the HNP is small, the particles are dispersed uniformly. Particle size distribution (figure 3) shows that the mean particle size of HNP is 77 nm, which is in agreement with the results of SEM.

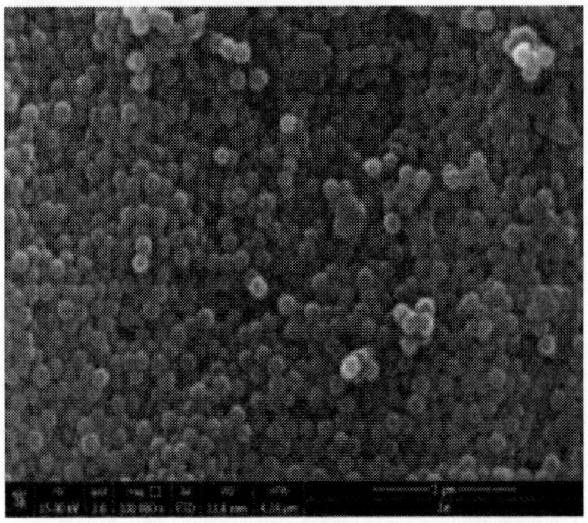

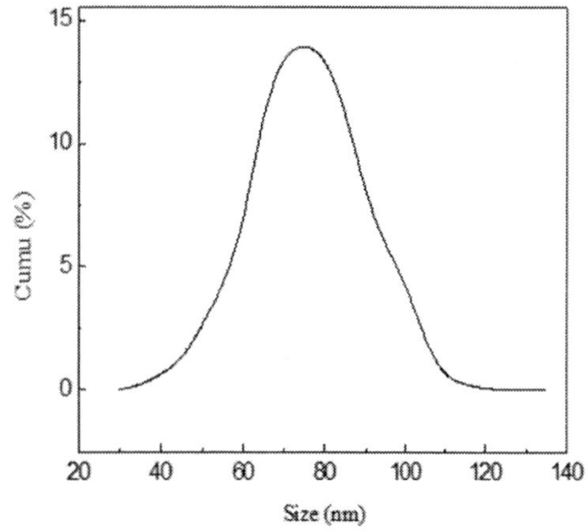

Figure 2. Scanning electron microscope for HNP. **Figure 3.** Particle size distribution for HNP.

3.1.2. Infrared spectrum analysis for HNP. Figure 4 shows the infrared spectra for the HNP. The absorptions at 3026 cm^{-1} and 698 cm^{-1} are attributed to the stretching vibration mode of —CH— and —CS— groups, respectively. The absorption at 1674 cm^{-1} which is assigned to the phenyl ring skeleton and the stretching vibration absorption at about 1600 cm^{-1}, 1500 cm^{-1}, 1450 cm-1 which are appeared in three different unsaturated —C=C— group suggest the existence of benzene ring. The absorption between 2925 and 2840 cm^{-1} are for the asymmetric stretching vibration of —CH_2— group. The absorption at 1027 cm^{-1} is stretching vibration absorption of —C-N— group. Absorption ranging from 1400 to 1430 cm^{-1} belong to NCS ring of MBT which are not observed indicates that —C=N— double bond in MBT cracks during the polymerization process. The absorption ranging from 880 to 680 cm^{-1} that change with the number and position of substituents on the phenyl ring are attributed to outside plane bending vibration of —CH— in alkene, which reveals the substituted degree and configuration on benzene ring. The above results confirm the reaction between styrene and MBT.

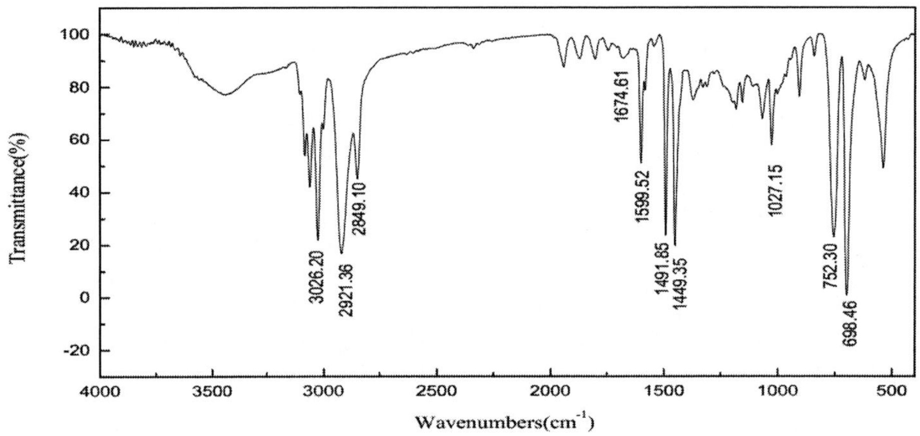

Figure 4. Infrared spectrum of HNP.

3.2. Flotation test

3.2.1. HNP dosage. Tests of the dosage of HNP at natural pH (pH=6) on the flotation efficiency of chalcopyrite, pyrite, quartz respectively were conducted and results are shown in figure 5.

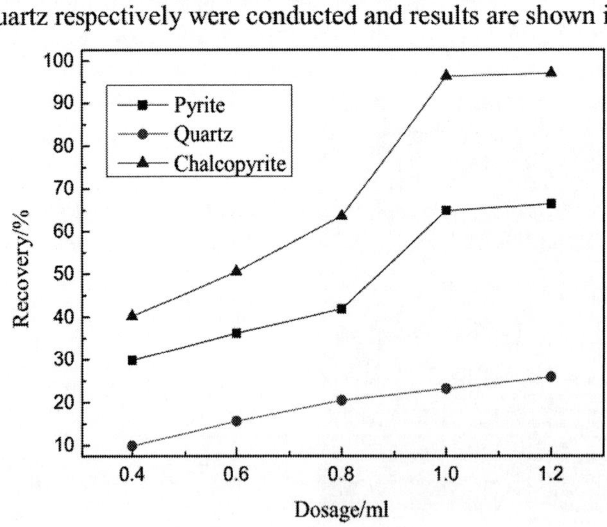

Figure 5. The relation between HNP dosage and recovery ratio.

It can be seen from figure 5 that in a certain range, with the increase of the dosage of HNP, the recovery of three kinds of mineral increased gradually, and recovery of chalcopyrite and pyrite grow much faster than that of quartz. Recovery of chalcopyrite increases from 40.21% to 96.32% rapidly with increasing dosage of HNP from 0.4 ml to 1.0 ml, and it doesn't continue to increase as the dosage of HNP goes up further. Recovery of pyrite increases from 29.87% to 66.44% with increasing dosage of HNP from 0.4 ml to 1.2 ml. Recovery of quartz increases from 9.89% to 25.88% slowly as the dosage of HNP rises up from0.4 ml to 1.2 ml. Comparison of the recoveries among the three kinds of minerals shows that the optimum concentration with a higher chalcopyrite recovery and a lower pyrite and quartz recovery appears in the dosage of HNP of 1.0ml so that good separation of chalcopyrite can be achieved.

3.2.2. pH value condition. The pulp pH value has a substantial influence on the mineral recovery rate in the sulfide minerals flotation. When amount of HNP is 1ml, the influence of pulp pH value on microfine particle chalcopyrite flotation efficiency was shown in figure 6. The pH values were adjusted with hydrochloric acid and sodium hydroxide respectively.

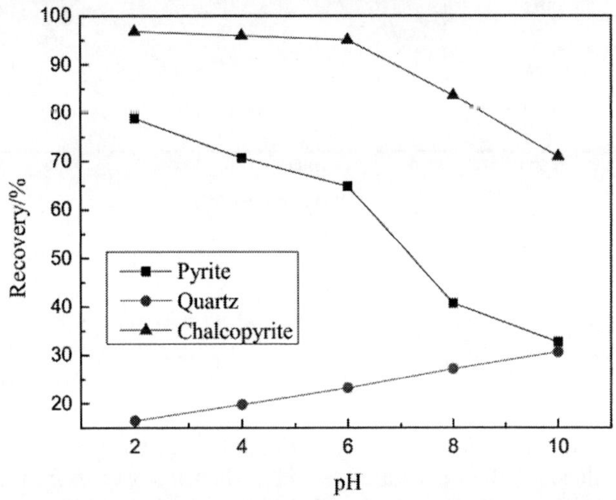

Figure 6. The influence of the pulp pH on recovery ratio.

It can be seen that, with the increase of pulp pH, chalcopyrite recovery decreases slowly at first and then dropped dramatically, and pyrite recovery decreases quickly in the pulp pH value set, while quartz recovery rate rise slowly. Chalcopyrite can be well floated in acid and neutral media, and the recovery amount are bigger than 95.0% when the pulp pH≤6. Recovery of pyrite reduce from 78.82% down to 32.70% with the pulp pH increases from 2 to 10. The maximum recovery ratioof quartz is 30.69% at pH=10. Then, pH 6.0 is the best result, giving 95.02% chalcopyrite recovery, 64.86% pyrite recovery and 23.20% quartz recovery.

The above flotation test results show that without depressant and frother, HNP shows well selectivity to chalcopyrite under neutral and weak acid conditions at dosage of 1 ml.

3.3. Interaction mechanism of HNP and minerals

3.3.1. Infrared spectra of chalcopyrite before and after interaction with HNP. The absorption mode between HNP and minerals can be analyzed by comparison of infrared spectrum of chalcopyrite before and after interaction with HNP. As shown figure 7 the infrared spectra of chalcopyrite treated with HNP changed obviously s. The characteristic peaks at 3025cm^{-1} and 697cm^{-1} of —CH— and —CS— stretching vibration absorption, respectively and at 2921.97cm^{-1} and 2851cm^{-1} of —CH_2— asymmetric stretching vibration as well as at 754cm^{-1} of —CH— bending vibration absorption of HNP appeared. It can also be seen that the wavenumber in 1516 cm^{-1} of —C=C— stretching vibration absorption peak shifts to 1492cm^{-1} leading to a 24 cm^{-1} displacement. The shift or disappear of corresponding peaks in the infrared spectra of chalcopyrite treated with HNP prove that chemical absorption occurs between HNP and chalcopyrite.

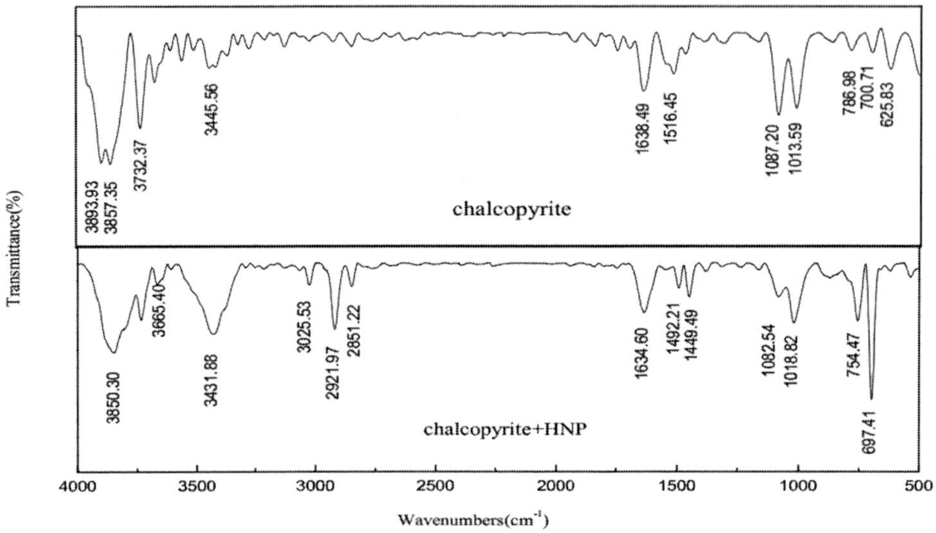

Figure 7. Infrared spectra of the chalcopyrite before and after interaction with HNP.

3.3.2. The zeta potential before and after chalcopyrite with HNP. Zeta potentials of chalcopyrite and chalcopyrite interaction with HNP with respect to pH value are shown in figure 8. The figure shows that chalcopyrite surface zeta potential showed a downward trend along with the solution pH increases. Isoelectric point of chalcopyrite changes from 5.1 to 4.3 after been adsorbed by HNP, which shows that the nanoparticles emulsion is a kind of anionic collector because of the anionic emulsifier SDS used in emulsion synthesis. When pH value is lower than isoelectric point, mineral charges positive and zeta potential of chalcopyrite adsorbed with HNP decreases, due to the electrostatic force. When the pH value is higher than isoelectric point, mineral turned to be negative and zeta potential of chalcopyrite adsorbed with HNP still decreases, which indicates that other effects such as hydrogen bond or chemical adsorption exist.

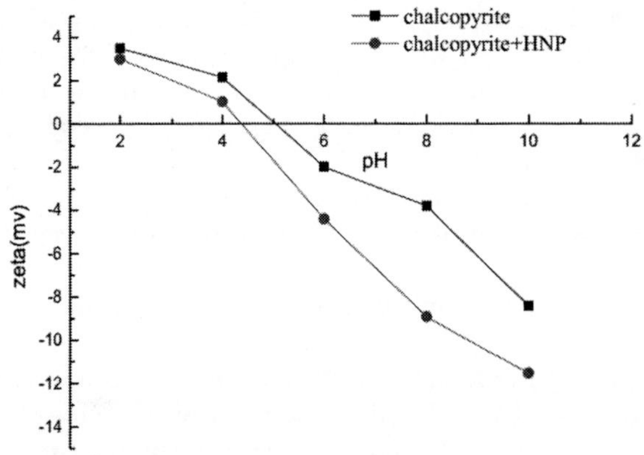

Figure 8. Relationship between zeta potential of chalcopyrite and pulp pH value.

3.3.3. Scanning electron micrographs of minerals with HNP. SEM photos of the mineral flotation concentrates and tailing in HNP system are shown in figure 9. It can be seen that there is a large number of HNP attached in chalcopyrite concentrate while very few in chalcopyrite tailings. The HNP attached in quartz concentrate can be ignored. There are some HNP in the pyrite concentrate but less than it in chalcopyrite concentrate. This shows that HNP display well adsorption with chalcopyrite which increase its flotability. The result of adsorption capacities of HNP on chalcopyrite, pyrite and quartz is proportional to the recovery ratioes in flotation.

a SEM of chalcopyrite concentrate with HNP b SEM of chalcopyrite tailing with HNP

c SEM of quartz concentrate with HNP d SEM of pyrite concentrate with HNP

Figure 9. SEM of the mineral with HNP.

4. Conclusions

In summary, the HNP prepared by emulsion polymerization showed that nanoparticle flotation collectors can promote microfine chalcopyrite flotation. The specific conclusions from this work are as following:

(1) Particle size and scanning electron microscope for HNP declared that during the emulsion polymerization, spherical nano particles were obtained with small size, uniform distribution and good dispersion.

(2) Infrared spectrum analysis for HNP illustrated that functional monomer MBT was bonded to styrene chemically.

(3) Flotation experiments showed that the HNP has well selectivity to chalcopyrite at acid and neutral media. With addition of 1 ml HNP, the recovery ratio of chalcopyrite is 96.32% at pH=6.

(4) Zeta potential experiments demonstrated that, after interaction with HNP, zeta potential values of chalcopyrite became more negative. The results indicated that the HNP belongs to anionic collector adhesive to minerals surface.

(5) Infrared spectrum demonstrates that the adsorption form of HNP onto chalcopyrite surface is chemical absorption, which result in the flotation selectivity of HNP to chalcopyrite.

(6) Scanning electron microscope confirmed that HNP has strong and selective adsorption on chalcopyrite.

Acknowledgements
The study was financially supported by the projects of Natural Science Foundation of China (No. 51364013).

References

[1] Yao W, Li Ml and Cui R 2015 Separation technology of fine minerals *Modern Mining* **1** 66-69
[2] Ai GH and Li XB 2011 Research status and prospect of fine wolframite beneficiation *Mining and Processing Equipment* **39(10)** 90-95
[3] Cai ZB and Xu HH 2015 Research status and Prospect of mineral processing technology of fine grain *Modern Mining* **11** 74-76,80
[4] Miettinen T, Ralston J 2010 The limits of fine particle flotation *Minerals Engineering* **23(5)** 420-437
[5] Nguyen A V, George P, Jameson G J 2006 Demonstration of a minimum in the recovery of nanopaticles by flotation: theory and experiment *Chemical Engineering Science* **61(8)** 2494-2509
[6] Joanne V and Stearnes B E 2000 Fine particle flotation and the influence of dissolved gas on interparticle interactions Australia:*Ian Wark Research Institute*
[7] Qiu Gz,Hu YH and Wang DZ 1993 Particle interaction and fine particle flotation Changsha:*Central South University of Technology Press* 234-345
[8] Zhang Y Y, Zhu Y Y and Zhang GF 2009 Autogenous carrier flotation of fine ilmenite *Chinese Journal of nonferrous metals* **19(3)** 554-560
[9] Guo JB 2003 Study on carrier flotation of hematite in East Anshan *Mining and metallurgical engineering* **23(3)** 29-31
[10] Zhang XP, Hu YH and Huang HJ 2007 Aggregation behavior of ultrafine kaolinite in aqueous media *Journal of China University of Mining and Technology* **36(4)** 514-517
[11] Espinoza O O, Song S and Lope'z VA 2003 Regrinding and floc-flotation of silver sulphide scavenger concentrate *Mineral Processing and Extractive Metallurgy* **112(2)** 90-94
[12] Song SX and Alejandro L V 2002 Parametric aspect of hydrophobic flocculation technology *Mineral Processing and Extractive Metallurgy Review* **23(2)** 101-127
[13] Zygmunt S and Izabela P 2004 Agglomerate flotation of fine oxide particles *International Journal of Mineral Processing* **74**(11) 85-90
[14] Liu J T, Wang Y T and Cao Y J 2006 Research status and development trend of flotation column technology *Coal Preparation Technology* **5** 25-28
[15] Huang G Y, Feng Q M and Ou L 2009 Recovery of fine Scheelite from flotation tailings with micro bubble flotation column *Journal of Central South University: Natural Science Edition* **40(2)** 263-267
[16] Wang X H, Gao B J and Lei H B 2012 Preparation of Functional Grafted Particles PVI-SiO2 and Preliminary Study of Their Adsorption Characters *Journal of Functional polymers* **25(2)** 127-134.
[17] Yang G, Yang G W and Xu H 2004 Flotation Mechanism and Molecular Design of Mercaptobenzo Flotation Reagents acta *chimica sinica* **629(2)** 153-159
[18] WANG YJ 2013 Functional polymeric materials for uranyl ions adsorption and determination Suzhou: *Soochow university*
[19] Songtao Y, Robert P and Adam R 2011 Nanoparticle flotation collectors: mechanisms behind a new technology *Langmuir* **27(17)** 10438–10446
[20] Songtao Y, Robert P and Carla A 2013 Towards nanoparticle flotation collectors for pentlandite separation *International Journal of Mineral Processing* **123** 137-144

IOP Conf. Series: Materials Science and Engineering **292** (2017) 012030 doi:10.1088/1757-899X/292/1/012030

Preparation of Ag superhydrophobic surface on metal substrates

J Y Li, S X Lu[1], W G Xu, Y Q Duan, X C Yang, Y Y Cheng, G He and S Cui

School of Chemistry and Chemical Engineering, Beijing Institute of Technology, Beijing 100081, P.R. China

[1] E-mail: shixianglu@bit.edu.cn

Abstract. In this work, the facile approaches are developed for preparation the Ag superhydrophobic surfaces (SHSs) on zinc (Zn), copper (Cu) and aluminium (Al) substrates. The water contact angles (WCAs) of the Ag SHSs on Zn, Cu and Al substrates are 167°, 165° and 154°, respectively. Furthermore, the water sliding angle (WSA) of each surface is less than 1°. The morphology and chemical composition of the samples are characterized using scanning electron microscopy (SEM) and X-ray diffraction pattern (XRD). The as-prepared three kinds of SHSs possess the self-cleaning performance, which can quickly take the chalk away when the water droplets fall down the SHSs. In addition, the superhydrophobicity of the SHSs can well maintain after exposure to the air for 6 months, indicating that the surfaces can sustain good stability.

1. Introduction

With the advent of new technologies, superhydrophobic surfaces (SHSs) have been constructed as interfacial materials, which exhibit high static water contact angles (WCAs) and ultra-low water sliding angle (WSA) [1]. Especially, SHSs have attracted tremendous attention due to their variety of special applications in the fields of self-cleaning, anti-icing, anti-corrosion, anti-biofouling, drug delivery and water–oil separation, *etc* [2-5]. There are numerous ways to construct superhydrophobic materials such as anodization [6], electrodeposition [7], sol-gel method [8], layer-by-layer deposition [9] and immersion process [10]. Generally speaking, on account of its facileness, low cost and suitability for industrial applications, immersion process is considered as an effective technique to prepare the SHS. Thus far, substrates, such as glass, polymers, metals, waxes, and semiconductors, have been investigated for SHSs [11, 12]. The fabrication of SHSs on various metal substrates has attracted increasing attentions due to the manageable morphology and technological applicability of the metallic SHSs [13]. Furthermore, as a widely used engineering material, silver (Ag) is attractive because of its high electrical and thermal conductivity, malleability and mechanical workability, stability and bactericidal property. Therefore, it is meaningful to construct Ag SHSs on metal substrates for practical application of metals [14].

In this article, we successfully fabricated the Ag SHSs on zinc (Zn), copper (Cu) and aluminum (Al) substrates via etching, followed on replacement deposition and then annealing. The WCA of each substrate could reach to 150° and the WSA of each sample was less than 1°. Moreover, the prepared SHSs exhibited self-cleaning property and long-term stability.

2. Materials and methods

2.1. Reagents and Materials

Silver nitrate ($AgNO_3$, 99.8%), sodium hydroxide (NaOH, 99.5%), nitric acid (HNO_3, 65~68 wt%), ammonia (NH_3•H_2O, 25~28%), acetone (C_3H_6O, 99.5%), ethanol (C_2H_5OH, 99.7%), sodium chloride (NaCl, 99.5%) were purchased from Beijing Fine Chemical Co. Ltd. China without further purification. Zn, Cu and Al sheets (99.9%, 1.0 cm×1.0 cm×0.1 cm) were obtained from Beijing Nonferrous Metal Research Institute.

2.2. Experimental sections

2.2.1. Fabrication of the Ag SHS on Zn substrate. Firstly, the Zn sheet was ultrasonically cleansed by immersion in ethanol and acetone for about 5 min, respectively, to remove the pollutants of the substrate. Then, the cleaned sheet was ultrasonically rinsed with deionized water and dried at ambient temperature. After that, the dried Zn sheet was immersed in 0.01 mol•L^{-1} $AgNO_3$ aqueous solution for 30 min, and finally, annealed at 200 °C for 1 h.

2.2.2. Fabrication of the Ag SHS on Cu substrate. In the first place, the Cu sheet was ultrasonically cleansed by immersion in ethanol and acetone for about 5 min, respectively, to remove the pollutants of the substrate. Then, in order to get rid of the oxide film, the cleaned sheet was ultrasonically cleansed in 26.7 wt% HNO_3 aqueous solution and rinsed with deionized water and ethanol. After dried at ambient temperature, the Cu sheet was immersed in 0.01 mol•L^{-1} $AgNO_3$ aqueous solution for 30 min, and finally, annealed at 180 °C for 12 min.

2.2.3. Fabrication of the Ag SHS on Al substrate. Firstly, the Al sheet was ultrasonically cleansed by immersion in ethanol and acetone for about 5 min, respectively, to remove the pollutants of the substrate. Then, the cleaned sheet was etched with 3 mol•L^{-1} NaOH aqueous solution for 1 min to remove the oxide film and then ultrasonically rinsed with deionized water for 2 min. After that, the dried Al sheet was immersed in the mixture of 100 mL 0.01 mol•L^{-1} $AgNO_3$ aqueous solution containing 3.6 mL NH_3•H_2O for 9 min. And NH_3•H_2O is used as a ligand combined with Ag^+ to change the potential of Ag^+ reduction as well as control the pH of the solution. Finally, annealed at 160 °C for 20 min.

2.3. Characterization

The WCAs and WSAs of the samples were measured with a remote computer-controlled goniometer system (FTÅ 200, Data physics Inc, USA). All WCAs of the samples were measured five times on different positions with 8 μL water droplets at ambient temperature. The measuring error of the WCA value was ± 1°. The surface morphologies of the samples were characterized by a scanning electron microscope (SEM, S-4800, Hitachi, Japan). The surface composition of the samples was measured by X-ray powder diffractometer (XRD, D8 Advance, Bruker, Germany) with Cu Kα radiation (40 Kv, 40 Ma, and $\lambda = 0.15418$ nm) at a scanning rate of 3° min^{-1}.

2.4. Self-cleaning performance test

The self-cleaning property of samples was studied by observing the phenomena of the water droplets falling down the sample surfaces covered with chalk.

3. Results and discussion

3.1. Surface morphology and wettability

The surface morphologies of the sample surfaces were investigated by SEM as shown in figure 1, and the insets are the WCAs of the relative samples. Figure 1a, b and c reflect the sample morphologies of the bare Zn, Cu and Al with the relatively smooth surfaces, even though some scratches exist and the

WCAs of them are 54°, 76° and 65°, respectively. Figure 1a', b' and c' display the SEM images of the Ag SHSs on Zn, Cu and Al substrates, respectively. It can be seen that there are numerous micro/nano-structures caused by the acid/alkaline cleaning, deposition of Ag from the replacement reaction and the annealing process. The WCAs of the corresponding Ag SHSs are 167°, 165° and 154°, respectively. Similarly, in ref [13], they measured the value of WCA of the Ag SHS is 164°. But, in ref [15], the Ag coating is superhydrophilic. The similarity or different are due to the roughness and surface composition are appropriate or not. In summary, the rough surfaces, coupled with the appropriate annealing process play the crucial roles in effecting wettability of the SHSs. Furthermore, in the true sense, a SHS should not only have a large static contact angle, but also a small sliding angle. Undoubtedly, the as-prepared each Ag SHS with the WSA less than 1° exhibits non-sticking behavior, which allows droplets to slide easily when the water droplets are placed on the surface.

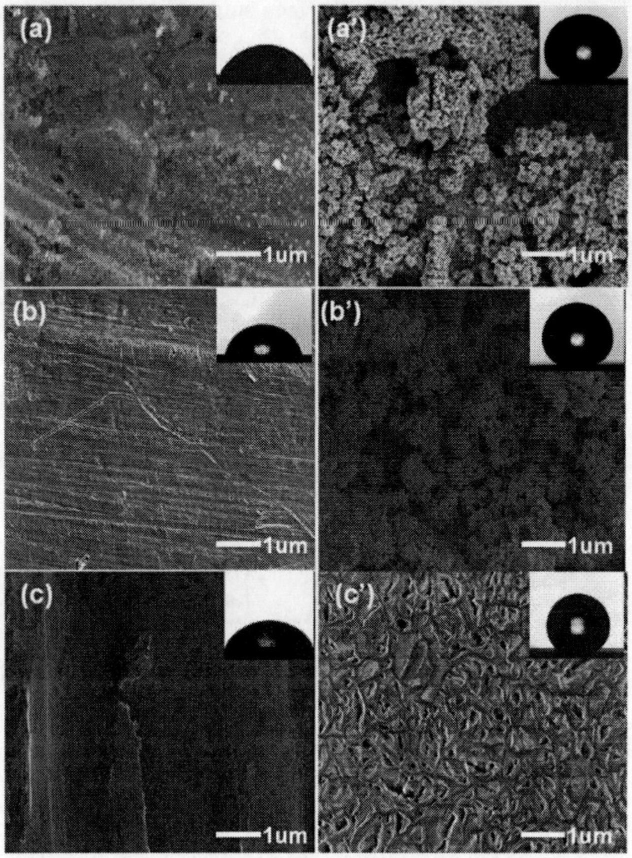

Figure 1. SEM morphologies of different samples: (a) pure Zn sheet; (a') Ag SHS on Zn substrate; (b) pure Cu sheet; (b') Ag SHS on Cu substrate; (c) pure Al sheet; (c') Ag SHS on Al substrate. The insets are the WCAs of relative samples, respectively.

Two models can account theoretically for the wetting properties. One is the Wenzel model [16], in which water drops will penetrate the grooves of the rough surface. However, Cassie *et al.,* through the studies of superhydrophobic phenomena in nature, found that when the water droplets are in contact with the solid surface, the water droplets cannot fully penetrate the grooves of the rough structure. Thus, the Cassie–Baxter model, where the SHS is regarded as a porous medium composed of air pockets, is always used as the calculation model [17]. The Cassie–Baxter state can be defined as follows:

$$\cos\theta^* = f_1 \cos\theta - f_2 \tag{1}$$

θ^* (the WCA of the SHS) is the apparent contact angle and θ (the WCA of pure metal substrate) is the intrinsic contact angle of the composite surface. f_1 is the fractional interfacial area of solid/water on the surface, f_2 is the fractional area of air/water and $f_1 + f_2 = 1$. The WCAs of the Ag SHSs on Zn, Cu and Al are all over 150° and the value of WCA of pure Ag surfaces is 79° [18]. According to Eq. (1), it can be obtained that the average value of f_1 is 0.112. This means that only less 11.2% of the area of water droplet directly contacts with the surface of the sample and the remaining more than 88.8% is in contact with air. Thus, the water droplets cannot penetrate the grooves, indicating the prepared SHSs are sufficiently rough and covered with the Ag micro/nano-structured.

3.2. XRD analysis

It is well-known that the wetting behavior of solid surfaces is governed by both the surface morphology and the chemical composition. The surface substance and crystal structure of samples are assessed by XRD spectra, as shown in figure 2.

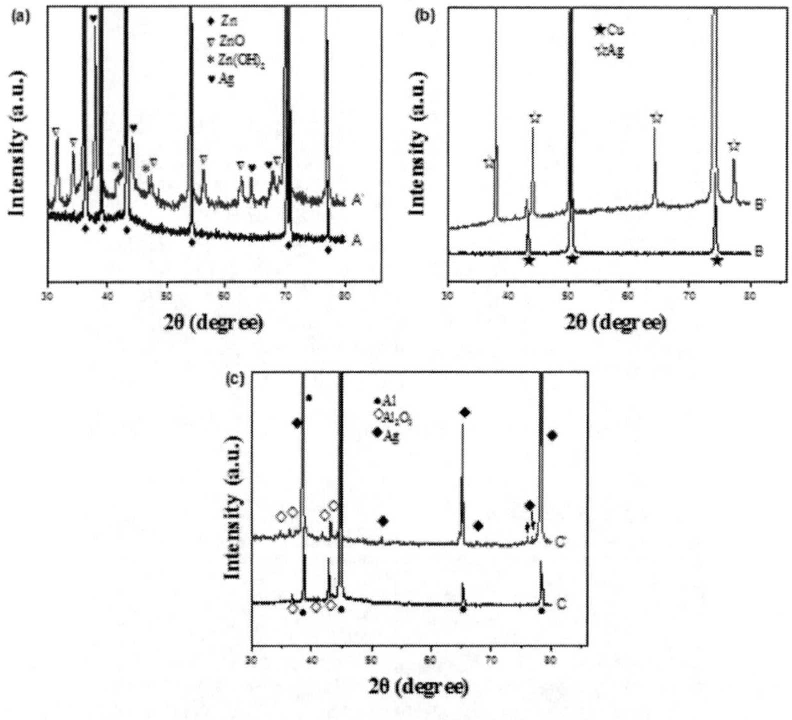

Figure 2. XRD spectra (a): pure Zn (A) and Ag SHS on Zn substrate (A'); XRD spectra (b): pure Cu (B) and Ag SHS on Cu substrate (B'); XRD spectra (c): pure Al (C) and Ag SHS on Al substrate (C').

The region of the 2θ is 30~80° and the scanning speed is 3°•min-1. The XRD spectra of pure Zn (A) and Ag SHS on Zn substrate (A') are shown in figure 2a. The symbols of • , , * and ♥ represent the characteristic peaks of Zn, ZnO, Zn(OH)2 and Ag, respectively. From curve (A), the sharp peaks mainly are attributed to the diffraction peaks of Zn. As shown in curve (A'), the peaks of ZnO and Zn(OH)2 appear, which indicates that some of Zn was oxidized in the reaction process. In addition, the sharp peaks showed at 2θ = 37.9°, 44.2°, 64.5° and 67.5° are attributed to the diffraction peaks of Ag(111), Ag(200), Ag(110), and Ag(106) (JCPDS card no.87-0598), which indicates that the Ag was successfully deposited on Zn substrate. In addition, the XRD spectra of pure Cu (B) and Ag SHS on Cu substrate (B') are shown in figure 2b, the symbols of ★ and ☆ represent the characteristic peaks of Cu and Ag, respectively. The three diffraction peaks at 2θ = 43.31°, 50.41° and 74.11° can be indexed to the Cu (JCPDS Card No. 04-0836) corresponding to Cu(111), Cu(200) and Cu(220), respectively from curve (B). And after the fabrication process as shown in curve (B'), the diffraction peaks of Ag show up, which indicates that the Ag was successfully deposited on Cu substrate. In the same way, the XRD spectra of pure Al (C) and Ag SHS on Al substrate (C') are shown in figure 2c. The symbols of ●, ◇ and ◆ represent the characteristic peaks of Al, Al2O3 and Ag, respectively. There are only some peaks of Al and Al2O3 on pure Al surface as shown in curve C. But after the reaction as shown in curve C', the peaks at $2\theta = 37.04°$, 51.67°, 64.52°, 67.5°, 76.01°, 76.79° and 78.89° are attributed to the diffraction peaks of Ag(101), Ag(104), Ag(110), Ag(106), Ag(200), Ag(201) and Ag(202) (JCPDS card no. 87-0598). In summary, the results indicate that the hierarchical micro/nano coatings are successfully prepared on the three substrates.

Through the analysis of XRD of the SHSs, the chemical reaction equations for fabricating the three SHSs can be obtained and showed in scheme 1-3, respectively.

$$2Zn + O_2 = 2ZnO \quad (1)$$
$$ZnO + H_2O = Zn(OH)_2 \quad (2)$$
$$Zn + 2AgNO_3 = 2Ag + Zn(NO_3)_2 \quad (3)$$

Scheme 1 The reactions for fabricating Ag SHS on Zn substrate.

$$2Cu + O_2 = 2CuO \quad (4)$$
$$CuO + 2HNO_3 = Cu(NO_3)_2 + H_2O \quad (5)$$
$$Cu + 2AgNO_3 = 2Ag + Cu(NO_3)_2 \quad (6)$$

Scheme 2 The reactions for fabricating Ag SHS on Cu substrate.

$$4Al + 3O_2 = 2Al_2O_3 \quad (7)$$
$$Al_2O_3 + 2NaOH = 2NaAlO_2 + H_2O \quad (8)$$
$$3AgNO_3 + Al = 3Ag + Al(NO_3)_3 \quad (9)$$

Scheme 3 The reactions for fabricating Ag SHS on Al substrate.

3.3. Self-cleaning property

The self-cleaning effect of SHS is a significant property due to the big WCA and the small WSA. In this experiment, the pure metal surfaces compared with the SHSs to study the self-cleaning property. As shown in figure 3, the text result of pure Al (A) and Ag SHS on Al substrate (B) are dusted with chalk powder. When the water droplets fall down the pure Al surface, the drops stick with chalk, and the dust still covers the Al surface finally. The same process is conducted for the Ag SHS on Al substrate, and the chalk can be promptly taken off the sample surface with rolling water droplets. In the same way, the tests results are consistent across the Ag SHS on Zn substrate and the Ag SHS on Cu substrate. These phenomena show that the SHSs are sufficiently rough because of the micro/nano structures, and this unique property makes the superhydrophobic materials have more widely application.

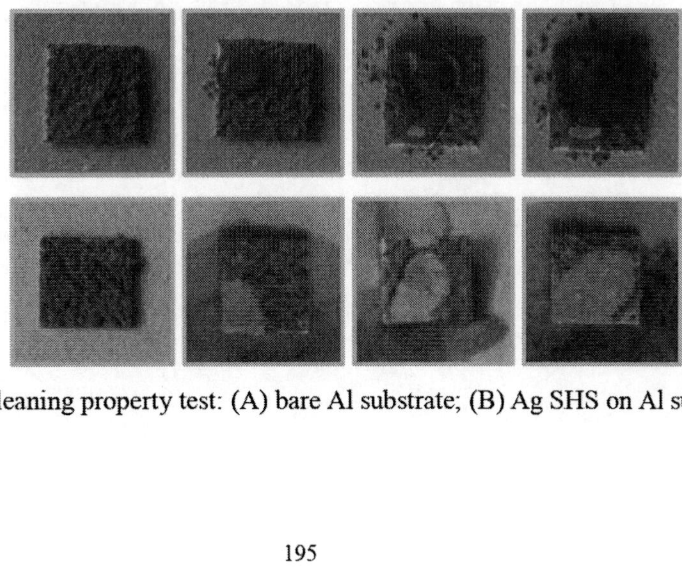

Figure 3. Self-cleaning property test: (A) bare Al substrate; (B) Ag SHS on Al substrate.

3.4. Stability

The stability of SHS is an important factor to determine the feasibility of the proposed method for the practical application. In our study, the stability of each SHS was evaluated by measuring the WCA after exposing to air in different period as shown in figure 4, and the error bar is 1°. We can see that the value of WCA of each sample is over 150° after 6 months, which had nearly no variation, indicating that the prepared SHSs display the good long-term stability in air.

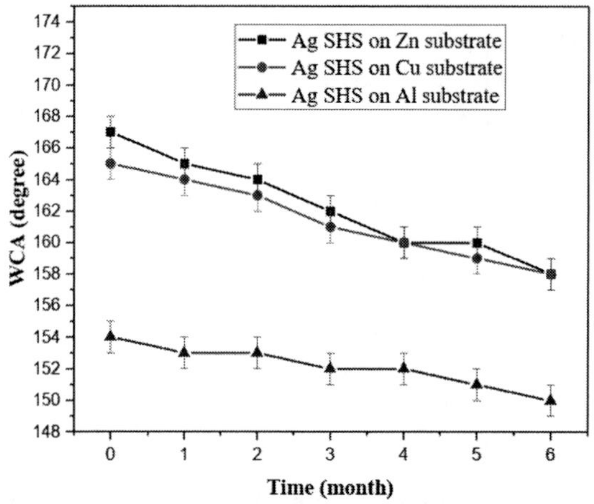

Figure 4. The WCAs of different samples after exposing to air in different period: (A) Ag SHS on Zn substrate; (B) Ag SHS on Cu substrate; (C) Ag SHS on Al substrate.

4. Conclusion

In this work, the facile approaches were developed for preparation the stable Ag superhydrophobic surfaces (SHSs) on zinc, copper and aluminium substrates. The water contact angle on each substrate could reach to 150° and the water sliding angle was less than 1°. The morphology and chemical composition of the surfaces were characterized using scanning electron microscopy (SEM) and X-ray diffraction pattern (XRD). The as-prepared three SHSs possessed the self-cleaning property and good stability, which had potential industrial applications.

Acknowledgements

We gratefully thank the National Natural Science Foundation of China (no. 21271027) for this work.

References

[1] Marmur A 2016 *Soft Matter* **8** 6867-6870
[2] Li J, Xu C C, Zhang Y, Wang R F, Zha F and She H 2016 *J. Mater. Chem.* A **4** 15546-15553
[3] Li J, Kang R M, Tang X H, She H, Yang Y X and Zha F 2016 *Nanoscale* **8** 7638
[4] Li J, Yan L, Li H Y, Li J P, Zha F and Lei Z Q 2015 *RSC Adv* **5** 53802-53808
[5] Li J, Yan L, Tang X H, Feng H, Hu D C and Zha F 2016 *Adv. Mater. Interfaces* **3** 1500770
[6] Wang D, Liu Y, Liu X, Zhou F, Liu W and Xue Q 2009 *Chemical communications* 7018-7020
[7] He G, Lu S, Xu W, Szunerits S, Boukherroub R and Zhang H 2015 *PCCP* **17** 10871-10880
[8] Guo P, Zhai S, Xiao Z and An Q 2015 *Journal of colloid and interface science* **446** 155-162
[9] Zhao J, Leng B, Shao Z, With G de and Ming W 2013 *RSC Advances* **3** 22332
[10] Wang Y, Liu X, Zhang H and Zhou Z 2015 *RSC Advances* **5** 18909-18914
[11] Zhang X., Shi F, Niu J, Jiang Y G and Wang Z Q 2008 *Journal of Materials Chemistry* **18** 621-633
[12] Zhou M, Li J, Wu C, Zhou X and Cai L 2011 *Soft Matter* **7** 4391-4396.
[13] Cui S, Lu S, Xu W, Wu B, Zhao N, He G, Hou X and Zhang H 2016 *New J. Chem.* **40** 8897-8904

[14] Liu Y, Cao H, Chen S and Wang D 2015 *The Journal of Physical Chemistry C* **119** 25449-25456
[15] Li D M, Wu R N, Li J, Li W J, Zhang Y and She H 2016 *RSC Adv.* **6** 77908–77912
[16] Wenzel R N 1936 *Industrial and Engineering Chemistry* **28** 988-994
[17] Cassie A B D and Baxter S 1944 *Transactions of the Faraday Society* **40** 0546-0550
[18] Safaee A, Sarkar D K and Farzaneh M 2008 *Applied Surface Science* **254** 2493–2498

IOP Conf. Series: Materials Science and Engineering **292** (2017) 012031 doi:10.1088/1757-899X/292/1/012031

Hydrogen production by sodium borohydride in NaOH aqueous solution

Q Wang[1], L F Zhang[1] and Z G Zhao[1, 2, 3]

[1] State Key Laboratory of Materials-Oriented Chemical Engineering, College of Chemical Engineering, Nanjing Tech University, Nanjing, China
[2] College of Renewable Energy & Advanced Materials, Nanjing Tech University Pujiang Institute, Nanjing, China

[3] E-mail: 35628489@qq.com

Abstract. The kinetics of hydrolysis reaction of $NaBH_4$ in NaOH aqueous solution is studied. The influence of pH of the NaOH aqueous solution on the rate of hydrogen production and the hydrogen production efficiency are studied for the hydrolysis reaction of $NaBH_4$. The results show that the activation energy of hydrolysis reaction of $NaBH_4$ increased with the increase of the initial pH of NaOH aqueous solution.With the increasing of the initial pH of NaOH aqueous solution, the rate of hydrogen production and hydrogen production efficiency of $NaBH_4$ hydrolysis decrease.

1. 1.Introduction

The intensive use of fossil fuels has caused great environmental problems [1]. As a promising energy, hydrogen energy is pollution free for its combustion only generating water without any byproduct [2-5]. At present, many countries around the world are studying hydrogen production both in large quantities and at low cost. People can produce hydrogen in different ways, for example, reforming, decomposition, and hydrolysis of compounds, water electrolysis or microbial fermentation or photosynthesis [6]. Boride hydride has high hydrogen storage density and it is solid state at room temperature and normal pressure. This makes it to be a good prospect as a miniaturized and mobile hydrogen source [7]. At present, hydrolysis and pyrolysis are two main ways to obtain hydrogen from boride hydride. Among the boride hydrides, $NaBH_4$ is the most studied used for hydrogen production by hydrolysis [8]. However, at room temperature, $NaBH_4$ reacts slowly in alkaline solution, and the hydrogen production efficiency is very low. Kreevoym proposed the following empirical equation [9]:

$$\log t_{1/2} = pH - (0.034T - 1.92) \tag{1}$$

$t_{1/2}$ is the half-life time, the unit is minute. T is the absolute temperature.

Hydrogen production by $NaBH_4$ in NaOH aqueous solution is investigated in this paper.

2. Experimental

Five glasses of 200 ml NaOH aqueous solution were prepared, their pH were 7.07, 7.59, 8.01, 8.49 and 9.00. In the five glasses of 200 ml NaOH aqueous solution, each of them was added 3 g $NaBH_4$ at 293.16 K. The hydrogen-production rate and the total amount of produced hydrogen of the hydrolysis of $NaBH_4$ were measured. After an hour of reaction, pH of the solution was measured.

The reaction facility is shown in figure 1. In a airtight room, the beaker contains NaOH aqueous solution, and 3 g $NaBH_4$ powder is put into the beaker by a lifting rod. The hydrogen-production rate is measured by the gas pipe and a glass rotameter. Temperature-control system is used to control the temperature.

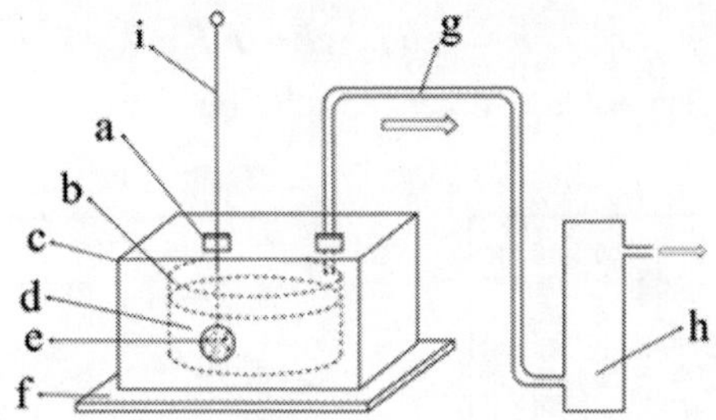

a: Seal; b: Beaker; c: Airtight room; d: NaOH aqueous solution; e: $NaBH_4$ powder; f: Temperature-control system; g: Gas pipe; h: Glass rotameter; i: Lifting rod

Figure 1. Schematic representation of the self-designed reaction facility.

3. Results and discussion

The relation between the rate and the pH of NaOH aqueous solution is shown in figure 2. The total production of hydrogen is obtained by integral of the curve in figure 2. Then the relation between hydrogen production and the pH of the NaOH aqueous solution is obtained, as shown in figure 3. The efficiency of hydrogen production of five glasses of 200 ml NaOH-$NaBH_4$ aqueous solution is 10.5 %, 9.6 %, 8.9 %, 8.4 % and 7.0 %, respectively.

One can see from figure 2 and figure 3, both of the rate of hydrogen production and the efficiency of hydrogen production decrease with the increase of pH of the NaOH aqueous solution.

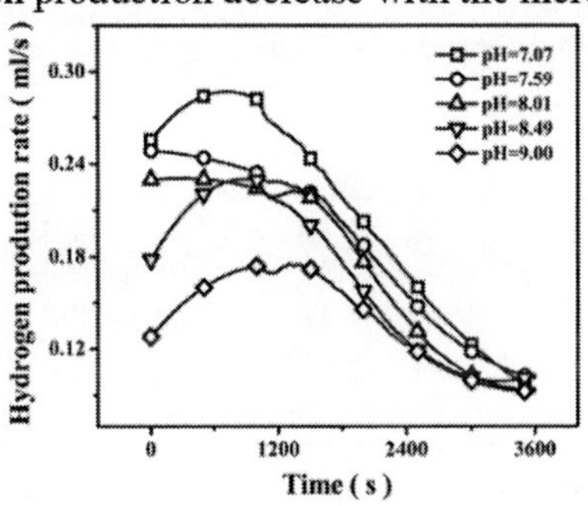

Figure 2. Dependence of hydrogen-production rate of $NaBH_4$ on the reaction time.

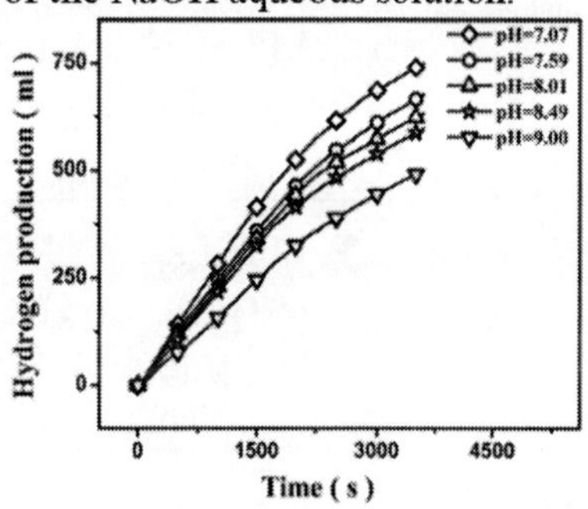

Figure 3. Dependence of the hydrogen-production of $NaBH_4$ on the reaction time.

The pH of the solution is measured with a pH meter 1 hours after the initiation of the reaction. It is found that the pH value of the final NaOH-$NaBH_4$ aqueous solution saturated at about 10.60.

The activation energy for hydrogenation is obtained by means of the Arrhenius equation:

$$Ea = -RT\ln(\frac{k}{ko}) \tag{2}$$

Ea is the activation energy, R is the gas constant, T is the absolute temperature, k_0 is a pre-exponential factor, k is the temperature dependent reaction-rate constant.

By the Avrami equation, one can obtain the constant k [10]:

$$f = 1 - \exp(-kt^n) \tag{3}$$

IOP Conf. Series: Materials Science and Engineering **292** (2017) 012031 doi:10.1088/1757-899X/292/1/012031

f is the reaction function (here it is the volume fraction of theoretical hydrogen production of 3 g $NaBH_4$), n is the order of reaction and t is the reaction time. People can get the following equation from equation (2) and (3) [11]:

$$\ln[-\ln(1-f)] = n\ln t + \ln k \quad (4)$$

$$Ea = -RT\ln k + RT\ln k_0 \quad (5)$$

From the data in the figure 3, people can get the relation between lnt and ln[-ln(1-f)], as shown in figure 4.

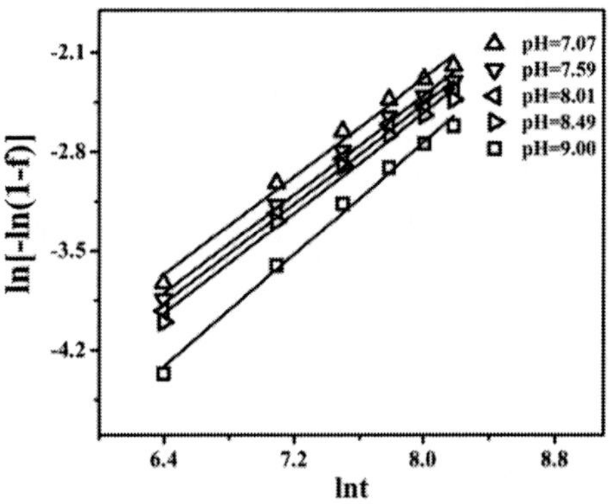

Figure 4. Relation between lnt and ln[-ln(1-f)].

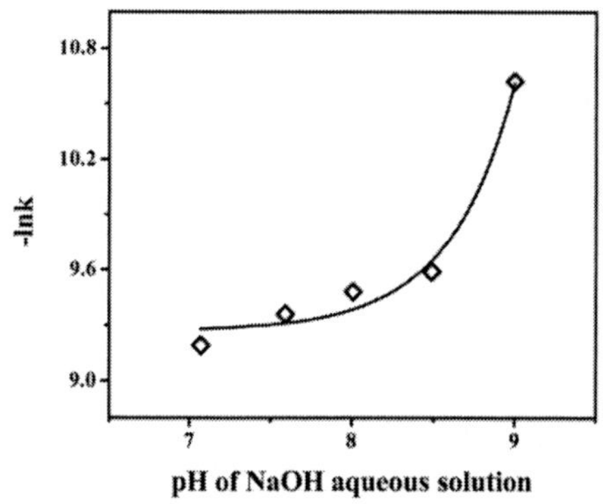

Figure 5. Relation between the pH of NaOH aqueous solution and –lnk.

People can get lnk and n of the $NaBH_4$ hydrolysis reaction in the different NaOH aqueous solution by figure 4, as shown in table 1:

Table 1. n and lnk for hydrogen production from $NaBH_4$ hydrolysis at different pH.

pH of NaOH(aq)	n	lnk
7.07	0.99	-9.19
7.59	0.87	-9.36
8.01	0.88	-9.48
8.49	0.88	-9.59
9.00	0.99	-10.62

In equation (5), k_0 is pre-exponential factor, it is a constant determined only by the nature of the reaction and independent of the temperature of the reaction and the concentration of matter in the system. So, the relation between the pH of NaOH aqueous solution and activation energy of $NaBH_4$ hydrolysis is similar to the relation between the pH of NaOH aqueous solution and -lnk, as shown in figure 5.

One can see from figure 5, the activation energy of $NaBH_4$ hydrolysis increases with the increase of the pH of NaOH aqueous solution.

4. Conclusions

With the increase of the pH of NaOH aqueous solution, both of the rate of hydrogen production and the efficiency of hydrogen production decreases. Increasing the pH of NaOH aqueous solution increases the activation energy of $NaBH_4$ hydrolysis.

Acknowledgments
This work was supported by the discovery project of State Key Laboratory of Materials-Oriented Chemical Engineering, Nanjing Tech University (38901159), The Hydrogen Energy Key Materials

R&D Center of Nanjing City (51201311) and the collaboration program between Nanjing Tech University and Jiangsu HY-Power Tech. Ltd.

References

[1] Sinha P, Roy S & Das D 2016 Genomic and proteomic approaches for dark fermentative biohydrogen production.*Renewable & Sustainable Energy Reviews* **56** 1308-21

[2] Rosales-Colunga L M, Rodríguez A D L 2015 Escherichia coli, and its application to biohydrogen production *Reviews in Environmental Science & Bio/technology* **14**(1) 123-35

[3] Steele B C, Heinzel A 2001 Materials for fuel-cell technologies *Nature* **414**(6861) 345-52

[4] Gorte R J 2015 Cooling down ceramic fuel cells *Science* **349**(6254) 1290-1290

[5] Costamagna P, Srinivasan S 2001 Srinivasan, s.: quantum jumps in the pemfc science and technology from the 1960s to the year 2000: part i. fundamental scientific aspects *Journal of Power Sources* **102**(s 1–2) 242–52

[6] Holladay J D, Hu J, King D L & Wang Y 2009 An overview of hydrogen production technologies *Catalysis Today* **139**(4) 244-60

[7] Satyapal S 2007 Annual Progress Report *Department of Energy*

[8] And Y S, Chen R 2006 Semiempirical hydrogen generation model using concentrated sodium borohydride solution *Energy & fuels* **20**(5) 2149-54

[9] Kreevoy M M, Jacobson R W 1979 The rate of decomposition of $NaBH_4$ in basic aqueous solutions *Ventron Alembic* **15**(2) 2-3

[10] Avrami M 1939 Kinetics of phase change. i general theory *Journal of Chemical Physics* **7**(12) 1103-12

[11] Xu E, Li H, You X, Bu C, Zhang L & Wang Q et al. 2017 Influence of micro-amount O_2 or N_2 on the hydrogenation/dehydrogenation kinetics of hydrogen-storage material MgH_2 *International Journal of Hydrogen Energy* **42**(12) 8057-62

IOP Conf. Series: Materials Science and Engineering **292** (2017) 012032 doi:10.1088/1757-899X/292/1/012032

Explosive performance of HMX/NTO co-crystal

J C Li[1, 3, 4], Q J Jiao[2], Y G Gong[3], Y Y Wang[1], T Liang[1] and J Sun[1]

[1] Institute of Chemical Defence, P.O. Box 1048, Beijing 102205, China
[2] National key laboratory of explosive science and technology, Beijing Institute of Technology, Beijing 100081, China
[3] State key laboratory of NBC Protection for Civilian, Research Institute of Chemical Defence, Beijing 102205, China

[4] E-mail: 510479079@qq.com

Abstract. A new co-crystal explosive of 1,3,5,7-tetranitro-1,3,5,7-tetrazocane (HMX) and 3-nitro-1, 2, 4-triazol-5-one (NTO) in a molar ratio of 1:1 has been prepared by solvent/anti-solvent method. The SEM photographs show that HMX/NTO co-crystals are distinctly different from HMX and NTO crystals. The co-crystals are prisms with well formed crystal surfaces. Thermal analysis results indicate the melting point of the co-crystal is 29.3℃higher than that of NTO. Moreover, the co-crystal exhibits a modified mechanical sensitivity. The characteristic height (H_{50}) of impact sensitivity increases 14.8cm, and the explosion percentage (*P*) of friction sensitivity decreases by 40% compared with HMX. The HMX/NTO co-crystals possess good thermal property and low sensitivity, which mean huge advantages in blasting engineering.

1. Introduction

The combination of high performance and low sensitivity has become a main challenge for further application and development of explosives. In the past, various approaches to modify the properties of explosives including crystallizing with solvents and coating with polymers have been adopted. Unfortunately, these traditional methods hardly produced remarkable improvements on the performances of existing explosives [1]. Aim of the co-crystallization technique is to include two or more different molecules in a single crystal lattice. This approach has been widely used in pharmaceutical field to improve the solubility and/or stability of either component of the co-crystal. In the area of energetic materials, co-crystallization is emerging as a new technology for modifying or enhancing the properties of existing energetic substances [2].Therefore co-crystallisation would potentially provide a way to achieve the combination of high energy and low sensitivity on the molecular level [3].

1,3,5,7-tetranitro-1,3,5,7-tetrazocane (HMX, 1, figure 1) is one of the most important energetic ingredients used in various propellants and explosives. HMX-based systems have many advantages, i.e. high melting point, high density, potentially high specific impulse, high detonation velocity, little smoke produced and, as compared to ammonium perchlorate, lower toxicity and lower corrosiveness [4]. However, its applications could potentially be enhanced by further decreasing its sensitivity. In an effort to reduce the sensitivity of highly energetic materials, Matzger and co-workers prepared a novel HMX/CL-20 co-crystal [5] and seven HMX co-crystals containing non-energetic materials [6]. In addition to these HMX co-crystals, HMX/N-methyl-2-pyrrolidone (NMP), HMX/*N-N* dimethylformamide (DMF) and HMX/2,4-dinitro-2,4-diazapentane co-crystals were reported [7]. Unfortunately, all these co-crystals possess low densities or high sensitivities.

The compound 5-nitro-2,4-dihydro-3*H*-1,2,4-triazol-3-one (NTO, **2**, figure 1) is a less sensitive energetic material. In previous studies [8], NTO is proposed to co-crystallize with HMX. The binding energies, heat of formations (HOFs),thermodynamic properties, atoms in molecules, and natural bond orbital analysis of four HMX/NTO complexes have been calculated using density functional theory methods. The calculation results show that HMX/NTO complexes are thermally stable and meet the thermal requirement of high energy density material [8].

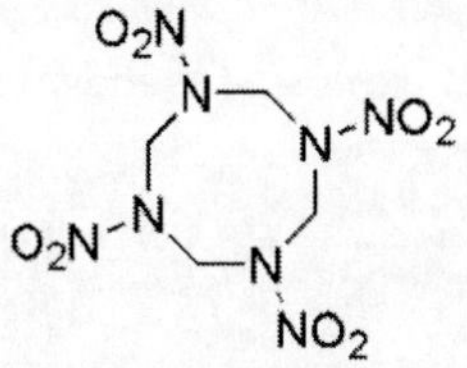

1,3,5,7-Tetranitro-1,3,5,7-tetrazocane (HMX)

1

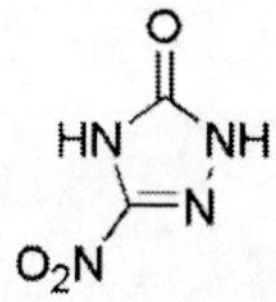

5-Nitro-2,4-dihydro-3H-1,2,4-triazol-3-one (NTO)

2

Figure 1. Chemical Structures of HMX (**1**) and NTO (**2**).

Herein, we prepared a HMX/NTO co-crystal in a 1:1 molar ratio by treatment with an appropriate mixture of solvent/anti-solvent. Scanning electron microscopy (SEM), differential scanning calorimetry (DSC) and mechanical sensitivity analyzer were employed to characterize of the samples.

2. Experimental

2.1. Material

β-HMX, E grade; HMX and NTO were obtained from North University of China. Acetone and trichloromethane were reagent grade and obtained from Guangdong Guanghua Chemical Reagent Chemical Technology Company.

2.2. Preparation method

A 1:1 molar ratio of HMX (1.48g) and NTO (0.65g) was dissolved into acetone (70ml) at 56℃ temperature, stir the solution. Then anti-solvent trichloromethane (50ml) was slowly dropped into the solution and filtered. The new co-crystal of HMX/NTO was formed.

2.3. Scanning electron microscopy (SEM)

Scanning electron microscopy of the samples were taken under the S-4700 SEM, Japan Hitachi.

2.4. Differential scanning calorimetry (DSC)

Differential scanning calorimetry (DSC) was carried out with a DSC-131 instrument, France Setaram. The sample (0.7±0.1) mg was heated from room temperature to 400℃ at a heating rate of 5℃/min in a nitrogen flow.

2.5. Impact sensitivity test

The impact sensitivity was conducted according to the GJB772A-97standard method 601.2 [9]. The test conditions are: drop weight, 5 kg; sample mass, 50 mg; trial quantity, 20 samples. For the varing of the impact energy in each trial, the ultimate impact energy is determined statistically with 50% probability of ignition.

2.6. Friction sensitivity test

The determination of the friction sensitivity was performed referring to the method of GJB772A-97 standard method 601.1 [9]. Sample amount is 20 mg, pendulum quality is 5 kg, the swing angle is 90° and gauge pressure is 3.92 MPa. The fraction sensitivity of each test sample was expressed by explosion probability (*P*).

3. Results and Discussion

3.1. Morphology

The crystal morphology of HMX, NTO and HMX / NTO co-crystal explosive is shown in figure 2.

(a) (b) (c)

Figure 2. Microscope images of HMX (a), NTO (b) and HMX / NTO co-crystal (c).

The SEM photographs show that HMX/NTO co-crystals are distinctly different from HMX and NTO crystals. HMX crystals have regular shape with smooth surface and uniform size. NTO with a rod-like morphology is easily agglomerated. The HMX/NTO co-crystals are prisms with well formed crystal surfaces.

The crystal structure of HMX/NTO co-crystal has been predicted using compass force field in Monte Carlo simulations, which are widely used to predict the molecular packing (crystal structure) of energetic materials. The HMX/NTO co-crystal is most likely to crystallize in triclinic system with *P*-1 space group, due to its lowest lattice energy [8].

3.2. Thermal analysis (DSC)

Differential Scanning Calorimeter (DSC) is helpful in studies on the thermal behaviour of the co-crystal, and the DSC curves of HMX/NTO co-crystal, HMX and NTO are depicted in figure 3.

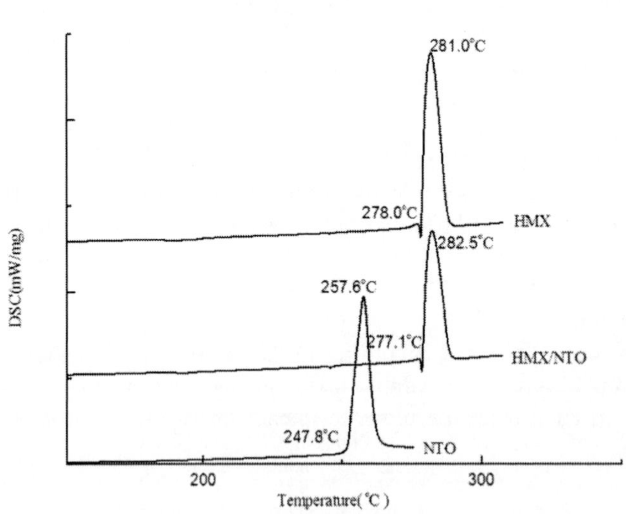

Figure 3. The DSC curves of HMX, NTO and HMX / NTO co-crystal.

It was evident from the curves that the thermal behavior of co-crystal differs from the crystals of their individual components, and the differences in thermal stability of these substances suggest the formation of a new crystal aggregate. The HMX/NTO co-crystal DSC curve shows a sharp and narrow endothermic peak at the temperature of 277.1℃, corresponding to the melting point of the co-crystal which is 29.3℃ higher than that of NTO (247.8℃). Besides, the co-crystal exhibits a strong exothermic peak in the temperature of 282.5℃ which is 1.5℃ higher than that of HMX (281.0℃). The shift may be caused by the change of lattice energy and crystal packing.

3.3. Sensitivity

To analyze the safety profile of these co-crystals, their impact and friction sensitivity were studied. The mechanical sensitivity tests of the HMX/NTO co-crystal, HMX and NTO were expressed by the drop height of 50% explosion probability (H_{50}) for impact sensitivity and explosion percentage (*P*) for the friction sensitivity. The results are presented in table1.

Table 1. The results of mechanical sensitivity for HMX, NTO and HMX / NTO co-crystal.

	H_{50}/cm	*P*/%
HMX	22.4	100
NTO	57.5	32
HMX/NTO	37.2	60

The co-crystal exhibits a modified mechanical sensitivity. The characteristic height (H_{50}) of impact sensitivity increases14.8 cm from 22.4cm (HMX) to37.2 cm (co-crystal).The explosion percentage (*P*) of friction sensitivity is 60%, which is 40% lower than HMX (100%). The HMX/NTO co-crystal shows a satisfactory safety than that of HMX. This phenomenon indicates that the formation of co-crystal can significantly increase the safety of the sensitive explosive. The intermolecular interactions between HMX and NTO molecules in the co-crystal structure make the molecules pack closer and decrease the volume and number of cavities which leads to a lower sensitivity of co-crystal than the mixture. Besides, the stability of co-crystal is also improved by hydrogen bond interactions.

4. Conclusion

A 1:1 molar ratio of HMX /NTO co-crystal explosive is prepared by treatment with a mixture of solvent/anti-solvent using acetone as the solvent and trichloromethane as the anti-solvent. The SEM photographs show that HMX/NTO co-crystals are distinctly different from HMX and NTO crystals. The co-crystals are prisms with well formed crystal surfaces. The HMX/NTO co-crystal is most likely to crystallize in triclinic crystal system with *P*-1 space group. Thermal analysis results indicate the HMX/NTO co-crystal shows good thermal property. The melting point of the co-crystal is 29.3℃ higher than that of NTO, and the exothermic peak temperature is 282.5℃ which is 1.5℃ higher than that of HMX (281.0℃).Moreover, the co-crystal exhibits a modified mechanical sensitivity. The characteristic height (H_{50}) of impact sensitivity increases 14.8cm, and the explosion percentage (*P*) of friction sensitivity decreases by 40% compared with HMX alone. The HMX/NTO co-crystals possess low sensitivity and good thermal property, which mean huge advantages in blasting engineering. Some other investigation on the properties of HMX/NTO co-crystal and researches on other co-crystal explosives are still ongoing.

References

[1] Yang Z W, Wang Y P, Zhou J H, Li H Z, Huang H and Nie F D 2014 *Propellants Explos. Pyrotech* **39** 9-13
[2] Aakeroy C B, Wijethunga T K and Desper J 2015 *Chem. Eur. J.* **21** 11029-11037
[3] Liu K, Zhang G, Luan J Y, Su P F and Shu Y J 2016 *J. Mol Struct* **1110** 91-96
[4] Ordzhonikidze O, Pivkina A, Frolov Yu, Muravyev N and Monogarov K.2011 *J Therm Anal Calorim* **105** 529–534

[5] Bolton O, Simke L R, Pagoria P F and Matzger A.J 2012 *Cryst. Growth Des.***12** 4311-14
[6] Landenberger K B, Matzger A J 2012 *Cryst. Growth Des.***12** 3603-09
[7] Lin H, Zhu S G, Li H Z and Peng X H 2013 *J. Mol Struct* **1048** 339–348
[8] Lin H, Zhu S G, Zhang L, Peng X H, Chen P Y and Li H Z 2013 *Int.J.Quantum Chem.***113** 1591-99
[9] National Military Standard of China 1997 Experimental Methods of Sensitivity and Safety GJB/772A-19 (in Chinese)

IOP Conf. Series: Materials Science and Engineering **292** (2017) 012033 doi:10.1088/1757-899X/292/1/012033

Manufacture of a ceramic paper for art applications

K Dölle[1, 3], A Honig[2], J Piatkowski[1] and C Kuempel[2]

[1] State University of New York, College for Environmental Science and Forestry, 1 Forestry Drive, Syracuse, New York, USA
[2] Munich University of Applied Sciences, Lothstr. 34, 80355 Munich, Germany

[3] E-mail: kdoelle@esf.edu

Abstract. Ceramic paper products are mostly used as high temperature ceramic insulation products. They offer an effective solution for most demanding heat management and insulation applications. The objective for this research project was to create a ceramic paper like product that combines the advantages of paper fibers, ceramic filler, and a clay product into one product, which can be produced on a continuous base with a paper machine. The produced ceramic paper product had a ceramic filler level between 59.68% and 78.8% with a basis weight between 322.9 g/m² and 693.7 g/m², and a final moisture content of 58.6% to 44.7% respectively. The wooden fiber served as a support medium for the ceramic filler material during production on the paper machine and during the conversion process into art pieces. During firing in a kiln, the fiber material combusted and the ceramic filler material mixture acts as common pottery clay, holding the desired shape of the art pieces produced.

1. Introduction

Since its invention by the Chinese Inventor, Ts'ai Lun, in 105 C.E., paper has made significant contributions to the advancement of society. Today paper products can be found everywhere in modern societies and its consumption is an indicator of the wealth and advancement of society. To save production cost filler material are added, because they are less expensive than fiber materials [1]. Paper products today range from newsprint, printing paper, tissue or hygiene paper, board and packaging paper, to specialty paper materials. Many paper products that are currently produced incorporate an inorganic material called fillers, designed to accommodate the individual customer needs. Filler materials are normally incorporated into the paper web during the formation process on the papermaking wire. This is achieved by mixing filler material and wooden fibers with water into a suspension during a preparation process. The water from the suspension is then drained out on the papermaking wire while suspended fiber and filler particles are retained in a fibrous web that is formed on the papermaking wire [1].

Today's primary filler materials in the paper industry are precipitated calcium carbonate (PCC) and ground calcium carbonate (GCC). Fillers are less expensive than fibers, allowing reduced production costs, improved optical paper properties, dimensional stability, better sheet formation and printability of the paper product. Filler content in today's paper products in North America is normally between 15 to 20%, but can reach up to 30% by weight in specialty paper products. Other filler materials such as titan dioxide, calcined clay and silica are incorporated in paper materials when specific properties are required such has abrasion, and hardness [1, 2]. The literature states that a paper can be produced with a filler content of over 70%. However, it is not being implemented because the paper product loses its integrity and required mechanical and physical properties if the filler contents is above 20%. [1-5].

The bases of ceramic materials are mineral fillers. These fillers are also the base component of graphic paper to increase the printability, surface and optical properties. Typical fillers are Kaolin, Talc, Ground Calcium Carbonate (GCC) or Titanium Dioxide. By incorporating these ceramic materials into a paper product results in reduced strength properties and run ability problems on a paper machine [6].

If a ceramic paper with a thickness below 2 mm and ceramic filler content above 40% can be manufactured, a paper-like composite material with a wide range of applications in industry and art products would result. At present time, a ceramic paper like product is available for insulation purpose [7, 8] but not for art or other applications.

Currently artists and sculptors add paper fibers to the clay body giving it important and unusual working characteristic such as enhanced surface and binding qualities in the green stage of the building process. It allows the production of large thin objects without cracking. [9, 10]. Paper fibers added to the clay body burn away early in the firing, leaving a durable ceramic product virtually unrecognizable from conventional ceramics, except for the lost weight of the fired paper ceramic [11].

The objective for this research project was to create a ceramic paper product that combines the advantages of paper fibers, ceramic filler, and clay product into one product that can be produced on a continuous base with a paper machine. The ceramic paper product should contain about 60% to 80% filler content with a moisture content level of 40 % to 50%, allowing the final product to be cut, rolled and molded and then finally fired in a kiln to produce a ceramic art piece.

2. Methodology

The methodology section describes the equipment, procedures and materials used for this project at the State University of New York (SUNY), College of Environmental Science and Forestry (ESF), department of Paper and Bioprocess Engineering (PBE). As part of this study several test methods were applied. These are described in detail in the following subsections. All tests were performed and reported according to TAPPI standards or as noted otherwise. Repeatability of the results stayed in between the allowable margins of the TAPPI testing standards.

2.1. Fiber materials

For the ceramic paper development paper machine run a total of 6 kg fibers were used. The pulp fiber mixture consisted of 2/3 Northern Bleached Softwood Kraft (NBSK) and 1/3 Eucalyptus Bleached Hardwood Kraft (EBHK). Both fiber fractions were used unrefined and had an initial Canadian Standard Freeness (CSF) of 600 ml and 400 ml respectively. The fiber mixture was pulped at a consistency of 5% and then diluted to 3% and stored in the paper machine chest. From the paper machine storage chest, the pulp fiber mixture was pumped to the headbox with prior dilution to 1.5% fiber content.

2.2. Ceramic filler materials

20.0 liter of a ceramic filler material suspension with a final solids content of 65.5% was prepared. The suspension had 32.0% Kaolin, 18.0% Feldspar, 18.0% Silica, 1.5% Ball Clay and 25.5% water. Content based on weight. 0.25% sodium silicate was added to improve dispersion. The ceramic filler suspension was stirred in the holding tank with a mixture to keep the solids in suspension while the suspension was pumped with a peristaltic pump into the headbox supply line.

2.3. Paper machine conditions

A 12 inches wide pilot Fourdrinier Paper Machine (FPM) at the PBE department (figure 1) was used for this research. The FPM was operated at a speed of 0.9 m/min. The fiber flow to the headbox was set at 5.6 l/min. at a consistency of 1% solids content. The ceramic filler flow started at set at 140 ml/min. and was increased during the run to 900 ml/min. To minimize ceramic filler loss, water removed from the fourdrinier section of the paper machine was used as dilution water for the paper fibers. The fourdrinier section was operated without a dandy roll. The vacuum boxes of the fourdrinier section were operated at maximum vacuum to remove as much water as possible from the suspension containing the fiber and ceramic filler material.

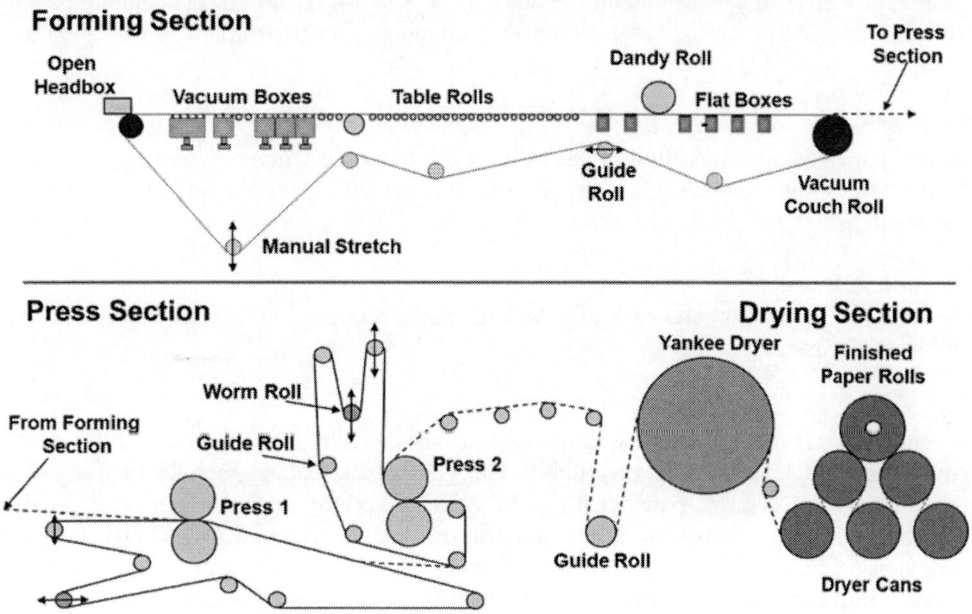

Figure 1. Fourdrinier pilot paper machine [12].

The press section was operated under no pressure. The Yankee dryer had an operation temperature of 190°C. The finished ceramic paper product was rolled up between the first two top drying cans as shown in figure 2.

Figure 2. Rolled up ceramic paper [13].

2.4. Testing

Testing needed for preparation of the pulp fibers and ceramic filler slurry, the paper machine run and for final testing of the ceramic paper product was carried out using TAPPI testing standards. Ceramic paper samples were conditioned according to T 402 sp-13, Standard conditioning and testing atmospheres for paper, board, pulp handsheets [14]. The grammage was determined by T 410 om-08 "Grammage of Paper and Paperboard (weight per unit area)" [15]. Moisture content of ceramic paper and ceramic filler slurry was determined by T412 om-06 "Moisture in pulp, paper and paperboard" [16], the freeness of pulp was measured as Canadian Standard Freeness (CSF) according to T 227 om-09 "Freeness of pulp (Canadian standard method)" [17]. The ash content of the ceramic paper product was determined by T-211 om-02 "Ash in wood, pulp, paper and paperboard: combustion at 525°C" [18].

2.5. Firing of ceramic paper product

The ceramic paper product was fired at gas kilns at the Syracuse University Art Facility in Syracuse New York. The bisque and the glaze firing was at cone 10 (1300°C/2380°F), typically used for proclaim products.

3. Discussion of Results

In this section, the production process of the ceramic paper product and its conversion into origami art applications is discussed.

3.1. Ceramic paper production

For the ceramic paper machine run the grammage or basis weight and the ceramic filler content were targeted as the ceramic paper specifications. The goal of the paper machine run was to produce various grammages of ceramic paper that can be used as working material to produce a ceramic art object. Earlier preliminary tests showed that a moisture content between 40% and 60% is desirable for the conversion of the ceramic paper product, allowing the artist enough time to convert the ceramic paper product into an art object before the ceramic paper product becomes to dry and stiff before it can be converted into art pieces.

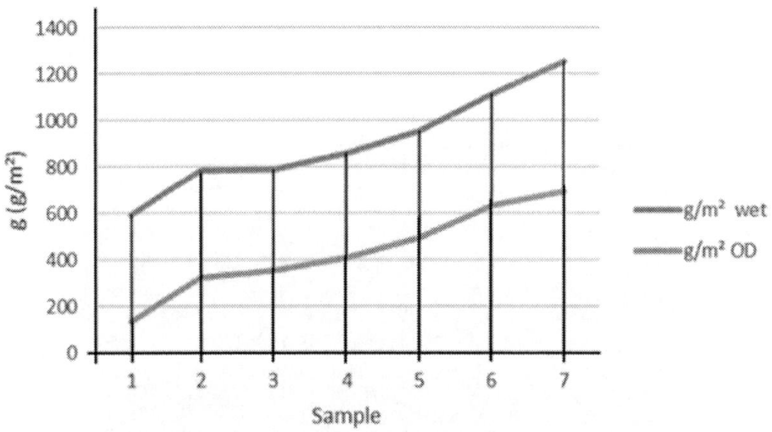

Figure 3. Ceramic paper product production sample.

Figure 3 represents the different ceramic paper products that were produced during the 3-hour paper machine run. Sample 1 was the paper sample with no ceramic filler, a dry basis weight of 130 g/m² and a moisture content of 22%. Sample 2 had a basis weight of 322.9 g/m², a ceramic filler content of 58.6% and a moisture content of 58.3%. Sample 3 showed a basis weight of 352.8 g/m², a ceramic filler content of 65.5% and a moisture content of 55.7%. Sample 4 had a basis weight of 322.9 g/m², a ceramic filler content of 59.6% and a moisture content of 58.3%. The production of sample 5 revealed a basis weight of 495.4 g/m², a filler content of 73.3% and a moisture content of 48.0%. The basis weight of sample 6 was 633.6 g/m², with a filler content of 76.2% and a moisture content of 43.1%. The produced sample 7 had a basis weight of 693.7 g/m², a filler content of 78.8% and a moisture content of 44.7%.

The produced ceramic paper product was rolled up and the specific sample marked with paper strips as shown in figure 2. After the production was finished the produced roll was wrapped in a plastic bag to prevent moisture to escape. The finished ceramic paper product is stored in a cold room till it is used.

3.2. Conversion into origami art

The finished ceramic paper product from sample 7 with a basis weight of 693.7 g/m² and a moisture content of 44.7% was first cut into smaller sheets of 20.0 cm by 20.0 cm. The cut sheets wee air dried for 30 minutes at room temperature achieve best folding results. The sheet was then folded into the origami structure shown in figure 4a. After air drying for 24h the green ware origami structure was bisque-burned in a gas fired kiln at 1300°C. After the bisque-burn a 20% reduction in size was noted (figure 4a). The origami ceramic product was glazed (figure 4b), followed by another burn at a 1300°C. After the second burn no shrinkage was observed (figure 4c).

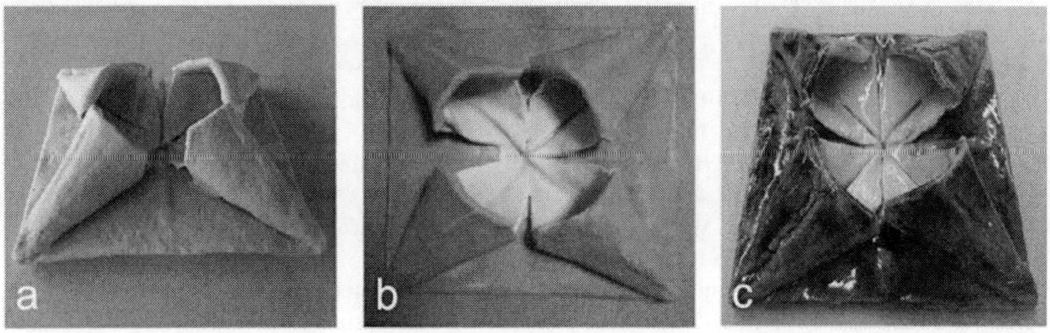

Figure 4. a) Ceramic paper origami after 1st firing, b) glazed, c) ceramic paper origami after 2nd firing with glaze [19].

The second approach using the ceramic paper sheet from sample 7 was the application as wall for a small drinking cup shown in figure 5a to c. As shown in figure 5b the wall thickness after glazing was about 1.6 mm. The production of the small cup included a bisque-burn in a gas fired kiln at 1300°C, followed by another burn after glazing at a 1300°C. A shrinkage of 20% after the first bisque-burn was observed. No shrinkage was observed after the second burn. The ceramic paper product showed excellent folding and forming capabilities during the creation of the two art objects shown in figure 4 a-c and figure 5 a-c. During firing of the ceramic paper art pieces in a kiln, the fiber material combusts and the ceramic filler material mixture acts as common pottery clay holding the desired 3-dimmensional shape of the art piece during the first bisque-firing and at its final state after the second firing for setting the glaze.

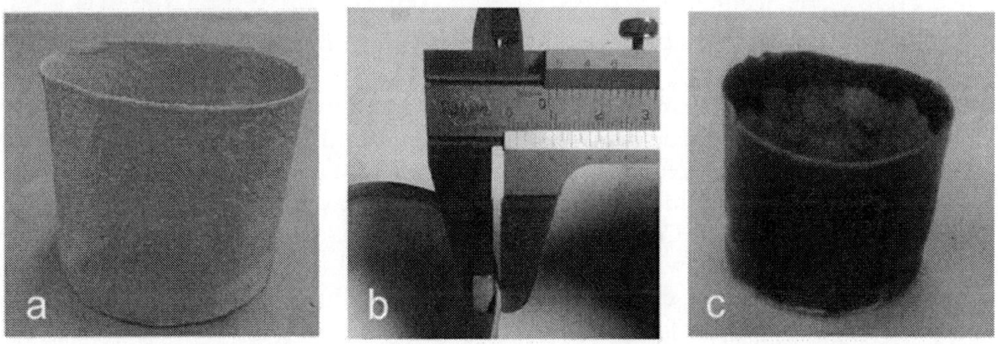

Figure 5. a) Ceramic paper cup after firing, b) wall thickness glazed 1.5 mm, c) paper clay cup after 2nd firing with glaze [20]

4. Conclusion

Production of a ceramic paper product for art applications using a paper machine is feasible. The ceramic paper product had a ceramic filler level between 59.68% and 78.80% with a basis weight of 322.90 g/m² and 693.70 g/m², and a moisture content of 58.6% to 44.7% respectively. The wooden fiber served as a support medium for the ceramic filler material during production on the pilot paper machine and the conversion process into art pieces using origami folding methods. During firing in a kiln, the fiber material combusts and the ceramic filler material mixture acts as common pottery clay holding the desired shape of the art piece during firing and at its final state after the first bisque firing. After the second firing for setting the glaze no changes in shape have been observed.

References

[1] Doelle K, Amaya J J 2012 Application of calcium carbonate for uncoated digital printing paper from 100% eucalyptus pulp *TAPPI JOURNAL* **11(1)** pp 41-49

[2] Doelle K., Heise O, Klungness J H, AbuBakr S 2000 Lightweight, High-Opacity Bible Paper by Fiber Loading *Intertech, Specialty & Technical Papers Europe* 2000 (Berlin Germany) pp1-9

[3] Fiedler M, Kretzschar J 2008 New Development in the field of highly filled papers *PTS NEWS (Munich Germany)* vol **1**

[4] Doelle K, Heise O 2005 Apparatus for loading fibers in a fiber suspension *United States Patent Office US Patent* 6,939,438

[5] Doelle K, Heise O 2002 Method of loading fiber suspension with calcium carbonate *United States Patent Office US Patent* 6,413,365

[6] Frank J S, Ashland Hercules Water Technologies, Wilmington, DE. The Influence of Filler Content and Process Additives on Wet Web Strength and Runnability Accessed June 15st 2017 Available: http://www.tappi.org/content/events/11papercon/documents/257.197%20doc.pdf

[7] Pacor Inc. , Ceramic Papers, Accessed June 15th, 2017 Available: http://www.pacorinc.com/ceramic-insulation-ceramic-papers

[8] Shandong Minye Refractory Fiber Co. ,Ltd. What is the usage of ceramic paper in automobile mufflers? Accessed June 20th 2017 Available: http://www.ceceramicfiber.com/Article/Whatistheusageofcera_1.html

[9] Gartside, B (1993) Mix what with clay?, *New Zealand Potter* vol **3** pp 32-33

[10] Juvonen, L (1995) Using paper fiber as a substitute in ceramic clays. *Proceedings of the Cimtec world ceramics congress (Finenze, Italy)* pp 193-200 Accessed July 20th 2017 Available: http://www.grahamhay.com.au/juvonen1997.html

[11] Gartside, B. (1994). The potential of paperclay. *Ceramic: Art and Perception* vol **18** pp 81-85

[12] Image by Klaus Dölle, Fourdrinier pilot paper machine jpg-file

[13] Photo by Klaus Dölle, Rolled up ceramic paper jpg-file

[14] TAPPI T 402 sp-13 "Standard conditioning and testing atmospheres for paper, board, pulp handsheets"

[15] TAPPI T 410 om-08 "Grammage of Paper and Paperboard (weight per unit area)"

[16] TAPPI T412 om-06 "Moisture in pulp, paper and paperboard".

[17] TAPPI T 227 om-09 "Freeness of pulp (Canadian standard method)"

[18] TAPPI T-211 om-02 "Ash in wood, pulp, paper and paperboard: combustion at 525°C"

[19] Photos by Klaus Dölle, a) Ceramic paper origami after 1st firing, b) glazed, c) ceramic paper origami after 2nd firing with glaze jpg-file

[20] Photo by Klaus Dölle, a) Ceramic paper cup after firing, b) wall thickness glazed 1.5 mm, c) paper clay cup after 2nd firing with glaze jpg-file

IOP Conf. Series: Materials Science and Engineering **292** (2017) 012034 doi:10.1088/1757-899X/292/1/012034

An electrochemical sensor based on nitrogen doped carbon material prepared from nitrogen-containing precursors

G Y Cui, C Y Wang, G Q Xiang and B Zhou[1]

Nanjing Normal University, Nanjing 210023, China

[1] E-mail: zhoubo@njnu.edu.cn

Abstract. In this work, a nitrogen doped carbon material (NDC) was prepared by using a copper adenine complex as precursor and applied to electrochemical sensing of Vitamin B2 (VB_2). The experimental results show that the nitrogen doped carbon material is obtained after calcination at 650 °C under argon atmosphere, afterwards, which were characterized by X-ray diffraction (XRD), transmission electron microscope (TEM), fourier transform infrared (FT-IR), and electrochemical method. According to the XRD data, the product was a carbon material, and infrared data demonstrates that there are two kinds of carbonyl nitrogen combination, respectively, C-N and C = N combination of ways. Importantly, we used NDC to construct electrochemical biosensor to detect VB_2 by differential pulse voltammetry (DPV). The linear range was 6×10^{-6} - 3.5×10^{-4} M (R=0.9962), the minimum detection limit was 5.4×10^{-6} M, and the sensitivity is better. Consequently, it has better sensing performance.

1. Introduction

Carbon-nitrogen compound materials, both in terms of theory, material properties and application development have made some achievements so far. After a small amount of nitrogen is introduced into the skeleton of the carbon material, the physical and chemical properties of the material are greatly optimized, and its application is further improved in electrochemistry. When carbon-nitrogen material is used as electrode material, it has gained remarkable achievements in the fuel cell, double-layer capacitors and lithium-ion batteries and other applications [1-3]. According to its different characteristics, it has been widely used in the aviation industry, optoelectronic industry [4] and other field.

In this work, Cu-Ade is used as a precursor to manufacture NDC by high temperature (about 400-1000 °C) and high pressure. Due to Cu-Ade has high nitrogen content, it does not need an additional nitrogen source. Metal adenine complex compounds containing carbon and nitrogen can be used to prepare carbon - nitrogen materials. But the metal adenine complex compounds which are as a precursor to synthesis carbonaceous materials have not been reported yet. Nitrogen atoms which have the lone pair of electrons can act as carriers, doping into the carbon material, and will increase the charge density of the carbon material, thereby the conductivity of carbon materials will increase [5]. The introduction of nitrogen atoms is bound to increase the defective position of the carbon material, so that results in an increase in activity in electrocatalytic or electrochemical reactions. Therefore, this paper uses carbon-nitrogen materials to catalyze VB_2 for the first time.

Vitamin B_2 (VB_2), also known as riboflavin, promoting metabolism is its main physiological function as a coenzyme. The yellow enzyme which is combined into riboflavin, phosphate and a molecule of protein plays a crucial role in metabolism of the sugar, fat and amino acid [6]. The growth of many animals and microbes is closely relevant to vitamin B_2. Scientific research found that, chemical carcinogens carcinogenic effect will be enhanced, provided that the lack of vitamin B_2. Therefore, everyone requires supplementing vitamin B_2 to reduce the incidence of cancer in the daily life. The current methods of detecting VB_2 are chemiluminescence [7], high performance liquid chromatography (HPLC) [8], fluorescence [9], ultraviolet spectroscopy [10], mass spectrometry [11] etc. These methods are more expensive, complicated and time consuming. The electrochemical method is simple, fast response, high sensitivity, cheap which has cause widespread attention.

2. Experimental section

2.1. Reagents

$CuSO_4{\cdot}5H_2O$, Adenine, sodium hydroxide, Glucose, Nitric acid were used as purchased without further purification. Riboflavin (VB_2) was obtained from Huixing Biochemical Reagent Co. (China). Ascorbic acid (AA), dopamine (DA) and uric acid (UA) were sigma reagents.

The 0.1 M phosphate buffer solution (PBS pH 7.0), which was made up from NaH_2PO_4, Na_2HPO_4, was employed as a supporting electrolyte, and all solutions were prepared with double distilled water.

2.2. Instruments

X-ray diffraction (XRD) measurements were carried out on a Shimadzu-XD-3AX-ray equipped with a Cu Kα radiation source ($\lambda = 0.15406$ nm), the scan rate is $10°min^{-1}$. Transmission Electron microscope (TEM) images were taken on Hitachi H-7650. Infrared spectrometers were taken on German brook Tensor 27. Tube furnace was purchased in Hefei Branch Crystal Material Technology Co., Ltd. Voltammetry experiments were performed on CHI660E electrochemical workstation (Shanghai Chenhua Instrument Company, China), which consisted of a bare or modified GCE (3.0 mm in diameter) working electrode, a platinum wire auxiliary electrode and a saturated calomel reference electrode (SCE).

2.3. Synthesis of complex

Firstly, Cu-Ade was prepared according to the reference [12]. Next, the precursor Cu-Ade was placed in a tube furnace, under the condition of argon protection, the temperature of the tube furnace was adjusted to 650 °C, and the calcination time was 6 h, the heating rate was 5 °C min^{-1}, and the black powder was obtained after the reaction was accomplished. Afterward, the final black powder was placed in a reaction kettle and sufficient concentrated nitric acid was added (But the volume of acid cannot exceed 60% of the volume of the reactor). After the reaction vessel was placed in an oven at 80 °C for 24 h, the copper nitrate solution was removed and the remaining black solid was washed with distilled water, centrifuged, filtered, and dried at room temperature, spared.

2.4. The preparation of NDC /GCE

2.5 mg of NDC was ultrasonically dispersed in 200 uL of deionized water. 6 μL of this suspension was cast onto the surface of a pre-polished glassy carbon electrode (GCE) and dried in the air. Then 2 μL of 1% nafion solution was dropped to cover the electrode surface. After drying in the air, NDC /GCE were obtained.

3. Results and discussion

3.1. Characterization of NDC

The black powder obtained by calcination of Cu_2 (N3, N9-Ade) $_4$ (H_2O) $_2{\cdot}5H_2O$ as precursor was characterized by XRD. As shown in figure 1, (a) is a simulate carbon material XRD pattern, the PDF number of standard card is 41-1487; (b) is the XRD pattern of NDC obtained experimentally. Compared with the two, the diffraction peak which is at 25 ° in the XRD pattern illustrates that the material obtained is a carbon material. The N element electronegativity is larger than C, the atomic radius is smaller than C. Therefore, the doped N element causes the crystal plane shrinking, and results interplanar spacing reducing, 2θ shifts to the right.

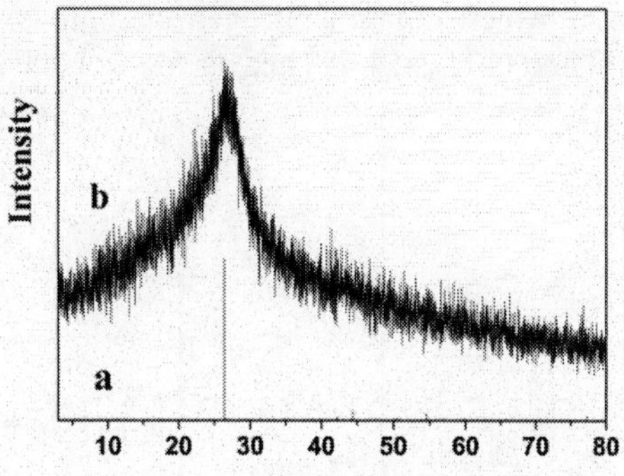

Figure 1. XRD patterns of JCPDS 41-1487 (a), and prepared NDC (b)

The morphology and size of the reaction product were characterized by TEM characterization, figure 2. It was seen from the figure that the NDC exhibits a layered structure. Its size is about 500 nm.

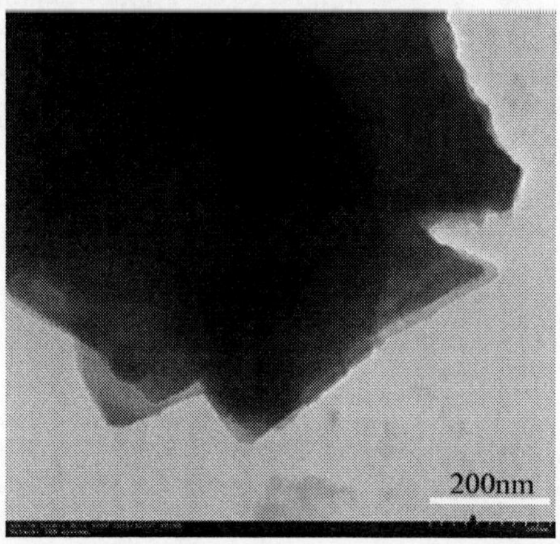

Figure 2. TEM images of the prepared NDC

In the experiment, the infrared peaks of different chemical bonds can be used to determine the type of bonding between different elements. Figure 3 is the infrared spectra of the samples prepared for the experiment. As shown in figure 3, the vibration peaks of C-N and C=N which are at 1378 cm^{-1} and 1623 cm^{-1} respectively were observed according to the reference [13].

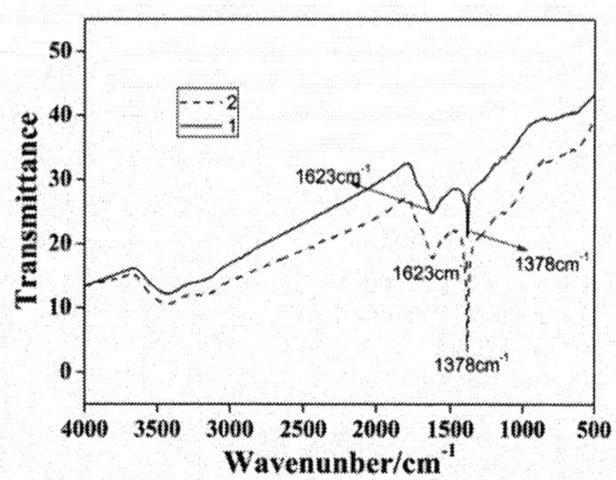

Figure 3. FTIR spectra of the prepared NDC

The elemental composition along with the N/C ratio is 46.4% as measured by XPS, Indicating that a part of nitrogen was lost in the synthetic process. But the content of nitrogen was still high. Clearly, there are three kinds of element which contain nitrogen, carbon, oxygen in the synthesized NDC (figure 4A). The content of C=N and C-N is 54.5% and 45.5% in NDC, respectively (figure 4B).

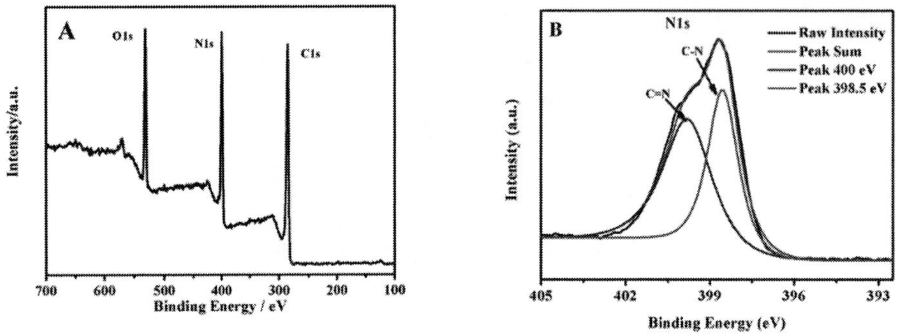

Figure 4. XPS spectrum of the synthesized NDC (A); XPS spectra of the N1s (B)

3.2. The electrochemical properties of NDC and electrochemical sensing of VB_2

Differential pulse voltammetry was carried out to investigate the electrochemistry of the NDC/GCE (figure 5). There was no significant redox peak in NDC/ GCE between -1.0 V and 0.2 V. When VB_2 was added, a reduction peak appeared at -0.45 V, which was attributed to the reduction peak of VB_2. The peak current of the reduction peak increased with the increase of the concentration of VB_2. From the DPV diagram can be seen that NDC has electrocatalytic ability towards VB_2.

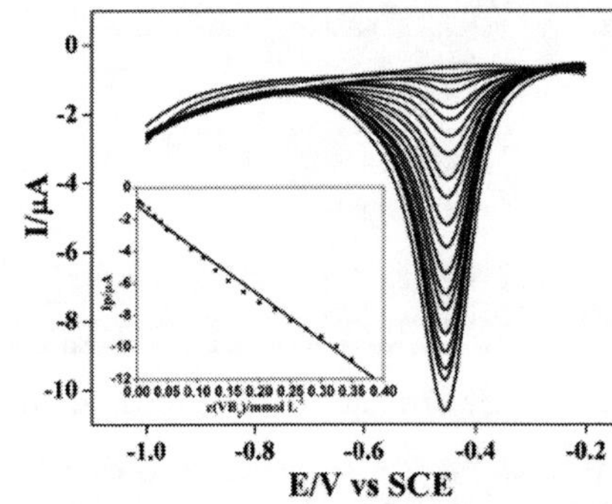

Figure 5. Differential pulse voltammetry for the interaction of VB_2 with NDC (from 0 to 0.350 mM) in 4 mL N_2-saturated PBS (0.1 mol L^{-1}, pH 7.0) Inset: calibration lot of the reduction peak currents as a function of VB_2 concentrations.

Inset in figure 5 is a linear relationship between response current and the concentration of VB_2 in the electrolyte. A linear dependence of the VB_2 reduction signals was observed for the VB_2 concentration in the range of 6×10^{-6} -3.5×10^{-4} M with a high sensitivity of 399 mA M^{-1} cm^{-2} (R=0.9962). Moreover, the detection limits (c_L) for investigated VB_2 was built according to the formula recommended by IUPAC: $c_L = 3\ s_b/S$ [14], where s_b is the standard deviation of the current density of the supporting electrolyte solution, S is the slope of the linear i_p vs c_{VB2} dependence. The detection limit is 5.4×10^{-6} M.

We examined the anti-jamming performance of NDC to detect hypoglycemic agents. According to the reference, dopamine (DA), hydrazine (N_2H_4), uric acid (UA) and sodium nitrite ($NaNO_2$) were selected as interfering substances in the detection of VB2. Figure 6 is the anti-jamming performance of the sensor obtained by adding the same concentration of interferences. The reduction signals are 1.508 μA, 1.529 μA, 1.510 μA, 1.367 μA, 1.538 μA, which are 1.39%, 0.133%, 9.35% and 1.99% of the current respond signal respectively. It can be seen that the addition of interfering substances has no effect on the detection of vitamin B_2 from the figure (the contribution of the disturbance to the current measurement is less than 10%), indicating that the sensor has good selectivity.

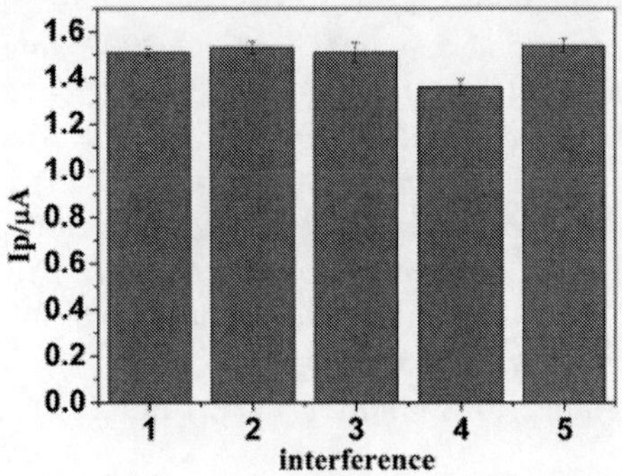

Figure 6. DPV peak currents of biosensor for 2.0×10^{-5} M VB2 in PBS (pH 7.0) before (1) and after adding 2.0×10^{-5} M, (2) DA, (3) N_2H_4, (4) UA and (5) $NaNO_2$

We used the NDC/ GCE sensor to detect the content of VB_2 in the VB_2 tablet (Nanjing Baijingyu Pharmaceutical Co., Ltd.) (table.1). VB2 was used as a standard solution to add into the electrolyte. The tablet was accurately weighed, dissolved in water, and filtered. Then, the filtrate was used as the medicine sample solution. As observed, the recovery of the spiked samples ranged between 93% and 101%, showing that the proposed method could be effectively used for the determination of VB_2 in real samples.

Table 1. Determination of VB_2 in the tablets samples and the recovery

sample	Sample concentration/mM	added/ mM	Found/mM	Recovery (%)	RSD (%)
		0.008	0.01742	93	9.30
1	0.01	0.010	0.02027	103	9.10
		0.015	0.02516	101	5.50

In the experiment, we also examined the reproducibility and stability of the sensor. The relative standard deviation (RSD) for 5 detections resulted from parallel electrodes was 9.2%. The modified electrode was stored at room temperature for 50 days and the current dropped to 90.3%. Reproducibility and stability of the experimental data show that we prepared NDC/GCE sensor has good reproducibility and stability.

4. Conclusions

In this paper, a carbon-nitrogen material was prepared from a copper-adenine complex which is as a precursor and used in electrochemistry to construct a sensor for detecting VB_2. The sensor has a wide linear range, high sensitivity and low detection limit. This work enriches the use of metal adenine complexes as precursors.

Acknowledgements

The authors thank National Natural Science Foundation of China (Grant No. 21175068 and 21335004) for financial support. We also thank the Priority Academic Program Development of Jiangsu Higher Education Institutions.

References

[1] Li Y, Li Z and Shen P K 2013 *Advanced Material* **25(17)** 2474-80
[2] Ma Y, Jiang S and Jian G et al 2009 *Energy & Environmental Science* **2(2)** 224-29

[3] Vivekananthan J, et al. 2015 *Electrochimica Acta* **182** 312-319
[4] Majeed S et al. 2013 *Nanotechnology Reviews* **2(6)** 615-635
[5] Patnaik S et al. 2016 *Inorganic Chemistry Frontiers* **3(3)** 336-347
[6] Song Z H and Wang L 2001 *Analyst.* **126(8)** 1393-1398
[7] Zandmeneghi M, Carnaro L, and Zandomeneghi G 2007 *Journal of Agricultural and Food Chemistry* **55(15)** 5990-5994
[8] Guo J, Lu Y and Dong H 2006 *Journal of Chromatographic Science* **44(9)** 552-556
[9] Kundu A, et al. 2013 *Acs Applied Materials & Interfaces* **5(15)** 7392-7399
[10] Barthus R C, Mazo L H and Poppi R J 2007 *Quimica Nova* **30(7)** 1638-1643
[11] Petteys B J and Frank E L 2011 *Clinica Chimica Acta* **412(1-2)** 38-43
[12] Gonzalez-Perez J M et al. 2006 *Inorganic Chemistry* **45(2)** 877-882
[13] Tian X L 2009 Hefei University of Technology
[14] Analytical Methods Committee 1987 *Analyst.* **112(2)** 199-204

IOP Conf. Series: Materials Science and Engineering **292** (2017) 012035 doi:10.1088/1757-899X/292/1/012035

Effect of synthesizing temperature on the electrochemical properties of nitrogen-doped Co-B alloys

Y L Liao, W Zhao, S J Qiu[1], H L Chu, Y J Zou, C L Xiang, H Z Zhang, F Xu, L X Sun

Guangxi Key Laboratory of Information Materials, Guangxi Collaborative Innovation Centre of Structure and Property for New Energy and Materials, and School of Materials Science and Engineering, Guilin University of Electronic Technology, Guilin 541004, PR China

[1] E-mail: qiushujun@guet.edu.cn

Abstract. Using urea as nitrogen source, nitrogen-doped Co-B alloys were prepared by chemical reduction method at different reaction temperatures (25, 50, and 75 °C). The structure and surface morphology were characterized by XRD and SEM. Then they were used as negative electrode materials for alkaline secondary batteries (ASBs). The electrochemical properties were measured by galvanostatic charge-discharge, electrochemical impedance spectroscopy (EIS), and anodic polarization (AP). The experimental results show that the lower the synthesizing temperature is, the better the electrochemical performance of nitrogen-doped Co-B alloy is. For example, nitrogen-doped CoB alloy prepared at 25 °C has an initial discharge capacity of 825.8 mAh/g. After 100 charge/discharge cycles, the discharge specific capacity has remained to be 369.5 mAh/g with capacity retention of 44.7%.

1. Introduction

With the development of social economy, the consumption of fossil energy is remarkably increased, which results in the coming of energy crisis and environment pollution. The development of green new energy can not only resolve energy problems, but also mitigate environment pollution. But new energies such as tidal energy, wind energy and sun energy, are needed to be converted into electrical energy to be stored in specific devices [1]. As a kind of energy storing device, alkaline secondary battery (ASB) employing alkaline aqueous solution as electrolyte is a general term. Among ASBs, nickel/metal hydride (Ni/MH) batteries are used as power supply for electronic products and portable electrical tools in recent years because of high specific energy, low cost, and stable functions in a wide temperature range [2-4]. Current ASB's negative electrode materials including Fe [5], Zn [6] and rare earth-based hydrogen storage alloys [7] cannot meet the ever-increasing requirements due to low capacity and fast capacity fading. Therefore, it is urgent to develop new materials used as negative electrodes.

Initially, Co-B alloy received an extensive research because it can accelerate hydrolysis of metal borohydride to produce hydrogen [8]. Later, some researchers found that it exhibited a reversible charge and discharge capacity in alkaline electrolyte, so they tried to improve its overall electrochemical properties [9]. As negative electrodes in ASBs, the theoretical electrochemical capacity and cycle stability of Co-B alloy have achieved a great progress compared to commercialized AB_5-type alloy and Mg_2Ni-type alloy [4]. Furthermore, Co-B alloy does not need any activation in charge/discharge cycles, which means that it can achieve its maximum discharge capacity at the first cycle. Therefore, Co-B alloy has great potential to be used as negative electrode in ASBs.

Herein, urea was introduced into the chemical reduction process to prepared nitrogen-doped Co-B alloy in order to further improve the electrochemical properties. Also the synthesizing temperature was systematically studied, especially its effect on the morphology, structure and electrochemical properties of Co-B alloys. There was a literature recording that urea can be broken down when the temperature reaches 90 °C [10]. In this study, the synthesizing temperatures should be lower than its decomposition temperature. Therefore, 25, 50, and 75 °C were selected as the synthesizing temperatures.

2. Experimental

2.1. Materials

$CoCl_2 \cdot 6H_2O$ (≥99%, Xilong Scientific), $NaBH_4$ (>98.0%, Tianjin Guangfu Fine Chemical Research Institute), $CO(NH_2)_2$ (>99.3%, Alfa Aesar), Ni powder (99.999%, Tianjin Jinbolan Fine Chemical Co., Ltd.), KOH (≥85%, Xilong Scientific). All raw materials were used without any further purification.

2.2. Sample preparation

First, 8.0 g $NaBH_4$ was dissolved in 30 mL distilled water to obtain solution A [11]. 10.0 g $CoCl_2 \cdot 6H_2O$ and 3.0 g $CO(NH_2)_2$ were dissolved in 60 mL distilled water to obtain solution B. Solution B was kept at 25 °C for 1 h using oil bath, and then solution A was dropwise added into solution B under moderate stirring. After the complete addition, this mixed solution was kept in oil bath at constant temperature of 25 °C until there was no gas bubbles produced. After naturally cooling to room temperature, the resulting sample was obtained through filtration and washing with distilled water and absolute ethanol for three times, and dried in vacuum at 60 °C for 12 h. Other samples were prepared under the same condition except for synthesizing temperatures. Here, these three samples were denoted by Y_{25}, Y_{50}, and Y_{75} according to the synthesizing temperatures of 25, 50, and 75 °C, respectively.

2.3. Characterization

The crystal structure and surface morphology of nitrogen-doped Co-B alloys were characterized by X-ray Diffraction (XRD, Bruker-D8 Advance, Germany) and scanning electron microscopy (SEM, QUANTA FEG450, USA). The composition was measured by an inductively coupled plasma atomic emission spectrometer (ICP-AES, Optima 8000, Perkin-Elmer, USA), and the content of N was measured by a CHNS elemental analyzer (EA, vario EL cube, Germany).

2.4. Electrochemical tests

About 100 mg of Co-B alloy powder and 300 mg of nickel powder was uniformly mixed, and then was pressed into a circular pellet of ø13 mm under 22 MPa pressure. Each side of the pellet was covered with a rounded foam nickel sheet of about ø25 mm, and then pressed at 4 MPa and spot-weld at the edge of foam nickel. The electrochemical tests were carried out in a standard three-electrode cell, including a Hg/HgO reference electrode, a $Ni(OH)_2$/NiOOH counter electrode, and a fabricated working electrode in 6 M KOH solution. Galvanostatic charge-discharge and high rate discharge performance tests were carried out on an automatic battery testing system (LAND-CT2001A, China). The electrochemical impedance spectroscopy (EIS) and anodic polarization (AP) were conducted on an electrochemical workstation (Zahner Elektrik IM6e, Germany). The testing temperature was held at 25 (±1) °C.

3. Results and discussion

3.1. Sample composition and structure

The composition of as-prepared Co-B alloys determined by ICP is $CoB_{0.40}$, $CoB_{0.33}$, and $CoB_{0.32}$ for Y_{25}, Y_{50}, and Y_{75}, respectively. Obviously, the content of B in these samples is decreased with the increase of synthesizing temperature. This result is consistent with the previous results reported by Lv et al., in which B content in Co-B alloy synthesized at 60 °C was higher than the sample prepared at 90 °C [4]. It has been reported that $CO(NH_2)_2$, as a precipitant, reacts with $CoCl_2{\cdot}6H_2O$ and then produces an intermediate product that has a catalytic effect for accelerating the hydrolysis of $NaBH_4$ [12]. Meanwhile, the increase of the synthesizing temperature could also accelerate the hydrolysis of $NaBH_4$ to a certain extent. These factors result in the decrease of BH_4^- concentration, which leads to the reduction of B content in the samples. EA results show that the nitrogen content is 0.645, 0.560, and 0.470 wt% for Y_{25}, Y_{50}, and Y_{75}. This result confirms the successful doping of nitrogen in Co-B alloys. Clearly, the higher synthesizing temperature gives rise to a lower nitrogen content. The variation trend of N content is the same as B in these samples. This may be resulting from the formation of ultrafine powder particles from the nucleation between Co atoms and Co-B alloys and it can be regarded as a catalyst, which further promotes the reaction between B and N to form BN precipitation. Then BN precipitation is coated on the surface of Co/CoB particles [13]. Under the combined action of B and N, the loss of active material could be reduced effectively, even in the following charge/discharge cycles.

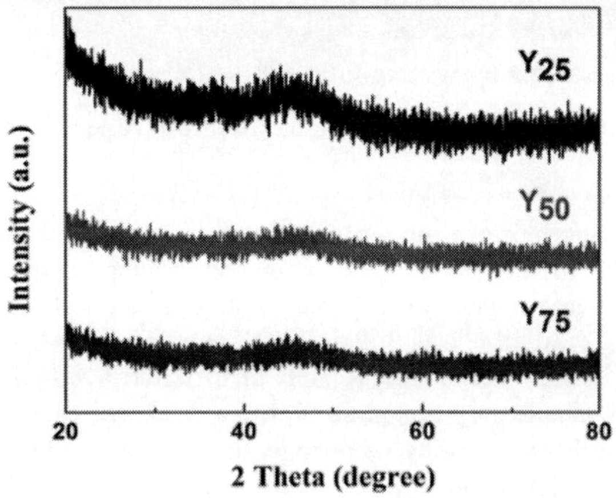

Figure 1. XRD patterns of as-prepared Co-B alloys.

Figure 1 shows the XRD patterns of three Co-B samples. No obvious crystal diffraction peaks are observed except for a broad peak at $2\theta = 45°$. This is consistent with the published results [14], indicating that nitrogen-doped Co-B alloys are in an amorphous state. SEM images and particle size distribution of Co-B alloys are shown in figure 2. It can be found that the three samples are free of agglomeration and the particles are evenly distributed with very fine particles. The average particle size is determined to be 59.7 nm for Y_{25}, 51.9 nm for Y_{50}, and 48.1 nm for Y_{75}. This phenomenon is in line with the reported result [4]: with the increase of reaction temperature, the particle size of the alloy is decreased accordingly. This can be attributed to the change of nucleation rate at different temperatures.

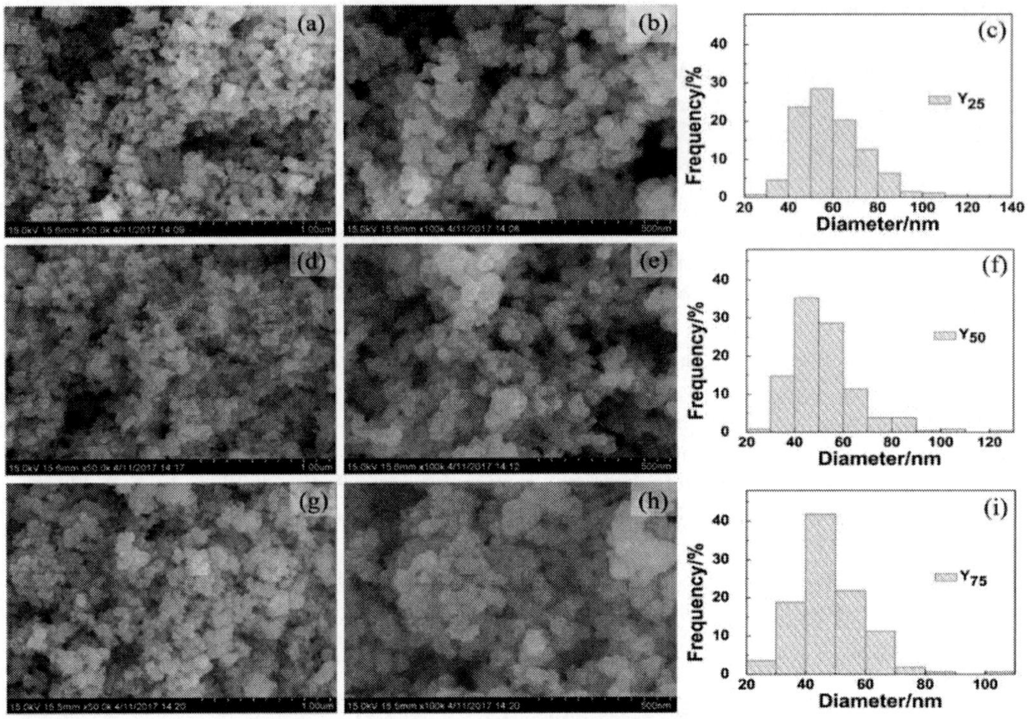

Figure 2. SEM and particle size distribution: (a, b, c) for Y_{25}, (d, e, f) for Y_{50}, and (g, h, i) for Y_{75}.

3.2. Electrochemical properties

Figure 3 shows the cycling performance of three alloy electrodes. The discharge capacity is significantly increased with the decrease of the synthesizing temperature. It must be mentioned that Y_{25} has a maximum discharge capacity of 825.8 mAh/g. After 100 charge/discharge cycles, the discharge capacity of Y_{25}, Y_{50}, and Y_{75} is 369.5, 339.8, and 315.2 mAh/g, respectively. Lv et al. [4] prepared CoB alloy with $CoSO_4{\cdot}7H_2O$ and $NaBH_4$ at different synthetic temperatures without the introduction of urea. Their sample prepared at 60 °C has a discharge capacity of 388.8 mAh/g after 60 cycles. In our case, the discharge capacity of nitrogen-doped Co-B alloys is 429.4, 387.9 and 364.4 mAh/g after 60 cycles for Y_{25}, Y_{50} and Y_{75}, respectively. Zhao et al. prepared CoB alloy through the reaction of $CoCl_2{\cdot}6H_2O$ and $NaBH_4$ under ultrasonication, which effectively improved the electrochemical performance of Co-B alloy with a maximum discharge capacity of 858.1 mAh/g. After 100 cycles, the discharge capacity is 322.6 mAh/g with capacity retention of 37.6% [11]. In our study, each sample achieves its maximum discharge capacity at the first cycle, but the capacity is decreased rapidly in the first few cycles. This can be attributed to the participation of B element during the first few cycles, which contributes to a high specific capacity of primary discharge. But its oxidation process is irreversible. In this case, the alloy electrode has the following reactions in the initial discharge process[14]:

$$CoB + 8OH^- \rightarrow Co(OH)_2 + BO_3^{3-} + 3H_2O + 5e^- \tag{3-1}$$

$$Co(OH)_2 + 2e^- \longleftrightarrow Co + 2OH^- \tag{3-2}$$

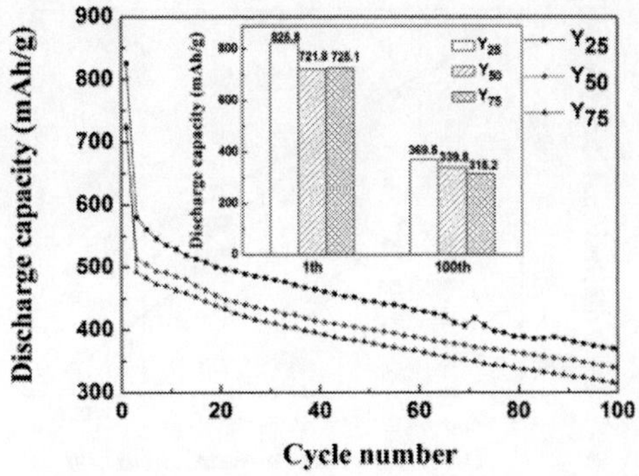

Figure 3. Cycling performance with discharge capacity at the 1st and 100th cycle shown in the inset.

At the same time, BO_3^{3-} is gradually dissolved into alkaline electrolyte, which thus increases the fresh surface of the alloy in this way. This indicates an increment of the electrochemical reaction area, which makes the reaction easier to take place and thereby improves the electrochemical performance [15-17]. As a result, the initial discharge capacity is greatly higher than that of the second one. After the second cycle, the discharge capacity has a slight decrease. Y_{25} has the best electrochemical performance among these samples probably due to more BN precipitation coated on Co/CoB particles, effectively preventing the dissolution of the active material.

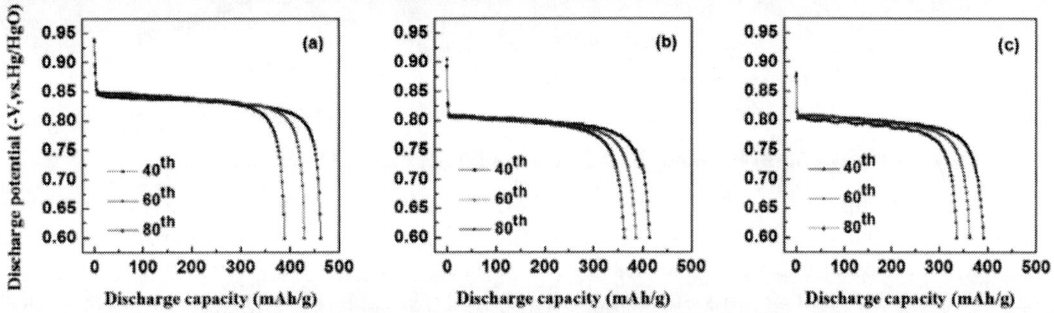

Figure 4. Discharge curves of (a) Y_{25}, (b) Y_{50}, and (c) Y_{75}.

In a general way, the more stable the discharge voltage plateau is, the more uniform the phase structure is and the better the consistency is [18]. As shown from figure 4, three electrodes show good stability for discharge voltage plateau. For example, the discharge voltage plateau of Y_{25} at 40th, 60th and 80th cycle is almost the same (about 0.83 V). However, the discharge voltage plateau of Y_{50} and Y_{75} has been attenuated in different degree. The discharge voltage plateau of Y_{75} is decreased to 0.8 V at the 80th cycle, indicating that the discharge stability is gradually deteriorated as the synthesizing temperature increases. Combined with figure 2, it is found that Y_{25} has larger particle size and more stable discharge voltage plateau. The result is in line with Zhao's report [11], in which larger particle size of the sample prepared under ultrasonication results in the improved electrochemical properties.

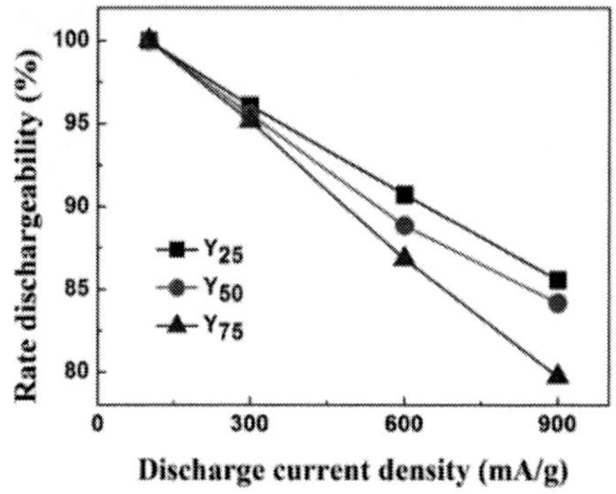

Figure 5. High-rate dischargeability (*HRD*) of as-prepared Co-B electrodes.

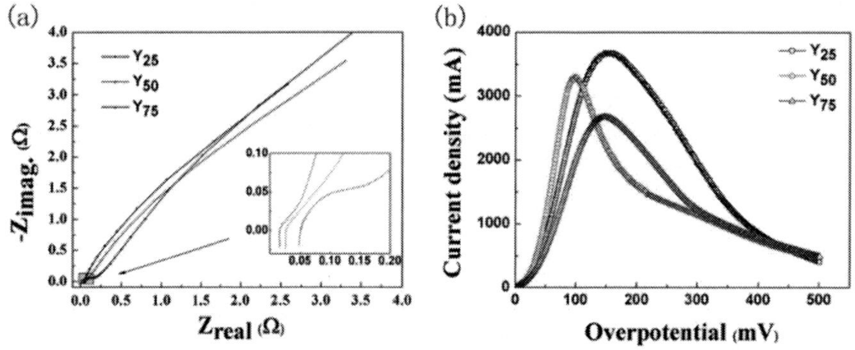

Figure 6. (a) EIS and (b) AP curves of as-prepared Co-B electrodes.

Figure 5 shows the high rate dischargeability (*HRD*) of the alloy electrodes, which is related to the electrochemical kinetics. It can be clearly seen that Y_{25} has superior *HRD* among these samples. Especially, when the discharge current density is 900 mA/g, HRD_{900} of Y_{25} can still keep above 85%, which is higher than that of CoB-30wt%$La_{0.7}Mg_{0.3}Ni_{3.5}$ (HRD_{900} = 82.5%) [14]. In order to further demonstrate the superiority of Y_{25}, we studied the influence of the synthesizing temperature on the kinetic properties. The electrochemical impedance spectroscopy (EIS) and anodic polarization (AP) tests were performed when the discharge depth of the electrodes was 50% after full activation. As can be seen from EIS in figure 6 (a), each curve consists of a semicircle and a straight line [14]. The resistance of the charge transfer reaction is reflected by the semicircle radius. It can be seen from the inset that the radius is increased with the increase of the synthesizing temperature, indicating that the charge transfer resistance of Y_{25} has a minimum value. This is consistent with *HRD* result. Figure 6 (b) showed AP curves of the alloy electrodes. Each curve has a peak that is defined as limit current density I_L. The larger I_L is, the better the diffusion properties in alloy electrode is [19]. It is found that I_L is decreased with the increase of the synthesizing temperature, indicating that the variation of the synthesizing temperature can significantly affect the diffusion behavior in the alloy electrodes. Y_{25} has the highest I_L from AP curves. Also as shown in the aforementioned data, it has the highest charge/discharge capacity, the most stable discharge voltage platform and the superior *HRD* performance. This indicates that the optimal synthesizing temperature is 25 °C in this study.

4. Conclusions

In this paper, urea was introduced into the chemical reduction process to prepare nitrogen-doped Co-B alloy. The synthesizing temperature was paid much attention. By comparison, we found that with the increase of synthesizing temperature, the electrochemical properties of Co-B alloy become gradually worse. When the synthesizing temperature is set at 25 °C, the initial discharge capacity of Co-B alloy is 825.8 mAh/g and after 100 cycles, the discharge specific capacity is still 369.5 mAh/g with capacity retention rate of 44.7%.

Acknowledgements
This research was financially supported by NSFC (51361006, 51401059, 51461010, 51361005, 51371060, U1501242, and 51461011) and the Innovative training program for College Students of Guangxi (No. 201710595140).

References

[1] Jamil M, Ahmad F and Jeon Y J 2016 *Renew. Sust. Energy Rev.* **55** 1181
[2] Liu Y F, Pan H G, Gao M X and Wang Q D 2011 *J. Mater. Chem.* **21(13)** 4743
[3] Taniguchi A, Fujioka N, Ikoma M and Ohta A 2001 *J. Power Sources* **100(1)** 117
[4] Lv D 2009 *Electrochemical study on ultrafine Co-B alloy* (Doctoral thesis from South China University of Technology) p13-23
[5] Li Y 2016 *The influence of battery additives on the performance of anode in Iron-Nickel battery* (Master thesis from Harbin Institute of Technology) p84
[6] Feng Z B, Yang Z H, Huang J H, Xie X E and Zhang Z 2015 *J. Power Sources* **276** 162
[7] Qiu S J, Huang J L, Chu H L, Zou Y J, Xiang C L, Zhang H Z, Xu F, Sun L X and Zhou H Y 2015 *J. Alloys Compd.* **648** 320
[8] Patel N and Miotello A 2015 *Int. J. Hydrogen Energy* **40(3)** 1429
[9] Wang Y D, Ai X P and Yang H X 2004 *Chem. Mater.* **16(24)** 5194
[10] Yang P P, Su M P, Yang X W, Liu G Z, Yu J F, Wu T H, Zhao D X, Zhang T S and Li D Q 2003 *Chinese J. Inorg. Chem.* **19(5)** 485
[11] Zhao W, Liao Y L, Huang J L, Chu H L, Qiu S J, Zou Y J, Xiang C L, Xu F and Sun L X 2017 *Key Eng. Mater.* **727** 751
[12] Durano M M, Tamboli A H and Kim H 2017 *Coll. Surf. A* **520** 355
[13] Shi G M, Dong Y, Huang Y and Zhang J H 2008 *Chinese J. Mater. Res.* **22(4)** 379
[14] Huang J L 2016 *Investigation on electrochemical performance of CoB-type alloys in a wide temperature range* (Master thesis from Guilin University of Electronic Technology) p7-17
[15] Wang Y D, Ai P, Cao Y L and Yang H X 2004 *Electrochem. Commun.* **6(8)** 780
[16] Yu X W and Licht S 2007 *Electrochim. Acta* **52(28)** 8138
[17] Yu X W and Licht S 2008 *J. Power Sources* **179(1)** 407
[18] Lai W H and Yu C Z 1996 *Battery Bimonthly* **26(4)** 189
[19] Kuriyama N, Sakai T, Miyamura H, Uehara I, Ishikawa H and Iwasaki T 1993 *J. Alloys Compd.* **202(1-2)** 183

IOP Conf. Series: Materials Science and Engineering **292** (2017) 012036 doi:10.1088/1757-899X/292/1/012036

The first solid state glycosylation of quercetin

D. V. Orlov, O. I. Lomovsky and I. O. Lomovskiy[1]

Institute of solid state chemistry and mechanochemistry Siberian branch Russian academy of sciences, 630090, Russia, Novosibirsk, Kutateladze str. 18

[1] E-mail: lomovsky@solid.nsc.ru

Abstract. This work presents first results of the study possibility of solid state glycosilation of quercetin. The main results were obtained by thermogravimetric and calorimetric analysis. It was shown that the product of glycosylation occurs in the presence of solid alkali as catalyst.

1. Introduction

Quercetin-3,3',4',5,7–pentahydroxyflavone, natural flavonoid that is in demand in medicine, preventive nutrition and animal husbandry. Quercetin has a high biological activity - it stabilizes cell membranes, passes antioxidants, anti-inflammatory properties, antihistamine action, antitumor, antithrombotic activity [1]. The main problem of quercetin bioavailability and the problem of isolation from plant raw materials is its small solubility in water. Mechanochemical methods for increasing the solubility and efficiency of extraction of flavonoids [2, 3] are based on the use of solid-phase chemical reactions with the formation of soluble forms of this compound during the joint mechanical treatment of mixtures of a solid reagent and flavonoid. Mechanochemical reactions use reagents that form soluble salts of polyphenols, complex compounds, inclusion compounds in branched soluble polymers. In nature, the main and soluble derivatives of quercetin are its glycosides - compounds with soluble sugars, in particular, rutin - a compound with rhamnose and glucose disaccharide.

The aim of the work is to elucidate the conditions under which the solid state mechanochemical reaction of quercetin and glucose occurs.

2. Experimental part

The main reagent used: quercetin (LTD "DiaM" Russia, > 99.3 %), glucose hydrate > 99.6 %. The other additives has purity higher than 99 %. Mechanochemical treatment of mixtures was carried out in planetary ball mill AGO-2 (Ltd. Novic, Russia) with balls acceleration 200 m/s^2, intensity treatment 1 W/g, time of treatment 5 minutes.

Thermoanalytical studies were carried out on differential scanning calorimeter DSC-550 (Instrument Specialist Inc.) in air atmosphere. The temperature range is from 20 to 380 °C, the heating rate is 10 deg / min.

X-ray phase analysis was carried out using Bruker D8 advance diffractometer at room temperature. IR-spectroscopic analysis was carried out using IR-Fourier spectrometer Infralum FT-801. Thin layer chromatography was carried out using Sorbfil CTX-1A plates with silicagel 50-120 mikrons and water/MeOH mixture as mobile phase.

3. Results and discussion

Usually mechanical treatment leads to the disappearance of crystalline phases of the initial reagents. The same effect is in our case - the broadening and disappearance of X-ray reflections is apparently associated with size reduction and amorphization of the components [4].

Figure 1 presents the results of gravimetric analysis of individual components (glucose and quercetin), a mixture of glucose and quercetin powders in a 1: 1 ratio by weight, and the same mixture treated under the conditions indicated above.

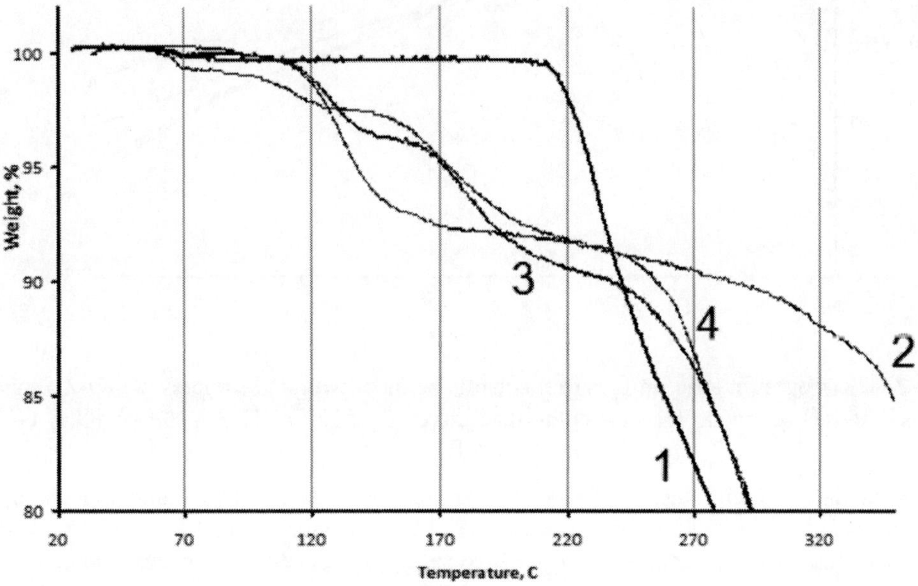

Figure 1. Thermogravimetric profiles of: 1 – glucose, 2 – quercetin, 3 – glucose : quercetin 1 : 1 mixture, 4 – glucose : quercetin mechanochemically treated mixture.

The results of glucose analysis correspond to those known [5]. The glucose is the most thermally stable component of the mixture. A slight decrease in weight at temperatures of 70 is associated with the removal of adsorbed water, dehydration with the formation of anhydrous glucose is observed at 210 °C. Further decomposition of anhydrous glucose occurs at temperatures above 310 °C.

Removal of water from quercetin is observed at 60 °C, the dehydration stage begins at 100 °C. Melting of quercetin occurs at 330 °C, which also corresponds to the literature data [6].

The mechanical mixture of powders of quercetin and glucose possesses stability corresponding to the stability of quercetin. The mixture of mechanically treated powders decomposes faster - at 100-190 °C. Thus, the preliminary mechanical activation of the powder mixtures accelerates the interaction of the components.

The glycosylation reactions of quercetin in the liquid phase provide acceptable rates of yield in aprotonic solvents and presence of catalysts, hydrochloric acid is usually used. Recently, it has been found that alkali can act as a catalyst in some glycosylation reactions [5]. The both types of catalyst were used in our work.

The effect of additives was studied in a mechanically treated mixtures by adding up to 5% by weight concentrated hydrochloric acid (with HCl concentration of 30%) and the addition of solid adipic acid in an ration of 5-20% by weight, but no catalytic effect on glycosylation reaction was observed.

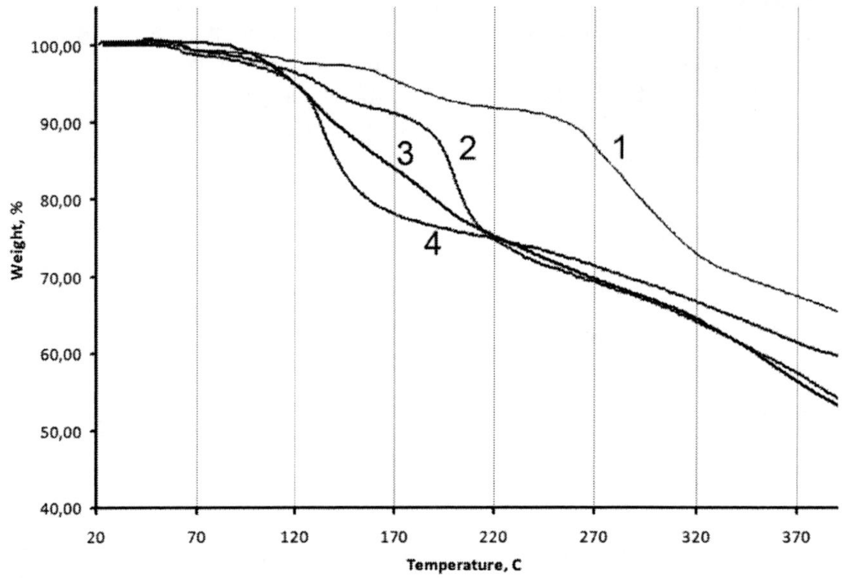

Figure 2. Thermogravimetric profiles of mechanochemically treated samples (AGO-2 5 min) of mixtures glucose : quercetin 1:1 : 1 – without additives, 2 - 5% NaOH, 3 – 10 % NaOH, 4 – 20% NaOH.

The solid alkali (Na_2CO_3, NaOH) was used as an alkaline catalyst. The significant changes were observed in samples treated in the presence of the strongest alkali - sodium hydroxide. Thermoanalytical data of mechanochemically treated samples depended on various amounts of sodium hydroxide are presented in figure 2. It can be seen that the reaction between the components proceeds in the range 120-200 °C is substantially accelerated by the introduction of increasing from 5 to 20% amounts of sodium hydroxide.

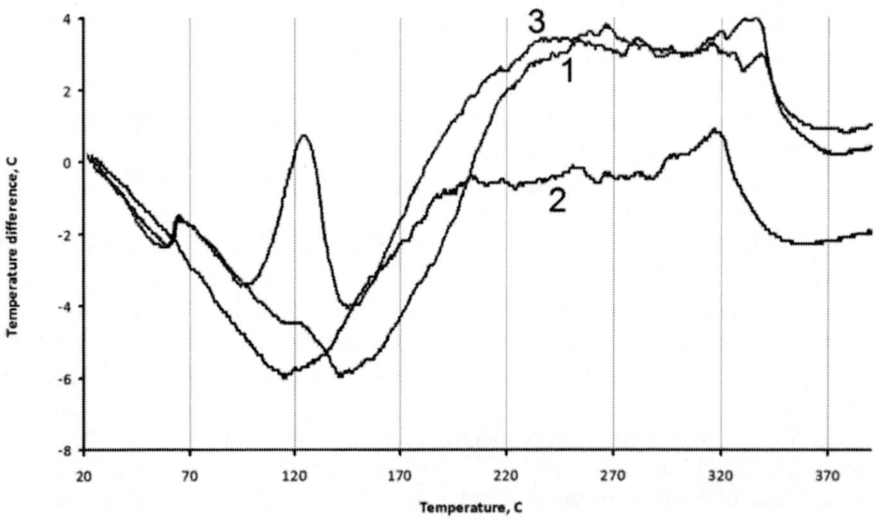

Figure 3. Differential thermal analysis of mechanochemically treated samples (AGO-2 5 min) of mixtures glucose : quercetin 1:1 : 1 - 5% NaOH, 2 – 10 % NaOH, 3 – 20% NaOH.

Figure 3 presents the results of differential thermal analysis of samples of the glucose-quercetin mixture treated mechanochemically in the presence of 5, 10 and 20% of sodium hydroxide. Curve of the sample with 10 % NaOH in the given temperature interval corresponds to the results of [6] for one of the samples of quercetin. The minimum of the endoeffect corresponding to the release of water, with an increase in the amount of alkali up to 10%, is shifted to 110 °C. With an addition amount of 20% against the background of a broad peak of the endoeffect, a relatively narrow peak of the exoeffect appears at 130 °C. This exoeffect could be a due to some reaction proceeds or recrystallization of components, The recrystallization process ware refuted by X-ray analysis. There is no decomposition of components at so low temperature so we assume that this exoeffect corresponds to glycosylation reaction.

The next exoeffect, apparently associated with the decomposition of the formed product, is already observed at a temperature of about 370 °C.

To confirm the nature of the endoeffect 124 ° C, the following additional experiments were carried out. Qualitative analysis by thin-layer chromatography showed that there are the compounds with retention times between reference samples - quercetin and rutine in the samples obtained in the presence of 20% alkali. These compounds can be interpreted as the glycosides of quercetin with 1 glucose molecule.

It is known that glycosylation of sugars in only one position requires special precautions for four of the five hydroxide groups of sugar capable of glycosylation [7]. Under our conditions, when the protection of the hydroxyl groups of the sugar was not carried out, it is also possible to form diglycosides, in which 2 glucose molecules are attached to two possible glycosylation sites of quercetin. This is indirectly indicated by the appearance of new substances with retention times close to the time of retention of rutin-glycoside quercetin with disaccharide ramnogalactose.

The IR spectra of samples obtained with 20% alkali and heated at 124 °C for 5 minutes, in the region characteristic for the manifestation of glycosidic bonds 950-1010 cm^{-1} [8], new signals appeared that were absent in the samples obtained without alkali.

4. Conclusions

The possibility of solid-phase mechanochemical glycosylation of quercetin is shown for the first time, the reaction proceeds in the presence of alkaline catalysts.

The formation of glycosides in absence catalysts is not observed, quercetin decomposes.

The formation of water-soluble glycosides with a different degree of glycosylation is observed after mechanochemical treatment of powder followed by heating of mixtures when stronger bases are used as catalyst.

Acknowledgment

The work was supported by grant of Russian scientific foundation N. 16-13-10200

References

[1] Russo M, Spagnuolo C, Tedesco I, Bilotto S, Russo G L 2012 The flavonoid quercetin in disease prevention and therapy: Facts and fancies *Biochem. Pharmacol.* **83 1** 6-15

[2] Lomovsky O I, Lomovsky I O 2012 *Mechanochemically Assisted Extraction, Contemporary Food Engineering Series. Enhancing Extraction Processes in the Food Industry*. ed. N Lebovka (CRS Press. Taylor & Francis Group Boca Raton London New York) 361 - 398

[3] Lomovsky O I, Lomovsky I O, Orlov D V 2017 Mechanochemical solid acid/base reactions for obtaining biological active preparations and extracting plant materials *Green Chemistry Letters and Reviews* **10 4** 171–85

[4] Orlov D V, Lomovskiy I O, Lomovskiy O I 2017 Quercetin stability under mechanical stress in the presence of plant-based polymers *Periodico Tche Quimica* **14 27** 83–90

[5] Netzsch thermal analysis, https://www.netzsch-thermal-analysis.com/us/materials-applications/us-applications/ thermal-decomposition-of-glucose

[6] da Costa E M, Filho J M B, do Nascumento T G, Macedo R O 2002 Thermal characterization of the quercetin and rutin flavanoids *Thermochim Acta* **392–393** 79 – 84
[7] Jensen K J 2002 O-Glycosylations under neutral or basic conditions *J Chem Soc Perkin Trans* **1** 2219-33
[8] Borghetti G S, Carini J P, Honorato S B, Ayala A P, Moreira J C F, Bassani V L 2012 Physicochemical properties and thermal stability of qurcetin hydrates in the solid statte *Thermochim Acta* **539** 109-14

IOP Conf. Series: Materials Science and Engineering **292** (2017) 012037 doi:10.1088/1757-899X/292/1/012037

Novel $Co(OH)_2$ with cotton-like structure as anode material for alkaline secondary batteries

W Zhao, Y L Liao, S J Qiu[1], H L Chu, Y J Zou, C L Xiang, H Z Zhang, F Xu, L X Sun

Guangxi Key Laboratory of Information Materials, Guangxi Collaborative Innovation Center of Structure and Property for New Energy and Materials, and School of Materials Science and Engineering, Guilin University of Electronic Technology, Guilin 541004, PR China

[1] E-mail: qiushujun@guet.edu.cn

Abstract. The cotton-like $Co(OH)_2$ (S-$Co(OH)_2$) was successfully synthesized and its electrochemical performance was systematically investigated. S-$Co(OH)_2$ was prepared through the "destruction" of the newly formed colloid $Co(OH)_2$ by the reduction using sodium borohydride. The crystal structure and surface morphology were characterized by X-ray diffraction (XRD), scanning electron microscopy (SEM) and energy dispersive spectrometry (EDS). Used as an anode material for alkaline secondary batteries, S-$Co(OH)_2$ sample exhibited better cycle stability, higher electrochemical capacity, and higher rate performance than those of conventional β-$Co(OH)_2$. At a discharge current density of 100 mA/g, the initial discharge capacity of S-$Co(OH)_2$ is 549.3 mAh/g and the discharge capacity is still sustained to be 329.2 mAh/g after 100 charge-discharge cycles with a capacity retention of 59.9%.

1. Introduction

Facing the global concerns of energy crisis and environmental risk, researchers have been trying their best to find new alternative energy sources and/or optimize the energy storage and conversion devices in these years [1-3]. As a device for electrical energy storage, alkaline secondary batteries have aroused extensive attention of many researchers due to their low cost, high specific energy, large power and safety performance [4-6]. However, with the ever-increasing requirement of the energy density and power density for all kinds of electronic equipments, alkaline secondary batteries are in the face of unprecedented challenges. Therefore, it is of great urgency to develop high-performance alkaline secondary batteries.

The anode materials for alkaline secondary batteries mainly including alloys or metals are crucial for their electrochemical properties. AB_5-type alloys, AB_2-type alloys, AB-type intermetallic compounds and Mg-based alloys have been developed to be used as the negative electrode materials for nickel/metal hydride (Ni/MH) secondary batteries [2, 7-9], which is a special kind of alkaline secondary batteries. Cobalt or cobalt-based alloys have been paid much attention thanks to high theoretical electrochemical capacity (909 mAh/g) [10]. They were also used for many devices, such as lithium ion batteries [11], alkaline secondary batteries [12], and supercapacitor [13]. Among Co-based compounds, $Co(OH)_2$ has received great interest owing to its layered structure with large inter-layer spacing, which makes it a promising electrode material. For example, the amorphous $Co(OH)_2$ nanorod arrays on Ni foam exhibited high specific capacity (172 mAh/g at 2 A/g), enhanced rate performance (83 mAh/g at 10 A/g), and superior cycling performance over 6000 cycles [14]. In addition, it can also be used as positive electrode materials in alkaline secondary batteries [14, 15]. As a negative electrode material in alkaline batteries, the electrode reaction is according to the following equation [5]:

$$Co(OH)_2 + 2e^- \leftrightarrow Co + 2OH^- \tag{1}$$

Feng et al. found that the main drawback of β-$Co(OH)_2$ was its higher discharge median voltage of 783 mV (vs. Hg/HgO) in comparison with AB_5-type alloy [16]. Also, the discharge capacity fluctuated during charge-discharge cycles. Toward these problems, we successfully synthesized a cotton-like β-$Co(OH)_2$ in this work. Its electrochemical properties were systematically investigated as anode materials for alkaline secondary batteries.

2. Experimental

2.1. Synthesis of β-Co(OH)₂

β-$Co(OH)_2$ was prepared by a simple hydrothermal method with some modification [17-19]. A typical synthesis process is as follows. 8.0 g cobalt chloride hexahydrate ($CoCl_2{\cdot}6H_2O$) was dissolved in 30 mL ultra-pure water and 3.0 g sodium hydroxide was dissolved in 40 mL ultra-pure water. Then these two solutions were transferred into a Teflon-lined stainless autoclave and homogenously mixed to form colloidal solution. It was maintained at 50 °C for 2 h and then cooled down to room temperature naturally. Finally, the pink precipitate was collected and washed with distilled water and absolute ethanol. β-$Co(OH)_2$ was obtained through drying the precipitate in vacuum at 60 °C for 24 h.

2.2. Synthesis of S-Co(OH)₂

S-$Co(OH)_2$ was synthesized by the "destruction" of the newly formed $Co(OH)_2$ by the reduction of sodium borohydride. 10.0 g cobalt chloride hexahydrate ($CoCl_2{\cdot}6H_2O$) was dissolved in 30 mL ultra-pure water and 3.4 g sodium hydroxide was dissolved in 30 mL ultra-pure water. These two kinds of solution were homogenously mixed and 30 mL sodium borohydride solution (6.9 mol/L) was slowly added at a rate of 1.5 mL/min with the assistance of ultrasonication. Finally, the gray precipitate was washed with distilled water and absolute ethyl alcohol. S-$Co(OH)_2$ was obtained through heating the precipitate under vacuum at 60 °C for 24 h.

2.3. Sample characterization

The crystal structure and surface morphology of the as-prepared samples were characterized by a Bruker D8 Advance diffractometer with Cu Kα radiation and scanning electron microscopy (SEM, QUANTA FEG450). In addition, the chemical composition of the samples was characterized by the energy dispersive spectrometer (EDS, X-Max20). The BET specific surface area was determined using a Quantachrome Autosorb-iQ2 sorptometer.

2.4. Electrochemical tests

The as-prepared $Co(OH)_2$ samples were used as negative electrode materials for alkaline secondary batteries. The working electrodes were fabricated according to our previous papers [5, 6, 20]. Typically, the as-prepared samples and Ni powder as conductive agent in a mass ratio of 1:3 were homogeneously mixed and then pressed into a wafer with a diameter of 13 mm under a pressure of 30 MPa. Each side of the wafer was coated with a rounded foam nickel sheet whose diameter is about 25 mm. Then the sandwich-like electrode was pressed under 6 MPa and spot-weld at the edge of foam nickel tightly. Finally, a nickel wire was attached to this pressed foam nickel sheet by spot welding. The charge/discharge cycles and high rate discharge-ability (*HRD*) were tested on an automatic instrument (LAND-CT2001A). The current density is 300 mA/g for charge and 100 mA/g for discharge in galvanostatic charge-discharge test. For *HRD* test, the charge current density is 300 mA/g and the discharge current density is 300, 600, and 900 mA/g. The electrochemical impedance spectroscopy (EIS) with an alternate current oscillation of 5 mV from 0.005 Hz to 10 kHz and anodic polarization (AP) at a scan rate of 1 mV/s from 0 to 0.5 V were measured on a Zahner Elektrik IM6e electrochemical workstation. All electrochemical tests were conducted in a three-electrode cell containing a $Ni(OH)_2$/NiOOH counter electrode, a Hg/HgO reference electrode, and a fabricated working electrode in 6 M KOH electrolyte. The testing temperature was set at 25 °C.

3. Results and Discussion

3.1. Sample structure

As can be seen from XRD patterns shown in figure 1, the diffraction peaks of two $Co(OH)_2$ samples match well with those of hexagonal phase $Co(OH)_2$ (JCPDS 74-1057) in space group *P-3m1* with the unit cell parameters of a = 0.3183 nm, c = 0.4652 nm, c/a = 1.461 for β-$Co(OH)_2$ and a = 0.3182 nm, c = 0.4656 nm, c/a = 1.463 for S-$Co(OH)_2$. There are no diffraction peaks for α-$Co(OH)_2$ at 22.22° [16] or any other impurity in the XRD patterns, suggesting high purity of the as-prepared samples. However, it is obvious that the peak intensity for S-$Co(OH)_2$ is decreased compared to that of β-$Co(OH)_2$, which may be attributed to the destruction of the crystallization process of $Co(OH)_2$ by $NaBH_4$. Specially, the destruction of crystallization process of the newly-formed colloid $Co(OH)_2$ resulting from $NaBH_4$ reduction may give rise to a low crystallite size. In addition, the effect of ultrasonication may exacerbate this destruction to the further formation of cotton-like structure [21, 22]. This is supported by crystallite size determined from Rietveld refinement. The crystallite size of S-$Co(OH)_2$ was decreased to be 23 nm from 68 nm for β-$Co(OH)_2$, which will be beneficial to the electrochemical reaction such as electron transport.

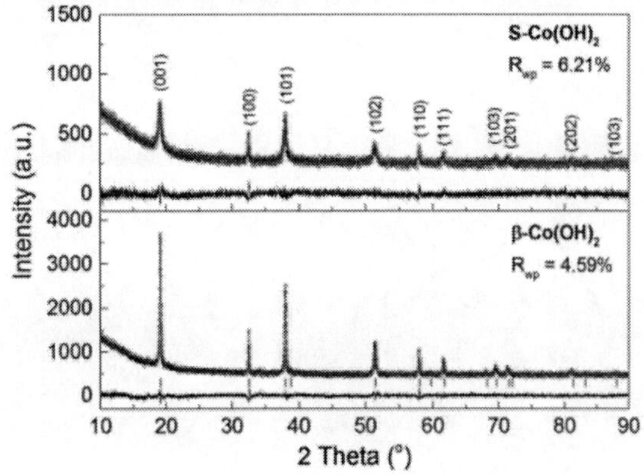

Figure 1. Rietveld refinement of XRD patterns for the as-prepared samples.

Figure 2 shows the SEM micrographs of the as-prepared $Co(OH)_2$ samples. Clearly, β-$Co(OH)_2$ shown in figure 2 (a) is observed to be consisting of agglomeration of nanosheets. As can be seen from magnified image in figure 2 (b), these sheets are in hexagonal shape with a thickness in the range of 50 ~ 200 nm. In contrast, S-$Co(OH)_2$ shown in figure 2 (c and d) has a cotton-like structure, which endows it a higher specific surface area of 42.3 m^2/g than β-$Co(OH)_2$ (18.2 m^2/g). The special structure may be attributed to the destruction of the newly prepared colloid $Co(OH)_2$. When $NaBH_4$ was added into the resultant solution, volume of newly formed colloid $Co(OH)_2$ was destructed by removing some oxygen atoms in its lattice during the reduction [23]. Meanwhile, both the bubbles produced from the reduction reaction of $NaBH_4$ and the effect of ultrasonication could exacerbate the self-assembled process of cotton-like structure [21, 22]. EDS analysis of S-$Co(OH)_2$ in figure 2 (e) shows that there is trace of boron element in this sample, which may come from the reduction of the newly formed $Co(OH)_2$ by $NaBH_4$. Upon the addition of $NaBH_4$, the newly formed colloid $Co(OH)_2$ may partly transform into Co and B with the induction of ultrasonication [21, 22]. Therefore, amorphous CoB alloy was suggested to be included into S-$Co(OH)_2$. However, the diffraction peak of amorphous

CoB alloy generally appearing at 2θ = 40-50 [5] was not observed, which is probably due to its smaller amount and/or stronger diffraction peaks of $Co(OH)_2$ in the XRD pattern.

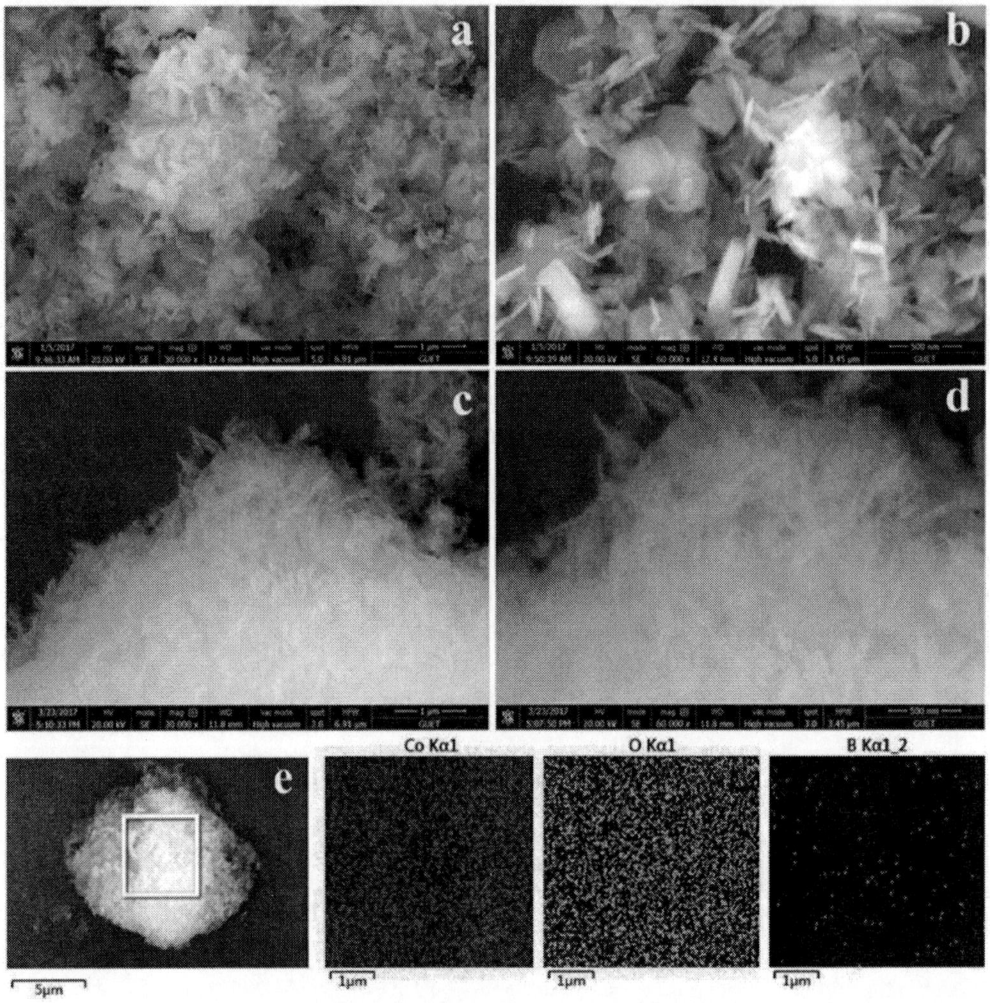

Figure 2. SEM micrographs: (a) and (b) for β-$Co(OH)_2$, (c) and (d) for S-$Co(OH)_2$. (e) Elemental distribution mappings of S-$Co(OH)_2$.

3.2. Electrochemical performance

For β-$Co(OH)_2$ electrode material, it is well accepted that the electron reaction can be expressed as equation (1) [16]. To determine their electrochemical properties as anode materials for alkaline secondary batteries, the charge-discharge capacities of these two samples were investigated and the related electrochemical properties were summarized in table 1. The galvanostatic charge-discharge curves in figure 3 (a) show that there is no activation process for both samples. Moreover, the maximum discharge capacity of S-$Co(OH)_2$ is 549.3 mAh/g, which is higher than that of β-$Co(OH)_2$. It is important to highlight that the discharge capacity of S-$Co(OH)_2$ is obviously higher than that of β-$Co(OH)_2$ at the 100th charge-discharge cycle. After 100 cycles, the discharge capacity of S-$Co(OH)_2$ is still sustained to be 329.2 mAh/g, which increased by about 4% than β-$Co(OH)_2$. Unlike the fluctuation in the curves of cycle life [16], they are very smooth in our case. As shown in figure 3 (b), the discharge voltage plateau for both $Co(OH)_2$ samples is almost the same at about 800 mV, which coincides with those observed in previous literatures [6, 23]. Note that it is higher than β-$Co(OH)_2$ (about 780 mV) synthesized by solid-state chemical reaction [16]. On the contrary, the charging voltage plateau of S-$Co(OH)_2$ sample is a little higher than that of β-$Co(OH)_2$, which indicates that the polarization of β-$Co(OH)_2$ electrode is more serious than S-$Co(OH)_2$. It is generally believed that the polarization has a negative effect on the electrochemical properties of the alloy electrodes [24], which is responsible for the improvement of electrochemical properties for S-$Co(OH)_2$. Figure 3 (c) shows the *HRD* of as-prepared $Co(OH)_2$ samples at 298 K. At a current density of 300, 600, and 900 mA/g, *HRD* of S-$Co(OH)_2$ and β-$Co(OH)_2$ are all higher than reported Co or Co-based alloys [12, 20, 24], which may be attributed to the high specific surface areas [24, 25].

Table 1. Electrochemical properties of the as-prepared samples.

Samples	Activation number	C_{max} (mAh/g)	C_{100} (mAh/g)	C_{100}/C_{max} (%)	HRD_{900}(%)
β-$Co(OH)_2$	1	533.1	316.6	59.4	96.6
S-$Co(OH)_2$	1	549.3	329.2	59.9	97.4

In order to deeply investigate the kinetic characteristics, we conducted the EIS and AP tests for both samples at the discharge depth of 50% after full activation of the electrodes, which are shown in figure 4. According to a previous report [24], there are a semicircle and a straight line in each EIS spectrum, and the semicircle is corresponding to the charge-transfer reaction resistance in the surface of alloy electrode. The big radius of the semicircle means higher charge-transfer resistance, which enhance the difficulty of charge-discharge process. That is to say, the bigger radius of the semicircle represents the higher charge-transfer reaction resistance of the electrode. In AP curves, the peak is related to the limiting current density I_L representing the diffusion rate in $Co(OH)_2$ electrode. The bigger the I_L is, the quicker the diffusion in the alloy electrode is [12, 20, 24]. Clearly we can see that S-$Co(OH)_2$ has a smaller radius of the semicircle in EIS curve and a higher peak in AP curve, which indicates that S-$Co(OH)_2$ sample exhibits better electrode reaction kinetic performance than β-$Co(OH)_2$ from the other point of view.

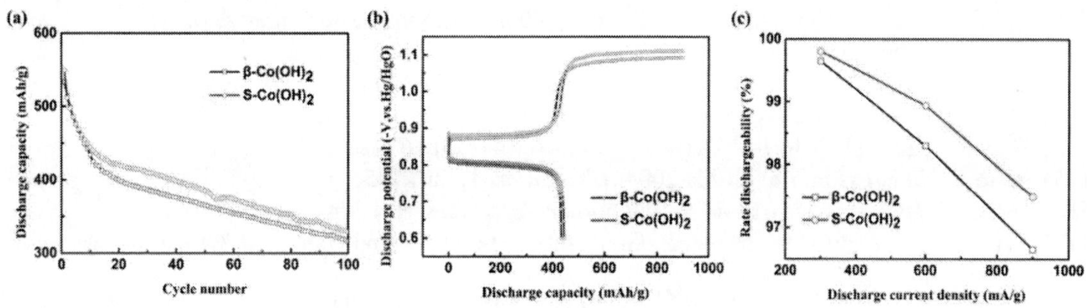

Figure 3. (a) Cycling performance curves, (b) Potential-capacity curves and (c) High rate dischargeability (*HRD*) curves of the as-prepared samples.

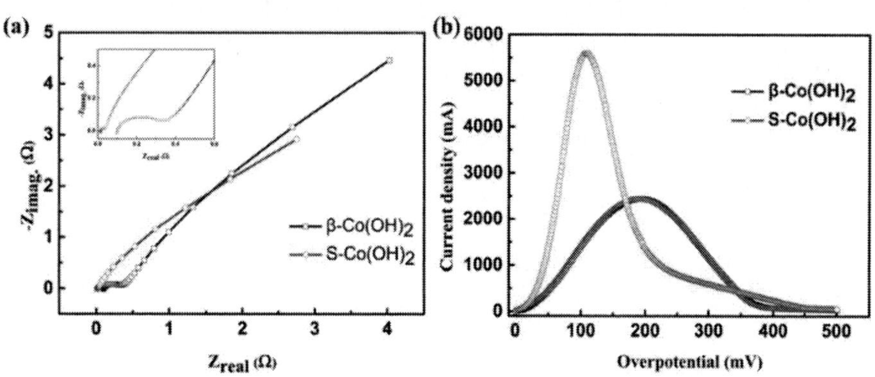

Figure 4. (a) EIS and (b) AP of the as-prepared samples.

4. Conclusions

In summary, the cotton-like $Co(OH)_2$ was successfully synthesized by a simple precipitation reaction by the destruction of the newly formed $Co(OH)_2$. As an anode material in alkaline secondary batteries, S-$Co(OH)_2$ sample has an improved electrochemical performance. After 100 charge/discharge cycles, it still has a discharge capacity of 329.2 mAh/g with a capacity retention of 59.93%, which are better than those of β-$Co(OH)_2$. Moreover, it has a higher rate dischargeability of HRD_{900} = 97.49% thanks to its improved electrochemical kinetic characteristics.

Acknowledgements

This research was financially supported by NSFC (51361006, 51401059, 51461010, 51361005, 51371060, U1501242, and 51461011), the Innovative training program for College Students of Guangxi (No. 201710595140) and the Program for Postgraduate Joint Training Base of GUET-CJYRE (No. 20160513-16-Z).

Reference

[1] Chu S, Cui Y and Liu N 2016 *Nat. Mater.* **16(1)** 16
[2] Liu Y, Cao Y, Huang L, Gao M and Pan H 2011 *J. Alloys Compd.* **509(3)** 675
[3] Reddy M, Subba Rao G and Chowdari B 2013 *Chem. Rev.* **113** 5364
[4] Taniguchi A, Fujioka N, Ikoma M and Ohta A 2001 *J. Power Sources* **100(1-2)** 117
[5] Zhao W, Liao Y, Huang J, Chu H, Qiu S, Zou Y, Xiang C, Xu F and Sun L 2017 *Key Eng. Mater.* **727** 751
[6] Qiu S, Huang J, Chu H, Zou Y, Xiang C, Yan E, Xu F and Sun L 2016 *Metals* **6(11)** 269
[7] Sun L, Wang D, Lin J, Liang F, Wu Y, Zheng W, Cao Z and Wang L 2016 *Diamond Relat. Mater.* **66** 10
[8] Zhang Y, Hou Z, Cai Y, Hu F, Qi Y and Zhao D 2016 *Int. J. Mater. Res.* **107(9)** 1
[9] Qiu S, Huang J, Chu H, Zou Y, Xiang C, Zhang H, Xu F, Sun L and Zhou H 2015 *J. Alloys Compd.* **648** 320
[10] Lv D 2009 *Doctoral thesis from South China University of Technology* 1-13
[11] Nithyadharseni P, Reddy M, Nalini B and Chowdari B 2015 *Mater. Lett.* **150** 24
[12] Qiu S J and et al. 2016 *Int. J. Hydrogen Energy* **41(36)** 16142
[13] Cao L, Xu F, Liang Y and Li H 2004 *Adv. Mater.* **16(20)** 1853
[14] Meng X and Deng D 2017 *ACS Appl. Mater. Interfaces* **9(6)** 5244
[15] Tronel F, Guerlou-Demourgues L, Goubault L, Bernard P and Delmas C 2008 *J. Power Sources* **179(2)** 837
[16] Feng Z, Han S, Li Y, Xu G and Liu Y 2008 *J. Yanshan Univ.* **32** 521
[17] Hou Y, Kondoh H, Shimojo M, Kogure T and Ohta T 2005 *J. Phys. Chem. B* **109(41)** 19094
[18] Xu Z and Zeng H 2000 *Int. J. Inorg. Mater.* **2(2-3)** 187
[19] Yao W, Yang J and Cheng H 2010 *Sci. China Chem.* **40** 1598
[20] Huang J L and et al. 2015 *Int. J. Hydrogen Energy* **40(41)** 14173
[21] Li H, Li H, Zhang J, Dai W and Qiao M 2007 *J. Catal.* **246(2)** 301
[22] Wang C, Li H, Chen S and Li H 2007 *J. Mol. Catal. (China)* **21(3)** 205
[23] Lee D, Kang M, Paek S and Jung H 2016 *Electrochim. Acta* **217** 132
[24] Qiu S J and et al. 2016 *Int. J. Hydrogen Energy* **41(6)** 3955
[25] Humana R, Thomas J, Ruiz F, Real S, Castro E and Visintin A 2012 *Int. J. Hydrogen Energy* **37(19)** 14966

IOP Conf. Series: Materials Science and Engineering **292** (2017) 012038 doi:10.1088/1757-899X/292/1/012038

Recent progress in the synthesis of thiazolo[3,2-a]pyrimidine compounds

F Y Wu, Y Luo and C B Hu[1]

School of Materials and Chemical Engineering, Chongqing University of Arts and Sciences, Chongqing 402160, China

[1] E-mail: bch1002@126.com

Abstract. In this paper, the progress in the synthesis of thiazole[3,2-a]pyrimidine compounds in the field of medicine and pesticide were reviewed. The main synthetic routes include: (i) synthesis of thiazolo[3,2-a]pyrimidines, spiro thiazolo[3,2-a]pyrimidines and pyrazolo[3,4-d]thiazolo[3,2-a]pyrimidines by multicomponent reactions (MCRs). (ii) synthesis of thiazolo[3,2-a]pyrimidines by condensation of pyrimidine-2-thiones, which were obtained by Biginelli reaction between aromatic aldehydes and thiourea, with substituted 2-bromo-1-phenylethanone or chloroacetic acid. (iii) synthesis of pyridothieno-fused thiazolo[3,2-a]pyrimidinones via Pictet–Spengler reaction. (iv) synthesis of pyrido[4,3-d]thiazolo[3,2-a]pyrimidine by reacting 2-aminothiazole with the α, β-unsaturated ketones.

1. Introduction

Systems including fused thiazole and pyrimidine units play important role in organisms because they have good biological potentials. The thiazolo[3,2-a]pyrimidine moiety as an kind of the most important structures, have been shown to display a great diversity of biological activities, including anti-inflammatory [1], antihypertensive [2], psychopharmacological [3], antimicrobial [4], antinociceptive [5], antibacterial [6], antitumor [7] and anti-HSV-1 [8]. They can act as calcium antagonists [9], diacylglycerol (DG) kinase inhibitors [10], HIV-1 reverse transcriptase inhibitors [11], group 2 metabotropic glutamate receptor antagonists [12]. and some thiazolo[3,2-a]pyrimidines have been reported as new acetylcholinesterase (AChE) inhibitors and CDC25B phosphatase inhibitors, used to treat Alzheimer's disease [13], cancer [14]. At present, the thiazolo[3,2-a]pyrimidine skeleton has been widely used as the core of the drug, which represents ritanserin and setoperone [15]. Therefore, the synthesis of thiazolo[3,2-a]pyrimidine compounds have been attracted wide attention of the chemical workers.

In the past decade, variety methods have been reported for the synthesis of thiazolo[3,2-a]pyrimidine derivatives. The main synthetic routes include: (i) synthesis of thiazolo[3,2-a]pyrimidines, spiro thiazolo[3,2-a]pyrimidines and pyrazolo[3,4-d]thiazolo[3,2-a]pyrimidines by multicomponent reactions (MCRs). (ii) synthesis of thiazolo[3,2-a]pyrimidines by condensation of pyrimidine-2-thiones, which were obtained by Biginelli reaction between aromatic aldehydes and thiourea, with substituted 2-bromo-1-phenylethanone or chloroacetic acid. (iii) synthesis of pyridothieno-fused thiazolo[3,2-a]pyrimidinones via Pictet–Spengler reaction. (iv) synthesis of pyrido[4,3-d]thiazolo[3,2-a]pyrimidine by reacting 2-aminothiazole with the α, β-unsaturated ketones.

IOP Conf. Series: Materials Science and Engineering **292** (2017) 012038 doi:10.1088/1757-899X/292/1/012038

2. Synthesis of thiazolo[3,2-a]pyrimidines by Multicomponent Reactions (MCRs)

Multicomponent reactions (MCRs) have become as one of the most efficient strategies for the assembly of complex structural frameworks through the combination of two or more distinct reactions into a one-pot transformation. The strategy is well suited for building complex molecules from readily available starting materials without the need for isolation and purification of intermediates, minimization of waste, labor, time, and cost, provides an eco-friendly protocol [16].

Adib et al. reported a one-pot three-component reaction for the synthesis of 5H-thiazolo[3,2-a]pyrimidines **1** with dialkyl acetylene-dicarboxylates, isocyanides and ethyl 2-oxo-2-(1,3-thiazol-2-ylamino)acetates as raw materials in 2007. The reaction was completed spontaneously in CH_2Cl_2 at room temperature within 24 h, and the yield could reach up to 90% [17].

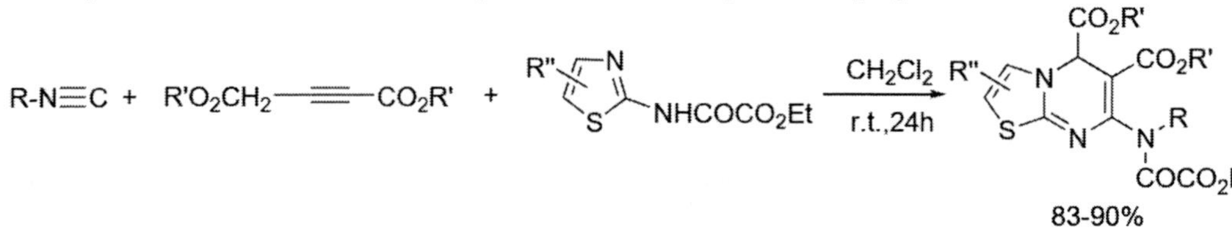

1

Scheme 1. Synthesis of 5H-thiazolo[3,2-a]pyrimidines.

In 2012, Esmaeili et al presented new polysubstituted 3-oxo-2,3-dihydro-5H-[1,3]thiazolo[3,2-a]pyrimidine compounds **2**, which was synthesized by the reaction of dialkyl acetylenedicarboxylates, isocyanides and 2-imino-1,3-thiazolidin-4-one via MCRs[18].

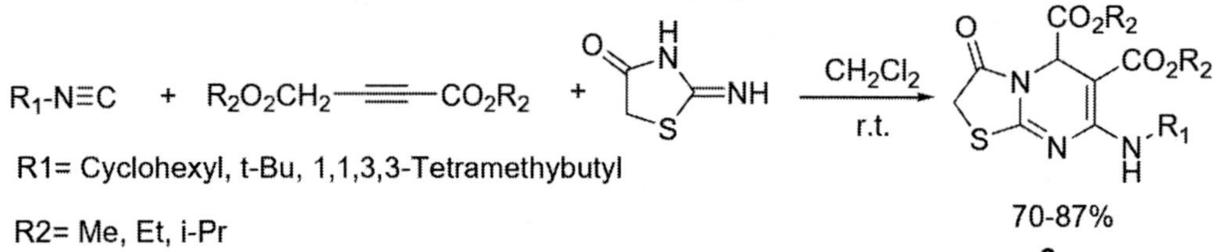

2

Scheme 2. Synthesis of 3-oxo-2,3-dihydro-5H-thiazolo[3,2-a]pyrimidine compounds.

Then a new series of spiro[indole-3,6'-pyrano[2,3-d][1,3]thiazolo[3,2-a]pyrimidine derivatives **3** were synthesized with the reaction of malononitrile, isatin, in the presence of diisopropylethylamine via MCRs by Esmaeili et al, too. This reaction could be completed in 14 to 40 minutes, with a yield of 89-95%, and this new spiro-system was expected to good biological activity because it contains pyranothiazolopyrimidine and oxindole scaffolds[19].

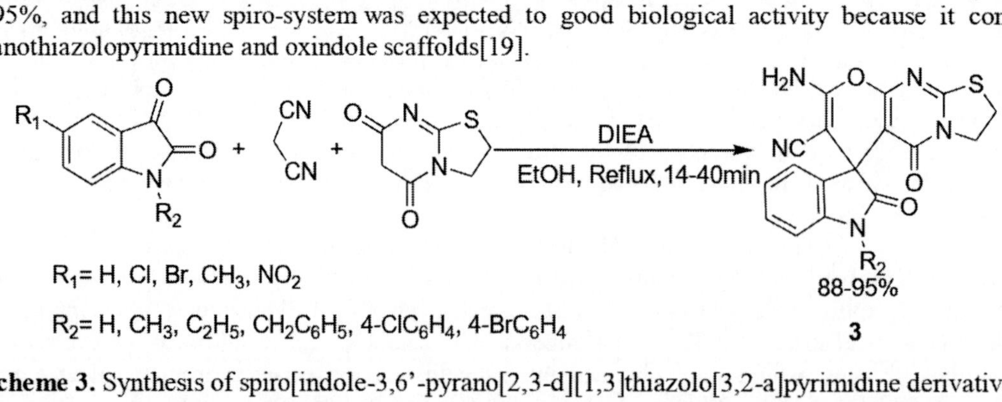

3

Scheme 3. Synthesis of spiro[indole-3,6'-pyrano[2,3-d][1,3]thiazolo[3,2-a]pyrimidine derivatives.

In 2016, a new multicomponent reaction strategy for the facile synthesis of densely functionalized pyrazolo[3,4-d]-thiazolo[3,2-a]pyrimidines **4** with easily obtained α-thiocyanate ketones, aryl aldehydes, and pyrazol-5-amines as raw materials was established by Jiang et al. The reaction could be completed within 20-30 minutes, with a yield of 56-82% in HOAc under microwave irradiation. The remarkable feature of the developed process was that it had been enabled cascade-type formation of C-N and C-C bonds through successive nucleophilic addition/5-exo-trig cyclization/6-endo-trig cyclization steps, this represented a new bicyclization strategy for the construction of N,S-containing heterocycles [20]. The compounds exhibit xanthine oxidase inhibitory activity because it incorporating both the thiazolo[3,2-a]pyrimidine and pyrazole motifs [21].

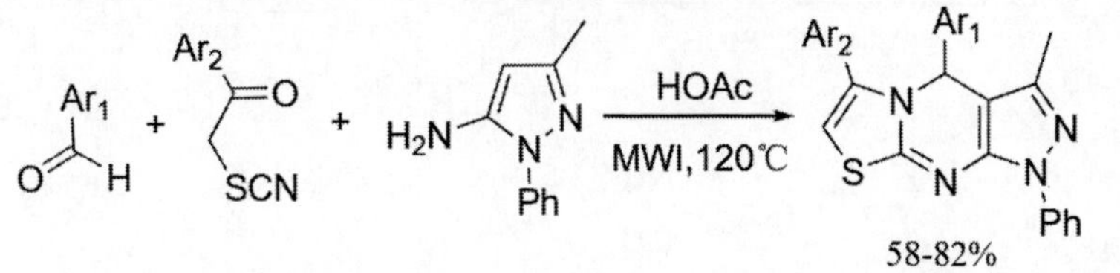

58-82%

4

Ar_1=C_6H_5, 4-ClC_6H_4, 4-BrC_6H_4, 4-FC_6H_4, 2-BrC_6H_4, 4-$CH_3C_6H_4$, 3,4,5-$OCH_3C_6H_2$, 2,3-$OCH_3C_6H_3$, 4-OH-3-$OCH_3C_6H_3$, 2,4-ClC_6H_3, 3-NO_2, 4-NO_2

Ar_2=3-$NO_2C_6H_4$, 4-$NO_2C_6H_4$, C_6H_5

Scheme 4. Synthesis of densely functionalized pyrazolo[3,4-d]-thiazolo[3,2-a]pyrimidines.

3. Synthesis of thiazolo[3,2-a]pyrimidines by condensation of pyrimidine-2-thiones, which were obtained by Biginelli reaction between aromatic aldehydes and thiourea, with substituted 2-bromo-1-phenylethanone or chloroacetic acid

Pyrimidine-2-thione is a kind of important Biginelli reaction products, because of its various pharmacological activities. In recent years, pyrimidine-2-thiones have received the widespread attention and research, and more and more thiazolo[3,2-a]pyrimidines were synthesized with pyrimidine-2-thiones as the raw materials. In 2014, eight 5H-thiazolo[3, 2-a]pyrimidine derivatives **5** were synthesized with Hantzsch condensation of pyrimidine-2-thione, which was obtained by the Biginelli reaction between aromatic aldehydes and thiourea, with substituted 2-bromo-1-phenylethanone by Jia et al. The antitumor activities were evaluated against MCF-7, A549 and HT-29 in vitro by MTT assay, and the results showed that antitumor activity of compound **5c** on HT-29 cells were similar to control drug [22].

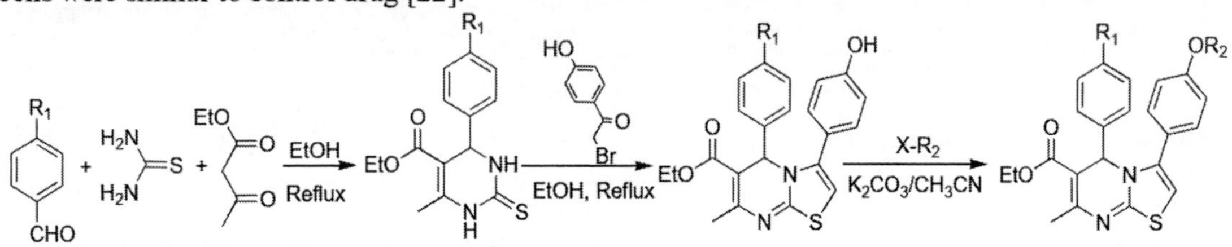

5

5a R_1= OCH_3, R_2= C_2H_5 **5b** R_1 = OCH_3, R_2= $PhCH_2$
5c R_1= Cl, R_2= C_3H_7 **5d** R_1 =Cl, R_2= C_4H_9
5e R_1= OC_2H_5, R_2= C_2H_5 **5f** R_1 = OC_3H_7, R_2= C_3H_7
5g R_1= OC_4H_9, R_2= C_4H_9 **5h** R_1 = OCH_2Ph, R_2= $PhCH_2$

Scheme 5. Synthesis of 5H-thiazolo[3, 2-a]pyrimidine derivatives.

Wang et al. synthesized five new thiazolo[3, 2-a]pyrimidine derivatives **6** containing pyrazole group by Knoevenagel reactions of 4-(5-chloro-3-methyl-1-phenyl-4-pyrazolyl)-3, 4-dihydronpyrimidin-2 (1H) -(thio)one which was prepared by Vilsmeier-Haack and Biginelli reaction using 3-methyl-1-phenyl-2-pyrazolin-5-one as the raw materials with arylaldehyde. The in-vitro anticanceractivities of the new thiazolo[3, 2-a]pyrimidines against human prostate cancer PC-3 cells were tested by MTT assay, and the results showed that **6a-6e** exhibited certain anticancer activities and **6a** exhibited the strongest anticancer activities with IC_{50} of 44.45 μM [23].

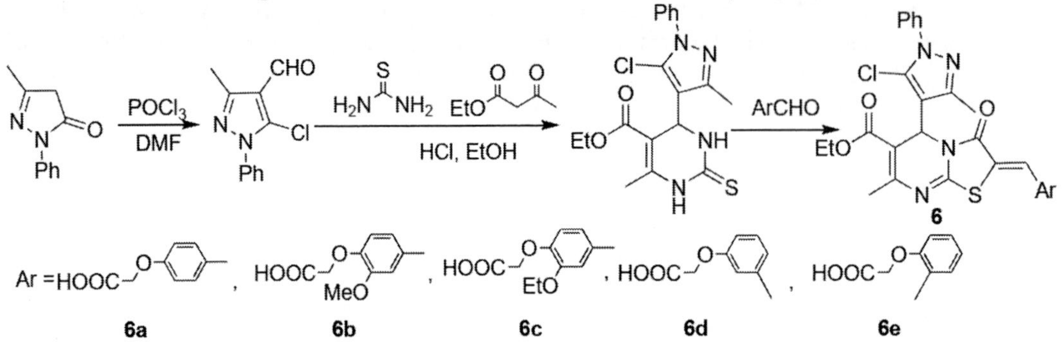

Scheme 6. Synthesis of thiazolo[3, 2-a]pyrimidines derivatives using 3-methyl-1-phenyl-2-pyrazolin-5-one as the raw materials.

Then, nine novel thiazolo[3,2-a]pyrimidine derivatives with pyrazole substituent **7** were synthesized from substituted phenylhydrazines via Vilsmeier-Haack, Biginelli and Knoevenagel reactions. The antiproliferative activities of target compounds against human prostate cancer PC-3 and human hepatoma HepG2 cell lines were evaluated by methyl thiazolyl tetrazolium (MTT) assay. The results displayed that several compounds showed moderate to potent antitumor activity [24].

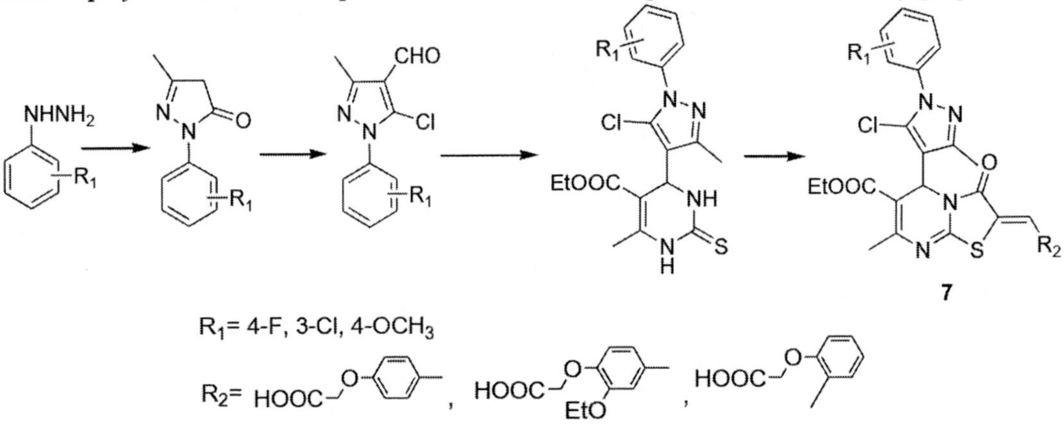

Scheme 7. Synthesis of thiazolo[3, 2-a]pyrimidine derivatives using substituted phenylhydrazines as the starting material.

4. Synthesis of pyridothieno-fused thiazolo[3,2-a]pyrimidinones via Pictet-Spengler reaction

The Pictet–Spengler reaction [25] is a kind of the most excellent strategies for C–C bond formation obtained considerable important for the synthesis of variety products and novel heterocycles of biological interest in synthetic organic chemistry [26], because of its outstanding functional group tolerance, stereo- and regio-selectivity. Wang et al. reported an effective method for the synthesis of pharmacologically excellent, functionalized fused pyrido[3", 2": 4', 5']thieno[3', 2': 2, 3]pyrido [4,5:d][1,3]thiazolo[3,2-a]pyrimidine-4-one derivatives **8** via Pictet–Spengler reaction between 2-(3-aminothieno[2,3-b]pyridin-2-yl)thiazolo[3,2-a]pyrimidin-5-one and aromatic aldehydes with NH_2SO_3H as the catalysis. 2-(3-Aminothieno[2,3-b]pyridin-2-yl)thiazolo[3,2-a]pyrimidin-5-ones were obtained by the condensation of 7-(chloromethyl)-5H-thiazolo[3,2-a]pyrimidin-5-one with 3-cyanopyridine-2-thione via Thorpe–Ziegler isomerization [27]. The advantages of this method were of easy available starting materials, mild reaction conditions, simple product isolation and good yields.

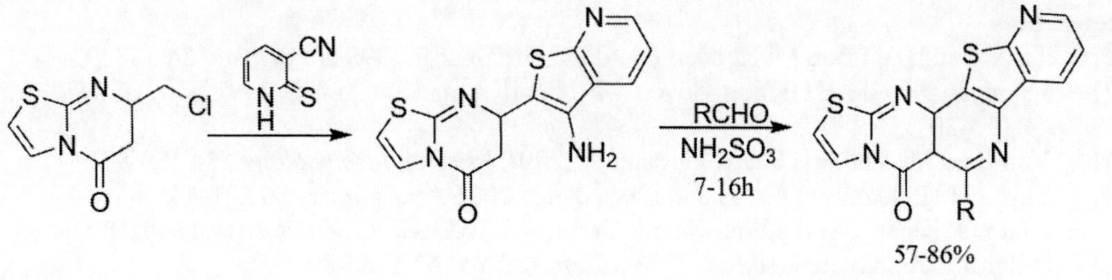

8

Scheme 8. Synthesis of pyridothieno-fused thiazolo[3,2-a]pyrimidinones via Pictet-Spengler reaction.

5. Synthesis of pyrido[4,3-d]thiazolo[3,2-a]pyrimidine by reacting 2-aminothiazole with the α, β-unsaturated ketones

Pyrido[4,3-d]thiazolo[3,2-a]pyrimidines **9** were designed and synthesized by reacting 2-aminothiazole with the α, β-unsaturated ketones which obtained by reacting N-ethyl-piperidone with various benzaldehyde analogues in ethanolic solution of sodium hydroxide. The in-vitro antitumor activity of the obtained compounds were evaluated at the National Cancer Institute (NCI) 60 cell lines panel assay. Some of the synthesized pyrido[4,3-d]thiazolo[3,2-a]pyrimidines showed outstanding broad-spectrum antitumor activity and could be considered as a useful template for future development to obtain more potent antitumor agent(s) [28].

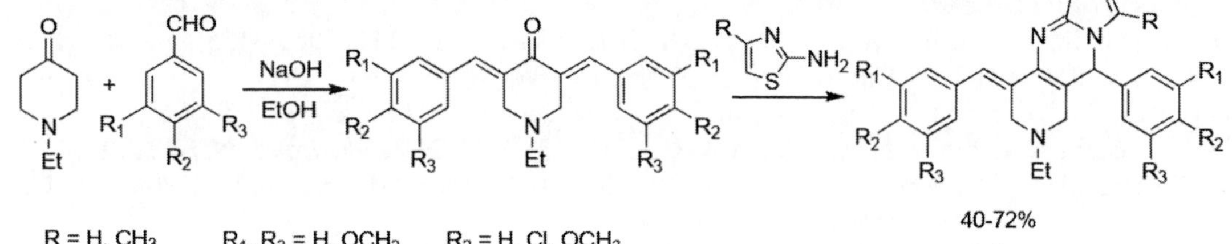

9

Scheme 9. Synthesis of pyrido[4,3-d]thiazolo[3,2-a]pyrimidine by reacting 2-aminothiazole with the α, β-unsaturated ketones.

6. Summary

As was mentioned above, the synthesis methods of thiazolo[3,2-a]pyrimidine compounds were numerous. Over the course of the study, the original synthesis of thiazolo[3,2-a]pyrimidines by condensation of pyrimidine-2-thiones, which were obtained by Biginelli reaction between aromatic aldehydes and thiourea, with substituted 2-bromo-1-phenylethanone reaction was improved and the new MCRs were constantly discovered. At the same time, many new thiazolo[3,2-a]pyrimidine compounds with high activity and high selectivity could be developed as medicine or pesticides. At present, the synthesis and research of thiazolo[3,2-a]pyrimidine compounds have attracted wide attention from the chemical industry at home and abroad. It is believed that in the near future, thiazolo[3,2-a]pyrimidine compounds will play an important role in the field of organic synthesis research and related applications.

Acknowledgements

This work was financially supported by Scientific and Technological Research Program of Chongqing Municipal Education Commission (KJ1501127, KJ1501128, KJ1401109), Natural Science Foundation Project of the Chongqing Science and Technology Commission (cstc2013jcyjA50022, cstc2015shmszx0962) and Chongqing University of Arts and Sciences (Y2015CH32).

References

[1] Tozkoparan B, Ertan M, Kelicen P and Demirdamar R 1999 *Il Farmaco* **54** 588–93

[2] Jeanneau-Nicolle E, Benoit-Guyod M, Namil A and Leclerc G 1992 *Eur. J. Med. Chem.* **27** 115-20

[3] Van Laar M, Volkerts E and Verbaten M 2001 *Psychopharmacology* **154** 189-97

[4] Sayed H H, Shamroukh A H and Rashad A E 2006 *Acta Pharm.* **56** 231-44

[5] Alam O, Khan S A, Siddiqui N and Ahsan W 2010 *Med. Chem. Res.* **19** 1245-58

[6] Coburn R A and Glennon R A 1973 *J. Pharm. Sci.* **62** 1785-89

[7] Abu-Hashem A A, Youssef M M and Hussein H A R 2011 *J. Chin. Chem. Soc.* **58** 41-48

[8] Mohamed S F, Flefel E M, Amr A and Abd EI-Shafy D N 2010 *Eur. J. Med. Chem.* **45** 1494-1501

[9] Balkan A, Uma S, Ertan M and Wiegrebe W 1992 *Pharmazie* **47** 687-88

[10] Nobe K, Miyatake M, Nobe H, Sakai Y, Takashima J and Momose K 2004 *Br. J. Pharmacol.* **143** 166-78

[11] Proudfoot J R 1998 *Expert Opin. Ther. Pat.* **8** 971-82

[12] Wichmann J, Adam G, Kolczewski S, Mutel V and Woltering T 1999 *Bioorg. Med. Chem. Lett.* **9** 1573-76

[13] Zhi H, Chen L, Zhang L, Liu S, Wan D C C, Lin H and Hu C 2008 *ARKIVOC* **xiii** 266-78

[14] Kolb S, Mondésert O, Goddard M, Jullien D, Villoutreix B O, Ducommun B, Garbay C and Braud E 2009 *ChemMedChem.* **4** 633-48

[15] Cornish J W, Maany I, Fudala P J, Ehrman R N, Robbins S J and O'Brien C P 2001 *Drug Alcohol Depend.* **61** 183-89

[16] a)Dömling A, Wang W and Wang K 2012 *Chem. Rev.* **112** 3083-3135; b)Zhu J P and Bienaymé H 2004 *Multicomponent Reactions* (Weinheim Wiley-VCH)

[17] Adib M, Nosrati M, Mahdavi M, Zhu L G and Mirzaei P 2007 *Synlett.* **17** 2703-06

[18] Esmaeili A A, Zangouei M, Fakhari A R and Habibi A 2012 *Tetrahedron Lett.* **53** 1351-53

[19] Esmaeili A A, Amini-Ghalandarabad S, Mesbah F, Tasmimi M, Izadyar M, Fakhari A R and Salimi A R 2015 *Tetrahedron* **71** 2458-62

[20] Hao W J, Zhou P, Wu F Y, Jiang B, Tu S J and Li G 2016 *Eur. J. Org. Chem.* **11** 1968-71

[21] Khobragade C N, Bodade R G, Dawane B S, Konda S G and Khandare N T 2010 *J. Enzyme Inhib. Med. Chem.* **25** 615-21

[22] Wang C H, Jia Y H and Cai D 2014 *Chemistry* **77** 174-77

[23] Wang J, Lin Y, Zhu X L, Nie J, Zhang L and Yao Q Z 2015 *Chinese J. Synth. Chem.* **23** 580-83

[24] Zhang L, Lin Y, Wang J, Zhu X L, Yao Q L and Yao Q Z 2015 *Chin. J. Org. Chem.* **35** 497-504

[25] Pictet A, Spengler T T 1911 *Ber. Dtsch. Chem. Ges.* **44** 2030-36.

[26] (a) Youn S W 2006 *Org. Prep. Proced. Int.* **38** 505–91; (b) Kundu B, Agarwal P K, Sharma S K 2012 *Curr. Org. Synth.* **9** 357-76

[27] Wang D L, Wang D, Yan L, Pan G Y and Yang J N 2016 *Chin. Chem. Lett.* **27** 953-56

[28] Al-Omary F A M, Hassan G S, El-Messery S M and El-Subbagh H I 2012 *Eur. J. Med. Chem.* **47** 65-72

IOP Conf. Series: Materials Science and Engineering **292** (2017) 012039 doi:10.1088/1757-899X/292/1/012039

Analysis of material parameter effects on fluidlastic isolators performance

Q Y Cheng[1], J H Deng, Z Z Feng and F Qian

Science and Technology on Rotorcraft Aeromechanics Laboratory, China Helicopter Research and Development Institute, Jingdezhen 333001, China

[1] E-mail: qy_cheng@163.com

Abstract. Control of vibration in helicopters has always been a complex and challenging task. The fluidlastic isolators become more and more widely used because the fluids are non-toxic, non-corrosive, nonflammable, and compatible with most elastomers and adhesives. In the field of the fluidlastic isolators design, the selection of design parameters of fluid and rubber is very important to obtain efficient vibration-suppressed. Aiming at getting the property of fluidlastic isolator to material design parameters, a dynamic equation is set up based on the dynamic theory. And the dynamic analysis is carried out. The influences of design parameters on the property of fluidlastic isolator are calculated. The material parameters examined are the properties of fluid and rubber. Analysis results showed that the design parameters such as density of fluid, viscosity coefficient of fluid, stiffness of rubber (K1) and loss coefficient of rubber have obvious influence on the performance of isolator. Base on the results of the study it is concluded that the efficient vibration-suppressed can be obtained by the selection of design parameters.

1. Introduction

Helicopter vibration is a critical aspect of helicopter design and a major reason for extended lead times during the aircraft development phase. Control of vibration in helicopters has always been a complex and challenging task. Increasing demands for expanding the flight envelop of helicopters, such as nap of earth flying, high speed, high maneuvers, coupled with the need to improved system reliability and reduce maintenance costs have resulted in more stringent vibration specifications.

Various methods have been applied to vibration control in the engineering field [1-6]. Traditionally, passive isolators and dampers are used to attenuate mechanical vibrations. The traditional approach to passive vibration isolation is to install relatively soft springs or elastomeric isolators to provide a low primary natural frequency [7-10]. These isolators would also incorporate sufficient damping to control resonant response. Soft systems with primary natural frequencies well below the N/rev exciting frequency are required to achieve isolation. Such systems result in large relative motion between the pylon and the airframe due to static loads. Natural frequencies low enough to isolate N/rev vibration would have static (1G) deflections up to 0.50 inches. Since flight controls and power transmission drive shafts cross this interface it is advantageous to keep the relative motion as small as practical. An effective method for isolating the N/rev vibration that did not allow large relative motions between the pylon and airframe was needed.

Fluidlastic products can also provide the long, predictable service life typical of elastomeric rotor bearings, because all of the relative motion across a Fludlastic device is accommodated by shear of the elastomer. There are no sliding seals, bushings, or bearings exposed to the environment [10-11]. Over a long service life, they typically show gradual degradation in appearance, and benign failure modes that allow for replacement long before their performance has been compromised. These characteristics also allow for on-condition replacement eliminating fixed service life intervals [12]. Many fluidlastic products cannot be distinguished from a conventional bonged rubber part by their external appearance.

There are many design parameters such as the density of fluid, viscosity coefficient of fluid, stiffness of rubber (K1) and loss coefficient of rubber that can affect the property of fluidlastic isolator. Each of these parameters is important considerations in the design of a fluidlastic isolator and affects property of fluidlastic isolator to a greater or lesser degree.

The present study is aimed at analyze the influence of material design parameters on property of fluidlastic isolator. The dynamic equation is set up based on the theory of dynamics. And the dynamic analysis is carried out. The influences of design parameters on the property of fluidlastic isolator are calculated.

2. Mathematical model

Figure 1 and figure 2 shows the cross-section and the mechanical model of Fluidlastic Isolator.

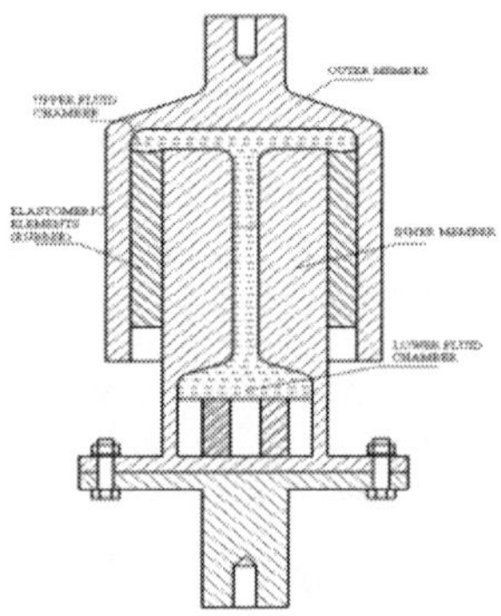

Figure 1. Cross-section of isolator

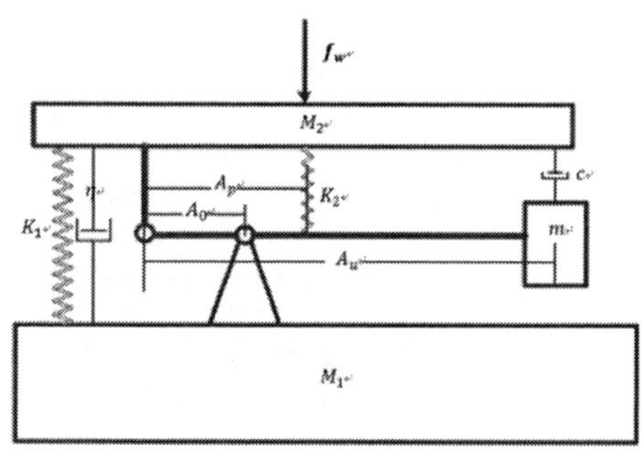

Figure 2. Mechanical model of isolator

According to the figure 1 and figure 2, the equations of motions (EOM) for the fluidlastic tuned isolator can be expressed as:

$$M_1\ddot{x}_1 = -K_1(1+i\eta)(x_1-x_2) - f - [m(\ddot{x}_0+\ddot{x}_2)+c\dot{x}_0] - K_2x_d \tag{1}$$

$$M_2\ddot{x}_2 = K_1(1+i\eta)(x_1-x_2) + f + K_2x_d + c\dot{x}_0 + f_w \tag{2}$$

$$[m(\ddot{x}_0+\ddot{x}_2)+c\dot{x}_0](A_u-A_0) + K_2x_d(A_p-A_0) = fA_0 \tag{3}$$

Where, f is the load that the M_2 imposed on the lever mechanism. And A_p is defined as:

$$A_p = A_uA_0 / A_d \tag{4}$$

The factors are defined as:

$$R_1 = \frac{A_u}{A_0} = \frac{x_0}{x_1-x_2} \tag{5}$$

$$R_2 = \frac{A_d}{A_0} = \frac{x_0}{x_d} \tag{6}$$

Taking Eq.(5) and Eq.(6) into the Eq.(3) ,We obtain the equation of f :

$$f = [mR_1\ddot{x}_1 - m(R_1-1)\ddot{x}_2 + cR_1(\dot{x}_1-\dot{x}_2)](R_1-1) + K_2\frac{R_1}{R_2}(x_1-x_2)\left(\frac{R_1}{R_2}-1\right) \tag{7}$$

Taking Eq.(7) into the Eq.(1) and Eq.(2) ,We obtain the following relations:

$$M_1\ddot{x}_1 + m\ddot{x}_1R_1^2 + c\dot{x}_1R_1^2 + K_1(1+i\eta)x_1 + K_2\frac{R_1^2}{R_2^2}x_1 = mR_1(R_1-1)\ddot{x}_2 + c\dot{x}_2R_1^2 + K_1(1+i\eta)x_2 + K_2\frac{R_1^2}{R_2^2}x_2 \quad (8)$$

$$M_2\ddot{x}_2 + m(R_1-1)^2\ddot{x}_2 + c\dot{x}_2R_1^2 + K_1(1+i\eta)x_2 + K_2\frac{R_1^2}{R_2^2}x_2 = mR_1(R_1-1)\ddot{x}_1 + c\dot{x}_1R_1^2 + K_1(1+i\eta)x_1 + K_2\frac{R_1^2}{R_2^2}x_1 \quad (9)$$

The Eq.(8) and Eq.(9) can be expressed as:

$$\begin{bmatrix} R_2^2M_1 + R_1^2R_2^2m & -R_2^2mR_1(R_1-1) \\ -mR_1(R_1-1)R_2^2 & R_2^2M_2 + R_2^2m(R_1-1)^2 \end{bmatrix}\begin{Bmatrix}\ddot{x}_1\\ \ddot{x}_2\end{Bmatrix} + \begin{bmatrix} R_1^2R_2^2c & -R_1^2R_2^2c \\ -R_1^2R_2^2c & R_1^2R_2^2c \end{bmatrix}\begin{Bmatrix}\dot{x}_1\\ \dot{x}_2\end{Bmatrix} + \begin{bmatrix} R_2^2K_1(1+i\eta)+R_1^2K_2 & -R_2^2K_1(1+i\eta)-R_1^2K_2 \\ -R_2^2K_1(1+i\eta)-R_1^2K_2 & R_2^2K_1(1+i\eta)+R_1^2K_2 \end{bmatrix}\begin{Bmatrix}x_1\\ x_2\end{Bmatrix} = \begin{Bmatrix}0\\ f_w\end{Bmatrix} \quad (10)$$

When the vibration load is a harmonic excitation, the load and displacement can be written as follows:

$$\left.\begin{aligned} f_w &= Fe^{j\omega t} \\ x_1 &= X_1e^{j\omega t} \\ x_2 &= X_2e^{j\omega t} \end{aligned}\right\} \quad (11)$$

The Eq.(11) is substituted into the Eqs.(10). We obtain the following relations:

$$\frac{X_1}{X_2} = \frac{\Delta_1}{\Delta_2} = \frac{\omega^2m(R_1-1)R_1R_2^2 - K_2R_1^2 - jc\omega R_1^2R_2^2 - K_1(1+i\eta)R_2^2}{\omega^2mR_1^2R_2^2 - K_2R_1^2 - jc\omega R_1^2R_2^2 + \omega^2M_1R_2^2 - K_1(1+i\eta)R_2^2} \quad (12)$$

From the equation (12), the Transmissibility of vibration can be written as:

$$T = \left|\frac{X_1}{X_2}\right| = \left|\frac{\Delta_1}{\Delta_2}\right| \quad (13)$$

When the X_1, Δ_1, c, η are all equal to zero, the frequency of undamped isolation system can be obtained:

$$F_r = \frac{1}{2\pi}\sqrt{\frac{K_1R_2^2 + K_2R_1^2}{(R_1-1)R_1R_2^2m}} \quad (14)$$

The amplitude of vibration can be written as:

$$X = \frac{R_1R_2^2F}{\left[K_1(1+i\eta)R_2^2 + K_2R_1^2\right] - m(R_1-1)^2R_2^2\omega^2 + jcR_1^2R_2^2\omega} \quad (15)$$

From the equation (15), the dynamic stiffness of isolation system can be obtained:

$$K^* = \frac{F}{X} = \frac{\left[K_1(1+i\eta)R_2^2 + K_2R_1^2\right] - m(R_1-1)^2R_2^2\omega^2 + jcR_1^2R_2^2\omega}{R_1R_2^2} \quad (16)$$

According to the static equation, the static stiffness of isolation system can be also obtained:

$$K = K_1 + K_2\frac{R_1^2}{R_2^2} \quad (17)$$

3. Results and Discussion

3.1. Stiffness of fluidlastic isolators

A civil helicopter is chosen for this study. The N/rev frequency of this helicopter is close to 25Hz.

The high loads and very small motions of high stiffness isolators are even more challenging. The effectiveness of a tuned isolator in a system can be estimated by measuring the dynamic stiffness (K*) over the frequency range of interest. The K* value at the tuned (N/rev) frequency is a good indicator of the effectiveness of the isolator in the system.

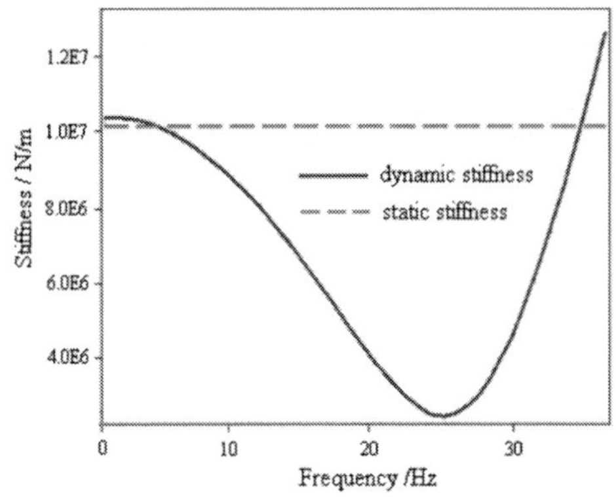

Figure 3. Fluid isolator stiffness versus frequency.

Fluidlastic isolators must also be designed to handle a specific range of input motions. The dynamic stiffness versus frequency curve is shown in figure 3. The dynamic stiffness varied with the frequency. The dynamic stiffness varies from nominally 1.03E7 N/m statically (Frequency is equal to 0 Hz) to approximately 1.91E6 N/m at 25 Hz. The effective dynamic stiffness at 25Hz is less than 20% of the static stiffness. So, the "rigid" isolator can still provide effective performance. However, the static stiffness is relatively high. It is advantageous to keep the relative motion between the pylon and airframe as small as practical.

3.2. Effect of fluid properties

The transmissibility versus fluid density curve is shown in figure 4. The results show that the alteration trend of transmissibility influenced by fluid density is similar to that affected by K1. There is also an optimal value exists for fluid density to get a good effect of vibration-suppressed.

According to the equation (14), the nature frequency of isolator is decreased with the increase of mass of fluid. When the nature frequency of isolator is varied from excitation frequency, the Transmissibility will be increased.

The influence of viscosity coefficient of fluid on the Transmissibility of vibration is shown in figure 5. The results show that the viscosity coefficient has obvious influence on the transmissibility of vibration. Compared to the loss coefficient, Compared the viscosity coefficient with the loss coefficient, the influence on Transmissibility of vibration is relatively higher.

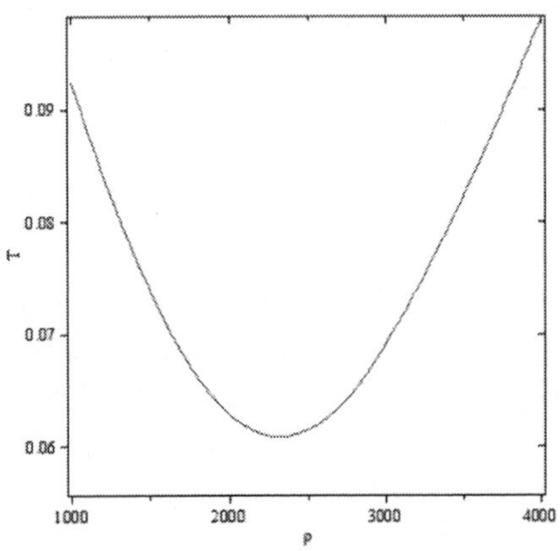

Figure 4. Density versus transmissibility.

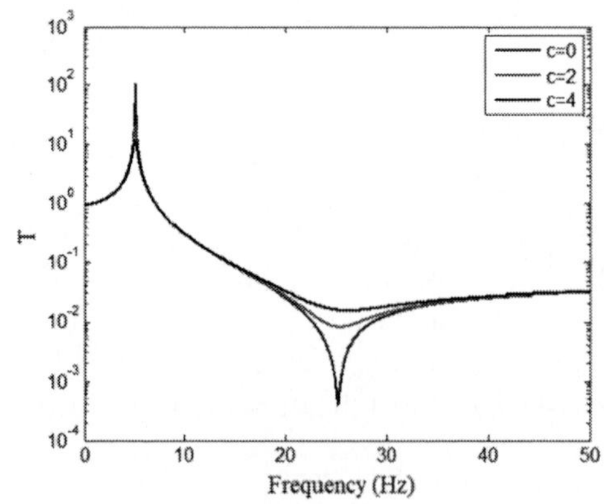

Figure 5. Viscosity coefficient versus transmissibility.

3.3. Effect of rubber properties

The transmissibility versus stiffness of rubber (K1) curve is shown in figure 6. It is observed that the K1 has obvious influence on the transmissibility. When the K1 increase to some value, the transmissibility will be attain to minimum value. This is means that the selection for K1 is very important to get a good effect of vibration-suppressed.

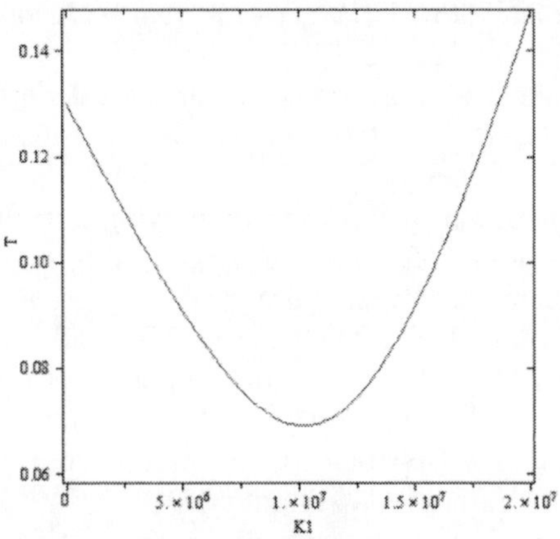

Figure 6. K1 versus Transmissibility.

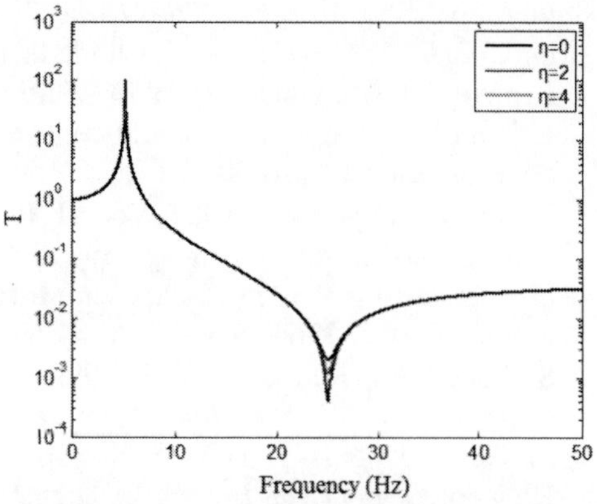

Figure 7. Loss coefficient versus Transmissibility.

The influence of loss coefficient of rubber on the Transmissibility of vibration is shown in figure 7. The results show that the loss coefficient has obvious influence on the transmissibility of vibration. With the increase of the loss coefficient, the transmissibility of vibration decreased. It is means that the efficiency of vibration-suppressed is improved. However, the result is also show that the loss coefficient has no influence on the resonance frequency of isolator.

4. Conclusions

The dynamic equation is set up based on the theory of dynamics. The dynamic analysis is carried out. The influences of design parameters on the property of fluidlastic isolator are calculated. Findings are listed below:

1) The design parameters such as the density of fluid, viscosity coefficient of fluid, stiffness of rubber (K1) and loss coefficient of rubber have significant influence on the performance of isolator. The efficient vibration-suppressed can be obtained by the selection of design parameters. The frequency of isolator can be also designed by the design parameters.

2) The properties of fluid have obvious influence on the performance of isolator. The nature frequency of isolator is affected by fluid density. And it is decreased with the increase of fluid mass. The viscosity coefficient and density of fluid are all have effect on the transmissibility of vibration. The selection for density and viscosity coefficient is very important to get a good effect of vibration-suppressed.

3) The rubber performance has some influence on properties of isolator. The nature frequency of isolator is increased with stiffness of rubber. The transmissibility of vibration is influenced by the stiffness (K1) and loss coefficient. And transmissibility is increased with loss coefficient. So the well vibration-suppressed efficiency can be obtained by selection of K1 and loss coefficient.

References

[1] Crews S T 1987 Rotorcraft Vibration Criteria A New Perspective *Proceedings of the 43rd Annual Forum of the American Helicopter Society* (St. Louis MO)

[2] Strehlow H, Mehlhose R, Znika P and Roth D 1990 Review of MBB's Passive and Active

Vibration Control Activities *The Royal Aeronautical Society in London*

[3] Heilmann J, Swanson D, Askari B A 2003 Vibration Attenuation Through the Use of Active Frahms *American Helicopter Society 59th Annual Forum* (Phoenix Arizona)

[4] Roget B, Chopra I 2004 Wind Tunnel Testing of an Individual Blade Controller for a Dissimilar Rotor *Presented at the American Helicopter Society 60th Annual Forum* (Baltimore, MD)

[5] C T Hugin, C Hatch 2006 Global control of helicopter vibrations using a semi-active vibration control system *47th AIAA/ASME/ASCE/AHS/ASC Structures, Structural Dynamics, and Materials Con* (Newport, Rhode Island)

[6] Brown T J, Mccloud J L 1980 Multicyclic control of a helicopter rotor considering the influence of vibration, load, and control motion *In: 36th annual National Forum of the AHS* (Washington. D C: AHS)

[7] Gupta B P, Logan A H 1984 Wood E R. Higher harmonic control for rotary wing aircraft *In: AIAA/AHS/ASE Aircraft Design Systems and Operations Metting* (San Diego, California)

[8] Halwes, Dennis R 1975 Nodal Beam Inertial Flexure U.S. Patent 3 858-831

[9] Flannelly, W G 1964 Dynamic Anti-resonant Vibration Isolator U.S. Patent 3,322,379

[10] Smith M R., Redinger W S 1999 The Model 427 Pylon Isolation System *American Helicopter Society 55th Annual Forum* (Montreal, Quebec, Canada)

[11] McGuire, Dennis P 1994 Fluidlastic Dampers and Isolators for Vibration Control in Helicopters *American Helicopter Society 50th Annual Forum* (Washington, DC)

[12] Dennis P McGuire 2006 Active Vibration Control Using Fluidlastic Pylon Struts *American Helicopter Society 62nd Annual Forum* (Phoenix, Arizona)

IOP Conf. Series: Materials Science and Engineering **292** (2017) 012040 doi:10.1088/1757-899X/292/1/012040

Effect of addition of Proline, ionic liquid [Choline][Pro] on CO_2 separation properties of poly(amidoamine) dendrimer / poly(ethylene glycol) hybrid membranes

S H Duan[1, 4], T Kai, F A Chowdhury, I Taniguchi[2] and S Kazama[3]

[1] Research Institute of Innovative Technology for the Earth (RITE), 9-2 Kizugawadai, Kizugawa-shi, Kyoto 619-0292, Japan
[2] International Institute for Carbon-Neutral Energy Research (WPI-I2CNER), Kyushu University, 744 Motooka, Nishi-ku, Fukuoka 819-0395, Japan
[3] Technical Development Bureau, Nippon Steel Corporation, 2-6-1 Marunouchi, Chiyoda-ku, Tokyo 100-8071, Japan

[4] E-mail: shduan@rite.or.jp

Abstract. Poly(amidoamine) (PAMAM) dendrimers were incorporated into cross-linked poly(ethylene glycol) (PEGDMA) matrix to improve carbon dioxide (CO_2) separation performance at elevated pressures. In our previous studies, PAMAM/PEGDMA hybrid membranes showed high CO_2 separation properties from CO_2/H_2 mixed gases. In this study, proline, choline and ionic liquid [Choline][Pro] compounds were selected as rate promoters that were used to prepare PAMAM/ PEGDMA hybrid membranes. The effect of addition of proline, choline, IL [Choline][Pro] on separation performance of PAMAM/PEGDMA) hybrid membranes for CO_2/H_2 separation was investigated. Amino acid proline, choline, and IL [Choline][Pro] were used to promote CO_2 and amine reaction. With the addition of [Choline][Pro] into PAMAM/PEG membrane, CO_2 permeance of PAMAM/PEG hybrid membranes are increased up to 46% without any change of selectivity of membrane for CO_2.

1. Introduction

A major reason of global warming is the emissions of CO_2 into the atmosphere, sea level change and extreme weather conditions [1]. CO_2 emission level in the atmosphere will still increase according to the current status of energy supply and demand such in industrial activity and energy though solar and nuclear energy increased. Therefore, reducing the emissions of CO_2 is crucial [2]. Many technologies are currently employed for separation and capture of CO_2 from gas streams [3], with membrane separation is being one of the promising solutions because of its energy efficiency and operation simplicity. Membrane based separation affords CO_2 separation from a pressurized gas stream, for example through the integrated coal gasification combined cycle (IGCC) process. The high-pressure difference between the feed and permeate sides of the membrane provides sufficient driving force to conduct membrane separation in the absence of additional compressors and vacuum pumps, thus affording reduced CO_2 separation costs [4]. In our research group, various CO_2 separation membranes containing PAMAM dendrimers have been developed for CO_2/H_2 separation [5]. Especially, the dendrimer can be readily immobilized in a crosslinked PEG by photopolymerization of PEGDMA in the presence of the dendrimer. The resulting polymeric PAMAM/PEGDMA/4GMAP membrane exhibits excellent CO_2 separation performance over H_2 under humidified conditions [6]. However,

IOP Conf. Series: Materials Science and Engineering **292** (2017) 012040 doi:10.1088/1757-899X/292/1/012040

permeability of CO_2 should be enhanced for practical use. A change in gas permeance with PAMAM/PEGDMA/4GMAP as a function of the membrane thickness was studied. It was found that thinner membranes give higher CO_2 permeation properties as expected, but the selectivity was decreased.

PAMAM dendrimer includes 4 primary amine groups at the terminal end is shown in figure 1. It is widely believed that carbamates could be formed by the reaction between primary/secondary amine and CO_2 in the absence of water [7]. The proposed reaction mechanism is as follows:

$$R^1R^2NH + CO_2 \rightleftarrows R^1R^2NH^+COO^- \tag{1}$$

$$R^1R^2NH^+COO^- + R^1R^2NH \rightleftarrows R^1R^2NCOO^- + R^1R^2NH_2^+ \tag{2}$$

where R^1 and R^2 represent functional groups including H or alkyl groups. The overall reaction is represented by the equation below.

$$2R^1R^2NH + CO_2 \rightleftarrows R^1R^2NCOO^- + R^1R^2NH_2^+ \tag{3}$$

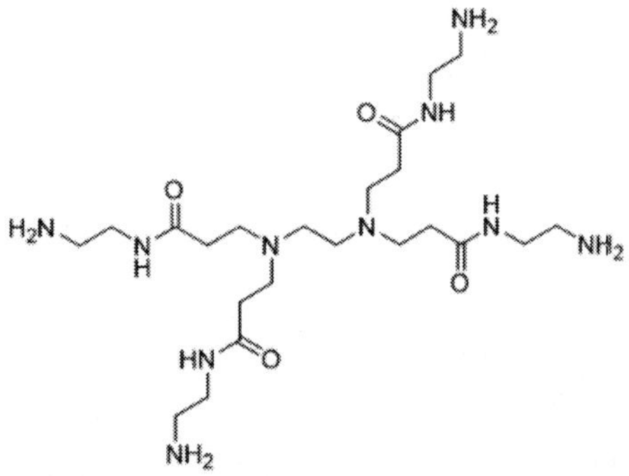

Figure 1. Polyamindoamine (PAMAM) dendrimer (G = 0), MW: 516.7.

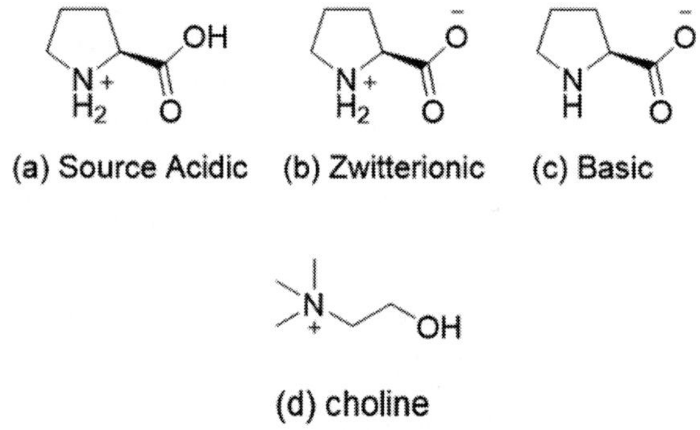

Figure 2. (a-c) Structures and pK of proline (d) Choline.

Proline is used as rate promoter in the reaction of primary amine with CO_2 reaction. With the addition of proline, the reaction mechanism can be different. Aqueous amino acid proline exist in three states: acidic, zwitterionic and basic or deprotonated (figure 2. (a ~ c)). The acidic state and zwitterionic state of the amino acids are much less reactive towards CO_2 than the deprotonated state (basic). Proline promotion mechanism can be summarized by equation (4). This reaction is followed by the removal of a proton from the zwitterionic carbamate by base, PAMAM, to form a neutral carbamate, changing chemical equilibrium of reaction of PAMAM and CO_2 shown in equation 3. It is resulting promotion of reaction of PAMAM and CO_2. Reaction between CO_2 and proline is represented by the equation below.

$$CO_2 + NHR^1R^2COO^- \underset{K_{-1}}{\overset{k_2}{\rightleftarrows}} {}^-OOCNH^+R^1R^2NCOO^- \tag{4}$$

In this work, proline, choline and IL [Choline][Pro] were added to PAMAM/PEG membrane to promote CO_2 and amine reaction. This addition can lead to high CO_2 flux transfer into membrane, as a result high CO_2 permeance Q_{CO2} and high selectivity will achieve. This way we can develop cost-effective technologies for CO_2 capture[8].

2. Experimental

2.1. Synthesis of [Choline] [Pro]

According to references of [9, 10], the route to prepare IL (2-hydroxyethyl)-trimethyl-ammonium (S)-2-pyrrolidinecarboxylic acid salt ([Choline][Pro]) is shown in figure 3. Choline chloride was first converted to choline hydroxide methanol solution. The solution was filtrated to remove KCl salt. The resulting solution was neutralized with equivalent amount of proline. After reaction complete, the evaporation of methanol was done at 40 ℃ under vacuum 10 mmHg to obtain the IL [Choline][Pro] - which is a light-yellow oil at room temperature.

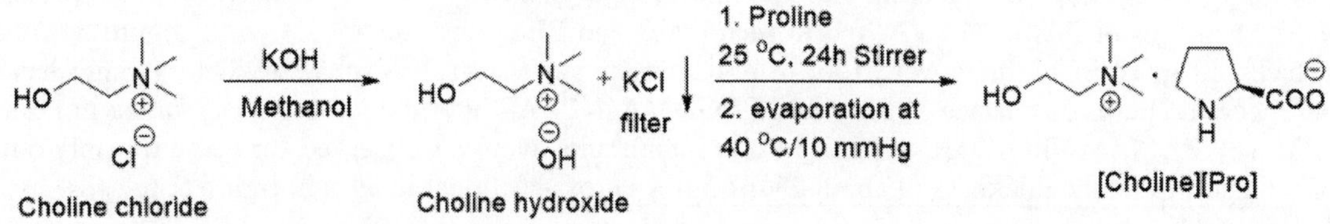

Figure 3. Schematic illustration to synthesized the IL [Choline][Pro]

2.2. Membrane preparation and SEM observation

A polymeric membrane was fabricated by photopolymerization of PEGDMA in the presence of PAMAM dendrimer in water [5]. The compositions of precursor solutions are (1) PAMAM (50 wt.%), PEGDMA (42.5 wt.%) and 4GMAP (7.5 wt.%) and addition (2) proline, (3) choline, (4) [Choline][Pro] with various mixture ratios (2) PAMAM/ proline/ PEGDMA/ 4GMAP = 47.5/2.5/42.5/7.5 (wt%), (3) PAMAM/ choline/ PEGDMA/ 4GMAP= 47.5/2.5/42.5/7.5 (wt%) and (4) PAMAM/ [Choline][Pro]/ PEGDMA/ 4GMAP = 47.5/2.5/42.5/7.5 (wt%) in water. The membrane thickness is controlled by sandwiching the reaction mixture with quartz plates and stainless steel spacers (50μm in thick) followed by the UV curing and transferred onto PES substrate membrane.

The morphologic analysis of composite membrane was performed using scanning electron microscopy (SEM, HitachiS-4800), at an accelerating voltage of 1.0 kV. Cross-sections of the composite membrane were obtained by fracturing the membrane in liquid N_2, and these sections were made conductive by coating with Pt/Pd.

2.2.1. Gas separation test

Test of gas membrane separation diagram setup used herein was provided in reference [5]. A CO_2/H_2 (80/20 by vol.) gas mixture was humidified at 80 % of relative humidity and then fed to a flat-type membrane cell at a flow rate of 100 ml/min. The CO_2 partial pressure under atmospheric condition was 80 kPa, while it was 560 kPa when the total pressure was 700 kPa. 80 % of relative humidity He was supplied to the permeate side of the cell as a sweep gas at a flow rate of 10 ml/min to collect the permeate gas. Operating temperature was 40 °C. CO_2 and H_2 concentration in both feed and permeate gas was measured by a gas chromatography with a pulsed discharge detector (GC-4000, GL Sciences Inc., Tokyo, Japan). In this experimental condition, the pressure ratio of the feed to the permeate side was sufficiently large, and the separation factor was thus accounted as an ideal separation factor (selectivity). Permeance, Q, and selectivity, $\alpha_{CO2/H2}$ were calculated according to equations (5) and (6):

$$\alpha(CO_2/H_2) = \frac{y_{CO_2}/y_{H_2}}{x_{CO_2}/x_{H_2}} \tag{5}$$

$$Q_i = \frac{N_i}{A{\cdot}\Delta p_i{\cdot}t} \tag{6}$$

Qi: permeance of gas I (m^3(STP)/(m^2 s Pa)); $\alpha CO_2/H_2$: CO_2/H_2 separation factor (selectivity); Ni: total flow rate of sweep gas(m^3 (STP)/s); x and y: mole fraction of feed and permeate side, respectively; Pf and Pp: Pressure of feed (Pf) and permeate side (Pa) respectively; A: Effective membrane area (m^2).

3. Results and discussion

3.1. Membrane characterization via photos and SEM images

The photographs of PAMAM/PEG hybrid membranes with various PAMAM dendrimer concentration were shown in figure 4. The solutions for prepared membranes became yellow with addition of choline and [Choline][Pro]. The resulting membranes were transparent as seen in figure 4. Membrane thickness was measured by scanning electron microscope (SEM).

SEM images of PAMAM/PVA hybrid membrane and PES substrate are shown in figure 5. From surfaces image of SEM, there were a lot of pores on the surface of PES substrate. On the other hand, there were no pores on surface of PAMAM/PEGDMA/4GMAP hybrid composite membrane in figure 5. (a), (b). PAMAM/PEGDMA/4GMAP hybrid membrane layer was observed formed uniformly onto PES substrate and the thickness of about 23.5μm from cross-sectional images in figure 5. (c), (d).

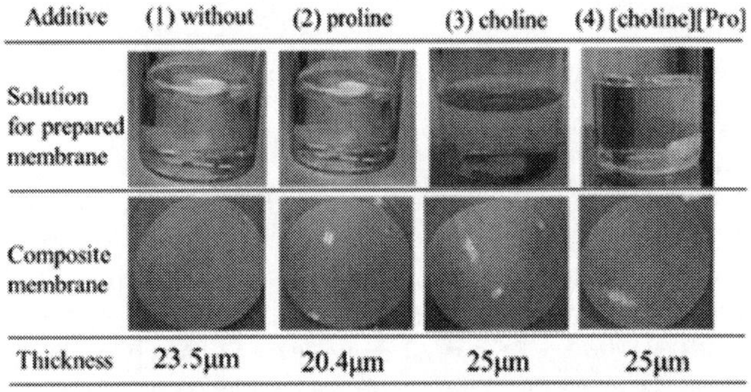

Figure 4. Preparation of solutions, membranes and membrane thickness.

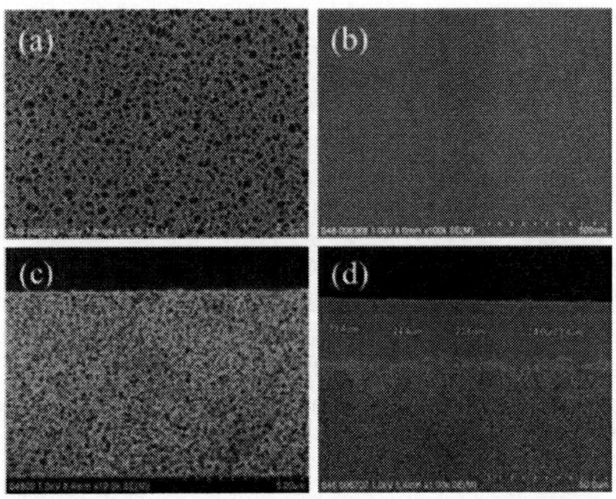

Figure 5. SEM images of (a) PES surface (b) PES cross-section (c) surface of membrane and (d) of PAMAM/PEGDMA/4GMAP = 50.0/42.5/7.5 (wt%) on PES substrate

3.2. Effect of CO_2 partial pressure in feed on CO_2 separation performance

Change in CO_2 separation properties as a function of CO_2 partial pressure were studied with (1)~(4) hybrid membranes and the results were shown in figure 6. As observed, Q_{CO2} decreased with increasing CO_2 partial pressures in the feed gas. Conversely, H_2 permeance was relatively constant and only decreased slightly as CO_2 partial pressure increased. A decrease in the permeance of CO_2 as pressure increases is typically observed when gas separation takes place via an improved transport mechanism process [11], as indicated by the CO_2 chemical sorption i.e., CO_2 sorption might reach near saturation at relatively low CO_2 partial pressure as indicated in (1) PAMAM/PEGDMA/4GMAP = 50.0/42.5/7.5 (wt%) hybrid membrane showed great decrease in Q_{CO2} at high applied CO_2 high pressure. Choline addition showed little decrease in Q_{CO2} at high applied CO_2 high pressure because of its space inhibitor as much reaction-inhibiting PAMAM with CO_2. [Choline][Pro] addition exhibited little decrease in Q_{CO2} and showed high Q_{CO2} at high applied CO_2 high pressure. [Choline][Pro] was found to absorb CO_2 by both chemical and physical sorption occurring at high applied CO_2 pressure [12]. Physical absorption was dominated by Van der Waals forces, and exhibit an easier desorption process resulting in transported CO_2 desorption fast from permeate side of membrane, so [Choline][Pro] addition membrane showed high Q_{CO2}.

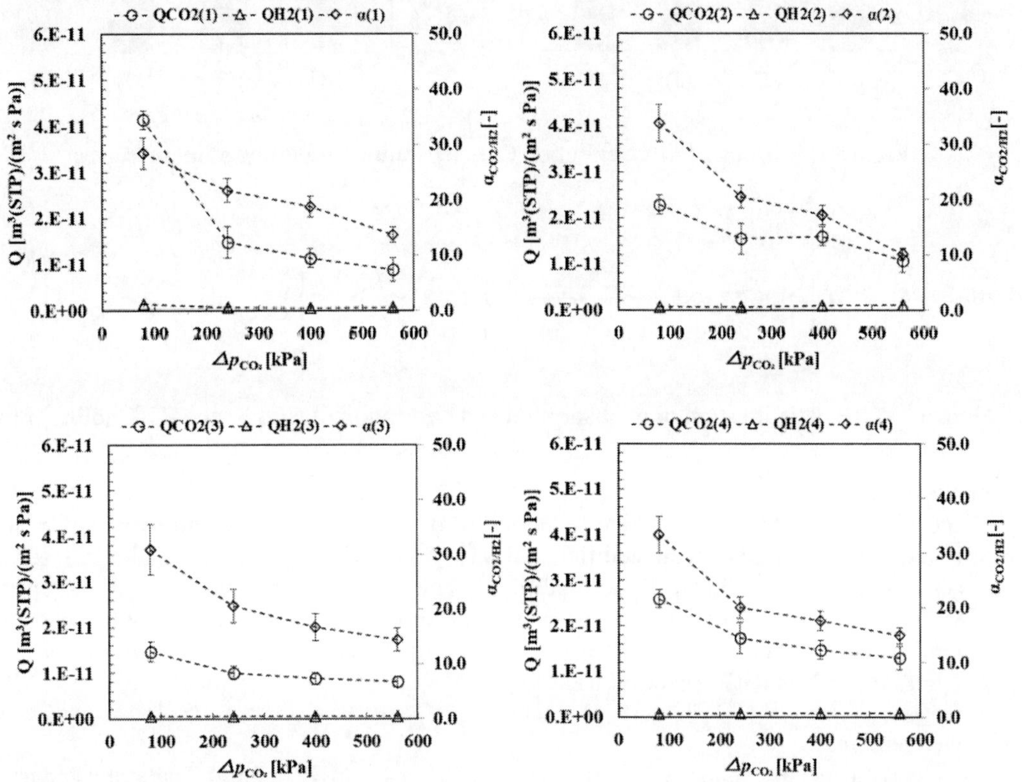

Figure 6. CO_2 separation properties of membranes as a function of CO_2 partial pressure.

3.3. Additive effect on CO_2 separation performance

The effect of various additive on CO_2 separation properties of membrane were shown in figure 7 at 0.56KPa of CO_2 partial pressure in feed. The membrane with addition [Choline][Pro] exhibit high CO_2 separation performance under higher pressurized conditions. The $\alpha(CO_2/H_2)$ and $Q(CO_2)$ are 14.9 and 1.30×10^{-11} $m^3(STP)\ m^{-2}\ s^{-1}\ Pa^{-1}$, respectively, at 0.56 MPa of $\Delta p(CO_2)$ and 0.70 MPa of total pressure. As result, addition of [Choline][Pro] into PAMAM/PEG membrane, Q_{CO2} of PAMAM/PEG hybrid membranes increased 46% without change of selectivity of membrane for CO_2. Addition of [Choline][Pro] in PAMAM/PEG membrane could enhance the rates of absorption and desorption of CO_2 [10] shown in figure 8. [Choline][Pro] are able to absorb CO_2 at low CO_2 partial pressure through a chemical reaction or chemical bonding. If the partial pressure of CO_2 is further increased, there will be a steady increasing as loading with an increase in pressure, providing evidence that both chemical and physical sorption are occurring at high applied CO_2 pressure with [Choline][Pro] [8, 12]. Choline][Pro] shows the effect facilitated transport of CO_2 though membrane by chemical and physical sorption.

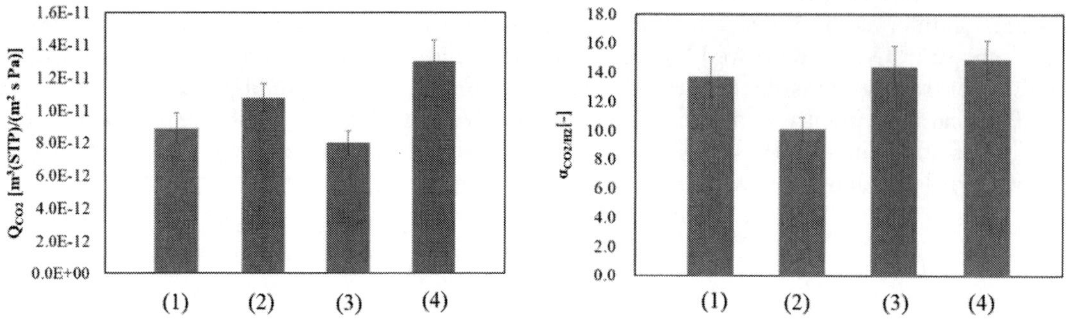

Figure 7. Various additive effect on CO_2 separation properties of membranes.

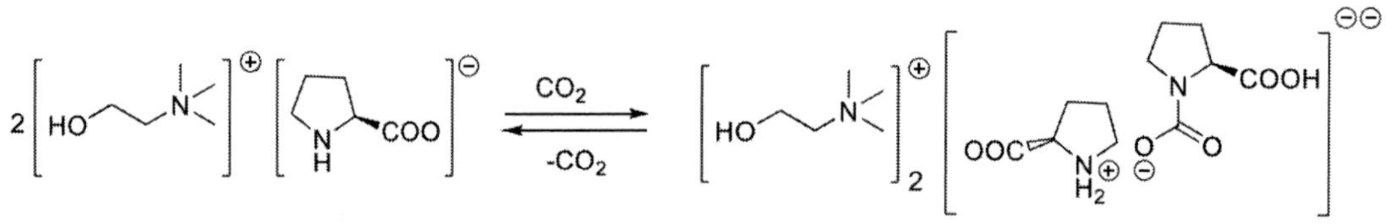

Figure 8. Schematic illustration of absorption and desorption of CO_2 by the IL [Choline][Pro]

4. Conclusions

The effect of addition of proline, choline, [Choline][Pro] on CO_2 separation properties of membrane were studied. The membrane with addition 2.5wt% [Choline][Pro] exhibits high CO_2 separation performance under higher pressurized conditions. The reason is addition of [Choline][Pro] in PAMAM/PEG membrane could enhance the rates of absorption and desorption of CO_2. The PAMAM dendrimer/PEG with addition of [Choline][Pro] shows a potential for CO_2 separation from H_2 at an elevated pressure such as IGCC practical use.

Acknowledgements

This work was financially supported by the Japanese Ministry of Economy, Trade and Industry and Nippon Steel& Sumikin Engineering Co. Ltd. The authors are thankful for the assistance of Ms. Keiko Ida, Ms. Kimiyo Kishi and Ms. Rie Sugimoto throughout the membrane fabrication processes and gas chromatography measurements.

References

[1] Liu A, H Liang 2011 Strategy for promoting low-carbon technology transfer to developing countries: The case of CCS *Energy Policy* **39** 3106–3116
[2] Hegerl G, Stott P 2014 Atmospheric science. From past to future warming Science **343** 844-845.
[3] Nagumo R, Kazama S and Fujioka Y 2009 Techno-economic evaluation of the coal-based integrated gasification combined cycle with CO_2 capture and storage technology *Energy Procedia* **1** 4089–4093
[4] Chu S, Majumdar A 2012 Opportunities and challenges for a sustainable energy future *Nature* **488** 294-303.
[5] Taniguchi I, Duan S, Kazama S and Fujioka Y 2008 Facile fabrication of a novel high performance CO_2 separation membrane: immobilization of poly(amidoamine) dendrimers in poly(ethyleneglycol) networks *Journal of Membrane Science* **322** 277–280
[6] Taniguchi I, Kai T, Duan S and Kazama S 2013 PAMAM Dendrimer Containing Polymeric Membrane for Preferential CO_2 Separation Over H_2—Interplay between CO_2 Separation Properties and Morphology *Energy Procedia* **37** 1067–1075

[7] Kovvali A S, Chen H and Sirkar K K 2000 Dendrimer membranes: A CO_2-selective molecular gate *Journal of the American Chemical Society* **122** 7594–7595

[8] Hu G P, Smith K, Wu Y, Kentish S and Stevens G 2017 Recent progress on the performance of different rate promoters in potassium carbonate solvents for CO_2 capture *Energy Procedia* **114** 2279 – 2286

[9] Hu S Q, Jiang T, Zhang Z F, Zhu A L, Han B X, Song JL, Xie Y and Li W J 2007 Functional ionic liquid from biorenewable materials: synthesis as a and application catalyst in direct aldol reactions *Tetrahedron Letters* **48** 5613-5617

[10] Li X Y, Hou M Q, Zhang Z F, Han B X, Yang G Y, Wang X L and Zou L Z 2008 Absorption of CO_2 by ionic liquid/polyethylene glycol mixture and the thermodynamic parameters, *Green Chemistry* **10** 879 – 884

[11] Zou J, Ho W S W 2006 CO_2-selective polymeric membranes containing amines in crosslinked poly(vinylalcohol) *Journal of Membrane Science* **286** 310–321

[12] Wang C M, Lou H M, Jiang D E, Li H R and Dai S 2010 Carbon dioxide capture by superbase-derived protic ionic liquids *Angewandte Chemie International Edition* **49** 5978–5981

IOP Conf. Series: Materials Science and Engineering **292** (2017) 012041 doi:10.1088/1757-899X/292/1/012041

Influence of heat treatment and hot extrusion on the microstructure and tensile properties of rare earth modified Mg-Zn based alloy

L Y Sheng[1], B J Wang, B N Du, C Lai and T F Xi

Shenzhen Institute, Peking University, southern 7 street, Hi-Tech Park, Shenzhen, China

[1] E-mail: lysheng@yeah.net

Abstract. In the present paper, the Mg-Zn-Y-Nd alloy was prepared by casting, heat treatment and hot extrusion. The microstructure and mechanical properties of the alloys were tested by OM, SEM, TEM and tensile test. The results showed that the $Mg_3Zn_2Y_3$ phase is the main strengthening phase and forms the eutectic structure with α-Mg matrix in the as cast alloy. The strengthening phases semi-continuously connect and separate the α-Mg matrix into cell structure. The average grain size of the as cast alloy is about 60 μm. The heat treatment promotes the solid solution of the strengthening phase and precipitation of small particles inside grain.Compared with the as cast alloy, the heat treatment increases grain size a little and mechanical properties more than 30%. The hot extrusion refines the grain and strengthening phase, which increase the mechanical properties significantly. Moreover, the great deformation by the hot extrusion results in the ultrafine structure and abundant of crystal defects. The intersection of micro-twins lead to the special region with nanometer size.

1. Introduction

Magnesium (Mg) alloys have attracted great attention as the light weight structural materials owing to their high specific strength, low density, ease of recycling and good damping, which can be applied in many fields, including implants, hand tools, sports equipment, automobiles, aerospace applications and electronic equipment [1-4]. Moreover, the recent research exhibited that the clinic trial of coronary stent made by rare earth doped Mg alloy exhibited good therapeutic effect without thrombosis, which demonstrated the good application prospect of Mg based implant [5]. However, the hexagonal close-packed (HCP) crystal structure of Mg limited the number of initiation slip systems during deformation at relative low temperature, which resulted in the poor deformability of Mg alloy [6]. The poor ductility and low strength of Mg alloy handicapped its wide application. To conquer these shortcomings of Mg alloy, many methods had been applied [7-9]. Alloying and thermal processing had been thought as the efficient method to improve the mechanical properties of Mg alloy at room and high temperature.

Recently, the investigations [10-12] on biomedical magnesium alloys revealed that the addition of Zn could improve its mechanical properties and corrosion resistance with little influence on its biocompatibility, because the Zn is the essential element of the human body. Then a series of new magnesium alloys for biomedical application were developed [13-15]. The research of Zhang et al. [10] exhibited that Zn addition in Mg could increases the mechanical performance and biocompatibility, and the Mg-6Zn (wt.%) alloy was the best choice. But the mechanical properties of Mg-Zn alloy is not high enough, because of the low proportion of strengthening precipitate and coarse grain [16,17]. Previous researches [18,19] exhibit that the rare earth could increase the strength of the alloy by enhancing grain boundary. The Cheng et al. studied the minor Y doped Mg alloy and exhibited that the minor Y addition could increase the impact toughness more than one times by changing the deformation mode from twinning to dislocation slip [20]. Kula et al. investigated the deformation behavior of the Mg-Y alloy and revealed that the Y addition could change the fracture mode by improving the dislocation movement [21]. The research of Li et al. investigated the Nd doped Mg-Zn-Zr alloy and exhibited that the Nd presence would increase the strength of the alloy obviously [22]. And moreover the precipitation by the aging treatment improved the strength of the alloy further. The recent investigations [23, 24] on the Y and Nd modified Mg based alloy exhibited that the formation of intermetallic phase was beneficial to the corrosion resistance and strength. Though the addition of Y and Nd could increase the strength of the Mg alloy, however their low solid solubility in Mg matrix promoted them to segregate along the grain boundary during solidification, which was detrimental to the ductility. The research of Liu et al. fabricated the WE54 alloy by severe plastic deformation and revealed that the great deformation could refine the microstructure and redistribute the precipitates, which improved the high temperature mechanical properties obviously [25]. The study of Li et al. exhibited that the great processing deformation and precipitates would improve the mechanical properties synergistically [26]. Though the severe thermal processing could refine microstructure and increase mechanical properties of Mg alloy, its application was limited due to its size limitation. Especially to the tube processing, the hot extrusion was still the main method. Based on the future application of Mg alloy in biomedical implant, the conventional extrusion processing was still needed to study accordingly. Therefore, in the present research, the Mg-Zn-Y-Nd alloy was fabricated by casting and hot extruded. Its microstructure evolution and mechanical properties were investigated simultaneously.

2. Experimental procedure

The Mg-4Zn-1.2Y-0.5Nd alloy (wt.%, Mg-Zn-Y-Nd for short) used in the present study was prepared from pure magnesium (99.9%), Zn (99.9%), Mg-25%Y and Mg-25%Nd master alloys using an electric resistance heating furnace in an SF_6 and CO_2 atmosphere with ratio of 99:1. The molten alloy was poured into a cylindrical metal mould with a diameter of 100 mm. Some ingots were investigated at as-cast state, the others were solid soluted at 720 K for 8 h and quenched in water. The sample for hot extrusion was heated at 700 K for 4 h and then hot extruded with an extrusion ratio of 8:1. Before the hot extrusion, the extrusion mould was heated to 500 K.

The specimen for microstructure observation and tensile test were cut from the alloys with different state, respectively. Microstructural characterizations of all samples were carried out on optical microscope (OM) and scanning electron microscopy (SEM). Phase analysis was carried out by transmission electron microscopy (TEM). The samples for transmission electron microscope (TEM) observation were cut from the alloys with different state and prepared by the conventional method. TEM observation was performed by a JEM-2010 transmission electron microscope operated at 200 kV. The tensile tests were conducted using the SANS-CMT5105 tensile testing machine at room temperature with a strain rate of 1×10^{-3} s^{-1}.

3. Results and discussion

Microstructure of the as-cast Mg-Zn-Y-Nd alloy is shown in figure 1. Clearly, it is mainly composed of α-Mg and strengthening phase along the grain boundary, as shown in figure 1 (a). The strengthening phases semi-continuously connect with each other and exhibits net shape, which is helpful to the strength. The α-Mg matrix is separated by the strengthening phase and forms the cell structure. The statistical analysis on the microstructrue shows that the average grain size is about 60 μm. Further observation on the strengthening phase exhibits that the coarse strengthening phase is composed of eutectic structure, as shown in figure 1 (b). The white strengthening phase and the black α-Mg phase align alternately. The EDS test reveals that the white strengthening phase is rich of Mg,

Zn and Y. According to the recent researches [27, 28], the white phase may be the I phase or W phase, whose morphology would influence the mechanical properties of the alloy obviously.

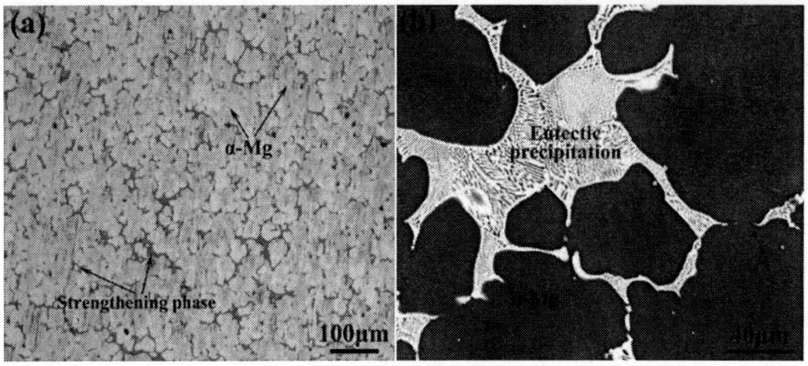

Figure 1. (a) Optical microscopy of the as cast alloy, (b) SEM back-scattered electron images showing the eutectic structure along the grain boundary.

In order to confirm the white strengthening phase, TEM has been applied to analyze the precipitates. TEM observation on the as cast Mg-Zn-Y-Nd alloy shows that there are fine eutectic structure, as shown in figure 2 (a). Two kinds of phases exhibit rod or lamella shape and package with each other. The selective area electron diffraction (SAED) pattern reveal that the strengthening phase is $Mg_3Zn_2Y_3$ phase, which has the face-centered cubic crystal structure with lattice parameter a = 0.6833 nm and the space group of Fm3m, as shown in figure 2 (b). Though no orientation relationship is found between the $Mg_3Zn_2Y_3$ phase and α-Mg matrix, the interface dislocation is found in the TEM observation. Based on the lattice parameters of $Mg_3Zn_2Y_3$ phase and α-Mg phase, there would be great difference between these phase. Moreover, the $Mg_3Zn_2Y_3$/α-Mg eutectic structure is the final solidification region, which would result in great element segregation and increase the crystal misfit along phase interface [29-31].

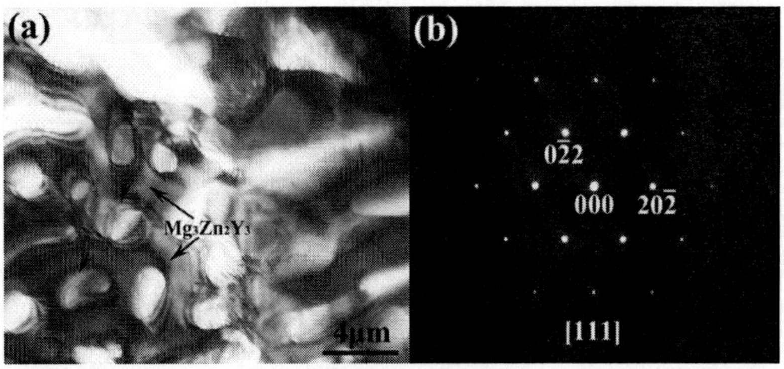

Figure 2. (a) Bright field TEM image of α -Mg and $Mg_3Zn_2Y_3$ eutectic structure (arrow indicates the interface dislocation), (b) SAED pattern of $Mg_3Zn_2Y_3$ phase.

Due to the detrimental effect of the coarse strengthening phase, the heat treatment is carried out to solid solute them. The microstructure of the heat treated Mg-Zn-Y-Nd alloy is exhibit in figure 3. Clearly, the heat treatment promote the solid solution and decrease the amount and size of the $Mg_3Zn_2Y_3$ strengthening phase, as shown in figure 3 (a). The coarse and semi-continuously distributed $Mg_3Zn_2Y_3$ strengthening phase has been change into small one and the distribution is irregular. The original coarse one results in relative big phase in the heat treated alloy. The grain boundary become clear and the strengthening phase is mainly distributed along grain boundary. In addition, there are a lot of small precipitates inside grain. TEM observations on the heat treated alloy reveal that there are two dual-scale precipitates, as shown in figure 3 (b). The big one with hundreds nanometer is mainly along the grain boundary and the small one with scores of namometers is mainly in grain. Further observation on the precipitate in the grain exhibits that small precipitates have two kinds of shape: lamella and rectangle, as shown figure 3 (c). These precipitates with different shape mingle with each other inside grain. The TEM observation on the precipitates adjacent to grain boundary reveals that small precipitates influence the crystal structure, as shown figure 3 (d). The edge dislocation can be observed on the interface of precipitates, which indicates the difference between the precipitate and the α-Mg matrix [32, 33].

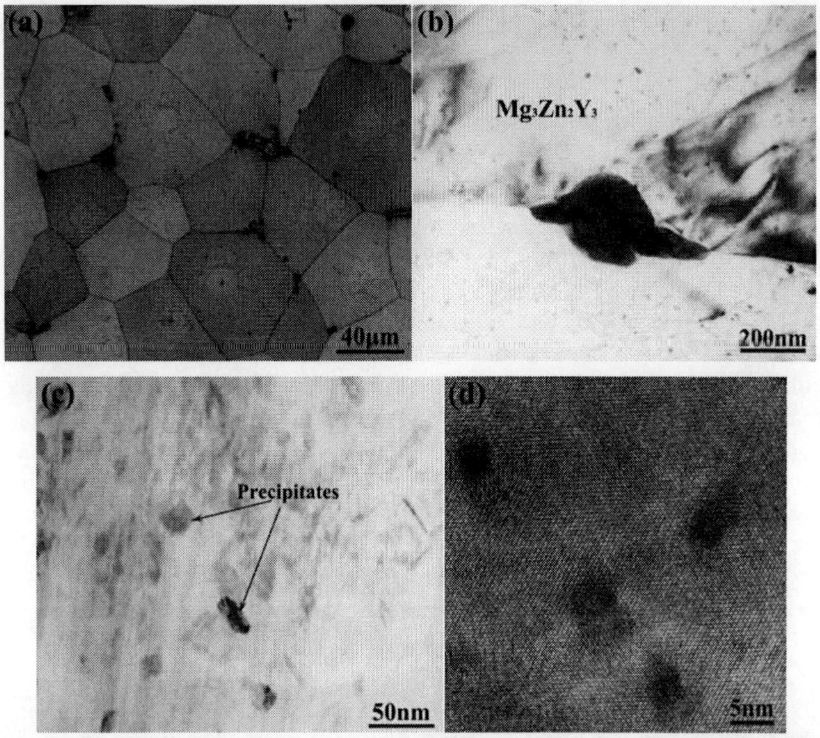

Figure 3. (a) Microstructure of the solid soluted alloy, (b) TEM image of residual precipitating phase, (c) High resolution TEM image showing the nanoscale precipitate along the phase boundary.

The microstructure of hot extruded Mg-Zn-Y-Nd alloy is shown in figure 4. It can be found that the hot extrusion changes the microstructure greatly. Compared with the as cast and heat treated alloy, the hot extrusion refine the grain obviously, as shown in figure 4 (a). The hot extruded alloy exhibits typical dual-scale crystal structure. The relative fine grains exhibits the strip distribution along the extrusion direction. The relative coarse grain is almost surrounded by the fine grains. Moreover, the $Mg_3Zn_2Y_3$ strengthening phases are fragmented into small ones and semi-continuously distributed along the extrusion direction as the strip shape. The cooperation of fragmented strengthening phase and extrusion promotes the refinement of the microstructure further [34-36]. The observation on the cross section of the hot extruded alloy exhibits that the $Mg_3Zn_2Y_3$ strengthening phases mainly distribute along the grain boundary and form the semi-continuous annular shape, as shown in figure 4 (b). Such a morphology should be ascribed to the original microstructure of as cast alloy. Though the heat treatment promotes the solid solution of the precipitates, it exceeds the solid solubility of the α-Mg matrix. The nonequilibrium state will be broken and the precipitation would be prior to occur in the region which has coarse $Mg_3Zn_2Y_3$ strengthening phases. The statistical analysis on the microstructure shows that the average grain size of the hot extruded alloy is about 20μm.

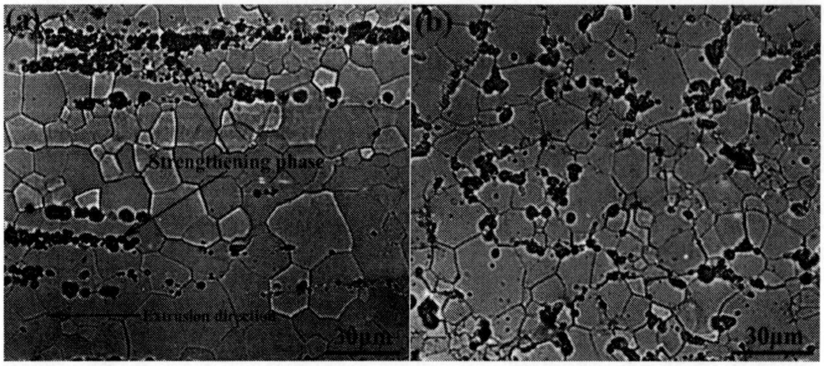

Figure 4. (a) Microstructure of the hot extruded alloy along the extrusion direction, (b) Microstructure of the hot extruded alloy perpendicular to the extrusion.

TEM observation on the hot extruded alloy reveals that the hot extrusion refines the microstructure greatly, as shown in figure 5. Along the fine grain region, the ultrafine structure with about hundreds nanometer is observed, as shown in figure 5 (a). Based on the Moreover, it also finds that there are great amount of crystal defects such as dislocations, micro-twins and stacking faults in the ultrafine structure. These crystal defects always traverse the whole ultrafine structure and some ones have intersected with each other. The detailed observation on the lamellar structure finds that they are composed of micro-twins and stacking faults, as shown in figure 5 (b). According to the recent investigations [37-39], the great deformation would lead to the coexistence of micro-twins and stacking faults. In addition, the formation of these crystal defects would generate the barrier to restrict the movement of dislocation, which is helpful to the strength. Moreover, the intersection of the micro-twins results in the special region, as the arrows indicated. The atom array is really different from the micro-twins and related with the thickness of the micro-twins.

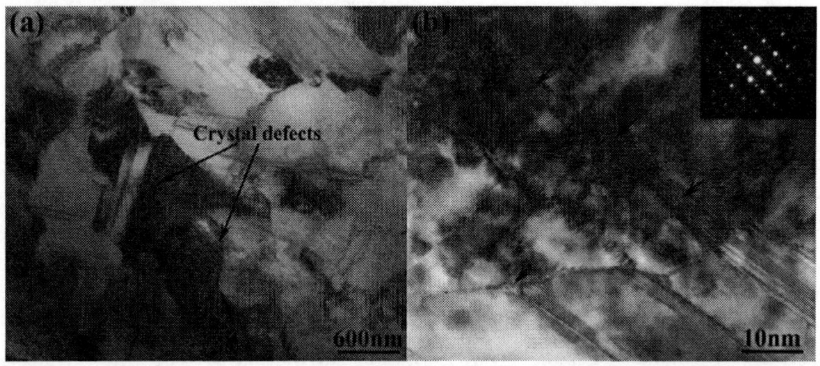

Figure 5. (a) Bright field TEM image of the ultrafine grain and crystal defects, (b) TEM image of the stacking faults and their intersection (Inset picture showing the SAED pattern).

The tensile properties of the Mg-Zn-Y-Nd alloy with different states are shown in figure 6. Clearly, with the heat treatment and hot extrusion, the mechanical properties of the alloy increase greatly. The strength and ductility of the heat treated alloy are more than 30% higher than that of the as cast alloy. However, the hot extrusion increases the strength and ductility about two times of the as cast alloy. According to the recent researches [23, 40], the mechanical properties of alloy is related with the microstructue greatly. The matrix with fine microstructure and well distributed strengthening phase would increase the strength and ductility simultaneously. In the present study, the heat treatment promotes the solid solution of strengthening phase and eliminates the bulk precipitate, which could contribute to the ductility. Moreover, the solid solution and ultrafine precipitate are also beneficial to the strength. Though the grain coarsening during heat treatment influences the mechanical properties, the heat treated alloy is still possesses better strength and ductility than the as cast alloy. In the hot extruded alloy, the extrusion refines the microstructure and the strengthening phase significantly, which contributes to the strength and ductility simultaneously [40, 41]. Though the strengthening phases still prefer to segregate along grain boundary, their small size could exert little influence on the synergistically deformability of the alloy. Therefore, the hot extrude alloy possesses the best mechanical properties.

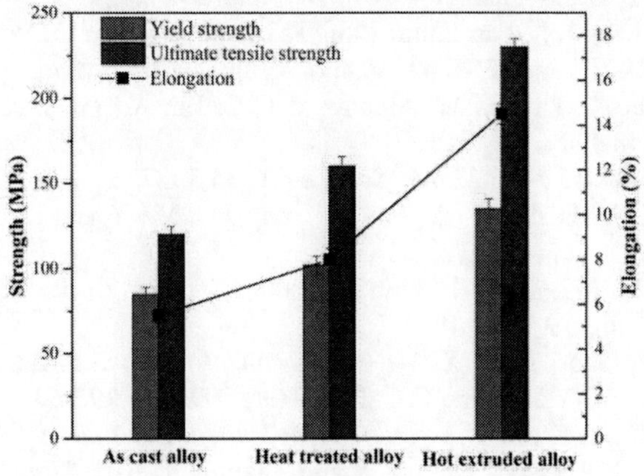

Figure 6. Mechanical properties of the Mg-Zn-Y-Nd alloy with different states at room temperature.

4. Conclusions

1) In the as cast Mg-Zn-Y-Nd alloy, the $Mg_3Zn_2Y_3$ phase is the main strengthening phase and forms the eutectic structure with α-Mg matrix. The strengthening phases semi-continuously connect and separate the α-Mg matrix into cell structure. The average grain size of the as cast alloy is about 60 μm.

2) The heat treatment promotes the solid solution of the strengthening phase and precipitation of small particles inside grain. Compared with the as cast alloy, the heat treatment increases grain size a little and mechanical properties more than 30%.

3) The hot extrusion refines the grain and strengthening phase, which increases the mechanical properties significantly. Moreover, the great deformation by the hot extrusion results in the ultrafine structure and abundant of crystal defects. The intersection of micro-twins lead to the special region with nanometer size.

Acknowledgements

The authors are grateful to the Shenzhen Technology Innovation Plan (CXZZ20140731091722497 and CXZZ20140419114548507) and the Strategic New Industry Development Special Foundation of Shenzhen (JCYJ20150529162228734, JCYJ20160427100211076, JCYJ20170306141749970, JCYJ20160329161539885, JCYJ20150625155931806 and JCYJ20160427170611414) for financial support.

References

[1] Li C Q, Xu D K, Wang B J, Sheng L Y and Han E H 2016 *J Mater. Sci. Technol.* **32** 1232
[2] Liu Z, Wang Y and Wang Z G 2000 *J. Mater. Res.* **14** 449
[3] Li Z J, Gu X N, Lou S Q and Zheng Y F 2008 *Biomaterials* **29** 1329
[4] Mordike B L and Ebert T 2001 *Mater. Sci. Eng. A* **302** 37
[5] Haude M, Ince H, Abizaid A, Toelg R, Lemos P A, Birgelen C, Christiansen E H, Wijns W, Neumann FJ, Kaiser C, Eeckhout E, Lim S T, Escaned J, Garcia-Garcia H M and Waksman R 2016 *Lancet* **387** 31
[6] Yi S B, Davies C H J, Brokmeier H G, Bolmaro R E, Kainer K U and Homeyer J 2006 *Acta Mater.* **54** 549
[7] Li C Q, Xu D K, Wang B J, Sheng L Y, Qiao Y X and Han E H 2017 *Sci. Rep.* **7** 40078
[8] Valle J A, Carreno F and Ruano O A 2006 *Acta Mater.* **54** 4247

[9] Feng B, Xin Y C, Guo F L, Yu H H and Liu Q 2016 *Acta Mater.* **120** 379
[10] Zhang S X, Zhang X N, Zhao C L, Li J N, Song Y, Xie C Y, Tao H R, Zhang Y, He Y H, Jiang Y and Bian Y J 2010 *Acta Biomater.* **6** 626
[11] Qin H, Yaochao Zhao Y C, An Z Q, Cheng M Q, Wang Q, Cheng T, Wang Q J, Wang J X, Jiang Y, Zhang X L and Yuan G Y 2015 *Biomaterials* **53** 211
[12] Xu L P, Ye G N, Zhang E L, Pan F and Yang K 2007 *J. Biomed. Mater. Res. Part A* **83** 703
[13] Wang H X, Guan S K, Wang X, Ren C X and Wang L G 2010 *Acta Biomater.* **6** 1743
[14] Cipriano A F, Sallee A, Tayoba M, Alcaraz M C C, Lin A, Guan R, Zhao Z and Liu H 2017 *Acta Biomater.* **48** 499
[15] Zhao X, Shi L L and Xu J 2013 *Mater. Sci. Eng. C* **33** 3627
[16] Zhang Y, Huang X F, Ma Z D, Li Y, Guo F, Yang J C, Ma Y and Hao Y 2017 *Mater. Sci. Eng. A* **686** 93
[17] Xu D K, Tang W N, Liu L, Xu Y B, Han E H 2007 *J. Alloys Comp.* **432** 129-134.
[18] Sheng L Y 2016 *Strength Mater.* **48** 107
[19] Sheng L Y, Yang F, Guo J T and Xi T F 2014 *Trans. Nonferrous Met. Soc. China* **24** 673
[20] Cheng J, Mu Y L, Zu G Y and Yao G C 2017 *Mater. Design* **123** 64
[21] Kula A, Jia X, Mishra R K and Niewczas M 2017 *Int. J. Plasticity* **92** 96
[22] Li H Z, Lv F, Liang X P, Qi Y L, Zhu Z X and Zhang K L 2016 *Mater. Sci. Engi. A* **667** 409
[23] Liu L, Yuan F, Zhao M, et al. 2017 *Mater.* **10** 477
[24] Shuai C, Yang Y, Peng S, et al. 2017 *J. Mater. Sci.-Mater. M.* **28** 130
[25] Liu X, Chen R and Han E 2008 *Mater. Sci. Eng. A* **497** 326
[26] Li C Q, Xu D K, Yu S, Sheng L Y and Han E H 2017 *J Mater. Sci. Technol.* **33** 475
[27] Wang S D, Xu D K, Wang B J, Sheng L Y, Han E H and Dong C 2016 *Sci. Rep.* **6** 29471
[28] Sheng L Y, Yang F and Xi T F 2015 *Appl. Mech. Mater.* **727** 111
[29] Sheng L Y, Du B N, Lai C, Guo J T and Xi T F 2017 *Strength Mater.* **49** 109
[30] Sheng L Y, Yang F, Xi T F, Zheng Y F and Guo J T 2012 *Intermetallics* **27** 14
[31] Sheng L Y, Yang F, Guo J T and Xi T F 2014 *Trans. Nonferrous Met. Soc. China* **24** 673
[32] Sheng L Y, Yang F, Xi T F, Lai C and Ye H Q 2011 *Compos. Part B-Eng.* **42** 1468
[33] Sheng L Y, Guo J T, Ren W L, Zhang Z X, Ren Z M and Ye H Q 2011 *Intermetallics* **19** 143
[34] Sheng L Y, Yang F, Xi T F, Guo J T and Ye H Q 2012 *Mater. Sci. Eng. A* 555 131
[35] Sheng L Y, Xi T F, Lai C, Guo J T and Zheng Y F 2012 *Trans. Nonferrous Met. Soc. China* **22** 489
[36] Sheng L Y, Yang F, Guo J T, Xi T F and Ye H Q 2013 *Compos. Part B: Eng.* **45** 785
[37] Zhang W, Du K, Chen X Q, Sheng L Y and Ye H Q 2016 *Philos. Mag.* **96** 58
[38] Sheng L Y, Guo J T, Xi T F, Zhang B C and Ye H Q 2012 *Prog. Nat. Sci.: Mater. Int.* **22** 231
[39] Sheng L Y, Zhang W, Guo J T, Wang Z S, Ovcharenko V E, Zhou L Z and Ye H Q 2009 *Intermetallics* **17** 572
[40] Li C Q, Xu D K, Zeng Z R, Wang B J, Sheng L Y, Chen X B and Han E H 2017 *Mater. Design* **121** 430
[41] Li C Q, Xu D K, Wang B J, Sheng L Y, Qiao Y X and Han E H 2017 *Sci. Rep.* **7** 40078

IOP Conf. Series: Materials Science and Engineering **292** (2017) 012042 doi:10.1088/1757-899X/292/1/012042

Simulation of the temperature field distribution in medium-voltage vacuum interrupter and experimental verification

W Sun[1,4], H L Liu[2], Y G Cai[2], Y L Li[3], H Y Zhou[1], Y Zhou[1]

[1] Electric Power Research Institute of State Grid Heilongjiang Electric Power Company Limited, Harbin 150030
[2] State Grid Heilongjiang Electric Power Company Limited, Harbin 150090
[3] State Grid Heilongjiang Electric Power Equipment and Materials Company, Harbin 150000)

[4] E-mail: sunwei19850525@163.com

Abstract: The temperature field distribution in the medium-voltage vacuum interrupter decides the thermal stability of it. In this paper, the simulation model of a kind of 12kV/3150A/40kA medium-voltage vacuum interrupter is constructed, and conductive bridge model is used. This paper simulates current contraction and Joule heating between contacts, and solves relevant problems using the function of the thermal-electrical coupling in the finite element software ANSYS. Steady-state temperature rise of vacuum interrupter at rated current and transient temperature rise of vacuum interrupter at short-time withstand current are calculated. Influence of the contact situation on vacuum interrupter temperature rise is analyzed. Steady-state temperature rise experiments for the interrupter are carried out, and experiment results verify the accuracy of simulation results. The results are useful in the designing and optimizing of medium-voltage vacuum interrupter.

1. Introduction

Electric current passes through a conductor of a circuit breaker, and heat is generated as a result during operation of the vacuum circuit breaker. A part of generated Joule thermal energy dissipates in the surrounding medium, while the remaining part heats the vacuum circuit breaker, thereby leading to temperature rise of the circuit breaker. Contact fusion welding may be caused in severe case. In the paper, the conductive bridge model is introduced to simulate current contraction and Joule heating among contacts. The thermal-electrical coupling function of finite element analysis software ANSYS is applied for solving temperature fields in the medium-voltage vacuum interrupter. The steady-state temperature rise of vacuum interrupter samples under uninterrupted duty at rated current of of 3150A is calculated. The influence of contacting condition among contacts on temperature rise of arc extinguish chambers is analyzed. The accuracy of the simulation results is verified through steady-state temperature rise experiments. Finally, the thermal stability of vacuum interrupter samples at short-time withstand current of 40000A and conduction time of 4s is simulated.

2. Establishment of temperature field simulation model

There were three heat transfer methods for medium-voltage vacuum circuit breakers: heat conduction, heat convection and heat radiation. Heat was transferred to external connectors of arc extinguish chamber and external medium thereof through heat conduction in the vacuum interrupter due to the

IOP Conf. Series: Materials Science and Engineering **292** (2017) 012042 doi:10.1088/1757-899X/292/1/012042

vacuum state in the arc extinguish chamber, no convection with outside air, and weak radiation[1]. Therefore, heat conduction dissipation was considered only in the vacuum interrupter when the temperature fields in the medium-voltage vacuum interrupter were simulated. However, three heat transfer modes of heat conduction, heat convection and heat radiation were considered aiming at external connectors and busbars of the arc extinguish chamber, wherein two heat transfer modes of heat convection and heat radiation were unified through surface coefficient of heat transfer, thereby reaching the purpose of simplifying calculation.

Therefore, internal 3D heat conduction equation of steady-state medium-voltage vacuum interrupter is shown in equation 1[2]:

$$\lambda(\frac{\partial^2 T}{\partial x^2}+\frac{\partial^2 T}{\partial y^2}+\frac{\partial^2 T}{\partial z^2})=-q \tag{1}$$

Internal 3D heat conduction equation of transient medium-voltage vacuum interrupter is shown in equation 2:

$$\rho c(\frac{\partial T}{\partial t})+\lambda(\frac{\partial^2 T}{\partial x^2}+\frac{\partial^2 T}{\partial y^2}+\frac{\partial^2 T}{\partial z^2})=-q \tag{2}$$

In the temperature：

T ——temperature /K;

λ ——heat conductivity coefficient/ $W \cdot m^{-1} \cdot K^{-1}$;

q ——energy generated in per unit volume/ $W \cdot m^{-3}$;

ρ —— material density/ $kg \cdot m^{-3}$;

c ——specific heat capacity/ $J \cdot kg^{-1} \cdot K^{-1}$;

t ——time/s;

x,y,z——rectangular coordinates;

Figure 1 shows the calculation model of temperature fields in the medium-voltage vacuum interrupter. In the paper, software Pro/Engineer was applied to establish a simulation model for calculating the temperature field in the vacuum interrupter according to the kind of 12kV/3150A/40kA medium-voltage vacuum interrupter in order to ensure the accuracy of the calculation results.

The temperature field calculation model was mainly composed of an incoming line terminal busbar, an incoming line terminal socket, a static terminal heat radiator, a static conducting rod, a static terminal contact, a dynamic conducting rod, a dynamic terminal contact, a dynamic terminal heat radiator, a leading-out terminal socket and a leading-out terminal busbar, wherein the incoming line terminal busbar and the leading-out terminal busbar were 3000mm long, 100mm wide and 30mm high.

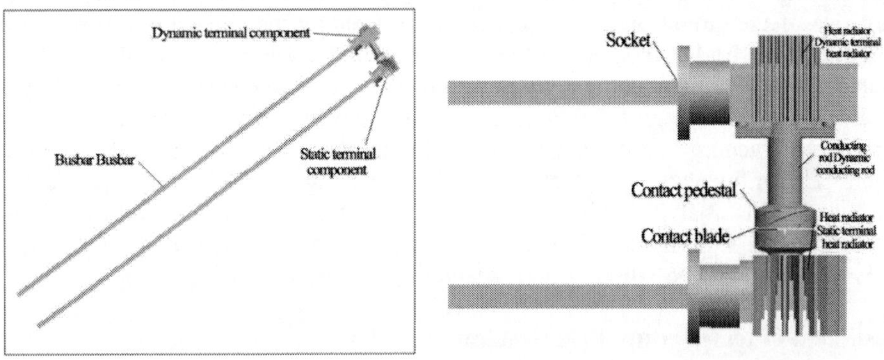

Figure 1. Calculation model of temperature fields in the medium-voltage vacuum interrupter.

Generally speaking, the seemingly smooth contact surface was actually bumpy. Dynamic and static contacts contacted actually in some protrusion points. The contact points were called conductive spots. When the electric current passed through the contact point of the contacts, the current line contracted, thereby increasing the current path near the conductive spot, decreasing the effective conductive interface, and producing contact resistance. It was usually assumed that there was only one conductive spot between the dynamic contact and the static contact, or all conductive spots formed one large conductive spot in the center position between the dynamic contact and the static contact aiming at general engineering application. In the paper, it was assumed that there was a cylinder in the center of the dynamic contact and the static contact (namely conductive bridge model) for connecting the dynamic contact and the static contact. Figure 2 shows the conductive bridge model.

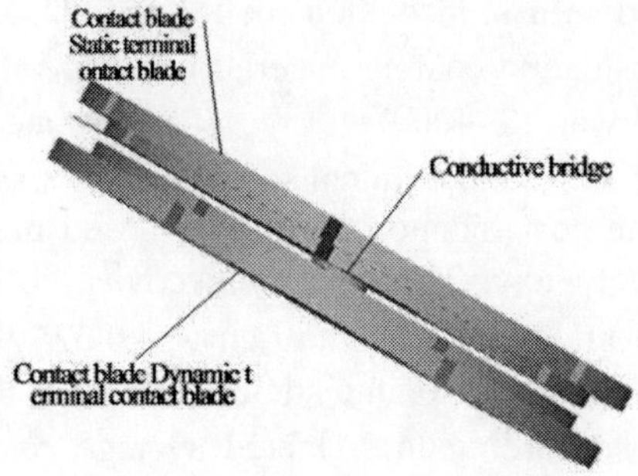

Figure 2. Conductive bridge model.

The material properties of the conductive bridge were the same as the contact material. The radius r of the conductive bridge was calculated from the Holm formula. In the paper, the contact force was 3858N, and the Brinell hardness of the contact material (CuCr50) was 128HB. The contact coefficient of the contact surface was 0.6, therefore the radius r of the conductive bridge was 4mm. The contacts were almostly closed, and the distance was little. Therefore, the height of the conductive bridge was 0.2 mm in the following simulation.

$$r = \sqrt{\frac{F}{\pi \xi H}} \tag{3}$$

In the formula:
F——contact force;
H—— material hardness;
ξ—— contact surface contact coefficient; it is generally 0.3~0.6.

3. Setting of material properties and boundary conditions

Corresponding material properties were set aiming at all parts after the establishment of the model, wherein the conducting rod, contact pedestal, busbar and socket were made of pure copper. The contact support was made of stainless steel. The contact blade was made of CuCr50. The heat radiator was made of aluminum. Table 1 shows the material properties for the model.

Table 1. Material properties of model.

Model name	Material name	Heat conductivity/ $W \cdot m^{-1} \cdot K^{-1}$	Material /Ω · m
Conducting rod	Copper (Cu)	400	$1/5.8 \times 10^{-7}$[3]
Contact pedestal	Copper (Cu)	400	$1/5.8 \times 10^{-7}$
Busbar	Copper (Cu)	400	$1/5.8 \times 10^{-7}$
Socket	Copper (Cu)	400	$1/5.8 \times 10^{-7}$
Heat radiator	Aluminum (Al)	217	2.7×10^{-8}
Contact support	Stainless steel	15.2	$1/1.1 \times 10^{-6}$
Contact blade	CuCr50	330	$1/1.41 \times 10^{-7}$ [4]

Setting of boundary conditions: two end faces of the incoming line terminal busbar and the leading-out terminal busbar were respectively selected as inlet surface and outlet surface of current. The experiment was carried out in a relatively closed temperature rise laboratory. No external airflow entered into the laboratory. The thermal radiation influence was very low. Therefore, heat conduction was the primary heat dissipation method. Busbar and a part of dynamic and static terminal sockets formed convection and heat radiation with outside air. The heat radiation coefficient was the highest. In addition, the busbar was coated with an insulation coating as thick as 0.3mm (alkyd paint). The heat conductivity coefficient of the insulation coating material was generally about 0.2 $W/m^2 \cdot K$. The heat conductivity coefficient of metal was 12~400 $W/m^2 \cdot K$. The integrated heat dissipation coefficient of the busbar and a part of dynamic and static terminal sockets was set as 0.3 $W/m^2 \cdot K$. The other part of the static terminal socket and the heat radiator were semi-closed in the insulating sleeve. The heat radiation coefficient was relatively low. The heat conductivity coefficient of plastic was 0.09 ~ 0.33 $W/m^2 \cdot K$, and the heat conductivity coefficient of air was 0.025~0.035 $W/m^2 \cdot K$. The integrated heat radiation coefficient of the other part of the static terminal socket and the static terminal heat radiator was set as 0.2 $W/m^2 \cdot K$ according to weighted average calculation. The other part of the dynamic terminal socket, the dynamic terminal heat radiator and a part of dynamic conducting rod were completely closed in the insulating sleeve. The heat radiation coefficient was the lowest. The integrated heat radiation coefficient of the three parts was set as 0.001 $W/m^2 \cdot K$ according to weighted average calculation. The remaining part of the calculation model was completely closed in the vacuum interrupter for adiabatic treatment. Steady state temperature field: the input current was the rated power frequency current, the effective value was 3150A, the environment initial temperature was set as 18 ℃. The input current of the transient temperature field was short-time withstand current: the effective value of the input current was 40000A. The initial temperature was the steady state temperature of the vacuum interrupter.

4. Calculation results of steady-state temperature field and experimental verification

4.1. Calculation results of steady-state temperature field

3D finite element analysis software ANSYS was used. Temperature fields of the medium-voltage vacuum interrupter were simulated through direct coupling of electric fields and thermal fields. The temperature field calculation model is calculated according to material properties and boundary conditions introduced in section 2. The current density distribution of the vacuum interrupter was obtained as shown in figure 3.

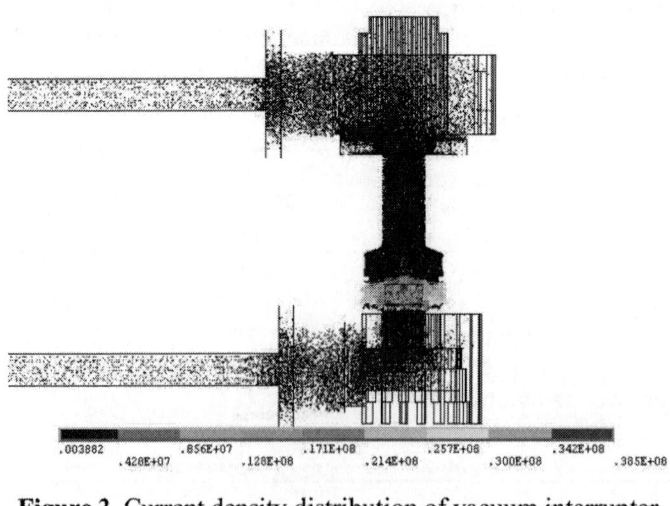

Figure 3. Current density distribution of vacuum interrupter.

Figure 3 shows that the current density distribution of the vacuum interrupter was uneven and there was greater difference among different parts. The current density of the conductive bridge was the largest due to current contraction, and the maximum value was $3.85e^9 \text{A/m}^2$. The current density of other parts was much smaller compared with the current density of the conductive bridge. It was obvious that the Joule heat caused by the contact resistance between the dynamic and static contacts was the main heat source of the vacuum interrupter.
The temperature field calculation model was calculated according to the previous conditions for obtaining the steady-state temperature field distribution of the vacuum interrupter shown as shown in figure 4.

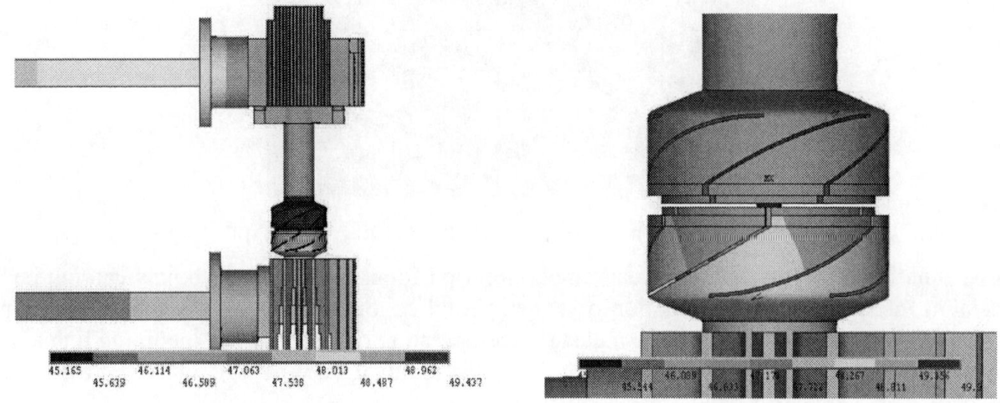

Figure 4. Nephogram of the steady-state temperature rise of the vacuum interrupter.

It is discovered in the figure that the maximum temperature rise value appeared in the conductive bridge part of the vacuum interrupter, which was followed by the temperature rise value of the dynamic contact and the static contact. The temperature rise value of the incoming line terminal busbar and the leading-out terminal busbar was the lowest. The reason is shown as follows: the contact resistance is the main heat source of the vacuum interrupter. The temperature rise value of the part was the highest. The incoming line terminal busbar and the leading-out terminal busbar were relatively far away from the conductive bridge. They formed convection heat dissipation with high convection coefficient. Therefore, its temperature rise value is lower than that of other parts.

The maximum allowable temperature is 90 ℃ in the contact of the AC high-voltage electric appliance under vacuum environment according to the national standards. The maximum temperature value of the vacuum interrupter during long-term work is 67.9 ℃ (ambient temperature of 18 ℃). Therefore, the steady-state temperature value of the vacuum interrupter is consistent with national requirements.

4.2. Experimental verification of steady-state temperature rise of medium-voltage vacuum interrupter

The simulation results should be experimented and tested in order to verify the accuracy of the simulation results. There were usually two methods for testing the temperature field distribution of the vacuum interrupter: the thermocouple method and the infrared camera method [5]. In the paper, a thermocouple was adopted for temperature rise experiment on the selected measurement points according to related national standard of high voltage switch electric appliances.

4.2.1. Experimental content. The experiment was carried out in the temperature rise laboratory. The temperature rise laboratory was not affected by sunlight or other heat radiation. No outside air flowed into the laboratory. The capacity of the temperature rise laboratory was sufficient to ensure normal natural airflow around experiment test object and uniform temperature.

IOP Conf. Series: Materials Science and Engineering **292** (2017) 012042 doi:10.1088/1757-899X/292/1/012042

Criteria of temperature rise experiment stability: it was determined that the test object has reached steady-state temperature rise when the temperature rise value change was smaller than 1℃ in one of five measurement points. Figure 5 shows the general circuit diagram of the temperature rise experiment.

Figure 5. General circuit diagram of temperature rise experiment.

Determination of thermocouple measurement point position: measurement points can not be set in the vacuum interrupter due to limitation of experimental conditions. Therefore, measurement points are respectively set in five positions for measuring temperatuer rise, including 'incoming line terminal', 'leading-out terminal', 'static terminal flange', 'static terminal cooling fin' and 'static leading-out terminal'.

4.2.2. Experiment results and analysis. Figure 6 shows the comparison diagram of calculated values and measured values in the measurement points. It is discovered in the figure that the calculated values of five measurement points were respectively 48.52℃, 47.761℃, 47.7℃, 47.833℃ and 48℃. The measured values were respectively 47.64℃, 47.6℃, 45.7℃, 45.02℃ and 48.47℃. The uncertainties between the calculated values and the measured values were respectively: 0.34%, 1.85%, 4.38%, 6.2% and 0.97%. The calculated values and the measured values of the measurement points were basically consistent mutually. However, there was still certain difference between the calculated values and the measured values in the measurement points mainly because the conductive bridge model was different from the actual contacting condition. Model simplification, comprehensive heat transfer coefficient and selection of material parameters also had certain influence on the simulation results.

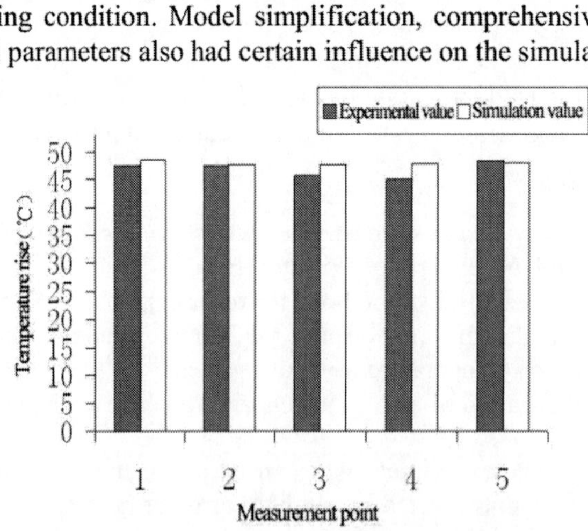

Figure 6. Calculated value and measured value of measurement point.

Another two groups of temperature rise experiments were carried out respectively for comparison: flexible coupling was respectively added in the incoming line terminal and leading-out terminal. Power frequency current with effective value of 3150A was input into the incoming line terminal. The experimental circuit was shown in figure 7. Flexible coupling was not added in the incoming line terminal and leading-out terminal. Power frequency current with effective value of 2000A was input into the incoming line terminal.

Figure 7. General circuit diagram when flexible coupling was added into terminal.

The experiment results of the above two groups of temperature rise experiments were compared with the steady-state temperature rise results of the test objects at rated current. The temperature rise experiment results of the measurement point in the 'leading-out terminal' was adopted as an example. The temperature rise change with time in the 'leading-out terminal' measurement point was concluded as figure 8.

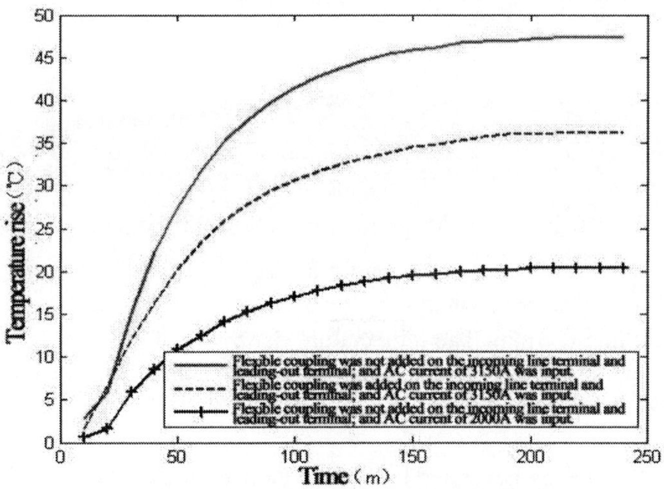

Figure 8. Temperature rise change with time in the 'leading-out terminal' measurement point.

The above figure shows that the temperature rise value of the test object was increased rapidly within short time after current was input into the experiment test object. However, the temperature rise value of the test object tended to be gentle. The test object reached the steady-state temperature rise value about 4 hours later. The steady-state temperature rise value of the experiment test object was lowered by 10℃ or so when flexible coupling was added on the incoming line terminal and leading-out terminal. The convection of the experiment test object and the outside world was enhanced since the contact area between the experiment test object and the external environment was increased by the flexible coupling, which was beneficial for heat dissipation of the experiment test object. The steady-state temperature rise value of the experiment test object was decreased greatly when power frequency current with effective value of 2000A was input into the incoming line terminal. The steady-state temperature rise value was decreased by 27.23 ℃. It was obvious that the experiment current has relatively high influence on steady-state temperature rise of the vacuum interrupter.

4.3. Influence of conductive bridge radius on temperature field distribution

The contacting condition among the contacts was mainly embodied in the dimension of the conductive bridge model, namely diameter and height. Changes of conductive bridge size meant changes of contacting conditions. The radius of the conductive bridge was mainly determined by contact contacted pressure, surface contacting condition and the Brinell hardness of contact materials [6]. Surface contacting condition and the Brinell hardness of contact materials were constant values. However the contacts were almostly closed, and the distance was little. Therefore, the influence of contact force on the steady-state temperature field in the medium-voltage vacuum interrupter can be concluded through changing the radius of the conductive bridge.

The height of the conductive bridge was hereby designed as 0.2 mm, and the radius was changed from 1mm to 10mm in order to study the influence of the conductive bridge radius on the temperature field distribution. The temperature rise change of all parts with the conductive bridge radius was obtained as shown in figure 9.

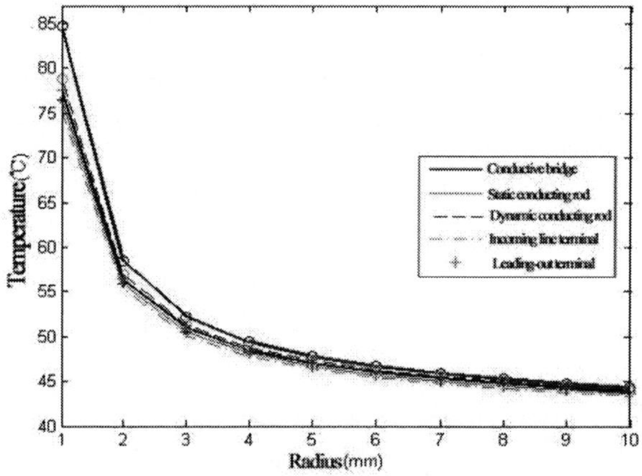

Figure 9. Temperature rise change of all parts with the conductive bridge radius.

Figure 9 shows that the maximum temperature in the arc extinguish chamber was increased from 62.32℃ to 102.78℃ under other constant conditions when the radius of the conductive bridge model was decreased from 10mm to 1mm. The temperature rise value of all parts in the vacuum interrupter was decreased with the increase of the conductive bridge radius. The temperature rise value was changed rapidly with conductive bridge radius at the beginning. The change tended to be slow subsequently because the energy generated by the conductive bridge was $p = I^2 R = I^2 \frac{\rho l}{s} = I^2 \frac{\rho l}{\pi r^2}$, and the change rate was $\frac{dp}{dr} = -\frac{2I^2 \rho l}{\pi r^3}$. In addition, the conductivity bridge was the main heat source, the change rate was relatively high when the radius was relatively small, and the change rate was relatively low when the radius was relatively large as a result. It was obvious that the mechanical burden of the contact was increased only if the contact force was increased merely after the contact force reached certain level. However, the steady-state temperature rise value of the vacuum interrupter can not be decreased prominently. Therefore, the heating problem of the vacuum interrupter can be controlled effectively if the contact force was controlled within certain scope.

5. Calculation results of transient temperature field

The transient temperature field was simulated for analyzing the thermal stability of the medium-voltage vacuum interrupter mainly, which was different from the simulation of the steady-state temperature field. Namely whether the contact of the arc extinguish chamber had fusion welding or not when short-time withstand current passed within short endurance time [7]. In the section, the temperature field calculation model was calculated according to the material properties and boundary conditions introduced in section 2, and the conductive bridge radius was designed as 4 mm. The transient temperature field distribution in vacuum interrupter was obtained. Moreover, the temperature rise value change with time at the maximum temperature point in the vacuum interrupter within 8s under short-time withstand current was studied in the section in order to study the temperature rise value change process with time at the maximum temperature point in the vacuum interrupter and the limit endurance time after short-time withstand current input into the vacuum interrupter. Since the conductive bridge model was relatively small, it can not be clearly displayed in the 3D transient temperature rise nephogram of the vacuum interrupter. Therefore, the transient temperature rise nephogram of the vacuum interrupter at the longitudinal profile was adopted for describing the transient temperature field distribution in the vacuum interrupter in order to describe the transient temperature rise value of the conductive bridge model and the transient temperature field distribution of the vacuum interrupter more clearly. Figure 10 shows the transient temperature rise nephogram of the vacuum interrupter at the longitudinal profile. Figure 11 shows the temperature rise value change with time at the maximum temperature point in the vacuum interrupter within 8s.

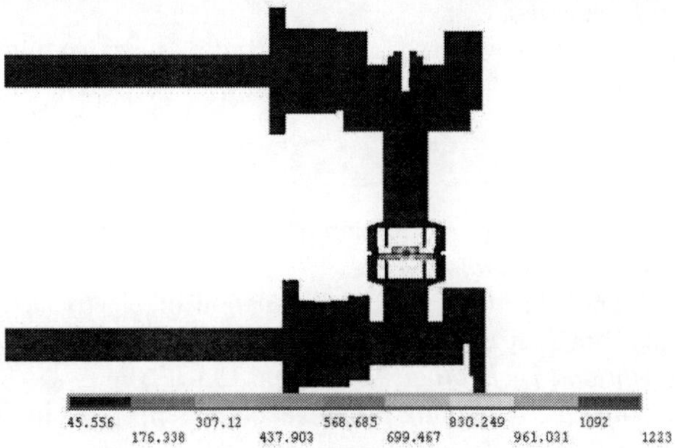

Figure 10. Transient temperature rise nephogram of the vacuum interrupter at the longitudinal profile.

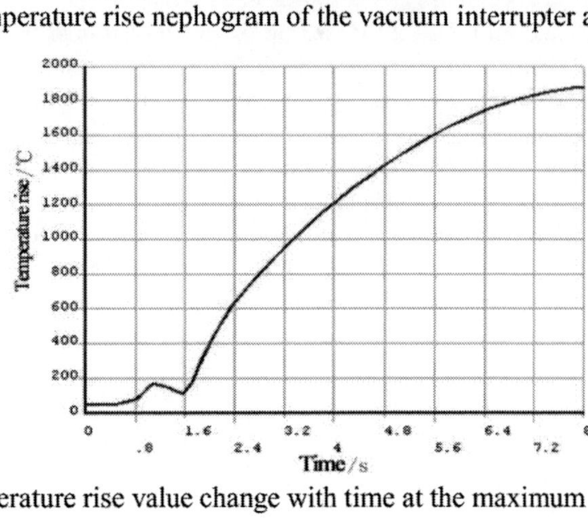

Figure 11. Temperature rise value change with time at the maximum temperature point.

Figure 10 and figure 11 show that the maximum temperature rise value of the temperature field model was 1223℃(environment temperature was set was 18℃) within the short endurance time of 4s when short-time withstand current (power frequency current with effective value of 40000A) was input into the incoming line terminal. Namely, the maximum temperature value of the vacuum interrupter was 1241 ℃. Since the contact blade was made of CuCr50, and the melting point of the CuCr50 was 1860℃, the temperature value was far less than the melting point of the contact materials. Fusion welding can not be produced, and thermal stability was satisfied.

Figure 12 also shows that the temperature rise value of the vacuum interrupter was increased rapidly at the beginning of the short circuit. However, the temperature tended to be stable gradually with time. The maximum temperature value of the vacuum interrupter was 1897 ℃. The temperature value has reached the melting point of the contact, and the contact underwent fusion welding.

6. Conclusion

In the paper, the conductive bridge model was utilized to simulate the current contraction and Joule heating among contacts. Steady-state simulation results showed that the maximum temperature in the arc extinguish chamber was increased from 62.32 ℃ to 102.78 ℃ when the conductive bridge radius was decreased from 10 mm to 1 mm. The temperature has been higher than the maximum allowable temperature of the contact of the AC high-voltage electrical appliance (90 ℃). However, the radius of the conductive bridge model was further increased, the temperature field change in the arc extinguish chamber tended to be slow.

Transient simulation results showed that the contact did not undergo fusion welding when short-time withstand current of 40kA passed through vacuum interrupter within the short endurance time of 4s. The thermal stability was satisfied.

In the paper, the steady-state temperature rise experiment was carried out. The results showed that the calculated value of the measurement points was in line with the measured value, and the maximum uncertainty was 6.2%.

Reference

[1] Yu L 2010 Research on development of large current vacuum circuit breaker and key issues *Shenyang University of Technology Press*

[2] Mateev V, Tanev R and Marinova I 2014 Simulation of electric and thermal fields of high voltage interrupter vacuum chamber *Proceedings of XVIII-th International Symposium of Electrical Apparatusand Technologies SIELA* pp 133-136

[3] Cadick J 2012 Predicting the remaining life of vacuum interrupters in the field *Group CBS Alan Seidel*

[4] Liang S H, Fan Z K and Hui R 2000 Research on optimized heat treatment process of CuCr50 contact materials *Transactions of Metal Heat Treatment* chapter **3** pp 66-70

[5] Chen D G, Li X W 2010 Virtual prototype technology of low voltage circuit breaker *Beijing: China Machine Press*

[6] Marinova I and Mateev V 2012 Inverse source problem for thermal fields *Compel International Journal of Computations and Mathematics in Electrical*

[7] Shemshadi A, Akbari A and Bathaee S 2013 A novel approach for reduction of electric field stress in vacuum interrupter chamber using advanced soft computing algorithms *IEEE Transactions on Dielectrics and Electrical Insulation*

Influence of stacking sequence and number of pin on load distribution of composite joints

F Liu[1, 2, 3], L Qi[1, 2] and S F Feng[1, 2]

[1] College of Civil Engineering, Xijing University, Xi'an 710123, China
[2] Shaanxi Key Laboratory of Safety and Durability of Concrete Structures, Xijing University

[3] E-mail: liufang_winter@163.com

Abstract. In this paper, finite element model of multi-pin composite joint is established. Matlab program is adopted to calculate equivalent material properties of composite laminate. Load distributions of double-lap composite joint are investigated, and simulation results show excellent agreement with experiments. By using equivalent material properties method, influence of stacking sequence on load distribution is studied. Equivalent material properties method could provide a simple and efficient method for optimizing load distribution of multi-pin composite joint.

1. Introduction

Advanced composite materials with fiber reinforcement are widely used in aerospace, shipbuilding and civil engineering due to their high stiffness and strength to weight ratios [1].

There are many methods for determining load distribution in bolted joints. Analytical and finite element methods are common means for load distribution investigation of multi-pin joint. Tate [2] developed a spring-based method for double-lap joints manufactured by isotropic materials. Nelson [3] extended spring-based method to composite joints. All these two mentioned methods are analytical ones. In recent years finite element method is wide used to determine load distribution of multi-pin joint. Fan [4] used a complex variable approach, while McCarthy [5] used three-dimensional finite element method.

Although three-dimensional finite element analysis is relatively simple, build of finite element (FE) model is complicated. Composite laminate material properties need to be given layer by layer, which need a lot of time. In this paper, a new FE model is established for load distribution investigation of multi-pin composite joint. In FE model, composite laminate is treated as one whole plate whose equivalent engineering material properties including modulus and Poisson's ratio are calculated by Matlab program developed by author. Matlab program is verified by experiments in literature [6]. To test and verify the new FE model, load distributions of double-lap and multi-bolt composite joints are investigated. At the same time FE model whose composite laminate material properties are given layer by layer is also created. Simulation results of two FE models and experimental results are compared with each other. At last, influence of stacking sequence and number of pin on load distribution of joint are investigated and load distribution of multi-pin composite joint is optimized.

2. Equivalent material properties of laminates

A new FE model is developed in this paper, composite laminate in FE model is treated as one whole plate. Equivalent material properties of laminate are calculated by Matlab program according to equations in literature [7].

Table 1. Unidirectional stiffness properties for HTA/6376.

E_{11} (GPa)	E_{22} (GPa)	E_{22} (GPa)	G_{12} (GPa)	G_{13} (GPa)	G_{23} (GPa)	υ_{12}	υ_{13}	υ_{23}
140	10	10	5.2	5.2	3.9	0.3	0.3	0.5

To verify program of equivalent material properties, a numerical example is chosen in literature [6]. In numerical example the composite laminate is manufactured from HTA/6376, a high-strength carbon fiber–epoxy material currently used in primary structures in the European aircraft industry. The unidirectional stiffness properties are shown in table 1. Equivalent material properties of laminate are shown in table 2.

Table 2. Comparison of equivalent material properties of laminates.

Equivalent material properties	E_x (GPa)	E_y (GPa)	E_z (GPa)	G_{xy} (GPa)	G_{xz} (GPa)	G_{yz} (GPa)	υ_{xy}	υ_{xz}	υ_{yz}
Experiments	54.25	54.25	12.59	20.72	4.55	4.55	0.309	0.332	0.332
Matlab	54.249	54.249	12.587	20.722	4.550	4.550	0.309	0.3326	0.3326
Error (%)	0.002	0.002	0.024	0.010	0.000	0.000	0.000	0.181	0.181

Table 2 indicates that the error of equivalent material engineering constants calculated by Matlab program is quite small, so that the program can be used to calculate the material properties of three-dimensional laminate.

3. Numerical study

In this section, load distributions of multi-bolt composite joints are investigated by equivalent material properties (EMP) FE model. To verify the FE model, double-lap joint model is established. Simulation results are compared with experiments in literature [5].

3.1. Geometry, FE model and boundary condition

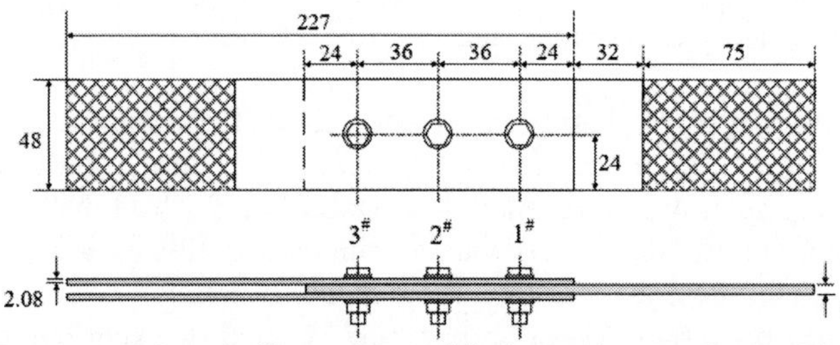

Figure 1. Joint configurations with three pins (mm): double-lap.

As shown in figure 1, the joint has three bolts and all the bolts are in single-column. Geometric parameters can be seen in figure 1. Diameter of bolt is 8mm. Composite laminates have stacking sequences [45° /0° /-45° /90°]$_{ns}$, thickness of each ply is 0.13mm. Unidirectional stiffness properties are shown in table 1. By using Matlab program, equivalent material properties of laminates can be obtained. The titanium bolt was modeled with isotropic material properties, with material constant E_x=110GPa, υ_b=0.29.

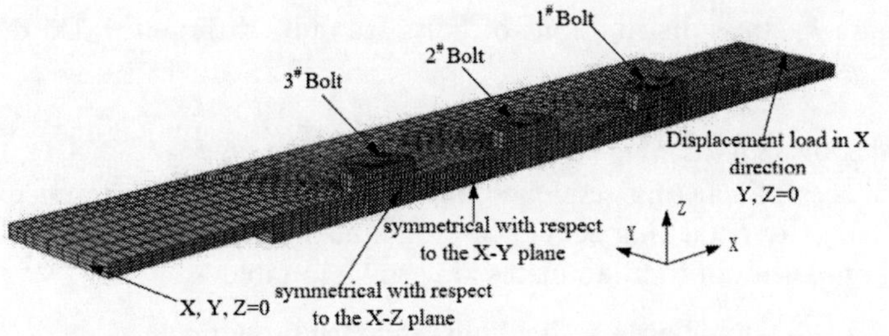

Figure 2. Mesh of FE model and boundary conditions: double-lap.

According to geometric parameters of joints, two FE model are established in ABAQUA shown in figure 2. In FE model, bolt and washer are simplified as one component. Bolt torque is 0.5Nm. To compare with experiments in literature [5], Bolt–hole clearances in FE model are 0. The laminates are modeled using three-dimensional solid elements with reduced integration which is C3D8R (8-node linear brick, reduced integration, hourglass control) in ABAQUS. Double-lap joint is symmetrical with respect to X-Z and X-Y plane, so one quarter of joint is modeled. Boundary conditions are shown in figure 2. One end of joint is fixed, and displacement load in X direction is carried out on the other end of joint.

3.2. Model validation

Composite joint is a contact FE problem, and material properties of three-dimensional (3D) composite laminate are usually given layer by layer which called as 3D FE method. To compare 3D FE method and EMP method, two FE models are established, and simulation results and experimental results [5] of load distribution are shown in figure 3. Simulation results show an excellent agreement with experiments.

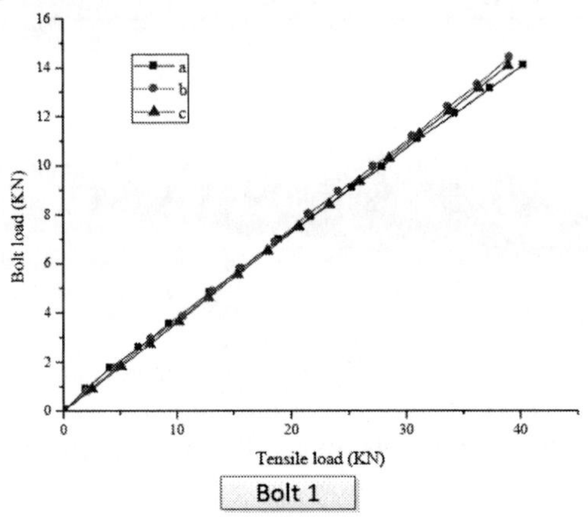

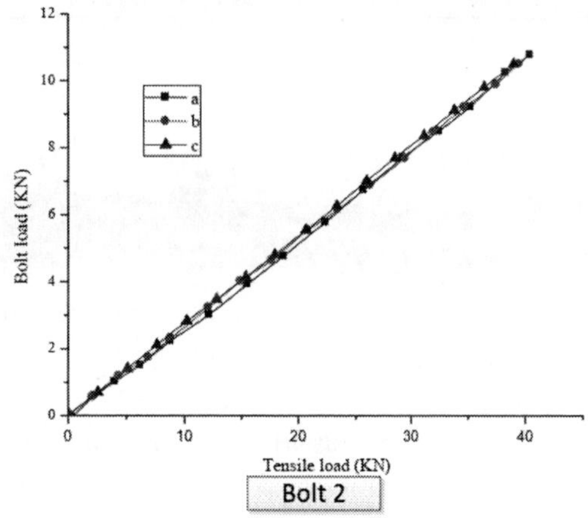

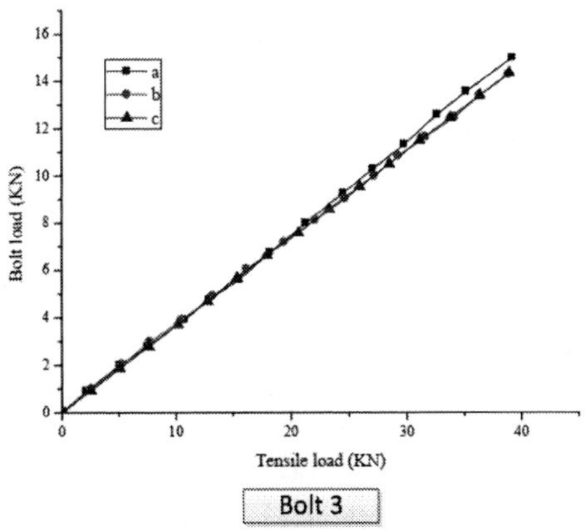

Double-lap joint

Figure 3 Load distribution comparison of experiments, 3D FE method and EMP method.
(a) Experiments [5] (b) 3D FE method (c) EMP method.

As shown in figure 3, stress distributions of bolts are quite different. Load distribution difference results in the reduction of joint efficiency.

3.3. Influence of stacking sequence, number of bolt and interference

In this section influences of stacking sequence, number of bolt and interference on load distribution are investigated. Firstly, four stacking sequence programs are design which can be found in table 3. Equivalent material properties of four laminates are shown in table 4.

Table 3. Stacking sequence program.

Number	Stacking sequence	Ratio (%)		
		0°	90°	±45°
Joint1	$[\pm45/90_2/\pm45/90_2/\pm45]_S$	0	40	60
Joint 2	$[\pm45/0/90/\pm45/0/90/\pm45]_S$	20	20	60
Joint 3	$[\pm45/0_2/90/\pm45/0_2/90]_S$	40	20	40
Joint 4	$[\pm45/0_3/\pm45/0_3]_S$	60	0	40

Table 4. Equivalent material properties of laminate for different stacking sequence program.

Number	Equivalent material properties								
	E_x (GPa)	E_y (GPa)	E_z (GPa)	G_{xy} (GPa)	G_{xz} (GPa)	G_{yz} (GPa)	υ_{xy}	υ_{xz}	υ_{yz}
Joint1	25.06	67.79	12.24	23.83	4.29	4.81	0.26	0.37	0.12
Joint 2	48.64	48.64	12.59	23.83	4.55	4.55	0.38	0.30	0.30
Joint 3	71.40	46.87	12.53	17.62	4.68	4.22	0.31	0.33	0.39
Joint 4	92.33	21.50	11.88	17.62	4.94	4.16	0.65	0.14	0.43

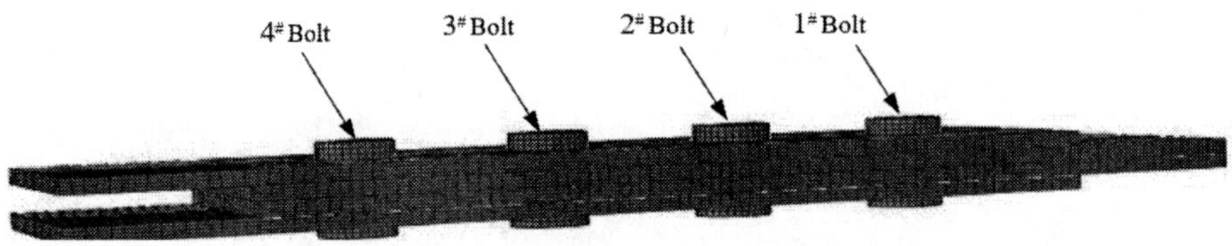

Figure 4. FE model of four-bolt joint.

FE model is shown in figure 4. Geometric parameters are similar to figure 1. Boundary condition is similar to section 3.1. Load bearing ratio of bolt is shown in figure 5. Here load bearing ratio bolt is defined as bolt load is divided by sum of all bolt load which represents joint efficiency. It can be seen from figure 5 that with increase of 0° ply, nonlinear of load bearing ration curve decreases.

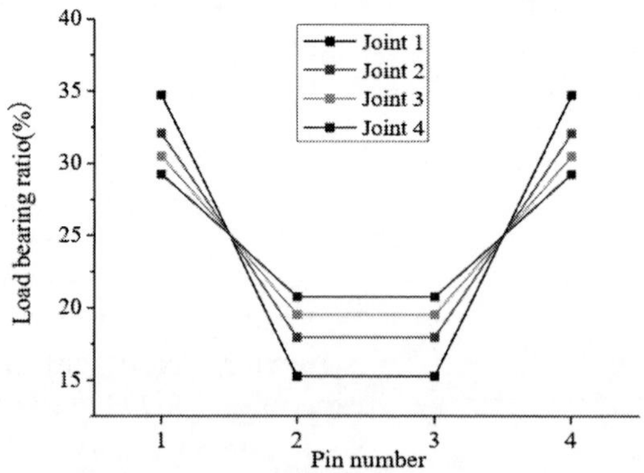

Figure 5. Load distribution of four-bolt joint.

Secondly, load distributions of two types of joint with four and five pins are investigated. Stacking sequence is same as joint 3. Load bearing ratios of every pin are shown in figure 6. Figure 6 shows that difference between maximum and minimum load bearing ratio in four-pin and five-pin joint are 10.98% and 15.79%, respectively. It presents that the more pins there are, the more nonuniform of load distribution is.

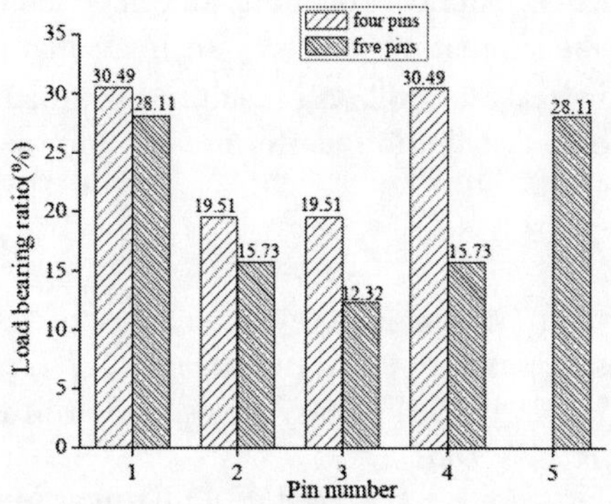

Figure 6. Comparison of load distribution of four-pin and five-pin.

At last, effect of interference between bolt and bolt-hole on load distribution is studied. It is known that proper interference could improve joint strength. 1% interference is the best choice for composite mechanical fastened joint. In this section four interference programs are shown in table 5. Stacking sequence of composite laminates in joint is according to Joint 3. Load bearing ration is shown in figure 7.

Table 5. Interference for four-pin double-lap joint.

Number	Interference			
	1#	2#	3#	4#
A	0	0	0	0
B	0	0.25%	0.25%	0
C	0	0.5%	0.5%	0
D	0	1.0%	1.0%	0

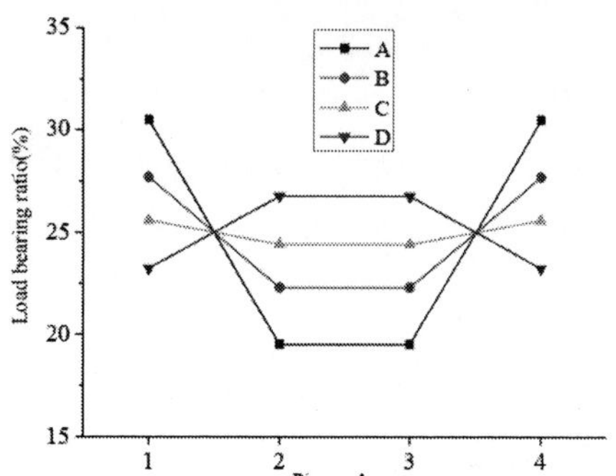

Figure 7. Load distribution of joint with different interference.

As shown in figure 7, uniformity of load distribution using interference C is the best, and nonlinear of interference A is the worst.

4. Discussion and conclusions

Composite mechanical fastened joint is widely used in various trades and industries. Load distribution investigation of multi-pin composite joint quite interesting.

Based on EMP method, FE model of double-lap is established, simulation results have an excellent agreement with experiments. Error of simulation result is caused by simplification of bolt and washer, which are simplified as one whole bolt. But the error is small, and the accuracy is good enough. Compared with 3D FE model, calculation accuracy based on EMP method is higher. Meanwhile, since laminate is treated as one whole plate, complication of model building is decreased greatly.

Influences of stacking sequence, number of bolt and interference on load distribution are investigated, respectively. From the simulation results we can see that increase of 0° ply could make load distribution more uniform; increase of bolt will result in load distribution more nonuniform. But 0.5% interference will made load distribution uniform, and joint efficiency could be improved effectively.

References

[1] Thoppul S D, Finegan J and Gibson F 2009 Mechanics of mechanically fastened joints in polymer-matrix composite structure-a review *Compos Sci Technol* **69** 301-29

[2] Tate MB, Rosenfeld SJ May 1946 Preliminary investigation of the loads carried by individual bolts in bolted joints *NACATN 1051*

[3] Nelson WD, Bunin BL, Hart-Smith LJ August 1983 Critical joints in large composite aircraft structure. *NASA CR-3710*

[4] Fan WX, Qiu CT 1993 Load distribution of multi-fastener laminated composite joints. *Int J Solids Struct* **30** 3013-23

[5] McCarthy MA, Lawlor VP and Stanley WF 2005 An experimental study of bolt-hole clearance effects in single-lap, multi-bolt composite joints *J Compos Mater* **39** 799-825

[6] Gray P J, McCarthy C T 2010 A global bolted joint model for finite element analysis of load distributions in multi-bolt composite joints *Compos Part B- Eng* **41** 317-325

[7] Sun C T, Sijian L 1988 Three-dimensional effective elastic constants for thick laminates *J Compos Mater* **22** 629-639

IOP Conf. Series: Materials Science and Engineering **292** (2017) 012044 doi:10.1088/1757-899X/292/1/012044

Breakdown characteristics of SF_6/N_2 in severely non-uniform electric fields at low temperatures

Y Wang[1], Z W Gao, G X Li, X C Zhu, C L Yu, J Q Liang and L Li

State Grid Heilongjiang Electric Power Company Limited Electric Power Research Institute, Harbin, Heilongjiang Province, China

[1] E-mail: wangyue_chn@163.com

Abstract. SF_6 has good electrical insulating properties, which is widely used as an insulating medium of GIS, GIL and other electrical equipment. However, the reliability of electrical equipments′ insulated gas is greatly challenged in cold areas, since SF_6 more readily liquefies. To solve the problem, SF_6 can be mixed with N_2 to maintain the insulating properties, and reduce its liquefaction temperature. Such practice has certain application prospect. In this paper, a breakdown experimental platform was built to study the insulating property of SF_6/N_2 at low temperature, wherein the temperature of the platform can be adjusted. A severely non-uniform electric field was generated by a rod-plate electrode. The breakdown characteristics of SF_6/N_2 with different mixing proportions at low pressures and low temperatures were measured. The result showed that the mixed gas was not liquefied within the temperature range. Temperature had insignificant influence on the insulating property thereof. The result in the paper has certain guiding significance for applying SF_6/N_2 mixed gas in high latitude areas.

1. Introduction

SF_6, as an electronegative gas with good insulating properties, is widely applied in GIS (Gas Insulated Switchgear), GIL (Gas Insulated Line), etc. However, SF_6 still has some disadvantages during application though it is widely used in electrical equipment, such as high molecular mass, easy liquefaction and insulating properties degradation at high pressure and low temperature. Therefore, the properties of SF_6 are affected in high altitude and alpine areas. High operation and maintenance cost is required for gas insulating electrical equipment in cold areas. N_2 has advantages of easy acquisition, low cost and low liquefaction temperature. SF_6 can be mixed with certain N_2 to maintain the most of the insulating properties of SF_6. For example, 50/50 SF_6/N_2 mixed gas still can maintain about 89% insulating properties of pure SF_6. The liquefaction temperature of the mixed gas is much lower than that of pure SF_6. Siemens and other companies have applied SF_6/N_2 mixed gas in electrical equipment practically [1-2]. Therefore, SF_6/N_2 mixed gas can be used as an insulating medium of electrical equipment for adapting to long-term low temperature operation environment at high altitude areas. It is of important practical significance to study the insulating properties of SF_6/N_2 mixed gas at low temperature.

The research methods of mixed gas insulating properties are divided into experimental research and theoretical research. The experimental research is applied widely with relatively fixed methods. The non-uniform electric field [3] generated by asymmetry electrodes is usually used for simulating the discharge structure in practical application environment. Power supply excitation includes DC, power frequency AC [4], lightning impulse [5], pulse [6], etc. High atmospheric pressure is mainly applied.

The breakdown characteristics of insulating gas with different mixing proportions are studied under various discharge condition [7-8]. Theoretical research is rarely applied with diversified methods, such as artificial neural network computation, FEM-FCT method [9], Boltzmann method [10], Monte Carlo method [11], etc. Existing research content is mainly concentrated at room temperature though it is of very important practical and theoretical significance to study the insulating properties of SF_6/N_2 mixed gas under extremely cold and highly cold conditions at room temperature to -60 ^{o}C [12]. The research on the insulating properties at low temperature is rarely reported [13], and related research is not comprehensive.

In the paper, a low-temperature breakdown experimental platform was built to study the breakdown characteristics of SF_6/N_2 mixed gas at low temperature. The temperature can be adjusted between 0 and -15 ^{o}C. A severely non-uniform electric field was generated by a rod-plate electrode. The breakdown voltage of mixed gas with different mixing proportions between 15 and 60kPa was measured, thereby the influence of low temperature on breakdown characteristics of SF_6/N_2 mixed gas was analyzed.

2. Experiment design and method

A breakdown experimental platform was built as shown in figure 1. It was used for measuring the breakdown voltage of SF_6/N_2 mixed gas at different temperatures, pressures and mixing proportions. The experimental platform was composed of a breakdown chamber, a cooling system, a gas distributing system and a measuring system. The breakdown chamber was cooled by liquid nitrogen. The temperature scope between 0 and -15 ^{o}C was selected in the breakdown experiment.

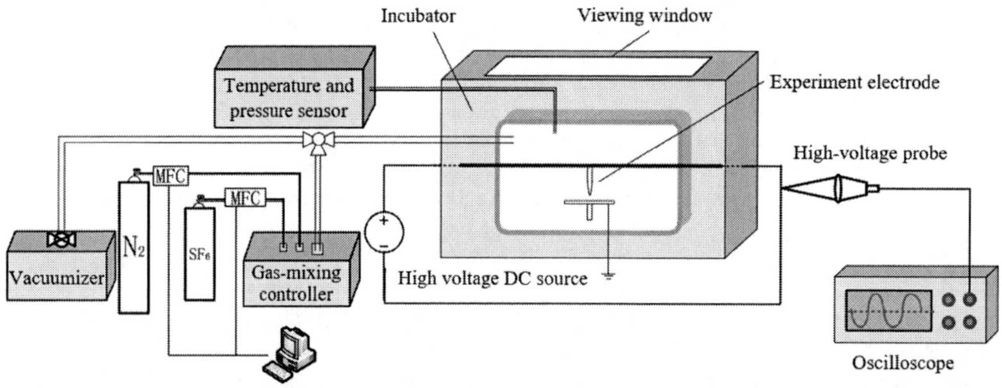

Figure 1. Diagram of devices for breakdown experiment.

Insulating gas discharge occurred between metal burr and metal plane usually in GIS. It belonged to point-to-plate discharge. To simulate the discharge condition, a rod-plate electrode was applied to generate the severely non-uniform electric field in the breakdown experiment. The electrode structure thereof was shown in figure 2(a). A copper electrode with tip radius of 2.5mm was applied as the rod electrode. A Rogowski electrode with radius of 100mm was selected as the plate electrode for eliminating the edge effect, wherein the electrode was made of copper. According to the definition of electric field non-uniformity, a severely non-uniform electric field was generated when the electric field non-uniform coefficient was greater than 4. The electric field non-uniform coefficient was 4.1265 when the electrode gap was set as 10 mm, and a severely non-uniform electric field was generated. The electric field distribution simulation results by Ansoft Maxwell thereof were shown in figure 2(b).

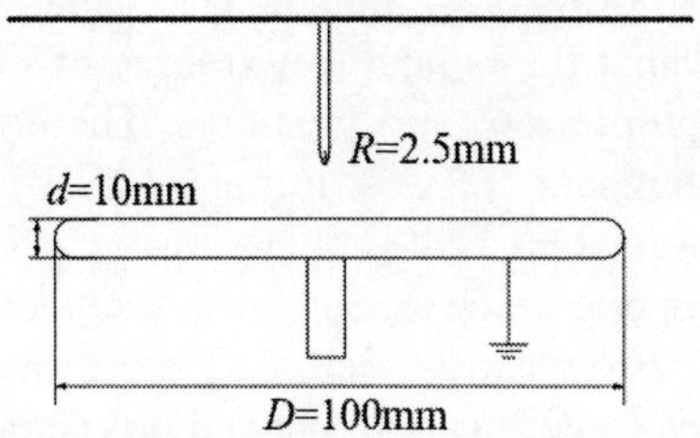

(a) Electrode structure diagram

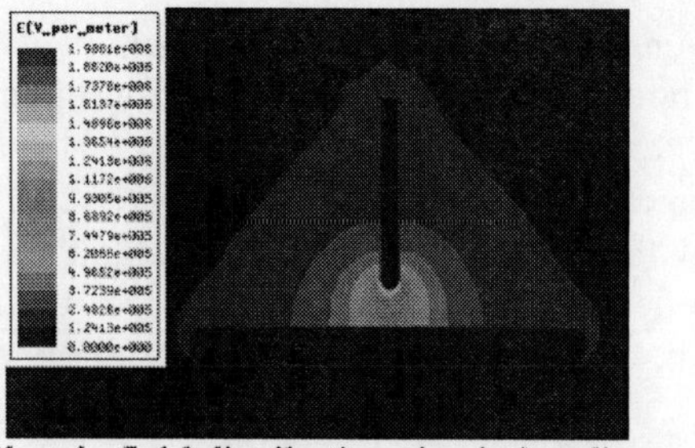

(b) Electric field distribution simulation diagram

Figure 2. Design of electrode for the experiment and electric field distribution simulation diagram.

Negative DC high voltage excitation was used in the experiment. Voltage was applied on the discharge electrode from zero and increased slowly according to 'High Voltage Experimental Technique' [14]. The voltage was increased according to the rate of 2% per second after the applied voltage reached about 75% of the estimated withstanding voltage. A sampling resistance was utilized for sampling the discharge current in the experimental circuit. The resistance was connected with the oscilloscope through a current probe. Breakdown was defined when the current waveform was distorted seriously. Corresponding voltage was defined as breakdown voltage.

3. Experimental results and analysis

3.1. Breakdown voltage at low temperature

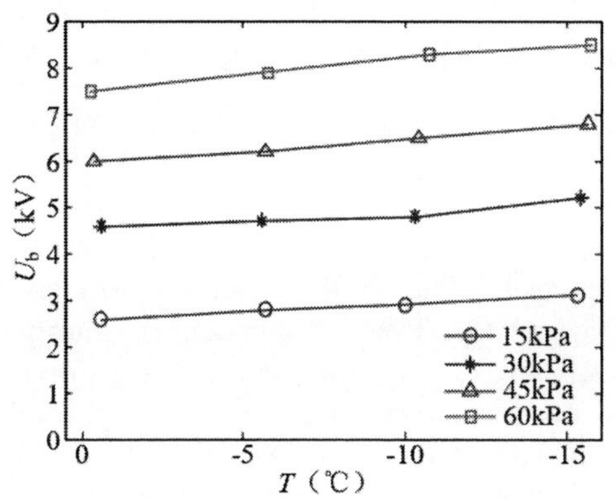

(a) SF_6/N_2:25/75

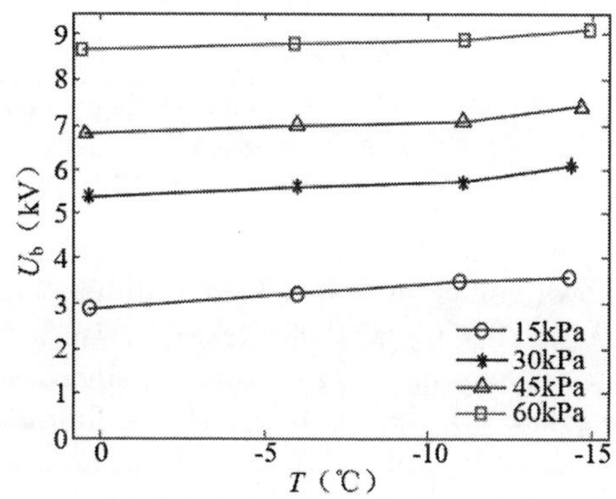

(b) SF_6/N_2:50/50

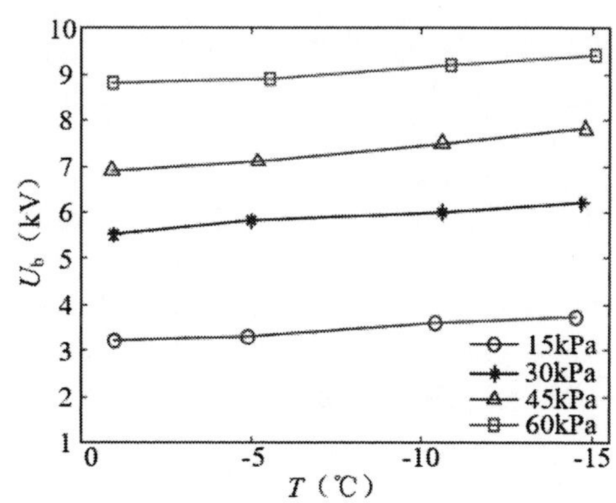

(c) SF_6/N_2:75/25

Figure 3. Breakdown voltage of SF_6/N_2 mixed gas in the severely non-uniform electric field.

Figure 3 indicated the measured breakdown voltages of SF_6/N_2 mixed gas under different conditions. The overall curve trend in the diagram showed that the breakdown voltage of mixed gas was changed slowly with temperature at different mixing proportions and pressures. The breakdown voltage was increased by about 0.75kV on average if the temperature was decreased by 15 °C. The breakdown voltage was increased slightly with temperature decrease because the mixed gas was not liquefied within the temperature scope. The influence of temperature on discharge was reflected in the microscopic process mainly, such as deceleration of particle Brownian motion [15], etc. However, the influence was very insignificant. Therefore, the breakdown voltage was not changed prominently with temperature decrease. In addition, relative distance among curves in the diagram indicated that the influence of gas pressure on the breakdown process was more significant than that of temperature in the scope.

It can be seen from figure 3 that the breakdown voltage of SF_6/N_2 mixed gas was linearly related with temperature and pressure. Therefore, corresponding linear fitting formula by polynomial fitting method and goodness of fit were obtained as shown in table 1. The goodness of fit of mixed gas breakdown voltage fitting formula was all above 0.98.

Table 1. Fitting formula of mixed gas breakdown voltage under severely non-uniform electric field.

Mixing proportion of mixed gas (SF_6/N_2)	Fitting formula	Goodness of fit
25/75	U_b=0.883+0.121p-0.048T	0.9869
50/50	U_b=1.474+0.122p-0.050T	0.9878
75/25	U_b=1.568+0.124p-0.051T	0.9879

[a] Wherein, U_b -- breakdown voltage of mixed gas, kV;
p-- pressure of mixed gas, kPa;
T -- temperature of mixed gas, °C.

The fitting formula in table 1 indicated that temperature coefficient was negative. The mixed gas breakdown voltage was increased with temperature decrease. The absolute values of the coefficient were 0.048, 0.050 and 0.051 respectively. It was obvious that when SF_6 content is increased, the temperature sensitivity was further enhanced slightly. However, the enhancement was insignificant. The pressure coefficient of the fitting formula was positive under three mixing proportions. The breakdown voltage of the mixed gas was increased with pressure increase. The value was 0.121, 0.122 and 0.124 respectively with little difference. The breakdown characteristics of mixed gas with three mixing proportions were affected by atmospheric pressure similarly.

3.2. Change law of SF_6/N_2 breakdown voltage with SF_6 content at low temperature

Figure 4 showed the change curve of SF_6/N_2 mixed gas breakdown voltage with SF_6 percentage content at different temperatures and different pressures in the severely non-uniform electric field. The sub-graphs in the diagram indicated that the breakdown voltage was linearly increased at the beginning with the increase of SF_6 percentage content in the mixed gas. The voltage tended to be saturated gradually after SF_6 was increased to certain percentage content. However, the saturation process lasted for relatively long time. The breakdown voltage of mixed gas containing 40% SF_6 was equal to about 80% voltage of mixed gas containing 75% SF_6. The curve increase trend was relatively similar. The change law of breakdown voltage with SF_6 percentage content was affected by temperature insignificantly.

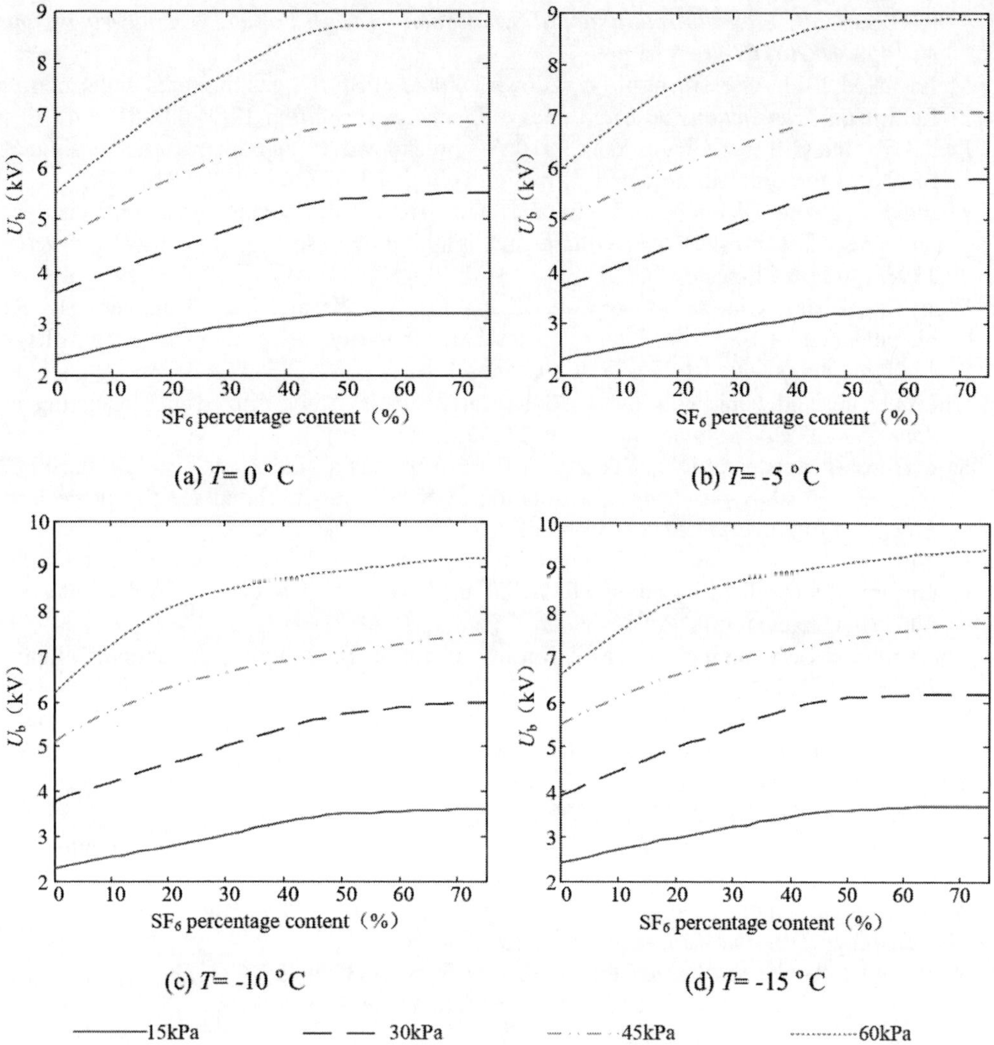

Figure 4. Change law of mixed gas breakdown voltage with SF_6 percentage content.

4. Conclusion

In the paper, a gas breakdown experimental platform was designed, wherein temperature can be adjusted. The breakdown voltage of SF_6/N_2 mixed gas at different temperatures, gas pressures and mixing proportions was measured. The following conclusions are obtained:

(1) The breakdown voltage of the SF_6/N_2 mixed gas is hardly affected by temperature regardless of SF_6 content thereof between 0 and 15 °C;

(2) The influence of gas pressure on SF_6/N_2 mixed gas breakdown voltage is more significant than that at 0~-15 °C and 15~60kPa;

(3) The insulating properties are increased firstly, which tends to be more and more saturate subsequently with increase of SF_6 content in the mixed gas. The change law of voltage with SF_6 percentage content was less affected by temperature.

References

[1] Xiao Dengming 2016 The development prospect of environment-friendly insulating gas *High Voltage Engineering* **42(4)** 1035-1046 (in Chinese)
[2] Dissentcr A, Finkel M, and Grund A et al. 1999 Dielectric properties of N_2/SF_6 mixtures for use

in GIS or GIL *Eleventh International Symposium on High Voltage Engineering Conf. Publ. No. 467, UK* **(3)** 67-70

[3] Mahdy A M 2011 Assessment of breakdown voltage of SF_6/N_2 gas mixtures under non-uniform field *IEEE Transactions on Dielectrics & Electrical Insulation* **18(2)** 607-612

[4] Lee B T, Huh C S and Chang Y M 2010 AC breakdown voltage characteristics simulation of SF_6/N_2 in non-uniform field and extra high voltage *PEOCO* **2010** 210-214

[5] Zhang Lu, Zhang Qiaogen, and Yin Yu, etc. 2013 SF_6 long-gap volt-second characteristics under very fast transient over-voltage and lightning impulse *High Voltage Engineering* **39(6)** 1396-1401 (in Chinese)

[6] Wen Tao, Zhang Qiaogen, Guo Can, Zhang Lu, Ma Yi and Tan Xiangyu 2015 Reversal phenomenon of SF_6 rod-plane gap discharge polarity effect under impulse voltage *High Voltage Engineering* **01** 275-281 (in Chinese)

[7] Qiu Yuchang and Feng Yunping 1993 Research on SF_6/N_2 mixed gas insulating medium *Journal of Xi'an Jiaotong University* **(1)** 1-6 (in Chinese)

[8] Sato Hitoshi, Morita Keiichi, Kikkawa Takashi,Hayakawa Naoki and Okubo Hitoshi 2002 A fundamental study on electrical insulation of N_2/SF_6 gas-insulated electric power apparatus *Electrical Engineering* **139(4)** 449-454

[9] Li Xintao, Lin Shen, and Xu Jianyuan, etc. 2014 SF_6 gas breakdown characteristics under uniform electric field based on FEM-FCT method *High Voltage Engineering* **40 (10)** 3046-3053 (in Chinese)

[10] Liu Zirui and Li Yanming 1997 Solution of gas discharge Boltzmann equation Computational *Physics* **(3)** 325-332

[11] Liu Xueli 2008 Monte Carlo simulation and experimental research on SF_6 alternative gas *Shanghai Jiaotong University* (in Chinese)

[12] Yang F, Chen Z, and Wu Y, etc. 2015 Two-temperature transport coefficients of SF_6-N_2 plasma *Physic s of Plasmas* **22(10)** 665-666

[13] Tang Xin, Liao Sijun and Yang Xin 2014 Research progress of SF_6 mixed/alternative gas insulating property *Insulation Materials* **06**18-22 (in Chinese)

[14] Zhang Renyu, Chen Changyu and Wang Changchang 2009 High voltage experimental technology *Tsinghua University Press* (in Chinese)

[15] Yang Jinji 1983 Gas discharge Beijing: *Science Press* (in Chinese)

IOP Conf. Series: Materials Science and Engineering **292** (2017) 012045 doi:10.1088/1757-899X/292/1/012045

The effect of material matching on the stress-induced power degradation for light-redirecting-ribbon-based silicon photovoltaic modules

X F Gou[1, 2], H Zhuang[2, 3], J Zhu[2] and X Y Li[1]

[1] Beijing University of Technology, 100124, Beijing, China
[2] CECEP Solar Energy Technology (Zhenjiang) Co. Ltd, 212132, Zhenjiang, China

[3] E-mail: zhuanghao@cecsec.cn

Abstract. Light redirecting ribbons (LRR) have recently been adopted in crystalline silicon modules in PV industry. The introduction of this LRR may bring additional stress at the boundary of the busbar/wafer contact, which may probably lead to higher power loss after aging. The thermal cycle and electroluminescence (EL) test are employed in this work to investigate this stress-induced power degradation for LRR-based crystalline silicon modules. The obtained results demonstrate that with increase in the thickness of the EVA encapsulant or the decrease in the thickness of the LRR, the power degradation induced by stress can be effectively diminished.

1. Introduction

Reliability and degradation of photovoltaic (PV) modules are an important concern to ensure the lifetime of PV modules longer than 25 years. PV module may degrade due to different environmental factors, such as irradiation, temperature, humidity and mechanical shock [1-4]. Several types of module degradation [5-11] can be triggered by each one of these various factors that induce corrosion, discoloration, delamination, cell breakage and cracking.

Stimulated by the strong demand for high-efficiency module products, light redirecting ribbons (LRR) have recently been adopted in high performance crystalline silicon PV modules to increase the power output. The light redirecting ribbon has delicately designed microstructures, which can reflect the incident light with an optimized angle to increase the chance of being reutilized by the cells. However, the introduction of this extra redirecting ribbon, which is usually over one hundred microns, can increase the risk of stress induced degradation of PV modules.

In this paper, thermal cycle test and electroluminescence measurement are adopted to investigate the stress induced power degradation of LRR-based modules. The influence of different thicknesses of the encapsulant and redirecting ribbon on the degradation level was systematically studied. The maximum output power and the electroluminescence were all measured after 200 thermal cycle test for each module. The electroluminescence images are in accordance with the level of module power loss. The obtained results demonstrate that with increase in the thickness of the EVA encapsulant or decrease in that of the LRR, the power degradation induced by stress can be effectively reduced. Our results can provide a guide for the thickness selection of encapsulant and LRR for LRR-based modules.

IOP Conf. Series: Materials Science and Engineering **292** (2017) 012045 doi:10.1088/1757-899X/292/1/012045

2. Experimental

Different module specimens were produced according to the industrial process at CECEP solar (Zhenjiang) with a standard composition of glass/EVA/cell/EVA/backsheet. Light redirecting ribbons of different thicknesses were assembled on the surface of the conventional interconnecting ribbon under heating using Autowell stringer. Thermal cycle test was performed according to the IEC 61215 standard (TC 200 cycles between -40°C and +85°C with a dwell time of 30 minutes at the temperature extremes). The power output and EL measurements were performed for each module after TC200 test. The maximum power output was measured under STC (25°C, 1000 W/m^2, AM 1.5).

3. Results and Discussion

3.1. Mechanism of stress

With the assistance of the bottom hot melt EVA, the light redirecting ribbon can be directly pasted to the welding ribbon under heating. For a conventional PV module, highly conductive ribbons (i.e., usually tin-coated copper ribbon) are soldered to the busbars from the front of one cell to the back of a neighbouring cell for current transport. It has already been noted that the coefficient of thermal expansion (CTE) is quite different between copper (17*10^{-6} K^{-1}), silicon (3*10^{-6} K^{-1}) and glass (9*10^{-6} K^{-1}) [12]. Their difference in CTE will cause a buildup of internal stresses as the materials cool down from high temperature. During the module lifetime of outdoor use, it may repeatedly undergo the high-temperature low-temperature cycle. Thus, stress often accumulates at the boundary of the busbar/wafer contact (figure 1). Furthermore, the introduction of this LRR layer on the welding ribbon will definitely increase the risk of stress induced degradation if no parameter of the materials is changed.

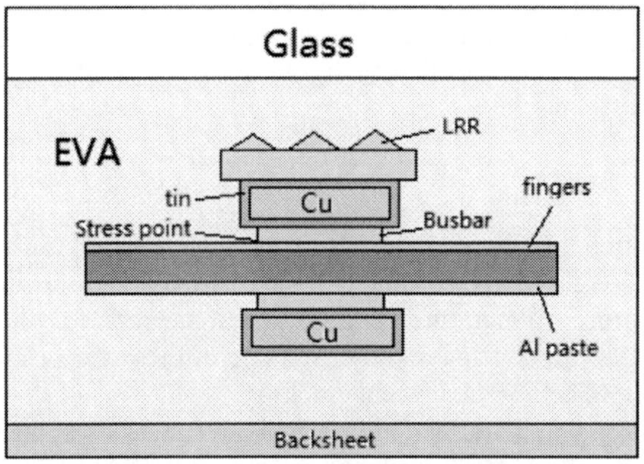

Figure 1. Scheme of the stress point of the LRR-based PV module.

3.2. EVA thickness effect on the power degradation

Currently, 0.45 mm thick EVA is widely adopted for PV module encapsulation. To investigate whether the introduction of LRR will cause much higher degradation when encapsulated with normal 0.45 mm thick EVA, LRR with different thicknesses, 0.115 mm and 0.15 mm, was employed for the LRR-based module specimen encapsulation. After encapsulation, the initial performance of the modules was measured. Then, the modules were put into the climate chambers to undergo the TC200 test. After the TC200 test, the typical EL images of the modules were measured and presented in figure 2. As can be seen, after the aging test, severe cell breakage and cracking problems occurred for both types of modules with different LRR thicknesses, suggesting that a much thicker EVA film should be used [13].

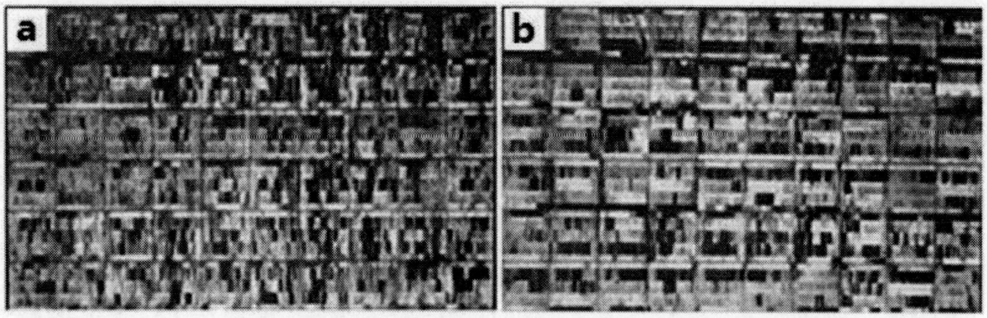

Figure 2. EL images of modules with 0.45 mm EVA after TC200 test: a) LRR 0.15 mm; b) LRR 0.115 mm.

Then the thickness of the front EVA film was increased to 0.65 mm, and modules were encapsulated with the same process. The LRR thickness in module A and B is 0.115 mm, and that in C and D is 0.15 mm. The electrical performance of the modules before and after the TC200 test was measured and listed in table 1. As can be seen, after the TC200 test, still high degradation level for the modules was observed. However, a LRR-thickness dependent degradation could be noted. The P_{max} of module A and B with thinner LRR degraded 5.96% and 4.85%, respectively. While the P_{max} of module C and D with thicker LRR degraded 6.21% and 5.17%, respectively.

Table 1. Performance degradation of modules with 0.65 mm EVA and different LRR thicknesses.

Module	LRR Thickness	Test Item	P_{max}	V_m	I_m	V_{oc}	I_{sc}	FF	CTM	P_{max} Degradation
A	0.115mm	Initial	274.01	31.21	8.78	38.24	9.27	77.30 %	100.90 %	5.96%
		After TC200	257.69	29.64	8.69	38.10	9.26	73.06 %	94.89%	
B		Initial	274.18	31.16	8.80	38.17	9.29	77.30 %	100.97 %	4.85%
		After TC200	260.88	29.88	8.73	38.05	9.30	73.69 %	96.07%	
C	0.15mm	Initial	272.79	31.11	8.77	38.16	9.35	76.40 %	100.45 %	6.21%
		After TC200	255.86	29.52	8.67	37.98	9.30	72.39 %	94.22%	
D		Initial	272.99	31.09	8.78	38.18	9.27	77.10 %	100.53 %	5.17%
		After TC200	258.87	29.82	8.68	38.00	9.31	73.17 %	95.33%	

The V_{oc}, I_{sc}, FF and P_{max} of modules with 0.65 mm EVA before and after TC200 test are depicted in figure 3. It shows that the degradation in the maximum output power is mainly driven by the loss in FF and the losses in Voc and Isc are quite small. Changes in electrical parameters can be explained below using electro-luminescence results. Figure 4 shows the EL measurement performed on the modules after the TC test. Compared with those using 0.45 mm EVA, cracks and breakage in the cells have been much improved. However, still evident in-active regions and finger defects can be noted. In-active regions of the cell may arise from the detachment of a part of the cell from the busbar [8]. These results demonstrate that increase in the EVA film thickness can effectively decrease the stress induced degradation level and thinner LRR is more desirable for its lower degradation level.

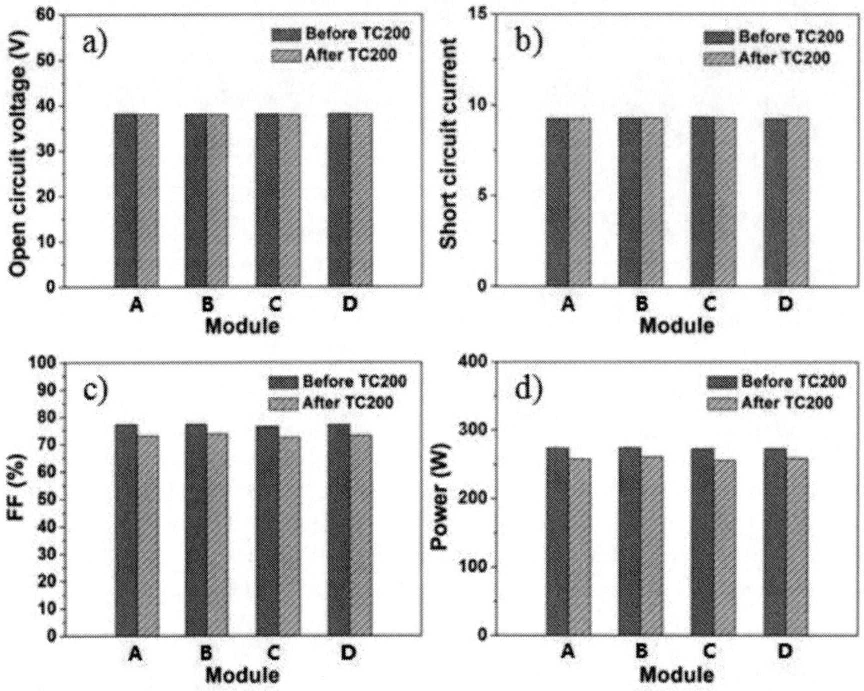

Figure 3. The V_{oc}, I_{sc}, FF and P_{max} of modules with 0.65 mm EVA before and after TC200 test.

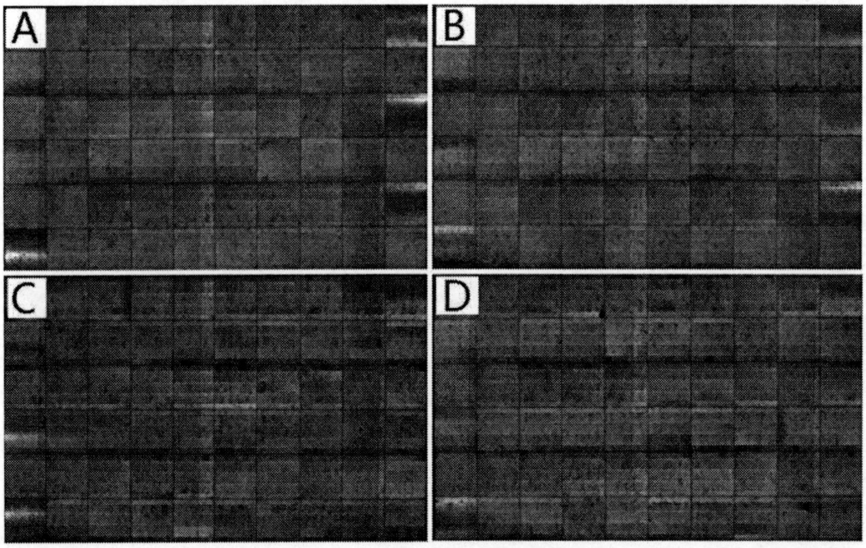

Figure 4. EL images of modules with 0.65 mm EVA after TC 200 test: A) and B) LRR 0.115 mm; C) and D) LRR 0.15 mm.

Since the power degradation level is still higher than the permitted maximum 5% degradation, an even thicker EVA was prepared and attempted. The electrical performance of the modules before and after the TC200 test was measured and summarized in table 2. The LRR thickness in module E is 0.15 mm, and that in F is 0.115 mm. After TC200 test, the P_{max} of module E with thicker LRR degraded 5.02% while the P_{max} of module F with thinner LRR degraded 3.52%, which is nearly 30% less than the module with 0.035 mm thicker LRR.

Table 2. Performance degradation of modules with 0.8 mm EVA and different LRR thicknesses.

Module	LRR Thicknes s	Test Item	P_{max}	V_m	I_m	V_{oc}	I_{sc}	FF	CTM	P_{max} Degradation
E	0.15mm	Initial	273	30.6	8.87	38.2	9.42	75.40%	99.98%	5.02%
		After TC200	259.3	29.8	8.69	38	9.30	73.40%	94.96%	
F	0.115mm	Initial	272.6	30.6	8.90	38.2	9.44	75.60%	99.83%	3.52%
		After TC200	263	30.1	8.74	37.9	9.28	74.70%	96.31%	

The V_{oc}, I_{sc}, FF and P_{max} of modules with 0.8 mm EVA before and after TC200 test are depicted in figure 5. It shows that the degradation in the maximum output power is mainly driven by the loss in FF and I_{sc}. Changes in electrical parameters can be explained below using electro-luminescence results. Figure 6 shows the EL measurement performed on the modules after the TC test. Compared with those using 0.65 mm EVA, cracks and in-active regions in the cells have been largely improved. However, still evident in-active regions and finger defects can be noted for the 0.15 mm LRR-based module. In contrast, for the 0.115 mm LRR-based module, few in-active regions of the cell due to detachment of a part of the cell from the busbar can be obviously observed, indicating reduced stress inside the module at the boundary of the busbar/wafer contact. These results demonstrate that further increase in the EVA film thickness to 0.8 mm can decrease the degradation of 0.115 mm LRR-based module to a reasonable level that meets the IEC 61215 standard.

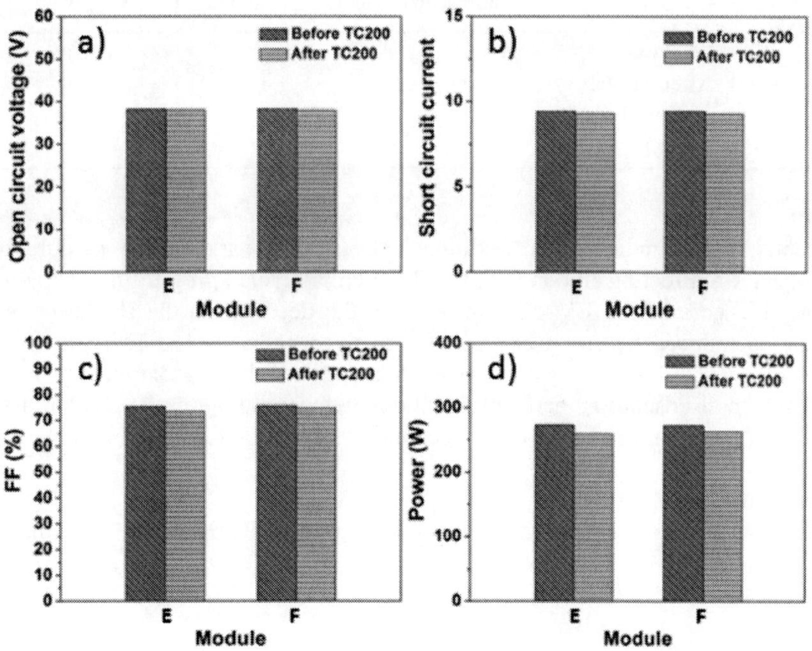

Figure 5. The V_{oc}, I_{sc}, FF and P_{max} of modules with 0.8 mm EVA before and after TC200 test.

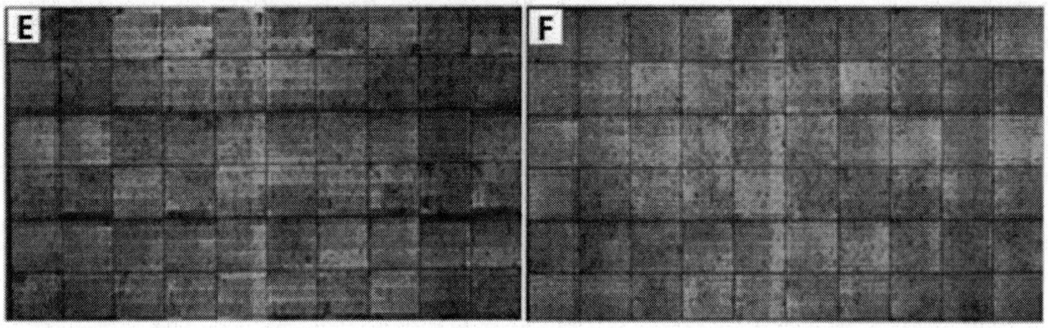

Figure 6. EL images of modules with 0.8 mm EVA after TC200 test: E) LRR 0.15 mm; F) LRR 0.115 mm.

3.3. Insulation performance

The insulation performance of the modules before and after TC200 test was also measured and summarized in table 3. The area of module samples is 1.63 m^2, which sets the pass criteria at 24.4 MΩ [14]. As can be seen, after the TC200 test, no obvious degradation in the insulation performance is observed. The insulation resistance for module E and F are both higher than 2000 MΩ before and after the TC200 aging test. All these obtained results suggest that 0.8 mm EVA is adequate to endure the stress induced degradation in 0.115 mm LRR-based modules.

Table 3. The insulation performance of the modules before and after TC200 test.

Module	State	Sample Area (m^2)	Standard (MΩ)	Experimental Value (MΩ)
E	Initial	1.63	>24.4	>2000
	After TC200			>2000
F	Initial	1.63	>24.4	>2000
	After TC200			>2000

4. Conclusion

The experimental results indicate that additional stress does exist in the module because of the introduction of light redirecting ribbon (LRR). The thermal cycle aging results demonstrate that with increase in the thickness of the EVA encapsulant or the decrease in the thickness of the LRR, the power degradation induced by stress can be effectively decreased. The EL images of the modules before and after the TC200 test are in correspondence with power degradation extent. The LRR-based modules showed good insulation performance both before and after the TC200 test. Our results provide a guide for the thickness selection of encapsulant and LRR for LRR-based modules.

Acknowledgements
This study was financially supported by the research program 17RD2 CECEP (Zhenjiang).

List of symbols and abbreviations

EVA: ethylene vinyl acetate copolymer
P_{max}: maximum power output
V_m: voltage at maximum power point
I_m: current at maximum power point
V_{oc}: open circuit voltage
I_{sc}: short circuit current
FF: fill factor
CTM: cell to module gain

IOP Conf. Series: Materials Science and Engineering **292** (2017) 012045 doi:10.1088/1757-899X/292/1/012045

References

[1] Quintana M A, King D L, McMahon T J and Osterwald C R 2002 Commonly observed degradation in field-aged photovoltaic modules *In: Proc. 29th IEEE Photovoltaic Specialists Conference* 1436-39

[2] Osterwald C R and McMahon T J 2009 History of accelerated and qualification testing of terrestrial photovoltaic modules: a literature review *Prog. Photovolt.: Res. Appl.* **17** 11-33

[3] Munoz M A, Alonso-Garcia M C, Nieves V and Chenlo F 2011 Early degradation of silicon PV modules and guaranty conditions *Sol. Energy* **85** 2264-74

[4] Vazquez M and Ignacio R S 2008 Photovoltaic module reliability model based on field degradation studies *Prog. Photovolt.: Res. Appl.* **16** 419-33

[5] Hoffmann S and Koehl M 2014 Effect of humidity and temperature on the potential-induced degradation *Prog. Photovolt.: Res. Appl.* **22** 173-79

[6] Duerr I, Bierbauma J, Metzgera J, Richtera J and Philippa D 2016 Silver grid finger corrosion on snail track affected PV modules-investigation on degradation products and mechanisms *Energy Procedia* **98** 74-85

[7] Zarmai M T, Ekere N N, Oduoza C F and Amalu E H 2017 Evaluation of thermo-mechanical damage and fatigue life of solar cell solder interconnections *Robot. Comput.-Integr. Manuf.* **47** 37-43

[8] Khatri R, Agarwal S, Saha I, Singh S K and Kumar B 2011 Study on long term reliability of photo-voltaic modules and analysis of power degradation using accelerated aging tests and electroluminescence technique *Energy Procedia* **8** 396-401

[9] Kempe M D, Dameron A A and Reese M O 2014 Evaluation of moisture ingress from the perimeter of photovoltaic modules *Prog. Photovolt.: Res. Appl.* **22** 1159-71

[10] Kempel M D, Miller D C and Wohlgemuth J H et al. 2015 Field testing of thermoplastic encapsulants in high-temperature installations *Energy Sci. Eng.* **3** 565-80

[11] Zhu J, Koehl M and Hoffmann S et al. 2016 Changes of solar cell parameters during damp-heat exposure *Prog. Photovolt.: Res. Appl.* **24** 1346-58

[12] Schneller E J, Brooker R P and Shiradkar N S et al. 2016 Manufacturing metrology for c-Si module reliability and durability Part III: Module manufacturing *Renew. Sustain. Energy. Rev.* **59** 992-1016

[13] Chiang S Y, Chou T L, Shih Z H, Hong H F and Chiang K N 2011 Life prediction of HCPV under thermal cycling test condition *Microelectron. Eng.* **88** 785-90

[14] Hoffmann S and Koehl M 2014 Effect of humidity and temperature on the potential-induced degradation *Prog. Photovolt.: Res. Appl.* **22** 173-79

IOP Conf. Series: Materials Science and Engineering **292** (2017) 012046 doi:10.1088/1757-899X/292/1/012046

Influence of B_2O_3 content on sintering behaviour and dielectric properties of La_2O_3-B_2O_3-CaO/Al_2O_3 glass-ceramic composites for LTCC applications

F L Wang[1], Y W Zhang[2], X Y Chen[1], H J Mao[1] and W J Zhang[1, 3]

[1] Department of Materials Science and Engineering, National University of Defense Technology, Changsha 410073, China
[2] Shandong Institute of Aerospace Electronic Technology, Yantai 264000, China

[3] E-mail: zhwjun@nudt.edu.cn

Abstract. La_2O_3-B_2O_3-CaO glasses with different B_2O_3 content were synthesized by melting method to produce glass/ceramic composites in this work. XRD and DSC results revealed that the diminution of B_2O_3 content was beneficial to increase the crystallization tendency of glass and improve the quality of crystalline phase, while decreasing the effect of glass during sintering process as sintering aids. The choice of glass/ceramic mass ratio was also influenced by the B_2O_3 content of glass. Dense samples sintered at 875 °C showed good dielectric properties which meet the requirement of LTCC applications: moderate dielectric constant (7.8-9.4) and low dielectric loss (2.0×10^{-3}).

1. Introduction

Low temperature co-fired ceramic (LTCC) technology has been widely used in microelectronic packaging to meet the requirements for miniaturization and high packaging density of highly integrated passive components [1-5]. Generally, the LTCC materials used for high frequency application should have a low sintering temperature ($<$900 °C), low dielectric constant and low dielectric loss [6-8].

Materials based on glass/ceramic systems are popular in LTCC applications for their outstanding designability characteristic. Al_2O_3 is one of the most common choices of ceramic filler in glass/ceramic systems, for it shows good dielectric properties [9, 10]. As for glass acting as sintering aids, it should have low softening temperature to ensure the sintering temperature of composites meet the requirement of LTCC applications. Several kinds of La_2O_3–B_2O_3-based (LB) glass have been studied, for they show good effects in reducing the sintering temperature of glass/ceramic composites. Another attractive character of LB glass is the appearance of some crystalline phases with excellent dielectric properties after the sintering of LB glass based ceramics. Chen et al. [11] found that the sintering temperature of $Ca_{0.6}La_{0.268}TiO_3$ (CLT) dielectric ceramic could be greatly reduced from 1350 °C to 900 °C when the La_2O_3-B_2O_3-ZnO (LBZ) glass was added. Xiang et al. [12] showed that the addition of La_2O_3 in CaO-B_2O_3-SiO_2 (CBS) glass could improve the glass-forming ability of the glass, as well as decrease its transition temperature. Tang et al. [13] reported that the sintering temperature of $Ba_{0.75}Sr_{0.25}(Nd_{0.75}Bi_{0.25})_2Ti_4O_{12}$ (BSNBT) ceramics doped with glass sintering aids of La_2O_3–B_2O_3–ZnO–CaO (LBZC) was decreased to 900 °C. Jo et al. [14] developed LB glass modified with CaO, MgO and ZnO, and the composites of glass and Al_2O_3 sintered at 850 °C showed good dielectric performance, i.e., dielectric constant around 8.1–8.3 and Q > 1,000 (at 17.1 GHz). It was also noticed that crystalline phases such as $LaBO_3$, $La(BO_2)_3$ and $LaAl_{2.03}(B_4O_{10})O_{0.54}$ were detected in the samples after sintering. Jung's group [15] and Seo's group [16] reported similar results. Chen et al. [17] researched the dielectric properties of $LaBO_3$, which was one of the main crystalline phases precipitated from LB glass and exhibited excellent dielectric properties with a dielectric constant around 11.8 and a quality factor value around 76,869 GHz (at 15 GHz).

In this work, La_2O_3-B_2O_3-CaO (LBC) glass with different content of B_2O_3 was synthesized. These glasses were used as sintering aids in a LBC glass/Al_2O_3 system. The sintering behavior and dielectric properties of the composites, and their correlation to the B_2O_3 content in LBC glass were discussed.

2. Experimental Procedure

LBC glasses were prepared by the melting method, and the chemical compositions were shown in table 1. The molar ratios of La_2O_3, B_2O_3 and CaO were set to be 1:2:1, 2:5:2 and 1:3:1 (noted as LBC121, LBC252 and LBC131, respectively). All raw materials were purchased from Sinopharm Chemical Reagent Co., Ltd. (Shanghai, China). The batches were mixed and melted in a platinum crucible at 1500 °C for 2h at a heating rate of 10 °C/min. The melt was rapidly quenched into deionized water, ball-milled with agate balls for 12h, and later dried at 120 °C.

Table 1. Chemical compositions of three types of La_2O_3 B_2O_3 CaO glass (wt%).

Sample	La_2O_3	B_2O_3	CaO
LBC121	62.45	26.82	10.73
LBC252	58.53	31.42	10.05
LBC131	55.07	35.47	9.46

The next step was to prepare glass/ceramic composites. High pure Al_2O_3 powder in the range from 30 to 50 wt% was mixed with three types of LBC glass by ball-milling in deionized water for 4 hours and dried at 120 °C. The powders were then admixed with 5 wt% polyvinyl alcohol (PVA) binder solution and pressed into disks (20 mm in diameter and 2 mm in height) at 20MPa. The samples were heat treated at 450 °C for 1h at a heating rate of 3 °C/min, then sintered at 875 °C for 2h. Disk samples only containing three types of glass were prepared using the same method.

The bulk densities of sintered samples were examined by the Archimedes method. The shrinkage values of samples were calculated by comparing the diameters of samples before and after sintering. Differential scanning calorimeter (DSC) analyses of glass samples were performed by a Netzsch STA 449F3 analyzer up to 1100 °C under 10 °C/min heating rate. Glass transition temperatures (T_g) and softening temperatures (T_s) of glass were obtained by thermo-mechanical analysis (TMA) via a dilatometer (Netzsch DIL 402C, Germany) with an alumina sample holder. The crystalline phases of samples were recorded using an X-ray diffractometer (XRD, Bruker AXS D8 Advance device, Germany), which employed Cu-Kα radiation (λ= 1.5418 Å). The microstructures of samples were observed by a field emission scanning electron microscopy (FE-SEM, Hitachi SU8010, Japan). The dielectric constant and dielectric loss of samples at 1 MHz were measured by a precision impedance analyzer (Wayne-Kerr 6500B, England).

3. Results and Discussion

3.1. The crystalline, microstructure and sintering behavior of glasses and composites

Figure 1a shows the XRD patterns of LBC glass samples before and after been heat treated at 875 °C. No obvious crystalline phase can be identified from the original LBC131 glass sample. However, the crystalline phase of $LaBO_3$ (JCPDS card no. 12-0762) is detected in the original LBC252 and LBC121 samples, while the original LBC121 sample also contains another kind of $LaBO_3$ phase ($LaBO_3^*$, also reported as low temperature forming $LaBO_3$ [18], JCPDS card no. 72-0074). In the samples been heat treated, the major crystalline phase is $LaBO_3$, and the secondary phase is CaB_2O_4 (JCPDS card no. 32-

0155). With the decrease of B_2O_3, the less crystalline phase of CaB_2O_4 can be observed. The results indicate that the reducing of B_2O_3 will improve the amount and purity of $LaBO_3$ phase, as well as increasing the crystallization tendency of LBC glass. The spontaneous crystallization of $LaBO_3$ and $LaBO_3^*$ phases before heat treatment will increase the viscosity of melted glass, which could have negative influence on the effect of glass as sintering aids.

The XRD patterns of composite samples containing 60 wt% glass and 40 wt% Al_2O_3 sintered at 875 °C are shown in figure 1b. The crystalline phases of $LaBO_3$ and Al_2O_3 (JCPDS card no. 10-0173) are identified as major ones in all three samples. And new crystalline phases of CaB_4O_7 (JCPDS card no. 31-0253) and $La_2CaB_{10}O_{19}$ (JCPDS card no. 54-0033) are found in LBC131 + Al_2O_3 sample. These two phases can be regarded as B-rich phases and their developments might be involved with the high content of B_2O_3 in LBC131 glass.

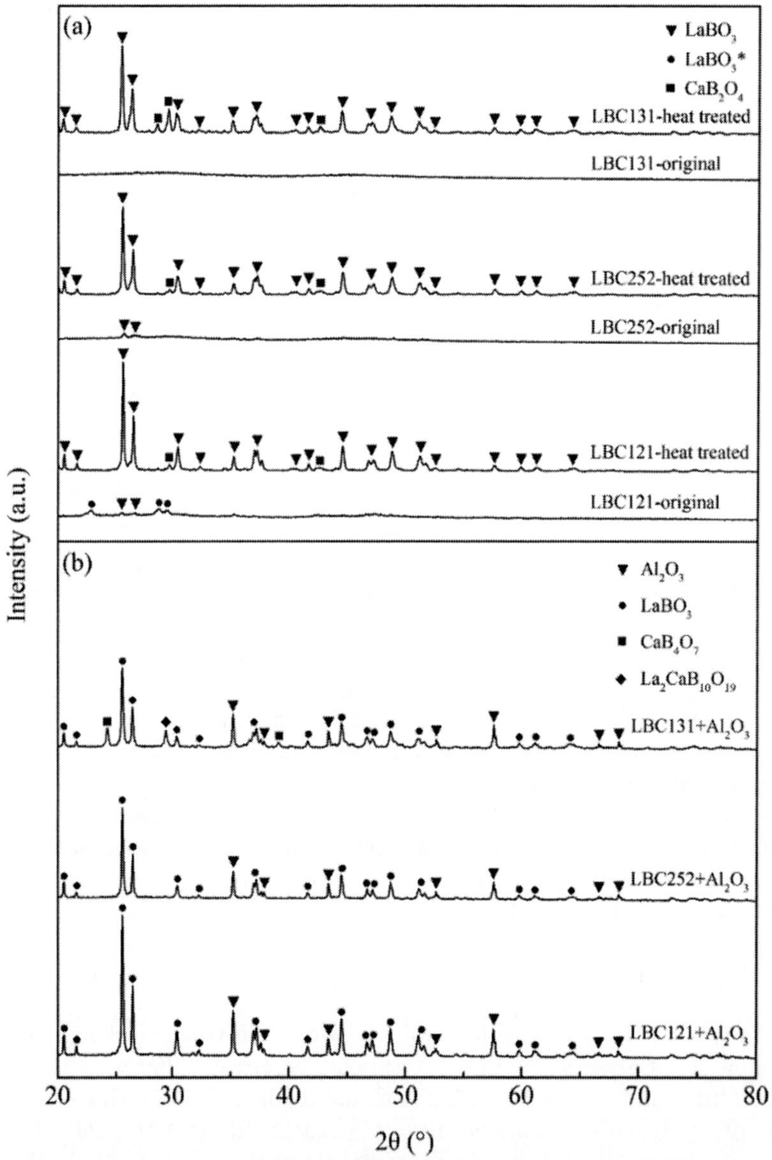

Figure 1. XRD patterns of (a) LBC glass samples before and after heat treatment, (b) 60 wt% LBC glass + 40 wt% Al_2O_3 composites sintering at 875 °C.

The DSC curves of three types of LBC glass are shown in figure 2. And table 2 summarizes the characteristic temperatures of glass samples based on the DSC and TMA results. The softening points of all kinds of LBC glass are around 650 °C, which just meet the requirement of LTCC applications. The change of peak positions in DSC curves suggests that the sintering behavior of LBC glass is influenced by the content of B_2O_3. For glass containing more B_2O_3, the T_g and T_s tend to be a bit higher. The major crystalline peak temperatures (T_{C1}) are around 700-740 °C, which correspond to the report of Jo et al. [14]. These peaks are presumably associated with the crystallization of the $LaBO_3$ phase. The secondary crystalline peak temperatures (T_{C2}) are around 800 °C, which are likely corresponding to the crystallization of CaB_2O_4. Moreover, an exothermic peak at 698 °C has shown in the DSC curve of LBC121 glass. According to the researches of Takada's group [19] and Jung's group [15], this peak could be attributed to the crystallization of a transient phase, such as $LaBO_3^*$ or $La(BO_2)_3$, which would transform into the $LaBO_3$ phase at higher temperature.

Table 2. Characteristic temperatures of LBC glass samples.

Glass	T_g (°C)	T_s (°C)	T_{C1} (°C)	T_{C2} (°C)
LBC121	636	642	737	809
LBC252	648	652	709	794
LBC131	656	669	740	803

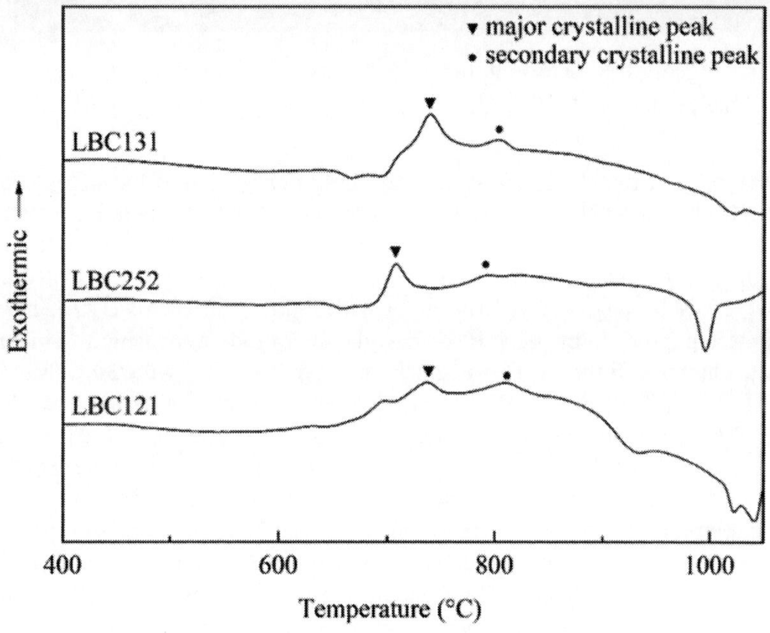

Figure 2. DSC curves of the LBC glass.

Figure 3 shows the SEM images of four selected samples, which performs the microstructures of their surface. As shown in figure 3, for samples having a LBC glass/Al_2O_3 ratio of 1:1, with the increase of B_2O_3 content in LBC glass, less pores can be observed. Figure 3d has clearly exhibited the porous structures of 70 wt%/30 wt% Al_2O_3 sample.

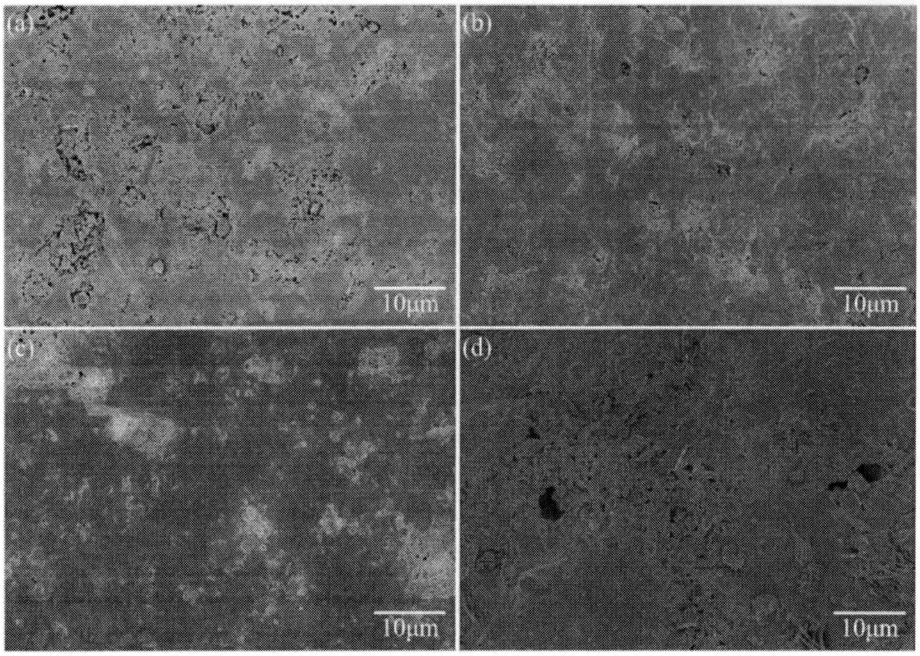

Figure 3. SEM images of samples containing (a) 50 wt% LBC121/50 wt% Al_2O_3, (b) 50 wt% LBC252/50 wt% Al_2O_3, (c) 50 wt% LBC131/50 wt% Al_2O_3 and (d) 70 wt% LBC131/30 wt% Al_2O_3.

Figure 4a and figure 4b present the shrinkage values and bulk densities of all samples containing three different kinds of LBC glass with 30 wt%, 40 wt% and 50 wt% Al_2O_3. Especially, the properties of samples containing 50 wt% Al_2O_3 are shown in table 3. Most samples show appropriate shrinkage values of about 12% with bulk densities around 3.2-3.3 g/cm^3. The shrinkage values of composite samples containing 50 wt% Al_2O_3 suggest that LBC glass with higher content of B_2O_3 shows more positive effect on the sintering process of sample at this glass/ceramic ratio, which could also be proved by the SEM results shown in figure 3. However, the composites of 70 wt%/30 wt% Al_2O_3, 70 wt% LBC252/30 wt% Al_2O_3 and 60% wt% LBC131/40 wt% Al_2O_3 show unusually low shrinkage values and bulk densities. Similar phenomenon has been reported by Thieme et al. [20]. Their glass-ceramic samples in the system BaO-SrO-ZnO-SiO_2 show obvious foaming at temperature above 900 °C, resulting in reducing shrinkage values and bulk densities. When the sintering temperature is higher than the crystalline temperature of low-melting-point glass in a high glass/ceramic ratio system, the release of gases inside the samples during the densification process could be blocked because of the viscosity rising of glass due to crystallization. Subsequently, these gases will lead to the formation of porous structures, just as figure 3d shows. Foaming is not observed in samples containing LBC121 glass, for the spontaneous crystallization in LBC121 glass could be regarded as a reducing in glass/ceramic ratio. However, this might have bad effect on the densification processes of composite samples, which could be proved by the decrease of shrinkage value and bulk density when the B_2O_3 content get lower in table 3.

Table 3. Characteristics of glass/ceramic compositions with 50 wt% Al_2O_3.

Glass	Shrinkage (%)	Density (g/cm^3)	Dielectric constant	Dielectric loss ($\times10^{-3}$)
LBC121	12.1	3.221	9.1	7.98
LBC252	12.3	3.247	7.9	1.92
LBC131	12.9	3.304	7.7	1.69

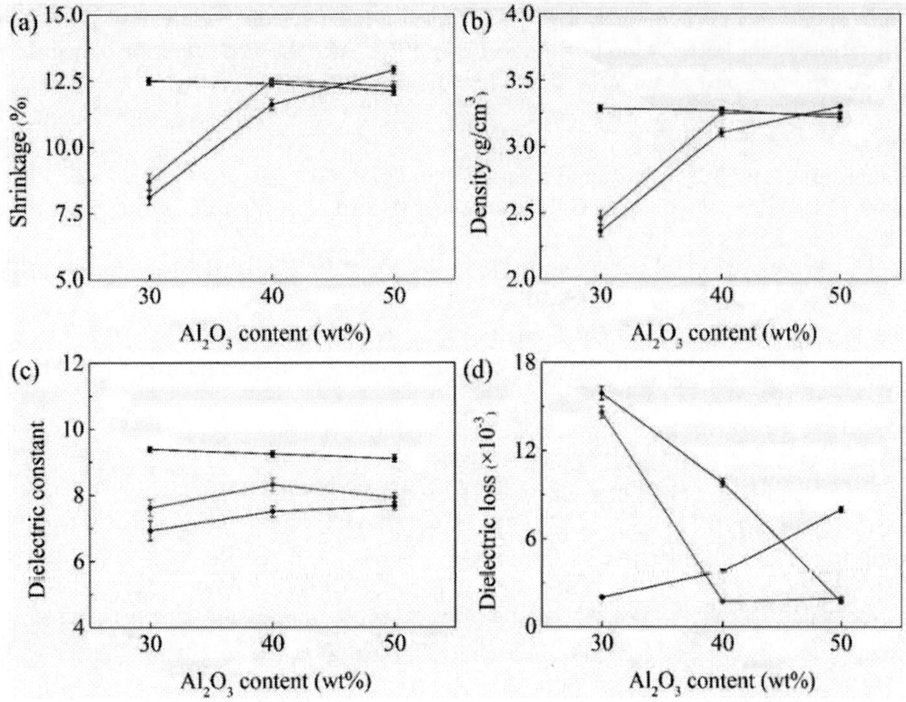

Figure 4. Variations in (a) shrinkage values, (b) bulk densities, (c) and (d) dielectric properties as a function of Al_2O_3 content for samples containing LBC121 (black), LBC252 (red) and LBC131 (blue) glass.

3.2. The dielectric properties of sintered glass/ceramic composites

The dielectric properties of samples containing 50 wt% Al_2O_3 measured at 1 MHz is shown in table 3. With the decrease of B_2O_3 content in LBC glass, the dielectric constant of samples increases gradually, which is associated with the increase of crystalline phase having high dielectric constant, eg. $LaBO_3$ phase shown in XRD results. And the dielectric loss shows an opposite variation tendency. This could be ascribed to the formation of more pores suggested by the reducing of shrinkage values and bulk densities.

Figure 4c and figure 4d show the dielectric properties of more samples with different Al_2O_3 content as well as different types of LBC glass. Samples containing LBC glass with high B_2O_3 content have lower dielectric constant. The varying tendency of dielectric loss of samples match well with that of shrinkage value and bulk density shown in figure 4a and figure 4b. Four samples with the contents of 70 wt% LBC121/30 wt% Al_2O_3, 60 wt% LBC252/40 wt% Al_2O_3, 50 wt% LBC252/50 wt% Al_2O_3 and 50 wt% LBC131/50 wt% Al_2O_3 show excellent combination dielectric properties, i.e., dielectric constant about 7.8-9.4 and dielectric loss about 2.0×10^{-3}.

Based on the above results, we believe that the LBC glass/Al_2O_3 composite has the potential to be chosen as the substrate material in LTCC applications. Our further study will concentrate on the thermal properties and mechanical properties of the composite, eg. coefficient of thermal expansion (CTE), thermal conductivity and bending strength, etc. In addition, we will also investigate the compatibilities of the composite with the low-melting and highly conductive silver and gold electrodes.

4. Conclusions

La_2O_3-B_2O_3-CaO glass with different B_2O_3 content and glass/ceramic composites based on LBC glass/Al_2O_3 had been fabricated. The decrease of B_2O_3 content would increase the crystallization tendency of LBC glass as well as improve the quality of crystalline $LaBO_3$ phase. But it would damage the effect of glass during sintering process of glass/ceramic composite. LBC glass with high

B_2O_3 content should be chosen for a low glass/ceramic ratio system, while low B_2O_3 content glass suited high ratio system better. Composites following this rule showed superior dielectric properties: moderate dielectric constant about 7.8-9.4 and low dielectric loss about 2.0×10^{-3}.

References

[1] Mohanram A, Lee S H, Messing G L and Green D J 2006 *J. Am. Ceram. Soc.* **89** 1923-29
[2] Valant M, Suvorov D, Pullar R C, Sarma K and Alford N M 2006 *J. Eur. Ceram. Soc.* **26** 2777-83
[3] Sayyadi-Shahraki A, Taheri-Nassaj E, Hassanzadeh-Tabrizi S A and Barzegar-Bafrooei H 2014 *J. Mater. Sci-Mater. El.* **25** 355-60
[4] Chen X, Zhang W, Bai S and Du Y 2013 *Ceram. Int.* **39** 6355-61
[5] Wilhelm S P, Kay R W, Mohammed M I, Lacrotte Y and Desmulliez M P 2013 *Microsyst. Technol.* **19** 801-7
[6] Tummala R R 1991 *J. Am. Ceram. Soc.* **74** 895-908
[7] Sebastian M T and Jantunen H 2008 *Int. Mater. Rev.* **53** 57-90
[8] Chang C R and Jean J H 1999 *J. Am. Ceram. Soc.* **82** 1725-32
[9] Lee Y H Aand Wei W C J 2005 *Key Eng. Mater.* **280** 935-40
[10] Müller R, Reinsch S, Eberstein M, Deubener J, Thiel A and Schiller W A 2008 *Adv. Mater. Res.* **39** 373-8
[11] Chen Y, Li E, Zou M, He H and Zhang S 2017 *J. Mater. Sci-Mater. El.* **28** 13132-8
[12] Xiang Y, Han J, Lai Y, Li S, Wu S, Xu Y, Zeng Y, Zhou L and Huang Z 2017 *J. Mater. Sci-Mater. El.* **28** 9911-8
[13] Tang B, Xiong Z, Li H, Long M and Zhang S 2015 *J. Mater. Sci-Mater. El.* **26** 8017-21
[14] Jo Y H, Kang M S, Chung K W and Cho Y S 2008 *Mater. Res. Bull.* **43** 361-9
[15] Jung B H, Hwang S J and Kim H S 2005 *J. Eur. Ceram. Soc.* **25** 3187-93
[16] Seo Y J, Shin D J and Cho Y S 2006 *J. Am. Ceram. Soc.* **89** 2352-55
[17] Chen X, Bai S, Li M and Zhang W 2013 *J. Eur. Ceram. Soc.* **33** 3001-6
[18] Hwang S J, Kim Y J and Kim H S 2007 *J. Electroceram.* **18** 121-8
[19] Takada T and Kageyama K 2005 *Jpn. J. Appl. Phys.* **44** 6629-35
[20] Thieme C, Schlesier M, Bocker C, Buzatto de Souza G and Rüssel C 2016 *ACS Appl. Mat. Inter.* **8** 20212-19

Interband optical transitions in Gd_2O_3: Er nanoparticles – prospective system for energy convertors

A F Zatsepin[1], Yu A Kuznetsova, M A Mashkovtsev and V N Rychkov

Institute of Physics and Technology, Ural Federal University, Yekaterinburg, Russia

[1] E-mail: a.f.zatsepin@urfu.ru

Abstract. The features of electron transitions in the region of fundamental absorption edge as well as vibrational characteristics are investigated in Gd_2O_3 nanoparticles doped with Er^{3+} ions at 0.25-8 mol% concentration. The analysis of initial reflection data is performed for cases of direct and indirect interband transitions. The dependences of corresponding energy gaps on activator concentration are obtained. The effect of strong decrease in the phonon frequency under doping of matrix with a small amount of Er^{3+} is found. The obtained data open new opportunities for improving the efficiency of energy conversion devices.

1. Introduction

Rare earth oxides are of interest as functional materials for energy conversion devices [1]. Gadolinium oxide with a C-type cubic structure attracts special attention as a host matrix due to the chemical and thermal stability and strong paramagnetic behavior [2]. Optically active intrinsic defects of cationic sublattice (irregular Gd^{3+} ions) demonstrate the possibility to donor of UV excitation energy to the activator ions with their subsequent emission in visible spectral region [3, 4]. As an activator ion, Er^{3+} is promising because of intense green and red luminescence and long-lived excited states (millisecond range) [5]. Moreover, for systems containing Er^{3+} ions the anti-Stokes (up-conversion) luminescence can be realized [6, 7]. These features determine prospects of Gd_2O_3:Er as material for UV and NIR energy convertors.

For optimization and improvement of characteristics of functional devices based on rare earth oxides the information about energy structure of material, electron states, mechanisms of radiative and nonradiative relaxations is needed. It is especially important to consider the features of the interaction between the electron and phonon subsystems because of it determines the key properties of material for optical applications such as absolute quantum and energy yield, kinetics of luminescence quenching, energy conversion efficiency, etc. In this work we investigate the features of interband electron transitions and phonon mode behavior in Gd_2O_3 nanoparticles under the introduction of Er^{3+} ions.

2. Experimental

Nanopowders of gadolinium oxide doped with Er^{3+} ions in 0.25-8 mol% concentration range were prepared by chemical precipitation method from a water-alcohol solutions of $Gd(NO_3)_2$ and $Er(NO_3)_2$, followed by annealing at 1000°C during 2 hours. For all investigated samples X-ray diffraction data obtained with the XPert Pro MPD diffractometer correspond to the cubic crystal structure of Gd_2O_3 without admixtures of any other phases. The unit cell parameter for undoped Gd_2O_3 is 10.813 Å. Introduction of Er^{3+} ions into the matrix causes to linear decreasing of lattice constant to a value of

10.793 Å for the Gd_2O_3:Er (8 mol%) sample with a maximum dopant concentration, which corresponds to the ideal Vegard law for the $Gd_{2-x}Er_xO_3$ system [8]. The average crystallite size is 55 nm for samples with a low Er^{3+} concentrations (up to 1 mol%) and decreases to 47 nm with a further increase of dopant concentration as was determined by the Debye-Scherrer's method [9].

Reflectance spectra for all Gd2O3:Er (0.25-8 mol%) samples were obtained using a Perkin Elmer Lambda 35 spectrophotometer equipped with an integrating sphere. For determining spectral dependences of optical density on the basis of primary experimental data we used the Kumar's analytical model [10]. Analysis of fundamental absorption edge in the regions of direct and indirect interband transitions was performed with use of power-law Tauc's expression [11].

3. Results and discussion

3.1. *Direct and indirect interband transitions*

Figure 1 shows the reflection spectra for nanoparticles of nominally pure and activated Gd_2O_3. A common characteristic of all spectra is the presence of dips in the near-UV region at 276 and 315 nm that correspond to optical absorption in Gd^{3+} ions. Such untypical optical activity for regular cations of matrix indicates the defective nature of Gd^{3+} absorption centers in Gd_2O_3 as discussed in more detail in our previous works [3, 4]. For activated samples, in reflection spectra there are a number of additional absorption bands in visible and near-IR regions that are identified as 4f-4f intra-configurational transitions in Er^{3+} activator ions. With the increase of concentration, the intensities of corresponding bands increase due to the growth of absorption centers quantity.

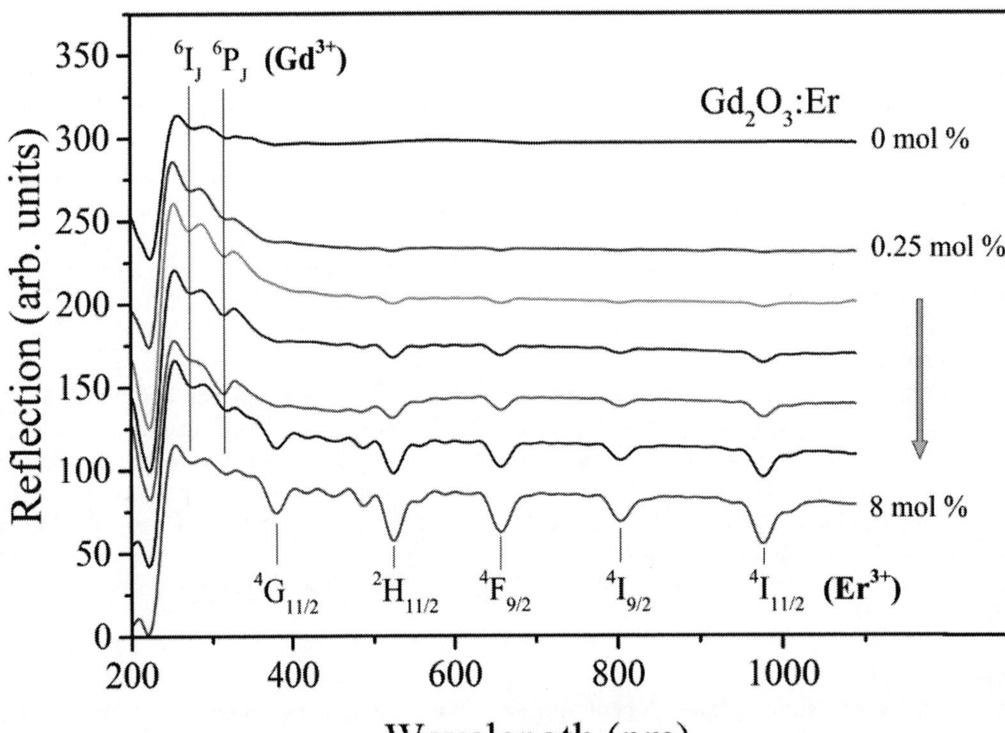

Figure 1. Reflection spectra of nominally pure and Er^{3+}-activated Gd_2O_3 nanoparticles. Absorption bands correspond to electron transitions in Gd^{3+} ($^8S_{7/2}\rightarrow{}^6I_J$, $^8S_{7/2}\rightarrow{}^6P_J$) and Er^{3+} ($^4I_{15/2}\rightarrow{}^4G_{11/2}$, $^4I_{15/2}\rightarrow{}^2H_{11/2}$, $^4I_{15/2}\rightarrow{}^4F_{9/2}$, $^4I_{15/2}\rightarrow{}^4I_{9/2}$, $^4I_{15/2}\rightarrow{}^4I_{11/2}$) ions. For clarity, spectra are shifted relative to each other.

The largest dip with a maximum at 220 nm in reflection spectra of all samples is due to the interband optical transitions in Gd_2O_3 matrix. For detailed analysis of behavior of fundamental absorption edge, we use fragments of reflection spectra corresponding to narrow range of wavelengths (225-250 nm) as initial data. Analytical processing of experimental $R(h\nu)$ reflection data is performed on the basis of Kumar's model in conjunction with the power-law expression [10, 11]:

$$[D(h\nu)\cdot h\nu]^n = \ln[(R_{max} - R_{min})/(R(h\nu) - R_{min})], \quad (1)$$

where $D(h\nu)$ is the optical density, R_{max} and R_{min} are the maximum and minimum values of reflection in the selected spectral range, n is the exponent determining the type of interband transitions (1/2 – for indirect and 2 – for direct transitions).

Since there is no reliable information on the band structure of Gd_2O_3 and it is reported about significantly different values of Gd_2O_3 energy gap (4.85-5.9 eV) in the literature [12], we use the equation (1) for the two most probable cases of direct and indirect interband transitions. The obtained results for Gd_2O_3:Er (1 mol%) is presented in the figure 2. Blue circles show the spectral dependence of fundamental absorption edge in the coordinates for direct allowed interband transitions (right scale). Approximation of 5.39-5.45 eV linear range to the intersection with the abscissa axis (as shown by the solid line) gives a value of 5.38 eV, which corresponds to the energy gap for direct band-to-band optical transitions in Gd_2O_3.

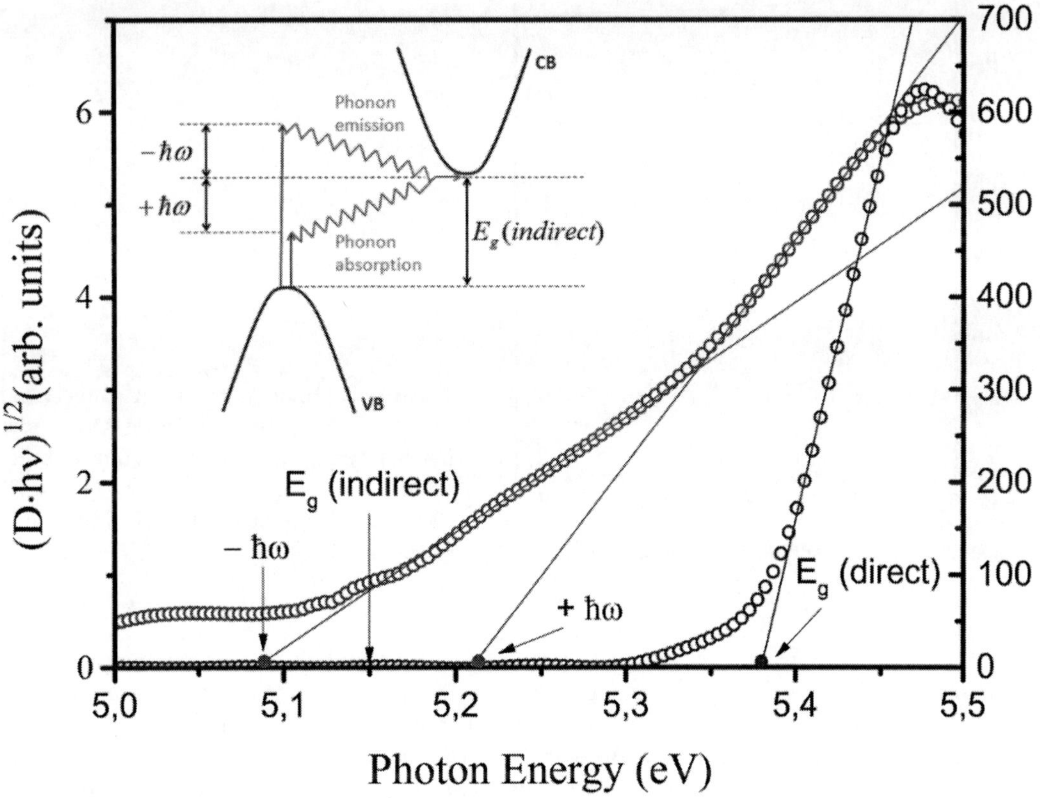

Figure 2. Optical absorption spectra of Gd_2O_3:Er (1 mol%) nanoparticles in the coordinates for direct (right scale) and indirect (left scale) interband transitions. The arrows show energy gap values for direct and indirect band-to-band transitions with absorption (+ ħω) and emission (- ħω) of phonons (see inset).

The case of indirect transitions (figure 2, red circles, and left scale) is more complicated. There are two linear regions in 5.15-5.44 eV energy range and, accordingly, two intersection points with the abscissa axis with values of 5.09 and 5.21 eV. It indicates the realization of two types of interband transitions – with absorption and emission of phonons, as schematically shown in the inset to figure 2. The midpoint of segment bounded by these points is the E_g (*indirect*) energy gap, equal to 5.15 eV. Phonon energy (ħω) absorbed and(or) emitted due to the indirect optical transitions in Gd_2O_3:Er (1 mol%) is 30 meV which corresponds to a frequency of 242 cm^{-1}.

Analysis of the fundamental absorption edge for all Gd_2O_3:Er (0.25-8 mol%) samples allows to reveal a number of regularities. Energy gap for direct interband transitions is invariant to the change of Er^{3+} concentration and equals 5.4 ± 0.02 eV (figure 3(*a*)). At the same time, small amounts of Er^{3+} ions cause an increase in the energy gap for indirect transitions from 4.93 eV for undoped sample to 5.16 eV for Gd_2O_3:Er (1 mol%). With further increase of Er^{3+} concentration up to 8 mol%, E_g (*indirect*) smoothly decreases to 5.08 eV.

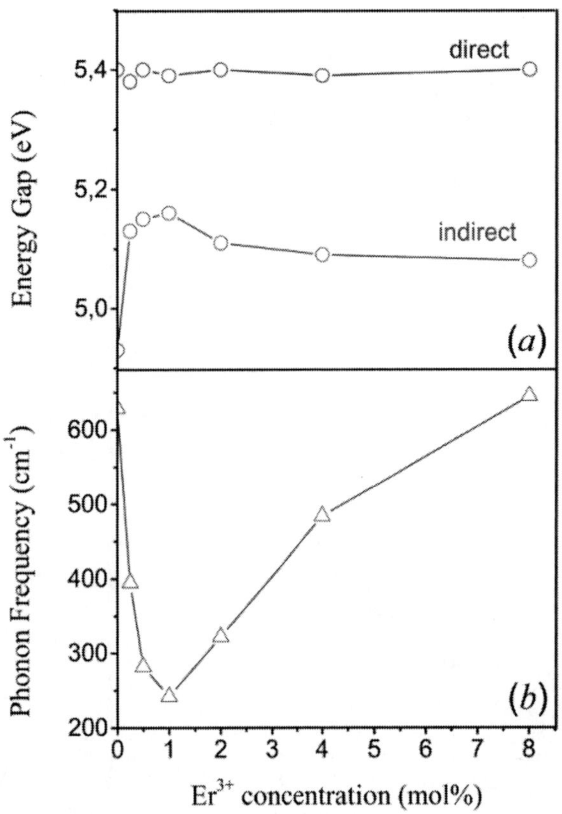

Figure 3. Energy gaps for direct and indirect interband transitions (*a*) and frequency of phonons involved in indirect optical absorption (*b*) for Gd_2O_3:Er (0-8 mol%) nanoparticles as a functions of Er^{3+} concentration.

3.2. *Giant phonon softening*

The frequency of phonons participating in indirect optical transitions is the most sensitive to the activator's introduction into the matrix. As can be seen from figure 3(*b*), introduction of even extremely small amount of Er^{3+} (0.25 mol%) leads to a strong decrease of phonon frequency by more than 1.5 times in comparison with the undoped matrix. The greatest softening of phonon mode is achieved at 1 mol% dopant concentration. Further increase of Er^{3+} amount causes the recovery of phonon frequency to a value of 645 cm^{-1} which is close to the data obtained for pure Gd_2O_3 matrix (629 cm^{-1}).

In the framework of the classical harmonic oscillator model it's known that frequency of lattice vibrations is directly proportional to the square root of the force constant of atom interaction divided by their reduced mass. Introduction of Er^{3+} ions into the Gd_2O_3 lattice results in two opposite effects. On the one hand, decrease in the unit cell parameter caused by the replacement of Gd^{3+} cationic positions by Er^{3+} ions with a smaller ionic radius should contribute to increase of vibrational frequency.

On the other hand, increase in the reduced mass resulted from introduction of heavier Er^{3+} ions into the lattice should cause a decrease in phonon frequency. Quantitative estimation of contributions from these two counteractive effects ($\Delta\omega_1$ and $\Delta\omega_2$) to the overall theoretical change of phonon frequency ($\Delta\omega_{th}$) can be performed using the expression:

$$\Delta\omega_{th} = \Delta\omega_1 + \Delta\omega_2 = -\omega_0\left(3\cdot\gamma\cdot\Delta a/a_0 + \Delta M/2M_0\right), \quad (2)$$

where ω_0, a_0 and M_0 - are the phonon frequency, unit cell parameter and reduced mass for undoped Gd_2O_3; Δa and ΔM - are the absolute changes in lattice constant and reduced mass of Gd_2O_3 caused by Er^{3+} introduction; γ - is the Grüneisen parameter (1.99 value was used [13]).

The results of calculation (table 1) show that the largest contribution to the phonon frequency change results from the $\Delta\omega_1$ component that is related to the lattice constant decreasing. The values of $\Delta\omega_2$ associated with increasing of reduced mass is by two order smaller in comparison with $\Delta\omega_1$. As a consequence, according to calculations, the vibrational frequency of Gd_2O_3 lattice permanently growths with the introduction of Er^{3+} ions and the increase in their concentration. The maximum value of phonon frequency shift is 7 cm^{-1} for sample with 8 mol% of Er^{3+}.

Theoretical calculations don't predict the effect of giant phonon softening that is experimentally observed. Probably, one reason is that calculations take into account behavior of lattice phonon modes only. Whereas experimental phonon softening may be related with the localized defective phonon modes near the structural disorder regions induced by Er^{3+} introduction.

Table 1. Phonon frequency changes in Gd_2O_3 nanoparticles induced by Er^{3+} introduction. $\Delta\omega_{th}$ - theoretical calculation data ($\Delta\omega_1$ - by decreasing of unit cell parameter, $\Delta\omega_2$ - by increasing of reduced mass); $\Delta\omega_{exp}$ - experimental data.

Er^{3+} concentration, mol%	0.25	0.5	1.0	2.0	4.0	8.0
$\Delta\omega_1$, cm^{-1}	0.475	0.537	1.078	1.944	3.536	7.137
$\Delta\omega_2$, cm^{-1}	-0.004	-0.009	-0.017	-0.035	-0.070	-0.139
$\Delta\omega_{th}$, cm^{-1}	0.471	0.528	1.061	1.909	3.466	6.998
$\Delta\omega_{exp}$, cm^{-1}	-234	-347	-387	-307	-145	16

4. Conclusion

The impact of Er^{3+} ions on the electronic-vibrational characteristics of Gd_2O_3 matrix was investigated probed by reflection spectroscopy. There are three types of interband optical transitions for both pure and activated Gd_2O_3 matrix: (i) – direct transitions; (ii) – indirect transitions with absorption of phonons and (iii) – indirect transitions with emission of phonons. Energy gap for direct transitions doesn't change with the introduction of Er^{3+} ions, while the energy gap for indirect absorption increases and achieves a maximum at 1 mol% of Er^{3+} concentration.

The phonon frequency is the most sensitive to the introduction of the activator into the lattice. The effect of giant phonon softening in Gd_2O_3 with a small amount of Er^{3+} ions was experimentally observed, however, theoretical calculations don't predict it. According to the preliminary forecast, this phenomenon can be associated with the local defective phonon modes induced by the Er^{3+} introduction into the Gd_2O_3 lattice. The obtained data indicate the possibility of thermal losses minimization during the radiation conversion. This allows improving the functional characteristics of energy convertors and increasing their efficiency.

Acknowledgments
The work has been funded by the Ministry of Education and Science (Government task №3.1485.2017/4.6) and by Act 211 Government of the Russian Federation, contract 02.A03.21.0006.

References

[1] Liu G K and Chen X Y 2007 *Handbook on the Physics and Chemistry of Rare Earths* vol 37, ed. K A Gschneidner Jr, J C G Bunzli and V K Pecharsky (North-Holland: Amsterdam, The Netherlands) 99
[2] Li H, Song S and Wang W 2015 *Dalton Trans.* **44(36)** 16081–16090
[3] Kuznetsova Yu A, Zatsepin A F, Tselybeev R A et all 2016 *J. Phys. Conf. Ser.* **741(1)**
[4] Zatsepin A, Kuznetsova Yu, Spallino L et al 2016 *Energy Proc.* **102** 144-151
[5] Chen X, Ma E, Liu G and Yin M 2007 *J. Phys. Chem. C* **111(27)** 9638–9643
[6] Mangiarini F, Naccache R, Speghini A et al 2010 *Mater. Res. Bull.* **45** 927-932
[7] Trofimova E S, Pustovarov V A, Kuznetsova Yu A and Zatsepin A F 2017 *AIP Conf. Proc.* **1886** 020024
[8] Vegard L 1921 *Zeitschrift für Physik* **5(1)** 17-26
[9] Klug P and Alexander L E 1954 *X-ray Diffraction Procedure* (New York: Wiley)
[10] Kumar V, Sharma T P and Singh V 1999 *Opt. Mater* **12** 115–119
[11] Tauc J 1974 *Amorphous and Liquid Semiconductors* (Plenum, New York)
[12] Gillen R, Clark S J and Robertson J 2013 *Phys. Rev. B* **87** 125116
[13] Sharma D N, Singh J, Vijay A et al 2016 *J. Phys. Chem. C* **120(21)** 11679-11689

IOP Conf. Series: Materials Science and Engineering **292** (2017) 012048 doi:10.1088/1757-899X/292/1/012048

Measurement and correlation of the solubility of gossypol acetic acid and gossypol acetic acid of optical activity in different solvents

B Zhang, H Tang[1], X Y Liu, X Zhai and X C Yao

School of Pharmaceutical Science, Shihezi University, Shihezi 832002, China

[1] E-mail: Th_pha@shzu.edu.cn

Abstract. The equilibrium method was used to measure the solubility of gossypol acetic acid and gossypol acetic acid of optical activity in isopropyl alcohol, ethanol, acetic acid and ethyl acetate at temperature from 288.15 to 315.15. The Empirical equation and the Apelblat equation model were adopted to correlate the experimental data. For gossypol acetic acid, the root-mean-square deviations (RMSD) were observed in the range of 0.023-4.979 and 0.0112-0.614 for the Empirical equation and the Apelblat equation, respectively. For gossypol acetic acid of optical activity, the RMSD were observed in the range of 0.021-2.211 and 0.021-2.243 for the Empirical equation and the Apelblat equation, individually. And the maximum relative average deviation was 7.5%. Both equations offered an accurate mathematical expression of the experimental results. The calculated solubility showed a good relationship with the experimental solubility for most of solvents. This study provided valuable datas not only for optimizing the process of purification of gossypol acetic acid of optical activity in industry but also for further theoretical studies.

1. Introduction

The Gossypol ((±)-GP, figure 1) is a natural polyphenol derived from cotton plant. Since the chemical bond connecting two naphthalenes has steric effect in the gossypol structure, there are two optical isomers of it [1]. Racemic gossypol ((±)-GP) is a mixture composed of (-)-GP and (+)-GP with different physiological activity [2]. (±)-GP is an effective natural antitumor and anticancer compound. Researchers have evaluated the anticancer property of gossypol against many types of cancer cell lines [3, 4]. The effect on cancer cells of (-)-GP is 2-5 times higher than racemic gossypol [5]. Furthermore, pharmacological researches showed that the side effects of (-)-GP is obviously lower than (+)-GP. In the process of deriving racemic gossypol into the gossypol acetic acid of optical activity, the solubility of (-)-GP in solvents is an important factor in the process of optimizing the conditions of derivatization. The solubility of gossypol acetic acid has been reported in some literatures. However, no experimental solubility data of gossypol acetic acid of optical activity is reported in the open literatures so far. What's more, it is necessary to know the solubility of the (-)-GP in different solvent systems to choose proper solvents.

In this work, the solubility datas of gossypol acetic acid of optical activity in pure isopropyl alcohol, ethanol, acetic acid and ethyl acetate were measured by the equilibrium method at temperature from 288.15 to 315.15 K under atmospheric pressure. The experimental datas were correlated by the Empirical equation and the Apelblat equation.

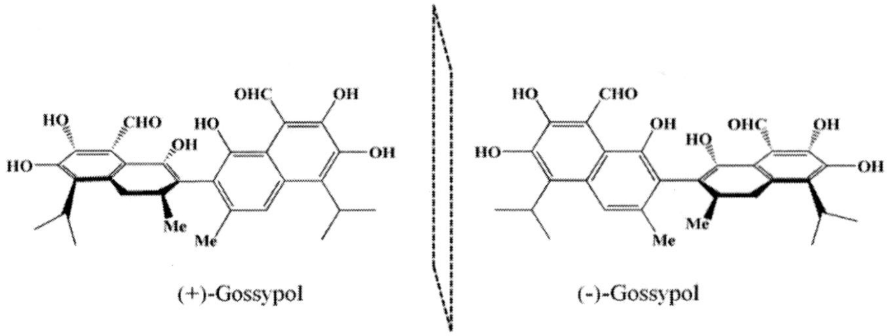

Figure 1. Chemical structure of the Gossypol.

2. Experimental

2.1. Materials and Apparatus

Gossypol acetic acid ((±)-GA) and gossypol acetic acid of optical activity ((-)-GA) with purity ≥ 99.0 %(determined by HPLC Chiral column) prepared in Key Laboratory of Xinjiang Phaytomedicine Resource. It was dried in vacuo at 323 K for 24 h and then stored in desiccator. All the organic solvents (HPLC grade) used for the experiment were purchased from Tianjin Fuyu Chemical Company, Ltd. The smart thermostatic bath (model: DC-2006) was provided by Ningbo Scientz Biotechnology Company, Ltd. Digital thermostat magnetic stirrer (model: 85-2), Japan Shimazu ultraviolet spectrophotometer (model: UV-2550), and distilled water was prepared in our laboratory.

2.2. Methods

The method of measuring the solubility of solid in solvents is analysis and synthesis [6-8]. In this work, we adopted an analytical method to determine the solubility of (-)-GA and (±)-GA in six selected solvents, with the temperature ranging from 288.15 to 315.15 K. Our analytical method is simple and reliable, and it allows the simultaneous measurement of a large number of samples.

The method of solubility measurement is similar to the method described in the literatures [9-11]. In the experiment, the equilibrium method was used to study the solubility of (-)-GA and (±)-GA in different organic solvents. The temperature was controlled by an intelligent thermostatic bath. The experiment was conducted in a closed small room with constant temperature and humidity and static air. (-)-GA and (±)-GA dissoluted in 10 mL organic solvent were placed into a 50 mL three-necked flask with a stopper, and stirred continuously by a magnetic stirrer at the temperature in the thermostatic bath. The precision of the temperature should be controlled within±0.01K. 12 h was conducted to ensure the solid–liquid equilibrium was established in the tube. At least 6 h was needed to settle the solution after turning off the stirring. Filtrated the supernatant by 0.45 μm microporous membrane and diluted it to appropriate concentration, and then determined it by UV. Each experiment was repeated three times to obtain the mean value. The saturated solution mole fraction solubility of solute(X) in each solvent can be calculated using the following equation:

$$X = \frac{m_A / M_A}{m_A / M_A + m_S / M_S} \tag{1}$$

Where m_A and m_s represent the mass of the solute and the solvent, and M_A and M_S are the molar mass of the solute and the solvent, respectively.

2.3. Test of the Apparatus

To prove the feasibility and demonstrate the uncertainties of the measurements, the solubility of NaCl in water was measured and compared with values reported in the literatures. The results agreed with the reported values with a mean relative deviation (RD) of 0.30%. The measured values were listed in table 1.

Table 1. Mole fraction of Nacl in water.

T (K)	X	X (NaCl) [12-13]	10^2RD
293.15	0.0996	0.0993±0.0004	0.30
303.15	0.1001	0.0999±0.0005	0.19
313.15	0.1009	0.1007±0.0006	0.20
323.15	0.1019	0.1017±0.0005	0.20

3. Results and Discussion

In order to use the UV method to calculate the concentrations, calibration curves were needed. The measured values were listed in table 2 and table 3.

Table 2. Standard curve equation of (-)-GA in four organic solvents.

Solvent	Calibration curves equation	R^2	The linear range
Isopropyl alcohol	A=0.0273*C+0.0051	0.9990	10.21μg.mL^{-1}~35.74μg.mL^{-1}
Ethanol	A=0.0285*C-0.0007	0.9998	9.99μg.mL^{-1}~34.97μg.mL^{-1}
Acetic acid	A=0.0301*C-0.0139	0.9992	10.05μg.mL^{-1}~35.18μg.mL^{-1}
Ethyl acetate	A=0.0305*C-0.0007	0.9999	10.08μg.mL^{-1}~35.28μg.mL^{-1}

Table 3. Standard curve equation of (±)-GA in four kinds of solvents.

Solvent	Calibration curves equation	R^2	The linear range
Isopropyl alcohol	A=0.0496*C-0.0677	0.9992	3.36μg.mL^{-1}~20.16μg.mL^{-1}
Ethanol	A=0.0550*C-0.0449	0.9995	5.15μg.mL^{-1}~25.75μg.mL^{-1}
Acetic acid	A=0.0513*C+0.015	0.9996	5.47μg.mL^{-1}~19.15μg.mL^{-1}
Ethyl acetate	A=0.0636*C+0.3452	0.9995	5.26μg.mL^{-1}~15.78μg.mL^{-1}

The measured mole fraction of (-)-GA in pure isopropyl alcohol , ethanol, acetic acid and ethyl acetate at the temperature from 288.15 to 315.15 K were presented in table 4 and the measured mole fraction solubility of (±)-GA in table 5. The mole fraction X values in table 4 and table 5 was the average value of three values taken from the same solvent. The RDs between the experimental values and the calculated values were also listed in table 4 and table 5.

IOP Conf. Series: Materials Science and Engineering **292** (2017) 012048 doi:10.1088/1757-899X/292/1/012048

Table 4. Mole fraction solubility data and correlation results of (-)-GA in four organic solvents.

Solvents	T/K	$10^{-4}x_{exp}$	Empirical equation		Apelblat equation	
			$10^{-4}x_{cal}$	10^2RD	$10^{-4}x_{cal}$	10^2RD
Isopropyl alcohol	288.15	11.223	11.489	2.37	11.533	2.76
	289.15	11.415	11.370	0.39	11.425	0.09
	293.15	11.607	11.113	4.26	11.208	3.44
	298.15	11.703	11.285	3.57	11.388	2.69
	303.15	11.990	12.006	0.13	12.063	0.61
	308.15	12.757	13.277	4.08	13.286	4.15
	313.15	15.153	15.096	0.38	15.176	0.15
	315.15	16.399	15.978	2.57	16.161	1.45
Ethanol	288.15	12.143	12.236	0.77	12.298	1.28
	289.15	12.252	12.204	0.39	12.277	0.20
	293.15	12.719	12.229	3.86	12.330	3.06
	298.15	12.737	12.602	1.06	12.705	0.25
	303.15	13.372	13.357	0.12	13.430	0.43
	308.15	14.502	14.492	0.07	14.539	0.26
	313.15	15.860	16.009	0.94	16.094	1.47
	315.15	17.149	16.722	2.49	16.862	1.68
Acetic acid	288.15	0.550	0.578	5.04	0.573	2.80
	289.15	0.570	0.574	0.72	0.572	0.85
	293.15	0.604	0.575	4.78	0.581	4.85
	298.15	0.604	0.612	1.37	0.619	1.28
	303.15	0.649	0.689	6.22	0.689	4.79
	308.15	0.807	0.807	0.04	0.798	2.29
	313.15	0.964	0.966	0.17	0.962	1.43
	315.15	1.023	1.040	1.76	1.047	1.19
Ethyl acetate	288.15	131.730	131.726	0.01	131.957	0.17
	289.15	132.046	132.878	0.63	133.188	0.86
	293.15	141.443	138.125	2.35	138.629	1.99
	298.15	144.200	146.126	1.34	146.651	1.70
	303.15	153.077	155.729	1.73	156.142	2.00
	308.15	170.964	166.935	2.36	167.248	2.17
	313.15	179.563	179.743	0.10	180.147	0.33
	315.15	184.660	185.315	0.35	185.855	0.65

The RDs were calculated according to:

$$RD = \frac{X_{exp} - X_{cal}}{X_{exp}} \tag{2}$$

Where X_{exp} and X_{cal} represent the experimental and calculated solubility values, respectively.
The relationship between temperature and mole fraction solubility in the different solvents was described by the Empirical equation and the Apelblat equation.

Table 5. Mole fraction solubility data and correlation results of (±)-GA in four organic solvents.

Solvents	T/K	$10^{-4}x_{exp}$	Empirical equation		Apelblat equation	
			$10^{-4}x_{cal}$	10^{2}RD	$10^{-4}x_{cal}$	10^{2}RD
Isopropyl alcohol	288.15	6.182	6.687	8.16	6.269	1.41
	289.15	6.278	6.653	5.97	6.241	0.59
	293.15	6.328	6.620	4.61	6.225	1.63
	298.15	6.459	6.815	5.51	6.410	0.76
	303.15	6.749	7.275	7.79	6.829	1.18
	308.15	7.487	7.993	6.76	7.510	0.30
	313.15	8.532	8.984	5.30	8.507	0.29
	315.15	9.031	9.452	4.66	9.013	0.20
Ethanol	288.15	11.629	11.714	0.73	11.593	0.31
	289.15	11.684	11.895	1.81	11.769	0.73
	293.15	12.616	12.612	0.03	12.473	1.13
	298.15	13.546	13.493	0.39	13.347	1.47
	303.15	14.192	14.357	1.16	14.211	0.14
	308.15	14.934	15.205	1.81	15.060	0.84
	313.15	15.992	16.035	0.33	15.887	0.66
	315.15	16.301	16.362	0.37	16.211	0.55
Acetic acid	288.15	0.147	0.169	14.97	0.141	4.08
	289.15	0.156	0.175	12.18	0.150	3.85
	293.15	0.206	0.212	2.91	0.194	5.83
	298.15	0.251	0.280	11.55	0.265	5.58
	303.15	0.340	0.374	10.00	0.356	4.71
	308.15	0.491	0.495	0.81	0.471	4.07
	313.15	0.618	0.641	3.72	0.617	0.16
	315.15	0.679	0.706	3.98	0.685	0.88
Ethyl acetate	288.15	100.376	96.450	3.91	100.726	0.35
	289.15	101.315	96.428	4.82	100.803	0.51
	293.15	102.694	97.274	5.28	101.953	0.72
	298.15	104.987	100.428	4.34	105.272	0.27
	303.15	110.080	105.912	3.79	110.739	0.60
	308.15	119.776	113.726	5.05	118.533	1.04
	313.15	128.851	123.870	3.87	128.952	0.08
	315.15	134.070	128.580	4.10	133.948	0.09

The effect of the temperature on solubility of (-)-GA in the selected solvents can be correlated by the Empirical equation [14].

$$X = A_1+A_2T+A_3T^2 \tag{3}$$

where X is the mole fraction solubility of (-)-GA or (±)-GA, T is the corresponding temperature in Kelvin, and A_1, A_2 and A_3 are the parameters of this equation. Values of the parameters A_1, A_2 and A_3 were listed in table 6 and table 7.

Table 6. Parameters of Empirical equation for (-)-GA in four organic solvents.

Solvents	A_1	A_2	A_3	R^2	RMSD	RAD
Isopropyl alcohol	960.688	-6.458	1.1×10^{-2}	0.9523	0.342	0.022
Ethanol	656.337	-4.431	7.621×10^{-3}	0.9776	0.243	0.012
Acetic acid	69.042	-0.471	8.083×10^{-4}	0.9828	0.021	0.025
Ethyl acetate	2470.29	-17.351	3.205×10^{-2}	0.9825	2.211	0.011

Table 7. Parametersof Empirical equation for(±)-GA in four organic solvents.

Solvents	A_1	A_2	A_3	R^2	RMSD	RAD
Isopropyl alcohol	455.755	-3.077	5.270×10^{-3}	0.9927	0.436	0.061
Ethanol	-68.877	0.378	-3.412×10^{-4}	0.9949	0.142	0.008
Acetic acid	41.382	-0.292	5.167×10^{-4}	0.9961	0.023	0.075
Ethyl acetate	3985.313	-26.923	4.660×10^{-2}	0.9969	4.979	0.044

The Apelblat equation is given as follows [15-17]:

$$\nabla X = A + \frac{B}{T} + C\nabla T \tag{4}$$

Where X is the mole fraction solubility of (-)-GA or (±)-GA, T is the experimental temperature in Kelvin, and A, B and C are regression parameters obtained by fitting to the experimental solubility data and are shown in table 8 and table 9.

Table 8. Parameters of Apelblat equation for (-)-GA in four organic solvents.

Solvents	A	B	C	R^2	RMSD	RAD
Isopropyl alcohol	-1016.901	44813.185	152.526	0.9626	0.295	0.019
Ethanol	-636.791	27810.049	95.840	0.9813	0.200	0.011
Acetic acid	-1115.635	48355.278	167.258	0.9837	0.021	0.024
Ethyl acetate	-192.569	7893.617	30.027	0.9821	2.243	0.012

Table 9. Parameters of Apelblat equation for (±)-GA in four organic solvents.

Solvents	A	B	C	R^2	RMSD	RAD
Isopropyl alcohol	-841.151	36821.309	126.283	0.9952	0.061	0.008
Ethanol	74.513	-4187.661	-10.158	0.9946	0.114	0.007
Acetic acid	105.003	-9301.701	-13.187	0.9953	0.0112	0.036
Ethyl acetate	-468.851	20450.330	71.068	0.9967	0.614	0.005

The root-mean-square deviation (RMSD), together with the relative average deviation (RAD) for the Empirical equation and the Apelblat equation, are also listed in table 6, and table 7, respectively. The RMSD are calculated according to:

$$RMSD = \sqrt{\frac{\sum_{i=1}^{N}\left(\frac{X^{i} - X_{cal}}{X_{exp}}\right)^2}{N}} \tag{5}$$

Where N is the number of experiment times and x_{cal} and x_{exp} represent the calculated and experimental solubility values. The RADs are calculated by:

$$RAD = \left(\frac{1}{N}\right)\sum_{=1}^{N}\left|\frac{X^{i} - X_{cal}}{X_{exp}}\right| \tag{6}$$

As we can see from tables 5, 6, 8, and 9, the calculated data of solubility of (-)-GA and (±)-GA in the four pure organic solvents showed good agreement with the experimental datas with small RMSD. Comparing the absolute values of the RD among all the values by two equations, the RD calculated by the Apelblat equation was lower than that calculated by the Empirical equation. The calculated datas of (-)-GA solubility were fitted by the Apelblat equation. The absolute values of the RD were lower than 4.85%, which indicated that the Apelblat equation was suitable for correlating the solubility data of (-)-GA in the selected pure solvents. The same conclusion can be drawn after analyzing the solubility data of the (±)-GA. For the (-)-GA, it can be seen from tables 5, 6 that the overall RMSD of calculated data for the (-)-GA in different solvents were 2.817 and 2.759, respectively. It can be observed that the regression results of the Apelblat equation is better than those of the Empirical equation. The same conclusion also applied to (±)-GA. Consequently, the Apelblat equation is more accurate than the Empirical equation for these systems. The results mentioned above suggested that the experimental datas and the correlations in this work are essential for modeling the purification process of (-)-GA

The graph of solubility of (-)-GA and (±)-GA in the selected solvent systems was shown in figure 2 and figure 3. It can be seen clearly from this figure that the solubility of (-)-GA and (±)-GA in each of the selected solvents would increase with the temperature range from 288.15 to 315.15 K. But the increment of solubility with temperature is different in each of the pure solvents. The solubility of (-)-GA in these solvents follows the order that Ethyl acetate > Ethanol>Isopropyl alcohol >Acetic acid. Among these solvents, the solubility of (-)-GA was highest in Ethyl acetate and lowest in Acetic acid. This result shows that the polarity of the solvents is not the only factor in determining the solubility of (-)-GA in the solvents. The selected four solvents are commonly used and essential for modeling the purification process.

In Ethyl acetate, (-)-GA had high solubility at the initial temperature studied (288.15K), and the increment of solubility with temperature is increasing quickly. Therefore, we can estimate that after recrystallization, (-)-GA can be obtained with higher yield in Ethyl acetate than in the other selected solvents. Thus, Ethyl acetate has the potential to be a solvent in purification and recrystallization process of (-)-GA.

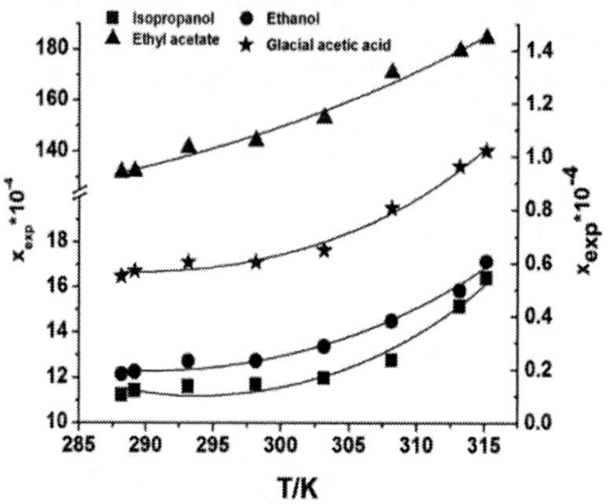

Figure 2. Solubility of (-)-GA in four organic solvents.

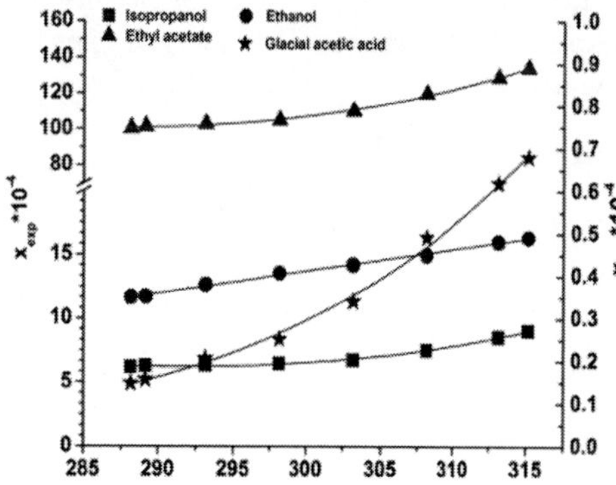

Figure 3. Solubility of (±)-GA in four organic solvents.

4. Conclusions

The equilibrium method was used to measure the solubility of gossypol acetic acid and gossypol acetic acid of optical activity in isopropyl alcohol, ethanol, acetic acid and ethyl acetate at temperature from 288.15 to 315.15 K. We can draw the following conclusions: (1) the solubility of gossypol acetic acid of optical activity in all of the selected pure solvents increases with temperature; (2) the solubility datas can be correlated using the Empirical equation and the Apelblat equation, and the Apelblat equation is more accurate than the Empirical equation; (3) the experimental solubility datas and the corresponding parameters can be used for optimizing the process of purification of gossypol acetic acid of optical activity in industry; (4) Ethyl acetate has the potential to be a solvent in recrystallization process of (-)-GA and (±)-GA.

Acknowledgments

This work was funded by the National Natural Science Foundation of China (No. 81260627 and No. 81660641) and the Science and technology attack plan project of Xinjiang Production and Construction Corps (No. 2016AD008)

References

[1] Kenar J A 2006 *Journal of the American Oil Chemists' Society* **83** 269-270
[2] Freedman T B, Cao X L, Oliveira R V, Cass Q B and Nafie LA 2003 *Chirality* **15** 196-200
[3] Vermel E M, Krugliak S A 1963 *Vopr Onkol* **21** 39-43
[4] Michael L L, Jennifer R. 2002 *Pharmacological Research*, 46:551-555.
[5] Blackstaffe L, Shelley M D, Fish R G 1997 *Melanoma Res* **7** 364- 372
[6] Brandani S, Brandani V, Flammi D 1994 *J Chem Eng* **39** 201–202
[7] Jiang Q, Gao G H, Yu Y X and Qin Y 2000 *J Chem Eng* **45** 292–294
[8] Guo S, Yang W, Hu Y, Wang K and Luan Y 2013 *J Solution Chem* **42** 1879–1887
[9] Liu M J, Fu H L, Yin D P, Zhang Y L, Lu C C, Cao H and Zhou J Y 2014 *J CHEM Eng* **59** 2070–2074
[10] Noubigh A, Aydi A, Mgaidi A and Abderrabba M 2013 *J MOL LIQ*, 187: 226–229.
[11] Eun S C, Hun Y S,Seong S Y and Ki P Y 1997 *J Chem Eng*,14: 292–296.
[12] Hu G Q, Li H, Wang X J, Zhang Y D 2010 *J Chem Eng Data*, 55:3969–3971
[13] Liu G Q, Ma LX, Liu J 2002 Data handbook of physical and chemical properties for chemistry and chemical engineering (inorganic volume) *Chemical Industry Press* Peking 2
[14] Tuo J M, jin Z H , Cen P L, and Wang M 2008 *J Chem Eng Chin Uni* **22** 733-738
[15] Apelblat A, Manzurola E 1999 *Journal of Chemical Thermodyn* **31** 85–91
[16] Manzurola E, Apelblat A 2002 *Journal of Chemical Thermodyn* **34** 1127–1136
[17] Apelblat A, Manzurola E 1997 *Journal of Chemical Thermodyn* **29** 1527–1533

IOP Conf. Series: Materials Science and Engineering **292** (2017) 012049 doi:10.1088/1757-899X/292/1/012049

Effect of GBCD on IGC susceptibility of solution treated 2024 Al alloy

G P Li, C L He[1], R Li, J M Shi, G F Ma, X Y Fu, H Z Chen and J M Wang

Liaoning Provincial Key Laboratory of Advanced Materials, Shenyang University, Shenyang 110044, China

[1] E-mail: ccllhhe@126.com

Abstract. Grain boundary character distribution (GBCD) of solution heat treated 2024 Al alloy was analyzed by electron backscattered diffraction, and intergranular corrosion (IGC) susceptibility and corrosion weight loss of the 2024 Al were investigated according to ASTM G110-92. The grain boundaries are composed of 12.17% low angle (∑1) boundaries, 10.87% low coincidence site lattice (∑3-∑29) boundaries and 76.96% random (R) boundaries. After corrosion for 6 h, the corroded boundary proportions of ∑1, ∑3, ∑5-∑29 and R boundaries were 10.71%, 11.11%, 21.95% and 25.14%, respectively. This shows that the GBCD has a great effect on IGC susceptibility of the solution treated 2024 Al, and both ∑1 and ∑3 boundaries exhibit much higher IGC resistance than ∑5-∑29 and R boundaries. Among all the boundaries, the R boundaries have the highest IGC susceptibility.

1. Introduction

In 1984, Watanabe [1] proposed the concept of grain boundary character distribution (GBCD) which is a parameter used to quantitatively describe the type and frequency of grain boundaries present in a microstructure. On the basis of coincidence site lattice (CSL) model, the grain boundaries are divided into low angle (∑1) boundaries, low ∑CSL grain boundaries ($\Sigma \leq 29$) and random grain boundaries (R). The R type boundaries refer to the general high-angle boundaries and the CSL grain boundaries of $\Sigma > 29$. Kim et al [2] found that IGC susceptibility of pure aluminum in HCl solution was strongly dependent on its GBCD. In 8%, 16% and 38% HCl, the IGC resistance of the ∑1 boundary was the best. In 8% and 16% HCl, the ∑3 grain boundary exhibited higher IGC resistance compared with the R boundaries. In NiAl intermetallic compound [3], the ∑1, ∑3 and ∑5 boundaries could obviously prevent the continuation of cracks, while ∑7, ∑11, ∑13, ∑21 and ∑23 grain boundaries had a relatively poor ability to prevent crack continuity. Meng et al [4] found that the selective features of IGC in Al alloy and Al matrix composites were mainly related to GBCD. High-angle boundaries were easy to be attacked while special grain boundary and low angle boundary had high IGC resistance [5].

In this paper, solution heat treated 2024 Al alloy was chosen as the research material and the relationship between GBCD and IGC susceptibility was studied in detail.

2. Materials and methods

The material used in this study was as rolled plates of 2024 Al alloy, its chemical composition was listed in table 1. Samples were solution heat treated in an air circulation furnace at 495 ±3 °C for 40 min and then cooled in air immediately. The sample dimensions were about 10 mm × 5 mm × 1 mm. The test samples were successively ground on emery papers from 400# to 2000#, and then polished by

1.0 μm diamond paste. After that, the samples were electropolished in an electrolyte containing 10 vol % $HClO_4$ + 90 vol % ethanol for 7 s under a DC voltage of 29 V, the electrolyte temperature was less than 10 °C. Stainless steel plate was selected as the cathode.

Table 1. The chemical composition of 2024 Al alloy (mass fraction, %).

Element	Cu	Mg	Mn	Fe	Si	Zn	Ti	Ni	Al
Content	3.8~4.9	1.2~1.8	0.3~0.9	0.5	0.5	0.3	0.1	0.1	Bal.

The GBCD data of the solution treated 2024 Al sample were obtained by using Hitachi S-4800 field emission scanning electron microscope equipped with EBSD detector and Oxford's CHANEL5 HKL. The tested area was automatically scanned with a step size of 2.5 μm. The distribution of grain size was obtained by the supporting software. The IGC experiments were carried out according to ASTM G 110-92, which involved degreasing in acetone and ethanol, immersion etching (1 min in 945 mL H_2O+50 mL HNO_3+5 mL HF at 93°C), rinsing in reagent water, immersion in concentrated nitric acid (70 %) for 1 min, and rinsing in reagent water and air dry, followed by 6 h immersion in solution (57 g NaCl and 10 mL H_2O_2 per litre) 30 °C. The corrosion morphology was observed by a metallographic microscope, and corrosion weight loss was measured by an electron balance with a precision of 0.1 mg.

3. Results and discussion

Figure 1a presents the grain size distribution of the solution heat treated 2024 Al (296 grains characterized), which shows that the distribution of grain sizes is non-uniform in the microstructure. The average grain size is 12.1 μm. The smallest grain size is 2.8 μm, and the largest is 81.1 μm. Thereinto, the fraction of grains with size between 2-4 μm is about 40% while grains greater than 40 μm is only about 7.1%. Figure 1b presents the misorientation angle distribution of the grain boundaries in the solution heat treated 2024 Al. In 460 grain boundaries characterized, the fractions of low angle (∑1) boundaries with misorientation angles less than 15° and R (∑>29) boundaries are 12.17% and 76.96%, respectively, whereas the total fraction of low ∑ CSL (∑3-∑29) boundaries is 10.87%. Except 1.96% ∑3, the fraction of the other type CSL boundaries (∑5-∑29) is less than 8.91%.

Figure 2a shows the EBSD map of the 2024 Al alloy before corrosion test, and figure 2b presents the corrosion morphology in the same area after 6 h corrosion. This figure clearly demonstrates that many random boundaries have been seriously attacked while ∑5-∑29 boundaries are only a small amount of corrosion. The ∑1 and ∑3 boundaries are resistant to corrosion. Among them, the two ∑1 boundaries are not corroded. Because the connectivity of random grain boundaries is interrupted by the low ∑CSL grain boundaries and low angle boundaries, the IGC is non-uniform. Additionally, figure 1b also clearly shows that serious pitting corrosion took place in the sample. The corrosion mechanism of IGC results from galvanic corrosion between grain boundary precipitates (e.g. Al_2Cu) and Cu-depleted zone, while pitting can be contributed to galvanic corrosion between secondary phase particles (AlCuFeMnSi phase [6]) and matrix Al.

Figure 3 shows the fractions of each type grain boundaries and corroded grain boundaries in the sample. After corrosion 6 h, the corroded proportions of ∑1, ∑3, ∑5-∑29 and R boundaries are 10.71%, 11.11%, 21.95% and 25.14%, respectively. Obviously, both the ∑1 and ∑3 grain boundaries reveal much lower IGC susceptibility, which is associated with their lower energies of grain boundaries, and less impurity segregation in the boundaries. Compared with the ∑1 and ∑3 grain boundaries, much more ∑5-∑29 grain boundaries are attacked. In all the boundaries, the number of R boundaries attacked is the most. This shows that the GBCD has a great effect on IGC susceptibility of the solution treated 2024 Al. In general, ∑1 and ∑3 (twin) boundaries possesses the lowest energy in all the grain boundaries, therefore, they should exhibit the best IGC resistance. Low ∑CSL grain boundaries have higher structural order, relative low grain boundary energy and less boundary precipitation [7]; therefore, their IGC resistance is relative good although it is much lower than ∑1 and ∑3 boundaries. Because R grain boundaries have the highest boundary energies, they present the lowest IGC resistance. The sequence of IGC susceptibility from low to high is: $\sum 1 \leq \sum 3 < \sum 5-\sum 29 <$ R. The proportion of corrosion boundaries in 460 grain boundaries characterized is 22.83%.

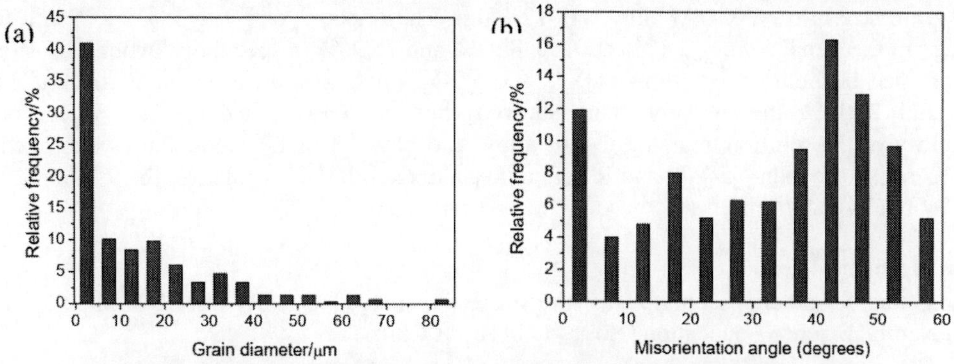

Figure 1. (a) Grain size distribution and (b) misorientation angle distribution for the 2024 Al alloy.

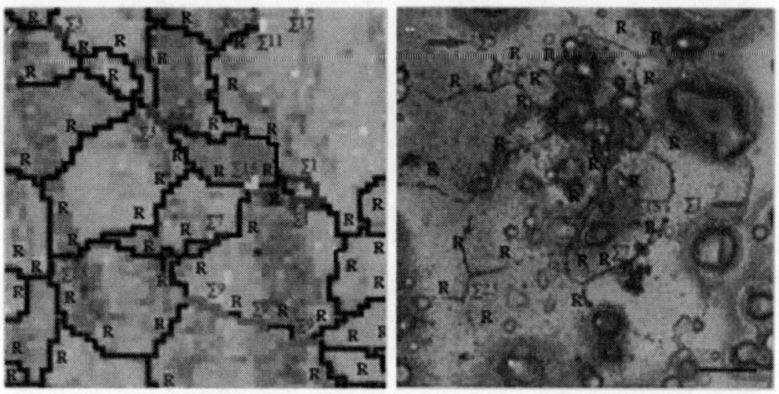

Figure 2. (a) EBSD map and (b) corrosion morphology of the sample before (a) and (b) after corrosion.

Figure 4 presents the change of weight loss with corrosion time. It is clearly shown that the weight loss is increased with corrosion time. However the corrosion rate is decreased with increasing time because the corrosion products covered on the sample surface, hindering the corrosion progress. When corrosion time increases from 1 h to 6 h, the corrosion rate of the 2024 Al alloy decreases from 18.91 mm/a to 17.86 mm/a.

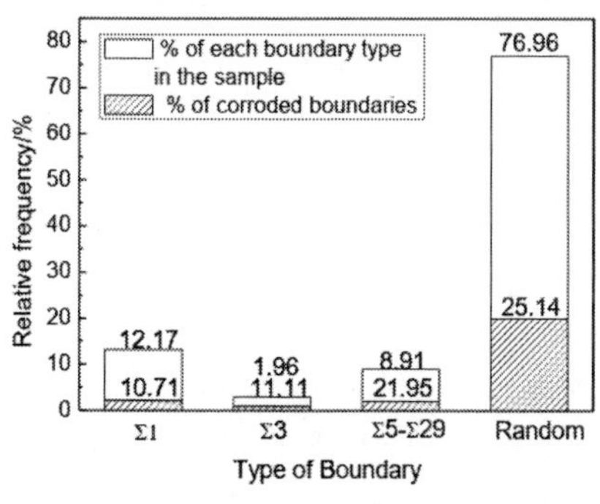

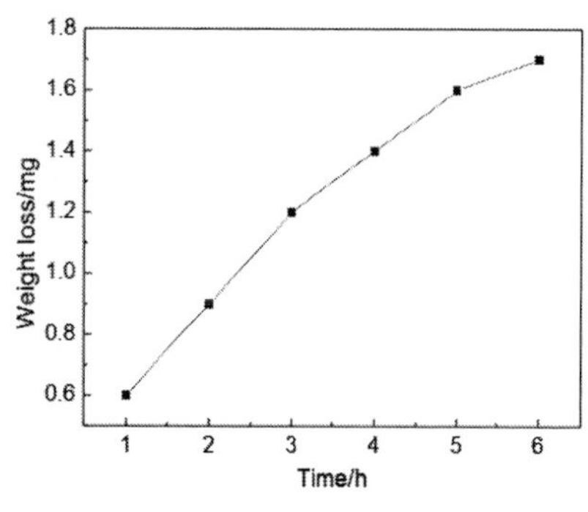

Figure 3. The relationship between GBCD and IGC susceptibility.

Figure 4. Weight loss vs corrosion time.

4. Conclusion

The GBCD of solution heat treated 2024 Al alloy was examined by EBSD, and IGC susceptibility and corrosion weight loss of the 2024 Al alloy were investigated according to ASTM G110-92. The average grain size of the 2024 Al alloy was 12.1 μm. The fractions of ∑1, ∑3, ∑5-∑29 and R grain boundaries in the sample were 12.17%, 1.96%, 8.91% and 76.96%, respectively. After corrosion for 6 h, the corroded boundary proportions of ∑1, ∑3, ∑5-∑29 and R boundaries were 10.71%, 11.11%, 21.95% and 25.14%, respectively. This showed that the GBCD had a great effect on IGC susceptibility of the solution treated 2024 Al alloy, and both ∑1 and ∑3 boundaries exhibited much better IGC resistance than ∑5-∑29 and R boundaries. Among all the boundaries, the R boundaries had the highest IGC susceptibility.

Acknowledgments

This work was financially supported by the Project-sponsored by SRF for ROCS, SEM and the National Natural Science Foundation (No. 51171118) of China.

References

[1] Watanabe T 1984 An approach to grain boundary design of strong and ductile polycrystals *Res Mechanica* **11** 47-84

[2] Kim S H, Erb U, Aust K T and Palumbo G 2001 Grain boundary character distribution and intergranular corrosion behavior in high purity aluminum *Scripta Materialia* **44** 835-839

[3] Kim T, Hong K T and Lee K S 2003 The relationship between the fracture toughness and grain boundary character distribution in polycrystalline *NiAl Intermetallics* **11** 33-39

[4] Meng C, Zhang D, Zhuang L and Zhang J 2016 Correlations between stress corrosion cracking grain boundary precipitates and Zn content of Al－Mg－Zn alloys *Journal of Alloys & Compounds* **655** 178-187

[5] Guérin M, Alexis J, Andrieu E, Laffont L, Lefebvre W and Odemer G 2016 Identification of the metallurgical parameters explaining the corrosion susceptibility in a 2050 aluminium alloy *Corrosion Science* **102** 291-300

[6] He C L, Lv S, Bai Y Y, Li R, Xie L P, Ma G F and Wang J M 2016 Effect of quenching methods on corrosion behavior of 2024 Al alloy *Journal of Shenyang University (Natural Science)* **28** 87-91

[7] Yuan X Y and Chen L Q 2016 Effect of grain and grain boundary features on anti-corrosion ability of a high manganese austenitic TWIP steel *Acta Metallurgica Sinica* **52** 1345-1352

IOP Conf. Series: Materials Science and Engineering **292** (2017) 012050 doi:10.1088/1757-899X/292/1/012050

Study on hydrothermal synthesis dynamics of xonotlite spherical particles

F Liu[1,2,3], S Chen[1,3], Q Lin[1,2], X D Wang[2] and J X Cao[1,2,3,4]

[1] School of Chemistry and Chemical Engineering, Guizhou University, Guiyang, Guizhou 550025, PR China
[2] Key Laboratory of Green Chemical and Clean Energy Technology, Guiyang, Guizhou 550025, PR China
[3] Key Laboratory of Efficient Utilization of Mineral and Green Chemical Technology, Guiyang, Guizhou 550025, PR China

[4] E-mail: jxcao@gzu.edu.cn

Abstract. The xonotlite crystals were synthesized via the hydrothermal synthesis manner from CaO and SiO_2 as the raw materials with their Si/Ca molar ratio of 1.0. The products were characterized by XRD and SEM techniques to investigate their crystalline phases and morphologies. Hydrothermal synthesis dynamics of xonotlite spherical particles was explored by masterly measuring the pH and the calcium concentrations of product slurries synthesized at various reaction temperature in this paper. The results indicated that the calculated values of the products' quality at various reaction temperatures were consistent with the measured values. Based on chemical reaction kinetic, using fourth-order Runge-Kutta method, spline interpolation and least-squares fitting method, the dynamic relationship of xonotlite spherical particles synthesized via hydrothermal synthesis process is of $-dc_A / dt = kc_A^{0.9}$.

1. Introduction

Xonotlite crystal belongs to monoclinic system (a=1.67nm, b=0.73nm, c=0.695nm), and it is a mineral of calcium silicate hydrate with the lowest crystalline water content, the best heat resistance and thermal stability (decomposition temperature of 1050-1100 ℃) [1-2]. Xonotlite has been widely utilized in the fields of thermal insulation, architecture and so on, for its distinct physical and chemical properties, such as thermal stability, biological activity and environmental friendliness [3-4]. In general, the xonotlite products used in industry were fabricated using dynamic process from CaO and SiO_2 as the raw materials with their Si/Ca molar ratio of 1.0 [5-17]. The previous researches revealed that raw materials posed a great impact on the growth of xonotlite crystals [14-18]. Due to the dissolution characteristics of siliceous material, the xonotlite crystals from H_2SiO_3 and K_2SiO_3 as siliceous materials could grow to form the various morphologies of spherical particles and spherical particles respectively [19]. Actually, it was difficult to synthesize xonotlite crystals because of the harsh conditions such as the high reaction temperature and pressure [14-17]. It can be confirmed by many scholars' research that the synthesis of xonotlite crystals was feasible based on the hydrothermal synthesis thermodynamics [21-22]. But since the invisibility of the reaction and the harsh conditions of xonotlites' formation, the dynamics data of xonotlite crystals was hardly possible to obtain so that few people took the hydrothermal synthesis dynamics into account [23-24]. In recent years, our group has completed the researches on the morphology and growth mechanism of xonotlite crystals

[4,6,7,13,19,20]. Moreover, by masterly measuring the electrical conductivities and the calcium concentrations of product slurries, our group achieved the dynamic test of xonotlite fibers and pointed out that the dynamic relationship of xonotlite fibers synthesized via hydrothermal synthesis process was of $-dc_A / dt = kc_A^{4/5}$. Therefore, based on the existing researches, hydrothermal synthesis dynamics of xonotlite spherical particles was explored by masterly measuring the pH and the calcium concentrations of product slurries in this paper, aiming to further reveal the dynamic relationship of xonotlite spherical particles.

2. Experiment

2.1. Materials and methods

The xonotlite spherical particles were synthesized via the dynamic hydrothermal synthesis method.

H_2SiO_3 solution and $Ca(OH)_2$ suspension were used as siliceous material and calcareous material respectively. $Ca(OH)_2$ suspension was prepared by mixing $CaCl_2$ with KOH $100g{\cdot}L^{-1}$ at the molar ratio of 1:2. The calcareous material, distilled water, and siliceous material were homogeneously mixed, and then the mixture was transferred into a reaction chamber with a stirrer at a heating rate of 1.5 ℃·min^{-1} and a stirring rate of 200 rpm. The hydrothermal treatment proceeded under the following conditions: the Ca/Si molar ratio of 1.0, the reaction temperature of 190 ℃, 210 ℃ and 230 ℃ respectively. After the hydrothermal treatment, the suspensions were filtrated and washed three times by distilled water. The resultant powders were dried at 75 ℃ for 12 h. Technological flow diagram of hydrothermally synthesizing xonotlite spherical particles was shown in figure 1.

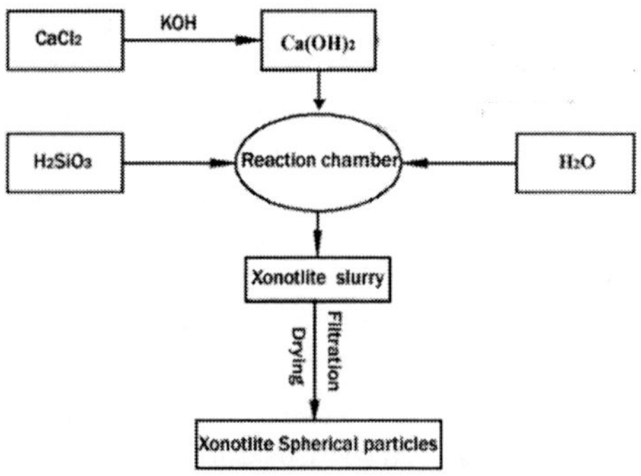

Figure 1. Technological flow diagram of hydrothermally synthesizing xonotlite spherical particles.

In this paper, the accuracy of the dynamic equation of xonotlite spherical particles synthesized via the dynamic hydrothermal synthesis method was verified by two methods, one was comparing the production rate of the xonotlite products calculated by the dynamic equation with the experimental values, and the other was comparing the apparent activation energies calculated by the dynamic equation with that calculated by the Kondo model ($[1-(1-\alpha)^{1/3}]^4 = k(t-t_0)$).

2.2. The determination method of samples

The reaction equation for the formation of the xonotlite spherical particles could be simplified as equation (1).

$$6Ca(OH)_2 + 6H_2SiO_3 + H_2O \rightarrow 6CaO \cdot 6SiO_2 \cdot H_2O + 12H_2O \quad (1)$$

The above equation (1) was only a simplified final reaction equation for the preparation of xonotlite spherical particles, but the whole reaction was actually made up of a series of complex chemical reaction steps. Due to the invisibility of the hydrothermal synthesis reaction, it was impossible to observe the reaction process directly. However, the above chemical reaction equation (1) was acid-base neutralization reaction, with increasing reaction time; the pH of the system would be close to neutral. Therefore, the dynamic test of the hydrothermal synthesis process of the xonotlite spherical particles can be achieved by measuring the change of the pH during the reaction process.

The product slurries synthesized by hydrothermal synthesis process at various reaction temperatures and time were kept at 25 ℃ in thermostat water bath cauldron, and the pH of various slurries was measured using a PHS-3C precise pH meter. The EDTA titration method was used to determine the calcium concentrations of various product slurries formed by hydrothermal synthesis process under various reaction temperatures and time. The product slurries were accurately weighed by FA2004 electronic balance after drying. After the induction time (t_0), the qualities of the products obtained under various reaction times can be defined as m(Total)=m(HCl)+m($Ca(OH)_2$)+m(C-S-H(I))+m(xonotlite).

The crystalline phase composition and microstructure characteristics of the samples were characterized by X-ray diffraction (XRD, Model D/max-2200, Japan). A scanning electron microscope (SEM, Model JSM-6490LV, operating at 25 KV, Japan) was used to observe the morphology of xonotlite crystals.

3. Results and discussion

Figure 2 shows the XRD patterns of the products synthesized hydrothermally at various temperatures. Clearly, the main crystalline phases prepared by hydrothermal synthesis method at 190 ℃, 210 ℃ and 230 ℃ respectively were xonotlite (corresponding to $Ca_6Si_6O_{17}(OH)_2$, standard JCPDS card No. 23-0125). From the analysis result of XRD patterns, the existence of a series of strong diffraction peaks showed that the xonotlite crystals synthesized at various temperatures were well-crystallized and almost no impurity.

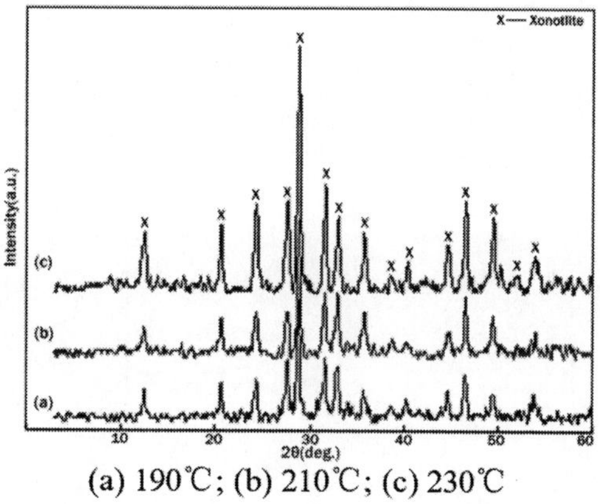

(a) 190℃; (b) 210℃; (c) 230℃

Figure 2. X-ray diffraction patterns of the products hydrothermally synthesized at various temperatures.

Figure 3 displays the SEM images of the products synthesized via the dynamic hydrothermal synthesis method at 190 ℃, 210 ℃ and 230 ℃ respectively. The products obtained by hydrothermal synthesis were xonotlite spherical particles with diameters of 10-50 μm.

The pH and calcium concentrations of the system were respectively measured after various reaction time, shown in figure 4 and figure 5 respectively. It can be seen from figure 4, with prolonging reaction time, the pH of hydrothermally synthesized products at various temperatures decreased, and its descending trend can be divided into three stages. The descending trend of these three curves in the first stage was fastest, in the second stage was relatively slow, obviously, the slowest of the descending trend existed in the third stage. And in the third stage, the pH tended to a certain value. The descending trend of these three curves indicated that the reaction rate in the first stage was fastest, then it became slightly slow in the second stage, and the slowest reaction speed appeared in the third stage.

(a) 230 ℃

(b) 210 ℃

(c) 190 ℃

Figure 3. SEM images of the products hydrothermally synthesized at various temperatures.

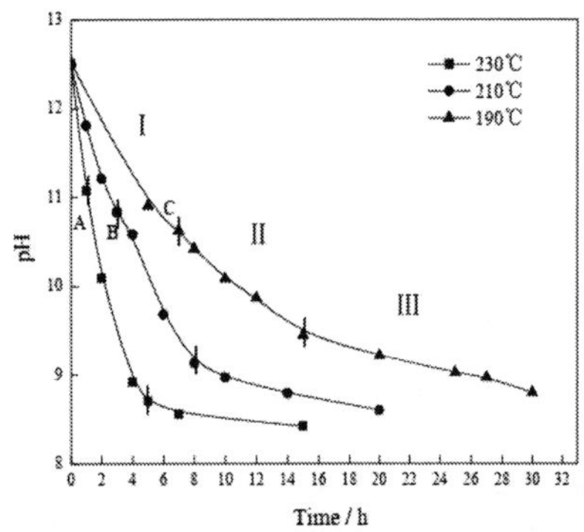

Figure 4. Time dependence of the pH of hydrothermally synthesized products at various temperatures.

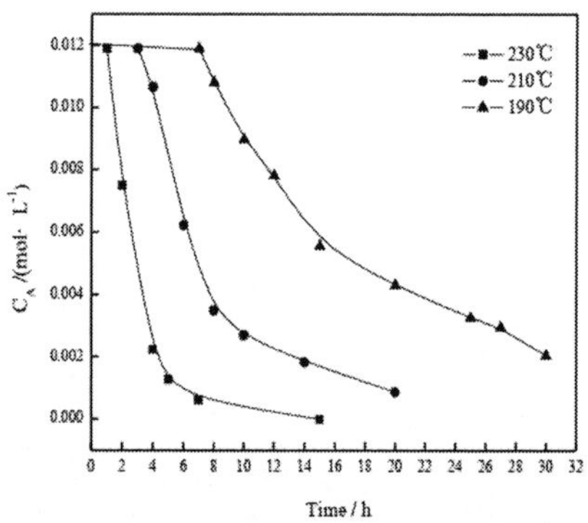

Figure 5. Time dependence of the calcium concentrations of the products hydrothermally synthesized at various temperatures.

In the hydrothermal synthesis process of xonotlite spherical particles, the reaction process can be expressed by the following reaction steps:

$$Ca(OH)_2 + H_2SiO_3 + H_2O \rightarrow C-S-H(gel) \quad (2)$$

$$C-S-H(gel) + Ca(OH)_2 + H_2SiO_3 + H_2O \rightarrow C-S-H(\text{II}) \quad (3)$$

$$C-S-H(\text{II}) + Ca(OH)_2 + H_2SiO_3 + H_2O \rightarrow C-S-H(\text{I})(5CaO\cdot 6SiO_2\cdot 5H_2O) \quad (4)$$

$$Ca(OH)_2 + C-S-H(\text{I})(5CaO\cdot 6SiO_2\cdot 5H_2O) \rightarrow 6CaO\cdot 6SiO_2\cdot H_2O + 5H_2O \quad (5)$$

Equation (2) represented the formation of amorphous C-S-H gels in the initial stage of the reaction. With the progress of the reaction, as shown in equation (3), C-S-H gels were transformed into calcium-rich hydrated calcium silicate C-S-H (II). Then due to the silicon-rich state in the reaction system, as shown in equation (4), C-S-H (II) would be transformed into the silicon-rich C-S-H(I) which owned a structure that similar to the tobermorite ($Ca_5Si_6O_{16}(OH)_2\cdot 4H_2O$). Finally, the silicon-rich C-S-H(I) directly transformed into the spherical xonotlite crystals, as shown in equation (5).

According to the reaction process of the xonotlite spherical particles and figure 4, it can be assumed that the above-mentioned chemical equation (2, 3 and 4) occurred in the first stage of figure 4 to form C-S-H. And there was no formation of the spherical xonotlite crystals in the first stage. Moreover, the reaction time of the first stage was assumed to be the induction time (t_0) required for the formation of spherical xonotlite crystals. Assuming that the chemical reaction of above equation (5) occurred in the second stage, at this moment, the nucleation and growth of xonotlite spherical particles took place on the surface of the C-S-H, and there was no C-S-H formation in this stage. As before, we also supposed that most of the xonotlite spherical particles had formed before the third stage. So in the third stage, the reaction rate decreased to the minimum, which led to the pH decreased slowly and tended to a certain value.

Based on the above assumptions and the data of calcium concentrations in figure 5, the qualities of the products under various reaction temperatures and time during the second stage can be calculated, as shown in tables 1, 2 and 3. According to the experimental method of 2.2, the quality-time values of the products at various reaction temperatures were measured, as shown in tables 4, 5 and 6. At the same reaction temperatures, the quality-time calculated values of the products were compared with the measured values, and the results were as shown in figure 6.

Table 1. Calculated data of quality-time of products hydrothermally synthesized at 230 ℃.

t/[h]	m(HCl)/[g]	m(Ca(OH)$_2$)/[g]	m(C-S-H(I))/[g]	m(Xonotlite)/[g]	m(Total)/[g]
1	7.4500	0.8806	6.0833	0.0000	14.4139
2	7.4500	0.5550	3.8343	2.1997	14.0391
4	7.4500	0.1654	1.1429	4.8321	13.5905
5	7.4500	0.0934	0.6452	5.3189	13.5075
7	7.4500	0.0454	0.3134	5.6435	13.4522
15	7.4500	0.0000	0.0000	5.9500	13.4000

Table 2. Calculated data of quality-time of products hydrothermally synthesized at 210 ℃.

t/[h]	m(HCl)/[g]	m(Ca(OH)$_2$)/[g]	m(C-S-H(I))/[g]	m(Xonotlite)/[g]	m(Total)/[g]
3	7.4500	0.8806	6.0833	0.0000	14.4139
4	7.4500	0.7896	5.4547	0.6148	14.3092
6	7.4500	0.4608	3.1836	2.8362	13.9306
8	7.4500	0.2583	1.7844	4.2047	13.6974
10	7.4500	0.1996	1.3789	4.6013	13.6298
14	7.4500	0.1350	0.9328	5.0377	13.5555
20	7.4500	0.0646	0.4461	5.5137	13.4744

Table 3. Calculated data of quality-time of products hydrothermally synthesized at 190 ℃.

t/[h]	m(HCl)/[g]	m(Ca(OH)$_2$)/[g]	m(C-S-H(I))/[g]	m(Xonotlite)/[g]	m(Total)/[g]
7	7.4500	0.8806	6.0833	0.0000	14.4139
8	7.4500	0.8003	5.5283	0.5429	14.3214
10	7.4500	0.6653	4.5958	1.4549	14.1660
12	7.4500	0.5785	3.9964	2.0412	14.0661
15	7.4500	0.4114	2.8418	3.1704	13.8737
20	7.4500	0.3182	2.1980	3.8002	13.7663
25	7.4500	0.2410	1.6651	4.3214	13.6775
27	7.4500	0.2185	1.5097	4.4734	13.6516
30	7.4500	0.1511	1.0435	4.9294	13.5739

Table 4. Experimental data of quality-time of products hydrothermally synthesized at 230℃.

t/[h]	0	1	2	4	5	7	15
m(Total)/[g]	15.05	14.42	14.05	13.58	13.50	13.45	13.41

Table 5. Experimental data of quality-time of products hydrothermally synthesized at 210℃.

t/[h]	0	1	2	3	4	6	8	10	14	20
m(Total) /[g]	15.05	14.73	14.58	14.42	14.31	13.92	13.71	13.62	13.55	13.48

Table 6. Experimental data of quality-time of products hydrothermally synthesized at 190℃.

t/[h]	0	5	7	8	10	12	15	20	25	27	30
m(Total)/ [g]	15.05	14.69	14.42	14.32	14.17	14.05	13.88	13.76	13.68	13.65	13.57

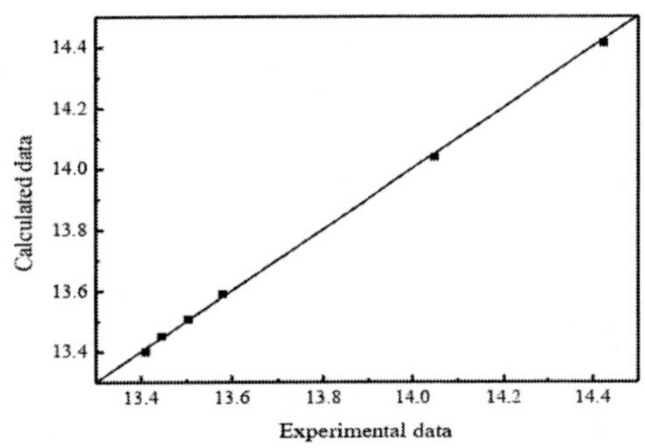

(a) 230 ℃

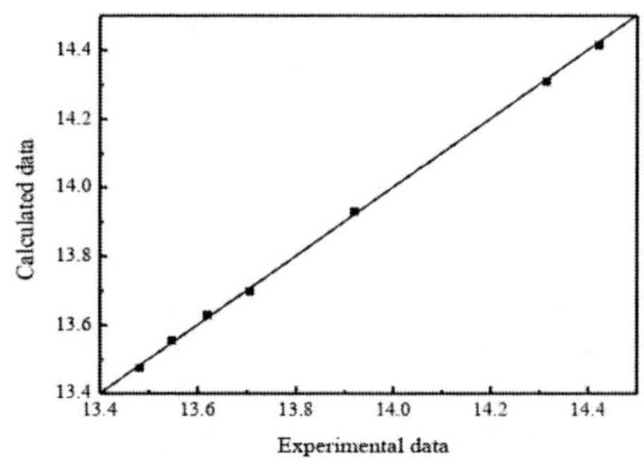

(b) 210 ℃

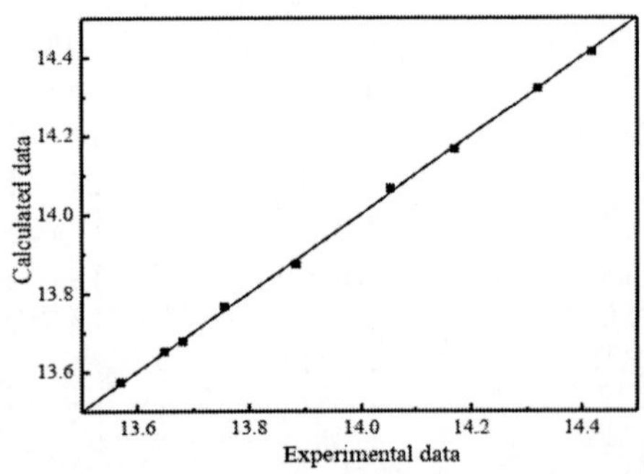

(c) 190 ℃

Figure 6. The experimental data comparison with calculated data of xonotlite spherical particles hydrothermally synthesized at various temperatures.

It can be seen from figure 6 that when the reaction temperature was 230, 210 and 190 ℃ respectively, the quality-time calculated values were consistent with the measured values after the reaction time of 1 h, 3 h and 7 h respectively. So the reaction time of 1 h, 3 h and 7 h can be defined as the induction time required for the formation of spherical xonotlite crystals at 230 ℃, 210 ℃ and 190 ℃ respectively, as indicated by points A, B and C in figure 4. There was no formation of spherical xonotlite crystals before the induction time. At the same time, the above analysis also showed that the above hypotheses were confirmed, that was, the second stage, beginning with the induction time, was the stage of the formation of xonotlite spherical particles, and there was no C-S-H produced in this stage.

According to the above analytical results, the equation (5) can be simplified as the following chemical reaction:

$$A+B \rightarrow C+5D \tag{6}$$

In the equation: A—$Ca(OH)_2$

B—$C-S-H(I)(Ca_5Si_6O_{16}(OH)_2 \cdot 4H_2O)$

C—$Ca_6Si_6O_{17}(OH)_2$

D—H_2O

According to the equation (6), the rate equation usually has the following form:

$$-dc_A/dt = k_A c_A^{n_A} c_B^{n_B} \tag{7}$$

Because $c_{A,0} = c_{B,0}$, that is, $c_A = c_B$. Thus, equation (7) can be simplified as:

$$-dc_A/dt = kc_A^n \tag{8}$$

In the equation, $k = k_A$ was the total rate constant of the reaction, $n = n_A + n_B$ was the total order of reaction.

In this equation, the kinetic parameters were only *k* and *n*, therefore, the rate equation can be established by determining the two parameters. Since there was only one constant in the equation, the form of the rate equation only depended on *n*, so the key to determine the rate equation was to determine the reaction order. In order to determine the reaction order, it was necessary to know the concentrations of the reactants (C_A) at different times (t) at a definite temperature, that was, to know the relationship of C_A-*t* in this chemical reaction.

In this paper, the internal performance function provided in Matlab were used. Then the reaction order (n) and the rate constant (k) were determined and calculated respectively by dealing with the dynamic data (C_A-t) corresponding to the various temperatures in figure 5 using the differential method. The values of *n* and *k* were used as the iterative initial values in the process of solving the fourth-order Runge-Kutta method, and the detailed solving process was shown in figure 7.

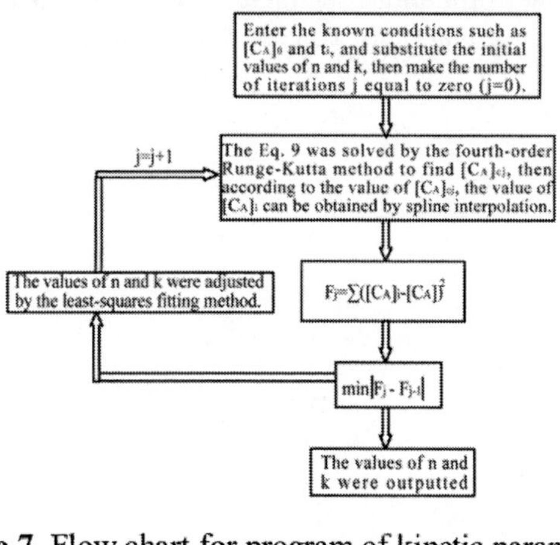

Figure 7. Flow chart for program of kinetic parameters.

It should be noted that a series of (t_i-$[c_A]_{ci}$) values obtained by the fourth-order Runge-Kutta method may not have a corresponding value ($[c_A]_i$). Therefore, it was necessary to use spline interpolation to solve the values of $[c_A]_i$. Firstly, the values of $[c_A]_i$ were calculated by using Matlab which completed the loop iteration of the fourth-order Runge-Kutta method and the spline interpolation. Then according to $\min F=\sum([c_A]_i-[c_A])^2$, the optimum solution of k were obtained by the least-squares fitting method.

$$\begin{cases} y_{n+1}=y_n+\frac{h}{6}[k_1+(2-2^{\frac{1}{2}})k_2+(2+2^{\frac{1}{2}})k_3+k_4 \\ k_1=f_n \\ k_2=f(t_n+\frac{1}{2}h,y_n+\frac{1}{2}hk_1) \\ k_3=f[t_n+\frac{1}{2}h,y_n+\frac{2^{\frac{1}{2}}-1}{2}hk_1+(1-\frac{2^{\frac{1}{2}}}{2})hk_2] \\ k_4=f[t_n+h,y_n-\frac{2^{\frac{1}{2}}}{2}hk_2+(1+\frac{2^{\frac{1}{2}}}{2})hk_3] \end{cases} \tag{9}$$

According to the above-mentioned method, the initial values of kinetic parameters n and k at hydrothermal synthesis temperature of 230 ℃ were 0.8765 and 0.2808$(mol\cdot m^{-3})^{1/5}\cdot s^{-1}$ respectively. Since the errors of experimental measurement and the errors of calculation, n=0.9 and k=0.2808$(mol\cdot m^{-3})^{1/5}\cdot s^{-1}$ were taken as the iterative initial values after rounded the reaction order. Then according to $\min F=\sum([c_A]_i-[c_A])^2$, the value of k that obtained by a least-squares fitting method was of 0.3147$(mol\cdot m^{-3})^{1/5}\cdot s^{-1}$.

Similarly, when the hydrothermal synthesis temperature was 210 ℃ or 190 ℃ respectively, according to $\min F=\sum([c_A]_i-[c_A])^2$, the values of k were 0.1330 $(mol\cdot m^{-3})^{1/5}\cdot s^{-1}$ and 0.0549$(mol\cdot m^{-3})^{1/5}\cdot s^{-1}$ respectively. When the hydrothermal synthesis temperature was of 230 ℃, 210 ℃ and 190 ℃ respectively, the value of n was of 0.9, and the values of k were of 0.3147$(mol\cdot m^{-3})^{1/5}\cdot s^{-1}$, 0.1330 $(mol\cdot m^{-3})^{1/5}\cdot s^{-1}$ and 0.0549$(mol\cdot m^{-3})^{1/5}\cdot s^{-1}$ respectively. The fitting results of the fourth-order Runge-Kutta method were shown in figure 8.

It can be seen from the figure 8 that the curves of reaction concentration-time fitted by fourth-order Runge-Kutta method at various temperatures were consistent with the experimental data. Obviously, the results showed that the values of n and k obtained by aforementioned computational method had a high accuracy. Therefore, the rate equation of xonotlite spherical particles hydrothermal reaction could be confirmed as $-dc_A/dt=kc_A^{0.9}$.

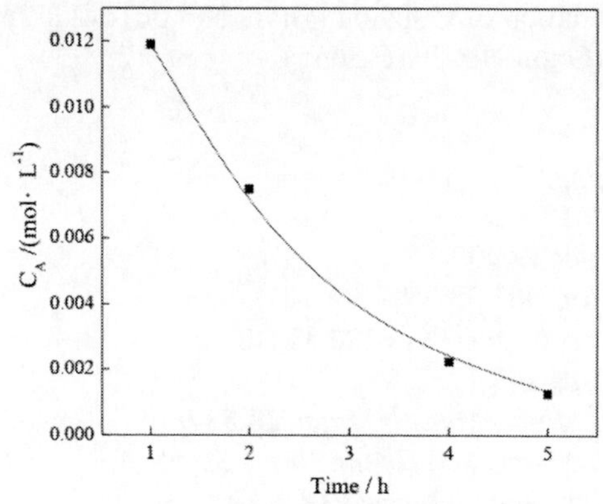

(a) 230 ℃

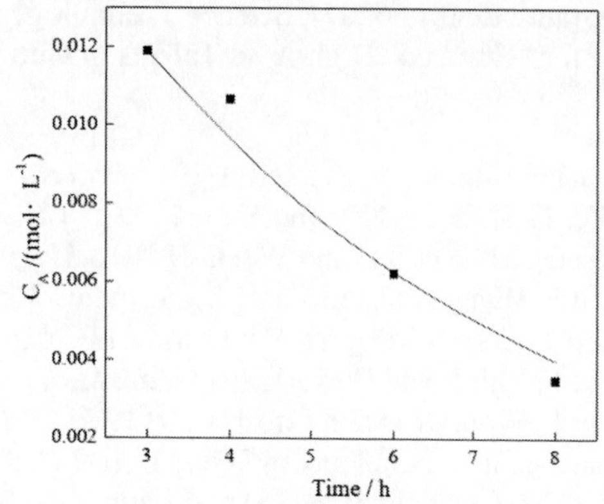

(b) 210 ℃

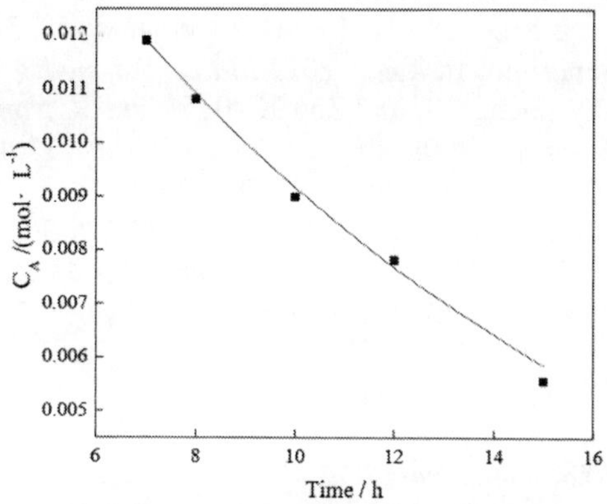

(c) 190 ℃

Figure 8. The curves of concentration-time fitted by fourth-order Runge-Kutta method at various temperatures.

4. Conclusions

(1) With H_2SiO_3 solution and $Ca(OH)_2$ suspension as siliceous material and calcareous material respectively, the dynamic test of the hydrothermal synthesis process of the xonotlite spherical particles can be achieved by measuring the change of the pH during the reaction process.

(2) By measuring the pH and the calcium concentrations of product slurries synthesized at various reaction temperature, it can be assumed that the hydrothermal synthesis process of xonotlite spherical particles can be divided into three stages, that was, the stage forming amorphous C-S-H gels, the stage forming the most of spherical xonotlite crystals, the stage forming a little xonotlite spherical particles.

(3) According to the data of calcium concentrations, the calculated values of the products' quality at various reaction temperatures were consistent with the measured values, showing that the hypotheses dividing the formation process of spherical xonotlite crystals into three stages were confirmed.

(4) Based on chemical reaction kinetic, using fourth-order Runge-Kutta method, spline interpolation and least-squares fitting method, the dynamic relationship of xonotlite spherical particles obtained by direct conversion of silica-rich C-S-H (I) is of $-dc_A / dt = kc_A^{0.9}$.

Acknowledgments
We gratefully acknowledge the supports of this work by the National Natural Science Foundation of China (Grant No. 21666007), Science Technology Foundation of Guizhou (Grant No. 2014-2007) and Program for "Hundred" High-level Talents in Guizhou (Grant No. 2016-5655).

References
[1] Dachowskia R and Stępień A 2011 *Procedia Engineering* **21** 1173-8
[2] Wei G S, Zhang X X and Yu F 2009 *J. Therm. Sci* **18** 142-9
[3] Hsiang H, Wei S C and Wei C H 2016 *Mater. Struct* **49** 905-15
[4] Liu F, Wang X D and Cao J X 2011 *Adv. Mater. Res* **382** 384-7
[5] Zou J, Guo C, Jiang Y, Wei C and Li F 2016 *Mater. Chem. Phys* **172** 121-8
[6] Liu F, Zhu B and Cao J X 2011 *Adv. Mater. Res* **148-149** 1755-8
[7] Liu F, Wang X D and Cao J X 2013 *Mater. Int. J. Miner. Metall. Mater* **20** 88-93
[8] Hartmann A, Schulenberg D and Buhl J C 2015 *J. Mater. Sci. Chem. Eng* **3** 39-55
[9] Zou J J, Guo C B, Wei C D and Jiang Y S 2015 *Res Chem Intermed* **18** 1-12
[10] Kurdowski W 2014 Cement and Concrete Chemistry (Dordrecht: Springer Netherlands) 205-77
[11] Shaw S, Clark S M and Henderson C M B 2000 *Chem. Geol* **167** 129-40
[12] Milestone N B and Ahari G K 2007 *Adv. Appl. Ceram* **106** 302-8
[13] Liu F, Chen S, Lin Q, Wang X D and Cao J X 2017 *Optoelectron. Lett* **13** 81-3
[14] Spudulis E, Šavareika V and Špokauskas A 2013 *Mater. Sci-medzg* **19** 190-6
[15] Guo X Y, Ma S H, Lü S Q, Zheng S L and Zou X 2015 *Chin. J. Nonferrous. Metals* **25** 534-44
[16] Black L, Garbev K and Stumm A 2009 *Adv. Appl. Ceram* **108** 137-44
[17] Liang H X and Li M Q 2002 *J. Chin. Ceram. Soc* **30** 294-9
[18] Baltakys K and Prichockiene E 2010 *Mater. Sci-Poland* **28** 295-304
[19] Liu F, Wang X D and Cao J X 2011 *Adv. Mater. Res* **382** 379-83
[20] Liu F, Zeng L K, Cao J X and Li J 2010 *J. Wuhan. Univ. Technol* **25** 295-7
[21] Clodic L and Meike A 1997 Thermodynamics of calcium silicate hydrates (California: Lawrence Livermore National Laboratory) 1-27
[22] Newman E S 1956 *J. Res.Nat. Bur. Stand* **57** 27-30
[23] Li M Q and Liang H X 2004 *China Part* **2** 124-7
[24] Moorehead D R and Mccartney E R 1965 *J. Am. Ceram. Soc* **48** 565-9

IOP Conf. Series: Materials Science and Engineering **292** (2017) 012051 doi:10.1088/1757-899X/292/1/012051

The CRG-PVA hydrogels study of properties with various nanoparticles and their application for cultivation of phototrophic microorganisms

K V Gorin[1, 3], G U Badranova[1], P M Gotovtsev[1], A Yu Shatalova[1], T E Grigoriev[1], S V Krasheninnikov[1], S A Tihomirov[2], O A Kondratev[2], M V Vishnevskaya[1] and R G Vasilov[1]

[1] Department of biotechnologies and bioenergy, NRC "Kurchatov institute", Moscow, 123182, 1 Akademic Kurchatov pl., Russia
[2] Resource center of laboratory X-ray methods, NRC "Kurchatov institute", Moscow, 123182, 1 Akademic Kurchatov pl., Russia

[3] E-mail: gkvbio@gmail.com

Abstract. In this work we are demonstrate results of researches of the hydrogels based on carrageenan and polyvinyl alcohols properties with various nanoparticles Al_2O_3 and ZrO_2 and their application for phototrophic microorganisms *Chlorella vulgaris* cultivation. X-ray diffraction of samples was carried out. Research on the cultivation of microalgae using the developed hydrogels has also been conducted, and the increase in productivity with the use of gels on average by 20% compared to the control sample has been shown. The highest productivity is observed with concentration 0.5% of ZrO_2 nanoparticles. We conclude that using of hydrogels in the developed photobioreactors possess a perspective.

1. Intoduction

At the present time in the world actively developing technologies based on using phototrophic microorganisms (PM) – microalgae and cyanobacteria. The reason of their utilization is that PM need only light, carbon dioxide and small amount of mineral additives in water for grow [1, 2]. At the same time, a wide range of products can be obtained from various PM, including biofuels, vitamins, substances for cosmetics or pharmaceuticals, feed and food additives and fertilizers [3, 4]. Also PM is the promising organisms for carbon dioxide fixation [3].

The reduction of biomass cost is one of the main challenges of microalgae based refinery development [5]. This cost strongly depends on the capital expenses for the photobioreactor itself, in which the cultivation are carried out and lighting expenses [6]. For elaboration a photobioreactor construction, the developers try for ensure the achieving highest as possible concentration of biomass. Although, in the later stages, the growth of biomass is strongly inhibited by the lightless in the deep of culture medium and the growth of biomass is limited only by surface layers [7, 8]. Thus, it is necessary to use either pipes of small diameter or shallow ponds, which leads to increase photobioreactor volume and rise cost of the equipment themselves [2, 7]. For another thing, these factors impose limitations on overall dimensions and construction solutions of photobioreactors, which limits their use in a number of areas, such as, for example, air purification with carbon dioxide fixation in large parking lots and buildings with a many people inside (shopping malls, train stations and etc.)

[9, 10]. For such applications, photobioreactors must fit into existing structural and technical solutions, so that on the one hand they avoid reconstruction, on the other hand, take into account social factors of people's behavior in these buildings [11]. Overcome of all these technical limitations is possible by use approaches that allow providing illumination in the depth of the culture medium [12]. One approach is to create conditions for the reflection of light within the cultivated environment, by placing various materials with a high albedo there [13]. The aim of this work is to analyze the possibility of application such materials with zero buoyancy based on polymer hydrogels consisting of carrageenan (CRG) and polyvinyl alcohol (PVA). The selection of such a system was due to its biocompatibility, high stability and the ability to immobilize various nanoparticles within it. For example, earlier it was shown the possibility of creating photocatalytic hydrogels based on CRG PVA with titanium dioxide nanoparticles immobilized therein [14]. The hydrogels remained stable, despite the photocatalytic activity of titanium dioxide. In our work we use hydrogels with various concentrations of nanoparticles. As nanoparticles Al_2O_3 and ZrO_2 were chosen due to its inertness and biocompatibility to microalgae strains. Also, those nanoparticles can gives strong white color to hydrogels. Our experiments in this stage will be provided by cultivation of PM in flasks with and without hydrogels presence, to demonstrate proof-of-principal. Future perspective is developing of new type of photobioreactors that can be applied in the existing buildings to make modern cities closer to CO_2 neutral status.

2. Materials and methods

2.1. Preparation of hydrogels

For preparing each of the gels, 10 ml CRG solution with a concentration of 2.5 g/l and 10 ml PVA solution with a concentration of 2.5 g/l too, were initially prepared at a temperature of 70 °C. The solutions were then mixed together and nanoparticles were added at the same time. Al_2O_3 and ZrO_2 from BDT company (Russian Federation) were used as nanoparticles. The dimensions of the nanoparticles were: 68±10 nm of aluminum dioxide; 74±10 nm of zirconium dioxide. Samples with concentrations of 0.25%, 0.50%, and 0.75% were prepared for each type of nanoparticles. The resulting mixture was stirred for an hour at 500-550 rpm at a temperature of 70°C. After this, the gels were cooled to room temperature and subjected to four cycles of freezing/thawing. The freezing temperature was -12 °C and the thawing temperature + 25 °C. The number of cycles and temperatures were chosen in accordance with the results of the studies presented in [14, 15].

2.2. Microalgae cultivation

The object of the study was the *Chlorella vulgaris* GKV1 culture from the collection of the National Research Center "Kurchatov Institute". *Chlorella vulgaris* GKV1 microalgae were grown on a Basal medium [16] with the following composition (g/l): KNO_3, 1.25; KH_2PO_4, 1.25; $MgSO_4 \cdot 7H_2O$, 1; $CaCl_2$, 0.0835; H_3BO_3, 0.1142; $FeSO_4 \cdot 7H_2O$, 0.0498; $ZnSO_4 \cdot 7H_2O$, 0.0882; $MnCl_2 \cdot 4H_2O$, 0.0144; MoO_3, 0.0071; $CuSO_4 \cdot 5H_2O$, 0.0157; $Co(NO_3)_2 \cdot 6H_2O$, 0.0049; EDTA·2Na, 0.5, the nutrient medium was prepared in filtered tap water, the pH of the nutrient medium was adjusted to 7. Cultivation was carried out in 250 ml in Erlenmeyer flasks. Gels (900 mm^2 in size) with aluminum and zirconium nanoparticles of different concentrations were placed in the flasks. The flask without gel was a control. Stirring was carried out by installing the flasks on a shaker with a frequency of 110 rpm. The culture temperature was maintained at 24 ± 2 °C by air conditioning in the room. The culture grew under conditions of constant illumination during the day and night at an intensity of 1100 lux. Growth control was performed by measuring the optical density of the culture at the same time each day on a Thermo Scientific Genesys 10S UV-Vis spectrophotometer at a wavelength of 680 nm.

2.3. Swelling study

The swelling study was carried out in according to [14]. We used the dried samples to determine the swelling capacity of the hydrogels and its dependency with time. The swelling capacity (SC) was

determined by immersing a weighted amount of dried hydrogel nanocomposite into 10 mL of nutrient medium and allowed to swell at room temperature for 5 min. Then, they were removed from aqueous solutions and blotted with filter paper to remove surface water and then weighed again. Then the swelling/weighting steps were repeated with the constant time interval of 5 min until the mass change in three consecutive measurements would remain constant with the accuracy of 0.003 g. The Ohaus PA64C analytical balance was used in those studies.
The degree of swelling $\alpha(t)$ at every given moment of time t was calculated by the formula

$$\alpha(t) = \frac{m(t)-m_0}{m_0} * 100\% \quad (1)$$

Where m_0 – is the mass of the initial (dry) sample in g, m(t) – is the mass of sample in swelling study in g.

The swelling of all samples of composite hydrogels with different concentrations (0.25%, 0.5%, 0.75%, and 1%) of nanoparticles (Al_2O_3, ZrO_2) in their composition was determined in a similar manner.

2.4. X-ray diffraction
The X-ray diffraction research was carried out on a Bruker Advance diffractometer using the characteristic line Cu Kα1 at a wavelength of 1,54Å. Samples of hydrogels were preliminarily placed in distilled water for 2 hours to ensure their swelling. The samples were then placed in a drying chamber and dried at 60° C for 24 hours. In the experiment, the correlation of the intensity of scattered X-ray radiation on the scattering angle in the θ/2θ scheme is measured (the angle of incidence is equal to the angle at which the scattered radiation is recorded). Diffraction maxima are observed at angles satisfying the Wolf-Bragg condition:

$$2d \sin\theta = n\lambda \quad (2)$$

where d – interplanar distance, θ – half the angle between the incident and recorded X-rays, n – diffraction degree, λ – X-ray length.

2.5. Measurement of mechanical characteristics of hydrogels
Measurements of mechanical characteristics are carried out at room temperature (23 ± 2° C) in a universal testing machine of Instron company (model 5965) in the regime of constant speed of clamping. For fixing samples in uniaxial tension tests, standard pneumatic clamps from a set of universal testing machine are used. Samples before the tests were conditioned at a temperature of 23 ± 2° C for at least 24 hours.

The procedure for testing samples under uniaxial tension is governed by GOST 11262-80. Dimensions of the working part of the sample – 10 x 3.5 x 0.5 mm. Speed of expansion of clamps - 1 mm/min.

Calculation of elastic moduli under uniaxial tension is carried out in accordance with GOST 9550-81. Calculation of the strength of deformation at break is carried out in accordance with GOST 11262-80.

The values of the mechanical parameters are obtained by averaging the measurement results in each series of at least 5 samples. The measurement error is defined as the average deviation from the mean values of the mechanical parameters.

3. Results and discussions

3.1. Research properties and characteristics of hydrogels
The presence of crystalline nanoparticles in the samples was determined by X-ray diffraction analysis. For both samples, the presence of crystal structures corresponding to those nanoparticles that were used in the synthesis process was recorded, which confirms their fixation in the gels (figure 1).

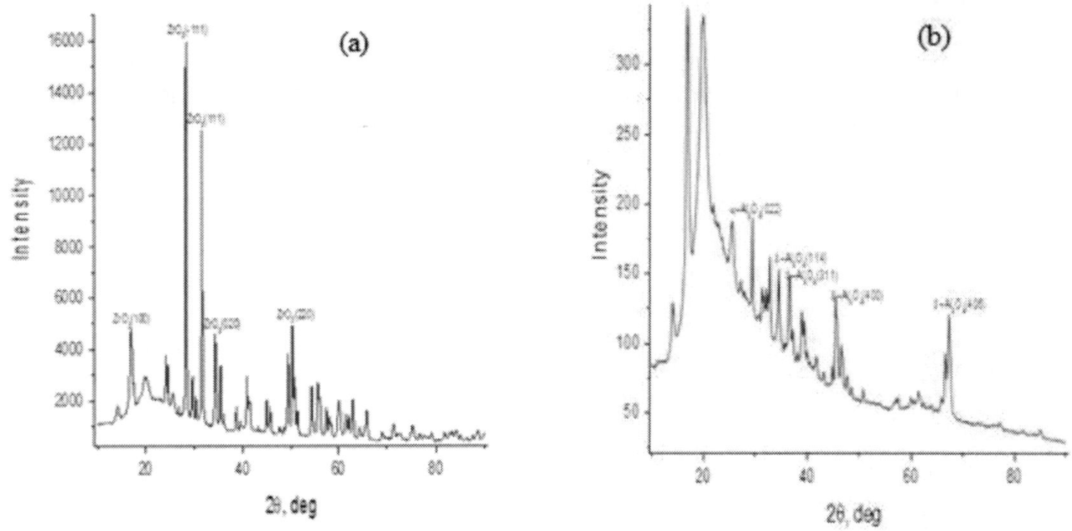

Figure 1. Diffractograms of the hydrogel samples under investigation, measured using Cu Kα radiation. Diffraction peaks: a) ZrO_2 and b) Al_2O_3 nanoparticles in the samples were noted.

Observation of swelling of the obtained hydrogels showed the following results: all hydrogels containing aluminum dioxide for half an hour absorbed 1500 - 1700% water from their weight, which is a high index of swelling. At the same time, gels containing zirconium dioxide swelled much worse, gaining 250-400% water. This difference in swelling indicates a greater hydrophilicity of hydrogels with aluminum dioxide nanoparticles.

The results of observation the mechanical properties of gels in a dry form are presented in table 1. From the presented results it can be seen that the gels (with aluminum dioxide) lose their mechanical strength as the content of nanoparticles increases therein. The same results was observed for gels with titanium dioxide nanoparticles [14]. Gel with 0.75% zirconium dioxide has more mechanical strength than gel with concentration 0.25% and 0.5%. A24, A34, A44 and are gels with a content of 0.25%, 0.5% and 0.75% of aluminum dioxide, respectively; Z24, Z34 and Z44 are gels with a content of 0.25%, 0.5% and 0.75% of zirconium dioxide, respectively; CRG + PVS is a gel without nanoparticles and K is a control flask in which the biomass was cultured without gels.

Table 1. Mechanical properties of hydrogels. Where: σ_{max} - maximum tensile stress, ε_b - tensile strain, E - Young's modulus.

Sample code	σ_{max} [MPa]	ε_b [%]	E [MPa]
Z24	17.3±4.0	0.6±0.1	3273±415
Z34	15.9±5.3	0.7±0.07	2867±300
Z44	38.7±6.9	1.3±0.03	3311±510
A24	46.8±2.34	1.9±0.09	3306±300
A34	27.5±2.1	0.9±0.2	3677±449
A44	24.9±1.24	0.9±0.045	3418±350
CRG+PVS	49.2±0.6	15.0±0.1	2055±60

Experimental data on the cultivation of microalgae with gel are presented in tables 2 and 3 for the first and second cultivation, respectively. It should be noted that the optical density measurements were carried out in a sample of the culture medium and this method did not take into account the fouling of gels by microalgae biomass.

Consider the results of the first cultivation. From table 2 it can be seen that all variants of gels allowed to provide more effective cultivation in comparison with a control sample without gel. In this case, the gels containing zirconia are largely fouling by biomass, so we can expect that a gel with 0.5% zirconia provides the greatest amount of biomass during cultivation under the conditions chosen in the experiment.

Table 2. Results of the first cultivation (gel designations are deciphered in the text).

Fermentation time, h	Sample code							
	A24	A34	A44	Z24	Z34	Z44	CRG+ PVS	K
0	0.142	0.142	0.142	0.142	0.142	0.142	0.142	0.142
24	0.12	0.162	0.105	0.121	0.12	0.11	0.13	0.143
48	0.181	0.213	0.168	0.188	0.2	0.189	0.334	0.173
96	0.237	0.268	0.228	0.234	0.249	0.267	0.375	0.2
168	0.369	0.417	0.337	0.377	0.375	0.383	0.422	0.271
192	0.422	0.453	0.367	0.435	0.423	0.398	0.473	0.319
Area with settled cells from the total area of the gel, %	5	10	1	85	90	90	5	-

Subsequently, using the same gels, repeated cultivation was carried out under the same conditions (table 3). At the same time, gels with zirconium dioxide were biomassed to a lesser extent, and at the same time the greatest amount of biomass was observed in the flasks with them. Gels with aluminum dioxide, on the contrary, were more prone to fouling during re-cultivation. This effect can be explained by the greater hydrophilicity of the gels and their ability to contain a larger volume of water and biomass that could remain after the first cultivation.

Table 3. Results of the second cultivation (gel designations are deciphered in the text).

Fermentation time, h	Sample code							
	A24	A34	A44	Z24	Z34	Z44	CRG+ PVS	K
0	0.05	0.05	0.048	0.053	0.05	0.049	0.048	0.053
24	0.052	0.066	0.065	0.064	0.057	0.056	0.072	0.066
48	0.035	0.034	0.069	0.053	0.057	0.062	0.109	0.079
72	0.06	0.026	0.116	0.087	0.109	0.096	0.174	0.101
144	0.049	0.028	0.242	0.230	0.214	0.202	0.296	0.136
192	0.059	0.019	0.283	0.28	0.288	0.279	0.375	0.072
312	0.525	0.2	0.51	0.368	0.581	0.403	0.506	0.123
Area with settled cells from the total area of the gel, %	60	100	30	20	20	20	10	-

4. Conclusions

The synthesis of novel CRG-PVA hydrogels with Al_2O_3 and ZrO_2 nanoparticles was achieved by proposed method. The presence of nanoparticles in the gels was determined by X-ray diffraction analysis. The tensile strength gel with aluminum dioxide lose as the content of nanoparticles increases therein, whereas gel with 0.75% zirconium dioxide has more mechanical strength than concentration 0.25% and 0.5%. The gels containing aluminum dioxide has swelling about for 4-6 fold higher than gels with zirconium dioxide. Research on the cultivation of microalgae using the developed hydrogels has also been conducted, and the increase in productivity with the use of gels on average by 20% compared to the control sample has been shown. The highest productivity of microalgae is observed with concentration 0.5% of ZrO_2 nanoparticles. The possible application of this work will include possess hydrogels with nanoparticles in the developing photobioreactors for increase productivity of microalgae.

References

[1] Wolf J, Stephens E, Steinbusch S, Yarnold J, Ross I L, Steinweg C, Doebbe A, Krolovitsch C, Müller S, Jakob G, Kruse O, Posten C and Hankamer B 2016 *Algal Res.* **15** 187–201
[2] Pruvost J, Borgne F Le, Artu A, Cornet J F and Legrand J 2015 *Advances in Chemical Engineering* **48** 257-310
[3] Valverde F, Romero-Campero F J, León R, Guerrero M G and Serrano A 2016 *Eur J Protistol* **55**, 95-101
[4] Singh P, Kumari S, Guldhe A, Misra R, Rawat I and Bux F 2016 *Renew. Sustain. Energy Rev.* **55** 1–16
[5] Quinn J C, Catton K, Wagner N and Bradley T H 2012 *BioEnergy Res.* **5** 49–60
[6] Yang Z, Cheng J, Yang W, Zhou J and Cen K 2016 *Bioresour. Technol.* **221** 492–497
[7] de Andrade G A, Berenguel M, Guzmán J L, Pagano D J and Acién F G 2016 *J. Process Control.* **37** 58–69
[8] Uduman N, Qi Y, Danquah M K, Forde G M and Hoadley A 2010 *J. Renew. Sustain. Energy* **2** 1-15
[9] oglu Huseynov E F 2011 *Procedia Eng.* **21** 534–542
[10] Pruvost J, Gouic B Le, Lepine O, Legrand J and Borgne F Le 2016 *Chem. Eng. J.* **284** 850–861
[11] Ragheb A, El-Shimy H and Ragheb G 2016 *Procedia - Soc. Behav. Sci.* **216** 778–787
[12] Hu J-Y and Sato T 2017 *Energy Convers. Manag.* **133** 558–565
[13] Gao X, Kong B and Vigil R D 2017 *Bioresour. Technol.* **224** 523–530
[14] Badranova G U, Gotovtsev P M, Zubavichus Y V, Staroselskiy I A, Vasiliev A L, Trunkin I N, and Fedorov M V 2016 *J. Mol. Liq.* **223** 16–20
[15] Guan Y, Bian J, Peng F, Zhang X-M and Sun R-C 2014 *Carbohydr. Polym.* **101** 272–280
[16] Gorin K V, Sergeeva Y E, Butylin V V, Komova A V, Pojidaev V M, Badranova G U, Shapovalova A A, Konova I A and P M Gotovtsev 2015 *Bioresour. Technol.* **193** 178–184

Life prediction for high temperature low cycle fatigue of two kinds of titanium alloys based on exponential function

G Y Mu[1,2,5], X Z Mi[3] and F Wang[4]

[1]Mechanical Engineering College, Dalian JiaoTong University, 794 Huanghe Road. Dalian, 116028, China
[2]Mechanical and Power Engineering College, Dalian Ocean University, 52 Heishijiao Street. Dalian, 116023, China
[3]Traffic and Transportation College, Dalian JiaoTong University, 794 Huanghe Road. Dalian, 116028, China
[4]Motor Train College, Dalian JiaoTong University, 794 Huanghe Road. Dalian, 116028, China

[5]E-mail: mugy1982@126.com

Abstract. The high temperature low cycle fatigue tests of TC4 titanium alloy and TC11 titanium alloy are carried out under strain controlled. The relationships between cyclic stress-life and strain-life are analyzed. The high temperature low cycle fatigue life prediction model of two kinds of titanium alloys is established by using Manson-Coffin method. The relationship between failure inverse number and plastic strain range presents nonlinear in the double logarithmic coordinates. Manson-Coffin method assumes that they have linear relation. Therefore, there is bound to be a certain prediction error by using the Manson-Coffin method. In order to solve this problem, a new method based on exponential function is proposed. The results show that the fatigue life of the two kinds of titanium alloys can be predicted accurately and effectively by using these two methods. Prediction accuracy is within ± 1.83 times scatter zone. The life prediction capability of new methods based on exponential function proves more effective and accurate than Manson-Coffin method for two kinds of titanium alloys. The new method based on exponential function can give better fatigue life prediction results with the smaller standard deviation and scatter zone than Manson-Coffin method. The life prediction results of two methods for TC4 titanium alloy prove better than TC11 titanium alloy.

1. Introduction

Titanium alloy has been widely used in the manufacture of aircraft engine compressor discs, blades and casing because of its high specific strength, good corrosion resistance, good heat resistance and good mechanical properties [1]. When titanium alloy is used for aircraft manufacturing materials, metal materials bearing the main form of the force structure failure is fatigue fracture, a serious threat to flight safety. Therefore, it is very important to study on the fatigue behavior of titanium alloy.

The high temperature low cycle fatigue (HTLCF) life prediction has been a hot research topic for many researchers. In order to effectively predict the HTLCF life of materials, many methods have been proposed in the past years. Manson and Coffin [2, 3] proposed Manson-Coffin equation. Manson-Coffin equation is widely used in fatigue life prediction. They noted that there was a linear relationship between plastic strain range and failure inverse number in the double logarithmic coordinates. Manson [4] considered that the fatigue life of the material can be estimated by alternative thermal fatigue test results when the thermal fatigue is superimposed by alternative mechanical loading. Halford and Manson [5] proposed the strain range partitioning method. They believe that high temperature and low cycle fatigue include fatigue damage and creep damage. The fatigue damage is mainly related to the deformation of the slip surface, and the creep damage is mainly related to the deformation of grain boundaries. Shengtian *et al.*[6] proposed a continuum damage mechanics method. They believed that with increment of damage the stress range decreases slowly and steady. This continuum damage stage is a main part of the whole life. The total continuum damage is not large. It means that the damage rate of TC4 titanium alloy is low but fatigue is sensitive to micro-crack damage. Yang *et al.*[7] proposed the fatigue life prediction method based on radial basis function neural network. Tisong *et al.*[8] proposed a life prediction model that describes the relation between damage energy density and fatigue life based on the formula of Three-Parameter Power Function. The results show that the proposed method is in good agreement with the experiment results. Weiping *et al.* [9] put forward a continuous miner method which combines stochastic load probability weighting, linear miner cumulative damage rule and fuzzy theory. The result shows that the improved fatigue prediction method is accurate within acceptable range. Ling *et al.* [10] proposed a new model for predicting low cycle fatigue life of magnesium alloy. The results show that the predicted results are in good agreement with the experimental results. Guodong *et al.* [11] proposed fatigue life prediction method based on power-exponent function. They considered that there was a power-exponent function relationship between the plastic strain range and inverse number in log-log scale coordinates. Although many models have been proposed for the HTLCF life prediction, each method has its limitations. Because the high temperature fatigue damage of materials is affected by creep-fatigue interaction and the other factors. It has been very difficult to accurately predict the high temperature fatigue life of the material. Yi *et al.* [12] considered that it is very difficult to analyze the behavior of fatigue creep interaction. So it becomes very important to study the high temperature fatigue life prediction model for specific materials.

In this paper, the HTLCF tests of TC4 titanium alloy and TC11 titanium alloy are carried out under total axial strain. The relationships of cyclic strain-life and stress-life are analyzed. The HTLCF life prediction models of two kinds of titanium alloys are established by Manson-Coffin method and new method based on exponential function. In addition, the fatigue life predictive ability of two methods was assessed. Related research can provide reference for component design and life evaluation.

2. Experimental methods

2.1. Materials

TC4 titanium alloy and TC11 titanium alloy are used as experimental materials. The Chemical composition of two kinds of titanium alloys are shown in table 1. The tensile property of two kinds of titanium alloys are shown in table 2 [13].

Table 1. The chemical composition of two kinds of titanium alloys (wt%).

Material	Al	C	Fe	Si	O	H	N	Ti
TC4	6.22	0.02	0.09	0.04	0.12	0.015	0.02	surplus
TC11	5.8~7.0	≤0.10	≤0.25	0.20~0.35	≤0.15	≤0.012	≤0.05	surplus

Table 2. The tensile property of two kinds of titanium alloys.

Material	Young's modulus E (GPa)	Fracture limit σ_b (MPa)	Yield strength $\sigma_{0.2}$ (MPa)	Elongation δ,%	Reduction of areaψ,%
TC4	109	969	948	16	45.5
TC11	83	806	641	20.8	63.5

2.2. *Experimental conditions and methods*

The HTLCF experimental conditions and methods of two kinds of titanium alloys are shown in table 3.

Table 3. The experimental conditions and methods of two kinds of titanium alloys.

Experimental conditions and methods	Material	
	TC4	TC11
Experimental temperature	350^0C	500^0C
Experimental control mode	Axial strain control	Axial strain control
Material specifications	d68 mm	d18mm
Loading Waves	Triangle wave	Triangle wave
Strain ratio	−1	−1
Loading frequency	0.033～0.33 Hz	0.167～0.333 Hz
Failure criterion	Crack	Crack
Heat treatment process	477^0C, 4.5 h hardening, 121^0C, 9h+177^0C, 9h aging	800^0C, 1 h, air-cooled

3. Results and discussion

3.1. *Low cycle fatigue experimental results and discussion of TC4titanium alloy*

The HTLCF experimental results of TC4 titanium alloy are shown in table 4.

Table 4. The HTLCF experimental results of TC4 titanium alloy.

Total strain ranges $\Delta\varepsilon_t/2$ (%)	Elastic strain range $\Delta\varepsilon_e/2$ (%)	Plastic strain range $\Delta\varepsilon_p/2$ (%)	Stress range $\Delta\sigma/2$ (MPa)	Fatigue inverse number $2N_f$
4.160	0.816	3.344	759	81
3.198	0.827	2.371	769	118
2.237	0.843	1.394	781	229
1.831	0.824	1.007	766	363
1.221	0.777	0.444	722	830
0.902	0.696	0.206	647	2176
0.682	0.617	0.065	574	5681

The linear relationship of $\Delta\varepsilon_t/2$ —$2N_f$, $\Delta\varepsilon_p/2$ —$2N_f$ and $\Delta\varepsilon_e/2$ —$2N_f$ in log-log coordinate is shown in figure 1. Regression formula (1) and formula (2) can be obtained by regression analysis:

$$\Delta\varepsilon_p/2=1.8919(2N_f)^{-0.9052} \quad (1)$$

$$\Delta\varepsilon_e/2=0.0117(2N_f)^{-0.0681} \quad (2)$$

The relationship curve of $\Delta\sigma/2$—$\Delta\varepsilon_p/2$ is shown in figure 2. Regression formula (3) can be obtained by regression analysis:

$$\Delta\sigma/2=1043.7(\Delta\varepsilon_p/2)^{0.0765} \quad (3)$$

3.2. *The HTLCF experimental results and discussion of TC11titanium alloy*

The linear relationship curve of $\Delta\varepsilon_t/2$ —$2N_f$, $\Delta\varepsilon_p/2$ —$2N_f$ and $\Delta\varepsilon_e/2$ —$2N_f$ in log-log coordinate are shown in figure 3. Regression formula (4) and formula (5) can be obtained by regression analysis:

$$\Delta\varepsilon_p/2=4.8798(2N_f)^{-0.9994} \quad (4)$$

$$\Delta\varepsilon_e/2=0.0151(2N_f)^{-0.1293} \quad (5)$$

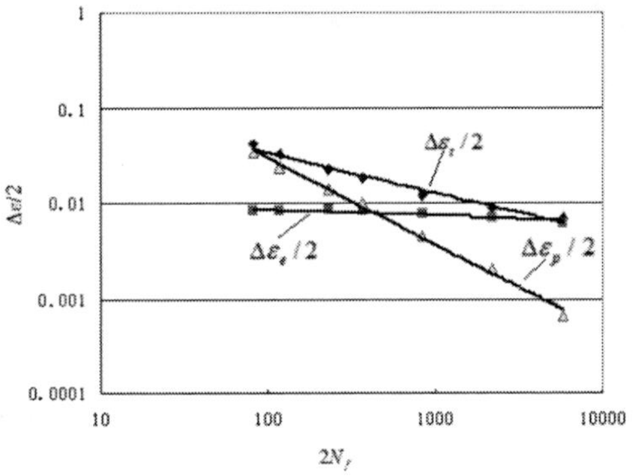

Figure 1. The relationship curve of $\Delta\varepsilon/2$—2N$_f$.

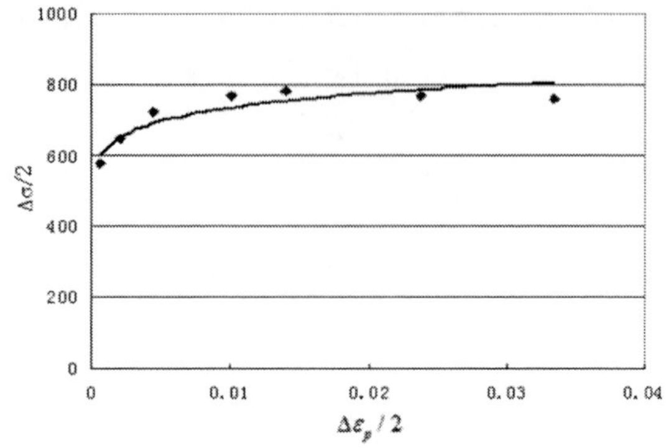

Figure 2. The relationship curve of $\Delta\sigma/2$ — $\Delta\varepsilon_p/2$.

The relationship curve of $\Delta\sigma/2$—$\Delta\varepsilon_p/2$ is shown in figure 4. Regression formula (6) can be obtained by regression analysis:

$$\Delta\sigma/2 = 1204.7(\Delta\varepsilon_p/2)^{0.1275} \tag{6}$$

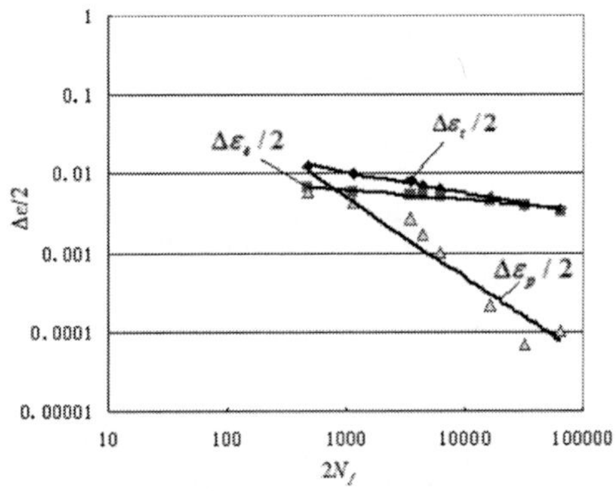

Figure 3. The linear relationship curve of $\Delta\varepsilon/2$—2N$_f$.

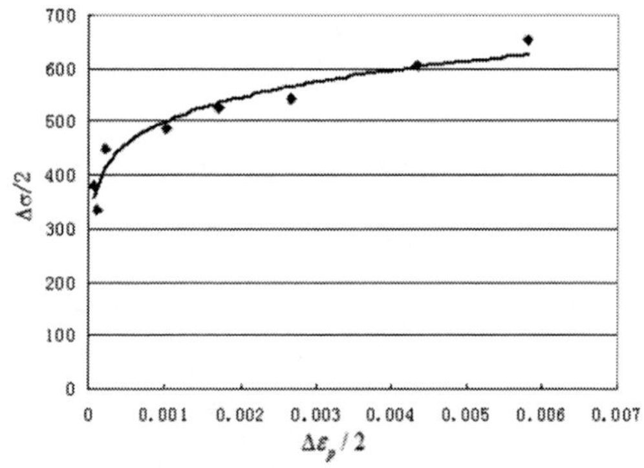

Figure 4. The relationship curve of $\Delta\sigma/2$ —$\Delta\varepsilon_p/2$.

4. Fatigue life prediction

4.1. Life prediction by the Manson-Coffin method

Manson-Coffin method is widely used in HTLCF life prediction. For HTLCF test under total strain ranges control, Manson-Coffin method predicts life using formula (7).

$$\frac{\Delta\varepsilon_t}{2} = \frac{\Delta\varepsilon_e}{2} + \frac{\Delta\varepsilon_p}{2} = \frac{\sigma_f'}{E}(2N_f)^b + \varepsilon_f'(2N_f)^c \tag{7}$$

where, $\frac{\Delta\varepsilon_t}{2}$ is total strain ranges, $\frac{\Delta\varepsilon_e}{2}$ is elastic strain range, $\frac{\Delta\varepsilon_p}{2}$ is plastic strain range, σ_f' is fatigue strength coefficient, *b* is fatigue strength exponent, ε_f' is fatigue ductility coefficient, *c* is fatigue ductility exponent. The $\frac{\sigma_f'}{E}$, *b*, ε_f' and *c* values can are obtained by regression analysis. The $\frac{\sigma_f'}{E}$, *b*, ε_f' and c values are shown in table 5.

Table5. Corresponding coefficients in formula (7) of two kinds of titanium alloys.

Material	$\frac{\sigma_f'}{E}$	b	ε_f'	c
TC4	0.0117	−0.0681	1.8919	−0.9052
TC11	0.0151	−0.1293	4.8798	−0.9994

4.2. Life prediction method based on exponential function

According to the HTLCF data analysis results, the relationship between plastic strain and failure inverse number presents nonlinear in log-log coordinate. It will lead to generating a prediction error using Manson-Coffin method. In order to solve this problem, a new method based on exponential function is proposed. Assuming a exponential relationship exists between $(-\ln\frac{\Delta\varepsilon_p}{2})$ and $\ln(2N_f)$, the relationship can be expressed as formula(8) :

$$(-\ln\frac{\Delta\varepsilon_p}{2}) = a_1 e^{a_2 \ln(2N_f)} \tag{8}$$

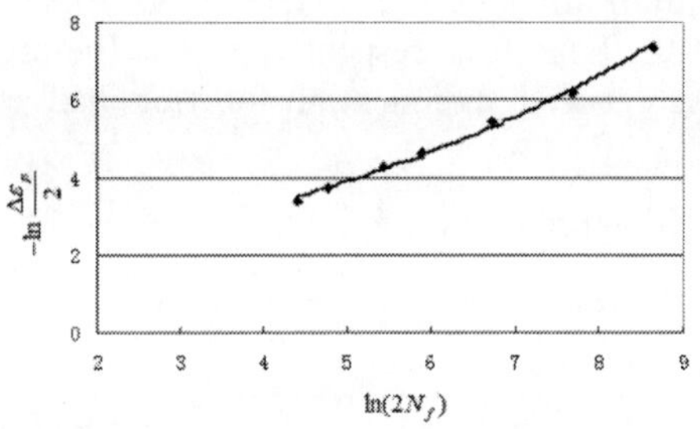

Figure 5. Curve of TC4 titanium alloy $(-\ln\frac{\Delta\varepsilon_p}{2})$-$\ln(2N_f)$.

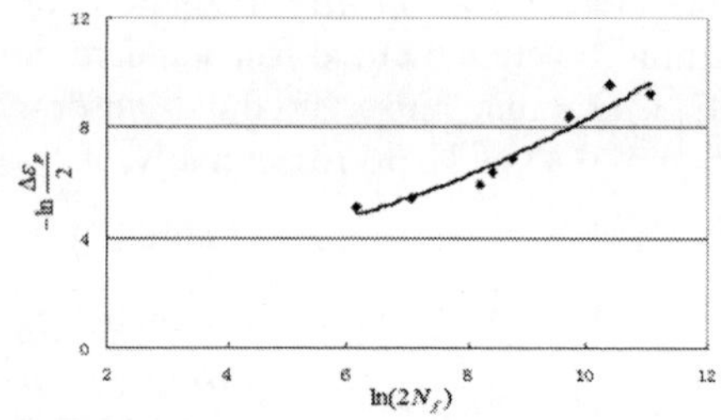

Figure 6. Curve of TC11 titanium alloy $(-\ln\frac{\Delta\varepsilon_p}{2})$-$\ln(2N_f)$.

Figure 5 and figure 6 present the exponent fit relationship of $(-\ln\frac{\Delta\varepsilon_p}{2})$ and $\ln(2N_f)$. a_1 and a_2 are obtained by regression for formula (8).

Formula (9) is obtained by transforming formula (8):

$$\frac{\Delta\varepsilon_p}{2} = e^{-a_1 e^{a_2 \ln(2N_f)}} \tag{9}$$

Therefore, the life prediction method based on exponential function can be expressed as the formula (10):

$$\frac{\Delta\varepsilon_t}{2} = \frac{\Delta\varepsilon_e}{2} + \frac{\Delta\varepsilon_p}{2} = \frac{\sigma_f'}{E}(2N_f)^b + e^{-a_1 e^{a_2 \ln(2N_f)}} \tag{10}$$

The $\frac{\sigma_f'}{E}$, b, a_1 and a_2 values are shown in table 6.

Table 6. Corresponding coefficients in formula (10).

Material	$\frac{\sigma_f'}{E}$	b	a1	a2
TC4	0.0117	−0.0681	1.6041	0.1772
TC11	0.0151	−0.1293	2.038	0.1407

5. Evaluation of life prediction ability

Shichao *et al.* [14] and Lijia *et al.* [15] believed that the life prediction ability of model can be quantitatively evaluated by scatter zone and standard deviation. The deviation degree between the experimental life and predicted life can be analyzed by scatter zone. The scatter zone (x_{max}) can be expressed as formula (11):

$$x_{max} = \max\left[\frac{N_f^p}{N_f^e}, \frac{N_f^e}{N_f^p}\right] \tag{11}$$

where, N_f^p is predicted life, N_f^e is experimental life.

The standard deviation is smaller; the life prediction ability of the model is better. The standard deviation(S) can be expressed as formula (12):

$$S = \left[\frac{\sum_1^n(\lg N_f^p - \lg N_f^e)^2}{(n-1)}\right]^{1/2} \tag{12}$$

where, n is total number of experimental samples.

Scatter zone and standard deviation of two life prediction models are shown in table 7. As seen from the table 7 that the life prediction for TC4 titanium alloy by the Manson-Coffin method and exponential function method had standard deviation of 0.20 and 0.12 respectively. The life prediction for TC11 titanium alloy by the Manson-Coffin method and exponential function had standard deviation of 0.43 and 0.36 respectively.

Table 7. Evaluation of life prediction.

Material	Prediction method	Scatter zone	Standard
TC4	Manson-Coffin	1.54	0.20
	New method	1.28	0.12
TC11	Manson-Coffin	1.83	0.43
	New method	1.62	0.36

6. Conclusions

In order to study the HTLCF behaviour of TC4 titanium alloy and TC11 titanium alloy, the HTLCF tests were carried out under total axial strain control. The relationships of cyclic strain-stress and strain-life are analyzed. Cyclic strain-stress and strain-life relationship curves of the two kinds of titanium alloys were obtained. According to analysis results, there was an exponential relationship between $\Delta\varepsilon_p/2$ and $2N_f$ in log-log coordinate.

The high temperature LCF life prediction models of two kinds of titanium alloys are established by Manson-Coffin method and new method based on exponential function.

Manson-Coffin method and exponential function method can effectively predict the HTLCF life of TC4 titanium alloy and TC11 titanium alloy. Prediction accuracy within ±1.83 times scatter zone.

The fatigue life prediction results of two kinds of titanium alloys by two prediction models showed new method based on exponential function proved more accurate and effective than Manson-Coffin method. The new method can give better life prediction results with the smaller scatter zone and standard deviation.

Manson-Coffin method and new method based on exponential function can effectively predict the HTLCF life of TC4 titanium alloy. The fatigue life prediction for TC4titanium alloy by the Manson-Coffin method and new method based on exponential function has scatter zone of 1.54 and 1.28 respectively and standard deviations of 0.20 and 0.12 respectively.

The life prediction capability of Manson-Coffin method and new method for TC4 titanium alloy prove more accurate than TC11 titanium alloy. Two methods can give better fatigue life prediction results for TC4 titanium alloy with the smaller standard deviation and scatter zone than TC11 titanium alloy.

IOP Conf. Series: Materials Science and Engineering **292** (2017) 012052 doi:10.1088/1757-899X/292/1/012052

Acknowledgements

This work was financially supported by Natural Science Foundation of Liaoning Province of China (Research Project20170540107).

References

[1] Huang Xinyue, Zhang Shichao, Lu Yuan & Yu Huichen 2011 Investigation on fatigue crack propagation behavior of TC11 and TC4 Ti alloys at room temperature and 400 ^{0}C*Journal of aeronautical materials* **31(5)** 82-85

[2] Manson S S 1954 Behavior of materials under conditions of thermal stress. National Advisory Commission on Aeronautics: Report 1170 *Lewis Flight Propulsion Laboratory, Cleveland*

[3] Coffin L F 1954 A study of the effects of cyclic thermal stresses on a ductile metal *Transactions of the American Society of Mechanical Engineers* **76** 931-50

[4] Manson S S 1974 Thermal stress and low-cycle fatigue *McGraw Hill Book company*

[5] Manson S S and Halford G 1967 A method of estimating high temperature and LCF behaviour of materials *Int. Conf. on thermal high-strain fatigue* 154-157

[6] Zhou Shentian, LiuJun & Huang Baozong 2008 Continuum damage mechanics study on low cycle fatigue damage of TC4titanium alloy *Journal of MechanicalStrength* **30(5)** 798–803

[7] Gao Yang, Bai Guangchen&YU Linchong 2009 Fatigue reliability analysis of turbinedisk based on RBFneural network*Journal of machine design* **26(5)** 8-14

[8] Shang Tisong, Zhao Ming&Chen Yanghui 2015 Low Cycle Fatigue Life Prediction Method Based on Three-Parameter Power Function*Journal ofPropulsion Technology* **36(6)** 907-911

[9] Li Weiping, Cao heli, Lilei & Kang Jin2017 Continuous Miner Method for Fatigue Life Estimation*Computer Simulation* **34(1)** 414-417

[10] Chen Ling, Zhang xianming & Liu Fei 2017 Discussion of Low Cycle Fatigue Life Prediction Models for Magnesium Alloys*China Mechanical Engineering* **28(5)** 512-517

[11] Gao Guodong, Duan Shulin, Zhang Wenxiao 2015 Low cycle fatigue life prediction for GH4133 at 550 ^{0}C based on power-exponent function*Journal of engineering research* **3(3)** 111–124

[12] Liu Yi, Chen Lijia& Wang Zhongguang2006 Fatigue-Creep behavior and deformation structures of GH4049 nickel-base superalloy *ActaMetallurgica Sinica* **35(9)** 955–959

[13] Xie Jizhou1992*Handbook of Low-cycle Fatigue*Beijing Institute of Materials Publications, Beijing

[14] Zhang Shichao, Yu Huichen&Li Ying 2013 Investigation on low cycle fatigue properties of superalloy GH3044 at elevated temperature*Journal of aeronautical materials* **33** 100–104

[15] Chen Lijia, Wu Wei&P.K.Liaw 2006 Creep-fatigue interaction behaviors and life predictions for three superalloys *ActaMetallurgica Sinica* **42** 952–958

IOP Conf. Series: Materials Science and Engineering **292** (2017) 012053 doi:10.1088/1757-899X/292/1/012053

Improvement in hydrogen storage properties of MgH_2 catalyzed with $BaTiO_3$ additive

J S Wang[1, 2, 3], W Zhang[1], S M Han[1, 2] and F Qin[2]

[1] State Key Laboratory of Metastable Materials Science and Technology, Yanshan University, Qinhuangdao 066004, PR China
[2] College of Environmental and Chemical Engineering, Yanshan University, Qinhuangdao 066004, PR China
[3] E-mail: bluceking@163.com

Abstract. The MgH_2+20 wt. % $BaTiO_3$ composites is prepared by ball milling MgH_2 with 20 wt. % $BaTiO_3$, and the effect of $BaTiO_3$ on the hydrogen storage properties of the MgH_2 is investigated. TPD (Temperature-programmed-desorption) results showed that with addition of $BaTiO_3$, the onset dehydrogenation temperature of the composite is 270 °C, which is 149 °C lower than that of pure MgH_2. And the hydrogen absorption / desorption rates of MgH_2 + 20 wt.% $BaTiO_3$ are 2.21 times and 2.12 times of pure MgH_2 at certain temperatures and hydrogen pressures, respectively. Furthermore, $BaTiO_3$ additives decreased the dehydriding activation energy of MgH_2 from 116 kJ /mol to 108 kJ /mol. $BaTiO_3$ plays an effective catalyst role in the hydrogenation / dehydrogenation process of MgH_2.

1. Introduction

Hydrogen as a high density energy carrier has been attached great importance on its clean and inexhaustible properties [1]. The application of hydrogen needs a support from high capacity storage materials, such as MgH_2, which shows a high hydrogen storage capacity of 7.6 wt.% [2, 3]. However, MgH_2 suffers from high dehydrogenation temperature and poor kinetics, which limit its commercial application [4, 5].

In order to improve the hydrogen storage properties of MgH_2, many kinds of additives has been reported to be doped into MgH_2, such as transitional elements (Ti, V, Nb et al.) [6, 7], transition metal oxides [8], borides [9], fluorides [10, 11] and sulfides [12-14]. Barkhordarian et al. [15] and Hanada et al. [16] reported the remarkable improvement of hydrogen sorption kinetics by ball milling certain contents of Nb_2O_5 and MgH_2, and the studies of Ma et al. [17, 18] further certified that the Nb crystals distributed in the composites can work as the gateway to facilitate hydrogen transportation, and result in an improvement of the hydrogen storage properties of MgH_2. In recent years, Ti-based catalysts, including Ti, TiO_2, $TiCl_3$, TiF_3, were employed to decrease the desorption temperature of MgH_2 [19, 20]. When ball milled with MgH_2, the titanium oxides were found to result in significantly effects on reducing the onset dehydrogenation temperature of MgH_2, and increased the hydrogenation and dehydrogenation rates [21].

More recently, metal composite oxides [22, 23], were used as catalysts to improve the hydrogen storage properties of MgH_2. Bariccoa et al. [24] investigated the effects of doping a perovskite-type $BaRuO_3$ into MgH_2, and reported that not only the RuO_6 octahedral structure could be penetrated by H_2, but the $BaRuO_3$ could destabilize MgH_2, which decreased the dehydrogenation temperature of

MgH_2. Besides, our group found that with 20 wt.% addition of $MgTiO_3$, the hydrogen storage performances of MgH_2 were remarkably improved [25].

In this paper, in consideration of the excellent catalytic effects of perovskite-type oxides and Ti-based additives, and the beneficial effect of perovskite-type structure compound on the diffusion of H_2 as mentioned above, a Ti-based perovskite-type $BaTiO_3$ was adopted as catalysts to further improve the hydrogenation/dehydrogenation properties of MgH_2. The effect of $BaTiO_3$ on the hydrogenation/dehydrogenation properties are studied, and more importantly, the effect mechanism is elucidated.

2. Experimental methods

The initial materials MgH_2 (98 %) and $BaTiO_3$ (99.5 %) were commercially purchased from Sigma-Aldrich. The MgH_2 powder was mixed with 20 wt.% $BaTiO_3$ powder which was mechanically milled for 2 h using a QM-ISP2 planetary, repeating 30 min milling and 15 min rest, with 500 $r \cdot min^{-1}$ under Ar atmosphere. The ball to powder weight ratio was about 40:1. All the sample manipulation was performed in an Ar-filled glove box to prevent getting in touch with oxygen and moisture.

The effect of $BaTiO_3$ addition on the dissociation temperature of the pure MgH_2 during the dehydrogenation was investigated by the temperature programmed desorption (TPD) apparatus. The hydrogen absorption/desorption kinetics of the samples was measured using a pressure-composition-temperature (P-C-T) characteristic measurement apparatus. Hydrogen absorption and desorption measurements were performed at desired temperature, with an initial hydrogen pressure of 3.0 MPa, and 0.1 MPa initial hydrogen pressure was given for the hydrogen desorption.

Samples after mechanical grinding and after hydriding/dehydriding cycling were characterized by X-ray diffraction (XRD) with Cu Kα radiation , using D/MAX-2500/PC, The 2θ angle was varied from 10° to 80°. Hydrogen desorption properties were investigated by a simultaneous thermal analyzer (DTG-60A) under flowing Ar at a rate of 100 $mL \cdot min^{-1}$, which were rising from room temperature to 500 °C with a heating rate of 5, 10, 15 and 20 $°C \cdot min^{-1}$.

3. Results and discussion

3.1. Hydrogen absorption/desorption performances

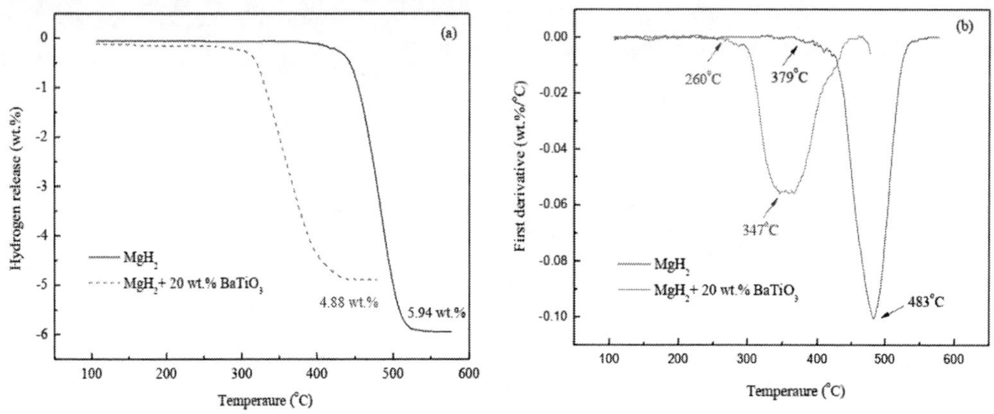

Figure 1. TPD curves of MgH_2 and MgH_2+20 wt. % $BaTiO_3$.
(a) Temperature-desorption contents curves; (b) The first derivative curves of (a).

Figure1 presents the dehydrogenation curves of the MgH_2+20 wt.% $BaTiO_3$ and the pristine MgH_2 when they are heated from 100 °C to 550 °C at a rate of 10 $°C \cdot min^{-1}$. The pristine MgH_2 starts to release hydrogen at about 389 °C, with a hydrogen desorption capacity of about 5.94 wt.%. In case of MgH_2+20 wt.% $BaTiO_3$, the onset dehydrogenation temperature is 260 °C, and the peak temperature is 347 °C, 119 °C and 136 °C lower than that of pure MgH_2, respectively, and its total dehydrogenation amount reaches 4.88 wt.% by 433 °C. Obviously, the addition is effective to improve the thermal decomposition characteristics of MgH_2 with results of decreasing desorption temperature. As for the maximum hydrogen release content, it seems that MgH_2+20 wt.% $BaTiO_3$ decreased almost 1wt.% than that of pure MgH_2. However, it should be noticed that the 20 wt.% $BaTiO_3$ additives were not hydrogen absorption materials. Thus, the maximum hydrogen desorption amount was shown relatively less in the diagram.

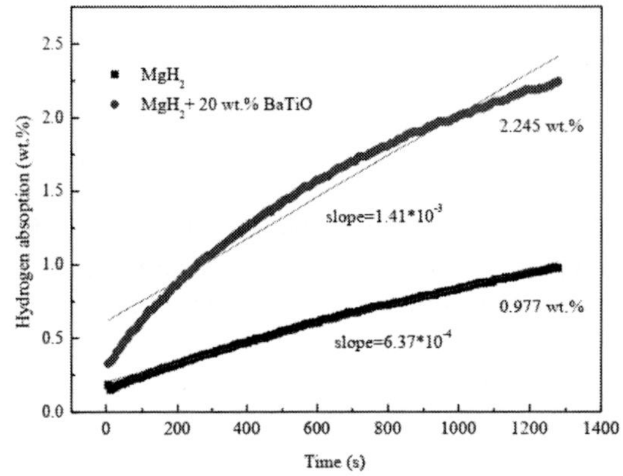

Figure 2. Comparison of the absorption kinetics of MgH_2 and MgH_2+20 wt. % $BaTiO_3$.

Figure 2 presents the hydrogen absorption curves of the received MgH_2 and the MgH_2+20 wt.% $BaTiO_3$ at 150 °C under 3 MPa hydrogen pressure. It should be noticed that all the samples are measured after two hydrogen absorption/desorption cycles. The representative curve shows that the received MgH_2 absorbs only 0.977 wt. % hydrogen within 21 min, whereas under identical conditions, MgH_2+20 wt. % $BaTiO_3$ absorbs 2.245 wt. % hydrogen, almost 2.3 times of pure MgH_2. Furthermore, the linear fitting of above isothermal desorption curves shows that for the MgH_2+20 wt. % $BaTiO_3$ sample, the average diffusion rate of H at 150 °C is $1.41*10^{-3}$ wt.%/s, and the value for pure MgH_2 is $6.37*10^{-4}$ wt.%/s. The results illustrate at 150 °C, the hydrogen absorption rate of MgH_2+20 wt. % $BaTiO_3$ is almost 2.21 times of pure MgH_2. It is obvious that the addition of $BaTiO_3$ significantly accelerates the hydrogen absorption rate of MgH_2.

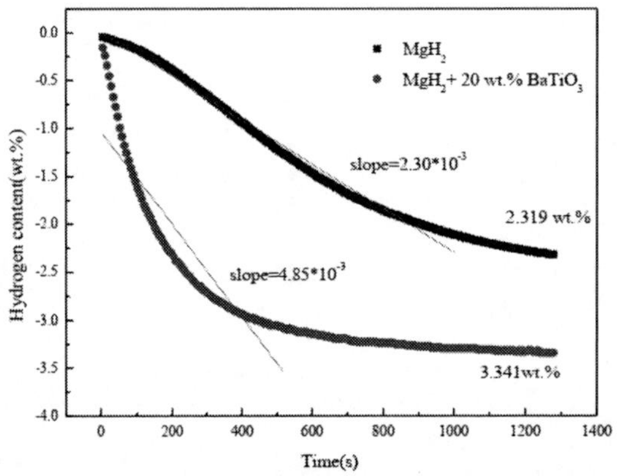

Figure 3. Comparison of the desorption kinetics of MgH_2 and MgH_2+20 wt. % $BaTiO_3$.

The isothermal dehydrogenation curve for MgH_2 samples with and without 20 wt. % $BaTiO_3$ at 350 °C are shown in figure 3. Within 21 min, 2.319 wt. % hydrogen was dehydrogenated for MgH_2. As for the MgH_2+20 wt. % $BaTiO_3$ composites, 3.341 wt. % hydrogen was dehydrogenated under the same condition, which was about 1.44 times than that of pure MgH_2. And for the MgH_2+20 wt. % $BaTiO_3$, releasing 90% hydrogen takes only 500 seconds, while it takes more than 1000s for pure MgH_2. According to the fitting curve, at 350 °C, the dehydrogenation rate of MgH_2+20 wt. % $BaTiO_3$ is $4.85*10^{-3}$ wt.%/s, almost 2.12 times of pure MgH_2. Therefore, with the introduction of 20 wt.%

$BaTiO_3$, not only the hydrogen absorption rate of MgH_2, but also the hydrogen desorption rate can be remarkably improved.

3.2. Activation energy for decomposition of the MgH_2-$BaTiO_3$ composite

The hydrogen desorption kinetics curves of pristine MgH_2 and MgH_2-$BaTiO_3$ composites at different heating rates are shown in figure 4. After doping $BaTiO_3$, the decomposition temperature of the samples is obviously decreased. At a heating rate of 10 °C/min, in compared with pure MgH_2, the dehydrogenation peak temperature is lowered by 43 °C for MgH_2+20 wt. % $BaTiO_3$.

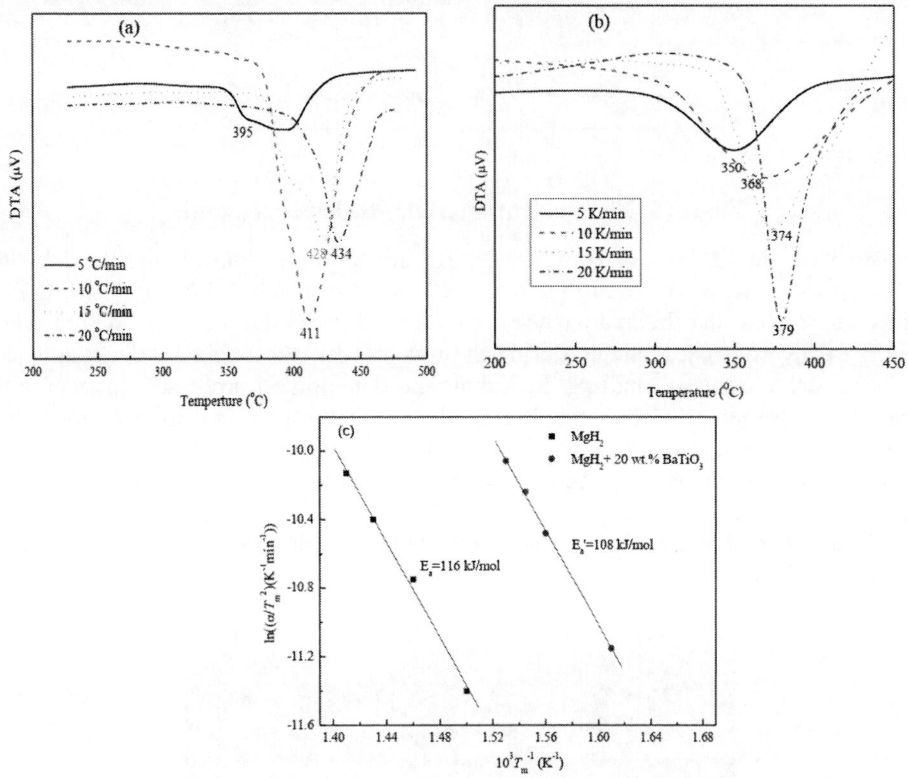

Figure 4. DTA curves of different sample at various heating rates (a) MgH_2; (b) MgH_2+20wt. % $BaTiO_3$; (c) Kissinger's plots.

In order to understand the fast hydrogen desorption kinetics, the activation energy of the dehydrogenation process for the pure MgH_2 and the MgH_2-$BaTiO_3$ was estimated with Kissinger's method [26].

$$\frac{d\left[Ln\left(\alpha/T_m^2\right)\right]}{d\left[1/T_m\right]} = -\frac{E_a}{R} \qquad (1)$$

Where α is the heating rate, Tm is the absolute temperature at the maximum desorption rate, and R is the ideal gas constant. The activation energies of different composites were evaluated by the DTA using Kissinger formula. Figure4 (c) shows Kissinger plots for the desorption reaction of MgH_2+20 wt.% $BaTiO_3$. The activation energy value is 116 kJ/mol and 108 kJ/mol for pristine MgH_2 and MgH_2+20 wt.% $BaTiO_3$ composite, respectively. The calculated results show that $BaTiO_3$ additives decrease activation energy during dehydriding, and therefore improve the dehydriding rate of the composite.

3.3. Structural characterization and effect mechanism of the MgH_2-$BaTiO_3$ composite

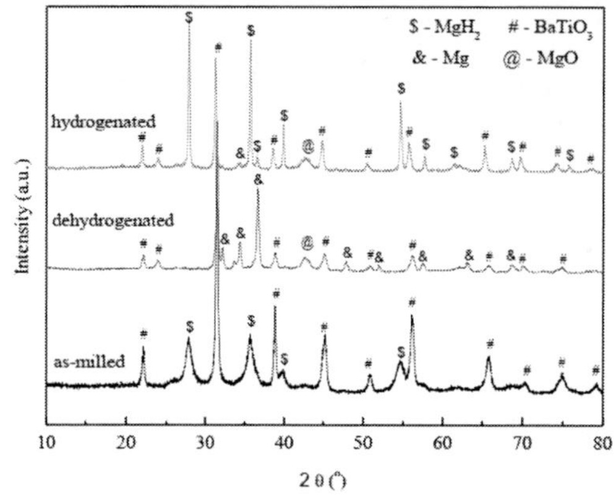

Figure 5. XRD patterns of MgH_2-$BaTiO_3$ composite.

The XRD patterns of the MgH_2 with 20 wt. % $BaTiO_3$ are shown in figure 5. For the dehydrogenated sample, MgH_2 phase is replaced by Mg phases, and a weak peak corresponding to MgO is found in both the dehydrogenated and the hydrogenated patterns, which is due to the oxidation of Mg when transferring the dehydrogenated sample during the test process. Furthermore, the results shows that $BaTiO_3$ remains stable after ball-milling, hydriding and dehydriding processes, indicating that there are no reaction between $BaTiO_3$ and MgH_2, but $BaTiO_3$ play a catalyst role during the hydriding/dehydriding process of MgH_2, which result in the decrease in desorption temperature and improvement in kinetics of MgH_2. According to the research results of Lin et al., the Ti-based catalysts owing to special electron distribution help in particle transfer in chemical reactions and have been applied in many chemical processing [27]. Therefore, $BaTiO_3$ play a catalytic role in the hydrogenation/dehydrogenation reactions of MgH_2, which helps to improve the hydrogen storage properties of MgH_2.

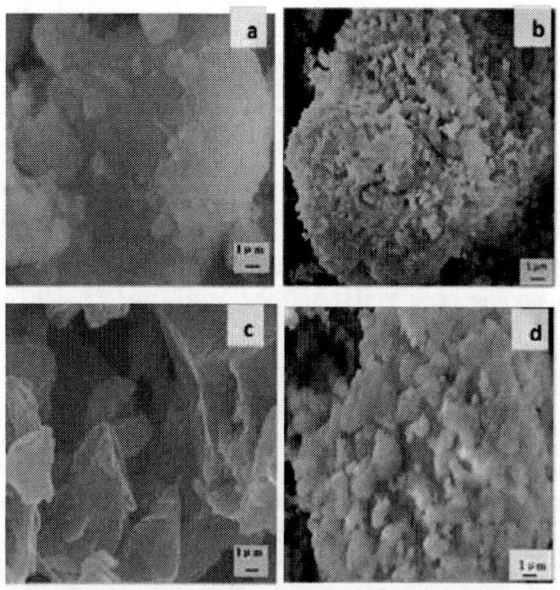

Figure 6. SEM morphology of the: (a) MgH_2 as-milled; (b) MgH_2 + 20 wt.% $BaTiO_3$ composite as-milled; (c) MgH_2 after hydrogenation; (d) MgH_2 + 20 wt.% $BaTiO_3$ composite after hydrogenation.

In order to futher study the microstructure and effect mechanism of the composite, Scanning Electron Microstructure is exploited to characterize the samples after ball-milling and hydrogenation process. As shown in figure 6, for the as-milled samples, the particle sizes of the MgH_2+20 wt.% $BaTiO_3$ shows a smaller and more homogeneous particle size compared to pure MgH_2. In addition, the average particle size of MgH_2+20 wt.% $BaTiO_3$ is also found to be smaller than that of pure MgH_2 after hydrogenation processes.That illustrates the addition of $BaTiO_3$ to MgH_2 can reduce the degree of aggregation of the particles during the ball-milling and the hydrogenation process. As reported by Peng et al.cx [28], smaller size particle and fewer amounts of clusters inside the hydrogen storage materials can facilitate hydrogenation/ dehydrogenation kinetics due to more H diffusion channels in the matrix. Therefore, by doping $BaTiO_3$ into the MgH_2, we're able to make uniform and homogeneous microstructures that have more remarkable hydrogen storage properties compared to pure MgH_2.

4. Conclusions

The thermodynamic and kinetic properties of MgH_2+ 20wt.% $BaTiO_3$ are investigated. In comparison with the pure MgH_2, the onset desorption temperature of the MgH_2+20 wt. % $BaTiO_3$ composite decreased about 119 °C, and the hydrogen absorption capacity increased by 2.3 times at 150 °C, the desorption capacity increased by 1.44 times at 350 °C within 21 min. According to DTA results measured at different heating rate, activation energy of composites decreased from 116 kJ/mol to 108 kJ/mol for MgH_2+20 wt. % $BaTiO_3$. $BaTiO_3$ play a catalytic role in the hydriding / dehydriding of MgH_2, which is responsible for the hydrogen storage properties improvement of MgH_2.

Acknowledgments
This work was financially supported by the National Natural Science Foundation of China (NOs.51571173 and 21606193), China Postdoctoral Science Foundation (2016M601281), and the Fundamental Research Funds for the Central Universities (Grant No. 3142015009).

References

[1] Pei L C, Han S M, Wang J S, Hu L, Zhao X and Liu B Z 2012 *Mater Sci Eng B* **177** 1589
[2] Kurko S, Milanović I, Grbović Novaković J, Ivanović N and Novaković N 2014 *Int. J Hydrogen Energy* **39** 862
[3] Floriano R, Leiva D R, Deledda S,Hauback B C and Botta W J 2014 *Int. J Hydrogen Energy* **39** 4404
[4] Shimada M, Higuchi E and Inoue H 2013 *J. Alloys Compd* **580** Supplement 1 S153
[5] Leng H Y, Pan Y B, Li Q and Chou K C 2014 *Int. J. Hydrogen Energy* **39** 13622
[6] Charbonnier J, Rango P D, Fruchart D et al. 2004 *J. Alloys Compd* **383** 205
[7] Rivoirard S, Rango P D, Fruchart D, Charbonnier J and Vempaire D 2003 *J. Alloys Compd* **Supplement 356–357** 622
[8] Floriano R, Leiva D R, Deledda S, Hauback B C and Botta W J 2013 *Int. J. Hydrogen Energy* **38** 16193
[9] Liu G,Qiu F, Li J, Wang Y,Li L, Yan C, Jiao L and Yuan H 2012 *Int. J. Hydrogen Energy* **37** 17111
[10] Jin S A, Shim J H, Cho Y W and Yi K W 2007 *J. Power Sources* **172** 859
[11] Zhang J, Huang Y N, Mao C and Peng P 2012 *J. Alloys Compd* **538** 205
[12] Jia Y H, Han S M, Zhang W, Zhao X, Sun P F, Liu YQ, Shi H and Wang J S 2013 *Int. J. Hydrogen Energy* **38** 2352
[13] Wang J S, Zhang W, Cheng Y, Ke D D and Han S M 2015 *J. Wuhan University of Technology-Mater. Sci. Ed* **30** 670
[14] Zhang W, Cheng Y, Han D and Han S M 2015 *Energy* **93** 625
[15] Barkhordarian G, Klassen T, Bormann R 2006 *J. Alloy. Compd* **407** 249
[16] Hanada N, Ichikawa T, Hino S and Fujii H 2006 *J. Alloy. Compd* **420** 46
[17] Ma T, Isobe S, Morita E, WangY, Hashimoto N, Ohnuki S, Kimura T, Ichikawa T and Kojima Y 2011 *Int. J. Hydrogen Energy* **36** 12319
[18] Ma T, Isobe S, Wang Y, Hashimoto N and Ohnuki S 2013 *J. Phys. Chem. C* **117** 10302

[19] Shahi R R, Bhatnagar A, Pandey S K, Dixit V and Srivastava O N 2014 *Int. J. Hydrogen Energy* **39** 14255
[20] Ren C, Fang Z Z, Zhou C, Lu J, RenY, Zhang X and Luo X 2014 *Int. J. Hydrogen Energy* **39** 5868
[21] Croston D L, Grant D M and Walker G S 2010 *J. Alloys Compd* **492** 251
[22] Xu G, Shen N, Chen L J, Chen Y and Zhang W 2017 *Mater. Res. Bull.* **89** 197
[23] Shan J W, Li P, Wan Q, Zhai F Q, Zhang J, Li Z L, Liu Z J, Volinsky A A and Qu X H 2014 *J. Power Sources* **268** 778
[24] Bariccoa M, Rahmana M W, Livraghia S, Castelleroa A, Enzob S and Giamello E 2012 *J. Alloy. Compd* **536** S216
[25] Wang J S, Han S M, Li Y, Shen N and Zhang W 2014 *Acta Physico-Chimica Sinica* **30** 2323
[26] He K 1957 *Anal. Chem,* **29** 1417
[27] Lin H J, Ouyang L Z, Wang H, Zhao D Q, Wang W H, Sun D L and Zhu M 2012 *Int. J. Hydrogen Energy* **37** 14329
[28] Peng S K, Xiao X Z, Xu R J, Li L, Wu F, Li S Q, Wang Q D and Chen L X 2010 *T Nonferr Metal Soc* **20** 1879

Thermodynamic calculation study on phase structures of Pt-Ir-M(Y, Zr, Mo)

S B Wang[1,2,3], Y Sun[1], J L Chen[2,3], S Wang[2], M J Peng[1], S Chen[2,3], MM Liu[2], J Q Hu[2], Y T Chen[2], J M Zhang[2], Y C Yang[2] and M Xie[1,2,3,4]

[1]School of Material Science and Engineering, Kunming University of Science and Technology, Kunming, Wenchang Road68, China
[2]State Key Laboratory of Advanced Technologies for Comprehensive Utilization of Platinum Metals, Sino-platinum Metals Co. Ltd., Kunming, Keji Road 988, China
[3]Kunming Institute of Precious Metals, Kunming, Keji Road 988, China

[4]E-mail: powder@ipm.com.cn

Abstract. The energy and electronic structure of the solid solutions of Pt-Ir-M (M=Mo,Y,Zr) alloy systems were investigated using the first-principle pseudo-potential plane-wave method. The results show that, adding the same percentage content, 1% Mo doped in Au-Pd solid solution is the most stable phase structure. When doped with Y and Zr, they are likely to react with Au-Pd to form the corresponding intermetallic compounds.

1. Introduction

Electrical contacts mainly undertake connected, points off circuit and load current. Its performance directly affects the safety running for switch apparatus, playing an important role in the service life of electrical appliance as well as reliability and security. In recent years, the requirement of high reliability of contacts which used in control and automobile engineering stimulates research on electrical contact phenomena of the noble metal contact materials. Pt-Ir, as one of them, has high melting point, high hardness, strong corrosion-resistance and low contact resistance. Which is widely used in aircraft engine ignitions, high sensitive relays, micro-motor relays and submarine gravity pendulum relays.

The existing literature more focused on microstructure, mechanical property and physical properties of Pt-Ir in high temperature environment[1–6]. However, as electrical contact material, problems still exist in composition segregation, poor processing performance and performance instability. Recently, it is one of hot spot that improving the above problems through doped other mental. In this work, the phase stability of Pt-Ir-M(M=Mo, Y, Zr) alloy were investigated in order to enhance the understanding of such alloy systems.

2. Calculation Procedure

Phase diagram shows that Pt-Ir alloy is continuous solid solution in high temperature zone under the solidus. Therefore, a solid solution model with Pt as solvent and Ir as solute was constructed, as shown in figure 1. The corresponding models were built with M(M=Mo,Y,Zr)content 1%, 2%and 5%(mass percent) according to the quality of composition proportion. Doping M could be achieved through modifying the atoms of digits.

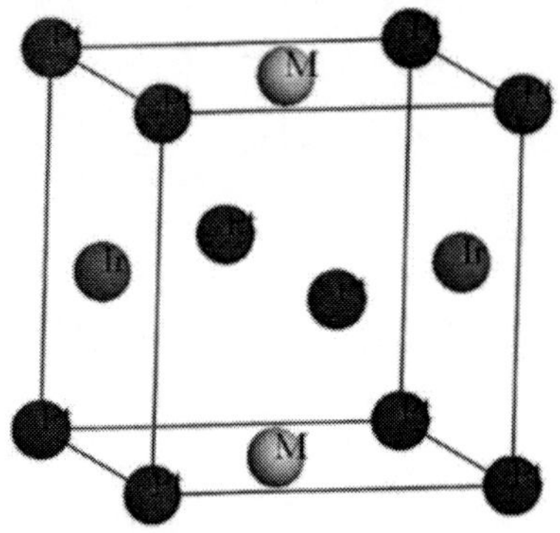

Figure1. Crystal structure of Pt-Ir-M.

All calculations were performed in CASTEP program which is based on density functional theory (DFT) of first-principles total energy pseudo potential plane wave method. CASTEP process was completed commonly used approximate energy density functional with the local density approximation(LDA)and generalized gradient approximation(GGA). Geometric optimization of lattice parameters was performed in original cell. The energy cut-off of the plane wave basis was all set as 340eV, which is sufficient for fully converge. The K-points were sampled by 6×6×6. Other parameters were all set as fine.

3. Result and discussion

3.1. Thermal stability of the solid solution model

Cohesive energy(E) and formation enthalpies(H) of Pt-Ir-M(M=Mo,Y,Zr) were calculated. The obtained results are presented in table 1. It can be seen from table 1 that all the H of these structures are negative, which means that the structures of these compounds can exist and are thermodynamically stable. With the increasing of doping element, E and H of Pt-Ir-M (M=Mo, Y, Zr) system are becomes higher, which implies that the lower content of doping element, the easier the solid solutions form, and the more stable of them.

Table 1. Cohesive energy and formation enthalpies of Pt-Ir-M(M=Mo, Y, Zr) with different composition ratio.

Percentage composition		E (eV/atom)	H (eV/atom)
	1%	-11.97	-14.30
Mo	2%	-11.74	-14.02
	5%	-11.04	-13.18
	1%	-10.56	-12.87
Y	2%	-10.36	-12.62
	5%	-9.71	-11.83
	1%	-11.78	-14.24
Zr	2%	-11.51	-13.92
	5%	-10.79	-13.05

In the comparison between calculated and experimental lattice parameter of Pt-Ir-M(M=Mo,Y,Zr) alloy phases presented in Section B, C and D, most of the calculated results are in good conformity with the experimental values, confirming that the calculation method and the model selected in this study are adequate.

Compared with three doping elements Mo, Y, Zr in Pt-Ir solid solution, we found that with the same percentage of doping, E and H of Pt-Ir-Mo are lower than the other two, which indicates that Pt-

Ir-Mo is formed more easily than Pt-Ir-Zr and Pt-Ir-Y, in other words, Pt-Ir-Mo is more stable than the other two.

3.2. The stability of Pt-Ir-Mo alloy phases

In Pt-Ir-Mo solid solution, there are four phase structures between Pt and Mo, which are Pt3Mo of P4/MMM, Pt2Mo of IMMM,PtMo of P63/MMC and PtMo3 of PM–3N. Similarly, there are three phase structures between Ir and Mo, which are Ir3Mo of P63/MMC, IrMo of PMMA and IrMo3 of Pm-3N. For these seven structures, the related theoretical calculation was done.

Table2 shows the calculated and experimentally determined lattice parameter of Pt-Ir-Mo alloy phases. Compared the calculated lattice parameters with the experimental values, it shows that the calculated results of lattice constants are in good conformity with the experimental values, confirming that computation parameters selected in this study are suitable.

Table 2. Lattice constants, cohesive energy and formation enthalpies of Pt-Ir-Mo.

Structure	Method	Space Group	Lattice constant, a (Å)	Lattice constant, b (Å)	Lattice constant, c (Å)	cell volume , V (Å3)	Cohesive energy, E (eV/atom)	Formation enthalpies,H (eV/atom)
Ir3Mo	Cal	P63/MMC	5.462	–	4.370	112.91	-11.73	-11.26
	Exp		5.487	–	4.385	114.33	–	–
IrMo	Cal	PMMA	3.863	3.854	3.859	57.44	-12.10	-11.62
	Exp		2.752	4.804	4.429	58.55	–	–
IrMo3	Cal	PM-3N	4.921	–	–	119.19	-12.22	-11.72
	Exp		4.968	–	–	122.62	–	–
Pt3Mo	Cal	P4/MMM	3.914	–	3.913	59.95	-8.96	-8.66
	Exp		3.897	–	3.936	59.77	–	–
Pt2Mo	Cal	IMMM	2.745	8.259	3.923	88.93	-9.65	-9.32
	Exp		2.765	8.296	3.938	90.33	–	–
PtMo	Cal	P63/MMC	4.450	2.713	4.903	59.20	-10.42	-10.04
	Exp		4.475	2.729	4.914	60.01	–	–
PtMo3	Cal	PM-3N	4.944	–	–	120.86	-11.39	-10.93
	Exp		4.987	–	–	124.03	–	–

The obtained E and H of Pt-Ir-Mo alloy phases are also presented in table 2. All H of these structures are negative, which means that the structures of these compounds can exist and are thermodynamically stable. The E and H of Ir-Mo alloy phases are higher than the other, which indicates that alloy phases of Pt-Mo system are formed more easily than Ir-Mo system at ambient conditions, in other words, Ir-Mo is more stable than the other. The H of IrMo3 is−11.72eV/atom, the lowest in the seven alloy phases and a similar situation is found of E. All this indicates that the IrMo3 has the strongest alloying ability among these alloy phases.

3.3. The stability of Pt-Ir-Y alloy phases

In Pt-Ir-Y solid solution, there are three phase structures between Pt and Y, which are Pt3Y of PM-3M, Pt4Y5 of PNMA and Pt2Y of FD-3M. Moreover, there are only two phase structures between Ir and Y, which are Ir2Y of FD-3M, IrY of PM-3M. For these five alloy phases, the related theoretical calculation was done.

The calculated lattice constants are listed in table 3, along with the experimental results. It can be seen that calculated lattice parameters are similar to experimental values. So a very good agreement is reached between calculated results and the experimental values.

Table 3. Lattice constants, cohesive energy and formation enthalpies of Pt-Ir-Y.

Structure	Method	Space Group	Lattice constant, a (Å)	Lattice constant, b (Å)	Lattice constant, c (Å)	cell volume , V (Å3)	Cohesive energy, E (eV/atom)	Formation enthalpies,H (eV/atom)
Pt3Y	Cal	PM-	4.108	–	–	69.34	-7.54	-9.56
	Exp	3M	4.075	–	–	67.67	–	–
Pt4Y5	Cal	PNMA	7.471	14.706	7.539	828.84	-7.23	-9.34
	Exp		7.458	14.546	7.519	815.69	–	–
Pt2Y	Cal	FD-	7.684	–	–	453.61	-7.50	-9.54
	Exp	3M	7.576	–	–	434.93	–	–
Ir2Y	Cal	FD-	7.538	–	–	428.37	-9.39	-10.74
	Exp	3M	7.524	–	–	425.94	–	–
IrY	Cal	PM-	3.399	–	–	39.28	-8.61	-10.21
	Exp	3M	3.400	–	–	39.30	–	–

Similar to Pt-Ir-Mo, all H of Pt-Ir-Y are negative, which also means that the structures of these compounds can existent and stable. The E and H of these five alloy phases are listed in table 3, too. The H of Pt-Y alloy phases in our calculation are −9.56, −9.34 and −9.54 eV/atom separately. Such results are all lower than those results of Ir-Y. It means that Ir-Y alloy phases are more stable than Pt-Y. The E were calculated to further estimate the stability of the above alloy phases. By comparing the value of E with the H, we can fine that the structural stability of five alloy phases based on the E has the same tendency as that based on the H. The E and H of Ir2Y are respectively −9.39 and −10.74 eV/atom, which are the highest among the five alloy phases. From the viewpoint of thermodynamics, it can be predicated that the Ir2Y structure has the highest stability than among these structures.

3.4. The stability of Pt-Ir-Zr alloy phases

Table 4. Lattice constants, cohesive energy and formation enthalpies of Pt-Ir-Zr.

Structure	Method	Space Group	Lattice constant, a (Å)	Lattice constant, b (Å)	Lattice constant, c (Å)	cell volume , V (Å3)	Cohesive energy, E (eV/atom)	Formation enthalpies,H (eV/atom)
Pt3Zr	Cal	PM-3M	3.983	–	–	63.20	-8.92	-8.66
	Exp		3.990	–	–	63.52	–	–
PtZr	Cal	PM-3M	3.340	–	–	37.26	-8.56	-10.87
	Exp		3.310	–	–	36.26	–	–
Pt11Zr9	Cal	I4/M	10.518	–	6.748	746.61	-8.60	-10.88
	Exp		10.297	–	6.888	730.32	–	–
Ir3Zr	Cal	PM-3M	3.913	–	–	59.90	-11.13	-10.72
	Exp		3.943	–	–	61.30	–	–
Ir2Zr	Cal	FD-3M	8.933	–	–	712.84	-7.85	-7.45
	Exp		7.359	–	–	398.53	–	–
IrZr	Cal	PM-3M	3.241	–	–	34.05	-10.43	-10.05
	Exp		3.318	–	–	36.53	–	–
IrZr2	Cal	I4/MCM	6.812	–	10.429	483.99	-9.42	-9.07
	Exp		6.512	–	5.673	240.57	–	–

There are three phase structures between Pt and Zr, in Pt-Ir-Zr alloy, which are Pt3Zr and PtZr of PM-3M, Pt11Zr9 ofI4/M. And there are four phase structures between Ir and Zr, which are Ir3Zr of PM-

3M, Ir2Zr of FD-3M, IrZr of PM-3M and IrZr2 of I4/MCM. For these seven alloy phases, the related theoretical calculation was done.

The calculated lattice parameters are listed in table 4, along with the experimental result. Similar to those of Pt-Ir-Mo and Pt-Ir-Y, the calculated lattice constant of Pt-Zr and Ir-Zr are a close approximation to the experiment results, confirming that the present calculations are reliable. However, it is worth mentioning that the calculated cell volume V of Ir2Zr is approximately twice than the experimental one which likely means that existing model will most likely need to be adjusted for Ir2Zr.

The value of the H and E of Ir3Zr in our calculation are −10.72eV/atom and −11.13eV/atom, both of which are lowest in seven alloy phases. This result implies Ir3Zr is the most stable, IrZr comes second, whose H= −10.05eV/atom and E is −10.43eV/atom.

4. Conclusions

The phase stability of the Pt-Ir-M(M=Mo, Y, Zr) have been investigated by the method of the first-principle pseudo-potential plane-wave. Comparison between theoretical results of E and H with Pt-Ir-M(M=Mo, Y, Zr) solid solutions show that the stability is aligned as Y<Zr<Mo with the same percentage of doping.

The H and E results show that the nineteen phase structures of Pt-Ir-M(M=Mo, Y, Zr) are all thermodynamically stable structures, among which the IrMo3 has the greatest alloying ability and the highest structural stability, and the Ir2Y is second.

Acknowledgment

The present work was supported by the National Natural Science Foundation of China (No. 515070575,U1602275, 51461023, U1302272), Yunnan Applied Basic Research Projects (2014FB164, 2015FA042)and the fund of the State Key Laboratory of Advanced Technologies for Comprehensive Utilization of Platinum Metals(No. SKL-SPM-201526).

References

[1] Ming X, Jiang C, Li C, Shiji F, Dekui N, Rongchun Z, Youcai Y, Ruming L, Yunxi D and Fuqian Z2004Investigation on the microstructure and properties of Pt-24Ir-RE alloys *Precious Metals*. **25** 25-28

[2] Zhongmin D, Jianlinag L, Xianyong L, An S, Weiming G and Li C2001Properties and microstructure of Pt-Ir -Ru alloys *Precious Metals*. **225** 1-3

[3] Weibiao Z, Zhengyan T, Zhizhen Y and Li S 1992Microstructure of platinum alloy with micro-amountof yttrium and zirconium *Precious Metals.* **40** 17-21

[4] FeifeiK, Yonghong G, Song C, Weiming G and Kunhua Z2011Investigation of internal oxidation on Pt-Ir-Zr high-temperature alloy *Rare. Metal. Mat. Eng.* **40** 585-589

[5] FeifeiK, Kunhua Z, Weiming G, Chen S and Jiming Z2012Effect of dispersion-strengthened phase on high temperature creep property of PtIr5 alloy *Rare. Metal. Mat. Eng.* **41** 241-245

[6] Weibiao Z, Zhizhen Y, Zhengyan T and Xiaoxia W1992Investigation ofhigh temperature strengthening ofplatinum alloy containing the trace YandZr *J. Mater. Eng.* **4** 17-19

IOP Conf. Series: Materials Science and Engineering **292** (2017) 012055 doi:10.1088/1757-899X/292/1/012055

Properties investigation and microstructures characterization of SiCp/6061Al composites produced by PM route

A Q Wang[1,3], H W Tian[1] and J P Xie[1,2]

[1] School of Materials Science and Engineering, Henan University of Science and Technology, Luoyang, 471023, China
[2] Non-ferrous metal generic technology collaborative innovation center of henan province, Luoyang, 471023, China

[3] E-mail: aiqin_wang888@163.com

Abstract. In this study, 35 vol.% SiC particles with different sizes reinforced 6061 aluminium alloy matrix composites were prepared by a powder metallurgy method. The Scanning Electron Microscope (SEM) images of composites were observed, the Coefficient of Thermal Expansion (CTE) and tensile strength of composites were examined, and the influences of SiC particle size on microstructures and properties of the composites were analyzed. Furthermore, the SiCp/6061Al composites with SiC particle size of 7.5 μm were selected to investigate the SiCp/Al interface microstructure and precipitated phases by the means of SEM, TEM and HRTEM. The study indicated that, with the increase of SiC particle size, the SiC particles distributed more uniformly in the matrix, the CTE of composites increased, but the tensile strength of composites decreased. The SiCp/Al interface in this experiment is clean and smooth, and the combination mechanism of SiC and Al is the formation of a half coherent interface by closely matching of atoms. Some micron-sized coarse intermetallic particles existed in the hot-pressed composites, such as random-shaped Mg_2Si, long stick shaped $Al_{15}(Mn, Fe, Cu)_3Si_2$. When the composites were solution treated at 510 ℃ for 2 h and then aging treated at 190 ℃ for 9 h, except long stick shaped $Al_{15}(Mn, Fe, Cu)_3Si_2$, numerous nano-sized precipitated phases (Mg_2Si) with diameters of 50-200 nm dispersively distributed in the matrix. After heat treatment, the tensile strength of composite with SiC particle size of 7.5 μm enhance from 298 MPa to 341 MPa.

1. Introduction

Owing to their high specific strength and elastic modulus, high wear resistance and low CTE values, SiC particles reinforced aluminium matrix composites have received significant attention over the past few decades and are attractive as candidate materials for aerospace and automotive applications [1-5]. The SiCp/Al composites can be prepared by the general fabricating methods, such as powder metallurgy (PM), squeeze casting, pressural or pressureless infiltration and spray forming, etc[6]. Powder metallurgy is considered as a good technique in producing aluminium matrix composite, due to its lower processing temperature compared to melting techniques. Therefore, interaction between the reinforcement and matrix is prevented. Moreover, powder metallurgy technique has the ability to manufacture near net shape product with low cost [7-8].

It is commonly recognized that both the particle amount and particle size have obvious effect on the properties of the composites [9-10]. The voids coexisted with the clustered SiC particles and around the large sized SiC particles can be treated as pre-existed cracks. These particles cannot transfer any load from soft matrix to the hard reinforcements, and thus lead to the degraded mechanical properties of the composites. Hardness, compression strength and wear resistance of the composites can be improved by proper addition of the reinforcement. For a certain particle volume fraction, there is a close relationship between the particle size and the properties of composite.

As the load transfers from the matrix to the reinforcements during the deformation process, the interfacial microstructures between SiC and Al are critical to control the composite mechanical properties, it is accepted that good bonding interfaces with coherency or semi-coherency are favorable to the mechanical properties, whereas interfaces with incoherency, especially those with the presence of brittle intermetallic phases at the interfaces, degrade these properties [11-12]. Hence, interface microstructures are of great scientific and technological interest. It is important to characterize the interface microstructure in order to understand the interaction between reinforcement and matrix during processing and to establish the relationship between microstructure and mechanical properties.

For 6xxx aluminium alloy matrix composites, the solid solution strengthening and precipitation hardening treatment are imparted to this alloy increase its strength. Mg and Si are the major alloying elements in this alloy, form precipitates through T6 treatment, which increases the strength of this alloy [13]. Edwards et al. [14] reported that the precipitation sequence in Al 6061 alloy is as follows: Supersaturated solid solution → GP Zones → β" → β' →β (Mg_2Si) phase. In addition, several kinds of constituent phases have been found in 6061Al alloy [15]. These include $Al_{15}(Fe, Mn)_3Si_2$phase (BCC structure with a = 1.23 nm), $Al_{13}(V, Fe)_4Si$ phase (BCC structure with a = 1.25 nm), which were proved that cannot dissolve during the solid solution.

In this study, 35 vol.% different sizes of SiC particles reinforced 6061 aluminium alloy matrix composite were produced using the PM technique. Our paper will be organized as follows. Section 2 will give the experimental materials and process. Section 3.1-3.3 will give a detailed research on the effect of particle sizes on the SiCp distribution, physical and mechanical properties. The microstructure and combination mechanism of the interface between SiCp and Al are investigated by the means of TEM and HRTEM and presented in the section 3.4. Finally, section 3.5 will research the shape, size and phase composition of alloy phases in the composites hot pressed and treated by a proper T6 heat treatment (solution treated at 510 ℃ for 2 h+ aging treatment at 190 ℃ for 9 h).

2. Experimental procedures

2.1. Experiment materials

SiC particles (mean particle size 40 μm、 25 μm、 15 μm、 7.5 μm) and gas atomized 6061 aluminium powders with average particle size of 10 μm were used as reinforcement and matrix, respectively. The chemical composition of 6061 aluminium was listed in table 1.The SEM images of SiC particle (mean particle size 7.5 μm) and 6061 aluminium matrix powder are respectively shown in figure 1(a) and figure 1(c). We can see that the SiC particles are angular in shape and 6061Al powders are close to spherical in shape. XRD patterns of SiC particle in figure 1(b) show that the SiC has predominantly the 6Hα poly-type and also contain few cubic β-SiC and 15R-SiC. It also can be obtained from the figure 1(d) that only Al diffraction peak exist in the pattern of 6061Al, which indicates that the alloy elements, like Mg, Si, Fe, are all supersaturated in the Al matrix in the process of gas atomization due to the extremely rapid cooling rate.

Table 1. Chemical compositions of 6061 aluminium matrix alloy (wt. %).

Alloy	Cu	Mg	Si	Fe	Al
6061Al	0.25	1.0	0.6	0.3	Bal.

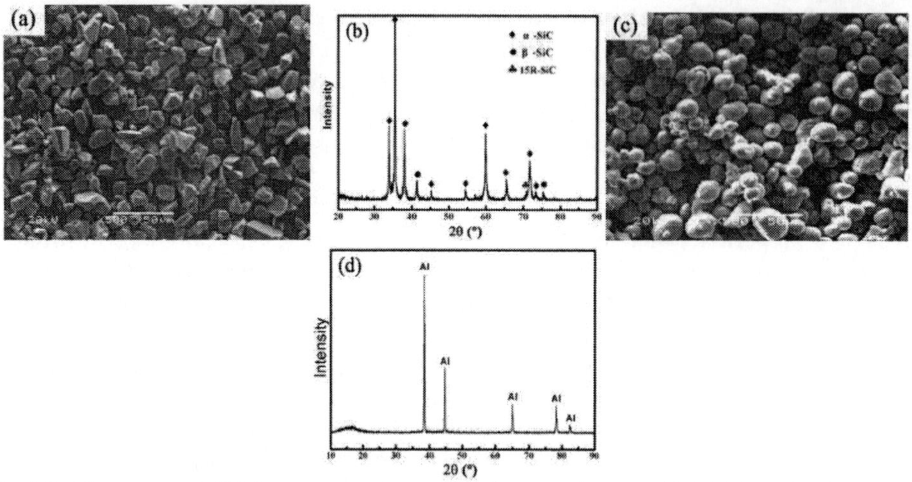

Figure 1. SEM images and XRD patterns of powders in experiment: (a) SiC powder image; (b) XRD pattern of SiC powder; (c) Al6061 alloy powder image; (d) XRD pattern of Al6061 alloy powder.

2.2. Experiment process

In this research, the aluminium matrix composite was manufactured by powder metallurgy technique. The powders (35 vol.% SiC particles) and balls with the ratio of 2:1 were blended in the Y style mixer for 24 h at the revolving speed of 50r/min and then the mixed powers were putted into the mould and hot-pressed up to 580 ℃ at 8 ℃/min in the VDBF-250 experiment machine with the vacuum pressure of 2.3×10^{-3} Pa. The stress of 80 MPa was applied in the powders for 3 h at 580 ℃ and then the powders were cooled in the furnace with the protection of vacuum. When the temperature drops to room temperature, the stress was removed and composite was acquired.

The tensile tests of composites were conducted using the Shimadzu AG-I250KN precision universal testing machine at a constant crosshead speed of 1 mm/min. The tensile property data for each condition was the average from the observations of four specimens. Figure 2 shows the sizes of tensile specimens. The CTE tests were carried out on a PCY-Ⅲdilatometer system from 25 to 100 ℃ with a heating rate of 5 ℃/ min. To eliminate system errors, the dilatometer was calibrated by measuring a standard alumina specimen under identical conditions. The metallographic samples were ground and polished following standard metallographic practices, then were etched using Keller's reagent. The microstructure was examined in scanning electron microscope (JSM-5610LV). The samples for TEM observations were machined into 0.5 mm by wire-electrode cutting and ground to 50 μm by Mechanical thinning, and then were cut into 3 mm diameter foils. After that, foils were prepared by argon ion milling using Gatan 691 Precision Ion Polishing System. Transmission electron microscopy (TEM) investigations were performed on JEM-2100 TEM microscope operated at 200 kV.

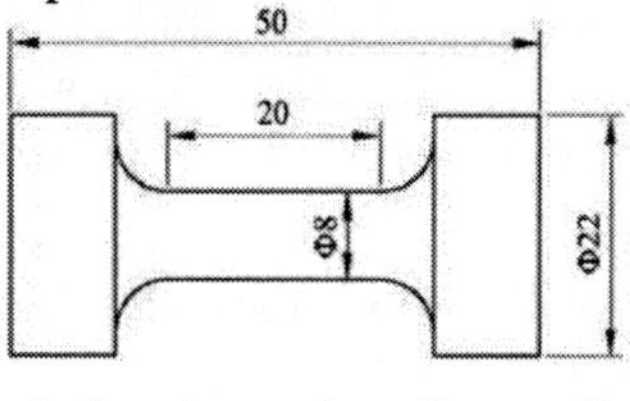

Figure 2. Specimen sizes for tensile test.

3. Experimental results

3.1. SEM microstructure of composite with different particle sizes

Figure 3 shows the SEM images of composites containing 35 vol.% SiC with different particle sizes of 40 μm, 25 μm, 15 μm, 7.5 μm. Grey regions imply Al matrix and dark grey and cornered particles imply the reinforcement SiC. It can be observed that the SiC particles with 40 μm are well dispersed and uniformly distributed in Al matrix, but the quantity of SiC in the composites become more and more as the decrease of SiC particles sizes and the distribution of SiC become more nonuniform. Slight SiC particle clustering can be observed in the composites with the particle size of 7.5 μm. It is well known that if the ratio of matrix/reinforcement particle sizes is much smaller, the reinforcements can cluster very easily in the matrix and result in an inhomogeneous distribution of reinforcements.

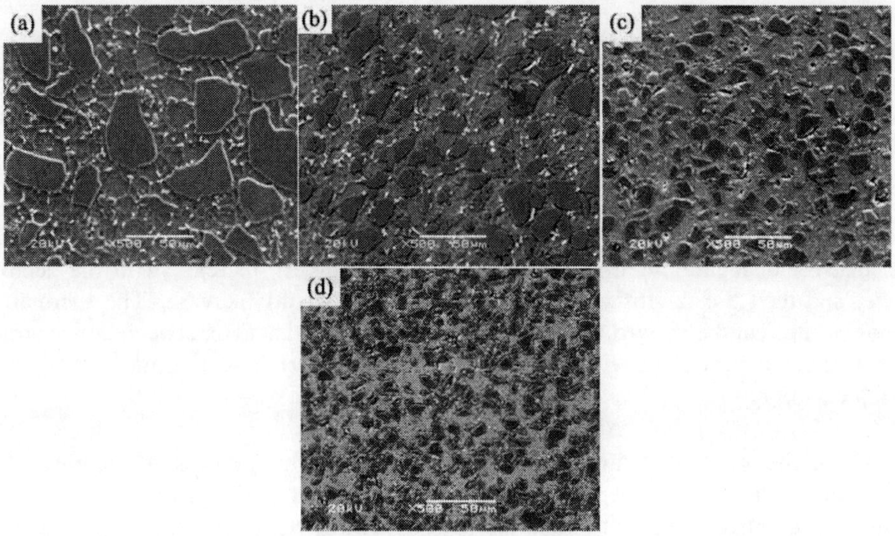

Figure 3. SEM images of SiC/6061Al composites with different SiC particle size: (a) 40 μm SiC; (b) 25 μm SiC; (c) 15 μm SiC; (d) 7.5 μm SiC.

3.2. Thermal expansion analysis of composite with different particle sizes

Figure 4 is the CTEs of SiCp/6061Al composite with four types of particle size measured between 25 ℃ and 100 ℃, it is seen that all the linear expansion coefficients of composites rise with the enhancement of temperature. Meanwhile, the decrease of SiC particle size can significantly reduce the CTE of composite. Furthermore, with the increase of temperature, composites with smaller particle size possess the smaller increase speed of linear coefficient expansion.

It can be easily seen from above that the particle size has a noticeable influence on the thermal expansion behavior. Because the big difference of CTEs between reinforcement and matrix (the CTE of SiC particle at room temperature is 4.7×10^{-6} ℃$^{-1}$, the CTE of pure Al at room temperature is 25.3×10^{-6} ℃$^{-1}$), the thermal residual stress will appear during the cooling process from the preparation to room temperature. When the value of thermal residual stress is higher than yield strength of matrix,

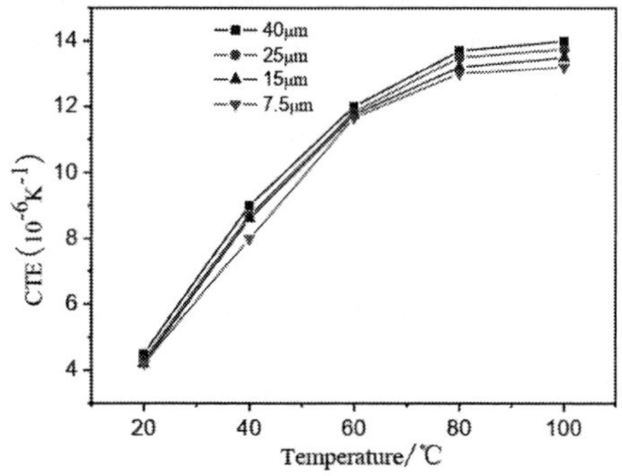

Figure 4. The CTEs of SiCp/6061Al with different particle sizes tested between 25 ℃ and 100 ℃.

it will relax in the form of matrix plastic flow and the particle dislocation density around the particle will thus increase. Arsenault and Shi [16] establish a dislocation density prediction model:

$$\rho = \frac{6f(\alpha_p - \alpha_M)\Delta T}{b(1-f)d} \quad (1)$$

Where b is the burgers vector of matrix, d and f is respectively the diameter and volume fraction of reinforcement, αp and αM is the CTE of reinforcement and matrix, respectively. We can obtain the inverse relationship between dislocation density and particle size from equation (1).

According to Taylor dislocation strengthening relations [17], the increase of dislocation density would cause the in-situ yield strength of matrix in the composite is greater than that of matrix without the reinforcement. The in-situ yield strength of matrix in the composite can be described as follows:

$$\sigma_{my} = \sigma_{m0} + \alpha\mu_m b\sqrt{\rho_{CTE}} \quad (2)$$

Where σ_{my} is the in-situ yield strength of matrix in the composite, σ_{m0} is the yield strength of pure matrix. α is a constant which is 1.25 for Al, μ_m is the shear modulus of matrix. From equation (2), we can obtain that the dislocation density of matrix would gradually increase with the decrease of the particle size, and then the in-situ yield strength of matrix would increase. The thermal expansion nature of composites can be viewed as the increases of average distance between lattice structure with temperature rise and the micro plastic relaxation in the matrix during heating process. So there will be the following equation:

$$\Delta L = \Delta L_\alpha + \Delta L_p \quad (3)$$

Where ΔL is the total elongation of composite during the process of heating, ΔL_α is the natural temperatures change volume, ΔL_p is the elongation caused by the plastic relaxation.

As we discussed above, the in-situ yield strength of matrix would increase as the decrease of the particle size, and thus the thermal mismatch stress is not easy to loose. As a result, the coefficient thermal expansion of composite decreases as the particle size decreases. When the particle is at a certain volume, the smaller the particle size is, the more the number of particle and the smaller the particle spacing are, and thus the expansion of the matrix can be effectively prevented under high temperature. So the smaller particle size can significantly reduce the increase rate of thermal expansion coefficient.

3.3. The mechanical properties analysis of composite with different particle sizes

The effect of the SiC particle size on the tensile strength of composites is shown in the figure 5. Considering this figure, by increasing the particle size of SiC from 7.5 μm to 40 μm, the tensile strength of composites decreased from 298 MPa to 212 MPa. Although the particle distribution of composites with the 40 μm SiC particle is the most uniformly, it is regretful that the tensile strength is the lowest. Apparently, the tensile strength of SiCp/6061Al composites has a close relationship with particle size.

Strengthening mechanisms of the metal matrix composites could be classified into direct and indirect mechanisms [10]. Direct strengthening is obtained by the load being transferred from the weaker matrix to the hard reinforcement during deformation, while indirect strengthening results from the variation of the matrix microstructures (such as Orowan strengthening mechanism, grain refinement, dislocation multiplication, etc.) in the composites by the addition of the reinforcements. In view of the above analysis, the influence mechanisms of particle size on the tensile strength of composites can be summed up as follows: Firstly, at a certain volume fraction of reinforcements, smaller size of SiC particle can bring a larger quantity of interface area, which can significantly enhance the capacity of transferring load from matrix to particle. Secondly, we have discussed in section 3.2 that composite with smaller particle size would possess a higher dislocation density, which can effectively strengthen the matrix; thereby the tensile strength is improved. Thirdly, the particle with larger size is more susceptible to contain defects. The cracked particles in the composites cannot transfer any load since the stress is released along the cracking surfaces. Furthermore, the cracked particles will act as the micro-crack initiators during deformation, leading to the decrease in the strength of the composites.

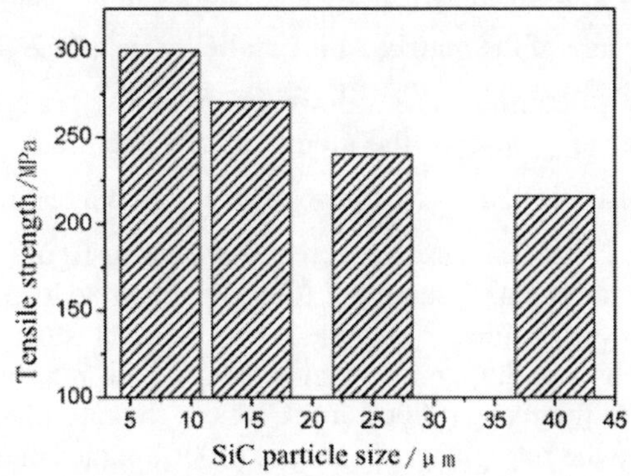

Figure 5. The tensile strength of SiCp/6061Al composites with four types of particle sizes.

3.4. Microstructure of the interface between SiC and Al

3.4.1. TEM characterization of SiC/Al interface. Figure 6(a) shows the TEM images of SiC/Al interfaces of the composites. A typical SiC/Al interface has been obtained. We can see from the figure that the interface is clean and smooth; there are no interface reactants and phenomenon of SiC particles dissolved. Previous studies [18] have shown that SiC particles are easy to react with Al to generate Al_4C_3 in the SiC particles reinforced aluminium alloy matrix composites. But we didn't find Al_4C_3 reactants in the composite materials in this work. It is mainly because the hot-press sintering technology was used in this experiment, the temperature throughout of which is too low to produce interface chemical reaction. So the interface of the composite prepared is clean, and the combining is good. It means that the external load can transfer from matrix to reinforcements and provides the prerequisites for the composite material with excellent performance. Moreover, figure 6(a) also displays the feature of high-density dislocations in the Al matrix adjacent to a very pointed SiC particle. The difference in the thermal expansion coefficients between the matrix and SiC particles necessitates the generation of dislocations in the matrix vicinal to SiC particles to accommodate thermal strain when change temperature. Figure 6(b) and figure 6(c) are the diffraction pattern of 6H-SiC in the zoneaxis $[1\bar{2}10]$ and the diffraction pattern of Al in the zoneaxis $[\bar{1}12]$.

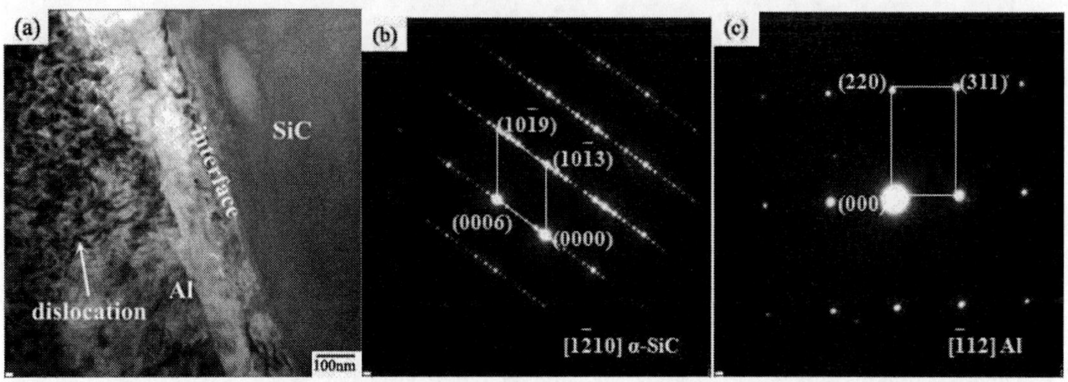

Figure 6. TEM images of the interface between SiC and Al: (a)the interface morphology of SiC/Al composites; (b) the diffraction pattern of SiC; (c) the diffraction pattern of Al.

3.4.2. HRTEM characterization of SiC/Al interface. Figure 7(a) shows an initial high resolution image of the interface between the reinforcement particle and the matrix. Figure 7(b) and (c) are corresponding SAED and indexed patterns. In figure 7(c), the circles represent the reflections from the matrix and the dots represent those from SiC particles. The incident beam is parallel with $[1\bar{2}1]$ Al and $[1\bar{2}10]$ 6Hα-SiC, the (111) plane of the matrix Al is parallel to the plane (0006) of the SiC phase. Thus, their orientation relationship is determined as follows: $[1\bar{2}10]$ SiC || $[1\bar{2}1]$ Al, (0006) SiC || (111) Al.

We can intuitively and clearly observe the atomic arrangement structure by the Inverse Discrete Fourier Transform (IFFT) of initial HRTEM image. The IFFT images of 6H SiC phase along $[1\bar{2}10]$ zone axis and Al along $[1\bar{2}1]$ zone axis are respectively shown in figure 7(d) and figure 7(e). We can see from figure 7(d) that the interplanar spacing of (0001) plane in 6Hα-SiC phase is 1.51 nm and each (0001) plane consist of six layers. Figure 7(f) is the IFFT image of square area 3 in figure 7(a) and we can intuitively and clearly observe the atomic arrangement of the interface between SiC and Al. The (111) plane of Al is parallel to the plane (0006) of SiC phase, the interplanar spacing of (111) planes in the matrix is 0.234 nm, while the interplanar spacing of (0006) planes in α-SiC phase is 0.251 nm. The lattice misfit between them is 0.07 and it is obviously that the combination mechanism of SiC and Al is the formation of a semi-coherent interface by closely matching of atoms. This semi-coherent interface can effectively transfer the load from matrix to SiC ceramic particles and provides the prerequisites for the composite material with excellent performance.

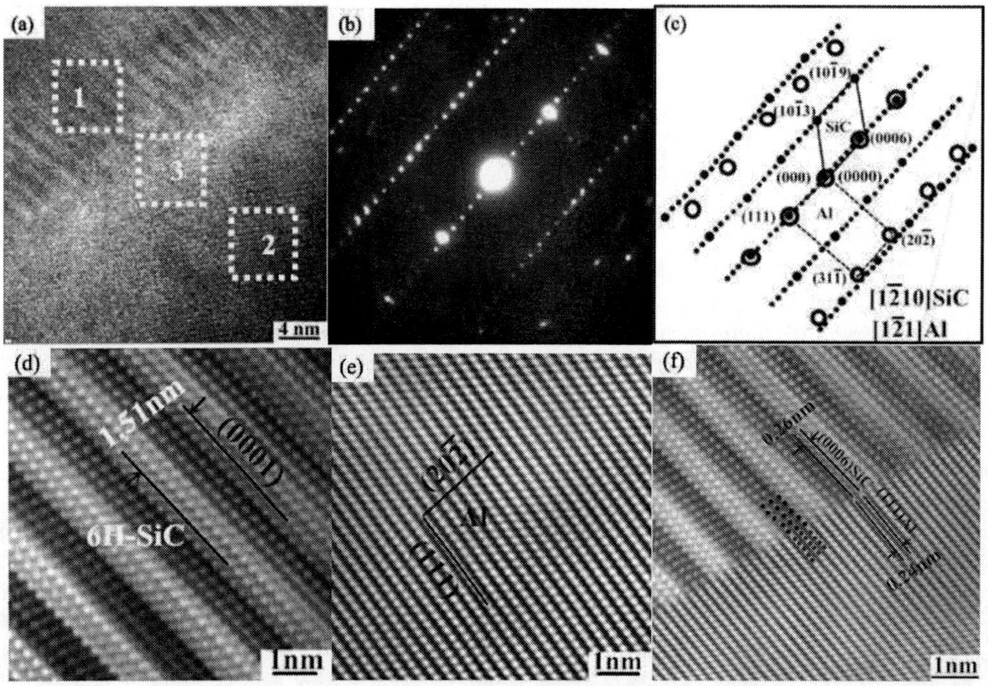

Figure 7. HREM images, corresponding SAED and IFFT images of the interface between SiC and Al: (a) The initial HREM image of interface between SiC and Al; (b) Corresponding SAED patterns of interface; (c) Indexed patterns; (d) IFFT image of square area 1 in Figure 6(a); (e) IFFT image of square area 2 in Figure 6(a); (f) IFFT image of square area 3 in Figure 6(a).

3.5. The evolution of alloy phases in the composite during heat treatment

3.5.1. Characterization of alloy phases in hot-pressed composites. When composites were prepared by vacuum hot pressing, the microstructure of alloy phases are especially complex by the reason that different kinds of alloying elements contained in the aluminium matrix can form various kinds of coarse precipitated phases at the coupled action of heat and stress. Considering the important role alloy phases played in the strengthening of composites, it is of practical importance to understand the microstructure of alloy phases in the hot-pressed and the evolution in the process of heat treatment. It can be seen from figure 8(a) that there is a considerable volume fraction of micron coarse intermetallic phases in the hot-pressed composite. The high brightness of these particles is due to the higher atomic number of Mg and Cu in comparison with that of Al, Si and C.

In order to clearly understand the morphological characteristics and phase composition of coarse intermetallic phases in the hot-pressed composites, TEM analysis was conducted, the TEM morphologies and corresponding diffraction patterns of coarse intermetallic particles were shown in figure 8(b)-(f). It can be found that two kinds of coarse intermetallic particle existed in the hot-pressed composites. The shape of coarse intermetallic particle in figure 8(b) is less regular and the size is about 3-5μm. Diffraction patterns from the coarse particle (see figure 8c) was indexed to be consistent with Mg_2Si(Space Group: Fm-3m, lattice parameters: a= 6.350 Å), which is considered the main strengthening phase in the 6000 series of Al alloys.

As shown in the figure 8(d), a spinel phase in the form of 100-300 nm size is present in the matrix, corresponding EDS, chemical composition and indexed patterns are presented in the figure 8(e) and figure 8(f). The EDS spectrum and chemical composition in figure 8(e) of the spinel phase reveal that they are Al-, Fe-, Mn-, Cu- and Si-rich with an atomic ration of Al:(Fe, Mn, Cu):Si≈15.1:3.26:2. By ignoring the experimental uncertainty, the phase can be written as $Al_{15}(Mn, Fe, Cu)_3Si_2$, which was confirmed to have a BCC structure with lattice parameter a=1.28 nm. This result was consisted with the research of Wang and Cheng [19-20].

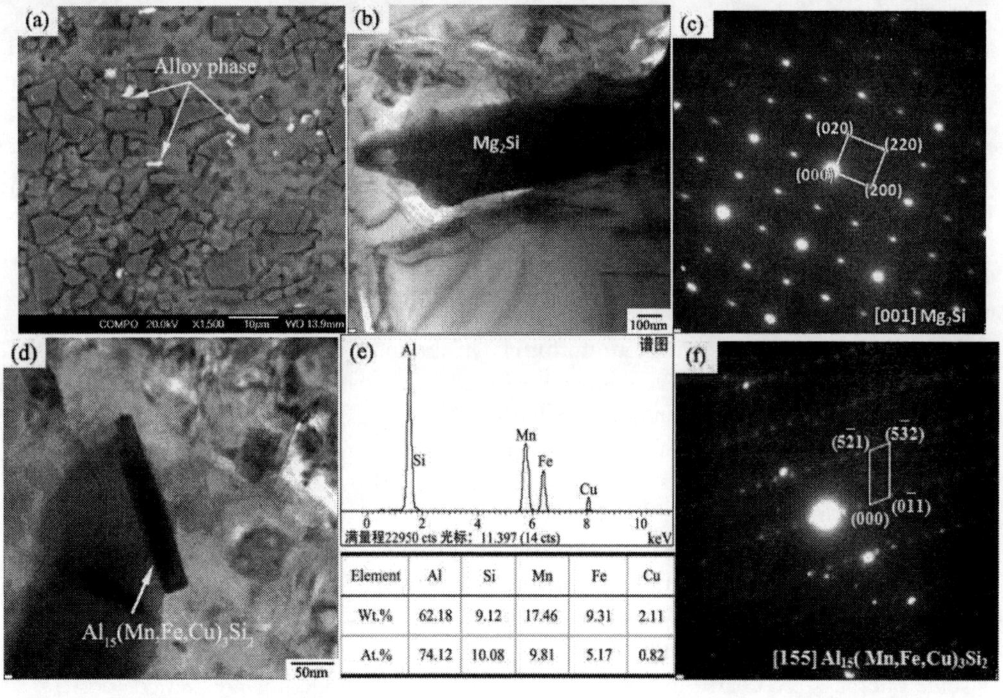

Figure 8. SEM and TEM analysis of alloy phases in hot-pressed SiCp/6061Al composites:
(a) SEM image of SiC/Al composites; (b) TEM image of coarse phase Mg_2Si; (c) Diffraction pattern of Mg_2Si; (d) TEM image of long stick phase; (e) EDS of long stick phase; (f) Diffraction pattern of long stick phase.

3.5.2. Characterization of alloy phases in composites after T6 heat treatment. When the composites were treated by a proper T6 heat treatment (solution treated at 510 ℃ for 2 h+ aging treatment at 190 ℃ for 9 h), the TEM images of alloy phases is shown in the figure 8. It can be seen from the figure 9(a) that a spinel phase about 200 nm appeared in the matrix, which is analogous to the phase shown in the figure 8(d). Further calibration of diffraction pattern shown in figure 9(b) confirmed the phase is

$Al_{15}(Mn, Fe, Cu)_3Si_2$. The shape and composition of phase in figure 9(a) are consonant with the result shown in figure 8(d). It can be obtained that the $Al_{15}(Mn, Fe, Cu)_3Si_2$ is mainly formed during the hot pressed sintering process and cannot dissolve in the solution process, due to the fairly low solubility of Mn and Fe elements in Al matrix. Figure 9(c) shows a TEM image of numerous nano-sized β'-Mg_2Si precipitated phases with diameters of 50~200 nm homogeneously distributed in the matrix.

As discussed above, the alloy phases in the hot-pressed composites are mainly micron coarse intermetallic particles, while in the aged composites are nano-sized precipitated phases. Compared with the coarse micron intermetallic phases in the hot-pressed composites, the precipitated phases in the aged composites are smaller in size and more in the quantity. Because the dislocation can be more effectively block by numerous nano-sized precipitated particles than slight micron coarse intermetallic particles, heat treatment is an effective way to enhance the mechanical properties of SiCp/6061Al composites. The strengthening effect is testified in this experiment that the tensile strength of composite with 7.5 μm particle size enhanced from 298 MPa to 341 MPa after heat treatment.

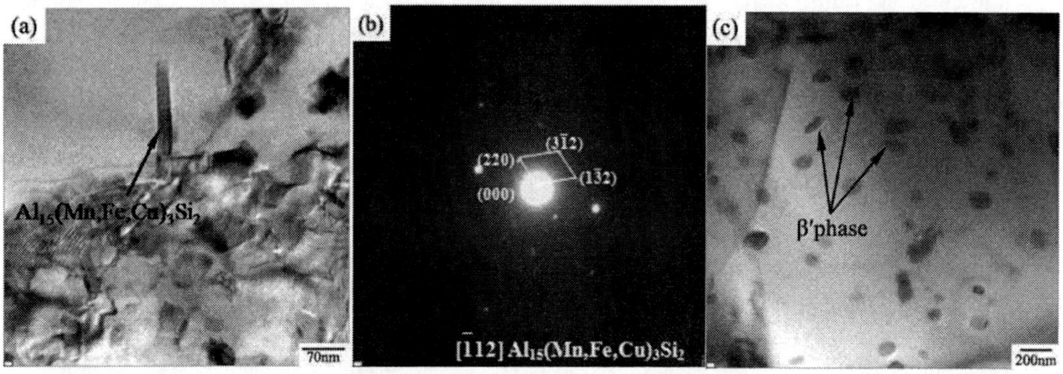

Figure 9. TEM analysis of alloy phases in the SiCp/6061Al composites after T6 heat treatment: (a) TEM image of long stick phase; (b) Diffraction pattern of long stick phase; (c) TEM image of dispersed phase.

4. Conclusions

Through this study, following conclusion could be achieved:

(1) The SiC particles distribute more uniformly in the matrix by increasing the particle size, but there is an inverse relation between tensile strength and particle size. The tensile strength of composites with 7.5 μm exhibited highest tensile strength 298 MPa.

(2) All the linear expansion coefficients of composites rise with the enhancement of temperature. Meanwhile, the decrease of SiC particle size can significantly reduce the expansion coefficients of composites. Furthermore, with the increase of temperature, composites with smaller particle size possess the lower increase speed of linear expansion coefficients.

(3) The SiCp/Al interface in this experiment is clean and smooth, and SiC particles are well combined with Al matrix. The combination mechanism of SiC and Al is the formation of a half coherent interface by closely matching of atoms.

(4) Some micron-sized coarse intermetallic particles existed in the hot-pressed composites, such as random-shaped Mg_2Si, long stick shaped $Al_{15}(Mn, Fe, Cu)_3Si_2$. After heat treatment, except long stick shaped $Al_{15}(Mn, Fe, Cu)_3Si_2$, numerous nano-sized precipitated phases (Mg_2Si) homogeneously distributed in the matrix. After heat treatment, the tensile strength of composite with SiC particle size of 7.5 μm enhance from 298 MPa to 341 MPa.

Acknowledgments

This research was financially supported by the National Natural Science Foundation of China (No. 51371077)

IOP Conf. Series: Materials Science and Engineering **292** (2017) 012055 doi:10.1088/1757-899X/292/1/012055

References

[1] H M Zakaria 2014 Microstructural and corrosion behavior of Al/SiC metal matrix composites *Ain Shams Engineering Journal* **5** 831-8

[2] X Du, T Gao, Y Y Wu and X F Liu 2014 Effect of copper addition on the in-situ synthesis of SiC particles in Al-Si-C-Cu system *Materials and Design* **63** 194-9

[3] M Hossein-Zadeh, O Mirzaee and P Saidi 2014 Structural and mechanical characterization of Al- based composite reinforced with heat treated Al_2O_3 particles *Materials and Design* **54** 245-50

[4] L Zhou, Y Wang, Z Y Ma and X L Yu 2014 Finite element and experimental studies of the formation mechanism of edge defects during machining of SiCp/Al composites *International Journal of Machine Tools & Manufacture* **84** 9-16

[5] D Mandal and S Viswanathan 2013 Effect of heat treatment on microstructure and interface of SiC particle reinforced 2124 Al matrix composite *Mater Characterization* **85** 73-81

[6] M S Arab, N EI Mahallawy, F Shehata and M A Agwa 2014 Refining SiCp in reinforced Al-SiC composites using equal-channel angular pressing *Materials and Design* **64** 280-6

[7] H L Wang, R Zhang, X Hu, C A Wang and Y Huang 2008 Characterization of a powder metallurgy SiC/Cu-Al composite *Journal of Materials Technology* **197** 43-8

[8] A Hassani, E Bagherpour and F Qods 2014 Influence of pores on workability of porous Al/SiC composites fabricated through powder metallurgy + mechanical alloying *Journal of Alloys and Compounds* **591** 132-42.

[9] O EI-Kady and A Fathy 2014 Effect of SiC particle size on the physical and mechanical properties of extruded Al matrix nanocomposites *Materials and Design* **54** 348-53

[10] Z W Wang, M Song, C Sun and Y H He 2011 Effect of particle size and distribution on the mechanical properties of SiC reinforced Al-Cu alloy composites *Materials Science and Engineering A* **528** 1131-7

[11] Z P Luo 2006 Crystallography of SiC/$MgAl_2O_4$/Al interfaces in a pre-oxidized SiC reinforced SiC/Al composite *Acta Materialia* **54** 47-58

[12] C Z Nie, J J Gu, J L Liu and D Zhang 2008 Investigation on microstructures and interface character of B_4C particles reinforced 2024Al matrix composites fabricated by mechanical alloying *Journal of Alloys and Compounds* **454** 118-22

[13] M Vaseghi, A K Taheri,S I Hong and H S Kim 2010 Dynamic ageing and mechanical response of Al-Mg-Si alloy through equal channel angular pressing *Materials and Design* **31** 4076-82.

[14] G A Edwards, K Stiller, G Dunlop and M J Couper 1998 The precipitation sequence in Al-Mg-Si alloys *Acta materialia* **46** 3893-904

[15] P Ravindran, K Manisekar and P Rathika 2013 Tribological properties of powder metallurgy – Processed aluminium self lubricating hybrid composites with SiC additions *Materials and Design* **45** 561-70

[16] P Jin, B L Xiao, Q Z Wang, Z Y Ma, Y Liu and S Li 2011 Effect of solution temperature on aging behavior and properties of SiCp/Al-Cu-Mg composites *Mater Sci Eng A* **528** 1504-11

[17] Y W Yan 2007 Effect of particle size on properties of SiCp/Al composites and its numerical simulation *Harbin: Harbin Institute of Technology*

[18] X D Sui, C P Luo and L Z Ouyang 2000 The interface behavior of SiC in SiCp/ZL109 composites *Acta Materiae Compositae Sinica* **01** 65-70

[19] R M Wang, M K Surappa, C H Tao, C Z Li and M G Yan 1998 Microstructure and interface structure studies of SiCp-reinforced Al (6061) metal-matrix composites *Meterials Science and Engineer A* **254** 219-26

[20] N P Cheng, S M Zeng and Z Y Liu 2008 Preparation microstructures and deformation behavior of SiCp/6066Al composites produced by PM route *Journal of materials processing technology* **202** 27-40

IOP Conf. Series: Materials Science and Engineering **292** (2017) 012056 doi:10.1088/1757-899X/292/1/012056

Research on connection structure of aluminumbody bus using multi-objective topology optimization

Q Peng[1, 4], X Ni[1], F Han[1], K Rhaman[1], C Ulianov[2] and X Fang[3]

[1]School of Mechanical and Automotive Engineering, Xiamen University of Technology, Xiamen, China
[2]NewRail Centre for Railway Research, Newcastle University, Newcastle upon Tyne, United Kingdom
[3]Department of Mechanical Engineering, Xiamen Technician College, Xiamen, China

[4]E-mail: pengqian603@163.com

Abstract. For connecting Aluminum Alloy bus body aluminum components often occur the problem of failure, a new aluminum alloy connection structure is designed based on multi-objective topology optimization method. Determining the shape of the outer contour of the connection structure with topography optimization, establishing a topology optimization model of connections based on SIMP density interpolation method, going on multi-objective topology optimization, and improving the design of the connecting piece according to the optimization results. The results show that the quality of the aluminum alloy connector after topology optimization is reduced by 18%, and the first six natural frequencies are improved and the strength performance and stiffness performance are obviously improved.

1. Introduction and background

Aluminum body is an ideal method to reduce weight in the development of a new energy bus. Owing to its low density and good mechanical properties, aluminum alloy could gradually replace steel as a dominant material of vehicles; however, due to the limited elastic modulus and poor welding characteristics, the reliability of aluminum body has been a great concern for the vehicle manufacturers.

In order to improve the reliability of aluminum structure, some replaceable connection methods were proposed and applied in the industry including riveted, bolted, bonded and composite connections [1]. Current aluminum body bus is mainly composed of extrusion parts and connected with mechanical joints, the strength, stiffness and longevity could hence strongly be related to the robustness of joint. Several academics and engineers proposed several types of joints to connect the extrusion components, Jia [2] invented a kind of frame connector structure, medium sized jointis used in the position of the bus frame below the window and the side door. Zheng [3] provided a kind of riveted structure of aluminum body. The shearing force of rivet is reduced by setting groove, so as to improve the service life and stability of rivets. Luo [4] released a skeleton connecting piece of aluminum body bus with simple process, high installation efficiency and good connection and fixation effect. However, the comprehensive performance of the joints with practical conditions of aluminum body bus need to be further investigated.

2. Multi objective topology optimization numerical simulation

The SIMP method based on isotropic materials, and no need to introduce microstructures and additional homogenization processes. The relative density of discrete element in the finite element model design space as the design variables, and assuming that the material parameter is constant inside the unit [5]. The material property of discrete element varies with the relative density of unit, and the relation form is shown in formula (1):

$$\begin{cases} E = (a_e)^{\Theta} E_0 \\ K = (a_e)^{\Theta} K_0 \\ M = a_e M_0 \end{cases} \tag{1}$$

In the formula, a_e is relative density of discrete element. E、E_0 is elastic modulus before and after unit optimization. K、$K_0$is the stiffness matrix before and after optimization. M、M_0 is the mass matrix before and after optimization. Θ is penalty coefficient.

2.1. objective function

The structural stiffness and the first 3 order low order natural frequencies of aluminum alloy structures under different working conditions are comprehensively considered. The multi-objective optimization problem is transformed into a single objective optimization problem by the compromise programming method.Then in order to transform the objective function into the minimum value problem, Υ(σ) = -Λ(σ).Finally, the optimization model which satisfies both the stiffness and the natural frequency optimization objective is obtained:

$$\begin{cases} \min P(\sigma) = \left[\psi^2 (\frac{C(\sigma) - C^{\min}}{C^{\max} - C^{\min}})^2 + (1-\psi^2)(\frac{\Upsilon(\sigma) - \Upsilon^{\min}}{\Upsilon^{\max} - \Upsilon^{\min}})^2 \right]^{\frac{1}{2}} \\ s.t. V(\sigma) - \bar{V} \le 0 \\ 0 < \sigma_j < 1 \\ j = 1, \dots, l \end{cases} \tag{2}$$

In the formula, $P(\sigma)$ is comprehensive objective function. $\Upsilon^{\max}$ 、 $\Upsilon^{\min}$ is the maximum and minimum value of $\Upsilon(\sigma)$. The meaning of other parameters is consistent with the previous definition.

2.2. Sub objective weight

The analytic hierarchy process combines qualitative and quantitative analysis, and solves the multi-objective problem based on strict mathematical theory [6, 7]. According to the problem that the designer wants to evaluate, the decision hierarchy is established. Then, the weights of the factors are solved by the method of paired comparison, and the importance degree between the working conditions is compared. Then the decision matrix is constructed. The weight value of 7 sub objectives is obtained.

$$\Psi = [0.27 \quad 0.13 \quad 0.14 \quad 0.20 \quad 0.09 \quad 0.09 \quad 0.08] \tag{3}$$

3. Multi objective optimization design of connectors

3.1. Connection structure and load analysis

In the initial stage of design, the connection structure of height and length is 110mm, width 35.2mm, thickness of 5mm. in the two corners transverse and longitudinal connecting plates respectively arranged on two bolt holes, two holes are 60mm, 25mm from the edge. In the two transverse and longitudinal connecting plates of the corner respectively arranged on two bolt holes. The two bolt holes are 60mm, 25mm from the edge. The material is Al6082, the density is 2.7g/cm3, the elastic modulus is 74Gpa, the Poisson's ratio is 0.3, and the yield strength is 260Mpa.

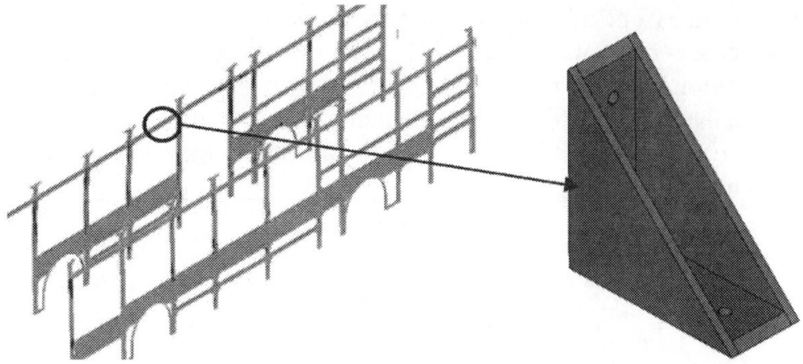

Figure 1. TheOriginal connect model.

The connection in the upper left corner of the first window after the middle door is located as the research object. The force condition is very complicated, and may be affected by several forces at the same time. Such as passenger, body, overhead air conditioning weight and vibration caused by road roughness, engine work, etc. It is difficult to accurately determine the load magnitude due to the influence of pavement condition, vehicle structural parameters and operating conditions. According to the deformation and force condition of the real vehicle and the reference [8], the load condition of the connector is determined. As shown in table 1, the load condition of the connection structure is shown.

Table 1. Finite element model load conditions.

Serial number	Load condition	Working condition description
1	Load condition 1	X force unit 1000N is applied to the model
2	Load condition 2	Y force unit 1000N is applied to the model
3	Load condition 3	Z torque 1×10^6N·m is applied to the model
4	Load condition 4	Y torque 1×10^6N·m is applied to the model

As shown in figure 2, the load constraint diagram of working condition 1 is shown. The remaining three operating conditions are the same as the condition 1, but the load directions is different.

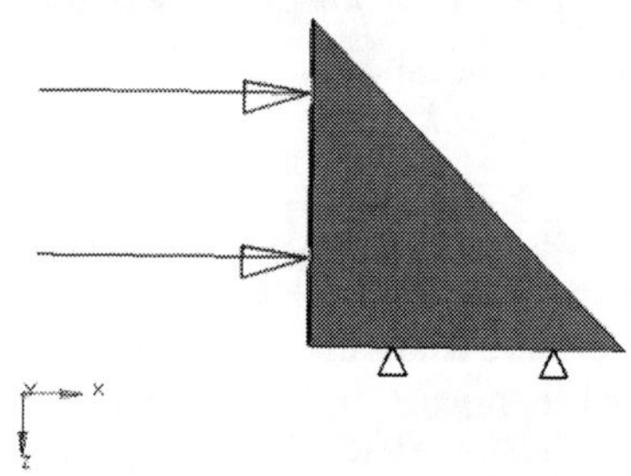

Figure 2. The load and constraint of condition 1.

3.2. Initial design of outer contour

The design of connector structure initial contour shape, is conducive to the follow-up of topology optimization, get connected topology optimum distribution. So the topography optimization identified contour connection is very necessary.

Using HyperMesh software, the compliance values and the first three order natural frequencies of the connectors under four conditions are normalized and weighted. Then the shape of the connection structure is optimized, and the position of the bolt hole is kept unchanged. The optimization results are shown in figure 3. The initial structure of the outer contour of the connector is designed as shown in figure 4. The length and height of the connecting structure are reduced by 10.5mm respectively, and the large blue redundant material in the original connection structure is removed.

Figure 3. Topography optimization density cloud map.

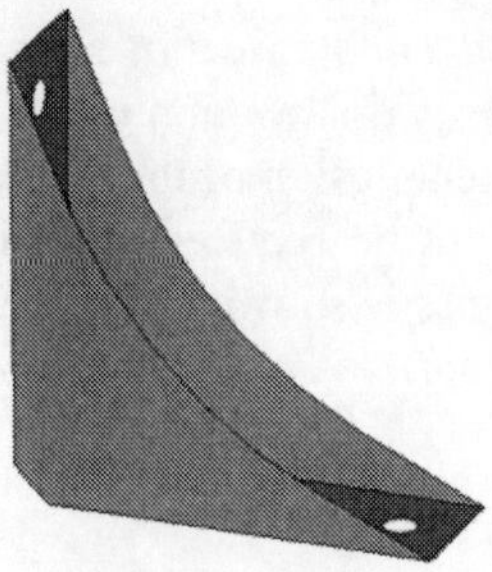

Figure 4. The Outer contourdesign.

3.3. Multi objective topology optimization

Before the topology optimization of the connector is made, the connection before connecting structure is divided into the design of regional and non regional design. non regional design to retain the original model, using shell elements are dispersed, and the design area is filled with solid unit. The optimum design space is shown in figure 5. The hollow area of the connecting piece is used as the topological design space, and the two plate like triangles and the horizontal and vertical connecting plates between the two triangular plates are used as the non optimized design areas.

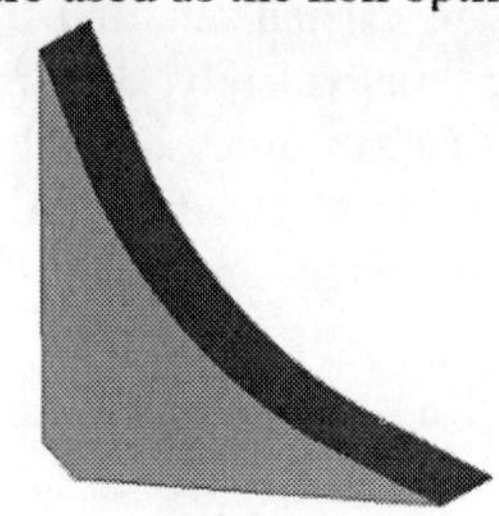

Figure 5. Topology optimization design space.

After defining the multi-objective topology optimization problem, the objective function is calculated, and the results are convergent after 32 iterations. As shown in figure 6, the density profile of the material after the optimization of the aluminum alloy connection structure is shown. It can be seen from the chart, the blue area topological space mainly concentrates in the middle of the connector and at both ends, the two parts should be removed, A area and B area for the Red areas show that the two areas need to consider strengthening member connections on the basis of the original.

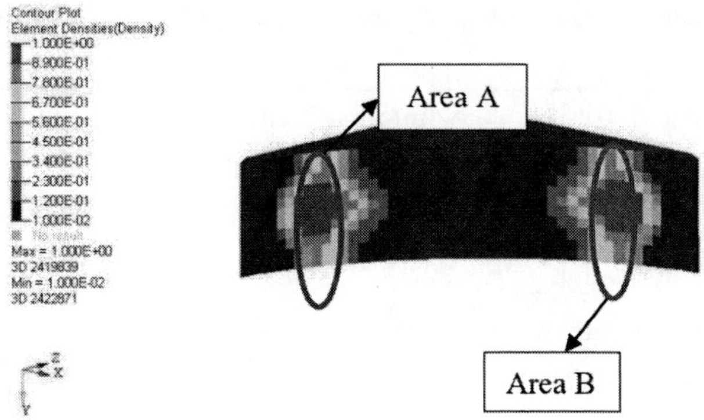

Figure 6. Topology optimization density cloud map.

3.4. Improvement and analysis of connection structure

The material density distribution nephogram of multi objective topology optimization is the reference for the improvement design of the connection structure. In order to ensure the strength of the improved connection will not be reduced and consider the actual structure of the connection. The improved connection structure is shown in figure 7, based on the original connections in the A area and B area were increased floor.

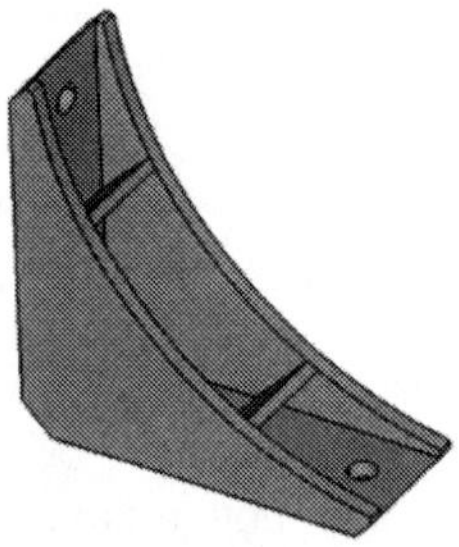

Figure 7. Connection improvement structure.

The quality of the improved connection structure is 0.175kg, which is reduced by 18% compared with the original structure. The inertia release analysis and free modal analysis of the improved aluminum alloy connecting structure are carried out, and the results are shown in table 2 and 3. In order to show the dynamic performance more clearly, the first six order free modal contrast is listed in table 2. According to the table, the first six order natural frequencies of the improved connection structure are slightly increased in addition to the fifth order, and the other five orders increase in varying degrees, and the dynamic characteristics are obviously improved.

Table 2. Comparison of natural frequencies before and after the improvement of Connect.

	1	2	3	4	5	6
Before	617.5	859.7	1008.1	1189.4	1299.2	1426.3
After	637.7	900.6	1137.7	1206.9	1263.7	1593.9

As shown in table 3, the deformation and stress of the improved connection structure under working conditions 1、2、3 are obviously reduced. The stress of the working condition 2 decreases, but the deformation of the working condition increases to some extent. But in general, the stress and deformation are more uniform under different conditions. The maximum stress is reduced from 189.4Mpa to 130.5Mpa. The maximum deformation decreased from 0.36mm to 0.22mm. The stiffness and strength properties are improved obviously after topology optimization.

Table 3. Comparison of the maximum stress before and after the improvement of Connect.

		Before	**After**	**Amplitude of change /%**
1	Maximum deformation /mm	0.39	0.3	-23.1
	Maximum stress /Mpa	159.8	113.01	-29.3
2	Maximum deformation /mm	0.24	0.27	12.5
	Maximum stress /Mpa	129.77	84.48	-34.9
3	Maximum deformation /mm	0.35	0.23	-34.3
	Maximum stress /Mpa	140.4	102.5	-27
4	Maximum deformation /mm	-0.49	-0.38	-22.4
	Maximum stress /Mpa	189.4	130.5	-31.1

4. The strength and fatigue performance of passengercar

The static strength performance of the car body structure has a direct impact on the fatigue life, crashworthiness and vibration characteristics of the aluminum alloy bus. According to the analysis results to verify the design is reasonable. In this paper, the stress distribution characteristics of aluminum alloy body are analyzed based on four typical conditions of bending, torsion, breaking and turning.

As shown in figure 8 the stress distribution of the aluminum alloy bus structure based on the bending condition is presented. From the diagram, the stress level of the front and rear sides and the side of the car body is lower. The maximum stress of the under frame structure is 169.02MPa, which is located at the main longitudinal beam at the center.The maximum stress value of the body structure is 123.91Mpa, located in the upper side of the connecting block. This result is less than the allowable stress of the material, indicating that the strength of the bus under bending condition is enough.

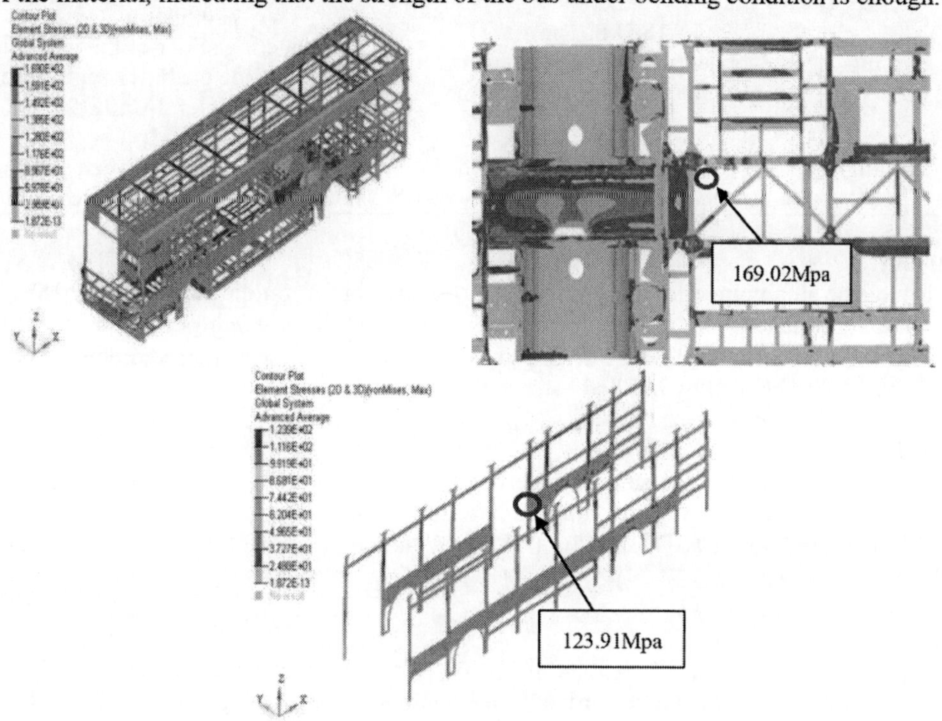

Figure 8. Calculation results of bending conditions.

The maximum stress calculated based on the four typical working conditions is shown in table 4. According to the design requirements of the body strength, the maximum stress of the structure under the limit condition should be less than the allowable stress of the material. From the table, it can be concluded that the topological optimization of the connector can meet the strength requirements of aluminum alloy body structure.

Table 4. Maximum stress distribution of skeleton structure under four typical working conditions.

working condition	Maximum stress and position of underframe	Maximum stress and position of bus body
Bend	**169.02Mpa** Central main longitudinal beam	**123.91Mpa** The connector on the upper middle door
Torsion	**253.82Mpa** Front suspension lower thrust rod	**149.68Mpa** The connector on the upper middle door
braking	**185.08Mpa** Connection between front and middle section of underframe	**150.6Mpa** Connecting slider of the upper skeleton of the posterior segment of the lateral wall
Turning	**238.75Mpa** Front suspension thrust rod support	**126.38Mpa** The connection between side longitudinal beam and side beam

5. Summary

The quality of the aluminum alloy connector after topology optimization is reduced by 18%, and the first six natural frequencies are improved to some extent, and the strength performance and stiffness performance are obviously improved.The improved connector can meet the requirements of strength and reliability of aluminum alloy bus body structure.

Acknowledgments

Authors wish to acknowledge the support received from 2011 Program of Bus and Specialty Vehicle Research& Design Collaborative Innovation Centre in Fujian Province (2016AYF004), the National High End Expert programme (GDT20153600065) and National Natural Science Foundation of China (51505403).

Reference

[1] Yongbing L, Yunwu M, Ming L et al. 2016 Advances in Welding and Joining Processes of Multi-material Lightweight Car Body *Journal of Mechanical Engineering* **52(24)** 1-23
[2] Jia B C 2010 A kind of frame connection structure[P] CN CN201020231168.0
[3] Zheng Y P, Lin X, Zhang D G 2016 A kind of riveted structure applied in Aluminum Space Frame[P] CN CN201610333157.5
[4] Luo Z H 2017 Aluminum bus body frame[P] CN CN201621112866
[5] Zhu X Y, Fang Z D, Shen S S, et al. 2011 Multi-objective Topology Optimization for the Control Arm of Vehicle Suspension[J] *Automotive Engineering* **33(02)** 138-141
[6] Yonetani H 2008 Laser-MIG hybrid welding to aluminum alloy carbody shell for railway vehicles *Weld. Int* **46(2)** 43-47
[7] Wu SC, Hu YN, Duan H, Yu C and Jiao HS 2016 On the fatigue performance of laser hybrid welded high Zn 7000 alloys for next generation railway components *Int. J. Fatigue.* **91** 1-10
[8] Hao S H, Hu R, Xu M L et al. 2016 Optimization Design of Roof Frame for Hybrid Aluminium Bus[J] *Automobile Technology* **2016(6)** 11-15

IOP Conf. Series: Materials Science and Engineering **292** (2017) 012057 doi:10.1088/1757-899X/292/1/012057

Infiltration sintering properties of Ni-4B-4Si(wt.%) alloy powders

Q Yang[1], X C Zhang, F L Wang and J T Zou

Faculty of Materials Science and Engineering, Xi' an University of Technology, Xi' an 710048, China

[1]E-mail: yangqing@xaut.edu.cn

Abstract. The Ni-4B-4Si(wt.%) alloy powders were infiltrated into the nickel skeletons, the effects of sintering temperatures (1050-1150 ° C) and skeletons (loose and compact nickel powders) on the microstructures and hardness of the sintered alloys were investigated. The Ni-B-Si alloy sintered at 1100 ° C consisted of γ -Ni and Ni_3B, and Si mainly solid soluted in the γ -Ni. The loose nickel powders favored to the infiltration of Ni-B-Si liquid alloy into the nickel skeletons, the sintered alloys exhibited dense microstructures and good interfacial bonding with Ni substrates. The interfacial hardness was equal to that of the sintered alloys and Ni substrates. Loose nickel powders ensured the density and interfacial bonding of the sintered alloys, the infiltration sintering process can be simplified and easily applied to practice.

1. Introduction

Nickel-based superalloys are widely used in modern aero engines, gas turbines and the hottest components of thermal power plants, such as working blades, turbine discs, combustion chambers, etc. [1]. After long-term use, the corrosion pits, cracks and other defects will appear on the hot components. The high cost of replacing these components promoted people to develop repair methods to extend the life of blades [2-4].

In 1990s, Liburdi proposed a powder metallurgy repair process called Liburdi Powder Metallurgy (LPM), the blades could be repaired through the infiltration of low-melting-point liquid alloy into high-melting-point porous skeleton [2]. Although various welding processes such as laser cladding techniques have been applied for blade repair [3,4], the LPM repair process can repair larger gaps with stable properties. If the alloy powders have the same composition as the matrix, high sintering temperature is required, which will degrade the microstructures and properties of the matrix. It is known that B and Si can effectively reduce the liquidus temperature of nickel alloy [5], which provide a possibility that the powder metallurgical repair process can be applied at lower temperatures that can reduce the influence of sintering temperature on the matrix.

In our previous work, the Ni-B-Si alloy powders with low melting point were prepared by mechanical alloying, and the initial melting temperature of Ni-4B-4Si(wt.%) alloy powders could be reduced to 1038 °C [6]. In this paper, the infiltration sintering properties of the Ni-4B-4Si alloy powders are investigated, the effects of sintering temperature and nickel skeleton on the microstructures and interfacial bonding properties of sintered alloys are analyzed.

2. Experimental

Carbonyl nickel powders (99%, 10 μm), silicon powders (99.9%, 3 μm) and boron powders (99.9%, 15 μm) were used as the source powders.The Ni-4B-4Si(wt.%) alloy powders were prepared by ball milling for 30 h on a planetary ball mill at a rotation speed of 380 rpm. The source powders, WC balls and dispersing liquid (ethanol) were sealed in a stainless steel jar with ball-to-powder weight ratio of 10:1. The preparation procedure and characterizations of Ni-4B-4Si alloy powders were reported in detail previously [6].

The as-prepared Ni-B-Si alloy powders were cold pressed and placed on the loose nickel powders (nickel powders on Ni substrate) and compact nickel powders (after cold press), respectively. The infiltration sintering was then performed at 1050-1150 ° C for 2 h.

The phase composition and microstructures of the sintered alloys were characterized by 7000S X-ray diffraction (XRD) and JSM-6700F scanning electron microscopy (SEM), respectively. The hardness near the interfacial regions between Ni substrates and sintered alloys was tested by Aglient-G200 nanoindentation.

3. Results and discussion

Figure 1(a-c) and (d-f) shows the SEM images of infiltration regions of sintered alloys prepared using loose and compact nickel powders, respectively. When the sintering temperature was 1050 °C, sintering necks were formed between the nickel powders during the sintering process, meanwhile, the Ni-B-Si alloy powders were melted and infiltrated into the as-formed nickel skeleton. However, the poor fluidity of the Ni-B-Si liquid alloy due to the low sintering temperature resulted in few Ni-B-Si alloy phases in the infiltration region. With the increase of sintering temperature, the fluidity of the liquid alloy was improved, the content of Ni-B-Si alloy phases in the infiltration region was obviously increased and the microstructures became dense.

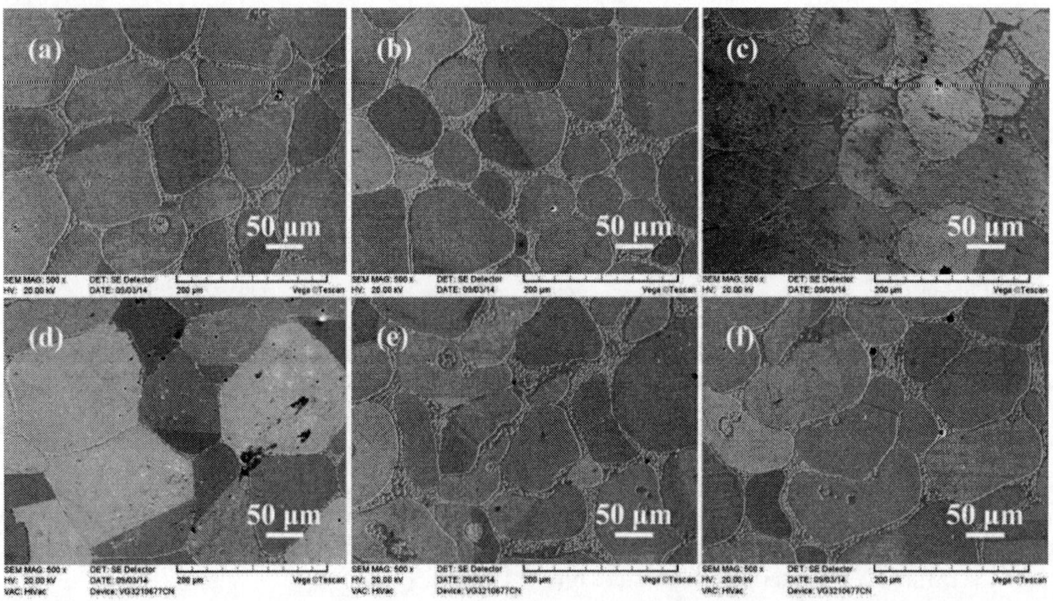

Figure 1. SEM images of infiltration regions of sintered alloys prepared using (a-c) loose and (d-f) compact nickel powders at temperatures of (a,d) 1050 °C, (b,e) 1100 °C, (c,f) 1150 °C.

As shown in figure 1(a-c), when the Ni-B-Si liquid alloy was infiltrated into the loose nickel powders, the capillary force and viscous flow of the liquid alloy between the nickel powders promoted the continuous infiltration of liquid alloy into the nickel skeleton. During the sintering process, sintering necks were formed between the nickel powders, the Ni-B-Si liquid alloy filled in the gaps of nickel skeletons and the compact sintered alloys were formed by metallurgical bonding. As shown in figure 1(d-f), the nickel powders deformed and became compact after cold press, which facilitated the formation of sintering necks between nickel powders during sintering process. During the sintering process, the elastic recovery of nickel powders occurred and a wide range of sintering necks were formed, which blocked the further infiltration of liquid alloy into the nickel skeleton, resulting in few Ni-B-Si alloy phases in the infiltration region.

Figure 2(a-c) and (d-f) shows the SEM images of interface regions between Ni substrates and sintered alloys prepared using loose and compact nickel powders, respectively. As shown in figure 2(a-c), after the infiltration of Ni-B-Si liquid alloy into the loose nickel powders, the interface between Ni substrate and sintered alloy exhibits good metallurgical bonding. When the sintering temperature increased to 1150 °C, the fluidity of the Ni-B-Si liquid alloy was enhanced, a large amount of Ni-B-Si alloy phases were formed near the interface, which weakened the interfacial bonding between Ni substrate and sintered alloy.

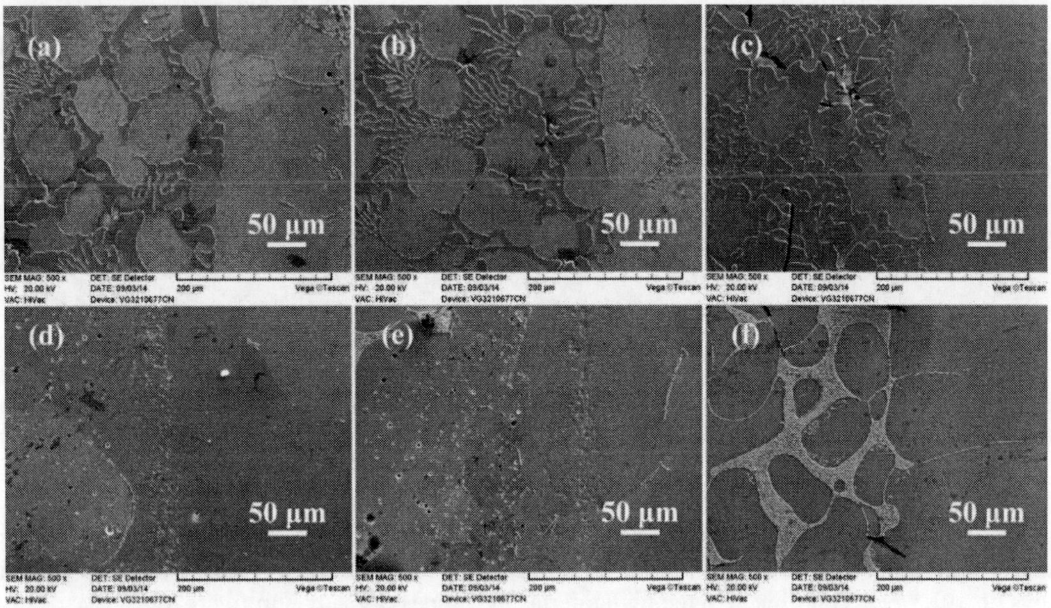

Figure 2. SEM images of interface regions between Ni substrates and sintered alloys prepared using (a-c) loose and (d-f) compact nickel powders at temperatures of (a,d) 1050 °C, (b,e) 1100 °C, (c,f) 1150 °C.

As shown in figure 2(d-f), the Ni-B-Si liquid alloy was infiltrated into the compact nickel powders. When the sintering temperature was 1050-1100 °C, the liquid alloy did not infiltrate to the interface, near which a large number of pores and sinters formed by the compact nickel powders are observed. During the sintering process, a lot of sintering necks were formed among the compact nickel powders, the gaps of the nickel skeleton became narrow, which was not helpful to the infiltration of Ni-B-Si liquid alloy. The fluidity of the liquid alloy was quite poor at low sintering temperature and the infiltration distance was limited. The increase of sintering temperature to 1150 °C facilitated the infiltration of Ni-B-Si liquid alloy, the pores near the interface disappeared and the compactness was obviously improved.

Figure 3 shows the hardness distribution near the interface regions between Ni substrates and sintered alloys prepared using loose and compact nickel powders at a sintering temperature of 1100 °C. After the infiltration of Ni-B-Si liquid alloy into the loose nickel powders, the interfacial hardness is equal to that of Ni substrate and sintered alloy, which reveals the good bonding property between Ni substrate and sintered alloy. As shown in figure 2(b), the Ni-B-Si liquid alloy infiltrated to the Ni substrate, enhancing the interfacial bonding of the sintered alloy and Ni substrate. After the infiltration of Ni-B-Si liquid alloy into the compact nickel powders, the interfacial hardness decreased sharply. As shown in figure 2(e), the Ni-B-Si liquid alloy did not infiltrate to the Ni substrate, the large number of pores distributing near the interface weakened the interfacial bonding between the sintered alloy and Ni substrate.

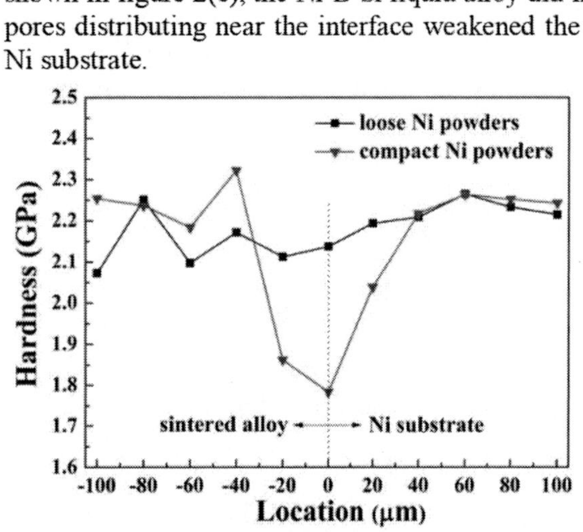

Figure 3. Hardness distribution near interface regions between Ni substrates and sintered alloys prepared using loose and compact nickel powders at a sintering temperature of 1100 °C.

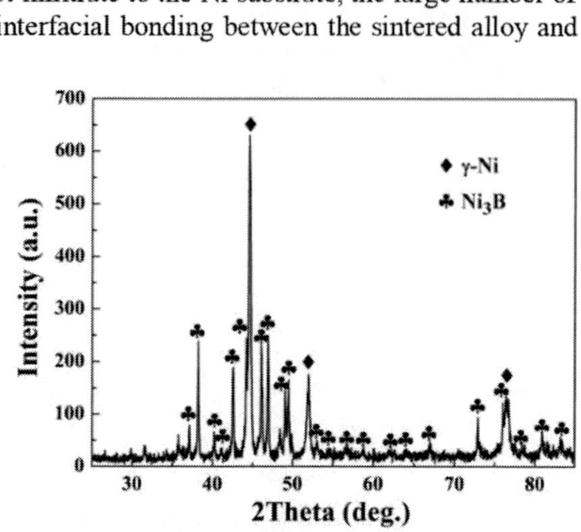

Figure 4. XRD pattern of Ni-B-Si alloy sintered at 1100 °C.

Figure 4 shows the XRD pattern of Ni-B-Si alloy sintered at 1100 °C. Ni-B-Si sintered alloy mainly consists of γ-Ni and Ni_3B. In combination of the microstructures of the sintered alloys shown in figures 1 and 2, it can be seen that the light phase is γ-Ni and the dark phase is Ni_3B. By comparing the Ni-B and Ni-Si phase diagrams [5], it can be seen that the solid solubility of B in Ni (0.3 at.%) at room temperature is much lower than that of Si in Ni (8 at.%). Meanwhile, the diffusion coefficient of B in Ni (6.22×10^{-11} m^2 s^{-1}) at 1100 °C is much higher than that of Si in Ni (3.09×10^{14} m^2 s^{-1}) [7]. Therefore, B easily diffused from the melt and precipitated in the form of Ni_3B [8]. When the melt temperature decreased to the eutectic temperature, γ-Ni and Ni_3B were formed by eutectic reaction, and Si mainly solid soluted in the γ-Ni. The formation of Ni_3B can improve the hardness of sintered alloys [9, 10], which contributed to the high interfacial hardness as shown in figure 3.

4. Conclusions

Ni-4B-4Si(wt.%) alloy powders were infiltrated into the nickel skeletons. γ-Ni and Ni_3B were formed in the infiltration regions and Si solid soluted in the γ-Ni. The loose nickel powders facilitated the infiltration of Ni-B-Si liquid alloy into the nickel skeleton as compared to the compact nickel powders, the sintered alloys became more compact and had better interfacial boding with Ni substrate. After infiltration of Ni-B-Si liquid alloy into the loose nickel powders, the interfacial hardness was equal to that of Ni substrate and sintered alloy, which was higher than that using compact nickel powders.

Acknowledgment
This work is supported by China Postdoctoral Science Foundation Funded Project (2013M532071).

References

[1] García Navas V, Arriola I, Gonzalo O and Leunda J 2013 *International Journal of Machine Tools and Manufacture* **74** 19
[2] Ellison K A, Lowden P and Liburdi J 1994 *Journal of Engineering for Gas Turbines and Power* **116** 237
[3] Sexton L, Lavin S, Byrne G and Kennedy A 2002 *Journal of Materials Processing Technology* **122** 63

IOP Conf. Series: Materials Science and Engineering **292** (2017) 012057 doi:10.1088/1757-899X/292/1/012057

[4] Shepeleva L, Medres B, Kaplan W D, Bamberger M and Weisheit A 2000 *Surface and Coatings Technology* **125** 45
[5] Okamoto H, 2000 *Desk Handbook: Phase Diagrams for Binary Alloys* ASM International, Materials Park, OH
[6] Wang F, Yang Q, Zou J and Liang S 2015 *Rare Metal Materials and Engineering* **44** 1985
[7] Pouranvari M, Ekrami A and Kokabi A H 2013 *Journal of Alloys and Compounds* **563** 143
[8] Binesh B and Gharehbagh A J 016 *Journal of Materials Science and Technology* **32** 1137
[9] Lopez J R, Mendez P F, Perez-Bueno J J, Trejo G, Stremsdoerfer G and Meas Y 2016 *International Journal of Electrochemical Science* **11** 4231
[10] Yuan X, Kang C Y and Kim M B 2009 *Materials Characterization* **60** 923

IOP Conf. Series: Materials Science and Engineering **292** (2017) 012058 doi:10.1088/1757-899X/292/1/012058

Enhancement of red emission intensity of $Ca_2Al_2SiO_7:Eu^{3+}$ phosphor by MoO_3 doping or excess SiO_2 addition for application to white LEDs

H Y Jiao[1], C R LiMao, Q Chen, P Y Wang and R C Cai

College of Electric Engineering , Key Laboratory for Electronic Materials of the State Ethnic Affairs Commission of PRC, Northwest University for Nationalities, Lanzhou, Gansu, 730030, PR China

[1]E-mail: dqjhy@xbmu.edu.cn

Abstract. $Ca_{1.86}Al_2(Si_{1-x}Mo_x)O_7$:0.14$Eu^{3+}$ and $Ca_{1.86}Al_2Si_{1+y}O_{7+2y}$:0.14 Eu^{3+} were synthesized by solid-state reaction. X-ray powder diffraction, excitation and emission spectra were used to investigate their structures and photoluminescence properties. The results shows that the phosphor $Ca_{1.86}Al_2SiO_7$:0.14Eu^{3+} cannot be excited efficiently by light of 393 nm. The introduced Mo ion does not change the position of the excitation peak, but increases both the absorption at 400nm and the emission intensity of Eu^{3+}. The intense red emitting phosphor $Ca_{1.86}Al_2(Si_{0.95}Mo_{0.05})O_7$:0.14$Eu^{3+}$ was obtained, which has 67% enhanced luminous intensity compared to that of the undoped sample $Ca_{1.86}Al_2SiO_7$:0.14Eu^{3+}. Otherwise, SiO_2 excess of non-stoichiometric phosphors $Ca_{1.86}Al_2Si_{1+y}O_{7+2y}$:0.14$Eu^{3+}$ showed the characteristic pattern of a tetragonal structure with a small SiO_2 concentration. The optimal phosphor of $Ca_{1.86}Al_2Si_{1.1}O_{7.2}$:0.14$Eu^{3+}$ has a luminous intensity about two times higher than that of the original stoichiometric phosphor $Ca_{1.86}Al_2SiO_7$:0.14Eu^{3+}. We confirmed that the photoluminescence intensity of the obtained phosphors is fairly enhanced by excessive SiO_2. The mechanism of this photoluminescence enhancement is discussed in this paper.

1. Introduction

After a decade of intense research, phosphor-converted light-emitting diodes (pc-LEDs) have attracted worldwide attention owing to their high luminescence efficiency and variety wide range of applications, such as in flashlights, display backlighting, traffic signals, and especially for solid-state lighting. The huge potential market in home lighting encourages the rapid development of pc-LEDs. White-light generation through the diode chips with the help of efficient phosphors has several advantages over the existing incandescent and fluorescent lamps in power efficiency, reliability, long lifetime, low energy consumption, and environmentally friendly characteristics [1-3]. At present, the most common method of preparing a pc-LED is to use the yellow-emitting YAG: Ce^{3+} phosphor with blue InGaN chips to produce the practical white light [4]. However, such white LEDs exhibits a high color temperature and low color-rendering index (Ra<80) because of their weak emission intensity in the red spectral region [5]. This drawback limits white LEDs' applications, such as for indoor illumination and screen display [6]. To warm the white light and improve the color-rendering index and color reproducibility, a promising strategy is to combine a separate red, green, and blue-emitting phosphors with a deep blue/near-UV LED chips. This requires the phosphor materials to have high absorption in the near-UV to blue spectral region. Unfortunately, the red phosphors $Y_2O_2S:Eu^{3+}$ has poor absorption in the near-UV to blue region and its efficiency is lower than that of the green and blue phosphors [7, 8]. Moreover, the red phosphor is chemically unstable and has a short working lifespan owing to its release of sulfide gas. Therefore, the lack of a proper red phosphor has become a bottleneck in the development of blue or near-UV LED-based white LEDs [9]. Hence, it is very essential to develop a stable and efficient red phosphor with high color purity that can be well coupled with blue and near-UV LEDs. As we know, Eu^{3+} ion is a promising red-emitting rare earth ion, and is a good place to start further discussion [10].

It is noted that the alkaline earth silicate-based inorganic material have been developed to be as the host materials for phosphors because of their various crystal structures, high physical-chemical stability, mild synthesis conditions, and lower production costs than nitrides[11, 12]. In particular, Eu^{3+}-doped $Ca_2Al_2SiO_7$ a compound with a melilite structure, exhibited superior red-emitting properties under near-UV LED excitation [13, 14]. However, in these two papers, two completely different strongest excitation peak positions were reported. One reported the strongest excitation peak for the broad charge transfer transition band (CTB) at around 240–340 nm, and the other reported the strongest excitation peak for the f–f transitions of Eu^{3+} around 394 nm (wavelength of near-UV LED chips), which is considered an important condition in LED. Therefore, a red-emitting phosphor $Ca_2Al_2SiO_7$:Eu^{3+} was newly synthesized in this work through high-temperature solid-state reaction. The strongest excitation peak was found at the broad CTB around 240–340 nm. In order to strengthen and broaden the absorption to around 350–400 nm to obtain more efficient phosphors for near-UV LED, one approach is doping with heterogeneous metal ions (e.g., W^{6+} or Mo^{6+} ions) for modification of the coordination environment of the Eu^{3+} luminescence center [15, 16]. Thus, we substituted some Si ions with Mo ions in this work. Some reports [17, 18] have focused on molybdate-doped Eu^{3+} because these phosphors have wide and intense charge-transfer absorption bands in the near-UV. Therefore, we synthesized phosphors of $Ca_{1.86}Al_2Si_{1+y}O_{7+2y}$:$0.14Eu^{3+}$ type ($0 \leq y \leq 0.3$) with excess SiO_2. We confirmed that the photoluminescence (PL) intensity of the obtained phosphors is fairly enhanced by SiO_2 excess addition. The mechanism of this PL enhancement is discussed in this paper.

2. Experimental

The phosphors $Ca_{1.86}Al_2SiO_7$:$0.14Eu^{3+}$, $Ca_{1.86}Al_2(Si_{1-x}Mo_x)O_7$:$0.14Eu^{3+}$ ($0.01 \leq x \leq 0.07$) and $Ca_{1.86}Al_2Si_{1+y}O_{7+2y}$:$0.14Eu^{3+}$ ($0 \leq y \leq 0.3$)were prepared by the solid-state reaction technique at high temperature. The starting materials were $CaCO_3$ (A.R.), Al_2O_3 (99%), H_2SiO_3, MoO_3 (99.5%) and Eu_2O_3 (99.99%). A small quantity of H_3BO_3 (about 5 mol%) was added as flux. The samples were prepared sintered at 900°C for 2 h in air. When the material cooled, put them in the tubular furnace and sintered at 1300°C for 8 h in air.

The phase assemblage of the product powders was analyzed using by a Rigaku D/Max-2400 X-ray diffractometer at a scanning rate of 2° min^{-1} and intervals of 0.02° in the 2θ range from 10° to 80° with Nifiltered Cu Ka radiation (λ = 0.15405 nm). The photoluminescence (PL) and photoluminescence excitation (PLE) spectra of the samples were measured by using an FLS-920T fluorescence spectrophotometer equipped with a 450 W Xe light source as the excitation source. All the measurements were performed at room temperature in the air and the spectral bandwidth of both excitation and emission spectrometer was adjusted to 0.50 nm ($\Delta\lambda_{ex} = \Delta\lambda_{em} = 0.50$ nm).

3. Results and discussion

3.1. XRD analysis

A series of XRD patterns of $Ca_{1.86}Al_2(Si_{1-x}Mo_x)O_7$:$0.14Eu^{3+}$ ($0.01 \leq x \leq 0.07$) and JCPDS-35-0755 are shown in figure 1. No impurity phase was detected when Eu^{3+} and Mo^{6+} ions were co-doped into the host lattice, indicating that the obtained samples are all of a single phase of $Ca_2Al_2SiO_7$ with tetragonal structure (space group $p\bar{4}2m$) and are consistent with JCPDS-35-0755 [19]. This indicates that the Ca^{2+} ions can be partly replaced by Eu^{3+} ions and Mo^{6+} can be substituted for Si^{4+} without a change of crystal structure. Based on the effective average ionic radius of the cations (R_{Eu3+} =1.066 Å, R_{Ca2+}

=1.12 Å, CN = 8 and R_{Mo6+} =0.41 Å, R_{Si4+} =0.26 Å, CN = 4) [20]. The results suggested that the samples form solid solutions. We have explained the imbalanced substitution using the Kröger - Vink defect notation [21], which has been reported in $CaSrAl_2SiO_7$:Eu^{3+} [22] .

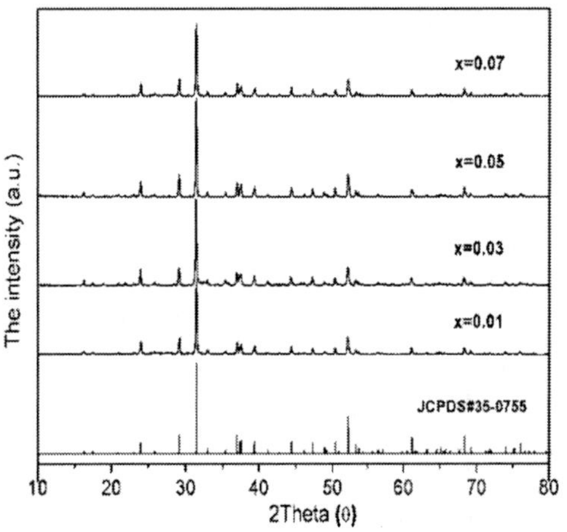

Figure 1. X-ray diffraction patterns of $Ca_{1.86}Al_2(Si_{1-x}Mo_x)O_7$:0.14$Eu^{3+}$ (0.01≤x≤0.07).

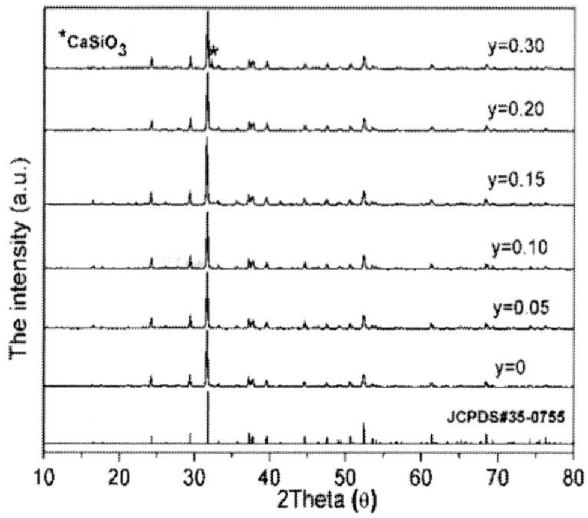

Figure 2. X-ray diffraction patterns of $Ca_{1.86}Al_2Si_{1+y}O_{7+2y}$:0.14$Eu^{3+}$ (0≤y≤0.3).

Figure 2 shows the XRD patterns of $Ca_{1.86}Al_2Si_{1+y}O_{7+2y}$:0.14$Eu^{3+}$ (0≤y≤0.3). The XRD patterns of the $Ca_{1.86}Al_2Si_{1+y}O_{7+2y}$:0.14$Eu^{3+}$ samples the tetragonal structure and are consistent with JCPDS-35-0755. However, with increasing in the SiO_2 content (y≥0.3), a small impurity peak appeared in the region 32-34°, which was assigned to a silicate impurity $CaSiO_3$. The impurity peak observed at larger SiO_2 concentration may be as a result of owing to the SiO_2 in $Ca_{1.86}Al_2SiO_7$ lattice. Therefore, we only investigated the photoluminescence properties of $Ca_{1.86}Al_2Si_{1+y}O_{7+2y}$:0.14$Eu^{3+}$ with a small SiO_2 concentration (y<0.3).

3.2. Luminescence properties

We first synthesized $Ca_{2-x}Al_2SiO_7$:xEu^{3+} phosphors. Owing to the similar photoluminescent character of $Ca_{2-x}Al_2SiO_7$:xEu^{3+} with different concentrations of Eu^{3+} (x), we only to investigated the luminescent properties of $Ca_{1.86}Al_2SiO_7$:0.14Eu^{3+} ; which the luminescent spectra are shown in figure 3. It can be clearly seen that the excitation spectrum is composed of an intense broad-band centered at about 254 nm and a group of sharp lines in the 300-550 nm longer wavelength rang. The broad-band is ascribed to the charge transfer transition from the 2*p* orbital of the O^{2-} ions to the 4f orbital of the Eu^{3+} ions (O^{2-}→Eu^{3+}), and a group of sharp lines in the 300-550 nm are assigned to the intra-configurational 4*f*-4*f* forbidden transitions of Eu^{3+}: $^7F_0$→$^5H_6$ (318 nm), $^7F_0$→$^5D_4$ (362 nm), $^5G_4$ (382 nm), $^5L_6$ (393 nm), $^5D_3$(414 nm), $^5D_2$ (464 nm), $^5D_1$ (525 nm) [23]. The emission spectrum of $Ca_{1.86}Al_2SiO_7$:0.14Eu^{3+}, which was excited by 393 nm, exhibited a well known characteristic Eu^{3+} emission. It consists of several emission lines, among which the main emission line is located at around 614 nm ($^5D_0$→$^7F_2$).

In order to strengthen the absorption in ~ 400 nm of the $Ca_{1.86}Al_2SiO_7$:0.14Eu^{3+} phosphor, the $Ca_{1.86}Al_2(Si_{1-x}Mo_x)O_7$:0.14$Eu^{3+}$ (0≤x≤0.07) series of samples was investigated. Figure 4 shows the excitation spectra (λ_{em} = 614 nm) of the $Ca_{1.86}Al_2(Si_{1-x}Mo_x)O_7$:0.14$Eu^{3+}$ series of samples. We can see that with increasing Mo^{6+} [24]. Significantly, when the Mo^{6+}-doped concentration is up to 0.05, the sharp peak at 393 nm is about four times higher than that of the original $Ca_{1.86}Al_2SiO_7$:0.14Eu^{3+}.

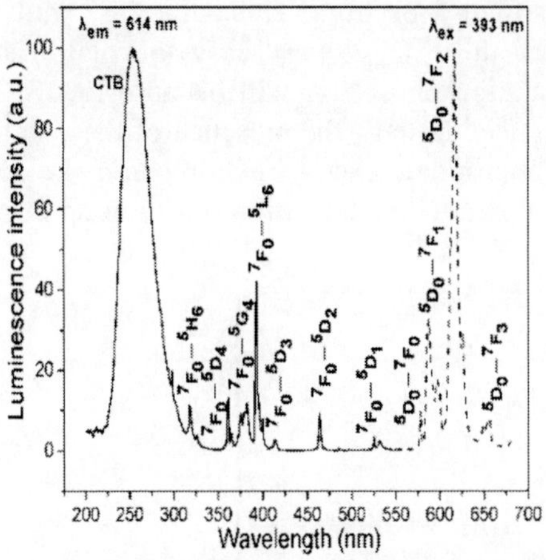

Figure 3. Excitation and emission spectra of $Ca_{1.86}Al_2SiO_7$:0.14Eu^{3+}.

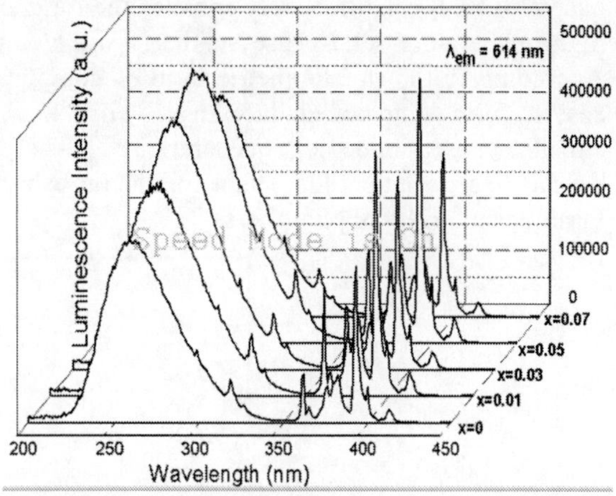

Figure 4. Excitation spectra of $Ca_{1.86}Al_2(Si_{1-x}Mo_x)O_7$:0.14$Eu^{3+}$ ($0 \leq x \leq 0.07$) with different doping concentration x of Mo^{6+}.

The comparison of emission intensity ($^5D_0 \rightarrow {}^7F_2$ transition) under 393 nm excitation for $Ca_{1.86}Al_2(Si_{1-x}Mo_x)O_7$: 0.14$Eu^{3+}$ ($0 \leq x \leq 0.07$) is shown in Table1. With the increase of the component of Mo^{6+}, the maximum emission intensity appears at x = 0.05. This can be primarily estimated that the presence of Mo, whose ionic radii is different from Si, results in somewhat diverse about the sub-lattice structure around the luminescent center ions, which may realize the optimum lattice environment for the energy transfer and luminescence of Eu^{3+} to some certain degree. The intensity for $Ca_{1.86}Al_2(Si_{0.95}Mo_{0.05})O_7$:0.14$Eu^{3+}$ was enhanced by 67% compared to that of the undoped sample $Ca_{1.86}Al_2SiO_7$:0.14Eu^{3+}. However, an increase in the concentration of Mo^{6+} ions lowers the crystallinity of the phosphors and results in the formation of more defects, which would increase the probability of nonradiative transitions.

Table 1. Comparison of emission intensity ($^5D_0 \rightarrow {}^7F_2$) of $Ca_{1.86}Al_2(Si_{1-x}Mo_x)O_7$: 0.14 Eu^{3+} ($0 \leq x \leq 0.07$) (λ_{ex} = 393 nm).

Composition	**Relative Intensity (100%)**
$Ca_{1.86}Al_2SiO_7$: 0.14 Eu^{3+}	100
$Ca_{1.86}Al_2(Si_{0.99}Mo_{0.01})O_7$: 0.14 Eu^{3+}	121
$Ca_{1.86}Al_2(Si_{0.97}Mo_{0.03})O_7$: 0.14 Eu^{3}	142
$Ca_{1.86}Al_2(Si_{0.95}Mo_{0.05})O_7$: 0.14 Eu^{3+}	167
$Ca_{1.86}Al_2(Si_{0.93}Mo_{0.07})O_7$: 0.14 Eu^{3+}	138

The luminescent properties of excess SiO_2 in the $Ca_{1.86}Al_2Si_{1+y}O_{7+2y}$:0.14$Eu^{3+}$ were researched by the luminescent spectra at room temperature. Owing of $Ca_{1.86}Al_2Si_{1+y}O_{7+2y}$:0.14$Eu^{3+}$ with different values of *y*, we just use the spectra of $Ca_{1.86}Al_2Si_{1.1}O_{7.2}$:0.14$Eu^{3+}$ for example. Figure 5 shows the excitation and emission spectra of $Ca_{1.86}Al_2SiO_7$:0.14Eu^{3+}. The strongest excitation peak is located at 393 nm, which is attributed to the characteristics of the transition absorption peak ($^7F_0$-$^5L_6$) of Eu^{3+}. The relationship of the relative intensity of 393 nm and the change in SiO_2 excess (y) is illustrated in the inset of figure 5. The relative intensities were increased with increasing excessive SiO_2 concentration (from y = 0.05 to 0.1) and also the non-stoichiometric phosphor $Ca_{1.86}Al_2Si_{1+y}O_{7+2y}$: 0.14$Eu^{3+}$ with y = 0.1 mol exhibited the strongest luminous intensity, which is about twice as high as that of the original stoichiometric phosphor $Ca_{1.86}Al_2SiO_7$:0.14Eu^{3+}. However, as y continues to increase, the luminous intensity gradually declines. As a result, we can consider that an appropriate excess of SiO_2 can effectively enhance the luminous intensity of phosphors. This is because while the high-temperature solid-phase synthesis silicates phosphors, the SiO_2 is very easy to volatilization. According to the stoichiometric ratio of component preparation phosphors will because low SiO_2 is easy to form some defects [25, 26]. Even at low doping concentrations, the presence of these defects can also cause luminescent quenching [27]. Therefore, an appropriate excess of SiO_2 can make up for defects to some extent [25], reducing the probability of the nonradiative transition so as to improve the luminous intensity of phosphors.

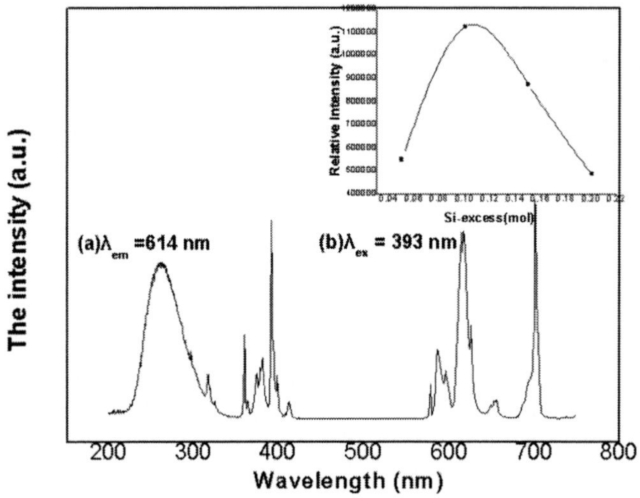

Figure 5. Excitation and Emission spectra of $Ca_{1.86}Al_2Si_{1.1}O_{7.2}$:0.14$Eu^{3+}$ (the inserted figure shows the relationship between relative intensity of 393 nm and the change of SiO_2 excess in $Ca_{1.86}Al_2Si_{1+y}O_{7+2y}$:0.14$Eu^{3+}$ ($0.05 \leq y \leq 0.2$)).

4. Conclusions

A new red-emitting phosphors $Ca_{1.86}Al_2(Si_{1-x}Mo_x)O_7$:0.14$Eu^{3+}$ ($0.01 \leq x \leq 0.07$) and $Ca_{1.86}Al_2Si_{1+y}O_{7+2y}$: 0.14$Eu^{3+}$ ($0 \leq y \leq 0.3$), was successfully synthesized. Our experimental data indicated that the phosphor $Ca_{1.86}Al_2SiO_7$:0.14Eu^{3+} cannot be excited 393 nm light. In order to strengthen the absorption of the phosphors at around 400 nm, Mo ion was introduced. The obtained phosphor $Ca_{1.86}Al_2(Si_{0.95}Mo_{0.05})O_7$:0.14$Eu^{3+}$ shows good absorption at 393 nm and enhanced emissions, which the luminous intensity was enhanced by 67% compared to that of the undoped sample $Ca_{1.86}Al_2SiO_7$:0.14Eu^{3+}. Moreover, non-stoichiometric phosphors $Ca_{1.86}Al_2Si_{1+y}O_{7+2y}$:0.14$Eu^{3+}$ ($0 \leq y \leq 0.3$) show the tetragonal structure (space group $p\bar{4}2m$) with a small SiO_2 concentration ($y<0.3$). An appropriate excess of SiO_2 can make up for the defects to a certain extent, reducing the probability of the nonradiative transition so as to improve the luminous intensity of phosphors. Therefore, these red phosphors $Ca_{1.86}Al_2(Si_{0.95}Mo_{0.05})O_7$:0.14$Eu^{3+}$ and $Ca_{1.86}Al_2Si_{1.1}O_{7.2}$:0.14 Eu^{3+} are considered as an prominent red-emitting phosphor for LEDs.

Acknowledgments

This work was supported by the National Natural Science Foundations of China (Grant No.51462031 and 21663026), the Fundamental Research Funds for the Central Universities (Grant no. 31920170005), and the introduction of excellent talent Research Funds for the Northwest University for Nationalities (Grant No. xbmuyjrc201125).

References

[1] Kim J S, Jeon P E, Choi J C, Park H L, Mho S I and Kim G C 2004 *Appl. Phys. Lett.* **84** 2931
[2] Xia Z G, Wang X M, Wang Y X, Liao L B and Jing X P 2011 *Inorg. Chem.* **50** 10134
[3] Mao N Z Y and Wang D J 2010 *Inorg. Chem.* **49** 4922

[4] Nakamura S and Fasol G 1997 *Springer* Berlin
[5] Ponce F A, Bour D P 1997 *Nature* **386** 351
[6] Piao X Q, Horikawa T, Hanzawa H and Machida K 2006 *Appl. Phys. Lett.* **88** 161903
[7] Zhang M, Wang J, Zhang Q H, Ding W J and Su Q 2007 *Mater. Res. Bull* .**42** 33
[8] Chien W C 2006 *J. Cryst. Growth* **290** 554
[9] Wang J G, Jing X P, Yan C H, Lin J H and Liao F H 2006 *J. Lumin.* **121** 57
[10] Miao S H, Xia Z G, Zhang J and Liu Q L 2014 *Inorg. Chem.* **53** 10386
[11] Tzou W C, Diao C C, Yang C F, Kuo C G and Huang C J 2011 *Ceram. Int.* **37** 1341
[12] Wang X M, Wang C H, Kuang X J, Zou R Q, Wang Y X and Jing X P 2012 *Inorg. Chem.* **51** 3540
[13] Chuai X H, Zhang H J, Li F S and Chou K C 2004 *Opt. Mater.* **25** 301
[14] Zhang Q H, Wang J, Zhang M, Ding W J and Su Q 2007 *Appl. Phys. A.* **88** 805
[15] Chiu C H, Liu C H, Huang S B andChen T M 2007 *J. Electrochem. Soc.* **154** J181
[16] Zhang Z H, Huang Q, Zhao X and Huang Z L 2009 *Phys. Status Solidi A* **206** 2839
[17] Li L L, Leng Z H, Zi W W and Gan S C 2014 *Journal of Electronic Materials* **43** 2588
[18] Geng , Yu C S, Wu H, Wu B J and Tian L H 2013 *J. Lumin.* **140** 71
[19] Kaminskii A A, Belokoneja E L, Mill B V and Sarkisov S E 1986 *Phys. Stat.Solidi A* **97** 279
[20] Su Q 1996 Science and Technology Publishing Company He Nan
[21] Kröger F A and Vink H J 1954 *Physica* **20** 950
[22] Jiao H Y and Wang Y H 2012 *Physica B* **407** 2729
[23] Pan Y X, Wu M M and Su Q 2003 *Mater. Res. Bull.* **38** 1537
[24] Wang Z L, Liang H B, Gong M L and Su Q 2007 *Opt. Mater.* **29** 896
[25] Kunimoto T, Yoshimatsu R and Ohmi K 2002 *IEICE Trans. Electron.* **E85-C(11)** 1888
[26] Kang Y C, Lenggoro W, Park S B and Okagama K 1999 *J. Solid State Chem.* **146** 168
[27] Furusho H, Hölsä J, Laamanen T, Lastusaari M, Niittykoski J, Okajima Y and Yamamoto A 2008 *J. Lumin.* **128** 881

IOP Conf. Series: Materials Science and Engineering **292** (2017) 012059 doi:10.1088/1757-899X/292/1/012059

Effect of gate bias sweep rate on the threshold voltage of in-plane gate nanowire transistor

H X Liu[1], J Li and R R Tan

The Faculty of Communication and Electronics, Jiangxi Science and Technology Normal University, Nanchang, 330013, People's Republic of China

[1] E-mail: liuhuixuan87@163.com

Abstract. In_2O_3 nanowire electric-double-layer (EDL) transistors with in-plane gate gated by SiO_2 solid-electrolyte are fabricated on transparent glass substrates. The gate voltage sweep rates can effectively modulate the threshold voltage (V_{th}) of nanowire device. Both depletion mode and enhancement mode are realized, and the V_{th} shift of the nanowire transistors is estimated to be 0.73V (without light). This phenomenon is due to increased adsorption of oxygen on the nanowire surface by the slower gate voltage sweep rates. Adsorbed oxygens capture electrons and cause a surface of nanowire channel was depleted. The operation voltage of transistor was 1.0 V, because the EDL gate dielectric can lead to high gate dielectric capacitance. These transparent in-plane gate nanowire transistors are promising for "see-through" nanoscale sensors.

1. Introduction

Nanowires have attracted intense research interest due to their potential applications in nanoscale optoelectronic devices, logic circuits, and sensors [1–3]. In_2O_3 nanwoire is of particular interest for potential. In particular, the electrical properties of nanowires have a strong influence by adsorption and desorption of ambient active elements onto and from the nanowire surface, due to the large surface area to volume ratio of nanowires [4]. Adsorbed oxygen is easily bound at the oxygen vacancy site of the In_2O_3 surface. The conductivity of the In_2O_3 nanowire channel is depleted because the oxygen adsorbed on the surface reacts with free electrons as oxygen ions in the forms of O^-, O_2^-, or O^{2-} [5]. Dayeh et al. reported that the transfer characteristics of InAs top-gated nanowire FETs were affected by gate bias sweep rate due to its influence on donor-type trap levels of the surface [6].

Simple device structure can reduce the preparation cost of the device, in-plane-gate transistor device with source/drain and gate electrodes on the same plane. However, conventional in-plane-gate transistors were fabricated by focused ion beam/electron beam lithography, which are generally complicated and expensive [7, 8]. Up to now, low-cost nanowire transistors with in-plane gate structure have little been reported. In other words, fabrication of these in-plane gate nanowire transistors is still rather challenging.

In this letter, we fabricated in-plane gate In_2O_3 nanowire transistors gated by SiO_2 solid electrolyte, by using a single nickel TEM grid shadow mask at room temperature. The operation voltage of the nanowire transistors is reduced to 1.5 V due to the extremely large EDL capacitance of SiO_2 solid electrolytes. The threshold voltage (V_{th}) can be effectively modulated by gate bias sweep rates. Both enhancement mode and depletion mode are realized, and the V_{th} shift of the nanowire transistors is estimated to be 0.73V (without light).

2. Experimental detail

In_2O_3 nanowires were synthesized by the vapor liquid solid (VLS) growth mechanism [9]. Figure 1(a) shows the low resolution scanning electron microscopy (SEM) image of the In_2O_3 nanowires. Notably the surfaces of In_2O_3 nanowires were clean and smooth. The length of the nanowires was in the range 1–20 mm, and the diameter was about 100 nm, as shown in the inset of figure 1 (a).

The entire process of in-plane-gate nanowire EDL transistors fabrication was performed on a Micromanipulator 1555 probe station at room temperature, the structure of the device is shown in figure 1(b). The fabrication process can be divided into four stages: (1) A 2.0-μm-thick SiO_2 solid-electrolyte film was deposited onto the clean ITO glass substrate by the plasma-enhanced chemical vapor deposition (PECVD) method, using SiH_4 and O_2 as reactive gases at room temperature. (2) A mechanical probe was used to transfer nanowires onto the SiO_2 solid-electrolyte film under an optical microscope [10]. (3) The TEM nickel grid mask was placed and fixed on the single In_2O_3 nanowire under the observation of an optical microscope. (4) After ITO (200 nm) electrodes have been deposited by radio frequency magnetron sputtering, the TEM nickel grid mask was moved from its position slightly. At last, a series of in-plane-gate In_2O_3 nanowire EDL transistors gated by SiO_2 solid-electrolyte solid electrolyte were fabricated. Scanning electron microscope (SEM) image of the nanowire EDL transistor was performed with a Hitachi S-4800 microscope. The electrical characteristics of the in-plane gate In_2O_3 nanowire EDL transistors were measured with a Keithley 4200 semiconductor parameter analyzer at room temperature in the dark. The capacitance-frequency measurement was performed using a WK 6500B precision impedance analyzer.

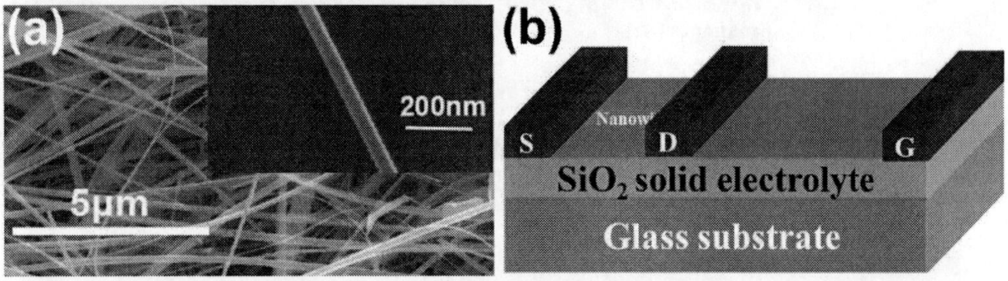

Figure 1. (a) SEM image of the In_2O_3 nanowires. (Inset) High-resolution SEM image of an individual In_2O_3 nanowire. (b) Schematic of the in-plane-gate nanowire transistor.

3. Results and discussion

Figure 2(a) shows the frequency-dependent capacitance of the SiO_2 solid electrolyte gate dielectric. A very high specific capacitance of 1.22 μF/cm^2 at 20 Hz due to EDL effect [11]. The figure 2(b) shows the optical transmission spectrum of the entire FETs arrays on the glass substrate in the wavelength range between 200 and 1000 nm. The average transmittance in the visible range (400–800 nm) is about 75%, indicating that the FETs are transparent to visible light.

The electronic properties of in-plane gate nanowire tansistors were measured as a function of gate bias sweep rate in different measurement environments. It can be seen from the figure 3 (a) and (b), the threshold voltage and I_{on} of device is different with light and without light. At the same time, a series of I_{ds}-V_{gs} curves measured at different gate bias sweep rates (25, 50, 100 and 500 mV/s) at a fixed drain voltage (V_{ds}) of 0.8 V under various environments. Since the gate bias sweep rate is the amount of change per unit time and the step of the voltage is independent of the time, the change in the gate bias sweep rate does not affect the step voltage. It is interesting to note that the transfer characteristics consistently shift in the negative direction as the gate bias sweep rates is changed from 25mV/s to 500mV/s. The operation mechanism of the threshold voltage change can be explained by the depletion of electrons by oxygen adsorption on the surface of In_2O_3 nanowire. Previous reports that the stable oxygen ions are O_2^-below 100 °C, O^- between 100 and 300 °C, and O^{2-} above 300 °C [12, 13]. The oxygen influence of the In_2O_3 nanowire transistors would be mainly attributed to adsorption of O_2- ions due to electrical properties of the nanowire transistor is tested at room temperature. Slower gate bias sweep rates imply longer gate biasing time. The oxygen ions would not be easily detached due to captured oxygen ions were strongly bonded by positive gate bias. Thus, slow gate bias sweep rates will cause more oxygen absorption than fast gate bias sweep rates. As a result, the current decreases and the threshold voltage shifts to the positive gate bias direction occur by slower gate bias sweep rates, due to longer gate biasing time. On the contrary, faster gate voltage sweep rates are applied in relatively shorter gate biasing time. Therefore, fast gate voltage sweep rates lead to less oxygen absorption. Thus, the current increases and the threshold voltage shifts to the negative gate voltage direction, as shown in figure 3. The figure 3(a) and (b) are the results of test with light and without light, the results show that the conductivity of the channel under light conditions is few stronger, there may be more carriers produced under with light than without light conditions.

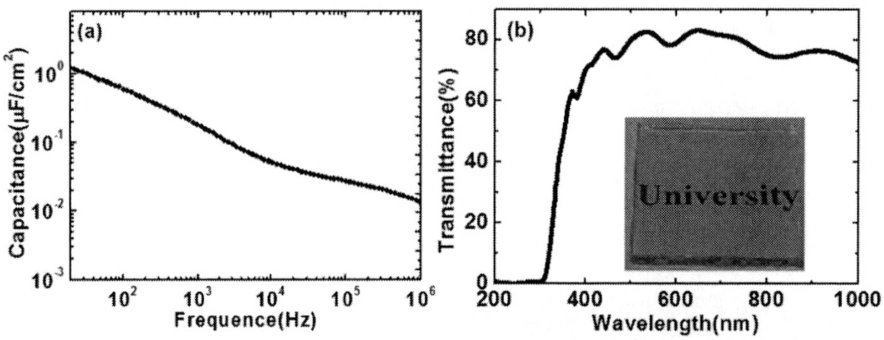

Figure 2. (a) Frequency-dependent specific capacitance of the SiO_2 solid-electrolyte. (b) Optical transmittance spectrum of the nanowire transistor. Inset: digital photograph of transistors fabricated on glass. The logo can be clearly seen through the devices.

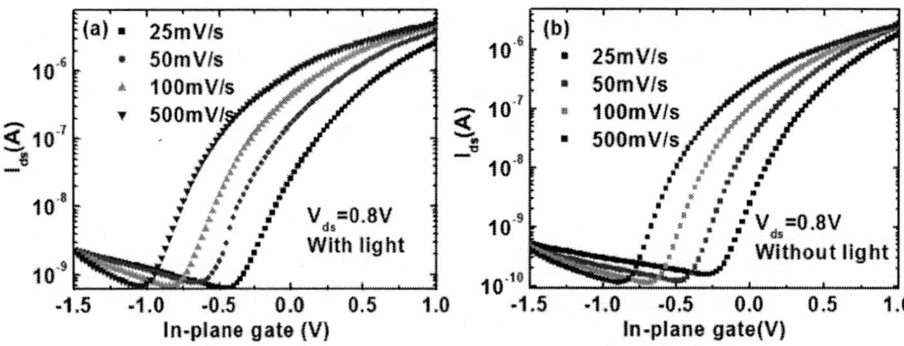

Figure 3. (a) and (b) transfer characteristics of the same device at V_{ds}=0.8V with light and without light.

Figure 4(a) shows the output characteristics (I_{ds}–V_{ds} curves) at V_{gs} varied from 0V to 1V in 0.2V steps (without light). A maximum on-current of 3.11 μA was measured at V_{ds} =0.8 V and V_{gs} = 1V. It indicates good current saturation behaviors at high V_{ds} and excellent linear characteristics of I_{ds} at low V_{ds}, this shows that a low resistance of the ohmic contact formed between the In_2O_3 nanowire and ITO source/drain electrodes.

We also measured the leakage current of the device, the leakage current is less than 5nA as shown in figure 4(b). The results strongly indicate that the SiO_2 solid-electrolyte is an electronically insulating but solid electrolyte. The EDL formed at the interface induces a rather large electric field on the order of megavolts per centimeter at the EDL region, resulting in a high density of charge carriers at the surface of the In_2O_3 channel. Such an electrostatic modulation process controls the conductance of the In_2O_3 channel layer, which is meaningful for low-voltage electrostatic modulation devices.

The field-effect mobilities of the nanowire EDL transistors are estimated by I_{ds}= (μC) $(V_{gs} - V_{th})^2/(2L^2)$, where L = 15 μm is the channel length. The electrolyte gate capacitance per unit length can be estimated as $C_g = C/L = 2\pi\varepsilon_0\varepsilon_r/\ln(1 + 2L_D/d)$, where $\varepsilon_r = 2$ is the dielectric constant of the SiO_2 solid-electrolyte medium, ε_0 is the permittivity of free space, and d = 100 nm is the diameter of In_2O_3 nanowire. Debye length L_D is simply defined as the typical distance required for screening the surplus charge by the mobile carriers present in a material [14]. According to the definition of the Debye length, we assume that the thickness of the EDL (1.0 nm) is the Debye length. Thus, L_D is assumed to be1.0nm [15, 16]. Based on this model, the saturation electron mobility of In_2O_3 nanowire transistors as gate bias sweep rate changes from 25 to 500mV/s are estimated to be from 141 to 95 cm^2/Vs with light, and from 175 to 66 cm^2/Vs without light, respectively.

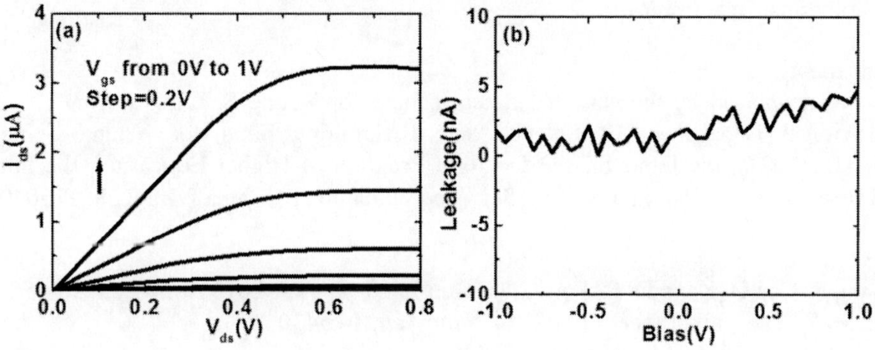

Figure 4. (a) Output characteristics of nanowire transistors. (b) The leakage current of the device.

Table 1. Summary of the electrical properties of in-plane gate In_2O_3 nanowire transistor with various voltage sweep rates with light and without light.

With Light				
voltage sweep rate (mV/s)	V_{th} (V)	S (mV/dec)	I_{ds} (μA)	μ (cm^2/Vs)
25	0.00	136	2.62	141.4
50	-0.23	133	3.78	135.0
100	-0.42	132	4.80	128.5
500	-0.74	130	5.34	95.2
Without Light				
gate voltage sweep rate (mV/s)	V_{th} (V)	S (mV/dec)	I_{ds} (μA)	μ (cm^2/Vs)
25	0.24	135	1.88	175.8
50	0.05	134	2.48	148.4
100	-0.23	136	2.62	93.5
500	-0.49	131	2.74	66.6

Table 1 summarizes the electrical performance of in-plane gate nanowire transistors with various gate bias sweep rate (25-500mV/s) with light or without light. The V_{th} of in plane-gate nanowire transistor with different gate bias sweep rate was calculated from x-axis intercept of the square root of I_{ds}-V_{gs} plot (Linear Extrapolation (LE) method [17]). Under light conditions, the V_{th} of the in-plane gate transistor shifts from 0 V to -0.74V when the gate voltage sweep rate changes from 25 to 500mV/s, and the shifts from 0.24 V to -0.49V without light, due to slow gate voltage sweep rates will cause more oxygen absorption than fast gate bias sweep rates, so the threshold voltage shifts to the positive gate bias direction occur by slower gate voltage sweep rates and longer gate biasing time. In the light conditions, the threshold voltage moves in the negative direction due to sensitivity of nanowires. The subthreshold slope S= $dV_{gs}/d(\log I_{ds})$, the smaller S, the easier it is to switch the transistor to an off-state. For this device, the S is almost identical (134mV/dec), due to S value is mainly related to the device structure and temperature, which was little affected by the ordinary light and gate voltage sweep rate. The mobility is decreased because the V_{th} and I_{ds} changes with various gate bias sweep rate. Since the mobility of the device is affected by the scattering of carriers, the carrier mobility is high when the density is lower. The change in threshold voltage in the case of light and without light is mainly due to sensitivity of nanowires.

4. Conclusions

We fabricated in-plane gate nanowire transistors using TEM grid mask, and studied the threshold voltage of In_2O_3 nanowire transistors as a function of gate bias sweep rate. The V_{th} of the transistors could be effectively tuned from 0.24 to -0.49V by ranging the gate voltage sweep rate from 25 to 500mV/s. These in-plane gate nanowire transistors had controllable V_{th}, simple fabrication, and low-voltage operation, making them very promising for next-generation low-cost portable "see- through" sensors and other nanoelectronics.

Acknowledgments

This work was supported by the National Science Fund for Young Scholars (Grant No. 61404062); Jiangxi Provinical Department of Science and Technology Fund for Young Scholars (Grant No.20151BAB217024), the Fund for the Doctoral Program of Higher Education; the Fund for Less Developed Regions of the National Natural Science Foundation of China (Grant No.11564015).

References

[1] Chuang S, Gao Q, Kapadia R, Ford A C, Guo J and Javey A 2013 *Nano Lett.* **13** 555
[2] Shim W Y, Yao J and Lieber C M 2014 *Nano Lett.* **14** 5430
[3] Yao S S and Zhu Y 2014 *Nanoscale* **6** 2345
[4] Zhang Y, Kolmakov A, Chretien S, Metiu H and Moskovits M 2004 *Nano Lett.* **4** 403
[5] Chaabouni F, Abaab M and Rezig B 2004 *Sens. Actuators B.* **100** 200
[6] Dayeh S A, Soci C, Yu P K L, Yu E T and Wang D 2007 *Appl. Phys. Lett.* **90** 162112
[7] Wieck A D and Ploog K 1990 *Appl. Phys. Lett.* **56** 928
[8] Nieder J, Wieck A D, Grambow P, Lage H, Heitmann D, Klitzing K V and Ploog K 1990 *Appl.Phys.Lett.* **57** 2695
[9] Wan Q, Dattoli E N, Fung W Y, Guo W, Chen Y B, X. Pan Q and Lu W 2006 *Nano Lett.* **6** 2909
[10] Li Q, Koo S M, Richter C A, Edelstein M D, Bonevich J E, Kopanski J J, Suehle J S and Vogel E M 2007 *IEEE Trans. Nanotechnol.* **6** 256
[11] Liu H X, Sun J, Tang Q X and Wan Q 2010 *J. Phys. Chem. C.* **114** 12316
[12] Takata M, Tsubone D and Yanagida H 1976 *J. Am. Ceram. Soc.* **59** 4
[13] Sadek A Z, Choopun S, Wlodarski W, Ippolito S J and Kalantar-Zadeh K 2007 *IEEE Sens. J.* **7** 919
[14] Debye P 1936 *Chem. Re.* **19** 171
[15] Panzer M J and Frisbie C D 2008 *Adv. Mater.* **20** 3177
[16] Pal B N, Dhar B M, See K C and Katz H E 2009 *Nature Mater.* **8** 898
[17] Ortiz-Conde A, García Sánchez F J, Liou J J, Cerdeira A, Estrada M and Yue Y 2002 *Microelectronics Reliability.* **42** 583

Effect of different fibers on mechanical properties and ductility of alkali-activated slag cementitious material

J Zhu[1, 3], W Z Zheng[2], C Z Qin[1], Z Z Xu[1] and Y Q Wu[1]

[1] College of Civil Engineering and Architecture, Harbin University of Science and Technology, Harbin 150080
[2] Key Lab of Structures Dynamic Behavior and Control of the Ministry Education, Harbin Institute of Technology, Harbin 150090

[3] E-mail: zhujing02@126.com

Abstract. The effect of different fibers on mechanical properties and ductility of alkali-activated slag cementitious material (AASCM) is studied. The research contents include: fiber type (plant fiber, polypropylene fiber), fiber content, mechanical property index, tensile stress-strain relationship curve, treating time. The test results showed that the compressive strength of two fibers reinforced AASCM was about 90 ~ 110MPa, and the tensile strength was about 3 ~ 5MPa. The reinforcement effect of polypropylene fiber is superior to that of plant fiber, and the mechanical properties of polypropylene fiber reinforced AASCM are superior to those of plant fiber, According to the comparison of SEM pictures, the plant fiber and polypropylene fiber are both closely bound with the matrix, and the transition zones are complete and close. Thus, it is proved that plant fiber and polypropylene fiber delay the crack extension and enhance the ductility of AASCM.

1. Introduction

Alkali-activated slag cementitious material (AASCM) is a type of cementitious material through the chemical reaction obtaining [1-3], with slag and alkaline activators such as sodium silicate, NaOH as the main raw materials, with or without a certain cold heat history of aluminate natural minerals containing silicon or industrial waste residue (e.g metakaolin, phosphorus slag, steel slag, coal gangue, fly ash, red mud, etc) and alkaline activators (such as alkali metal hydroxide, alkali metal carbonate, alkali metal sulfate, etc). It uses industrial waste slag as the raw material and has the advantage of high temperature resistance, corrosion resistance, energy conservation and environmental protection, and has aroused the wide concern of scholars both at home and abroad [4-8]. Since the 1980s in China, AASCM has begun to be studied and also made a lot of research results [9, 10]. AASCM has excellent compressive properties, heavy metal ion solidification, and thermostability [11], but AASCM belongs to brittle material, so its tensile properties are relatively low. To overcome the brittleness of AASCM, Li [12] studies steel fiber reinforced alkali-activated slag panels, witch has good reinforced effect. This essay used plant fiber and polypropylene fiber (PPF) in increasing the ductility of AASCM.

2. Materials and Formulation

2.1. Raw materials and mix proportion

S95 slag is proposed by mass fraction of 36.9% SiO_2, mass fraction of 15.66% Al_2O_3, mass fraction of 37.57% CaO, mass fraction of 9.3% MgO, mass fraction of 0.36% Fe_2O_3. The activity indexes of slag are mainly: the quality coefficient 1.69, the alkaline coefficient 0.97, and the activity coefficient 0.42.The main parameters of potassium silicate solution are shown in table 1.The mass fraction of sodium hydroxide is not less than 96.0%.

Table 1. Parameters of potassium silicate solution

Baume degree	Density	Modulus	mass fraction, %	
			K_2O	SiO_2
46.3	1.465	2.76	15.98	28.15

[a] Note: Baume degree for potassium silicate is the unit of density for liquids measured by hydrometer.

Test uses two relatively optimal W32, W35 of AASCM and main parameters are shown in table 2.

Table 2. The proportion parameters used for text

Number	Slag	Modulus of potassium silicate	potassium silicate amount, %	Sodium hydroxide,%	Water amount %
W32	1	1.0	14	4.58	32
W35	1	1.0	12	5.35	35

The plant fiber and PPF were selected for the test fiber type, and wheat straw is chosen as plant fiber. Its length is 2 cm ~ 3 cm. To improve the moisture content of wheat straw and the mechanical interaction of wheat straw and matrix, the wheat straws were treated by mass fraction of 10% sodium hydroxide solution to soak for 20, 60 min, then soaked in clean water for 2 min. After wheat straws were placed in the shade to dry, the final moisture content of wheat straws was controlled about 15%. In addition, the length of PPF is 1cm ~ 2cm and its elastic modulus is 42GPa. The fiber pictures are shown in figure 1.

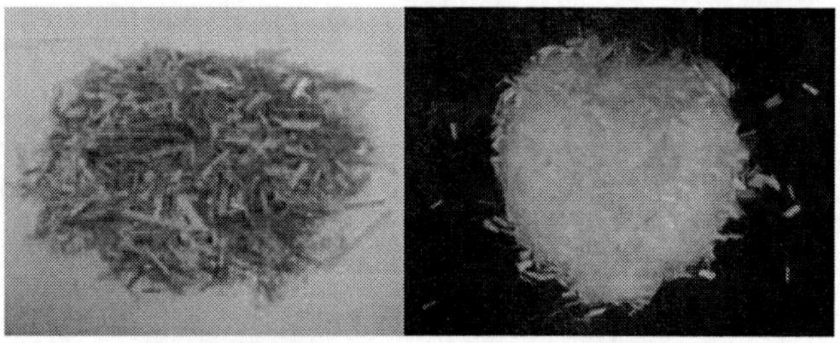

(a) Plant fiber (b) PPF

Figure 1. Fiber pictures

2.2. Production and maintenance of specimens

The samples were prepared by mixing raw materials--slag and alkaline activators in a 5 liter pan mixer. The sequence of mixing was important to homogenize the fresh paste. Thus, the raw materials were firstly mixed for about 30 s. the alkaline activators and water were then poured into the pan, and stirred at low speed for approximately 6 min until the mixture was glossy and well combined. Then, the plant fibers (The percentage of plant fiber is 1%) or polypropylene fibers (The percentage of polypropylene fibers is 0.7% and 1%) were mixed and stirred for approximately 1 min. The fresh paste was cast in various shapes of moulds for measurement of strength. They were maintained in a humidistat at 20 ± 2℃ and 95% relative humidity. The samples were demolded after 24 h of humidity maintaining and then stored in a humidistat at 20 ± 2℃ and 50% relative humidity until testing commenced.

3. The strength varies with age

3.1. Compressive strength of cementing materials

Figure 2 shows compressive strength of cementing materials of two fibers reinforced AASCM with the age,

By comparison, the compressive strength of two fibers reinforced AASCM was 90~105MPa, witch are better than AASCM matrix. After observing the failure of the specimen, it is found that plant fiber and polypropylene fiber not only improve the ductility of AASCM, but also improve compressive strength of AASCM. The reason is that the wheat stalk and PPF are small and tightly coupled with the matrix, which can be distributed evenly in AASCM to achieve enhanced mechanical properties. The incorporation of straw and PPF can obviously delay the crack extension then enhance the ductility of AASCM. Thus, the enhancement of PPF to AASCM is better than that of plant fiber.

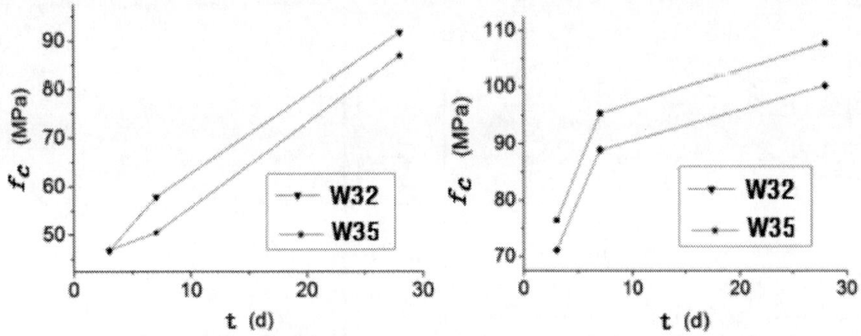

(a) Plant fiber reinforced AASCM (b) PPF reinforced AASCM

Figure 2. Compressive strength of cementing materials

3.2. Axial tensile strength

The axial tensile strengthes of the two kinds of fibers reinforced AASCM are shown in figure 3.

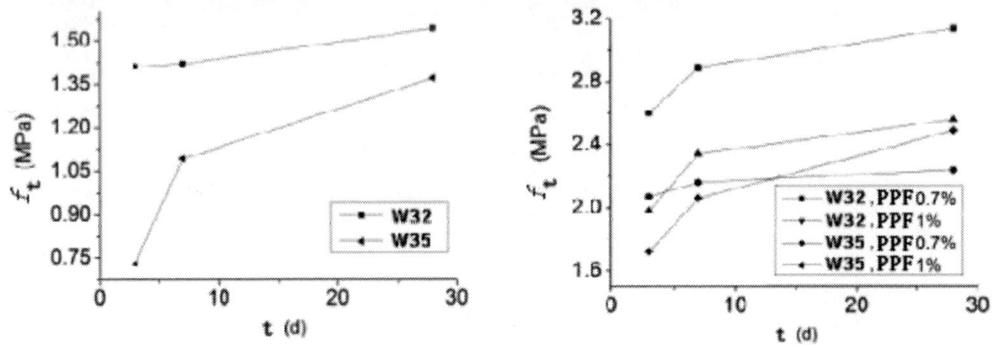

(a) Plant fiber reinforced AASCM (b) PPF reinforced AASCM

Figure 3. Effect of different fibers on Axial tensile strength of dumbbell type specimen

It can be seen from figure 3 that the development of tensile strength of AASCM is similar to compressive strength and flexural strength, which increase with the increase of age. The axial tensile strength of W32 specimen is maximum. The 28d AASCM flexural strength is about 1/10 of compressive strength of cementing materials, and axis tensile strength is about 1/40 of compressive

IOP Conf. Series: Materials Science and Engineering **292** (2017) 012060 doi:10.1088/1757-899X/292/1/012060

strength of cementing materials, and the tensile strength of PPF reinforced AASCM is superior than that of of plant fiber reinforced AASCM.

The axial tensile strength reflects the enhancement of the ductility of AASCM, which shows that the fundamental reason for the change of the macroscopic mechanical properties of fibers reinforced AASCM is that the fiber makes the AASCM microstructure more compact, and The tensile effect of fibers enhance the toughness of the AASCM matrix.

3.3. Comparison of tensile stress-strain relationship curve

The comparison of tensile stress-strain relationship curve is shown in figure 4. The percentage of plant fiber is 1%, and the percentage of polypropylene fibers is 0.7%.

According to the Figure 4, it can be seen that with the addition of two kinds of fibers, the area under the tensile stress and strain curve increases, which indicates that the ductility and toughness of the fibers reinforced AASCM are better than matrix. The PPF is superior than plant fiber on improving the ductility of AASCM.

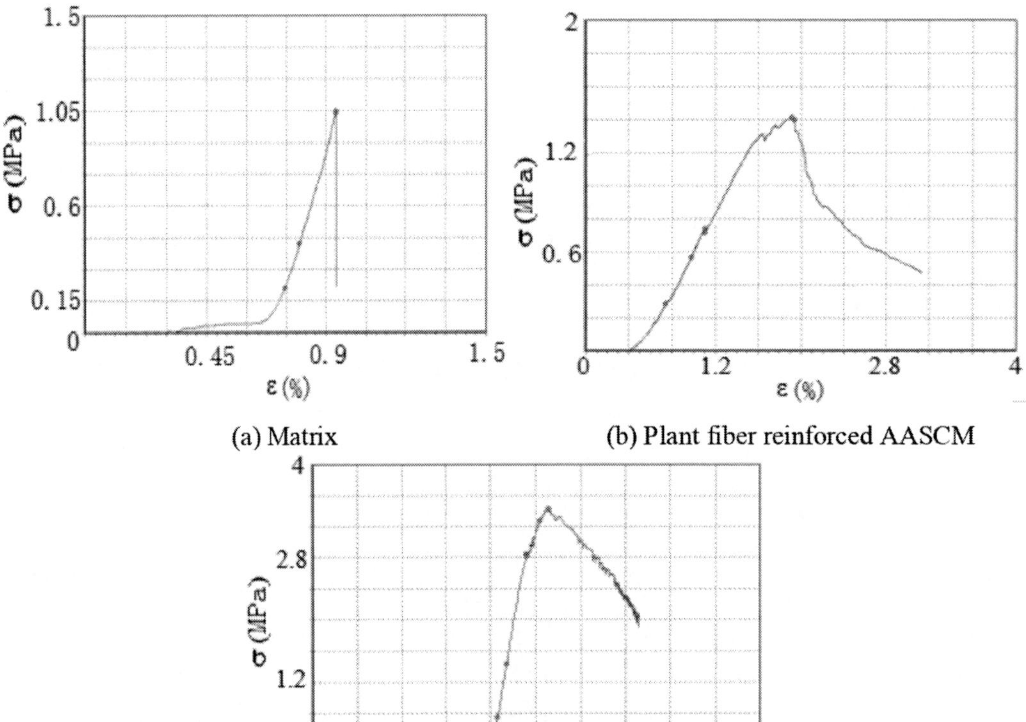

(a) Matrix (b) Plant fiber reinforced AASCM

(c) PPF reinforced AASCM

Figure 4. tensile stress-strain relationship curve

When the AASCM is not reinforced with plant fiber, the descending curve of AASCM matrix is very steeper, which indicates that the failure process is rapid and the brittleness of AASCM is obvious. When the AASCM is reinforced with plant fiber, the descending curve is flatter and longer, which indicates that the plastic deformation capacity of AASCM has improved. When the AASCM is reinforced with PPF, the descending curve is the flattest and longest, which indicates that the plastic deformation capacity of PPF reinforced AASCM has further improved, and the PPF is superior than plant fiber on improving the ductility of AASCM, just as mentioned in 3.1 and 3.2.

IOP Conf. Series: Materials Science and Engineering **292** (2017) 012060 doi:10.1088/1757-899X/292/1/012060

4. Compressive strength of different fibers reinforced AASCM block

The comprehensive strengths of different fibers reinforced AASCM block are shown in figure 5, figure 6 and figure 7. The percentage of plant fiber is 1%, and the percentage of polypropylene fibers is 0.7%. The fracture forms of the single row of holes block compressive strength, single-hole block compressive strength and standard solid brick compressive strength are carefully analyzed.

The test results showed that the compressive strength of single row of holes block of matrix AASCM is equivalent to that of MU10 (***MU means m***asonry unit, and 10 means the compression strength of block is 10MPa) concrete block. Moreover, the compressive strength of standard solid brick of matrix AASCM is in the middle of that of MU20 and MU25 concrete block. Which basically meet engineering requirements. In addition, the compressive strength of the PPF reinforced AASCM is highest, and that of plant fiber reinforced AASCM is better, then that of matrix is lowest.

The research contents include: fiber type: plant fiber, polypropylene fiber; mechanical index: single row of holes block compressive strength, single-hole block compressive strength and standard solid brick compressive strength. The vertical cracks are crushed at the corners. Compared with the matrix, the plant fiber and PPF have good tensile properties because of those high fiber strength. Thus the fibers reinforced AASCM have ideal ductility.

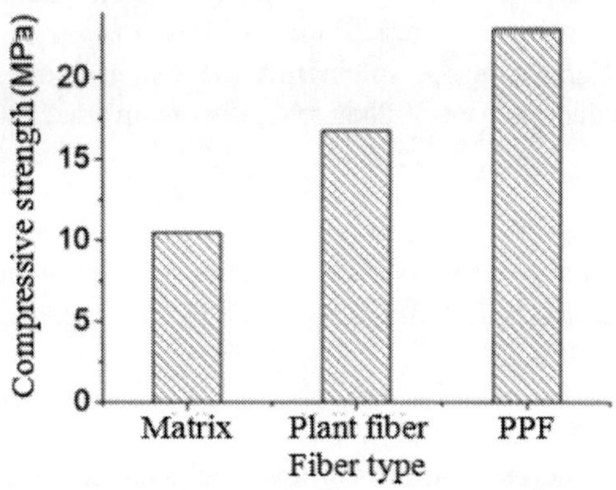

Figure 5 Compressive strength of single-row hole block

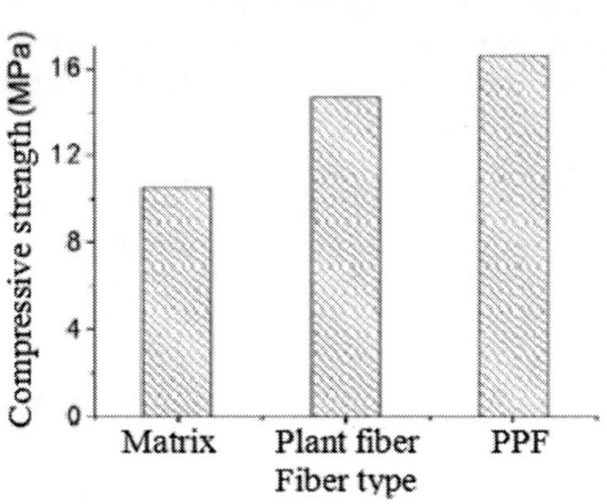

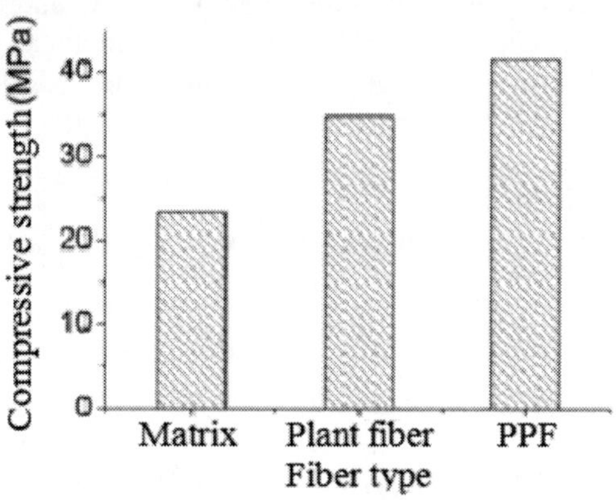

Figure 6 Compressive strength of single-hole block **Figure 7** Compressive strength of solid brick

5. SEM analysis

The different fibers reinforced AASCM are analyzed by SEM, and the SEM pictures are shown in figure 8. According to SEM pictures of figure 8, it can be seen that the plant fiber and PPF are both closely bound with the matrix, and the transition zones are complete and close. Thus, it is further proved that plant fiber and PPF can delay the crack extension and enhance the ductility of AASCM.

(a) Plant fiber reinforced AASCM (b) PPF reinforced AASCM

Figure 8. SEM pictures of fibers reinforced AASCM

According to scanning electron microscopy (SEM) test, it can be seen a series of changes; the combined with plant fiber had good biocompatibility, interface morphology complete. Polypropylene fiber and plant fiber are different degree intensifies the material internal structure defects. Thus, Microstructural changes revealed the root of changes in the macro mechanical properties.

6. Conclusions

(1) Test proves that plant fiber and PPF reinforced AASCM have excellent mechanical properties and these two kinds of fibers can improve the mechanical properties and enhance the ductility of AASCM matrix. Compared with the matrix, the plant fiber and PPF have good tensile properties because of those high fiber strength.

(2) The compressive strength of the PPF reinforced AASCM is highest, and that of plant fiber reinforced AASCM is better, then that of matrix is lowest. The compressive damage of the AASCM block is similar to that of the concrete, and the vertical cracks are crushed at the corners. Single row of holes block compressive strength, single-hole block compressive strength and standard solid brick compressive strength basically meet engineering requirements.

(3) According to SEM pictures, the plant fiber and PPF are both closely bound with the matrix, and the transition zones are complete and close, which can delay the crack extension and enhance the ductility of AASCM. Polypropylene fiber and plant fiber are different degree intensifies the material internal structure defects.

Acknowledgements

This work was financially supported by the National Natural Science Foundation of China (NO. 51508140).

References

[1] Ri-On Oh, Sang-Sun Cha, Seong-Yong Park, et al 2014 Mechanical properties and water purification characteristics of natural jute fiber-reinforced non-cement alkali-activated porous vegetation blocks *Paddy and Water Environment* **12(1)** 149-156

[2] Zheng W. Z, Zhu J 2013 The effect of elevated temperature on bond performance of alkali-activated GGBFS paste *Journal of Wuhan University of Technology (Materials Science Edition)* **28(4)** 721-725

[3] Atis C D, Bilim C, Celik Ö, et al 2009 Influence of activator on the strength and drying shrinkage of alkali-activated slag mortar *Construction and Building Materials* **23(1)** 548-555

[4] Palomo A, Alonso S, Fernández-Jiménez A 2004 Alkaline activation of fly ashes, a NMR study of the reaction products *Journal of the Amerian Ceramic Society* **87(6)** 1141-1145

IOP Conf. Series: Materials Science and Engineering **292** (2017) 012060 doi:10.1088/1757-899X/292/1/012060

[5] Zheng W. Z, Zhu J 2015 Application foundation of alkali-activated slag cementitious material in structural engineering *Harbin Institute of Technology Press*

[6] Pimraksa K, Chindaprasirt P, Rungchet A,et al 2011 Lightweight Geopolymer Made of Highly Porous Siliceous Materials with Various Na_2O/Al_2O_3 and SiO_2/Al_2O_3 Ratios *Materials Science and engineering A-Structural materials properties microstructure and processing* **528(21)** 6616-6623

[7] Zhu J, Zheng W Z 2012 Shear performance of alkali-activated slag cementitious material used for FRP-strengthened concrete structures *Journal of Southeast University* **42(5)** 962-969

[8] Won J P, Kang H B, Lee S J, et al 2012 Eco-Friendly Fireproof High-Strength Polymer Cementitious Composites *Cement and Concrete Composites* **30(4)** 406-412

[9] Davidovits J 2005 Geopolymers, green chemistry and sustainable development solutions *Saint Quentin: Geopolymer Institute* 145-232

[10] Dai J G, Gao W Y, Teng J G 2012 Bond-Slip Model for FRP Laminates Externally Bonded to Concrete at Elevated Temperature *ASCE-Journal of Composites for Construction* **3(2)** 1-32

[11] Muniz-Villarreal M S, Manzano-Ramirez A, Sampieri-Bulbarela S, *et al* 2011 The effect of temperature on the geopolymerization process of a metakaolin-based geopolymer *Materials Letters* **65(6)** 995-998

[12] Li T, Zhang Y M and Dai J G 2017 Flexural behavior and microstructure of hybrid basalt textile and steel fiber reinforced alkali-activated slag panels exposed to elevated temperatures *Construction and Building Materials* **152(15)** 651-660

IOP Conf. Series: Materials Science and Engineering **292** (2017) 012061 doi:10.1088/1757-899X/292/1/012061

The design of wideband metamaterial absorber at E band based on defect

L S Wang[1, 3], D Y Xia[2], X Y Ding[1] and Y Wang[1]

[1] Science and technology department, Sanya University, Sanya, Hainan, China, 572022

[2] Finance and economics department, Sanya University, Sanya, Hainan, China, 572022

[3] E-mail: wlswls1982@126.com

Abstract. A kind of wideband metamaterial absorber at E band is designed in this paper; it is composed of round metal cells with defect, dielectric substrate and metal film. The electromagnetic parameters of unit cell are calculated by using the finite element method. The results show that the wideband metamaterial absorber presents nearly perfect absorption above 90% with absorption ranging from 65.38GHz to 67.86GHz; the reason of wideband absorption is the overlap of different absorption frequency which is caused by electromagnetic resonance; the size parameters and position of defect has important effect on its absorption property. It has many advantages, such as simply, easy to preparation and so on. It has potential application on aerospace measurement and control, remote data communication, LTE wideband mobile communication and other fields.

1. Introduction

The research of microwave absorber is always the front of military struggle. However, the traditional microwave absorber cannot meet the practical requirement with the development of radar detection technology. The metamaterial absorber provides new way to improve the stealth performance of military equipment. The metamaterial absorber essentially is a kind of resonant structure which is composed of resonant metal unit, dielectric substrate and metal film. In 2008, Landy et al firstly proposed the metamaterial absorber based on split ring resonator and short metal wire, it achieved perfect absorption at simple microwave frequency [1]. Since then, the design of metamaterial absorber is expanded from microwave band [2, 3] further to THz band [4, 5], infrared band [6, 7] and light band [8-10]; the absorption frequency is expanded from simple frequency further to dual band [11-13], tri band [14-16], wideband [17] and tunable absorption [18]. The metamaterial absorber is mainly applied on thermal radiation detection [19], thermal radiation imaging [20], and biochemical substance detection [21] and so on.

The microwave at E band with the frequency at the range of 60-90GHz is mainly used on aerospace measurement and control, remote data communication and LTE wideband mobile communication, so the research of metamaterial absorber at E band has great application prospect. A kind of wideband metamaterial absorber at E band based on round metal cells with defect is designed in this paper, its absorption remains above 90% between 65.38GHz t- 67.86GHz, its unit cell has the advantages of simple, easy to prepare and so on.

2. Model design

The designed unit cell of wideband metamaterial absorber at E band in this paper is shown in figure 1. It is composed of round metal cells with defect, dielectric substrate and metal film; the round metal cells with defect is made of copper with the thickness of 0.05mm, its dimensions are: a=b=4.4mm, r=2mm, w1=0.8mm, w2=0.7mm; the dielectric substrate is FR4 dielectric with the thickness of 0.5mm; the metal film is also made of copper with the thickness of 0.1mm.

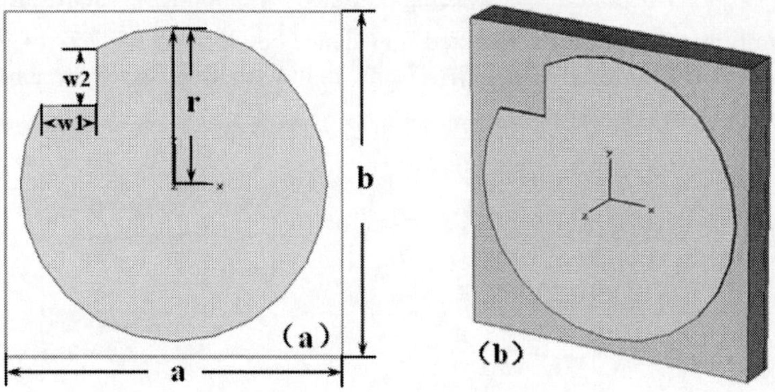

Figure 1. The schematic diagram of unit cell (a) the front view; (b) the perspective view.

The unit cell is designed and simulated by commercial three-dimensional electromagnetic simulation software--Microwave studio CST. In the simulation process, the x and y directions are set as unit cell boundary condition, the z direction is set as open, the all+floquet ports was used to simulate the incoming and outgoing waves. The electromagnetic parameters are calculated by frequency domain electromagnetic solver.

3. Results and discuss

The metal film of unit cell makes the incident wave cannot penetrate, the transmittance is $T(w)=|S_{12}|^2=0$. The defect configuration can convert the incident wave partially into the cross-polarization wave, so the absorption should be calculated by $A(w)=1-R_{x.x}(y.y)-R_{x.y}(y.x)$ with $R=|S_{11}|^2$.

The calculated absorption is shown in figure 2. We can see from figure 2 that the unit cell presents nearly perfect absorption above 90% with absorption ranging from 65.38GHz to 67.86GHz; its width is 2.48GHz.

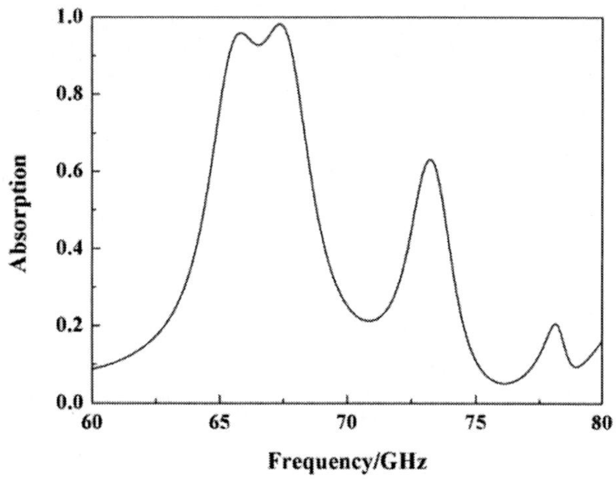

Figure 2. The absorption curve of unit cell.

According to the calculation formula of absorption $A(w) = 1 - R(w) = 1 - \left|\frac{Z(w) - Z_0}{Z(w) + Z_0}\right|^2$, the perfect absorption of incident wave is achieved when the input impedance of metamaterial absorber with free space is equal [22].According to the parameters extracted by simulation, the normalized impedance ($Z(w)/Z_0$) is calculated by scattering parameter method [23], the result is shown in figure 3. We can see from figure 3 that the normalized impedance between 65.38- 67.86GHz is close to one, it indicates that the good impedance match of unit cell with free space is achieved; the perfect absorption can be realized in this case.

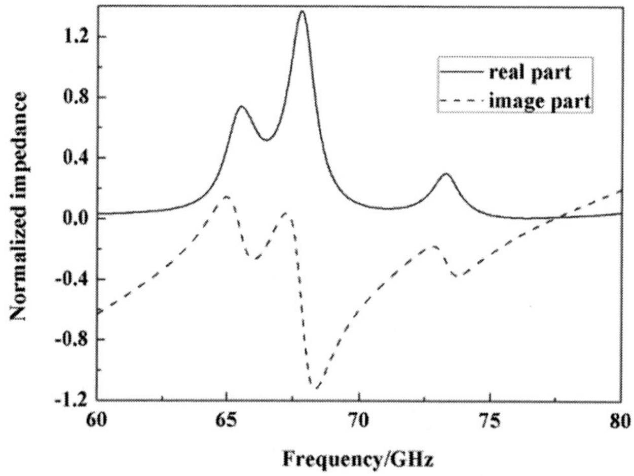

Figure 3. The relative input impedance of unit cell with free space.

In order to explore how the absorption peak is generated, we simulate the surface current of unit cell at 65.823GHz and 67.38GHz, the results are shown in figure 4 and figure 5. It can be seen from figure 4 and figure 5 that the surface current at upper resonant unit is inverse with the surface current at metal film, they form current loop, it indicates that the unit cell couples strongly to the incident magnetic field which lead to an intense magnetic resonance at the absorption peak and the absorption is generated from the loss of dielectric. The overlap of different resonant frequency leads to wideband absorption.

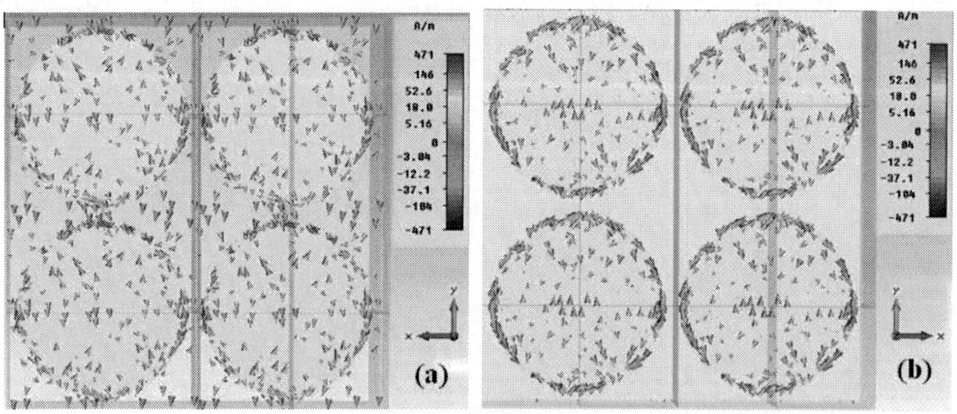

Figure 4. The surface current distribution at 65.823GHz, (a) the upper resonant unit; (b) the metal film.

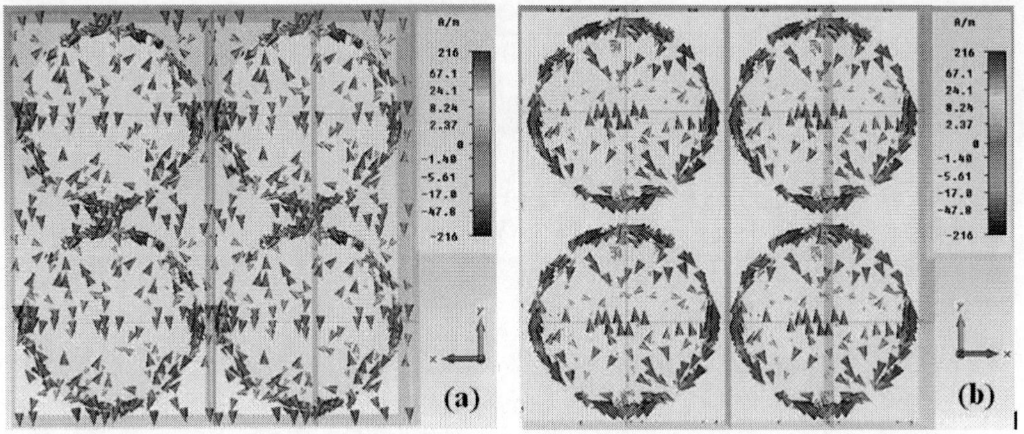

Figure 5. The surface current distribution at 67.38GHz, (a) the upper resonant unit; (b) the metal film.

Here, we discuss the effect of parameters w1 and w2 on the absorption property, respectively. First, in figure 6, we change the parameter w1 from 0.6 mm to 0.8mm with other parameters fixed. There are two group peaks (P1, P2) in the picture. We can see that the relative position of two peaks is sensitive to the parameter w1. When changing the parameter w1 from 0.6 mm to 0.8mm, P1 come close to P2. When the parameter w1 is equal to 0.8mm, the two peaks form a continuous absorption spectrum between 65.38- 67.86GHz with absorptions all above 90%. Figure 7 shows the absorptions with the parameter w2 changing from 0.5mm to 0.7mm.From the picture, the relative position of the absorption peaks is also sensitive to the parameter w2. The result is same with figure 6.

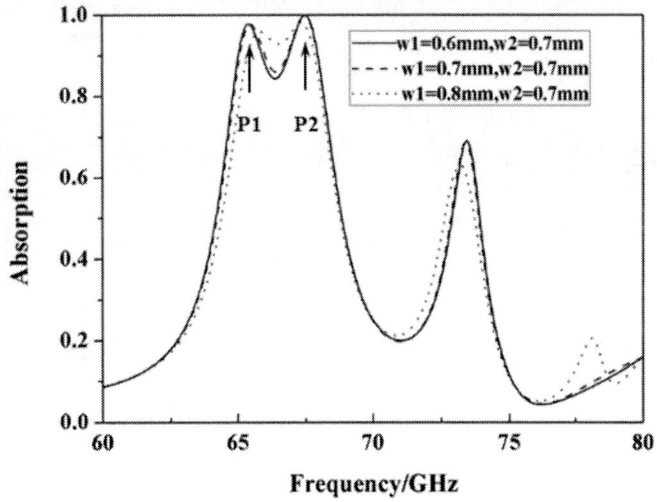

Figure 6. The curve of absorption property with different w1.

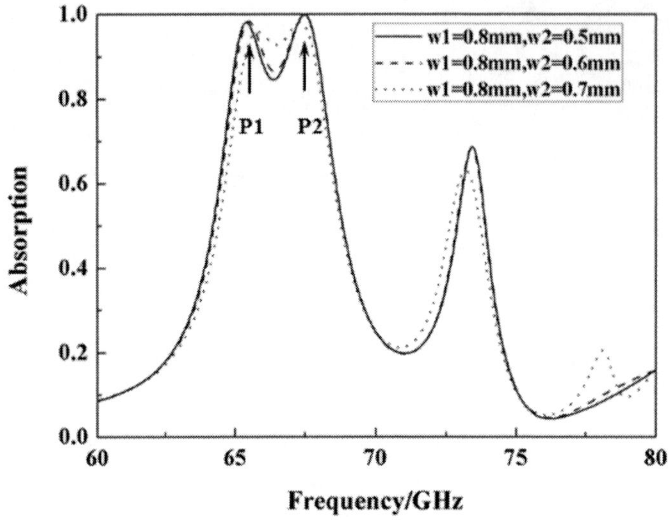

Figure 7. The curve of absorption property with different w2.

Otherwise, we discuss the effect of defect position on its absorption property. In figure 8, we rotate the resonant unit cell along Z axis counterclockwise from 0 degree to 120 degree with other parameters fixed. We can see from figure 8 that the continuous absorption spectrum between 65.38-67.86 GHz with absorptions all above 90% become two absorption peaks (P1, P2) and the P1 absorption peak shift to lower frequency with the increasing of rotating angle.

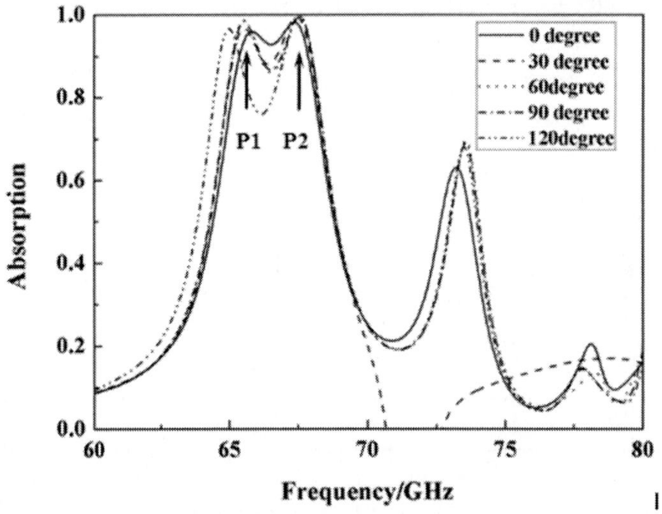

Figure 8. The curve of absorption with different rotation angles.

4. Conclusions

In summary, we propose a wideband metamaterial absorber at E band in this paper which is consisted of round metal cells with defect, dielectric substrate and metal film. The simulation results show that the wideband metamaterial absorber presents nearly perfect absorption above 90% with absorption ranging from 65.38GHz to 67.86GHz. The analysis of the surface current has been performed to better understand the resonant mechanism. Otherwise, we discuss the effect of parameters and position of defect on its absorption property. Therefore, the wideband absorber has potential application in the design and the development of high-performance absorbers for use in aerospace measurement and control, remote data communication, LTE wideband mobile communication and other fields.

Acknowledgments
This paper is financially supported by Hainan natural science foundation of china (No.20165200).

Reference

[1] Landy N I, Sauuyigbe S, Mock J J, Smith D R and Padilla W J 2008 *Phys.Rev.Lett* **100(20)** 207402
[2] Cheng Y Z, Yang H L, Cheng Z Z and Wu N 2011 *Appl.Phys. A* **102(1)** 99-103
[3] Wang B N, Koschny T and Soukoulis C M 2009 *Phys. Rev. B* **80** 033108
[4] Tao H, Bingham C M, Strikwerda A C, Pilon D, Shrekenhamer D, Landy N I, Fan K, Zhang X, Padilla W J and Averitt R D 2008 *Phys. Rev. B* **78 (24)** 241103
[5] Grant J, Ma Y, Saha S, Lok L B, Khalid A and Cumming D R S 2011 *Opt.Lett* **36(8)** 1524-1526
[6] Liu N, Mesch M, Weiss T and Giessen H 2010 *Nano Lett* **10(7)** 2342-2348
[7] Liu X, Starr T, Starr A F and Padilla W J 2010 *Phys. Rev. Lett* **104 (20)** 207403
[8] Aydin K, Ferry V E, Briggs R M and Atwater H A 2011 *Nat. Commun***57 (2)** 517-528
[9] Wang Y, Sun T, Paudel T, Zhang Y, Ren Z and Kempa K 2012 *Nano Lett* **12(1)** 440
[10] Wang J Q, Fan C Z, Ding P, He J, Cheng Y, Hu W, Cai G, Liang E and Xue Q 2012 *Opt. Express* **20 (14)** 14871-14878
[11] Wen Q Y, Zhang H W, Xie Y S, Yang Q H and Liu Y L2009 *Appl. Phys. Lett* **95** 241111
[12] Lee J and Lim S 2011 *Electron. Lett* **47(8)** 8-9
[13] Shen X P, Cui T J and Ye J X 2012 *Acta. Phys.Sin* **61(5)** 058101
[14] Shen X P, Cui T J, Zhao J M, Ma H F, Jiang W X and Li H 2011 *Opt. Express* **19(10)** 9401-9407
[15] Li L, Yang Y and Liang C 2011 *J. Appl. Phys* **110** 063702
[16] Shen X P, Yang Y, Zang Y Z, Gu J Q,Han J G, Zhang W L and Cui T J 2012 *Appl. Phys. Lett* **101** 154102
[17] Bao S, Luo C R, Zhang Y P and Zhao X P 2010 *Acta Phys.Sin* **59(5)** 3187-3191
[18] Yang Y J, Huang Y J, Wen G J, Zhong J P, Sun H B and Oghenemuero Gordon 2012 *Chin. Phys. B* **21(3)** 038501
[19] Kuznetsov S A, Paulish A G, Gelfand A V, Lazorskiy P A and Fedorinin V N 2012 *Progress in Electro-magnetics Research* **122** 93-103
[20] Liu X, Starr T, Starr A F and Padilla WJ 2010 *Phys. Rev. Lett* **104** 207403
[21] Kuznesov S A，Gelfand A V，Lazorskiy P A, Lazorskiy P A and Fedorinin V N 2011 *Appl. Phys.Lett* **99** 023501
[22] Claire M W, Liu X L and Willie J P 2012 *Adv.Mater* **24** 98-120
[23] Smith D Rand Schultz S 2002 *Phys.Rev.B* **65** 195104

IOP Conf. Series: Materials Science and Engineering **292** (2017) 012062 doi:10.1088/1757-899X/292/1/012062

Synthesis and microwave absorbing properties of Cobalt ferrite

H Y Liu[1] and Y S Li

Department of Physics and Electronic Engineering, Heze University, Heze 274015, China

[1] E-mail: liuhongyan507@163.com

Abstract. Cobalt ferrite power $CoFe_2O_4$ was synthesized through the chemical co-precipitation method. The structure, morphology and microwave absorbing properties were studied by changing raw materials, annealing temperature and experimental steps. The measurements of X-ray diffraction and scanning electron micrograph suggest that annealed $CoFe_2O_4$ sample is still a spinel structure. Moreover, the crystalline and grain sizes become large with the enhancement of annealing temperature. The measurements of microwave absorbing properties show that the reflection loss decreases continuously, and the wavelength of maximum absorption loss shift to short-wave limit as the sample thickness increases.

1. Introduction

Ferrite is a new type of non-metallic magnetic materials, which is quickly developed since 1940s. Currently, it is widely applied to not only in military, but also in the civil aspect, with the development of science and technology, and has become a hot topic in the context of function materials [1-3]. As the composition and structure of ferrite material are different, then their permeability, coercive force and loss of intrinsic magnetism are also various. Therefore, one can prepare ferrite materials with excellent performances by controlling their wide composition and structure.

It is generally known that Cobalt ferrite is a typical ferrite material, which has the unique physical, chemical and magnetic properties. For example, its coercive force and resistivity can achieve the level of dozens of times compared with magnetic alloys [4-6]. The magneto-crystalline anisotropy constant of unit Cobalt ferrite at room temperature can reach $2.7*10^5 J/m^3$. The chemical stability is better, abrasion resistant and corrosion stability. Additionally, Cobalt ferrite can still keep high real and imaginary parts of complex permeability at C and Ku wave bands under microwave frequency, which indicates it can produce dielectric loss and magnetic hysteresis loss simultaneously. Therefore, Cobalt ferrite can be used as an important and conventional microwave absorbent [7-9].

In this paper, $CoFe_2O_4$ particle is synthesized by the chemical co-precipitation method by changing the raw materials, annealing temperatures and experimental steps. Magnetic field annealing is used to controlling and improving the microwave absorption properties. Its composition, phase structure and morphology are characterized by X-ray diffraction (XRD: DX-2000, Cu target, Wavelength 1.54184 Å) and scanning electron micrograph (SEM: JMS-6610 LV). The reflection loss is measured by vector network analyzer (AV3629D).

2. Experiments

$CoFe_2O_4$ particle is synthesized by the chemical co-precipitation method based on the following two steps, which are cited as method one and method two for convenience.

Method one: Firstly, $Co(CH_3COO)_2{\cdot}4H_2O$ and $Fe(NO_3)_3{\cdot}9H_2O$ were stoichiometric dissolved in deionized water with Fe/Co=2. Continuously heating when the temperature came to 90 ℃, and then controlling the thermostat and stirred for several minutes. Subsequently, poured into NaOH solution slowly and immediately generated precipitation. After the reaction stopped, and then continuously retained thermal insulation for 40 minutes. Filtrated the gel sediment, washed by deionized water several times until the solution became neutral pH = 7. Finally, it was dried at (100 ±10) ℃ in air and grinded to obtain the precursor of $CoFe_2O_4$, the mixture of $Co(OH)_2$ and $Fe(OH)_3$, which was respectively heated to 500 ℃ and 700 ℃ for 2h , and cooled to room temperature.

Method two: $CoFe_2O_4$ ferrite nanoparticle was prepared by improved chemical co-precipitation method through boiled in water. Firstly, $Co(CH_3COO)_2{\cdot}4H_2O$ and $Fe(NO_3)_3{\cdot}9H_2O$ were stoichiometric dissolved in deionized water and gently stirred. Secondly, poured the mixture into NaOH solution slowly, stirred for several minutes, stopped heating when the temperature of mixture came to 110 ℃ and lasted this temperature for 2 hours. Thirdly, filtrated the gel sediment and washed by deionized water several times until the solution became neutral. Finally, the mixture was dried at 80 ℃ in air and grinded. In order to prepare monodisperse $CoFe_2O_4$ ferrite nanoparticles, $CoFe_2O_4$ powder was washed in alcohol, acetone and distilled water, undergone neultrasonic and centrifugal treatment, dried and grined in air.

3. Results and discussion

3.1. Phase structure and morphology

The XRD patterns of $CoFe_2O_4$ sample synthesized by the method one with annealing temperature 800 K (denoted by A1) and 1000 K (denoted by A2) are plotted in figure 1. It shows that the samples are spinel structures. Moreover, the strongest diffraction peak appears at crystal plane (311). Therefore, the average grain size of A1 sample calculated by Scherrer formula is about 8.54 nm, and that is about 8.57 nm which corresponds to A2 sample. According to low angle (311) diffraction peak and high angle (400) diffraction peak, one can obtain the lattice constant 8.385 Å by Bragg equation, which well coincides with 8.377 Å reported by Ref. [10].

The XRD patterns of $CoFe_2O_4$ sample synthesized by the method two with annealing temperature 400 K (denoted by B1) and 500 K (denoted by B2) are plotted in figure 2. It suggests that diffraction peaks are weaker at lower annealing temperature because of disorderly inter-crystalline structure and nanometer crystal defect result in the continuous change of the lattice space. This indicates that grain is tiny, crystal growth is not complete, and it contains amorphous component. As annealing temperature increases, diffraction peaks become stronger, and $CoFe_2O_4$ grain tends to more complete.

Figure 3 (a) and figure 3 (b) plot the SEM images of $CoFe_2O_4$ sample synthesized by the method one with annealing temperature 800 K and 1000 K, respectively. Figure 4 (a) and figure 4 (b) also plot the SEM images of $CoFe_2O_4$ sample synthesized by method two with annealing temperature 400 K and 500 K, respectively. The elaborate comparisons between different methods and annealing temperature are rather obviously. It can be seen that the morphology of pure $CoFe_2O_4$ is relatively homogeneous rod with the diameter in micron scale. Moreover, the average grain size of $CoFe_2O_4$ nanoparticles synthesized by the same method continuously increases as annealing temperature increases. It shows that there exists obvious aggregation. As the average grain size of $CoFe_2O_4$ nanoparticles is too small and the distance between particles is short, and then the van der Waals force is greater than the gravity, which leads to an aggregation, even crowds or masses into a dense cluster [11, 12], as shown in figure 4.

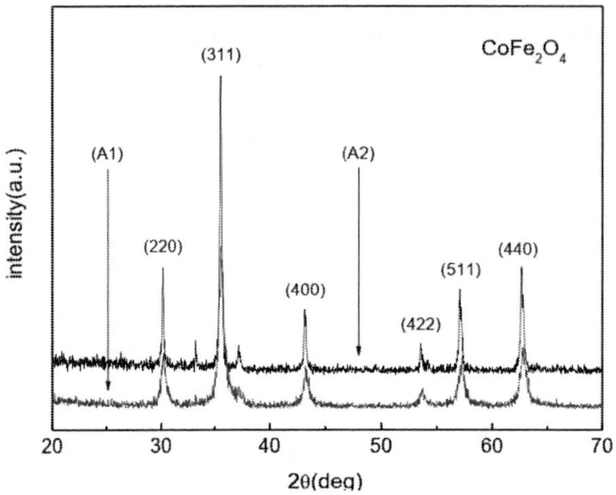

Figure 1. XRD patterns of $CoFe_2O_4$ synthesized by the method one by annealing temperature 800 K (denoted by A1) and 1000 K (denoted by A2).

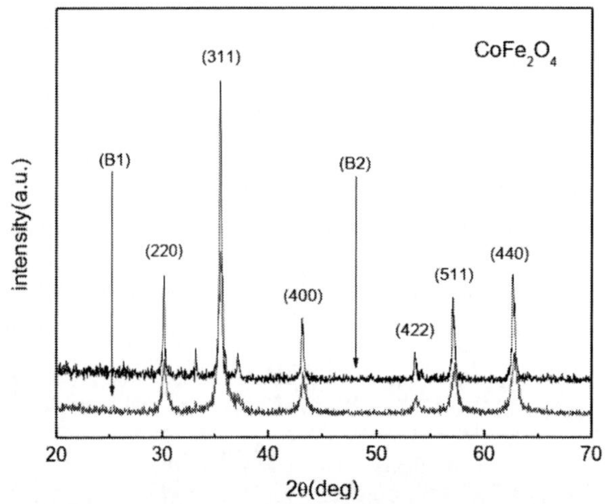

Figure 2. XRD patterns of $CoFe_2O_4$ synthesized by the method two by annealing temperature 400 K (denoted by B1) and 500 K (denoted by B2).

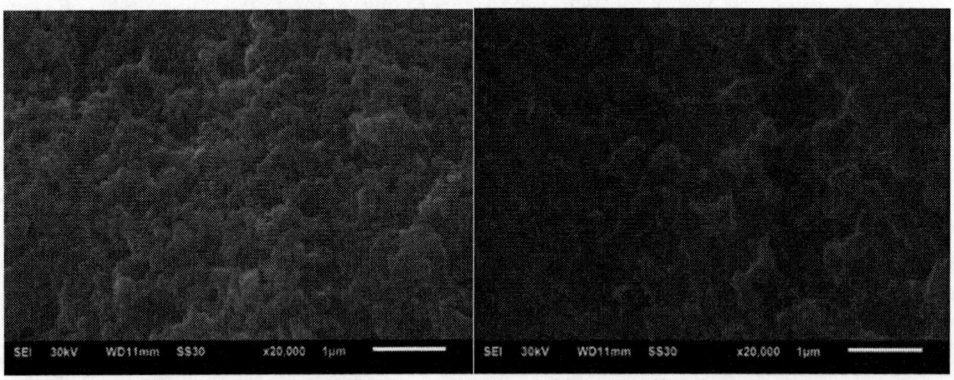

(a) Annealing temperature 800 K (b) Annealing temperature 1000 K

Figure 3. SEM images of $CoFe_2O_4$ synthesized by the method one at different annealing temperatures.

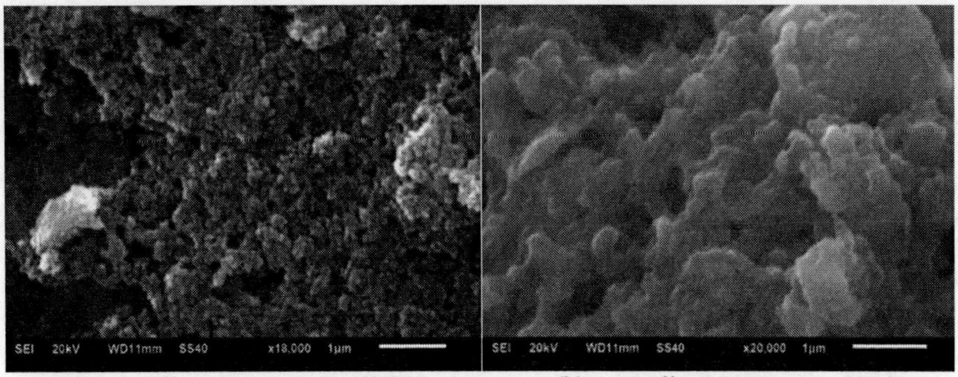

(a) Annealing temperature 400 K (b) Annealing temperature 500 K

Figure 4. SEM images of CoFe2O4 synthesized by the method two at different annealing temperatures.

3.2. *Microwave absorbing properties*

To obtain the reflection loss, one should mix the samples and paraffin wax with ratio 1: 1, poured the mixtures with a beaker into oven, heated to 80 ℃ for several minutes, put them into a coaxial cylinder, pressed into annulations with inner diameter 3.04 mm, outside diameter 7.00 mm, and thickness 1-3 mm.

Figure 5(a) and figure 5(b) are the curves of reflection loss versus frequency of $CoFe_2O_4$ and paraffin wax mixtures synthesized by the method one with different thicknesses at annealing temperature 800 K and 1000 K, respectively. As shown in figure 5, the change trends and reflection loss degree of $CoFe_2O_4$ sample with same thickness and different annealing temperatures are basically similar. The main reason of reflection loss lies in that the change in magnetization lags behind the magnetic field. When the responses of magnetic domain rotation or domain wall displacement lag behind the variation under external magnetic field annealing process, it will produce a large loss. It can be seen that the degree of reflection loss decreases as the sample thickness (2-3mm) increases. Moreover, the wavelength of maximum absorption loss tends to the direction of low frequency and short-wave limit [10]. The reflection loss changes significantly and the maximum absorption peak of the samples can reach to -48 dB, which appears at about 14.5 GHz with a thickness of 2 mm. These results suggest that it is easier to improve microwave absorbing performance through magnetic field annealing. Further experiments demonstrate that the same conclusions can also been obtained by the sample synthesized by the method two.

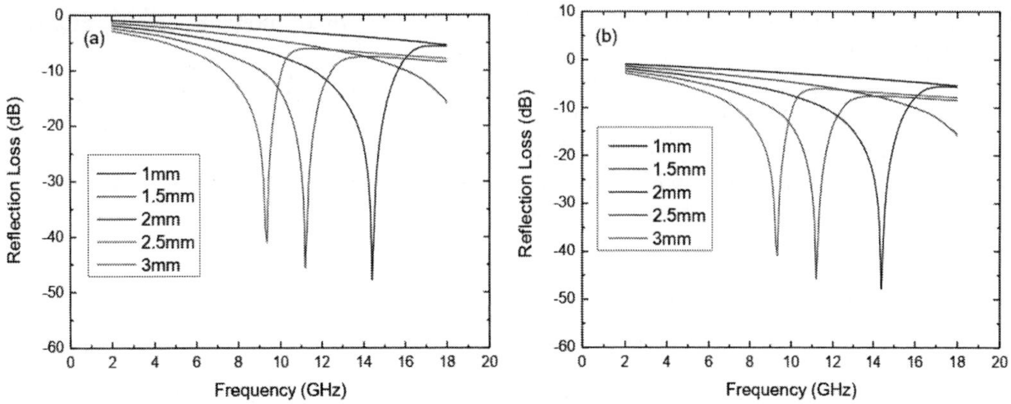

Figure 5. Reflection loss versus frequency of $CoFe_2O_4$ synthesized by the method one with different thicknesses: (a) annealing temperature 800 K, (b) annealing temperature 1000 K.

IOP Conf. Series: Materials Science and Engineering **292** (2017) 012062 doi:10.1088/1757-899X/292/1/012062

4. Conclusions

We have successfully prepared the light-weight and strong absorption microwave absorber $CoFe_2O_4$ by using chemical co-precipitation method. It is suggested that $CoFe_2O_4$ ferrite is still a spinel structure based on different annealing temperatures. The crystalline and grain sizes become large as annealing temperature increases. Our results also show that the reflection loss decreases along with the enhancement of sample thickness. Moreover, the wavelength of maximum absorption loss tends to the direction of short-wave limit. Annealing is advantageous to the improvement of absorbing properties.

Acknowledgements

This work is supported by the natural science foundation of Shandong province (No. ZR2017MEM012) and the technology program for doctor of Heze University (No. XY16BS32).

References

[1] Wu M Z, He H H, Zhao Z S and Yao X 2000 Electromagnetic and microwave absorbing properties of iron fibre–epoxy resin composites *J Phys D: Appl Phys* **33** 2398-2401

[2] Wei J, Liu J H and Li S M 2007 Electromagnetic and microwave absorption properties of Fe_3O_4 magnetic films plated on hollow glass spheres *J Magn Magn Mater* **312** 414-417

[3] Ohlan A, Singh K, Chandra A and Dhawan S K 2008 Microwave absorption properties of conducting polymer composite with barium ferrite nanoparticles in 12.4–18 GHz *Appl Phys Lett* **93** 053114

[4] Rajendran M, Pullar. C, Bhattacharya A K and Majumdar C K 2001 Magnetic properties of nanocrystalline $CoFe_2O_4$ powders prepared at room temperature: variation with crystallite size *J Magn Magn Mater* **232** 71-83

[5] Zhou P H, Deng L J, Xie J L and Liang D F 2008 Effects of particle morphology and crystal structure on the microwave properties of flake-like nanocrystalline Fe_3Co_2 particles *J Alloy Compd* **448** 303-307

[6] Mohamed R M, Rashad M M, Haraz F A and Sigmund W 2010 Structure and magnetic properties of nanocrystalline cobalt ferrite powders synthesized using organic acid precursor method *J Magn Magn Mater* **322** 2058-2064

[7] Zhang X F, Dong X L, Huang H, Lv B, Lei J P and Choi C J 2007 Microstructure and microwave absorption properties of carbon-coated iron nanocapsules *J Phys D: Appl Phys* **40** 5383–5387

[8] Che R C, Zhi C Y, Liang C Y and Zhou X G 2006 Fabrication and microwave absorption of carbon nanotubes/$CoFe_2O_4$ spinel nanocomposite *Appl Phys Lett* **88** 033105

[9] Li Y S and Liu H Y 2015 Preparation and microwave absorbing properties of $CoFe_2O_4$/cenosphere composites *J At Mol Phys* **32** 147-152

[10] Park J, Privman V and Matijevic E 2001 Model of formation of monodispersed colloids *J Phys Chem B* **105** 11630-11635

[11] Wu S H, Zhang S M, Huang W P, Shi J and Li B Q 2003 Mechano-synthesis Characterization and Magnetic Properties of Nanoparticles of Cobalt Ferrite $CoFe_2O_4$ orepared by metal vapor tevhniques *J Mater Sci Technol* **19** 422-424

[12] Mooney K E, Nelson J A and Wagner M J 2004 Superparamagnetic cobalt ferrite nanocrystals synthesized by alkalide reduction *Chem Mater* **16** 3155-3161

IOP Conf. Series: Materials Science and Engineering **292** (2017) 012063 doi:10.1088/1757-899X/292/1/012063

An experimental investigation of a thermoelectric power generation system with different cold-side heat dissipation

Y H Li[1], Z H Wu[1, 3], H Q Xie[1], J J Xing[1], J H Mao[1], Y Y Wang[1] and Z Li[2]

[1] School of Environmental and Materials Engineering, College of Engineering, Shanghai Polytechnic University, Shanghai 201209, China
[2] School of Environmental and Chemical Engineering, Shanghai University, Shanghai 200444, China

[3] E-mail: wuzihua@sspu.edu.cn

Abstract. Thermoelectric generation technology has attracted increasing attention because of its promising applications. In this work, the heat transfer characteristics and the performance of a thermoelectric generator (TEG) with different cold-side heat dissipation intensity has been studied. By fixing the hot-side temperature of TEG, the effects of various external conditions including the flow rate and the inlet temperature of the cooling water flowing through the cold-sided heat sink have been investigated detailedly. It was showed that the output power and the efficiency of TEG increased with temperature different enlarged, whereas the efficiency of TEG reduced with flow rate increased. It is proposed that more heat taken by the cooling water is attributed to the efficiency decrease when the flow rate of the cooling water is increased. This study would provide fundamental understanding for the design of more refined thermoelectric generation systems.

1. Introduction

Thermoelectric generators (TEGs) are solid state devices that use the Seebeck effect to convert heat (driven by a temperature difference) directly into electrical energy [1-3]. The basic structures of TEGs are formed by p- and n-type semiconductor thermoelements (TEs). TEs are connected electrically in a serial manner to increase the output voltage of the TEG and connected thermally in parallel to decrease the thermal resistance of the TEG [4]. Compared to the conventional generator, thermoelectric devices (TEDs) have numerous advantages such as low-grade thermal energy source required, environmental friendly, longer lifetime, high reliability, compactness and robustness [5-7]. Generally, a thermoelectric generation system mainly includes a heat source, a cooler, heat exchanger and thermoelectric device that are thermally and electrically combined together as a whole. The design of a good thermoelectric generation system involves integrate consideration of the structure, heat source, type of thermoelectric device, installation, etc.[8]

Over the past 30 years, there has been growing interest in applying TE technology to improve the efficiency of waste heat recovery, using various heat sources such as automotive exhaust systems [9, 10], cooling systems[11], solar thermal systems [12], domestic water heaters [13, 14]. Gou et al [15] investigated the viability and performance of the TEG for low-temperature waste heat recovery in industry area and showed a promising potential of using TEG in industrial fields. Lu et al. [16] studied two types of heat enhancement processes for exhaust heat exchanger on the performance of TEG. The result showed there was an optimal fin transverse spacing or fin thickness at the operating conditions for rectangular offset-strip fin and metal foam with low porosity which made the TEG got maximum net power output. Hasani et al. [17] experimentally inspected that thermoelectric cooler as a power generator applied in waste heat recovery from a Proton Exchange Membrance (PEM) fuel cell and proved a suitable solution for recovering waste heat from a PEM fuel cell.

However, low conversion efficiency and power density are crucial factors restricting the development of TEGs. Numerous studies have been conducted on improving the conversion efficiency of TEGs [9, 18-20]. Liu et al [21] studied the influences of the hot and cold side temperature, flow rates and the load resistance on the power output and voltage of TEG. Through adjusting the flow rates, the maximum power output was 183.24W with copper heat exchanger and engine revolution of 3200 r/min. Chen et al.[22] theoretically built a model of a two-stage TEG structure and conducted the performance of the model on the output power and thermal efficiency by different working electric-current. Casano and Piva [23] built an experimental rig to evaluate the characteristic of the thermoelectric modules . They presented the power output and the conversion efficiency of the thermoelectric modules under different load resistances. Researchers have proposed a number of TEG systems with different features, structures and properties to improve the output power and conversion efficiency. For a commercial TEG system, the physical properties such as Seebeck coefficient and thermal conductivity were always taken as constant to consider the external heat exchanger performance affecting the performance of a TEG. In this study, the output power and conversion efficiency of TEGs at various external parameters, such as cooling water flow rate and temperature of copper heat sink to change the cold-side temperature of TEG, were conducted by self-designed devices. The results of the experiments could provide a fundamental understanding for the optimization of more refined thermoelectric generation systems.

2. Experimentation

A schematic diagram of the experimental apparatus used in the present work was shown in figure1. The main components of the experimental apparatus include a peristaltic pump (Masterflex L/S, 77202-50), thermostatic bath (NingBo Scientz Biotechnology co., LTD, DCW-2008), copper heat sink (shown in figure 2(d)), thermoelectric device (shown in figure 2(b and c)), heating plate (IKA, C-MAG-HP4) and liquid cabin.

In the thermostatic bath, the working medium, water, was guided through the copper tube with a radius of 2.4mm and total length of 3.50m. Three K-type thermocouples (Omega, KMTXL-040G-6) were used to test the initial temperature (T_1), terminal temperature (T_2) of the flowing water and water bath temperature (T_w), respectively. The liquid cabin with 1000ml accommodate of water was properly insulated for reducing heat transfer to the environment. Bi_2Te_3-based thermoelectric devices (TEDs) were supplied by Shanghai Institute of Ceramics, Chinese Academy of Sciences. The appearance and specifications of the TEDs are shown in figure2 (b), figure2 (c) and table 1, respectively. One side of the TED was attached to the heating plate and the other side was attached to copper heat sink in vertical orientation. The temperature of the heating plate was fixed at 100°C in the present work. The peristaltic pump with a maximum capacity of 1000 ml/min was employed to accurately control the flow of water through the copper heat sink. The pipe connecting the various components was platinum-cured silicone tubing (Masterflex, 96410) with internal diameter of 0.48cm.

The thermocouples (omega, K-type, KMTXL-040G-6) were fixed at appropriate points as shown in figure1. The thermocouples were connected to a data acquisition device (34970A, Agilent), which was connected to a computer to measure and record the detailed data (open circuit voltage, temperature) for further processing.

The resolutions of flow meter, voltage measurements, and thermocouple were $0.01L \cdot min^{-1}$, 1.25mV, and 0.1°C, respectively. The uncertainty of the experiments was smaller than 6% from references [7, 24] so that the experiments in the current study are reliable. Moreover, each case has been repeated twice at least and the data of experiments has been ensured.

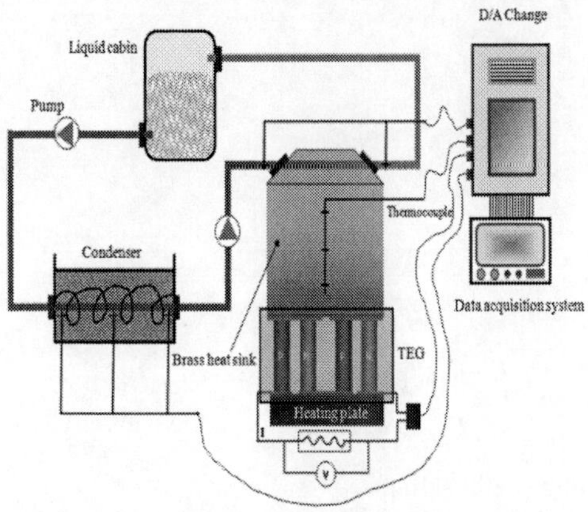

Figure 1. Schematic of the experimental test setup

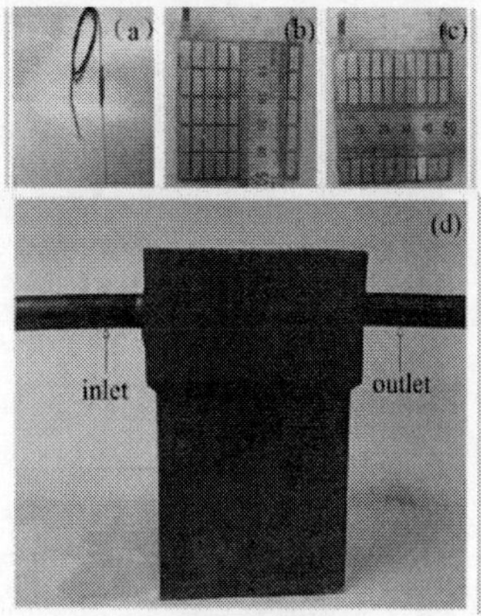

Figure 2. Photographs: (a) K-type thermocouples; (b) and (c) thermoelectric device; (d) copper heat sink

Table 1. Specifications of the thermoelectric device

Specifications	Value
Type of material	Bi_2Te_3
Maximum current (A)	7
Maximum voltage (V)	3.5
Maximum power (W)	7.6
T_{max} (°C)	270
Dimensions(L×W×H) (mm^3)	50*50*3.5
Number of couples	49

3. Methodology

3.1. Thermostatic bath

The role of thermostatic bath was to adjust the fluid temperature of copper cube via heat dissipation. In the system, the thermostatic bath ensured the terminal temperature (T_2) (seen from figure 3) of the flowing water close to the set temperature of water bath (T_w) and made the fluid entering into the heat sink at the designed experimental temperature.

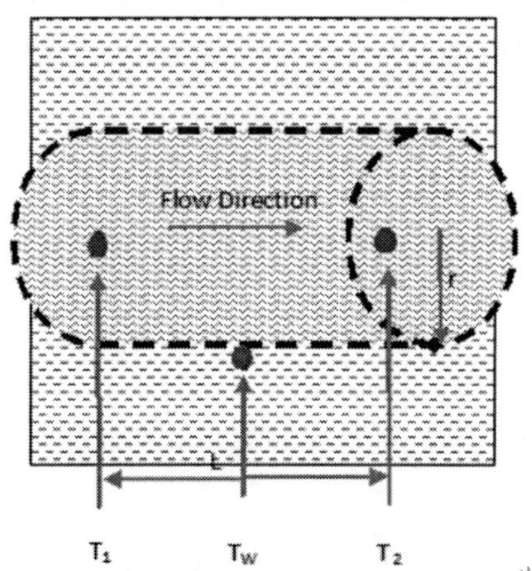

Figure 3. Schematic of fluid through thermostatic bath

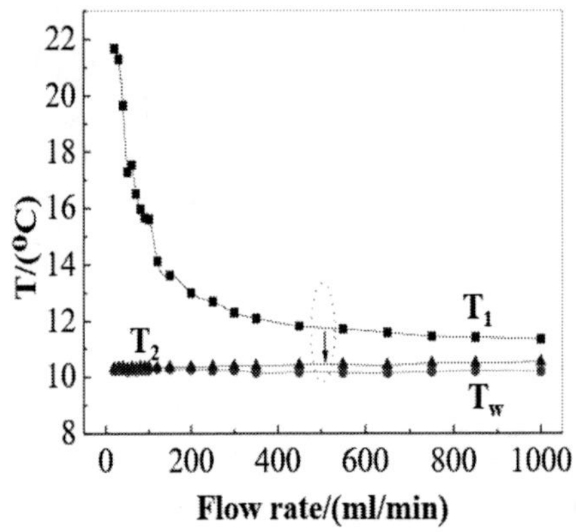

Figure 4. Distributions of T_1, T_2, and T_w with the thermostatic bath temperature set at 10 °C

3.2. Copper heat sink

Deionized water was used as coolant in this study. The inlet temperature (T_{in}) and outlet temperature (T_{out}) of the heat sink were measured by two thermocouples, the average temperature and the heat taken away by water was calculated by the following expressions.

$$T_C \approx T_{average} = \frac{T_{in} + T_{out}}{2} \tag{1}$$

$$Q_W = M \times C_P \times (T_{out} - T_{in}) \tag{2}$$

where M is mass flow rate, C_p is the fluid (water) heat capacity, T_{in} and T_{out} are the inlet temperature and the outlet temperature, respectively, as showed in Figure2 (d).

3.3. Thermoelectric generator

The temperature difference between the hot and cold sides is got as the following.

$$\Delta T = T_H - T_C \tag{3}$$

where T_H and T_C are the hot-side temperature and the cold-side temperature of TEG, respectively. The output power (W) can be calculated as

$$P_{out} = {U^2}/{4R} \tag{4}$$

where U is the output voltage (V) which is measured from the electronic load, R is the internal resistance of the TEG and is measured to be 0.254Ω.

The efficiency of the thermoelectric generation power is calculated by

$$\varphi = \frac{P_{out}}{P_{out} + Q_W} \times 100\% \tag{5}$$

4. Results and discussion

4.1. Thermostatic bath

Figures 4-6 show the distributions of T_1, T_2, and T_w at different flow rate with the thermostatic bath temperature set at 10, 20, and 30 °C, respectively. It can be seen that there is an obvious change from initial to terminal temperature of the fluid at low flow rates and the difference between T_1 and T_2 becomes smaller with the flow rate increased. The terminal temperature (T_2) was close to the designed water bath temperature (T_W) which made the fluid entering into the heat sink at the designed experimental temperature.

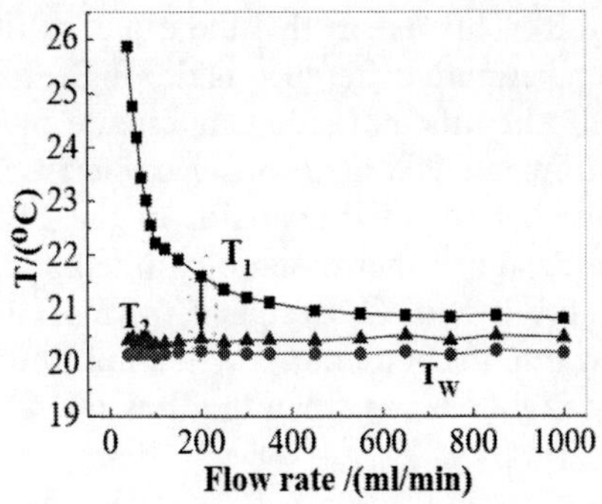

Figure 5. Distributions of T_1, T_2, and T_w with the thermostatic bath temperature set at 20°C

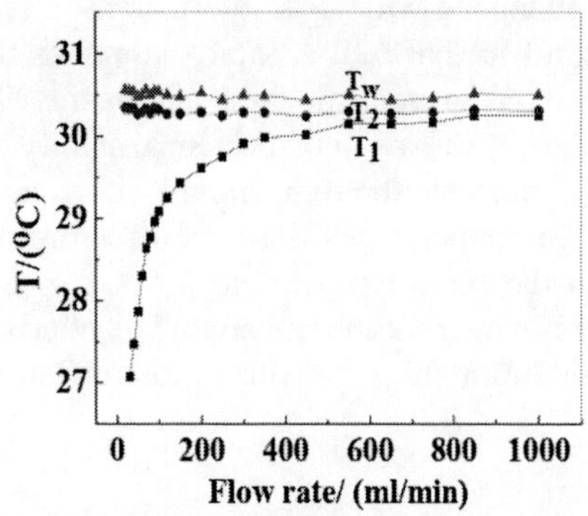

Figure 6. Distributions of T_1, T_2, and T_w with the thermostatic bath temperature set at 30°C

4.2. Copper heat sink

To test the influence of cold-side heat dissipation on the performance of TEG, cooling water with different initial temperature (from the thermostatic bath) was flow through the copper heat sink at different flow rate. Figure7 presents the temperature difference of outlet and inlet of the heat sink at different flow rate and thermostatic bath temperature. It can be seen from figure 7 that temperature difference of outlet and inlet of heat sink reduced rapidly at the lower flow rate (≤100ml/min). This means that water can effectively transfer heat from the cold side of TEG. When the flow rate was over 100ml/min, the temperature difference only changed slightly with the flow rate increase. It indicates that further increasing flow rate would not have obvious effect on reducing the average temperature of the cold side of the TEG system when the flow rate was over 100ml/min. That is to say, the output power will not improve obviously by increasing flow rate when the flow rate was over 100ml/min. Further discussion will be presented in the following part. When the flow rate of cooling water is increased, more pump power would be consumed. Also, the cooling water may take away more heat from the cold side of the TEDs. Therefore, increasing the flow rate of the cooling water is not always helpful to the TEG system, especially at high flow rates.

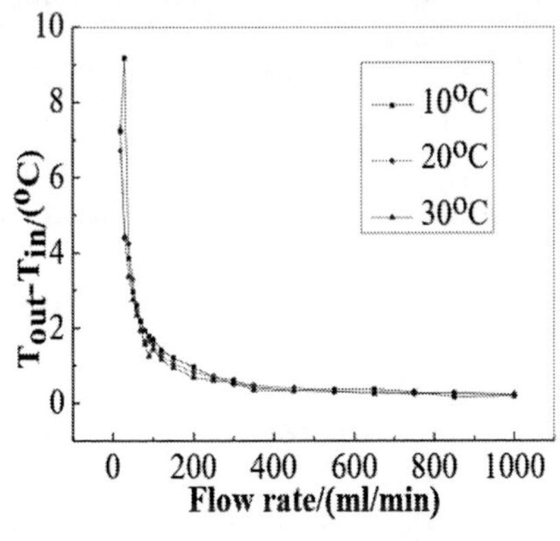

Figure 7. Temperature difference of the outlet and inlet of the heat sink at different flow rate and thermostatic bath temperature

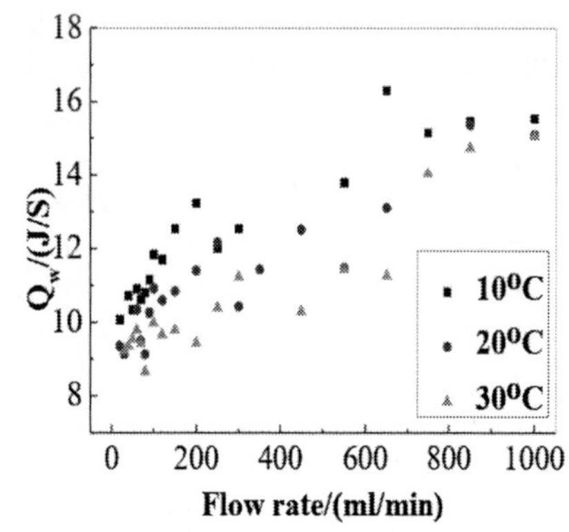

Figure 8. Variation of the heat flow of the cold side of the TED with the flow rate

Figure 8 shows the variation of the heat flow taken away by the cooling water at different flow rate and different temperature conditions of thermostatic bath. From Eq.(2), it can be seen that the heat flow is related to the flow rate, water specific heat capacity and the temperature difference of outlet and inlet of the heat sink. Although the temperature difference from Figure7 was reduced, the heat flow was enhanced with flow rate increased (seen in Figure8). This means that the effect of the flow rate and fluid special heat capacity surpasses that of the temperature difference under high flow rate. More heat is taken from the cold side of the TEDs. That is, when the hot-side temperature of TEG is fixed, the heat taken away by heat sink mainly depends on flow rate and fluid self-characteristic.

Figure 9 presents the temperature difference between the hot-side and cold-side of the TED. It is seen that the temperature difference rises rapidly when the designed thermostatic bath temperature is lower than the room temperature (25 °C), but reduces slightly when the designed thermostatic bath temperature is over room temperature at relatively low flow rate (≤100ml/min). When the flow rate is higher than 100ml/min, the temperature difference has only slight change when the flow rate is further increased.

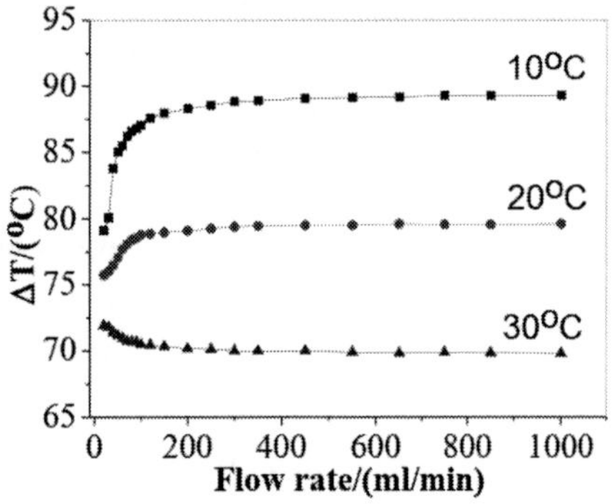

Figure 9. Temperature difference between the hot and cold sides of the TED at different flow rates

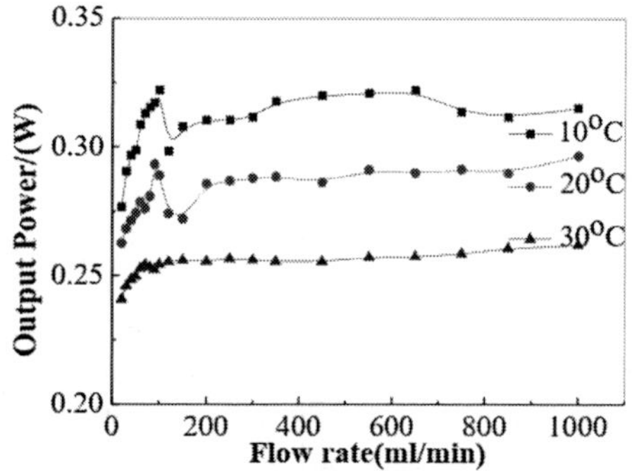

Figure 10. Distribution of the output power of the TEG system at different flow rate

4.3. Thermoelectric generator

The influence of temperature difference and flow rate on the performance of TEG is shown in figure10. The cold-side temperature was varied while the hot-side temperature of TEG was fixed at 100°C in this study. The temperature difference between the hot-side and cold-side of the TED is shown in figure 9. It can be seen that the temperature difference has a great effect on the output power. When the flow rate was 0 (namely natural convection cooling), the output power was 0.08152W, 0.07102W and 0.07069W when the thermostatic bath was set at 10°C, 20°C and 30°C, respectively. When the cooling water flow rate was 20ml/min, the output power was 0.27669V, 0.26267V and 0.24078V when the thermostatic bath was set at 10 °C, 20 °C and 30 °C, which was increased by about 239%, 270% and 241% compared to natural convection cooling, respectively. Cold side heat dissipation has significant effect on the output power of the TEG system.

At the condition of the same inlet temperature of the cold side heat sink and fixed temperature of the hot-side heat sink, more heat would be taken away by the cooling water when the flow rate is increased. The average temperature of the cold-side will be decreased. The open circuit voltage and the output power will be increased. Practically, when the flow rate was lower than 100 ml/min, the temperature difference could not reach the ideal temperature difference (the temperature difference between the hot side and the set thermostatic temperature), as showed in figure 9. The output power could be increased obviously when the flow rate of the cooling water was increased. However, from figure 10, it can be observed that the output power was not change accordingly to the temperature difference shown in figure 9. When flow rate was greater than 100 ml/min, the effect of flow rate increase was not so significant. Similar behavior has been reported in the theoretical and practical results of Esarte et al. [7, 25] and Chen et al. [7, 25] .

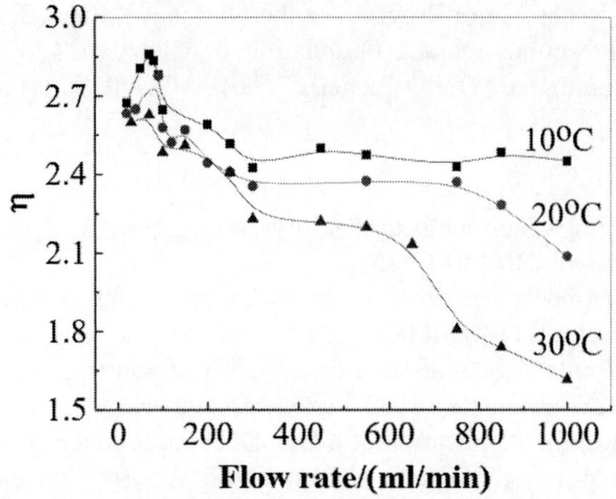

Figure 11. Conversion efficiency of the TEG system at different flow rate

Based on the heat flow taken away by the cooling water (figure 8) and the output power (figure 10), the conversion efficiency of the TEG system can be obtained by using Eq. (5). Figure 11 shows the variation of the conversion efficiency at different flow rate and different temperature conditions of the thermostatic bath. The conversion efficiency of the TEG system was enhanced with temperature difference increased and reduced with flow rate increased for the same thermostatic bath temperature when the hot-side temperature fixed. As expressed in Eq. (5), the efficiency of the TEG system was inversely proportional to Q_w/P_{out}. With increasing the flow rate, the output power increased slightly. At the same time, more heat was taken away from the cold side. Therefore, the conversion efficiency would not be raised with increasing flow rate accordingly. However, in spite of the efficiency of the TEG is not increased with the flow rate of the cooling water, it is desirable to enhance the capacity of the cold side cooling. Because many types of the heat sources for the TEG systems, e.g. the waste heat such as automobile exhaust heat and solar heat are continuous. If the heat is not used or recovered, it will be lost totally. Therefore, more power generation is utmost consideration and the optimal efficiency is not always required.

5. Conclusions

The paper presented the experimental results of the performance of a model thermoelectric generator for heat recovery and power generation at various external parameters. Water as flowing fluid could absorb and release heat to adjust the whole system temperature. The primary conclusions from this study are summarized as follows.

(1) The thermostatic bath was employed to adjust the cold-side temperature of the TEG by controlling the set temperature. It was found that the terminal temperature (T_2) of the copper cube was close to the set water bath temperature (T_W), making the cooling water to enter into the cold-side heat sink with the designed temperature.

(2) For the cold-side heat sink, although more heat (Q) was taken away by the cooling water with flow rate increased, the temperature difference ($\triangle$T) between hot-side and cold-side of TEG was not always increased. When the flow rate was over 100ml/min, $\triangle$T only changed slightly because the heat transfer capacity could overpass the heat transfer capacity from the hot side to cold side.

(3) For the TEG, the temperature difference was the most important factor for the output voltage and the power. Although the efficiency of the TEG was enhanced with the temperature difference between the hot and cold sides increased when the hot-side temperature was fixed, it was interesting that the efficiency of the TEG was reduced with the cooling water flow rate increased under the condition of the same inlet temperature. It is proposed that more heat taken by the cooling water is attributed to the efficiency decrease.

Acknowledgment
This work was supported by the Major Program of the National Natural Science Foundation of China (51590902), the National Natural Science Foundation of China (51476095), and the Program for Professor of Special Appointment (Young Eastern Scholar, QD2015052) at Shanghai Institutions of Higher learning.

References
[1] Riffat SB , Ma XL 2003 Thermoelectrics: a review of present and potential applications *Applied Thermal Engineering* **23**(8) 913-935
[2] Tritt TM 2011 Thermoelectric Phenomena, Materials, and Applications *Annual Review of Materials Research* **41**(1) 433-448
[3] Bjrk R, Sarhadi A, Pryds N, et al 2016 A thermoelectric power generating heat exchanger: Part I – Experimental realization *Energy Conversion and Management* **119** 473-480
[4] Mamur H, Ahiska R 2015 Application of a DC–DC boost converter with maximum power point tracking for low power thermoelectric generators *Energy Conversion and Management* **97** 265-272
[5] Aanguren P, Astrain D 2014 Computational and experimental study of a complete heat dissipation system using water as heat carrier placed on a thermoelectric generator *Energy,* **74** 346-358
[6] Al-nimr M, Tashtoush B, Jaradat A 2015 Modeling and simulation of thermoelectric device working as a heat pump and an electric generator under Mediterranean climate *Energy* **90 Part 2** 1239-1250
[7] Chen W, Liao CY, Hung CI, et al 2012 Experimental study on thermoelectric modules for power generation at various operating conditions *Energy* **45**(1) 874-881
[8] Ashby SP, Garc AJ, Min G, et al 2013 Measurement of Thermoelectric Properties of Phenylacetylene-Capped Silicon Nanoparticles and Their Potential in Fabrication of Thermoelectric Materials *Journal of Electronic Materials* **42**(7) 1495-1498
[9] Liu X, Deng YD, Zhang K, et al 2014 Experiments and simulations on heat exchangers in thermoelectric generator for automotive application *Applied Thermal Engineering* **71**(1) 364-370
[10] Liu X, Deng YD, Chen S, et al 2014 A case study on compatibility of automotive exhaust thermoelectric generation system, catalytic converter and muffler *Case Studies in Thermal Engineering* **2** 62-66
[11] Jeng LY, Teng TP 2013 Performance evaluation of a hybrid cooling system for electronic chips *Experimental Thermal and Fluid Science* **45** 155-162
[12] Zheng XF, Yan YY, Simpson K 2013 A potential candidate for the sustainable and reliable domestic energy generation–Thermoelectric cogeneration system *Applied Thermal Engineering* **53**(2) 305-311
[13] Qiu K, Hayden ACS 2008 Development of a thermoelectric self-powered residential heating system *Journal of Power Sources* **180**(2) 884-889
[14] Qiu K, Hayden ACS 2009 A Natural-Gas-Fired Thermoelectric Power Generation System *Journal of Electronic Materials* **38**(7) 1315-1319
[15] Gou XL, Xiao H, Yang SW 2010 Modeling, experimental study and optimization on low-temperature waste heat thermoelectric generator system *Applied Energy* **87**(10) 3131-3136
[16] Lu C, Wang SX, Chen C, et al 2015 Effects of heat enhancement for exhaust heat exchanger on the performance of thermoelectric generator *Applied Thermal Engineering* **89** 270-279
[17] Hasani M, Rahbar N 2015 Application of thermoelectric cooler as a power generator in waste heat recovery from a PEM fuel cell – An experimental study *International Journal of Hydrogen Energy* **40**(43) 15040-15051
[18] Zhang GB, Fan LH, Niu ZQ, et al 2015 A comprehensive design method for segmented thermoelectric generator *Energy Conversion and Management* **106** 510-519

[19] Jin SY, Ziabari A, Koh YR, et al. Enhanced thermoelectric performance of P-type nanowires with pulsed laser assisted electrochemical deposition, *Extreme Mechanics Letters*
[20] Attia SI 2015 The influence of condenser cooling water temperature on the thermal efficiency of a nuclear power plant *Annals of Nuclear Energy* **80** 371-378
[21] Liu X, Li C, Deng YD, et al 2015 An energy-harvesting system using thermoelectric power generation for automotive application *International Journal of Electrical Power & Energy Systems* **67** 510-516
[22] Chen LG, Li J, Sun FG, et al 2005 Performance optimization of a two-stage semiconductor thermoelectric-generator *Applied Energy* **82(4)** 300-312
[23] Casano G, Piva S 2011 Experimental investigation of the performance of a thermoelectric generator based on Peltier cells *Experimental Thermal and Fluid Science* **35(4)** 660-669
[24] Chen WH, Liao CY, Wang CC, et al 2015 Evaluation of power generation from thermoelectric cooler at normal and low-temperature cooling conditions *Energy for Sustainable Development* **25** 8-16
[25] Esartea J, Minbroweb DM 2001 Modelling heat exchangers for thermoelectric generators *Journal of Power Sources* **93** 72-76

IOP Conf. Series: Materials Science and Engineering **292** (2017) 012064 doi:10.1088/1757-899X/292/1/012064

Microstructure analysis of snail trails in photovoltaic modules

J Tao[1], Z X Shi, S Y Chen, J X Kong, Q S Huang and X Fg Gou

CECEP Solar Energy Technology (Zhenjiang) Co., Ltd.,Bei Shan Road 9 , Zhenjiang 212132, People's Republic of China

[1]E-mail: waitflying@126.com

Abstract. Snail trails on photovoltaic modules are a source of enormous concern to the solar industry as few scientific reports on the mechanisms producing this global phenomenon were previously available. The samples surface were treated with CH_3OH /CH_2Cl_2 and used the SEM and Raman for material analysis. The size of the discoloration silver grid is about 80-200 nm. From the Raman spectroscopy can be seen snail trails and the surrounding discoloration of silver on the Ag_2CO_3 generation.

1. Introduction

The investigation of defects and degradation in photovoltaic modules has become an important subject since reliability and product lifetime are keys to system performance [1, 2]. One of the degradation effects that has been observed increasingly during recent years is "snail trails" or "earth worms", thus called because of their visual and power appearance [3, 4]. However, long-term negative influences [5] cannot be excluded and require further study.

The main goal of the work presented is to uncover the chemical and microstructural nature of discoloured grid fingers. It can be investigated thoroughly several PV modules affected by snail trails after outdoor exposure, as well as mini modules with snail trails induced after treatment in the damp-heat chamber. It can be shown that the same chemical mechanism is underlying the grid finger discoloration in both the real modules and the model system. This enables us to set up a model for the formation of snail trails and to develop strategies for their prevention.

2. Material and methods

Samples of EVA encapsulation about 400×400 mm^2 in glass size were incubated under accelerated conditions. PV modules with snail trails were disassembled mechanically to obtain the cell and encapsulation foil surface after earthworms appear. As shown in figure 1, select the snail trails section or around the gate line discoloration of the module, and heat on the heating table for 5 minutes at 80 $^{\circ}C$ temperature. After heating, the EVA film with cell use the blade directly to the battery chip with shovel. Using of tweezers on the surface of the battery chip stripped of EVA, and then cut into small 1×1cm^2 with scissors. Raman test samples no longer need other treatment, direct focus test can be tested. The microstructure snail trails analysis were prepared for field emission scanning electron microscope (SEM, JSM-7001F) and Laser Raman Spectrometer (USA, ThermoFisher, DXR).

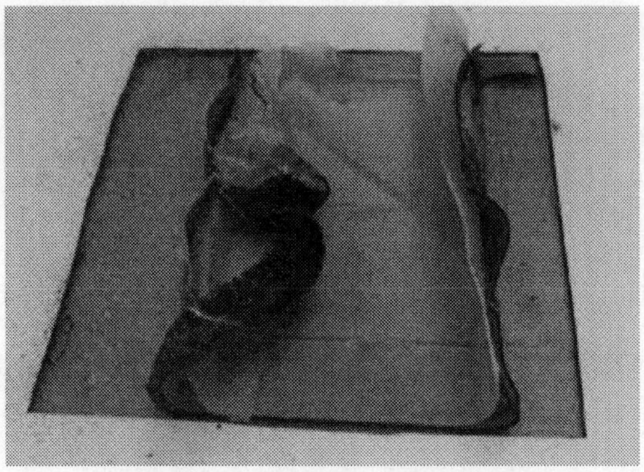

Figure 1. Sample drawing.

3. Results and discussion

3.1. Sample preparation

Figure 2 shows the snail trails appear in the physical map and EL (Electro Luminescence) under the test picture. Figure 2(a) is an enlarged view of figure 2(b). The small gray particles is the snail trails in figure 2(a). The Snail trails are generally appear next to the silver gate line. This is a familiar phenomenon in the industry. It can be seen from the figure 2, snail trails appear similar to the yellow floral color of small particles. The samples were treated separately using water and CH_3OH/CH_2Cl_2, respectively. And figure 3 shows the SEM image before and after CH_3OH/CH_2Cl_2 washing. Samples for SEM testing require spray treatment. In view of the test found that Ag oxide generation control below 80 °C to peel off, the sampling temperature of the heating platform can not be too high, otherwise it will form silver oxide. As shown in figure 3, an oxygen peak occurs at 145 °C and no at 80 °C [6, 7]. EVA stripping material on the test results will have a great impact [8, 9]. The CH_3OH/CH_2Cl_2 of snail trails on the sample surface treatment, organic matter is dissolved and the surface nanoparticles still exist, which as a method of testing the following particles. It can be seen that the particle size does not change.

a b

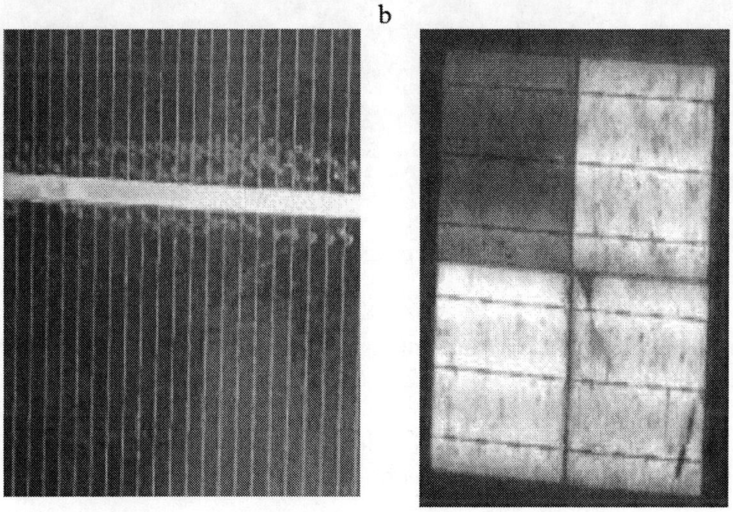

Figure 2. (a) appear snail trails; (b) EL test.

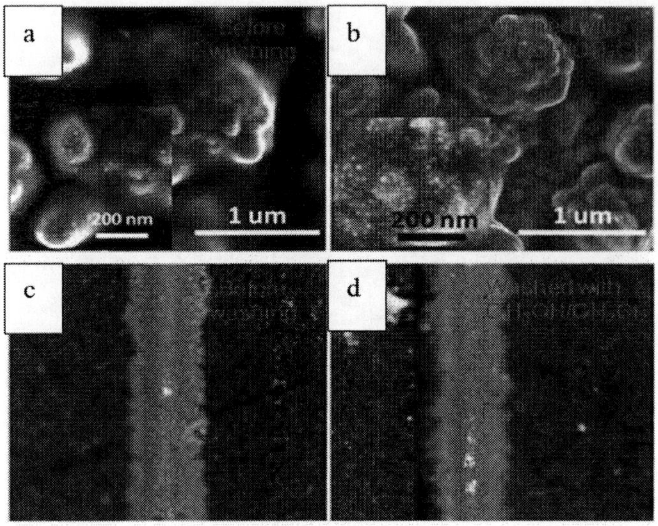

Figure 3. (a)(c)SEM of snail trails before washing ;
(b)(d) SEM of snail trails after CH_3OH/CH_2Cl_2 washing.
(80 °C stripping)

3.2. SEM and EDS

It can be seen from figure 4 that there are a large number of particles on the snail trails grain, and the particle size is about 80 nm. There is also a small amount of particles on the normal grid lines at the discoloration. The particles are larger in relation to the snail trails, which is about 100-200 nm.

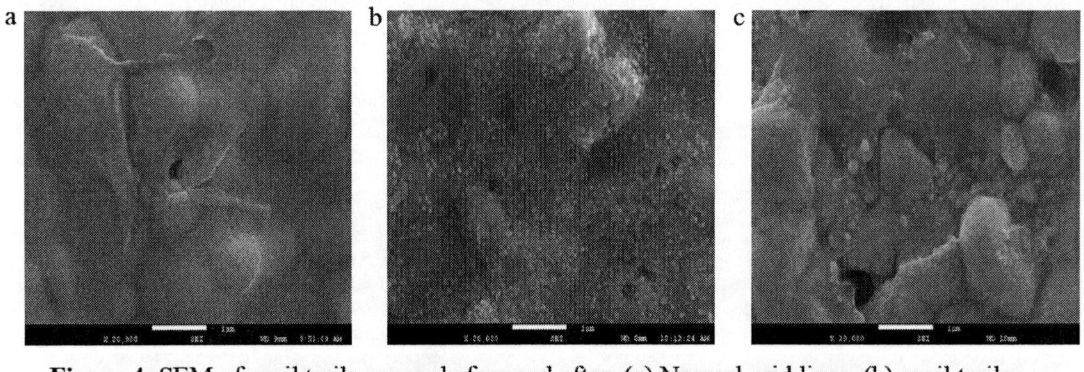

Figure 4. SEM of snail trails appear before and after. (a) Normal grid lines, (b) snail trails, (c) surrounding color lines.

3.3. Raman spectroscopy

Figure 5 is the Raman spectra of normal grating line and the discolored line. It can be seen from the figure 5 that there is a significant $\nu_{O\text{-}C\text{-}O}$ absorption vibration peak at 1375cm^{-1} and 1128cm^{-1}. Ag-O characteristic vibration peak appears in the 462cm^{-1}, 583 cm^{-1} and 966 cm^{-1}. And the peak of the bending of the CO_3^{2-} appears at 1052cm^{-1}, 1438 cm^{-1} and 709 cm^{-1} [10] peaks. From the Raman spectroscopy can be seen snail trails and the surrounding discoloration of silver on the Ag_2CO_3 generation.

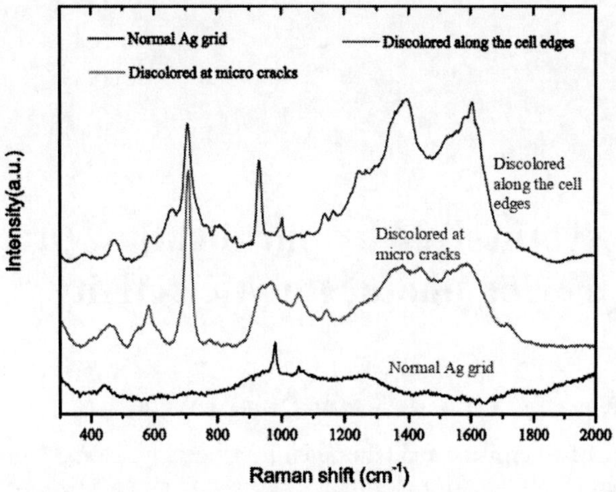

Figure 5. Raman spectra of silver gate lines with normal Ag grid and discolored.

4. Conclusion

At present, the emergence of snail trails can be reproduced through different experimental conditions. The CH_3OH/CH_2Cl_2 of snail trails on the sample surface treatment, organic matter is dissolved and the surface nanoparticles still exist, which as a method of testing the following particles and the particles will not be destroyed by oxidation. The particle size of snail trails and discoloration lines is about 80-200 nm. From the Raman spectroscopy can be seen snail trails and the surrounding discoloration of silver on the Ag_2CO_3 generation. In this paper, the specific material of snail trails has been analyzed. According to the generated material in this paper, and the matching material performance can prevent the emergence of snail trails. With the reduction of snail trails phenomenon, the modules power attenuation is reduced. And the rigid quality control would be executed and therefore optimize the energy output of the PV module over 25 years lifetime.

Acknowledgements

This work has been supported by the CECEP project cecep-kj-2014-012 and 14RD5. The authors thank Jiang Su University for test instrument.

References

[1] Peng P, Hu A, Zheng W and Su P 2012 *RSC ADV* **2** pp 11359-11365
[2] Meyer S, Timmel S, Richter S, Werner M, Gl Ser M, Swatek S, Braun U and Hagendorf C 2014 *Solar Energy Materials & Solar Cells* **121** pp 171-175
[3] Kontges M, Kunze I, Naumann V, Richter S and Hagendorf C 2011 *TUV Rheinland, Modulworkshop* **8**
[4] Meyer S, Richter S, Timmel S, Glaser M, Werner M, Swatek S and Hagendorf C 2013 *Energy Procedia* **38** pp 498-505
[5] Liu H C, Huang C T, Lee W K, Yan S S and Lin F M 2015 *Energy and Power Engineering* **7** pp 348-353
[6] Stark W and Jaunich M 2011 *Polymer Testing* **30** pp 236-242.
[7] Isarankura S, Ayutthaya Na and Wootthikanokkhan J 2008 *Journal of Applied Polymer Science* **107** pp 3853-3863
[8] Czanderna AW and Pern FJ 1996 *Sol. Energy Mater Sol. Cells* **43** pp 101-181
[9] Zweifel H, Maier RD and Schiller M 2009 *Plastic Additive Handbook*
[10] Meyera S, Timmela S and Gläsera 2014 *Solar Energy Materials and Solar Cells* **130** pp 64-70

IOP Conf. Series: Materials Science and Engineering **292** (2017) 012065 doi:10.1088/1757-899X/292/1/012065

Fe-doped ZnO synthesized by parallel flow precipitation process for improving photocatalytic activity

Q M Meng[1, 3], Q L Lu[1], L X Wang[1] and J Wang[2]

[1] College of Chemistry and Chemical Engineering, Bohai University, Jinzhou 121013, P.R. China
[2] College of Chemistry and Chemical Engineering, Northeast Petroleum University, Daqing 163318, P.R. China

[3] E-mail: qingmingmeng@163.com

Abstract. ZnO and Fe-doped ZnO photocatalysts with different molar ratio of Fe/Zn (0.25%, 0.50%, 0.75% and 1.0%) were synthesized by a parallel flow precipitation method. The structures and morphologies of as-synthesized samples were tested by the X-ray diffraction (XRD), field emission scanning electron microscopy (FE-SEM), Brunauer-Emmett-Teller (BET) surface area, respectively. The results reveal that Fe-doped ZnO has higher BET surface area than pure ZnO. XRD spectra of Fe-doped ZnO photocatalysts display merely the characteristic peaks of hexagonal wurtzite structure. The morphologies of ZnO vary from the regular hexagonal prism rod to irregular short rod and flake when Fe is introduced into ZnO particles. For degradation of RhB dye, the 0.50%Fe photocatalyst possesses the most outstanding photocatalytic performance under visible light radiation, which is more than ten times as high as the pure ZnO.

1. Introduction

ZnO is a promising photocatalyst due to its high catalytic efficiency, strong oxidative activity, and no toxicity [1]. However, during the photodegradation process of organic dye pollutants, photoexcited electrons (e^-) and holes (h^+) fast recombination and photocorrosion make ZnO exhibit poor photocatalytic performance. To promote the photocatalytic activity of ZnO, varieties of metal ions have been doped into ZnO, such as Co^{2+}, La^{3+}, Ce^{3+}, etc [2]. Modification of ZnO-based photocatalyst with these metal ions can extend response range of light and promote photo-induced charge carriers to separate, which may improve its photocatalytic activity effectively [3]. Fe-doped ZnO had been synthesized by various approaches, among of which, the parallel flow precipitation technology is an extremely attractive one for mass production in industry and low cost, removing the need for complication equipments [4, 5]. Furthermore, the influence of Fe-doping on photodegradation rate of ZnO photocatalyst synthesized using a parallel flow precipitation method has been seldom concerned. The primary driving force in this study is to concentrate on the influence of Fe-doping on morphology, structure, crystal grain size as well as the relationship between them and photodegradation rate of ZnO samples synthesized using the parallel flow precipitation process. The photocatalytic activity was tested by degradation of Rhodamine B (RhB).

2. Experimental section

All chemical reagents of analytical grade were obtained from Damao Chemical Reagent Factory and used without further purification. Fe-doped ZnO were synthesized in the following parallel flow precipitation process. 7.35 g $Zn(NO_3)_2 \cdot 6H_2O$ and different amount of $Fe(NO_3)_3 \cdot 9H_2O$ were completely dissolved in 30 mL of deionized water and marked A (molar ratio of Fe/Zn was 0%, 0.25%, 0.50%, 0.75% and 1.0%). With two constant flow pumps, solution A and $NH_3 \cdot H_2O$ (1 mol/L) were simultaneously infused into the same reactor under constant stirring and the pH was controlled at 6.8. After being washed, precipitates obtained were dried by spray and then calcined for 2 h at 673 K in the muffle oven. The photocatalysts with various molar ratio of Fe/Zn were labeled as ZnO, 0.25%Fe, 0.50%Fe, 0.75%Fe and 1.0%Fe, respectively. The Brunauer-Emmett-Teller (BET) surface area was tested using a NOVA 2000e automatic surface analyzer (Quantachrome). The morphologies and dimensions of samples were studied by a Hitachi S-4800 field emission scanning electron microscopy (FE-SEM). X-ray diffraction (XRD) patterns were carried out by a D/max2200VPC diffractometer with Cu-Kα (wavelength 0.15406 nm, 30 mA and 40 kV). The photocatalytic activity of photocatalysts (50 mg) prepared was assessed using the degradation of 50 mL RhB solution (10 mg/L).

3. Results and discussion

The XRD patterns of the pure and Fe-doped ZnO photocatalysts are shown in figure 1. All samples obtained have a hexagonal wurtzite structure according to the standard spectrum (JCPDS no. 36-1451). Additionally, no other peaks were observed and the peak positions were not changed greatly with the adding of Fe. However, the full width at half maximum (FWHM) of pure ZnO was narrower than that of Fe-doped ZnO, illustrating the crystal size of ZnO decreased as introducing of Fe. This is because the ionic radius of Fe (0.064 nm) is different from Zn (0.074 nm), and a finite solid solution can be formed when Fe substituted for Zn [6]. Thus, the lattice shrinkage was increased and crystal growth was impeded with the increasing of Fe content. According to the Scherrer equation, the average crystallite size of ZnO, 0.25%Fe, 0.50%Fe, 0.75%Fe and 1.0%Fe is 97 nm, 91 nm, 76 nm, 85 nm and 89 nm, respectively.

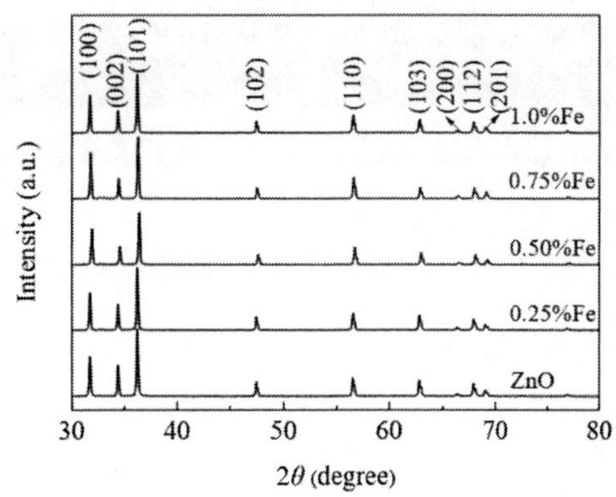

Figure.1 XRD patterns of ZnO and Fe-doped ZnO.

Table 1. Specific surface area of the synthesized photocatalysts.

Photocatalysts	ZnO	0.25%Fe	0.50%Fe	0.75%Fe	1.0%Fe
S_{BET} (m^2/g)	16.9	19.7	30.3	23.6	20.4

The BET surface area of Fe-doped ZnO samples is summarized in table 1. The specific surface area tends to increase firstly and then decline with the adding of Fe, the maximum is 30.3 m^2/g of 0.50%Fe, which is almost double of that of the pure ZnO. This may be attributed to Fe substituted for Zn into the crystal lattice of ZnO and restrained the growth of ZnO. It is generally accepted that the photocatalytic process is mainly related to the adsorption and desorption of pollutant molecules on the surface of photocatalyst [7]. The larger specific surface area can supply the more reactive adsorption and desorption sites for the photocatalytic degradation of organic pollutants.

FE-SEM photographs of the pure and Fe-doped ZnO photocatalysts are shown in figure 2. The images display that undoped ZnO is composed of typical hexagonal prism and average grain size is 2 μm in length and 1 μm in diameter approximately. Compared with undoped ZnO, the crystal size of Fe-doped ZnO decreased and their morphologies varied from regular hexagonal prism rod to irregular short rod and flake when Fe was introduced into ZnO particles. Among these Fe-doped ZnO, average crystal size of the 0.5%Fe was minimum, which was in accordance with the result of XRD.

Figure 2. FE-SEM images of ZnO and Fe-doped ZnO.

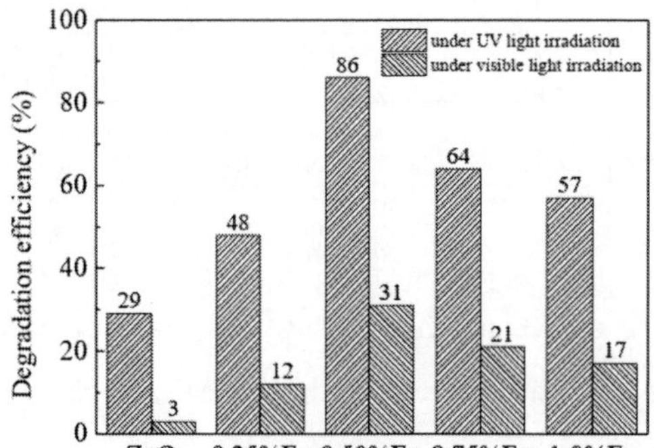

Figure 3. Photocatalytic performance of the synthesized photocatalysts for degradation of RhB.

The photodegradation rate of the pure and Fe-doped ZnO is presented in figure 3. As described in figure 3, all Fe-doped ZnO photocatalysts displayed better photocatalytic activity than that of pure ZnO under both UV and visible light illumination. Furthermore, the 0.50%Fe photocatalyst exhibited the highest photocatalytic performance under visible light radiation, which was more than ten times as high as the undoped ZnO. It is well known that ZnO-based photocatalysts are exposed to the light of energy higher than its band gap, the photoexcited e^-/h^+ pairs can be produced, then further react with O_2 and H_2O to produce the highly reactive superoxide radical anions ($\cdot O_2^-$) and hydroxyl radicals ($\cdot OH$) [8, 9]. These active species (h^+, $\cdot O_2^-$ and $\cdot OH$) can react with the organic pollutants near or on the surface of photocatalyst, causing the decomposition of pollutants into CO_2 and H_2O. Fe into the ZnO lattice introduced inter-band trap site, which captured photoexcited e^- and reduced the recombination speed of h^+ and e^-, and the charge separation efficiency was improved. High charge separation efficiency enhanced the formation of these active species, which benefited for the photodegradation of RhB dye. Here, the improved photodegradation rate of Fe-doped ZnO material is attributed to the decreased crystal size, the increased specific surface area and the enlarged separating rate of photogenerated charge carriers.

4. Conclusions

Fe-doped ZnO were successfully synthesized by a parallel flow precipitation process. The as-synthesized Fe-doped ZnO photocatalysts were merely comprised of hexagonal wurtzite structure and exhibited much higher photocatalytic performance in degradation of RhB solution under both UV and visible light radiation than pure ZnO. The 0.50%Fe photocatalyst possessed the best photocatalytic activity, which could be ascribed to the decreased crystal size, the increased specific surface area and the enlarged separating rate of photogenerated charge carriers. Doping ZnO with Fe is an effective way for enhancing its photocatalytic activity.

References

[1] B İkizler and S M Peker 2014 *Thin Solid Films* **558** 149
[2] W Raza, M M Haque and M Muneer 2014 *Appl Surf Sci* **322** 215
[3] R Saleh and N F Djaja 2014 *Superlattices Microstruct* **74** 217
[4] X H Yu, D W Meng, C Z Liu, X W He, Y Q Wang and J Xie 2012 *Mater Lett* **86** 112
[5] S S Yi, J B Cui, S Li, L J Zhang, D J Wang and Y H Lin 2014 *Appl Surf Sci* **319** 230
[6] B P Kafle, S Acharya, S Thapa and S Poudel 2016 *Ceram Lnt* **42** 1133
[7] J Z Li, J B Zhong, X Y He, S T Huang, J Zeng, J J He and W L Shi 2013 *Appl Surf Sci* **284** 527
[8] S S Xiao, L Zhao and J S Lian 2014 *Catal Lett* **144** 347
[9] M Rezapour and N Talebian 2014 *Ceram Int* **40** 3453

IOP Conf. Series: Materials Science and Engineering **292** (2017) 012066 doi:10.1088/1757-899X/292/1/012066

Development of material formula and structure property indicators for low cold-resistant characterization of Cables' Material

W Sun[1,4], Y G Cai[2], Y M Feng[2], Y L Li[3], H Y Zhou[1] and Y Zhou[1]

[1] Electric Power Research Institute of State Grid Heilongjiang Electric Power Company Limited, Harbin 150030
[2] State Grid Heilongjiang Electric Power Company Limited, Harbin 150090
[3] State Grid Heilongjiang Electric Power Equipment and Materials Company, Harbin 150000

[4] E-mail: sunwei19850525@163.com

Abstract. Alpine regions account for about 27.9% of total land area in China. Northeast China, Inner Mongolia, Northwest China and other regions are located in alpine regions, wherein the above regions are rich in energy. However, the low-temperature impact embrittlement temperature of traditional PVC cable materials is between -15℃ and -20℃, which is far lower than actual operation requirements. Cable insulation and sheath are always damaged during cable laying in alpine regions. Therefore, it is urgent to develop low-temperature-resistant cables applicable to low-temperature environment in alpine regions, and safe and stable operation of power grids in the alpine regions can be guaranteed. In the paper, cold-resistant PVC formula systems were mainly trial-manufactured and studied. Appropriate production technologies and formulas were determined through selecting raw materials and modified materials. The low-temperature impact embrittlement temperature was adjusted below -50℃ under the precondition that PVC cable materials met national standard property requirements. Cold-resistant PVC cable materials were prepared, which were characterized by excellent physical and mechanical properties, and sound extrusion process, and cold-resistant PVC cable materials can meet production requirements of low-temperature-resistant cables. Meanwhile, the prepared cold-resistant cable material was used for extruding finished product cables and trial-manufacturing sample cables. Type tests of low temperature elongation ratio, 15min withstand voltage, etc. were completed for 35kV and lower sample cables in Mohe Low-temperature Test Site. All properties were consistent with standard requirements.

1. Introduction

Alpine regions account for about 27.9% of total land area in China. Cable insulation and sheath are always damaged during cable laying due to influence of high-cold climate. Cable cracking and strand breaking are available sometimes, thereby directly threatening safe and stable operation of the power grid. with implementation of global energy interconnection and 'Arctic Circle and Equator' strategy. UHV project goes through alpine region inevitably. Low temperature environment of alpine region seriously affects safe operation of cable and other power transmission and transformation equipment as well as the reliability of the power grid. However, low-temperature impact embrittlement temperature of traditional PVC cable material is between -15℃ and -20℃, which is far less than actual operation requirements [1].

In the paper, cold-resistant PVC formula system was trial-manufactured and studied. The low-temperature embrittlement temperature was dropped below -50℃. Cold-resistant PVC cable materials with excellent physical and mechanical properties and sound extrusion process were prepared, wherein the production requirements of low-temperature-resistant cables were satisfied. The optimum cold-resistant cable processing technology meeting actual operation environment in the alpine regions were determined. 35kV and lower cold-resistant power cables and control cables were completed. Low temperature type tests of sample cables were completed in Mohe Low Temperature Test Site. All properties were consistent with standard requirements.

2. Research on cold-resistant cable material formula

2.1. Determination of research method and plan

Since there were no national standards about low-temperature-resistant PVC cable materials, related property indicators of low-temperature-resistant PVC cable materials were formulated in the paper according to GB/T8815-2008 'Soft Polyvinyl Chloride Plastic for Wires and Cables', wherein low-temperature impact embrittlement temperature was lower than PVC common cable materials by 30℃.

Strong interaction force among ordinary PVC molecules will lead to relatively poor low-temperature impact properties of cable materials. In addition, it is difficult for PVC products to produce relative displacement change in the case of external shock. External shock cannot be converted to internal friction heat, thereby leading to structure fracture of material ontology [2]. There are many methods to reduce PVC intermolecular forces. For example, a large amount of plasticizer, chlorinated polyethylene (CPE), etc. can be added. However, plasticizer is composed of low molecular weight organic compounds, which can precipitate out easily due to excessive addition. The plasticizer can be volatilized, thereby leading to anti-plasticization. The properties of the material are not substantially improved [3]. In comparison, chlorinated polyethylene is composed of high- molecular polymers. It is not necessary to consider precipitation, evaporation and similar problems. However, the vitrification temperature is above -10℃. Therefore, low-temperature impact properties also cannot be improved. In the paper, high molecular weight modified plastification material and cold-resistant plasticizer were selected for replacing a part of traditional plasticizer, thereby reaching the purposes of permanent plastification, no precipitation, migration resistance, etc. PVC intermolecular forces can be significantly reduced. Vitrification temperature can be greatly reduced. The following defects can be overcome, for example CPE has high vitrification temperature, plasticizer is prone to anti-plasticization, etc. Therefore, the technical key in the paper is shown as follows: the low-temperature impact embrittlement temperature is reduced to -50℃ or lower level on the basis of guaranteeing conventional properties of PVC cable materials.

2.2. Selection of base stock and additives

2.2.1 Screening of PVC resin. Currently, polyvinyl chloride resin with high polymerization degree has been widely used in cable material. However, it should be matched with low polymerization degree resin and modified material. Sheathing material: SH-200 (polyvinyl chloride resin with polymerization degree of 2500) was separately used as matrix resin. It was difficult to process the material although all property indicators reached the indicator requirements. The extruder load was relatively high in actual production, and materials even cannot be plasticized. S-1000 and SH-200 were mixed in use. The low-temperature impact embrittlement property of sheathing material can be as low as -40℃, while the low-temperature impact embrittlement property of insulation material can reach -35℃. Therefore, the requirements cannot be satisfied. WS-1300 and SH-200 were mixed. The low-temperature impact embrittlement properties of sheathing material and insulation material can meet requirements with excellent plasticization. They can be extruded for production. Insulation material: S-800 and S-1000 (polyvinyl chloride resin with polymerization degree of 800 and 1000) were separately used as matrix resin, and the low temperature property was unqualified. It can be matched with WS-1300 resin (polyvinyl chloride resin with polymerization degree of 1300), and the low temperature property was still unqualified. WS-1300 was separately used as matrix resin, and its low-temperature impact embrittlement property satisfied the requirements.

2.2.2 Screening of plasticizer. Plasticizer is a volatile substance. The plasticity can be improved, and the softness thereof can be improved by adding it into plastic. Plasticizer was added into high polymer for reducing the interaction force among resin molecules, thereby reducing resin softening temperature and melting temperature. The plasticizer cold-resistant property was determined by the structure composition. The plasticizer with excellent intermiscibility had relatively poor cold-resistant property. Especially when the plasticizer contained annular structure, cold - resistance was significantly reduced. On the contrary, aliphatic esters with straight chain methylene (-CH2-) as main body had better ability to resist coldness. The plasticizer with straight chain alkyl had strong ability to resist coldness. In addition, the alkyl chain was longer, the ability to resist coldness was stronger. If the alkyl branched chain was increased, the ability to resist coldness also can be worsened correspondingly [4]. The optimum plan was selected after comprehensive consideration. Dioctyl sebacate (DOS), dioctyl phthalate (DOP) was used asplasticizer of the sheathing material, and dioctyl sebacate (DOS), di-iso-decylphthalate (DIDP) and dioctyl terephthalate (DOTP) were used as plasticizer of the insulation material.

2.2.3 Screening of heat stabilizer. The addition of calcium/zinc compound stabilizer had certain influence on thermal stability time of sheathing material and insulation material. More calcium/zinc compound stabilizer was added, the thermal stability time of sheathing material and insulation material was longer, and the property was better. Main reason is shown as follows: calcium/zinc compound stabilizer can be added for inhibiting the decomposition of PVC effectively and allowing production proceed normally. If less stabilizer was added, the thermal stability time of the sheathing material and the insulation material was very short, thereby burning material during processing. It was impossible to continue production normally. In the formula, calcium/zinc compound stabilizer 8890 was finally selected as the formula system heat stabilizer of sheathing material and insulation material through rigorous screening.

2.2.4 Screening of lubricant. Lubricant is mainly used for improving the liquidity of the plastic melt. However, the shear friction effect of the compound can be greatly reduced if it is excessively used, thereby prolonging the plasticizing time prominently. It even cannot be melted and plasticized completely. Polyethylene wax has excellent thermostability and chemical resistance as lubricant. Its molecular structure belongs to a very long non-polar carbon chain with very stronger polar center. In the experiment, some parts in the polyethylene wax structure, compatible with PVC polarity, can play a role of internal lubrication. The parts with different polarity from PVC can play a role of external lubrication. Therefore, it can be used as lubricant of PVC pipes. Its lubrication principle is shown as follows: the friction of polymer melt and mold surface can be improved, thereby forming a lubrication film on the surface of finished products. Therefore, polyethylene wax was adopted as lubricant of the formula system during design of the formula.

2.2.5 Selection of low-temperature-resistant modified material. Selection of low-temperature-resistant modified material was the key to develop the product. Whether materials were selected correctly or not directly affected final properties of the material. Low-temperature-resistant modifier Elvaloy 4924 (ethylene-vinyl acetate carbonyl (E/VA/CO) copolymer) was used as modifier and dispersed in polyvinyl chloride matrix aiming at cold-resistant sheathing material. A net structure was formed. When the material suffered from external force shock, most energy was absorbed by the elastic

network structure, thereby further improving the material toughness [5]. The low temperature flexibility of polyvinyl chloride can be improved, and the embrittlement temperature can be lowered especially in low temperature environment, therefore excellent property thereof can be kept in continuous low temperature environment. Semi-crosslinked nitrile powder was selected with model P83. When P83 elastomer scattered in polyvinyl chloride matrix, P83 wrapped polyvinyl chloride, and P83 elastomer further formed a continuous net structure. When the material was subjected to strong external shock, the continuous net structure was acted, thereby absorbing most impact energy, preventing materials from serious damage, and protecting the material itself. Small molecule liquid plasticizer can easily migrate and precipitate especially in low temperature environment. However, high polymer solid plasticizer did not migrate and precipitate, thereby guaranteeing the low temperature flexibility of polyvinyl chloride and, improving the impact strength at low temperature.
Low-temperature-resistant modifier thermoplastic polyether urethane elastomer was added in cold-resistant insulation material for improving hardness, tensile property and low-temperature property of cold-resistant insulation material. Therefore, Polyether TPU elastomer was selected as modified material of cold-resistant insulation material suitably.

2.2.6 Selection of antioxidant and protective agent. Antioxidant and protective agent were introduced to inhibit or delay the oxidation of resin and elastomer. Meanwhile, oxidation of plasticizer also can be inhibited, thereby obtaining ideal anti-aging effect [6]. Bisphenol A was adopted in the formula system as antioxidant of the formula system. 4010NA was used as protective agent. After bisphenol A and 4010NA were added, the tensile strength change rate and breaking elongation change rate of the material were improved prominently according to experiment contrast. The material properties were better after aging. It indicates that the aging of PVC sheathing material and insulation material caused by oxygen effect also can be inhibited effectively by adding bisphenol A and 4010NA.

2.2.7 Screening of modified filler. Filler can reduce cost and improve some physical and mechanical properties aiming at PVC cable material. For example: calcium carbonate can improve thermostability and dimensional stability; pottery clay filler can improve electrical insulation property; antimony oxide filler can improve flame retardant property [7], etc. Commonly-used fillers of polyvinyl chloride cable materials include the follows: calcium carbonate, talcum powder, antimony oxide, pottery clay, magnesium hydroxide, aluminium hydroxide, glass fiber, barium sulfate, magnesium silicate, etc. In the paper, Nano calcium carbonate with surface treatment by silane coupling agent was adopted in the formula plan. Plasticizing behavior, mechanical property, etc. of PVC sheathing material and insulation material were improved greatly after addition of Nano calcium carbonate according to experiment contrast.

3. Determination of cold-resistant cable material formula and property test

3.1. Determination of formula system

The following materials were selected for test in research on PVC sheathing material and insulation material formula with low-temperature-resistance of -50℃ through analysis as well as material screening and comparison for many times.

Sheathing material formula plan: polyvinyl chloride resin (model SH-200) with polymerization degree of 2500, polyvinyl chloride resin with polymerization degree of 1300, dioctyl sebacate (DOS), dioctyl phthalate (DOP), calcium/zinc compound stabilizer 8890, polyethylene wax, low-temperature-resistant modifier Elvaloy 4924 (ethylene-vinyl acetate carbonyl (E/VA/CO) copolymer), nitrile rubber powder P83, bisphenol A, anti-aging agent 4010NA and modified Nano calcium carbonate.

3.2. Formula component

The optimum formula was selected through test and comparative analysis on a series of formula properties. The results are show in table 1 and table 2.

Table 1. Optimum formula of PVC sheathing material with low-temperature resistance of -50℃.

Raw material	Name	Formula
PVC resin 1	SH-200 (polymerization degree 2500)	50
PVC resin 2	WS-1300 (polymerization degree 1300)	50
DOS	Dioctyl sebacate	15
DOP	Dioctyl phthalate	45
Stabilizer	Calcium/zinc compound stabilizer 8890	8
Lubricant	Polyethylene wax	1
Modifier 1	Elvaloy 4924	5
Modifier 2	P83	5
Antioxidant	Bisphenol A	1
Anti-aging agent	4010NA	1
Filler	Nano calcium carbonate	15

Table 2. Optimum formula of insulation material with low-temperature resistance of -50℃.

Raw material	Name	Formula
PVC resin	WS-1300 (polymerization degree 1300)	100
DOS	Dioctyl sebacate	15
DIDP	Di-iso-decylphthalate	10
DOTP	Dioctyl terephthalate	35
stabilizer	Calcium/zinc compound stabilizer 8890	7
Lubricant	Polyethylene wax	1
Modifier	Polyether TPU elastomer	10
Antioxidant	Bisphenol A	1
Anti-aging agent	4010NA	1
Filler	Nano calcium carbonate	15

3.3. property test

Cold-resistant polyvinyl chloride novel sheathing material formula and cold-resistant insulation material formula were determined through comprehensive study and screening of the above formula. The properties thereof were measured. The results are shown in table 3 and table 4.

Table 3. Measured characteristics of cold-resistant PVC sheathing material.

Content	Standard requirement	Experimental result
Tensile strength /MP	≥16.0	18.3
Elongation at break /%	≥180	306
Thermal deformation /%	≤50	22.7
Impact embrittlement property (test temperature -50℃)	Qualified	Qualified
Thermal stability time at 200℃ /min	≥80	116
Volume resistivity at 620℃ /(Ω·m)	$\geq 1.0\times10^{9}$	3.8×10^{10}

Table 4. Measured characteristics of cold-resistant PVC insulation material.

Content	Standard requirement	Experimental result
Tensile strength /MP	≥15.0	17.9
Elongation at break /%	≥150	324
Thermal deformation /%	≤50	19.6
Impact embrittlement property (test temperature -50℃)	Qualified	Qualified
Thermal stability time at 200℃/min	≥60	89
Volume resistivity at 20℃ /(Ω·m)	$\geq 1.0\times10^{12}$	2.8×10^{12}

4. Research on structure property of cold-resistant cables

In the paper, the influence factors of cable sheath cracking phenomena during cable laying were analyzed. The relationship among mechanical and electrical properties, cable structure and material additives of cold-resistant cable at low temperature was summarized. The structures of 35kV and lower cold-resistant power cables and control cables were designed. Sample cables were trial-manufactured. Their property indicators were verified and analyzed. The low-temperature-resistant property assessment of cold-resistant cable at -50℃ was perfected. Product test standards were perfected, and cold-resistant cable type test was completed.

4.1. Trial-manufacturing of cold-resistant cable samples

In the paper, four sample cables were prepared, including power cables with rated voltage of 26/35kV, power cables with rated voltage of 8.7/10kV, power cables with rated voltage of 0.6/1kV and control cable with rated voltage 450/750kV. Specific structure and dimension are shown in tables 5-8.

Table 5. Structure and dimension of power cable samples with rated voltage of 26/35kV.

Model and specification		YJV-26/35 1×300		Remark
No.	Structure	Thickness(mm)	Outer diameter(mm)	
1	Conductor	60/2.68	20.5 (-0.2,+0.2)	Copper conductor
2	Conductor screen	0.9	22.3	
3	Insulation	10.5	43.6	
4	Insulation screen	0.7	45.0	
5	Copper tape screen	1×0.12	45.4	
6	Wrapping tape	1×0.15	45.9	Non-woven fabrics
7	Outer sheath	2.5	50.9±1.5	Sheathing material with cold resistance of -50℃

Table 6. Structure and dimension of power cable samples with rated voltage of 8.7/10kV.

Model and specification		YJV22-8.7/10 3×300		Remark
No.	Structure	Thickness(mm)	Outer diameter(mm)	
1	Conductor	60/2.68	20.5 (-0.2,+0.2)	Copper conductor
2	Conductor screen	0.7	21.9	
3	Insulation	4.5	31.0	
4	Insulation screen	0.7	32.4	
5	Copper tape screen	1×0.10	32.7	
6	Cable		71.6	Filled with rock wool rope
7	Wrapping tape	2×0.15	72.5	Non-woven fabrics

Table 7. Structure and dimension of power cable samples with rated voltage of 0.6/1kV.

Model and specification		YJV22-0.6/1 3×240+2×120		Remark
No.	Structure	Thickness(mm)	Outer diameter(mm)	
1	Conductor	36/3.08 24/2.56	18.4 (-0.2,+0.2) (phase conductor) 12.9 (-0.2,+0.2) (neutral conductor)	Copper conductor
2	Insulation	1.7 1.2	22.0 15.5	
3	Cable		53.9	
4	Wrapping tape	2×0.15	55.1	Non-woven fabrics
5	Inside liner	1.6	58.3	Sheathing material with cold resistance of -50℃
6	Armour	2×0.5	60.3	
7	Outer sheath	2.9	66.1	Sheathing material with cold resistance of -50℃

Table 8. Structure and dimension of power cable samples with rated voltage of 450/750kV.

Model and specification		KVVP2-450/750 14×6		Remark
No.	Structure	Thickness(mm)	Outer diameter(mm)	
1	conductor	1/2.74	2.74±0.02	Copper conductor
2	Insulation	0.8	4.5	Insulation material (yellow) with cold resistance of -50℃
3	Cable		19.9	Filled with rock wool rope
4	Wrapping tape	2×0.05	20.2	Polyester tape
5	Copper tape screen	1×0.06	20.4	
6	Wrapping tape	1×0.15	20.8	Non-woven fabrics
7	Outer sheath	1.7	22.2	Sheathing material with cold resistance of -50℃

4.2. Low temperature type test of cold-resistant cable sample

Low temperature elongation ratio, low-temperature impact, 15min withstand voltage and similar tests were implemented aiming at four sample cables (respectively 80m) at outdoor natural environment of -50℃~-45℃ in Mohe Low Temperature Test Site according to related requirements in GB/T 12706.1-2008, GB/T 12706.2-2008, GB/T 12706.3-2008, GB/T 9330.2-2008 and other standards. Test photos are shown in figure 1, figure 2 and figure 3.

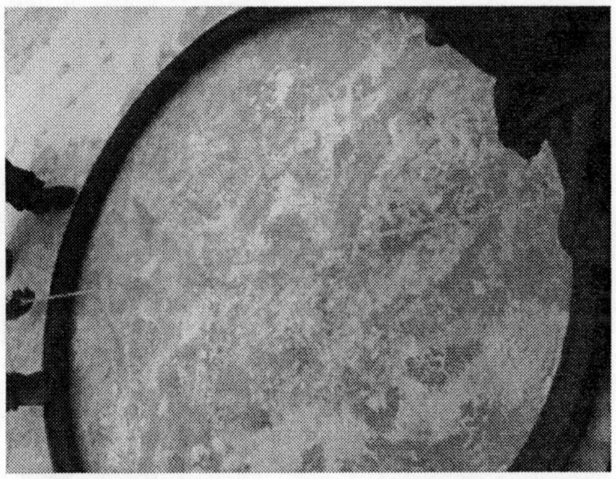

Figure 1. The Experimental field of cable low temperature elongation ratio

Figure 2. Wiring diagram of cable low temperature 15min withstand voltage test

Figure 3. The Experimental field of cable low temperature 15min withstand voltage

The Experimental results of four sample cables are shown in table 9, table 10, table 11 and table 12. All properties of four sample cables were consistent with standard requirements according to actual tests.

Table 9. The experimental results of power cable samples with rated voltage of 26/35kV.

Experimental project	Unit	Standard requirement	Experimental result	Evaluation
Bending experiment	Time	3	Carry out six times experiments on a cylinder with the diameter of 1450mm	Meet the requirements
15min withstand voltage	kV	91	Pass the experiment	Meet the requirements
Low temperature elongation ratio	%	≧20	185	Meet the requirements
Low-temperature impact	-	Crack-free	Crack-free	Meet the requirements

Table 10. The experimental results of power cable samples with rated voltage of 8.7/10kV.

Experimental project	Unit	Standard requirement	Experimental result	Evaluation
Bending experiment	Time	3	Carry out six times experiments on a cylinder with the diameter of 1700mm	Meet the requirements
15min withstand voltage	kV	30.5	Pass the experiment	Meet the requirements
Low temperature elongation ratio	%	≧20	185	Meet the requirements
Low-temperature impact	-	Crack-free	Crack-free	Meet the requirements

Table 11. The experimental results of power cable samples with rated voltage of 0.6/1kV.

Experimental project	Unit	Standard requirement	Experimental result	Evaluation
Low temperature elongation ratio	%	≧20	192	Meet the requirements
Low-temperature impact	-	Crack-free	Crack-free	Meet the requirements

Table 12. The experimental results of power cable samples with rated voltage of 450/750kV.

Experimental project	Unit	Standard requirement	Experimental result	Evaluation
Low temperature elongation ratio	%	≧20	187	Meet the requirements
Low-temperature impact	-	Crack-free	Crack-free	Meet the requirements

5. Conclusion

(1) Property indicators of cold-resistant cable insulation and sheath materials were set. The optimum cold-resistant cable design formula was determined. Cold-resistant PVC cable material with excellent physical and mechanical properties and sound extrusion technology was prepared, wherein the material met production requirements of low-temperature-resistant cables.

(2) The low-temperature property of the prepared cold-resistant sheathing material and cold-resistant insulation material was qualified in -50℃ low-temperature impact embrittlement experiment. The heavy metal content in the formula system was consistent with requirements of green environment protection design and manufacturing.

(3) The processing technology of cold-resistant cable insulation and sheath materials were mastered through formula difference contrast and process combination design. Cold-resistant cable insulation and sheath properties were analyzed. The optimum cold-resistant cable processing technology meeting actual operation environment in the alpine region was determined. In addition, processing technology temperature of cold-resistant (below -50℃) cable PVC plastic was determined through material trial-manufacturing and test for many times. The developed cold-resistant sheathing material and cold-resistant insulation material had stable extrusion technology properties.

(4) The relationship among mechanical and electrical properties, cable structure and material additives of cold-resistant cables at low temperature was summarized. The structures of 35kV and lower cold-resistant power cables and control cables were designed. Meanwhile, the developed cold-resistant sheath and insulation materials were used for finished product cable extrusion experiment and trial-manufacturing sample cables. Type tests of low temperature elongation ratio, low-temperature impact, 15 min withstand voltage, etc. were completed for 35kV and lower sample cables in Mohe Low-temperature Test Site. All properties were consistent with standard requirements.

(5) The cold-resistant cable developed in the paper was applicable to -50℃ low temperature operation environment. Seasonality of cable laying in alpine regions was broken, thereby effectively guaranteeing operation reliability of cable lines and filling the blank of domestic and foreign related technology fields.

Reference

[1] Wang C J 2002 Electric wire and cable manual *Beijing: China Machine Press* **1**
[2] Wang G H 1995 Principle of polymer material shaping processing *Beijing: Chemical Industry Press*
[3] Coran A Y and Patel R 1983 Nitrile rubber polyolefin blends with technological com-

patibilization *Rubber Chemistry and Technology* chapter **5** pp 1045-1060
[4] Han J B, Tian H C and Xu X 2005 Influence of plasticizer DOP on the property of NBR/PP blends *Rubber Industry* chapter **8** pp 472-475
[5] Luo S W, Xi Y J and Mao J H 2006 Analysis on electric properties of polyvinyl chloride cable material *Polyvinyl Chloride* chapter 8 pp 9-12
[6] Wang J J, Ma X Y and Wang Y 2003 Research progress of polyvinyl chloride blend modification *Insulation Material* chapter **3** pp 40-45
[7] Ellul M D 1998 Plasticization of polyolefin elastomers semicrystal-line plastics and blends crosslinked in situ during melt mixing *Rubber Chemistry and Technology* chapter **2** pp 244-274

IOP Conf. Series: Materials Science and Engineering **292** (2017) 012067 doi:10.1088/1757-899X/292/1/012067

Diagnostics of the thickness of a plasma electrolytic oxidation coating on a nanostructured Mg-Sr alloy

V R Mukaeva[1, 5], O B Kulyasova[2], R G Farrakhov[1], E V Parfenov[1, 3] and Y F Zheng[4]

[1] Department of Theoretical Basis of Electrical Engineering, Ufa State Aviation Technical University, 12 Karl Marx Street, Ufa, 450008, Russia
[2] Institute of Physics of Advanced Materials, Ufa State Aviation Technical University, 12 Karl Marx Street, Ufa, 450008, Russia
[3] Academy of Sciences of Republic of Bashkortostan, 15 Kirov Street, Ufa, Russia
[4] College of Engineering, Peking University, 5 Yi-He-Yuan Road, Hai-Dian District, Beijing, China.

[5] E-mail: veta_mr@mail.ru

Abstract. This study is focused on the development of a method for controlling the thickness of a plasma electrolytic oxidation (PEO) coating on a new nanostructured Mg-2wt%Sr alloy. The process of high pressure torsion of a solution-treated Mg-2wt%Sr magnesium alloy was investigated. A significant reduction in the grain size to 500 nm was attained. The PEO coating process on the new nanostructured Mg-2wt%Sr alloy was studied. The produced coating was compared with a PEO coating with a coarse-grained structure. The waveforms of unipolar voltage pulses of a power source were studied, and electrical transients that can be described by an exponential of the second order were revealed. The relationship between the time constant of the electrical transient exponent and the thickness of the PEO coating was justified. A method for estimating the thickness of the PEO coating on the surface of the Mg-2wt%Sr alloy in terms of the parameters of the electric transient was proposed.

1. Introduction

Magnesium and magnesium alloys are of a great interest for biomedical applications, since they have a number of advantages compared to other materials [1, 2]. In particular, magnesium is a biodegradable material, and it has a Young's modulus close to that of a human bone [3]. A biodegradable magnesium implant enables avoiding a second surgery on a patient after a fracture unites [4]. Also, magnesium, as a chemical element, is vital for the human body, and it can be considered as a medical material that encourages bone tissue restoration.

The general disadvantage of all magnesium alloys is low strength and considerable corrosion rate; therefore, studies on the enhancement of the mechanical and corrosion properties of magnesium implants are actively carried out globally at the cutting edge of chemical and material science. Several technological treatments for a magnesium alloy and its surface can be applied in order to solve this problem: (i) alloying using a biocompatible Sr element; (ii) refinement of a grain structure via methods of severe plastic deformation, such as high pressure torsion (HPT) and equal channel angular pressing [5, 6]; (iii) applying a biocompatible coating by plasma electrolytic oxidation (PEO) [7, 8].

Properties of coatings on biodegradable implant surface have a significant influence on the dissolution rate. The PEO process helps to achieve the required surface properties better than other methods, and it has advantages in terms of environmental safety and effectiveness [9]. Coating thickness is its most important property. In-situ diagnostics and accurate achievement of the coating thickness will help to control the dissolution rate of the coated implant.

One of the in-situ ways to obtain information about the coating thickness is to study the electrical characteristics of the electrolyzer in response to the input pulses generated by the power supply. Therefore, the aim of this study is to develop a method for controlling the thickness of the PEO coating on a new nanostructured Mg-2wt%Sr alloy during the treatment.

2. Experimental

2.1. Material and surface characterization

A Mg-2wt%Sr magnesium alloy was chosen as the initial material. The samples were subjected to solution treatment at a temperature of 450 °C for 20 hours, followed by quenching. The thermal treatment of the Mg-2wt%Sr samples was carried out in an electric furnace Nabertherm (in air atmosphere).

The microstructure of the Mg-2wt%Sr sample after the solution treatment is shown in figure 1. The microstructure of the alloy is a solid solution (α-Mg) with a separated eutectic (α-Mg + $Mg_{17}Sr_2$) at the grain boundaries. The average grain size of α-Mg was 85 μm. The volume fraction of the eutectic phase was 20.9%.

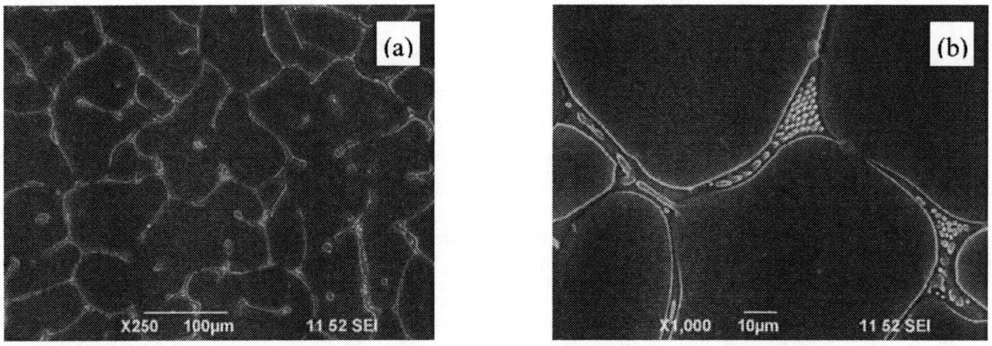

Figure 1. Microstructure of Mg-2%Sr alloy in the solution-treated state.

The solution-treated samples of the alloy under investigation in the form of disks with a diameter of 20 mm and a thickness of 2 mm were subjected to high pressure torsion for structure refinement [5]. The alloy disks under investigation were deformed at room temperature, at a speed of 1 rpm, with an applied pressure of 6 GPa; the number of revolutions was 10. During HPT the sample temperature reached 50-80 °C.

Analysis of the microstructure was carried out with a JEM6390 scanning electron microscope (SEM) with an accelerating voltage of 10 kV. To reveal the alloy microstructure, the cross-sections were etched with the following solution: 2.5 mg picric acid; 2.5 ml. acetic acid; 5 ml. distilled water; 50 ml. ethyl alcohol. The average grain size was calculated using the linear intercept method from the SEM images.

Study of the nanostructure was performed on a JEOL JEM-2100 transmission electron microscope (TEM) at an accelerating voltage of 200 kV. The foils for the TEM investigation were prepared using a Tenupol-5 double-jet electro-polishing machine with the following electrolyte: 30% HNO_3, 70% methanol. The average grain size was estimated by averaging the maximum and minimum grain sizes, according to the dark-field images. The number of the analyzed grains was 300.

2.2. Plasma electrolytic oxidation

The following Mg-2wt%Sr samples were used for the plasma electrolytic oxidation: solution-treated samples, with the initial coarse-grained structure (CG); samples with a nanostructure after HPT processing (nano).

The PEO treatment was performed using a unipolar voltage-controlled power supply. Square waveform voltage was applied to the sample connected as an anode; and the electrolyzer connected as a cathode. The processing parameters are given in table 1. The electrolyte composition which enables producing biocompatible Ca- and P- compounds in the PEO coating was chosen as elsewhere [10]. The coating thickness was measured by a Defelsko Positector 6000 eddy current gauge.

Table 1. Parameters of the PEO treatments and coating thickness.

Sample No.	Mg-2%Sr structure	Coating thickness (μm)	Electrolyte (g/L)	Temperature (°C)	Voltage pulse amplitude (V)	Frequency (Hz)	Duty cycle (%)	Time (min)
1	CG	11.5	12 g/L $Na_7PO_4 \cdot 12H_2O$ 2 g/L $Ca(OH)_2$	20	470	500	10	10
2	nano	14.3						

During the PEO treatment, the instantaneous values and DC components of current and voltage were recorded at a sampling frequency of 1 MHz, and at a rate of 1 s^{-1} by an automated data acquisition system L502 (L-Card, Russia) running under the LabVIEW control software.

3. Results and discussion

3.1. Microstructure of CG and nanostructured Mg-2wt%Sr alloy and PEO coating

As a result of the HPT processing, a new structure was produced in the Mg-2wt%Sr alloy, as shown in figure 2. During the HPT processing, the eutectic was destroyed and partially dissolved; its volume fraction became 17.5% (figure 2a). TEM studies revealed a strong refinement of the average grain size down to 500±37 nm (figure 2b).

Figure 3 shows the surface images of the PEO coating on CG and nanostructured Mg-2wt%Sr alloy. The coating thickness *h* on the nanostructured sample is 14.3±1.5 μm, and on the CG sample – 11.5±1.4 μm. The thicker coating on the nanostructured samples can be explained by the fact that this structure of this alloy has much higher total length of the grain boundaries which provide favorable conditions for the appearance of microdischarges [11].

Study of the SEM images using ImageJ software revealed the porosity of the coatings. It was found that the porosity of the coatings on CG and nanostructured samples is the same: 22.0±0.1%. Therefore, the electrical properties of the coatings that are discussed below do not depend on porosity.

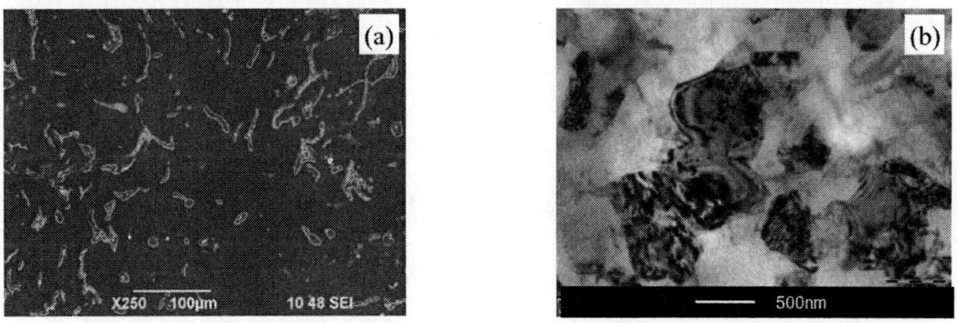

Figure 2. Microstructure of Mg-2%Sr alloy after HPT: (a) SEM, (b) TEM.

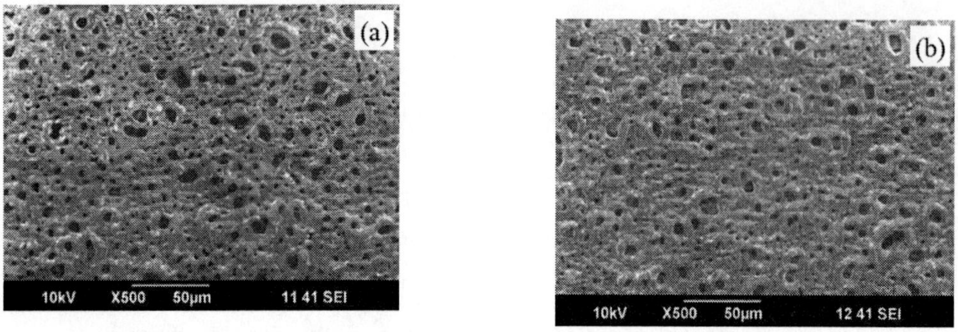

Figure 3. SEM images of the resulting surface after the PEO of CG (a) and nano (b) Mg-2wt%Sr.

3.2. Electrical characteristics of the PEO process of CG and nanostructured Mg-2wt%Sr alloy

For analysis of the PEO electrical characteristics, the instantaneous values of voltage were represented as waveforms. The voltage waveforms at the beginning of the treatment and after 10 minutes are shown in figure 4. The evolution of the DC components of the current and voltage during the PEO are presented in figure 5. As it can be seen from the waveforms, the shape of the electrical signal changes in long time scale (curves 1 and 2 in figure 4). It is known that in the equivalent circuit of the PEO electrolyzer, the coating can be represented as two series branches containing a resistor and a capacitor in parallel [12]. Therefore, electrical transients seen in figure 4 appear as a reaction to switching off the power supply current pulses. Within the PEO process, the duration of electrical transients increases, which is associated with an increase in the coating thickness as shown elsewhere for CG Mg and CG Al [10, 12]. This study extends this method to a nanostructured substrate.

The curve fitting for the electrical transients was obtained using the following equation:

$$u(t) = A_1 e^{-\frac{t-t_0}{\tau_1}} + A_2 e^{-\frac{t-t_0}{\tau_2}} + C, \qquad (1)$$

where A_1, A_2, C are integration constants; τ_1 and τ_2 are time constants; t_0 is the starting time of the voltage fall, t is time.

Evolution of time constants τ_1 and τ_2 during the PEO treatment is shown in figure 6. According to figure 6, the value of the time constant τ_1 is comparable with the value of the voltage pulse cycle. The value of τ_1 increases exponentially and tends to approach a constant value. The values of the time constant τ_2 are an order of magnitude smaller than that of τ_1; this time constant grows linearly. The difference in τ_2 between CG and nanostructured substrate is negligible.

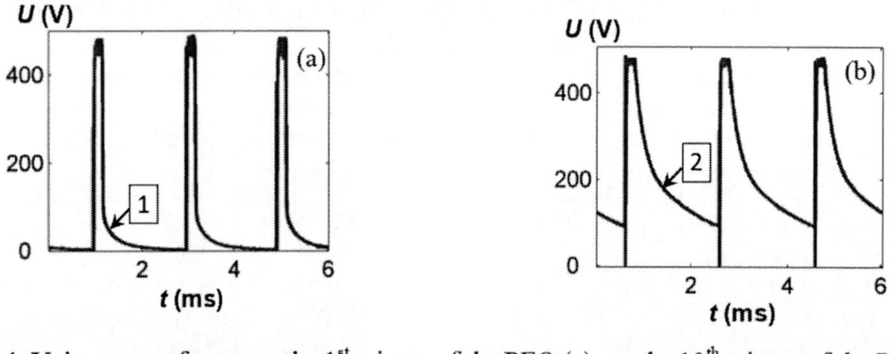

Figure 4. Voltage waveforms: on the 1st minute of the PEO (a); on the 10th minute of the PEO (b).

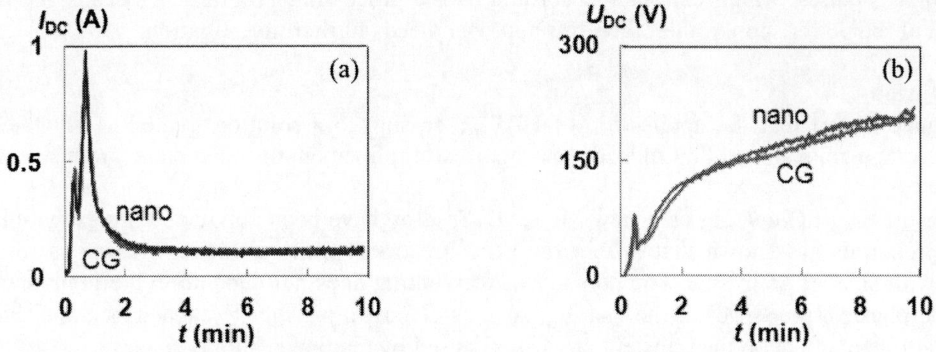

Figure 5. Evolution of DC current (a) and voltage (b) components during PEO of CG and nanostructured Mg-2wt%Sr alloy.

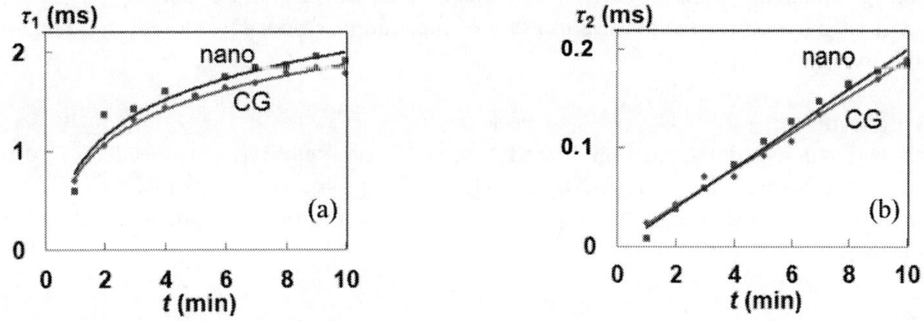

Figure 6. Evolution of the time constants τ_1 (a) and τ_2 (b) during PEO of CG and nanostructured Mg-2wt%Sr alloy.

Usually, a PEO coating on Mg alloy has a porous outer and a very thin compact inner layer [13]. This is reflected in two distinct time constants standing for the two layers of the coating structure both in the electrochemical impedance spectroscopy of PEO coatings on Mg [14] and in this study. These time constants characterize dielectric relaxation processes within those layers [10] as

$$\tau_1 = R_1 C_1, \tau_2 = R_2 C_2. \tag{2}$$

where R_1, R_2 – coating layer resistance, C_1, C_2 – coating layer capacitance.

Therefore, considering similar porosity of the outer layer for both CG and nanostructured samples, the outer layer capacitances should be the same; consequently, the resistances and time constants reflect the coating thickness.

As a result, taking into account the influence of the coating growth on electrical transients during the PEO, the coating thickness h can be evaluated as:

$$h = k_1 \cdot \tau_1 \tag{3}$$

where k_1 is a calibration constant. The coefficient k_1 is calculated from experimental data; for this study it is 7.0±0.3 [μm/ms].

Moreover, since there is a functional relationship between the duration of the electrical transients defined by the time constants, and the area under the waveform graph of the voltage pulses, the coating thickness can also be determined from the DC voltage component U_{DC} (figure 5) as

$$h = k_2 \cdot U_{DC} \tag{4}$$

The benefits of this method include simpler hardware and software implementation. The disadvantage is that the result of calculation of the coating thickness will be affected by the amplitude of the voltage pulses, which can vary according to the processing program or because of unstable operation of the power supply; therefore, this approach needs further investigation.

4. Conclusion

Studies have shown that the application of HPT processing to a solution-treated Mg-2wt%Sr alloy with an average grain size of 85 μm leads to a significant refinement of the average grain size down to 500 nm.

Studies of the PEO coating on a new Mg-2wt%Sr alloy have been conducted. Investigation of the coating properties has shown that within the same treatment time, a thicker coating is formed on samples with smaller grain size. The porosity of the coating does not depend on the grain size of the alloy. The physical properties of the coating, such as electrical resistance, capacitance and thickness, have a significant effect on the transient processes caused by the power supply pulses.

A diagnostic method for determining the coating thickness during the PEO process for both coarse grained and nanostructured Mg-2wt%Sr alloy has been developed; this helps to carry out measurements without interrupting the technological process.

Accurate attainment of the specified thickness of a biocompatible coating on the surface of implants will help to control the dissolution rate of the future temporary implant in a human body for successful bone repair.

Acknowledgments

This work was supported by Russian Foundation of Basic Research (Project No. 16-53-53039 in cooperation with NSFC, China, and No. 16-38-60062 for young scientists (V.R.M.)) and by the Ministry of Education and Science of the Russian Federation through a platform grant in the field of basic research (No. 16.7268.2017).

References

[1] Li H, Zheng Y and Qin L 2014 *Progress in Natural Science* **24** 414
[2] Tian P and Liu X 2015 *Regenerative Biomaterials* **2** 135
[3] Staiger M P, Pietak A M, Huadmai J and Dias G 2006 *Biomaterials* **27** 1728
[4] Gu X N, Xie X H, Lia N, Zheng Y F and Qin L 2012 *Acta Biomaterialia* **8** 2360
[5] Valiev R Z, Islamgaliev R K and Alexandrov I V 2000 **45** 103
[6] Kulyasova O B, Islamgaliev R K, Zhao Y and Valiev R Z 2015 *Advanced Engineering Materials* **17** 1738
[7] Hornberger H, Virtanen S and Boccaccini A R 2012 *Acta Biomaterialia* **8** 2442
[8] Bagheri H R, Aliofkhazraei M, Gheytani M., Masiha H R, Sabour Rouhaghdam A and Shahrabi T 2015 *Surface and Coatings Technology* **283** 1
[9] Parfenov E V, Yerokhin A, Nevyantseva R R, Gorbatkov M V, Liang C J and Matthews A 2015 *Surface and Coatings Technology* **269**
[10] Gao Y, Yerokhin A, Parfenov E and Matthews A 2014 *Electrochimica Acta* **149** 218
[11] Martin J, Melhem A, Shchedrina I, Duchanoy, Nomine A, Henrion G, Czerwiec T and Belmonte T 2013 *Surface and Coatings Technology* **221** 70
[12] Fatkullin A R, Parfenov E V, Yerokhin A L, Lazarev D M, Matthews A 2015 *Surface & Coatings Technology* **284** 427
[13] Mohedano M, Luthringer B J C, Mingo B, Feyerabend F, Arrabal R, Sanchez-Egio P J, Blawert C, Willumeit-Römer R, Zheludkevich M L and Matykina E 2017 *Surface and Coatings Technology* **315** 454
[14] Esmaily M, Svensson J E, Fajardo S, Birbilis N, Frankel G S, Virtanen S, Arrabal R, Thomas S, Johansson L G 2017 *Progress in Materials Science* **89** 92

IOP Conf. Series: Materials Science and Engineering **292** (2017) 012068 doi:10.1088/1757-899X/292/1/012068

A study of low cycle fatigue life prediction method for 1Cr11Ni2W2MoV at 360℃

G D Gao[1, 2, 5], W X Zhang[3] and D Chen[4]

[1] Nautical and Marine Engineering College, Dalian Ocean University, 52 Heishijiao Street. Dalian, 116023, China
[2] Marine Engineering College, Dalian Maritime University, 1 Linghai Road.Dalian, 116026, China
[3] Mechanical Engineering College, Dalian Ocean University, 52 Heishijiao Street. Dalian, 116023, China
[4] COSCO (Dalian) Shipyard CO.,LTD, 80 Zhongyuan Road, Dalian, 116113, China

[5] E-mail: gmc2022@163.com

Abstract. The high temperature low cycle fatigue tests of 1Cr11Ni2W2MoV are conducted under total axial straincontrol. The specimens are tested under fully reversed strain. The relationships of cyclic strain-life, stress-strain and stress-life are analyzed. Manson-Coffin method and a new method based on power transformation theory are evaluated by the low cycle fatigue data. The results showed that two methods can effective predict low cycle fatigue life of 1Cr11Ni2W2MoV at 360 ℃. Prediction accuracy within ± 1.65 times scatter band. The life prediction capability of new methods proves more effective and accurate than Manson-Coffin method. The new method has smaller standard deviation than Manson-Coffin method. Manson-Coffin method and new method have standard deviations of 0.32 and 0.18 respectively.

1. Introduction

1Cr11Ni2W2MoV is a martensitic stainless steel, mainly for the manufacture of load-bearing components which work in wet conditions at below 550℃. It can be used to manufacture rotor blade plate and other rotating components in aero engine compressor. GuoXiaoguang*et al.* proposed that these components bear the complex alternating loads at high temperatures during the service, which may be the cause of early fatigue damage [1].

The high temperature low cycle fatigue (LCF) life prediction is an important basis for safety assessment on the project, and has long been concerned by domestic and foreign scholars [2, 3]. In order to accurately predict the high temperature LCF life prediction, various methods have been proposed over the years. Manson and Coffin established Manson-Coffin method [4]. This method is widely used in LCF life prediction. They noted that in some cases, relationship of plastic strain amplitude and failure reverse number presents linear relationship in the double logarithmic coordinates. Chen Lijie*et al.* proposed quadratic characters of plastic strain amplitude vs. failure reverse number in log-log scale coordinates. Power transformation method can be applied to construct low-cycle fatigue life prediction model [5]. Fan Zhichao*et al.* proposed the life prediction method based on ductility exhaustion theory [6]. Young Il Kwon &ByeongSoo Limproposed artificial neural network method [7]. Chen Hong and JiangHongde proposed the low cycle fatigue life prediction method which can consider the effect of average stress, strain ratio and crystal orientation on fatigue life. The method is validated using low cycle fatigue test data of DD3 and PWA1840 nickelbasesingle crystal superalloy, and the results show that the proposed method is acceptable [8]. Chen Ling *et al.* proposed the life prediction model for low cycle fatigue based on continuum damage mechanics. Results show that the predicted results which are obtained by the sampling data of different life periods are in good agreement with the experiment results [9].

In this paper, the high temperature LCF tests of 1Cr11Ni2W2MoV are conducted under total axial straincontrol. The specimens are tested under fully reversed strain. The relationships of cyclic strain-life, stress-strain and stress-life are analyzed. Manson-Coffin method and a new method based on power transformation theory are evaluated by the LCF data. Related research can provide reference for component design and life evaluation.

2. Experimental methods

2.1. Material

The testing material was 1Cr11Ni2W2MoV bar. The base material was given a normalizing treatment at 1000 to 1020℃ for 1 hour followed by oil quenching and subsequently tempering at 540 to 600℃ for 1.5 hour followed by air cooling, before machining the standard fatigue specimens. The chemical composition of 1Cr11Ni2W2MoV is presented in table 1. The tensile properties for 1Cr11Ni2W2MoV at 20℃ and at 360℃ are shown in table 2.

Table 1. The chemical composition of 1Cr11Ni2W2MoV (weight, %)

C	Cr	Mn	Si	Ni	W	Mo	P	S	Nb
0.10~0.16	10.50~12.00	<0.60	<0.60	1.40~1.80	1.50~2.00	0.35~0.50	<0.030	<0.020	0.75~1.25

Table 2. The tensile property for 1Cr11Ni2W2MoV

Temperature T(℃)	Young's modulus E (MPa)	Fracture limit σ_b (MPa)	Yield strength $\sigma_{0.2}$ (MPa)	Elongation δ,%(weight)	Reduction of area ψ,%(weight)
20℃	206000	1183	993	8	68
360℃	163000	989	825	14	71

2.2. Test conditions and methods

The LCF tests were conducted under total axial strain control at 360℃. The specimens are tested under fully reversed strain. The temperature variation did not exceed ±2℃. The experimental frequency was 0.333HZ for all tests. Experimental waveform was a triangular waveform. Three specimens are tested under each amplitude. The average of similar data is taken as the experimental result.

3. Results and discussion

3.1. Low cycle fatigue test results and discussion of 1Cr11Ni2W2MoV

The linear relationship of $\Delta\varepsilon_t/2$ —$2N_f$, $\Delta\varepsilon_p/2$ —$2N_f$and $\Delta\varepsilon_e/2$ —$2N_f$ in double logarithmic coordinates is presented figure 1. Fitting coefficient (R^2) of$\Delta\varepsilon_t/2$—$2N_f$,$\Delta\varepsilon_p/2$—$2N_f$and $\Delta\varepsilon_e/2$—$2N_f$is 0.914, 0.9704 and 0.975 respectively. Regression equation (1) and equation (2) can be obtained by regression analysis.

$$\Delta\varepsilon_p/2=6.9776(2N_f)^{-1.1448} \tag{1}$$

$$\Delta\varepsilon_e/2=0.0076(2N_f)^{-0.0878} \tag{2}$$

The curve of $\Delta\sigma/2$—$\Delta\varepsilon_p/2$ is shown in figure 2. Regression equation (3) can be obtained by regression analysis.

$$\Delta\sigma/2 = 1216.9(\Delta\varepsilon_p/2)^{0.0764} \tag{3}$$

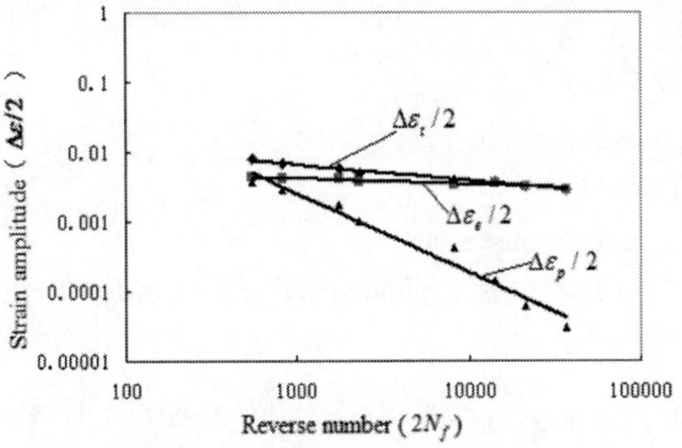

Figure 1. Curve of $\Delta\varepsilon/2$—$2N_f$

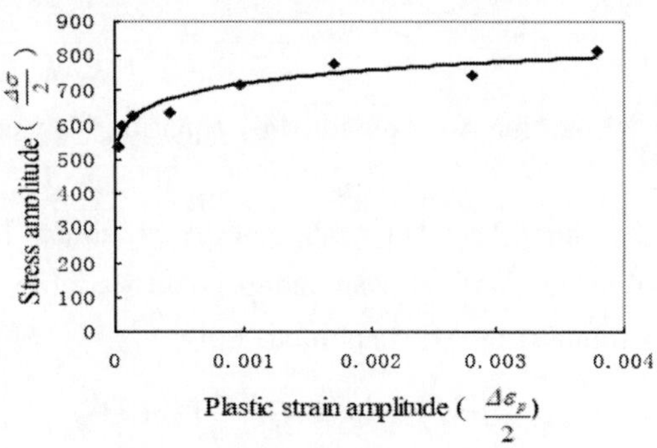

Figure 2. Curve of $\Delta\sigma/2$—$\Delta\varepsilon_p/2$

3.2. Fatigue life prediction

3.2.1. Life prediction by the Manson-Coffin method. Manson-Coffin method is widely used in LCF life prediction. For LCF test under total strain amplitude control, Manson-Coffin method predicts life using equation (4).

$$\frac{\Delta\varepsilon_t}{2} = \frac{\Delta\varepsilon_e}{2} + \frac{\Delta\varepsilon_p}{2} = \frac{\sigma_f'}{E}(2N_f)^b + \varepsilon_f'(2N_f)^c \tag{4}$$

where, $\frac{\Delta\varepsilon_t}{2}$ is total strain amplitude, $\frac{\Delta\varepsilon_e}{2}$ is elastic strain amplitude, $\frac{\Delta\varepsilon_p}{2}$ is plastic strain amplitude, σ_f' is fatigue strength coefficient, b is fatigue strength exponent, ε_f' is fatigue ductility coefficient, c is fatigue ductility exponent. The $\frac{\sigma_f'}{E}$, b, ε_f' and c values are obtained by regression analysis. Therefore, Manson-Coffin method predicts the low cycle fatigue life of 1Cr11Ni2W2MoV at 360℃ thorough equation(5).

$$\frac{\Delta\varepsilon_t}{2} = 0.0076(2N_f)^{-0.0878} + 6.9776(2N_f)^{-1.1448} \tag{5}$$

3.2.2. Life prediction based on the power transformation theory. According to LCF data analysis results, the relationship between plastic strain and failure reverse number presents nonlinear in the double logarithmic coordinates. It will lead to generating a prediction error using Manson-Coffin method. In order to solve the problem, a new method based on power transformation theory is proposed.

3.2.3. Power transformation theory. Box and Cox proposed the power transformation method as follows [10]:

$$T_p(x) = \begin{cases} \frac{x^p - 1}{p}, p \neq 0 \\ \ln x, p = 0 \end{cases} \tag{6}$$

Order M_x and M_y are the median of the variable x, y respectively. Assuming the real model data as follows:

$$\frac{y^p-1}{p}-\frac{M_y^p-1}{p}=K(x-M_x) \tag{7}$$

Order $z=\frac{y^p-1}{p}$, if p≠0,there:

$$M_y \approx (pM_Z+1)^{1/p} \tag{8}$$

Equation (9) is obtained by equation (6) variable substitution:

$$z-M_z \approx K(x-M_x) \tag{9}$$

Where, M_z is the median of the variable z, K is the model constants.

Make (pz+1) as an independent variable, $y=(pz+1)^{1/p}$ is expanded by second-order Taylor formulas in (pM_z+1) point as follows:

$$y \approx (pM_Z+1)^{1/p}+(pM_Z+1)^{(1/p-1)}(z-M_Z)+\frac{1-p}{2}(pM_Z+1)^{(1/p-2)}(z-M_Z)^2 \tag{10}$$

The equation (8) and equation (9) is substituted into equation (10). Equation (11) is obtained.

$$y \approx M_y+\frac{KM_y}{pM_Z+1}(x-M_x)+\frac{1-p}{2M_y}(\frac{KM_y}{pM_Z+1})^2(x-M_x)^2 \tag{11}$$

Order $C=\frac{KM_y}{pM_Z+1}$, the equation (11) is simplified to equation (12):

$$y \approx M_y+C(x-M_x)+\frac{1-p}{2M_y}C^2(x-M_x)^2 \tag{12}$$

Order $L=y-M_y-C(x-M_x)$, $F=\frac{C^2(x-M_x)^2}{2M_y}$, the equation (12) is simplified to equation (13):

$$L=(1-p)F \tag{13}$$

Where, C is obtained by linear regression for equation $y-M_y=C(x-M_x)$; p is obtained by linear regression for equation (13).

3.2.4. Life prediction model based on the power transformation theory. For metal material whose the plastic strain amplitude is less than the metal material 1, take (-ln $\frac{\Delta\varepsilon_p}{2}$) as intermediate variable. Equation (14) is obtained by transforming equation (6).

$$\frac{(-\ln\frac{\Delta\varepsilon_p}{2})^p-1}{p}-\frac{M_{y'}^p-1}{p}=K_0[\ln(2N_f)-M_{x'}] \tag{14}$$

Where, $M_{x'}$、$M_{y'}$ are the median of the variable 2Nf and (-ln$\frac{\Delta\varepsilon_p}{2}$) respectively. K_0 is the model constants.

Order $C_0=\frac{M_{y'}^p}{p}-K_0M_{x'}$, the equation (14) is simplified to equation (15):

$$(-\ln\frac{\Delta\varepsilon_p}{2})^p = K_0 \ln(2N_f) + C_0 \tag{15}$$

Figure 3 presents the linear relationship of $(-\ln\frac{\Delta\varepsilon_p}{2})^p$ and $\ln(2N_f)$. Linear correlation has reached more than 0.99. C_0 and K_0 are obtained by linear regression for equation (15).

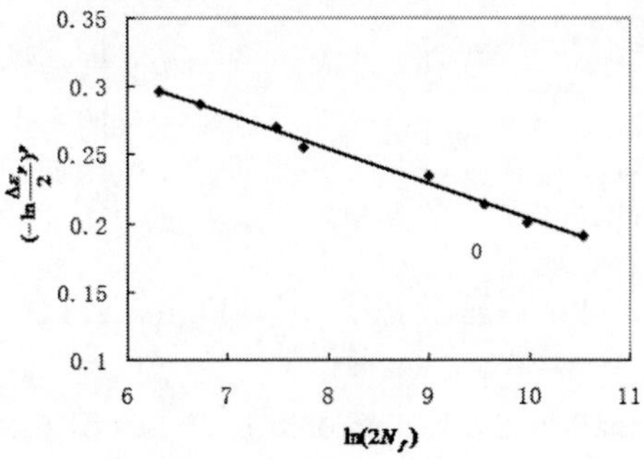

Figure 3. Curve of $(-\ln\frac{\Delta\varepsilon_p}{2})^p$ — $\ln(2N_f)$

Equation (16) is obtained by transforming equation (15).

$$\frac{\Delta\varepsilon_p}{2} = e^{-[C_0 + K_0 \ln(2N_f)]^{1/p}} \tag{16}$$

Therefore, the life prediction method based on power transformation theory can be expressed as the equation (17):

$$\frac{\Delta\varepsilon_t}{2} = \frac{\Delta\varepsilon_e}{2} + \frac{\Delta\varepsilon_p}{2} = \frac{\sigma_f'}{E}(2N_f)^b + e^{-[C_0 + K_0 \ln(2N_f)]^{1/p}} \tag{17}$$

The $\frac{\sigma_f'}{E}$, b, p, C_0 and K_0 values are obtained by regression analysis. Therefore, the new method predicts the low cycle fatigue life of 1Cr11Ni2W2MoV at 360℃ thorough equation(18).

$$\frac{\Delta\varepsilon_t}{2} = 0.0076(2N_f)^{-0.0878} + e^{-[0.4574 - 0.0254\ln(2N_f)]^{1/(-0.7061)}} \tag{18}$$

3.3. *Quantitative evaluation of life prediction ability*

Scatter band and standard deviation can be used to evaluate quantitatively the life prediction ability of model. Scatter band shows that the degree of deviation from the predicted life and experimental life [11, 12]. The scatter band (s') can be expressed as equation (19).

$$s' = \max\left[\frac{N_f^p}{N_f^m}, \frac{N_f^m}{N_f^p}\right] \tag{19}$$

Where, N_f^p is predicted life, N_f^m is experimental life.

Figure 4 presents all of life prediction value fall within 1.65 times scatter band by Manson-Coffin method. Figure 5 presents all of life prediction value fall within 1.57 times scatter bandby the new method.

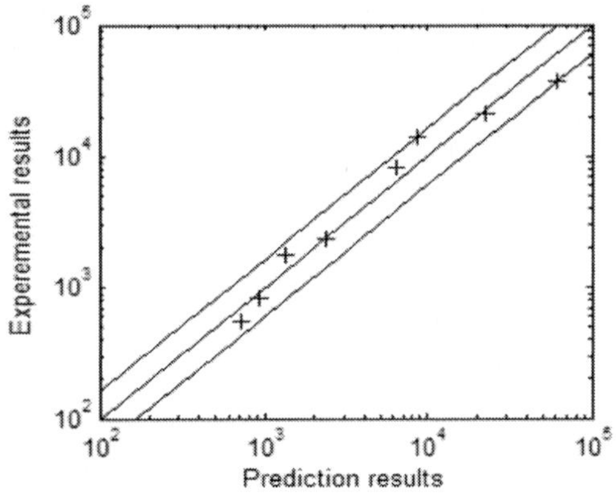

Figure 4. Results predicted by Manson-Coffin method

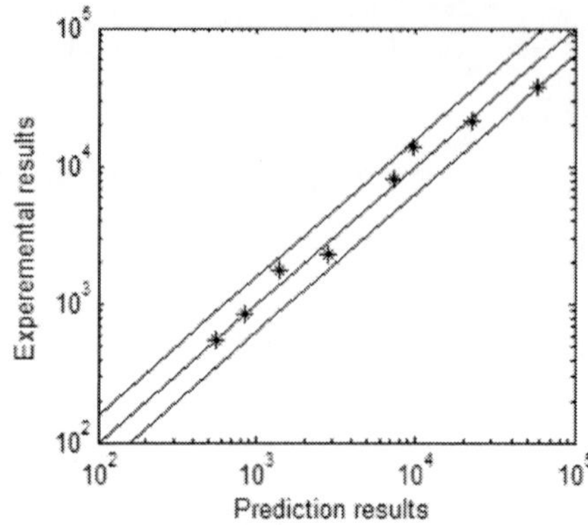

Figure 5. Life prediction result predicted by new method

The standard deviation is smaller; the life prediction ability of the model is better. The standard deviation(s) can be expressed as equation (20).

$$s = \left[\frac{\sum_{1}^{n} (\lg N_f^p - \lg N_f^m)^2}{(n-1)} \right]^{1/2} \tag{20}$$

Where, n is number of data points.

The fatigue life prediction for 1Cr11Ni2W2MoV by the Manson-Coffin method and new method had standard deviation of 0.32 and 0.18 respectively. The fatigue life prediction for 1Cr11Ni2W2MoV by the Manson-Coffin method and new method had Scatter bands of 1.65 and 1.57 respectively.

The evaluation of life prediction methods is shown in figure 6. The new method based on power transformation theory has smaller standard deviation and scatter band than Manson-Coffin method.Therefore, the life prediction capability of new methods proves more effective than Manson-Coffin method.

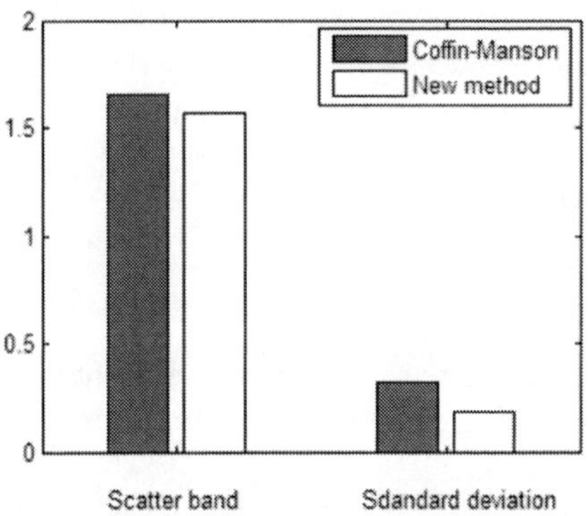

Figure 6. Evaluation of life prediction methods

4. Conclusions

The LCF tests of 1Cr11Ni2W2MoV were carried out under total axial strain control at 360℃. The relationships of cyclic strain-life, stress-strain and stress-life are analyzed. Cyclic stress-strain and strain-life curves of the 1Cr11Ni2W2MoV were obtained. There was a nonlinear relationship between the plastic strain amplitude ($\Delta\varepsilon_p / 2$) and reverse number ($2N_f$) in double logarithmic coordinates.

The LCF life prediction model of 1Cr11Ni2W2MoV at 360℃ is established by Manson-Coffin method and power transformation method.

Manson-Coffin method and new method can effective predict LCF life of 1Cr11Ni2W2MoV at 360 ℃. Prediction accuracy within ±1.65 times scatter band.

The life prediction capability of new methods proves more accurate than Manson-Coffin method. The results of life prediction showed new method has smaller standard deviation and scatter band than Manson-Coffin method. The Manson-Coffin method and new method have scatter bands of 1.65 and 1.57 respectively and have standard deviations of 0.32 and 0.18 respectively.

Acknowledgements
This work was financially supported by Natural Science Foundation of Liaoning Province of China(Research Project20170540107).

References

[1] Guo X G, Guo J T, Zhou L Z, et al. 2007 Rotational bending fatigue behavior of ni-based casting superalloy K435 at room temperature *Journal of Northeastern University* **28** 357-360
[2] Humayun K S M, Tae-in Y 2014 Evaluation of an energy-based fatigue approach considering mean stress effects *Journal of Mechanical Science and Technolog* **28(4)** 1265–1276
[3] Ozaltun H, Shen M H H, George T et al. 2011 An energy based fatigue life prediction framework for in-service structural components *Experimental Mechanics* **51(5)** 707–718
[4] Manson, SS 1954 Behavior of materials under conditions of thermal stress*National Advisory Commission on Aeronautics: Report 1170* Lewis Flight Propulsion Laboratory Cleveland
[5] Chen L J, Gang T Q, Xie L Y 2006 Low cycle fatigue life prediction method based on power-exponent function model *Journal of Mechanical Strength* **28** 761-765
[6] Fan Z C, Chen X D, Chen L and Jiang J L 2006 Prediction method of fatigue-creep interaction life based on ductility exhaustion theory *Acta Metallurgica Sinica.* **42** 415-420
[7] Young Il Kwon and Byeong Soo Lim 2011 A Study of Creep-Fatigue Life Prediction Using an Artificial Neural Network *Metal and Materials International* **4** 311–317
[8] Chen H and Jiang H D 2015 Low cycle fatigue life prediction methodfor single crystal nickel-base superalloysat high temperature *Journal of Mechanical Strength* **37(5)** 857–862
[9] Chen L, Zhang X M and OuyangP 2015 A life prediction model for low cycle fatigue based on continuum damage mechanics *Engineering* **26(18)** 2506-2517
[10] Box G E P and Cox DR 1964 An analysis of transformations*Journal of the Royal Statistical Society* **B26** 211-252
[11] Zhang S C, Yu H C and Li Y 2013 Investigation on low cycle fatigue properties of superalloy GH3044 at elevated temperature*Journal of aeronautical materials* **33** 100–104
[12] Zhang G D, Liu S L, He Y H and Zhou B Z 2004 Life predication of thermomechanical fatigue in DS superalloy DZ125 *Journal of aerospace power* **19(1)** 17–22

IOP Conf. Series: Materials Science and Engineering **292** (2017) 012069 doi:10.1088/1757-899X/292/1/012069

Effect of physicochemical action on the aggregative properties of detonation-synthesized nanodiamonds

Z W Fan[1], H Ilnitska[2], V Lysakovskyi[2], S Ivakhnenko[2] and T Kovalenko[2, 3]

[1] Beijing Longshengfu International Culture Exchange Center, China, China 100078 Beijing, Fengtai District, Fangzhuang Fangqunyuan, Shidaifangqun 3-302
[2] V.Bakul Institute for Superhard Materials National Academy of Sciences of Ukraine, Ukraine, 04074, Kyiv, Avtozavodskaya, 2

[3] E-mail: tetiana.v.kovalenko@gmail.com

Abstract. The results of researches of physicochemical action on aggregate properties of nanodiamond are presented. The kinetics of aggregation of nanodiamond powder was studied as a function of time, temperature, and pH of the solution. The effect of the sp2–sp3 hybridization ratio of carbon in nanodiamond powders on their aggregation was studied. It is shown that the presence of non-diamond carbon in detonation synthesis nanodiamond powders leads to the increase of the mean diameters of particles, i.e., their agglomeration. The theoretical justification of the aggregation mechanism is proposed. It is shown that it is possible to control aggregative properties of nanodiamond powders by physicochemical influences, e.g., gas-phase thermal treatment to reduce the size of agglomerates and to create a well-developed reconstructed surface of diamond particles with a low content of functional groups on their surface.

1. Introduction

At present the world industry has aroused great interest in nanosized materials, especially in detonation-synthesized nanodiamond powders [1-3]. An essential obstruction in manufacturing various products based on nanodiamond powders is their high tendency to the aggregation, which is seen both in air and in liquid medium. The nanodiamond aggregation in liquid media is a complex physicochemical process. For the last ten years at the Bakul Institute for Superhard Materials physicochemical and surface properties of submicron and nanodisperse diamond powders of static and detonation syntheses are studied in detail. It was defined that physicochemical properties of nanodisperse diamond powders depend on the chemical composition of their surface and the nature of their functional groups [4-7]. Nanodiamond powders are characterized by an essential aggregation of grains, which does not allow steady suspensions to be prepared. The aggregation of nanodiamond particles consists of three basic stages [8, 9]: the integration of the initial particles, the formation of coarser aggregates from the initial particles, and the consolidation of aggregates through the redistribution of separate particles and initial aggregates under the action of viscous stress, which acts at the expense of the liquid medium. The aggregation of nanodiamond particles depends on the physicochemical properties of their surface. Physicochemical characteristics of the diamond functional cover are responsible for many diamond properties. In this case the portion of the surface atoms (C_{surf}) in the total amount of the carbon atoms, which is responsible for their contribution into the diamond properties, depends on the sizes of the diamond macromolecules. The contribution of the detonation-

IOP Conf. Series: Materials Science and Engineering **292** (2017) 012069 doi:10.1088/1757-899X/292/1/012069

synthesized nanosized diamond particles is especially high, according to the calculation, the weight of the functional groups is commensurable with the weight of a polycarbonic nucleus [9-11]. Any action on a diamond starts with the action on its functional cover; therefore the chemical state of the diamond surface may affect both the features of the action and the final result. The disordering of the surface layer may be considered as a structural effect of the surface strong deforming action on the neighbouring volumetric layers of a diamond. The main method to decrease the rate of the aggregation of nanodisperse diamond powders is a decrease of the amount of inclusions, in particular, metal and carbon, and the neutralization of energy charges of their surfaces [6, 8, 10]. As the powder dispersion increases, the aggregative and sedimentary stabilities increase. Accordingly, the free and interfacial energy increases, and according to the Gibbs–Helmholtz principle this energy in disperse systems tends to a spontaneous decrease. A decrease of the free interfacial energy occurs through a coagulation–coarsening of particles. Thus nanodiamond powders, whose single grain is 4–6 nm of size, in suspensions exhibit sizes from 3 to 20 μm. As it is known, in the real conditions suspensions of nanodiamond powders belong to thin suspensions, whose sizes are from 100 to 0.5 μm [6, 7]. The nature of a liquid phase, its viscosity, temperature, and pH value are the fundamental factors in the creation of stable suspensions. The prerequisite to the agglomeration of the nanodiamond powders is their chemical composition and adsorption properties. Therefore, the aim of our work was to study the effect of various physicochemical actions of the medium on the aggregative properties of the detonation-synthesized nanodiamond powders.

2. Methods of the investigations

The investigations were performed on nanodiamond powders of the following grades: ACUD80, ACUD90, ACUD95, and ACUD99 with the different contents of carbon phases (by the normative documents of Ukraine) in the initial state and with the reconstructed surface. Aggregative properties of nanodiamond powders have been evaluated by histograms of particle-size distribution. The histograms have been plotted by the instrumentality of a Seishin LMS-30 device. Mean diameter of nanodiamond agglomerates has been defined from the histogram data by special programs. In addition, aggregative properties of nanodiamond powders have been estimated by the rate of their sedimentation in solutions of different types. We studied the effects of the temperature, pH of the medium, contents of different forms of carbon, thermal gas-phase reconstruction of the surface on the diamond aggregative properties. The effects of the temperature and pH of suspensions on the aggregation kinetics of nanodiamond powders were studied on nanodiamond powders of the ACUD-99 grade at temperature from 25 to 75°C in solutions with pH from 3.5 to 8.9 for the time from 5 s to 1 h. Tested were 0.2% suspensions of nanodiamond powders in 1l of distilled water with pH = 3.5 and in the solution of 9% sodium chloride with pH = 5.2 and pH = 8.9. The acidation of the suspensions of nanodiamond powders to pH = 3.5 was carried out by an addition of the solution of hydrochloric acid and the alkalization of suspensions of nanodiamond powders to pH of 8.9 was achieved by the addition of sodium hydroxide. A thermal gas-phase reconstruction of the surface was performed in two variants. According to the first variant (sample 1), samples of nanodiamond powder were annealed at 1000°C in the argon atmosphere and then cooled in argon. According to the second variant (sample 2), after the annealing of nanodiamond powders in the argon atmosphere at 1000°C the formed graphitized layer was oxidized by the air oxygen at 450°C [12]. The thermal gas-phase reconstruction of the surface was performed in two stages: annealing of nanodiamond powders at 1000°C in the argon atmosphere and then oxidation of the formed graphitized layer by air oxygen at 450°C. The rate of the nanodiamond particles aggregation during the deposition was defined by photometric method from a change of the suspension optical density and a change of the concentration of the nanodiamond particles in the suspension [13, 14]. For the first time by changing of optical density the stability of nanodiamond powders water suspension were evaluated. Thermal desorption spectra were taken using a MI 1201 mass-spectrometer in the temperature interval 20–600°C. Adsorption-structural and physicochemical studies of the powders surfaces were carried out by the classical method of the gas adsorption at lower temperatures in concordance with the BET principle using a NOVA 2200 gas-adsorption analyzer (Quantachrome, USA), which are intend for the measurement of the porosity, surface specific area, radii of pores, etc. The principle of operation is a static manometric method. Studies of the metal ions adsorption process on the diamond surface were performed using the photocolorimetric method from solutions of same-named salts in concentrations of 1:10.

3. Results and discussion

The nature of the liquid phase, its viscosity, pH values and temperature are fundamental factors at the creation of stable suspensions. In figure 1 the kinetics of the change of the ASUD 99 nanodiamond powders sedimentation rate in time is shown, and in figure 2 the variations of the sedimentation rate of nanodiamond powders vs. pH of solutions.

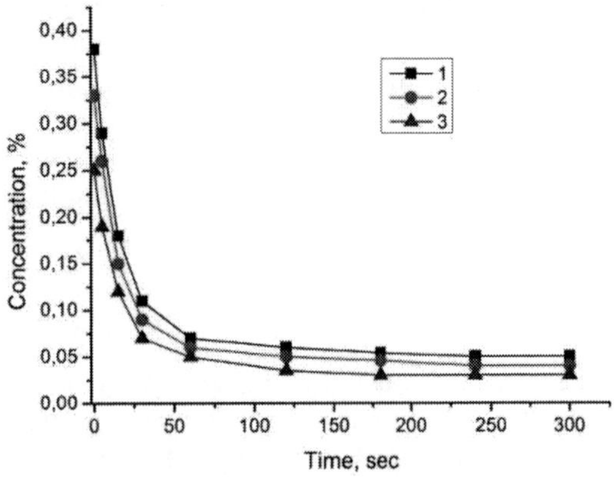

Figure 1. Dependence of the sedimentation rate of the ASUD 99 nanodiamond powder in distilled water on the temperature and exposure time:T = 25°C (curve 1), T = 50°C (curve 2), T = 75°C (curve 3).

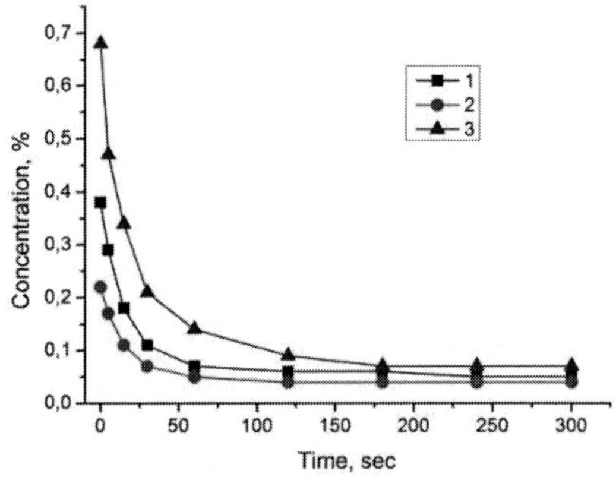

Figure 2. Dependence of the sedimentation rate of the ASUD 99 nanodiamond powder on pH of a solution (suspension) and exposure time: pH = 8.9 (curve 1), pH = 6.5 (curve 2), pH = 3.5 (curve 3).

It follows from figure 1 that for the first ten seconds of the deposition of diamond particles aggregates a rectilinear section is observed that characterizes a decrease in the sedimentation rate of diamond particles. After the first 10 s the sedimentation rate decreases and after 70 s becomes virtually constant, i.e., the process of aggregation is completed for 70 s.

In studying the effect of heating the suspensions of nanodiamond powders in distilled water at temperatures from 25 to 55°C, and 75°C on the aggregation kinetics of nanodiamond particles (figure 1), it was found that with increasing temperature of a suspension of nanodiamond powders in a distilled water the concentration of particles in the interval under study decreases, i.e., with increasing temperature the aggregation rate increases.

The aggregation rate of nanodiamond powders depends on the pH of the suspension solution. The aggregation kinetics of nanodiamond powders in solutions based on distilled water at the room temperature at different pH values of the solution is given in figure 2. It is seen that as the pH value of the solution increases, the concentration of diamond particles in the suspension in the interval under study increases, thus decreasing the sedimentation rate, i.e., in the alkaline medium the aggregation process becomes slower. Thus, the alkalization of suspensions essentially reduces the sedimentation rate of nanodiamond particles.

Of interest was to compare the change of the mean diameter of particles of nanodiamond powders of different grades with the change of the carbon phase concentration. The following grades of nanodiamond powders were studied: ASUD 80, ASUD 90, ASUD 95, ASUD 99, which differed by the fixed relation of carbon phases of the sp^2- and sp^3- hybridization.

The table 1 gives the basic adsorption-and-structural characteristics of ASUD 99 nanodiamond powders after the two-staged gas-phase reconstruction of the surface.

Table 1. Physicochemical properties of the surfaces of the initial and reconstructed ASUD 99 powders.

Characteristic	ASUD 99 init.	Sample 1	Sample 2
Specific surface area S_{BET}, m^2/g	242	245.0	256.4
Pore volume, V, ml/g	1.045	1.0	1.816
Mean radius of pores, R, Å	86	80.0	141
Mean diameter of particles, d_{mean}, μm	12.6	2.28	4.764
Adsorption potential, A, J/g	12.2	12	14
Parameter K, %	12.6	12	10

In our studies it was defined that with increasing concentration of non-diamond carbon in nanodiamond powders the mean diameter of particles increases, thus indicating the intensification of the aggregation process (figure 3).

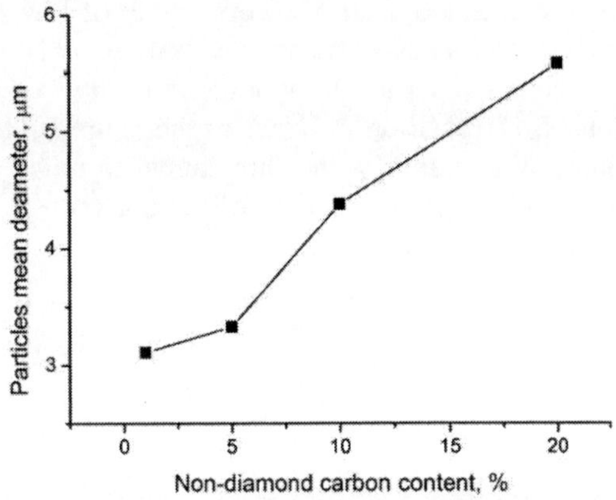

Figure 3. Dependence of the change of the particles mean diameter of the nanodiamond powder on the concentration of non-diamond carbon

Based on the data obtained the following mechanism of the aggregation is proposed. The mechanism implies that the aggregation is realized in several stages:

– adsorption of oxygen-containing groups from solutions and thus formation a functional cover;

– establishing the equilibrium between the solution and functional cover,

agglomeration through the Van der Waals forces;

– agglomeration through covalent (chemical) forces.

The nanodiamond powder is characterized by a polycrystalline structure. The nanodiamond particle consists of a nucleus, a shell of non-diamond carbon and functional groups of different composition. As a rule, agglomerates of nanodiamond powders consist of several diamond clusters, in which the bond between diamond nuclei is realized with the help of functional groups [4, 15].

It follows from the above that in many cases oxygen-containing groups are responsible for aggregation properties of powders. Because of this we studied the influence of the thermal gas-phase reconstruction on a changing the functional cover and on aggregative properties of nanodiamond powders. It was defined that the samples of nanodiamond powders subjected to the thermal annealing at 1000°C in the argon atmosphere and cooled in the same atmosphere (sample 1) exhibited the minimum size of agglomerates – 2.8 μm. Obviously, the annealing of powders at 1000°C and cooling them in the argon atmosphere made it possible to clear the surface from oxygen-containing groups, graphitize the powder surface, and reduce the aggregation rate. The oxidation of powders by the air oxygen (sample 2) allowed us to remove a graphitized layer of nanodiamond powder, but at the same time new oxygen-containing compounds appeared on the surface and the mean diameter of nanodiamond powders increased by a factor of 2. The mean diameter of the agglomerates are presented in the table 1.

It follows from the data given in the table that after a two-stage gas–phase reconstruction of the surface of nanodiamons powder according to the described above scheme, the specific surface of powders (sample 2) and the adsorption potential of them somewhat increased (no more than by 10%), but at the same time the volume of pores increased considerably by a factor of 1.7.

Figure 4 shows the IR spectra of the initial and after thermal gas-phase reconstruction of nanodiamond powders were recorded in the near and middle infrared range on the Nexus IR Fourier spectrometer. The broad band at 3410 cm^{-1} corresponds to valence O-H vibrations of absorbed water on the surface (or N-H oscillations). Peaks 2890 and 2950 cm^{-1} are symmetrical and asymmetric valence vibrations C-H. Peak 1750 cm^{-1} corresponds to valence C=O vibrations in the carbonyl and carboxyl groups, 1635 cm^{-1} responds to the deformation O-H vibration (or / and N-H and C-N valence vibrations).

In the frequency range 1000-1500 cm^{-1} there is a large number of bands (with a complex contour), among which the main interest connected with 1380 cm^{-1} peak of low intensity (a small component compared to others) corresponds to the N-O valence vibration of NO_x groups (a consequence of treatment with nitric acid), or corresponds to OH vibrations in the COOH group. Peak 1260 cm^{-1} is associated with C-N vibrations. Superposition of bands in the range 1000-1200 cm^{-1} correspond to C-O vibrations =C–O–C= groups. Wide bands with a low signal-to-noise ratio in the range 1900-2700 cm^{-1} are artifacts (absorption of the diamond element of the apparatus), which arose when measured by the method of fully deformed total reflection.

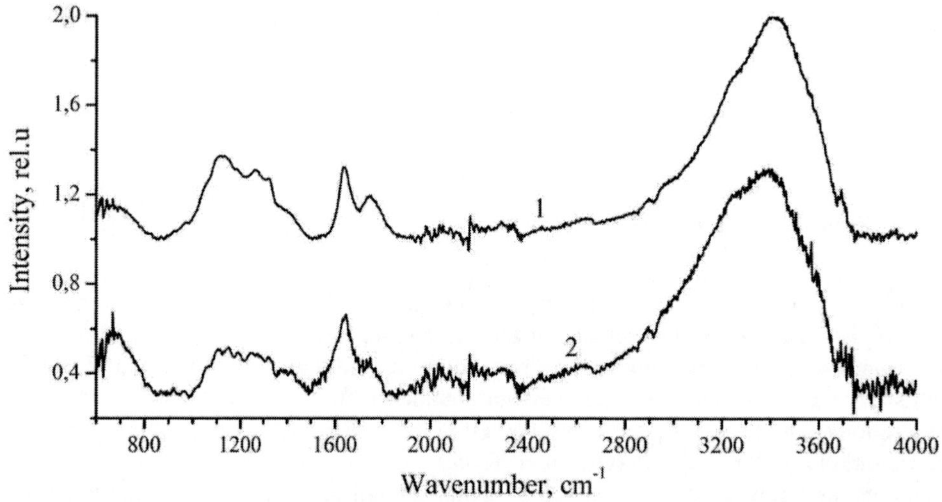

Figure 4. IR spectra of initial nanodiamond powders (1) and after gas-phase treatment (2).

IR-spectrum of sample 2 (table 1) after thermal gas-phase reconstruction (curve 2) in comparison with the spectrum of the initial sample (curve 1) contains bands at the frequency range 1000-1750 cm^{-1} of significantly lower intensity, what means a lower content of carbonyl and carboxyl groups.

Thermodesorption spectra (figure 5) of vapors, carbon monoxide and dioxide for the initial nanodiamonds and after the gas-phase reconstruction of the surface (for sample 2 in table 1) verify that on the surface of nanodiamond powders with the reconstructed surface (curve 2) the amount of physical and chemical adsorbed vapors is much below than on the initial surface, the content of CO is insignificant, and CO_2 is absent. These studies confirm the obtaining of nanodiamond powders surface, which is purer from functional groups.

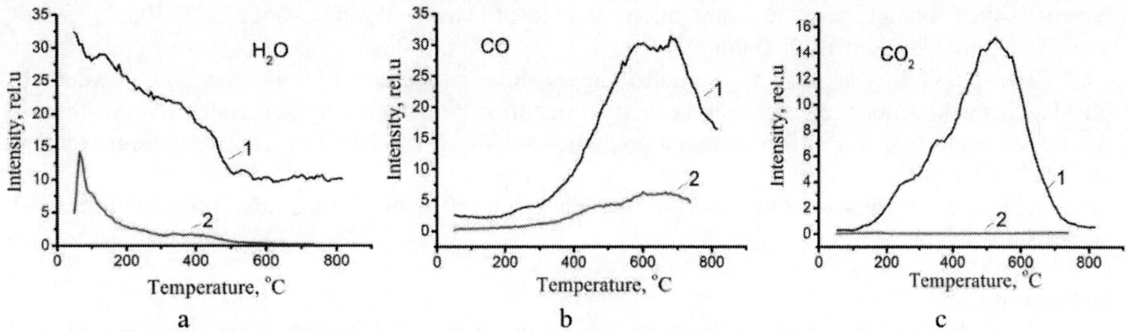

Figure 5. Thermodesorption spectra of vapors (a), carbon monoxide (b) and dioxide (c) on the initial diamonds (curve 1) and after thermal gas-phase reconstruction of the surface (curve 2).

Thus, it is shown that, by physicochemical influences, it is possible to control the aggregative properties of nanodiamond powders, for example by gas-phase heat treatment, to reduce the size of agglomerates and to create an elaborate reconstructed surface of diamond particles with a low content of functional groups on their surface.

It was of interest to study the sedimentation stability of produced nanodiamond powders treated according to two-stage gas-phase treatment. From the ASUD 99 nanodiamond powders (sample 2) a water suspension with the concentration 0.64% and volume of 250 ml was prepared.

The sedimentation stability of this suspension with time was studied. The suspension was stable for 3 months. In this case the optimal density of the suspension D = 5.5 B was kept for all the period of time. The distribution of agglomerates of diamond particles and their mean diameter was defined in the finished suspension. The mean diameter of agglomerates in a finished suspension was 0.403 μm.

We studied adsorption properties of a nanodiamond powder with a reconstructed surface relative to chromium ions (VI) [16].

The efficiency of the heavy metal ions recovery from the modal solution exceeds 70%, the kinetic parameters of the process are defined. It has been established that nanodiamonds rapidly and efficiently recover chromium ions (VI) at the static conditions: for first 5 min the concentration of them decreases more than by 50%.

Based on our investigations, a novel composite filter material on the base of basalt, lavsan, and cellulose fibers with the addition of detonation-synthezed diamond nanopowder for purification of water from heavy metals ions was developed. The testing of the developed material after additional (final) drains after the treatment of the synthesis product during the diamond recovery shows that the drains correspond to the maximum permissible values for a drinking water.

4. Conclusions

1. The kinetics of aggregation of nanodiamond powder depending on time, temperature, and pH solution was studied. It has been established that the aggregation rate of the freshly made solution is so much large that the aggregation of almost 90% of nanodiamond powder occurs for 10 s. With the temperature increase the aggregation rate also increases. It was defined that the alkalization of the solution results in a decrease of the aggregation rate because due to the dissociation of oxygen-containing compounds of the surface layer.

2. The effect of the ratio of the sp^2–sp^3 hybridization of carbon in nanodiamond powders on their aggregation was studied. It is shown that the presence of non-diamond carbon in nanodiamond powders of detonation synthesis leads to the increase of the mean diameters of particles, i.e., their agglomeration.

3. The theoretical justification is proposed of the aggregation mechanism, which implies that the aggregation process takes place in several stages that include the adsorption of oxygen-containing groups from the solution and thus formation of the functional cover; establishing of the equilibrium

IOP Conf. Series: Materials Science and Engineering **292** (2017) 012069 doi:10.1088/1757-899X/292/1/012069

between the solution and agglomeration functional cover through Van der Waals forces; agglomeration due to covalent (chemical) forces.

4. The possibility is shown to control aggregative properties of nanodiamond powders by physicochemical actions, e.g., gas-phase thermal treatment to decrease agglomerates and create well-developed reconstructedn surface of diamond particles with a low content of functional groups on their surface.

5. The use of nanodiamonds having the developed surface ensures the producing a stable suspensions.

References

[1] Gruen D M 2005 *Synthesis, Properties and Applications of Ultrananocrystalline Diamond* Ed. by Gruen D M, Shenderova O A and Vul A Ya (NATO Science Series – Dordrecht The Netherlands: Springer) p 401
[2] Novikov N V, Bogatyreva G P and Voloshin M N 2004 *Physics of Solids* **46**(4) 585
[3] Danilenko V V 2003 *Synthesis and sintering of diamond by explosion* (Moskow: Energoatomizdat) p 272
[4] Novikov N V, Bogatyreva G P 2008 *J. Superhard Materials* **2** 3
[5] Bogatyreva G P, Marinich M A 2009 *Diamond and Related. Materials* **9(12)** 2002
[6] Bogatyreva G P, Ishchenko E V, Gvyazdovskaya V L, Bazaliy G A and Oleynik N A 2004 *Structure of properties of SHM. Methods of studies* (Kyiv) p 126 (in rus)
[7] Bogatyreva G P, Marinich M A, Bazaliy G A, Ilnitskaya G D, Bilochenko V A and Tsyba N N 2010 *Rock destruction and metal-working tools – engineering and technology of its production and application*. Proceeding, issue 13. (ISM, NASU, Kyiv) p 326-331 (in rus.)
[8] Goncharuk O V, Zarko V I, Bogatyrev V M and Terpilowski K 2014 *Chemistry, Physics & Technology of Surface* **5(2)** 210
[9] Tsubota T, Tanii S, Ida S, Nagata M and Matsumoto Y 2004 *Diamond Relat. Mater.* **13** 1093.
[10] Krueger A J 2008 *Mater. Chem.* **18**, 1485
[11] Zhu Y W, Xu F, Shen J L, Wang B C, Xu X Y and Shao J B 2007 *J. Mater. Sci. Technol.* **23**, 599
[12] Ilnitska H, Antonenko L, Tymoshenko V 2014 Method of producing a filter material Pat UA96472 U C01B31/06 (2006.01)
[13] Tymoshenko V V *2013 Nano and Advanced Materials Workshop and Fair NAMF* (Warsaw, Poland) 3
[14] Ilnitska H, Oleynik N, Zaitseva I and Tymoshenko V 2013 *Scientific notes of LNTU* **41**(1) 106 (in rus.)
[15] Chukhaeva S I 2004 *Physics of the Solid State* **46**(4) 610
[16] Ilnitska H, Antonenko L, Tymoshenko V 2015 Method of producing a filter material Pat UA96472 U C01B31/06 (2006.01)

IOP Conf. Series: Materials Science and Engineering **292** (2017) 012070 doi:10.1088/1757-899X/292/1/012070

Application of P.O and R-SAC mortar for 3D printing in construction

J C Lin[1], X Wu, W Yang, R X Zhao and L G Qiao

National Enterprise Technology Center, China State Construction Ready Mix Concrete CO., LTD, Wuhan, China

[1] E-mail: Sillybear@yeah.net

Abstract. Compared with traditional properties of building materials, 3D printing materials should have controllable setting time, appropriate workability and superior mechanical properties. Adaptability of two types of cementitious system on the 3D printing were investigated in this paper. Results showed that both of them can match of 3D printing process with the working window time required (10-90min) by using the compound adjustable solidification agents, which have the high mechanical properties (3d≥30MPa, 28d> 63MPa), and appropriate suitable parameter range of fluidity (R-SAC: 210-240mm,P.O:200-220mm).

1. Introduction

3D printing technology, also known as the additive manufacturing technology. Nowadays 3D printing contain much more multidisciplinary advanced technology such as digital modeling technology, material science, computer technology, sensors technology, etc. This technology is called one of the samples of the third industrial revolution.

3D printing materials are designed for the conditions of 3D process and requirements. There are four general forms of the materials: powder, filament, lamellar, liquid. The major varieties include engineering plastics, photosensitive resin, metals and ceramics, and it is widely used in the mold manufacturing, medical treatment, aerospace and other fields. But because of the issue of cost, technology, practicability, etc., the 3D printing materials mentioned above are not appropriate for the construction building. However, the cementitious a large class of building materials has the advantages of high strength, high durability, self-coagulation and low cost, Combined with the technology of "contour process", the One-piece building elements of 3D-printing is achievable with cementitious materials.

1.1. Analysis of performance requirements of 3D printing cementitious materials

3D printing with cementitious materials is an innovative construction process for fabricating concrete components employing an additive, layer-based, manufacturing technique, also called freeform construction [1, 2].The material needs to have good flow performance, but also provides a certain degree of strength to support the subsequent accumulation after the extrusion of the printing head. With the current technique of building manufacturing process, cementitious materials are achieved long the setting time and excellent working performance Therefore, different from the traditional construction of the material properties, 3D printing material performance should be designed to meet the following requirements:

(1) Appropriate work performance

3D printing materials need to have an acceptable degree of extrudability to be extruded through an extrusion printing head containing nozzles to form small materials filaments. The filaments must bond together to form each layer, as the fresh materials is continuously extruded to form consecutive filaments layered on the previous ones to build complete 3D components.

(2) Controllable setting time

The materials should not only have small amount of deformation with high printing speed, but also must provide sufficient support strength in a short period of time, so the material's setting should be relatively rapid. At the same time, the material working window time is not fixed to avoid the material setting too fast or loosely bound of layers. And the material should provide a controlled adjustment of condensation hardening range from 10-90min.

(3)Superior mechanical properties and durability

The layered structure of 3D printing is likely to congenital defects of compactness, the structural strength of 3D printing construction is less than the prefabricated in order to ensure the safety and integrity of the component of 3D print structure, 3D printing materials require higher mechanical strength and good durability.

Therefore, compared with traditional high performance development of concrete materials, the performance design of 3D printing materials is not convergence, and how to choose the 3D printing cementitious materials is very important. With this, the matching of the materials under R-SAC cement and P.O system with 3D printing is explored.

2. Raw materials and experimental methods

2.1. Raw materials

(1) Cement:

A) 42.5R sulphoaluminate cement produced by Wang Lou Cement Industry Co., Ltd., which basic performance is shown in table 1.

Table 1. Physical properties of sulphoaluminate cement.

Standard consistency water consumption (%)	Setting time (h:min)		Compressive Strength/MPa			Flexural Strength/MPa		
	Initial setting time	Final setting time	1d	3d	28d	1d	3d	28d
27	10	18	30.0	41.0	45.0	5.3	5.9	6.2

B) P·O42.5 cement produced by Hubei Yadong Cement Plant, whose basic erformance is shown in table 2.

Table 2. Basic performance of concrete.

Standard Consistency/g	Density/ $g\cdot cm^{-3}$	Stability	Compressive strength/MPa		Flexural Strength/MPa	
			3d	28d	3d	28d
127	3.20	Qualified	23.30	48.10	5.23	8.72

(2) Coagulant: Calcium formate (industrial pure); Promoting powder; Liquid low-alkali accelerator; $AlSO_4$ (chemical pure concentration of 20% of the saturated solution), NaF.

(3) Retarder: Boric acid (industrial pure); Sodium gluconate (pure industrial); Sucrose.

(4) Sand:River sand(fineness modulus is 3.0).

2.2. Test methods

Setting time test is performed according to the Standard of test methods for water requirement of normal consistency, setting time and soundness of the P.O (GB/T 1346-2011).

Compressive strength test is performed according to the Standard for Method of testing cements-determination of strength (GB/T 17671-1999).

IOP Conf. Series: Materials Science and Engineering **292** (2017) 012070 doi:10.1088/1757-899X/292/1/012070

Mortar fluidity test is performed according to the Standard for Methods of testing uniformity of concrete admixture (GB8077-2012),

3D printing test is performed according to construction process of project developed equipment (figure 1) with the component quality of the mortar, Cement: Microsphere: Slag Powder: Sand: Water=40:9:5:74:15.

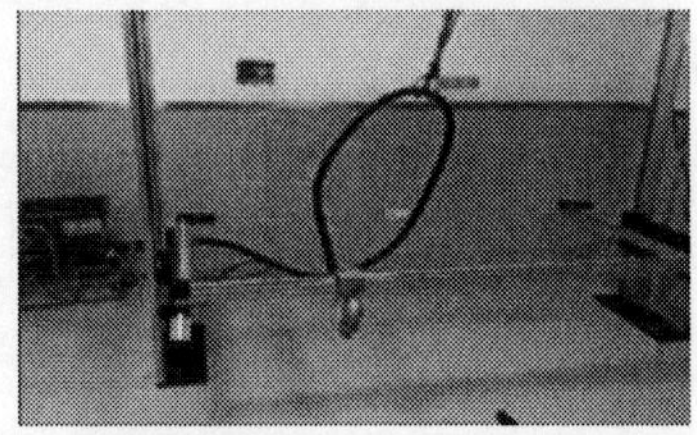

Figure 1. 3D printer.

3. Experimental results and discussion

3.1. Applicability of sulphoaluminate cement

3.1.1. Controllability of setting time. Sulphoaluminate cement is used for building quick repair materials, and its fast settling speed and higher early strength are more matched for 3D printing construction. The initial setting time of the sulphoaluminate cement used in the trial is about 10min, it is necessary to use retarders to reduce the rate of hydration, the test selected three kinds of common retarder to control the setting time.

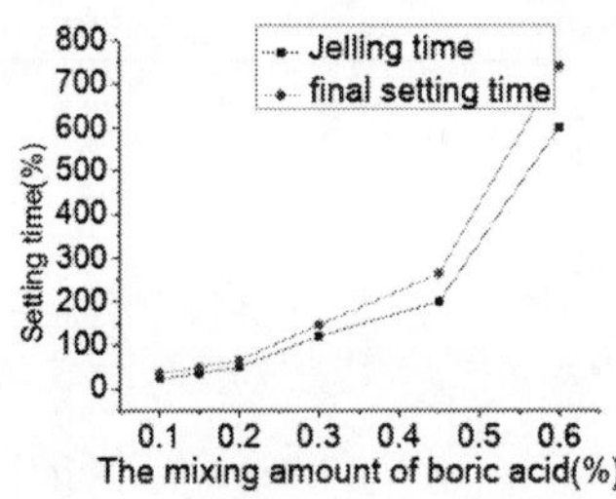

Figure 2. Effect of boric acid on setting time

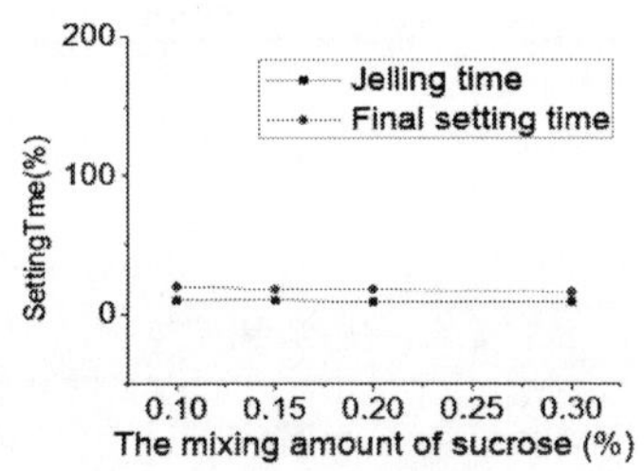

Figure 3. Effect of sucrose on setting time

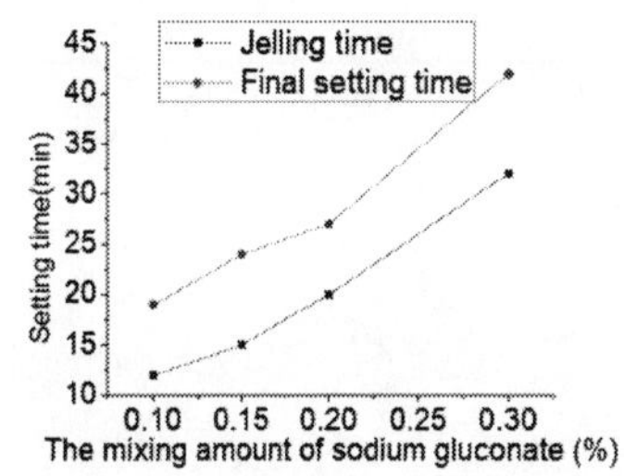

Figure 4. Effect of sodium gluconate on setting time

The results showed that the effect of different retarders on the setting time of R-SAC is different. Its retarding effect order is boric acid > sodium gluconate > sucrose. The setting time of R-SAC is prolonged with the increasing content of boric acid. When the amount is 0.1%, the initial setting time can be increased by 14% to 24min, but when the amount is 0.3%, the setting time is increased by more than 144%, excessive increase in setting time is not conducive to control, which will increase the use of risk (figure 2). Both sodium gluconate and sucrose are common retarders for P.O, the retardation effect of sodium gluconate is more linear. When the amount of 0.3%, initial setting time is increased by 20 minutes, the retardation effect is more stable(figure 4). With the increase of sucrose amount, the retardation effect of R-SAC is not improved, so it cannot be used as R-SAC retarder(figure 3). It is also found that when the boronic acid amount is 0.15%, the setting time of R-SAC reached the standard of initial setting (>30 min). Therefore, with 0.15% boric acid, the preparation of R-SAC with stable settling time gradient is carried out by using retarder and quick-setting agent.

As shown in the figure 5, the test group combined with boric acid as a retarder, only the sodium gluconate group has a retardation effect, sucrose group still no retardation effect. In the sodium group, with the increase of the amount of sodium gluconate from 0.1% to 0.4%, the setting time average increase is less than 30%, and the initial setting time is fixed in the range of 75-160 min.

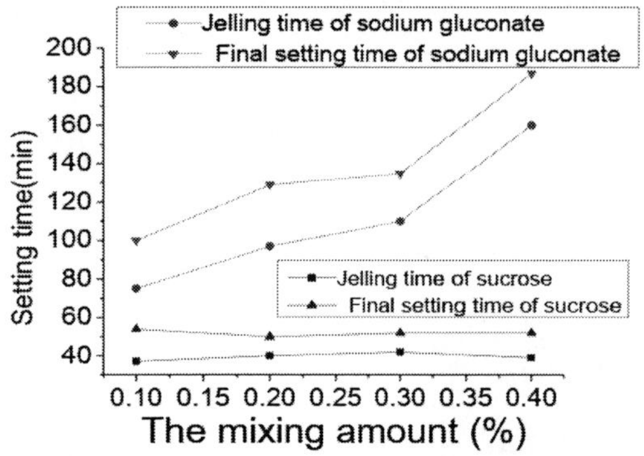

Figure 5. Effect of retarders on setting time.

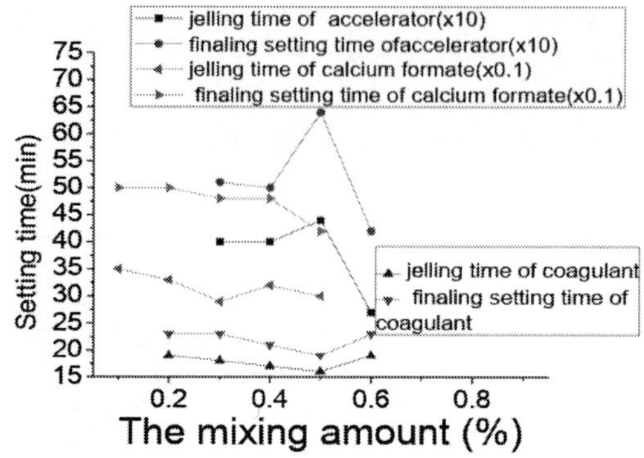

Figure 6. Effect of coagulant on setting time.

The figure 6 shows that the composite coagulation effect of promoting powder group is the best, When the amount of 0.2%, initial setting time is reduced to 20 minutes, but after that, with the increase in the amount of promoting powder ,setting time is essentially flat. The Liquid low-alkali accelerator has a retarding effect before 5%, when proportion up to 6%, the setting time is shortened to 27min. And calcium formate compound with boric acid has no obvious coagulation effect. In summary, When R-SAC is used as 3D printing material, the R-SAC setting time control formula is shown in the table 3:

Table 3. R-SAC setting time control formula

NO.	Setting time /min	promoting powder /%	retarder		initial setting time /min	final setting time/min	Time window /min
			variety	mixing amount /%			
1	10-20	0.2	boric acid	0.15	19	26	7
2	20-30	/	boric acid	0.1	24	36	12
3	30-40	/	Sodium gluconate	0.3	32	49	17
4	50-60	/	boric acid	0.2	60	75	15
5	70-90	/	boric acid + sodium gluconate	0.15+0.1	75	100	25

From the table we can see that the R-SAC setting time can be controlled in 10 ~ 90minto match the different construction conditions of 3D printing.

3.1.2. Work performance range. In this experiment, the adjustable solidification agent is chosen as the fifth group in 3.1.1, and the material's setting time is set to 75min. According to the different amount of water reducing agent, fluidity of 150-260mm of the mortar is tested for 3D printing effect, the experimental data is shown in table 4.

Table 4. 3D printing performance of R-SAC mortar.

NO.	amount of water reducing agent/%	Fluidity/mm	layer width/cm	The layer number	Printing status
6	0.15	150	4.1	/	/
7	0.24	175	4.1	22	pipe blocking
8	0.33	195	4.3	22	pipe blocking
9	0.42	210	4.2	30	well
10	0.50	240	4.5	35	well
11	0.55	260	5.4	15	expanding

With the increase of the amount of water reducing agent, the setting time does not change much, the fluidity of cement mortar will gradually increase. But when its amount of more than 0.5%, the mortar is bleeding. When the mortar is super-stiff (fluidity ≤ 150mm), it does not have the extrusion property, and also can't be connected with the substrate, the print performance is extremely poor; and when the fluidity of mortar is better than250mm, its extrusion performance is excellent, but the 3D printing maximum number of layers is less than 15, and the component engender deformation. This shows that the fluidity of the mortar has a great influence on the 3D printing performance, and there is suitable parameter range of R-SAC fluidity is 210-240mm.

3.1.3. Mechanical properties. The mechanical properties of mortar with adjustable solidification agents in 3.1.1 are verified, the experimental data is shown in table 5.

Table 5. Mechanical properties of R-SAC.

NO.	promoting powder/%	adjustable solidification agents		initial setting time /min	The compressive strength/MPa		
		variety	amount /%		3d	7d	28d
0	/	/	/	12	43.2	54.4	67.9
1	0.2	boric acid	0.15	29	48.7	56.7	64.5
2	/	boric acid	0.1	21	42.6	53.2	66.6
3	/	Sodium gluconate	0.3	29	41.2	54.2	68.2
4	/	boric acid	0.2	67	37.1	51.85	63.1
5	/	boric acid + sodium gluconate	0.15+0.1	85	41.0	52.9	65.2

It was found that the addition of retarder had a slight decrease in the strength of 3d, but it did not affect the strength of a late stage. Only when the boric acid amount exceeds 0.2% 3d strength decreased by about 14%.

3.2. Applicability of P.O System

3.2.1. Controllability of setting time. The setting time of P.O material is generally more than 90min. Therefore, three kinds of common accelerators were selected to control the setting time.

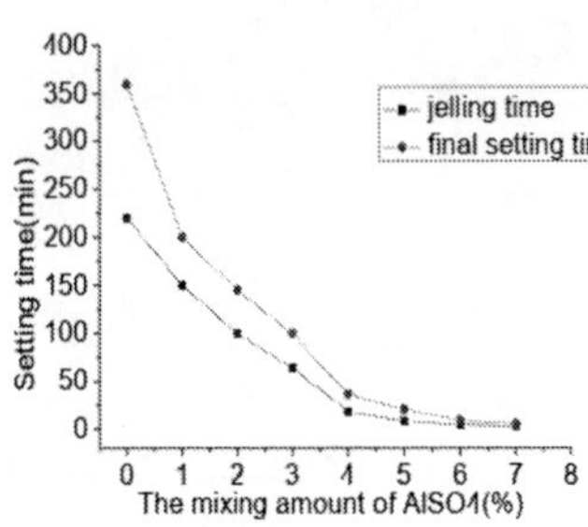

Figure 7. Effect of $AlSO_4$ on setting time.

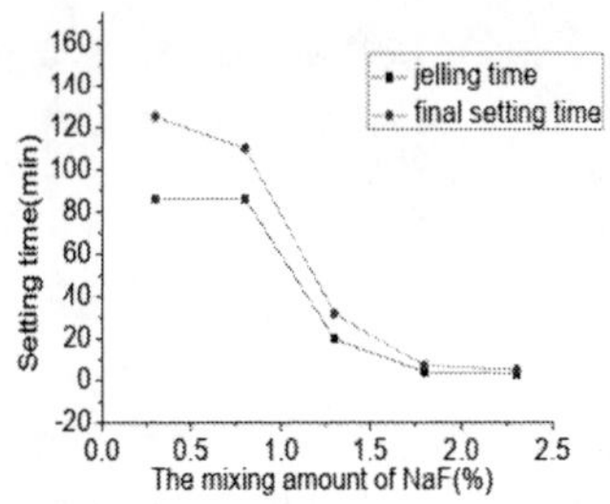

Figure 8. Effect of NaF on setting time.

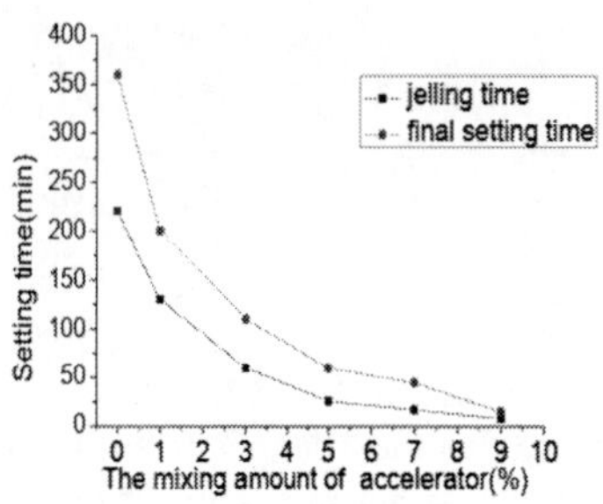

Figure 9. Effect of Liquid low-alkali acceleratoron setting time.

The results shows that the effect of different accelerators on the setting time are similar, and the effect order is NaF> AlSO4> liquid low-alkali accelerator. It can be seen from figure 7 that with the increase of the amount of $AlSO_4$, the setting time are shortened rapidly, When the amount of $AlSO_4$ is 4% ~ 7%, the effect of coagulation is the most significant, and the initial setting time is less than 20min.This is the AFt crystal overlap with each other to form a skeleton [3, 4], but when the amount of $AlSO_4$ is more than 5%, the setting time tends to be stable. It can be seen from figure 8 that the amount of NaF is more than 1.3%, the initial setting time is less than 20min, and when the amount is more than 1.8%, the setting time is not change greatly. It should be noted that NaF contains alkali metal ions, its dosage should be controlled to avoid the alkali aggregate reaction. As shown in figure 9, the effect of the liquid low-alkali accelerator on setting time is relatively stable increases. When the amount is over 3%, the initial setting time is shortened to 60min, when the amount is 9%, the initial setting time is less than 10min.

Therefore, When P.O material is used as 3D printing material, the setting time control formula is shown in the table 6:

Table 6. Setting time control formula of P.O.

NO	Setting time/min	accelerators		initial setting time /min	final setting time/min	Time window /min
		variety	mixing mount %			
12	10-20	$AlSO_4$	5	8	20	12
13	20-30	NaF	1.3	20	32	12
14	30-40	liquid low-alkali accelerator	5	26	60	34
15	50-60	liquid low-alkali accelerator	3	60	110	50
16	70-90	liquid low-alkali accelerator	2	99	158	59

From the table we can see that the setting time can be controlled in 10 ~ 90minto match the different construction conditions of 3D printing.

3.2.2. Applicability of work performance. In this experiment, the adjustable solidification agent is chosen as the group 16 in 3.2.1, and the material s ' setting time is set to 80 min. According to the different amount of water reducing agent, fluidity of 150-260 mm of the mortar is tested for 3D printing effect, the experimental data is shown in table 7.

Table 7. 3D printing performance of P.O mortar.

NO	amount of water reducing agent /%	Fluidity/mm	layer width/cm	The layer number	Printing status
17	0.84	150	4.1	/	/
18	0.89	170	4.1	/	/
19	0.95	190	4.1	22	pipe blocking
20	1.08	205	4.2	25	well
21	1.10	220	4.4	20	well
22	1.19	245	5.0	12	expanding
23	1.22	260	5.6	8	expanding

The results show that with the increase of the water reducer, the fluidity of the mortar gradually increases, and the 3D printing layers stability layers increased first and then decreased. When the mortar is super-stiff (fluidity ≤ 190mm), its extrusion performance is barely, the 3D printing effect is extremely poor as shown in figure 10. And when the fluidity of mortar is better than 245 mm, its extrusion performance is excellent, but it is easy to produce large deformation of the single layer in the printing process shown in figure 11.This shows that the suitable parameter range of fluidity of P.O mortar is 200-220mm.

Figure 10. 3D Printing Performance of low fluidity.

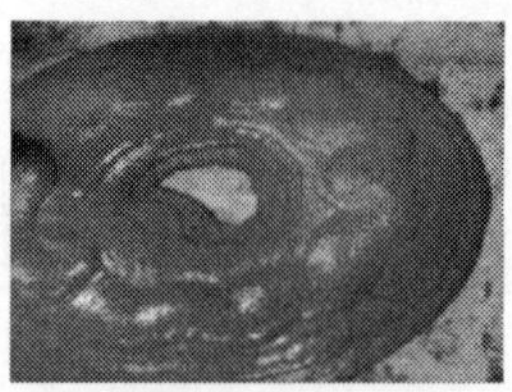

Figure 11. 3D Printing Performance of high fluidity.

3.2.3. Mechanical properties. The mechanical properties of mortar with adjustable solidification agents in 3.2.1 are verified, the experimental data is shown in table 8.

Table 8. Mechanical properties of P.O mortar.

NO	adjustable solidification agents		initial setting time /min	The compressive strength/MPa	
	variety	amount /%		3d	28d
	/	/	300	30.3	83.2
12	$AlSO_4$	5	8	44.5	76.6
13	NaF	1.3	20	42.5	78.8
14	liquid low-alkali accelerator	5	26	31.5	75.2
15	liquid low-alkali accelerator	7	60	42.4	77.1
16	liquid low-alkali accelerator	2	99	43.2	81.0

It is found that the adjustable solidification agents in the suitable amount can be more accurate to control the setting time of the material, and the early mechanical properties of the material can be improved greatly, and the influence on the later performance is less than 10%.

4. Conclusion

In summary, 3D printing materials can choose R-SAC system and P.O system as a cementing material. Both of them can match the working window time required (10-90min) for 3D printing. At the same time, the materials also have excellent mechanical properties, 3d strength ≥ 30MPa, 28d strength > 63MPa. In the actual 3D printing process, the R-SAC mortar has large viscosity and high printing effect than P.O mortar. So the R-SAC suitable parameter range of fluidity is 210-240 mm, the P.O is 200-220mm.

Acknowledgements

This paper is supported by CSCEC Fund Project (cscec-2014-z-33).

References

[1] Buswell R, Soar RC, Gibb A and Thorpe T 2007 Freeformconstruction: mega-scale rapid manufacturing for con-struction *Autom Constr* **16** 224–231

[2] Lim S, Le T, Webster J, Buswell R, Austin S, Gibb A and Thorpe T 2009 Fabricating construction componentsusing layer manufacturing technology *Proc Int Conf GlobInnov Constr Loughborough* pp 512–520

[3] Paglia C, Wombacher F and Böhni H 2001 The influence of alkali-free and alkaline shotcrete accelerators within cement systems –I Characterization of the setting behavior *Cement & Concrete Research* **31(6)** 913-918

[4] Zhang L,W Dongmin, Zhang, etc. 2014 Compatibility and Hydration-accelerating Mechanism Analysis of Alkali-free Liquid Accelerating Agent *Bulletin of The Chinese Ceramic Society* **06** 1365-1371

Use B-spline interpolation fitting baseline for low concentration 2, 6-di-tertbutyl p-cresol determination in jet fuels by differential pulse voltammetry

D S Wen[1], H Wen[2], Y G Shi[1,4], B Su[3], Z C Li[1] and G Z Fan[1]

[1]Department of Training, Army Logistics University of PLA, Chongqing 401311, PRC
[2]Department of Oil Quality and Technical Supervision, Navy 91458, Hainan 572000, PRC
[3]Oil Application Group, Beijing Oil Research Institute, Beijing 102310, PRC

[4]E-mail: ygshi@126.com

Abstract. The B-spline interpolation fitting baseline in electrochemical analysis by differential pulse voltammetry was established for determining the lower concentration 2,6-di-tert-butyl p-cresol(BHT) in Jet Fuel that was less than 5.0 mg/L in the condition of the presence of the 6-tert-butyl-2,4-xylenol.The experimental results has shown that the relative errors are less than 2.22%, the sum of standard deviations less than 0.134mg/L, the correlation coefficient more than 0.9851. If the 2,6-ditert-butyl p-cresol concentration is higher than 5.0mg/L, linear fitting baseline method would be more applicable and simpler.

1. Introduction

Taking characteristic peak of a substance as the shoulder peak of another is a common way to analyze overlapping peaks in spectral analysis [1]. For accurately picking up the shoulder peak and determining its height or area, the primary task is to the baseline definition of the overlapped shoulder peak, which directly makes a great impact on the accuracy and precision of the determination [2]. For the obvious shoulder peak, the baseline is determined by the line passing through the starting point and end point of shoulder peak. But if the concentration of the substance concerned is lower, its peak will be seriously overlapped, it is impossible to determine the linear baseline because the starting and end point of the peak could not be acquire daccurately [3].The phenomenon was also found in examining the adaptability of determining 2, 6-di-tert-butyl-p-cresol (BHT) in jet fuel by differential pulse voltammetry in the condition of the presence of the more than one antioxidants [4]. The mixed antioxidant is composed of BHT, 6-tert-butyl-2, 4-xylenol and p-cresol. If the BHT in the mixed antioxidant is as low to 15%~20% and the 24.0 mg/L mixed antioxidants are added in jet fuel, the BHT oxidation peak will be concealed by the oxidation peak of 6-tert-butyl-2,4-xylenol and the baseline cannot be determined. So a new baseline determination method must be established.

2. Basic Principle

2.1. B-spline Interpolation Fitting

B-spline which suggested by Schoenberg in 1946, essentially is a method of segmented curve fitting and is applied in free curve and surface modeling widely for its characteristic of low order smooth[4-6].

$\{(x_i, y_i)\}_{i=1}^{n}$ is the electrochemical data point range in two-dimensional space and$P$ represented the nodes in$x$ axis ,x_i is increasing sequence:

$$P_0 = minx_i = x_0 = a,\ P_{N+1} = maxx_i = x_n = b \quad (1)$$

The number of internal nodes in $B = [a, b]$ is N, and P_0, P_{N+1} are as the boundary nodes [7]. After defining the sequences of nodes,$B_j^k(x)$ is can be computed by DeBoor-Cox [4], which is k or $k-1$order B-spline function on x.

$$B_j^k(x) = \begin{cases} B_j^1(x) = \begin{cases} 1, & P_j < x < P_{j+1} \\ 0, & \text{other} \end{cases} \\ B_j^k(x) = \frac{x - P_j}{P_{j+k-1} - P_j} B_j^{k-1}(x) + \frac{P_{j+k} - x}{P_{j+k} - P_{j+1}} B_{j+1}^{k-1}(x) \end{cases} \quad (2)$$

The B-spline curve equation on $\{(x_i, y_i)\}_{i=1}^{n}$is written as follow: $y = \sum_{j=-k+1}^{N} c_j B_j^k(x)$, where$\{c_j\}$ is column vector of control factor.

If $B_j^k(x_i) = t_{i,j}$,$\{(x_i, y_i)\}_{i=1}^{n}$ could be expressed by the equations (3):

$$Y_n = T_n C_n + E_n \quad (3)$$

Where$Y_n = \begin{bmatrix} y_1 \\ y_2 \\ \vdots \\ y_n \end{bmatrix}$,$T_n = \begin{bmatrix} t_{1,-k+1} & \cdots & t_{1,N} \\ \vdots & \ddots & \vdots \\ t_{n,-k+1} & \cdots & t_{n,N} \end{bmatrix}$, $C_n = \begin{bmatrix} c_{-k+1} \\ c_{-k+2} \\ \vdots \\ c_{-k+n} \end{bmatrix}$, fittingerror $E_n = \begin{bmatrix} e_1 \\ e_2 \\ \vdots \\ e_n \end{bmatrix}$.

Here it's used for determining the baseline of the electrochemical characteristic curve of BHT. The key is to determine the peak width of BHT. The electrochemical characteristic curve of antioxidant mixture described above is shown in figure 1.Where the concentration ratio of three main additives which are BHT, 6-tert-butyl-2, 4-xylenol and 6-tert-butyl-2,4-xylenol is 3:15:2.

The BHT peak width is determined by the following method. Combining $\{(x_i, y_i)\}_{i=1}^{n}$ and $R_{j,x_i} = \Delta \frac{d_{y_i}}{d_{x_i}}$, the sequence is obtained:

$$R_{j,x_i} = \{R_{j,0} = \Delta_a \frac{d_{y_i}}{d_{x_i}}, R_{j,N+1} = \Delta_b \frac{d_{y_i}}{d_{x_i}} / j \in Z^*\} \quad (4)$$

Where R_{j,x_i}candetermine the peak width,j is the sequence of $R_{j,x_i} = 0$. According to figure 1and figure 2, the BHT peak width node could be determined by the following criteria:

$$\left[x_{R_{1,x_i}=0}, x_{maxR_{j,x_i}}\right], \text{ where } x_i \in \left[x_{R_{1,x_i}=0}, x_{R_{2,x_i}=0}\right]$$

2.2. Signal processing algorithm

(1)Input the original data, the smooth curve by Savitzky-Golay smooth in figure 1 is indicated in red. Here using the 27 points Savitzky-Golay filter will offer a better fit, because it can retain the more information as much as possible.

(2)Determine the peak width of BHT by 1st derivative curve of smooth curve and the blue line area, $\left[x_{R_{1,x_i}=0}, x_{maxR_{j,x_i}}\right]$ is the peak width, shown in feigure 2.

(3) Fit the curve of 6-tert-butyl-2, 4-xylenol based on B-spline interpolation method, then taking it as the baseline of BHT, shown in figure 3.

(4) Fit with GuassAmp after subtracting baseline, obtain the peak height of the oxidation peak, shown in figure 4.

(5) Construct a standard addition calibration plot and calculate BHT concentration in jet fuel.

Here the standard addition curve is calculated by following equation: $i_p(\times 10^{-7}A) = \text{kc} + \text{b}$, c as the BHT standard addition concentration c(mg/L) and $i_p(\times 10^{-7}A)$ as the oxidation peak height.

The BHT concentration in jet fuels is calculated by equation (5):

$$c_0 = \frac{b}{k} \times \frac{V}{V_0}. \tag{5}$$

Here c_0(mg/L) is the BHT concentration in sample, V is volume of volumetric flask,V_0 is sampling volume.

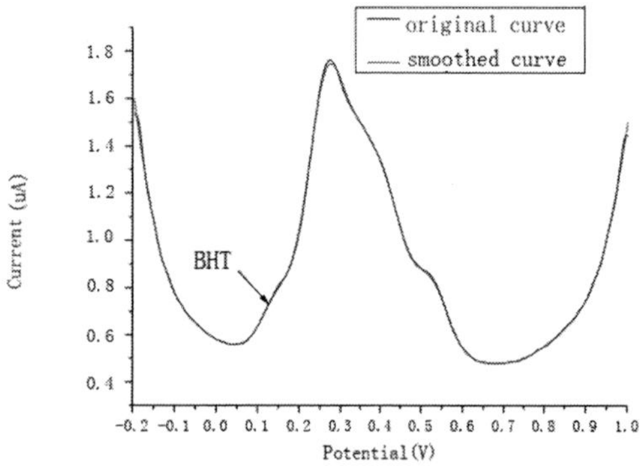

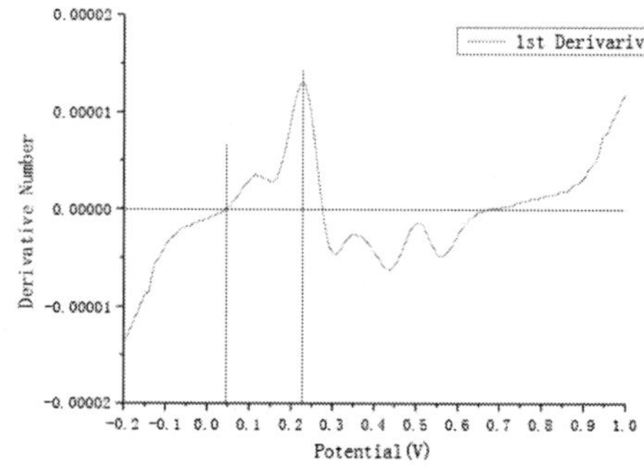

Figure. 1 The characteristic electrochemical curve. **Figure. 2** The BHT peak width.

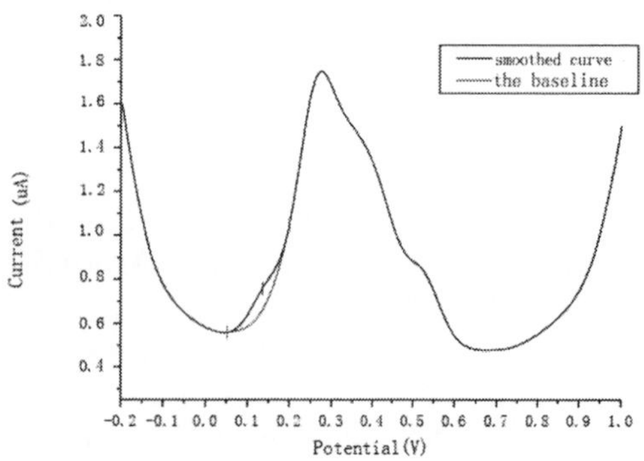

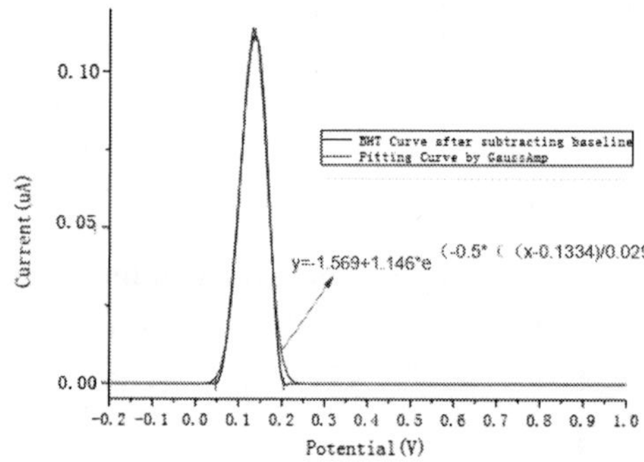

Figure. 3 The baseline difined by b-spline. **Figure. 4** The BHT GuassAmp curve.

3. Experiments

3.1. Materials and instruments

3.1.1. Materials. Absolute ethyl alcohol (AR, Chongqing Chuandong Chemical Co.),KOH (AR, Chongqing Chuandong Chemical Co. Ltd.), n-heptane (AR, Tianjin Kermel Reagent Co.Ltd.), 2,6-di-tertbutylp-cresol(BHT,AR, Chengdu Kelong Chemical plant),p-cresol and 6-tert-butyl-2,4-xylenol (AR, Shanghai Ziyi Reagent Factory),qualitative filter paper, volumetric flasks (100mL,50 mL and10mL),pipets, beakersand weighing bottles (25mm×40mm).

Absolute alcohol saturated with KOH, BHT in n-heptane (500mg/L), p-cresol in n-heptane solution(500mg/L), 6-tert-butyl-2,4-xylenol in n-heptane solution(500mg/L) and Jet fuel containing BHT 8.0mg/L. The seven Jet-fuel samples have been formulated as shown in table 1.

3.1.2. Instruments. Electrochemical station(CHI1230, Shanghai Chenhua instrument company, China) with the composite electrode consisted of glassy carbon work electrode, platinum wire auxiliary electrode and reference electrode. The Absolute ethyl alcohol saturated with KOH was used as supporting electrolyte. The electrochemical characteristic curve of the samples was determined by differential pulse voltammetry.

Instrument conditions: the voltage scan range0.0V to1.0V, the voltage increment 0.005V, the pulse amplitude0.05V, pulse width 0.05s, sampling width 0.01s, silence time 2.0s and sensitivity 1×10^{-5}A/V.

Table 1. Main additives concentration of prepared Jet fuel samples.

Samples	Antioxidant Concentration(mg/L)		
	BHT	6-tert-butyl-2,4-xylenol	p-cresol
A	3.6	20.4	0.0
B	3.6	13.2	7.2
C	5.0	13.2	5.8
D	5.0	18.0	1.0
E	5.0	19.0	0.0
F	10.0	14.0	0.0
G	10.0	13.2	0.8

3.2. Electrochemical analysis

All standard addition solutions were prepared with 10mL volumetric flask at room temperature. The addition concentrations of BHT in samples were 0.0mg/L, 5.0mg/L, 15.0mg/L and 25.0mg/L respectively. The standard addition solutions have been formulated according to the following steps: get the four 10mL volumetric flasks, fill 9.0mL samples in each flask, then add 0.00mL, 0.10mL, 0.30mLand 0.50mL BHT standard addition solution respectively and finally append n-Heptane in each flask to 10mL.

2mL standard addition solution and 2mL supporting electrolyte were added in electrolytic cell respectively. In order to make the solution mixed completely, it is imperative to shake the cell at least10 seconds, and then the composite electrode is inserted in electrolytic cell. The electrochemical analysis was conducted according to the given conditions. Each standard addition solution was measured at least 5 times.

4. Results and discussion

The relative error Δc_0 and the sum of standard deviations were used for expressing accuracy and precision of the method.The correlation coefficient also named Pearson's R is used as criterion of linearity. They could be calculated as following:

$$s=\sum_{c=0}^{c=25}\delta_i,\ \delta_i=\sqrt{\frac{\sum_{c=0}^{c=25}(x_i-\bar{x})^2}{n-1}}(mg/L) \quad (6)$$

$$\Delta c_0=\frac{c_0-c}{c}\times 100\% \quad .(7)$$

$$R(X,Y)=\frac{Cov(X,Y)}{\sqrt{Var[X]Var[Y]}} \quad (8)$$

Where $Cov(X,Y)$ is the covariance of variable X and Y, $Var[X]$ means the variance about variable X.

The results of sample A to E calculated by B-spline Fitting baseline are shown in table2 and figure 5(a). When the concentration is 10.0mg/L, the results of sample F to G calculated by B-spline Fitting Baseline and Linear Fitting Baseline are shown in table 3 and figure 5(b).

When the BHT level is 3.6~5.0mg/L, the method of linear fitting baseline cannot achieve any effective information about the peak of BHT. However the method of B-spline fitting baseline could get the datas shown in the table 2 and figure 5(a), the relative errors are less than 2.22%, the sum of standard deviations less than 0.134mg/L and R greater than 0.9851. When the BHT concentration is higher than 5.0mg/L, the two baseline definition methods will have the similar ability shown in the table 3 and figure5 (b). Therefore the b-spline fitting baseline is helpful to gain more test information, no matter in high or low concentration of BHT conditions. And it is more suitable especially in the low concentration. For the linear fitting baseline method, only the concentration is higher than 5.0mg/L, it is more applicable and simpler, because the peak of BHT in the characteristic electrochemical curve is more obvious that the peak height can be obtained.

Table 2. The results of sample A to E with b-spline fitting baseline.

peak height/10^{-7}A / sample and times		standard addition concentration/mg/L				BHT concentration/mg/L
		0	5	15	25	
sample A	1	0.1224	0.5269	1.028	1.547	3.68
	2	0.1237	0.5259	1.035	1.551	
	3	0.1271	0.5278	1.041	1.541	
	4	0.1296	0.5252	1.030	1.550	
	5	0.1248	0.5238	1.029	1.548	
average		0.1255	0.5259	1.033	1.547	
precision		s=0.014 , Δc_0 =2.22% , R=0.9963				
sample B	1	0.1389	0.4121	0.8766	1.434	3.52
	2	0.1416	0.4223	0.8175	1.336	
	3	0.1336	0.4105	0.9214	1.369	
	4	0.1372	0.4102	0.8732	1.322	
	5	0.1336	0.4107	0.9023	1.313	
average		0.1370	0.4132	0.8782	1.355	
precision		s=0.097 , Δc_0 = -2.22% , R=0.9994				
sample C	1	0.6941	1.446	2.416	4.221	5.05
	2	0.6756	1.445	2.428	4.332	
	3	0.6675	1.451	2.433	4.232	
	4	0.6691	1.442	2.419	4.444	
	5	0.6828	1.443	2.421	4.344	
average		0.6778	1.445	2.423	4.315	
precision		s=0.113 , Δc_0 = 0.60% , R=0.9903				
sample D	1	0.3960	0.6213	1.203	2.083	5.10
	2	0.3975	0.5988	1.217	2.097	
	3	0.3950	0.5992	1.214	2.094	
	4	0.3910	0.6024	1.211	2.086	
	5	0.3966	0.6136	1.221	2.099	
average		0.3952	0.6071	1.213	2.092	
precision		s=0.026 , Δc_0 = 2.00% , R=0.9915				
sample E	1	0.5372	1.036	1.695	3.162	5.05
	2	0.5198	1.048	1.693	3.163	
	3	0.5284	1.010	1.697	3.162	
	4	0.5360	1.012	1.711	3.169	
	5	0.5297	1.029	1.702	3.159	
average		0.5302	1.027	1.699	3.163	
precision		s=0.034 , Δc_0 = 1.00% , R=0.9851				

Define the limits of quantification (LOD) as the smallest amount of analysis that can be tested with full certainty, then the LOD can be represented with the amount of analysis that average is $\mu_B + 6\delta_B$, here δ_B is standard deviation. The IUPAC defined that the LOD is the amount of analysis that average based on $\mu_B + 3\delta_B$, which is a conditional concept depended on the repeated times and risk of test. The limits of quantification (LOQ) is defined as the minimum concentration of reliable determination and quantification of analyte in a particular degree of confidence. Typically, the LOD for an analysis method is defined as $3\delta_B$ and LOQ is defined as $10\delta_B$.

Here, δB were obtained under the conditions which tested 5 times repeatedly for a particular concentration of antioxidants (3.6~10.0 mg/L) by the antioxidants determination method and analyzed the information to obtain the results by the presented method. The largest standard deviation was

0.303 mg/L from table 3 and table 4, thus the LOD and LOQ should be 0.909mg/L and 3.03mg/L respectively.

Table 3. The results of sample F to G with B-spline Fitting Baseline and Linear Fitting Baseline.

sample and times \ peak height/10^{-7}A			standard addition concentration/mg/L 0	5	15	25	BHT concentration/mg/L
linear fitting baseline	sample F	1	1.124	1.456	3.034	4.177	10.11
		2	1.139	1.842	3.110	4.187	
		3	1.147	1.556	3.207	4.122	
		4	1.104	1.664	2.924	4.107	
		5	1.147	1.655	3.327	3.969	
	average		1.132	1.635	3.120	4.112	
	precision		s=0.404 , $\Delta c_0 = 1.10\%$, R=0.9963				
bspline fitting baseline	sample F	1	1.224	1.733	3.225	4.351	10.29
		2	1.229	1.725	3.218	4.359	
		3	1.227	1.745	3.205	4.353	
		4	1.224	1.724	3.231	4.357	
		5	1.227	1.734	3.119	4.352	
	average		1.226	1.732	3.120	4.354	
	precision		s=0.060 , $\Delta c_0 = 2.90\%$, R=0.9992				
linear fitting baseline	sample G	1	0.8022	1.313	2.325	2.987	10.82
		2	0.8061	1.325	2.258	2.988	
		3	0.8094	1.345	2.245	2.997	
		4	0.8089	1.324	2.246	2.998	
		5	0.8052	1.314	2.259	3.075	
	average		0.8064	1.324	2.267	3.009	
	precision		s=0.086 , $\Delta c_0 = 8.20\%$, R=0.9976				
bspline fitting baseline	sample G	1	0.8112	1.313	2.325	2.972	10.93
		2	0.8181	1.325	2.258	2.981	
		3	0.8114	1.345	2.245	2.987	
		4	0.8119	1.324	2.246	2.998	
		5	0.8112	1.314	2.259	3.075	
	average		0.8128	1.324	2.267	3.003	
	precision		s=0.090 , $\Delta c_0 = 9.30\%$, R=0.9976				

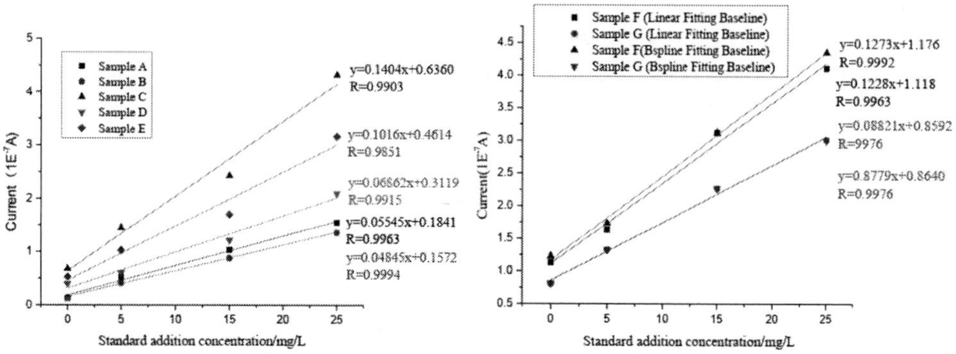

(a) b-spline fitting baseline (b) b-spline fitting baseline and linear fitting baseline

Figure 5. The curve of standard addition concentrations vs peak height.

Table 4. The BHT concentration test results of sample A to G.

sample	BHT actual concentration /mg/L	BHT test concentration /mg/L						relative error Δc_0 /%	standard deviations /mg/L
		1	2	3	4	5	average		
A	3.6	3.64	3.64	3.78	3.72	3.64	3.68	2.22%	0.064
B	3.6	3.03	3.68	3.44	3.7	3.77	3.52	-2.22%	0.303
C	5.0	5.39	4.98	5.23	4.63	5.02	5.05	1.00%	0.287
D	5.0	5.26	5.03	5.02	5.03	5.14	5.10	2.00%	0.104
E	5.0	5.15	5.07	4.94	5.01	5.08	5.05	1.00%	0.079
F	10.0	10.3	10.26	10.37	10.28	10.23	10.29	2.90%	0.053
G	10.0	11.07	11.12	10.88	10.80	10.80	10.93	9.30%	0.152

5. Conclusion

As shown above, the baseline fitting method based on b-spline interpolation was established for determining the low BTH in jet fuel interfered by other antioxidants with electrochemical method. When the BHT concentration was as low as 3.6mg/L-5.0mg/L in the presence of the 6-tert-butyl-2, 4-xylenol, it was difficult to define the baseline of BHT peak. The baseline fitting method based on b-spline interpolation could solve this problem. With the method determining BHT in jet fuel at low concentration(3.6~5.0mg/L), its relative errors are less than 2.22%, the sum of standard deviations less than 0.113mg/L and R more than 0.9851.

Acknowledgments

This work is supported by national standard program 20090226-T-469.

References

[1] Hu Y G, Zhang X X, Zhao Z Y et al 2012 Strategy and implementation of resolve overlapping spectra based on curve fitting *Journal of Chongqing University* 76-81

[2] Zhang Lei, Wang Yuming, Wang Yongjun et al. 2009 Recognition Algorithm for the peak of Chromatograph *Computer Application and Software* **8** 256-258

[3] Shi Yonggang, Liu Shaopu, Li Zicun et al. 2006 The Assay of AntioxidantIn Jet Fuel Based on Differentiating pulse Voltammetry.*Petroleum processing and Petrochemicals* **7** 54-57

[4] Renner G, Weiß V 2004 Exact and approximate computation of B-spline curves on surfaces.*Computer-Aided Design* **36(4)** 351-362

[5] GuoJianfeng, Zhu Changqing 2000 Spline Least Squares with Two Variables and Its Application in Cartography).*Engineering of Surveying and Mapping* **9(4)** 25

[6] Wang Renhong, Li Chongjun, Zhu Chungang et al. 2008 Computation Geometry.Beijing: *Sciencepublishing company(press)*

[7] Wang Xin, Fan Xianguang, Xu Yingjie et al 2014 Baseline Correction Method of Raman Spectra Based on B-spline.*Spectroscopy and Spectral Analysis* **8** 2117-21

IOP Conf. Series: Materials Science and Engineering **292** (2017) 012072 doi:10.1088/1757-899X/292/1/012072

Analysis and selection of materials for the design of lightweight railway vehicles

C Ulianov[1,3], A Önder[1] and Q Peng[2]

[1] NewRail Centre for Railway Research, Newcastle University, Newcastle upon Tyne, United Kingdom
[2] School of Mechanical and Automotive Engineering, Xiamen University of Technology, Xiamen, China

[3] E-mail: cristian.ulianov@ncl.ac.uk

Abstract. Railway vehicle manufacturers are focusing nowadays on building lighter vehicles, which are more environmental- and track-friendly. The article presents a methodology that was developed and is being proposed for the assessment and selection of materials for lightweight railway vehicles. The analysis methodology consists of a two level stages that consider different categories of key criteria. Example of analyses that were carried out on the basis of some of the most relevant criteria are presented along with specific recommendations regarding the potential use of various classes of materials in different applications, and subsequent conclusions.

1. Introduction and background

Lightweight materials have largely been developed for and further deployed in aerospace and automotive sectors. In the recent past, an increasing uptake in the rail industry for both passengers and freight vehicles was also noticed. This has been mostly driven by consideration for reduced emissions and impact on the infrastructure. Lightweighting also provides a feasible mean to achieve higher dynamic performance with either the same or reduced traction power. The main method to achieve lightweight construction is to replace materials of high specific weight with lower density materials, without reducing stiffness and durability. Common lightweight materials are, for example, metals such as aluminium, magnesium, high strengthened steels or various types of unreinforced and reinforced composites.

Although some of these materials tend to be expensive, their use in railway applications is justified with respect to the overall life cycle operational costs. Therefore, recent published studies, e.g. [1-6], reported significant applications of new generation lightweight materials into both passengers and freight rail vehicle designs. These reported studies focused on using different categories of materials for specific applications; however, the methodology applied for the selection of specific materials seems to be different, or not even explained in most of the cases. This gap was noticed by engineers in industry, which need an efficient and rapid method for the assessment and selection of novel lightweight materials to be used in the design of modern railway vehicles.

2. Methodology for the assessment and selection of materials for lightweight railway vehicles

A methodology for the assessment and selection of the most feasible candidate materials for the construction of lightweight railway vehicles was developed and further tested for various vehicle designs and classes of materials.

A preliminary analysis focuses on materials that have been previously validated in other transport applications for lightweighting purposes. The next step should consider separately the relevant vehicle components and sub-assemblies, such as the bogie, structural frames, carbody, etc., as well as the requirements related to environmental impacts, investment and overall life-cycle-costs, etc. However, as the number of the criteria increases, the selection process becomes more and more complicated and impractical. Therefore, prioritisation and compromises should be made. The proposed methodology assesses the candidate materials in two steps, according to different criteria that are subsequently clustered on two levels (Level 1 and Level 2), with respect to two main aspects:

- The importance/effect on the behaviour and overall properties of the rail vehicle;
- The practicality in using them in the selection methodology.

The ***Level 1 criteria*** that were considered for the first step of the assessment include:

- Specific elasticity modulus (stiffness divided by density);
- Specific tensile strength (strength divided by density);
- Fatigue behaviour/strength;
- Material cost (Euro/kilogram);
- Applicability to different vehicle components and sub-assemblies, with respect to:
    - Manufacturing processes (including joining);
    - Specific stresses on different vehicle components and/or sub-assemblies (flexural, tensile, compressive, etc.)

The first two criteria above (specific modulus and strength) are related to the potential lightweighting achievable by using a lower density material such as composites, since the densities of, e.g., different steel grades are nearly the same. Therefore, these criteria are effective when there is a comparison between steel and composite materials. The material cost criterion is used to provide the relative price of each material in Euro/kg. It is not an indication of the cost effectiveness, because, considering potential mass savings, using a more expensive material may be cheaper for the whole product.

The applicability of the material type for the intended use can be accounted as a more complex criterion, which takes into account the type of stresses and loads, the available processing techniques to manufacture suitable parts and/or sub-assemblies by using the assessed materials, and the required joining techniques.

Level 2 criteria include a more diverse range, which can impact on the final product, i.e.:

- Life-cycle-costs (LCC);
- Environmental impact, with respect to: *Recyclability, CO2 footprint, Noise reduction, Energy save/use, etc.;*
- Resistance to degrading factors, including: *Impact resistance, Fire-smoke-toxicity resistance, Abrasion resistance, Resistance against factors such as chemicals, humidity, temperature, etc.*

Although these requirements are also important to assess the vehicle's value, the way to take into account each of their effect is very difficult and not straightaway compared to the Level 1 criteria. Most of the lightweight solutions for railway vehicles are very innovative and have a low technology readiness level, therefore, it is still extremely challenging to use accurate data or predictions regarding quantitative indicators related to the above criteria. The reason is mainly the lack of information on many elements that determine the 'value' of these criteria as quantitative indicators (e.g., no info on maintenance costs, disposal, recycling, etc.). Therefore, using general information available in literature, as well as in similar applications (in other industries), the proposed methodology uses qualitative indicators.

3. Assessment of materials and discussion of potential applications

3.1. Level 1 assessment and pre-selection of candidate materials

The identification of the potential lightweight materials from various material families was carried out by using Granta CES EduPack 2017 software [7]. This software is an extensive database of engineering materials with their primary properties, such as mechanical, thermal, electrical, etc., as well as other characteristics, such as cost, manufacturing, joining, life-cycle, environmental impact, etc. The important specifications for rail vehicle parts include the tensile strength (and yield strength for metallic materials), elongation at failure, density, Young's modulus, and their Price. The relevant material families are conventional metallic materials, lightweight composite materials, plastics, and foams which have important applications in sandwich composite configurations. Some relevant examples of candidate material types from different families are listed below in table 1.

Table 1. Examples of candidate materials for rail vehicle applications.

Material Family	Material type	Examples
Steels	Conventional	S275, S355
	High strength steel	S890, S960, TWIP steels, TRIP steels
Composites	Monolithic	Carbon and Glass fibre reinforced laminated polymers
	Sandwich	Fibre reinforced laminates with foam or honeycomb core
Aluminium		Al6160-T6, Al6082,-T6, Al7020-T6

A general comparison between the potential material families is presented in figure 1.

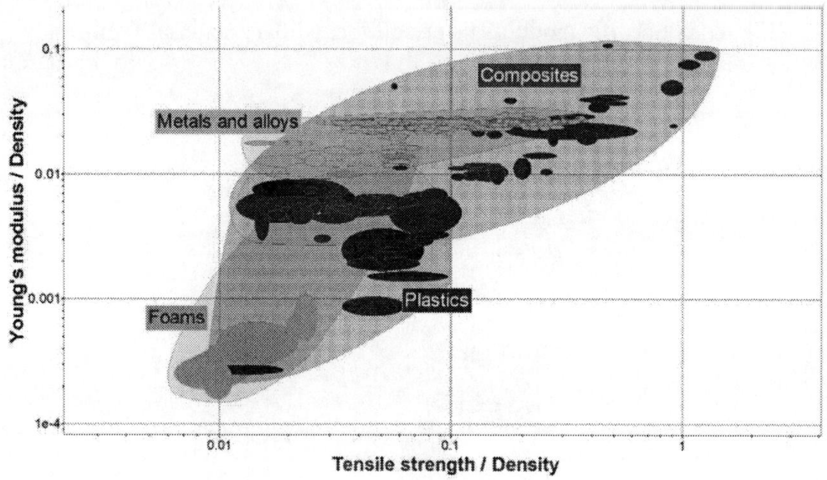

Figure 1. Specific modulus vs specific tensile strength for candidate material families.

Considering: i) the steel grades available on the current market; ii) outcomes of similar previous work [e.g., 5, 6]; and iii) relevant information in literature, the types of steels considered for further analysis include the conventional structural steels, low alloy high strength steels, and 1st and 2nd generation high-strength-steels (HSS) such as twining induced plasticity (TWIP) and transformation induced plasticity (TRIP), which have higher elongation.

The main candidate types of composite materials are the fibre reinforced polymer composites (FRP). Two different configurations of FRPs can be considered, i.e.: monolithic (laminated) and sandwich composites, which incorporate a core material in-between two laminates. Glass and carbon fibres have been widely used in many transport industries. Although FRPs paved the way for lightweighting in aeronautics sector with the extensive use of carbon fibres, the cost is a key criterion in novel vehicle design. Unfortunately, despite their excellent strength and stiffness properties, carbon fibres remain one of the most expensive reinforcing material. Therefore, in order to compete with the reasonable price of steels, carbon fibres were excluded from the candidate material selection. As shown in figure 2 and figure 3, composites have lower density, however, their current price is still high.

The price indicated in the figures is just the bulk material cost, without taking into account other factors such as manufacturing methods.

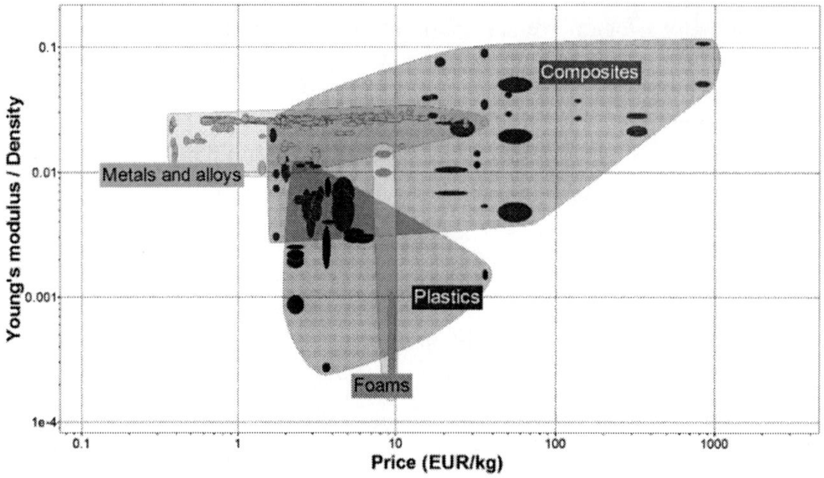

Figure 2. Specific modulus vs price for candidate material families.

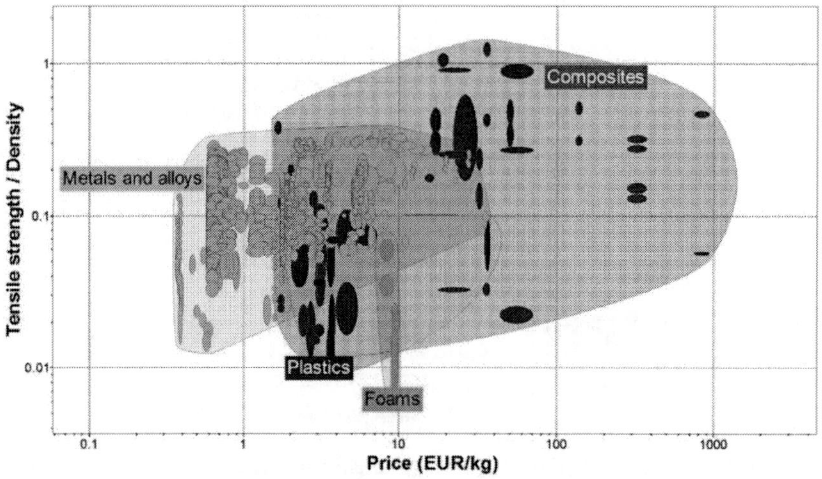

Figure 3. Specific tensile strength vs price for candidate material families.

According to the above considerations and information, some relevant example materials from each material family were listed in table 2 below. In this table, the benchmark material for each family was highlighted in yellow, and the properties of other materials were assessed relatively to these benchmark materials. According to the assessment, the green colour indicates superior properties (or lower price), while the red coloured cells indicates inferior properties or higher price, relative to the base material.

"Structural steel S355" was selected as base material, being the commonly used steel for rail vehicle structures. The reason to select "Polyester/E-glass fibre, pultruded rod" is its cost effectiveness, i.e., being one of the cheapest composite material considered. It can be noticed that TRIP, TWIP, and S960QL high strength steels have better properties in terms of strength, however, they are more expensive. On the other hand, aluminium grades have lower density compared to steels, but have lower strength and are more expensive. Clearly, the cost factor is not an exact indication due to the ignorance of other effects such as manufacturing, volume, size, etc., but it can give an idea in terms of comparison. The situation is similar for the composites family as well: E-glass/polyester pultruded rod outperforms other configurations. However, caution should be given in this case, as the performance of composites depends predominantly on the properties of the final product such as the geometry, the thickness, fibre/resin type, material configuration details, the type of loading etc. Therefore, different configurations should be considered for the intended application.

Table 2. Shortlist of example selected materials.

Fam	Material type	Yield S (MPa)	Tensile S (MPa)	Elongation (%)	Price (EUR/kg)
Metallic Materials	TRIP (Transformation induced plasticity) steel	47.88%	52.72%	20.45%	-131.11%
	TWIP (Twinning induced plasticity) steel	40.84%	78.18%	150.00%	-170.00%
	Structural steel, S960QL	170.42%	93.63%	-54.54%	n/a
	Structural steel, S275N	-32.39%	-21.81%	6.81%	-19.22%
	Aluminium - Al6106-T6	-43.66%	-54.54%	-63.63%	-840%
	Aluminium - Al6005-T6	-30.98%	-46.72%	-55.90%	-750%
	Aluminium - Al6082-T6	-26.76%	-41.81%	-55.90%	-895%
	Structural steel, S355JR	355	470 - 630	22	0.4 - 0.5
Composites	Polyester/E-glass fibre, woven fabric, QI lay-up	-74.70%	74.70%	-32.50%	-23.89%
	Epoxy/E-glass fibre, woven prepreg, biaxial lay-up	-41.30%	-41.30%	-11%	-1817%
	Epoxy/E-glass fibre, UD prepreg, UD lay-up	-7.77%	-7.77%	25%	-1487%
	Polyester/E-glass fibre, pultruded rod	690 - 828	690 - 828	2	1.6 - 1.8

3.2. Level 2 assessment and final selection of candidate materials

The candidate materials that pass the Level 1 criteria should be further assessed according to the Level 2 criteria, in order to refine the list of selected materials. As mentioned earlier, due to the lack of specific information on many elements and to the complexity of the parameters, Level 2 criteria will be evaluated qualitatively. Therefore, some material properties can be assessed as *"excellent, good, acceptable, unacceptable, etc."* in comparison to the other candidate materials. This approach could be more dependable and efficient if the relevant vehicle body part is known. For example, in the case of a freight wagon carrying liquids or chemicals, the resistance of the wagon body/tank to the effects of relevant chemicals should be sufficient to be operated safely. Table 3 shows an example comparison for such case, where the resistance properties of two HSS is compared against a polymer composite. These three materials were selected on the basis of their performance vs the Level 1 criteria (table 2).

3.3. High strength steels in railway vehicle applications

Having improved mechanical properties compared to conventional steels, high strength steels (HSS) are sophisticated materials with carefully selected chemical compositions and multiphase microstructures resulting from precisely controlled heating and cooling processes. HSS can be considered a feasible option for lightweight design, as they have been widely used in automotive sector for body chassis frame and for crashworthiness applications. Railway industry has recently made use of HSS in both passengers and freight wagon applications. The structural frame of the prototype vehicle in SUSTRAIL project [5] was manufactured using the RQT-701 grade supplied by TATA Steel, and mass reduction was 26%. Therefore, it is estimated that an overall wagon mass reduction of 30-35% can be achieved by extending the use of HSS to other structural parts of the vehicle (e.g., the bogie frame, wagon body, etc.). Besides being capable to enable significant mass savings, HSS can be cost effective as well, and it is 100% recyclable, therefore, more environmental friendly than composites. A drawback of late generations of HSS is the availability on the market and the acceptance by the rail sector.

Table 3. Example comparison of selected materials with respect to some relevant Level 2 criteria.

Property (Resistance to)	Candidate Material		
	TWIP steel	TRIP steel	Polyester/E-glass fibre, pultruded rod
Water	Acceptable	Acceptable	Excellent
Weak acids	Limited use	Limited use	Acceptable
Strong acids	Unacceptable	Unacceptable	Acceptable
Weak alkalis	Acceptable	Acceptable	Unacceptable
Strong alkalis	Limited use	Limited use	Unacceptable
Organic solvents	Excellent	Excellent	Unacceptable
UV radiation (sunlight)	Excellent	Excellent	Good
Flammability	Non-flammable	Non-flammable	Highly flammable
Galling resistance (adhesive wear)	Acceptable	Acceptable	-

3.4. Polymer composites in railway vehicles applications
An increasing trend of using lightweight composites in railway vehicles for mass and cost savings was noticed in the last 20 years. Their application was more successful so far in passenger vehicles compared to freight wagons. Successful applications (prototype/production) since late 1970's include full scale bodyshells, interior elements such as tables, toilets, etc., and component and sub-assembly level such as bogie frames, crashworthy front cab, and fairings. Some of these examples are feasible for different parts of freight wagons and this has already been considered by INNOWAG project [8]. Past studies [4] demonstrated mass savings of up to 35-40% by using lightweight composite materials. The flexibility in manufacturing of 3D complex shaped parts is also a significant advantage of composites, which also reduces the number of parts and the amount of labour, leading to cost savings. The major barrier to the application of composite materials into rail vehicle solutions is the lack of relevant standards to take into account composite intrinsic properties. In addition, FRPs have different inner structures and failure mechanisms compared to metallic materials, which can limit their use in certain parts of the vehicle. For example, FRPs are sensitive to impact loads and, therefore, special caution should be taken when applying such materials in key structural vehicle configuration.

3.5. Aluminium in railway vehicle applications
A vast range of aluminium alloy grades applications in railway vehicle becomes an inevitable tendency no matter in EU and China for the purpose of lightweight and good reliability. Although the cost of aluminum alloy is higher, the capability of mass reduction for carbodies could achieve 50% compared to traditional steel bodies; furthermore, potential 14% power savings could be achieved due to the lower consumption. The relative higher cost of aluminium components could be compensated from the lower consumption within 2-3 years, while the time for individual maintenance of aluminium alloy body is typically shorter than that of the steel by 15%-45%; the recycle rate of aluminium alloy is up to 80% for when the vehicle comes to the end of life cycle.

3.6. Other potential materials in railway vehicle applications
In addition to the materials discussed above, some other novel material solutions exist and could be potentially applied to railway vehicles. Steel foams are relatively new forms of steel materials that are designed to achieve lighter steel structures. These types of steel incorporate a porous structure which can be produced by different manufacturing processes such as powder metallurgical, polymer/oxide ceramic foam precursor, lotus-type, etc. [9]. As a result, they have lower density compared to the parent metal. However, not all of these methods are standardised and, most importantly, the physical properties of the manufactured structures depend predominantly on the processing method. This situation creates another difficulty for selection, in addition to the lack of numerical data regarding different material and property types. Nevertheless, steel foams can be used to replace, for example,

bulky cast or forged parts in the running gear like housings/covers, supports, etc., which do not require to withstand intense loads/stresses, however can achieve significant mass savings.

Another potential lightweight material is the natural or recyclable fibre reinforced composites. At the moment, their current capabilities, manufacturing processes, and long-life-cycle characteristics are fairly new. Therefore, the knowledge is very limited and invalidated. One of the drawbacks of carbon and glass FRPs in use today is that they are either non-recyclable or very difficult and costly to recycle. Even though, the novel natural/recyclable FRPs requires more practice to be implemented in railway vehicle applications; they are clearly promising materials with unique advantageous properties and can replace the non-recyclable FRPs in, e.g., secondary structures such as interior elements.

4. Conclusions

A practical method for the assessment and selection of novel materials for lightweight rail vehicle design solutions was developed and tested for different case studies. Some relevant example results are presented in this article. In-depth analysis and testing is required for validating the feasibility of using the selected materials in railway applications. Preliminary structural strength analyses validated the concepts using HSS, however, further analyses and/or testing are required for composite materials and for validating critical issues such as joints, fatigue behaviour, resistance to degrading factors, etc.

Acknowledgments

Authors wish to acknowledge the support received from the National High End Expert programme (GDT20153600065) and Xiamen University of Technology Programme (2016AYF004).

References

[1] Lee W-G, Kim J-S, Sun S-J and Lim J-Y 2016 The next generation material for lightweight railway car body structures: Magnesium alloys *Proc IMechE Part F: J Rail and Rapid Transit* **0(0)** 1–18

[2] Matsika E, O'Neill C, Grasso M and De Iorio A 2016 Selection and ranking of the main beam geometry of a freight wagon for lightweighting *Proc IMechE Part F: J Rail and Rapid Transit* **0(0)** 1–19

[3] Ulianov C, Shaltout R and Balan A 2015 Lightweight Vehicle Structural Design with Advanced Steel Grades and Profiles *Applied Mechanics and Materials* vol. **809-810** p 1199-1204

[4] Wennberg D 2013 Multi-Functional Composite Design Concepts for Rail Vehicle Car Bodies *KTH Royal Institute of Technology* PhD thesis

[5] SUSTRAIL project 2015 Outline Design of a Novel Innovative Lightweight High Performance Freight Wagon Body; Deliverable D3.4. EU FP7 project SST.2010.5.2-2. Grant no. 265740

[6] SPECTRUM project 2014 Detailed Structural Design Concept for a High Performance Freight Wagon; Deliverable 3.1. EU FP7 project SST.2010.5.2-3. Grant no. 266192

[7] CES EduPack 2017 Granta Design Limited

[8] INNOWAG project 2017 Benchmark and market drivers for an integrated intelligent and lightweight wagon solution; Deliverable D1.1. H2020 Shift2 Rail project S2R-OC-IP5-03-2015 Grant no. 730863

[9] Smith B H, Szyniszewski S, Hajjar J F, Schafer B W and Arwade S R 2012 Steel foam for structures: A review of applications, manufacturing and material properties *Journal of Constructional Steel Research* vol **71** p 1-10

IOP Conf. Series: Materials Science and Engineering **292** (2017) 012073 doi:10.1088/1757-899X/292/1/012073

Performance characteristics of rubber seed oil biodiesel

P Liu, M Qin, J Wu and B S Chen[1]

Dept. of Army Logistical University of PLA, Chongqing 401311, P.R China

[1] Email: boshuichen@163.com

Abstract. The lubricity, ignition quality, oxidative stability, low temperature flow property and elastomeric compatibility of rubber seed oil biodiesel(RSM) were evaluated and compared with conventional petro-diesel. The results indicated that RSM and its blends with petro-diesel possessed outstanding lubricity manifested by sharp decrease in wear scar diameters in the high-frequency reciprocating rig(HFRR) testing. They also provided acceptable flammability and cold flow property,although the cetane numbers (CN) and cold filter plugging points(CFPP) of biodiesel blends slightly decreased with increasing contents of petro-diesel. However, RSM proved to be very susceptible to oxidation at elevated temperatures during prolonged oxidation durations, characterized by increased peroxide values, viscosity, acid values and isooctane insolubles. The oxidation stability of RSM could be significantly improved by antioxidants such as BD100, a phenol antioxidant produced by Ciba corporation. Furthermore, RSM provided poor compatibility with some elastomeric rubbers such as polyacrylate, nitrile-butadiene and chloroprene, but was well compatible with the hydrogenated nitrile-butadiene elastomer.

1. Introduction

The study of biodiesel is not a new topic. Defined as the mono-alkyl esters of long chain fatty acids derived from renewable lipids such as vegetable oils and animal fats, biodiesel has proved to be a competent, environmentally friendly substitute for conventional petro-diesel since it provides very similar properties to those of mineral diesel fuels. In fact, the overall properties of biodiesel are determined predominantly by the sources of feedstocks, or in turn the structural features of the various individual fatty acids, especially chain length, degree of unsaturation, and branching of the chain [1-3]. Although in the past decades, vast amount of studies on the performances of biodiesel have been performed and reported [4-7], it is still important to fully understand how the various fatty acid profiles of different sources can influence biodiesel properties.

Rubber seed oil, a non-edible oil extracted from the seeds of rubber trees and also aco-product of natural rubber industry, with properties similar to linseed and soybean oil, has great potential for supplementing conventional biodiesel feedstocks since, because of reduced food availability, diversion of biodiesel feedstocks from traditional edible oils such as soybean, corn, sunflower, peanut, and rapeseed, to non-edible sources such as palm, castor, rubber seed and cotton seed, has in recent years gained continuously increasing interest [8-12]. But unfortunately, there are until now only a few studies concentrating on performance characteristics of rubber seed oil based biodiesel [13-14]. Further investigations on properties of rubber seed oil biodiesel are as such of important economical and practical significances. In the present investigation, some typical performances characteristics including lubricity, ignition quality, oxidation stability, low-temperature flow property and elastomeric compatibility of rubber seed oil biodiesel were evaluated and compared in relation to those of a conventional petro-diesel.

2. Experimental

2.1. Materials

Rubber seed oil biodiesel (RSM): a biodiesel fuel of rubber seed oil origin, provided by Prof. Dr. Yang Ru of Beijing University of Chemical Technology. The mass percentage of the main fatty acid compositions of RSM were 12.3% for palmitic acid, 2.9% for stearic acid, 29.6% for oleic acid, 37.5% for linoleic acid, and 17.1% for linolenic acid, respectively. Some typical physicochemical indices of RSM are presented in table 1.

Table 1. Typical physicochemical indices of RSM and petro-diesel.

parameter	RSM	Petro-diesel
density(20 °C)/(kg/m^3)	881	835
Kinematic viscosity/(mm^2/s)	4.06(40°C)	4.60(20°C)
flash point(closed cup)/°C	150	65
sulfur / (m/m%)	0	0.119
acid value / (mgKOH/g)	0.22	-
free glycerin/ (m/m %)	0.016	-
total glycerin/ (m/m %)	0.02	-

Petro-diesel: No. 0 commercial petroleum-based diesel fuel obtained from a filling station of Sinopec Group. For comparison purpose, some typical physicochemical properties of the petro-diesel are also listed in table 1.

Biodiesel-petrodiesel blends: uniform blendings of a proper amount of RSM with petro-diesel on a volume basis, termed B×, where × is a number denoting the volume percent of RSM incorporated in a finished blend, e.g.B20， B50 and B80.

Elastomers: polyacrylate rubber (ACM, 60% of rubber content), hydrogenated nitrile-butadiene rubber (HNBR, 55% ofrubber content), nitrile-butadiene rubber (NBR, 50% of rubber content), chloroprene rubber (CR, 52% of rubber content).

Anti-oxidants： BD50 and BD100were commercially available from Ciba corporation, while1406 (phenol antioxidant) and 1407 (amine antioxidant) and MOA (phenol- amine antioxidant) were provided by Sinopec research institute of petroleum processing.

2.2. Lubricity test

Fuel lubricity affects durability of vehicle components such as fuel injector and fuel pump. The reduced lubricity of a fuel can be damaging to the engine and fuel systems.In this work, the lubricities of biodiesel blends, viz. B0, B20, B40, B60, B80 and B100, were evaluated respectively on a high-frequency reciprocating rig (HFRR) based on the ISO12156-1 method.The HFRR test involved a weighted steel ball and a stationary steel disk which was completely immersed in the test fuel.The ball and disk were heated to 60℃ and brought into contact with each other under the load of 10N and the entire apparatus was vibrated at 50 Hz for 75 min. There after the wear scar diameter (WSD) of the steel ball was measured and reported as the lubricity of the test fuel.

2.3. Ignition quality measurement

Cetane number (CN) is a prime indicator to specify the ignition quality of a diesel fuel, and is thus one of the most significant properties for determination of a fuel performance. Reflecting on the delay durations of ignition, cetane number ensures the readiness of a diesel fuel to autoignite when injected into the combustion chamber of an engine. The CN is generally dependent on the chemical compositions of a fucl. Thc higher the cetane values of a fuel, the shorter the ignition delay and thus better ignition quality. However,a cetane number that is too high or too low can lead to engine problems such as poor startability, high noise level and exhaust emissions. According to the international regulationon ASTM D613, the cetane number of a diesel fuel is commonly determined on a standardized mono-cylinder engine by comparing it to reference fuels of *n*-hexadecane (with an assigned CN=100) and 2,2,4,4,6,8,8-heptamethylnonane (with an assigned CN =15). In this work, the CNs of neat RSM and its blends with No. 0 conventional petro-diesel, namely, B0, B20, B40, B60, B80 and B100, were measured following the procedures of the Chinese method GB/T 386-2010, which well corresponds to the ASTM D613 in the CN determination.

2.4. Low-temperature fluidity test
Cold filter plugging point (CFPP), the temperature at which a fuel jams the filter due to the formation of agglomerates of crystals, is one of the most important indices related to low-temperature operability of diesel fuels. In the present work, CFPPs of various biodiesel blends were determined on a low-temperature flow tester following the Chinese SH/T0248 procedures which well corresponds to EN 116 method.

2.5. Oxidation stability test
Due to polyunsaturated fatty acid content, high oxidative instability of biodiesel is usually a special concern. Oxidation stability is therefore an important quality parameter for biodiesel since oxidation products originated with biodiesel affect storage life and contribute to deposit formation in tanks, fuel systems and filters. In the present study, the oxidation stability of RSM was tested based on the Chinese method SH/T0690, a method equivalent to ASTM D4625which is the most widely accepted method for assessing the storage stability of fuels. Specifically, RSM was oxidized at 43°C over an extended period of time up to 20 weeks. During oxidation experimentation, typical parameters such as peroxide values, kinematic viscosity, acid values and isooctane insolubles, of the aged RSM were determined every four weeks to characterize its oxidation stability.To further understand the oxidation characteristics of RSM, effects of some antioxidants on RSM oxidation stability were also evaluated (by determining peroxide values, kinematic viscosity, acid values and isooctane insoluble) based on the Chinese method SH/T 0175 which well corresponds to the ASTM D2274, and by determining oxidation induction periods based on the Rancimat method EN14112.

2.6. Elastomeric compatibility test
Some rubber elements in engines such as gaskets and seals can be exposed to oil and fuel. The physical, chemical and mechanical properties of the elastomeric materials can deteriorate during exposure to these fluids. The use of biodiesel as the diet of diesel engines may create unforeseenable problems with some elastomeric components,and resistant plastic materials are required.For this reason, the compatibility of different RSM biodiesel blends with four elastomeric elements which have been commonly used in diesel engines was tested in this study. The laboratory trials were conducted by immersing the elastomers with RSM biodiesel blends under atmospheric temperatures for 72 h. Thereafter, the changes in mass, volume and hardness of the elastomers before and after testing were determined to characterize the elastomeric compatibility of RSM.

3. Results and Discussion

3.1. Lubricity
Figure 1 shows the HFRR wear scar diameters tested with B0, B20, B40, B60, B80 and B100, respectively.

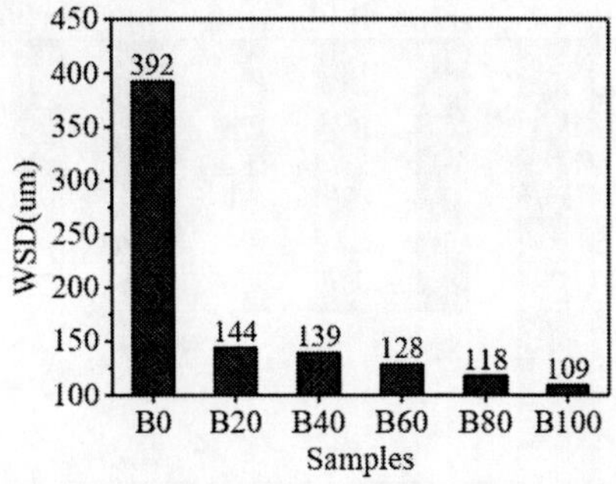

Figure 1. Wear scar diameters tested with biodiesel blends.

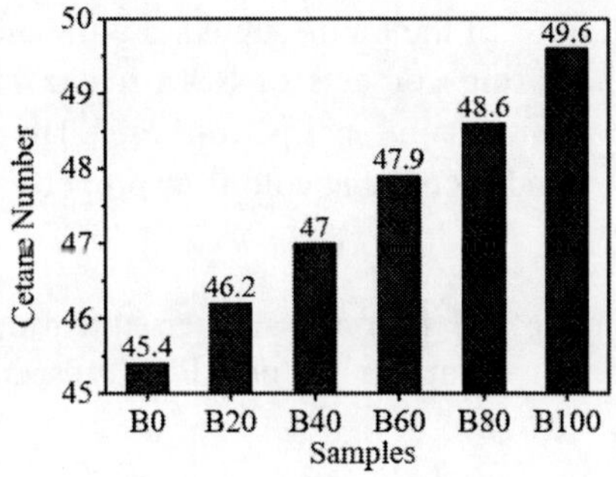

Figure 2. Cetane numbers of biodiesel blends.

It is clear from figure.1 that neat RSM and its blends with proper amount of petro-diesel provided outstanding lubricity with respect to neat petro-diesel. The lubricity of RSM biodiesel decreased with increasing contents of petro-diesel in a blend, indicating that petro-diesel to some extent impaired the lubricity of RSM. From practical point of view, excellent lubricity of RSM is very important to reduce engine wear and to protect the injection system. It can also be drawn from figure 1 that incorporation of RSM into petro-diesel markedly improved the lubricity of petro-diesel, manifested by sharp decrease in WSD. The excellent lubricity of RSM may be attributed to its unique FAME (fatty acid methyl esters) molecules, which adsorbed on or tribo-reacted with the friction surfaces to form a lubricious boundary protection film [15-16].

3.2. Ignition quality

Shown in figure 2 are the cetane numbers of B0, B20, B40, B60, B80 and B100, respectively. It can be found from figure 2 that the CN of neat RSM (B100) is higher than that of neat petro-diesel(B0), demonstrating that the flammability of RSM biodiesel is better than that of conventional No. 0 petro-diesel. The CNs of biodiesel blends from B80 to B20 slightly decreased with the increase of petro-diesel contents. According to the ASTM D6751 biodiesel standard, which prescribes a minimum of 47 for biodiesel cetane numbers in the United States, the cetane numbers of the RSM biodiesel blends are all acceptable, although in some European countries,higher cetane values of biodiesel are usually required.

3.3. Low-temperature Flow Property

The CFPPs of different RSM biodiesel blends, viz. B0, B20, B40, B60, B80 and B100, are listed in figure 3.

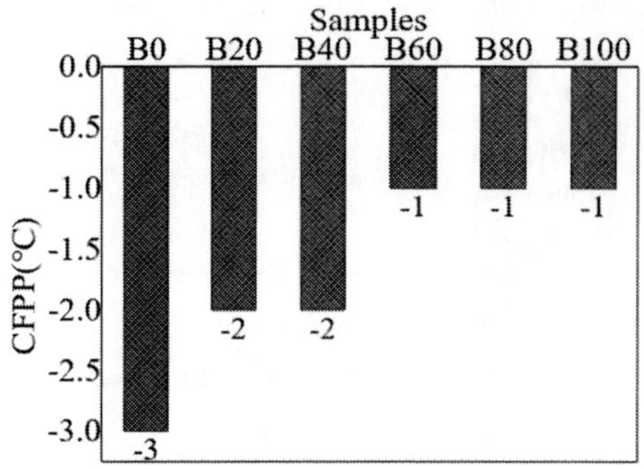

Figure 3. CFPPs of biodiesel blends.

We can see from figure 3 that CFPPs of the biodiesel blends, all below 0℃, were slightly different, although CFPPs increased with increasing contents of RSM in the blends. Virtually, CFPP of neat RSM was only 2℃ higher than that of conventional petro-diesel. The results demonstrated that RSM and its blends with petro-diesel provided acceptable cold flow property.

3.4. Oxidation Stability of RSM

Shown in figure 4, figure 5, figure 6, and figure 7 are peroxide values, viscosity, acid values and isooctane insolubles against oxidation durations for neat RSM, respectively.

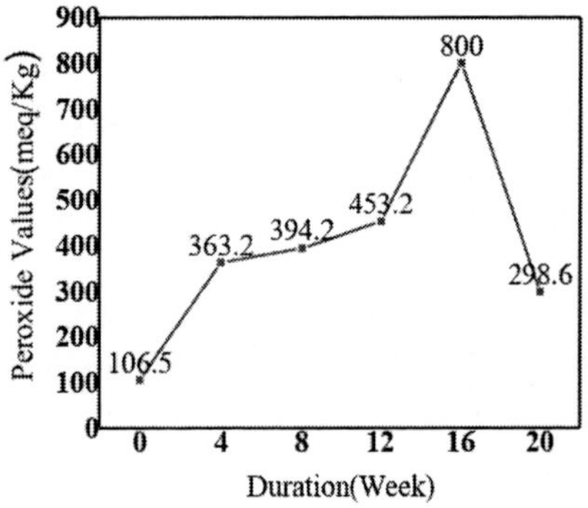

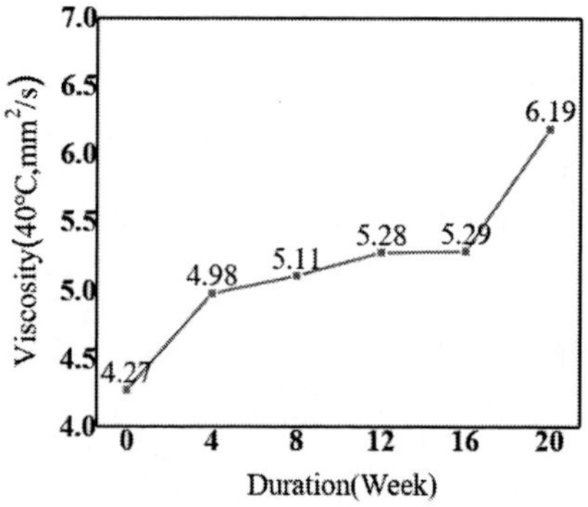

Figure 4. Peroxide values *vs*. test durations. **Figure 5.** Viscosity *vs.*test durations.

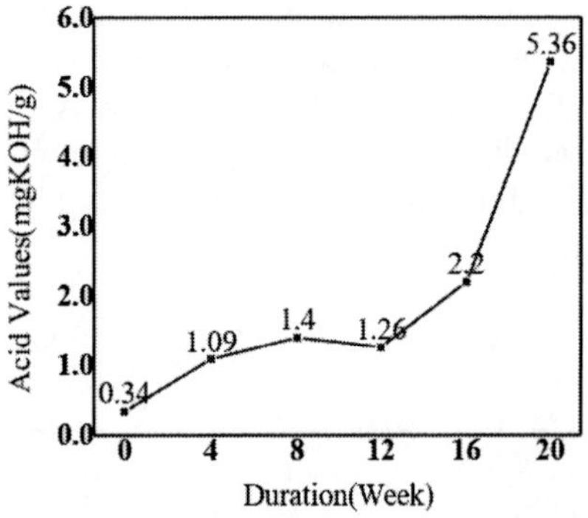

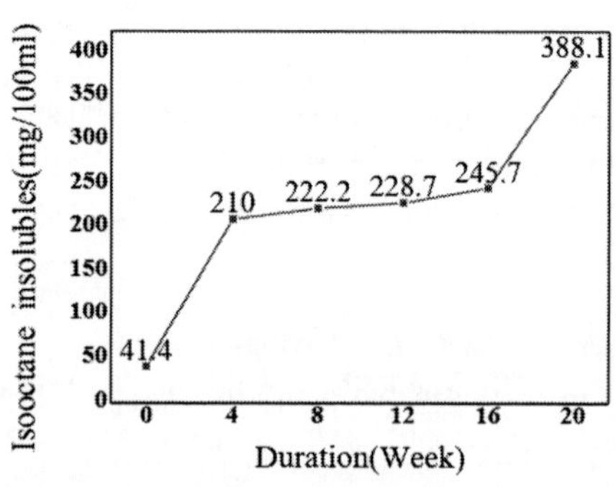

Figure 6. Acid values *vs.*test durations. **Figure 7.** Isooctane insoluble *vs.*test durations.

From figure 4 it can be clearly observed that, from the beginning of oxidation test onwards, the peroxide values of RSM increased with increasing test durations, followed by a sharp decrease after 16 weeks. The sharp decrease of peroxide values after 16 weeks may be ascribed to dramatic elimination of peroxides due to transformation to secondary oxidizers [17-18]. In figure 5, the viscosity of neat RSM increased gradually as the oxidation time increased because of oxidation polymerization and thus formation of viscous high molecular polymers [19]. In figure 6 shows that the acid values of RSM increased mildly from the beginning of the oxidation test up to 16 weeks, but markedly increased at 20 weeks. This well coincided with the sharp increase of peroxide values at 16 weeks shown in figure 4, and may be attributed to the substantial production of acidic oxidation products [20]. Concurrently, characterized by an all-along increase towards oxidation, the variation of isooctane insoluble with oxidation durations shown in figure 7 exhibited a very similar trend to those of peroxide values, viscosity and acid values. This well corresponded to continual oxidation polymerization and thus formation of polar polymeric materials as a result.

Table 2 gives the oxidation test results for RSM and its formulations with 1.0 wt.% of different antioxidants, in which the peroxide values, viscosity, acid values and isooctane insolubles were determined based on SH/T0175, while the induction periods determined based on EN14112.The results presented in table 2 show that the oxidation stability of RSM was significantly improved by the antioxidants, manifested by markedly increased induction periods and decreased peroxide values, viscosity, acid values, as well as insolubles.It can also be observed from table 2 that BD100 was the most effective antioxidizer. The mechanisms of antioxidants in inhibiting oxidation of RSM need further investigation.

Table 2. Oxidation results of RSM doped with oxidation inhibitors.

Test sample	Acid value (mgKOH/g)	Peroxide value (mg/kg)	Viscosity (40℃, mm^2/s)	Iso-octane insoluble (mg/100ml)	Induction period (*h*)
RSM	7.18	750.6	7.91	349.5	0.6
RSM+BD50	3.45	360.4	4.46	135.3	5.0
RSM+BD100	3.12	300.3	4.38	78.9	6.2
RSM+1407	4.09	561.5	6.57	304.2	3.1
RSM+1406	4.88	603.7	7.10	311.1	2.2
RSM+MOA	3.06	393.1	4.85	227.2	4.8

3.5. Elastomeric compatibility

Figure 8 and figure 9 show the changes in mass, volume and hardness of four elastomers ACM, CR, NBR and HNBR before and after immersion with RSM, respectively.

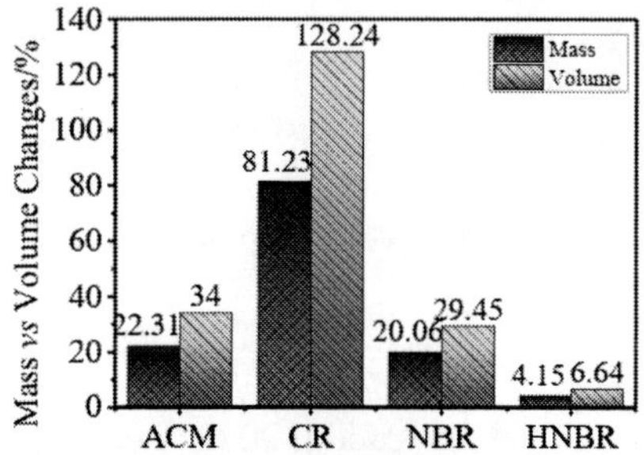

Figure 8. Mass *vs* Volume changes of elastomers after immersion with RSM.

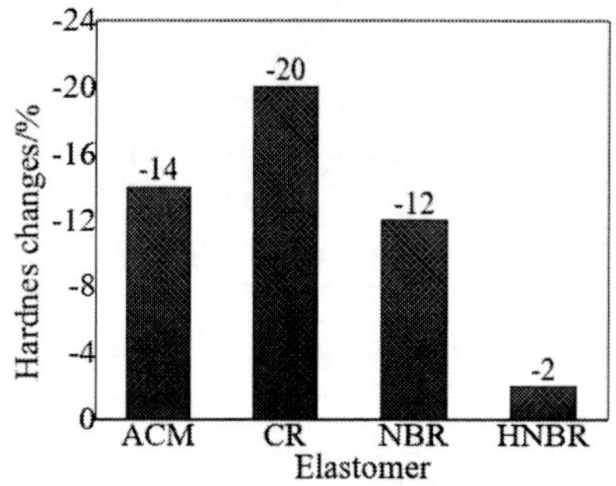

Figure 9. Hardness changes of elastomers after immersion with RSM.

It can be observed from figure 8 and figure 9 that, characterized by dramatic increases in mass, volume and hardness, the elastomers ACM, NBR and CR were to great extent degraded by RSM biodiesel blends, emonstrating that they were poorly compatible with RSM. Furthermore, it is also clear from figure 8 and figure 9 that ACM, NBR and CR were more severely deteriorated with increasing contents of RSM in the blends. Otherwise, the mechanical properties of HNBR were almost un-affected by RSM blends at whatever concentrations of RSM. This indicated that HNBR seemed to be an excellent RSM-resistant material, and is probably a promising elastomeric candidate for RSM-fueled engines.

4. Conclusions

The results of the present study revealed that rubber seed oil biodiesel and its blends with petro-diesel provided excellent lubricity, acceptable flammability and cold flow property, with respect to neat petro-diesel. However, RSM were prone to oxidation although the oxidation stability of RSM could be markedly enhanced by specific antioxidants such as BD100. Furthermore, RSM exhibited poor compatibility with many elastomers with the exception of hydrogenated nitrile-butadiene, which might be potential as a perspective elastomeric material for RSM-powered engines. The experimental results also provided the basis for future work regarding the performances and applicability of RSM/petro-diesel blends.

Acknowledgement

The authors gratefully acknowledge financial support of National Natural Science Foundation of China (project1 No. 51375491; project2 No. cstc2017jcyjAX0058)

References

[1] Gerhard K 2005 Dependence of biodiesel fuel properties on the structure of fatty acid alkyl esters *Fuel Processing Technology* **86** 1059-1070
[2] C A W Allen, K C Watts and R G Ackman 1999 Predicting the viscosity of biodiesel fuels from their fatty acid ester composition *Fuel* **78** 1319-1326
[3] Daniel P G and John W G 2004 Effects of specific fatty acid methyl esters on diesel fuel lubricity *Fuel* **83** 2351-2356
[4] Laforgia D and Ardito V 1995 Biodiesel fueled IDI engines: Performances, emissions and heat release investigation *Bio resource Technology* **51**(1) 53-59
[5] A E Atabani, A S Silitonga and I A Badruddin 2012 A comprehensive review on biodiesel as an alternative energy resource andits characteristics *Renewable and Sustainable Energy Reviews* **16** 2070-2093
[6] S K Hoekman, Amber B and Curtis Robbins 2012 Review of biodiesel composition, properties, and specifications *Renewable and Sustainable Energy Reviews* **16** 143-169
[7] Mustafa C, Ahmet E and Erol A 2006 Performance and exhaust emissions of a biodiesel engine

Applied Energy **83** 594–605

[8] H Sharon,K Karuppasamy and D R Soban Kumar 2012 A test on DI diesel engine fueled with methyl esters of used palm oil *Renewable Energy* **47** 160-166

[9] Masjuki Hj Hassan, Md Abul Kalam 2013 An overview of biofuel as a renewable energy source: development and challenges *Procedia Engineering* **56** 39-53

[10] D Huang,H Zhou and L Lin 2012 Biodiesel an Alternative to Conventional Fuel *Energy Procedia* **16** 1874-1885

[11] L Ranganathan, G Lakshmi Narayana Rao and S Sampath **2011** Experimental Investigation of a Diesel Engine Fuelled With Optimum Biodiesel Produced From Cotton Seed Oil *European Journal of Scientific Research* **62(1)** 101-115

[12] M A Fazal, A S M A Haseeb, H H Masjuki **2010** Comparative corrosive characteristics of petroleum diesel and palm biodiesel for automotive materials *Fuel Processing Technology* **(91)** 1308-1315

[13] WW Liu etc, 2005 Preparation of biodiesel fuel with rubber seed oil rubber seed oil *China oils and fats* **30(10)** 63-66

[14] Ramadhas etc, 2005 Performance and emission evaluation of a diesel engine fueled with methyl esters of rubber seed oil *Renewable Energy* **30** 1789-1800

[15] Gerhard K and K R Steidley 2005 Lubricity of Components of Biodiesel and Petrodiesel. The Origin of Biodiesel Lubricity *Energy & Fuels* **19** 1192-1200

[16] D P Geller and J W Goodrum 2004 Effects of specific fatty acid methyl esters on diesel fuel lubricity *Fuel* **83** 2351-2356

[17] N Kapilan, T P Ashok Babu and R P Reddy 2009 Technical Aspects of Biodiesel and its Oxidation Stability *International Journal of ChemTech Research* **1(2)** 278-282

[18] A Bouaid, M Martinez and J Aracil 2007 Long storage stability of biodiesel from vegetable and used frying oils J *Fuel* **86(16)** 2596-2602

[19] Knothe G 2007 Some aspects of biodiesel oxidative stability *Fuel Processing Technology* **88(7)** 669-77

[20] M C RL,Ratcliff M, Moens L and Lawrence R 2007 Several factors affecting the stability of biodiesel in standard accelerated tests *Fuel Processing Technology* **88(7)** 651-7

IOP Conf. Series: Materials Science and Engineering **292** (2017) 012074 doi:10.1088/1757-899X/292/1/012074

Load distributions analysis of muti-pin composite joints: single-lap

F Liu[1, 2, 3], L Qi[1, 2] and S F Feng[1, 2]

[1] College of Civil Engineering, Xijing University, Xi'an 710123
[2] China Shaanxi Key Laboratory of Safety and Durability of Concrete Structures, Xijing University

[3] E-mail: liufang_winter@163.com

Abstract. In this paper, finite element model of muti-pin composite joint is established. Composite laminate of joint is treated as one whole plate. Matlab program is adopted to calculate equivalent material properties of composite laminate. Load distributions of single-lap composite joint are investigated, and simulation results show excellent agreement with experiments. Compared with FE model whose composite laminate material properties are given layer by layer, new FE model not only ensured calculation accuracy but also reduced model building time.

1. Introduction

With wide application of composites, composite components are usually fasted to other structural components by mechanical means. The current investigation [1-3] shows that failure of single-bolt, single-lap composite joint firstly occurs at the maximum stress point. But in practical application, composite laminates are fastened by muti-pin/bolt, and load distribution of pin usually is different. Large load distribution difference results in reduction of joint efficiencies [4]. So load distribution investigation of muti-pin composite joint is quite important.

Three-dimensional finite element analysis is rather time-consuming and simpler method compared with the spring-based method. A simplified method for determining the effects of bolt-hole clearance on the load distribution in composite multi-bolt joints is presented in literature [5]. The method is validated against three-dimensional finite element models.

In this paper, a new FE model is established for load distribution investigation of muti-pin composite joint. In FE model, composite laminate is treated as one whole plate whose equivalent engineering material properties including modulus and Poisson's ratio are calculated by Matlab program developed by author. Matlab program is verified by experiments in literature [6].

To test and verify the new FE model, load distributions of single-lap and multi-bolt composite joints are investigated. At the same time FE model whose composite laminate material properties are given layer by layer is also created. Simulation results of two FE models and experimental results are compared with each other.

2. Equivalent material properties of laminates

A new FE model is developed in this paper, composite laminate in FE model is treated as one whole plate. Equivalent material properties of laminate are calculated by Matlab program according to equations in literature [7].

To verify program of equivalent material properties, a numerical example is chosen in literature [4]. In numerical example the composite laminate is manufactured from HTA/6376, a high-strength carbon fiber–epoxy material currently used in primary structures in the European aircraft industry. The unidirectional stiffness properties are shown in table 1. Stacking sequence of laminate is [45° /0° /-45° /90°]$_{5s}$. Equivalent material properties of laminate are shown in table 2.

Table 1. Unidirectional stiffness properties for HTA/6376.

E_{11} (GPa)	E_{22} (GPa)	E_{22} (GPa)	G_{12} (GPa)	G_{13} (GPa)	G_{23} (GPa)	υ_{12}	υ_{13}	υ_{23}
140	10	10	5.2	5.2	3.9	0.3	0.3	0.5

Table 2. Comparison of equivalent material properties of laminates.

Equivalent material properties	E_x (GPa)	E_y (GPa)	E_z (GPa)	G_{xy} (GPa)	G_{xz} (GPa)	G_{yz} (GPa)	υ_{xy}	υ_{xz}	υ_{yz}
Reference value	54.25	54.25	12.59	20.72	4.55	4.55	0.309	0.332	0.332
Matlab	54.249	54.249	12.587	20.722	4.550	4.550	0.309	0.3326	0.3326
Error (%)	0.002	0.002	0.024	0.010	0.000	0.000	0.000	0.181	0.181

Table 2 indicates that the error of equivalent material engineering constants calculated by Matlab program is quite small, so that the program can be used to calculate the material properties of three-dimensional laminate.

3. Numerical study

In this section, load distributions of muti-bolt composite joints are investigated by equivalent material properties (EMP) FE model. Simulation results are compared with experiments in literature [8].

3.1. Geometry, FE model and boundary condition

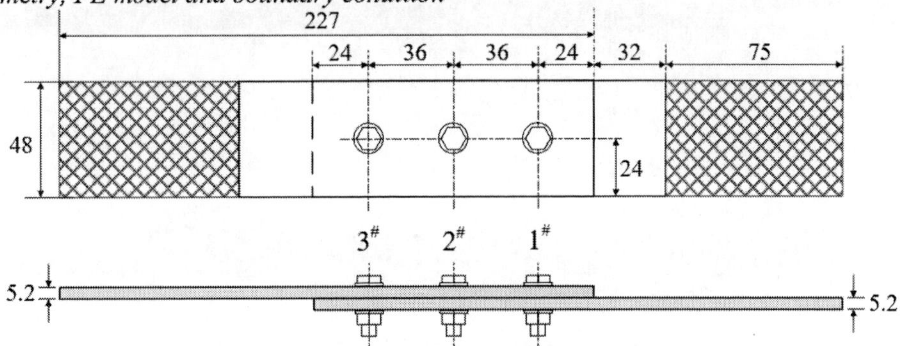

Figure 1. Joint configurations with three pins (mm): single-lap.

As shown in figure 1, the joint has three bolts and all the bolts are in single-column. Geometric parameters can be seen in figure 1. Diameter of bolt is 8mm. Composite laminates have stacking sequences [45° /0° /-45° /90°]$_{ns}$, thickness of each ply is 0.13mm. Unidirectional stiffness properties are shown in table 1. By using Matlab program, equivalent material properties of laminates can be obtained. The titanium bolt was modeled with isotropic material properties, with material constant E_x=110GPa, υ_b=0.29.

IOP Conf. Series: Materials Science and Engineering **292** (2017) 012074 doi:10.1088/1757-899X/292/1/012074

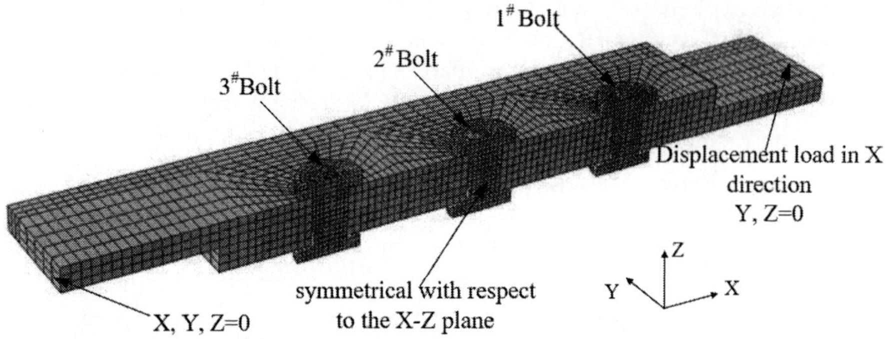

Figure 2. Mesh of FE model and boundary conditions: single-lap.

According to geometric parameters of joints, the FE model is established in ABAQUS shown in figure 2. In this model, bolt and washer are simplified as one component. Bolt torque is 0.5Nm. To compare with experiments in literature [8], Bolt–hole clearances in FE model are 0. The laminates are modeled using three-dimensional solid elements with reduced integration which is C3D8R in ABAQUS. Since single-lap joint is symmetrical about X-Z plane, only one half of joint is modeled. One end of joint is fixed, and displacement load in X direction is carried out on the other end of joint.

3.2. Model validation

FE simulation stress contour by EMP method is shown in figure 3.

To compare 3D FE method and EMP method, two FE models are established, and simulation results and experimental results [8] of load distribution are shown in figure 4. Simulation results show an excellent agreement with experiments.

As shown in figure 3, stress distributions of bolts are quite different, which can be seen from figure 4, too. For example, maximum bolt loads of bolt 1 and 3 in single-lap joint are 4.6KN and 5.5KN.

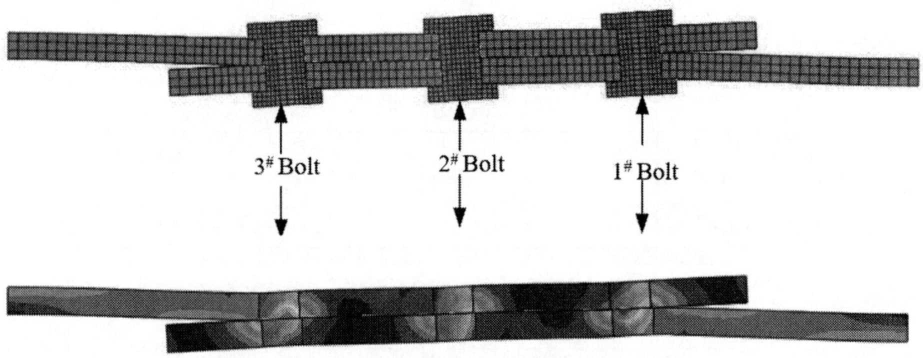

Figure 3. Stress contour of single-lap joint.

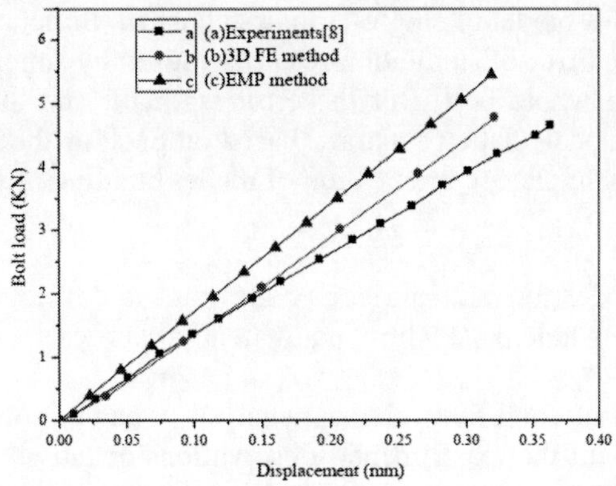

Bolt 1

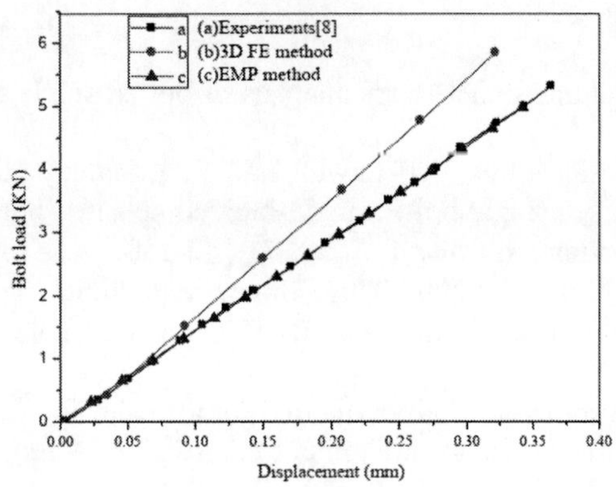

Bolt 2

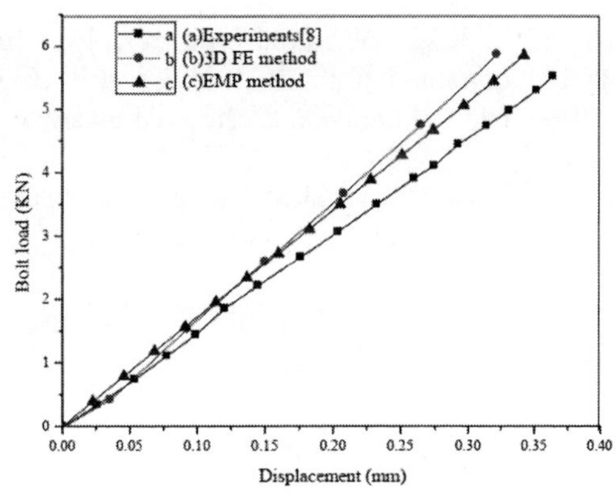

Bolt 3

Single-lap joint

Figure 4. Load distribution comparison of experiments, 3D FE method and EMP method
(a) Experiments[8] (b) 3D FE method (c) EMP method.

4. Discussion

Composite laminates are modeled layer by layer in traditional FE method. In this paper, composite laminates are modeled as one whole plate. Equivalent material properties of laminates are calculated by Matlab program which is verified by experiments. Error of Matlab program is quite small.

Based on EMP method, FE model of single-lap is established. Simulation results have an excellent agreement with experiments. Error of simulation result is caused by simplification of bolt and washer, which are simplified as one whole bolt. But the error is small, and the accuracy is good enough. Compared with 3D FE model, calculation accuracy based on EMP method is higher. Meanwhile, since laminate is treated as one whole plate, complication of model building is decreased greatly.

5. Conclusions

In this paper, a new FE model for muti-pin composite joint is developed. In FE model, composite laminate is simplified as one whole plate whose equivalent properties are calculated by Matlab written by author.

From the experimental results and finite element analysis, it can be concluded that:

EMP FE model agrees with the experimental observations quantitatively. Compared with 3D FE model, since composite laminates are treated as one plate, model building time decreases obviously. EMP FE model is a rather time-consuming and simpler method for load distribution investigation.

References

[1] Ireman T 1998 Three-dimensional stress analysis of bolted single-lap composite joints *Compos Struct* **43** 195-216

[2] M A McCarthy, C T McCarthy, V P Lawlor and W F Stanley 2005 Three-dimensional finite element analysis of single-bolt, single-lap composite bolted joints: part I -model development and validation *Compos Struct* **71** 140-158

[3] C T McCarthy, M A McCarthy 2005 Three-dimensional finite element analysis of single-bolt, single-lap composite bolted joints: Part II-effects of bolt-hole clearance *Compos Struct* **71** 159-175

[4] P J Gray, C T McCarthy 2010 A global bolted joint model for finite element analysis of load distributions in multi-bolt composite joints *Compos Part B- Eng* **41** 317-325

[5] M A McCarthy, C T McCarthy and G S Padhi 2006 A simple method for determining the effects of bolt-hole clearance on load distribution in single-column multi-bolt composite joints *Compos Struct* **73** 78-87

[6] P J Gray, C T McCarthy 2010 A global bolted joint model for finite element analysis of load distributions in multi-bolt composite joints *Compos Part B- Eng* **41** 317-325

[7] Sun C T and Sijian L 1988 Three-dimensional effective elastic constants for thick laminates *J Compos Mater* **22** 629-639

[8] McCarthy M, Lawlor V and Stanley W 2005 An experimental study of bolt-hole clearance effects in single-lap, multi-bolt composite joints *J Compos Mater* **39** 799-825

Influences of the current density on the performances of the chrome-plated layer in deterministic electroplating repair

H Xia, X M Shen[1], X C Yang, Y Xiong and G L Jiang

College of Field Engineering, PLA Army Engineering University, No. 1 Haifu Street, Nanjing, Jiangsu 210007, P. R. China

[1] E-mail: shenxmjfjlgdx2014@163.com

Abstract. Deterministic electroplating repair is a novel method for rapidly repairing the attrited parts. By the qualitative contrast and quantitative comparison, influences of the current density on performances of the chrome-plated layer were concluded in this study. The chrome-plated layers were fabricated under different current densities when the other parameters were kept constant. Hardnesses, thicknesses and components, surface morphologies and roughnesses, and wearability of the chrome-plated layers were detected by the Vickers hardness tester, scanning electron microscope / energy dispersive X-ray detector, digital microscope in the 3D imaging mode, and the ball-milling instrument with profilograph, respectively. In order to scientifically evaluate each factor, the experimental data was normalized. A comprehensive evaluation model was founded to quantitative analyse influence of the current density based on analytic hierarchy process method and the weighted evaluation method. The calculated comprehensive evaluation indexes corresponding to current density of $40\,A/dm^2$, $45\,A/dm^2$, $50\,A/dm^2$, $55\,A/dm^2$, $60\,A/dm^2$, and $65\,A/dm^2$ were 0.2246, 0.4850, 0.4799, 0.4922, 0.8672, and 0.1381, respectively. Experimental results indicate that final optimal option was $60\,A/dm^2$, and the priority orders were $60\,A/dm^2$, $55\,A/dm^2$, $45\,A/dm^2$, $50\,A/dm^2$, $40\,A/dm^2$, and $65\,A/dm^2$.

1. Introduction

Deterministic electroplating repair is a new and practical method for rapidly repairing the attrited parts, which is based on the deterministic surfacing theory [1] and electroplating technique [2, 3]. Schematic diagram of the deterministic electroplating repair system is shown in the figure 1. Original profile of the attrited part is obtained by three-coordinate measuring machine. Through subtracting the expected shape from the original profile, theoretical distribution of the optimal deposition layer is calculated. The electroplating solution is sucked from the container to the nozzle head by the peristaltic pump and is absorbed back to the container by the atmospheric pressure generated from the vacuum ejector. Through controlling concentration of the electroplating solution and the current density between the anode and the attrited part, shape of the deposition function and its efficiency can be kept steady. Distribution of the dwelling time can be achieved by the Fourier transform and convolution operation based on the theoretical distribution of the optimal deposition layer and the deposition function [4]. By controlling movement of the nozzle head which is installed on the three-dimension (3D) motion platform according to distribution of dwelling time, the attrited part can be deterministically repaired. Different from electro-brush plating [5] and electro-bath plating [6], parameters in the deterministic electroplating repair should be accurately controlled, because stability and veracity of the deposition function is the foundation of the deterministic electroplating method. Among these parameters, current density is an important factor, which affects efficiency of the deposition function and performance of the chrome-plated layer. Thus, investigation of the influences of current density was conducted in this study. The chrome-plated layers were fabricated under the different current densities when the other parameters were kept constant. Then, hardnesses, thicknesses and components, surface morphologies and roughnesses, and wearability of the chrome-plated layers were detected respectively.

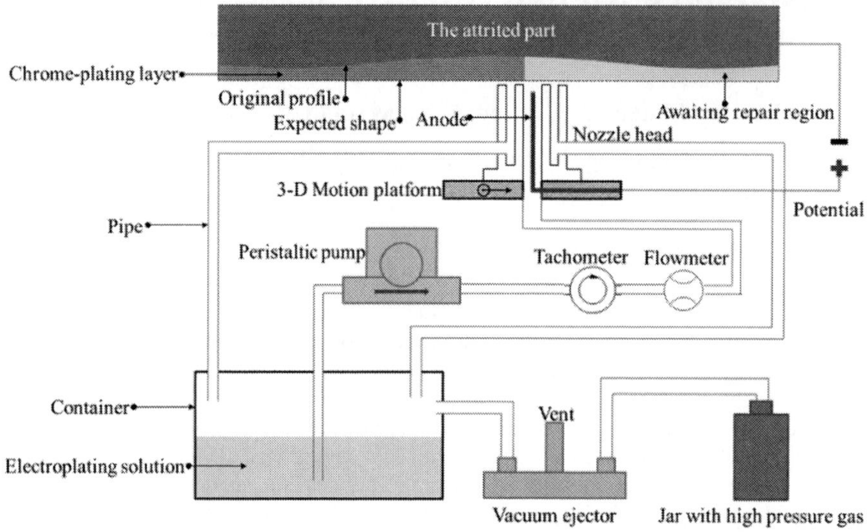

Figure 1. Schematic diagram of the deterministic electroplating repair system.

2. Experimental parameters

The chrome-plated layer was deposited on the specimen 1045 steel. The pressure, concentration, and time of the wet sand blasting were 0.48 MPa, 10-15%, and 15 min, respectively. The activations before formal plating consisted of the dipping and rinse for 5 min, warm-up for 60 s, countercurrent at $26\,\mathrm{A/dm^2}$ for 80 s, and primary plating at $6\,\mathrm{A/dm^2}$. The current gradually raised $70\,\mathrm{A/dm^2}$ in 10 min, which started the concussive plating. The concussive plating stood for 4 min. Afterwards, the formal plating was conducted for 116 min, which meant that the total effective plating time was 120 min, and the investigated current densities were $40\,\mathrm{A/dm^2}$, $45\,\mathrm{A/dm^2}$, $50\,\mathrm{A/dm^2}$, $55\,\mathrm{A/dm^2}$, $60\,\mathrm{A/dm^2}$,and 65 $\mathrm{A/dm^2}$, respectively. When preparations of the chrome-plated layers were finished, performances of the layers were studied, including the hardness detection, thickness measurement, component analysis, surface morphology investigation, roughness examination, and wearability testing.

3. Results and discussions

3.1. Hardness detection

Hardnesses of the chrome-plated layers were detected by the Vickers hardness tester. The values of hardnesses of the chrome-plated layers corresponding to the current densities $40\,\mathrm{A/dm^2}$, $45\,\mathrm{A/dm^2}$, $50\,\mathrm{A/dm^2}$, $55\,\mathrm{A/dm^2}$, $60\,\mathrm{A/dm^2}$, and $65\,\mathrm{A/dm^2}$ were 884 HV0.1, 845 HV0.1, 855 HV0.1, 855 HV0.1, 905 HV0.1, and 874 HV0.1, respectively, which were obviously larger than hardness of the substrate (specimen 1045 steel) 192 HV0.1. Hardness of the chrome-plated layer did not increase or decrease by degrees along with increase of the current density. The best option was $60\,\mathrm{A/dm^2}$ and the worst option was $45\,\mathrm{A/dm^2}$. Target of deterministic electroplating repair was the attrited part, and its usability can be improved by fabricating the high-hardness chrome-plated layer at the worn region.

3.2. *Thickness measurement and component analysis*

Thicknesses and components of the chrome-plated layers were measured by the scanning electron microscope and energy dispersive X-ray detector [7, 8], respectively, as shown in the figure 2, and the results were summarized in table 1. It could be found that concentrations of the Cr element ranged from 97.8% to 98.36% and the average was 98.09%, which indicated that component consistencies of the chrome-plated layers in the six experiments were excellent. Influence of the current density to the plating efficiency was fluctuant. The best option was $60\ A/dm^2$ and the worst option was $65\ A/dm^2$. The major reason was that the increase of current density would raise the over-potential for the Hydrogen (H) evolution and that for the Cr evolution [9]. The average thickness of the chrome-plated layer was 157 μm when the current density was $60\ A/dm^2$.

Table 1. The summarized results of thickness measurement and component analysis.

Current density	Thickness measurement			Components analysis	
	Position 1	Position 2	Average	Cr (wt %)	C (wt %)
$40\ A/dm^2$	78.4 μm	70.8 μm	74.6 μm	98.08%	1.92%
$45\ A/dm^2$	120 μm	119 μm	119.5 μm	98.36%	1.64%
$50\ A/dm^2$	141 μm	131 μm	136 μm	98.15%	1.85%
$55\ A/dm^2$	82.5 μm	90.3 μm	86.4 μm	97.97%	2.03%
$60\ A/dm^2$	160 μm	154 μm	157 μm	97.8%	2.2%
$65\ A/dm^2$	61.4 μm	65 μm	63.2 μm	98.15%	1.85%

3.3. Surface morphology investigation and roughness examination

Surface morphologies and roughnesses of the chrome-plated layers were investigated by the digital microscope in the 3D imaging mode [10], as shown in the figure 3. The calculated roughness average (Ra) and the selection of ten-point height of irregularities (Rz) was summarized in table 2.

It could be observed that the surface roughnesses were improved when current density rose from $40\ A/dm^2$ to $55\ A/dm^2$ in the first place, and were deteriorated when the current density further rose from $55\ A/dm^2$ to $65\ A/dm^2$ later. Therefore, the optimal option was $55\ A/dm^2$. Surface roughness of the chrome-plated layer was also a critical factor in deterministic electroplating repair, because precision of the repaired part affected its application. It had been reported that surface roughnesses of the attrited part had little change or changed worse after the electroplating repair. Thus, Pretreatment of the attrited part before repair was necessary. Surface roughness can be improved by fine polishing of the attrited part, which was propitious to obtain high surface quality after electroplating repair.

Table 2. Summary of the calculated Ra and Rz of the chrome-plated layers.

Current density	Surface roughness of the cross-sectional line	
	Ra	Rz
$40\ A/dm^2$	8.572 μm	31.88 μm
$45\ A/dm^2$	6.315 μm	27.68 μm
$50\ A/dm^2$	6.182 μm	25.58 μm
$55\ A/dm^2$	5.659 μm	20.23 μm
$60\ A/dm^2$	7.114 μm	29.89 μm
$65\ A/dm^2$	7.375 μm	31.83 μm

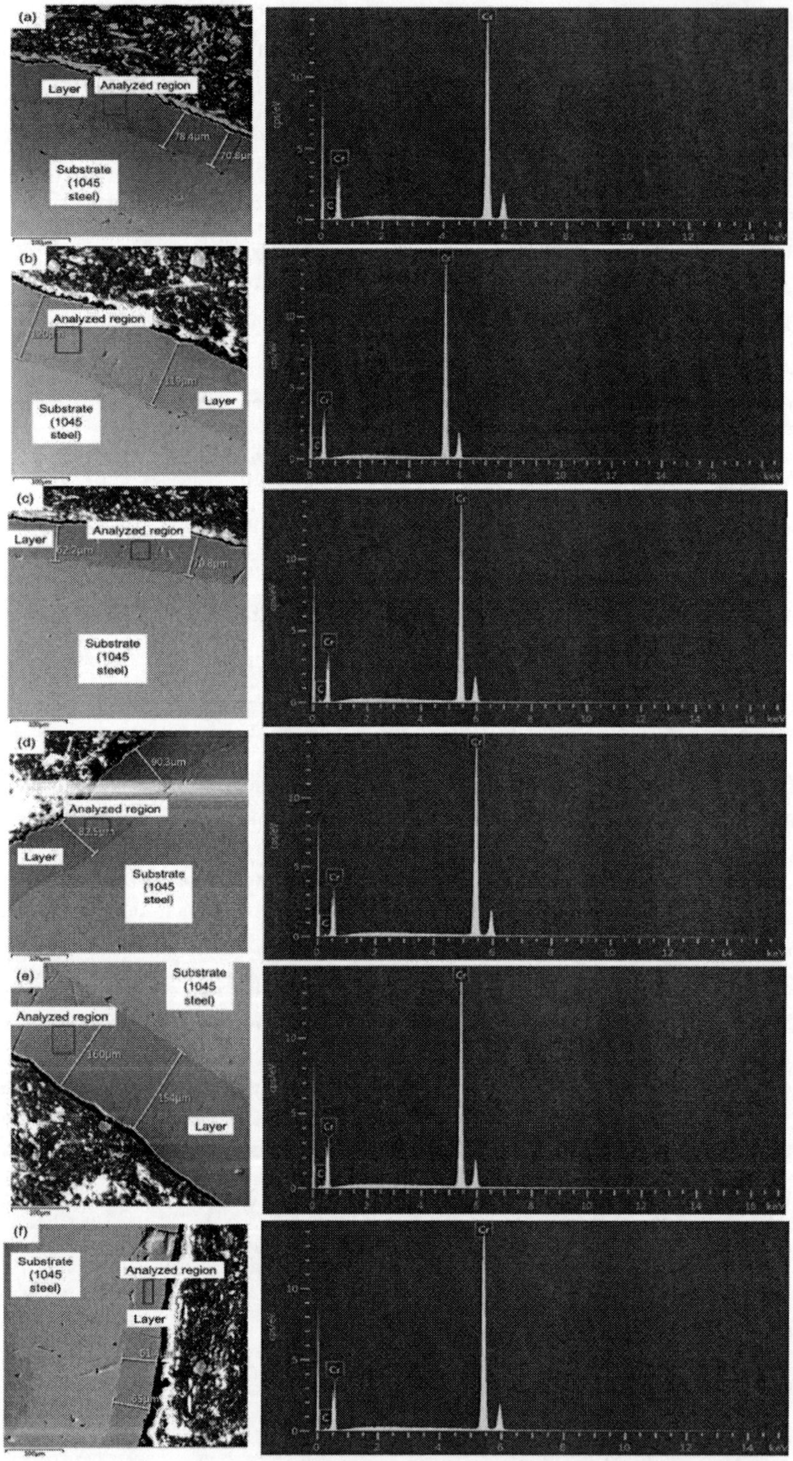

Figure 2. Thickness and component analysis of chrome-plated layer. (a) current density was 40 A/dm^2. (b) current density was 45 A/dm^2. (c) current density was 50 A/dm^2. (d) current density was 55 A/dm^2. (e) current density was 60 A/dm^2. (f) current density was 65 A/dm^2.

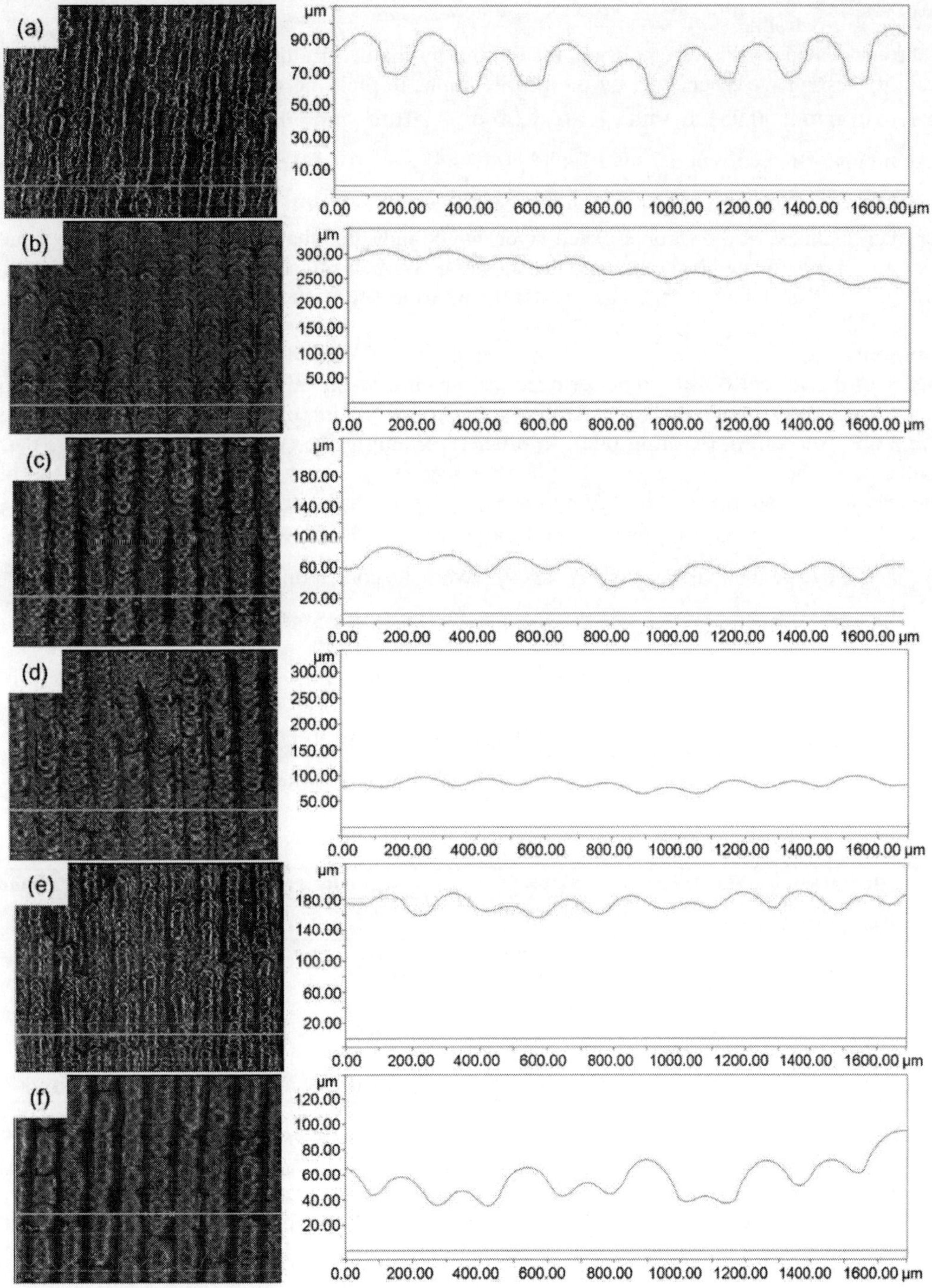

Figure 3. Surface morphologies and roughnesses of the chrome-plated layers. (a) current density was 40 A/dm^2. (b) current density was 45 A/dm^2. (c) current density was 50 A/dm^2. (d) current density was 55 A/dm^2. (e) current density was 60 A/dm^2. (f) current density was 65 A/dm^2.

3.4. Wearability testing

The chrome-plated layers were polished for 60 min by the ball-milling instrument, and results of the wearability testing were achieved by the profilograph. It could be calculated that the wearing rates were 1.10 μm/min, 0.95 μm/min, 1.18 μm/min, 0.90 μm/min, 0.83 μm/min, and 1.18 μm/min corresponding to the current density of 40 A/dm^2, 45 A/dm^2, 50 A/dm^2, 55 A/dm^2, 60 A/dm^2, and 65 A/dm^2. Especially when the current density was 65 A/dm^2, depth of the polished spot was larger than thickness of the chrome-plated layer, which indicated that the polishing process penetrated through the chrome-plated layer and into the substrate. Wearability of the chrome-plated layer affected the useful life of the repaired part. The smaller the wearing rate is, the larger the useful life is.

4. Summaries

Influence of the current density to performance of the chrome-plated layer was evaluated based on the obtained experimental data, including hardness, thickness (plating efficiency), surface roughness (Ra and Rz), and wearability (wearing rate). In order to scientifically evaluate each factor, the data was normalized, as shown in the table 3. For the positive factors, such as the hardness and the plating efficiency, normalization was achieved by the Eq. 1. Meanwhile, for the negative factors, such as the surface roughness and wearing rate, normalization was obtained by the Eq. 2. Here a_i and b_i both represented the experimental data, and α_i and β_i were the corresponding normalized results.

$$\alpha_i = \frac{a_i - min\left(\left\{a_i \middle| i=1,2,3,4,5,6\right\}\right)}{max\left(\left\{a_i \middle| i=1,2,3,4,5,6\right\}\right) - min\left(\left\{a_i \middle| i=1,2,3,4,5,6\right\}\right)} \tag{1}$$

$$\beta_i = \frac{max\left(\left\{b_i \middle| i=1,2,3,4,5,6\right\}\right) - b_i}{max\left(\left\{b_i \middle| i=1,2,3,4,5,6\right\}\right) - min\left(\left\{b_i \middle| i=1,2,3,4,5,6\right\}\right)} \tag{2}$$

Table. 3 Summary of the normalized data of the evaluation factors.

Current density(A/dm^2)	Hardness	Plating efficiency	Surface roughness		Wearing rate
			Ra	Rz	
40	0.650	0.122	0.000	0.000	0.229
45	0.000	0.600	0.775	0.361	0.657
50	0.167	0.776	0.820	0.541	0.000
55	0.167	0.247	1.000	1.000	0.800
60	1.000	1.000	0.501	0.171	1.000
65	0.483	0.000	0.411	0.004	0.000

A comprehensive evaluation model was founded to quantitative analyze influence of the current density based on the analytic hierarchy process (AHP) method and weighted evaluation method [11]. Therefore, it would be found that relationship between Ra and Rz was highly linear correlation, which indicated that the five evaluation factors could be divided into four groups. Fine surface roughness of the repaired part would avoid the secondary machining of ultra-hard chrome-plated layer, which was the remarkable advantage of deterministic electroplating repair process. However, surface roughness Ra and Rz of the chrome-plated layers were mainly decided by pretreatment of the attrited part before repair, and the electroplating process had few influences. Thus, weight of evaluation factors for Ra and Rz were both 0.1. Meanwhile, the plating efficiency was the most important factors, which decided the final repairing efficiency. Therefore, weight for the plating efficiency was 0.4. Furthermore, the hardness determined the usability of the repaired part and the wearing rate decided its useful life, so they shared the rest of weight 0.4. According to the above analysis, the final comprehensive evaluation model was obtained, as shown in Eq. 3. Here E was the comprehensive evaluation index. $n_{hardness}$, $n_{plating\_efficiency}$, n_{Ra}, n_{Rz}, and $n_{wearing\_rate}$ were the normalized data in table 3, respectively.

$$E = 0.2 \cdot n_{hardness} + 0.4 \cdot n_{plating\_efficiency} + 0.1 \cdot n_{Ra} + 0.1 \cdot n_{Rz} + 0.2 \cdot n_{wearing\_rate} \quad (3)$$

The calculated comprehensive evaluation indexes corresponding to current density of 40 A/dm^2, 45 A/dm^2, 50 A/dm^2, 55 A/dm^2, 60 A/dm^2, and 65 A/dm^2 were 0.2246, 0.4850, 0.4799, 0.4922, 0.8672, and 0.1381, respectively. It could be observed that the final optimal option was 60 A/dm^2, and the priority orders were 55 A/dm^2, 45 A/dm^2, 50 A/dm^2, 40 A/dm^2, and 65 A/dm^2.

5. Conclusions

Investigation of the influences of current density was conducted in this study. Hardnesses, thicknesses and components, surface morphologies and roughnesses, and wearability of chrome-plated layers were detected respectively. Based on the analytic hierarchy process (AHP) method and the weighted evaluation method, a comprehensive evaluation model was founded to quantitative analyze influence of the current density. Experimental results indicate that the final optimal option was 60 A/dm^2.

Acknowledgments

This work was supported by a grant from National Natural Science Foundation of China (Grant No. 51505498), a grant from Natural Science Foundation of Jiangsu Province (Grant No. BK20150714), and a grant from National Key Research & Development Program (Grant No. 2016YFC0802903).

References

[1] Dai Y F, Song C, Peng X Q and Shi F 2010 Calibration and prediction of removal function in magnetorheological finishing *Applied Optics* **49** 298-305

[2] Bocking C, Cameron B 1994 Use of high speed selective jet electrodeposition of gold for the plating of connectors *Transactions of the Institute of Metal Finishing* **72(1)** 33-40

[3] Kunieda M, Katoh R and Mori Y 1998 Rapid prototying by selective electrodeposition using electrolyte jet *CIRP Annals-Manufacturing Technology* **47(1)** 161-164

[4] Yamamura K 2007 Fabrication of ultra-precision optics by numerically controlled local wet etching *Ann. CIRP* **56** 541-544

[5] Fukunaka Y, Yamamoto T and Kondo Y 1989 Electrodeposition of silver under direct and pulsed current *Journal of the Eleetrochemical* **136(11)** 3278-3283

[6] Yang H, Dong S Y and Xu B S 2008 Microstructure and properties of brush electroplated nano-SiC-Al2O3/Ni composite coating *Key Engineering Materials* **373-374** 285-288

[7] Shen X M, Dai Y F, Deng H, Guan C L and Yamamura K 2013 Ultrasmooth reaction-sintered silicon carbide surface resulting from combination of thermal oxidation and ceria slurry polishing *Optics Express* **12(21)** 14780-14788

[8] Shen X M, Dai Y F, Deng H, Guan C L and Yamamura K 2013 Comparative analysis of oxidation methods of reaction-sintered silicon carbide for optimization of oxidation-assisted polishing *Optics Express* **22(21)** 26123-26135

[9] Ben-Jacob E, Garik P 1990 The formation of patterns in nonequilibrium growth *Nature* **343** 523-530.

[10] Shen X M, Tu Q Z, Deng H, Jiang G L and Yamamura K 2015 Mechanism analysis on finishing of reaction-sintered silicon carbide by combination of water vapor plasma oxidation and ceria slurry polishing *Optical Engineering* **54(5)** 055106

[11] Zhang H Q, Zhan S F 2004 Comprehensive evaluation model for residential quarters *Journal of Hohai University(Natural Sciences)* **32(6)** 716-718

Preparation and mechanical properties of carbon nanotube-silicon nitride nano-ceramic matrix composites

C Y Tian[1] and H Jiang

Department of Mechanical Engineering, Hefei University, Hefei, Anhui, China

[1] E-mail: tianchy2007@126.com

Abstract. Carbon nanotube–silicon nitride nano-ceramic matrix composites were fabricated by hot-pressing nano-sized Si_3N_4 powders and carbon nanotubes. The effect of CNTs on the mechanical properties of silicon nitride was researched. The phase compositions and the microstructure characteristics of the samples as well as the distribution of carbon nanotube in the silicon nitride ceramic were analyzed by X-ray diffraction and scanning electron microscope. The results show that the microstructure of composites consists mainly of α-Si_3N_4, β-Si_3N_4, Si_2N_2O and carbon natubes. The addition of proper amount of carbon nanotubes can improve the fracture toughness and the flexural strength, and the optimal amount of carbon nanotube are both 3wt.%. However the Vickers hardness values decrease with the increase of carbon nanotubes content.

1. Introduction

Silicon nitride (Si_3N_4) ceramic has been regarded as one of the promising engineering ceramic materials because of its high strength, high hardness, high corrosion and wear resistance [1, 2]. However, silicon nitride ceramics still have some inadequacy, such as insufficient resistance to brittle fracture, insufficient thermal-shock resistance and fracture toughness. Therefore, improving the mechanical properties and wear behaviour of silicon nitride ceramic is an important object in developing Si_3N_4-based ceramics [3, 4].

In the late 80's Niihara proposed the nanocomposite concept. Ceramic nanocomposite has attracted a lot of scientific interest because of its high mechanical properties, wear resistance, chemical inertness, and corrosion resistance [5, 6]. It is an effective way to incorporate a secondary phase such as nanoparticles, whisker into Si_3N_4 matrix to improve the mechanical properties [7, 8].

In recent years, carbon nanotubes (CNTs) has been considered as the most promising reinforcement for composite materials because of its exceptional mechanical ($E \approx 1.8TPa$) and electrical properties [9, 10]. The potential for nanocomposites reinforced with CNTs which have extraordinary specific stiffness and strength represents tremendous opportunities for application in the 21st century. It is expected that the addition of carbon nanotubes will significantly improve the quality of different matrixes including metal, polymer or ceramic. Therefore there have been several attempts to fabricated CNT/ceramic composites with improved flexural strength and fracture toughness [11, 12]. Zhan et al. fabricated CNT/alumina nanocomposite by spark-plasma-sintering the mixture of single-wall nanotubes and nano-alumina powders, and their experimental results show that the facture toughness of the nanocomposite is three times as much as that of monolithic alumina ceramic [13]. Other research results show that the addition of CNT into alumina results in a rather low increment in fracture toughness, while a significant increase in the contact damage and wear resistance [14]. But carbon nanotubes reinforced silicon carbide and silicon nitride matrix composites only show modest improvements [15, 16], which may be related to the bonding between CNT and ceramic matrix as well as the distribution of CNT in the ceramic matrix. So it is crucial to achieve a homogeneous distribution of CNT and a good bonding between CNT and ceramic matrix.

Synthesis of dense ceramics based on silicon nitride nano-powders is a promising route towards ceramic engineering materials. An advantage of nano-powders is their ultra-diffusivity and large specific surfaces. In this study, carbon nanotube–silicon nitride nano-ceramic matrix composites were fabricated by hot-pressing nano-sized Si_3N_4 powders and carbon nanotubes. The homogeneous distribution of CNTs on the silicon nitride matrix was studied to achieve a good bonding between CNT and ceramic matrix. The effect of carbon nanotubes on properties of silicon nitride nano-ceramics and the strengthening mechanism were researched.

2. Experimental procedure

2.1. Charateristics of the original powders

Nanosized Si_3N_4 and multi-wall carbon nanotubes were used to prepare silicon nitride matrix ceramic composites. The silicon Nitride powders are mixture of amorphous Si_3N_4 and α-phase Si_3N_4, and the amount ratio of amorphous Si_3N_4 to α-phase Si_3N_4 is 4. Nanosized yttrium oxide and alumina powders were used as sintering aids and both of their amounts were 5wt%. Details about composition of the original powder mixtures are shown in table 1.

Table 1. Composition of the original powder mixtures in all systems (in wt%).

Specimen	Si_3N_4	Al_2O_3	Y_2O_3	CNT
A	90	5	5	0
B	89	5	5	1
C	87	5	5	3
D	85	5	5	5

The morphology of silicon nitride powders and carbon nanotubes is determined by TEM (H–800 Transmission Electron Microscopy) analysis, as shown in figure1.

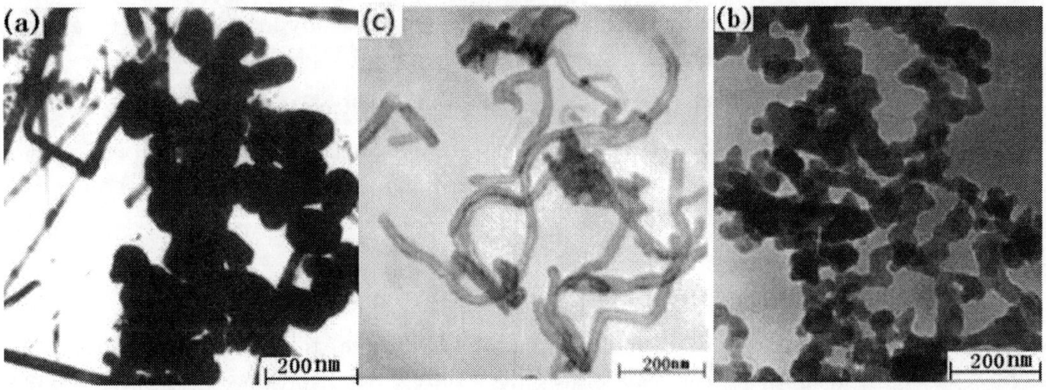

Figure 1. TEM images of original Si_3N_4 powders and CNTs (a) α-Si_3N_4 particles, (b) amorphous Si_3N_4 particles, (c) CNTs

It can be seen from the TEM images that the amorphous Si_3N_4 powders are spherical particles (20~50nm), while the α-Si_3N_4 powders consist of hexagonal particles(50~100nm) as well as some whisker-like particles with diameter of 20nm or so. The original CNTs are not straight and tangle together, and their diameter is between 20nm~40nm and length is 2~5um. The CNTs used in the experiments were acid treated with nitric acid in order to remove catalysis particles and improve the surface properties, which contributes to a good bonding between CNT and ceramic matrix. In addition, the walls of CNTs become straight and length becomes shorter after the acid treatment that contributes to the homogeneous distribution of CNTs on the silicon nitride matrix.

2.2. Preparation of samples

First, the Si_3N_4, Al_2O_3 and Y_2O_3 powders mixtures were supersonically dispersed for 1h and ball milled with agate balls in anhydrous ethanol for 16h, then the CNTs were added into the powder mixtures. Second, the CNTs and powder mixtures were supersonically dispersed for 1h and ball-milled in ethanol for 2h so as to get a better homogenous dispersion, then dried at 80℃ in a vacuum furnace and sieved with 200-mesh sieve. The dried mixtures were sintered at 1700℃ in a hot-press furnace which is full of nitride gas, and the applied pressure is 30MPa and holding time is 1h.

Testing specimens with dimension of 3mm×4mm×36mm were prepared by diamond cutting from the sintered materials. All surfaces of the samples were finely ground with a diamond wheel and polished with 1um diamond pastes.

2.3. Structure observation and properties testing

The relative density of the sintered samples was measured by the Archimedes method in distilled water. The phase compositions of the sintered samples were identified by X-ray diffractometer (Rigaku, D/max-rB x-ray diffractometer). The microstructure and fracture surface morphology of the samples were observed by a scanning electron microscope (SEM, LEO-1530VP) after coating gold.

The flexural strength was measured on a universal testing machine (MTS tester) using a 3-point bending method, and the span length and cross-head speed were 30mm and 0.2 mm/min respectively. The hardness and the fracture toughness were measured on a HV120 hardness tester by the Vickers indentation technique, with a load of 3kgf and 5kgf respectively.

3. Results and discussion

3.1. Microstructure and phase

X-ray diffraction (XRD) patterns of the samples are shown in figure 2. A phase transformation from α-Si_3N_4 to β-Si_3N_4 takes place during sintering. The crystalline phases of carbon nanotube–silicon nitride nano-ceramic matrix composites are α-Si_3N_4, β-Si_3N_4 and a little amount of Si_2N_2O. No characteristic peak of CNT is observed. The existence of Si_2N_2O can be explained by the following reaction which occurs during sintering process when the temperature reaches 1600 ℃ [17]:

$$2Si_3N_4\ (s) + 1.5O_2\ (g) = 3\ Si_2N_2O\ (s) + N_2\ (g) \quad (1)$$

In reaction (1), the exit of O_2 may due to the insufficient vacuum of sintering equipment. In addition, the oxygen may interfuse into the powder mixtures during the dispersing and drying process.

When the temperature is higher than 1650 ℃, another reaction happens and the amount of Si_2N_2O gradually decreases:

$$2Si_2N_2O\ (s) = Si_3N_4\ (s) + SiO\ (g) + 0.5O_2\ (g) \quad (2)$$

The morphology of fracture surface indicates that the CNTs are still present in the structure after sintering, as shown in figure 3. The monolithic silicon nitride nano-ceramic shows an inter-granular fracture mode and there is no obvious β-Si_3N_4 grains pullout phenomenon in the fracture surface, whereas all composites show a different degree of CNTs pullout effect. The most CNTs are vertical to the fracture surface and parallel to the pressed surface which is beneficial to the improvement of mechanical properties. The CNTs homogeneously disperse in silicon nitride matrix when its amount is lower, but the agglomerated CNTs can be seen from the fracture surface when the amount of CNTs is 5wt.%.

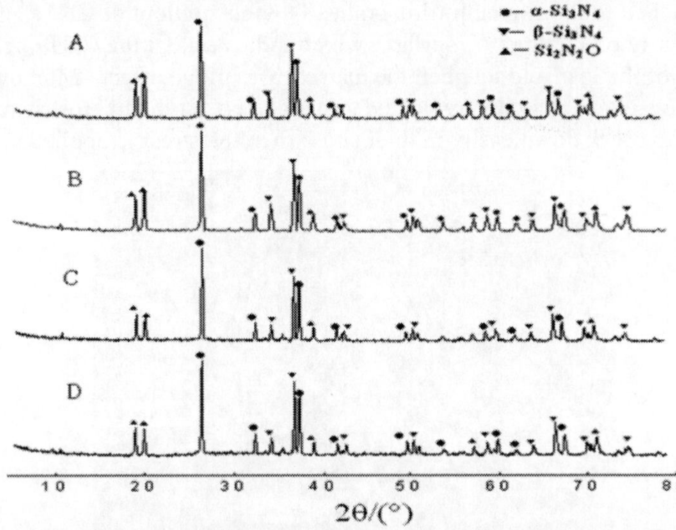

Figure 2. X-ray diffraction patterns of sample A, B, C and D.

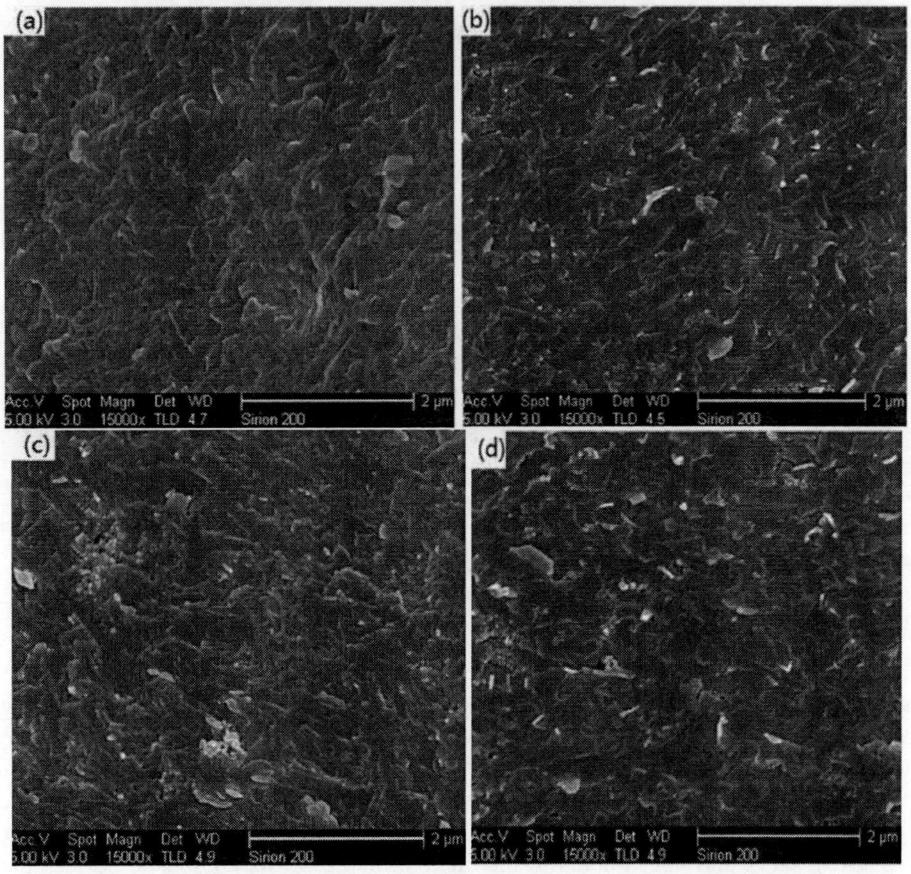

Figure 3. SEM micrographs of fracture surfaces at low magnification
(a) sample A, (b) sample B, (c) sample C, (d) sample D

Figure 4 shows the fracture surface micrograph of sample C at high magnification and EDX spectrum of selected areas. The CNTs are mainly located at grain boundaries of silicon nitride matrix and they are well attached to the silicon nitride grains. Obvious pullout of CNTs and short outcrops of CNTs can be seen clearly on the fracture surface, which indicates that the CNTs bear significant stress by shearing a portion of the load and toughen the matrix by a bridge effect. Moreover, the degradation of CNTs also can be observed, which may be due to the defects on the microstructure of CNTs. These defects make the CNTs break down easily in their sites when the stress is applied.

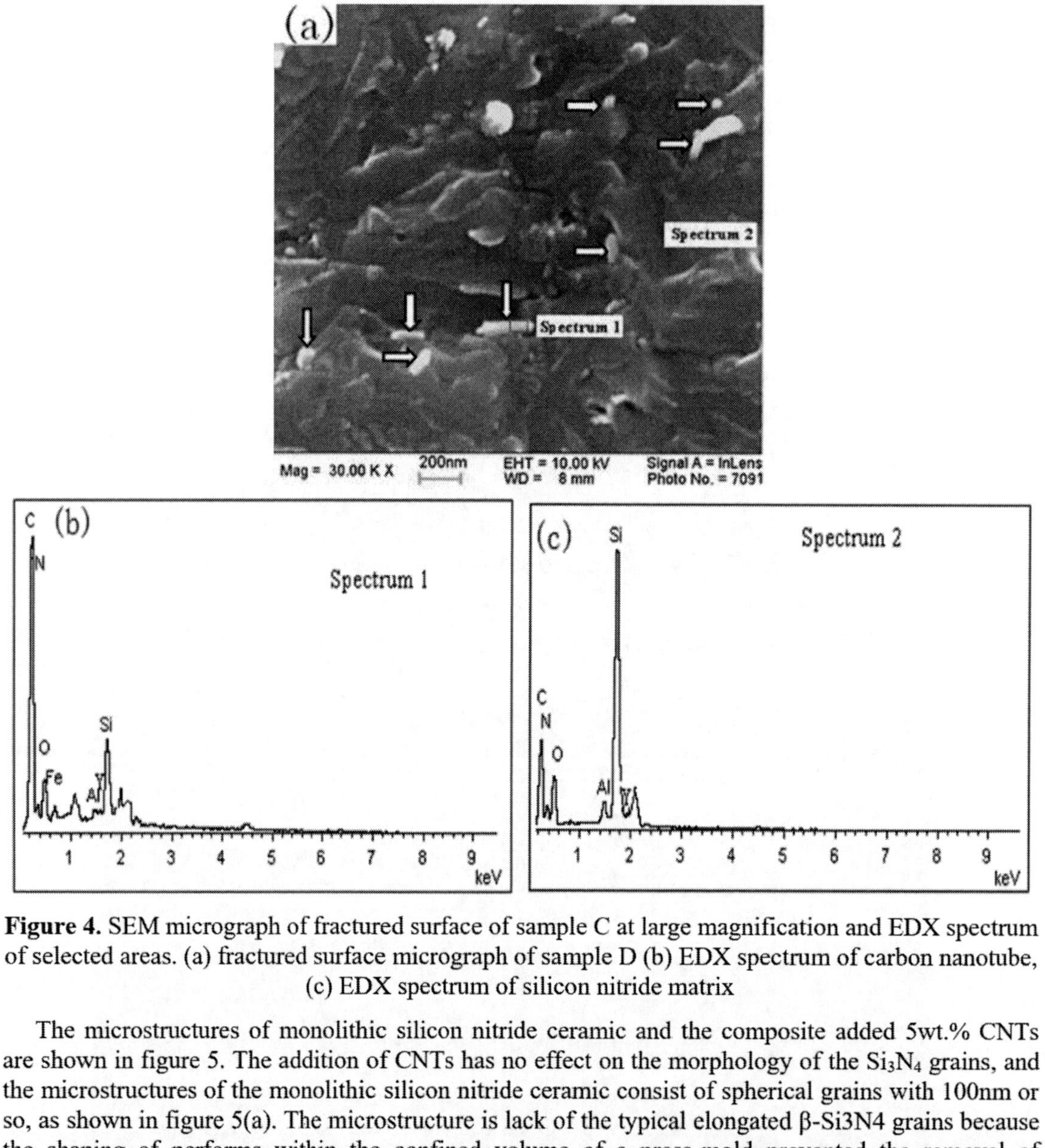

Figure 4. SEM micrograph of fractured surface of sample C at large magnification and EDX spectrum of selected areas. (a) fractured surface micrograph of sample D (b) EDX spectrum of carbon nanotube, (c) EDX spectrum of silicon nitride matrix

The microstructures of monolithic silicon nitride ceramic and the composite added 5wt.% CNTs are shown in figure 5. The addition of CNTs has no effect on the morphology of the Si_3N_4 grains, and the microstructures of the monolithic silicon nitride ceramic consist of spherical grains with 100nm or so, as shown in figure 5(a). The microstructure is lack of the typical elongated β-Si3N4 grains because the shaping of performs within the confined volume of a press-mold prevented the removal of monoxide from the starting powders [18]. From figure 5(b) it can be seen that there are agglomerated CNTs, separated CNTs and micro-pores in the microstructure of the composite besides the spherical grains.

IOP Conf. Series: Materials Science and Engineering **292** (2017) 012076 doi:10.1088/1757-899X/292/1/012076

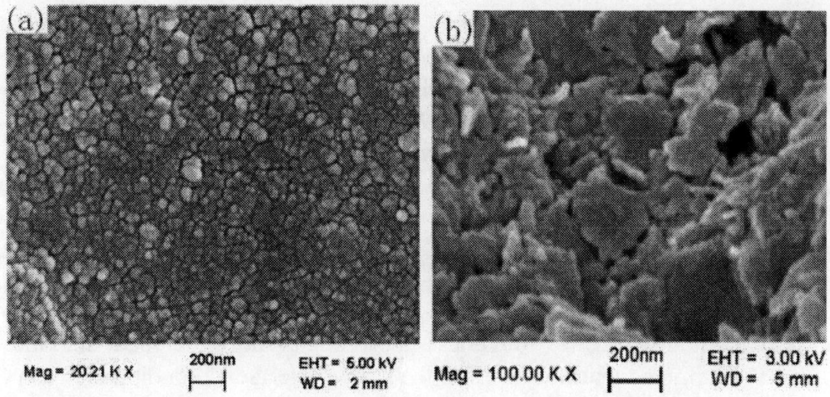

Figure 5. SEM microstructure micrographs of sample A and sample D (a) sample A (b) sample D.

3.2. Mechanical properties

Flexural strength, Vickers hardness and fracture toughness of the samples were tested and the effects of CNTs amount on the mechanical properties were researched, as shown in figure 6. The flexural strength and fracture toughness of carbon nanotube-silicon nitride nano-ceramic matrix composites both increase initially and then decrease with the increase in CNTs amount. The flexural strength and the fracture toughness both reach a maximum when CNTs content is 3wt.%.

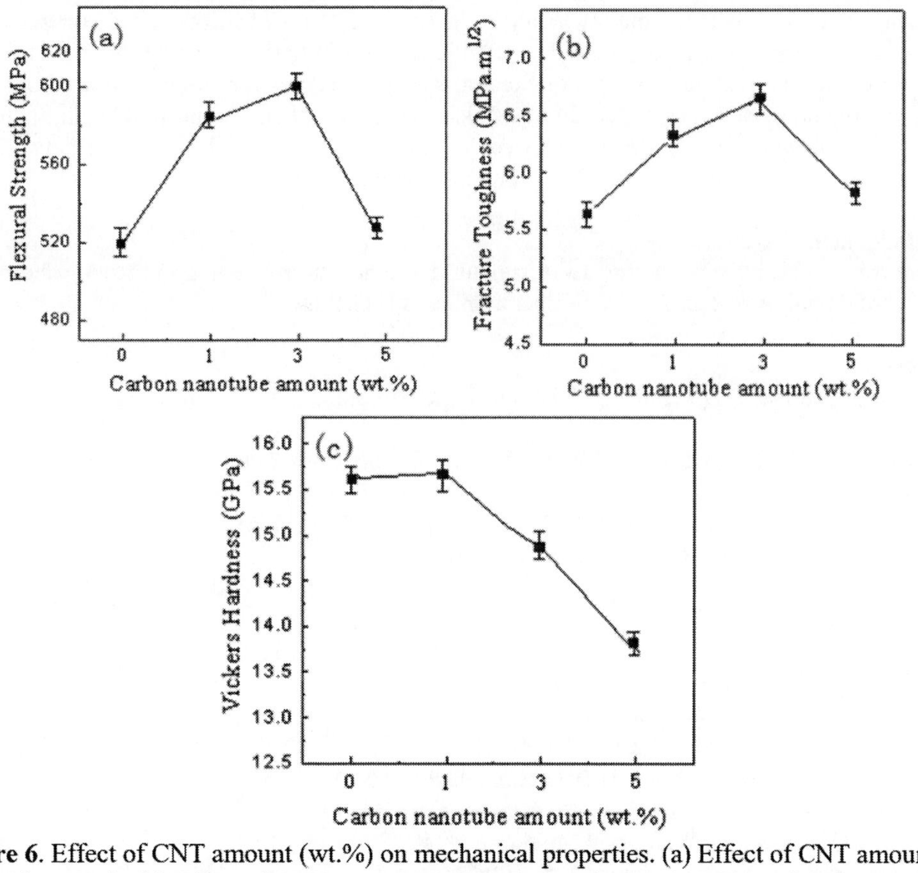

Figure 6. Effect of CNT amount (wt.%) on mechanical properties. (a) Effect of CNT amount on flexural strength, (b) Effect of CNT amount on fracture toughness, (c) Effect of CNT amount on Vickers hardness.

The increase in strength is mainly related to the strengthening effect caused by the CNTs which can bear significant stress by shearing a portion of the load though pullout on the fracture surface. In fiber reinforced composites, a principal source of increased fracture toughness is crack deflection caused by reinforcing fiber ahead of a propagating crack [19]. The crack deflection caused by CNTs prolonged the propagation path of the crack and more energy will be dissipated, which increase the fracture toughness of silicon nitride nanocomposites. In addition, because of the different thermal expansion coefficients between the Si_3N_4 and CNTs, the internal residual stress generated during cooling process will induce the crack to propagate toward the CNTs and diverge, which is also contributed to the increase of the fracture toughness. But when the CNTs content is high, the agglomeration of CNTs causes the decrease in relative density and the micro-pores in microstructure that lead to the decrease in the flexural strength and fracture toughness. So it can be concluded that the homogenous distribution of CNTs within the matrix as well as the strong interfacial bonding between CNTs and the matrix are important factors affecting the strengthening effect of the CNTs. The Vickers hardness initially has little change, but it start to gradually decrease when the CNTs content is above 1wt.%. This is mainly due to the decrease in the relative density and the agglomeration of CNTs.

4. Conclusion

Carbon nanotube–silicon nitride nano-ceramic matrix composites were fabricated by hot-pressing nano-sized Si_3N_4 powders and carbon nanotubes at 1700℃. The optimal manufacturing process has been performed to homogenously disperse the carbon nanotubes in the silicon nitride matrix. The phase composition, microstructure and mechanical properties of the sintered materials are investigated.

The microstructure of sintered composites is mainly constituted of α-Si_3N_4, β-Si_3N_4, Si_2N_2O and carbon nanotubes. The CNTs are mainly located at grain boundaries of silicon nitride matrix.

The flexural strength and fracture toughness both increase initially then decrease with the increase of CNTs amount. The maximum values of flexural strength and fracture toughness are both obtained in silicon nitride nano-ceramics added 3wt.% CNTs. The strengthening and toughening mechanisms are CNTs pullout and crack deflection caused by the CNTs. The Vickers hardness values of the composites start to decrease when the CNTs content is above 1wt.%.

Acknowledgement

This work was financially supported by the natural science foundation provided by the education department of Anhui province, and the project number is KJ2014A215.

References

[1] Shuzo K, Manuel E B, Cecilia V, Kiyoshi H and Motohiro T 1997 *J. Eur. Ceram. Soc.* **17** 1841-47
[2] Hirano T, Niihara K, Ohji T and Wakai F 1996 *J. Mater. Sci. Lett.* **15** 505-07
[3] Wang Y X, Wang L P and Xue Q J 2011 *Surf. Coat. Tech.* **205(8/9)** 2770-77
[4] Tapasztó O, Kun P and Wé ber F 2011 *Ceram. Int.* **37(8)** 3457-61
[5] An Y M, Xu X H and Gui K X 2016 *Cera. Int.* **42** 14066-70
[6] Karolina P S T, Elisa P, Claudio B, Sara B, Matteo P and Paolo F 2016 *Mater. Sci. & Eng. A* **659** 158-64
[7] Yang W Y, Gao F G and Xu C G 2010 *Solid State Sci.* **12(9)** 1692-95
[8] Ahmad N and Sueyoshi 2010 *Ceram. Int.* **36(2)** 491- 96
[9] Rochie S Ann 2000 *Chim. Sci. Mater.* **25** 529-32
[10] Sark S and Das P K 2014 *Ceram. Int.* **40(5)** 7449-58
[11] Kothari A, Hu S H and Xia Z H 2012 *Acta. Mater.* **60(8)** 3333-38
[12] Nima N, Saeed Z, and Sara A 2012 *Mater. & Desi.* **34** 1-14
[13] Zhan G D, Kuntz J D, Wan J and Mukherjee A K 2003 *Nat. Mater.* **2** 38-42
[14] Deng C F, Wang D Z and Zhang X X, 2007 *Mater. Sci. & Eng. A* **444** 138-45
[15] Ma R Z, Wu J, Wei B Q, Liang J and Wu D H 1998 *J. Mater. Sci.* **33** 5243-63

[16] Liu X J, Huang Z Y, Xiang C S and Pan Y B 2006 *J. Chin. Ceram. Soc.* **2** 1363-66
[17] Szépvölgyi J and Mohai I 1999 *Ceram. Int.* **25** 717-21
[18] Paranosenkov V P , Kelina I Yu, Plyasunkova L A and Bykov Yu V 2003 *Ref. Ind. Ceram* **44(4)** 223–26
[19] Ohji T, Jeong Y K, Choa Y H and Niihara K 1998 *J. Am. Ceram. Soc.* **81(6)** 1453-60

IOP Conf. Series: Materials Science and Engineering **292** (2017) 012077 doi:10.1088/1757-899X/292/1/012077

Study the effect of $Ni_{0.5}Zn_{0.5}Fe_2O_4$ doped electromagnetic and microwave absorbing properties on RGO based composites

B Mei, X J Su, S Bi[1], G L Hou and J Li

Xi'an Research Institute of High-Tech, Xi'an, 710025, China

[1] E-mail: bisong5011@163.com

Abstract. $Ni_{0.5}Zn_{0.5}Fe_2O_4$ is one of the traditional microwave absorbing materials, and it has attract the focus of the whole world in the past years. The composites behave well in absorbing areas due to its coexistence of dielectric property and ferrimagnetism. The graphene is one of the new absorbing materials and it has been used in varies areas due to its excellent electrical conductivity, thermal conductivity and mechanical properties. In the work, the $Ni_{0.5}Zn_{0.5}Fe_2O_4$ nanoparticles were surficiently compounded with the reduced graphene (RGO), and the microstructure and microwave absorbing performance of the composites were characterized. Consequently, the results indicated that the as-prepared materials perform well in microwave absorption area.

1. Introduction

The electromagnetic pollution has turned out to be one of the new environmental pollution in the pasted years duo to the development of the electronics technique. The existence of the weaponry have faced the great challenge due to the continuous renewal of the radar detection technique. Therefore, the study of the microwave absorbing materials that with excellent microwave absorbing properties is of great importance [1-6]. The magnetic loss, resistance loss and dielectric loss characteristics account for the major role in microwave absorption abilities [7-9], then it is a point to combine the different kinds of microwave absorbing materials together to prepare the composites that is of better absorbing properties.

Reduced Graphene (RGO) have been widely applied in microwave absorption due to its excellent electrical conductivity and mechanical properties, and once it were compouned with other absorbing materials, the microwave absorption performace of the composites may be strengthened [10-12]. For example, once oxidized graphene is doped into the barium ferrite, thus the absorbing frequency band is widened and the absorbing abilities is increased [13-15]. Moreover, ferrite serve as a traditional microwave absorption due to its coexistence of dielectric property and ferrimagnetism. And the ferrite possess excellent microwave absorption characteristics when the Nickel and Zinc were added into the composites. Besides, once the material is prepared in nano-sized, the microwave absorbing ability would be greatly enhanced [16]. For example, the $BaFe_{11}(Ti_{0.5}Mn_{0.5})_4O_{19}$ nanoparticles have been prepared by coprecipitation method, and the result indicates that the particles were nano-sized, and the qualified absoption bandwidth (<-10dB) coverd 9.2~12.2GHz. In the work by Chung and his coworkers, hydrazine was applied in the fabrication of Fe_3O_4/RGO/polyaniline composites and the results suggest that the composites of electrical conductivity up to 2.6 S/cm exhibited absorption shielding ~26 dB when the composite thickness was 2.5 mm.

In this work, Ni-Zn ferrite were compounded with RGO by sufficient ultrasonic agitation method. And the microstructure of the as-prepared samples were characterized by XRD and SEM. Then, the electromagnetic parameters of the composites were tested by VNA. Consequently, the microwave absorption performance was discussed.

2. Exprimental

The RGO were prepared by hummers method, the mixture of 1 g $NaNO_3$ and 2 g graphite powders were added into concentrated 120 ml H_2SO_4 under stirring at 0 °C for 0.5 h in an ice bath. Then stirred the solution at 0°C for 1 h. Subsequently, 6 g $KMnO_4$ powders were slowly added into the mixture. Until stirring for 2h, the mixture was settled at 30 °C for 0.5 h and 150 ml deionized water was slowly added. After the mixture was heated and kept at 98 °C for 15 min, 50 ml 5% H_2O_2 was slowly added into the solution. Then, the resulting solution was washed with 5% HCl and deionized water for 5 cycles and the final solution was the GO aqueous solution.

The Ni-Zn ferrite were prepared by sol-gel method, citric acid monohydrate were added to the mixed solution of zinc nitrate and nichel nitrate (quantitative weigh) and regulated the pH to 6 by ammonia solution, kept heated reaction until the solution turned to sol rubber, then keep heating the solution at 80 °C until the samples turn into xerogel, finally, the samples were calcined in muffle for two hours under 650 °C, then would get the Ni-Zn ferrite powder. The RGO and the as-prepred ferrite were sufficient ultrasounded for 50 min, and then add the ferrite into the RGO solution. Expecially, in experiment, the mass ratio of the RGO in the composites is 5%. Followed by fully stirred method. Then the mixture were freeze drying, and the RGO/$Ni_{0.5}Zn_{0.5}Fe_2O_4$ were gained.

The microstructure of the as-prepared samples were investigated by X-ray diffraction (XRD, Rigaku Ultima IV, Cu-Ka) in the range of 5°~90°, with a scan speed of 3 s and a step size of 0.02° in 2θ. The morphologies of the as-prepared samples were investigated by scanning electron microscope (Hitachi S-4700 SEM). And the electromagnetic properties of the samples were tested by the vector network analysis (VNA, KEYSIGHT E5071C) via coaxial method.

3. Results and discussion

The XRD patterns of the RGO/$Ni_{0.5}Zn_{0.5}Fe_2O_4$ nanoparticles were shown in figure 1. Apparently, the wide peak between 8° and 20° indicated the existence of RGO in the nanoparticles, and all the peaks that were shown between 20° and 8° indicated the existence of $Ni_{0.5}Zn_{0.5}Fe_2O_4$. Therefore, the results indicate the $Ni_{0.5}Zn_{0.5}Fe_2O_4$ nanoparticles were successfully grown with the RGO, and the RGO/ $Ni_{0.5}Zn_{0.5}Fe_2O_4$ nanoparticles were gained.

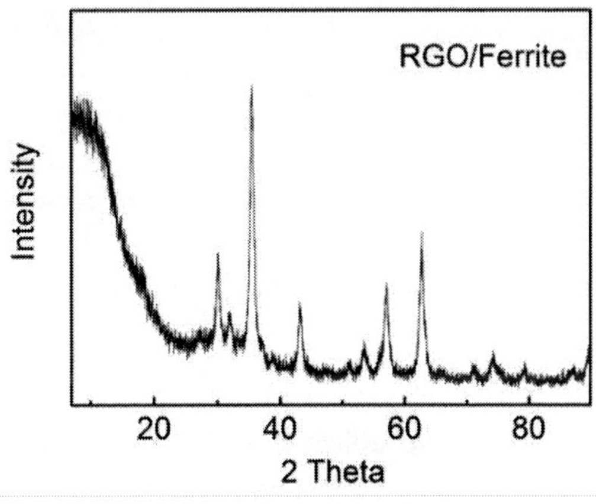

Figure 1. XRD patterns of the RGO/$Ni_{0.5}Zn_{0.5}Fe_2O_4$ nanoparticles.

As is shown in figure 2, it is obvious that the size of the $Ni_{0.5}Zn_{0.5}Fe_2O_4$ is within 100 nm, indicating the nano-sized $Ni_{0.5}Zn_{0.5}Fe_2O_4$ particles were gained. Moreover, the RGO is apparently to see in the picture, and the $Ni_{0.5}Zn_{0.5}Fe_2O_4$ nanoparticles is well grown on the surface of RGO, indicating the RGO and Ni-Zn ferrite are well growed together by sufficient ultrasonic agitation method.

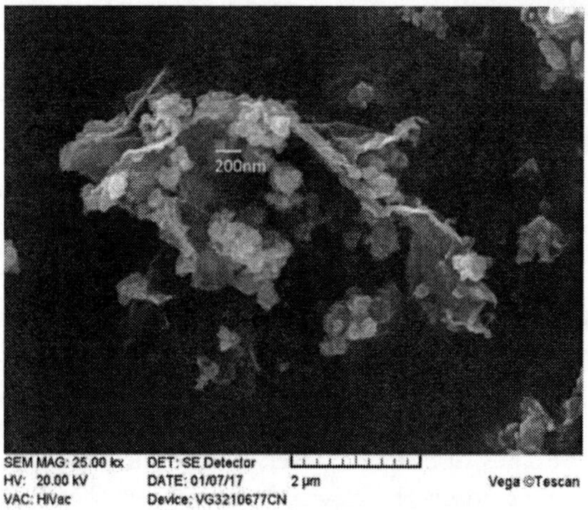

Figure 2. SEM image of the RGO/$Ni_{0.5}Zn_{0.5}Fe_2O_4$ nanoparticles.

In the preparation of the samples, paraffin was used as the matrices, and the mass ratio of the composites in the samples were 90%. The inner diameter of the specimens is 3 mm, and the external diameter is 6.9 mm. Consequently, the permittivity and permeability of the RGO/$Ni_{0.5}Zn_{0.5}Fe_2O_4$ nanoparticles were gained in the range of 2~12.4 GHz (figure 3). As is shown in figure 3a, the real permittivity of the composites is around 4.8, and the imaginary permittivity of the composite is around 0.1. Besides, the real permeability of the composites is around 1.2, and the imaginary permeability of the composites is around 0.2.

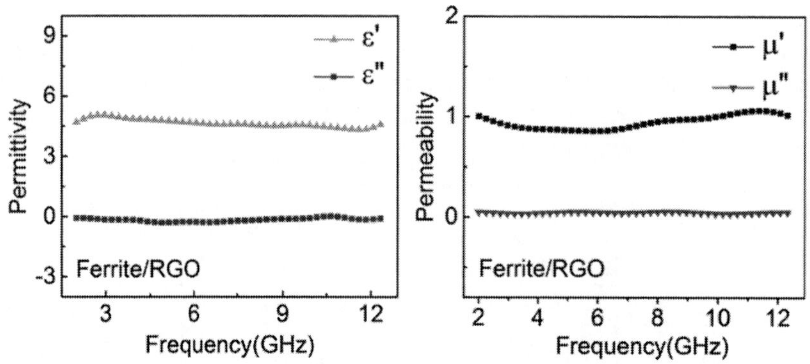

Figure 3. Permittivity and permeability of the RGO/$Ni_{0.5}Zn_{0.5}Fe_2O_4$ nanoparticles (a: permittivity; b: permeability).

The reflection loss (R_L), which could be a reflection of the ability of the microwave absorption performance of the absorbing layers on the ideal metal plane, was achieved by the relation

$$R_L = 20\log(|Z_{in}-1|/|Z_{in}+1|). \quad (1)$$

Moreover, the normalized input impedance (Z_{in}) of absorption layer is given as

$$Z_{in} = (\mu_r/\varepsilon_r)^{1/2}\tan h[j2\pi fd(\mu_r\varepsilon_r)^{1/2}/c] \quad (2)$$

where c is the light velocity, f the microwave frequency, d the thickness of the absorber, μ_r and ε_r the complex permittivity and permeability of the specimens, respectively [17]. Therefore, the microwave absorption performance of the as-prepared samples were characterized.

Figure 4 exhibits the microwave absorption performance of the RGO/$Ni_{0.5}Zn_{0.5}Fe_2O_4$ nanoparticles. The microwave absorbing abilities at different thickness are shown in figure 4a, it is obvious that the composites possess best microwave absorption performance at the thickness of 6 mm. And the absorption intension reaches -19.5dB at 5.68GHz, and the qualified bandwith of the sample reaches 1.6 GHz. Besides, the three-dimensional microwave absorption performance is shown in figure 4b.

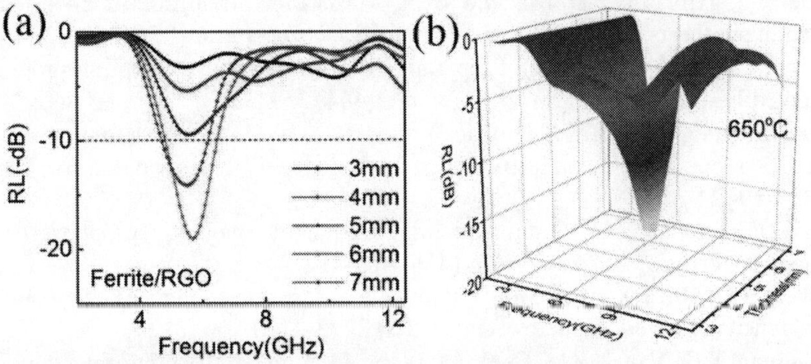

Figure 4. Microwave absorption performance of the RGO/$Ni_{0.5}Zn_{0.5}Fe_2O_4$ nanoparticles.

According to the discussion of the electromagnetic property and microwave absorbing ability of the as-prepared RGO/$Ni_{0.5}Zn_{0.5}Fe_2O_4$ nanoparticles, the results indicate that it is a potential method to compound the RGO with the $Ni_{0.5}Zn_{0.5}Fe_2O_4$ composite to synthesis a new microwave absorption material. The SEM results indicate that the $Ni_{0.5}Zn_{0.5}Fe_2O_4$ nanoparticles are well compounded with the RGO. Therefore, the as-synthesised RGO/$Ni_{0.5}Zn_{0.5}Fe_2O_4$ nanoparticles might serve as a potential material in microwave absorbing area.

4. Conclusion

The sufficient ultrasounded method were conducted to synthesis the RGO/$Ni_{0.5}Zn_{0.5}Fe_2O_4$ nanoparticles. The microstructure of the as-prepared composites were characteriazed. The results indicate that the $Ni_{0.5}Zn_{0.5}Fe_2O_4$ is well compounded with the RGO, and the as-prepared samples possess well electromagnetic properties and microwave absoption performance. The composites show best wave absorption performance at the thickness of 6mm, and the qualified bandwith of the bandwith reaches 1.6 GHz. Moreover, the absorption intension of the composites reaches -19.5 dB at 5.68 GHz. Thus the study provide a guide for preparing the composites that possess excellent microwave absorption performance.

5. References

[1] Bi S, Ma L, Mei B, Tian Q, Liu C H, Zhong C R, Xiao Y D, et al. 2014 Silicon carbide/carbon nanotube heterostructures: Controllable synthesis, dielectric properties and microwave absorption *Adv. Powder Technol.* **25** 1273-1279

[2] Moayad H F, Sahrim H A, Mustaffa H A, Rozaidi R, Abdul H S, Ayman A, Sivanesan A, et al 2014 Preparation, thermal, magnetic and microwave absorption properties of thermoplastic natural rubber matrix impregnated with NiZn ferrite nanoparticles *J. Alloys Compd.* **(96)** 103-108

[3] Khairy M 2014 Synthesis, characterization, magnetic and electrical properties of polyaniline/$NiFe_2O_4$ nanocomposite *Synth. Metals* **(189)** 34-41

[4] Zhao Y, Shen M Y, Li Z X, SUN R, Xia A L, Liu X G, et al 2016 Green synthesis and enhanced microwave absorption property of reduced graphene oxide-$SrFe_{12}O_{19}$

nanocomposites *J. Alloys Compd.* (**689**) 1037-1043

[5] Zhang H, Hong M, Chen P, Xie A J, Shen Y H, et al 2016 3D and ternary rGO/MCNTs/Fe_3O_4 composite hydrogels: Synthesis characterization and their electromagnetic wave absorption properties *J. Alloys Compd.* (**665**) 381-387

[6] Khadijenh D, Elham Y, Farshid N A, Seyed A M, Naeimeh B L, et al 2015 Radar Absorption Properties of $Ni_{0.5}Zn_{0.5}Fe_2O_4$/PANI/epoxy Nanocomposites *J. Chin. Chem. Soc.* (**62**) 826-831

[7] Ghazzawy E H, Amer M A 2016 Structural, elastic and magnetic studies of the as-synthesized $Co_{1-x}Sr_xFe_2O_4$ nanoparticles *J. Alloys Compd.* (**135**) 293-303

[8] Albuquerque A S, Ardisson J D, Macedo W A, Maria C M, et al 2000 Nanosized powders of NiZn ferrite: synthesis, structure, and magnetism *J. Appl. Phys.* (**87**) 4352-4356

[9] Li X A, Hah X J, Tah Y J, et al 2008 Preparation and microwave absorption properties of Ni-B al-loy-coated Fe_3O_4 particles *J. Alloys Compd.* (**464**) 352-356

[10] Li Z J, Hou Z L, Song W L, Liu X D, Cao W Q et al 2016 Unusual continuous dual absorption peaks in Ca-doped $BiFeO_3$ nanostructures for broadened microwave absorption *Nanoscale* (**19**) 10415-10424

[11] Ghaemi A, Morisako K 2008 Static and high frequency magnetic properties of Mn-Co-Zr substituted Ba-ferrite *J. Alloys Compd.* (**456**) 485-491

[12] Mu G H, Shen H G, Qiu J X, et al 2006 Microwave absorption properties of composite powders with low density *Applied surface science* (**253**) 2278-2281

[13] Song W L, Guan X T, Fan L Z, Cao W Q, Wang C Y, et al 2015 Turning three-dimensional textures with graphene aerogels for ultra-light flexible graphene/texture composites of effective electromagnetic shielding *Carbon* (**93**):151-160

[14] Liu J, Cao M S, Luo Q, Shi H L, Wang W Z, Yuan J 2016 Electromagnetic Property and Tunable Microwave Absorption of 3D Nets from Nickel Chains at Elevated Temperature *Acs Appl. Mater. Interfaces* **8** 22615-22622

[15] Tang H, Jian X, Wu B, Liu S Y 2016 Fe_3C/helical carbon nanotube hybrid: Facile synthesis and spin induced enhancement in microwave absorbing properties, Compos. Part B **107** 51-58

[16] Dong C S, Wang X, Zhou P H, Liu T, Xie J L 2014 Microwave magnetic and absorption properties of M-type ferrite $BaCo_xTi_xFe_{12-2x}O_{19}$ in the Ka band *J. Magn. Magn. Mater.* **354** 340-344

[17] Wang Y P, Peng Z, Jiang W 2016 Controlled synthesis of Fe_3O_4@SnO_2/RGO nano composite for microwave absorption enhancement *Ceram. Int.* **42** 10682-10689

IOP Conf. Series: Materials Science and Engineering **292** (2017) 012078 doi:10.1088/1757-899X/292/1/012078

Numerical analysis of temperature field in the high speed rotary dry-milling process

N X Wu, L J Deng and D H Liao[1]

School of Mechanical and Electronic Engineering, Jingdezhen Ceramic University Jingdezhen, 333403

[1] E-mail: 13320149686@163.com

Abstract. For the effect of the temperature field in the ceramic dry granulation. Based on the Euler Euler mathematical model, at the same time, made ceramic dry granulation experiment equipment more simplify and established physical model, the temperature of the dry granulation process was simulated with the granulation time. The relationship between the granulation temperature and granulation effect in dry granulation process was analyzed, at the same time, the correctness of numerical simulation was verified by measuring the fluidity index of ceramic bodies. Numerical simulation and experimental results showed that when granulation time was 4min, 5min, 6min, maximum temperature inside the granulation chamber was: 70℃, 85℃, 95℃. And the equilibrium of the temperature in the granulation chamber was weakened, the fluidity index of the billet particles was: 56.4. 89.7. 81.6. Results of the research showed that when granulation time was 5min, the granulation effect was best. When the granulation chamber temperature was more than 85℃, the fluidity index and the effective particles quantity of the billet particles were reduced.

1. Introduction

At present, the raw material of building ceramic bodies were mainly made by wet ball-milling technology, the high energy consumption, serious pollution and other problems do not conform to the environment protection policy which was made by the country [1]. Building ceramic billets dry granulation technology could solve such problems fundamentally, but the building ceramic bodies still had some problems, such as irregular surface shape, high hardness and other defects. These problems will lead to some bad consequences, such as ceramic body forming process easily stratified, low intensity, surface roughness and so on [2-3]. Temperature will affect the plasticity of ceramic bodies in a certain extent. Especially, when the temperature exceeded 80℃, the plasticity of ceramic bodies would be greatly damaged [4].

The author established the dry granulation test platform and found that: the temperature of the granulation chamber's wall rose quickly in the dry-milling process. In order to find the relationship between temperature field and granulation effect in dry granulation process, the author combined experimental data to establish the three-dimensional physical model of dry-milling granulation chamber which based on computational fluid dynamics method [5] to simulated and analyzed the pattern between the temperature field and granulation time. It effectively controlled the dry granulation time to prevent the damage of billet particles, improved the dry milling granulation effect and provided theoretical guidance for the further promotion of dry milling technology in building ceramics industry.

2. Model establishment

2.1. Mathematical model

The size of the building ceramic materials before granulation was the um level, it could be assumed that the building ceramic materials, as the same as quasi-fluid phase, when the temperature field mainly rose in the mixing stage, only the internal granulation chamber had air phase and powder particle phase. The initial particle size distribution of the powder particles accounted for about 1/4 of the volume in the granulation chamber, therefore, the air phase was selected as the main phase and the powder particle phase was selected as the secondary phase. The mathematical model of the temperature field of dry milling process was established by Euler - Euler model [6-7]. In the iterative calculation process, air phase and powder particles coexisted and permeated each other. Their speed, volume distributions were relatively independent. The air-phase and powder-particle phases had their continuity equations, momentum conservation equation and the energy conservation equation [8-9], specific theoretical deduction equation as follows:

(1) Continuity equation

Continuity equation of air phase：

$$\frac{\partial}{\partial t}\left(\alpha_g \rho_g\right)+\nabla\cdot\left(\alpha_g \rho_g \vec{v}_g\right)=\sum_{g=1}^{n}\dot{m}_{gs} \tag{1}$$

Continuity equation of powder particle：

$$\frac{\partial}{\partial t}\left(\alpha_s \rho_s\right)+\nabla\cdot\left(\alpha_s \rho_s \vec{v}_s\right)=\sum_{s=1}^{n}\dot{m}_{sg} \tag{2}$$

Wherein: α_g, α_s - Air phase, powder particle phase volume distribution percentage; ρ_g, ρ_s- the density of air phase and the powdery particle phase; $\vec{v}_g$, $\vec{v}_s$ - Air phase, and power particle phase velocity; $\dot{m}_{gs}$, $\dot{m}_{sg}$ - Mass transfer enthalpy of air phase and powder particle phase($\dot{m}_{sg}$=-$\dot{m}_{gs}$).

(2) Momentum conservation equation

The momentum equation of the air phase:

$$\frac{\partial}{\partial t}\left(\alpha_g \rho_g \vec{v}_g\right)+\nabla\cdot\left(\alpha_g \rho_g \vec{v}_g \vec{v}_g\right)=-\alpha_g \nabla p+\nabla\cdot\bar{\bar{\tau}}_g+\sum_{g=1}^{n}\left(\vec{R}_{gs}+\dot{m}_{gs}\vec{v}_{gs}\right)+\alpha_g \rho_g\left(\vec{F}_g+\vec{F}_{lif,g}+\vec{F}_{Vm,g}\right) \tag{3}$$

Wherein, τ_g- Strain tensor of air phase.

$$\bar{\bar{\tau}}_g=\alpha_g \mu_g\left(\nabla\vec{v}_g+\nabla\vec{v}_g^T\right)+\alpha_g\left(\lambda_g-\frac{2}{3}\mu_g\right)\nabla\cdot\vec{v}_g\bar{\bar{I}} \tag{4}$$

The momentum equation of powder particle phase:

$$\frac{\partial}{\partial t}\left(\alpha_s \rho_s \vec{v}_s\right)+\nabla\cdot\left(\alpha_s \rho_s \vec{v}_s \vec{v}_s\right)=-\alpha_s \nabla p+\nabla\cdot\bar{\bar{\tau}}_s+\sum_{s=1}^{n}\left(\vec{R}_{sg}+\dot{m}_{sg}\vec{v}_{sg}\right)+\alpha_s \rho_s\left(\vec{F}_s+\vec{F}_{lif,s}+\vec{F}_{Vm,s}\right) \tag{5}$$

Wherein, τ_s- Strain tensor of powder particle Phase.

$$\bar{\bar{\tau}}_s=\alpha_s \mu_s\left(\nabla\vec{v}_s+\nabla\vec{v}_s^T\right)+\alpha_s\left(\lambda_s-\frac{2}{3}\mu_s\right)\nabla\cdot\vec{v}_s\bar{\bar{I}} \tag{6}$$

Wherein: μ_g、λ_g- The molecular viscosity and the bulk viscosity of the air phase,，μ_s、λ_s、- The molecular viscosities and bulk viscosities of the powdery particles; $\vec{F}_g$ - The volume force of the air phase, $\vec{F}_s$ - The Volume Force of Powder Particle Phase; $\vec{F}_{lif,g}$ - The lift of the air phase, $\vec{F}_{lif,s}$ - The lift of particle phase; $\vec{F}_{vmg}$ - The virtual mass force of the air phase, $\vec{F}_{vms}$ - The Virtual Mass Force of Particle Phase; $\vec{R}_{gs}$、$\vec{R}_{sg}$ - Interphase interaction forces, which were closed to each other ($\vec{R}_{gs}$= -$\vec{R}_{sg}$); p- two-phases sharing of pressure; $\bar{I}$- Unit tensor.

(3) Energy conservation equation

The energy conservation equation of air phase:

$$\frac{3}{2}\left[\frac{\partial}{\partial t}\left(\rho_g\alpha_g T_g\right)+\nabla\cdot\left(\rho_g\alpha_g\vec{v}T_g\right)\right]=\left(-p_g\bar{\bar{I}}+\bar{\bar{\tau}}\right):\nabla\vec{v}+\nabla\cdot\left(k_{T_g}\nabla T_g\right)-\gamma T_g+E_{gs} \quad (7)$$

Energy conservation equation of powder particle:

$$\frac{3}{2}\left[\frac{\partial}{\partial t}\left(\rho_s\alpha_s T_s\right)+\nabla\cdot\left(\rho_s\alpha_s\vec{v}T_s\right)\right]=\left(-p_s\bar{\bar{I}}+\bar{\bar{\tau}}\right):\nabla\vec{v}+\nabla\cdot\left(k_{T_s}\nabla T_s\right)-\gamma T_s+E_{sg} \quad (8)$$

Wherein: T_g、T_s - temperature of air phase and the particle; k_{Tg}、k_{Ts} - Air phase, powder particle phase diffusion coefficient; γ- Energy dissipation coefficient of the collision; E_{gs}、E_{sg} - Energy exchange between two phases(E_{gs}=E_{sg}).

2.2. *Physical Model*

According to the specific nature of the agitation, the physical model of ceramic dry granulation process can be divided into two regions. An area was divided into a moving area by a reamer, blade and its vicinity. The remaining regions were divided into static regions. The entire granulation process was in a closed area. The granulation chamber wall was set as a wall. The walls of the reamers and blades were also set as walls. The intersection of the active and static regions was set as the interface.

Specific physical model and boundary conditions are shown in figure 1.

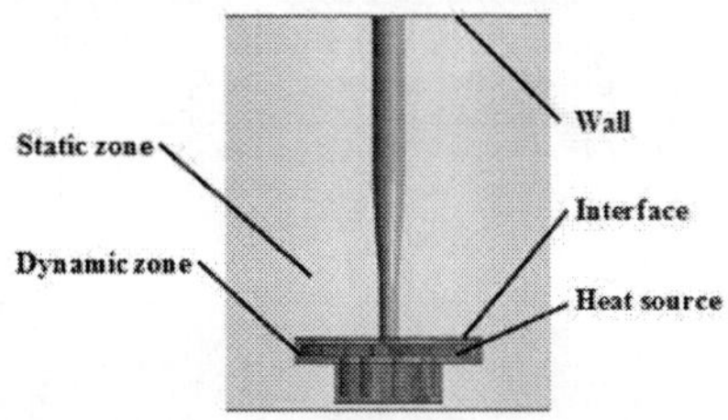

Figure 1. Simplified diagram of physical model and boundary conditions.

3. Simulation results and analysis

Analysis figure 2 granulation chamber sectional view of the axial temperature field: when granulation time t_1=4min, the maximum temperature inside the granulation chamber was about 70 ℃, the minimum value was about 20℃, the area temperature near reamers, leaves was significantly higher, but powder particle distribution area temperature in the whole granulation chamber was relatively uniform; when granulation time t_1=5min, the maximum temperature inside the granulation chamber was about 85℃, the minimum value was about 20℃, the area near the reamer and blade's temperature was significantly higher, the temperature of power particles distribution area in granulation chamber was relatively uniform. There had a small number of regional temperature was too high; when granulation time t_1=6min, the maximum temperature inside the granulation chamber was about 95℃, the minimum value was about 20 ℃, the area near the reamer and blade's temperature was significantly higher, the temperature distribution of the powder granules in the whole granulation chamber showed a great difference. There had some areas' temperature was too high.

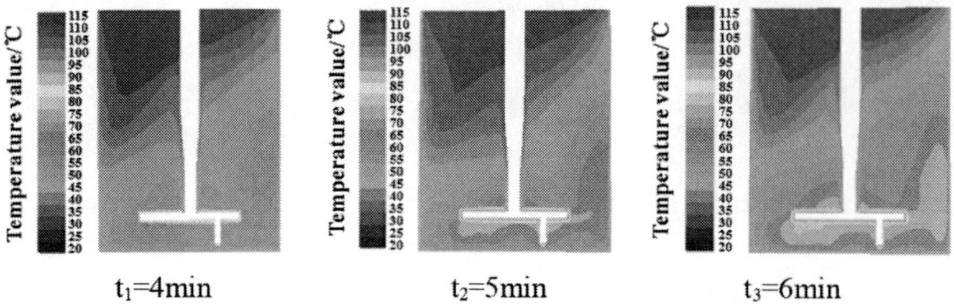

Figure 2. Sectional view of the axial temperature field.

4. Experimental results and analysis

Adopted YTN-1001-type powder comprehensive properties tester measured billet particles' angle of repose, compression ratio, uniformity, plate angle, analyzed the fluidity of billet particles. From the experimental values showed in table 1 different granulation time of the billet particles fluidity index: when granulation time was 4min, the angle of repose was 49.51°, the compression ratio was 21.47%, uniformity was 18.69, plate angle was 54.68°; when granulation time was 5min, the angle of repose of the billet particles was 29.64°, compression ratio was 9.29%, uniformity was 6.92, plate angle was 31.76°, the fluidity evaluation index of the billet particles was 89.7; when granulation time was 6min, the angle of repose of the particles was 30.58°, compression ratio was 12.68%, uniformity was 8.32, plate angle was 32.11°, particle fluidity evaluation index was 81.6.

Table 1. Ceramic body particles liquidity index of different granulating time.

Experiment numbers	Granulating time /min	Angle of repose /°	Compressibility /%	Uniformity	Angle of plate /°	Liquidity index
1	4	45.27	20.54	17.72	52.57	56.4
2	5	26.53	9.52	6.49	32.06	89.7
3	6	29.32	11.13	7.87	30.11	81.6

5. Conclusion

1. Numerical simulation and experimental results showed that: when granulation time was 5min, the maximum temperature inside the granulation chamber was about 85℃, the minimum value was about 20℃, the region near the reamer and blade's temperature was relatively high, there are small number of regional temperature was too high, but temperature was relatively uniform of particle distribution area in granulation chamber. The fluidity index of the billet particles was 89.7 at this time, the effective billet particles were 90.66%, which was the optimum value of the billet particles, when the temperature was more than 85℃, the fluidity of the billet particles and the number of effective particles was reduced.

2. The author combines Euler's two-fluid theory, the mathematical model of temperature field in dry milling process was established, which provided the theoretical support for the study of granulation chamber temperature field in dry granulation process and provided a theoretical basis to promote dry milling technology in the building ceramic industry.

References

[1] Li S, Yungang C 2011 Analysis on energy saving measures of spray drying tower *Modern Manufacturing* **9** 82-83

[2] NanXing W, DaHai L and Tiantian Z 2014 Effect of Mixer Shaft Eccentricity of Ceramics Dry Granulating Machine on Particles Dispersion *Bulletin of The Chinese Ceramic Society* **12** 3300-3303

[3] Fengjiao L 2015 Application of Dry Powder Process in the Production of Ceramic Outer Wall

Tiles *South China University of Technology*

[4] Chen M, Xu Z, Chu R, et al. Y_2O_3-modified $Ba(Ti_{0.96}Sn_{0.04})O_3$, ceramics with improved piezoelectricity and raised Curie temperature *Materials Research Bulletin* **59(16)** 305-310

[5] Chi D, Yaping Z and Chengjia G 2003 Numerical Simulation of Temperature Field for Bioceramic Coating Cladded by Laser *Journal of Materials Science and Engineering* **21(4)** 503-506

[6] Hassan N M S, Khan M M K and Rasul M G 2013 Temperature Monitoring and CFD Analysis of Data Centre *Procedia Engineering* **56(6)** 551-559

[7] Meier H F, Vegini A A and Mori M 2011 Four-Phase Eulerian-Eulerian Model for Prediction of Multiphase Flow in Cyclones *Journal of Computational Multiphase Flows* **3(2)** 93-106

[8] Jianping W, Lin H and Huai Z 2001 Modeling and Simulation of Gas-Liquid-Solid three-phase Turbulent Flow Based On E/E/L Model *Journal of Chemical Industry* **52(04)** 343-348

[9] Li Z, Hong Z and Hongxiang Y 2012 Phenolic Wastewater Treatment by Combination of Oxidation Combined with Adsorption and in a Gas-liquid-solid Three-phase Fluidized Bed Reactor *Chemical Industry and Engineering Progress* **31(02)** 458-461.

IOP Conf. Series: Materials Science and Engineering **292** (2017) 012079 doi:10.1088/1757-899X/292/1/012079

The influence of fine aggregates on the 3D printing performance

J C Lin[1], X Wu, W Yang, R X Zhao and L G Qiao

National Enterprise Technology Center ,China State Construction Ready Mix Concrete CO., LTD, Wuhan, China

[1] E-mail: Sillybear@yeah.net

Abstract. Influences of nature Particle, size, grain shape and fineness modulus of fine aggregates on the 3D printing performance of cement-based mortar were investigated. Results showed that the working performance of the mortar is not only dependent on the fineness of the aggregate, but also the gradation and grain size of the aggregate. And the mechanical properties of the mortar are increasing with the increase of Mx in the same test condition. The research shows that it is effective to choose different properties of materials for different design requirements, and the fluidity of mortar must be decreased under assuring construction quality and the pumpability of 3D printing materials.

1. Introduction

3D printing technology, as a rapid prototyping technology, fabricates three-dimensional entity structure using layer-by-layer superimposed method based on 3D digital model. As one of the most promising manufacturing technology in the 21st century, it is known as the core of the "third industrial revolution".

The application of 3D printing originated from 20th centrury in US in the building field. A new construction method of free form component that accumulative building structures layer by layer with selective solidification of cementitious materials was proposed by Joseph Pegnain1997 [1].

According to the diversity of building materials, the current 3D printing construction methods can be divided into three categories: 1st,the construction method of stratified squirting superimposed with cement-based material, 2nd,the construction method based on gravel powder Layered bonding,3rd,Three-dimensional construction method with large manipulator [2].

3D printing technology puts forward more special requirements for the workability of building materials. At the same time, the printing process also has great influence on the structural performance and material hardening performance. Therefore, the effect of 3D printing is dependent on not only the rheological properties, plasticity and self-condensation properties of cement-based materials, but also with matching between the 3D printing equipment and construction process. For now, 3D printing test and application in the building field mostly use pumping technique with mortar, the aggregate accounted for more than 70% of the total volume in the cement-based materials, so the performance of the aggregate has a significant impact on the extrusion performance, printing performance, mechanical property, durability of 3D printing.

Therefore, this paper explored the major characteristics of 3D printing materials and their applications based on the research of nature Particle, size and grain shape of micro-aggregates on workability, mechanical property and the 3D printing performance of the cement-based mortar.

2. Raw materials

2.1. Raw materials

2.1.1. Cementitious material. P·O42.5 cement produced by Hubei Yadong Cement Plant, whose basic performance is shown in table 1. S95 mineral powder is produced in Xinzhou, Wuhan. microspheres produced by Tianjin ZHUXIN New Material Technology Co., Ltd. was used. Their basic performance indexes and particle size distribution are shown in table 2 and table 3.

Table 1. Basic performance of concrete.

Standard Consistency/g	Flexural Strength/MPa		Compressive Strength/MPa		Density/g·cm^{-3}	Stability
	3d	28d	3d	28d		
127	5.23	8.72	23.30	48.10	3.20	Qualified

Table 2. Physical properties of cement and mineral admixtures.

Raw Materials	Density/g·cm^{-3}	Specific Surface Area/m^2·kg^{-1}	WaterRe-quirement/%	Fluidity Ratio/%	28d Compressive Strength Ratio/%
Cement	3.20	350	100	100	100.0
S95 Slag	2.80	425	105	100	90.1
Microsphere	2.73	710	95	105	83.5

Table 3. Particle size distribution of raw materials.

RawMaterials	D_{50}/μm	D_{10}/μm	D_{25}/μm	D_{75}/μm	D_{90}/μm
Cement	15.00	2.63	5.85	21.42	37.47
S95 Slag Powder	16.00	2.44	5.55	25.06	41.52
Microsphere	1.00	0.61	0.78	1.42	1.99

2.1.2. Aggregate. The test uses river sand 1# with 3.1 fineness modulus; river sands 2# with 2.84 fineness modulus; river sand 3# with 1.84 fineness modulus; river sand 4# with1.68fineness modulus; quartz sand 7# with 1.0 fineness modulus; manufactured sand 7# with 3.5 fineness modulus. Their clay lump content, silt content are less than 1%,apparent density are 2600±30kg/m^3;

2.1.3. Admixture. Polycarboxylate superplasticizer was used, whose solid content of 25% and water-reducing rate of 28%.

3. Test methods

3.1. Mortar compressive strength test
Concrete strength test is performed according to the Standard for Method of testing cements-determination of strength(GB/T 17671-1999).The mortar specimen dimensions in this paper are 40mm×40mm×160mm.

IOP Conf. Series: Materials Science and Engineering **292** (2017) 012079 doi:10.1088/1757-899X/292/1/012079

3.2. Mortar workability test

Concrete fluidity test is performed according to the Standard for Methods of testing uniformity of concrete admixture (GB8077-2012), Test method for fluidity of cement mortar (GB / T 2419-2005) test.

3.3. 3D printing test

3D printing Test process: 3Dprinter builds φ30cm circular components based on the medium-sized 3D printer, print layer height 1.4cm, flat print speed 8cm / s.

4. Results and Discussion

Based on the design of C70 concrete mixing proportion, the research analyses workability, mechanical property and the 3D printing performance with the component quality of the mortar, Cement: Microsphere: Slag Powder: Sand: Water=40:9:5:74:15.

4.1. Workability of 3D print material

In the mixing proportion invariable situation, the fluidity of fresh mortar usually increase and the viscosity decrease with the increasing of the fineness modulus (M_x as following) of fine aggregate [3]. Figure 1 presents the result is in accordance with the expected except the trial group of manufactured sand. When M_x of sand less than 2.2, the fresh mortar has poor fluidity and high viscosity, the reason is that the specific surface area increased significantly, the sands require more cement slurry to fill the space. When M_x of sand more than 2.8, with decreasing of the percentage of voids in fine aggregate, flowable slurry in the mortar is more opulent because of the accumulation effect of sand particles, the workability of fresh mortar is effectively improved. Meanwhile that segregation and weeping of fresh mortar is serious .And with the increased angularity index of the manufactured sand which has a high fineness modulus, the fluidity and workability rentention of the fresh mortar is becoming bad relatively. Also by contrast it find that the group of river sand has more excellent fluidity and thixotropy properties than the group of quartz sand in the 1.68~1.84 range of M_x .This is because the river sands have continuous gradation and less surface area.

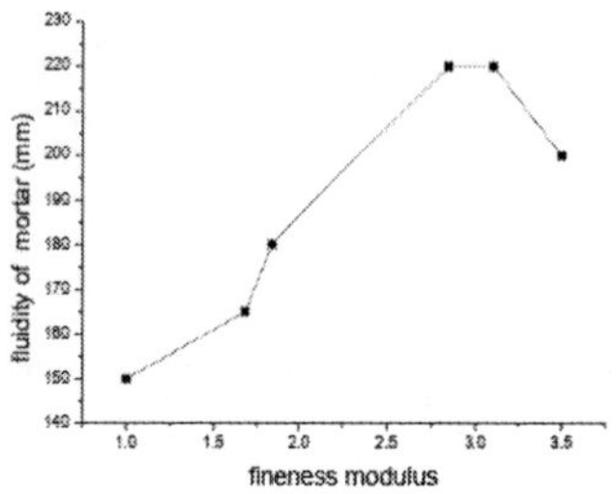

Figure 1. Effect of fineness modulus on Fluidity.

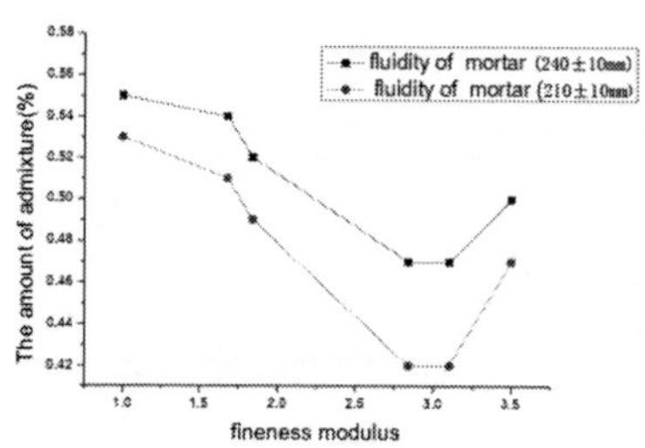

Figure 2. Effect of fineness modulus on amount of additives.

As shown in figure 2, it can also be found than the change trend of amount of additives gradually decreases with the increase of the fineness modulus of the sand. But in the same Mx range, poor granulation or poor gradation of the fine aggregate increases significantly the amount of additives. Therefore, the working performance of the mortar is not only dependent on the fineness of the aggregate, but also the gradation and grain size of the aggregate.

4.2. Mechanical property of 3D print material

The results are shown in figure 3, the compressive strength of the mortar are increasing with the increase of M_x in the same test condition. Compared with Mx 1.0 group, the 28d strength of Mx 3.5 group increase 8% .This is due to the mortar with a larger fineness modulus offers lower viscosity and higher uniformity, the strength is improved with decreasing porosity and surface area.

IOP Conf. Series: Materials Science and Engineering **292** (2017) 012079 doi:10.1088/1757-899X/292/1/012079

But the Mechanical performance of the mortar leveled off after M_x more than 2.2, there is a 5% increase of strength of manufactured sand group only. And compared with natural one, manufactured sand has the characteristics of poor graduation, high fine modulus, rugged surface and sharp particle, so the intensity and contractibility of the mortar is improving after the hardening process [4-6].

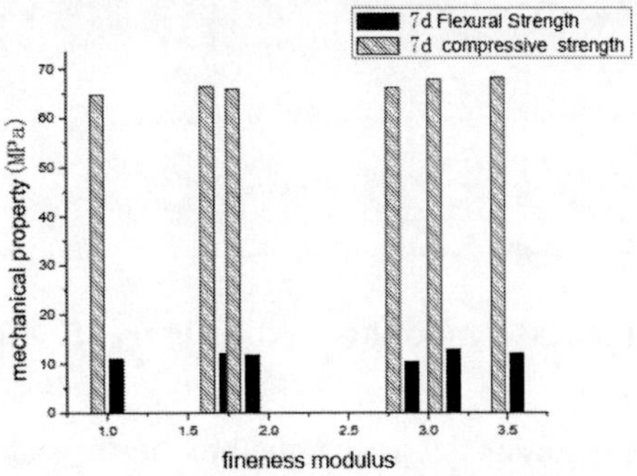

Figure 3. The Mechanical Properties of Mortar.

At the same time, in the experimental groups with the ratio of Mx between 1.68 ~ 1.84, it is found that the strength of quartz sand group is about 2% higher than that of continuous graded sand group, and that's related to the higher hardness and angular shape of quartz sand.

4.3. 3D printing performance of 3D print material

According to the 3D printing process(3Dprinter builds with diameter of 30cm circular components with the fluidity of 220 ± 10mm and 240 ± 10mm of the mortar , print layer height 1.4cm, flat print speed 8cm / s),the experiments test the effect of 3D print quality on the various mortars workability. As shown in table 4. It is found that there is no significant difference in the extrusion volume of the mortar under the same pumping conditions. In addition to manufactured sand group which has higher resistance of pipeline because of the heat generated by the screw is with rugged surface and sharp particle of fine aggregate. The extrusion force of the other groups is gradually reduced with the fine aggregate increased, but at the same time the loss of working performance is gradually increased, that is good agreement with the peculiarity of poor bleeding are consistent and lower viscosity of the manufactured mortar.

Table 4. 3D Printing Performance of Mortar.

NO.	The Variety of Sand	Fineness Modulus	Number of Layers	Layer Height/ cm	Single Layer Width/ cm	Difficulty of Transport	Effect of Surface	Fluidity/ mm
1	river sand	3.1	26	1.4	4	medium	good	220
2	river sand	3.1	20	1.4	4	medium	excellent	230
3	river sand	2.84	25	1.4	4	medium	good	220
4	river sand	2.84	15	1.4	4.2	low	excellent	240
5	river sand	1.84	22	1.4	4.7	low	excellent	210
6	river sand	1.84	13	1.4	5.6	low	excellent	230
7	quartz sand	1.68	16	1.4	5.2	low	excellent	210
8	quartz sand	1.68	12	1.2	5.6	low	excellent	235
9	quartz sand	1	15	1.4	5.2	low	excellent	210
10	quartz sand	1	10	1.1	6	low	excellent	230
11	manufactured sand	3.5	42	1.4	4	hard	tiny crack	210
12	manufactured sand	3.5	32	1.4	4	medium	excellent	230

IOP Conf. Series: Materials Science and Engineering **292** (2017) 012079 doi:10.1088/1757-899X/292/1/012079

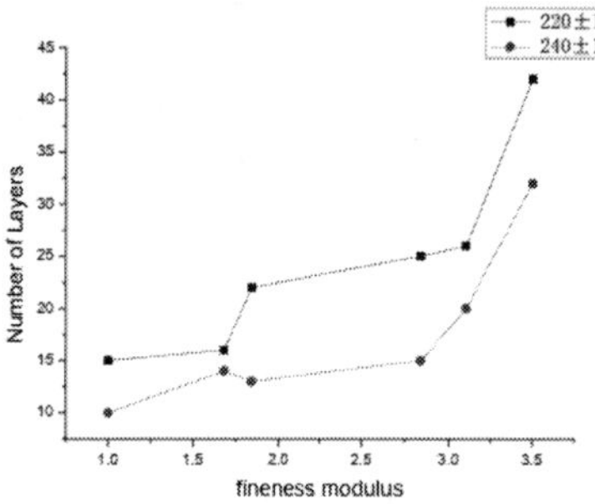

Figure 4. Effect of fineness modulus on Number of Layers.

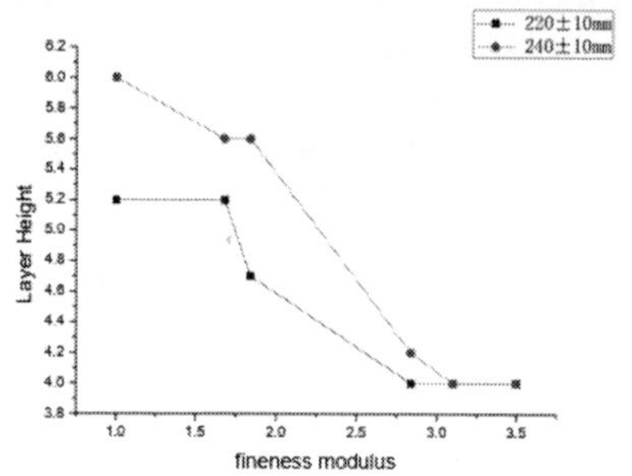

Figure 5. Effect of fineness modulus on Single Layer Width.

Moreover, as the fineness modulus decreases, the number of stable print layers decreases and the single-layer printing width gradually increases ,at same time the apparent effect of the member is improved (figure 5-figure 6).That is due to the reduction of Mx improve the mortar thixotropic performance, the structure has the higher tendency to deformation when 3D printing. When the Mx of the test group is less than 1.68, the single layer height of the mortar is less than the set value with the same condition. After forming only several layers, the structure begin lateral collapseing as shown in figure 7, it is relative to mismatch between the default parameters and material properties. When the Mx of the test group is higher than 2.86, the print layers and the single layer width of the mortar are stable, and the average number of layers is more than 15 layers, and the maximum is more than 40 layers, as shown in figure 8.

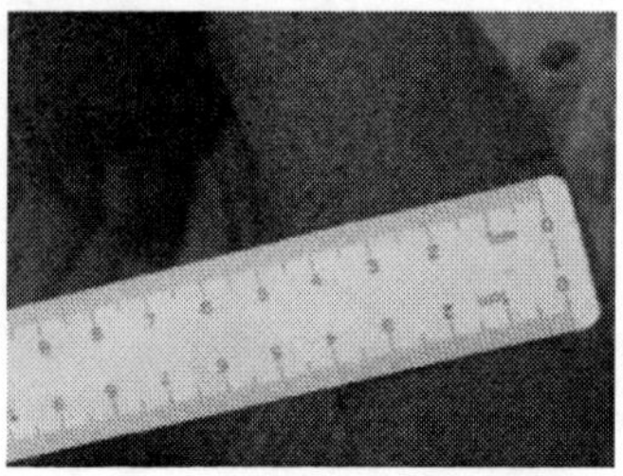

Figure 6a. Single Layer Width of Mx 1.84.

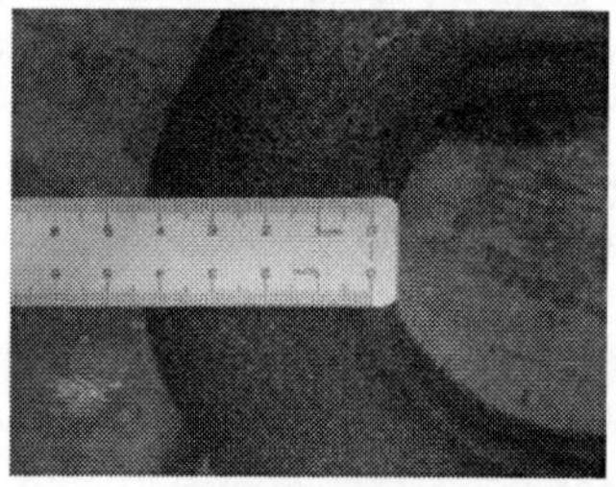

Figure 6b. Single Layer Width of Mx 2.84.

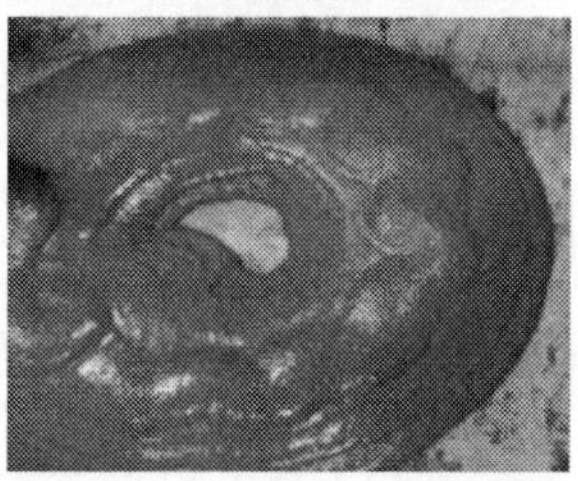

Figure 7. The phenomenon of Lateral Collapseing.

Figure 8. Maximum Layers of 3D Printing.

Meanwhile, all that mortar pumping takes a lot of effort with larger aggregate, due to the operation of the friction between the metal screw and rubber bushing will produce a certain degree of friction heat. So for the avoidance of rigidity impact of mortar pump in movement, it is necessary to optimize slurry viscosity and water retention to counteract the effects of rough particle and angular.

5. Summary

In summary, according to the different 3D printing construction conditions and design requirements, it is necessary to choose different specifications of the fine aggregate: for the excellent appearance effect of small components with low requirement of construction speed, the fineness modulus of sands less than 1.68 of 3D printing materials is an appropriate choice; for medium-sized components, the fine aggregate with fineness modulus during 1.6 ~ 2.86should be selected for 3D printing materials; for large volume components, it is suggested to select the mortar with mechanism sand which the fineness modulus is more than 2.86 to ensure the building speed and stability .

Acknowledgements
This paper is supported by CSCEC Fund Project (cscec-2014-z-33).

References

[1] Pegna J 1997 Exploratory investigation of solid freeform construction *Automation in Construction* **5(5)** 427-437
[2] Lieyun D, Jie X and Yawe T 2015 Research and Application Review of the Digital Construction Technology of 3D Printing for Construction *Journal of Civil Engineering and Management* **03** 1-10
[3] Yang J J, Yin S H, Wang H C, Wen Z Y, Lin Y Q and Li W 2008 Influence of sand quality on performance of C80 HPC *Concrete* **11** 58-61
[4] Ou W H, Yang Y B, Zheng W J, Lin J S etc. 2010 Experimental study of high strength rock chips concrete *Concrete* **3** 89-91
[5] Wu S M, Yang Y B, Peng Y X, etc. 2010 Research and development of chip concrete *Architecture Technology* **41 (1)** 67-70
[6] Yuan T H, Liu Z M, Sun K P, etc. 2010 Stone chips in the product research and application of concrete *Concrete* **8** 103-105

IOP Conf. Series: Materials Science and Engineering **292** (2017) 012080 doi:10.1088/1757-899X/292/1/012080

The research of penetration testing in the FRP pipe applications

M L Xin[1], Y Chen, M D Li, Z G Wang and G H Gu

Guangzhou Special Pressure Equipment Inspection and Research Institute, Guangzhou 510663

[1] E-mail: xdmingliang@126.com

Abstract. In this paper, for the first time will be introduced to the penetrant metal nondestructive testing technology in FRP pipe non-destructive testing, glass fiber reinforced plastic are verified through the SEM surface sine qua non of penetrant testing. Through the prefabricated crack, glass fiber reinforced plastic materials available to the nondestructive testing methods for crack detection. By penetrant testing of glass fiber reinforced plastic, partial fluorine polyethylene and rigid polyvinyl chloride (PVC) for inspection, to find obvious crack.

1. Introduction The first section in your paper

Glass Fiber Reinforced plastic products, commonly known as glass Fiber Reinforced plastic abbreviations FRP (Fiber Reinforced Plastics), is a kind of Fiber Reinforced plastic is the general, common glass Fiber after dipping unsaturated resin after curing polymer Reinforced material. According to the different kinds of unsaturated resins, it can be divided into: polyester fiberglass, epoxy glass steel and phenolic glass steel. According to different fiber, it can be divided into glass fiber reinforced composite plastic [1] (GFRP), carbon fiber reinforced composite plastic [2] (CFRP), boron fiber reinforced composite plastic, etc. Glass fiber reinforced plastic materials, the diameter of the fiber is very small, plays a role of stress in the whole is responsible for the materials, the vast majority of the outside world under load, because the fiber is a brittle material, the fracture strain is about less than thirty over one thousand, so easy to damage, fracture and from corrosion. The base body mainly ACTS as a bonding fiber, and its strength and modulus are much lower than the fiber, but the matrix can withstand large deformation and have good viscoelasticity and elastoplastic properties for ductile material. The combination makes the composite materials have the advantages of light and hard, non-conducting, stable performance, high mechanical strength and corrosion resistance. Markova and other [3] have studied the glass fiber board in the natural environment and acid environment treatment, the longitudinal bending performance of the sample, the glass steel has replaced the steel in many fields, such as automobile, ship shell, etc.

Glass steel pressure pipe due to the nature itself materials prone to aging, show the impact resistance performance is poor, eventually leading to the composite material structure performance is decreasing, even may cause great economic losses. Researchers to study for a long time, glass steel pressure vessel combined with practical experience summed up the various damage forms of the glass fiber reinforced plastic, such as matrix cracking, interface separation, fiber debonding and fiber breakage, etc. [4]. In order not to cause material damage, nondestructive testing techniques are used to test equipment performance.

NDT (Non-destructive Testing) is defined as: on the premise of not damage the specimen, by means of physical or chemical method, with the help of advanced technology and equipment, the structure of the internal and surface of the specimens, nature and the state inspection and test methods.

Nondestructive testing is based on the development of modern science and technology. During the development of nondestructive testing technology, there are three stages: discovery stage, detection stage and evaluation stage. Nondestructive testing techniques have been developed over the years, and there are many relatively mature nondestructive testing methods at home and abroad, such as the four major nondestructive testing techniques: radiation, ultrasonic, magnetic powder and penetrating detection. Later, visual inspection, eddy current detection and magnetic leakage detection were introduced, and the detection of sound emission and phased array detection were developed with the development of technology. It is the four major conventional nondestructive testing techniques for the manufacturing quality of the pressure metal special equipment and the most commonly used nondestructive testing methods. RT and UT are mainly used to detect internal defects of specimens, and MT and PT are mainly used to detect surface defects of specimens. The non-destructive testing method of plastic pipe and equipment is mainly visual inspection, ultrasonic detection and ray detection.

Glass fiber reinforced plastic equipment due to the characteristics of material itself, its products performance difference is bigger, the factory inspection qualified products will appear leakage phenomenon in the process of practical application, and appear leakage often has reached the point of irreparable. Early prevention of leakage will need to find micro cracks, micro cracks and testing technology is commonly used in penetrant testing, liquid penetrant will try to introduce FRP equipment, and explore the characteristics of glass surface is suitable for penetrant testing is particularly important.

Based on FRP pipe, FRP lining partial fluorine ethylene and glass fiber reinforced plastic lining rigid PVC pipe as the research object, using penetrant testing, metal materials commonly used nondestructive testing technology is introduced into the glass fiber reinforced plastic material of surface crack detection test, and two kinds of glass fiber reinforced plastic and observed by a scanning electron microscopy (sem) plastic surface morphology and analysis, at the same time, the prefabricated crack on the glass fiber reinforced plastic penetration test feasibility in FRP nondestructive testing, and puts forward requirements for glass fiber reinforced plastic products processing.

2. Experimental

2.1. Materials

FRP pipeline, FRP - PVDF pipeline for fiberglass reinforced plastics, frp-upvc pipeline, yade chemical equipment (Shanghai) co., LTD.

Cleaning agent, osmotic agent, imaging agent, Shanghai xinmeida flaw detection equipment co. LTD.

2.2. Equipment and instruments

Scanning electron microscope, Quanta 650, Czech, FEI;

Slicer, CUT 5062, SLEE, Germany.

2.3. Sample preparation

FRP, FRP - PVDF and FRP - UPVC tubing. Solid samples using slice mechanism for 20 μ m thick sheet, set aside.

2.4. Penetration test

The test method is based on the NB/T 47013.5-2015 pressure equipment nondestructive testing part 5: penetration detection. The pipe pretreatment is used to clean the surface of the surface with cleaning agent. Apply osmotic agent, the osmotic temperature range of 5°C ~ 50°C, lasts for no less than 10min. Dry treatment, remove excess penetrant, apply imaging agent, observe.

3. Results and discussion

3.1. Pipeline surface analysis

FRP early NDT research using X-ray, ultrasonic testing method, later will be introduced to the acoustic emission detection of glass fiber reinforced plastic resin cracking of pressure vessel, fiber breakage and acoustic signal recognition made some breakthrough. In the 1990s, the technology of ultrasonic scanning of composite materials was taken seriously, and it was widely used in the field of outdoor rapid detection. In the new century, computerized tomography is applied in FRP materials, which can be used to detect cracks, inclusions, stomata and stratified [5]. Infiltration failure is one of the reasons for the failure of polymer materials. In order to check whether there is an infiltration phenomenon that needs to be permeated, the surface of the material is selective.

No defects were found in the appearance of the three selected pipes. Select FRP piping for radial sampling, as shown in figure 1. The exterior observation of FRP pipe is two kinds of compound pipe of different materials. Through the SEM, the left of the pipe is shown in figure 1, the outer material of the pipe is shown, the middle image is the transition section, and the most right is the inner tube. Can clearly see from the picture the composite pipe of FRP pipe differences for the adding amount of glass fiber is different, the outer pipe main mechanical force to add a lot of fiber, plays a role of anti-corrosion lining the main fiber need not too much.

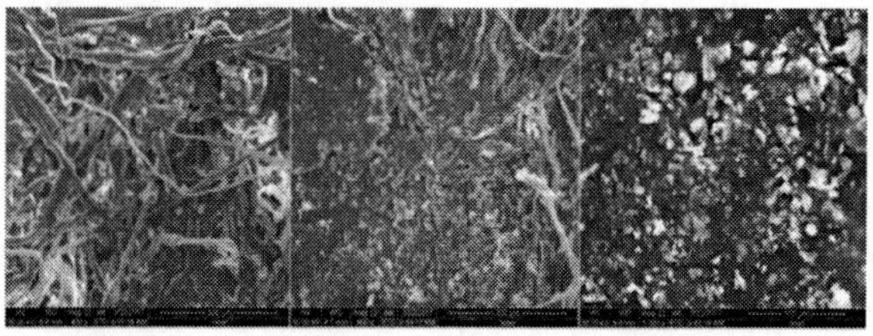

Figure 1. Radial SEM image of FRP piping.

In order to verify the applicability of seepage detection in FRP piping, it is necessary to confirm whether the material of such pipeline and composite pipeline meets the conditions of penetration testing. The surface morphology of FRP, FRP-UPVC and FRP-PVDF were observed by SEM, and the characteristics of the three materials were to be prepared. The surface images of the three kinds of tubes were shown in figure 2. It can be seen from the figure that the surface of FRP is remarkably smooth and smooth than the surface of the other two kinds of materials. From the SEM images, the FRP surface can be judged to meet the requirements of penetration detection.

Figure 2. SEM images of FRP (left), FRP - UPVC (middle) and FRP-PVDF (right).

The surface of the FRP-PVDF shows the surface of most plastics, and some of the mustard structure should be formed in the process of pipeline processing. Because of its molecular structure, FRP-PVDF has a relatively flat surface and a small uniform shallow pit, which is caused by shrinkage of materials and stress relaxation in the process of material processing.

FRP material surface is smooth, when the surface crack of material damage or will be introduced to the internal fiber layer, porosity will further increase, at the same time, also facilitate the defect amplification of penetration test more intuitive display defect position improve the sensitivity of permeability test, FRP lining structure as shown in figure 3.

Can be seen from the figure 3 cellulose in FRP surface enlarge figure can clearly see the distribution, can be seen from the diagram of cellulose in glass fiber reinforced plastic distribution more uniform and no winding and crosslinking. From the fiberglass scan, the fiberglass diameter is 23 μ m, and the fiber is smooth and smooth.

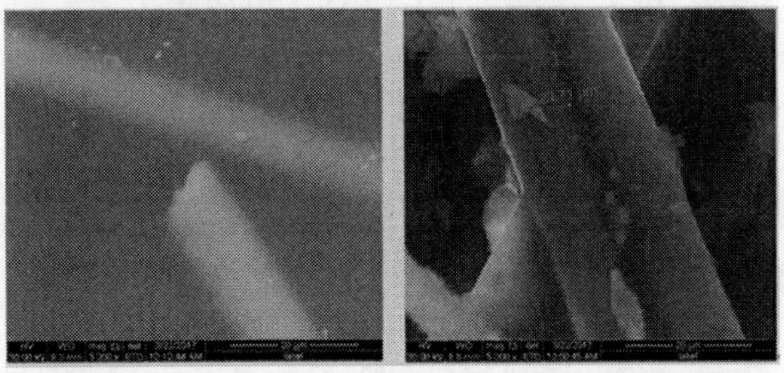

Figure 3. FRP surface and internal SEM images.

3.2. Penetration detection

Before penetration testing, the surface of the sample must be processed to remove impurities or liquid contaminants that affect the experiment. The wet properties on the surface of non-metal specimens are different from the wet properties of the metal specimen surface. Before penetration testing of non-metal specimens, it is necessary to prefabricate the crack verification penetration test to satisfy this non-metal specimen [6].

Since all kinds of non-destructive testing methods have their own scope and limitations, new nondestructive testing methods have been continuously developed and applied. Generally, any physical, chemical, or other possible technical means may be developed into a nondestructive testing method, as long as the basic definition of nondestructive testing is met.

The FPR specimen was compressed, so that the surface of the surface was obviously cracked, and the permeation test was carried out, as shown in figure 4.

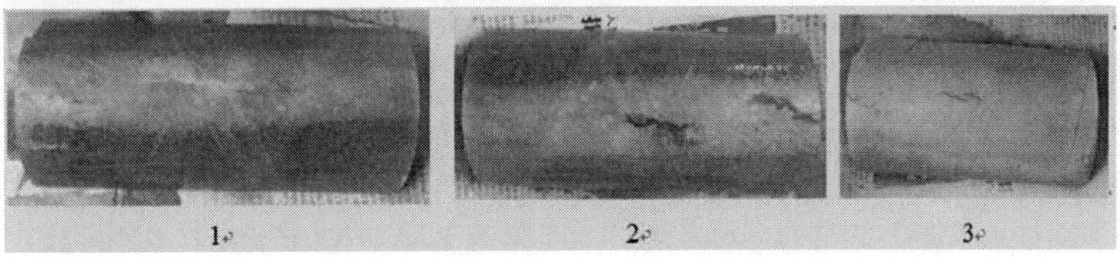

Figure 4. Prefabricated crack FPR specimen penetration detection 1-initial sample, 2-permeation process, 3-imaging process.

Can be seen from the graph, through the prefabricated crack penetration testing process can be clearly detected crack defects, can support penetrant test can be used in FRP material near the surface of the crack detection. In order to better verify the detection effect of osmotic detection in pipe, the permeation test of polymer materials of three kinds of materials is carried out, and the results are shown in figure 5. It can be seen from the figure that there are no obvious cracks in the three materials. The red stripe of the FRP image in the figure is not a crack in the process. Penetrant testing can effectively detect the defects on the surface of the sample and potholes, you can see from the picture the reflect on the crack of the penetrant test is quite sensitive and can see in the FRP - UPVC imaging figure some red dot, this is Nick, processing or the installation process pipeline penetration test can be obviously visible reaction material form of near surface defects. When the pipe is processed or used in the process of scratches or cutting, the penetration detection can be found better, providing an intuitive data support for early warning and precautionary prevention. From FRP diagram in figure 5 shows that there are some red dots appear, this means there are some holes, usually by appearance test, may be difficult to find a small hole should be the cause of the curing process, these holes will be one of the causes of leakage, it is only by penetrant test found holes, and carries on the repair will reduce the possibility of leakage failure.

The penetration test can not only detect cracks in the surface of glass steel, but also have some applicability to the plastic pipe. It can be seen from FRP-UPVC that some scratches also show that the test results of the permeation test are verified because of the fragile installation process of UPVC materials. It can be seen from FRP-PDFE materials that there are no obvious cracks and defects in surface formation.

Figure 5. Permeation test of FRP, FRP-UPVC and FRP-PVDF.

4. Conclusions

Penetration testing is the most frequently used conventional nondestructive testing method, it is introduced into the polymer down especially of glass fiber reinforced plastic equipment, the use of glass fiber reinforced plastic equipment and safety protection will play a role in promoting further. In terms of high polymer material non-destructive testing technology also has a certain reference value, develop metal nondestructive testing technology out into the polymer materials by technical means the corresponding detection, in order to achieve the online non-destructive testing of polymer materials, to improve the detection and maintenance cost driven.

The surface morphology of three kinds of polymer materials was analyzed by scanning electron microscopy, and the determination of surface cracks applied to three kinds of materials was determined by the test of penetration test. Through the prefabricated cracks, the reagent of the metal penetration test is proved to be able to directly reflect the defect position. Finally, the penetration test was used to detect the near surface of three specimens without obvious cracks. The next step will be to study the research and penetration detection sensitivity of macromolecule material.

Acknowledgments

This study is supported by the Science and Technology Plan of Guangzhou (201607010113) and Administration of Quality and Technology Supervision of Guangdong Province (2015CT04).

References

[1] Abbasi A, Hogg PJ 2005 Temperature and environmental effects on glass fibre rebar: modulus, strength and interfacial bond strength with concrete *Composites: Part B: engineering* **36(5)** 394-404

[2] Santos TFA, Vasconcelos GC, Souza WA, et al. 2015 Suitability of carbon fiber-reinforced polymers as power cable cores: Glavanic corrosion and thermal stabilit evaluation *Materials and Design* **65(1)** 780-788

[3] Markova A V, Savin V F, Yu B, et al. 2011 Corrosion resistance tests of strained rods from polymer composite materials *Inorganic Materials* **47(15)** 1713-1716

[4] Wang G Y 2013 Molding process and failure prediction research for high pressure FRP pipe

Harbin: Northeast Forestry University

[5] Hu Y ZH 2014 Review of standardization and detection technology of FRP/composite materials *Fiber Reinforced Plastics/Composites* **9** 63-70

[6] Yang B, Hu X ZH, Gong G, et al. 2016 The influence and countermeasures of the surface state of workpiece on penetration detection *Nondestructive Testing* **38(8)** 55-59

IOP Conf. Series: Materials Science and Engineering **292** (2017) 012081 doi:10.1088/1757-899X/292/1/012081

Research progress of self-healing intelligent composite materials

P Li[1], G Z Liu and J Y Huang

School of Mechatronics & Vehicle Engineering, East China Jiaotong University, Nanchang, China

[1] E-mail: ecjtulipeng@126.com

Abstract. Self-healing composite material is an important intelligent composite material. In this paper, according to the recent research progress of the self-healing of composite materials, introduced the two kinds of typical extrinsic self-healing methods, which are microcapsule self-healing and hollow fiber self-healing, the paper expounds the repair mechanism of the two methods of self-healing, self-healing system characteristics and research status. This paper introduces some problems existing in the field of self-healing, and looks forward to the application prospect and development direction of self-repairing intelligent composite materials.

1. Introduction

With the rapid development of aviation and aerospace industry, composite materials are widely used in the industry field, but the composites will be damaged to a certain extent in the process of using, the damage is mainly the macroscopic and microscopic damage the crack. Macroscopic damage is usually easier to spot and solve in time, while microscopic damage is not easy to find and cannot be dealt with in a timely manner. The most common failure mode is the formation and propagation of microcracks that reduce the material's structural capabilities. Damage may be fixed through traditional bolted or bonded repair methods, but such repair requires temporary decommission of a part, collection of repair materials, and employee time and effort to enact the repair [1]. Therefore, when people study and solve this problem, the self -healing method is proposed to imitate the biological system, and the development of self-healing materials has also attracted the attention of many scholars [2].

In 1980s, the concept of intelligent biomimetic self-healing composite was proposed and widely studied [3, 4]. Self-healing composite materials can repair itself without the need for external material supplies when it is damaged. The main core of the self-healing technology is the supply of energy and material to the damage. With the development of the intelligent biomimetic composite technology, the self-healing technology of materials has been paid more and more attention and developed rapidly. Self-healing material is a kind of intelligent bionic material. When the material produces tiny cracks, the ability of the intelligent biomimetic self-healing material to repair itself is achieved by mimicking the self-healing features of the organism. Self-healing of composite material means that the material can be repaired without any external damage. Usually these tiny cracks are invisible to the human eyes and it will expand after the formation and with the expansion of the crack, the bearing capacity of the matrix material will decline, and the direct fracture of the material. To prevent the cracks from expanding, the scientists developed a new kind of material; this material will identify the presence of damage and make the appropriate response to self-healing. The self-healing function will greatly improve the use life of the composite material and prolong the service life of the material [5-9].

At present, various methods have been used to restore the mechanical properties of composite materials. In accordance with the need for additional repair agents, the self-healing system can be divided into extrinsic and intrinsic [10, 11]. Among them, the self-healing system of extrinsic can realize self-healing function by adding repair agent into the material system, mainly including microcapsule type and hollow fiber type [12]. The intrinsic self-healing is achieved by the physical or chemical effects of the polymer material itself. It mainly includes reversible covalent self-healing and reversible noncovalent self-healing [11, 13, 14]. Intrinsic self-healing must be performed under specific external conditions, the preparation process is more complicated, and the chemical structure of the resin molecule is strict, the repair effect is poor, and it cannot be applied to commonly used composites. In this paper, two typical extrinsic self-healing methods are summarized for self-healing composites. The repair mechanisms and characteristics are introduced, and analysis its advantages and disadvantages. Finally, the development direction and application prospect of self-repairing composite materials are prospected.

2. Hollow fiber type

The repairing mechanism of hollow fiber self-healing method is to insert hollow fiber into the matrix material. The hollow fiber is equipped with a repair agent fluid, when the material is damaged, the damaged area is repaired by releasing the repair fluid in the hollow fiber and bonding the crack. The method was originally proposed by Dry and Sottos, which was originally used in concrete materials, and then extended to the field of polymer materials. In the process of forming polymer-based composites, hollow glass fibers containing one-component cyanoacrylate or two-component epoxy resin adhesives are embedded in the composite matrix by Dry and others [15]. As shown in figure 1 [11], when the composite material is damaged, the crack propagation force causes the liquid core fiber to burst and the repair agent to bond the crack and restore the mechanical properties of the material. In 1999, Motuku [16] planted liquid core fibers containing vinyl resin or epoxy resin restorers in fiber glass /epoxy composite laminates, through the low speed impact test, they explored the materials of hollow fibers, quantity, spatial distribution, repair agent and other factors on the performance of composite materials. In 2001, Bleay [17] diluted the repairing agent by adding acetone, and filled the repairing agent to a hollow glass fiber (outer diameter is 15 mm, inner diameter is 5 mm) by means of vacuum technology. The experimental results show that the existence of the hollow fiber does not significantly reduce the impact properties of the composite material, but the repair efficiency is low, only about 10%. The utility model has the advantages that not only the material can be self repaired, but also the hollow fiber with small diameter can be embedded into the composite material as an reinforcing body, the disadvantage is that it is difficult to inject the fluid into the fiber. In recent years, a large number of hollow fiber self-healing composite materials have been developed by the Bond IP group in the UK [18~20], they embedded the hollow fiber with two component epoxy resin repair agent and UV dye vertically into the resin matrix composite, and achieved structural enhancement, damage detection and self-healing. The results of four point bending test show that the flexural strength of the material can be restored to about 97% before the impact damage. At the same time, the addition of hollow fibers reduced the initial bending strength of the material by 16%. Kouaourakis [21] discussed the effect of the addition of hollow fibers on the mechanical properties of carbon fiber/epoxy composites. They use a fiber diameter of up to 680μm, the influence of the arrangement and diameter of the hollow fiber on the mechanical properties (modulus of elasticity, tensile strength, etc.) of the base material is discussed. For hollow fiber self-repair system, single component repair agent generally use cyanoacrylate, but its repair reaction speed is fast, the repair agent that has not completely flow into the damage has solidified, so the repair effect is poor, the two-component repair usually using epoxy resin and its curing agent as a repair system, but the epoxy resin viscosity, will also affect the repair effect to a certain extent [22].

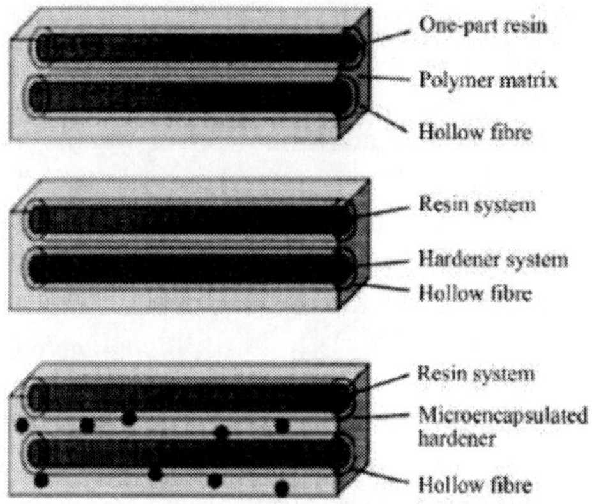

Figure 1. Diagram of hollow-fiber self-healing polymer materials.

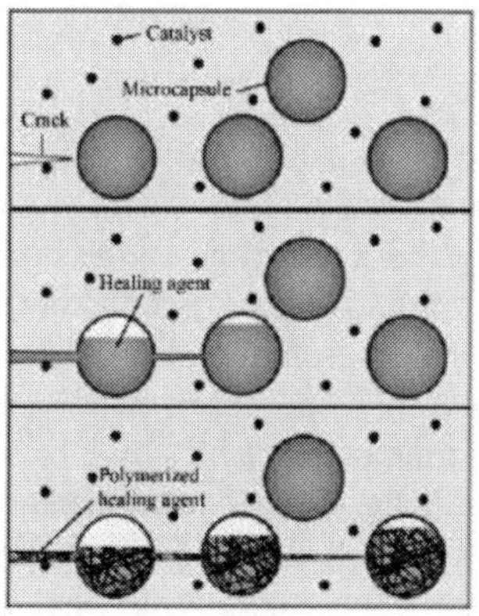

Figure 2. Diagram of microcapsule self-healing mechanism.

For hollow fiber self-repair system, single component repair agent generally use cyanoacrylate, but its repair reaction speed, the repair agent has not completely flow into the damage that occurs at the curing, so the repair effect is poor, two-component repair usually using epoxy resin and its curing agent as a repair system, but the epoxy resin viscosity, will also affect the repair effect to a certain extent. In addition, because the hollow fiber diameter is relatively large, this system is not suitable for self-repair of smooth surface, fiber diameter, fiber arrangement and so on have a great impact on self-repairing composite materials. To make this self-healing system applied to the industry, method need to be further improved. The specific measures include: to explore a better method of filling and sealing hollow fiber; explore other hollow fiber materials such as carbon nanotubes and apply them in practice; effectively filled hollow fiber in large scale application.

3. Microcapsule type

The microcapsule self-healing polymer material was first proposed by White [23] in 2001. It has become the focus of scientific researchers in the following ten years, and has become the most important method of self-healing at present. The self-healing mechanism is shown in figure 2 [23]. The microcapsules containing the repairing agent are embedded in the polymer matrix material while the catalyst is embedded in the matrix. After the material cracks, the crack growth leads to the rupture of microcapsule, and the release of the repair agent spreads to the damage area under the action of siphonage, and the polymerization occurs after the catalyst, and the crack is repaired. In recent years, research has been carried out on the selection of remediation agents, the selection of catalysts and the corresponding microencapsulation technologies.

3.1. Dicyclopentadiene (DCPD) self-healing agent system

The earliest repair system proposed by White [24]is dicyclopentadiene (DCPD) and metal ruthenium based (Grubbs) catalyst. DCPD can react with Grubbs catalyst to initiate ring opening metathesis polymerization, and play a role in repairing cracks, core material which was prepared by in situ polymerization of DCPD microcapsules, and the microcapsules with catalyst embedded in resin matrix, repair efficiency can reach 90%. Kessler et al. [25, 26] adopted the system to study its repair process of e-glass fiber/epoxy resin composites, the fracture toughness of the system was reduced after the microcapsule was buried in the middle of the 16 laminated plates, the fracture toughness test of the double cantilever beam is used to crack the material. Then manually injected DCPD with the catalyst mixture after the recovery efficiency was 67%, while the manual injection of DCPD after embedded catalyst in the composite material was only 19 percent. This indicates that the good contact between restorer and catalyst is the key to realize self-healing performance. Brown et al. [27, 28]also used the system to study the expansion and self-healing properties of microcapsules on the fatigue crack of materials. The disadvantage of this system is that Grubbs catalyst is expensive, the catalyst is easy to reunite in the matrix and the catalyst is easily lost in contact with the amine curing agent in epoxy resin. To solve the problem of reunion and deactivation of catalyst features, Rule, etc. [29] using paraffin wax formed outside the Grubbs catalyst layer packages, when the catalyst mass fraction is 0.25% and 0.75%, the epoxy resin matrix in the repair efficiency of 75% and 93% respectively. Wilson [30] with epoxy ethyl ester as substrate, Grubbs catalyst DCPD microcapsule and paraffin coated matrix, the epoxy lipid peroxidation agent ethyl amine on the catalyst performance, adding 15% (mass fraction) of DCPD microcapsules and 15% (mass fraction) Grubbs catalyst paraffin-coated, since the repair efficiency is 30%. Jackson et al. [31]encapsulated the Grubbs catalyst with polymethyl methacrylate or polypropylene, and then coated the SiO_2 layer in situ on the surface of the coated DCPD urea-formaldehyde resin microcapsules, which, to a certain extent, allowed the microcapsules and the catalyst in the matrix played a uniform role in the dispersion. Jones and other [32, 33] investigated the effects of the morphology, size and dissolution kinetics of Grubbs catalysts on the self-healing properties and fatigue crack growth. Wilson et al. [34] discussed the repair effect of the second generation Grubbs catalyst, which increased in chemical and thermal stability compared with the first-generation Grubbs, and improved the initial polymerization rate and repair efficiency. Jin et al. [35] added 6.6% (mass fraction) of DCPD microcapsules and 10 mg of Grubbs catalyst to the epoxy resin patch to achieve a maximum repair efficiency of 58%. The initial fracture toughness of the film material was reduced by adding microcapsules with a diameter of 117 μm in the 750 μm thick film material, and the uniform distribution of the microcapsules could alleviate this effect to a certain extent. In order to solve the expensive Grubbs catalyst problem, Kamphaus et al. [36] replaced the Grubbs catalyst with WCl6 catalyst, which also enabled the open-loop metathesis polymerization of DCPD. The self-healing performance of the epoxy resin matrix showed that the catalyst efficiency is lower than Grubbs catalyst.

3.2. Epoxy resin self-healing agent system

At present, most researchers use epoxy resin as self-repairing agent. A large number of studies have been carried out by zhang mingqiu of sun yat-sen university [37-38]. They used melamine-formaldehyde resin as wall materials to prepare microcapsules containing epoxy resin. In the meantime, the microcapsules of the curing agent were prepared by the mixed system of liquid polysulfide pentaerythritol tetramine and benzyl dimethylamine catalyst, the two kinds of microcapsules added to the bisphenol A type epoxy resin matrix, under normal temperature the repair efficiency can reach 82% ~ 88%, under low temperature (-10°C), the repair efficiency can reach 86%. The system has been applied to glass fiber / epoxy resin laminates. After the impact of the laminates, the damage area of the 6 h can be recovered by 85%, and the damage is basically repaired after 12 h. But the system is not suitable for high temperature self-repairing materials, in order to solve this problem, they bisphenol A epoxy resin EPON 828 as the core material preparation of microcapsules, curing agent is still polythiol system, the catalyst using tertiary amine [Tris (dimethylaminomethyl) phenol], the system even at higher than 250°C ambient temperature, the repair efficiency is still up to 72% to 86% [39].

In order to solve the problem that the self-healing can happen rapidly at room temperature or low temperature to avoid further crack propagation, Zhang et al. [40] also prepared microcapsules of trifluorinated boron ether complex, as the curing agent of epoxy resin restorer. The system can quickly cure adhesive cracks at low temperatures. Experiments show that, by adding the epoxy resin microcapsules with the mass fraction higher than 5% to the epoxy resin matrix and the microcapsules with the boron trifluoride ether complex with the mass fraction higher than 1%, the material repair efficiency can reach 88%. However, because of the large polarity and high activity of three boron fluoride, it is difficult to direct microencapsulation. For this purpose, hollow microcapsules are prepared, and then three fluorinated boron ether complexes are encapsulated by osmosis.

All the above methods belong to the double capsule self-healing system. Zhang also studied the self-healing system of single capsule. Its curing agent USES imidazoles type latent curing agent brominated copper with 2 - methyl imidazole complex [CuBr2 (2 - MeIm) 4], and applied to the epoxy self-healing system, by adding the mass fraction of 10% of epoxy resin microcapsule and 2% [CuBr2 (2 - MeIm) 4], the repair efficiency up to a maximum, repair efficiency up to 111%, this is because the fracture toughness is higher than the fracture toughness of the substrate material after curing.

4. Summary and Prospect

At present, there are still some problems to be solved for the microcapsule self-healing composite materials, for example, how to prepare microcapsules with proper mechanical properties, how to improve the stability of microcapsule and the compatibility of microcapsule and substrate etc. For the hollow fiber type self-healing composite material, it is necessary to solve the problems of how to ensure the arrangement of fibers without changing in the process of processing. The future development trend is as follows: to optimize and develop new self-healing system, improve repair efficiency and self-healing cycle times; the self-healing function does not affect the properties of the matrix material; Self-healing has self-diagnostic function at the same time. It can make self-healing polymer material achieve real bionic material; from laboratory to engineering application.

There is still a gap between China's research on self-healing biomimetic materials and the advanced level of the world, the research of theory and experiment is not carried out deeply and extensively, mostly on the basis of foreign research. Therefore, it is necessary to further intensify the research and strengthen the technology. From raw materials to preparation methods, there should be some breakthroughs. I believe that with the development and improvement of the self-healing composite material technology, self-healing and repair of composite materials will be closer to the organism, its application will be further expanded, the self-healing function will also be more perfect.

Funding

This work is supported by the National Natural Science Foundation of China (Grant No. 51365012).

References

[1] Scheiner M, Dickens T J and Okoli O 2016 Progress towards self-healing polymers for composite structural applications *Polymer* **83** 260-282
[2] Kim H J, Lee K J 1996 Healing of fractured polymers by inter diffusion *Polymer* **37** 4593~4597
[3] Hu J F, Chen H Q and Zhang Z B 2009 Mechanical properties of melamine formaldehyde microcapsules for self-healing materials *Materials Chemistry and Physics* **118(1)** 63-70
[4] Miao Q J, Fang Z P and Cai G P 2004 Research progress of self-healing composite materials *Journal of materials science and Engineering* **22(2)** 301-303
[5] Rule J D, Brown E N, Sottos N R 2005 Wax-protected catalyst microsphere for efficient self-healing materials *Advanced Materials* **17(2)** 205-208
[6] Beres W, Koul A K, Thamburaj R A 1997 Tapered double cantilever beam specimen designed for constant testing at elevated temperatures *Journal of Testing and Evaluation* **25** 536 -542

[7] Yang H T, lv Q, Fang Z P 2009 Research progress of polymer - based self-healing materials *Journal of materials science and Engineering* **27(5)** 798-803.
[8] Wang R G, Li H Y, Hu H L 2009 Preparation and characterization of self-healing microcapsules with poly (urea-formaldehyde) grafted epoxy functional group shell *Journal of Applied Polymer Science* **113(3)** 1501-1506
[9] Brown E N, White S R, Sottos N Rm 2004 Microcapsule induced toughening in a self-healing polymer composite *Journal of Materials Science* **39(5)** 1703-1710
[10] Haiping Wang, Zhimin Rong, Mingqiu Zhang 2010 Microcapsule Filling Self - repairing Polymers and Their Composites *Chemical Progress* **22(12)** 2397-2407
[11] Li G, Meng H 2015 Recent advances in smart self-healing polymers and composites.
[12] Tee C K, Wang C, Allen R 2012 An electrically and mechanically self-healing composite with pressure- and flexion-sensitive properties for electronic skin applications *Nature Nanotechnology* **7(12)** 825-32
[13] Li C H, Wang C, Keplinger C 2016 A highly stretchable autonomous self-healing elastomer *Nature Chemistry* **8(6)** 618-624
[14] Wojtecki R J, Meador M A, Rowan S J 2011 Using the dynamic bond to access macroscopically responsive structurally dynamic polymers *Nature materials* **10(1)** 14-27
[15] Dry C 1996 The study of self healing ability for glass micro-bead filling Epoxy resin composites *Computer Structure* **35** 263
[16] Motuku M, Vaidya U K, Janowski G M 1996 Parametric studies on self-repairing approaches for resin infused composites subjected to low velocity impactb*Smart Materials & Structures* **8(8)** 623
[17] Bleay S M, Loader C B, Hawyes V J 2001 A smart repair system for polymer matrix composites *Composites Part A: Applied Science and Manufacturing* **32(12)** 1767-1776
[18] Trask RS, Bond IP 2006 Biomimetic self-healing of advanced composite structures using hollow glass fibres *Smart Mater Struct* **15** 704-10
[19] Pang JWC, Bond IP 2005 A hollow fibre reinforced polymer composite encompassing self-healing and enhanced damage visibility *Compos Sci Technol* **65(11-12)** 1791-1799
[20] C J Norris, I P Bond, R S Trask 2011 Interactions between propagating cracks and bioinspired self-healing vascules embedded in glass fibre reinforced composites *Composites Science and Technology* **71(6)** 847-853
[21] Qi H Z, Zhao H Y, Zhu K Y 2011 Research progress of self-healing polymer materials *Chemical Progress* **23(12)** 2560-2567
[22] Lee J Y, Buxton G A, Balazs A C 2004 Using nanoparticles to create self-healing composites *J. Chem. Phys* **121**5531-5540
[23] Kessler M R,White S R 2001 Self-activated healing of delamination damage in woven composites *Compos. Part A:Appl. S* **32(5)** 683-699
[24] White S R, Sottos N R, Geubelle P H 2001 Autonomic healing of polymer composite *Nature* **409** 794-797
[25] Kessler M R, Cottos N R, White S R 2003 Self-healing structural composite materials *Compos Part A:Appl. S* **34(8)** 743-753
[26] Glogowski E, Tangirala R, Russell T P 2006 Functionalization of nanoparticles for dispersion in polymers and assembly in fluids *J.Polym. Sci. Pol. Chem* **44** 5076-5086
[27] Brown E N, White S R, Sottos N R 2005 Retardation and repair of fatigue cracks in a microcapsule toughened epoxy composite part :Manual infiltration *Compos. Sci. Technol* **65** 2466-2473
[28] Brown E N, White S R, Sottos N R 2005 Retardation and repair of fatigue cracks in a microcapsule toughened epoxy composite part:In situ self- healing *Compos.Sci. Technol* **65** 2474-2480
[29] Rule J D, Brown E N, Sottos N R 2005 Wax-protected catalyst microspheres for efficient self-healing materials *Adv. Mater* **17(2)**205-208

[30] Wilson G O, Moore J S, White S R 2008 Autonomic healing of epoxy vinyl esters via ring opening metathesis polymerization *Adv. Funct. Mater* **18** 44-52
[31] Jackson A C, Bartelt J A, Marczewski K 2010 Silica-protected micron and sub-micron capsules and particles for self-healing at the microscale *Macromol. Rapid Comm* **31** 1-7
[32] Jones A S, Rule J D, Moore J S 2006 Catalyst morphology and dissolution kinetics for self-healing polymers *Chem. Mater* **18** 1312-1317
[33] Jones A S, Rule J D, Moore J S 2007 Life extension of self-healing polymers with rapidly growing fatigue cracks *J. R. Soc. Interface* **4** 395-403
[34] Wilson G O, Caruso M M, Reimer N T 2008 Evaluation of ruthenium catalysts for ring-opening metathesis polymerization-based self-healing applications *Chem. Mater* **20** 3288-3297
[35] Jin H H, Miller G M, Pety S J 2013 Fracture behavior of a self-healing,toughened epoxy adhesive *Int. J. Adhe. Adhes* **44** 157-165
[36] Kamphaus J M, Rule J D, Moore J S 2008 A new self-healing epoxy with tungsten (Ⅵ) chloride catalyst *J. R. Soc. Interface* **(5)** 95-103
[37] Yuan Y C, Rong M Z, Zhang M Q 2009 Study of factors related to performance improvement of self-healing epoxy based on dual encapsulated healant *Polymer* **50** 5771-5781.
[38] Yuan Y C, Rong M Z, Zhang M Q 2008 Self-healing polymeric materials using epoxy/mercaptan as the healant *Macromolecules* **4**15197-5202
[39] Yuan Y C,Ye X J, Rong M Z 2011 Self-healing epoxy composite with heat-resistant healant *ACS Appl. Mater. Inte*r **3(11)** 4487-4495
[40] Xiao D S, Rong M Z, Zhang M Q 2007 A novel method for preparing epoxy-containing microcapsules via UV irradiation-induced interfacial copolymerization in emulsions *Polymer* **48** 4765-4776

IOP Conf. Series: Materials Science and Engineering **292** (2017) 012082 doi:10.1088/1757-899X/292/1/012082

Influences of glass content on the microstructure and properties of BaO-CaO-Al_2O_3-SiO_2 glass/alumina composite for LTCC applications

Z X Li[1], H J Mao[1], Y W Zhang[2] and W J Zhang[1,3]

[1] Department of Material Science and Engineering, National University of Defense Technology, Changsha 410073, China
[2] Shandong Aerospace Electronic Technology Research Institute, Yantai 264000, China

[3] E-mail: zhwjun@nudt.edu.cn

Abstract. In this paper, a new amorphous Ba-Ca-Al-Si glass was prepared and the composites with different glass/alumina ratios, which ranged from 40 wt. % glass to 60 wt. % glass, have been studied. According to XRD patterns, the main crystalline phases of the composites are Al_2O_3 and $BaAl_2Si_2O_8$, and $BaAl_2Si_2O_8$ is generated by the reaction between glass and alumina. With regard to dielectric and mechanical properties, both the dielectric constant and the bending strength of the samples show positive relationships with glass content, and then the values tend to decline when the glass content exceeds 55 wt. %, which can be explained by the results of sintering shrinkage curves, bulk densities and SEM images. The optimal composition of 55 wt. % glass-45 wt. % alumina was developed, exhibiting good dielectric properties (ε_r=5.882, $tan\delta$=2.75× 10^{-3}) and high bending strength (σ=169 MPa). Finally, the 55 wt. % glass/alumina composite was chosen for tape casting, showing a well chemical compatibility with silver.

1. Introduction

Electronic packing is a method of enclosing, protecting or providing physical structure to electronic components and finished electronic devices [1]. In order to get electronic components with smaller size, higher reliability and higher frequency, multilayer cofired ceramics is popular in recent years [2]. Low temperature cofired ceramics (LTCC) adjusts the ceramics with glass to be sintered less than 900 °C, which can be cofired with Au, Ag, Cu and other low melting metal [3]. Because the use of higher conductivity metal gives LTCC substrate materials better performance, LTCC shows good application prospects in computers, aerospace and military communication field.

Generally, LTCC materials can be divided into two categories: glass-ceramic and glass/ceramic [4]. Glass-ceramic is obtained by sintering amorphous glass to make it crystallize [5]. To meet the mechanical and electrical requirements, the crystallization always needs to be accurately manipulated to a certain degree, which makes the sintering process complicated [6]. Glass/ceramic systems are made up of low softening point glass and crystalline ceramics, in which the glass phases wet and connect the ceramic particles to lower the sintering temperature, meanwhile, the ceramic-fillers provide excellent mechanical and electrical properties to the system [7]. Therefore, required performance of the materials can be achieved by designing the composites and the synthetic process is relatively simple and controllable. On account of these advantages, glass/ceramic materials have been widely used.

In the glass/ceramic systems, alumina is one of the most commonly used ceramics because of its outstanding mechanical and electrical properties. In terms of the glass, lead borosilicate glass with low-softening temperature, low-dielectric constant and low thermal conductivity has attracted much attention, and related commercial products have been developed by the companies of Dupont, Fujitsu and NEC. However, in these systems, the large amounts of residual glassy phases are detrimental to the high-frequency dielectric properties, and lead-based products are not environment-friendly. Theses disadvantages seriously limit the further application of glass/ceramic systems in the field of LTCC substrate materials. Therefore, based on alumina, a lot of other glass systems have been developed. *Induja et al.* reported [8] a new LTCC composition consisting of Bi_2O_3-ZnO-B_2O_3-SiO_2 and Al_2O_3, which had ε_r=10.9 and tanδ=0.009 (at 5 GHz) after sintered at 850 °C. In the research of *Zeng et al.*, the lead-free, low-softening point ZnO-Bi_2O_3-B_2O_3 glass +Al_2O_3 system were prepared below 850 °C. This system exhibited good dielectric properties: small permittivity of 5.73 and low loss tangent of 8.94×10^{-4} at 1 MHz, and its bending strength was 129.9 MPa [9]. *Xia et al.* obtained a K_2O-B_2O_3-SiO_2-Al_2O_3 glass/alumina + silica composite with the dielectric constant of 6.1 and dielectric loss of 2×10^{-3} at 1MHZ, and flexural strength of 168 MPa [10]. The experiment described here attempts to reduce the residual glassy phase in the glass/ceramic system through the avenue of making the glass phases react with the Al_2O_3 ceramics during the sintering process, and then the high-frequency dielectric properties will be improved. The composition of lead-free BaO-CaO-Al_2O_3-SiO_2 (BCAS) glass + Al_2O_3 ceramic is selected, the reaction between the two phases is investigated, the effects of the glass/ceramic ratio on the dielectric and mechanical properties are studied, and the microstructure and compatibility of co-firing with Ag electrode are also analyzed.

2. Experimental procedures

The starting materials were high purity powders (>99%, Sinopharm Chemical Reagent Co., Ltd.) of $BaCO_3$, CaO, Al_2O_3, SiO_2, B_2O_3, ZrO_2 and TiO_2. The designed composition of Ba-Ca-Al-Si glass is shown in table 1. After the homogeneous mixture process, the powder was melted in a platinum crucible at 1500°C for 2 hours. Then the frits were obtained by pouring the liquid directly into deionized water for quenching. The glass powder was finally obtained after milling the frits in agate jars with absolute ethanol at the speed of 500 r/h for 12 hours.

Table 1. Composition of Ba-Ca-Al-Si glass (mol%).

BaO	CaO	SiO_2	$Al2O_3$	B_2O_3	ZrO_2	TiO_2
15	11	54	7	10	2	1

Different glass/ceramic samples were prepared by mixing 40 wt. %, 45 wt. %, 50 wt. %, 55 wt. % and 60 wt. % glass (D50=2 μm) with Al_2O_3 (D50=5 μm). For homogeneous mixing, the samples were ball milled with absolute ethanol at 300 r/h for 3 hours. After drying the slurries up, the powders were well mixed with 5 wt. % polyvinyl alcohol (PVA) separately and pressed into pellets and bars. Finally the samples were continuously sintered at 450 °C for 2 h, 750 °C for 1 h and 875 °C for 1 h with a heating rate of 2 °C /min.

The identification of crystalline phases was carried out by X-ray diffraction analysis (XRD, Bruker AXS D8 Advance device, Germany). The microstructures were studied by a scanning electron microscopy (SEM, Quanta FEG 250, US). The sintering shrinkage curves were determined in the temperature range from 10 °C to 1000 °C at 5 °C /min (Netzsch DIL 402C, Germany). The densities of the sintered samples were measured using Archimedes method. The dielectric constants ε_r and dielectric losses *tanδ* of the composites were tested using a network analyser (N5230C, Agilent Co.).

IOP Conf. Series: Materials Science and Engineering **292** (2017) 012082 doi:10.1088/1757-899X/292/1/012082

3. Results and discussion

Figure 1(a) shows the XRD patterns of original BCAS glass, sintered glass and the composite of 50 wt. % glass-50 wt. % alumina. It indicates that the designed BCAS glass has amorphous nature with small amount of quartz, regardless of whether the glass is sintered, while the composition with alumina generates new crystalline phases. According to the number and intensity of the peaks, we can determine that the main crystalline phases of the composites are Al_2O_3 and $BaAl_2Si_2O_8$. Moreover, the disappearance of bread peak with the addition of Al_2O_3 into the BCAS glass suggests the reduction of the glass phases. Therefore, it can be concluded that the emerging $BaAl_2Si_2O_8$ phase is generated from the reaction between the BCAS glass and Al_2O_3. Further analyses of the samples with different amounts of glass content are demonstrated in figure 1(b). It can be seen that the content of glass from 40 wt. % to 60 wt. % doesn't affect the main type of crystalline phases.

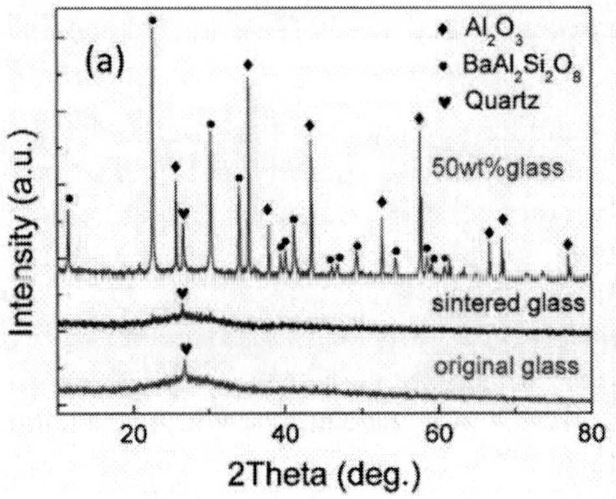

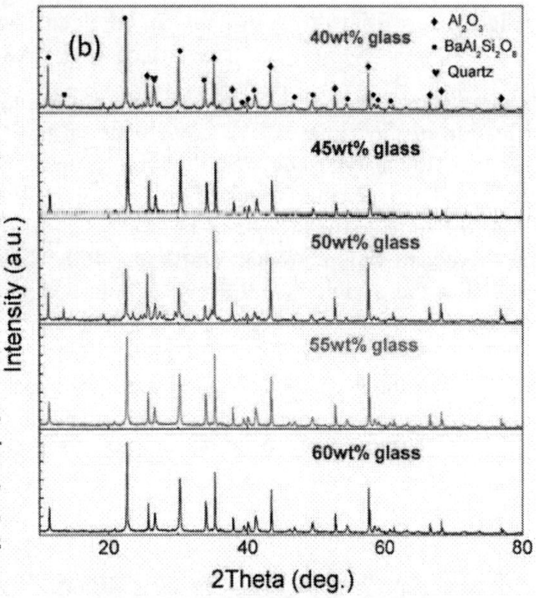

Figure 1. XRD patterns of (a) BCAS glass powder, sintered glass powder and 50 wt. % glass/alumina composite, (b) samples with 40 wt. %, 45 wt. %, 50 wt. %, 55 wt. % and 60 wt. % glass content.

We now turn towards the dielectric and mechanical properties of the series of samples. For the design of advanced electronic packaging in integrated circuits, signal transmission quality is one of the most important aspects, and dielectric constant can be a direct measure for it. For substrates, the dielectric constant and the delay time t_d has the following relationship:

$$t_d = l \times (\varepsilon_r)^{1/2}/c \tag{1}$$

Where l is the wiring length, ε_r is the dielectric constant and c is the velocity of light. The equation indicates that reducing the dielectric constant makes the signal transmission speed up. Therefore, a low dielectric constant is an essential prerequisite for the commercial application of an LTCC substrate material. Figure 2(a) shows the dielectric data of the samples. It can be found that all the samples have a dielectric constant lower than 6 at ~10GHz. Furthermore, the dielectric constant increases with the increase of glass content. When it comes to a critical value, the dielectric constant begins to decrease if the glass content continues to increase. The maximum dielectric constant 5.882 of the samples can be achieved when the composite contains 55 wt. % glass. Besides, the absorption of electric signal by the dielectric material is inevitable during its propagation process, thus, in order to get better performance, low dielectric loss is also an essential element. As shown in figure 2(a), all the samples display a dielectric loss less than 0.003 at ~10 GHz. Bending strength is another key performance indicator for the LTCC substrate materials, determining the service life of the related commercial products. The variation of the bending strength illustrating in figure 2(b) displays the similar tendency as the dielectric constant. The optimal composite is 55 wt. % glass/alumina and the bending strength achieves 169 MPa.

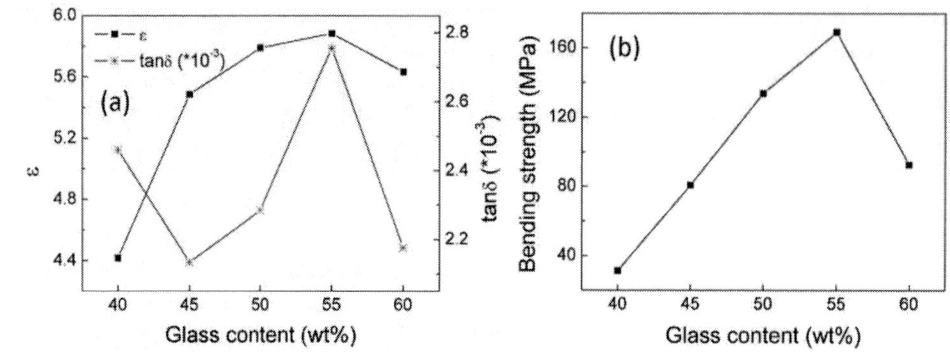

Figure 2. (a)Dielectric constants and dielectric losses measured at ~10 GHz and (b) bending strength of the BCAS /Al_2O_3 systems.

It is well known that the dielectric constant and bending strength are fairly interrelated to the densification of the samples. We now focus on the shrinkage process, density and microstructure of the samples to analyze the densification process and unravel the reason of the aforementioned phenomena. The sintering curves of different glass-content samples are shown in figure 3(a). It can be seen that the temperature range of shrinkage in all the samples is from ~700 °C to ~875 °C, which means the densification process starts at ~700 °C and ends at~ 875 °C. Moreover, it is obviously detected that the shrinkage intensity is enlarged as the glass content elevates. And the shrinkage percentages are ~8 %, ~10 %, ~11.3 %, ~12.6 % and ~16.6 % for the samples with glass content of 40 %, 45 %, 50 %, 55 % and 60 %, respectively. This phenomenon can be explained by the role of glass phases in the composites. In glass/ceramic systems, glass phases wet and connect the ceramic particles, so that the more glass is contained, the larger the shrinkage percentage will be. In addition, we can find that the slope of the shrinkage curve gradually increases with the increase of the glass content, implying that the samples with larger amount of glass appear to have a high shrinkage speed. The differential curves in figure 3(a) also prove this view. It should be noted that too low shrinkage rate may be detrimental to the densification of the sample, and excessively high shrinkage speed may lead to the formation of more pores. The experiment results of the sample density and microstructure below provide evidences to this deducing.

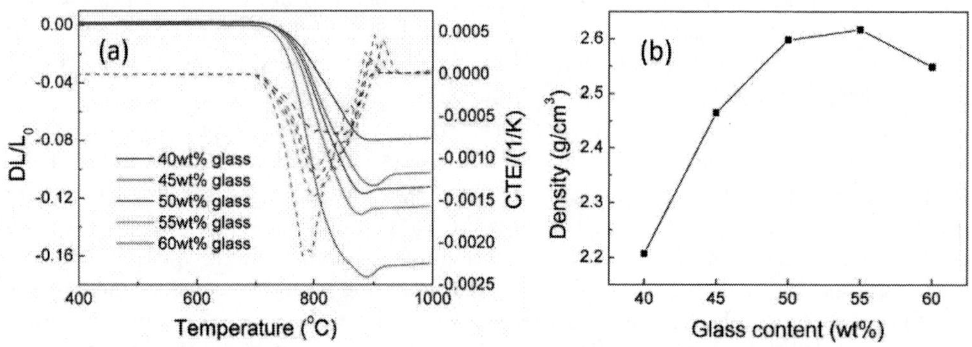

Figure 3. (a) sintering shrinkage curves, and (b) densities of the BCAS /Al_2O_3 systems.

The measured density data in figure 3(b) reveal that the maximum value of 2.617 is obtained when the glass content reaches 55 wt. %, and further increase of the glass content induces the decline of the density. This feature is nearly consistent with the variation of the dielectric constant and bending strength. As mentioned before, the increase of the glass content is accompanied by enlarged shrinkage intensity, which directly results in the enhancement of the density and then improves the dielectric and mechanical properties. As to the lower density of the sample with 60 % glass, the too fast shrinkage speed bringing about more pores is the critical factor, which will be discussed later. More visualized and detailed proofs are given by the detection of the cross-section morphology.

Figure 4 shows the cross-section images of the samples with different content of glass measured by SEM.

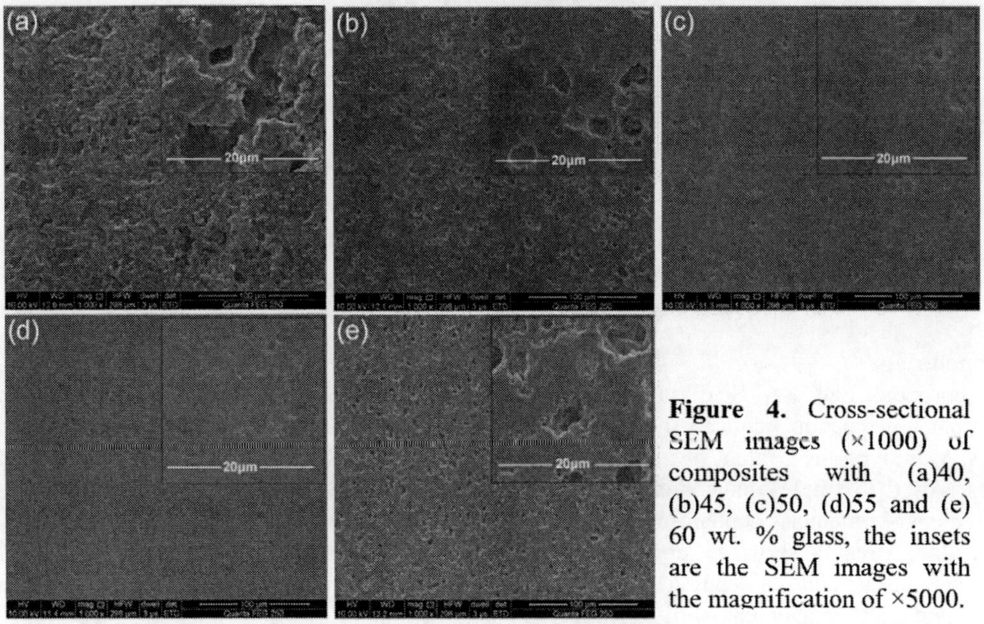

Figure 4. Cross-sectional SEM images (×1000) of composites with (a)40, (b)45, (c)50, (d)55 and (e) 60 wt. % glass, the insets are the SEM images with the magnification of ×5000.

A typical glass/ceramic behavior, with large Al_2O_3 grains embedded within an amorphous glass matrix, is observed in all the samples. However, the details in the microstructure images are distinct. For the 40 wt. % glass sample, there are plenty of uncoated Al_2O_3 phases and large pores displayed in figure 4(a), resulting in lower density, dielectric constant and bending strength as mentioned before. As the increase of glass content, the size of the pores is scaled down and the structure becomes more homogenous, and the most dense morphology can be obtained when the glass content is 55 wt. %, seen from figure 4(b) to (d). Whereas, figure 4(e) indicates that excess glass has negative effect, leading to the emergence of numerous large pores. This is ascribed to the excessively fast shrinkage speed analyzed in figure 3(a). So it is easy to understand that the 55 wt. % glass sample possesses the optimal dielectric and mechanical properties.

In order to confirm whether the BCAS glass /Al_2O_3 ceramic is suitable for LTCC substrate application, the properties of the samples laminated with green tapes and the chemical compatibility of this composition and silver are evaluated. The 55 wt. % glass/alumina composite was chosen for tape casting. From figure 5(a), we can see that the laminated sample after the same sinter process as aforementioned samples demonstrates a denser and more homogeneous microstructure, but inevitable pores still exist. Additionally, the density is enhanced to 2.74, the dielectric constant increases to 6.53, and the bending strength is improved to 210 Mpa. Figure 5(b) shows the backscattered-electron diffraction (BSED) image of the samples with the printing of silver paste on to the layered green tapes. After co-fired at 875 °C, a silver layer with the thickness of ~10 μm is formed, accompanied with the formation of a ~50 μm buffer layer. The EDS results in figure 5(c) and (d) reveal that the bright white layer is pure silver, and the dark gray part in the buffer layer is abundant with the element of Al, Si and Ba. These results imply that the reaction between the Ag and BCAS glass /Al_2O_3 ceramic is not taken place, showing a well chemical compatibility. The segregation of the phases with the Al, Si and Ba elements and the pores at the buffer layer may be due to the mismatching between the silver paste and green tape, and our later work will devote to the development of the better matched silver paste.

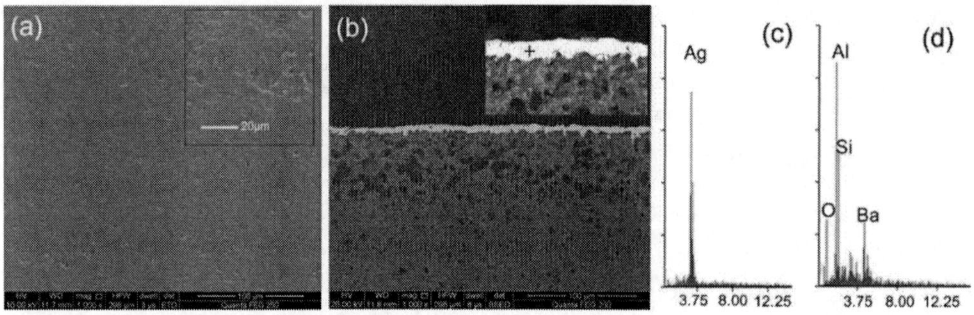

Figure 5. Cross-sectional SEM images (×1000) of (a) the laminated tape and (b) the tape with silver paste, (c) and (d) SEM-EDS results of the tape with silver paste, the blue and red crosses in the inset of (b) correspond to (c) and (d), respectively. Both the inset in (a) and (b) are the SEM images with the magnification of ×5000.

4. Conclusions

The amorphous Ba-Ca-Al-Si glass was prepared and synthesized with alumina. The XRD patterns show that the reaction between the BCAS glass and alumina can generate $BaAl_2Si_2O_8$ crystalline phases. Moreover, the content of glass from 40 % to 60 % doesn't affect the main type of crystalline phases. All the samples have dielectric losses less than 0.003, which can meet the commercial requirement. As the glass content increases from 40 wt. % to 55 wt. %, both the dielectric constant and the bending strength show an upward tendency. When the glass content exceeds 55 wt. %, the dielectric constant and the bending strength begin to decline. The results of sintering shrinkage curves, density and SEM images can well explain this phenomenon. With the increase of glass content, the shrinkage intensity and the shrinkage speed are elevated. Low shrinkage rate is detrimental to the densification process, while high shrinkage speed may lead to the formation of more pores. The optimal composite, which consists of 55 wt. % glass and 45 wt. % alumina, can achieve the dielectric of 5.882, the bending strength of 169 MPa. The optimal composition was chosen for tape casting. The tape has better properties and more homogeneous microstructures, and shows chemical compatibility with silver. Finally, further research needs to be carried out to develop suitable silver paste matching better with the tape.

Reference

[1] Wessely H, Fritz O, Horn M, Klimke P, Koschnick W and Schmidt K H 1991 Electronic packaging in the 1990s: the perspective from europe *IEEE Transactions on Components Hybrids & Manufacturing Technology* **14** 272-84

[2] Sebastian M T and Jantunen H 2008 Low loss dielectric materials for LTCC applications: a review *International Materials Reviews* **53** 57-90

[3] Berglund M, Persson A and Thornell G 2015 Evaluation of Dielectric Properties of HTCC Alumina for Realization of Plasma Sources *Journal of Electronic Materials* **44** 3654-60

[4] Shimada Y, Utsumi K, Suzuki M, Takamizawa H, Nitta M and Watari T 2003 Low firing temperature multilayer glass-ceramic substrate *IEEE Transactions on Components Hybrids & Manufacturing Technology* **6** 382-8

[5] Shimada Y, Yamashita Y and Takamizawa H 2002 Low dielectric constant multilayer glass-ceramic substrate with Ag-Pd wiring for VLSI package *Micronanoelectronic Technology* **11** 163-70

[6] Li B, Qing Z, Li Y, Li H and Zhang S 2016 Crystallization, microstructures and properties of low temperature co-fired CaO-Al_2O_3-SiO_2 glass-ceramic *Journal of Electroceramics* 1-6

[7] Induja I J, Surendran K P, Varma M R and Sebastian M T 2017 Low κ, low loss alumina-glass composite with low CTE for LTCC microelectronic applications *Ceramics International* **43** 736-40

[8] Induja I J, Abhilash P, Arun S, Surendran K P and Sebastian M T 2015 LTCC tapes based on Al_2O_3-BBSZ glass with improved thermal conductivity *Ceramics International* **41** 13572-81
[9] Zeng D, Xu J, Chen Y, Chen W, Yu Y, Wang H and Zeng R 2015 Novel lead-free glass/ceramics system with low permittivity, low loss for ltcc application *International Journal of Applied Ceramic Technology* **12** E112-6
[10] Xia Q, Zhong C W and Luo J 2014 Low temperature sintering and characteristics of K_2O-B_2O_3-SiO_2-Al_2O_3, glass /ceramic composites for LTCC applications *Journal of Materials Science Materials in Electronics* **25** 4187-92

IOP Conf. Series: Materials Science and Engineering **292** (2017) 012083 doi:10.1088/1757-899X/292/1/012083

Adsorption equilibrium and dynamics of toluene vapors onto three kinds of silica gels

K L Yan[1, 2, 3], Q Wang[1, 2]

[1] SINOPEC Research Institute of Safety Engineering, 339 Songling Road, Qingdao 266000, P. R. China
[2] State Key Laboratory of Safety and Control for Chemicals, 218 Yanansan Road, Qingdao 266071, P. R. China

[3] E-mail: yankele214@163.com

Abstract. The benzene is the representative of VOCs and widely exists in the industrial waste gas. In this study, adsorption equilibrium and dynamics of toluene vapors at five initial concentrations (1.39 g·m^{-3}, 5.12 g·m^{-3}, 8.38 g·m^{-3}, 15.6 g·m^{-3}, 21.3 g·m^{-3}) onto three kinds of silica gels (GA, GB and GC) were investigated and compared. The experimental results showed that GA has the rich microporous and mesoporous distributions, and the larger surface area and microporous volume than GB and GC. It can be clearly seen that the order of the adsorption rate of adsorbents on the silica gels samples is GA, GB and GC. Due to the suitable pore distribution in the region of micropore and mesopore (1–4 nm), GA exhibits the comparable breakthrough adsorption capacities with GB and GC for a given initial concentration. Moreover, the experimental data were fitted to the Langmuir and Freundlich models, respectively. The Freundlich isotherms correlated with the experimental data presented a better fitting than Langmuir model. Taken together, it is expected that GA silica gel would be a promising adsorbent for the removal of toluene vapors from gas streams.

1. Introduction

Nowadays, the emission of volatile organic compounds (VOCs) has gained more and more attention not only for their damage to human health and environment but for economic interests as well. Adsorption technology has been recognized as a preferred method for the removal of VOCs because it provides additional benefits from the recovery of the valuable VOCs for reuse [1]. The heart of an adsorption process is usually a porous solid adsorbent. The activated carbon considered as the major commercial adsorbent provides excellent adsorption capacity, but it has been recognized that it owns some disadvantages such as fire risk, pore blocking and hygroscopicity [2–5]. In addition, the activated carbons are predominantly consisted of micropores (< 2 nm in size) with the specific surface area and pore volume, so it is very difficult to regenerate adsorbed VOCs due to the strong dispersive force resulting from the overlap of attractive forces from opposite pore walls of micropore [6, 7].

As a result, alternative adsorbents such as hydrophobic zeolite and silica gels have attracted considerable attention to control VOCs. Hydrophobic zeolites have been found to be effective in VOCs removal [8]. However, serious diffusion restrictions imposed by the micropores (<2 nm) tend to inhibit its ability to adsorb large VOC molecules [9]. Silica gels with high surface areas have attracted a great deal of attention due to their wide range of application as adsorbents for environmentally hazardous chemicals, reaction catalysts, catalyst supports, chemical sensors, and electrical and optical devices [10-13]. Recently, Sui et al. [14] demonstrated that the silica gel would be a good candidate for the removal and recovery of high concentrations of toluene vapor from gas stream through atmospheric pressure adsorption and vacuum desorption. However, studies on the dynamic adsorption performance of silica gels in regards to VOCs have been very limited in comparison to those on their adsorption equilibria [15].

In this study, three kinds of silica gels, were chosen and considered as representatives of silica gels. The textural properties of the above three silica gels were systematically characterized. In addition, the adsorption behavior of the toluene vapors onto the silica gels at five initial concentrations were investigated. Meanwhile, the experimental data were correlated with Langmuir and Freundlich adsorption models.

2. Experimental

2.1. Materials

Three kinds of silica gels with different pore structures were used as the adsorbents, and named GA, GB and GC, respectively. The silica gel samples were supplied from Qingdao Meigao Co. Ltd.. All samples were desiccated in vacuum at 423.15 K for 12 h and preserved in a vacuum sealed bag as the samples were cooled to room temperature in vacuum. Toluene was supplied from Tianjin Guoyao Co. Ltd. with a purity of 99.5% by mass.

2.2. Characterization

Dried silica gels were characterized for its structure properties, such as surface area, pore volume and mean pore diameter. The nitrogen adsorption/desorption isotherms of three silica gels were determined using an adsorption analyzer ASAP 2020 (Micromeritics Instrument Co., USA) operated at 77 K. The experimental procedure was described briefly as follows: The silica gel particles were weighed (0.1 g) and poured into a sample bottle. The sample bottle was circulated in a heating jacket, heated to 523 K and vacuumed below 3 μm Hg to remove the moisture within the silica gel particles. After the silica gel sample was completely dried, it was weighed again and placed into liquid nitrogen to keep the sample temperature at 77 K. The freezing silica gel sample bottle was filled with nitrogen gas to determine the surface area and pore volume. The specific surface area was determined using the N_2 isotherms data by means of the Brunauer-Emmett-Teller (BET) method. The total pore volume (V_{total}) was estimated to be the liquid volume of nitrogen at a relative pressure of about 0.98. The micropore volume (V_{micro}) was calculated from the N_2 isotherms data by Barrett-Joyner-Halenda (BJH) method. The pore size distributions were calculated by applying the density functional theory (DFT) to N_2 isotherm data.

2.3. Adsorption Experiment

The experimental setup is shown in figure 1. It basically consisted of a toluene vapor generator, an adsorption column (10 mm in inner diameter and 150 mm in height) where silica gel samples were held, and a gas analysis system. Firstly, the nitrogen steam passed through a dryer bed in order to remove water, and was divided into two streams using two mass flows. One nitrogen flow was conducted to the bubble saturator containing pure toluene, which was held at a constant temperature by means of a thermostatic water bath. Secondly, the third nitrogen stream was used to diluted and attain a given toluene concentration. Then, the given concentration toluene vapor was passed through the adsorption column, and the flow rate of toluene vapor was fixed at 50 ml/min at room temperature. At the outlet of the adsorption column, a gas chromatograph (HP 6890, Agilent, Japan) with a flame ionization detector (FID) was used to measure the concentrations of toluene. The exhaust was discharged into the atmosphere after several rounds of purification.

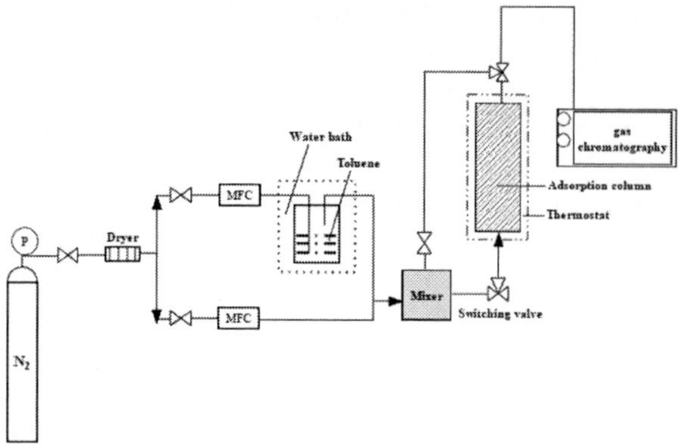

Figure 1. Schematic diagram of the experimental apparatus.

2.4. *Isotherm Models*

Adsorption experimental data were correlated with two different isotherms, namely, Langmuir and Freundlich models.

Langmuir model. The well-known Langmuir isotherm can be expressed by equation (1):.

$$\frac{C_e}{q_e} = \frac{1}{q_m b} + \frac{1}{q_m} C_e \tag{1}$$

where q_e is the amount adsorbed (mg·g^{-1}), q_m is the saturated amount adsorbed mg·g^{-1}, C_e is the equilibrium concentration (g·m^{-3}), and b is the adsorption affinity (m^{-3}·g).

Freundlich model. The Freundlich isotherm is expressed by equation (2):

$$q_e = K_f C_e^{n_f} \tag{2}$$

where q_e is the amount adsorbed (mg·g^{-1}), K_f and n_f are empirical constants that are generally temperature-dependent, and C_e is the equilibrium concentration (g·m^{-3}). The Freundlich equation is very popularly used in the description of adsorption VOCs from gas stream.

3. Results and Discussion

3.1. *Characteristics of Silica gels*

In order to understand more clearly the pore structure of adsorbents, the N_2 adsorption-desorption isotherms and pore size distributions of three kinds of silica gels were shown in figure 2; the textural properties of all the adsorbents studied were included in table 1. It can be seen from figure 2 that at lower relative pressure (p/p_0) below 0.05, the nitrogen uptake increases sharply with the increment of relative pressure, proving the existence of micropore structure in the GA and GB. Meanwhile, visible hysteresis loop means that GB and GC contain mesopore distribution. According to IUPAC classification, the adsorption–desorption isotherms of GA are close to type I, reflecting the domination of micropores in the pore structure. And GB is typical of adsorbents with mixed micro- and mesoporous structure. Also, it is significantly learned from the pore size distributions of three silica gels that GA is also typical of adsorbents with mixed micro- and mesoporous structure, the pore size is mainly distributed in the regions of 1-4 nm. GB is a typical mesopore adsorbent with a small amount of micropore (< 2 nm). In contrast, the pore sizes of GC are distributed in the regions of 10-40 nm near the macropore. It can be clearly seen from table 1 that three adsorbents have obvious different textural properties. GA has the largest surface area (700.5961 m^2·g^{-1}) and micropore volume V_{micro} (0.088194 cm^3·g^{-1}).

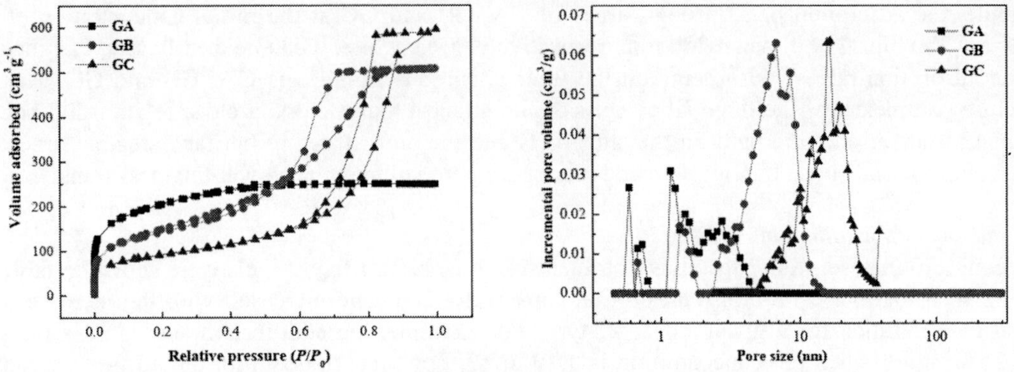

(c) GC silica gel

Figure 2. The N_2 adsorption-desorption isotherms at 77 K and pore size distributions of three kinds of silica gels.

Table 1. Textural properties of three kinds of silica gels.

Adsorbent	BET ($m^2 \cdot g^{-1}$)	Average pore diameter (nm)	V_{total} ($cm^3 \cdot g^{-1}$)	V_{micro} ($cm^3 \cdot g^{-1}$)
GA	701	2.2	0.3855	0.088194
GB	523	6.0	0.7861	0.001273
GC	322	11.4	0.9223	0.00012

3.2. Dynamic adsorption properties

In this study, five initial toluene concentrations including 1.39 $g \cdot m^{-3}$, 5.12 $g \cdot m^{-3}$, 8.38 $g \cdot m^{-3}$, 15.6 $g \cdot m^{-3}$, 21.3 $g \cdot m^{-3}$, were chosen to investigate the adsorption dynamic adsorption behavior on the different kinds of silica gels. Figure 3 just shows the breakthrough curves of toluene vapors adsorbed on three silica gels at 21.3 $g \cdot m^{-3}$, where C_0 is the inlet toluene concentration ($g \cdot m^{-3}$), C_t is the outlet concentration ($g \cdot m^{-3}$) and t is the adsorption time (min).

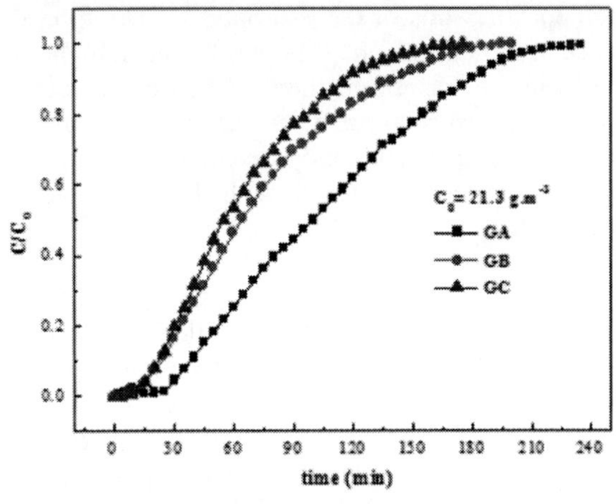

Figure 3. Breakthrough curves of toluene on three kinds of silica gels.

Evidently, under a given concentration, the breakthrough curves of toluene vapors onto three silica gels are similar, however, GA shows longer adsorption breakthrough time (the outlet concentration is equal to 95 percent of the inlet concentration of toluene) for toluene vapors than GB and GC. For example, the adsorption breakthrough times of GA, GB and GC at the initial concentration of 8.38 $g \cdot m^{-3}$ are 280 min, 240 min and 200 min, respectively. Meanwhile, it can be clearly seen that the order of the adsorption rate of adsorbents on the three silica gels samples is GA, GB and GC. The pore structure parameters of the three silica gel samples suggest that there is a close relationship between the mass transfer rate internally in the silica gels and the pore structure (surface area and micropore pore volume). The bigger surface area and micropore pore volume, the slower the mass transfer rate.

3.3. Adsorption equilibrium

The equilibrium adsorption capacities of toluene on three kinds of silica gels were shown in table 2. It can be seen that the equilibrium adsorption capacities of toluene increased with the increase of the initial concentration for a given silica gel type. For example, the equilibrium adsorption capacity is just 21.0 $mg \cdot g^{-1}$ when the concentration is 1.39 $g \cdot m^{-3}$, however, the equilibrium adsorption capacity increases to 220.5 mg/g at the initial concentration of 21.3 $g \cdot m^{-3}$. Moreover, the order of the equilibrium adsorption capacity of toluene is GA, GB and GC for a given initial concentration. For example, the equilibrium adsorption capacities of toluene for GA, GB and GC are 113.5 $mg \cdot g^{-1}$, 51 $mg \cdot g^{-1}$ and 26 $mg \cdot g^{-1}$ at the initial concentration of 5.12 $g \cdot m^{-3}$.

Table 2. Equilibrium adsorption capacities of toluene on three kinds of silica gels.

Concentration	Adsorption capacities（$mg \cdot g^{-1}$）		
（$g \cdot m^{-3}$）	GA	GB	GC
1.39	21	14.5	12
5.12	113.5	51	26
8.38	155.4	71.1	37.5
15.6	201	98	54
21.3	220.5	113.5	60

It is clearly that the adsorption capacities of the GA and GB increased sharply with the increase of toluene vapors concentrations at lower concentrations, manifesting the existence of microporous structure inside the two adsorbents while those on GC slowly increased with the increment of concentration, which is typical of adsorption in a mesoporous adsorbent. It is well-known that the adsorption energy in the micropore is much larger than in the mesopores due to the overlapping of adsorption forces from the opposite walls of the micropores. Therefore, the adsorption capacities of GA and GB for the low concentration of toluene vapors were much higher than GC. By comparison of the adsorption capacities of toluene vapors on three adsorbents in the experimental concentration range, it is found that, GA has the best adsorption capacities. The results were mainly due to the fact that GA has the largest BET surface area and micropore volume. Among the three adsorbents, GC has the lowest adsorption capacities for toluene vapors, which is consistent with its lowest values of micropore volume and surface area.

3.4. Isotherm models

The adsorption equilibrium data of toluene vapors at five initial concentrations onto GA, GB and GC at 298 K were obtained. Langmuir and Freundlich equations are used to correlate the experimental equilibrium data of toluene vapors. A non-linear optimization routine is used to optimize the parameters of the models to fit the experimental data. The fitting parameters and correlation coefficients of Langmuir and Freundlich equations for the adsorption of toluene onto GA, GB and GC are presented in table 3. It is found that all of the equilibrium data were correlated by Langmuir and Freundlich equations with large correlation coefficient. From table 3, it is clear that the Freundlich equation has a better correlation with experimental data than the Langmuir equation. The curves of Freundlich fitting on three silica gels were shown in figure 4. The parameters K_F and n_f varied with different adsorbents; however, there was no apparent correlation between K_F, n_f, and the adsorbent properties (surface area and porosity) within the examined concentration ranges. In the present work, the adsorption isotherms are used mainly for comparing adsorption capacities between different adsorbents.

Table 3. Fitting parameters of Langmuir and Freundlich models for toluene on three silica gels.

Adsorbent	Langmuir Model			Freundlich Model		
	q_m (mg·g^{-1})	b (m^3·g^{-1})	R^2	K_f(m^{3nf}·mg·g^{nf+1})	n_f	R^2
GA	490.196	0.04359	0.9175	41.5025	0.5644	0.9478
GB	201.082	0.06017	0.9743	18.2349	0.6069	0.9803
GC	89.686	0.09247	0.9633	10.6588	0.5751	0.9891

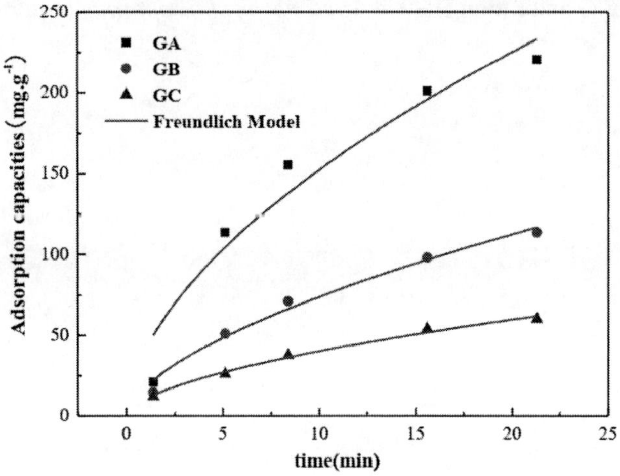

Figure 4. Curves of Freundlich fitting on three silica gels.

4. Conclusions

Adsorption equilibrium and dynamics of toluene vapors onto three kinds of silica gels with different textural properties were evaluated at five initial concentrations. It is concluded that GA will be an efficient and competitive adsorbent for toluene vapors adsorption. Due to the bigger surface area and micropore volume, the order of the adsorption rate of adsorbents on the three silica gels samples is GA, GB and GC. Compared with GB and GC, the results of adsorption equilibrium experiments confirmed that the equilibrium adsorption capacities of toluene increased with the increase of the initial concentration for a given silica gel type, and GA had the best breakthrough adsorption capacity for a given initial concentration. In addition, it also can be seen that the Freundlich model has a better correlation with experimental data than the Langmuir model for toluene adsorption onto silica gels.

Acknowledgments

This research was financially funded by Project of SINOPEC Science and Technology Department (Grant No 316049), National Natural Science Foundation of China (Grant No E080301), and Project of SINOPEC Research Institute of Safety Engineering (Grant No Y-208).

References

[1] Ghoshal A K, Manjare S D 2002 *J. Loss Prevent. Proc. Ind.* **15** 413
[2] Zerbonia R A, Brockmann C M, Peterson P R and Housley D 2001 *J. Air Waste Manage. Assoc.* **51** 1617
[3] Brennan J K, Bandosz T J, Thomson K T and Gubbins K E 2001 *Colloid Surf. A: Physicochem. Eng. Aspects* **187** 539

[4] Cosnier F, Celzard A, Furdin G, Bégin D and Marêché J F 2006 *Adsorpt. Sci. Technol.* **24** 215
[5] Bradley R H 2011 *Adsorpt. Sci. Technol.* **29** 1
[6] Inagaki M 2009 *New Carbon Mater.* **24** 193
[7] Yun J H, Choi D K and Moon H 2000 *Chem. Eng. Sci.* **55** 5857
[8] Brosillon S, Manero M H and Foussard J N 2001 *Environ. Sci. Technol.* **35** 3571
[9] Christensen C H, Johannsen K and Schmidt I 2003 *J. Am. Chem. Soc.* **125** 13370
[10] Kosuge K, Kubo S, Kikukawa N and Takemori M 2007 *Langmuir* **23** 3095
[11] Qiao S Z, Bhatia S K and Nicholson D 2004 *Langmuir* **20** 389–395
[12] Hartmann M, Bischof C 1999 *J. Phys. Chem. B* **103** 6230
[13] Stein A 2003 *Adv. Mater.* **15** 763
[14] Sui H, Liu H X, An P, He L, Li X G and Cong S 2017 *J. Taiwan Inst. Chem. E.* **74** 218
[15] Lee J W, Shim W G and Moon H 2004 *Microporous Mesoporous Mater.* **73** 109

IOP Conf. Series: Materials Science and Engineering **292** (2017) 012084 doi:10.1088/1757-899X/292/1/012084

Comparison of sound absorbing performances of copper foam and iron foam with the same parameters

X C Yang, X M Shen[1], P J Xu, X N Zhang, P F Bai, K Peng, Q Yin and D Wang

College of Field Engineering, PLA University of Science and Technology, No. 1 Haifu Street, Nanjing, Jiangsu 210007, P. R. China

[1] E-mail: shenxmjfjlgdx2014@163.com

Abstract. Sound absorbing performances of the copper foam and the iron foam with the same parameters were investigated by the AWA6128A detector according to standing wave method. Two modes were investigated, which included the pure metal foam mode and the combination mode with the settled thickness of metal foam. In order to legibly compare the sound absorbing coefficients of the two metal foams, the detected sound frequency points were divided into the low frequency range (100 Hz ~ 1000 Hz), the middle frequency range (1000 Hz ~ 3200 Hz), and the high frequency range (3500 Hz ~ 6000 Hz). Sound absorbing performances of the two metal foams in the two modes were discussed within the three frequency ranges in detail. It would be calculated that the average sound absorbing coefficients of copper foam in the pure metal foam mode were 12.6%, 22.7%, 34.6%, 43.6%, 51.1%, and 56.2% when the thickness was 5 mm, 10 mm, 15 mm, 20 mm, 25 mm, and 30 mm. meanwhile, in the combination mode, the average sound absorbing coefficients of copper foam with the thickness of 10 mm were 30.6%, 34.8%, 36.3%, and 35.8% when the cavity was 5 mm, 10 mm, 15 mm, and 20 mm. In addition, those of iron foam in the pure metal foam mode were 13.4%, 20.1%, 34.4%, 43.1%, 49.6%, and 56.1%, and in the combination mode were 25.6%, 30.5%, 34.3%, and 33.4%.

1. Introduction

Metal foam has been gradually introduced in the noise reduction of large-scale mechanical equipment, because it has the advantages of high sound absorbing coefficients, high strength, easy processability, fine associativity, outstanding permeability, excellent fire resistance, and a long service life [1-3]. Among these known metal foams, copper foam and iron foam are two promising materials for the sound absorption [4, 5]. In different ranges of sound frequency, the sound absorbing performances are different [6, 7]. Therefore, in order to establish the optimal option for sound absorption with different ranges of sound frequency, comparison of sound absorbing coefficients of copper foam and iron foam with same parameters were conducted in this study.

The investigated sound frequency could be divided into three ranges, which included the 100 Hz ~ 1000 Hz, 1000 Hz ~ 3200 Hz, and 3500 Hz ~ 6000 Hz. For different ranges of the sound frequency, the sound absorbing coefficients of copper foam and that of iron foam with the same parameters were detected. Meanwhile, for the purpose of thoroughly comparing the sound absorbing performances of copper foam and iron foam, metal foams with different thickness were investigated. In addition, it had been reported that the sound absorbing performances of metal foam would be obviously improved by combining with cavity [8-10]. Therefore, sound absorbing coefficients of metal foam with different cavity were also studied. Finally, the synthetical sound absorbing performances of copper foam and that of iron foam with same parameters were primarily given. Research on the comparison of sound absorbing performances of the copper foam and the iron foam with same parameters would promote application of the metal foam and ameliorate the noise reduction method.

2. Experimental setup and parameters

Detection of the sound absorbing coefficients of metal foam was obtained by the AWA6128A detector, as shown in the figure 1, which was realized according to the standing wave method [11, 12]. The detected metal foam was fixed in the standing wave tube, and the back cavity could be adjusted by the adjustment. The input audio power signal was generated by the computer, and the received sound signal was obtained by the pickup with acoustic probe. The detection process was tracked by the computer and displayed on the monitor.

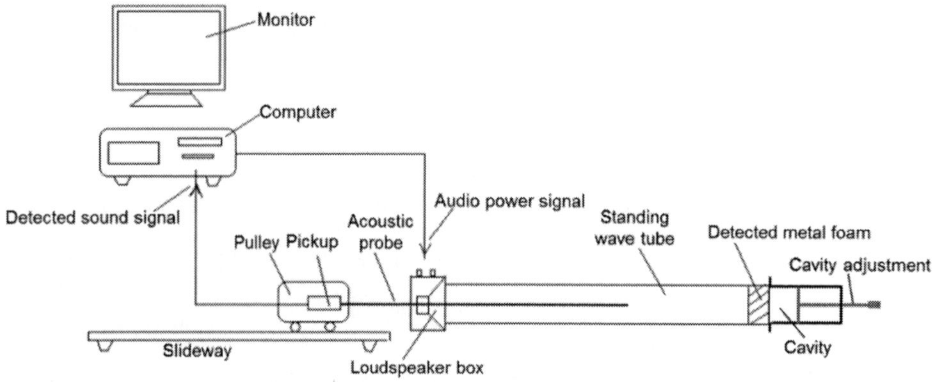

Figure 1. Schematic diagram of the experimental system for detection of sound absorbing coefficients.

Parameters of the copper foam and those of the iron foam were summarized in table 1. It could be found that the porosity, pore-per-inch (PPI), bulk density, open-cell rate, and aperture of the two metal foams were same, which indicated that the comparisons of sound absorbing performances of the two metal foams were conducted with the same parameters.

Table 1. Summaries of parameters of copper foam and iron foam.

Material	Porosity	Pore-per-inch (PPI)	Bulk density	Open-cell rate	Aperture
Copper foam	75%	90	0.3 g/cm^3	98%	0.2 mm
Iron foam	75%	90	0.3 g/cm^3	98%	0.2 mm

Two series of experiments were conducted by experimental system shown in figure 1 to compare sound absorbing performances of the two metal foams, as shown in table 2. On the basement of these experiments, the sound absorbing performances of copper foam and that of the iron foam with the same parameters were compared respectively. In the pure metal foam mode, the investigated thickness of the metal foam was 5 mm, 10 mm, 15 mm, 20 mm, 25 mm, and 30 mm, respectively. Meanwhile, in the combination mode, thickness of the metal foam was settled as 10 mm, and dimension of the cavity was set as 5 mm, 10 mm, 15 mm, and 20 mm, respectively.

Table 2. The designed experiments for comparison of sound absorbing performances of metal foam.

The designed experiments	Thickness of the metal foam	Dimension of the cavity
Pure metal foam mode	5 mm, 10 mm, 15 mm, 20 mm, 25 mm, 30 mm.	0 mm
Combination mode with the settled thickness of metal foam	10 mm	5 mm, 10 mm, 15 mm, 20 mm.

3. Results and discussions

According to the characteristics of the experimental system, the detected sound frequency in the range of 100 Hz ~ 1000 Hz was 100 Hz, 200 Hz, 300 Hz, 400 Hz, 500 Hz, 600 Hz, 700 Hz, 800 Hz, and 950 Hz. Similarly, the detected sound frequency in the range of 1000 Hz ~ 3200 Hz was 1100 Hz, 1300 Hz, 1500 Hz, 1800 Hz, 2000 Hz, 2300 Hz, 2600 Hz, 2900 Hz, and 3200 Hz, and that in the range of 3500 Hz ~ 6000 Hz was 3500 Hz, 3800 Hz, 4100 Hz, 4400 Hz, 4700 Hz, 5000 Hz, 5300 Hz, 5600 Hz, and 6000 Hz. In each frequency range, the detected sound frequency point was 9.

3.1. Pure metal foam mode

In the pure metal foam mode, the investigated thickness of the metal foam was 5 mm, 10 mm, 15 mm, 20 mm, 25 mm, and 30 mm, respectively. The obtained sound absorbing coefficients of the two metal foams in the frequency range of 100 Hz ~ 1000 Hz were shown in figure 2. It could be found that for the same material of metal foam, no matter copper foam or iron foam, the sound absorbing coefficients increased along with the increase of thickness from 5 mm to 30 mm in most detected frequency points, which was coincident with the results reported in the research of sound absorbing performances of the porous materials [13-15]. Meanwhile, it could be observed that the sound absorbing coefficients of the two metal foams in the low frequency range were low, and the maximum was no more than 40%. The major reason was supposed that wavelength of the sound signal was long in the low frequency range, so the sound wave diffracted the metal foam and the sound energy could not be absorbed. In addition, the experimental results indicated that the sound absorbing coefficients of copper foam were lower than that of the iron foam with the same parameters in most detected frequency points when the thickness was 5 mm, 10 mm, and 30 mm. When the thickness was 15 mm, 20 mm, and 25 mm, the synthetical sound absorbing performances of the two metal foams were almost same. Thus, it could be concluded that in the low frequency range, the sound absorbing performances of copper foam was worse than that of the iron foam with the same parameters.

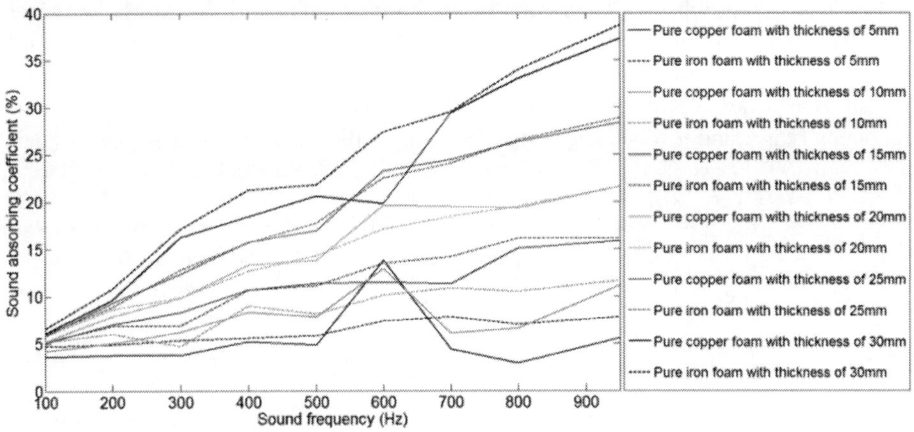

Figure 2. The obtained sound absorbing coefficients of the two metal foams in the frequency range of 100 Hz ~ 1000 Hz in the pure metal foam mode.

The obtained sound absorbing coefficients of the two metal foams in the frequency range of 1000 Hz ~ 3200 Hz were shown in figure 3. It could be observed that the sound absorbing coefficients were almost linear with the sound frequency for the metal foam with a certain thickness. Meanwhile, sound absorbing coefficients were also near linear with thickness of the metal foam for a certain frequency point. It was interesting to notice that the two lines had intersection. For example, when the thickness was 30 mm, sound absorbing coefficients of the copper foam were worse than that of the iron foam in the frequency range of 1000 Hz ~ 2000 Hz, and be better than that of the iron foam in the frequency range of 2000 Hz ~ 3200 Hz. There also existed an intersection frequency 2000 Hz when the thickness was 15 mm. It could be found that except the thickness of 20 mm, sound absorbing performances of the copper foam was better than that of the iron foam along with the increase of sound frequency. Thus, sound absorbing performances of the copper foam was better than that of the iron foam with the same parameters in the middle frequency range, especially when the sound frequency was increased.

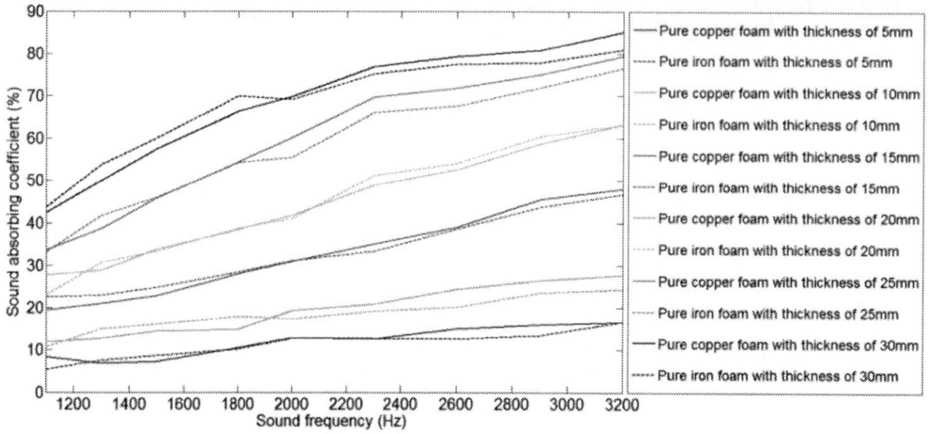

Figure 3. The obtained sound absorbing coefficients of the two metal foams in the frequency range of 1000 Hz ~ 3200 Hz in the pure metal foam mode.

The obtained sound absorbing coefficients of the two metal foams in the frequency range of 3500 Hz ~ 6000 Hz were shown in figure 4. It would be observed that sound absorbing coefficients of the two metal foams increased along with the increase of thickness. When the thickness increased from 5 mm to 20 mm, improvement of the sound absorbing performances of the two metal foams was obvious, and the further increase of thickness had no obvious improvement to the sound absorption. It could be found that except when the thickness was 5 mm, sound absorbing coefficients of the copper foam were better than those of the iron foam with the same parameters in the high frequency range. Judging from evolvements of the two metal foams it could be primarily concluded that sound absorbing performances of copper foam was worse than that of the iron foam with same parameters in the low frequency range, and it was changing better along with the increase of sound frequency.

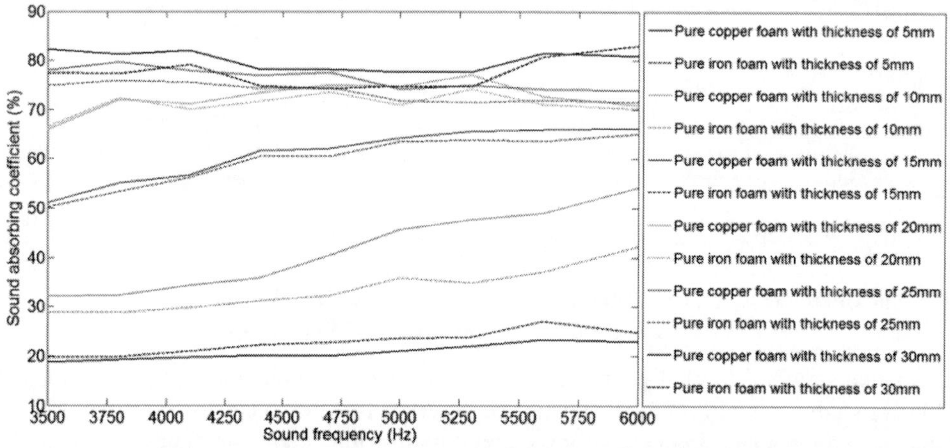

Figure 4. The obtained sound absorbing coefficients of the two metal foams in the frequency range of 3500 Hz ~ 6000 Hz in the pure metal foam mode.

3.2. Combination mode with the settled thickness of metal foam

The achieved sound absorbing coefficients of the two metal foams in the frequency range of 100 Hz ~ 1000 Hz were shown in figure 5. It could be found that when dimension of the cavity was low, sound absorbing coefficients of the copper foam and those of the iron foam were almost same. When the cavity increased to 15 mm and 20 mm, it could be observed that sound absorbing coefficients of the copper foam were better than those of the iron foam with the same parameters. Relative to the sound absorbing performances of the pure metal foam with thickness of 10 mm in figure 2, introduction of cavity would be propitious to improve sound absorbing performances of the metal foam, although the improvement was no remarkable in the low frequency range.

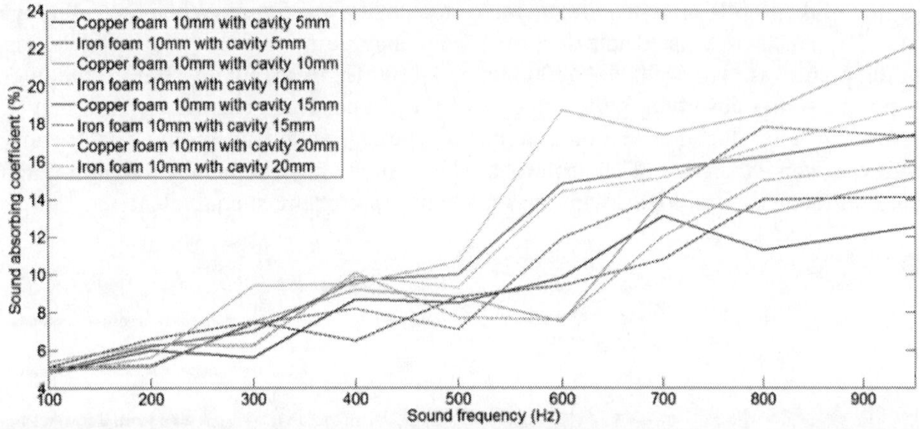

Figure 5. The achieved sound absorbing coefficients of the two metal foams in the frequency range of 100 Hz ~ 1000 Hz in the combination mode with the settled thickness of metal foam.

The achieved sound absorbing coefficients of the two metal foams in the middle frequency range were shown in figure 6. It would be observed that sound absorbing coefficients of the copper foam were obviously better than those of the iron foam with same cavity. Meanwhile, relative to sound absorbing performances of the pure metal foam in figure 3, it could be found that the improvements of the sound absorbing coefficients were remarkable by the introduction of cavity, and the improvement level of the copper foam was larger than that of the iron foam.

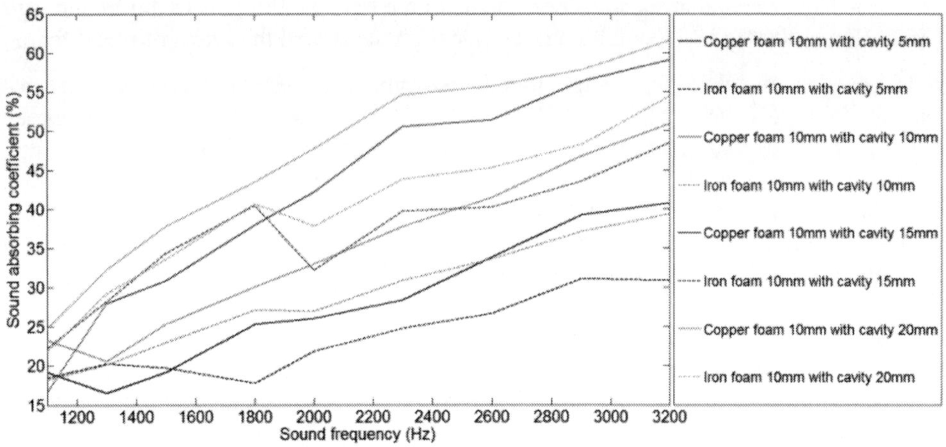

Figure 6. The achieved sound absorbing coefficients of the two metal foams in the frequency range of 1000 Hz ~ 3200 Hz in the combination mode with the settled thickness of metal foam.

The achieved sound absorbing coefficients of the two metal foams in the middle frequency range were shown in figure 7. Firstly, it could be found that introduction of the cavity would be obvious to improve the sound absorbing coefficients of the metal foam. In figure 4, the average sound absorbing coefficients of the metal foam with the thickness of 10 mm were 35%. Through introducing the cavity, sound absorbing coefficients were 1.5 times even when the cavity was only 5 mm. Secondly, it could be observed that evolvement of the sound absorbing coefficients along with increase of the cavity were disordered for the two metal foams [16-18], which was quite different from that in the pure metal foam mode, and also be different from that in low or middle frequency range in the combination mode. It was supposed that the major reason was that there existed sound absorption by the sympathetic vibration in the cavity, and the efficient of the sympathetic vibration was not in the positive correlation with dimension of the cavity. Thirdly, sound absorbing performances of the copper foam was always better than that of the iron foam when dimension of the cavity was 5 mm. When the cavity was 10 mm, 15 mm, and 20 mm, there existed intersectional frequency points for the two metal foams, and the corresponding point was near 5600 Hz, 4700 Hz, and 4100 Hz, respectively. Before the intersectional frequency points, sound absorbing performances of the copper foam was better than that of the iron foam. Meanwhile, after the intersectional frequency points, sound absorbing performances of the copper foam was worse than that of the iron foam. Thus, in the high frequency range, sound absorbing performances of the two metal foams with the same cavity were almost equivalent.

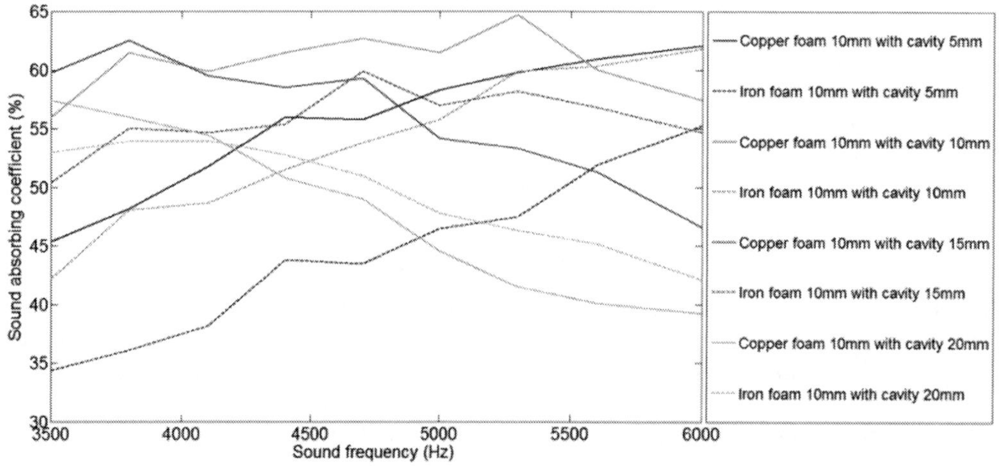

Figure 7. The achieved sound absorbing coefficients of the two metal foams in the frequency range of 3500 Hz ~ 6000 Hz in the combination mode with the settled thickness of metal foam.

It would be calculated that the average sound absorbing coefficients of copper foam in the pure metal foam mode were 12.6%, 22.7%, 34.6%, 43.6%, 51.1%, and 56.2% when the thickness was 5 mm, 10 mm, 15 mm, 20 mm, 25 mm, and 30 mm. meanwhile, in the combination mode, the average sound absorbing coefficients of copper foam with the thickness of 10 mm were 30.6%, 34.8%, 36.3%, and 35.8% when the cavity was 5 mm, 10 mm, 15 mm, and 20 mm. In addition, those of iron foam in the pure metal foam mode were 13.4%, 20.1%, 34.4%, 43.1%, 49.6%, and 56.1%, and in the combination mode were 25.6%, 30.5%, 34.3%, and 33.4%.

4. Conclusions

Comparison of sound absorbing performances of the copper foam and iron foam with same parameters were conducted in this study. The pure metal foam mode and the combination mode with the settled thickness of metal foam were investigated. The investigated frequency was ranged into three parts. The experimental result indicated that sound absorbing coefficients of the metal foam had relationship with the sound frequency and the metal parameters, which meant that the optimal option for the best sound absorbing performances should be determined by the concrete conditions.

Acknowledgments
This work was supported by a grant from National Natural Science Foundation of China (Grant No. 51505498), a grant from Natural Science Foundation of Jiangsu Province (Grant No. BK20150714), and a grant from National Key Research & Development Program (Grant No. 2016YFC0802903).

References

[1] Lu T J, Hess A and Ashby M F 1999 Sound absorption in metallic foams *Journal of Applied Physics* **85(11)** 7528-7539

[2] Sun J X, Duan C Y and Liu P S 2016 Sound absorption characterization of aluminum foam made by press infiltration casting *Multidiscipline Modeling in Materials and Structures* **12(4)** 737-747

[3] Liu P S, Qing H B and Hou H L 2015 Primary investigation on sound absorption performance of highly porous titanium foams *Materials & Design* **85** 275-281

[4] Ru J M, Kong B, Liu Y G, Wang X L, Fan T X and Zhang D 2015 Microstructure and sound absorption of porous copper prepared by resin curing and foaming method *Materials Letters* **139** 318-321

[5] Xie Z K, Ikeda T, Okuda Y and Nakajima H 2004 Sound absorption characteristics of lotus-type porous copper fabricated by unidirectional solidification *Materials Science and Engineering A* **386(1-2)** 390-395

[6] Wang X L 2007 Porous metal absorbers for underwater sound *The Journal of the Acoustical Society of America* **122(5)** 2626-2635

[7] Wang X L, Peng F and Chang B J 2009 Sound absorption of porous metals at high sound pressure levels *The Journal of the Acoustical Society of America* **126(2)** EL55-EL61

[8] Ning J F, Zhao G P 2016 Sound absorption characteristics of multilayer porous metal materials backed with an air gap *Journal of Vibration and Control* **22(12)** 2861-2872

[9] Yairi M, Minemura A 2006 Effect of honeycomb structure in the back cavity on the absorption characteristics of microperforated panel absorbers *The Journal of the Acoustical Society of America* **119(5)** 3250

[10] Sakagami K, Nakajima K, Morimoto M 2006 Sound absorption characteristics of a honeycomb-backed microperforated panel (MPP) absorber *The Journal of the Acoustical Society of America* **120(5)** 3146

[11] Emanov A F, Krasnikov A A 2016 Use of the standing wave method to study seismically insulated buildings *Seismic Instruments* **52(4)** 323-349

[12] Ning F L, Li X F 2013 Numerical simulation of finite amplitude standing waves in acoustic resonators using finite volume method *Wave Motion* **50(2)** 135-145

[13] Liu Z Q, Zhan J X, Fard M and Davy J L 2016 Acoustic properties of a porous polycarbonate material produced by additive manufacturing *Materials Letters* **181** 296-299

[14] Wang W, Liu H T and Gu W D 2017 A novel fabrication approach for improving the mechanical and sound absorbing properties of porous sound-absorbing ceramics *Journal of Alloys and Compounds* **695** 2477-2482

[15] Chen W H, Chen T N, Wang X P, Wu J H and Li S B 2016 Fractal Model for Acoustic Absorbing of Porous Fibrous Metal Materials *Shock and Vibration* **2016** 1-17

[16] Yahya I, Kusuma J I, Harjana, Kristiani R and Hanina R 2016 Laboratory investigation on the role of tubular shaped micro resonators phononic crystal insertion on the absorption coefficient of profiled sound absorber *IOP Conference Series: Materials Science and Engineering* **107(Conference 1)** 012046

[17] Bernard C, Alexei M, Denis L and Mohamed S 2008 Low frequency in situ metrology of absorption and dispersion of sound absorbing porous materials based on high power ultrasonic non-linearly demodulated waves *Applied Acoustics* **69(7)** 634-648

[18] Supitcha T, Pornlada P, Wanlop H and Kantima C 2016 Enhancing sound absorption properties of open-cell natural rubber foams with treated bagasse and oil palm fibers *Journal of Reinforced Plastics and Composites* **35(8)** 688-697

IOP Conf. Series: Materials Science and Engineering **292** (2017) 012085 doi:10.1088/1757-899X/292/1/012085

Study on bamboo gluing performance numerical simulation

Z R Zhao[1, 3], W H Sun[1], X M Sui[2, 4] and X F Zhang[1]

[1] Department of Logistics, Beijing WUZI University, Beijing, 101149, China
[2] Department of electronic and Information Engineering, North China Institute of Science &Technology, Sanhe, 065201, China
[3] National Natural Science Foundation of China （No.31470588） Science and technology project of Hebei Province（No.15211830）

[4] E-mail: Sunny7801@sina.com

Abstract. Bamboo gluing timber is a green building materials, can be widely used as modern building beams and columns. The existing bamboo gluing timber is usually produced by bamboo columns or bamboo bundle rolled into by bamboo columns. The performance of new bamboo gluing timber is decided by bamboo adhesion character. Based on this, the cohesive damage model of bamboo gluing is created, experiment results are used to validate the model. The model proposed in the work is agreed on the experimental results. Different bamboo bonding length and bamboo gluing performance is analysed. The model is helpful to bamboo integrated timber application.

1. Introduction

Bamboo is similar to a gradient material for fiber longitudinal distribution characteristics, so it has high tensile strength, excellent elastic properties, it is good for engineering materials. On the other hand, as the growth time is a short, it is a sustainable green material. It is widely used in modified bamboo-based panels, laminated material. At present, there are mainly bamboo reorganization methods such as, the use of the original bamboo multi-party reorganization using for building beams proposed by Wansi Fu [1].The bamboo unit of the same diameter is processing a regular polygonization ,laying at cross-section side by side, extending the bamboo material along longitudinal by finger.Yushun Li [2] proposes the bamboo composite plate made of two bamboo plywood at top and steel profiles in the middle. Qisheng Zhang [3] proposes a bamboo strips laminated material manufacturing method (CN100999092A), the use of bamboo strips applied adhesive hot pressing. Yongfu Yang [4] proposes natural hollow structure bamboo/wood composite material manufacturing method patent, the use of the same diameter of the bamboo section processed into four prism which horizontal arbitrary into the honeycomb shape, the upper and lower surface pressure veneer to form a natural hollow structure of bamboo composite sheet. The use of bamboo in the prior art can be a reorganization of bamboo bundles or the use of whole bamboo for reorganization.

The property of the bamboo reorganization material depends on the performance of the bamboo units and the performance of the bamboo gluing. So the gluing property of the whole bamboo and bamboo bundles is key factor. The experimental method can be used to optimize the design of this new type of bamboo reorganization material, but the experiment takes a long time, at the same time to achieve a variety of parameters of high cost of experimental high cost. It is a better choice to study the bamboo reorganization material by numerical simulation. To achieve the digital design of bamboo composite bamboo bundles, need to solve the mathematical properties of bamboo gluing simulation. To solve the problem, the research group studied the numerical simulation of bamboo gluing, and studied the bamboo slabs lap peeling. Many scholars have done research on the influence of process parameters on the properties of bamboo reorganization. Shan Huang [5] studied the relation between different phenolic resin amount of bamboo bundles and time under various states. Chungui Du [6] studied the effect of bamboo beam shape on its dipping, secondary drying process and the performance of bamboo. For the lap joint mechanics analysis, some scholars use analytical methods to solve the problems [7-8]. Lap joint numerical calculation aspects [9-11] are mostly concentrated in the adhesive layer stress distribution, optimization the adhesive layer stress distribution to improve the glue strength. It is difficult to obtain accurate solution by using the analytical method. The experimental data are very limited due to the limitation of the experimental conditions and the experimental time and cost reasons. Numerical simulation is a good method to intuitively get bamboo sticking model and get a reasonable answer. From the current literature, there is little research on the damage process and the influencing factors of bamboo sticks. In this paper, the finite element simulation method is used to study the shear failure process of the bamboo slab under the loading condition.

2. Mthodology of modelling of lap bamboo peeling

2.1. Gluing lap bamboo slab stress analysis

In this paper, the properties of bamboo reorganization material are influenced by the properties of bamboo gluing, and the bonding of bamboo is the key factor of bamboo reorganization. Therefore, the study of bamboo gluing is an important point to realize the bamboo aggregate. Bamboo shape and gluing properties determine the overall performance of the material. In order to facilitate the establishment of the numerical simulation model of bamboo gluing capacity, the simplified analysis model was taken as bamboo lap, and the two bamboo slabs were glued together by adhesive. Through studying the various influencing factors of gluing failure of lap joint. Thickness and gluing can affect the gluing performance. The size of the bamboo slab is 200mm × 22mm × 6mm and the length of cement is 50mm. The model is shown in figure 1.

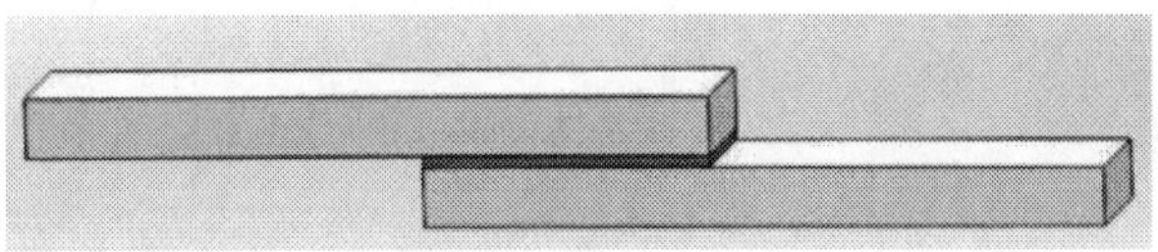

Figure 1. Model of bamboo adhesive lap.

One end of the bamboo slab is fixed, the displacement load is applied at the other end to analyze the damage of the adhesive layer. Under the load the elastic deformation of bamboo slab appears. A linear elastic mathematical model modeling is created. The failure of the gluing layer mainly appears in the shear failure of the adhesive layer, the disruption of the interface stripping, and the tensile failure of the bamboo slab. Through the experimental test, it is found that the bonding of the lap bamboo is broken due to the occurrence of shear failure in the adhesive layer. The numerical analysis of the lap bamboo is mainly analyzed in the separation of the adhesive layer under tensile stress and shear stress. Due to the crack tip stress singularity and the mesh size effects on crack tip unit integration point stress, a linear elastic fracture mechanics analysis of crack propagation is not suitable for calculating the crack tip a greater yield situation.

Virtual crack closure technology based on fracture mechanics (VCCT) and cohesion model based on the damage mechanics are used for calculating the fracture. Virtual crack closure technology based on fracture mechanics (VCCT) assumes that when the strain energy release rate reaches the critical value, the fracturing crack is developing. It is necessary to set the initial crack, but in most of the actual situation the crack cannot be preset where the occurrence of cracks, it is only solve the crack start time but the crack direction is not resolved. In addition, it is more dependent on the thickness of the grid. Based on the cohesive force model of damage mechanics, the cohesive model is used to simulate the interface failure, which is consistent with the actual physical mechanism. This method is still effective when the plastic deformation occurs in the cracked area. Compared with the fracture mechanics virtual crack closure technique (VCCT) method, the cohesive model based on the damage mechanics is not dependent on the grid. Based on the above reasons, this paper uses the damage mechanics cohesive force model to analyze the cracking of glued lap bamboo.

2.2. *Glue layer cohesion model*

Using the cohesive model analysis on the gluing layer, the tensile force increases with the increase of the cracked displacement at the beginning. When the force reach a point, the crack appears in gluing layer, the tensile force decreases with the increase of the crack displacement, the force will be zero until gluing layer completely cracked. There is also a peak between the shear stress and the cracking displacement. When the displacement reaches the cracking, the shear stress reaches the maximum value, and the glue layer begins to crack until the entire adhesive layer is completely broken. The relationship between force and cracking displacement in cohesive force model is:

$$T=\begin{cases}\dfrac{T_c}{u_c}u & (u\le u_c)\\ \dfrac{u_f-u}{u_f-u_c}T_c & (u>u_c)\end{cases} \tag{1}$$

$$\tau=\begin{cases}\dfrac{\tau_c}{u_\tau}u & (u\le u_\tau)\\ \dfrac{u_\tau^f-u}{u_\tau^f-u_\tau}\tau_c & (u>u_\tau)\end{cases} \tag{2}$$

In the above equation, T_c、u_c、u_f、τ_c、u_τ、u_τ^f are the tensile stress, the tensile stress under the action of tensile stress, the crack propagation under tensile stress, the shear stress intensity, the shear displacement under shear stress, Stress propagation under the crack, respectively. Under the action of internal tensile stress and shear stress, the bond strength of the adhesive layer is:

$$G_{Tc}=\int T du_c \tag{3}$$

$$G_{\tau c}=\int \tau du_\tau \tag{4}$$

When the internal tensile stress and shear stress reach to T_c and τ_c respectively, the adhesive layer is cracking. The above two kinds of stress will play a role in the cracking of the adhesive layer. In this paper, the B-k model is adopted considering the mixing of the internal tensile stress and the shear stress.

3. Results of model analysis

3.1. *Calculation and verification of the model*

In order to test the bamboo plywood bonding performance, the research group made lab bamboo samples, bamboo size is selected 200mm × 22mm × 6mm, bonding length of 50 mm, the use of urea-formaldehyde resin, the amount of coating 200g / m² ~ 260g / m² evenly is applied to the lap. Under external force and cold curing 6 hours, a lap bamboo is made. The lap bamboo sample is placed in hydraulic universal test machine WE-300A of Jinan Test Group Co. Ltd.. One the holder of the holdings of test machine hold a lab bamboo in one end, the next holder hold a lap bamboo at the other end, a strain gauge and a force sensor are provided to read the displacement and force readings by applying a force to the two ends of the upper and lower retainers until the t lap bamboo is peeling. The broken lap bamboo specimen is as shown in figure 2.

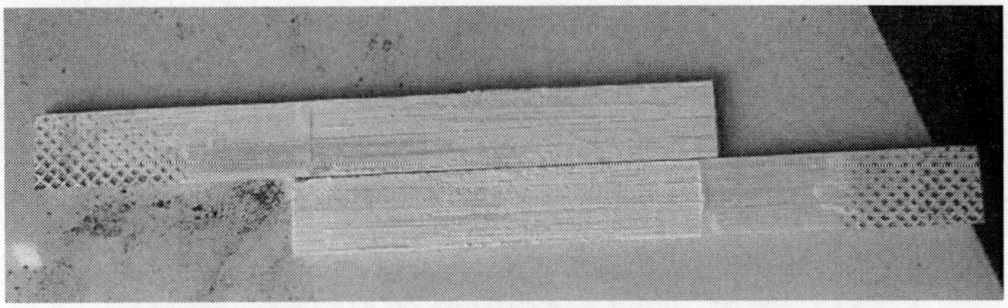

Figure 2. Bamboo piece after opening lap.

According to the experimental sample, a three-dimensional lap bamboo model is created in Abaqus. The property of bamboo is set to be elasticity, elastic modulus and poison's ratio is 0.65GPa, 0.36 respectively. The property of gluing layer is such as: the tensile strength is 74.6MPa, the shear stress intensity is 68MPa, the energy release of the adhesive layer is $G_{\Pi c}$=2.1KJ/m², G_{Ic}=0.3KJ/m². The part of bamboo is used hexahedral mesh C3D8I, glued layer is used hexahedral bonding unit COH3D8.The grid of lap is encrypted, the same structure is built by experiment. A lap bamboo length 50mm cracking deformation is shown in figure 3, the lab bamboo cracking comparison between experimental and finite element simulation of the is shown in figure 4.

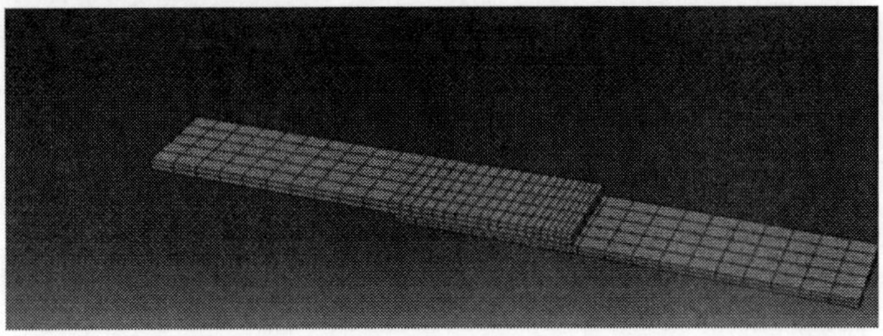

Figure 3. Gluing bamboo piece deformation after opening lap.

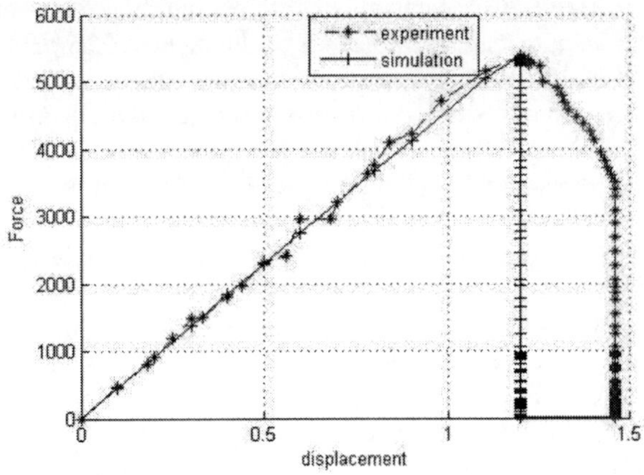

Figure 4. Comparison between experiment and simulation.

It can be seen from figure 5 that under the displacement load, the force of the adhesive layer increases with the increase of cracking displacement, and cracks occur when the cracking force reaches 5.4KN. It can be seen that the experimental value and the simulated values are consistent. But the experimental value in the fracture moment did not drop sharply to 0, there is a slow process of change, this shows that the adhesive layer began to crack is a process, to a certain stage to achieve the dissipation of the adhesive layer, cracking occurred violently. The stress cloud of the lap is shown in As can be seen from figure 5, the stress occurs at the maximum near the boundary of the glue layer.

IOP Conf. Series: Materials Science and Engineering **292** (2017) 012085 doi:10.1088/1757-899X/292/1/012085

Figure 5. Bamboo lap stress cloud map.

4. Conclusion

In this paper, in order to improve the performance of bamboo reorganization material, this paper focuses on the numerical modeling of bamboo gluing performance. The fracture peeling strength of the adhesive layer is 5.4kN when the lap length is 50mm, and the proposed model was verified by experiments.

In this paper, the finite element model of bamboo gluing performance is established, and the influence of some gluing parameters on the performance is analyzed. The established model has a certain guiding effect on the numerical optimization design of bamboo reorganization.

References

[1] Wansi Fu 2008 Bamboo original multi-reorganization material and its manufacturing method, China Patent NO. 200710179001.7

[2] Yusun Li 2010 Profiled steel plate - bamboo plywood composite floor and its making method, China Patent NO. 200710106252.2

[3] Qisheng Zhang 2008 A method of manufacturing bamboo products,China Patent NO. 200710019222.8

[4] Yongfu Yang 2011 Natural hollow structure bamboo / wood composite board and manufacturing method,China Patent NO. 200910260365.7

[5] Huang S, Zhao M and He Q 2011 Analysis of influencing factors of bamboo bundle and wood bundle *J. Forestry Science and Technology Development* vol **25** p 93-96

[6] Du C G, Wei J G and Yu H W 2015 Effects of bamboo shapes on preparation technology and performance of recombinant bamboo *J.Forestry Science and Technology* vol **40** p 44-47

[7] Da Silva L F M, Das Neves P J C, Adams R and Spelt J K 2009 Analytical models of adhesively bonded joints-part I: Literature survey *J. International Journal of Adhesion and Adhesives* vol **29** p 319-30

[8] Luo Q T,Tong L Y 2009 Analytical solutions for nonlinear analysis of composite single-lap adhesive joints *J. International Journal of Adhesion and Adhesives* vol **29** p 244-54

[9] Rudawsks A 2010 Adhesive joint strength of hyrid assemblies:Titanium sheet-composites and aluminium sheet-composites experimental and numerical verification *J. International of Adhesion and Adhesives* vol **30** p 574-82

[10] De Monra M F S F, Goncalves J P M, Chousal J A G and Campilho R D S G 2008 Cohesiveandcontinum mixed-mode damage models applied to the simulation of the mechanical behavior of bonded Joints *J. International Journal of Adhesion and Adhesives* vol **28** p 419-26

[11] Krueger R, Cvitkovich M K, O'Brien T K and Minguet P J 2003 Testing and analysis of composite skin/stringer debonding under multi-axial loading *J. Composite Materials* vol **4** p 1263-300

Fabrication of graphene/polydopamine/copper foam composite material and its application as supercapacitor electrode

Y Zheng[1], S X Lu[1, 3], W G Xu[1], G He[1], Y Y Cheng[1], F Y Xiao[1] and Y Zhang[2]

[1] School of Chemistry and Chemical Engineering, Beijing Institute of Technology, Beijing 100081, P.R. China
[2] State Gride Shandong Electric Power Research Institute, Jinan 250003 P.R. China

[3] E-mail: shixianglu@bit.edu.cn

Abstract. In this work, a composite electrode was fabricated by chemical deposition of polydopamine (PDA) and graphene oxide (GO) on the copper foam (CF) surface, followed by annealing treatment. Owing to the cohesive effect of the PDA middle film, GO was coated on CF surface successfully, and then reduced simultaneously while annealing. The resulted rGO/PDA/CF composite electrode was directly used as a supercapacitor electrode and exhibited excellent electrochemical performance, with a high specific capacitance of 1250 F g^{-1} at 2 A g^{-1} and favorable cycle stability.

1. Introduction

Supercapacitors have drawn great interest in energy device because of their fast charge and discharge rate, long cycle lifetime and high reliability [1, 2]. Due to the difference of their storage mechanisms, supercapacitors can be classified as electronic double layer capacitors (EDLCs) and pseudocapacitors. Compared with EDLCs who store energy through reversible ions at electrode / electrolyte interface, pseudocapacitors possess higher charge-storage capacity, induced by rapid Faradaic redox reactions [3, 4].

Carbon materials are widely used in electrodes because of high specific surface area and easy accessibility. Among them, graphene has been extensively explored due to its superior conductivity, pore size distribution and excellent electrochemical performance [5-7]. However, the self-aggregation and restacking of graphene sheets occurred during the preparation process, leading to an actual specific capacitance far below its theoretical value. In order to prevent the stacking of graphene effectively, we try to reduce the graphene oxide (GO) via annealing.

Dopamine can self-polymerize and spontaneously deposit conformal polydopamine (PDA) films, which could be utilized to functionalize various material surfaces [8]. In this work, we use PDA as a middle film, because (1) the secondary amine groups on PDA layer react with hydroxyl groups on GO, and then form amide groups. Thus GO could be coated more tightly on the surface of copper foam. (2) PDA contributes to conductivity. Liu et al. found that carbon nanoparticles obtained after PDA carbonization had better conductivity [8]. Hence, PDA is a good choice to be the middle film between copper foam and GO.

In this study, copper foam (CF) acts as the electrode substrate to carry active substances and the current collector for its high conductivity. To prevent the stacking of graphene sheets effectively, GO was first deposited on the CF surface uniformly with the assistant of PDA film and then annealed to be reduced graphene oxide (rGO), and we named the sample as rGO/PDA/CF composite electrode. It can be used as the supercapacitor electrode directly, when tested in 1 mol L^{-1} Na_2SO_4 electrolyte, it exhibited typical pseudocapacitance characteristics and excellent performance.

2. Experimental section

2.1. Preparation of composite electrode rGO/PDA/CF

GO was prepared by the oxidation of natural graphite powder according to a modified Hummers' method [9, 10]. Typically, Cu foam (2.0 cm × 1.0 cm × 0.1 cm) were ultrasonically cleaned with acetone, alcohol and doubly distilled water respectively, then dried under vacuum for 24 h. The preparation of rGO/PDA/CF electrode includes the following sections. First, the cleaned CF sheet was immersed into 20 mg mL^{-1} PDA solution whose pH was about 8.5 for 1.5 h at room temperature. Second, the as-synthesized GO was diluted to a concentration of 0.5 mg mL^{-1} suspension. The Cu foam decorated by PDA (PDA/CF) was then placed into the GO suspension for 40 min at room temperature. Thus the GO/PDA/CF composite was obtained. Finally, it was annealed at 200 °C for 1.5 h. In this way, the final electrode material, rGO/PDA/CF composite sheet was synthetized successfully.

2.2. Characterizations

Structural and phase characterizations of the electrode were done by XRD using a Bruker D8 Advance Phaser diffractometer with Cu K_α radiation at a continuous scanning mode (40 kV, 40 mA, and λ = 0.15418 nm) at a scanning rate of 3° min^{-1}. Scanning electron microscopy (SEM) images was determined using a scanning electron microscope (SEM, S-4800, Hitachi, Japan). Raman spectroscopy investigations were performed by using Witec confocal Raman microscopy (LabRAM HR Evolution) with laser wavelength of 532 nm.

2.3. Electrochemical measurement

The electrochemical performance of the GO/PDA/CF and rGO/PDA/CF composite electrodes were measured by using a three-electrode configuration with an electrochemical potentiostat (CHI 760E, CH Instruments) in the potential range of -0.03 V to -0.83 V. A 1 mol L^{-1} Na_2SO_4 aqueous solution was used as the electrolyte. Pt sheet and saturated calomel electrode were used as a counter and reference electrode, respectively. The capacitive performance of rGO/PDA/CF was characterized by cyclic voltammetry (CV), galvanostatic charge and discharge (GCD). The gravimetric specific capacitance (C) of the electrode was calculated from the following equation:

$$C = \frac{I \times \Delta t}{m \times \Delta V} \tag{1}$$

where I represents the constant discharge current, Δt represents the discharging time, m refers to the mass of active substance on the electrode and ΔV is the operating voltage.

3. Results and discussion

3.1. Morphology and composition analysis

Figure 1a confirms that CF possesses smooth surface. From figure 1b, we can see that after immersing in PDA solution, small particles are appeared on CF surface, which provide good attachment points for GO. It can be seen clearly in figure 1c, that GO was successfully deposited on the surface of PDA/CF. After annealing, GO film fits more compactly on the surface and more wrinkles can be observed, as shown in figure 1d.

Figure 2 shows the Raman spectra of GO/PDA/CF and rGO/PDA/CF composite electrodes, in which two typical peaks belonged to GO and rGO can be observed within the range of 500 to 2500 cm^{-1}. The peak at around 1600 cm^{-1} and 1350 cm^{-1} are assigned to the E_{2g} mode, i.e. G band and D band, respectively [11]. The I_D/I_G value of GO/PDA/CF was calculated to be 0.92, while that of rGO/PDA/CF was 0.84, the decrease after annealing indicates that GO was partly reduced to rGO [12]. The XRD spectra of CF, GO/PDA/CF and rGO/PDA/CF composites are showed in figure 3a. From the figure, we can see that the sharp peaks are located at 43.6°, 50.6° and 74.3°, corresponding to metal Cu (JCPDS card no. 04-0836). In figure 3b, there are no peaks can be observed in the intercept part of the spectra in the range of $2\theta = 3$ - 35°. Therefore, GO/PDA/CF and rGO/PDA/CF composites synthesized in the experiment may be all amorphous phase.

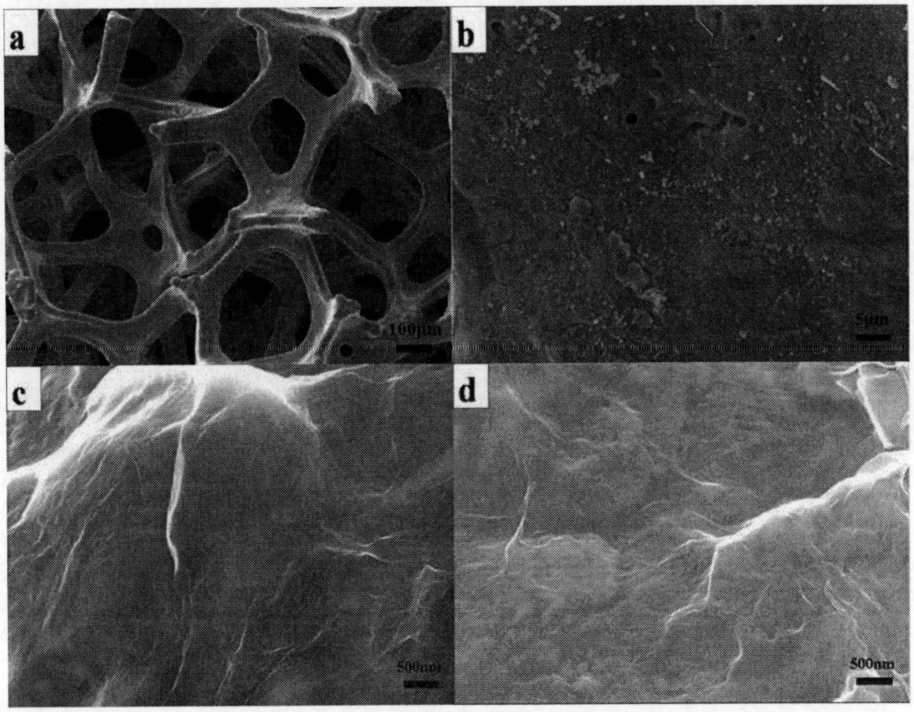

Figure 1. SEM images of different sample surfaces: (a) CF, (b) PDA/CF sheet, (c) GO/PDA/CF sheet, (d) rGO/PDA/CF sheet.

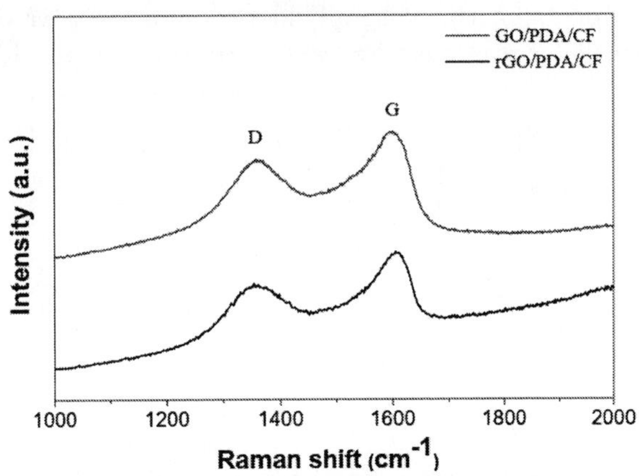

Figure 2. Raman spectra of GO/PDA/CF and rGO/PDA/CF composites.

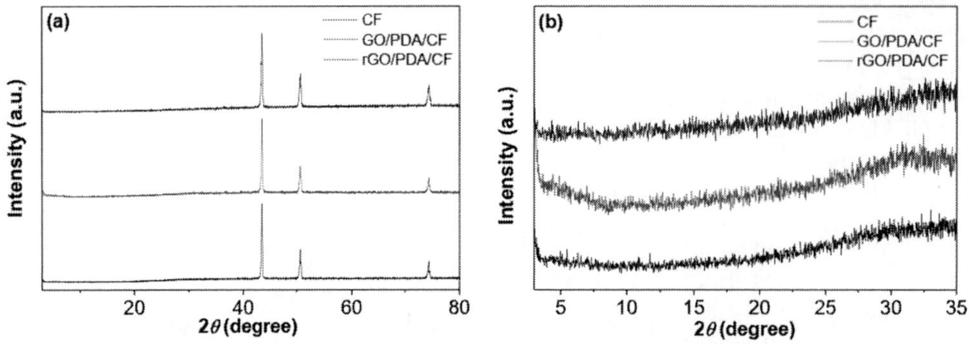

Figure 3. (a) XRD patterns of CF, GO/PDA/CF and rGO/PDA/CF composites. (b) The magnified part of the region $2\theta = 3\text{-}35°$.

3.2. Electrochemical performance

The electrochemical performances of GO/PDA/CF and rGO/PDA/CF composite electrodes were investigated by galvanostatic charge-discharge (GCD) tests, as shown in figure 4. The slightly nonlinear and asymmetric charging/discharging curves validate the dominant pseudocapacitance derived from Faradiac reactions [13]. On the other hand, the specific capacitance value of rGO/PDA/CF electrode is rather better than GO/PDA/CF electrode by contrast of the discharge time. This could be due to rGO reduced from GO, which can also be supported from Raman result.

The CV curves of rGO/PDA/CF composite electrode at different scan rates ranging from 5 to 100 mV s^{-1} are showed in figure 5a. All CV curves displayed broad redox peaks, suggesting the typical pseudocapacitance characteristics [14, 15]. We propose that a small amount of elemental copper was oxidized during annealing process, which results in the increase of capacitance value. The GCD curves of rGO/PDA/CF composite electrode in the current density range of 2 - 32 A g^{-1} are shown in the figure 4b. The specific capacitance of rGO/PDA/CF electrode was about 1250 F g^{-1} at current density of 2 A g^{-1}. When the current density was 4 A g^{-1}, the specific capacitance was about 760 F g^{-1}, which is much higher than 248 F g^{-1} at current density of 2 A g^{-1} in previous paper [16].

Figure 6a shows the cycling performance at 4 A g^{-1} for 1000 cycles, as illustrated that, the rGO/PDA/CF electrode showed 74.5% retention, displaying favorable cycle stability. Figure 6b illustrates cyclic voltammograms of the rGO/PDA/CF electrode under different scan rates after 1000 cycles. It is obviously that the CV curves were all close to rectangular which showed typical EDLCs behavior. This may be exactly the reason for the reduction of specific capacitance.

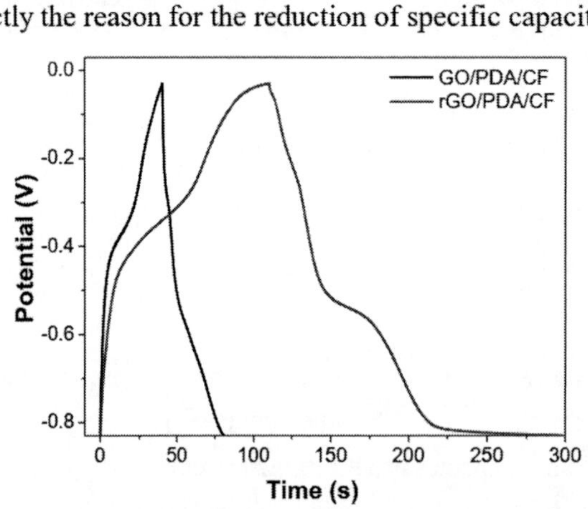

Figure 4. Discharge curves of GO/PDA/CF and rGO/PDA/CF electrodes at 4 A g^{-1}.

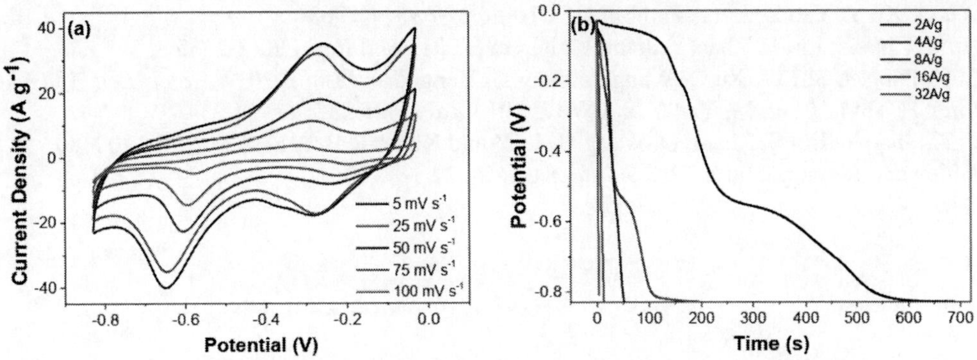

Figure 5. (a) Cyclic voltammograms of the rGO/PDA/CF electrode under different scan rates; (b) Discharge curves of rGO/PDA/CF electrode at different applied currents.

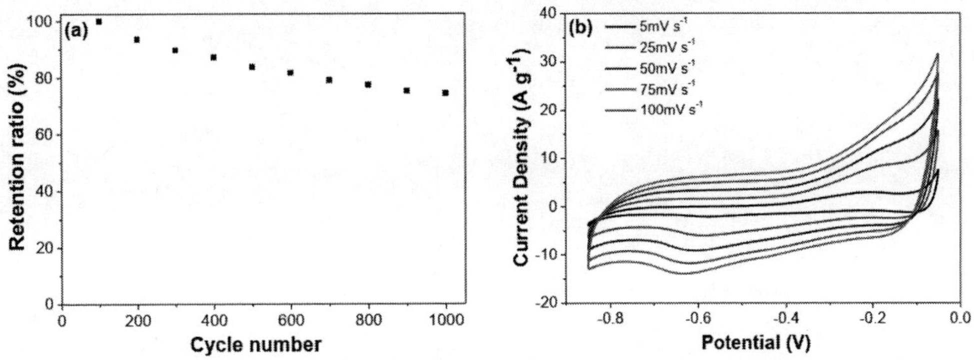

Figure 6. (a) Capacity retention versus cycle numbers for rGO/PDA/CF electrode at the current density of 4 A g^{-1}; (b) Cyclic voltammograms of the rGO/PDA/CF electrode under different scan rates after 1000 cycles.

4. Conclusion

In summary, the rGO/PDA/CF composite supercapacitor electrode has been fabricated successfully. In the experiment, GO was attached on the Cu foam by simple chemical immerse using PDA as the middle film and was reduced by annealing method. The final electrode exhibits superior specific capacitance (1250 F g^{-1} at 2 A g^{-1}) and excellent cycling stability with the capacitance retention of 74.5% after 1000 cycles.

Acknowledgements

We gratefully thank the National Natural Science Foundation of China (no. 21271027) for this work.

References

[1] M Winter and R J Brodd 2005 *Chem. Rev.* **105** 1021
[2] Simon P and Gogotsi Y 2008 *Nat. Mater.* **7** 845
[3] Zhao B T, Deng X, Ran R, Liu M L and Shao Z P 2016 Adv. Energy. Mater. **6** 495
[4] Cao F F, et al. 2016 *J. Am. Chem. Soc.* **138** 6924
[5] Mahmood N, Zhang C, Yin H and Hou Y 2014 *J. Mater. Chem.* A **2** 15
[6] Yang X, Cheng C, Wang Y, Qiu L and Li D 2013 *Science.* **341** 534
[7] Stoller M D, Park S, Zhu Y, An J and Ruoff R S 2008 *Nano. Lett.* **8** 3498
[8] Liu Y, Ai K and Lu L 2014 *Chem. Rev.* **114** 5057
[9] Xu Y, Zhao L, Bai H, Hong W, Li C and Shi G 2009 *J. Am. Chem. Soc.* **131** 13490
[10] Hummers W S and Offeman R E 1958 *J. Am. Chem. Soc.* **80** 1339

IOP Conf. Series: Materials Science and Engineering **292** (2017) 012086 doi:10.1088/1757-899X/292/1/012086

[11] Wu Q, Xu Y, Yao Z, Liu A and Shi G 2010 *Acs. Nano.* **4** 1963
[12] Hu C, Zhai X, Liu L, Zhao Y, Jiang L and Qu L 2013 *Sci. Rep-UK.* **3** 2065
[13] Li X, Ding R, Shi W, Xu Q, Wang L, Jiang H, Yang Z and Liu E 2017 *Mater. Lett.* **187** 144
[14] Chen H, Hu L, Chen M, Yan Y and Wu L 2014 *Adv. Func. Mater.* **24** 934
[15] Li K, Shao Y, Liu S, Zhang Q, Wang H, Li Y and Kaner R B 2017 *Small.* **13** 1700380
[16] Jaidev and Ramaprabhu S 2012 *J. Mater. Chem.* **22** 18775

Research on influence of water-cement ratio on workability and mechanical properties of geopolymer grouting material

X Q Wang[1], P H Wen[2], Z W Gao[2, 3] and C H Wang[2, 4]

[1] Tianjin Municipal Engineering Design & Research Institute, Tianjin 300392, China
[2] School of Highway, Chang'an University, Xi'an 710064, Shaanxi Province, China
[3] Xizang Minzu University, Xianyang 712082, Shaanxi Province, China

[4] E-mail: 2539526581@qq.com

Abstract. To investigate the influence of water-cement ratio on the fluidity and mechanical strength of geopolymer grouting materials, the change of fluidity, compressive strength and flexural strength of road geopolymer grouting material with different water-cement ratio were tested. The results showed that the fluidity of geopolymer grouting material was enhanced while the compressive and flexural strength decreased with the increase of water - cement ratio. The optimum value of water-cement ratio was in range (0.39-0.42). The road geopolymer grouting material had the better fluidity and the strength requirement of reinforcement if the main constituent materials was not changed.

1. Introduction

As a kind of alkaline activated cementitious material, geopolymer grouting material is an inorganic polymer with amorphous three-dimensional network and mainly composed of ionic bond and covalent bond [1]. It has good corrosion resistance, high temperature performance, impermeability and volume stability [2-3]. At present, the production of geopolymer grouting material mainly is prepared using industrial wastes such as fly ash and blast furnace slag as main raw materials [4]. The resulting CO_2 is 80% less than Portland cement. In view of its environmental pleasant aspect, low prices, excellent performance and other advantages of geopolymer grouting material, it is more in line with the green concept, through research on its performance is necessary. Yu Q et al. found that the addition of $BaCl_2$ in alkali activated material can prolong the coagulation time and revealed the main mechanism [5]. Cristelo N et al. studied the relationship among the mechanical properties of fly ash-based geopolymer materials with alkali-activators, capital and environment, and found that using fly ash-based geopolymer material can reduce CO_2 emissions and construction costs significantly [6]. Khademi et al. used some data-driven models to predict the compressive strength of environment-friendly road materials [7]. Nie Qiong obtained the optimum proportion of slag-based grouting material in different ages through mean square design optimization process and compared the material with cement slurry [8]. Chen Lihua determined the optimum proportion of alkali-activated superfine slag powder grouting material, and the compressive strength was up to 13.82 MPa [9]. Ibrahim et al. ascertain compressive strength and water absorption of geopolymer bricks in different aging time [10]. Al Bakri Abdullah MM et al. investigated on the compressive strength of geopolymer material and the ratios of fly ash/alkaline activator and Na_2SiO_3 solution/NaOH solution [11]. But the effect of water-cement ratio on the workability and mechanical properties of the blast furnace slag-based geopolymer grouting material remains unclear.

This paper used blast furnace slag as the basic raw material to prepare geopolymer grouting material with alkali-activators, revealed the influence of water-cement ratio on the workability and mechanical properties of the geopolymer based on fluidity, compressive and flexural test.

2. Materials

It is an effective method to prepare the geopolymer grouting material by alkali-activated blast furnace slag at normal temperature. In this study, the main materials include blast furnace slag (SL), sodium silicate (SS), sodium hydroxide (SH), water-reducing agent and so on. Blast furnace slag (superfine powder, produced by a plant in Hebei province, average grain diameter is equal 0.05mm, activity index⩾95%, 35% of silicon dioxide, 14% of aluminum oxide, 41% of calcium oxide, and 8% of magnesium oxide), alkali-activators including sodium silicate solution (9% of Na_2O, 30% of SiO_2, density is equal 1.39 g/ml, industrial grade, modulus is equal 3.3) and sodium hydroxide solution (purity⩾99%, white, flaky solid, industrial grade), water-reducing agent (the main component is sulfonated melamine（SM）, white powder, no chloride, bulk density is equal 450-750 kg/m^3, less than 4% of weight loss on drying, PH about 9-11), water.

3. Test method

The geopolymer grouting material can be prepared by using alkaline activator to stimulate the aluminum source materials including the potential active silicon or industrial wastes such as fly ash, blast furnace slag, steel slag. The geopolymer grouting material was prepared with a certain proportion of blast furnace slag, sodium silicate, sodium hydroxide, water-reducing agent and water. Table 1 shows the mixture proportion of geopolymer grouting material.

Table 1. The mixture proportion.

Series number	m_W/m_{SL}	m_{SS}/m_{SL}	m_{SH}/m_{SL}	m_{SM}/m_{SL}
1	0.36	0.09	0.07	0.006
2	0.39	0.09	0.07	0.006
3	0.42	0.09	0.07	0.006
4	0.45	0.09	0.07	0.006
5	0.47	0.09	0.07	0.006

A certain amount of blast furnace slag, sodium silicate solution and NaOH were weighed out according to the mix ratio described in table 1. Sodium silicate and water was put into the beaker to mix, and NaOH was added in the mixing process, and the beaker was sealed with a plastic wrap to avoid the evaporation of water after the mixing was finished, until NaOH was dissolved and stirred evenly, stoped stir and cooled it down to ambient temperature. After the solution was completely cooled, the geopolymer slurry was prepared according to "Test methods for water requirement of normal consistency setting time and soundness of the Portland cement" (GBT1346-2011 of China).

The fluidity of the geopolymer grouting material is an important index to study its practical applicability in road engineering, and determined its effective diffusion radius.

The fluidity test mold was selected according to "cementitious grout" (JC/T 986-2005 of China). The truncated cone is required as follows: The inner wall is smooth, the height is 60mm ± 0.5mm, the inner diameter is 70mm ± 0.5mm, the inner diameter of the lower mouth is 100mm ± 0.5mm, the outer diameter is 120mm, and the wall thickness is more than 5mm. Test method reference to "Cade for concrete admixture application" (GB 50119-2013 of China), i.e., the 500mm×500mm glass plate was placed horizontally, glass plate, truncated cone mold and mixer internal surface were wetted, and the mold was covered with a damp cloth, the geopolymer slurry was poured into the expansion cylinder for mortar fluidity test, the maximum diameter of the two perpendicular directions were measured after 10 seconds, and the results were represented by averages.

When the mechanical strength of geopolymer grouting material was tested, prepared geopolymer grouting material according to the design ratio and formed 40mm×40mm×160mm specimens using the way of removing the bubble by artificial patted. Six samples were prepared for each mixture ratio. They were sealed with plastic wrap to prevent rapid evaporation and precipitation of water, the specimen were removed from mold after curing for 24 hours at ambient temperature of 20℃±5℃ and room humidity about 50%, then specimens were cured in water until reached the test age. The flexural strength and compressive strength were measured according to "Method of testing cements-Determination of strength"(GB/T 17671-1991 of China), and the loading speed should be strictly controlled within 2200 N/s-2600 N/s during the process of determination.

4. Test results and discussion

4.1. Workability

The fluidity of geopolymer grouting material with different water-cement ratio was analyzed, the results of the analyses are shown in figure 1.

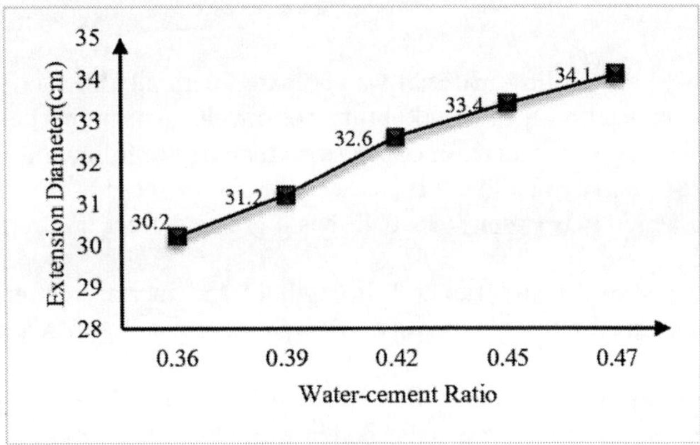

Figure 1. The fluidity.

From figure 1, it is noted that the fluidity increases as the water-cement ratio increases, when the water-cement ratio is between 0.39-0.42, the extension diameter of the mortar increase larger, but the increasing of the extension diameter is reduced when the water-cement ratio exceeds 0.42. In view of the results, it is indicating that the water-cement ratio is more influential to the fluidity when the water-cement ratio is between 0.39-0.42.

4.2. Mechanical properties

According to the test results, the compressive strength and the flexural strength of geopolymer grouting material of 7 days curing ages related with different water-cement ratio is shown in figure 2.

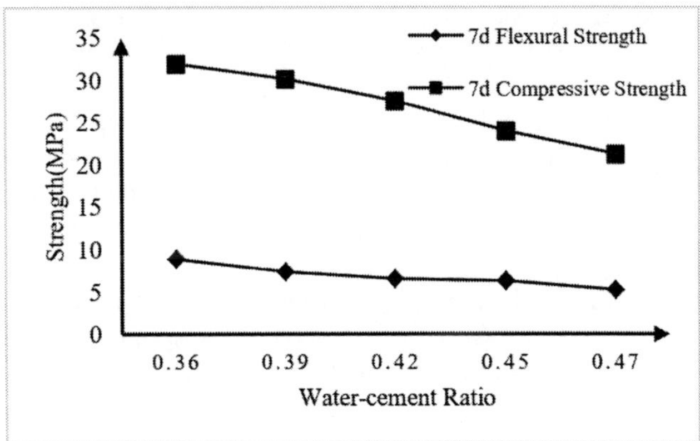

Figure 2. Mechanical properties.

The following conclusions can be drawn according to the results shown in figure 2.

(1) The flexural strength and compressive strength of the specimen decrease gradually with the increasing water-cement ratio, the decrease of compressive strength is larger than that of flexural strength.

(2) Water-cement ratio is an important factor that affects the strength of the geopolymer grouting material, the remaining water of the material is higher with the increase of the water-cement ratio after the polymerization reaction, the excess water remain in the interior of the geopolymer grouting material that become blisters or produce bubbles will affect the strength properties.

Considering the influence of water-cement ratio on the workability and mechanical properties of the geopolymer grouting material, it is suggested that the water-cement ratio can be controlled between 0.39 and 0.42 in the preparation process. The optimum water-cement ratio can make the geopolymer grouting material have better working performance and mechanical properties in the process of road reinforcement.

5. Conclusions

In this study the geopolymer grouting material was prepared with alkaline activators and investigated the effects of water-cement ratio on the workability and mechanical properties based on the fluidity test, compressive and flexural test. The main conclusions include the following aspects:

(1) The fluidity of the geopolymer slurry increases with the increase of the water-cement ratio, and the water-cement ratio which is between 0.39-0.42 has a greater effect on the fluidity of the polymer grouting material.

(2) The compressive strength and flexural strength of the mortar material decreases with the increasing of the water-cement ratio, compressive strength decreases significantly with the increasing of the water-cement ratio.

(3) The water-cement ratio can be controlled between 0.39 and 0.42 in the preparation process to ensure the geopolymer grouting material have better workability and mechanical properties in the process of road reinforcement.

References

[1] Shi H S, Lin M S and Guo X L 2011 Research advance on fly ash based geopolymer material *Fly Ash Comprehensive Utilization* **(3)** 42-46

[2] Tao W H, Fu X H, Sun F J and Yang Z X 2008 Studies on properties and mechanisms of geopolymer cementitious material *Bulletin of The Chinese Ceramic Society* **(4)** 730-735

[3] Jian J C, Liu Z, Yang H B, Liu L R and Yu A P 2014 Research on preparation and application status of geopolymers *Multipurpose Utilization of Mineral Resources* **(3)** 18–22

[4] Faris M A, Abdullah M M A B, Sandu A V, Ismail K N, Moga L M, Neculai O and Muniandy R 2017 Assessment of alkali activated geopolymer binders as an alternative of portlant cement *Materiale Plastice* **54** 145-154

[5] Yu Q J, Zhao S Y, Huang J Q, Qiao F, Guo W Y, Yin S H, Wen Z Y and Gu G B 2005 Retardation of gelation and improvement of flowability for alkali-activated carbonatite-slag cementitious-grouting material *Journal of the Chinese Ceramic Society* **33** 871-875

[6] Cristelo N, Miranda T, Oliveira D V,Rosa I, Soares E, Coelho P and Fernandes L 2015 Assessing the production of jet mix columns using alkali activated waste based on mechanical and financial performance and CO_2 (eq) emissions *Journal of Cleaner Production* **102** 447-460

[7] Khademi F, Jamal S M, Deshpande N and Londhe S 2016 Predicting strength of recycled aggregate concrete using artificial neural network, adaptive neuro-fuzzy inference system and multiple linear regression *International Journal of Sustainable Built Environment* **5** 355-369

[8] Nie Q 2010 Development and test research of new grouting materials on slag powder[D] AnHui *University of Science and Technology(An-hui)*

IOP Conf. Series: Materials Science and Engineering **292** (2017) 012087 doi:10.1088/1757-899X/292/1/012087

[9] Chen L H, Chen G M, Xiao M and Fang L B 2013 Test research of new superfine slag powder slurry grouting materials *Journal of Hunan City University (Natural Science)* **22** 23-26
[10] Ibrahim W M W, Al Bakri Abdullah M M, Sandu A V, Hussin K, Sandu I G, Ismail K N, Radir A A and Binhussain M 2014 Processing and characterization of fly ash-based geopolymer bricks *Revista de Chimie* **65** 1340-1345
[11] Abdullah M M A, Tahir M F M, Hussin K, Binhussain M, Sandu I G, Yahya Z and Sandu A V 2015 Fly ash based lightweight geopolymer concrete using foaming agent technology *Revista de Chimie* **66** 1001-1003

IOP Conf. Series: Materials Science and Engineering **292** (2017) 012088 doi:10.1088/1757-899X/292/1/012088

Influence of temperature and moisture joint action on oil-paper insulation compound electric field distribution

H D Zhang[1], L Li[1, 3], M J Wu[1], Y Zhou[1], H Zhang[1] and Y M Feng[2]

[1] State Grid Heilongjiang Electric Power Company Limited Electric Power Research Institute, Harbin, Heilongjiang, 150030, China
[2] State Grid Heilongjiang Electric Power Company Limited, Harbin, Heilongjiang, 150090, China

[3] E-mail: lilin77@yeah.net

Abstract. In the paper, the electric field change in transformer oil and oil-impregnated insulation paperboard within wide temperature range (20℃~100℃) was studied. The influence law of temperature on oil-paper insulation electric fields was summarized. Oil-paper insulation electric field distribution under the action of different wave-form voltages was simulated for analysis. The electric field distribution condition during moisture balance steady state and transient state was obtained. The reasons of electric field distribution change were analyzed.

1. Introduction

EHV/UHV transformer is an important device in the power transmission system. Its operating reliability has direct influence on safe operation of the system. Existing operation experience shows that moisture dynamic balance process and existence form between oil and insulating paper in the transformer has important influence on oil paper insulation breakdown characteristics and surface flashover characteristics. Temperature is an important factor affecting the above-mentioned process [1-4]. In recent years, transformer insulation faults occurred in Northeast China and Northwest China for several times, which were related with changes of transformer operating temperature. Domestic electric power operation departments attach great importance to related research. The space charge accumulation characteristics of transformer oil-paper insulation under the action of compound electric field are mostly studied at room temperature and narrow temperature scope. It is necessary to study space charge accumulation of oil-paper insulation and its influence on electric field distribution characteristics in the possible operation temperature scope in view of actual working condition of oil-paper insulation.

2. Specimen pre-treatment and testing

2.1. Treatment of transformer oil

KIX45 transformer oil produced by Kunlun Company was selected in the experiment according to the temperature requirements of the project. The dielectric properties of the transformer oil were reduced since moisture and gas must be mixed in the transformer oil during delivery, carbon black was produced in transformer oil after breakdown test, and moisture was absorbed during refrigeration temperature-reversion. Transformer oil underwent standardized treatment before test, therefore it can meet various requirements regulated in national standards.

Oil filter was selected for treating transformer oil. The solubility of moisture and air in oil was reduced under vacuum and high temperature conditions, and the principle was utilized for dehydrating and degassing oil in the treatment equipment. The oil filter was used for filtering impurity particles with tiny particle size in oil. The transformer oil treated by the oil filter met the national standard 'GB 2535-90 Transformer Oil'. The treatment process was shown in figure 1.

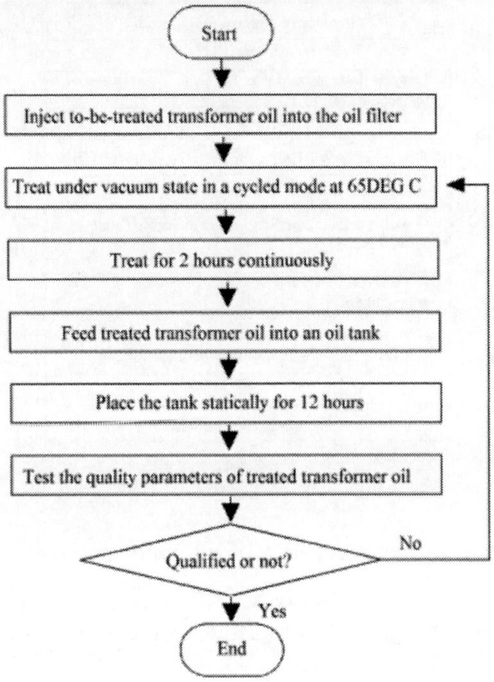

Figure 1. Treatment flow of transformer oil.

The moisture content was measured firstly for the treated transformer oil, which was measured again at certain interval, thereby ensuring no change in the moisture content. The transformer oil was finally stored in a sealed oil tank.

2.2. Treatment of oil impregnated paperboard

The vacuum drying oven transformed in the laboratory was utilized for treating paperboard according to the national standard 'GB10580-89'. The treatment process included vacuum drying and oil immersion. Related parameters of treated oil impregnated paperboard were tested. The operation flow was shown in figure 2.

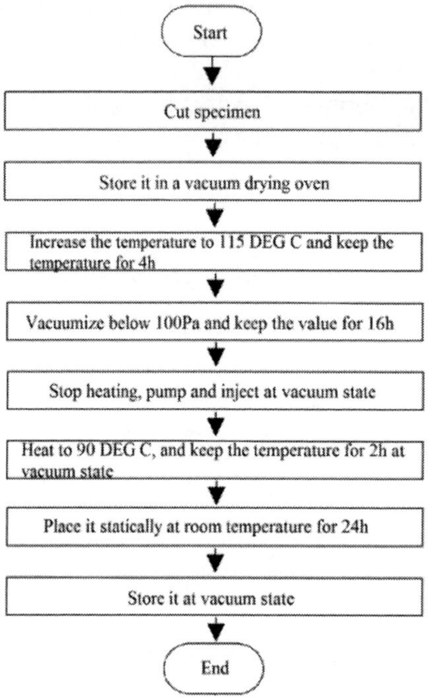

Figure 2. Treatment flow of oil impregnated paperboard.

The moisture content was measured firstly after treatment, which was measured again at certain interval, thereby ensuring no change in the moisture content. The treated paperboard was stored in transformer oil which was qualified in the test.

3. Influence of temperature and moisture joint action on oil-paper insulation compound electric field distribution

3.1. Influence of temperature on electric field distribution during moisture balance

Moisture of oil and paperboard was estimated through the moisture balance equation given by ABB Company [5]. The relation equation of oil-paperboard electric field distribution and moisture content under moisture balance state was obtained:

$$\begin{cases} E_1 = \dfrac{\rho_1}{\rho_1 d_1 + \rho_2 d_2}(1\text{-}\eta)U + \dfrac{\varepsilon_2}{\varepsilon_2 d_1 + \varepsilon_1 d_2}\eta U \\ E_2 = \dfrac{\rho_2}{\rho_1 d_1 + \rho_2 d_2}(1\text{-}\eta)U + \dfrac{\varepsilon_1}{\varepsilon_2 d_1 + \varepsilon_1 d_2}\eta U \\ \rho_1 = \mathrm{A}_1 \times e^{-\left(\frac{C_{paper}}{\alpha_1} + \frac{T}{\beta_1}\right)} + \mathrm{B}_1 \\ \rho_2 = \mathrm{A}_2 \times e^{-\left(\frac{C_{\mathrm{oil}}}{\alpha_2} + \frac{T}{\beta_2}\right)} + \mathrm{B}_2 \\ \varepsilon_1 = \mathrm{C}_1 + \mathrm{D}_1 \times e^{\theta_1 \times C_{\mathrm{paper}}} \\ \varepsilon_2 = \mathrm{C}_2 + \mathrm{D}_2 \times e^{\theta_2 \times T} \\ C_{\mathrm{paper}} = 2.06915 e^{-0.02970T} \times C_{\mathrm{oil}}^{\;0.40489T^{0.09733}} \end{cases} \tag{1}$$

η=the content of AC voltage in compound voltage, ε=dielectric constant ρ=volume resistivity, d_1=2mm, d_2=4mm and U=150kV were set in the equation (1). The electric field intensity change trend in transformer oil and its corresponding oil impregnated paperboard under AC-DC superposition voltage was shown in figure 3 and figure 4, wherein different temperatures and different moisture contents during moisture balance were considered.

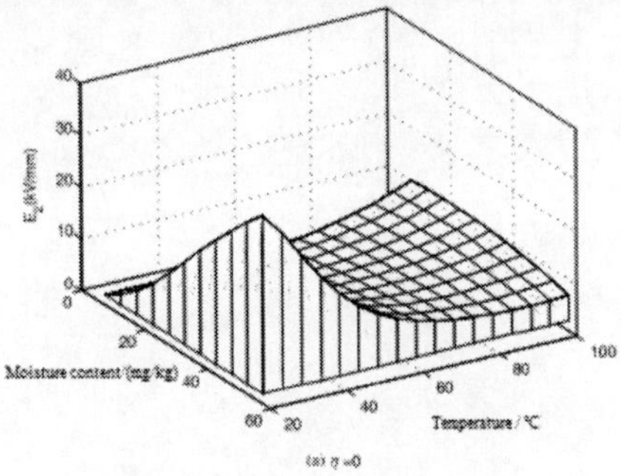

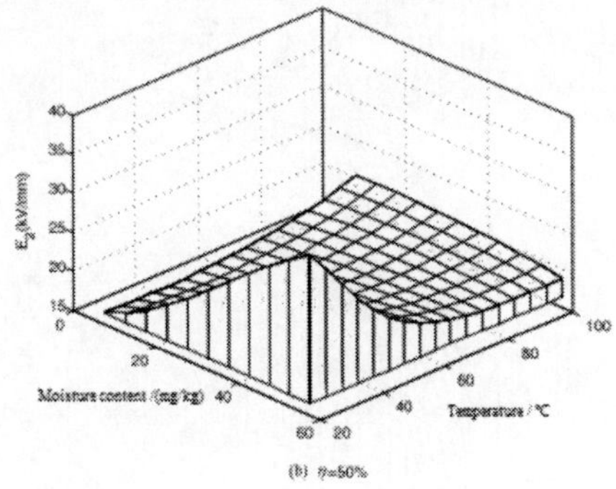

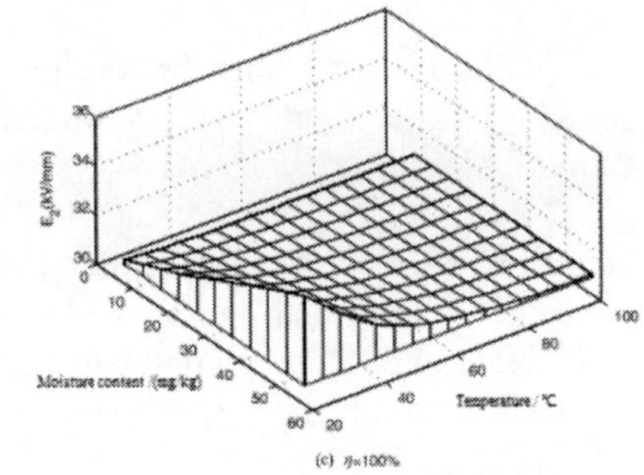

Figure 3. Field intensity distribution in oil at different temperatures and moisture contents.

Figure 3 showed that the field intensity in transformer oil was the lowest at low temperature and low moisture content when AC content was low (namely η=0), and it was the highest at low temperature and high moisture content. However, the lowest electric field value was discovered at high temperature and low moisture content when AC content was high (namely η=100). The field intensity was reduced according to exponential law with temperature rising or moisture content reduction. The AC content was lower, the electric field change amplitude was larger.

Figure 4 showed that the field intensity change trend in the oil impregnated paperboard was just opposite to that in the transformer oil. The field strength in the oil impregnated paperboard was the lowest at low temperature and high moisture content, and it was the highest at low temperature and low moisture content. The field intensity was increased according to exponential law with temperature rising or moisture content reduction. The AC content was lower, the electric field change amplitude was larger.

Moisture contents-temperature planes in figure 3 and figure 4 were compared. It was obvious that oil projection on the plane was rectangle, the projection of corresponding paperboard on the plane was irregular graphics, namely: the paperboard moisture content corresponding to oil with high moisture content was not high at high temperature.

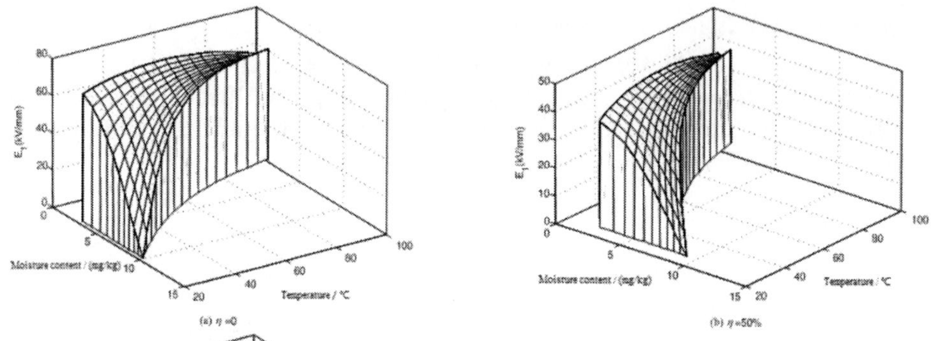

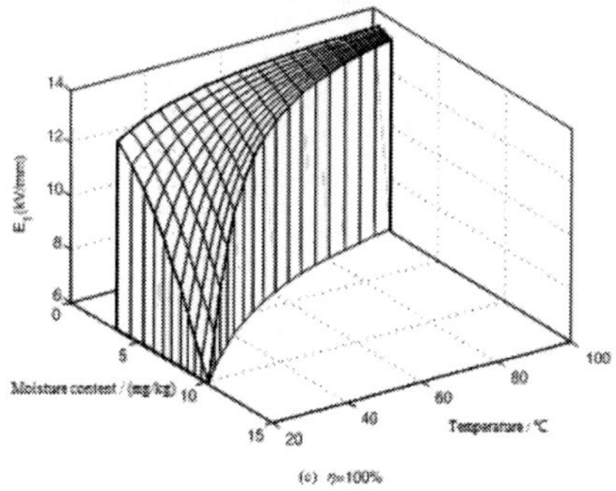

Figure 4. Field intensity distribution in oil impregnated paperboard at different temperatures and moisture contents.

T=20℃ and Coil=7.7mg/kg during moisture balance were regarded as the initial states to analyze the influence of temperature on oil-paper insulation electric field distribution under AC-DC superposition electric field concretely. The change of electric fields in oil impregnated paperboard and oil with temperature rising was analyzed. The change of electric field intensity in paperboard and oil with temperature at moisture balance was shown in figure 5 and figure 6.

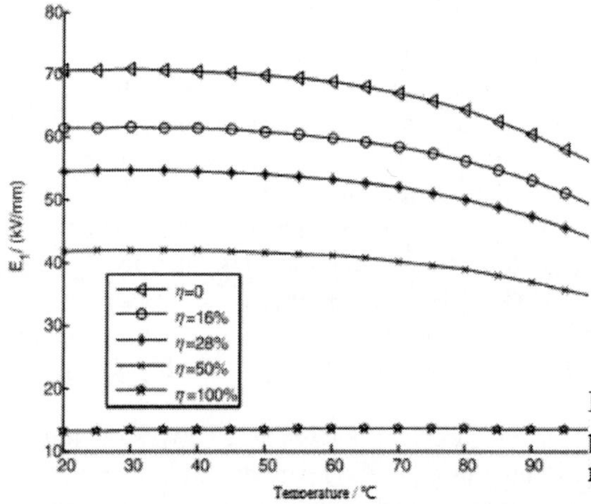

Figure 5. Field strength in oil impregnated paperboard at different temperatures during moisture balance.

Figure 5 showed that the field intensity increase with temperature in the paperboard under compound electric field was decreased gradually during moisture balance. The temperature was higher, the field intensity decrease speed was faster. AC content η was lower, the phenomenon was more prominent. The electric field was not changed basically with temperature when AC content η was equal to 100% (namely pure AC).

The field intensity change in transformer oil at different temperatures was shown in figure 6 during moisture balance. Figure 6 showed that the field intensity change condition in oil was just opposite to that in the paper under compound electric field during moisture balance. The field intensity in the oil was gradually increased with temperature rising. The temperature was higher, the field intensity rising was faster. Similarly, AC content η was lower, and the phenomenon was more prominent.

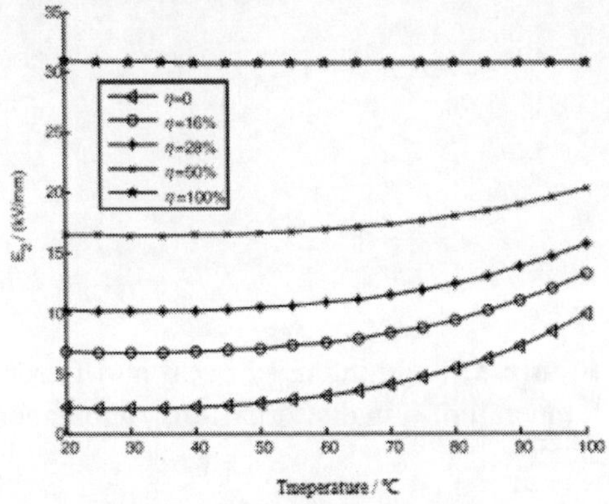

Figure 6. Field intensity in oil at different temperatures during moisture balance.

3.2. *Influence of temperature on electric field distribution during moisture unbalance*

When the temperature of oil-paper insulation suddenly rose, it was believed that resistivity and relative dielectric constant were changed due to temperature change only rather than moisture content because of slow moisture movemen.

Similarly, T=20℃ and C_{oil}=7.7mg/kg were set as initial states. The moisture content C_{paper} in the paper during T=20℃ was obtained through calculation, namely 3.45%. Therefore, the change trend of respective electric field strength with temperature was shown in figure 7 and figure 8 when moisture content in paper and water was constant.

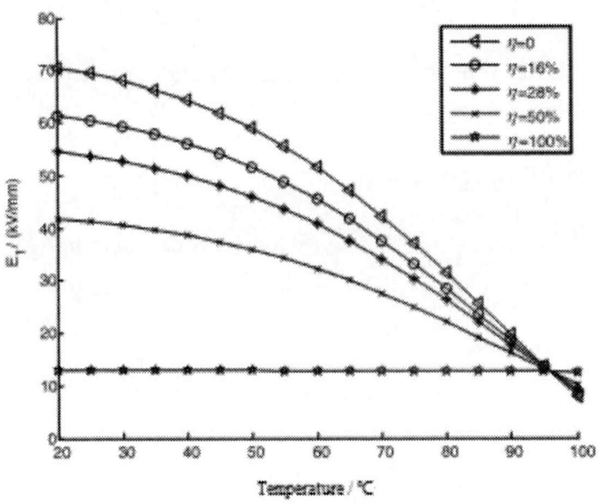

Figure 7. Field intensity of oil impregnated paperboard at different temperatures under moisture unbalance state.

Figure 7 showed that the change trend of electric field intensity in paperboard under compound electric field with temperature was basically the same as that under moisture balance, but the change amplitude value was far higher than that under moisture balance when the moisture content of the paperboard was constant. When the AC content was high (such as η=100%, namely pure AC), the field intensity was not changed with temperature change basically. When AC content was low, the field intensity in the paperboard was gradually decreased with temperature rising. When the temperature rose to 95 ℃ or so, the field strength under compound electric field was almost equal to the field intensity under pure AC, the temperature rose continuously, and the field intensity under compound electric field was lower than that under pure AC.

Figure 8 showed the change of field intensity in transformer oil with temperature under moisture unbalance state.

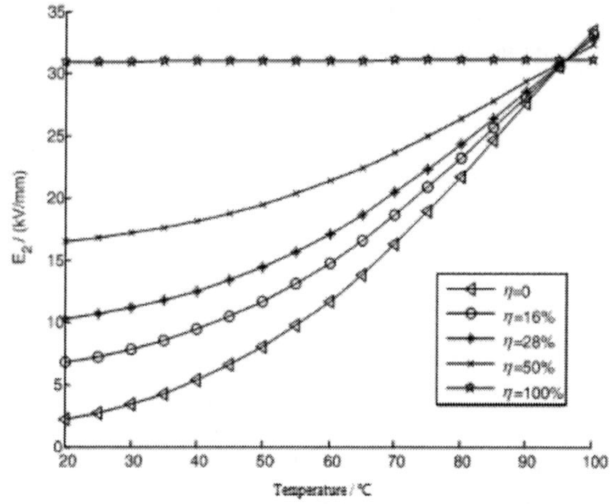

Figure 8. Field intensity in oil at different temperatures under moisture unbalance state.

Figure 7 showed that the change law of field intensity in transformer oil was just opposite to that in the paperboard. The electric field intensity in oil was gradually increased with temperature rising under compound electric field. The temperature was higher, the change speed was faster. The AC content η was lower, the change amplitude value was larger. When the AC content was the highest (such as η =100%, namely pure AC), the field intensity was not changed with temperature change basically. When AC content was lower, the field intensity in the oil was gradually increased with temperature rising. When the temperature rose to 95 ℃ or so, the field strength under compound electric field was almost equal to the field intensity under pure AC. The temperature rose continuously, and the field intensity under compound electric field was lower than that under pure AC.

3.3. Analysis on simulation results

Figure 5 and figure 8 were compared. It was discovered that the field intensity change amplitude was large at low AC content, and the field intensity change was small at high AC content with temperature rising because electric field distribution mainly depended on the ratio of relative dielectric constant at high AC content. The electric field distribution mainly depended on the resistivity ratio at low AC content.

The change of relative dielectric constant ratio of paper and oil with temperature was shown in figure 9.

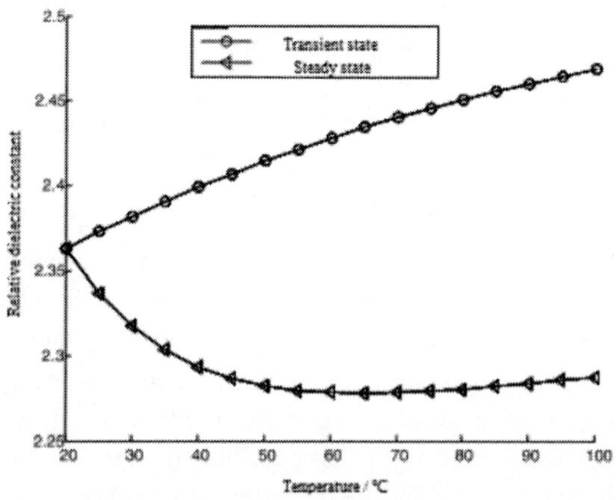

Figure 9. Relative dielectric constant ratio of paper and oil at different temperatures.

Figure 9 showed that the relative dielectric constant ratio of paper and oil under moisture unbalance state was increased with temperature rising. However, the ratio was increased from 2.36 to 2.47 only with amplitude of 4.6%. The ratio change law during moisture balance was different from that under short-time influence. Its ratio was decreased with temperature rising. The highest value was 2.36, and the lowest value was 2.28. The amplitude was only 3.3%. It was obvious that relative dielectric constant had small change amplitude, and it was the main reason why electric field change was small during high AC content.

Figure 10 showed the change of resistivity ratio between oil and paper at different temperatures.

Figure 10 showed that the resistivity ratio under moisture unbalance state was slightly different from that during moisture balance with temperature increase. However, it showed a decrease trend as a whole. The resistivity ratio was decreased gradually with temperature rising under water unbalance state, namely it was decreased from the highest ratio 32.34 at 20℃ to the lowest ratio 0.23 at 99℃. The decrease amplitude was 99.28%. The ratio was increased firstly and decreased then during water balance. It was decreased from the highest ratio 33.71 at 30℃ to the lowest ratio 5.60 at 99℃. The amplitude of fluctation was 83.39%. Resistivity ratio was changed greatly because the decrease speed of paper was much higher than that of oil though the volume resistivity of oil and paper was decreased exponentially with temperature rising. It was obvious that oil and paper resistivity was not proportionally changed with temperature, and it was the main reason why electric field was greatly changed with temperature at low AC content.

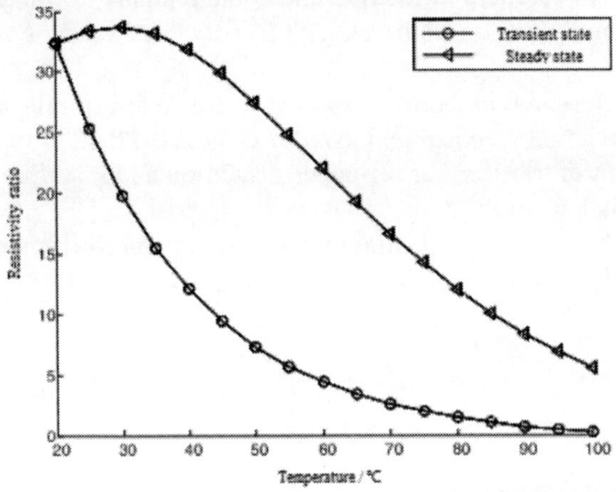

Figure 10. Resistivity ratio of paper and oil volume at different temperatures.

Figure 6 and figure 8 were compared. It was obvious that the electric field change amplitude in transformer oil was much larger than that during moisture balance at low AC content under moisture unbalance state. Field intensity in oil was gradually increased with temperature rising during moisture balance. However, the highest value of compound electric field was not higher than 2/3 under its AC condition. The value under pure DC condition (η=0) was only 29% of AC condition. However, the electricity field in oil was changed significantly under moisture unbalance state, and the field intensity in oil was even higher than that under AC at high temperature. Pure DC (η=0) was adopted as an example. The electric field was increased from 2.18kV/mm to 9.86kV/mm when the temperature rose from 20℃ to 99℃ during moisture balance. The electric field was increased from 2.18kV/mm to 33.55kV/mm under moisture unbalance state. The highest value was nearly 4 times of that during moisture balance. However, the AC field intensity (η=100%) was always kept at 30.7 kV/mm or so during the process.

4. Conclusion

Temperature change affects the resistivity and relative dielectric constant of oil and paper on the one hand, and the moisture balance point between oil and paper is moved on the other hand. Therefore the resistivity relative dielectric constant of oil and paper is changed, and electric field distribution of oil-paper insulation under compound electric field is further affected.

IOP Conf. Series: Materials Science and Engineering **292** (2017) 012088 doi:10.1088/1757-899X/292/1/012088

Electric field is changed with temperature slightly during moisture balance due to the following reasons: the moisture balance point is moved with temperature rising during moisture balance, and moisture in paper is moved to oil, thereby lowering the moisture content of paper and further reducing the decrease speed of the resistivity. The resistivity ratio decrease amplitude is much smaller than that under the moisture unbalance state, and many electric fields are concentrated in paper. Time is limited for moisture movement with temperature rising under the moisture unbalance state, and many electric fields are concentrated in oil as a result.

In conclusion, temperature has more prominent influence on oil-paper insulation in compound electric field compared with AC electric field. Oil test of oil-paper insulation in compound electric field during sudden temperature rising is more stringent than that during high temperature moisture balance.

Reference

[1] Lu Xiaode, Chen Shikun and Fang Zhiye, etc. 1998 Analysis on characteristics of winding end electric field on the side of converter transformer valve J *Transformer* **35(6)** 16-22

[2] Wen K C, Zhou Y B and Fu J, et al. 1993 A calculation method and some features of transient field under polarity reversal voltage in HVDC insulation J *IEEE Transactions on Power Delivery* **8(1)** 223-230

[3] Zhao Y G, Wei X L and Wang Y H 2014 Research on breakdown characteristic test of oil-paper insulation system under AC-DC superposition electric field J *Smart Grid* **2(10)** 8-12

[4] Liu Y X 2002 Analysis on moisture content of transformer oil-paper insulation and its influence on operating safety J *Transformer* **39(5)** 1-6

[5] Du Y, Zahn M, Lesieutre B C, et al. 1999 Moisture equilibrium in transformer paper-oil systems J *IEEE Electric Insulation Magazine* **15(1)** 11-19

IOP Conf. Series: Materials Science and Engineering **292** (2017) 012089 doi:10.1088/1757-899X/292/1/012089

Raman spectroscopic study on structure and phase transition of zirconium tungstate synthesized with a CO_2 laser

J P wang, Q D chen[1]

Faculty of Aerospace Engineering, Binzhou University, Binzhou 256603, China

[1]E-mail: chenqingdong-88@163.com

Abstract. Negative thermal expansion material of ZrW_2O_8 has been successfully synthesized by rapid solidification with a CO_2 laser. The materials solidified in densely packed blocks with both smooth surface and dark green color. Raman spectroscopic and XRD analysis reveal that the samples solidified in an orthorhombic structure. It is analyzed that the compressive stress can be effectively reduced by lowering the laser scan speed. ZrW_2O_8 samples with cubic structure were produced when the laser scan speed was less than 2mm/s. The samples cooled by water are mainly dominated by α phase which demonstrates that a phase transition from γ to α has occurred after annealing.

1. Introduction

Materials with negative thermal expansion (NTE) have attracted significant attention in recent years due to scientific curiosity and technological interest [1-7]. Among these materials, ZrW_2O_8 is of considerable interest due to their large isotropic NTE over a wide range of temperature. It exhibits a framework structure that consists of the corner-sharing WO_4 tetrahedra and ZrO_6 octahedra. The polyhedra in ZrW_2O_8 are connected at almost all vertices, the exception being one terminal oxygen atom in each WO_4 tetrahedron. The NTE in ZrW_2O_8 was explained in terms of the rigid unit mode (RUM) model [5].

Several methods have been developed for the production of ZrW_2O_8 such as solid state reactions [8], sol-gel method [9] and combustion route. The methods are not only tedious, but also highly time and energy wasting. Another difficulty associated with this procedure is the volatility of WO_3 at elevated temperatures. In this paper, ZrW_2O_8 bocks have been successfully synthesized by rapid solidification with a CO_2 laser and Raman spectroscopic study the structure and phase transition of zirconium tungstate. With this technique a sample can be produced within a few seconds. Laser synthetic route is a rapid and green manufacturing technology with which a sample can be produced within a few seconds. It is easy to realize industrial mass production by continuous feeding of raw materials.

2. Experimental procedure

The commercial chemicals of ZrO_2 and WO_3 (99.9% purity) were mixed according to the stoichiometric ratios of destination materials of ZrO_2 and WO_3. The mixtures were ground in a mortar for 2 hours and then pressed into cylindrical pellets by uni-axial cold press with a steel mould at a pressure of 10 MPa.The mixtures were ground in a mortar for 2 hours and then pressed into cylindrical pellets by uni-axial cold press with a steel mould at a pressure of 10 MPa. The synthesis was performed by using a 5 KW continuous-wave CO_2 laser. The laser beam was striking onto the pellets which were placed in a distance where the defocus length was set to 120 mm. In this case, the beam spot on the samples was approximately 10.4 mm in diameter. During the synthesis, either the pellets or the laser beam run straight at a scan speed of 1-5 mm/s. The laser power was varied from 400W to 800 W.

All the samples were analyzed by Raman spectroscopy with a Renishaw MR-2000 Raman spectrometer. Laser excitation wavelengths of 633nm and 532nm were used (the laser power was 3mW).X-ray diffraction (XRD) measurements were carried with an X'Pert PRO X-Ray Diffractometer.

3. Results and discussion

Figure 1a shows the Raman spectrum of zirconium tungstate synthesized by a CO_2 laser with 800 W laser power and 5 mm/s scan speed. For comparison, the Raman spectrum of ZrW_2O_8 prepared by solid state reaction is shown in figure 1c.It is obvious that the Raman spectrum of ZrW_2O_8 synthesized by a CO_2 laser is very different from that by solid state reaction [8]. The Raman modes at 1031, 1015, 974 and 932 cm^{-1} can be assigned to symmetric stretching(ν_s)while those at 903, 889, 870, 868, 802, 777, 745, 722, 707, 683 and 654 cm^{-1} to asymmetric stretching vibrations (ν_{as}) of the WO_4tetrahedra. The peaks at 432, 413, 392,375,351 and 329 cm^{-1} are ascribed to the asymmetric bending (δ_{as}) where as those at 313, 301 and 268 cm^{-1} is from the symmetric bending (δ_s) modes. The other modes observed at 242, 235, 202, 183, 141, 137, 92, 84, 75 and 64 cm^{-1} originate from lattice modes arising from Zr atom motion and translational and librational (hindered rotation) modes of WO_4. As compared to the Raman spectrum of ZrW_2O_8 by solid state reaction, the most distinctive features of the Raman spectrum of ZrW_2O_8 synthesized by a CO_2 laser are (i) appearance of more Raman peak, which have strong peak at 840,745,707,683,654,432,413,392,351,242,202,137 and 75 cm^{-1}; (ii) disappearance of modes at 931,902 and 888 cm^{-1}; (iii) the characteristic Raman spectra of ZrW_2O_8 become weak and wide. These features suggest that the ZrW_2O_8 synthesized by CO_2 laser solidified in orthorhombic structure (γ phase) instead of cubic (α phase).

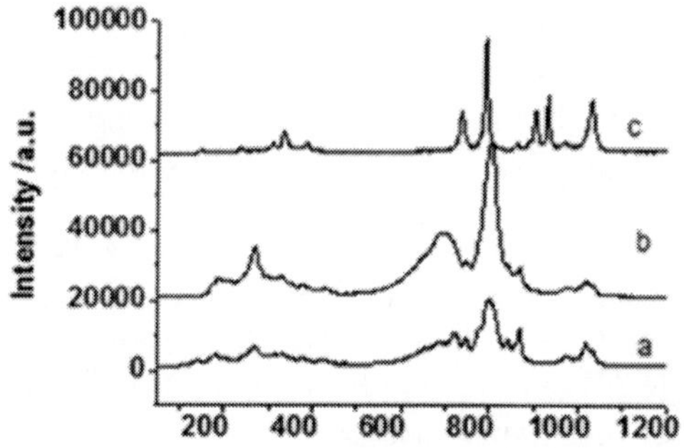

Figure 1. Raman spectra of ZrW_2O_8: (a) synthesized by CO_2 laser; (b) synthesized by laser and annealed; (c) synthesized by solid state reaction.

Figure 1b shows the Raman spectrum of zirconium tungstate synthesized by a CO_2 laser after annealing. When the sample after annealing, the Raman modes at 903 and 932 cm^{-1} vanished and the ν_{as} modes split into a number of peaks (see spectrum in figure 1b). Besides, the mode at 1015 cm^{-1} regained its intensity. The restoration for the δ_{as} and δ_s modes occurred as well. This suggests that a reverse phase transition from cubic to orthorhombic took place when the sample was cooled. Since phase transition from α to γ and the disappearance of the bands at 899 and 936 cm^{-1} were only observed under pressures above 0.21 and 0.5 GPa previously, it can be inferred that the ZrW_2O_8 block synthesized by a CO_2 laser still encounters a compressive stress after the temperature circle. It is the compressive stress that accounts for the recovery of the γ phase after annealing.

Through literature survey found that ZrW_2O_8 crystallizes are cubic phase with the space group $P2_13$ by all the previously reported synthetic methods [8]. But there is some γ phase ZrW_2O_8 crystallizes in the laser synthesis which suggests that the sample under the certain pressure during the solidification process, which is because the γ phase belongs to high pressure phase only under the effect of pressure α phase shift to γ phase. Sudden drop of temperature from the molten pool to the ambience during the solidification should account for the compressive stress generation. High pressure neutron diffraction studies showed that ZrW_2O_8 undergo an irreversible structural transition from α (cubic) to the γ (orthorhombic) phase at 0.2 GPa and finishes at about 0.5 GPa [10]. Ravindran, et al have carried out high pressure Raman spectroscopic study of ZrW_2O_8[11].They observed a dramatic splitting of the degenerate v_3 modes of the WO_4 into a large number of components at pressures above 0.21 GPa and the disappearance of the Raman modes at 904 and 932 cm^{-1} when the pressure was increased to above 0.4 ~ 0.5 GPa.

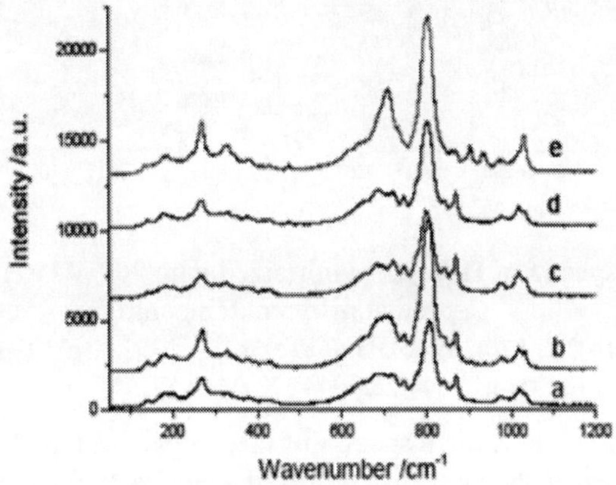

Figure 2. Raman spectra of ZrW_2O_8 synthesized with different scan speeds at 500W laser power:(a) 5mm/s; (b) 4 mm/s; (c) 3 mm/s; (d)2 mm/s and (e)1 mm/s.

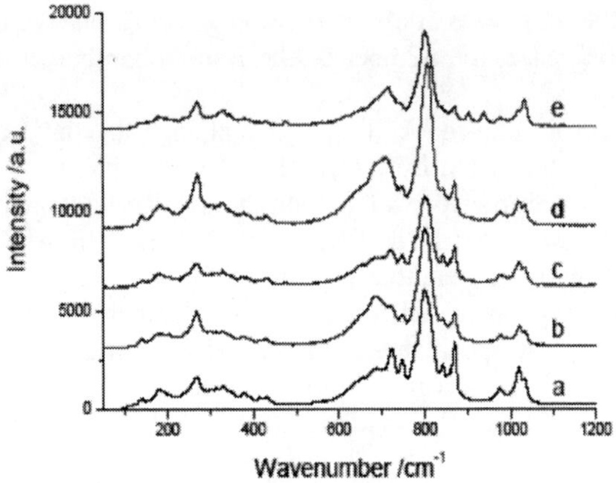

Figure 3. Raman spectra of ZrW_2O_8 synthesized with 800W laser power at different speed:(a)5mm/s(b)4 mm/s(c)3 mm/s(d)2 mm/s(e)1 mm/s.

Figures 2a ~2e and figures 3a ~3e show the Raman spectra of ZrW_2O_8 synthesized with 500 W and 800 W at different scan speeds. The prerequisite condition for α phase production is the compressive stress under 0.2 GPa. The strength of the compressive stress is proportional to the cooling rate which is proportional to the temperature difference between the molten pool and the ambient, and inversely proportional to the time for the sample to cool. It is analyzed and confirmed that the compressive stress can be effectively reduced by lowering the laser scan speed. It is obvious that the γ phase dominates when the scan speed is more than 2 mm/s. However, with lower scan speed, α phase is produced as indicated by the degeneration of the v_{as} modes, and the disappearance of the Raman modes at 899, 936 and 1015 cm^{-1}.These results show that the compressive stress can be tuned by laser synthetic parameters and it can be used to the production of different phases of ZrW_2O_8.

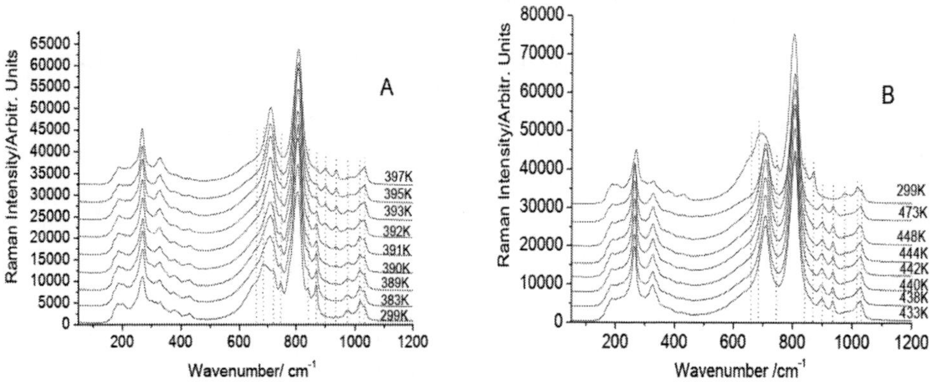

Figure 4. Raman spectra of HfW_2O_8 synthesized with 700W laser power and 1mm/s scan speed at different temperature.

A: (a)299 K(b)383 K(c)389 K(d)390 K(e)391 K(f)392 K(g)393 K(h)395 K(i)397 K

B: (a)433 K(b)438 K(c)440K(d)442 K(e)444 K(f)448 K(g)473 K and(h)cooled back to 299 K

Figures 4A and 4B show the Raman spectra of the sample synthesized by a CO_2 laser with 500 W laser power and 4 mm/s scan speed at different temperatures from 299 K to 473 K and then cooled back to 299 K, respectively. With temperature increase, distinct changes in the Raman spectra are observed: (i) The Raman modes at 903 and 931 cm^{-1} appear at about 390 K and increase in intensity till to 397 K; (ii) The Raman band at 745,840,861,974 and 1015 cm^{-1} vanishes progressively above 390 K; (iii) Only two of the v_{as} modes at about 707 and 798cm^{-1} survive and others disappear. As the temperature continues to increase, the number of the Raman bands remains unchanged. Most of the Raman bands shift to red.

It is obvious that several Raman mode frequencies change discontinuously around 390 K. These changes indicate that a phase transition from orthorhombic to cubic occur at about 390 K. In α phase, the structure of ZrW_2O_8 is a framework of corner-sharing ZrO_6octahedra and WO_4tetrahedra. The octahedra share all their six corners with six different WO_4 tetrahedra while WO_4 tetrahedra share only three of their four oxygens with the neighboring ZrO_6 octahedra. The remaining oxygen in each WO_4 tetrahedron is formally singly coordinated. An isostructural order-disorder transition to another cubic Pa_3 structure is expected at about 430 K due to the disordering of the WO_4 tetrahedra. The disordered phase has higher oxygen mobility. However, no obvious changes in the Raman spectra can be seen in this temperature region. These give the evidence that the ZrW_2O_8 synthesized by CO_2 laser is γ-ZrW_2O_8.

As can be seen from figures 4 some of the high-frequency stretching modes, particularly the asymmetric stretching modes, give rise also to negative slopes vs temperature and hence contribute also to the negative thermal expansion. This confirms that the correlated motions of translation and libration with the asymmetric stretching vibrations of the polyhedra are responsible for the negative thermal expansion in the low-temperature phase. Above 440 K, some of the correlated motions may be destroyed as evidenced by the disappearance of the coupled librational motion around 40 cm^{-1} with the marked weakening of the out-of-phase asymmetric stretching vibration across the order-disorder phase transition and cause a smaller negative thermal expansion of the β phase.From figure 2B we can see that when the sample cooled to room temperature, the Raman modes at 899 and 936 cm^{-1}vanished and the v_{as} modes split into a number of peaks.Besides, the v_S mode at 1015 cm^{-1} restored its original intensity. The restoration for the δ_{as} and δ_s modes occurred as well. This suggests that a reverse phase transition from cubic to orthorhombic took place when the sample was cooled. Since phase transition from α to γ and the disappearance of the bands at 899 and 936 cm^{-1} were only observed under pressures above 0.21 and 0.5 GPa previously, it can be inferred that the ZrW_2O_8 block synthesized by a CO_2 laser still encounters a compressive stress after the temperature circle. It is the compressive stress that accounts for the recovery of the phase after annealing.

Through the mechanism of laser synthesis route we can understand the reasons for the formation of γ phase [11, 12]. In the laser synthesis, laser energy is absorbed by ZrO_2 and WO_3 efficiently, and translated into heat on its own where the chemical reaction took place. When the laser beam to move forward the reaction products can fast curing, which is very similar to the process of laser cladding. For most of the laser cladding alloy coating, very big tensile stress in the process of rapid solidification so often leads to the cladding layer crack. The pressure is unavoidable in the process of rapid solidification of ZrW_2O_8 samples. Since ZrW_2O_8 is negative expansion material, making a compressive stress and not tensile stress. It is the compressive stress led to ZrW_2O_8 exists in orthogonal structure. In fact, the laser power and speed are key technological parameter whether the sample can directly sinter into the shape. The laser power and speed can denote with the laser energy of unit time. When the laser power is higher and scan speed is slower, more heat transfer into the sample. It is favorable to solidification level sand sinter levels to connect well. But if the unit time laser energy excessively high causc the commixture powder be prone to gasify which leads to pit, cave and even sink and the sinter level have the warp distortion and so on. And if the unit time laser energy excessively smaller which influence the conjunction between solidification level and sinter level even unable to take shape.

In order to confirm the above analysis, the sample was characterized by XRD which is shown in figure 5. It shows the sample is in accordance with that of an orthorhombic structure ZrW_2O_8 with space group $P2_12_12_1$, indicating that the sample which is natural cooled by air is indeed dominated by the γ phase. Figure 4b shows the XRD of a sample for comparison. The sample is cooled by water with space group $P2_13$, which is almost α-ZrW_2O_8. It demonstrates that a phase transition from γ to α has occurred after annealing. The water cooling to reduce the γ- ZrW_2O_8, show that the compressive stress decreases in the samples. It is noticed that the XRD and Raman peaks of the sample synthesized by laser are relatively broad. This may indicate that the sample is composed of nano-crystallines and amorphous phases. The samples which are just sintered by laser and were quickly put into the water. Due to the water has a large heat capacity and the high thermal conductivity, the heat will be quickly absorbed by the water in the laser sintering samples. So the temperature field have been rapidly eroded and this will decrease the compressive stress in the sample, thereby reducing the γ- ZrW_2O_8 generated.

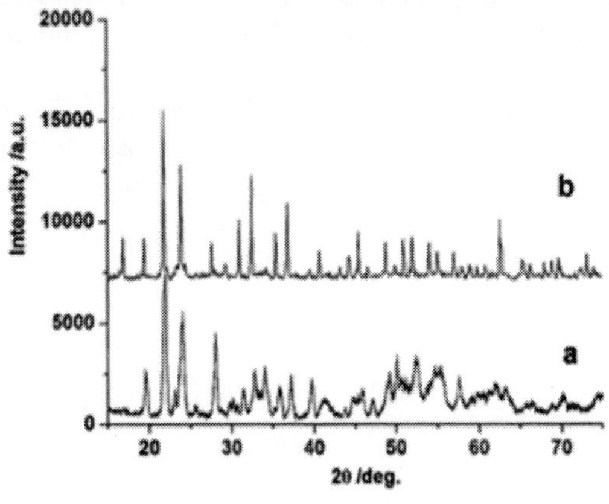

Figure 5. XRD patterns of ZrW_2Osynthesized with 500 W laser power and 2mm/s scan speed; (a) naturally cooled by air (b) cooled by water.

4. Conclusion

ZrW_2O_8 has successfully been synthesized with a high power CO_2 laser. It is a rapid process in which a sample can be produced in few seconds. The sample of laser synthesis compare with the samples which are synthesized by solid state reaction or other chemistry methods, there is γ phase in the sample. The γ phase formation gives a definite indication that a compressive stress was induced during the solidification process. Analyzing by the mechanisms of the laser synthetic route, the strength of the compressive stress is proportional to the cooling rate which is proportional to the temperature difference between the molten pool and the ambient, and inversely proportional to the time for the sample to cool.A recovery of the γ phase has been observed when the samples were cooled to room temperature. This suggests that cubic structure is only metastable at temperatures above 390 K and at room temperature the γ phase is preferred.It is analyzed and confirmed that the compressive stress can be effectively reduced by lowering the laser scan speed. ZrW_2O_8 samples with cubic structure were produced when the laser scan speed was less than 2mm/s. The samples cooled by water are mainly dominated by α-ZrW_2O_8. We also attempt to avoid the appearance of γ phase by another method which improves the resist compression of material.

Acknowledgement

This work was supported by National Natural Science Foundation(41401384, U1731121); A Project of Shandong Province Higher Educational Science and Technology Program (J17KB127);Binzhou City Science and Technology Development Plans (2014ZC0307);Binzhou City Science and Technology Development Plans(2015ZC0210);Binzhou University Research Fund Project (BZXYG1513)and Binzhou University Research Fund Project (BZXYG1706).

References

[1] Mary T A,Evans J S O,Vogt T &SleightA W 1996 *Science* **272** (5) 90-92
[2] Evans J S O, Zhu Z, Jorgenson J D, Argyrious D N, Shortand S& Sleight A W 1997*Science* **275** 61
[3] Wang J P, Chen Q D, Feng W W& Liang E J 2016*Optik* **127**(5) 2837-2839
[4] Ding P, Liang E J, Jia Y & Du Z Y 2008 *J.Phys.: Condens. Matter*. **20** 275224
[5] Liang E J, Liang Y, Zhao Y, Liu J, Jiang Y J 2008 *J. Phys Chem A* **112** 12582-12587
[6] Wang ZP,Song WB,Zhao Y,Jiang YJ&Liang EJ2011*Chin. J. Light Scatt.* **23** 250–255
[7] GeX,MaoY,LiuX,ChengY,YuanB, M Chao&Liang EJ 2016 *Scientific Report* 24832 **6**
[8] Yamamura Y, Nakajima N and Tsuji T 2001 *Phys.Rev.B*. **64** 184109 **.**
[9] Kameswari U, Sleight A W and Evans S O 2000 *International Journal of Inorganic Material*. **2** 333-337
[10] Mittal R, Chaplot S L, Kolesnikov A I, Loong C-K and Mary T A 2003 *Phys.Rev.B.* **68** 054302
[11] Ravindran T R, Arora A K and Mary T A 2001 *J Phys: CondensMatter* **13** 11537–11588
[12] Chao M J and Liang E J 2004 *Surf. Coat.Techn*. 179265

IOP Conf. Series: Materials Science and Engineering **292** (2017) 012090 doi:10.1088/1757-899X/292/1/012090

Magnetic effect on oxide-scale growth of Fe-5Cr alloy

C H Zhou[1, 3], X W Li[1], S H Wang[1] and H T Ma[2]

[1] School of Materials Science and Engineering, Heilongjiang University of Science and Technology, Harbin 150022, P. R. China
[2] School of Materials Science and Engineering, Dalian University of Technology, Dalian 116085, P .R. China

[3] E-mail: dlut2325@163.com

Abstract. The oxidation behaviour of Fe-5Cr alloy was investigated at 650℃ in the presence of magnetic field. Results indicated that the oxide scales were both consisted of an outer Fe-oxide scale and an inner mixed-oxide scale in the presence or absence of magnetic field. The oxide-scale growth of Fe-5Cr alloy, gained by measuring the oxide-scale thickness, was verified to follow parabolic lawyer. And the oxidation kinetics showed that the applied magnetic field retarded the oxide-scale growth of Fe-5Cr alloy.

1. Introduction

The development of superconducting magnet techniques promote the material researches in a high magnetic field [1]. Many investigations have focused on the magnetic effect on the annealing [2], grain growth [3, 4], phase transformation [5, 6], texture [7], and interfacial reaction [8-14]. Most of these studies are related to the atom diffusion behaviour in a magnetic field. Earlier, Youdelis [8] reported that the diffusion of Cu in Al was retarded by a 3T magnetic field. And many subsequent investigations about the magnetic effect on the diffusion behaviour in the diffusion couple systems were in agreement with the result [9-11]. However, some other studies have found different magnetic effect on the diffusion behaviour. Nakajima [12] pointed out that the diffusion of Ni in Ti was not affected by an applied 4T magnetic field. Zhao [13, 14] reported that the growths of intermetallic compounds at Sn-3Ag-0.5Cu/Cu and Sn1.5Cu/Cu interface were increased by the applied magnetic field. Therefore, it can be seen that the magnetic effect on the diffusion behaviour is still unclear.

Difference with atom diffusion in a diffusion couple system, the diffusion behaviour during oxidation/corrosion of alloys is closely related to the ion diffusion. Lu [15] reported that the electrochemical corrosion behaviour of metal in aqueous solutions was significantly affected by an applied magnetic field. Zhou [16] pointed out that the subscale growth of Fe-20Ni alloy was retarded by an applied magnetic field. Nevertheless, it is hard to explain the magnetic field effect on the ion diffusion behaviour based on the recently investigations More investigations are still needed to explore the magnetic effect mechanism on the ion diffusion. The oxide-scale growth of Fe-Cr alloys is a typical case which is controlled by ion diffusion [17]. And many investigations have been carried out to study the ion diffusion behaviour during oxidation of Fe-Cr alloys [18-20]. In the present work, the authors investigated the oxidation behaviour of Fe-5Cr alloy in a series of magnetic fields, the objective was to give some evidences on the magnetic effect on the ion diffusion behaviour during the oxide-scale growth of Fe-5Cr alloy.

2. Experimental materials and procedures

The alloy with nominal composition of Fe-5Cr was used. It was prepared by vacuum arc-melting using appropriate mixtures of high purity metals (Fe 99.97 wt.%, Cr 99.7 wt.%). And the actual composition determined by energy dispersive X-ray spectroscopy (EDX) was Fe-5.45Cr (in wt.%). The samples with dimensions about 1 × 6 × 12 mm^3 were ground with SiC papers down to 2000 polishing grit (2.5 μ m), and then cleaned with acetone and ethanol. Discontinuous oxidation tests were carried out, which means that the samples were exposed to the conditions of high temperature and magnetic field for a given time, and then were fetched out from the furnace. And the average oxide-scale thicknesses oxidized at different time were measured by using Q500IW image analysis meter to plot a fully oxidation kinetics curve. And at least 5 SEM images were used. The mean thickness of oxide scales was the mean value measured by different SEM images.

The magnetic field apparatus was introduced by many investigations [9-11]. During the oxidation in the uniform magnetic field, the samples were placed in a flat quartz capsule, and the main oxidation surface was parallel to the magnetic field direction, as shown in figure 1. The oxidation test was carried out in air at 650 ℃ and the applied magnetic field intensity was 0 T, 2 T and 6 T. After the tests, the samples were mounted in resin and the oxide-scale cross-sections were examined using a JSM-5600LV scanning electronic microscopy.

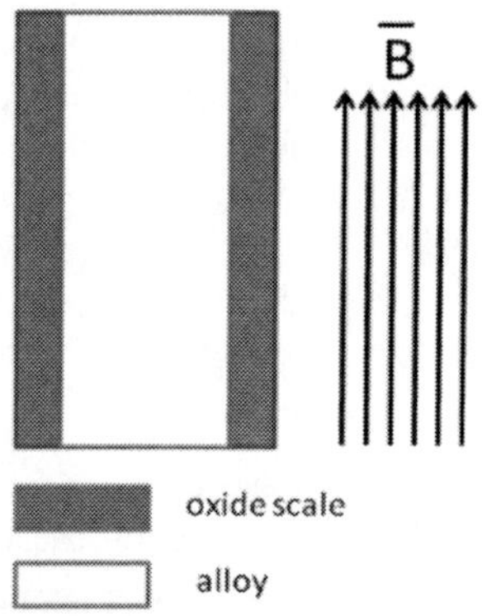

Figure 1. Schematic diagram of the oxidized samples in the magnetic field.

3. Experimental materials and procedures

3.1. Cross-sectional morphologies

Figure 2 shows the cross-sectional morphology of Fe-5Cr alloy oxidized at 650 ℃for 7 h. It can be seen that the oxide scale was composed of three distinguishable zones, i.e., the outmost dark layer, the intermediate gray layer and the inner dark layer. The element composition of every layer was detected by EDX, as indicated in table 1. It is indicated that the outmost layer marked by number 1 and 2 in figure 2 only contained Fe and O elements and the ratio of Fe/O is approximately 1, which implies the outmost layer is FeO. Though Cr element existed in the intermediate layer marked by number 3, its content is lower and the ratio Fe(Cr)/O is also about 1. It is implied that the main oxide of intermediate layer is also FeO. However, the Cr content at the location 4 and 5 was up to 6.86% and 6.61% which is somewhat higher than that of the alloy substrate. It may be FeO and dispersed Fe-Cr-oxide. The microstructures of oxide scales on an Fe-5Cr alloy were in accordance with previous investigations [17]. In order to clear the oxidation mechanism of Fe-5Cr alloy under a high magnetic field, the oxide scale was divided to two parts. The location numbered by 1-3 was defined as Fe-oxide scale and the location numbered by 4 and 5 was called mixed-oxide scale.

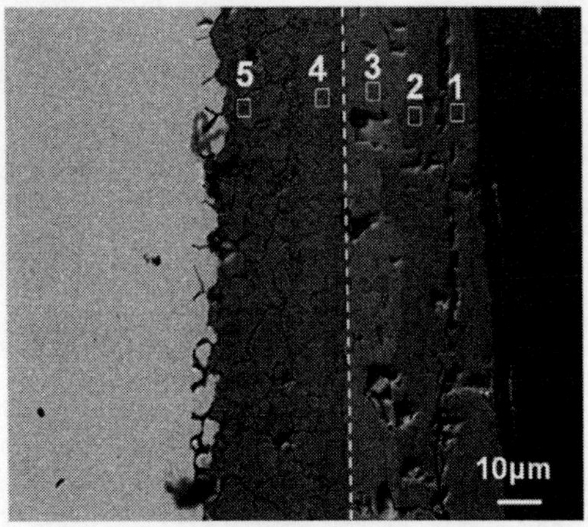

Figure 2. Cross-sectional morphologies of Fe-5Cr alloy oxidized at 650 ℃ for 7 h.

Figure 3 displays the cross-sectional morphologies of Fe-5Cr alloy oxidized at 650 ℃ for 5 h under a series of magnetic field intensities. It can be seen that the cross-sectional morphologies of samples under different magnetic field are nearly same, which means that all of them were composed of an Fe-oxide layer and a mixed-oxide layer. The difference is that the oxide-scale thickness in the absence of magnetic field is thicker than those in the presence of magnetic field, which implies that the applied magnetic field induced a decrease of oxide-scale growth. Moreover, it can be found that the difference of oxide-scale thicknesses for 2 T and 6 T are almost indistinguishable in this magnification.

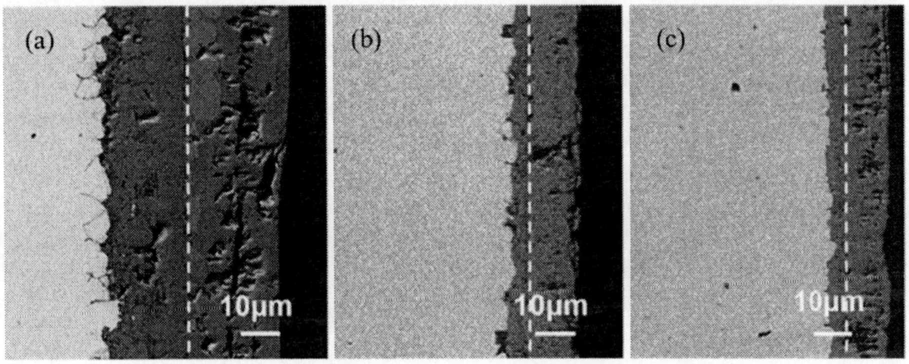

Figure 3. Cross-sectional morphologies of Fe-5Cr alloy oxidized at 650 ℃ for 5 h under a magnetic field of (a) 0 T; (b) 2 T; (c) 6 T.

3.2. Growth of oxide scale

Figure 4(a) represents the oxide-scale thickness as a function of oxidation time for Fe-5Cr alloy oxidized at 650 ℃ with and without magnetic field. It can be obviously gained that the application of magnetic field decreases the oxide-scale growth, which is in agreement with the current investigations about the magnetic effect on the diffusion in diffusion couple systems and high temperature oxidation process [8-11, 16]. The variation of oxidation kinetics affected by magnetic field is attributed to a change of the growths of Fe-oxide layer and mixed-oxide layer.

The relation between square of oxide-scale thickness and time is shown in figure 4(b), which is implied that it followed a parabolic law. Calculations of quantifying the oxide rate constant were performed. And the oxide rate constants for Fe-5Cr alloy oxidized at 0 T, 2T and 6 T were 1.43×10^{-7} $cm^2 \cdot s^{-1}$, 1.98×10^{-8} $cm^2 \cdot s^{-1}$ and 1.67×10^{-8} $cm^2 \cdot s^{-1}$, respectively. It can be seen that the decreased effect of 2 T magnetic intensity on the oxidation kinetics constant of oxide scale by a factor of approximately

7. Under 6 T, the decreased effect can be by a factor of more than 9. It is confirmed that the application of magnetic field strongly affects the oxide-scale growth of Fe-5Cr alloy.

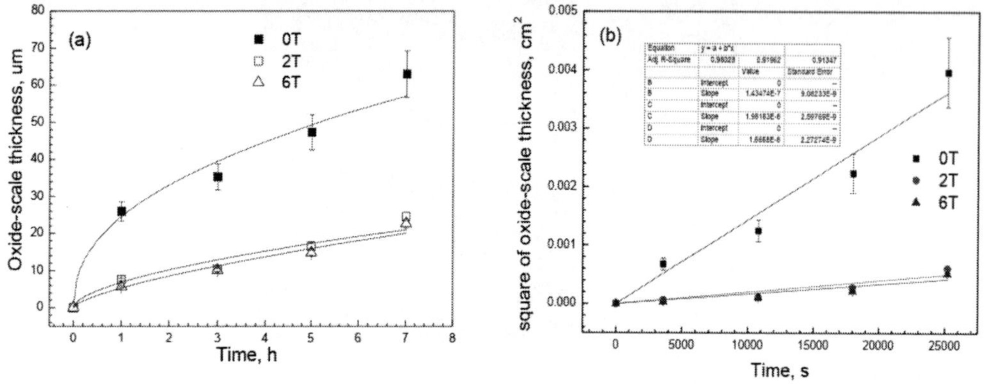

Figure 4. The oxidation kinetics of the Fe-5Cr alloy.
(a) oxide-scale thicknesses as a function of time. (b) square of oxide-scale thickness as a function of time

3.3. Growth of mixed-oxide scale

As indicated in figure 3, the thicknesses of mixed-oxide scales oxidized under magnetic field were somewhat similar. To further clear the magnetic effect on the growth of mixed-oxide scale, the thicknesses of mixed-oxide scales were measured and curves of the relation between thicknesses of the mixed-oxide scale and time were plotted in figure 5. It can be seen that the growth of mixed-oxide scale approximatively obeyed a linear law, which is related to the lower Cr content in Fe-5Cr alloys. At low Cr contents, some Cr will enter solution in the FeO phase but, due to the stability of spinel, the solubility is limited, so Cr-rich oxides dispersed in FeO layer [17]. Since the diffusion distance for solution formation is short, Cr-rich oxides can be fast generated, then the formation of mixed-oxide scale is also fast, consequently the growth kinetics displays a linear law. As for the magnetic effect on the growth of mixed-oxide scale, it can be attributed to the restrained effect of magnetic field on the diffusion of iron ions. The rate constants of mixed-oxide scale were calculated. And the oxide rate constants at 0 T, 2T and 6 T were 1.28×10^{-6} cm•s^{-1}, 3.11×10^{-7} cm•s^{-1} and 2.42×10^{-7} cm•s^{-1}, respectively. It can be seen that the decreased effects of 2T, 6T magnetic intensities on the rate constants of mixed-oxide scales by a factor of more than 4 and 6, respectively.

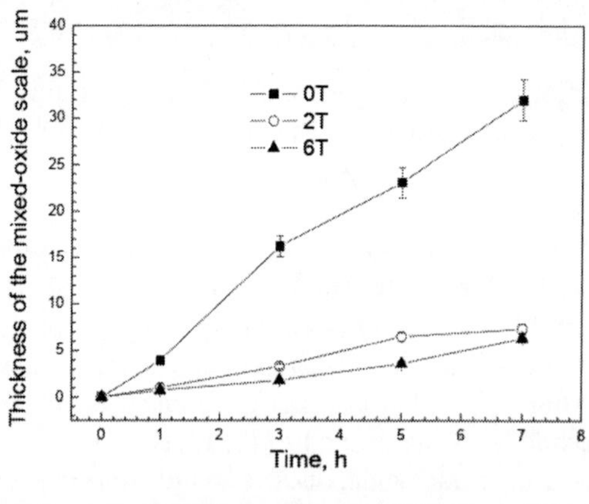

Figure 5. Thickness of the mixed-oxide scales as a function of time for samples oxidized with and without magnetic field.

4. Discussion

The magnetic effects on the diffusion process, such as grain growth, interfacial reaction, were intensively investigated [3, 4, 8-14]. Under magnetic field, a physical explanation of the magnetic effect on the diffusion coefficient of charged particle is usually identified as [8, 21, 22]:

$$D_{\perp} \approx \frac{1}{1+\omega_C^2\tau^2} D \text{ and } D_{//} = D \tag{1}$$

where τ is the mean collision time of the charged particle, D is the diffusion coefficient without magnetic field, $\omega_c = qB/m$, q/m is the charge-mass ratio of the charged particle, subscripts ⊥ and // indicate the direction perpendicular and parallel to the magnetic field, respectively. Based on equation (1), it is indicated that the diffusion is suppressed in the direction perpendicular to the magnetic field, by which a large volume of reports about magnetic field retarded effect on diffusion were demonstrated [8-11, 16]. During the oxide-scale growth of Fe-5Cr alloy, the diffusion of Fe cations is perpendicular to the magnetic field. Such process must be retarded by the magnetic field according to the equation (1).

Moreover, the diffusion coefficient D is depended on the lattice parameter and the jump rate of atom, as indicated in equation (2).

$$D = n\beta d^2\Gamma \tag{2}$$

where n is the number of the nearest neighbor interstitial positions, 2 β the probability when an interstitial atom jumps into a new interstitial position, and d the projected length onto the diffusion direction. Symbol Γ in Equation 2 represents the jump rate [23]. When the anions or cations diffuse in the high magnetic field, they are affected by the magnetic field in the form of Lorentz force, and the Lorentz force $\vec{f}$ is expressed by [24, 25]:

$$\vec{f} = q\vec{V} \times \vec{B} \tag{3}$$

where $\vec{V}$ is the average velocity of ions in the absence of magnetic field, $\vec{B}$ is the magnetic field intensity. During the growth of oxide scale of Fe-5Cr alloy, the diffusion of Fe cations perpendicular to the magnetic field is retarded by the Lorentz force, and then the probability when an atom jumps into a new interstitial position is decreased. Meanwhile, the jump rate for Fe cations perpendicular to the magnetic field is also decreased by the Lorentz force. Those magnetic effects on the diffusion behavior of Fe cations result in the decreased diffusion coefficient of high temperature oxidation of Fe-5Cr alloy.

As an oxidation process, the oxidation rate is controlled by the cations and anions diffusion. Once the ion diffusion behavior is changed, the oxidation rate is also changed. Wagner proposed the high temperature oxidation model based on the ions-electron theory, as indicated in equation [17].

$$k_p = \frac{2M_{OX}E\sigma n_e\left(n_c + n_a\right)}{nF\rho_{OX}} \tag{4}$$

where k_p the parabolic oxidation rate constant, M_{OX} molar mass of oxide, E electric field intensity, σ oxide scale conductivity, n_e electron transport number, n_c cation transport number, n_a anion transport number, n valence, F faraday's constant, ρ_{OX} oxide density, n_e+n_c+n_a =1. According to aforementioned analysis on the effect of magnetic field on the ion diffusion in equation (1), (2), (3), the effective n_e, n_c, n_a along the oxide-scale growth direction are all retarded by the applied magnetic field during the oxidation of Fe-5Cr alloy. In addition, the oxides formed in all experimental conditions are not changed as indicated in figure 2 and figure 3, i.e. M_{OX}, ρ_{OX} and σ are not changed by the applied magnetic field. Then the decreased n_e, n_c, n_a result in oxidation parabolic rate constant decreasing, as shown in figure 4.

5. Conclusion

Fe-5Cr alloy was subjected to oxidation under magnetic field at 650 ℃. The cross-sectional morphologies and the oxide-scale growth were examined. The major conclusions from this investigation are as follows:

- The oxide-scale structures, consisted of outer Fe-oxide layer and inner mixed-oxide layer, were not changed by the applied magnetic field.
- The application of magnetic field retarded the oxide-scale growth.
- The applied magnetic field did not change the oxidation kinetics law.

Acknowledgements

The work was supported by National Natural Science Foundation of China (No. 51501058) and Scientific Research Fund of Heilongjiang Provincial Education Department (No. 12543069).

References

[1] Asai S, Sassa K S and Tahashi M 2003 *Sci. Tech. Adv. Mater.* **4** 455
[2] Ghasemi A 2016 *J. Supercond. Nov. Magn.* **29** 1
[3] Harada K, Tsurekawa S, Watanabe T, Wantanabe T and Palumbo G 2003 *Scripta Mater.* **49** 367
[4] Molodov D A, Bollmann C, Konijnenberg P J, Barrales-Mora L A and Mohles V 2007 *Mater. Trans.* **48** 2800
[5] Kustov S, Corró M L, Pons J and Cesari E 2009 *App. Phys. Lett.* **94** 191901
[6] Ludtka G M, Jaramillo R A, Kisner R A, Nicholson D M, Wilgen J B, Mackiewicz-Ludtka G and Kalu P N 2004 *Scripta Mater.* **51** 171
[7] Molodov D A and Sheikh-Ali A D 2004 *Acta Mater.* **52** 4377
[8] Youdelis W V, Colton D R and Cahoon J 1964 *Can. J. Phys.* **42** 2217
[9] Liu Z F, Dong J, Zeng X Q, Lu C, Ding W J and Ren Z M 2007*J. Alloys Compd.* **440** 132
[10] Ren X, Chen G Q, Zhou W L, Wu C W and Zhang J S 2009 *J. Alloys Compd.* **472** 525
[11] Li D G, Wang Q, Li G J, Lv X, Nakajima K and He J C 2008 *Mater. Sci. Eng.* A **495** 244
[12] Nakajima H, Maxkawa S, Aoki Y and Masahiro K 1985 *Trans. Jpn. Ins. Met.* **26** 1
[13] Zhao J, Yang P, Zhu F and Cheng C Q 2006 *Scripta Mater.* **54** 1077
[14] Cheng C Q, Zhao J, Xu Y, Xu F M and Huang M L 2009 *Soldering Surf. Mount. Technol.* **21** 14
[15] Lu Z P and Yang W 2008 *Corros. Sci.* **50** 510
[16] Zhou C H, Ma H T, Wang L and Li X W 2013 *Corros. Sci.* **77** **292**
[17] Birks N, Meier G H and Pettit F S 2006 *Introduction to the High-Temperature Oxidation of Metals* (Cambridge: Cambridge Press)
[18] Horita T, Yamaji K, Xiong Y P, Kishimoto H, Sakai N and Yokokawa H 2004 *Solid State Ionics* **175** 157
[19] Ostwald C and Grabke H J 2004 *Corros. Sci.* **46** 1113
[20] Briickman A, Emmerich R and Mrowect S 1972 *Oxid. Metal.* **5** 137
[21] Hu X W 2006 *Theoretical Basis of Plasma* (Beijing: Beijing Press)
[22] Wang C J, Wang Q, Wang Z Y, Li H T, Nakajima K and He J C 2008 *J. Crystal. Growth* **310** 1256
[23] Smith W F and Hashemi J 2011 *Foundations of Materials Science and Engineering* (Beijing: China Machine Press)
[24] Han Y, Ban C Y, Ba Q X, Guo S J, Wang S H and Cui J Z 2006 *Mater. Lett.* **60** 1884
[25] Li D G, Wang Q, Li G J, Liu T and He J C 2009 *Mater. Lett.* **63** 890

Long term mechanical properties of alkali activated slag

J Zhu[1, 3], W Z Zheng[2], Z Z Xu[1], Y F Leng[1] and C Z Qin[1]

[1] College of Civil Engineering and Architecture, Harbin University of Science and Technology, Harbin 150080
[2] Key Lab of Structures Dynamic Behavior and Control of the Ministry Education, Harbin Institute of Technology, Harbin 150090

[3] E-mail: zhujing02@126.com

Abstract. This article reports a study on the microstructural and long-term mechanical properties of the alkali activated slag up to 180 days, and cement paste is studied as the comparison. The mechanical properties including compressive strength, flexural strength, axis tensile strength and splitting tensile strength are analyzed. The results showed that the alkali activated slag had higher compressive and tensile strength, Slag is activated by potassium silicate (K_2SiO_3) and sodium hydroxide (NaOH) solutions for attaining silicate modulus of 1 using 12 potassium silicate and 5.35%sodium hydroxide. The volume dosage of water is 35% and 42%. The results indicate that alkali activated slag is a kind of rapid hardening and early strength cementitious material with excellent long-term mechanical properties. Single row of holes block compressive strength, single-hole block compressive strength and standard solid brick compressive strength basically meet engineering requirements. The microstructures of alkali activated slag are studied by X-ray diffraction (XRD). The hydration products of alkali-activated slag are assured as hydrated calcium silicate and hydrated calcium aluminate.

1. Introduction

Alkali activated slag is a type of cementitious material through the chemical reaction obtaining [1-3], with slag and alkaline activators such as sodium silicate, NaOH as the main raw materials, with or without a certain cold heat history of aluminate natural minerals containing silicon or industrial waste residue (e.g metakaolin, phosphorus slag, steel slag, coal gangue, fly ash, red mud, etc) and alkaline activators (such as alkali metal hydroxide, alkali metal carbonate, alkali metal sulfate, etc). It uses industrial waste slag as the raw material and has the advantage of high temperature resistance, corrosion resistance, energy conservation and environmental protection, and has aroused the wide concern of scholars both at home and abroad [4-6]. Since the 1980s in China, alkali activated slag has begun to be studied and also made a lot of research results [7-10]. But its long-term mechanical properties still hasn't been published, and the guide specifications and standardsis on alkali activated slag are rarely seen at home and abroad. Moreover, the lack of detailed and reliable technical indicators greatly hinders its popularization and application. Therefore, this essay studies on the long-term mechanical properties of the alkali-activated slag. The results indicate that alkali activated slag is a kind of rapid hardening and early strength cementitious material with excellent long-term mechanical properties. The microstructures of alkali activated slag and cement paste are studied by X-ray diffraction (XRD). The hydration products of alkali-activated slag are assured as hydrated calcium silicate and hydrated calcium aluminate.

2. Materials and Formulation

2.1. Raw materials and mix proportion

S95 slag is proposed by mass fraction of 36.9% SiO_2, mass fraction of 15.66% Al_2O_3, mass fraction of 37.57% CaO, mass fraction of 9.3% MgO, mass fraction of 0.36% Fe_2O_3.The activity indexes of slag are mainly: the quality coefficient 1.69, the alkaline coefficient 0.97, and the activity coefficient 0.42.The main parameters of potassium silicate solution are shown in table 1.The mass fraction of sodium hydroxide is not less than 96.0%.

Table 1. Parameters of potassium silicate solution.

Baume degree	Density	Modulus	mass fraction, %	
			K_2O	SiO_2
46.3	1.465	2.76	15.98	28.15

[a] Note: Baume degree for potassium silicate is the unit of density for liquids measured by hydrometer.

P.O42.5 ordinary Portland cement with mass fraction of 94.5% SiO_2 is used to prepare cement paste (OPC), which is analyzed as counterpart. Test uses two relatively optimal mixing ratio (mixing ratio is various mass fractions of raw materials) W35, W42 of alkali activated slag, and main parameters are shown in table 2.

Table 2. The proportion parameters used for text.

Number	Slag	Cement	Modulus of potassium silicate	potassium silicate amount, %	Sodium hydroxide,%	Water amount %
W35	1	0	1.0	12	5.35	35
W42	1	0	1.0	12	5.35	42
OPC	0	1	0	0	0	25

2.2. Production and maintenance of specimens

The samples were prepared by mixing raw materials--slag and alkaline activators in a 5 liter pan mixer. The sequence of mixing was important to homogenize the fresh paste. Thus, the raw materials were firstly mixed for about 30 s. the alkaline activators and water were then poured into the pan, and stirred at low speed for approximately 6 min until the mixture was glossy and well combined. The fresh paste was cast in various shapes of moulds for measurement of strength. The OPC control samples with a water/binder ratio of 0.25 were machined to the same size for comparison. They were maintained in a humidistat at 20 ± 2℃ and 95% relative humidity. The samples were demolded after 24 h of humidity maintaining and then stored in a humidistat at 20 ± 2℃ and 50% relative humidity until testing commenced.

3. The strength varies with time

3.1. Flexural strength

Figure 1 shows the test results of flexural strength of paste with time, it can be seen that with the increase of the age, the flexural strength of alkali activated slag is increasing. Before 14d, the flexural strength developed rapidly, and after 28d, the flexural strength developed gradually. The flexural strength of the W35 specimen was significantly higher than that of W42 and OPC. The damage process of alkali-activated slag specimen is significant. With the increase of load, a clear crack is appeared in the edge of the test piece. When the ultimate load is reached, the specimens are damaged.

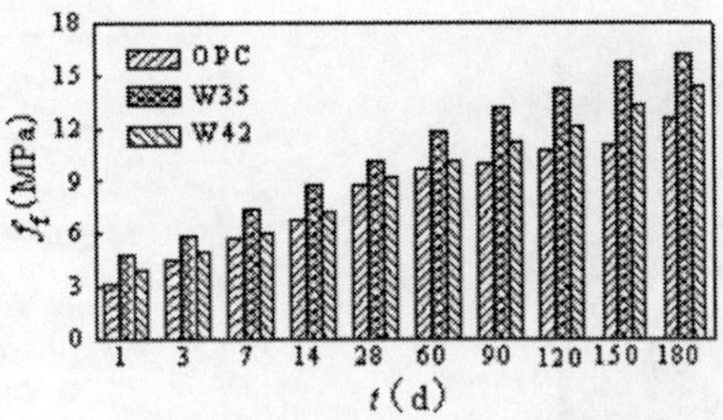

Figure 1. Flexural strength of cementing materials.

3.2. Compressive strength of cementing materials

Table 3 and table 4 show the compressive strength of W35 and W42. Figure 2 shows test apparatus and variation curves of compressive strength of cementing materials with time. It can be seen from figure 2 that at 1d ~ 3d, the development of compressive strength of alkali-activated slag is fastest. Compared with the compressive strength of 28d, the compressive strength of 1d and 3d can reach 54% and 72%, respectively, and its compressive strength at 3d ~ 7d was slow, and its compressive strength of 7d was 88%. After 7d, its compressive strength of was slower, but the strength continued to increase with time. In addition, its strength tended to stabilize until 120d ~ 180d. Under the same curing condition, the compressive strength of alkali-activated slag is significantly better than that of OPC. Test proves that alkali-activated slag is a kind of cementitious material with fast hardening speed and high early strength.

Table 3. Compressive strength of W35.

Age (d)	Compressive strength of W35 (MPa)						$f_{l.40}$ (MPa)
	A01	A02	B01	B02	C01	C02	
1	50.42	52.90	54.25	52.59	51.38	52.66	52.37
3	67.06	67.31	72.82	77.35	73.63	77.88	72.67
7	77.25	79.31	80.25	82.56	82.13	82.65	80.69
14	82.55	84.63	81.75	83.54	85.67	84.98	83.85
28	85.50	90.48	89.87	90.91	92.82	91.38	90.16
60	101.40	103.81	105.12	104.38	104.82	105.49	104.17
90	111.75	113.86	112.89	114.03	115.64	114.80	113.82
120	116.39	118.57	118.88	117.81	119.36	118.63	118.27
150	118.14	119.23	120.75	121.62	118.77	119.84	119.73
180	120.59	122.65	119.44	120.86	122.24	121.31	121.18

Table 4. Compressive strength of W42.

Age (d)	Compressive strength of W42 (MPa)						$f_{l.40}$ (MPa)
	D01	D02	E01	E02	F01	F02	
1	35.86	34.33	35.61	32.35	35.02	38.24	35.23
3	49.41	42.35	48.84	48.59	51.89	50.05	48.52
7	66.60	69.78	68.48	67.03	71.87	69.32	68.84
14	76.25	75.12	80.38	80.36	78.09	76.64	77.80
28	80.81	79.88	79.56	78.94	85.97	80.13	80.88
60	89.96	90.52	92.04	93.27	91.17	90.62	91.26
90	98.77	100.31	99.84	100.19	98.90	99.64	99.61
120	101.22	103.79	102.27	104.38	103.95	102.48	103.02
150	104.56	105.25	107.57	106.15	105.56	104.31	105.57
180	105.84	106.08	107.27	106.13	104.75	106.15	106.04

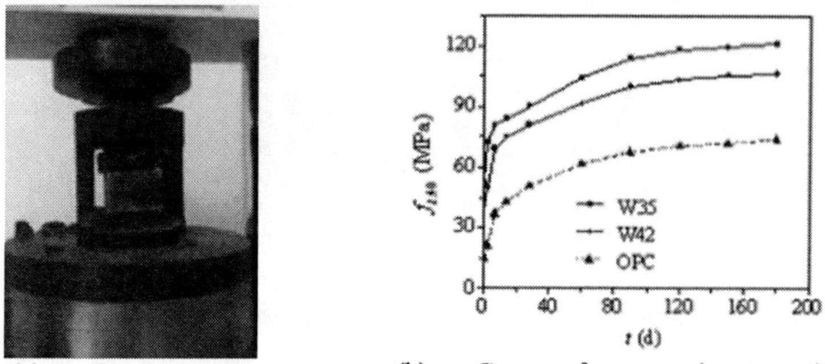

(a) Test apparatus (b) Curves of compressive strength

Figure 2. Compressive strength of cementing materials.

3.3. Compressive strength of cube

Table 5 shows the test results of compressive strength of cube with time. From table 3, it can be seen that the cube compression strength of W35 is slightly higher than that of W42. During the determination of the compressive strength of the cube specimens, the test phenomenon is that when the ultimate load is near the limit load, the specimen will make a large, brittle sound and crack. alkali-activated slag is similar to the ordinary concrete compressive test pieces, and the destruction form of alkali-activated slag specimen has two opposite cones, and the details are shown in figure 3. Curing age for 180d, the highest compressive strength of cubic specimens is 96.43MPa of W35 specimens.

Table 5. Test results of compressive strength of cube.

Age (d)	Compressive strength of W35			$f_{cu.70.7}$ (MPa)	Compressive strength of W42			$f_{cu.70.7}$ (MPa)
	G0	H0	I0		J0	K0	L0	
1	40.68	42.55	41.79	41.67	27.37	28.87	28.31	28.18
3	57.92	58.23	57.33	57.83	38.61	39.16	38.68	38.82
7	63.65	64.69	64.29	64.21	54.01	53.95	55.08	54.35
14	65.91	66.42	67.84	66.72	61.17	62.69	61.85	61.91
28	71.69	72.24	71.31	71.75	63.73	65.24	64.58	64.53
60	81.92	82.75	84.01	82.89	75.95	76.40	75.53	73.01
90	89.56	90.21	91.95	90.57	78.22	81.01	79.83	79.69
120	93.76	94.52	94.07	94.12	81.97	82.70	82.58	82.43
150	94.52	95.76	95.55	95.28	83.25	85.39	84.73	84.46
180	96.53	95.11	97.65	96.43	84.07	85.27	85.16	84.83

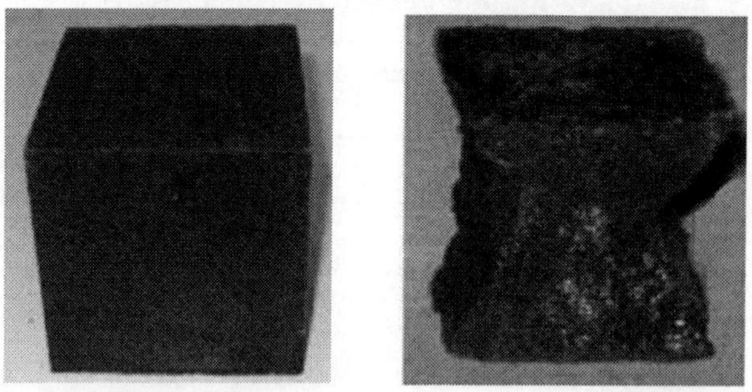

(a) Before the test (b) After the test

Figure 3. Compressive strength of cube.

3.4. Splitting tensile strength

The splitting tensile strengthes of the two kinds of ratio W35 and W42 are shown in table 6.

Table 6. Test results of splitting strength of cube.

Age (d)	Splitting strength of W35			f_{ts} (MPa)	Splitting strength of W42			f_{ts} (MPa)
	PA0	PB0	PC0		PD0	PE0	PF0	
1	1.47	1.28	1.36	1.37	0.89	0.92	1.10	0.97
3	2.12	2.26	2.18	2.19	1.43	1.48	1.51	1.47
7	2.63	2.75	2.59	2.66	2.25	2.18	2.37	2.27
14	2.97	2.84	2.90	2.91	2.63	2.75	2.71	2.70
28	3.41	3.25	3.04	3.23	2.96	3.12	2.99	3.02
60	3.65	3.59	3.84	3.69	3.27	3.35	3.10	3.24
90	3.91	4.03	3.73	3.89	3.36	3.47	3.39	3.41
120	4.02	3.97	4.15	4.06	3.51	3.49	3.58	3.53
150	4.08	4.16	4,.14	4.13	3.55	3.59	3.78	3.64
180	4.22	4.19	4.04	4.15	3.64	3.81	3.80	3.75

The fracture form of the alkali-activated slag specimen is shown in figure 4. Along the middle line, the crack emerges, and the fracture interface is clear and plain. By comparing the axial tensile strength and splitting tensile strength of the test pieces of W35 and W42, these conversion coefficient of the two are basically the same, and the average value is 0.933.

Figure 4. Fracture form of dumbbell type specimen.

3.5. Axial tensile strength

The axial tensile strengthes of the two kinds of ratio W35 and W42 are shown in figure 5.

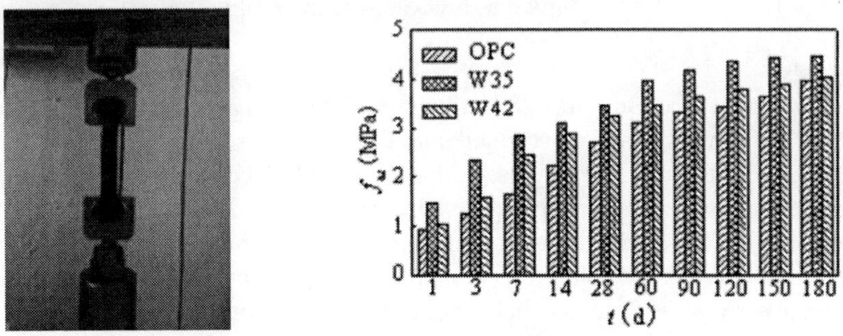

(a) Test apparatus (b) Curves of axial tensile strength

Figure 5. Axial tensile strength of dumbbell type specimen.

It can be seen from figure 5 that the development of tensile strength of alkali-activated slag is similar to compressive strength and flexural strength, which increase with the increase of age. The axial tensile strength of W35 specimen is highest, the second of W42 and the lowest of OPC. The tensile strength of the alkali-activated slag specimens 1d, 3d and 7d at normal temperature can reach 33%, 51% and 78% of tensile strength at 28d. After 60d, the development of tensile strength of alkali-activated slag slowed down. At 180d, the tensile strength of alkali-activated slag increased by 126%

compared with 28d. By comparison, the 28 d alkali-activated slag flexural strength is about 1/23 ~ 1/9 of compressive strength of cementing materials, and axis tensile strength is about 1/34 ~ 1/25 of compressive strength of cementing materials, which show that with the increase of age, the growth of axis tensile strength is slow, and its strength is lower than compressive and flexural strength.

4. Compressive strength of block

The block compressive strength of the ratio W35 are shown in table 7. The fracture forms of the single row of holes block compressive strength, single-hole block compressive strength and standard solid brick compressive strength are shown in figure 6. The results showed that the compressive strength of single row of holes block of alkali activated slag is equivalent to that of MU10 (*MU means m*asonry unit, and 10 means the compression strength of block is 10MPa) concrete block. Moreover, the compressive strength of standard solid brick of alkali activated slag is in the middle of that of MU20 and MU25 (*MU means m*asonry unit, and 10 means the compression strength of block is 20MPa and 25MPa) concrete block. Which basically meet engineering requirements.

Table 7. Test results of compressive strength of block.

Block type	1	2	3	4	5	Average, MPa
Single row of holes block	8.88	10.97	10.3	10.11	10.46	10.46
single-hole block	14.78	14.67	13.98	15.24	15.17	14.75
standard solid brick	23.13	23.04	23.45	23.67	22.98	23.25

(a) Single row of holes block (b) Single-hole block (c) Standard solid brick

Figure 6. Fracture forms of block.

5. XRD analysis

The standard maintenance 3d, 7d, 28d alkali-activated slag cementing materials and standard maintenance of 28d cement paste are scanned, and the diffraction pattern is shown in figure 7.

Through analyzing figure 7, it can be seen that for the OPC control sample, the XRD pattern shows some sharper reflection peaks corresponding to ettringite [$Ca_6[Al(OH)_6]_2(SO_4)_3 \cdot 26H_2O$], portlandite [$Ca(OH)_2$], calcite ($CaCO_3$), and also to calcium silicate hydrate (CSH) gel [$(CaO)_x \cdot SiO_2 \cdot (H_2O)_y$]. The XRD patterns of alkali-activated slag samples have been detected the presence of CSH-type hydration products using a peak of ~30° 2θ, which overlapped the amorphous hump of the slag. Peaks of a hydrotalcite-type phase similar to $Mg_6Al_2CO_3(OH)_{16} \cdot 4H_2O$ were also identified. Peaks of tetracalcium aluminate hydrate $(C,M)_4AH_{13}$ were also observed, and some of them overlapping those of hydrotalcite. Some carbonation seemed to have taken place, because all sample preparation was performed in the open laboratory. However, the peaks were weak and overlapped with other phases. It has also reported that the cement paste 28d hydration products have a large amount of calcium hydroxide, calcium alum and calcium carbonate and other crystal substances, Which is obviously different from the hydration product of alkali-activated slag with hydrated calcium silicate and hydrated calcium aluminate. In addition to the hump in the region between 17 :35 degree which reflect the amorphous composition of the geopolymer structure, where this hump increased in broadness with time reflecting the increased content of the formed geopolymer structure with time.

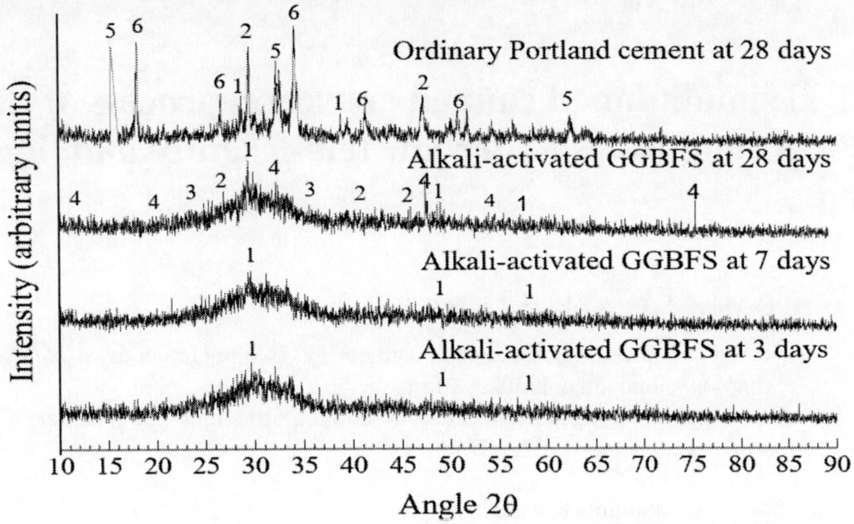

Figure 7. XRD patterns at room temperature.

6. Conclusions

(1)Test proves that alkali-activated slag has excellent long-term mechanical properties and is a kind of rapid hardening and early strength cementitious material.

(2) Single row of holes block compressive strength, single-hole block compressive strength and standard solid brick compressive strength basically meet engineering requirements.

(3) The hydration products of alkali-activated slag are assured as hydrated calcium silicate and hydrated calcium aluminate by XRD analysis, which is obviously different from the hydration products of OPC.

Acknowledgements

This work was financially supported by the National Natural Science Foundation of China (NO. 51508140).

References

[1] Atis C D, Bilim C and Celik Ö, et al. 2009 *Construction and Building Materials* **23(1)** 548-555
[2] Ri-On Oh, Sang-Sun Cha and Seong-Yong Park, et al. 2014 *Paddy Water Environ* **12(1)** 149-156
[3] Zheng W Z, Zhu J 2013 *J Wuhan Univ Technol (Materials Science Edition)* **28(4)** 721-725
[4] Palomo A, Alonso S and Fernández-Jiménez A 2004 Journal of the Amerian Ceramic Society **87(6)** 1141-1145
[5] Zheng W Z, Zhu J 2015 Harbin Institute of Technology Press
[6] Zhu X.H, Kang X.J, Yang et al, 2017 *Constr Build Mater* **132(1)** 290-295
[7] Pimraksa K, Chindaprasirt P, Rungchet A, et al. 2011 A-Structural materials properties microstructure and processing *Materials Science and engineering* **528(21)** 6616-6623
[8] Zhu J, Zheng W Z 2012 *Journal of Southeast University* **42(5)** 962-969
[9] Won J P, Kang H B, Lee S J, et al. 2012 *Cement & Concrete Composites* **30(4)** 406-412
[10] Wang W C, Wang H Y and Lo M H 2014 *Constr Build Mater* **68(15)** 409-415

Numerical simulation of canned extrusion process of the W-25wt. %Cu composites under low temperature and liquid sintering

D Q Ma[1, 2, 3], J P Xie[1, 2] and Y H Song[1]

[1] School of Materials Science and Engineering, Henan University of Science and Technology, Luoyang, 471023, China
[2] Non-ferrous metal generic technology collaborative innovation center of henan province, Luoyang, 471023, China

[3] E-mail: madouqin@haust.edu.cn

Abstract. The W-25wt. % Cu compacts prepared by low temperature vacuum sintering (sintering at 1090℃), and then hot extrusion under low temperature and canning backward extrusion (extrusion temperature is 1090℃, extrusion ratio is 7.7), finally annealing at 900℃ have better properties. The defects such as fracture and canning cracking and the low utilization rate of materials due to the uneven distribution of the W-Cu materials along the longitude were solved by the numerical simulation of the hot extrusion process of the billet and defect analysis in the canning backward extrusion process. The good coincidence of the distribution of the W-Cu materials and the loading force in the hot extrusion experiment with the numerical simulation results proved the accuracy of the model and parameter selection in numerical simulation. So the simulation results have practical significance. The results show that canning cracking was mainly due to the difference strain and flow speed between the canning and the W-Cu billet, while the flow of the W-Cu billet was affected by the structure and material of the canning. It is concluded that when the extrusion temperature was near the melting point of copper phase, the extrusion speed was in the range of 5 to 10 mm/s, extrusion ratio was in the range of 7.7 to 11, the canning thickness was 5 mm and material was AISI-1016 can not only improve the temperature and stress and strain distribution of billet and canning but also enhanced the flow uniformity of billet.

1. Introduction

The rapid development of high-end area of electronic communication, aerospace and national defense industry has made the tungsten-copper composites with more excellent properties, such as higher intensity and better air tightness (which ask for the relative density is not less than 98%), more uniform composition and microstructure, more excellent thermoelectric properties and lower thermal expansion coefficient. It is necessary to promote the development of new preparation and synthetic process technology of the tungsten-copper composites [1-5].

The most common method of producing tungsten-copper composites is the presureless infiltration technique and liquid-phase sintering method [6]. The presureless infiltration technique is always used to prepared the tungsten-10~40 wt.% copper composites with macrograin [7, 8]. As to the liquid-phase sintering method, the densification process is mainly through the rearrangement of tungsten particles and solid phase sintering of tungsten skeleton owing to the poor intersolubility of tungsten and copper element [9]. When the content of liquid copper up to 40 wt.% [10], the W skeleton is still difficult to deform [11, 12]. It is not easy to obtain compact tungsten and copper composites. Moreover the huge difference of melting point between tungsten and copper element caused to the composition uncontrollability of the tungsten copper sintered compacts. According to the research of Guo Shu, the content of liquid copper is 30wt.% the copper begin to loss. When the content of liquid copper is up to 50wt.% , the loss of the copper up to 10wt.% [13]. Therefore, it is of little significance to improve the densification degree of tungsten-copper composites by simply using higher sintering temperature.

In recent years, to solve the densification of tungsten-copper composites, scholars around the world have taken a large number of basic and theoretical studies on the structure, synthesis/preparation and processing of tungsten-copper composites and their relationship with their performance. The results indicated that sub-micron tungsten copper powders with great activity and advantage, can improve the densification process of tungsten-copper composite material [14-18]. And the higher content of tungsten materials, the more dependent on the particle size [9].

To obtain the high quality of tungsten-copper sintered compacts, in this paper the ultra-fine tungsten copper powder is sintered by low temperature liquid phase sintering, and then canned hot extrusion [19]. While in the process of the experiment, there are many problems such as cracking, fracture of the steel jacket (as shown in figure 1) and low material utilization owing to the uneven distribution of copper along the axis. Therefore, the canned extrusion process of the W-25wt. %Cu composites under low temperature and liquid sintering is numerical simulated using the DEFORM software simulation to shorten the experimental process [20].

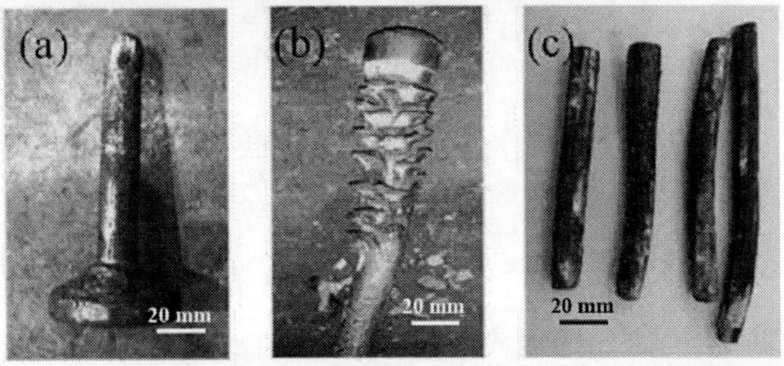

Figure 1. The defects in the process of extrusion: (a) fracture of the extrusion billet; (b) crack of the 45 steel jacket and (c) tungsten copper material extrusion out first and then fracture.

2. Experimental procedures

2.1. The raw materials

First, nano-sized tungsten copper composite powder contains 25 wt. % copper was synthesized by hydrothermal method; and then was processed by low temperature liquid phase sintering and canning backward extrusion; finally, the high density ultrafine W-Cu composites were obtained. It is concluded that when the extrusion temperature and the preheat temperature was near the melting point of copper phase(1030, 1050, 1070, 1090 or 1120℃), the extrusion speed was in the range of 2 to 10 mm/s(2, 5,10 or 15 mm/s), extrusion ratio was in the range of 5.2 to 17.4(5.2, 7.7, 11 or 17.4) [21].

2.2. The parameters of the numerical simulation process

In order to ensure the accuracy of the simulation results and the consistency with the experimental results, the appropriate flow stress, thermal physical property parameters, interface heat transfer coefficient and friction factor of tungsten copper material should be used.

The real stress-strain curves of W-25wt. %Cu sintering billet were measured by Gleeble-1500 thermal simulator. The constitutive equation of W-25wt. %Cu composite material was calculated as the Equation 1 [21].

$$\dot{\varepsilon} = 1.29\times10^{15}[\sinh(0.00488\sigma)]\exp[-317.15\times10^{3}/(RT)] \quad (1)$$

Because the heat capacity is not affected by the microstructure of the material, the heat capacity C_{comp} of the tungsten copper material can be calculated according to the law of the mixture as the Equation 2 [22, 23]:

$$C_{comp} = W_{Cu}C_{Cu} + W_{W}C_{W} \quad (2)$$

Where the W_{Cu} , W_W is the weight percent of copper and tungsten respectively ; C_{Cu} , C_W is the specific thermal capacity of copper and tungsten respectively, and the C_{Cu},C_W is 385,136 J/(kg·K) respectively [21]. So the specific thermal capacity of W-25wt.%Cu is 198.25 J/(kg·K).

According to the theoretical calculation formula of thermal expansion coefficient (α_T) proposed by German [22, 23].

$$\alpha_T = \alpha_{Cu} + \frac{\alpha_W - \alpha_{Cu}}{1+\frac{B_{Cu}}{B_W}\left[\frac{8R-8R^2}{1-2R^2} - \frac{(1-2R)\pi R^2}{1-\pi R^2}\right]} \quad (3)$$

$$R = 0.0113 + 1.58V_{Cu} - 1.83V_{Cu}^{3/2} + 1.06V_{Cu}^{3} \quad (4)$$

$$B_i = \frac{E_i}{3\times(1-2\gamma_i)} \quad (i=\text{Cu, W}) \quad (5)$$

Where the α_{Cu}, α_W is the coefficient of thermal expansion of copper and tungsten respectively; B_i is the volume modulus of each component; γ_i is the Poisson's ratio of each component; E_i is the elasticity modulus of each component; V_{Cu} is the volume fraction of copper. So the coefficient of thermal expansion of W-25wt.%Cu is 9.7 ppm/K.

According to the theoretical calculation formula of thermal conductivity (Q_T) proposed by German[22, 23].

$$Q_T = \pi R^2 Q_{Cu} + (1-2R)^2 Q_W + \frac{Q_W Q_{Cu} A_L}{3/2RQ_W + (1-3/2R)Q_{Cu}} \quad (6)$$

$$A_L = 4R(1-R) - \pi R^2 \quad (7)$$

Where Q_W, Q_{Cu} is the thermal conductivity of copper and tungsten respectively; R is calculated by Equation 7. So the thermal conductivity of W-25wt.%Cu is 284W/(m·K).

German assumes the material without strain, porosity, and with the ideal interface combining. And in fact, the residual stress and porosity in the material could affect the thermal conductivity. Kohn and Fortini estimate the influence of porosity on the thermal conductivity as the Equation 8 [24].

$$Q_r = \frac{1-\varepsilon}{1+11\varepsilon^2} Q_c \quad (8)$$

Where Q_r, Q_c is the actual and theoretical thermal conductivity of the material respectively, ε is the porosity of the material. ε is calculated by Equation 9.

$$\varepsilon = (1-\frac{\rho_0}{\rho})\times100\% \quad (9)$$

Where ρ_0, ρ is the actual and theoretical density of the material respectively.

The density of W-25wt.% Cu sintered compacts is 92%, so ε is 8%. After correction, the thermal conductivity of W-25wt.% Cu is 244W/(m·K).

The heat transfer coefficient between the W-25 wt. % Cu and air is 0.021 N/(s·mm·℃), and the heat transfer coefficient of the mould is 11 N/(s·mm·℃) [22]. The temperature at the time of extrusion is measured with an infrared thermometer. When the melt temperature is 1030℃, 1050℃, 1070℃, 1090℃and 1120℃; the coating temperature is 790℃, 804℃, 818℃, 830℃ and 845℃ respectively.

2.3. Numerical simulation model of canned extrusion process

In the process of simulation, the geometrical shape of the upper die (extrusion die); the lower die (container) and the billets are determined; especially for extrusion die Angle (which is set as 60 °). The model is simplified to 2D and the one-half structure owing to the extruded blank is symmetric about the z axis. The process of the extrusion and the sketches of the billet were shown in figure 2. The bottom and wall thickness of the steel sleeve is fixed to 10mm and 5 mm. Tungsten-copper materials are defined as porous material. The package is defined as a plastic body, and the upper and lower die are defined as rigid bodies. The top of the package is set to 2, 5 or 10 mm. The package material is an important factor in the extrusion process considering the cost of the experiment; the steel sleeve is made of the common carbon structural steel (AISI-1016, AISI-1045 or AISI-1060).

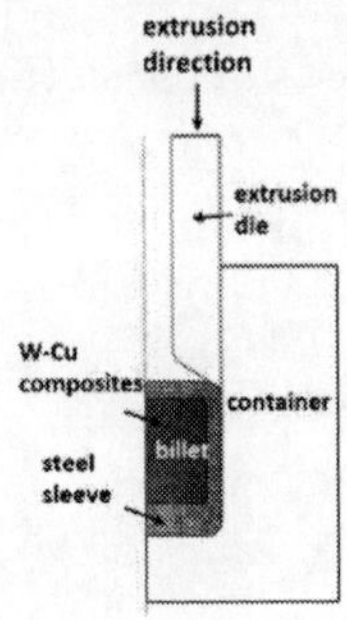

Figure 2. The process of the extrusion and the sketch of the billet.

3. Numerical simulation of canned extrusion process

The phenomenon crack of the steel sleeve occurred in the process of heat extrusion. When change the thickness of the steel sleeve, the "big head" and "small head" phenomenon occurred due to the uneven distribution of copper along the axis. So the material utilization is low. Figure 3 indicated the effects of top thickness and materials of steel jackets on the hot extrusion simulation results and the billet shape. When the top thickness of the steel sleeve is 2mm, the steel sleeve cracked and the "big head" occurred (figure 3 a~c). When the top thickness of the steel sleeve is 5mm, the steel sleeve cracked is slightly in AISI-1045 steel jackets, but the distribution of copper along the axis is even (figure 3 d~f). When the top thickness of the steel sleeve is 10mm, the steel sleeve is not cracked and the "small head" occurred (figure 3 g~i).In order to describe the crack of the steel sleeve more accurately, such as the fracture position and after fracture the continuous of the deformation. The formation processes of the steel jackets (2 mm, AISI-1016) was analysed (figure 4). The steel sleeve first breaks at the die gate (figure 4 c). And then the material below the die gate will continue to deform until the simulation is over. Figure 5 shows the changes of damage coefficient, temperature, stress-strain and velocity field with the extrusion time of the inside and outside of the billet and steel jacket. According to the strain and velocity curve and the damage coefficient, it can be seen that the crack of the steel sleeve is mainly related to the strain and velocity difference of billet and steel jacket (figure 5 e, f). Therefore, in order to ensure the good surface quality of the compacts, the uniform strain and velocity field is important.

The effects of the jacket thickness and materials on the temperature, stress, strain and velocity field were shown in figure 6 and figure 7. The results show that the canning thickness was 5 mm and material was AISI-1016 can not only improve the temperature and stress and strain distribution of billet and

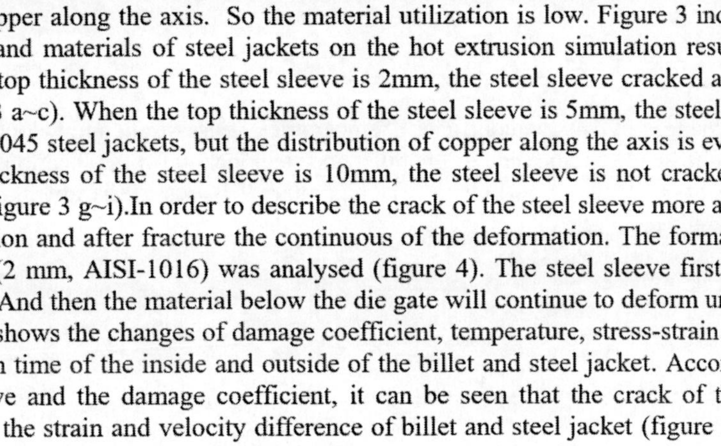

Figure 3. Effects of top thickness and materials of steel jackets on the simulation results: (a) 2 mm, 1016; (b) 2 mm, 1045; (c) 2 mm, 1060; (d) 5 mm, 1016; (e) 5 mm, 1045; (f) 5 mm, 1060; (g) 10mm, 1016; (h) 10 mm, 1045; (i) 10 mm, 1060.

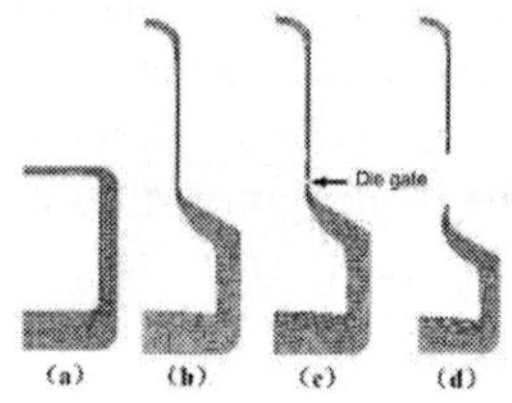

Figure 4. The formation processes of the steel jackets in the extrusion process.

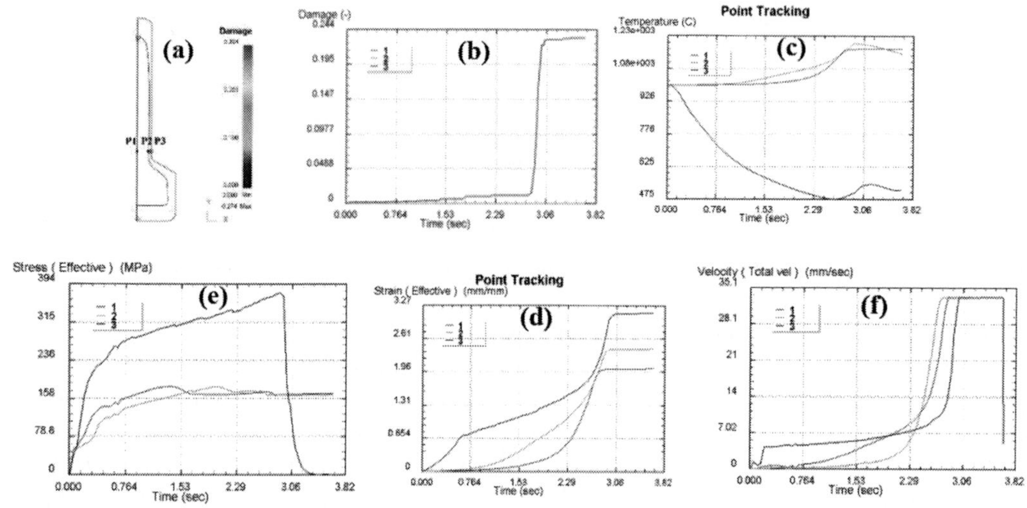

Figure 5. The changes of damage coefficient, temperature, stress-strain and velocity field with the extrusion time of the inside and outside of the billet and steel jacket: (a) the locations of point tracking point; the curves of (b) damage coefficient; (c) temperature; (d)stress; (e)strain and (f)velocity field.

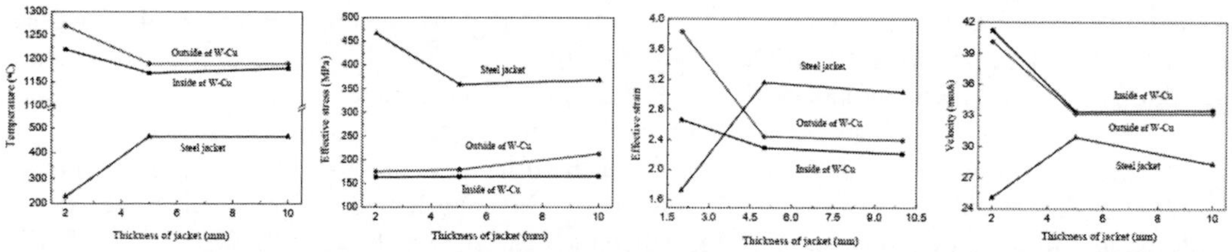

Figure 6. The effects of the jacket thickness on the temperature, stress, strain and velocity field.

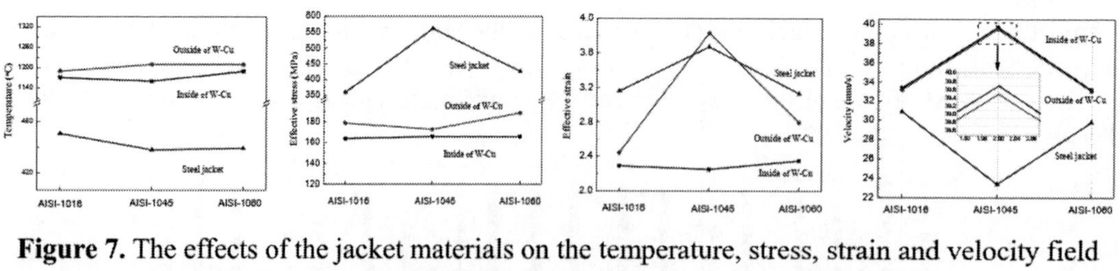

Figure 7. The effects of the jacket materials on the temperature, stress, strain and velocity field

canning but also enhanced the flow uniformity of billet. Which is consistent with results hot extrusion simulation results and the billet shape (figure 3).

4. Result validations

Figure 8 shows the products and the shape of the numerical simulation of the extrusion billet, when the coating thickness is 5 mm, material was AISI-1016, extrusion temperature is 1090℃, extrusion ratio

7.7, the extrusion speed of 5 mm/s. It can be seen that the copper distributed uniformly along the axis and are consistent with the numerical simulation.

Figure 9 shows the comparison of the load between the simulation and experimental value at different extrusion temperatures and extrusion ratios. The difference of the load between the simulation and experimental value are not more than 5%. So the results can be determined the the accuracy of numerical simulation. While the trend of load value between the simulation and experimental value is different, and the inflection point is when the extrusion temperature is 1080℃. When the extrusion temperature higher than the melting point of copper, the actual load force less than that of numerical simulation, this is mainly because the copper melting promoted the rearrangement of tungsten particle. The actual load force is higher than that of numerical simulation, this is mainly due to the time remove from heat treatment furnace billet to begin to extrusion in the actual process of hot extrusion needed slightly longer than the set value. So the practical extrusion temperature is lower than the temperature of the parameters set by the numerical simulation.

The good coincidence of the distribution of the W-Cu materials and the loading force in the hot extrusion experiment with the numerical simulation results proved the accuracy of the model and parameter selection in numerical simulation. So the simulation results have practical significance.

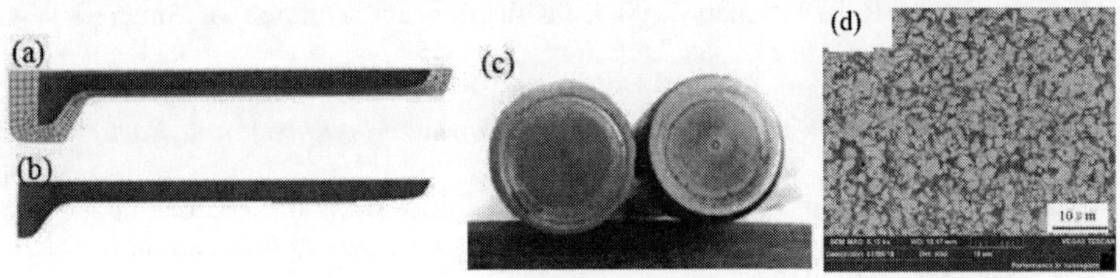

Figure 8. W-Cu extrusion billet: (a and b) simulation; (c) experimental products (d) microstructure of the W-Cu composite.

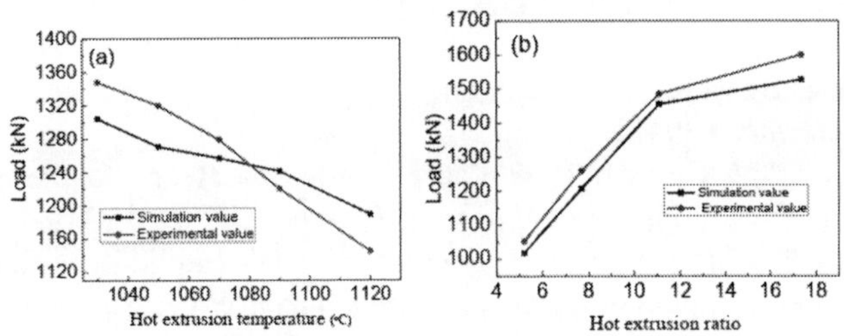

Figure 9. Comparison of the load between the simulation and experimental value: (a) at different extrusion temperatures and (b) at different extrusion ratio.

5. Conclusions

(1) The crack of the steel sleeve is mainly related to the strain and velocity difference of billet and steel jacket. The uniformity of velocity was mainly affected by the structure and material of the steel jacket.

(2) It is concluded that when the extrusion temperature was near the melting point of copper phase, the extrusion speed was in the range of 5 to 10 mm/s, extrusion ratio was in the range of 7.7 to 11, the canning thickness was 5 mm and material was AISI-1016 can not only improve the temperature and stress and strain distribution of billet and canning but also enhanced the flow uniformity of billet.

References

[1] A Elsayed, W Li and O A El Kady 2015 Experimental investigations on the synthesis of W-Cu nanocomposite through spark plasma sintering *Journal of Alloys and Compounds* **63** 373-80

[2] S Liang and L Chen 2015 Infiltrated W-Cu composites with combined architecture of

hierarchical particulate tungsten and tungsten fibers *Materials Characterization* **1** 33-8

[3] X Wei, J Tang and N Ye 2016 A novel preparation method for W-Cu composite powders *Journal of Alloys and Compounds* **66** 471-475

[4] Q Zhou and P Chen 2016 Fabrication of W-Cu composite by shock consolidation of Cu-coated W powders *Journal of Alloys and Compounds* **65** 215223

[5] A A Bothate 2010 Advances in W-Cu: New Powder Systems [D] [Doctoral Dissertation] *California; San Diego State University*

[6] P W Ho, Q F Li and J Y H Fuh 2008 *Materials Science and Engineering: A* **485** 657-63

[7] H Ibrahim, A Aziz and A Rahmat 2013 Comparison of liquid phase sintering and Cu-melt infiltration methods to consolidate 80W-Cu composite using Nickel as sintering activator *Advanced Materials Design and Mechanics* **37** 34-40

[8] W S Wang 1998 The Effect of Tungsten Particle Size on the Processing and Properties of Infiltrated W-Cu Compacts *Metallurgical and Materials Transactions A* **29A** 1509-16

[9] J L Johnson, J J Brezovsky and R M German 2005 Effects of Tungsten Particle Size and Copper Content on Densification of Liquid-Phase-Sintered W-Cu *Metallurgical and Materials Transactions A* **36A** 2807-14

[10] A Upadhyaya and R M German 1998 Densification and Dilation of Sintered W-Cu Alloys *International Journal of Powder Metallurgy* **34** 43-5

[11] J L Johnson, A Upadhyaya and R M German 1995 Effect of Solubility on Shape Retention during Liquid Phase Sintering *Advances in Powder Metallurgy and Particulate Materials* **1**:219-28.

[12] A G Hamidi, H Arabi and S Rastegari 2011 A feasibility study of W-Cu composites production by high pressure compression of tungsten powder *International Journal of Refractory Metals and Hard Materials* **29**123-7

[13] G Shu 2008 Study on the densification process of tungsten copper composites with different components [D] [Master Dissertation] *Harbin: Harbin Institute of Technology*

[14] W F Wang 1997 Effect on Tungsten Particle Size and Copper Content on Working Behavior of W-Cu Alloy Electrodes during Electrodischarge Machining *Powder Metallurgy* **40** 295-300

[15] M Ahangarkani, S Borgi and H Abbaszadeh 2012 *International Journal of Refractory Metals and Hard Materials* **3** 39-44

[16] J L Johnson and R M German 1993 Phase equilibria effects on the enhanced liquid phase sintering of tungsten- copper *Metallurgical Transactions A* **24** 2369-77

[17] Y V Naidich, I A Lavrinenko and V A Evdokimov 1977 Liquid phase sintering under pressure of tungsten-nickel-copper composites *Translated from Poroshkovaya Metallurgiya* **172** 43-9

[18] V V Panichkina, M M Sirotyuk and V V Skorokhod 1982 Liquid-phase sintering of very fine tungsten-copper powder mixtures [J] *Translated from Poroshkovaya Metallurgiya* **234** 27-31

[19] G Gusmano, A Bianco and R Polini 2001 Chemical synthesis and sintering behaviour of highly dispersed W/Cu composite powders [J] *Journal of Materials Science* **3** 901-7

[20] T Nitulescu and S Talu 2001 Applications of descriptive geometry and computer aided design in engineering graphics, Cluj-Napoca, Romania, Risoprint Publishing house ISBN 973-656-102-X

[21] D Q Ma 2016 Study on processing and electrical contact propertied of high dense ultrafined W-Cu composites [D] [Doctoral Dissertation] *Zhengzhou: Zhengzhou University*

[22] R M German 1993 A Model for the Thermal Properties of Liquid-Phase Sintered Composites *Metallurgical Transactions A* **24A** 1475-1752

[23] Y X Cai, B W Liu and L X 1997 Tan Thermal and physical properties model and calculation of tungsten-copper alloys *Science and engineering of powder metallurgy materials* **2** 11-4

[24] D R Li 2009 Study on the densty process and plastic deformation of W-Cu powder heat extrusion [D] [Doctoral Dissertation] *Harbin: Harbin Institute of Technology*

IOP Conf. Series: Materials Science and Engineering **292** (2017) 012093 doi:10.1088/1757-899X/292/1/012093

Preparation and characterization of flexibility and lightweight CF-RGO composites

J Li, S Bi[1], X J Su, G L Hou, B Mei and W Q Ma

Xi'an Research Institute of High-Tech, Xi'an, 710025, China

[1] E-mail: bisong5011@163.com

Abstract. Absorbing materials are developed with the concept of "thin, light, wide and strong", and various methods and materials have been attempted to satisfy the demand. In this work, the in situ growth method were conducted to grow reduced graphene (RGO) into carbon fiber (CF) to prepare CF-RGO composites. Moreover, the phase, microstructure and electromagnetic performance of the flexibility and lightweight composites were characterized. Consequently, the XRD and SEM results indicate the RGO is well grown on the surface of CF. The microwave absorption performance of the as-prepared samples were calculated, and the results show that the effective absorption band of the composites move towards low frequency once the RGO were introduced into the CF.

1. Introduction

Wave-absorbing materials that have been widely applied in electromagnetic controlling and communication technology are considerably explored in the recent years [1-5]. Among the pursued nanoscale fillers, carbon nanomaterials, one-dimensional (1D) carbon nanotubes (CNTs) and two dimensional (2D) graphene nanosheets (GNs) in particular, have received great attention coupled with extensive development effort due to their exclusive chemical structures and tunable electrical properties [6-9]. In the past decades, CNT- and GN-based nano-fillers with low percolation threshold and appropriated electrical loss have already demonstrated dramatically enhanced microwave absorption performance in the pursed composites [10-14]. Among various strategies, combination of electrical loss and magnetic loss has shown great potential owing to the excellent flexibility in manipulating complex permittivity and permeability, and thus has been extensively explored. In the early examples, Che and coworkers have encapsulated iron into the tubular space of the multiwalled CNTs (MWCNTs). The as-prepared Fe@MWCNT heterostructures (filler loading of 20 wt%) were further fabricated into epoxy-based composites and exhibited a broad effective microwave absorption region with the peak absorption up to 25 dB in the span of 2-18 GHz [15-17].

With presence of wave-absorbing materials, the electromagnetic wave energy would be weakened via absorption patterns and reflection patterns. Reduced graphene oxides (RGOs) have drawn significant attention since it posssess large specific surface area, high young's modulus, high thermal conductivity, high optical transmittance and good electrical conductivity. In this work, the RGO were in situ created into the CF network by electrochemical deposition technology. similar explorative studied have also been attempted in the fabrication of graphene-heterostructure filler with combined electrical loss/magnetic loss [18-22]. For instance, Chen et al. have recently converted graphene oxide (GO) and Ni precursors into reduced graphene oxide (RGO)/Ni heterostructures. The composites incorporated with such heterostructures at filler loading of 60 wt% possessed a maximum reflection loss ~17 dB. In the report by Sun et al, concentration of Fe_3O_4 was tuned in the process of converting GO into RGO, and the wax-based composites with Fe3O4/RGO heterostructures exhibited a reflection loss up to 27 dB at 40 wt% filler loading. Likewise, Fu et al. synthesized $NiFe_2O_4$ nanorods on the RGO and fabricated the heterostructures into wax-based composites, showing the best absorption performance up to 29.2 dB at 60 wt% filler loading [23].

In the preparation, graphene oxide (GO) aqueous solution was prepared by modified hummers method. And the in situ growth method were conducted to grow reduced graphene (RGO) into carbon fiber (CF) to prepare CF-RGO composites.

2. Results and discussion

X-Ray diffraction (XRD) spectras of the as-prepared composite textiles are shown in figure 1. The structure and morphology of the samples were investigated by X-ray diffraction (XRD, Rigaku Ultima IV, Cu-Ka) in the range of 10°~90° with a scan speed of 3 s and a step size of 0.02° in 2θ. As exhibited in the spectrum of the CF, there is a broad peak around 21~25°, indicating the existence of the carbon of amorphous CF. Meanwhile, there is a distinct peak around 25~26° in the spectrum of the CF-RGO, referring to the growth of the RGO in the CF. the results indicate the RGO were successfully introduced into CF.

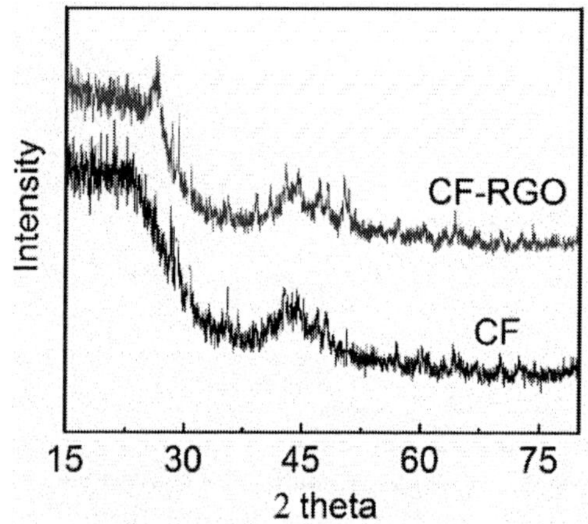

Figure 1. XRD spectras of the as-prepared samples.

FESEM images of CF textile and CF-RGO textile are demonstrated in figure 2. As is shown in figure 2a, the CF textile possesses a smooth surface. Figure 2b indicates that the RGO layers are successfully deposited on the surface of the CF, and the separated CF wire are bonded together by the graphene nanosheets. Consequently, as is, the conductive network were gained.

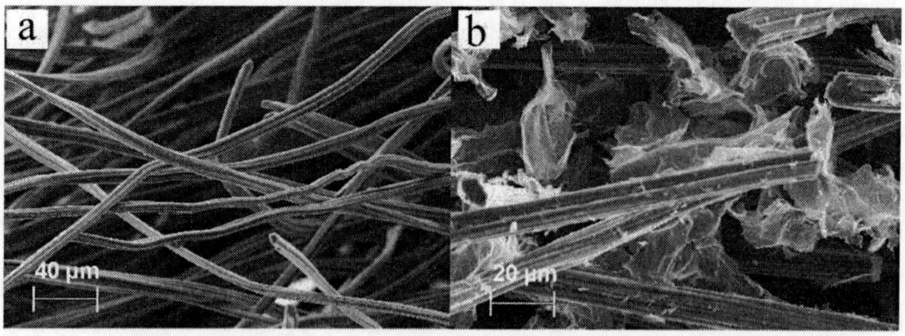

Figure 2. SEM images of the neat CF (a) and CF-RGO (b).

Coaxial method was conducted to measure electromagnetic properties and dielectric properties of the wax-based composites in the range of 2~18 GHz. To explore the influence of the filler loading in the wax-based composites, the samples with different proportions (10%, 15%, 20%, and 30%) have been prepared. Consequently, the complex permittivity of the samples (figure 3) are given as marked in the related figures. According to figure 3, the real permittivity of CF-RGO is much higher than that of CF, indicating the growth of RGO in the CF would obviously improve the real permittivity of the composites. Such increase is associated with the introduction of RGO interfaces.

The reflection loss (R_L), which could be a direct reflection of the ability of the microwave absorption performance of the absorber, is achieved by the relation

$$R_L = 20\log(|Z_{in}-1|/|Z_{in}+1|). \quad (1)$$

Here the Z_{in} refers to the input impedance of the absorption layer, which could be achieved by

$$Z_{in} = (\mu_r/\varepsilon_r)^{1/2}\tanh[j2\pi fd(\mu_r\varepsilon_r)^{1/2}/c]. \quad (2)$$

where, μ_r and ε_r is the complex permittivity and permeability of the samples, respectively. f is frequency. d is thickness of the absorber, and c is the light velocity.

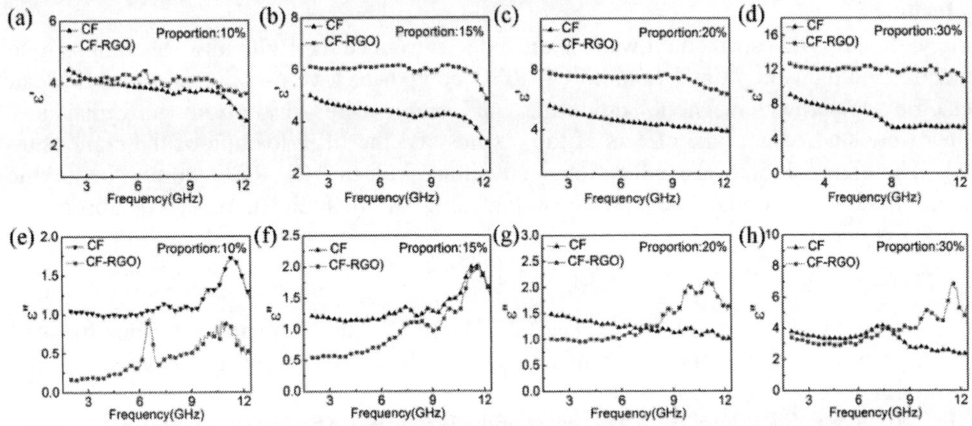

Figure 3. Complex permittivity of the wax-based composites embedded with the samples in different proportion.

Figure 4 show the microwave absorption performance of the as-prepared samples at different thickness. It is obvious that the three composites all possess best microwave absorbing abilities at the proportion of 30%. And the maximum reflection loss values of CF and CF-RGO are up to 36.62 and 27.78 dB, respectively. It is obvious that the CF-RGO composites possess better absorbing performance than that of CF composites at lower frequency. The results suggest that the effective microwave absorption band would move towards low frequency with the introduce of two-dimensional RGO interfaces, which might be a potential method to enlarge the microwave absorption band towards low frequency.

IOP Conf. Series: Materials Science and Engineering **292** (2017) 012093 doi:10.1088/1757-899X/292/1/012093

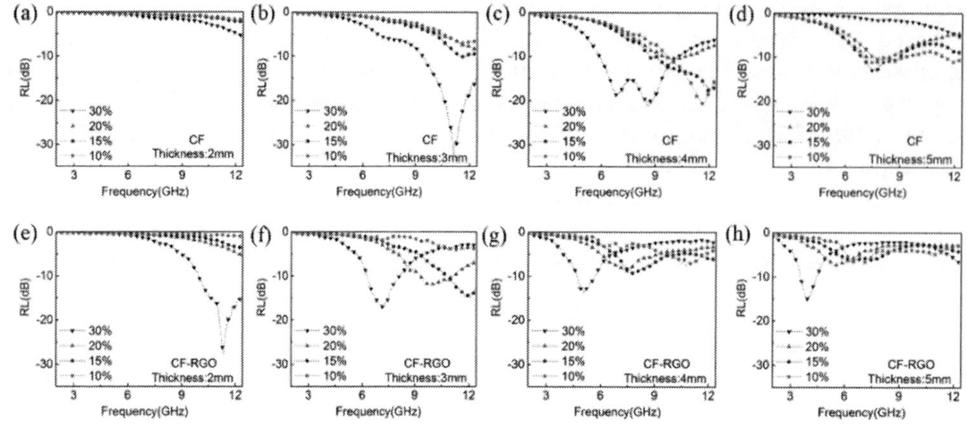

Figure 4. Microwave absorption performance of the as-prepared samples.

3. Conclusions

In summary, in-situ synthesis method was conducted to introduce the RGO into CF composite textiles. Consequently, the lightweight and flexible CF-RGO composite textiles were fabricated. Meanwhile, the dielectric properties, magnetic properties and microwave absorption performance of the composites were studied, and the effects of the textiles and the filler loading of the composites were discussed. Therefore, due to the exceptional advantages of the CF-RGO composite textiles, the composites could serve as the potential absorbing materials to broaden the wave absorption band towards low frequency.

References

[1] WL Song, MS Cao, MM Lu, S Bi, CY Wang, J Liu, J Yuan, LZ Fan 2014 Flexible grapheme/polymer composite films in sandwich structures for effective electromagnetic interference shielding *Carbon* **66** 67-76

[2] CJ Li, B Wang, JN Wang 2012 Magnetic and Microwave Absorbing Properties of Electrospun $Ba_{(1-x)}La_xFe_{12}O_{19}$ Nanofibers *J. Magn. Magn. Mater.* **324** 1305-1311

[3] ZJ Li, ZL Hou, WL Song, XD Liu, WQ Cao, XH Shao, MS Cao 2016 Unusual continuous dual absorption peaks in Ca-doped $BiFeO_3$ nanostructures for broadened microwave absorption *Nanoscale* **19** 10415-10424

[4] YJ Kim, SS Kim 2010 Magnetic and microwave absorbing properties of Ti and Co substituted M-hexaferrites in Ka-band frequencies (26.5~40 GHz) *J. Electrochem* **24** 314-318

[5] S Bi, L Ma, B Mei, Q Tian, CH Liu, CR Zhong, YD Xiao 2014 Silicon carbide/carbon nanotube heterostructures: Controllable synthesis, dielectric properties and microwave absorption *Adv. Powder Technol.* **25** 1273-1279

[6] HS Chen, WY Wei, JX Liu, DN Fang 2012 Propagation of a mode-III interfacial crack in a piezoelectric-piezomagnetic bi-material *Int. J. Solids Struct.* 49 2547-2558

[7] GM Shi, YF Li, L Ai, FN Shi 2016 Two step synthesis and enhanced microwave absorption properties of polycrystalline $BaTiO_3$ coated Ni nanocomposites *J. Alloys Compd.* **680** 735-743

[8] HS Chen, J Ma, YM Pei, DN Fang 2013 Anti-plane Yoffe-type crack in ferroelectric materials *Int. J. Fract.* **179** 35-43

[9] WL Song, XT Guan, LZ Fan, YB Zhao, WW Cao, CY Wang, MS Cao 2016 Strong and thermostable polymeric graphene/silica textile for lightweight practical microwave absorption composites *Carbon* **100** 109-117

[10] CS Dong, X Wang, PH Zhou, T Liu, JL Xie, LJ Deng 2014 Microwave magnetic and absorption properties of M-type ferrite $BaCo_xTi_xFe_{12-2x}O_{19}$ in the Ka band *J. Magn. Magn.*

Mater. **354** *340-344*

[11] S Bi, XJ Su, GL Hou, GQ Gu, Z Xiao 2013 Microstructure characterization of alumina/carbon nanotubes composite coating *Ceram. Int.* **39** 5979-5983

[12] HS Chen, WY Wei, JX Liu, DN Fang 2014 Propagation of a semi-infinite conducting crack in piezoelectric materials: Mode-I problem *J. Mech. Phys. Solids* **68** 77-92

[13] X Ding, SP Li, N Zhang, JG Wang 2016 $FeNi_3$ nanoalloy decorated on 3D architecture composite of reduced graphene oxide/molybdenum disulfide giving excellent electromagnetic wave absorption properties *J. Alloys Compd.* **689** 208-217

[14] CY Zhao, MY Shen, ZX Li, R Sun, AL Xia, XG Liu 2016 Green synthesis and enhanced microwave absorption property of reduced graphene oxide-$SrFe_{12}O_{19}$ nanocomposites *J. Alloys Compd.* **689** 1037-1043

[15] YB Feng, T Qiu 2012 Preparation, characterization and microwave absorbing properties of FeNi alloy prepared by gas atomization method *J. Alloys Compd.* **513** 455-459

[16] S Gholami, AS Rad, A Heydarinasab, M Ardjmand 2016 Adsorption of adenine on the surface of nickel-decorated graphene; A DFT study *J. Alloys Compd.* **686** 662-668

[17] AS Rad 2016 Adsorption of mercaptopyridine on the surface of Al- and B-doped graphenes: Theoretical study *J. Alloys Compd.* **682** 345-351

[18] B Wang, Y Wang, J Park, H Ahn, G Wang 2011 In situ synthesis of Co_3O_4/graphene nanocomposite material for lithium-ion batteries and supercapacitors with high capacity and supercapacitance *J. Alloys Compd.* **509** 7778-7783

[19] SL Zhang, QZ Jiao, CX Wang, H Yu, Y Zhao, Q Wu 2016 In situ synthesis of Mg/FeLDO/ carbon nanohelix composites as absorbing materials *J. Alloys Compd.* **658** 505-512

[20] WL Song, XT Guan, LZ Fan, WQ Cao, CY Wang 2015 Tuning three-dimensional textures with graphene aerogels for ultra-light flexible graphene/texture composites of effective electromagnetic shielding *Carbon* **93** 151-160

[21] WL Song, J Wang, LZ Fan, Y Li, CY Wang, MS Cao 2014 Interfacial engineering of carbon nanofiber-graphene-carbon nanofiber heterojunctions in flexible lightweight electromagnetic shielding networks *ACS Appl. Mater. Interface* **6** 10516-10523

[22] I Villalpando, P John, S Porro, JIB Wilson 2016 Carbon fibre production during hydrogen plasma etching of diamond films *RSC Adv.* **6** 64421-64427

[23] L Guadagno, et al. 2015 Effective formulation and processing of nanofilled carbon fiber reinforced composites *RSC Adv.* **5** 6033-6042

IOP Conf. Series: Materials Science and Engineering **292** (2017) 012094 doi:10.1088/1757-899X/292/1/012094

The study of size and stability of n-butylcyanoacrylate nanocapsule suspensions encapsulating green grass fragrance

G Y Zhu[1,4], C T Lin[2], J M Chen[2], D M Lei[2] and G X Zhu[3]

[1] Shanghai Institute of Technology, Shanghai, China
[2] Guangzhou Levon Flavor & Fragrance Technology Co., Ltd, Guangdong, China
[3] 14846, 43 Ave. Edmonton, Alberta, T6H 5S1, Canada

[4] E-mail: zgy012@126.com

Abstract. Green grass fragrance has been widely used in many fields. However, fragrances are volatile compounds that do not last long. In order to prolong its odor, nanocapsules encapsulated green grass fragrance were prepared. The paper deals with the preparation of green grass fragrance nanocapsules by emulsion polymerization. N-butylcyanoacrylate (BCA) with excellent biocompatibility and biodegradability was used as encapsulant. The nanocapsule suspension systems were characterized and its stability was investigated. The physicochemical properties of polymeric nanocapsules (average diameter and polydispersity) were evaluated as a function of time to assess the system stability. The result showed that the system (containing 0.8% of green grass fragrance, with a polydispersity index (PDI) near 0.1 and an average diameter in the range of 20-30 nm) was an ideal state and relatively stable. Besides, the distinction of stability of three nanocapsule suspensions with different green grass fragrance content was also obvious from scanning electron microscopy (SEM).

1. Introduction

Fragrance is a mixture of fragrant essential oils, aroma compounds, fixatives and solvents. It is used to give the human body, objects, and living-spaces a pleasant scent. Green grass fragrance has an intense, fresh grassy-green, leafy odor. It has been widespread used in perfuming, daily chemical and other industries as a pleasant fragrance of covering up the bad odor and purifying the air [1]. However, applications of green grass fragrance have turned out to be particularly complicated due to its chemical and physical properties, such as low stability, high volatilization and release in the presence of light, heat and oxygen [2]. Hence, finding technologies to solve these problems is much desired. Among the existing approaches (e.g., nanoparticles, liposomes, cyclodextrin inclusions) [3], it is obvious that nanoencapsulation has attracted increasing attention.

In general, nanoencapsulation is one of nanotechnology, which can improve the stability, solubility and bioavailability of encapsulated species, increase their protection, reduce evaporation, promote easier handling and control the release during prolonged storage and applications [4-7]. It is already today an important processing method in industry and bears great potential for a wide range of applications in numerous fields [8], including foods, pharmaceutical, chemical, cosmetics, perfumes, textiles, agriculture, etc.[9] And it is a processing by which one material or a mixture of materials, such as volatile and flavor being coated with or entrapped within another material or system, producing particulate dispersions or solid particles, with sizes ranging from 10 nm to 1 μm [4]. The coated material is called the core, active material or nucleus, and the coating material is called the wall material, shell, carrier or encapsulant [10, 11].

An important step in developing microcapsules is the selection of a wall material that meets the required criteria [12]. N-butylcyanoacrylate (BCA), which has been widely used in the medical field, is the most attractive for emulsion polymerization system owing to its biodegradability, availability, biocompatibility and low toxicity. And it has several biomedical applications, such as tissue adhesives [13-17] that were based on the fact that the anionic polymerization could be easily initiated by traces of nucleophiles like water, amines, alcohols, or phosphines [18]. Polybutylcyanoacrylate (PBCA) nanoparticles have been extensively investigated for 30 years to be used as drug carrier for their excellent properties [19-24]. So BCA was chosen as the coating material to build the shell on the green grass fragrance.

Additionally, an important issue to be considered for nanocapsules encapsulated perfume materials is their mechanical properties. For example, the use of suspension systems of polymeric nanocapsules is often limited due to problem of stability over prolonged storage. A colloidal suspension does not normally show phase separation, since both sedimentation and aggregate formation are slow. However, storage for many months can sometimes result in aggregation and consequently phase separation [25, 26].

Hence, this work was aimed to prepare a series of PBCA nanocapsule suspensions encapsulated green grass fragrance and the emphasis was on the relationship between the stability and the particle size and polydispersity of them.

2. Materials and methods

2.1. Materials

Butylcyanoacrylate (BCA) was purchased from Zhejing Jinpeng Chemical Company. Castor oil polyoxyethylene ether 40 (EL-40) was provided by Shanghai Haojiong Reagent Co. Ltd. Hydrochloric acid and sodium hydroxide were purchased from Sinopharm Chemical Reagent Co. Ltd. Green grass fragrance was blended by Guangzhou Levon Flavor & Fragrance Technology Co., Ltd. Deionized water was produced in laboratory.

2.2. Preparation of nanocapsules

Polybutylcyanoacrylate (PBCA) nanocapsules were prepared by emulsion polymerization as described in literatures [27, 28]. A glass beaker was equipped with a magnetic stirrer (Shanghai Mei Yingpu instrument and Meter Manufacturing Co., Ltd.). Deionized water, emulsifying agent and core material were added in the beaker, and emulsified at 1100 rpm for 10 min. Then the pH of the system was adjusted to 2.0 with an aqueous solution of HCl (1M). BCA was added to the beaker at the speed of 3d/min. After that, the reaction was carried out for 1 h in room temperature, using a stirring rate of 1100 rpm, too. At last, the pH of this system was adjusted to 7.0 using an aqueous solution of NaOH (1M), and the reaction would go on for another 0.5 h.

2.3. Particle size and polydispersity measurements

The mean size and the size distribution of the BCA encapsulated green grass fragrance nanocapsule suspensions were measured using a dynamic light scattering instrument (Zetasizer Nano ZS, Malvern Instruments Ltd, UK) in three replicates. The size distribution was evaluated for all samples according to the polydispersity index (PDI). PDI is an indicator of the nanocapsule size distribution in the range 0-1 and the values of it < 0.2 indicate a monodisperse emulsion, whereas values > 0.5 show larger distributions [27].

2.4. Nanocapsule suspensions stability measurements

The stability of the PBCA nanocapsule suspensions encapsulated green grass fragrance was evaluated by the size and PDI of the suspensions stored at room temperature for a period of time (24 h, 15 d and

30 d). The morphology and structure of nanocapsules were examined using a scanning electron microscopy (Hitachi S-3400 N, Hitachi High-Technologies, Japan) at an accelerated voltage of 15.0 kV and at 90 d after the samples preparation. The nanocapsule samples were fixed on metal stubs with double sided tape and coated with gold by a gold sputter coater in a high-vacuum evaporator.

3. Results and discussion

3.1. Effects of green grass fragrance concentration on size and PDI of nanocapsules

The particle size and the size distribution of the green grass fragrance nanocapsule suspensions were measured by Zetasizer Nano ZS (Malvern Instruments Ltd, UK). The average diameter and the PDI were recorded. All measurements were performed in triplicate and the average values were obtained. Figure 1 shows the particle size and distribution of nanocapsules changing with core material concentration. Data are expressed as mean ± standard deviation with error bars as shown in figure 1.

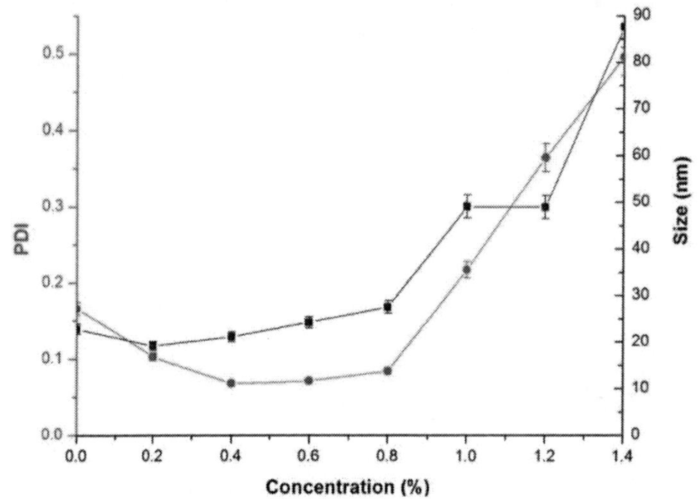

Figure 1. The average size and PDI of PBCA nanocapsule suspensions as a function of green grass fragrance concentration for storage time 24± 0.1 h.

Fixing emulsifier concentration at 1.6% and changing the core content, nanocapsule systems of various particle size and PDI were gained. Making the adjustment is to improve the proportion of core material in the system, hoping to get higher stability and loading capacity. As showed in figure 1, it also ensured that the particle size and distribution of nanocapsule were relatively ideal when the ratio of emulsifiers to core material decreased compared with the previous work [27]. The average diameter of PBCA nanocapsules encapsulated green grass fragrance enhanced with green grass fragrance increased, whereas PDI just increased sharply when green grass fragrance concentration more than 0.8%. An excess of core material would be attached on the outside of the capsule resulting in particle size increased and distribution unevenly. From results of Zetasizer Nano ZS, it can be concluded that the size distribution of the nanocapsules (containing 0.8% of green grass fragrance) was monodisperse, and had a narrow peak.

3.2. Storage stability analysis of green grass fragrance nanocapsule suspensions

Figure 2 and figure 3 show the changes of the mean sizes and PDI of PBCA nanocapsule suspensions along with storage time, under the concentration of green grass fragrance in range of 0-1.4%. With the increasing of storage time, there are different changing trends in the size and PDI of sample systems.

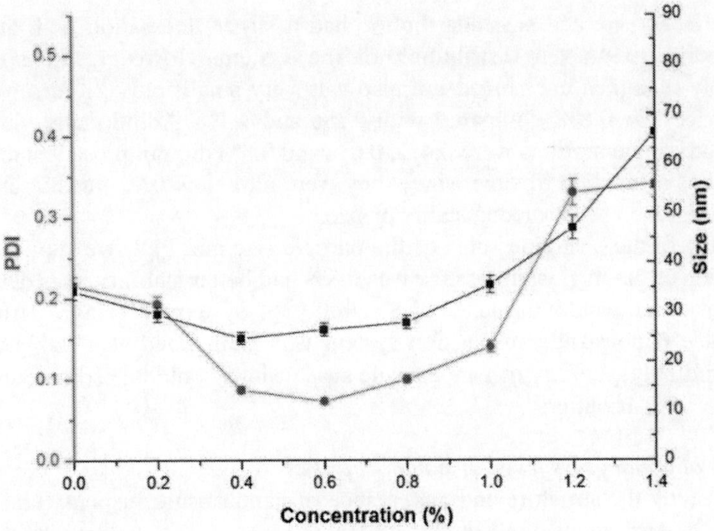

Figure 2. The average size and PDI of PBCA nanocapsule suspensions as a function of green grass fragrance concentration for storage time 15 ± 0.01 d.

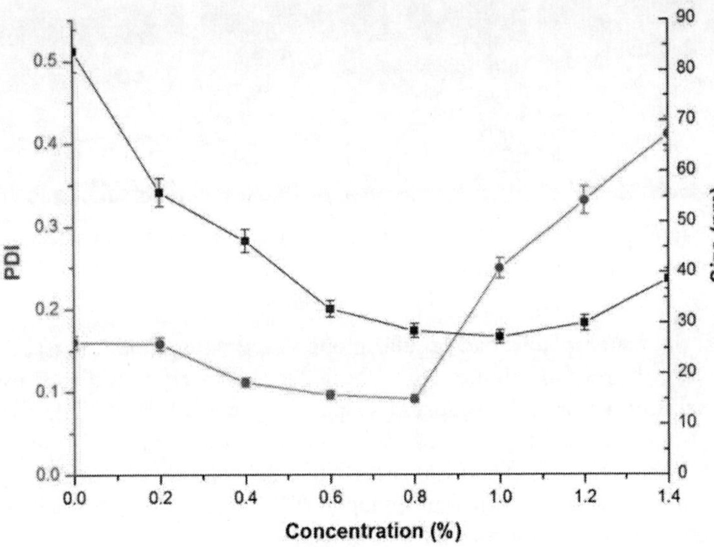

Figure 3. The average size and PDI of PBCA nanocapsule suspensions as a function of green grass fragrance concentration for storage time 30 ± 0.01 d.

The particle size of PBCA nanocapsule suspensions increased as the storage time increased at the lower green grass fragrance concentration (0, 0.2% and 0.4%); besides, PDI changing along with the storage time. It may occur flocculation firstly, and then coalescence and both of them result in increase of the mean particle size [29]. While the system had a high concentration (1.0%, 1.2% and 1.4%), the change rule of the particle size and PDI was just on the contrary with the law at low core material concentration. There were macroscopic particle in freshly prepared nanocapsule suspensions with higher green grass fragrance concentration, which may be the main reason that the mean particle size was big initially. After the macroscopic particle subsided, the average particle size of the nanocapsules decreased. Another reason that suspension systems were unstable at lower green grass fragrance concentration and mean particle size increased sharply in storage for 30 d. So size change of PBCA nanocapsules had a decrease trend when stored for 30 d, compared to 24 h and 15 d.

In a word, the mean size and size distribution had a larger fluctuation both at lower and higher concentration, leading to uneven distribution of the system. However, the size of nanoparticle suspension not only remained unchanged but also was very small, only 28 nm, when the content of green grass fragrance was 0.8%, compared with 0.2% and 1.4%. Additionally, the polydispersity of the samples that core concentrations were 0.4%, 0.6% and 0.8% did not present statistically significant differences in PDI as a function of time, since they were all below 0.2, but the samples of 0.4% and 0.6% green grass fragrance performed unstable in size.

Hence, according to the changing rules of the particle size and PDI, we supposed that the system which the proportion of green grass fragrance was 0.8% had better stability. It is demonstrated that the content of core material would influence the stability of system. Average particle size and size distribution of stable nanocapsule suspension system was maintained at a stable level within given storage time. Accordingly, stability of nanocapsule suspensions could be estimated by the changing of mean particle size and distribution.

3.3. SEM analysis of green grass fragrance nanocapsules

SEM was used to study the structure and appearance of nanocapsule suspensions. The morphologies of nanocapsules at 90 d are showed in figure 4.

Figure 4. Scanning electron microscopy micrographs of nanocapsules containing green grass fragrance in various concentrations at 90 d after sample preparation, A (0.2%), B (0.8%), C (1.4%).

It confirmed that the nanocapsules had regular globe shape with smooth surface and without holes and cracks, but varying degrees of reunion and adhesion were observed too as shown in figure4. The distinction of the stability of three nanocapsule suspensions was obvious. The average particle size showed a significant change varying in the range 500-900 nm as storage time increased to 90 d, and were larger than the values obtained from Zetasizer Nano ZS. Moreover, combining with distribution morphology of the nanocapsule suspensions loading 0.2% and 1.4% of green grass fragrance, it indicated that flocculation and coalescence existed indeed.

4. Conclusions

This study describes production of PBCA nanocapsules briefly and investigates its storage stability. After 24 h, 15 d and 30 d of storage, even 90 d, no layer separation was observed of all nanocapsule suspensions, but their transparency was different in surface appearance. Unstable system became turbid seriously over time. Besides, the distinction of the stability of three nanocapsule suspensions (with 0.2%, 0.8% and 1.4% of green grass fragrance) was also obvious from SEM.

In respect of mean size and size distribution, relative stable system like containing 0.8% of green grass fragrance nearly did not changed as the storage time increased in a given time range, however, the systems in which percentage composition of green grass fragrance was lower or higher changed greatly, that is, the PBCA nanocapsule system with various physical stability could be achieved by optimizing of process formulation, and stability of it could be estimated by the changing of mean particle size and distribution, and SEM micrographs.

Reference

[1] Zhu G, Xiao Z, Zhou R and Feng N 2015 *J. Food Sci. Technol.* **52** 4607
[2] Zhu G, Feng N, Xiao Z, Zhou R and Niu Y 2015 *J. Therm. Anal. Calorim.* ***120*** *1811*
[3] Lee S Y, Rim Y, McPherson D D, Huang S L and Kim H 2014 *Bio-Med. Mater. Eng.* **24** 61
[4] Sousa Lobato K B D, Paese K, Forgearini J C, Guterres S S, Jablonski A and Oliveira Rios A D2013 *Food Chem.***141** 3906
[5] Baranauskiene R, Bylaite E, Zukauskaite J and Venskutonis R P 2007 *J. Agric. Food Chem.* **55** 3027
[6] Peña B, Panisello C, ArestéG, Garcia-Valls R and Gumí T 2012 *Chem. Eng. J.* **179** 394
[7] Martins I M, Rodrigues S N, Barreiro M F and Rodrigues A E 2012 *Ind. Eng. Chem. Res.* **51** 11565
[8] Neubauer M P, Poehlmann M and Fery A 2014 *Adv. Colloid Interface Sci.* **207** 65
[9] Xiao Z B, Liu W L, Zhu G Y, Zhou R J and Niu Y W 2014 *Flavour Fragr. J.* **29** 166
[10] Madene A, Jacquot M, Scher J and Desobry S 2006 *Int. J. Food. Tech.* **41** 1
[11] Zhu G Y, Xiao Z B, Zhou R J and Yi F P 2012 *Adv. Mater. Res.***535** 440
[12] Calvo P, Hernandez T, Lozano M and Gonzalez-Gomez D 2010 *Eur. J. Lipid Sci. Technol.***112** 852
[13] Arias J L, Gallardo V, Linares-Molinero F and Delgado A V 2006 *J. Colloid Interf. Sci.* **299** 599
[14] Carriles Y R, Brito R A, Sánchez R M, Acevedo E S, Domínguez P R and Mueller W D 2014 *Molecules* **19** 6220
[15] Dadas B, Alkan S, Cifci M and Basak T 2007 *Eur. Arch. Otorhinolaryngol.* **264** 539
[16] Voon L W, Chua C N and Hanson R 2004 *Arch. Ophthalmol.* **122** 279
[17] Wang Y N, Wei Q H, Pan F L, Yang M M and Wei S M 2014 *Bio-Med. Mater. Eng.* **24**8 25
[18] Zhang Y, Zhu S Y, Yin L H, Qian F, Tang C and Yin C H 2008 *Eur. Polym. J*.**44** 1654
[19] Vauthier C, Dubernet C, Fattal E, Pinto-Alphandry H and Couvreur P 2003 *Adv. Drug Deliv. Rev.* **55** 519
[20] Vauthier C, Dubernet C, Chauvierre C, Brigger I and Couvreur P 2003 *J. Control. Rel.* **93** 151
[21] Moghimi S M, Hunter A C and Murray J C 2001 *Pharmacol. Rev.***53** 283
[22] Vonarbourg A, Passinari C, Saulnier P and Benoit J P 2006 *Biomaterials* **27** 43
[23] Couvreur Pand Vauthier C 2006 *Pharm. Res.* **23** 1417
[24] Wu M, Dellacherie E, Durand A and Marie E 2009 *Colloid. Surface. B* **69** 141
[25] Moraes C M, Paula E D, Rosa A H and Fraceto L F 2010*J. Braz. Chem. Soc. Rev.* **21** 995
[26] Schaffazick S R, Guterres S S, Freitas L L and Pohlmann A R 2003 *Quim. Nova.* **26** 726
[27] Hu J, Xiao Z B, Zhou R J, Li Z, Wang M X and Ma S S 2011 *Flavour Fragr. J.* **26** 162
[28] Xiao Z, Lei D, Zhu G and Niu Y 2015 *J. Polym. Res.* *22* 10
[29] Tadros T 2004 *Adv. Colloid Interface Sci.* **108-109** 227

Study on the quality of FRP fishing vessel based on improved Fishbone Chart

J H Sui[1, 3], Y F Yu[1], Q F Du[1, 2] and D W Jiang[2]

[1] Navigation and Ship Engineering College, Dalian Ocean University, Dalian 116023, China

[2] Dalian Global Yachting Manufacture (int' l) Co., Ltd., Dalian 116046, China

[3] Email: 18018989368@163.com

Abstract. The construction quality of FRP fishing vessels influences their production, use and industry development. In order to explore the factors that affect the construction quality of FRP fishing vessels, key factors affecting the construction quality of FRP fishing vessels are determined based on the quality problems of FRP fishing vessels constructed. The improved Fishbone Chart is used to analyze the eight factors of "human, machine, material, process, environment, inspection, design and information". Taking the factors that affect the construction quality of FRP fishing vessels as the central target, the eight influencing factors were condensed into five aspects and a composite Fishbone Chart is drawn. The Fishbone Chart is used as the basic model, the influencing factors are sorted, screened and discriminated, and the system model convenient for construction site management and control is established. Finally, the causes of poor construction of FRP fishing vessels are analyzed and discussed, and relevant suggestions are put forward.

1. Introduction

FRP is a kind of composite material made of glass fiber and its products as reinforcing material and synthetic resin as matrix material. FRP has the characteristics of small relative density and high specific strength. Its advantages include shock resistance, corrosion resistance, heat insulation and electrical insulation. The FRP fishing vessel constructed with this material has beautiful appearance, smooth finish and long service life. The Chinese FRP fishing vessel began construction in 1970s, and it has undergone three ups and three falls. The service life of FRP fishing vessels can be up to 50 years. In 1985, China developed 16.88 m FRP trawler. It has a good hull form and a high level of hull structure design. Due to the imperfect construction technology, FRP fishing vessels have been scrapped in less than 10 years. During the Ninth Five Year Plan period of China, a serious defect in the design and construction of a 33 meter trawler led to abandonment. In 2004, during the operation at Fiji fishing grounds, there was a quality accident in the FRP tuna fishing vessel built by the Shandong West harbor ship repair factory. These problems have challenged the fishing community to the quality of domestic FRP fishing vessels [1-2]. Therefore, the improvement of FRP fishing vessel construction quality has become the focus of attention at the moment.

2. Improvement of Fishbone Chart analysis

Fishbone Chart, also called causal analysis chart or characteristic factor diagram, is an important method of accident analysis [3]. The factors that influence product quality is divided into human, machine, material, process and environment by traditional Fishbone Chart. However, the quality of the products has been beyond the scope of the traditional Fishbone Chart. The improved Fishbone Chart combined the ideas of Japanese famous expert Mr. Taguchi Xuanichi about the quality of products. It will increase the original five major factors to eight major factors, namely, " human, machine, material, process, environment, inspection, design and information" (Shown in figure 1) [4].

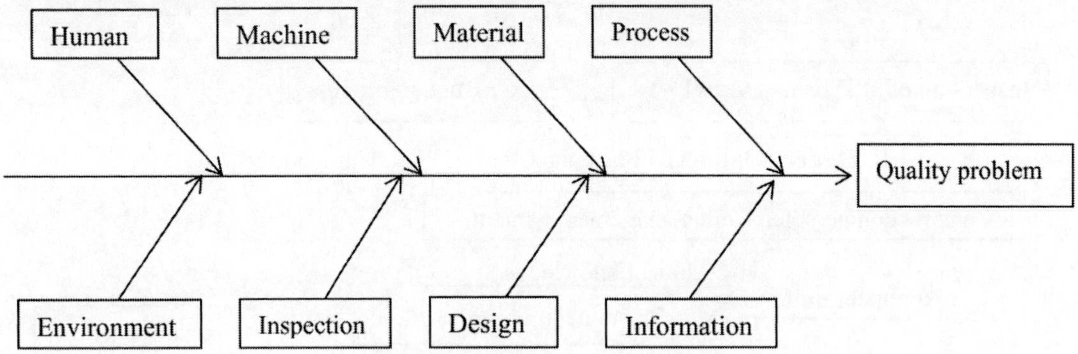

Figure 1. Improved Fishbone Chart.

After the improved Fishbone Chart has been applied, we can systematically and comprehensively summarize the influencing factors and put forward reasonable preventive measures. The analysis steps of improving Fishbone Chart are as follows [5]:

- Survey: a comprehensive understanding of accidents and awareness and analysis to prepare for accurate graphics.
- Fix the problem: the accident, the problem or the object of study as the "result", draw on the right hand side of the picture, and draw the trunk and arrow.
- Analyse the causes and categories: collect, analyse and list the causes, organize and classify them according to the importance and cause and effect.
- Mapping: fill in the drawings one by one.

3. Construction quality analysis of improved Fishbone Chart

3.1. Ideas for building improved Fishbone Charts

In the design and construction process of FRP fishing vessels, the design of the product quality is the first step, followed by the construction and inspection. The improved Fishbone Chart was established by combining the quality of FRP fishing vessels with the formation of the real situation and based on the concept of total quality management and source quality management. The eight factors are divided into five major areas, namely, construction personnel and environment, hull design, raw materials, construction process, inspection of FRP fishing vessel. Based on the complexity and complexity of the influencing factors, a composite Fishbone Chart of five aspects is established (Shown in figure 2).

3.2. The induction and analysis of influencing factors

3.2.1. Construction personnel and environment. At present, the moulding method of FRP fishing vessel in China is hand lay up moulding. Its mode of operation is mainly manual operation. Although the operation is simple, it requires higher operating techniques. Therefore, not only the construction staff needs to have a serious work attitude and rich experience, but also managers need to supervise and manage the work. The construction of the building capacity of shipyard should stand the test and environmental control in the workshop temperature of less than 32℃, relative humidity is not more than 85% (Shown in figure 3).

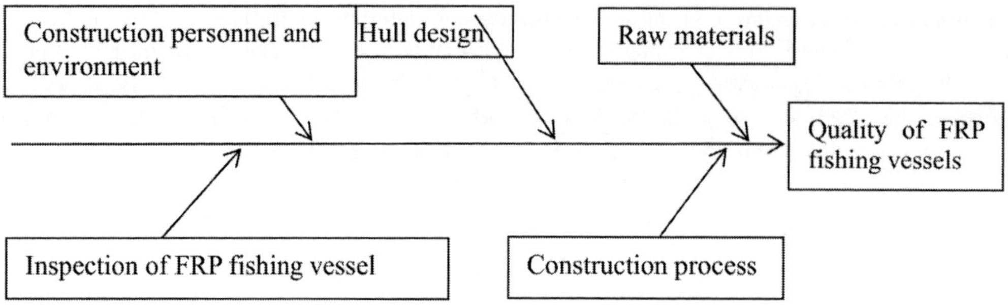

Figure 2. Improved Fishbone Chart of FRP fishing vessel.

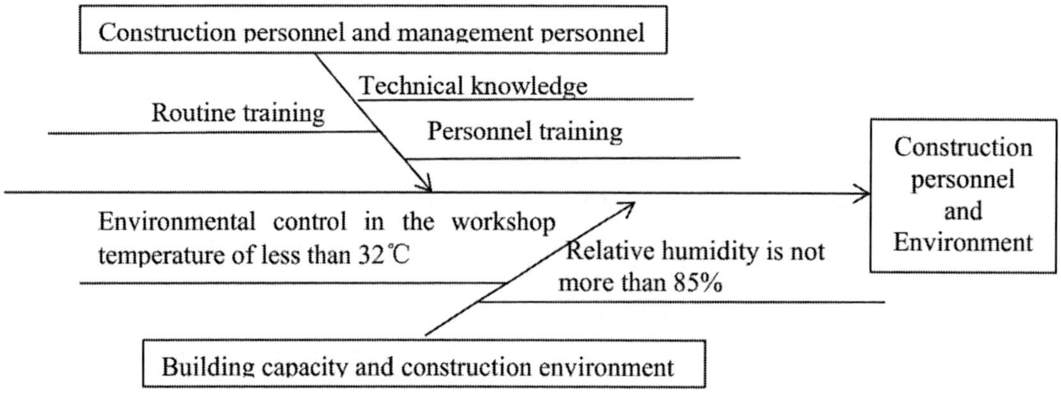

Figure 3.Construction personnel and environmental factors Fishbone Chart.

3.2.2. Hull design. Hull design can be divided into structural design, alignment design and local design. The structural design factors of FRP fishing vessels are the connection of the important nodes, the form of the frame and the choice of the plate structure. Linear design is the mother ship type selection and bilge line selection. The local design is mainly the choice of sandwich structure and the manufacture of bulbous bow (Shown in figure 4).

3.2.3. Material factors. Selection of raw materials. The choice of raw materials is mainly divided into glass fiber base material and synthetic resin. The roving is made of fiberglass from a FRP fishing vessel and can be woven into either a twist or a chopped fiber felt. Compared with short cut fiber felt, twist - free coarse gauze is not only of high strength but also of high cost. Therefore, it is necessary to combine the purposes of increasing the strength of the material and reducing the use cost. Raw material treatment is mainly glass fiber surface treatment, that is, glass fiber pretreatment. The treatment can improve the state and physical properties of the glass fiber surface. Glass fiber surface treatment methods are generally three kinds: pretreatment, post-processing, and direct blending with resin [6]. The mechanical properties, electrical properties, weather resistance and water resistance of the FRP were improved after the treatment of the surface of the glass fiber (Shown in figure 5).

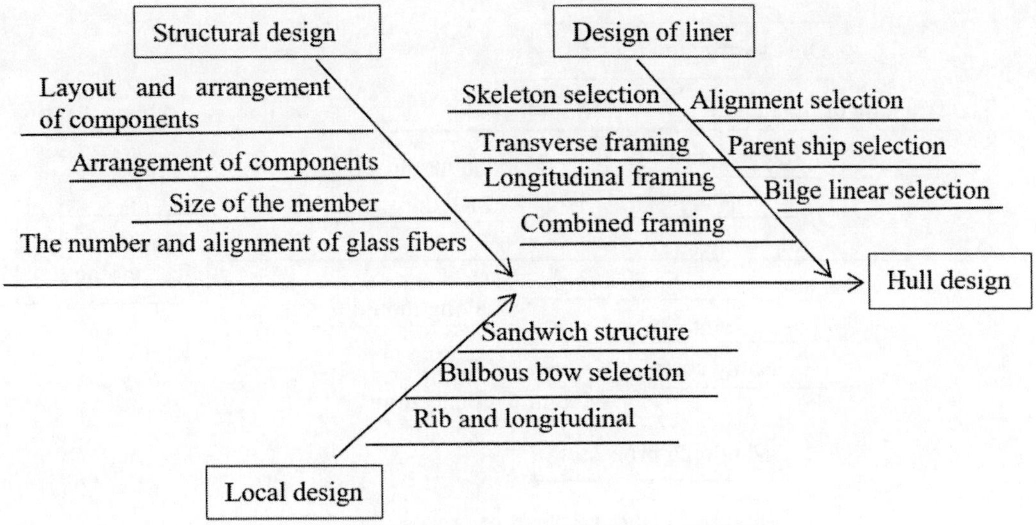

Figure 4. Design factor Fishbone Chart.

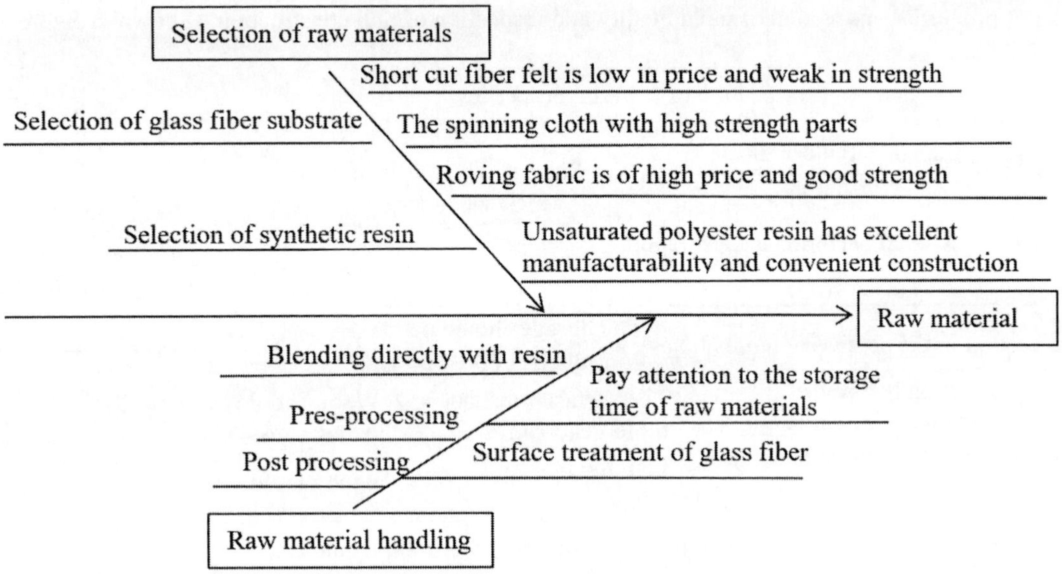

Figure 5. Raw material classification Fishbone Chart.

3.2.4. Construction technology. The construction process of FRP fishing vessel consists of mold construction process and forming process. The mold construction process includes the material and structure of the mold. The construction of the mold affects its quality, ease of operation, and ease of film removal. FRP fishing vessels are relatively large products in FRP products. The appropriate mold form is selected through the construction of quantities, plant conditions, and economic benefits. FRP fishing vessels are mainly made of mold, including wooden mold and glass steel mould and the structure is composed of two parts: integral die and assembled die [7].

The forming process of FRP mold: cleaning, paint remover, spray coating, laying molding, stripping and mounting connection. In order to build a high quality FRP fishing vessel, it is necessary to strictly control every link of the molding process and formulate a strict technological process (Shown in figure 6) [8].

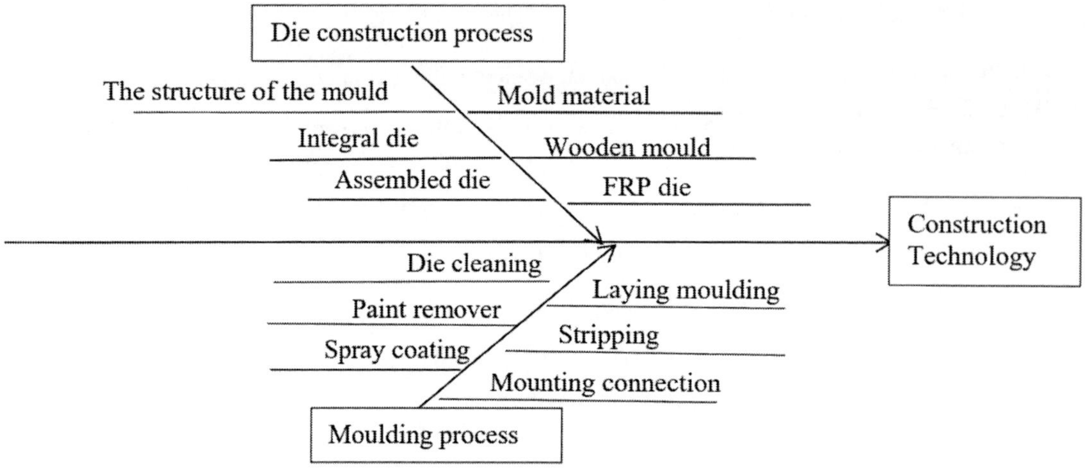

Figure 6. Fishbone Chart of process factors.

3.2.5. Inspection of FRP fishing vessel. Inspection of FRP fishing vessels includes inspection of material properties, inspection of mold quality and inspection of hull construction. (Shown in figure 7).

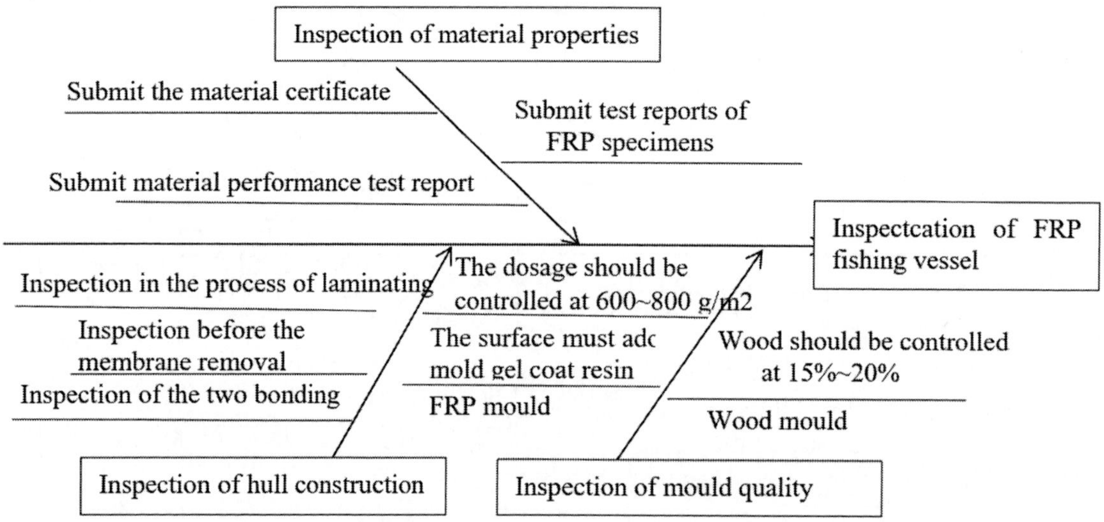

Figure 7. Fishbone Chart of Inspection of FRP fishing vessel.

- Inspection of material properties requires the construction of resins and fiberglass reinforced materials for FRP fishing vessels to be approved by the vessel inspection division. The shipyard shall submit the material certificate and material performance test report to the inspection department. In addition, FRP fishing vessels in the production phase shall submit test reports of FRP specimens to the inspection departments.
- The mould test includes inspection during manufacture and inspection before laminating. The mould is mainly made of wood mould and FRP mould. The moisture content of wood should be controlled at 15%~20%. The FRP mould surface must adopt mould gel coat resin; the dosage should be controlled at 600~800 g/m^2. The surface of the mould shall be smooth and the joint shall be smooth.

- The inspection of hull construction is divided into the inspection in the process of laminating, the inspection before the membrane removal and the inspection of the two bonding. The surface should be smooth, uniform color, gel coat layer do not have wrinkles, cracks, bubbles and other defects.

3.3. Measures to improve Fishbone Chart to determine the influencing factors
Through the improvement of Fishbone Chart, the unsafe factors in the construction process of FRP fishing vessels are analyzed. Through the above analysis, we formulate measures for different factors:

- Train the construction staff to achieve the required technical level. Enterprise qualification has been certified and irregularly checked.
- The skeleton forms of FRP fishing vessels are mostly mixed skeleton structure. It is characterized by vertical and horizontal spacing is basically the same aggregate can ensure the longitudinal strength and transverse strength of fishing vessels [9-10]. The plate structure has a trapezoidal foam structure, which is subjected to greater failure stress [11]. Bilge linear selection of angle than round bilge rolling performance [12]. Choose the angle of FRP bilge not only overcome poor rigidity and do not need to install the vertical fin or increase the breadth.
- In the coating, the permeability of the fabric will affect the beauty and quality of the FRP. Therefore, the first and last layers of paste are made of fiber felt. The synthetic resin adopts unsaturated polyester resin. The advantages of these resins are good curability, easy construction and easy quality assurance [13-14]. In addition, the storage time of unsaturated polyester resin is generally half a year. During construction, the environment and storage time affect the properties and material of the resin, so the storage of the material needs to be noticed.
- The wooden die was the most widely used in the construction of FRP fishing vessels. From the point of view of accuracy and benefit, the FRP die reinforced plastic will be adopted. In combination with the economic benefits and construction quality, most of the fishing vessels are made of wooden dies and the parts with higher accuracy (such as bulb bow) is made of FRP dies.
- The fiber width is not wide, so it is necessary to Ply Splice process. The process is divided into lap joint and butt joint, which reduces the loss of the strength of FRP and guarantees the appearance quality and dimensional accuracy [6]. In the process of paste making of FRP fishing vessels, the form of lap joint is mainly adopted. This form is to stagger the joints in one direction and form a trapezoidal joint. The form of the lap joint increases the strength of the joint area and the retention ratio of the elastic modulus.

4. Analysis of the advantages and disadvantages of Fishbone Chart

4.1. Advantage
Fishbone Chart is a method of finding the root cause of the problem. It has the characteristics of intuition and simplicity [15-16]. On the one hand, Fishbone Charts can give full consideration to the influencing factors, identify symptoms, analyze the causes, and find solutions. On the other hand, by brainstorming, there is unlimited free association and discussion. The relevant risk factors are summarized and the causal relationship is clarified in an orderly and easy to read way. In addition, the method is easy to master, and can be used to guide field safety construction after short-term training.

4.2. Disadvantages
Compared with the fault tree analysis method, the fault tree can be used for quantitative analysis and systematic evaluation of the influencing factors [17-18]. According to the probability of occurrence of each factor, the probability of accident is calculated and a specific quantity concept is provided to achieve the purpose of quantifying the index of influencing factors. Compared with the method of

arrangement, arrangement is the influence degree of the factors affecting the histogram sequence, so as to find out the main factors [19]. According to the principle of "critical minority" and "secondary majority", the analysis of product quality is realized. But the Fishbone Chart is the qualitative analysis of the influencing factors and the probability of each factor cannot be accurately calculated. Fishbone Chart is a tool to break the big concept into some small concepts with motivation, which reflects the influence and logical relation among various factors. It is not convenient to establish the chain of cause and effect between each influence factor.

5. Summary and outlook

The factors that affect the construction quality of FRP fishing vessels come from every aspect. In the construction process, we need to learn from the mature design of developed countries and regions, combine the characteristics of raw materials in China, improve the construction process, reduce construction costs, and facilitate the fishermen to accept. The influencing factors were divided into 3 types: design, material and process. The different types of factors were studied from different angles, and the basic model of Fishbone Chart was established. With the advantages of simple and practical, intuitive and easy to understand, Fishbone Chart model was applied to every link of the construction process.

The structure design, raw material and molding process of FRP fishing vessel interact with each other, but Fishbone Chart is not convenient to establish the chain of three aspects. The basic model of Fishbone Chart should be weighted and arranged according to the time sequence. The logical relationship between the factors is found, and the structure analysis chart is built. Finally, the perfect basic model is built and the relevant database is built. Through further practice, the applicability of this method in the construction of FRP fishing vessels is studied, and the model is more truly reflected in the construction process of FRP fishing vessels.

Reference

[1] Zhao S P, Wang W Y, and Yao L 2013 *J. Chinese Fisheries Economics* **31** 28-47
[2] Zheng J L and Gao C 2016 *J. Chinese Fisheries Economics* **34** 42-47
[3] Chen J M and Wen R S 2009 *J. Shanxi Architecture* **35** 40-41
[4] Wang J L Discussion about "Fishbone Chart" Improvement J. Standard Science JHEP02 (2010)
[5] Wang S P 2011 *J. Safety in Coal Mines* **42** 113-116
[6] National Building Materials Bureau Shanghai FRP Research Institute 1984 *FRP hand lay up molding* (Beijing: China Building Industry Press)
[7] Takashi F, Kasahara K and Kanayama Y 1989 FRP Fishing vessel (Dalian: Dalian University of Technology Press)
[8] Zhang W G 2001 Research Report on construction technology of FRP fishing vessel *J. Fiber Composites* JHEP02 (2001)
[9] Qiu J X 2007 *J. China Water Transport* **5** 8-9
[10] Song X F Wan R and Gao Q L 2003 Study on FRP fishing vessel connection technology *J. Shandong Fisheries* JHEP03 (2003)
[11] Qiu T X and Li X M 2000 *J. Journal of Ocean University of China* **30** 62-67
[12] Qiu T X, Li W T, Song X F and Gao Q L 2000 *J. Journal of Fishery Sciences of China* **7** 87-90
[13] Chinese Society for corrosion and protection 1995 *Synthetic Resin and FRP* (Beijing: Chemical Industry Press)
[14] Zheng Z C, Li Y Z and Tang Y 2008 *J. Sichuan Architecture* **28** 183-186
[15] Wang J Y 2011 *J. Liaoning Chemical Industry* **40** 1270-1271
[16] Deng X W, Chen Q H and Jiang D X 2013 *J. Guangdong Electric Power* **26** 73-77
[17] Hao C X, Xu Y and Gong S W 2012 *J. Journal of Safety Science and Technology* **8** 154-159
[18] Yang H X and Sun Y X 2011 *J. Journal of Safety Science and Technology* **7** 203-206
[19] Liu Z, Nie X T, Wu S S and Gao J Y 2016 *J. Shanxi Architecture* **42** 229-230

IOP Conf. Series: Materials Science and Engineering **292** (2017) 012096 doi:10.1088/1757-899X/292/1/012096

Ultra-thin carbon-fiber paper fabrication and carbon-fiber distribution homogeneity evaluation method

L F Zhang[1], D Y Chen[2], Q Wang[1], H Li[1, 2] and Z G Zhao[1, 2, 3]

[1] State Key Laboratory of Materials-Oriented Chemical Engineering, College of Chemistry and Chemical Engineering, Nanjing Tech University, Nanjing, China
[2] College of Renewable Energy and Advanced Materials, Nanjing Tech University Pujiang Institute, Nanjing, China

[3] E-mail: 35628489@qq.com

Abstract. A preparation technology of ultra-thin Carbon-fiber paper is reported. Carbon fiber distribution homogeneity has a great influence on the properties of ultra-thin Carbon-fiber paper. In this paper, a self-developed homogeneity analysis system is introduced to assist users to evaluate the distribution homogeneity of Carbon fiber among two or more two-value images of carbon-fiber paper. A relative-uniformity factor W/H is introduced. The experimental results show that the smaller the W/H factor, the higher uniformity of the distribution of Carbon fiber is. The new uniformity-evaluation method provides a practical and reliable tool for analyzing homogeneity of materials.

1. Introduction

Carbon-fiber paper has been recognized as a primary material for MEA preparation of fuel cell due to its many attractive characteristics, such as high gas-permeability, high conductivity, high corrosion resistance and low density etc. As an extension material of Carbon fiber, carbon-fiber paper has also been applied for making super capacitor, lithium battery sandwich, electromagnetic shielding materials, aerospace materials, heating materials, functional materials, etc [1-6].

Many properties of carbon-fiber paper are strongly related with the thickness of carbon-fiber paper and plane-distribution uniformity, especially for the applications in fuel cell [7]. The distribution uniformity of Carbon fiber directly or indirectly affects the application of carbon-fiber paper [8]. For ultra-thin carbon-fiber paper, the distribution uniformity will play even more important role.

The preparation methods of carbon-fiber paper are mainly classified as dry-method and wet-method. Air-forming technique is generally used in dry-method preparation of carbon-fiber paper [9, 10]. In this technique, Carbon fiber are randomly overlapped each other to form a carbon-fiber net with uniformly thickness. Due to the electrostatic effect and many other characteristics, Carbon fiber are often mutual entanglement which results in an uneven-distribution. Therefore, the dry-method preparation of carbon-fiber paper is not yet widely applied. Wet-method paper-making is the main method at present to produce high quality carbon-fiber paper. The fabrication process mainly contents short cut Carbon fiber, fiber impregnated resin, dissociation, drying, hot pressing, high temperature carbonization and graphitization [11, 12]. The fabrication process of ultra-thin carbon-fiber paper has not reported yet.

Some research groups have reported their research work on the uniformity of fibers in carbon-fiber paper, but mainly in the initial stage. Hu [13] has reported the influence of dispersant on the dispersion

IOP Conf. Series: Materials Science and Engineering **292** (2017) 012096 doi:10.1088/1757-899X/292/1/012096

of Carbon fiber and paper evenness through analysis on the absorbent, resistance and tensile strength of carbon-fiber paper. The results obtained from the three methods are merely the same.

Su [14] reported that by using computer technology, the information of the vertical and the horizontal arrangement of Carbon fiber in the plane was obtained and analyzed. The influence of fiber length and the distribution on the twine phenomenon and on the pore structure of carbon-fiber paper has been studied. The results show that the large variation of Carbon fiber arrangement between vertical and horizontal direction in the carbon-fiber paper will reduce the uniformity of porosity distribution. Pan [15] reported fiber uniformity evaluation method for nano-fiber non-wovens based on time-frequency transform technology. Song [16] has also reported on uniformity evaluation method based on luminance uniformity of LED display screen. By obtaining the brightness information of CCD image through sensors, LED pixels are analyzed statistically, with the size of the standard deviation, the uniformity of brightness is evaluated.

In this paper, we will report a fabrication process of ultra-thin carbon-fiber paper and introduce a carbon-fiber plane-distribution uniformity evaluation method.

2. Experimental

The ultra-thin carbon-fiber paper fabrication process is presented in figure 1.

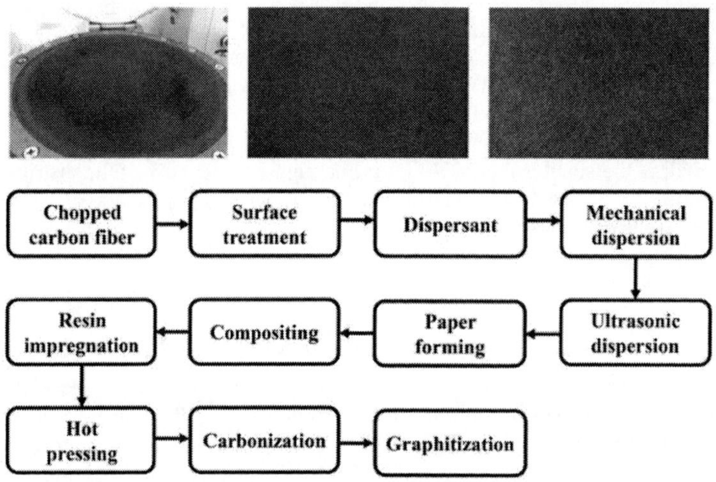

Figure 1. Schematic diagrams for preparation technology of ultra-thin carbon-fiber paper.

Firstly, Carbon fiber are cut to 6 mm long, and then, chopped fibers are put into a muffle furnace for a 90-minutes sintering at about 400℃. After sintering, the Carbon fiber are dispersed by using sodium carboxymethyl cellulose solution. Then, this carbon-fiber solution is put into ultrasonic cleaner for further disperse. After the ultrasonic treatment, the carbon-fiber solution is poured into a pattern forming device to form carbon-fiber paper blank. The carbon-fiber blank is then deposited on the filter and dry. The carbon-fiber paper blank is taken off from the filter by a positive-negative transfer method for further desiccation. The semi-finished carbon-fiber paper is then dipped into phenolic resin for completely impregnation. The naturally dried and impregnated semi-finished carbon-fiber paper will be hot pressed for 20 minutes at about 300°C. The pressure used in this process is 100 kg/cm^2. Finally, a carbonization process is carried out at 1450℃ for 3 hours in a vacuum furnace under high purity nitrogen gas atmosphere.

The gas permeability of the carbon-fiber paper is measured by YG-461E permeability tester and plane resistivity is measured by four-probe resistivity measurement method using WSP-22 four-probe resistivity tester.

The carbon-fiber distribution homogeneity is evaluated by a self-developed homogeneity analysis system based on average-pixel-filling statistics method. This system is based on the analysis of filling/un-filling pixel point distribution for a two value picture. This homogeneity analysis technique is applicable to multi-fibers or similar materials those distributed randomly in a plane. With this system, an object picture is converted into a two-value picture first. The vertical and horizontal pixels of the picture are counted line by line. The ratio of filling to un-filling for each line is assigned to a variable "r". The number of lines those with equal ratio of filling to un-filling are accumulated and assigned to function "f(r)". For each f(r) - r curve, the half-height width "W" over the height "H" is defined as W/H. By comparing the W/H value of each sample, the plane-distribution homogeneity of ultra-thin carbon-fiber paper is evaluated.

3. Results and Discussions

3.1. Ultra-thin carbon-paper fabrications

The ultra-thin carbon-fiber paper before dipped in phenolic resin is shown in figure 2a and 2b. The density of Carbon fiber for this sample is 20 g/m^2 and the plane density of the sample is 0.0022 g/cm^2. The picture is taken in front of the computer screen. One can see that the sample is almost transparency due to its low plane density and ultra-thin structure. The gas permeability of this sample reaches 610 mm/s. A relatively low plane-density is obtained in comparing with commercial carbon-fiber paper such as Toray TPG-H-060. The plane density for TPG-H-060 is 0.0082 g/cm^2 which is about 3 times higher than present result. A little banding force (tensile 2.8 N) is tested which is attributed to the low plane-density.

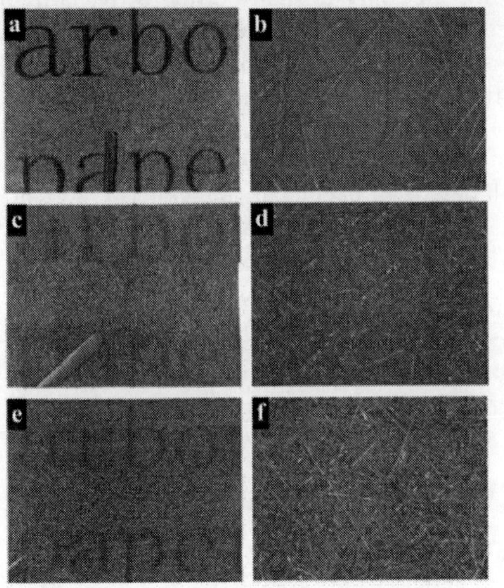

Figure 2. Conventional photos and optical images (40X) Carbon fiber paper at various pressing steps. a and b: before dipped in phenolic resin; c and d: after hot pressed; e and f: after 1450℃ and 3 hours carbonization.

With this semi-finished paper sample, an optical microscope photograph is shown in figure 2b. It shows that Carbon fiber are aligned randomly and no sodium-carboxymethyl-cellulose solution solidified substance and volatile residue exists around the crossing area of Carbon fiber. From figure 2b one can also see that the Carbon fiber are piled up randomly and almost no broken fibers. The region of blurred image indicates that the loose phenomenon of carbon-fiber superposition. This is because that no external pressure applied yet. Figure 2c presents a conventional photo of semi-finished carbon-fiber paper that took after dipped into phenolic resin and hot pressed. The effect of light transmission is obviously weakened due to the phenolic resin existence. Figure 2d shows the morphology of carbon-fiber paper after the hot pressed. Little solidified phenolic resin can be clearly seen in figure 2d. This may reduce the property of the carbon-fiber paper as it may cause a un-uniformity of surface density and the surface resistivity. From figure 2d one can also see that some broken Carbon fiber on the surface. This is directly related to the effect of hot pressing. Figure 2d

IOP Conf. Series: Materials Science and Engineering **292** (2017) 012096 doi:10.1088/1757-899X/292/1/012096

indicates that the pressure we used in hot-pressed procedure is acceptable. In this paper, about 100 kg/cm^2 pressure is applied and the thickness of the semi-finished carbon-fiber paper at this stage is 0.093 mm.

The phenolic resin existence induces a remarkable enhancement of the plan density (from 0.0022 g/cm^2 to 0.0040 g/cm^2) and the tensile (from 2.8 N to 12 N) comparing with the semi-finished carbon-fiber paper that before dipped into phenolic resin. In the meantime, decreasing of the plane resistance (from 160 mΩ•cm to 140 mΩ•cm) as well as the gas permeability from 610 mm/s to 330 mm/s are found. From figure 2d one can also see that due to the increasing of the density caused by hot pressing, the blurred phenomena that can be seen from figure 2b disappeared, the number of visible carbon-fiber increased significantly and the fluffy phenomena decreased. Figure 2e shows the conventional optical photo of carbon-fiber paper after 1450 ℃ carbonization. Comparing figure 2e to figure 2c, the light transmission effect is slightly difference. Relatively sharp characters show in figure 2e proofs that the influence of carbonization on the light transmission of the carbon-fiber paper is positive. The thickness of the carbon-fiber paper is also decreased due to the carbonization process (from 0.093 mm to 0.090 mm), while, the plane density and the plane resistivity decrease remarkably, that means the carbonization process cause the carbon-fiber paper shrink a little. This shrink is mainly related with the mass evaporation: almost 14% resin volatilized during the carbonization process. As results, the gas permeability of the carbon-fiber paper increases from 330 mm/s to 410 mm/s. Further higher gas permeability can be expected after higher temperature graphitization. Figure 2f shows morphology picture after carbonization. From figure 2f one can see that only a small portion of phenolic resin remains between Carbon fiber and porosity increased significantly. Tensile of the sample increases from 12 N (before carbonization) to 19N after carbonization which indicates that the Carbon fiber bond each other even tightly. From figure 2f one can find that the Carbon fiber are mostly remain their original length without broken. An enhanced porosity can be expected after graphitization treatment. The plane resistivity of the sample after the carbonization is equal to 30 mΩ•cm, much lower than one after the hot-pressed (140 mΩ•cm). However, it is still very high in comparison with Toray's TPG-H-060 which is about 5.8 mΩ•cm (graphitized). This is because the relatively low carbonization temperature we used in this paper.

Table 1. Some physical parameters for ultra-thin Carbon fiber paper with different states.

	Before dipped into phenolic resin	After hot pressed	After carbonization	Toray TPG-H-060
Plane density	0.0022 g/cm^2	0.0040 g/cm^2	0.0033 g/cm^2	0.0082 g/cm^2
Thickness	0.120 mm	0.093 mm	0.090 mm	0.185 mm
Plane resistivity	160 mΩ•cm	140 mΩ•cm	30 mΩ•cm	5.8 mΩ•cm
Tensile strength	2.8 N	12 N	19 N	34 N
Carbon fiber content	20 g/m^2	20 g/m^2	20 g/m^2	/
Permeability	610 mm/s	330 mm/s	410 mm/s	500 mm/s

Table 1 presents some parameters of ultra-thin carbon-fiber paper at different preparation states. The plane resistivity will be decreased and the tensile strength will be increased if graphitization finished.

3.2. *Evaluation of Carbon fiber distribution homogeneity*

One of important factors that affect the property of Carbon fiber paper is homogeneity of carbon-fiber distribution. In this paper, we introduce a self-developed homogeneity analysis system for evaluating carbon-fiber distribution uniformity of carbon-fiber paper.

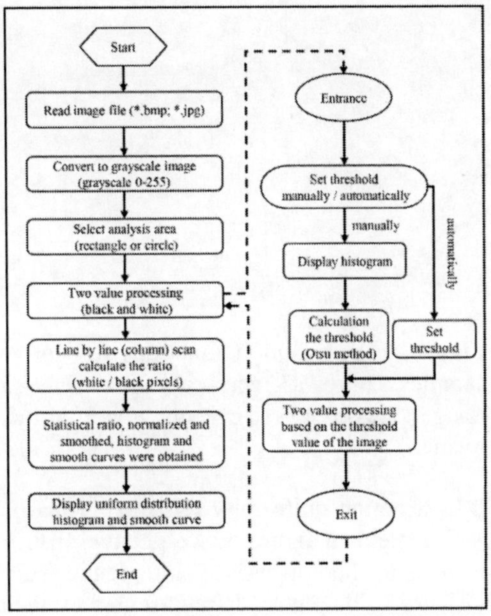

Figure 3. Flow charts of the self-developed homogeneity analysis system.

Figure 3 represents the flow chart of the analysis model used in the self-developed system. To evaluate the Carbon fiber distribution of one or several carbon-fiber paper samples, one can take images for each sample with the same magnification and size. The analysis program works as following: firstly, converting an image of object sample into a two-value image, then, the ratio of black and white pixel on each horizontal and vertical lines of the two-value image is calculated. The lines those with same ratio (r) are accumulated and labeled with f(r). A graph can then be drawn by taking f(r) as y-axis and (r) as x-axis. For a sample that the Carbon fiber are randomly distributed, this program will generate a single smoothly and regularly peak. This method is applicable only to analysis those images with multi fibers or particles and randomly plane distribution. There are some extreme cases, for example, un-randomly distributed situation; figure 4 presents the analysis results.

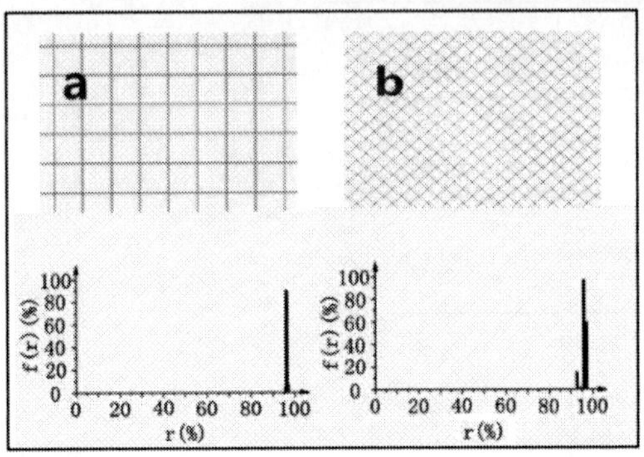

Figure 4. Distribution functions f(r) for two ideal uniformly (un-randomly) distributions of fiber material.

By applying this model, a graph with ideal uniformly (un-randomly) distribution of fiber material should be one or more straight line perpendicular to the horizontal axis. Figure 4a and 4b showed this case. Figure 4 represents the case that for a homogeneous distribution, there is only one or few ratio of filling pixel to un-filling pixel exist throughout the whole picture. The distribution function will be a few vertical lines only.

For an image with a randomly distribution structure, the homogeneity distribution is evaluated by the factor (W/H), while, W is the half height width of fitting curve f(r) and H is the peak height of fitting curve f(r). The better the distribution uniformity, the smaller the W/H value is.

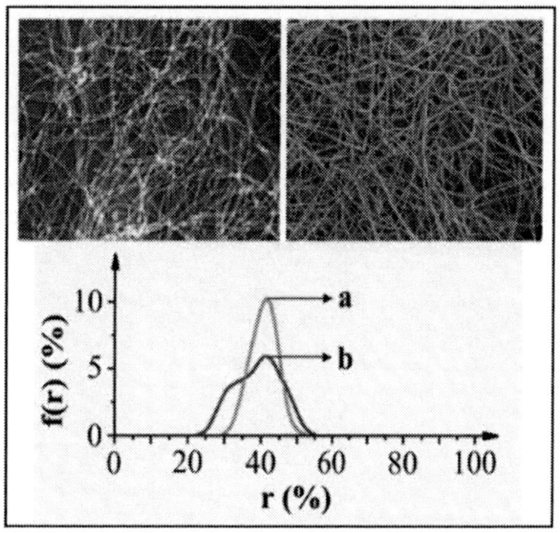

Figure 5. Distribution functions f(r) for two samples: curve "a" represents upper right-side image, curve "b" represents upper left-side image.

In figure 5, we chose two samples with different distribution uniformity remarkably that can be easily distinguished visually. By applying this analysis system to figure 5, we get two distribution curves. In figure 5, the curve "a" (W/H = 0.9) corresponds to the upper-right side picture which has better plane distribution than the upper-left side one. The left side picture is represented by the curve "b" (W/H = 3.1) in the figure 5. One can see that the distribution function f(r) is strongly related with the distribution homogeneous of the sample.

To applying this model to the homogeneous analysis on the ultra-thin carbon-fiber paper presented in this paper, we choose figure 2b, 2d, and 2f as analysis objects and study the relationship between the fiber distribution uniformity and the distribution function (curves) of these three images. The results are shown in figure 6. The curve "a" represents figure 2b, sample before dipped into phenolic resin. The curve "b" represents figure 2d, sample after hot-pressed and the curve "c" represents figure 2f, sample after carbonization; Table 2 shows the W/H values for each sample.

Table 2. W/H values for various samples.

Sample	Curve	W/H
Sample before dipped into phenolic resin	"a"	0.16
Sample after hot-pressed	"b"	0.23
Sample after carbonization	"c"	0.22

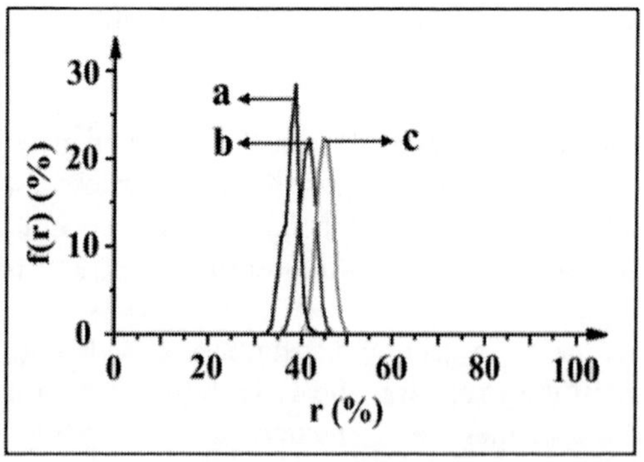

Figure 6. Distribution functions f(r) for three samples of carbon-fiber paper. Analysis is done based on figure 2b represented by curve "a", figure 2d represented by curve "b" and figure 2f represented by curve "c".

In figure 6, curve "a" shows a small different with curve "b" and "c". This is because that there is no resin added at this stage. Foreign substance, in this case is resin, contribute negatively to the analysis result. Curve "b" and "c" are nearly the same indicating that the carbon-fiber distribution ie same at these two steps. This is understandable because that both these two samples are resin added, the hot-press and carbonization treatment hardly can influence the carbon-fiber distribution. Figure 6 implies that this analyst technique is applicable only under certain condition: objects' image should be comparable.

4. Conclusions

Ultra-thin carbon-fiber paper (0.09 mm thickness) has been fabricated. The distribution uniformity of the Carbon fiber play a key role in identifies the quality and properties of ultra-thin carbon-fiber paper. By applying the self-developed distribution homogeneity analysis system introduced in this work, the homogeneity distribution of Carbon fiber in carbon-fiber paper has been analyzed. W/H can be used as a factor to evaluate the uniformity of randomly distributed carbon-fiber paper.

Acknowledgments

This work is supported by the discovery project of State Key Laboratory of Materials-Oriented Chemical Engineering, Nanjing Tech University (38901159); Nanjing Engineering Research Center for Hydrogen-Energy Key Materials (51201311).

References

[1] Ra E J, Raymundo-Piñero E Lee, Y H and Béguin F 2009 *CARBON* **47**(13) 2984-2992
[2] Su Y S, and Manthiram A 2012 *NAT COMMUN* **3** 1166
[3] Yang X P,Rong H M and Shen Z M 2000 *Hi-Tech Fiber & Application* **3** 007
[4] Zou T, Zhao N, Shi C and Li J 2011 *B MATER SCI* **34**(1) 75-79
[5] Lin R H, Luo X P, Shen Z C, Gong D, Ou Z X, Wu Y, Wu J F, Zhu Y M and Wang A S 2004 *Paper Science &Technology* **01** 8-11
[6] Mathur R B, Maheshwari P H, Dhami T L, Sharma R K and Sharma C P 2006 *J POWER SOURCES* **161**(2) 790-798
[7] Hung C H, Chiu C H, Wang S P, Chiang I L and Yang H 2012 *INT J HYDROGEN ENERG* **37**(17) 12805-12812
[8] Amirnasr E, Shim E, Yeom B Y and Pourdeyhimi B 2014 *J TEXT I* **105**(4) 444-453
[9] Xiao Y, Huang Q Z, Chang X, Wang X F and Guo J 2008 *J CENT SOUTH UNIV (Science and Technology)* **6** 017
[10] Yan H P 2000 *Nonwovens* **8**(3) 18–24
[11] Teruaki K,Tadaaki Yand lwaki 1976 U.S. Patent 3 998 689[P]
[12] Zhang X and Shen Z 2002 *FUEL* **81**(17) 2199-2201
[13] Hu Z J, Hu G L and Ye Y 2013 *Paper and paper making* **11** 011
[14] Su F, Gao J, Wen Y and Yang Y 2009 *Materials Review* **14** 012
[15] Pan R R, Liu J H, Gao W D, Wang H B and Sun Y Y 2012 Chinese Patent CN102888747A
[16] Song X L, Zheng X F, Ling L Q and Hao Y R 2009 *Chinese Journal of Liquid Crystals and Displays* **01** 140-144

IOP Conf. Series: Materials Science and Engineering **292** (2017) 012097 doi:10.1088/1757-899X/292/1/012097

Investigation on photoluminescence emission of (reduced) graphene oxide paper

J J Ding[1], H X Chen, D Q Feng and H W Fu

College of Science, Xi'an Shiyou University, Xi'an, Shaanxi 710065, China

[1] E-mail: dingjj303@163.com

Abstract. In order to contrastively investigate optical properties of graphene oxide (GO) and reduced graphene oxide (rGO) paper, GO is prepared by improved Hummer method and controlled reduced using hydration hydrazine to obtain good dispersive rGO in organic solvent. Finally, GO and rGO paper are obtained by vacuum filtration method. Samples morphology and optical properties are analyzed by scanning electron microscopy (SEM) images, Raman spectra, absorbance spectra and photoluminescence (PL) spectra. Results indicate that there are large numbers of localized states in both GO and rGO paper, and optical gaps of two samples are 0.62 eV. In PL spectra of GO paper, we observe three emission peaks at 565, 578 and 608 nm, respectively whose intensity decreases evidently after reduced, which is due to the decrease of oxide functionalized groups and expansion of sp^2 clusters. PL emission will gradually decrease during GO are reduced.

1. Introduction

Graphene has attracted a lot of research interest since its discovery in 2004 [1]. As a semiconductor material with zero band gaps, it has excellent mechanical, thermal, optoelectronic property, superior electrical conductivity and special electron transport property. It possesses many important applications in field effect transistor, photovoltaic cells, liquid crystal display, nanocomposite materials, field emitter, sensors, and save energy materials *etc* [2]. For optical properties, Bao et al. [3] reported that, compared with the traditional semiconductor material, single-layer graphene has lower light absorbance saturation and higher light carrier density. This means that it is very easy to attain saturation under illumination from visible to infrared light wave band, which can be used as saturated absorption layer of fiber laser device to produce ultra-quick laser.

Graphene oxide (GO) is a graphene sheet modified with oxygen functional groups [4], it exhibits interesting steady-state photoluminescence (PL) properties [5-7]. Such as, Luo et al. [8] reported the visible light band from solid GO and discussed its possible band gap mechanism. Eda et al. [9] observed a narrow peak at 390 nm and a similar peak at 440 nm in PL spectra of GO prepared by chemical method. Mei and Zhang [10] synthesized GO sheets with strong PL and proposed a fluorescence on-off mechanism. In addition, PL can also be observed in chemical decoration GO [11].

Generally, optical properties of materials are closely related to their structures and component. Both GO and reduced graphene oxide (rGO) mainly contained carbon functional groups related to sp^2 (graphite) and sp^3 (oxide functional groups). At the same time, the sp^3 carbon regions divide the graphite into sp^2 carbon clusters with different sizes and shapes, and wraps the sp^2 clusters to make them localized. GO has similar structures with amorphous carbon, while the atoms used to form hybrid sp^3functional groups are different. In GO and rGO, each carbon atom in the lattice has a π

orbital that contributes to a delocalized network of electrons. At present, properties of GO and rGO raw materials have been gradually mastered. So we want to comparatively investigate optical properties of GO and rGO paper. GO and rGO paper are synthetic material based on graphite raw materials. They have light weight, good strength and high hardness. In addition, they have better flexibility than steel, and they are also environmentally friendly materials.

In this paper, GO is prepared by improved Hummer method and controlled reduced using hydration hydrazine to obtain good dispersive rGO in organic solvent. Finally, GO and rGO paper are obtained by vacuum filtration method. The morphology and optical properties are analyzed by using scanning electron microscopy (SEM) images, Raman spectra, absorption spectra and PL spectra.

2. Experimental details

Native graphite powder (3 g) is mixed with concentrated H_2SO_4 (12 mL), $K_2S_2O_8$ (2.5 g), and P_2O_5 (2.5 g), and then incubated at 80 °C for 5 h. The product is cooled to room temperature and diluted with deionized water and then set aside for one day. The mixture is filtered using 0.45 μm microspores to remove the remaining acid and dried in vacuum at room temperature to get a preoxidized graphite powder. The preoxidized graphite powder is placed in concentrated H_2SO_4 (120 mL) at 0 °C and stirred quickly. $KMnO_4$ (15 g) is added gradually with stirring while keeping the temperature of the mixture below 20 °C. The mixture is dispersed into 250 ml of water and then stirred at 35 °C for 2 h, followed by the addition of water (700 mL) and 30% H_2O_2 (20 mL), and stir for 2 h. The GO brilliant-yellow mixture is filtered and washed with 10 wt.% HCl aqueous solution (1000 mL) to remove metal ions and wash repeatedly with H_2O to remove the acid until the pH of the filtrate is neutral. The resulting GO slurry is dried in a vacuum oven at 60 ℃. Then, 100 mg of the as-prepared GO powers are dispersed into 200 mL of water under mild ultrasound for 1 h. Then, the GO solvents are centrifuged with 4800 rpm/min for 30 min to obtain even and stabilized GO with transparent yellow-brown suspension. The above solution (20 mL) is mixed with water (20 mL), and then add hydrazine hydrate (17.5 μL) and ammonia (313.6 μL) and stir 10 min. The mixture is transferred into a 100 mL flask equipped with a condenser and stirred for 2 h at 95 °C, and then cooled to room temperature to obtain rGO solution, then filtered repeatedly, and dried in vacuum at 60 °C to obtain rGO. Finally, GO paper and rGO paper are prepared by vacuum suction filtration.

SEM images are characterized by a FEI Quanta 250 scanning electron microscope under 10K magnification. Raman spectra are investigated using a JY LabRAM HR800 laser Raman spectrometer. The absorption spectra are measured by using a Jasco V-570 UV/VIS/NIR spectrophotometer.PL spectra are carried out by a SENS-9000 fluoroSENS fluorescence spectrometer under the 278 nm excitation wavelength. All spectra are measured at room temperature.

3. Results and discussion

Figure 1 shows the photos of GO (left) and rGO paper (right). As prepared GO paper and rGO paper are circle with a diameter of 3 cm. The surface is flat and brittle.

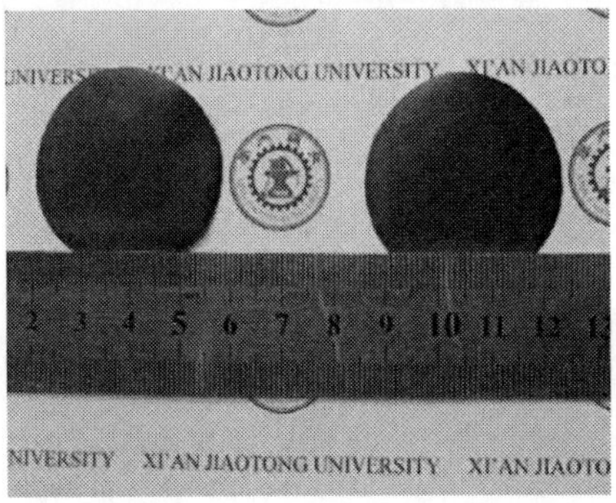

Figure 1. Photos of GO (left) and rGO paper (right).

IOP Conf. Series: Materials Science and Engineering **292** (2017) 012097 doi:10.1088/1757-899X/292/1/012097

Figure 2 presents cross-sectional SEM images of GO (a) and rGO (b) paper. The GO paper exhibits a curly graphene layer. The cross-sectional view of rGO paper clearly shows the superposition of the lamellar structure.

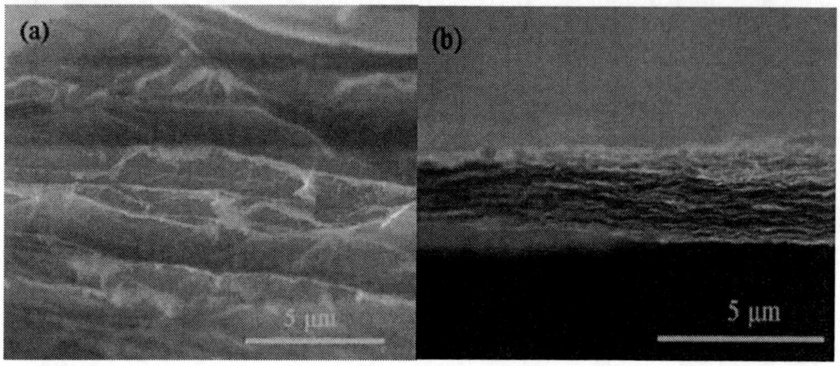

Figure 2. Cross-sectional SEM images of GO (a) and rGO (b) paper.

Raman Spectra is used to widely investigate the bonding structure of various carbon materials. Disordered carbon material exhibit the regular two peaks at 1350 and 1587 cm^{-1} corresponding to the D and G peaks, respectively. Figure 3 presents Raman spectra of GO and rGO paper. Two strong D and G peaks located at 1329 and 1593cm^{-1}, respectively, are observed, which indicates that both GO and rGO paper have high disorder structures. The D peak is due to the presence of structural disorders in graphene sheets. The G peak attributes to optical E_{2g} phonons at the Brillouin zone center, whereas the ratio of the intensity of the G-band to the D-band is related to the in-plane crystallite size [12]. Besides, the intensity ratio of I_G/I_D is widely used to characterize the defect quantity in graphene and a low ratio indicates a great disorder arising from structural defects [13]. As shown in figure 3, the intensity ratio I_G/I_D of rGO (0.79) is larger than that of GO (0.70) paper, indicating that structural defects in GO paper are partially repaired.

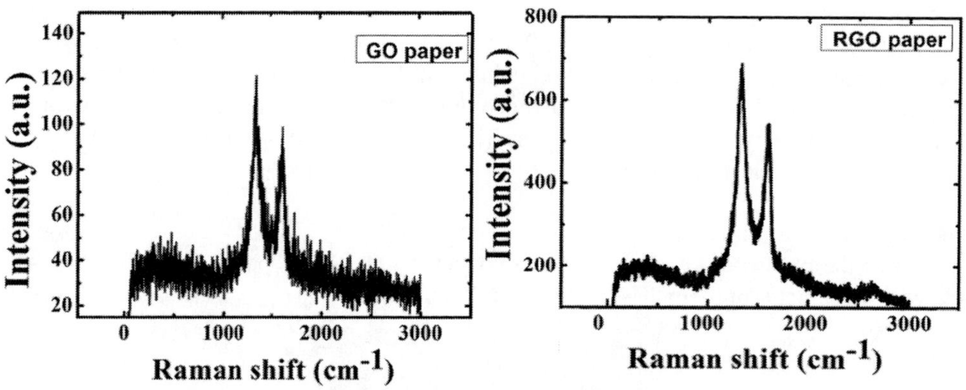

Figure 3. Raman spectra of GO (left) and rGO paper (right).

In order to calculate optical band gap of GO and rGO paper, we use the following formula [14]:

$$\alpha h\nu = A(h\nu - E_g)^n \tag{1}$$

Where, α is absorption coefficient, A is a constant, h is Planck's constant, ν is the photon frequency, E_g is the optical band gap, n is an exponent which have values of 1/2 for direct-band semiconductors. As can be seen from figure 4, both plots are basically same and E_g values are about 0.62 eV.

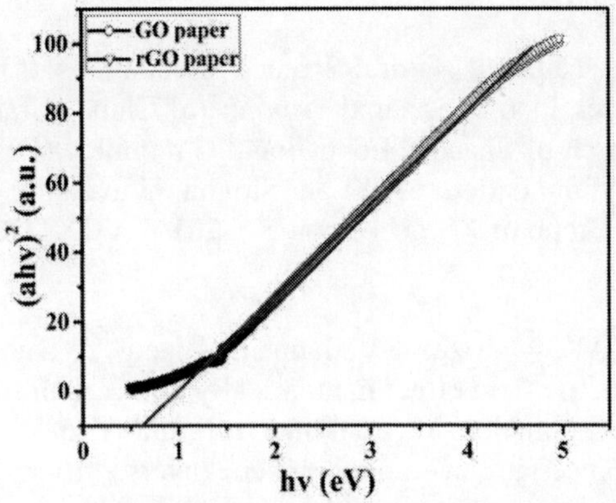

Figure 4. $(ahv)^2$ vs. hv plots of GO and rGO paper.

Figure 5 shows PL spectra of GO and rGO paper. Graphite is semiconductor material with zero band gap and do not own luminescence properties. However, as its size is decreased to nano-scale, band gap will be opened due to quantum effect. In the GO and rGO sheets, oxide functional group and carbon vacancies will cut the graphene to form some nano-carbon cluster, which makes them show semiconductor properties and occur luminescence phenomenon. Three emission peaks located at 565, 578 and 608 nm, respectively, are observed in GO paper and they are attributed to the recombination of electron-hole pairs in local state of sp^2 carbon cluster embedded in sp^3 matrix [15]. Compared with GO paper, PL intensity of rGO decreases rapidly. The peak at 578 nm shows blue shift. As GO is reduced, oxide functional groups decrease and sp^2 carbon cluster expand. GO gradually transform into graphene structures and PL emissions decrease and even disappear.

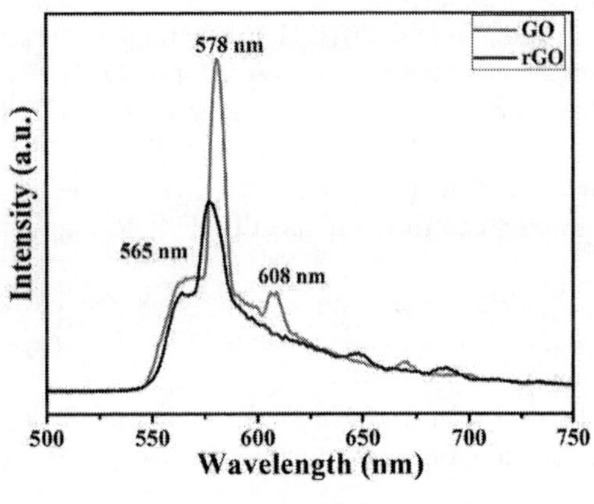

Figure 5. PL spectra of GO and rGO paper.

4. Conclusions

GO and rGO paper are prepared by Hummer method and vacuum suction filtration. Surface micrograph and optical properties are characterized using SEM images, Raman spectra, absorption spectra and PL spectra. Results indicated that rGO shows separated, flake-like graphene layer, and both samples have high disorder. Three emission peaks located at 565, 578 and 608 nm, respectively, are observed in GO paper and they are attributed to the recombination of electron-hole pairs in local state of sp^2 carbon cluster embedded in sp^3 matrix. Compared with GO paper, PL intensity of rGO decreases rapidly. The peak at 578 nm shows blue shift. As GO is reduced to become rGO, oxide functional groups decrease and sp^2 carbon cluster expand. GO gradually transform into graphene structures and PL emissions decrease and even disappear.

Acknowledgment
This work is supported by the National Natural Science Foundations of China (Grant No. 11447116), Natural Science Basic Research Plan in Shaanxi Province of China (Grant No. 2016JQ5037), Special Program for Scientific Research of Shaanxi Educational Committee (Grant No. 16JK1601), Doctoral Scientific Research Startup Foundation of Xi'an Shiyou University (Grant No. 2016BS12) and Creative Scientific Research Group of XSYU (Grant No. 2014KYCXTD02).

References

[1] Novoselov KS, Geim AK, Morozov SV, Jiang D, Zhang Y, Dubonos SV, Grigorieva IV and Firsov AA 2004 Electric field effect in atomically thin carbon films *Science* **306** 666-669
[2] Tombros N, Jozsa C, Popinciuc M, Jonkman HT and Van Wees BJ 2007 Electronic spin transport and spin precession in single graphene layers at room temperature *Nature* **448** 571-574
[3] Bao QL, Zhang H, Wang Y, Ni ZH, Yan YL, Shen ZX, Loh KP and Tang DY 2009 Atomic-layer graphene as a saturable absorber for ultrafast pulsed lasers *Adv. Funct. Mater.* **19** 3077-3083
[4] Cai WW, Piner RD, Stadermann FJ, Park S, Shaibat MA, Ishii Y, Yang DX, Velamakanni A, Sung JA, Stoller M, An J, Chen DM and Ruoff RS 2008 Synthesis and solid-state NMR structural characterization of ^{13}C-labeled graphite oxide *Science* **321** 1815-1817
[5] Gokus T, Nair RR, Bonetti A, Bohmler M, Lombardo A, Novoselov KS, Geim AK, Ferrari AC and Hartschuh A 2009 Making graphene luminescent by oxygen plasma treatment *ACS Nano* **3** 3963-3968
[6] Lu J, Yang JX, Wang IZ, Lim A, Wang S and Loh KP 2009 One-pot synthesis of fluorescent carbon nanoribbons, nanoparticles, and graphene by the exfoliation of graphite in ionic liquids *ACS Nano* **3** 2367-2375.
[7] Pan DY, Zhang JC, Li Z and Wu MD 2010 Hydrothermal route for cutting graphene sheets into blue-luminescent graphene quantum dots *Adv. Mater.* **22** 734-738
[8] Luo ZT,Vora PM,Mele EJ,Johnson ATC and Kikkawa JM 2009 Photoluminescence and band gap modulation in graphene oxide *Appl. Phys. Lett.* **94** 111909
[9] Eda G, Lin YY, Mattevi C, Yamaguchi H, Chen HA, Chen IS, Chen CW and Chhowalla M 2010 Blue photoluminescence from chemically derived graphene oxide *Adv. Mater.* **22** 505-509
[10] Mei QS and Zhang ZP 2012 Photoluminescent graphene oxide ink to print sensors onto microporous membranes for versatile visualization bioassays *Angew. Chem. Int. Ed.* **51** 5602-5606
[11] Mei QS, Zhang K, Guan GJ, Liu BH, Wang SH and Zhang ZP 2010 Highly efficient photoluminescentgrapheneoxide with tunable surface properties *Chem. Commun.* **46** 7319-7321
[12] Subrahmanyam KS, Vivekchand SRC, Govindaraj A and Rao CNR 2008 A study of graphenes prepared by different methods: characterization, properties and solubilization *J. Mater. Chem.* **18** 1517-1523
[13] Pimenta MA, Dresselhaus G, Dresselhaus MS, Cancado LG, Jorio A and Saito R 2007 Studying disorder in graphite-based systems by Raman spectroscopy *Phys. Chem. Chem. Phys.* **9** 1276-1290
[14] Girtan M and Folcher G 2003 Structural and optical properties of indium oxide thin films prepared by an ultrasonic spray CVD process *Surf. Coat. Technol.* **172** 242-250
[15] Zhu YW, Murali S, Cai WW, Li XS, Suk JW, Potts JR, and Ruoff RS 2010 Graphene and graphene oxide: synthesis, properties and applications *Adv. Mater.* **22** 3906-3924

IOP Conf. Series: Materials Science and Engineering **292** (2017) 012098 doi:10.1088/1757-899X/292/1/012098

High-temperature creep properties and life predictions for T91 and T92 steels

J P Pan[1, 5], S H Tu[1], G L Sun[2], X W Zhu[1], L J Tan[3] and B Hu[4]

[1] Jiaxing Special Equipment Inspection Institute, Jiaxing 314051, China
[2] Jiaxing Nanhu District Market Supervision Administration, Jiaxing 314001, China
[3] Shanghai Jiao Tong University, Shanghai 200240, China
[4] China Special Equipment Inspection and Research Institute, Beijing 100029, China

[5] E-mail: jxlpjp@foxmail.com

Abstract. 9-11%Cr heat-resistant steels are widely used in high-temperature and high-pressure boilers of advanced power plants. In the current paper, high-temperature creep behaviors of T91 and T92 steels have been investigated. Creep tests were performed for both steels at varied temperatures. The creep mechanisms of T91 and T92 steels were elucidated by analyzing the creep rupture data of the two steels. In addition, Manson-Haferd model was employed to predict the creep life of T91 and T92 steels, the results of which indicate that the Manson-Haferd model works well for the two steels.

1. Introduction

Construction of advanced power plants requires materials with improved high-temperature strength and superior resistance to high temperature corrosion. 9-11%Cr tempered martensite ferritic steels are imperative high-temperature materials frequently used for critical components of fossil-fired power plants that operate in the temperature range of 500-650 °C [1, 2]. Usually, service lives exceeding 100,000 h are expected for power plant boilers. In order to guarantee safe and economical operations, life prediction is always a major concern for the materials serviced in power plants [3, 4]. In the present work, we took samples from new boilers of a power station that belong to T91 and T92 steels, respectively. The creep mechanisms of the two 9-11%Cr steels were elucidated through analyzing the data from creep tests. Moreover, Manson-Haferd (M-H) model was utilized to predict the life of the steels to be serviced in boilers.

2. Experimental

2.1. Materials

The T91 and T92 steels sampled from newly-installed boilers of a power station were studied. The chemical compositions of the two steels were determined using an Optical Emission Spectrometer (ARL 4460, Thermo Scientific), and the results are listed in table 1.

Table 1. Chemical compositions of T91 and T92 steels (%).

Steel	C	Si	Mn	P	S	Cr	Mo	Nb	V	N	Ni	W	Ti
T91	0.083	0.314	0.403	0.009	0.002	8.59	0.918	0.074	0.202	/	/	/	/
T92	0.09	0.378	0.4	0.012	0.003	8.68	0.374	0.049	0.19	0.048	0.187	1.54	0.003

IOP Conf. Series: Materials Science and Engineering **292** (2017) 012098 doi:10.1088/1757-899X/292/1/012098

2.2. Creep tests
Tensile creep tests in air at 500 °C, 550 °C, 600 °C and 650 °C with preset load were conducted using an electronic high temperature creep & rupture testing machine (GWT2105, MTS) to determine the creep rupture times. Standard cylindrical samples according to DIN50125 B 4×20 were used.

3. Results and discussion

3.1. Creep mechanisms
Compared with heat resistant steels with lower Cr contents, more second-phase particles exist in microstructures of 9-11 %Cr steels [5, 6], which play a pivotal role in providing superior creep properties of these steels. In minimum creep rate versus applied stress plots of second-phase particles strengthened materials, three separable stress regions can be divided according to the stress exponent n values [7]. At low stress levels usually occurring at elevated temperatures, the stress exponent n is found at ~1. At higher stress levels, however, the n values are larger than 1, and the creep is dominated by the glide of dislocations rather than the migration of vacancies. Two possible mechanisms have been proposed for dislocations overcoming the second-phase particles in steels, i.e., climbing and Orowan mechanisms [8]. When dislocations bypass second-phase particles by the climbing mechanism, the n is typically between 3 and 5. When the second-phase particles are bypassed by Orowan mechanism, the region is called a "power-law-breakdown" (PLB) region and the n exceeds 7.

As shown in figure 1, both T91 and T92 steels were increasingly sensitive to the applied stress with decreasing temperature, as reflected by the growing n values. In the data range, the n values of both the steels are larger than 7 or at least close to 7 at the testing temperatures, indicating that Orowan mechanism dominates the second-phase particles by-pass and is thus the dominant creep mechanism. It is also found that the n values of the T92 steels were larger than those of the T91 steels under the same conditions, showing that the creep rate of the T92 steels is more sensitive to the applied stress.

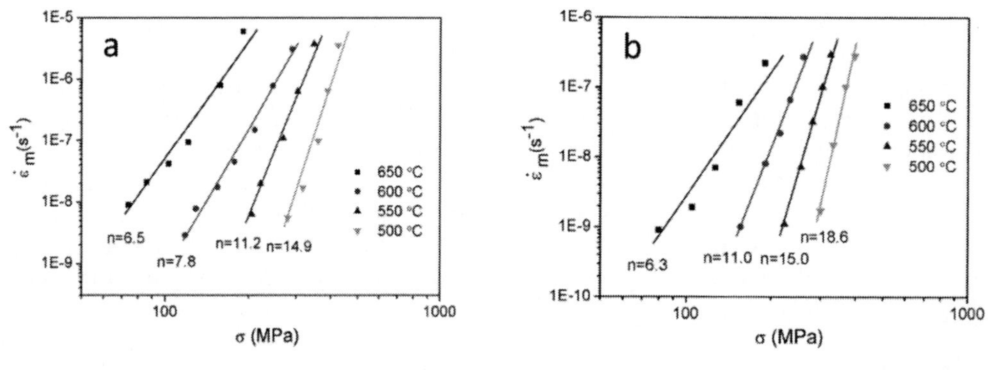

Figure 1. Minimum creep rate versus applied stress for (a) T91 and (b) T92 at different temperatures. The data are linearly fitted and the n values represent the slopes of the fitting lines.

3.2. Manson-Haferd model
As well known, Larson-Miller (L-M) model is popular in life prediction of conventional heat-resistant steels. For the life prediction of 9-11%Cr steels, however, the L-M model is not an ideal choice. The Manson-Haferd (M-H) model is expected to be more effective in predicting the creep rupture life of T91 and T92 steels, which can be described by the following equation:

$$P_{MH}(\sigma) = (lgt_r - lgt_a)/(T - T_a) \tag{1}$$

where σ is the applied stress (MPa); lgt_a and T_a are characteristic parameters of the model. In M-H model, the relationship between lgt_r and T is linear when the σ is fixed. To obtain the optimal values of lgt_r and T is critical for creep rupture life prediction. Here the lgt_r and T were calculated by determining the point whose distance from the isostress lines is at a minimum. The following equation was used for calculation:

$$d = \min\left(\sum_{i=1}^{8}(a_i x + b_i y + c_i)/\sqrt{a_i^2 + b_i^2}\right) \quad (2)$$

where d is the distance of the optimal point from the isostress lines; a_i, b_i and c_i are the coefficients of isostress lines; (x,y) is the coordinate values of the optimal point. We limited the x value in the range of 500-800 and the y value in the range of 10-20. MatLab was used for calculation and fitting, and the obtained optimal parameters are listed in table 2.

Table 2. Manson-Haferd model parameters.

a_0	a_1	a_2	a_3	lgt_a	T_a
0.2518	-0.4992	0.2933	-0.06522	14.02	665.89

3.3. Creep curves

The changes of stress with creep time and the M-H predictions for T91 and T92 steels at different temperatures over a period of more than 10000 h are shown in figure 2. It can be seen that the M-H prediction curves fit the experimental data well for both the steels at the three selected temperatures, indicating that the M-H model has high prediction accuracy for 9-11%Cr heat-resistant steels. Besides, the T92 steel exhibited a little better mechanical performance than the T91 steel during the high-temperature creep tests, as reflected by higher stress under the same testing conditions.

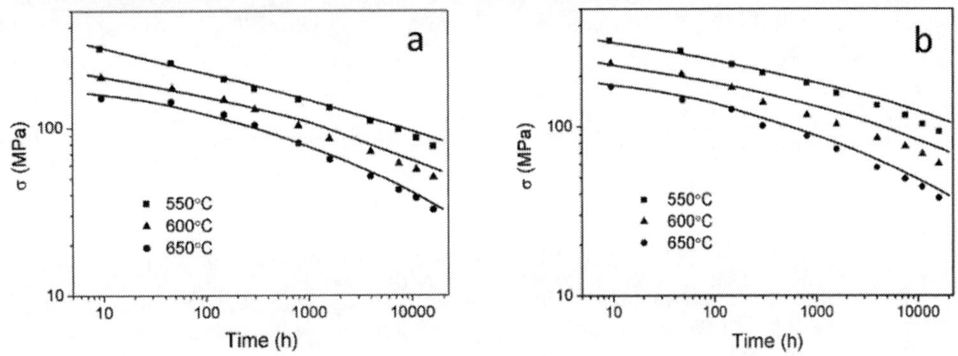

Figure 2. Changes of stress with creep time (scatter) along with M-H predictions (curve) for (a) T91 steel and (b) T92 steel at different temperatures.

4. Conclusions

In conclusion, we have studied the high-temperature creep behaviors of T91 and T92 steels. Creep tests were conducted at selected temperatures, the results of which indicate that Orowan mechanism is the dominant creep mechanism of both two steels at high temperatures. Manson-Haferd model has been utilized to predict the creep life of the steels. The results of calculation and experiments demonstrate that the Manson-Haferd model is suitable for the two steels.

Acknowledgment

This work is financially supported by the Quality and Technology Supervision of Zhejiang Province (Grant No. 20170251) and The National Key Research and Development Program of China (Grant No. 2017YFF0210702).

References

[1] Aghajani A, Somsen Ch and Eggeler G 2009 On the effect of long-term creep on the microstructure of a 12% chromium tempered martensite ferritic steel *Acta Mater.* **57** 5093-

IOP Conf. Series: Materials Science and Engineering **292** (2017) 012098 doi:10.1088/1757-899X/292/1/012098

5106
[2] Auerkari P, Holmstrom S, Veivo J and Salonen J 2007 Creep damage and expected creep life for welded 9-11% Cr steels *Int. J. Pres. Ves. Pip.* **84** 69-74
[3] Vallant J C, Vandenberghe B, Hahn B, Heuser H and Jochum C 2008 T/P23, 24, 911 and 92: New grades for advanced coal-fired power plants-properties and experience *Int. J. Pres. Ves. Pip.* **85** 38-46.
[4] Cerjak H, Hofer P and Schaffernak B 1999 The influence of microstructural aspects on the service behaviour of advanced power plant steels *ISIJ Int.* **39** 874-888
[5] Armaki H G, Chen R, Maruyama K and Igarashi M 2011 Creep Behavior and Degradation of Subgrain Structures Pinned by Nanoscale Precipitates in Strength-Enhanced 5 to 12 Pct Cr Ferritic Steels *Metall. Mater. Trans. A* **42** 3084
[6] Kimura K, Kushima H and Sawada K 2009 Long-term creep deformation property of modified 9Cr-1Mo steel *Mater. Sci. Eng. A* **510-511** 58-63
[7] Yavari P, Langdon T G 1982 An examination of the breakdown in creep by viscous glide in solid solution alloys at high stress levels *Acta Mater.* **30** 2181-2196
[8] Chen Y X, Yan W, Wang W, Shan Y Y and Yang K 2012 Constitutive equations of the minimum creep rate for 9% Cr heat resistant steels *Mater. Sci. Eng. A* **534** 649-653

IOP Conf. Series: Materials Science and Engineering **292** (2017) 012099 doi:10.1088/1757-899X/292/1/012099

Study on the temperature field of large-sized sapphire single crystal furnace

J P Zhai[1, 4], J W Jiang[2], K G Liu[3], X B Peng[3], D L Jian[1] and I L Li[1]

[1] Shenzhen Key Laboratory of Laser Engineering, College of Optoelectronic Engineering, Shenzhen University, Shenzhen 518060, P. R. China
[2] College of Electronic Science and Technology, Shenzhen University, Shenzhen 518060, P. R. China
[3] College of Mechatronics and Control Engineering, Shenzhen University, Shenzhen, 518060, P. R. China

[4] E-mail: jpzhai@szu.edu.cn

Abstract. In this paper, the temperature field of large-sized (120kg, 200kg and 300kg grade) sapphire single crystal furnace was simulated. By keeping the crucible diameter ratio and the insulation system unchanged, the power consumption, axial and radial temperature gradient, solid-liquid surface shape, stress distribution and melt flow were studied. The simulation results showed that with the increase of the single crystal furnace size, the power consumption increased, the temperature field insulation effect became worse, the growth stress value increased and the stress concentration phenomenon occurred. To solve these problems, the middle and bottom insulation system should be enhanced during designing the large-sized sapphire single crystal furnace. The appropriate radial and axial temperature gradient was favorable to reduce the crystal stress and prevent the occurrence of cracking. Expanding the interface between the seed and crystal was propitious to avoid the stress accumulation phenomenon.

1. Introduction

Due to the good optical properties, stable chemical structure and strong mechanical strength of sapphire crystals, it is widely used in satellite space technology, infrared military devices, high-intensity laser window and other military fields [1–4]. Because of its lattice constant close to GaN and the growth process is mature, sapphire single crystal is widely used in SOS microelectronic circuits, LED lamp beads and other substrate industry [5, 6]. Owing to its high hardness, dielectric constant in line with the capacitive screen requirements, sapphire crystal can be used in the field of wearable equipment window [7-9]. Because of its special characteristics of the atomic level transition, titanium-doped sapphire single crystal was used in high-power laser gain media [10, 11]. With the rapid development of these industries, the growth of large-sized and high-quality sapphire single crystal has become particularly urgent [12-16]. Therefore, the temperature field study of the large-size sapphire single crystal furnace has great application value. It has been reported that 300 kg single sapphire crystal is successfully synthesized in Monocrystal Company. However, the research of large-size sapphire single crystal furnace thermal field had not been reported at present. Therefore, it had great application value that studied the large-size sapphire single crystal furnace temperature field.

In this paper, the 120 kg, 200 kg and 300 kg grade single crystal furnace temperature field have been designed and simulated. The power consumption, axial and radial temperature gradient, solid-liquid surface shape, stress distribution and flow of these three temperature fields were compared.

2. Simulation

120 kg, 200 kg and 300 kg grade temperature fields with the same of insulation system structure, material and crucible ratio of height to diameter have been designed. The temperature field structure is shown in figure 1. The middle and bottom insulation system both include the 10 layers of molybdenum screen inside and zircon outside. The temperature field diagram has been introduced into the simulation software CGSim. The power consumption, axial and radial temperature gradient, solid-liquid surface shape, stress distribution and flow of these three temperature fields have been comprehensive studied.

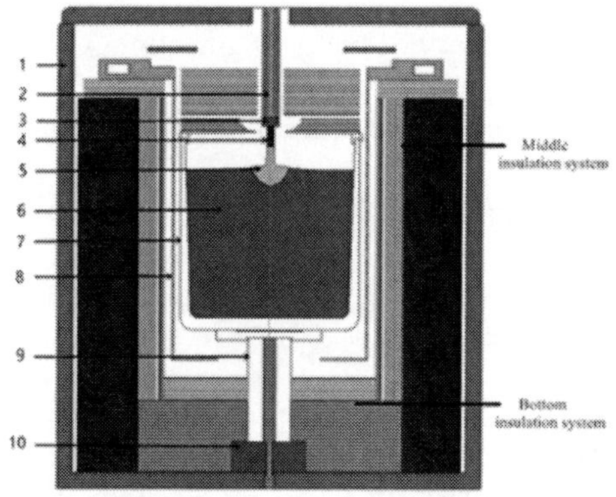

Figure 1. The structure illustration of Single crystal furnace temperature field: 1 furnace shell; 2 seed bar; 3 seed chuck; 4 seed; 5 crystal; 6 melt; 7 crucible; 8 heating; 9 tungsten column; 10 tungsten column base care.

3. Results and discussion

3.1. Temperature field insulation

In order to study the thermal insulation effect of various grades (120kg, 200kg and 300kg) thermal field, 13 points marked by A', B', C', D', E', F', G', H', I', J', K', L' and M' (as shown in figure 2) are taken from the upper, middle, and bottom insulation system. The temperature of these points is measured and drawn as shown in figure 3.

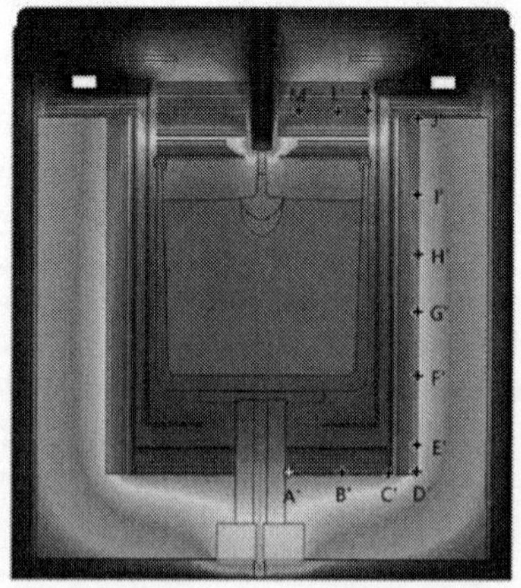

Figure 2. Schematic diagram of temperature field test.

As can be seen from figure 3, in the upper and bottom region of temperature field, the temperatures at the same position (A', B', C', M', L' and K') of the 120kg, 200 kg and 300 kg remain unchanged. However, for the same position (E', F', G', H', I' and J') of the side insulation system, the temperature of the 120 kg single crystal furnace is the lowest, followed by 200 kg and the highest is 300 kg. From the temperature curve, it can be seen that the 120 kg grade had the largest temperature gap difference between with 200 kg grade at the F 'point, reaching 69.9 K, 200 kg grade had the largest temperature gap difference between with 300 kg grade at the D' point, reaching 47.34 K. Thus, as the size of the crucible increases, it can be concluded that the heat dissipation of the single crystal furnace middle insulation system increased, and more heat is taken away by the cooling water inside the furnace shell. The insulation effect of the temperature field is gradually deteriorated and the power loss of the single crystal furnace increased. Table 1 shows the power consumption, unit power consumption and maximum temperature, respectively. Not surprisingly, the 200 kg grade single crystal furnace power consumption increased 26 kw than that of 120 kg, and the 300 kg grade single crystal furnace power consumption increased 47 kw than that of 120 kg grade. These results are consistent with the results of figure 3. With the increase in crystal size, the power consumption is increase, dramatically. However, the unit power consumption decrease with increasing crystal size, which would be favorable to decrease the unit cost of sapphire products. The maximum temperature of 200 kg grade single crystal furnace is close to the 120 kg, but the maximum temperature of 300 kg grade single crystal furnace decrease 8.52 K than that of 120 kg, which would result in a decrease of the temperature gradient in the axial and radial temperature curves

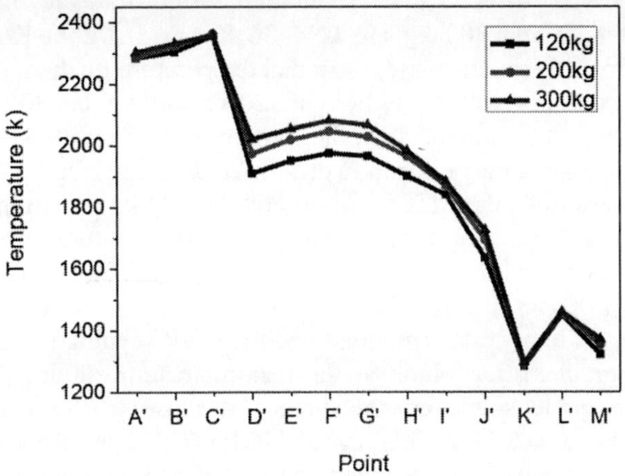

Figure 3. The temperature curve of the point taken.

Table 1. The power consumption, unit power consumption and the maximum temperature comparison of 120kg, 200kg and 300kg single crystal furnace temperature field.

	power consumption (kw)	unit power consumption (kw/kg)	maximum temperature (K)
120kg	71.675	0.5973	2555.1
200kg	97.0344	0.4851	2555.97
300kg	117.948	0.39316	2546.58

3.2. Axial and radial temperature gradients

In order to verify the axial and radial temperature gradient of 120 kg, 200 kg and 300 kg single crystal furnace, the axial and radial temperature curves are plotted. The measured curve position is shown in figure 4. The measurement of the axial temperature curve is from point A to point B and the radial temperature curve is from point C to point D.

IOP Conf. Series: Materials Science and Engineering **292** (2017) 012099 doi:10.1088/1757-899X/292/1/012099

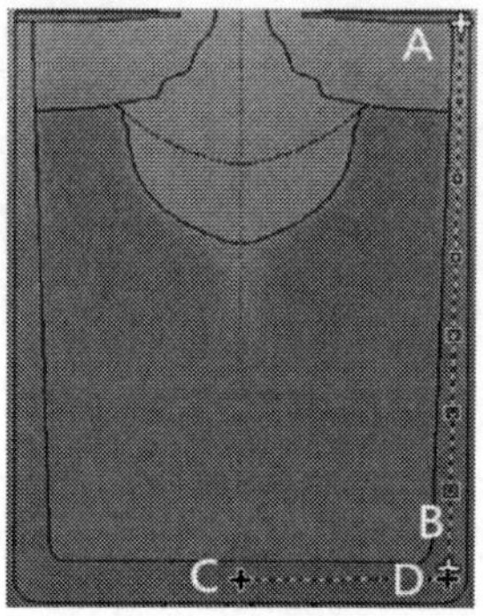

Figure 4. The measurement illustration of axial and radial temperature curve. (Point A to point B for the axial temperature curve measurement; Point C to point D for the radial temperature curve measurement).

Figure 5 shows the radial temperature gradient of 120 kg, 200 kg and 300 kg single crystal furnace. These three temperature curves are similar to each other. The starting point C of these curves is almost coincident. With the increasing of the distance from C to the D, the temperature curve began to appear differences. All the corresponding point temperatures of the 120 kg are the highest, the 200 kg is followed and the 300 kg is minimum. Due to the different crucible size, the ending position of these radial temperature curves is different. The simulation results indicate that the radial temperature gradient of 120 kg, 200 kg and 300 kg are 0.16436 K/mm, 0.14990 K/mm and 0.10389 K/mm, respectively. With the crucible size increase, the radial temperature gradient decrease.

Figure 6 shows the axial temperature gradient of 120 kg, 200 kg and 300 kg single crystal furnace. It can be seen that the three curves coincide at the starting point A. With the increase of distance from point starting A, the temperature curve began to differ. All the corresponding point temperatures of the 120 kg also are the highest, the 200 kg is followed and the 300 kg is minimum. However, due to the different crucible size, the ending position of these axial temperature curves is also different. The simulation results show that axial temperature gradient of 120 kg, 200 kg and 300 kg respectively are 0.45622 K/mm, 0.39511 K/mm and 0.33707 K/mm, respectively. As the size of the single crystal furnace increases, the axial temperature gradient decrease. By keeping in mind the decrease of radial and axial temperature gradient, let's look at the maximum temperature (Table 1). The maximum temperature of 200 kg grade single crystal furnace is close to the 120 kg, but the maximum temperature of 300 kg decrease 8.52 K than that of 120 kg. These result in the decrease of unit power dissipation and the axial temperature gradient, with the size of the sapphire single crystal size increasing.

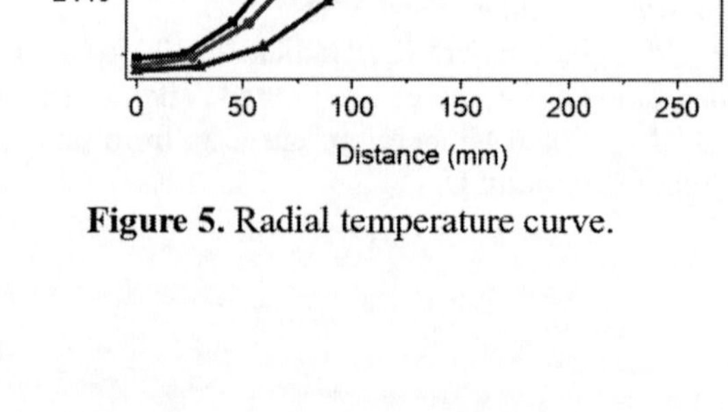

Figure 5. Radial temperature curve.

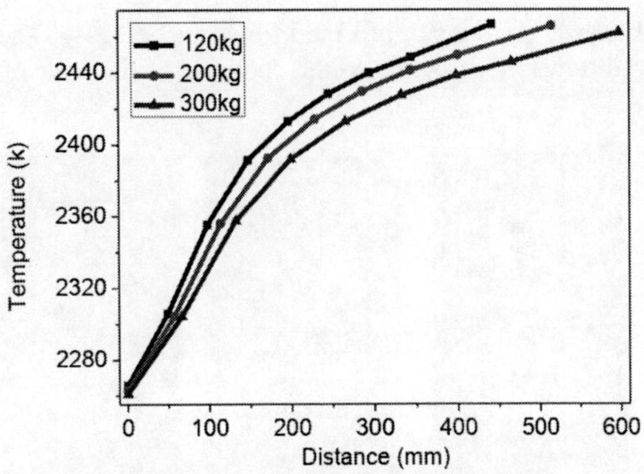

Figure 6. Axial temperature curve.

3.3. Solid-liquid surface shape and stress distribution

The change of radial and axial temperature gradient would cause the growth angle of solid-liquid surface and crystal internal stress to change. It is important to analyse the shape of the solid-liquid surface and its stress distribution. Figure 7 shows that Solid-liquid surface shape and stress distribution of 120 kg, 200 kg and 300 kg. With the increase of crystal size, the growth angle between the solid and liquid surface increase, this inevitably led to the change of the stress distribution inside the crystal and the stress value.

It is well known that the maximum stress distribution of the crystal is at the junction of the seed and crystal and the junction of the crystal and melt. From the crystal local stress amplification diagram (figure 7 right), it can be seen that no stress enrichment phenomenon occurs at the junction of seed and crystal for 120 kg grade temperature field. With the increase of crystal size, the stress enrichment phenomenon appears and becomes serious. Very serious stress enrichment is observed when the crystal size reaches to 300 kg grade. Therefore, the cracking probability of the 300 kg grade single crystal becomes larger during the process of crystal growth. Expanding the interface between the seed and crystal was propitious to smooth the junction of the angle and avoid the stress accumulation phenomenon.

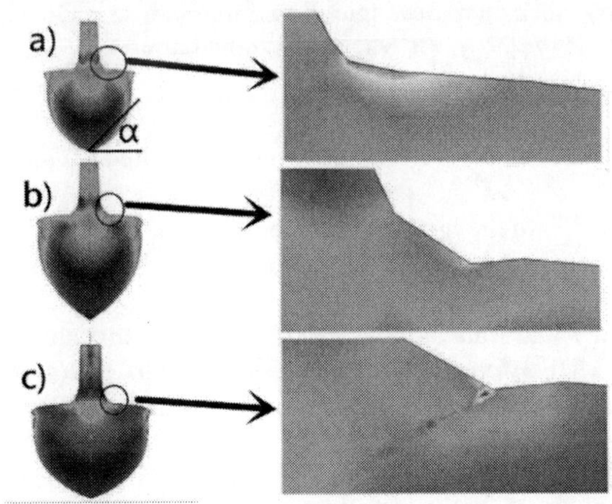

Figure 7. Solid-liquid surface shape and stress distribution: (a) 120 kg; (b) 200 kg and (c) 300 kg.

3.4. The melt flow conditions

Figure 8 shows that the melt flow condition of 120 kg, 200 kg and 300 kg single crystal furnace. The liquid flow of these three temperature fields is similar, only one liquid vortex exists. However, with the increase of the crucible size, the density of liquid flow line decrease. These results indicate that the flow rate gradually slowed down. This is due to the decrease of the melt internal temperature and the expansion of the crucible.

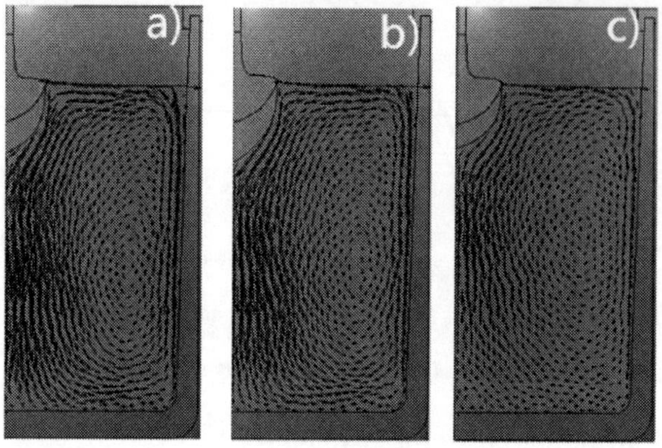

Figure 8. The melt flow conditions: (a) 120 kg; (b) 200 kg and (c) 300 kg.

4. Conclusions

In this paper, the power consumption, axial and radial temperature gradient, solid-liquid surface shape, stress distribution and melt flow of 120 kg, 200 kg and 300 kg grade single-crystal furnace have been studied. The simulation results show that with the increase of the single crystal furnace size, the power consumption increase, the temperature field insulation effect become worse, the growth stress value increase and the stress enrichment phenomenon occurs. Therefore, during the design process of large-size sapphire single crystal furnace, the middle and bottom insulation system, insulation effect should be enhanced. The appropriate radial and axial temperature gradient can effectively reduce the crystal stress and prevent the occurrence of cracking. During seeding process, expanding the interface between the seed and the crystal can slow down the angle between the seed and the crystal and avoid the stress accumulation phenomenon.

Acknowledgment

This work was financed by Shenzhen Strategic Emerging Industry Development Special Fund of Knowledge (JCYJ20150525092941038, JCYJ20150525092940969)

References

[1] Nassau K Dr A V L Verneuil J 1972 The man and the method J. *Journal of Crystal Growth* **13-14** 12-18
[2] Harris D C 2003 A peek into the history of sapphire crystal growth *Proc Spie* **5078** 1-11
[3] Akasaki I 2007 Key inventions in the history of nitride-based blue LED and LD *Journal of Crystal Growth* **300**(1) 2-10
[4] Komaki H, Nakamura T and Katayama R J 2007 Growth of In-rich InGaN films on sapphire via GaN layer by RF-MBE J *Journal of Crystal Growth* s **301-302**(4) 473-477
[5] Foxon C T, Campion R P and Grant V A J 2007 Use of band-gap thermometry to investigate the growth of GaN on sapphire and GaAs J *Journal of Crystal Growth* **301**(4) 482-485
[6] Yamada M, Tao C J 2000 Microscopic observation of strain induced in heteroepitaxial layers with reflection type of infrared polariscope J. *Journal of Crystal Growth* **210**(1) 102-106
[7] Dobrovinskaya E R, Lytvynov L A and Pishchik V M 2009 Sapphire: Material, Manufacturing, Applications M *Springer Publishing Company, Incorporated*,
[8] Schmid F, Khattak C P and Rogers H H J 1999 Current status of very large sapphire crystal growth for optical applications J *Proceedings of SPIE - The International Society for Optical Engineering* 70-76

[9] Song C, Hang Y, Xia C, et al J 2005 Growth of composite sapphire/Ti:sapphire by the hydrothermal method J *Journal of Crystal Growth* **277**(1-4) 200-204
[10] Schmid F and Viechnicki D J P 1972 Apparatus and method for unidirectionally solidifying high temperature material *U.S.* Patent 3,653,432, [P] 1972-4-4
[11] Khattak C P and Schmid F J 2001 Growth of the world's largest sapphire crystals J *Journal of Crystal Growth* **225**(2-4) 572-579
[12] Labelle H E and EFG J 1980 the invention and application to sapphire growth J *Journal of Crystal Growth* **50**(1) 8-17
[13] Akselrod M S and Bruni F J J 2012 Modern trends in crystal growth and new applications of sapphire J *Journal of Crystal Growth* **360** 134-145
[14] Budenkova O N, Vasiliev M G and Yuferev V S J 2004 Simulation of global heat transfer in the Czochralski process for BGO sillenite crystals J *Journal of Crystal Growth* **266**(1-3) 103-108
[15] Bunoiu O M, Nicoara I and Santailler J J 2005 On the void distribution and size in shaped sapphire crystals J *Crystal Research & Technology* **40**(9) 852-859
[16] Nicoara I, Bunoiu O M and Vizman D J 2005 Voids engulfment in shaped sapphire crystals J *Journal of Crystal Growth* **287**(2) 291-295

Experimental exploration of the hydrodynamic effect polishing machinability for different types of material

W Q Peng[1, 3], Y Li[1], Z Wang[2] and S Y Li[2]

[1] College of Basic Education, National University of Defense Technology, Changsha, Hunan, China, 410073
[2] College of Mechatronic and Automation, National University of Defense Technology, Changsha, Hunan, China, 410073

[3] E-mail: plxhaz@126.com

Abstract. Hydrodynamic effect polishing (HEP), in which the material removal relies on the chemisorption between nanoparticles and the workpiece surface in elastic mode, can realize automatic level smooth surface without surface/subsurface damage. The machinability of different types of optical material (such as monocrystalline silicon and crystalline quartz, amorphous silicate glass, Zerodur and so on) were investigated experimentally. The workpiece surfaces before and after being polished by HEP was observed by atomic force microscopy. The experimental results show the surface roughness of monocrystalline silicon and quartz, amorphous silicate glass have decreased from Rms 0.737nm to Rms 0.175nm, Rms 0.490nm to Rms 0.187nm, Rms 0.469nm to Rms 0.157nm respectively, and meanwhile all the defects and bumpy structures have been removed clearly. However the surface roughness has increased from Rms 0.213nm to Rms 0.321nm with the obvious increment of micro unevenness. By comparison, we can conclude that excellent performance is shown when HEP is applied on the optical material structure with a single monocrystalline or amorphous component. However the ultrasmooth surface cannot be obtained when HEP was applied on the combinational materials such as Zerodur glass. The micro unevenness increases gradually along with polishing process due to the different material removal of the monocrystalline and amorphous component.

1. Introduction

The development of modern optical technology has put high requirement on the optical component surface quality. Taking the extreme ultraviolet lithography (EUVL) for example, the mirror surface roughness should be fabricated at atomic level [1-2]. It also requires that there is no defects and crystalline destroy on the processed surface. Although the surface roughness can be restricted at a relative low level by using the ultra-smooth polishing method with the material removal in plastic mode, surface/subsurface damage and residual stress can't be avoided. Therefore the traditional ultrasmooth polishing method can't meet the demand of ultrasmooth surface for modern optical system. How to fabricate the ultra-smooth and defect-free surface has become a hot research point in ultra-precision machining area. Different optical systems need different optical components. While different optical material is composed of different unit. How to obtain an ultrasmooth surface of different optical component is a difficulty problem in the polishing area.

Elastic emission machining, an ultra-smooth polishing method presented by Mori [3, 4], has been considered as the highest precise polishing method so far. The good machinability on semiconductor materials such as silicon and silicon carbide has been validated through experiment research.

Hydrodynamic effect polishing (HEP), a non-contact polishing method, was present based on the theory of float polishing [5-7] and elastic emission machining. The atomic-level material removal rate of HEP is realized by the chemisorption between nanoparticle and workpiece. Defect-free surface with surface roughness at atomic-level can be obtained [8-11]. As the chemisorption relies on the workpiece material construction, which types of material can be polished by HEP is still unknown. Up to now the reports about the machinability for different types of optical materials are very limited. In this paper, the machinability of different types of material was explored experimentally, and the structure characteristic of machinability material by HEP has been summarized.

2. Experiment procedure

For the sake of the exploration becoming universal and pertinent, four typical workpiece made up by different optical materials were selected as experimental sample. The four different types of optical materials were silicon, monocrystalline silicon and crystalline quartz, amorphous silicate glass and Zerodur glass. The silicon and quartz are crystalline material while the silicate glass is amorphous material. Zerodur glass is composite material which is composed by crystalline and amorphous material. Therefore three different types of construction are involved. The optimal process parameters are set for polishing experiment. The polishing slurry is a mixture of deionized water, some dispersant and silicon-oxide nanoparticles with average diameter of 20nm. The uniform raster scan with the scan step size of 10μm was selected as the polishing path. After HEP process, all the samples were rinsed with the deionized water under irradiation with a 1.3M supersonic wave to remove the adsorbed nanoparticle on the surface. At last, the surface before and after polishing was observed by atomic force microscopy (AFM, Bruker's Dimension Icon) with scanning area of 10μm×10μm at a resolution of 512 × 512 pixels. By contrast of surface morphology and surface roughness before and after HEP, the machinability can be concluded.

A monocrystalline silicon sample with the thickness of 3mm was selected as the sample, and the polishing area was 15mm×8mm. As a treatment before HEP, the sample was dipped into hydrofluoric acid (HF) with concentration of 5% for 30min to remove the oxide layer. Then the sample was rinsed by deionized water for 1 minute to remove the residual solution. Then HEP was used for polishing the sample with the removal depth about 140nm.

A circular crystalline quartz (Beijing Brightcrystals Techonlogy INC.) with crystal surface direction of (1 0 0) was selected as the sample. The diameter of sample is 50mm, and the size of 5mm×5mm on the sample surface was choose as the polishing area with removal depth of 200nm for the HEP treatment process.

As the amorphous silicate was mainly consisted of silica, we chose the silicate glass as the polishing sample representing the typical amorphous silicate material. The machinability of silicate glass by HEP was then conducted. The size of 5mm×5mm was chosen as the process area, and the removal depth is about 160nm.

The Zerodur glass was produced by Schottag Corporation in German. The diameter of sample is 50mm, and the size of 5mm×5mm on the sample surface was choose as the polishing area with removal depth of 160nm for the HEP treatment process.

3. Results and discussion

3.1. Monocrystalline Silicon

Figure 1 shows the AFM images of 10μm×10μm of the silicon sample surface before and after HEP process. There are lots of un-uniform plastic pits and bumpy structure on the initial surface, as shown in figure 1(a). After removal depth of 140nm, the initial defects have been removed clearly, and the surface looks very smooth, as shown in figure 1(b). Meanwhile surface rms roughness has decreased from 0.737nm to 0.175nm. By comparing with the section profiles before and after HEP process, the profile fluctuation has been greatly depressed when the surface was polished by HEP. The surface height was almost at the same level with the micro-unevenness within ±0.3nm. By comparison we can see HEP has good smooth ability on the monocrystalline silicon material surface. The silicon material is composed of uniform silicon unit cell. Each removal rate of the unit cell is the same. Experiment results also demonstrate the machinability of silicon by HEP is feasible.

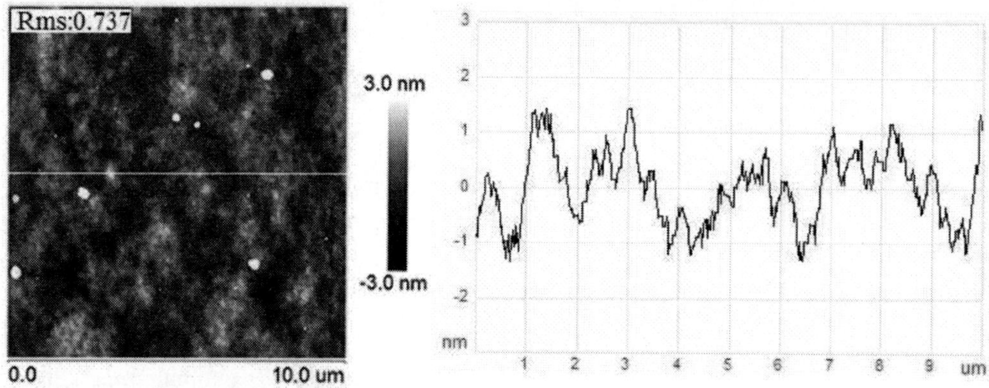

(a) Before polishing process

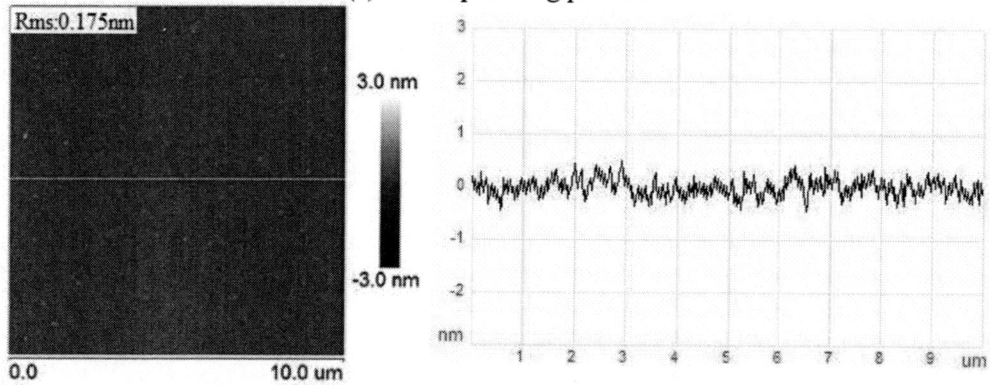

(b) After polishing process

Figure 1. AFM images of silicon surface before and after polishing process.

3.2. *Crystalline quartz*

Figure 2 shows the AFM images of crystalline quartz before and after HEP process. There are many micro plastic scratches on the initial surface, as shown in figure 2(b). The initial surface profile fluctuation was restricted within ±1nm, and the surface rms roughness was 0.490nm. When the sample was polished by HEP, all the plastic scratches and bumpy structures have been removed clearly. The processed surface looks much smooth, and the profile fluctuation has been depressed within ±0.5nm, as shown in figure 2(b). The surface rms roughness has decreased to 0.187nm. It demonstrates that the feasibility of HEP machinability of crystalline quartz is very effective. For crystalline quartz glass is composed of uniform crystalline unit cell, each unit cell has the same removal rate.

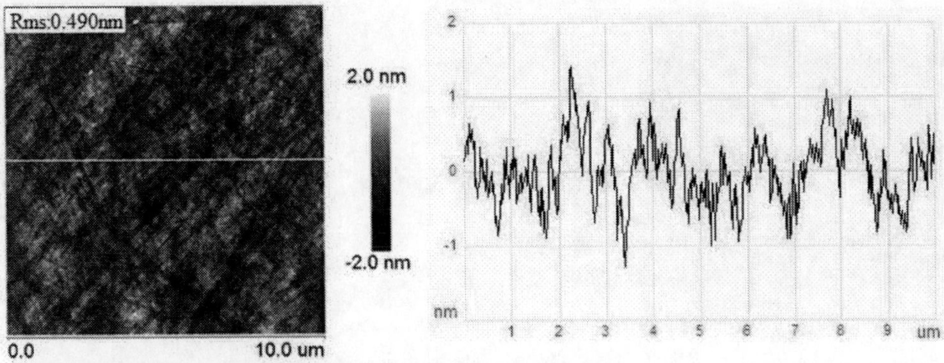

(a) Before polishing process

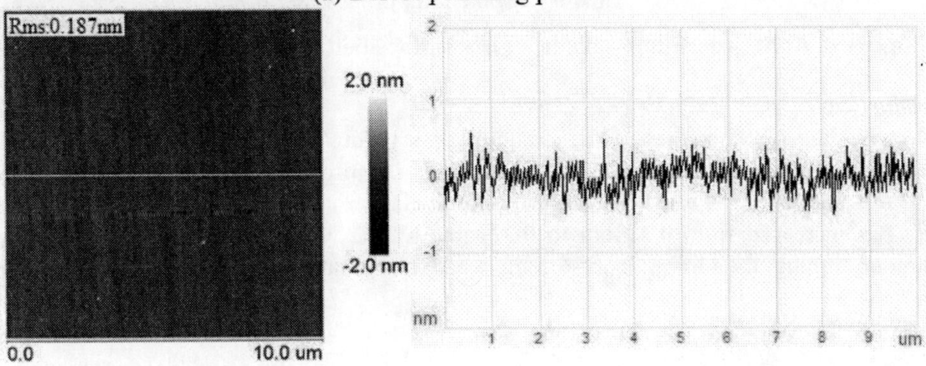

(b) After polishing process

Figure 2. AFM images of quartz surface before and after polishing process.

3.3. *Silicate glass*

The AFM observation of processed surface of silica glass is shown in figure 3. The mechanical damages such as the scratches and cracks on the initial surface have been removed clearly when the surface was processed by HEP. Meanwhile the degree of micro-unevenness has greatly depressed, and the surface looks much smooth with the surface rms roughness decreasing from 0.469nm to 0.157nm. It demonstrates that HEP has great effect on removal of mechanical damages and improvement of the surface quality. As the silicate glass has the similar compositions with the silica glass, we can conclude HEP has good smooth ability on the silicate glass. It also means the machinability of amorphous structure is very well. Silicate glass is amorphous material, and the material of each amorphous phase is the same.

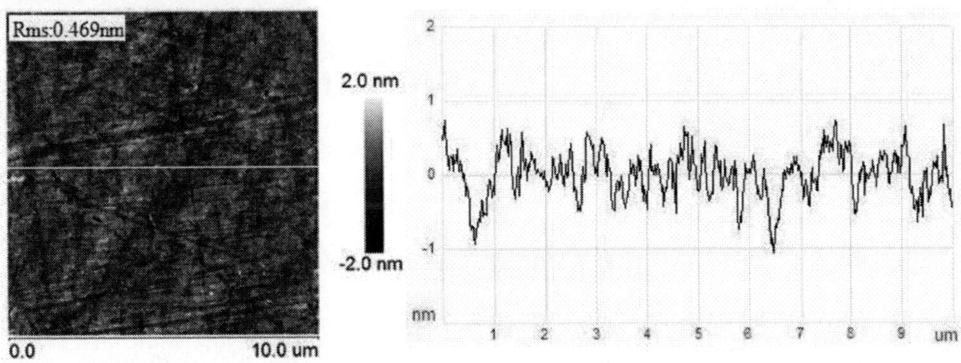

(a) Before polishing process

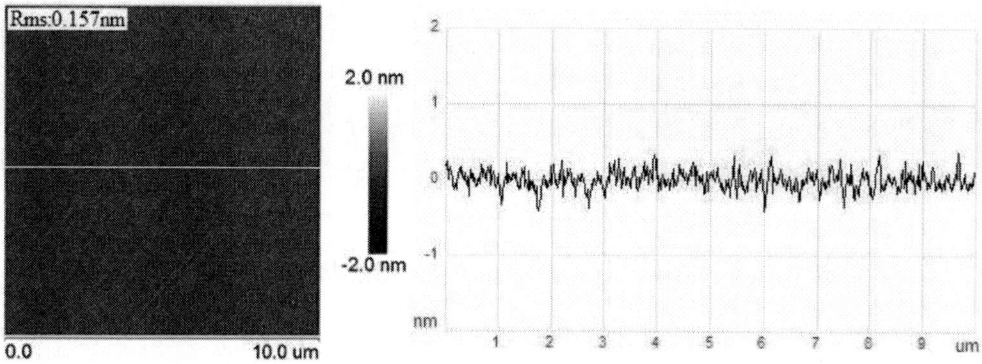

(b) After polishing process

Figure 3. AFM observations of silica glass surface before and after polishing process.

3.4. *Zerodur glass*

Figure 4 shows the AFM images of the Zerodur glass before and after HEP process. Although the scratches on the initial surface have been removed, the profile fluctuation is not depressed but increased at some extent. From the value of the surface roughness, we can see the surface rms roughness has increased from 0.213nm to 0.321nm. As the wheel was manufactured by the single point diamond turning, the turning marks on the wheel has greatly duplicated on the processed surface [6].

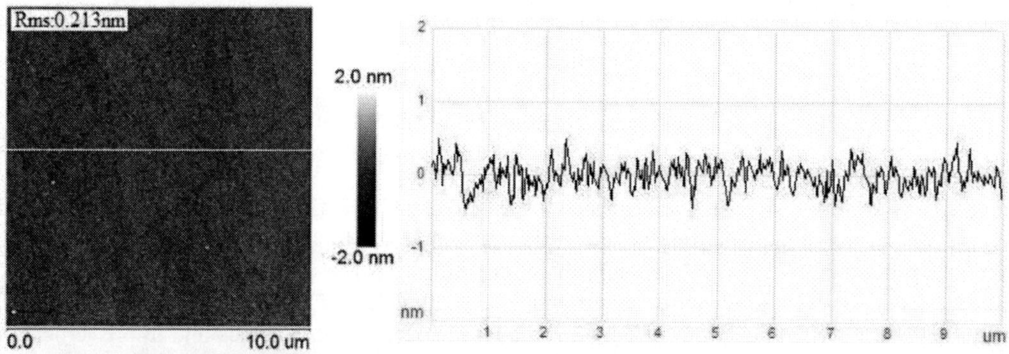

(a) Before polishing process

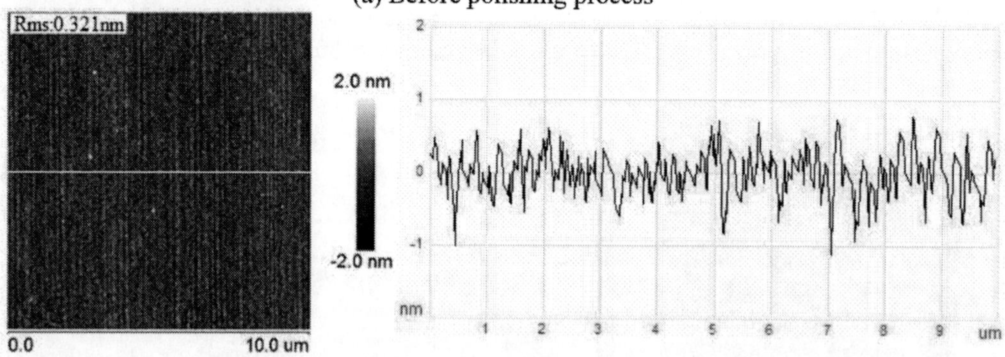

(b) After polishing process

Figure 4. AFM observations of Zerodur glass surface before and after polishing process.

The Zerodur glass is a composite phase structure made up of crystalline and amorphous phase material. The crystalline phase, which is formed by micro crystalline particles, is uniformly distributed in the amorphous structure, as shown in figure 5. Because of the difference of manufacturing process, the size of the crystalline particle is different but normally about 50nm [9]. Crystalline phase is negative thermal expansion material, while the amorphous phase is positive thermal expansion material. By controlling the components of these two materials at a proper ratio, the Zerodur glass is zero thermal expansion. That is why it has good thermal stability and has wide application in modern optical system such as EUVL and spatial telescope mirrors and so on.

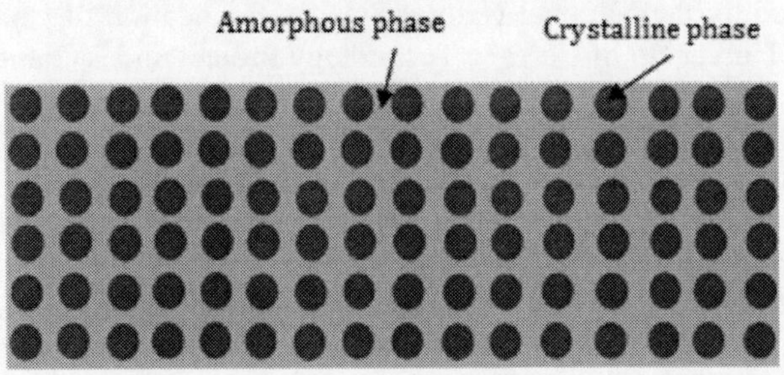

Figure 5. Structure mode of Zerodur glass.

As the different structure of crystalline and amorphous phase, the average surface atom bonding energy is incongruent. So the material removal rate of these two phases is different even if the process is at the same condition. That is mean the material removal rate of one phase is higher than that of the other. Therefore the surface will become coarse and the profile fluctuation will increase as the HEP process progresses. That is why the bumpy structure on processed surface and the surface roughness increased, as show in figure 4. It demonstrates that HEP process is not suitable for polishing the composite phase structure such as Zerodur glass. From the above analysis we can know the machinability of the silicon, quartz and silica is very well. Although they have different structures, they have the same characteristic that is they are all consisted of (crystalline or amorphous phase). Thereby if there is a post-treatment method that can change the thin surface layer on Zerodur glass into a single phase, then ultra-smooth fabrication can be realized by HEP process. One method is electron beam irradiation technology. It is reported that *Mori* [12] has successfully changed the crystalline structure on Zerodur glass into amorphous structure by electron beam irradiation technology, as shown in figure 6(a). Another way is adding technology that is a thin amorphous phase film is coated on the Zerodur glass, as shown in figure 6(b). *Liao ea tl.* [13] has obtained a ultra-smooth surface by ion beam figuring when the Zerodur glass surface was coated a thin silicon or silica film. Therefore, the ultra-smooth surface can be polished when the surface has been changed into a single phase.

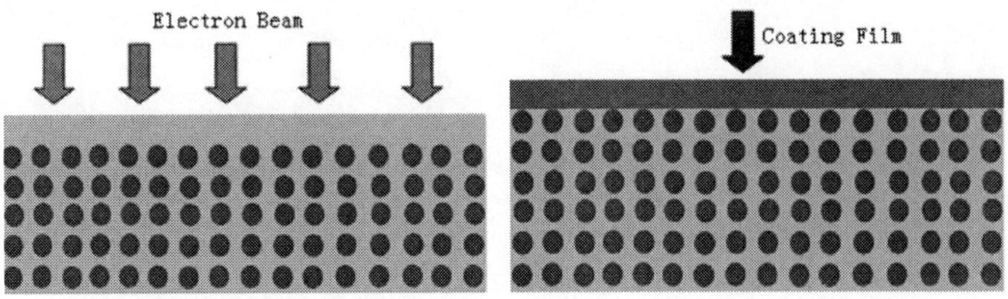

(a) Electron beam irradiation technology (b) Adding technology

Figure 6. Surface post-treatment technology for Zerodur glass.

4. Conclusions

The HEP machinability of different types of optical material was investigated experimentally in this paper. It demonstrated that HEP has good smooth ability on single phase structure such as monocrystalline silicon, crystalline quartz and amorphous silica glass. The surface defects on these materials can be effectively removed and the surface roughness can be reached at atomic level. As the material removal rate of different phase is different, the HEP can't be directly applied on the composite phase structure material such as Zerodur glass. However, ultra-smooth surface can be obtained when the surface was changed into a single phase.

Acknowledgements
This work was supported by the National Natural Science Foundation of China (NSFC) under Grant 61505259 and National University of Defense Technology Project under Grant JC14-03-03.

References
[1] Murakami K, Oshino T and Kondo H 2007 *Proc. of SPIE* **6517** 65170J
[2] Miura T, Murakami K and Suzuki K 2007 *Proc. of SPIE* **6517** 651707
[3] Mori Y, Yamauchi K and Endo K 1987 *Precision engineering* **9(3)** 123-128
[4] Mori Y, Yamauchi K and Endo K 1988 *Precision engineering* **10(1)** 24-28
[5] Namba Y, Tsuwa H 1987 *Annals of CIRP* **36(1)** 211-214
[6] Bennett M, Shaffer J 1987 *Applied Optics* **26(4)** 687-703
[7] Namba Y, Ohnishi N, Yoshida S, Harada K and Yoshida K 2004 *Annals of CIRP* **53(1)** 459-462
[8] Peng W Q, Guan C L and Li S Y 2013 *Applied Optics* **52(25)** 6411-6416
[9] Peng W Q, Guan C L and Li S Y 2014 *Optics Express* **22(11)** 13951-13961
[10] Peng W Q, Guan C L, Li S Y 2014 *Applied Optics* **53(29)** 6913-6919
[11] Peng W Q, Guan C L and Li S Y 2012 *Proceeding of SPIE* **9281** 92810B
[12] Mori 2012 United States **US8235769B2**
[13] W Liao, Y Dai and X Xie 2014 *Optics Express* **22(1)** 377-386

Galvanealed coating evolution for hot forming steel

F Fang[1], X F Du and Y M Chen

Research and Development Center, Wuhan Iron and Steel Co., Wuhan, 430080, China

[1] E-mail: fangfang_82@baosteel.com

Abstract. The aim of the galvanealing experiments with three parameters of dew point, protective atmosphere and galvanealing time was to monitor the changes in coating quality and the evolution of coating after hot forming. The results show that, the three parameters have little effect on coating weight (thickness), and the average iron amount was within 9.51 to 14.9 %. The ζ, δ and Γ phase for the as-received galvanealed samples transformed to Γ-$Fe_4Zn_9$and α-Fe (Zn) after reheating. The existence of a thin film of Γ-Fe_4Zn_9 increases the electrochemical potential to the base steel by a good amount and will not generate LME.

1. Introduction

Corrosion resistance and the avoidance of scale formation during the heating of the material are the main drivers for the development of new coatings for hot stamping [1].Currently there are three industrial relevant coating systems available in producing hot forming steel: Zinc, Aluminum/Silicon and x-tec. The friction coefficient of Al-Si coated steel was generally higher and unstable due to galling of the Al-Si coating on the tool [2]. Moreover, it is not sufficient in corrosive protection for Al-Si coated steel sheets to be applied to wet under body areas, compared with galvanized or galvannealed steel sheets [3, 4].X-tec is a paint system based on nanotechnology which avoids scale formation during heating but offers only a limited corrosion resistance for the formed component. The coating can be used for both process variants of hot stamping [5].

Zinc coated hot forming steels offer the chance to have cathodic corrosion protection as an advantage compared to today's standard applications with uncoated and AlSi coated materials. To utilize all the potential of the coating it is favourable to have at least a little amount of Γ-Fe4Zn9 in the coating. This increases the electrochemical potential to the base steel by a good amount [6]. In which, galvanealing (Zn-Fe coating) is a well knownalloying process whereby zinc from the galvanizecoating and iron from the steel substrate combine in various zinc-iron intermetallic phases.Previous research work [7-9] have mentioned the primary factors appear to be alloyingtime, alloying temperature, steel chemistry and pot aluminum. But the effect of dew point and protective atmosphere on the coating thickness and structure change also needs more attention. It is essential to permitthe alloying process to achieve desired coating thickness with proper alloying phases in avoiding "powdering" or flaking when the finished product is subjected to forming operations.

Despite of the industrial application of hot stamping for almost 20 years, there isstill only one base material that is used for this application, the boron alloyed steel 22MnB5. Therefore, by changing dew point, protective atmosphere and alloying time, the steel of 22MnB5 was used in the paper to study their effect on the thickness and structure of Zn-Fe coating.

2. Experimental procedure

The hot forming steel used in the paper was industrially produced with a regular cold-rolled mill finish to thickness of 1.5mm. The material chemistry was 22MnB5 (0.22C-0.25Si-1.35Mn-0.003B, wt%) with a small amount of Cr and Ti to achieve the desirable mechanical properties.

The galvannealed samples were conducted on Iwatani-Surtec HDPS simulator. The coupons comprised 110mm × 220mm with the long axis parallel to both the rolling and dipping directions, cleaned and dried per standard practice [10]. The samples were assembled to perform the following processes: heating by rate of 5 ℃/s to 800 ℃, with slow cooling rate of 4 ℃/s and fast cooling rate of 30 ℃/s to 475 ℃, dipping in the zinc bath with temperature stabilized at 460 ℃, after cooling by wiping knife, the hot dipped samples were reheated to 500 ℃ and cooling by rate of 10 ℃/s to room temperature. Changing the dew points from -50 to 30 ℃, protective atmosphere contents from 5 to 30 %H_2-N_2 and galvanealing time from 5 to 15 s to study their effects on coating quality. The galvanealed samples will be reheated on 910 ℃ for 5 min, and hot formed after that to analyze the coating evolution.

Surface state (zinc coating weight and thickness), average iron content of samples galvanealed by the simulator were analyzed by Glow Discharge Optical Emissivity Spectroscopy (GDOES) using a Horiba Jobin- Yvon Profiler. Coating quality and reactive wetting were assessed by the counting of bare spots and bare spot area in the uniform 90mm ×90mm coating area of the coated panel [11]. Samples were also taken from this area for SEM (HITACHI-S-4300) examination of the coating/substrate interfacial layer.

3. Results and Discussions

3.1. *Effect of galvanealing parameters on coating weight (thickness)*

It is widely accepted that dew point affect greatly the selective oxidation formed on the steel surface during annealing, the increase of dew point will transfer the external oxidation to inner oxidation which will contribute to galvanizing [12]. However, the results obtained in the experiment displayed that the change of dew point has no significant effect on coating weight, as shown in figure 1a. When dew point changed from dry to moist (-60 to 30 ℃), the coating weight was among 46.2 and 59.2 g/m^2, and coating thickness was among 6.66 and 8.25 μm. Although there is a minute difference of coating weight and thickness, the trend is not regular.

However, when controlling the H_2 content from 5 to 30%, as shown in figure 1b, the coating weight and thickness decreased from 84.15 to 53 g/m^2 and from 12 to 7.63 μm, respectively. Especially, the trend become obvious when H_2 content was above 20%, which enlarge the Hydrogen reduction ability on steel surface to suppress the diffusion of Fe and alloying elements from steel to coat.

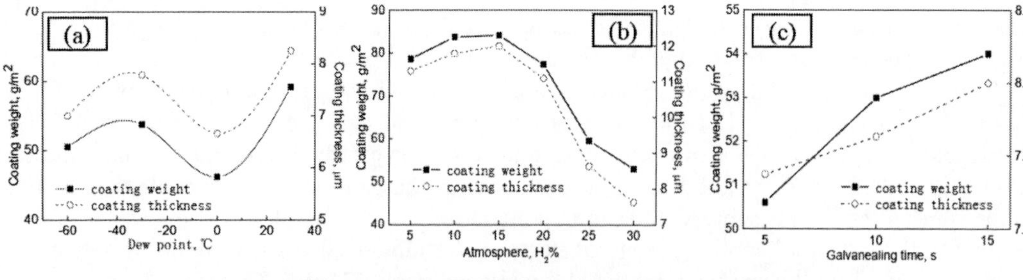

Figure 1. Coating weight and thickness controlled by:
(a) dew point, (b) H_2 content, (c) galvanealing time detected by GDOES.

Generally, the coating weight increase with longer dipping and reheating (galvanealing) time, as shown in figure 1c, the trend is obvious. The specific data showed that, the coating weight and thickness decreased from 50.6 to 54 g/m^2 and from 7.37 to 8 μm, respectively. Although the trend is obvious, the differences were not large enough to specify if there is a relationship of galvanealing time with coating weight, but the phase structure may differentiate with the time, which will be discussed next.

3.2. Effect of galvanealing parameters on iron content

The iron amount has a close relationship with phase structure for galvanealed coating. The average content of iron amount was measured under different galvanealing parameters, as shown in figure 2. When dew point changed from -60 to 30 ℃, the iron amount was within 9.67 and 10.9%; when H_2 content changed from 5 to 30 %, the iron amount was within 9.51 and 10.3%; when galvanealing time changed from 5 to 15 s, the iron amount was within 10.2 and 14.9%. All of the iron contents correspond to the existence of ζ and δ phases in the coat, and the existence of Γ phase need further analysis which will be shown later. The iron contents were stable by changing of dew point and H_2 content, but increase with galvanealing time. So that, if the time increase further, the iron content will reach another range and Γ phase will appear.

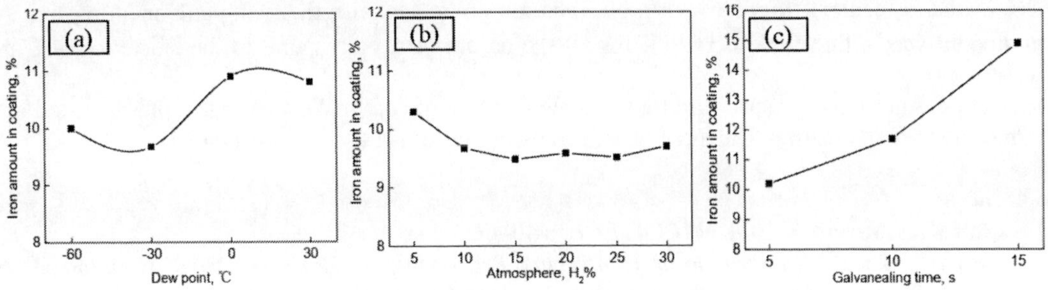

Figure 2. Iron amount controlled by:
(a) dew point, (b) H_2 content, (c) galvanealing time detected by GDOES.

3.3. Structure changes before and after hot forming process

The morphology of zinc coating before and after hot forming was analyzed by SEM shown in figure 3. As the average iron content of coating was shown in figure 2, the detailed energy spectrum were measured to make sure if there exists Γ phase. As shown in figure 3a, Spectrum 1 to 3 displayed iron content as 5.87, 8.48 and 25.89 wt %, which correspond to ζ, δ and Γ phase. Combined figure 2 and figure 3a, the amount of Γ phase was very little, which will not damage the whole coating affiliation with steel matrix.

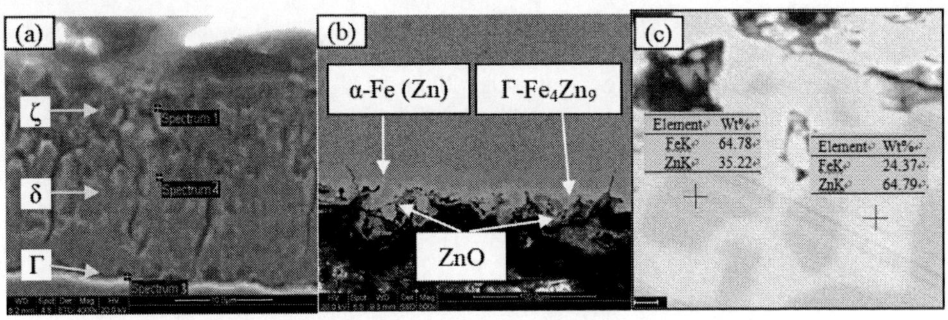

Figure3.SEM Morphology of zinc coating before and after hot forming:
(a) beforeand (b) after hot forming, (c) elements distribution after hot forming.

The coating morphology and elements distribution after hot forming were shown in figure 3b and 3c. The coating generally contained two distinctive phases, i.e. the Zn-rich gamma phase (Γ-Fe_4Zn_9) and Fe-rich alpha phase (α-Fe (Zn)). Surface oxidation occurred during annealing, leading to a surface covered with ZnO (figure 3b). Coatings contained greater than 30 % Zn in α-Fe (Zn) which verified cathodic protection potential through corrosion testing and electrochemical work. The coating become thicker significantlyafter hot forming, caused by the elements in the matrix diffuse into the coating. There is a thin film of Γ-Fe_4Zn_9distributed above α-Fe (Zn) phase, the little amount of Zn in the Γ phase can not generate liquid metal brittlement (LME) in the later welding process. Also, as it was mantioned in the beginning of the paper, the Γ phase increases the electrochemical potential to the base steel by a good amount. However, the Zn coating can easily be oxidized in the reheating procedure, the oxides need to be removed by shot blasting.

4. Conclusions

(1) The coating weight (thickness) measured by GDOES displayed that: the change of dew point has no significant effect; the more H_2 content, the thinner of coating thickness, especially when H_2 was above 20%; when galvanealing time was limited within 5 to 15 s, the differences were not large enough to specify if there is a relationship of galvanealing time with coating weight;

(2) The average contents of iron amount were almost the same when changing dew point from -60 to 30 ℃ and H_2 content from 5 to 30%, but increase obviously with the rising galvanealing time. The iron amount was within 9.51 to 14.9 % for all the parameters, correspond to the existence of ζ and δ phases;

(3) The result of energy spectrum for the different area of coat make sure that Γ phase exited on the interface of coat and matrix. The three phases transformed to Γ-$Fe_4Zn_9$and α-Fe (Zn).

References

[1] Ralf Kand Robert V 2011 *GALVATECH* p 40-47
[2] Sergej M 2015 *5th International Conference Proceeding on Hot Sheet Metal Forming of High-Performance Steel* ed.: Kurt Steinhoff, Mats Oldenburg, Braham Prakash, Toronto, Ont., Canada p 167-175
[3] Berto F 2016 *International Journal of Fatigue* **93(41)** pp168-172
[4] Berto F 2015 *Strength of Materials* **47(5)** pp 719-727
[5] Lenze F-J 2008 *Proceedings of the 1st International Conference in Hot Sheet Metal Formingof High-performance Steel (CHS2)* Kassel Germany pp 15-21
[6] Thomas K 2015 *5th International Conference Proceeding on Hot Sheet Metal Forming of High-Performance Steel* ed.: Kurt Steinhoff, Mats Oldenburg, Braham Prakash, Toronto, Ont. Canada pp 345-353
[7] Zapico D 2014 *Metall. Res. Technol* **111** p 9–15
[8] Chakraborty A 2010 *Surface & Coatings Technology* **204** p 3481–3489
[9] Moon-Hi H 2005 *ISIJ International* **45(6)** pp 896–902
[10] Joseph R 2013 *GALVATECH* pp 127
[11] Liu H 2011 *Surface and CoatingsTechnology* **206** p 1237
[12] Lee S 2011 *Metals and Materials International* **17** p 251

IOP Conf. Series: Materials Science and Engineering **292** (2017) 012102 doi:10.1088/1757-899X/292/1/012102

Structure and Raman scattering of chained carbon films on copper substrate: ab initio approach

E A Buntov[1, 4], A F Zatsepin[1], M B Guseva[2], D A Boqizoda[1] and B L Oksengendler[3]

[1] Institute of Physics and Technology, Ural Federal University, Ekaterinburg, 620002, Russia
[2] Faculty of Physics, Moscow State University, Moscow, 119991, Russia
[3] Institute of Ion-Plasma Technologies, Tashkent, 100125, Uzbekistan

[4] E-mail: e.a.buntov@urfu.ru

Abstract. During the past decades one-dimensional carbon allotropes were attracting much interest. One of such modifications consists of the sp^1 chains, ordered in a hexagonal array with nearly 5 angstroms interchain spacing – linear-chained carbon (LCC). The technological problem of LCC synthesis is the lack of characterization technique confirming its structure along chains. In this work we try to improve the simple structural model of LCC by adding copper substrate and dopants. We compare the experimental Raman spectra with ones modelled by DFT. The calculation results explain the low-frequency features seen in experimental Raman spectra by the substrate-enhanced interaction of carbon chains.

1. Introduction

A plenty of carbon nanomaterials have been studied in the last few decades [1, 2]. They already have a large variety of applications in daily life and industry. In 1985 the discovery of C_{60} molecules made major contribution to medical applications [1]. Another type of carbon material, carbon nanotubes, was proven to give impressive elastic moduli [2]. More recently discovered carbyne [2] is predicted to be a strong material as well [3]. Theoretical study of Young's modulus of carbyne, a parallel carbon chains with kinks, gives a breakthrough at over 1TPa which attracts material scientists [3, 4]. Other unique properties of carbyne and its modifications, such as linear-chained carbon (LCC), provide wide application areas ranging from nanoelectronics to medicine.

Though a pure carbyne crystal of macroscopic size has never been synthesized, the LCC films with 5 angstrom chain-to-chain distance have been produced since 1990s [5]. One of the problems that arise in the synthesis of LCC is the lack of rapid characterization technology, which confirms the structure of the new material. At the same time, the Raman scattering method is able to detect carbon chains and reflects their structural features [6]. The shape of the Raman spectrum of LCC films differs significantly from that of isolated carbon chains [7] due to the Coulomb and Van der Waals interaction of chains in the array. The lines in the spectrum are largely broadened, their interpretation is difficult and cannot be performed within the framework of a simple structural model like one proposed by M.B. Guseva and V.G. Babayev [5]. The reason may be in the model factors not taken into account: the effect of the substrate, impurity atoms, and the degree of structural disorder.

The purpose of this paper is to modify the structural model of a LCC film on a copper substrate (see figure 1) with further comparison of the experimental and calculated Raman spectra.

2. Numerical techniques

The modified structural model used in this paper (2D surface slab, see figure 1) includes three near-surface atomic copper layers and a chain of 12 carbon atoms with a kink. The impurity atoms were placed on the kinks and ends of the chains. The interatomic distances were: C-C 1.275 Å, C-H 1.09 Å, Cu-C 1.86 Å. To calculate the structure, phonon modes and the intensity of the corresponding Raman lines of carbon chains on the crystalline copper substrate (111) and (001) (figure 1), the density functional theory (DFT) method was used with the use of the relativistic LDA pseudopotential HGH [8] and a cutoff energy of 45 Ha in the ABINIT package [9]. SCF convergence was achieved with 10^{-12} Ha tolerance for absolute differences of total energy. Substrate atoms were fixed, positions of carbon and impurity atoms were relaxed by Broyden-Fletcher-Goldfarb-Shanno minimization down to a maximal absolute force tolerance of $5 \cdot 10^{-6}$ Ha/Bohr. We used the dispersion correction method vdw-DF2 (LMKLL) [10]. Integration over the Brillouin zone was carried out using 7 x 7 x 1 Monkhhorst and Pack grids [11].

Figure 1. Schematic diagram of the LCC structural model on a copper substrate used for ab initio calculations.

3. Results and discussion

The results of structural optimization with the minimization of the total energy (figure 2, left) show that, in the absence of impurities, the chain on the copper surface straightens, the kinks disappear. Hydrogen atoms provide a bending angle close to 45 degrees (figure 2, right).

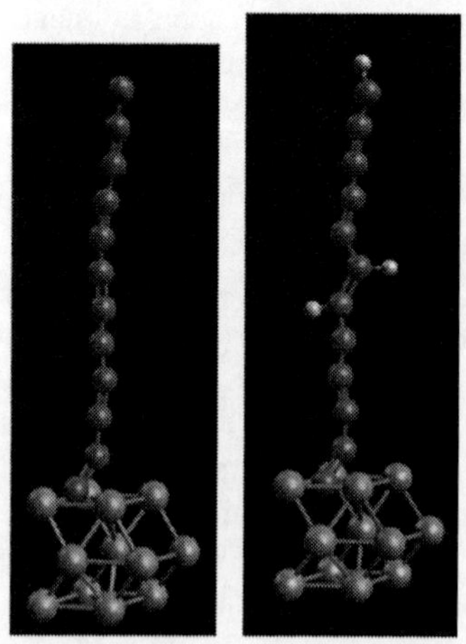

Figure 2. Structural relaxation results for carbon chains on copper substrate.

DFT modelling of the crystal of carbon chains with 6 atoms straight section and 60 degree kink allows to identify the types of normal vibration modes involved into the Raman scattering. The corresponding eigenvectors are shown in figure 3. One can see that the low-frequency modes (375 cm^{-1} and 750 cm^{-1}) correspond to the chain bending. The so-called D and G modes are common for the Raman spectra of different carbon allotropes. In our case the D mode corresponds to the atoms displacements away from chain axis, while G and BLA (bond length alternation) vibrations occur along this axis.

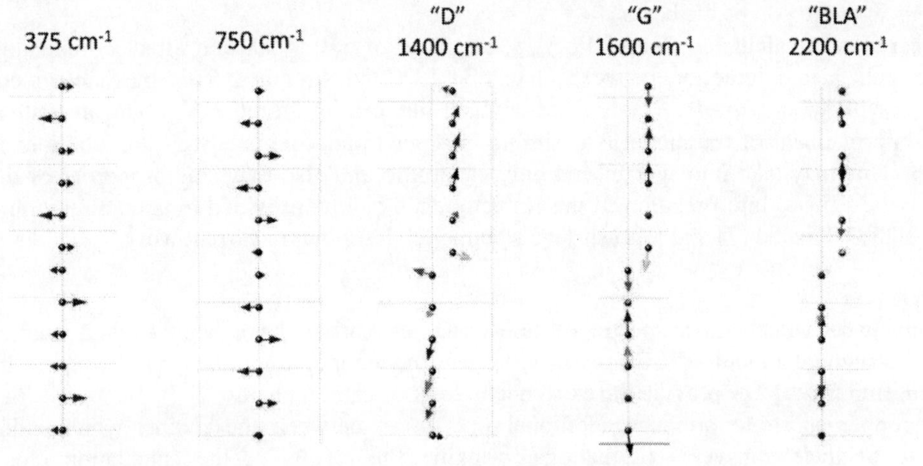

Figure 3. Eigenvectors of the Raman-active vibrational modes of kinked carbon chains. Kinks are not observable for the first two chains due to view angle rotation.

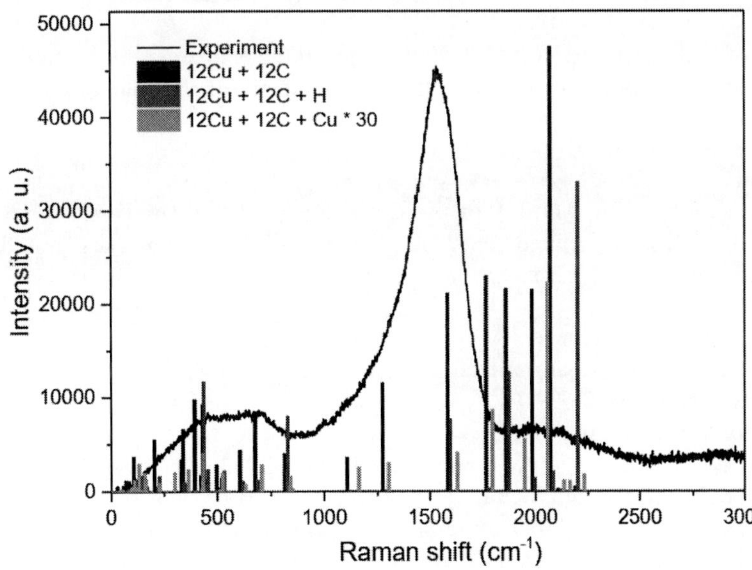

Figure 4. Comparison of the experimental and calculated spectra of carbon chains on the copper surface (111). The Raman intensity for copper-doped chains is multiplied by a factor of 30.

The calculated Raman spectra of LCC structures on a copper substrate are shown in figure 4: pure, hydrogen- and copper-doped chains on Cu(111) plane. For comparison, the experimental spectrum of the LCC film on a polycrystalline copper substrate is presented at the background [7]. Compared with the previous model of linear-chain carbon, they better describe the transverse vibrations of the chains (500-1300 cm^{-1}), and the longitudinal ones are worse.

The copper impurity reduces the intensity of the Raman spectra by a factor of 30. A small number of bends, and therefore sp^2-bonds, in comparison with the bulk LCC crystal, reduces the intensity of the 1500 cm^{-1} peak and unnecessarily increases the maxima in the 2000 cm^{-1} region.

The presence of hydrogen leads to a shift in the lines in the Raman spectrum to the high-frequency region. The admixture of copper sharply reduces the intensity of the Raman scattering by damping the vibrations in the region of the bends.

Appearance of calculated Raman lines in the range of 550 – 900 cm^{-1} may be explained by the substrate-enhanced interaction between chains in LCC/Cu structure. This mechanical coupling is combined with weak Van-der-Waals forces, which induces low-frequency modes at tenths of cm-1. Further improvement of the model is desirable in order to approximate well the whole experimental spectrum. One may take into account not only electronic, but also magnetic properties of the system. However the general interpretation of the Raman bands is now provided by a combination of simple infinite 3D LCC model [7] and a 2D surface slab model described in current work.

4. Conclusion

In this paper, the Raman spectra of linear-chained carbon have been studied numerically. A modified structural model of LCC with hydrogen and copper impurities on a copper substrate is proposed. Impurity atoms provide the existence of kink of carbon chains.

The copper substrate provides additional interaction between the chains, which enhances the amplitude of their transverse oscillations, bringing the results of the calculation closer to the experimental data.

The small number of kinks and, consequently sp^2-bonds, compared with the bulk crystal, reduces the intensity of the peak 1500 cm^{-1} and unnecessarily increases the maxima in the vicinity of 2000cm^{-1}.

The presence of hydrogen leads to a shift in the lines in the Raman spectrum to the high-frequency region. The admixture of copper suddenly reduces the intensity of the Raman scattering by damping the vibrations in the region of the kinks.

Acknowledgments

The research was supported by Act 211 of the Government of the Russian Federation, contract no. 02.A03.21.0006, as well as by government task 3.1485.2017/4.6. Authors are grateful to Prof. V.D. Kochakov from Chuvash State University for the synthesis of LCC films.

References

[1] Bakry R, Vallant R M, Najam-ul-Haq M, Rainer M, Szabo Z, Huck C W and Bonn G K 2007 *Int. J. Nanomedicine.* **2** 639
[2] Yao N, Lordi V 1998 *J. Appl. Phys.* **84** 1939
[3] Liu M, Artyukhov V I, Lee H, Xu F and Yakobson B I 2013 *ACS Nano* **7** 10075
[4] Castelli I E, Salvestrini P and Manini N 2012 *Phys. Rev.* B **85** 214110
[5] Babaev V G, Cuseva M B 1999 Ion-assisted condensation of carbon *Carbyne and Carbynoid Structures* ed R B Heimann, S E Evsyukov, L Kavan. (Dordrecht/Boston/London:Kluwer Academic Publishers) p 159
[6] Ferrari A C and Robertson J 2004 *Phil. Trans. R. Soc. Lond. A* **362** 2477
[7] Buntov E A, Zatsepin A F, Guseva M B and Ponosov Yu S 2017 *Carbon* **117** 271
[8] Hartwigsen C, Goedecker S and Hutter J 1998 *Phys. Rev. B* **58** 3641
[9] Gonze X, Amadon B, Anglade P M et al. 2009 *Computer Physics Communications* **180** 2582
[10] Lee K, Murray E D, Kong L et al. 2010 *Phys. Rev. B* **82** 081101
[11] Monkhorst H J, Pack J D 1976 *Phys. Rev. B* **13** 5188

Enhanced photoluminescence properties of Al doped ZnO films

H X Chen[1], J J Ding

College of Science, Xi'an Shiyou University, Xi'an, Shaanxi 710065, China

[1] E-mail: chxia8154@163.com

Abstract. Al doped ZnO films are fabricated by radio frequency magnetron sputtering. In general, visible emission is related to various defects in ZnO films. However, too much defects will cause light emission quench. So it is still a controversial issue to control appropriate defect concentrations. In this paper, based on our previous results, appropriate Al doping concentration is chosen to introduce more both interstitial Zn and O vacancy defects, which is responsible for main visible emission of ZnO films. A strong emission band located at 405 nm and a long tail peak is observed in the samples. As Al is doped in ZnO films, the intensity of emission peaks increases. Zn interstitial might increase with the increasing Al^{3+} substitute because ZnO was a self-assembled oxide compound. So Zn interstitial defect concentration in Al doped ZnO films will increase greatly, which results in the intensity of emission peaks increases.

1. Introduction

ZnO materials with both wide band gap (~3.37 eV) and large exciton binding energy (~60 meV) is an ideal material for optoelectronic devices [1]. It can also be applied to ultraviolet lasers, solar cells, laser diodes, transparent conductive contacts, thin film transistors and emitters [2]. Optical properties of ZnO have been investigated for a long time. Rao et al. [3] explained the role of defects in photoluminescence (PL) properties of ZnO. Hwang et al. [4] elucidated the surface plasmon resonance effects on enhancing the UV response in p-Si/n-ZnO heterojunction photodetectors. Tabib et al. [5] revealed a high density of defects in ZnO nanocrystals which are mainly oxygen vacancies. Rao et al. [6] integrated different ZnO nanostructures into white light-emitting diodes to improve quality in angular CCT uniformity. Shatnawi et al. [7] observed a strong correlation between the magnetic and optical behavior of the Co-doped ZnO nanoparticles.

In general, visible emission is related to various defects in ZnO films. As above mentioned, different methods are used to introduce defects to improve visible emission of ZnO. However, too much defects will cause light emission quench. So it is still a controversial issue to control appropriate defect concentrations. In this paper, based on our previous results [8], appropriate Al doping concentration is chosen to introduce more both interstitial Zn and O vacancy defects, which be responsible for main visible emission of ZnO films.

2. Experimental details

The samples are deposited on glass substrates by radio frequency and direct current magnetron co-sputtering. High-purity Al and ZnO targets (≥ 99.995%) are used in the experiment. The sputtering chamber is evacuated to 6×10^{-5} Pa before introducing Ar and O_2 mixed gas. Note that the substrate rotation rate during sputtering process is maintained at 10 rpm. ZnO film is carried out in the growth ambient with O_2:Ar=15:25 (sccm) at room temperature (RT) and the sputtering power of 100 W is used during sputtering for 1 h. In addition, Al target is direct current co-sputtered at voltage of 150V and current of 0.15 A. To further improve the crystal quality and photoluminescence properties, the samples are annealed in 2×10^{-4} Pa vacuum at 350℃ for 1 h. Then the heated system is suddenly switched off and cooled down to room temperature.

X-ray diffraction (XRD) patterns of the samples are studied using a D/Max-2400 X-ray diffractometer. Scanning electron microscopy (SEM) images of the samples are characterized by a JEOL JSM-7600F field emission scanning electron microscope. Transmittance and absorption spectra are measured by a Jasco V-570 UV/VIS/NIR spectrophotometer. PL properties of the samples are measured by a FluoroMax-3 fluorescence spectrometer. All spectra are measured at room temperature.

3. Results and discussion

Figure 1 shows XRD patterns of ZnO (a) and Al doped ZnO films (b). A strong ZnO (002) diffraction peak centered at 34.38°, which is smaller than that of standard ZnO powder (2θ=34.42° [9]), is observed in ZnO films. This indicates that compressive stress is introduced in the ZnO film. As Al is doped in ZnO films, ZnO (002) peak shifts to 34.60°, as shown in the inset of figure 1. A tensile stress is formed in Al doped ZnO film. This indicates that Al atoms may incorporate interstitially or replace Zn atomic sites substitutionally in the ZnO hexagonal lattices. Compared with ZnO film, the full width at half maxima (FWHM) of the (002) peak in Al doped ZnO increased, implying that crystal quality is deteriorated after Al incorporation.

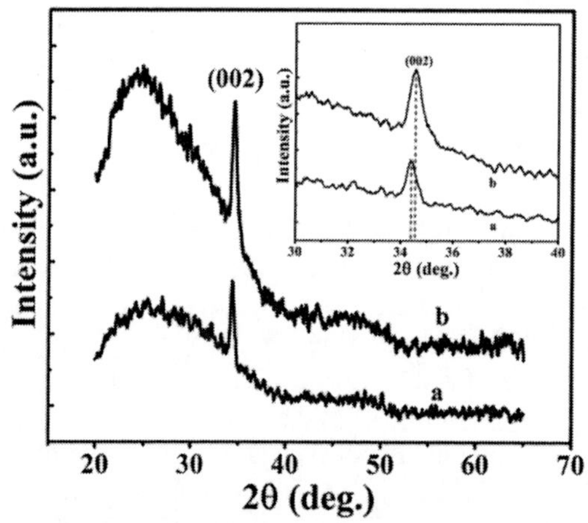

Figure 1. XRD patterns of ZnO (a) and Al doped ZnO films (b). The inset is the corresponding enlarged view.

Figure 2 presents SEM images of ZnO (a) and Al doped ZnO films (b). Dense and irregular ZnO nano particles are observed in ZnO film. As Al is doped in ZnO films, ZnO nano particles become regular and even. This can be ascribed to compressive stress is introduced in the ZnO film, while a tensile stress is formed in Al doped ZnO film.

Figure 3 displays transmission spectra of ZnO (a) and Al doped ZnO films (b). It is clear that a transmittance of over 80% in the visible range of all the samples is obtained. It shows an excellent shoulder peak at 354 nm in ZnO film. As Al is doped in ZnO films, average transmittance increases and shoulder peak disappears.

Figure 4 illustrates $(ahv)^2$ vs hv plots of ZnO (a) and Al doped ZnO films (b). In order to calculate the direct optical band gap of samples, we used the Tauc relationships follows [10]:

$$\alpha h\nu = A(h\nu - E_g)^{\frac{1}{2}} \tag{1}$$

Where α is the absorption coefficient, A is a constant, h is Planck's constant, v is the photon frequency, E_g is the optical band gap. E_g values are 4.00 and 4.05 eV for ZnO and Al doped ZnO films, respectively.

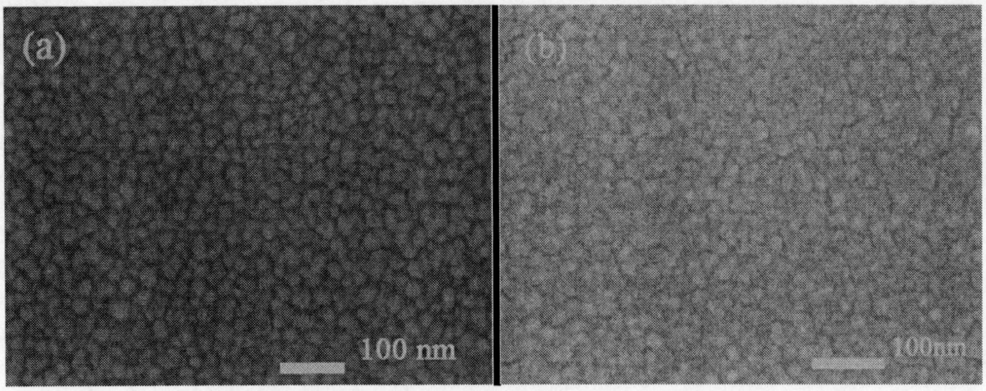

Figure 2. SEM images of ZnO (a) and Al doped ZnO films (b).

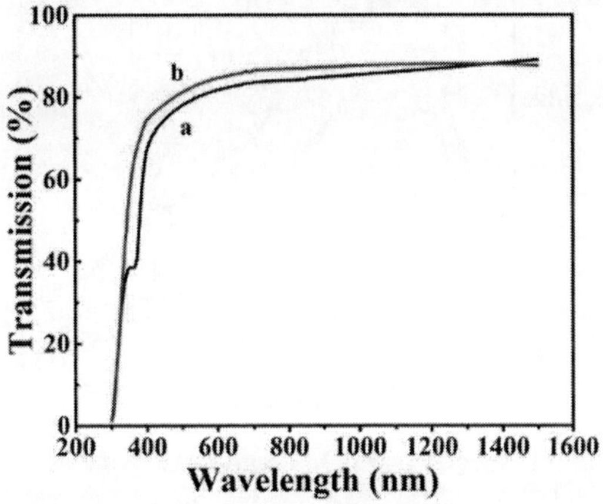

Figure 3. Transmission spectra of ZnO (a) and Al doped ZnO films (b).

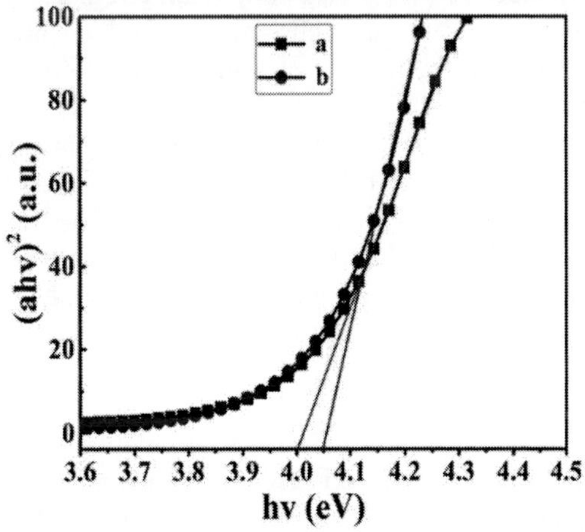

Figure 4. $(ahv)^2$ vs hv plots of ZnO (a) and Al doped ZnO films (b).

Figure 5 depicts PL spectra of ZnO (a) and Al doped ZnO films (b). A strong emission band located at 405 nm and a long tail peak is observed in the samples. As Al is doped in ZnO films, the intensity of emission peaks increases.

The origin of the defect-related PL emission has been investigated for a long time. Several different hypotheses have been proposed, however, they are still under investigation. As Al is incorporated in ZnO, Al^{3+} may replace Zn^{2+} ions sites substitutionally in the ZnO hexagonal lattices because the radii of Al^{3+} (0.053 nm) is less than that of Zn^{2+} ions (0.074 nm). Zn interstitial might increase with the increasing Al^{3+} substitute because ZnO is a self-assembled oxide compound [11]. So Zn interstitial defect concentration in Al doped ZnO films will increase greatly, which results in the intensity of emission peaks increases. Interstitials Zn and oxygen vacancies are main donor defects while Zn vacancies and interstitials oxygen are main acceptor defects in intrinsic ZnO [12]. Kohan et al. [13] have calculated the energy levels of defects in ZnO thin films by the full-potential linear muffin-tin orbit method. The luminescence center observed in this work agrees quite well with Kohan's theoretical calculation. Therefore, the blue emission band is assigned to the electron transition from the Zn interstitial levels to the top of the valence band.

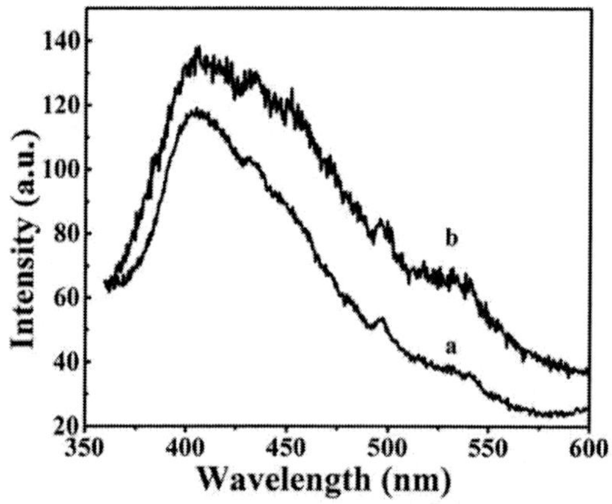

Figure 5. PL spectra of ZnO (a) and Al doped ZnO films (b).

4. Conclusions

Al doped ZnO films are fabricated by radio frequency magnetron sputtering. Based on our previous results, appropriate Al doping concentration is chosen to introduce more both interstitial Zn and O vacancy defects, which is responsible for main visible emission of ZnO films. As Al is doped in ZnO films, crystal quality is deteriorated, average transmittance increases and optical band gap decreases. In addition, the intensity of emission peaks increases. In ZnO film, a strong emission band located at 405 nm and a long tail peak is observed in the samples. As Al is doped in ZnO films, the intensity of emission peaks increases. The enhanced mechanism is discussed. The luminescence center observed in this work agrees quite well with previous theoretical calculation. Therefore, the blue emission band is assigned to the electron transition from the Zn interstitial levels to the top of the valence band. This work may help the development of advanced photoelectric devices based on ZnO materials.

Acknowledgement

This work is supported by the National Natural Science Foundations of China (Grant No. 11447116), Natural Science Basic Research Plan in Shaanxi Province of China (Grant No. 2016JQ5037), Special Program for Scientific Research of Shaanxi Educational Committee (Grant No. 16JK1601), Doctoral Scientific Research Startup Foundation of Xi'an Shiyou University (Grant No. 2016BS12).

References

[1] Godlewski M, Guziewicz E, Kopalko K, Luka G, Lukasiewicz MI, Krajewski T, Witkowski BS

and Gierałtowska S 2011 *Low Temp. Phys.* **37** 235-240
[2] Ding J J, Wang M Q, Zhang X Y, Ran C X, Shao J Y and Ding Y C 2014 *Appl. Phys. Lett.* **105** 233106
[3] Rao T P, Goswami G K and Nanda K K 2014 *J. Appl. Phys.* **115** 213513
[4] Hwang J D, Wang F H, Kung C Y, Lai M J and Chan M C 2014 *J. Appl. Phys.* **115** 173110
[5] Tabib A, Bouslama W, Sieber B, Addad A, Elhouichet H, Férid M and Boukherroub R 2017 *Appl. Surf. Sci.* **396** 1528-1538
[6] Rao L S, Tang Y, Li Z T, Ding X R, Li J S, Yu S D, Yan C M and Lu H G 2017 *Optics Express* **25** A432-A433
[7] Shatnawi M, Alsmadi A M, Bsoul I, Salameh B, Alnawashi G A, Al-Dweri F and El Akkad F 2016 *J. Alloys Compd.* **655** 244-252
[8] Chen H X, Ding J J, Wang X M, Wang X J, Chen G X and Ma L 2016 *Optical Materials* **62** 505-511
[9] Chen Y, Bagnall D M, Koh H J, Park K, Hiraga K, Zhu Z and Yao T 1998 *J. Appl. Phys.* **84** 3912-3918
[10] Girtan M and Folcher G 2003 *Surf. Coat. Technol.* **172** 242-250
[11] Ding J J, Ma S Y, Chen H X, Shi X F, Zhou T T and Mao L M 2009 *Physica B* **404** 2439-2443
[12] Zhang S B, Wei S H and Zunger A 2001 *Phys. Rev. B* **63** 075205
[13] Kohan A F, Ceder G and Morgan D 2000 *Phys. Rev. D* **61** 15019

IOP Conf. Series: Materials Science and Engineering **292** (2017) 012104 doi:10.1088/1757-899X/292/1/012104

Design of transmissive lens with nearly 100% efficiency by using Huygens metasurface

B C Lin[1], G M Wang, T Cai, H S Hou, W L Guo, P Xie, S Y Shuai, Z H Liao, K Li, L Zhu and X J Gao

Air and Missile Defense College, Air Force Engineering University, Xi'an 710051, China

[1] E-mail: 18740495294@163.com

Abstract. We proposed a high-efficiency meta-surface with a parabolic phase distribution that can focus a plane wave to a point image in transmission geometry. For realizing the predicted effects, we use the meta-atoms with 100% transmission to make it. Then we operate the simulation in the CST Microwave Studio to perform the focusing characteristic. The results show that the maximum efficiency is up to 90.6% at 10GHz and the half-power bandwidth is 31.8%, which is better than other previous works.

1. Introduction

Until now, Meta-surfaces (MS) have been a research spot for their electromagnetic (EM) control, EM integration, low cost and easy fabrication [1-5]. Typically, one of the most important applications for meta-surfaces is to realize the super lens which could transform the plane wavefronts into a focusing spherical wavefronts, inspired by Pendry's work [2]. Besides, what utilize the phenomenon of reversible transmission of light or EM wave to design the lens antenna with high gain has been successful in many researches [6-10]. Consequently, there has been more expectation for designing a lens with high performance.

Therefore, many scientists and engineers are endeavored for improving the performance of lens [11-21]. For instance, ultrathin flat MS without impedance mismatch problems in reflection geometry, working in the microwave regime, have been engineered, which makes the lens can focus efficiently [11]. What's more, it is designed and experimented that a lens in reflection geometry with 91% efficiency [12]. However, a lens with very high efficiency in transmission geometry is more useful in practice. Unfortunately, there have been minority of designs to achieve the goal in recent years, for others' focusing efficiency used to be less than 70% [16-20].

In this paper, we use Huygens meta-surface (HMS) to improve the focusing efficiency of transmissive lens. Huygens meta-surface can exhibit both electric and magnetic responses, and it was succeeded in obtaining anomalous refraction with high-efficiency in a recently paper, nevertheless, it did not apply to improve the performance of lens in previous works [21, 22]. Thus, for application to lens, we rigorously prove that achieving the 100% transmission coefficient with arbitrary phase should satisfy the certain criterions, and the theoretical result coincide well with the result by using finite-difference time-domain (FDTD) simulation in Section II. Then the transmissive lens is engineered by such meta-atoms with 100% transmission. Obviously, these meta-atoms must meet the parabolic-phase profile in Section III. Finally section IV concludes the whole paper.

2. The 100% transmission mechanisms of the proposed meta-atom

It can be easy to achieve 100% transmission when using Huygens matematerial which can exhibit both electric and magnetic responses [21]. In this paper, we utilize three types of meta-atoms to achieve the phase coverage up to 360. For demonstrating the 100% transmission, we select two types of meta-atoms to illustrate shown in figure 1 (a) and figure 1 (b). The substrate is F4B with ε_r=2.65+0.001*i*, 1.5mm thickness. We can see from figure 1 (c) and figure 1 (d) that the transmission amplitude of two selected meta-atoms are 100% and the phase is 21.8 deg. and -143.8 deg. (or 216.2 deg.), respectively, which satisfy the parabolic transmission-phase distribution. The serial number of the meta-atoms in figure 1 can be seen in figure 2 (d).

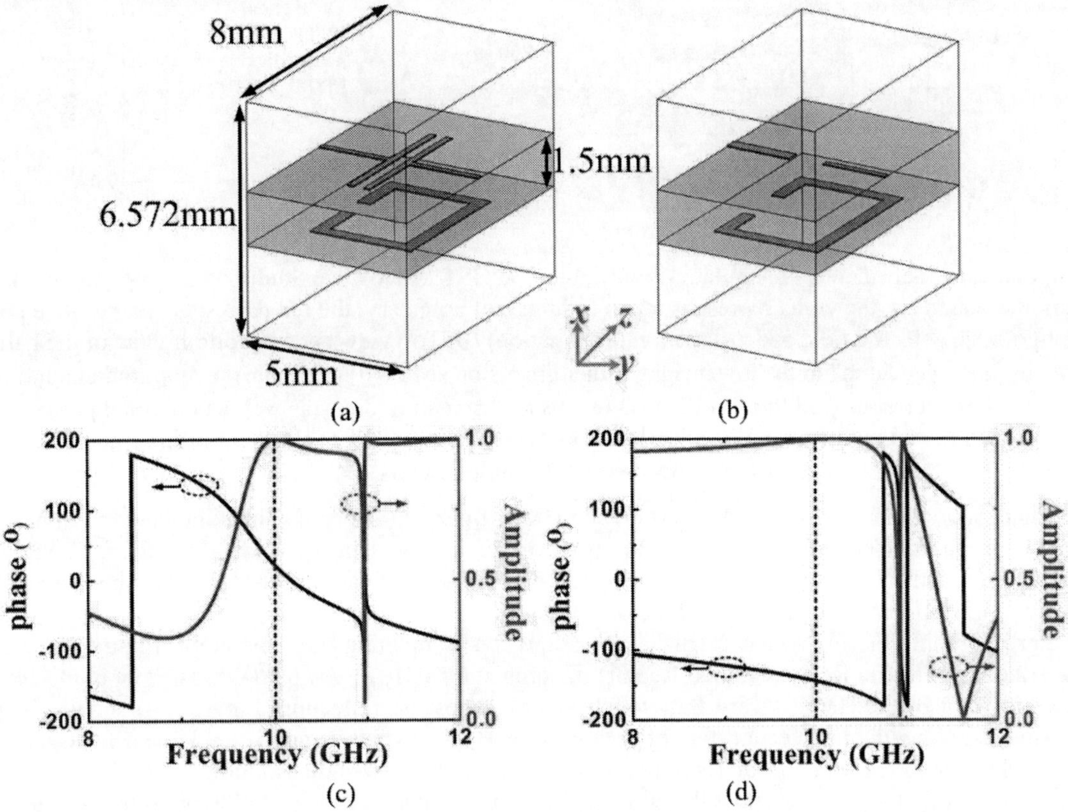

Figure 1. (a) Schematics of meta-atom 19, (b) schematics of meta-atom 26, (c) transmission amplitude and phase distribution of meta-atom 19, (d) transmission amplitude and phase distribution of meta-atom 26.

3. Design and Analysis of the nearly 100% efficiency lens

In this part, we use the 100% transmission coefficient meta-atoms for designing the nearly 100% efficiency lens. Thus, the HMS should satisfy the following parabolic refraction-phase profile for the incident electric and magnetic (EM) wave with propagation along z-direction and E field parallel to y-direction.

$$\varphi(x,y) = k_0\left(\sqrt{x^2+f^2}-f\right)+\varphi_0 \qquad (1)$$

Where $k0=2\pi/\lambda 0$ is the propagation constant in free space, f is the focal distance, $\varphi 0$ is reference phase, $\lambda 0$ is the wavelength at the 10GHz. It is known that the focusing point can be arbitrary, thus, the focal distance f=2λ0=60mm is selected for generality.

Figure 2 (d) shows the transmission amplitude of each meta-atom and the transmitting phase distribution of the proposed HMS. Besides, the simulated one is calculated by FDTD and the theoretically one is obtained by MATLAB and we can see that all transmission amplitudes are more than 0.9, even to 1, which coincide well with the theoretical one. The refracted-phase distribution also agrees very well.

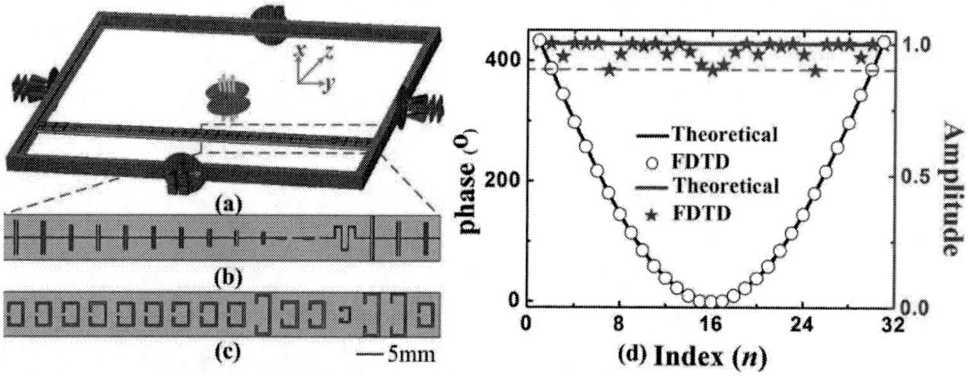

Figure 2. (a) Schematics of simulated environment in CST Microwave Studio, the orange represents periodic boundary, the violet represents open (add space) boundary, the red represents plane wave port (y-polarized electric field and z-direction propagation) (b) Top view and (c) bottom view of the half structures. "—— 5mm" in the lower right is the dimension scale, (d) transmission amplitude including theoretical results (red line) and FDTD results (red star structures), as well as refracted phase distribution including theoretical results (black line) and FDTD results (black circle structures), the red dash line present the amplitude is 0.9.

Then, we can complete our 100% efficiency HMS. figure 2 (a) depicts the simulated environment of the proposed meta-surface in CST Microwave Studio by using periodic boundary. Here, we intercept the half of the structures to describe the detailed size of each meta-atom, just showed as the figure 2 (b) (top view) and figure 2 (c) (bottom view).

For exhibiting the focal characteristic, we have been completing the simulation. Figure 3 depicts the simulated electric field amplitude Re (E_y) patterns at 10 GHz in *yoz*-plane. Most importantly, EM wave in front of the HMS is very flat, which means almost no reflection. Figure 4(a) shows power distribution along the intersection line of *yoz*-plane at 10GHz which is good supplement for figure 3. For making sure the location of focal point, we retrieve the power distribution along line y=0 just shown in figure 4 (b), which show the simulated maximal power converge to the focal line located at z=63mm, which can be explained by the finite-size effect.

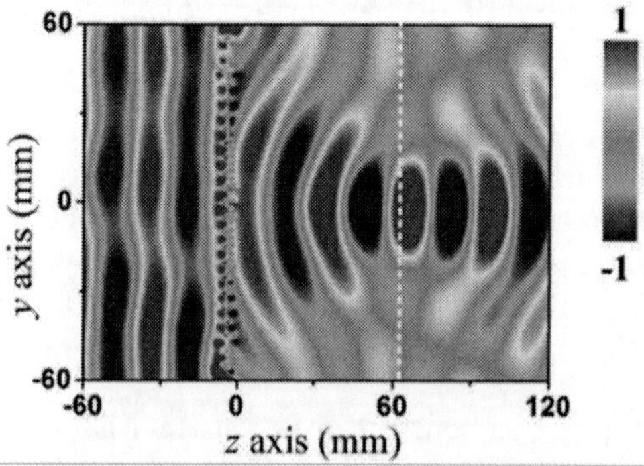

Figure 3. Simulated normalized electric field amplitude Re (E_y) patterns at 10 GHz in *yoz*-plane, the white dash line present the focal line.

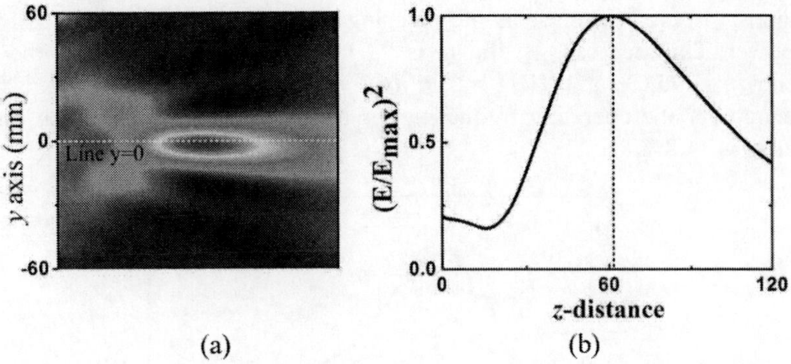

Figure 4. (a) Simulated power distribution at 10GHz, the white dash line present line *y*=0, (b) Normalized power distribution along line *y*=0.

For calculating the focusing efficiency, we define that P is the total power, P_f is the focusing power, P_s is the scatter power, P_r is the reflection power, P_t is the transmission power, t_i represent the transmission coefficient of meta-atom (i is the serial numbers can be seen in figure 2 (d)). Thus, we can obtain their relationships as following.

$$P = P_f + P_s + P_r \tag{2}$$

$$P_t = P_f + P_s \tag{3}$$

$$\eta_1 = \frac{P_f}{P_t} \tag{4}$$

$$\eta_2 = \frac{P_t}{P} = \frac{\sum_{i=1}^{n} t_i}{n} \tag{5}$$

Where the η_1 is the focusing efficiency of transmission part, η_2 is the transmission efficiency which is equal to the average of all meta-atoms' transmission coefficient. We can know that the reflection power P_r is negligibly small for the proposed HMS has almost no reflection. So power P_r can ignore. Here we define that η is the total focusing efficiency. So we can immediately calculate η by

$$\eta = \frac{P_f}{P} = \frac{P_f}{P_t} \times \frac{P_t}{P} = \eta_1 \times \eta_2 \tag{6}$$

Equation (6) tells us that we can divide into two parts to get the efficiency η. One part is calculating η_1, which we can retrieve the power distribution of the focal line to obtain, just shown in figure 6 (b). Another part is calculating η_2, which we can immediately get from figure 2, and the numerical value is 0.97 according to Eq. (5).

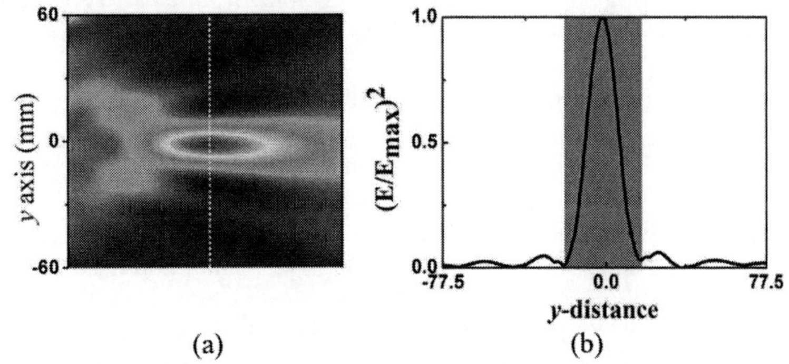

Figure 5. (a) Simulated power distribution at 10GHz, the white dash line present focal line, (b) normalized power distribution along corresponding focal line.

IOP Conf. Series: Materials Science and Engineering **292** (2017) 012104 doi:10.1088/1757-899X/292/1/012104

The orange part in figure5 (b) presents the focusing power in the transmission part, others present the scattering power. Thus we can get the η_1 is 93.4% by Eq. (4). Therefore, the total simulated focusing efficiency η is 90.6% at 10GHz by Eq. (6), which is higher than others [16-20]. The total focusing efficiency over the operating frequencies is shown in figure 6. We can see that the half-powered bandwidth is 31.8%.

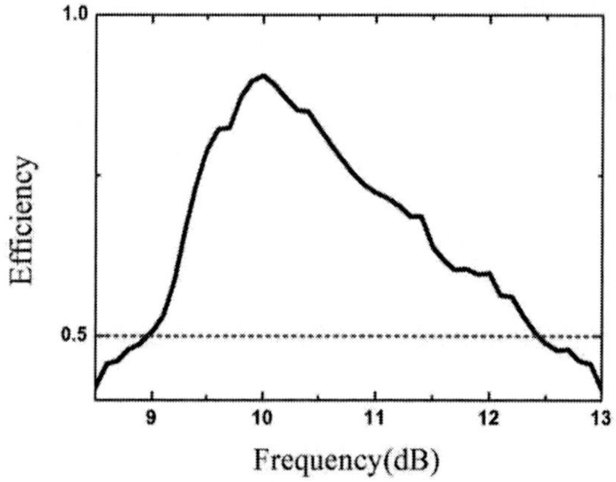

Figure 6. (a) Simulated efficiency of the focal HMS against the frequency. The red dash line presents that the efficiency is 0.5.

4. Conclusion

In summary, we show that the microwave lens with nearly 100% efficiency can be realized by using Huygens meta-materials with 100% transmission characteristic. The results show that the maximal focusing efficiency reaches 90.6% at 10GHz higher than other previous work, and half-power bandwidth reaches 31.8%. In general, this strategy to achieve the high efficiency of lens is proposed and demonstrated.

References

[1] T Cai, G M Wang, H X Xu, S W Tang and J G Liang 2016 *Opt. express* **24** 22606-15
[2] J B Pendry 2000 Phys. Rev. Lett. **85** 3966
[3] H X Xu, G M Wang, T Cai, J Xiao and Y Q Zhuang 2016 *Opt. express* **24** 27836-48
[4] N Yu, P Genevet, M A Kats, F Aieta, J P Tetienne, F Capasso and Z Gaburro 2011 *Science* **334** (6054) 333–7
[5] L H Gao, et al. 2015 *Light Sci. Appl.* **4**(9) e324
[6] Cai T, Wang G M and Liang J G 2017 *Appl. Phys. A* **766** (7) 399
[7] An W X, Xu S h, Yang F and Gao J F 2013 *IEEE Trans. Antennas Propag.* **62** (11) 5539–46
[8] Abdelrahman A H, Elsherbeni A Z and Yang F 2014 *IEEE Antennas Wireless Propag. Lett.* **13** 1288–91
[9] Liu G, Wang H J, Jiang J S, Xue F and Yi M 2015 *IEEE antennas wireless Propag. Lett.* **14** 1415-18
[10] Li H P, Wang G M, Xu H X, Cai T and Liang J G 2015 *IEEE Trans. Antennas Propag.* **63** 5144-49
[11] Li X, Xiao S Y, Cai B G, He Q, Cui T J and Zhou L 2012 *Opt. Lett.* **37** (23) 4940-42
[12] Cai T, Tang S W, Wang G M, Xu H X, Sun S L, He Q and Zhou L 2017 *Adv. Optical Mater.* **5** 1600506
[13] Xu H X, Ma S, Luo W, Cai T, Sun S, He Q and Zhou L 2016 *Appl. Phys. Lett.* **109**, 193506
[14] Xu H X, Wang G M, Qi M Q, Li L and Cui T J 2013 *Adv. Opt. Mater.* **1** (7) 495-502
[15] Xu H X, Wang G M, Qi M Q, Lv Y Y and Gao X 2013 *Appl. Phys. Lett.* **102** 193502
[16] Pfeiffer C and Grbic A 2013 *Appl. Phys. Lett.* **102** 231116

[17] J Luo, H L Yu, M W Song and Z J Zhang, 2014 *Opt. Lett.* **39** 2229-2231
[18] Pfeiffer C, Emani N K, Shaltout A M, Boltasseva A, Shalaev V M and Grbic A 2014 *Nano Lett.* **14** 2491-2497
[19] West P R, Stewart J L, Kildishev A V et al. 2014 *Opt. express* **22** (21) 26212-21
[20] Arbabi A, Horie Y, Bagheri M et al. 2015 *Nature nanotechnology* **10** (11) 937-43
[21] Pfeiffer C and Grbic A 2013 *Phys. Rev. Lett.* **110** 197401
[22] Wong J P S, Selvanayagam M and Eleftheriades G V 2015 *IEEE Trans. Microw. Theory Techn.* **63** (3) 913-24

Effect of Zn doping on the microwave absorption of BFO multiferroic materials

S Bi[1], J Li, B Mei, X J Su, C Z Ying and P H Li

Xi'an Research Institute of High-Tech, Xi'an, 710025, China

[1] E-mail: bisong5011@163.com

Abstract. The microwave absorbing materials were firstly used in the Second World War. And the $BiFeO_3$ (BFO) based microwave absorbers have been widely applied into the microwave absorbing area due to its possession of excellent electromagnetic properties. Various methods have been conducted to improve the microwave absorption performance of the BFO based materials. In the work, the sol-gel method were used to prepare the BFO, and the Zn were doped into the BFO to prepare the $Bi_{1-x}Zn_xFeO_3$ nanoparticles. The X-ray diffraction, scanning electron microscope, and vector network analysis (VNA) were conducted to characterize the microstructure and electromagnetic properties of the as-prepared samples. The results indicate that the $Bi_{1-x}Zn_xFeO_3$ nanoparticles were successfully gained and the as-prepared samples possess excellent microwave absorption properties.

1. Introduction

With the development of the electronic industry, including electronic communication and computer office, the electromagnetic pollution becomes worse and worse in our daily life.With the development of the radar technique, the demand of the microwave absorbing materials that possess excellent wave absorption performance is urgency. $BiFeO_3$ based microwave absorbing material is the traditional microwave absorber due to the long-range spiral spin modulation (period length of 64 nm), and the materials are one of most used multiferroic materials. Significant efforts have been dedicated to understanding their intrinsic properties and various methods were conducted to improve the microwave absorbing properties of the composites [1-7]. However, the macroscopic magnetization of BFO is quite small, thus greatly restricting its practical applications [8-9]. When the substitution concentration reaches a high level, the ferroelectricity is often decreased, which is originated from the space group transformation or the decrease in the stereochemical activity of Bi lone electron pairs [10-15].

Intensive research has been focused on tuning magnetization of the BFO based materials to achieve effective microwave absorption. In particular, cation doping has shown unique advantages among all the different approaches [16-21]. For example, Li et al. [22] have investigated Ca-doped BFO nanoparticles, and the qualified bandwidth in X-band was successfully increased to 3.4 GHz. Besides, in the La-substituted BFO investigated by Cao and his coworkers, the minimum reflection loss was elevated from -12 dB to -30dB [23]. In the work, we utilized the sol-gel method to prepare the Zn-doped BFO $Bi_{1-x}Zn_xFeO_3$ (x=0.1, 0.15, 0.2) samples, and the dielectric properties of the as-prepared samples are discussed.

2. Experimental

The sol-gel method was used to prepare the $Bi_{1-x}Zn_xFeO_3$ (x=0.1, 0.15, 0.2) samples, and ethylene glycol was used as the solvent. Firstly, stoichiometric amounts of $Bi(NO_3)_3 \cdot 5H_2O$, $Fe(NO_3)_3 \cdot 9H_2O$ and $Zn(NO_3)_2 \cdot 6H_2O$ were dissolved in ethylene glycol and then mixed together, the mixture was heated to 50 °C until the sol solution were obtained. Then the xerogel was obtained by continually stirring the solution at 90 °C. Subsequently, the powders were put into the tube furnace to heat at 120 °C for 1 h followed by a treatment at 300 °C for 2 h. Then, the samples were finely ground into powders. Finally, the powders were calcined in tube furnace at 650 °C for 1 h for cooled quickly, and the powders washed by deionized water.

3. Results and discussion

As is shown in figure 1, the XRD patterns of the Zn-substituted BFO (x=0.1, 0.15, 0.2) were characterized by X-ray diffraction (XRD, Rigaku Ultima IV, Cu-Ka) in the range of 15°~90° with a scan speed of 3 s per step, and a step size of 0.02° in 2θ. Comparing the X patterns of the Zn-substituted BFO (x=0.1, 0.15, 0.2) with the pure BFO, it is obvious that the distinct peaks around 22°, 32°, 38°, 22°, 46.5° and 53° are all presented in the three patterns, indicating that the Zn^{2+} have replaced the Bi^{3+}. Therefore, in the work, the $Bi_{1-x}Zn_xFeO_3$ (x=0.1, 0.15, 0.2) nanoparticles were successfully obtained.

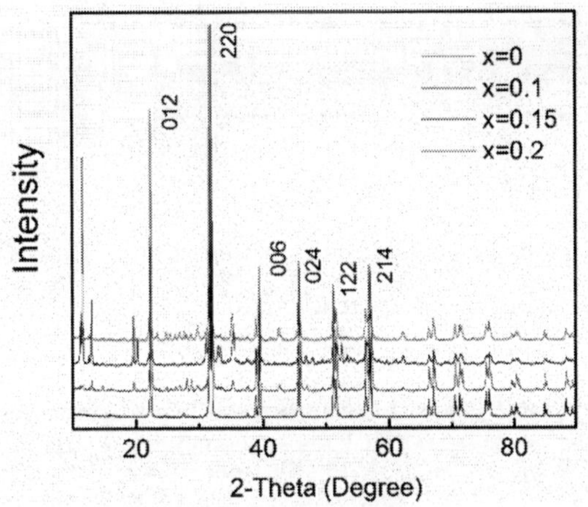

Figure 1. XRD patterns of the $Bi_{1-x}Zn_xFeO_3$ (x=0.1, 0.15 and 0.2) nanoparticles.

The microstructure of the as-prepared $Bi_{1-x}Zn_xFeO_3$ (x=0.1, 0.15 and 0.2) nanoparticles was demonstrated in figure 2. It is obvious that the $Bi_{1-x}Zn_xFeO_3$ samples are well-crystallized and have uniform particle size distribution. The average particle sizes are roughly 130, 110 and 100 nm for the $Bi_{0.9}Zn_{0.1}FeO_3$, $Bi_{0.85}Zn_{0.15}FeO_3$ and $Bi_{0.8}Zn_{0.2}FeO_3$ samples, respectively. Therefore, it can be concluded that the particle sizes do not have a significant impact on the electromagnetic performance of the samples, and the well crystallized samples were gained.

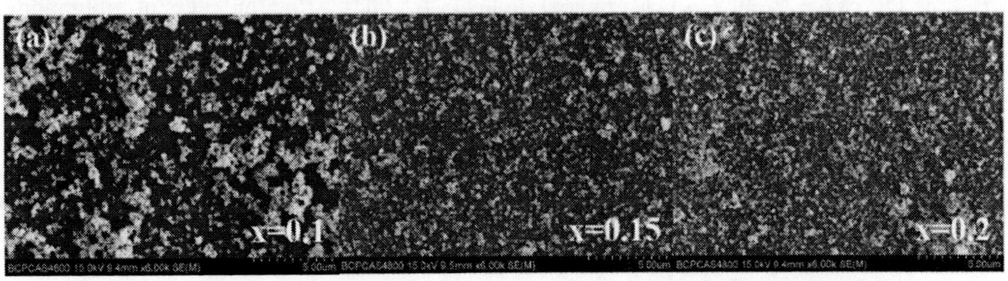

Figure 2. SEM images of the $Bi_{1-x}Zn_xFeO_3$ (x=0.1, 0.15 and 0.2) nanoparticles.

The vector network analyzer (VNA, E5071C KEYSIGHT) was used to characterize the permittivity and permeability of the as-prepared samples (8.2~12.4 GHz) via coaxial method. In the preparation of the samples, paraffin was used for the matrices. The inner diameter of the specimens is 3 mm, and the external diameter is 6.9 mm. The frequency-dependent complex permittivity (ε) and permeability (μ) of the samples are exhibited on figure 3. It is obvious that with the increase of frequency, the permittivity of the samples all have a slight decline trend. The real permittivities of the $Bi_{0.9}Zn_{0.1}FeO_3$, $Bi_{0.85}Zn_{0.15}FeO_3$ and $Bi_{0.8}Zn_{0.2}FeO_3$ are roughly up to 10.3, 13.2 and 8.6, respectively.

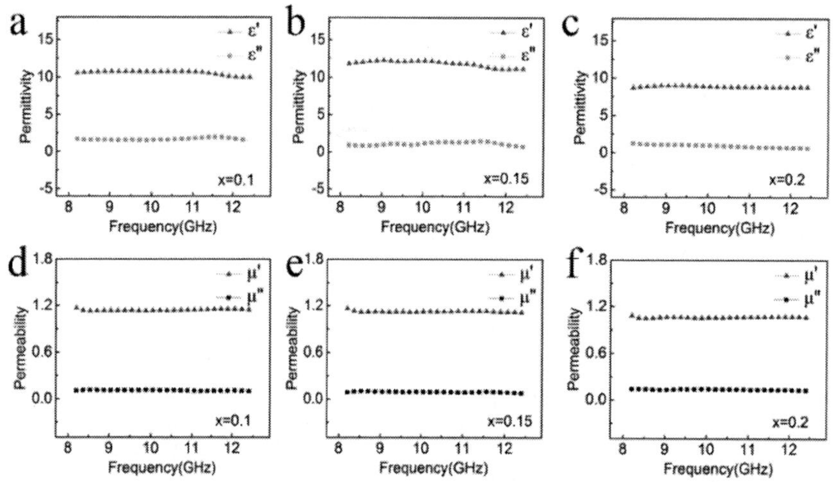

Figure 3. Frequency-dependent permittivity and permeability of the $Bi_{1-x}Zn_xFeO_3$.

The reflection loss (R_L) is achieved by the relation:

$$R_L = 20\log(|Z_{in}-1|/|Z_{in}+1|) \quad (1)$$

Here Z_{in} refers to the input impedance of the absorption layer, which could be achieved by:

$$Z_{in} = (\mu_r/\varepsilon_r)^{1/2}\tanh[j2\pi fd(\mu_r\varepsilon_r)^{1/2}/c] \quad (2)$$

where, μ_r and ε_r are the complex permittivity and permeability of the samples, respectively, f is frequency, d is thickness of the absorber, and c is the light velocity. Meanwhile, it could be the reflection the ability of the microwave absorption performance of the absorber,

The microwave absorption performance of the as-prepared samples at different frequency were calculated by the relation (1 and 2), and the microwave absorption performance of the as-prepared samples at different thickness (4mm, 5mm, 6mm) were shown in figure 4. Apparently, the qualified bandwidth (QB, RL < -10dB) of the $Bi_{0.9}Zn_{0.1}FeO_3$ nanoparticles is 0.8 GHz at the thickness of 4mm. The QB reaches 1 GHz at the thickness of 6mm, and the QB is 0.6 GHz at the thickness of 6mm. Likewise, the QB of the $Bi_{0.85}Zn_{0.15}FeO_3$ nanoparticles is 1.1 GHz at the thickness of 5mm, and that of the $Bi_{0.85}Zn_{0.15}FeO_3$ nanoparticles at the thickness of 6mm reaches 1.4 GHz. Besides, the QB of the $Bi_{0.8}Zn_{0.2}FeO_3$ nanoparticles is 1.1 GHz at the thickness of 4mm. The QB is 1 GHz at the thickness of 5mm, and that of the $Bi_{0.85}Zn_{0.15}FeO_3$ nanoparticles at the thickness of 6mm is 0.6 GHz.

According to figure 4, the minimum reflection loss (RL_{min}) of the $Bi_{0.9}Zn_{0.1}FeO_3$ nanoparticles reaches -22.12dB, and the RL_{min} of $Bi_{0.8}Zn_{0.2}FeO_3$ nanoparticles is -18.75dB. And the minimum reflection loss of the $Bi_{0.85}Zn_{0.15}FeO_3$ nanoparticles reaches -29.89dB. Consequently, it is obvious that the composites possess best microwave absorption performance when the Zn doping ratio is 15%.

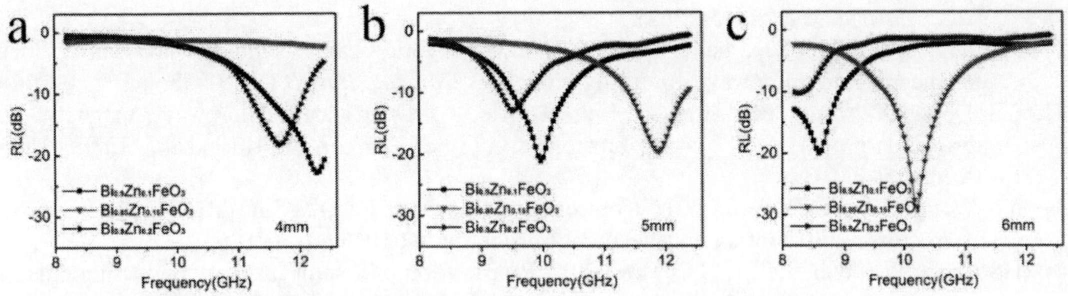

Figure 4. Microwave absorption performance of the as-prepared samples at different thickness.

Therefore, the as-prepared $Bi_{1-x}Zn_xFeO_3$ nanoparticles were characterized and the microwave absorption performance of the composites were discussed. Consequently, the composites perform well in electromagnetic loss area, especially when the doping ratio is 15%. Moreover, the density of the BFO based composites is relatively low compared with the traditional absorbing materials, thus the study provides a guide for preparing microwave absorbing materials that possess excellent electromagnetic loss performance.

4. Conclusion

In summary, the $Bi_{1-x}Zn_xFeO_3$ composites are successfully prepared by sol-gel method. The microstructure and the electromagnetic properties of the as-prepared samples were characterized. The results show that the $Bi_{1-x}Zn_xFeO_3$ samples are crystallize well, have uniform particle size distribution and have the lowest dielectric losses when the doping ratio of Zn^{2+} is 15%. The as-prepared $Bi_{0.85}Zn_{0.15}FeO_3$ nanoparticles exhibit excellent microwave absorbing abilities at the thickness of 6mm. The qualified bandwidth of the samples is 1.4 GHz, and the minimum reflection loss of the samples reaches -29.89dB. Consequently, the study may provide a guide to explore a material that performs well in microwave absorption area.

References

[1] CJ Li, B Wang, JN Wang 2012 Magnetic and Microwave Absorbing Properties of Electrospun $Ba_{(1-x)}La_xFe_{12}O_{19}$ Nanofibers *J. Magn. Magn. Mater.* **324** 1305-1311

[2] ZN Yang, F Luo, Y Hu, DM Zhu, WC Zhou 2016 Dielectric and microwave absorption properties of TiAlCo ceramic fabricated by atmospheric plasma spraying *Ceram. Int.* **42** 8525-8530

[3] YJ Kim, SS Kim 2010 Magnetic and microwave absorbing properties of Ti and Co substituted M-hexaferrites in Ka-band frequencies (26.5~40 GHz) *J. Electrochem* 24 314-318

[4] PJ Liu, ZJ Yao, JT Zhou 2016 Fabrication and microwave absorption of reduced graphene oxide/$Ni_{0.4}Zn_{0.4}Co_{0.2}Fe_2O_4$ nanocomposites *Ceram. Int.* **42** 9241-9249

[5] HS Chen, WY Wei, JX Liu, DN Fang 2012 Propagation of a mode-III interfacial crack in a piezoelectric-piezomagnetic bi-material *Int. J. Solids Struct.* **49** 2547-2558

[6] HS Chen, J Ma, YM Pei, DN Fang 2013 Anti-plane Yoffe-type crack in ferroelectric materials *Int. J. Fract.* 179 35-43

[7] HS Chen, HL Wang, YM Pei, YJ Wei, B Lin, DN Fang 2015 Crack instability of ferroelectric solids under alternative elec-tric loading *J. Mech. Phys. Solids.* **81** 75-90

[8] L Zhou, S Cui, Y Zhai, F Luo, YY Dong 2015 Dielectric and microwave absorption properties of plasma sprayed Cr/Al_2O_3 composite coatings *Ceram. Int.* **41** 14908-14914

[9] Z Amouzegar, R Naghizadeh, HR Rezaiea, M Ghahari, M Aminzare 2015 Cubic $ZnWO_4$ nano-photocatalysts synthesized by the microwave-assisted precipitation technique *Ceram. Int.* **41** 1743-1747

[10] J Liu, M. Cao, Q Luo, HL Shi, WZ Wang, J Yuan 2016 Electromagnetic Property and Tunable Microwave Absorption of 3D Nets from Nickel Chains at Elevated Temperature *Acs Appl.*

Mater. Interfaces **8** 22615-22622

[11] H Tang, et al 2016 Fe_3C/helical carbon nanotube hybrid: Facile synthesis and spin induced enhancement in microwave absorbing properties *Compos. Part B* **107** 51-58

[12] CS Dong, X Wang, PH Zhou, T Liu, JL Xie, LJ Deng 2014 Microwave magnetic and absorption properties of M-type ferrite $BaCo_xTi_xFe_{12-2x}O_{19}$ in the Ka band *J. Magn. Magn. Mater.* **354** 340-344

[13] YP Wang, Z Peng, W Jiang 2016 Controlled synthesis of Fe_3O_4@SnO_2/RGO nano composite for microwave absorption enhancement *Ceram. Int.* **42** 10682-10689

[14] HS Chen, WY Wei, JX Liu, DN Fang 2014 Propagation of a semi-infinite conducting crack in piezoelectric materials: Mode-I problem *J. Mech. Phys. Solids.* **68** 77-92

[15] Li Yong, Cao Mao-sheng, Wang Da-wei, Yuan Jie 2015 High-efficiency and dynamic stable electromagnetic wave attenuation for La doped bismuth ferrite at elevated temperature and gigahertz frequency *RSC Adv.* **5** 77184-77191

[16] N S Sowmya, A Srinivas, P Suresh, A Shukla and S V Kamat 2015 Synthesis and study of structural, magnetic and microwave absorption properties in multiferroic $BiFeO_3$ electroceramic *J. Mater. Sci.: Mater. Electron.* **26** 5368-5379

[17] ZL Hou, HF Zhou, LB Kong, HB Jin, X Qi, MS Cao 2012 Enhanced ferromagnetism and microwave absorption properties of $BiFeO_3$ nanocrystals with Ho substitution *Mater. Lett.* **84** 110-113

[18] FS Wen, N Wang, F Zhang 2010 Enhanced microwave absorption properties in $BiFeO_3$ ceramics prepared by high-pressure synthesis *Solid State Commun.* **150** 1888-1891

[19] HJ Yang, WQ Cao, DQ Zhang, TJ Su, HL Shi, WZ Wang, J Yuan,. MS Cao 2015 NiO Hierarchical Nanorings on SiC: Enhancing Relaxation to Tune Microwave Absorption at Elevated Temperatures *Acs Appl. Mater. Interfaces* **13** 70-73

[20] E Bahraoui, M Taibi, AM ET S Tlemc, AA Albassam, MA Lefdil, IV Kityk, NS AlZayed 2015 Multiferroic Eu doped $BiFeO_3$ microparticle polymer composites as materials for laser induced gratings *Journal Materiasl Science: Materials in Electronics* **26** 9949-9954

[21] CY Quan, Y Han, N Gao, WW Mao, J Zhang, JP Yang, XA Li, W Huang 2016 Comparative studies of pure, Ca-doped, Co-doped and co-doped $BiFeO_3$ nanoparticles *Ceram. Int.* **42** 537-544

[22] ZJ Li, ZL Hou, WL Song, XD Liu, WQ Cao, XH Shao, MS Cao 2016 Unusual continuous dual absorption peaks in Ca-doped $BiFeO_3$ nanostructures for broadened microwave absorption *Nanoscale* (**19**) 10415-10424

[23] Y Li, HJ Yang, WG Yang, ZL Hou, JB Li, HB Jin, J Yuan MS Cao 2013 Structure, ferromagnetism and microwave absorption properties of La substituted $BiFeO_3$ nanoparticles *Mater. Lett.* **111** 130-137

IOP Conf. Series: Materials Science and Engineering **292** (2017) 012106 doi:10.1088/1757-899X/292/1/012106

Study on the effect of polymeric rheology modifier on the rheological properties of oil-based drilling fluids

C Ma[1,5], L Li[2], Y P Yang[2], W W Hao[2], Q Zhang[3] and J Lv[4]

[1] Jiangxi Science & Technology Normal University, Nanchang 330013, China
[2] CNPC Drilling Research Institute, Beijing 100083, China
[3] Petrochina Tarim Oilfield Company, Kuerle 841000, China
[4] CNPC Guangzhou Training Center, Guangzhou 510000, China

[5] E-mail: macha001@126.com

Abstract. A new type of polymeric rheology modifier was synthesized by suspension polymerization, and the effect of rheology modifier on the rheological properties of oil-based drilling fluids was investigated. The results indicated that the obtained polymer had good capacity of improvement of shearing force of oil-based drilling fluids under high temperature and high pressure conditions. Moreover, the obtained polymer can improve the stability of oil-based drilling fluids greatly. As a result, the obtained polymer is a good rheology modifier for oil-based drilling fluids, and it can optimize oil-based drilling fluid system with good rheological properties, good static suspension ability for cuttings and environmental protection function. It can play an essential role in safe drilling jobs and improvement of drilling efficiency.

1. Introduction

Due to the contribution from shale gas plays, estimated gas reserves in South China have risen greatly in the last few years. These new shale plays have been developed in recent years by the application of advanced horizontal drilling techniques. In order to maintain wellbore stability of the horizontal section, oil-based drilling fluids are often the first choice of drilling fluids in the drilling operation of unconventional shale plays [1-5]. Despite the differences amongst shale plays, oil-based drilling fluids are flexible enough to be used across the majority of the unconventional shale gas reserves. Excellent shale formation stabilization, lubricity, and anti-contamination ability are just a few of the benefits of oil-based drilling fluid system [6-10]. However, compared with water-based drilling fluids, the rheological properties of oil-based drilling fluids are more difficult to control, and there is a lack of suitable additive to adjust the rheological properties of oil-based drilling fluids [11-16].

In order to meet the requirement of drilling operation in the deep formations, a new type of polymeric rheology modifier for oil-based drilling fluids is developed, and the effect of obtained polymer on the rheological properties of oil-based drilling fluids is tested in details. The experimental results indicate that the obtained polymer could improve the shearing force of oil-based drilling fluids greatly. In addition, it has good temperature resistance property.

The rheology modifier has good capacity of improving the shearing force and viscosity of oil-based drilling fluids, which is conducive to improving the capacity of carrying cuttings of drilling fluids and wellbore stability.

IOP Conf. Series: Materials Science and Engineering **292** (2017) 012106 doi:10.1088/1757-899X/292/1/012106

2. Experiment

To a 500 mL, four-necked flask equipped with a nitrogen inlet and a magnetic stirrer, methyl methacrylate, methyl styrene, octadecyl acrylate, dimethyl azobis isobutyrate and 300 mL THF are added. The mixture is stirred at 80 °C for 16 h in nitrogen.

After the solvent is removed at reduced pressure, the product is washed by distilled water for several times to remove residual THF, dried in vacuum at 100 °C for 24 h, pulverized through a 80~100 mesh sieve, and then the polymeric rheology modifier (PRM) is obtained as a white powder. The chemical structure of PRM polymer is showen in figure 1.

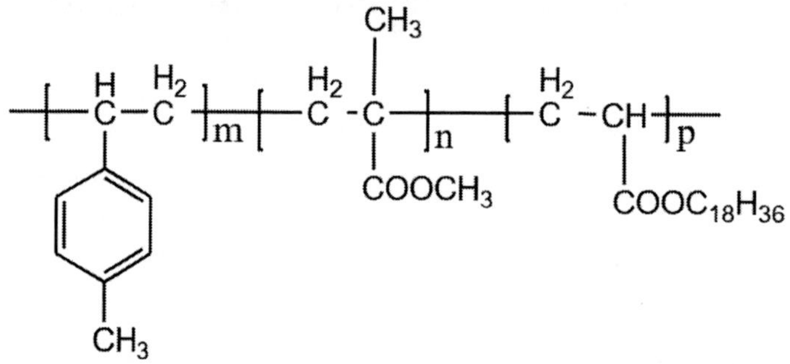

Figure 1. The structural formula of polymeric rheology modifier.

3. Results and discussion

3.1. Mechanism of the effect of PRM on the rheology of oil-based drilling fluids

Based on the similarity-intermiscibility theory, due to its long alkyl chain, PRM has good oil solubility, so PRM molecular chain could spread freely in the oil phase. PRM has two kinds of strong polar groups, including ester group and ether group. So the space grid structure with certain intensity could be formed through the hydrogen bonding interaction by the combination of PRM molecular and water drop in invert oil-emulsion drilling fluids, which could play an important role in improving the shearing force of oil-based drilling fluids.

As the shear value of drilling fluids is depended on the number of structure chain in unit volume fluid and the strength of single structure chain, the space grid structure will increase the strength of structure chain, which can improve the static suspension property of drilling fluids.

3.2. The effect of dosage of PRM on the rheological properties of oil-based drilling fluids before and after rolling at 120°C

Table 1 illustrates the influence of the dosage of PRM on the rheological properties of oil-based drilling fluid formula before and after rolling 16 h at 120 °C. The results indicate that PRM has good capacity of improving the shearing force and viscosity of oil-based drilling fluids at 120 °C, and the capacity of improving shearing force of oil-based drilling fluids is better and better with the increase of dosage of PRM.

In addition, the shear strength of oil-based drilling fluids becomes better and better, too. PRM has very certain capacity of depressing fluid loss of oil-based drilling fluids. The rheological properties of oil-based mud change little before and after rolling 16 h even when the dosage is 3%.

PRM has good effect of thickening and improving shear strength. All things considered, the optimum dosage of PRM at 120 °C is 1%. In addition, PRM could increase the emulsion-breaking voltage of drilling fluids after rolling 16 h at 120 °C, which could improve the stability of oil-based drilling fluids.

Table 1. Influence of dosage of PRM on the properties of oil-based drilling fluids before and after rolling 16 h at 120 °C.

Dosage of PRM/%	Aging condition	AV/mPa·s	PV/mPa·s	YP/Pa	GEL/Pa/Pa	FL_{API}/mL	FL_{HTHP}/mL	ES/V
0	before roll	23.5	19	4.5	3.5/4	1.2	-	1280
	after roll	22	19	3	2.5/3	-	5.0	603
1%	before roll	27	20	7	4/4.5	0.7	-	1123
	after roll	28.5	23	5.5	3.5/4	-	3.0	936
2%	before roll	32.5	24	8.5	4.5/5	0.5	-	1301
	after roll	39	31	8	4.5/5	-	2.9	960
3%	before roll	48	33	15	7/7.5	0.3	-	1268
	after roll	51.5	39	12.5	5.5/6	-	3.2	1160

[a] Formula for preparing oil-based drilling fluids: 240 mL diesel oil + 60 mL calcium chloride aqueous solution with concentration of 20% + 2% organic clay + 2% primary emulsifier + 2% auxiliary emulsifier + 1% wetting agent + 0-3% PRM + 1% calcium oxide + barite (The density of mud is 1.20 g/cm^3).
[b] The rolling condition is 120 °C × 16 h, HTHP filtration condition is 120 °C × 3.5 MPa.
[c] AV—Apparent viscosity.
[d] PV—Plastic viscosity.
[e] YP—Yield point.
[f] GEL—Initial gel strength/Final gel strength.
[g] FL_{API}—API fluid-loss value.
[h] FL_{HTHP}—HTHP fluid-loss value.
[i] ES—Emulsion-breaking voltage.

3.3. The effect of dosage of PRM on the rheological properties of oil-based drilling fluids before and after rolling at 150 °C

Table 2 illustrates the influence of the dosage of PRM on the rheological properties of oil-based mud. From table 2, the experimental results indicate that PRM has good capacity of improving the shearing force and viscosity at 150 °C. The effect of thickening and improving shearing force and becomes better and better with the increase of dosage of PRM. Moreover, the rheological properties of oil-based mud change little before and after rolling.

All things considered, mainly on account of the cost and the capacity of improving shear strength, the optimum dosage of PRM at 150 °C is 1%. Moreover, PRM can increase the emulsion-breaking voltage of oil-based drilling fluids after rolling at 150 °C, so the stability of drilling fluids could be improved greatly.

Table 2. Influence of dosage of PRM on the properties of oil-based drilling fluids before and after rolling 16 h at 150 °C.

Dosage of PRM /%	Aging condition	AV/mPa·s	PV/mPa·s	YP/Pa	GEL/Pa/Pa	FL_{API}/mL	FL_{HTHP}/ mL	ES/V
0	before roll	23.5	19	4.5	3.5/4	1.2	-	1280
	after roll	16	16	0	0.25/0.5	-	37	420
1%	before roll	27	20	7	4/4.5	0.7	-	1123
	after roll	28.5	24	4.5	2.5/3	-	18.4	698
2%	before roll	32.5	24	8.5	4.5/5	0.5		1301
	after roll	36.5	30	6.5	3/3.5	-	40	960
3%	before roll	48	33	15	7/7.5	0.3	-	1268
	after roll	47.5	36	11.5	4/4	-	20	1000

[a] Formula for preparing oil-based drilling fluids: 240 mL diesel oil + 60 mL calcium chloride aqueous solution with concentration of 20% + 2% organic clay + 2% primary emulsifier + 2% auxiliary emulsifier + 1% wetting agent + 0-3% PRM + 1% calcium oxide + barite (The density of mud is 1.20 g/cm^3).
[b] The rolling condition is 150 °C × 16 h, HTHP filtration condition is 150 °C × 3.5 MPa.

3.4. The effect of dosage of PRM on the rheological properties of oil-based drilling fluids before and after rolling at 180 °C

Table 3 indicates the influence of the dosage of PRM on the rheological properties of oil-based mud before and after rolling 16 h at 180 °C. From table 3, the results indicate that PRM has good capacity of improving the shearing force and viscosity at 180 °C. PRM has very certain capacity of depressing fluid loss of oil-based drilling fluids. The effect of improving the shearing force and viscosity becomes better and better with the increase of dosage of PRM.

Table 3. Influence of dosage of PRM on the properties of oil-based drilling fluids before and after rolling 16 h at 180 °C

Dosage of PRM /%	Aging condition	AV/mPa·s	PV/mPa·s	YP/Pa	GEL/Pa/Pa	FL_{API}/mL	FL_{HTHP}/ mL	ES/V
0	before roll	23.5	19	4.5	3.5/4	1.2	-	1280
	after roll	15	15	0	0/0.25	-	84	406
1%	before roll	27	20	7	4/4.5	0.7	-	1123
	after roll	20.5	20	0.5	1/1.5	-	64	698
2%	before roll	32.5	24	8.5	4.5/5	0.5	-	1301
	after roll	33	28	5	2.5/3	-	66	816
3%	before roll	48	33	15	7/7.5	0.3	-	1268
	after roll	38	22	16	3/3	-	72	900

[a] Formula for preparing oil-based drilling fluids: 240 mL diesel oil + 60 mL calcium chloride aqueous solution with concentration of 20% + 2% organic clay + 2% primary emulsifier + 2% auxiliary emulsifier + 1% wetting agent + 0-3% PRM + 1% calcium oxide + barite (The density of mud is 1.20 g/cm^3).
[b] The rolling condition is 180 °C × 16 h, HTHP filtration condition is 180 °C × 3.5 MPa.

The rheological properties of oil-based mud have little change before and after rolling at 16h at 180 °C. All things considered, mainly on account of the cost and the capacity of thickening and improving shear strength, the optimum dosage of PRM at 180 °C is 2%. In addition, PRM can increase the emulsion-breaking voltage of drilling fluids after rolling at 180 °C, so the stability of drilling fluids could be improved.

4. Conclusions

1) PRM has good capacity of thickening and improving the shearing force of oil-based drilling fluids at HTHP conditions.

2) PRM has little influence on the properties of oil-based drilling fluids, and it can optimizate the rheological properties of oil-based drilling fluids. PRM can solve the problem that the rheological properties of oil-based mud are difficult to adjust.

3) PRM can improve the stability of oil-based drilling fluids greatly under HTHP conditions.

4) PRM has good temperature resistance properties under 180 °C. The optimum dosage of PRM for oil-based drilling fluids is less than 1% under 150 °C.

PRM could play an essential role in improving hole cleaning, maintaining borehole stability, and avoiding severe drilling problems. It is a good rheology modifier for oil-based mud, and it could meet the requirement of drilling operation under extreme condition. PRM is expected to possess a wide prospect of application in drilling operations.

Acknowledgments

This study is supported by the National High Technology Research and Development Program of China (863 Program) (No. 2012AA091502) and CNPC Project (No. 2014E-2102).

References

[1] Messler D, Kippie D and Webb T 2002 Improved techniques of deepwater SBM displacements: A case history *Soc. Pet. Eng.* **1** 731

[2] Neff J M, Hart A D, Ray J P, Limia J and Purcell T 2005 An assessment of seabed impacts of synthetic-based drilling-mud cuttings in the Gulf of Mexico *Soc. Pet. Eng.* **1** 94

[3] Li L, Xu X G, Zhu J Z, *et al* 2016 Application of innovative high-temperature high-density oil-based drilling fluid technology in the efficient exploration and development of ultra-deep natural gas resources in West China *Soc. Pet. Eng.* **1** 10-22

[4] Yin D, Li L, X. G. Xu, *et al* 2016 Application of high density non-aqueous fluid technology in the efficient development and production of super-deep complicated formations in the Tian Mountain Front Block *Soc. Pet. Eng.* **1** 33-39

[5] Zhu J Z, Li L, Li L and Zhang Z 2015 Application of UDM-2 drilling fluid technology in the development of upper-deep oil and gas resources in Tarim Basin *Soc. Pet. Eng.* **1** 240-254

[6] Nasr-El-Din H A, Al-Otaibi M B, Al-Qahtani A A and Samuel M 2007 An effective fluid formulation to remove drilling fluid mud cake in horizontal and multi-lateral wells *SPE Drill. Completion* **22** 26-32

[7] Ma C, Li L, Zhang J, *et al* 2015 Research on performance of high-temperature high-density oil-based drilling fluids International Conference on Materials Chemistry and Envionmental Protection **21** 1-4

[8] Li L, Xu X G, Sun J S, Yuan X B and Li Y M 2012 Vital role of nanomaterials in drilling fluid and reservoir protection applications *Soc. Pet. Eng.* **1** 640-647

[9] Zhu J Z, Li L, Yin D, *et al* 2017 Application of comprehensive harmless waste treatment technology to treat drilling cuttings in the Tian Mountain Front Block in Tarim Basin *Soc. Pet. Eng.* **1** 32-43

[10] Li L, Yin D, Li L, Yang Y P, Xu X G and Zhang J 2017 Application of innovative leaching and resource utilization technology to treat oily drilling cuttings on the harmlessization handle in the Tian Mountain Front Block in West China *Soc. Pet. Eng.* **1** 688-699

[11] Li L, Yin D, Li L, *et al* 2018 Application of innovative high density high-performance water-based drilling fluid technology in the efficient development and production of ultra-deep complicated formations in the Tian Mountain Front Block *Soc. Pet. Eng.* **1** 618-629

[12] Li L, Yuan X B and Sun J S. 2013 Vital role of nanotechnology and nanomaterials in the field of oilfield chemistry *Soc. Pet. Eng.* **1** 85-91

[13] Oyler K W, Burrows K J, West G C, Siems D, and Kirsner J 2010 Diesel oil-based invert emulsion drilling fluids and methods of drilling boreholes US Patent 7696131

[14] Nabhani N, Emami M and Moghadam T A B 2011 Application of nanotechnology and nanomaterials in oil and gas industry *AIP Conf. Proc.* **1415** 128-131

[15] Li L, Yuan X B, Ma C, Cheng R C and Yang Y P 2014 Study on the effect of humic acid acetamide on the rheological properties of gas-to-liquid based drilling fluids *Appl. Mech. Mater.* **641-642** 447-450

[16] Ma C, Li L, Yuan X B, Cheng R C and Wang G 2014 Analysis of fluid loss-reducing mechanism of humic acid acetamide compound with lipophilic property *Adv. Mater. Res.* **1004-1005** 891-894

IOP Conf. Series: Materials Science and Engineering **292** (2017) 012107 doi:10.1088/1757-899X/292/1/012107

Cobalt-cadmium bimetallic porphyrin coordination polymers for electrochemistry application

C Y Wang, G Y Cui, D Ding and B Zhou[1]

Nanjing Normal University, Nanjing 210023, China

[1] E-mail: zhoubo@njnu.edu.cn

Abstract. In this paper, we used tetra (4-carboxyphenyl) porphyrin (H_2TCPP) and metal cadmium, cobalt as reactants to synthesize metal porphyrin coordination polymers that they had different metal ratio. They were expressed as Co1Cd3TCPP, Co1Cd1TCPP, Co3Cd1TCPP, respectively. The results were characterized by X-ray diffraction (XRD), transmission electron microscopy (TEM) and inductively coupled plasma atomic emission spectrometer (ICP). Herein, a series of metal porphyrin coordination polymers has multiple metal active centers and constructs electrochemistry sensors. In order to increase the conductivity, multi-walled carbon nanotubes (MWCNTs) can be used to modify the electrodes. The polymer/MWCNTs/GCE electrode was studied by cyclic voltammetry and chronoamperometry as sensor for sodium nitrite. The performance of Co1Cd1TCPP/MWCNTs/GCE electrode is best, the sensitivity for sodium nitrite is 350.95 mA M^{-1} cm^{-2} and the. The results indicate that metal porphyrin coordination polymers have excellent performance. It also enriches the application of metal porphyrin coordination polymer in electrochemistry sensor.

1. Introduce

The metal-organic frameworks (MOFs) are assembled by metal ions and organic ligands linked together through a coordination interaction and become a new kind of highly porous crystalline material [1]. Compared with the traditional inorganic porous materials, MOFs have a larger surface area, higher porosity, richer structure and function [2]. Therefore, metal-organic frameworks materials are widely used in gas adsorption and separation [3], catalytic reactions [4] and chemical sensing [5]. These applications typically require the incorporation of coordinative unsaturated metal ions in the framework to act as functional sites. The metal ligand has two different metal-binding sites, which can easily synthesize heteronuclear metal skeletons containing coordination unsaturated metal ions. Square planar building blocks based on tetraarylporphyrin, such as meso-tetra (4-pyridyl) porphyrin (TPyP) and meso-tetra (4-carboxyphenyl) porphyrin (TCPP), have been used extensively as metal ligands in the synthesis of MOFs, with the goal of preparing catalytic materials [6]. Porphyrin ligands are used to synthesize heterogeneous metal frameworks and provide two different metal binding sites. The metal can be coordinated with the carboxyl or pyridine surrounding the porphyrin ring or the nitrogen of the porphyrin cavity. Cd (II) can be preferentially coordinated with carboxyl or pyridine around porphyrin due to the relatively large metal ion radius. This strategy has made some achievements in the synthesis of some mixed-metal frameworks using TPyP [7]. Co (II) has an empty d orbitals, making the transition metal prone to coordination. At the same time, the valence state of the metal cobalt changes as a result of electron transfer processes, and the metal cobalt has excellent electrochemical performance.

Many articles have been published on the detection of small molecules using metal active centers, but there are less reportes that through different proportions of bimetallic synthesis of metal coordination polymer to detect small molecules. The coordination polymers can provide two metal active centers, and through experimental studies can find the best metal proportion of the metal coordination polymers for electrochemical sensing. In this paper, Co (II) and Cd (II) are coordinated with metal porphyrins which form a frame material to detect sodium nitrite.

2. Experiment section

2.1. Experimental materials

The cadmium chloride（$CdCl_2 \cdot 2.5H_2O$, AR）was purchased from Sinopharm Chemical Reagent Co., Ltd (Shanghai, China). The cobalt chloride ($CoCl_2 \cdot 6H_2O$, AR) was purchased from Shanghai Reagent Plant (Shanghai, China). Multi-walled carbon nanotubes (MWCNTs, 97%) were purchased from Shenzhen Nanotechnology Technology Co., Ltd. (Shenzhen, China). Sodium nitrite ($NaNO_2$, AR) was purchased from Nanjing Chemical Reagent Factory (Nanjing, China). The disodium hydrogen phosphate (Na_2HPO_4, AR) and the sodium dihydrogen phosphate (NaH_2PO_4, AR) were purchased from Sinopharm Group Chemical Reagent Co., Ltd. The 0.1 M phosphate buffer solution (PBS) was obtained from 0.1 M NaH_2PO_4 and 0.1 M Na_2HPO_4. All solutions were prepared with ultra-pure water purified by a Millipore-QSystem ($\rho \geq 18.2$ MΩ cm).

The polymers were characterized by X-ray diffraction (XRD, the Germany Brooke production diffractometer with Cu Kα radiation λ=1.5406Å, υ=4s/step at room temperature), transmission electron microscopy (TEM, H-7650, resolution: 0.204nm, magnification: high resolution mode: x200 ~ x600,000 and low magnification mode: x50 ~ x1,000) and inductively coupled plasma atomic emission spectrometer (ICP, US LEEMANLABS production, analytical elements: 73 elements, wavelength range 178 ~ 800nm, resolution: 0.006nm). The electrochemical measurements were carried out on a conventional three electrodes system with the glassy carbon electrodeh as the working electrode, a platinum wire as the counter electrode, and a saturated calomel electrode (SCE) as the reference electrode on a CHI660E electrochemistry workstation (Shanghai, Chenhua).

2.2. Synthesis of metal porphyrin coordination polymer

H_2TCPP were prepared according to the literature [8]. Then, H_2TCPP (80 mg, 0.1 mmol), $CdCl_2 \cdot 2.5H_2O$ (X_1 mmol) and $CoCl_2 \cdot 6H_2O$ (X_2 mmol) (X_1: X_2= 3: 1, 1: 1 and 1: 3) were dissolved in N, N-dimethylformamide (DMF) (6 mL) and H_2O (0.3 mL). The mixture was transformed into a Teflon-lined stainless steel autoclave having a capacity of 25 mL and heated to 150 ℃ for 48 h. Finally, the resulting precipitate was washed with DMF, alcohol and H_2O several times, and the product was dried at 50 ℃ in a vacuum.

2.3. Improved glassy carbon electrode

The glassy carbon electrode (GCE) was polished with alumina slurry and then ultrasonically cleaned in secondary distilled water to obtain mirror-like surface. Modified electrodes were fabricated as follows: First 6μL of suspension (5mg/mL) of MWCNTs was cast onto the GCE surface and dried at room temperature. Then 6μL of suspension (5mg/mL) of metal porphyrin coordination polymers were cast onto the surface of the GCE and dried in the air. Finally 2 μL of 1% nafion solution was dropped to cover the electrode surface. After drying in the air, the metal porphyrin coordination polymers/MWCNTs/GCE were obtained.

3. Results and discussion

3.1. XRD characterization of metal porphyrin coordination polymer

As it can be seen from figure 1, the XRD pattern of a series of metal porphyrin coordination polymers are compared with that of 1-Co (CCDC-865037) synthesized in literature. It is found that the peaks of (a) and (d) are in good agreement with each other. Through the comparison of (a), (b) and (c), it can be seen that the general position of the peak does not change, indicating that when the put into ratio is different, but we synthesized the structure of the metal porphyrin coordination polymers are same. It follows that the polymer composition changes when the metal is incorporated, but the spatial structure of the polymer is unchanged. So, we synthesized a series of metal porphyrin coordination polymers have the same structure. The diffraction peak of the metal coordination polymers were mainly due to their weak crystallinity.

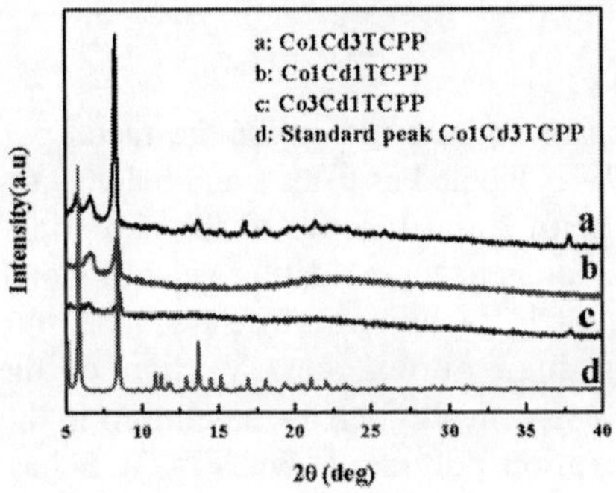

Figure 1. XRD pattern of metal porphyrin coordination polymer.

3.2. TEM characterization of metal porphyrin coordination polymer

The morphology and size of the metal porphyrin coordination polymers were characterized by TEM. Figure 2 (a), (b), (c) represents the Co1Cd3TCPP, Co1Cd1TCPP and Co3Cd1TCPP polymers, respectively. It can be seen from the figure that the metal porphyrin coordination polymers present irregular particles with a width of 2 μm and a length of about 4 μm. It can be considered that the synthetized polymers do not present a specific morphology.

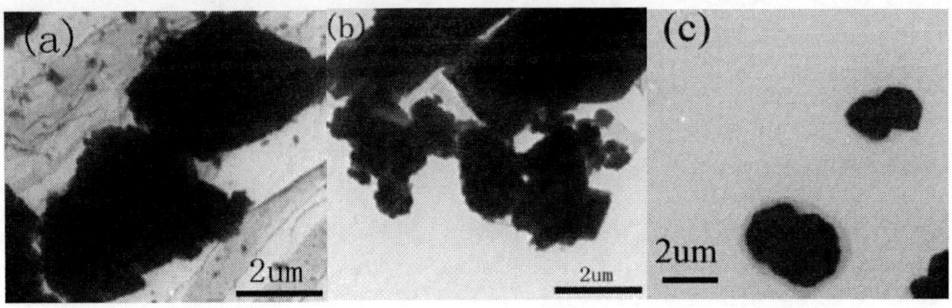

Figure 2. TEM (Resolution: 0.204nm, magnification: high resolution mode: x200 ~ x600, 000 and low magnification mode: x50 ~ x1,000) diagram of metal porphyrin coordination polymers a: Co1Cd3TCPP, b: Co1Cd1TCPP and c: Co3Cd1TCPP).

3.3. ICP characterization of metal porphyrin coordination polymer

In order to analyze the proportion of metal elements in the metal porphyrin coordination polymers, we tested the three kinds of metal porphyrin coordination polymers by ICP. The samples were pre-treated by dry ashing method and a small amount of samples were placed in a porcelain crucible and then placed in a muffle furnace. The temperature was gradually increased (100 ℃/min) to 540 ℃ for 8 h in

IOP Conf. Series: Materials Science and Engineering **292** (2017) 012107 doi:10.1088/1757-899X/292/1/012107

the air atmosphere. The residue was dissolved with nitric acid at 7%. The cobalt and chromium were determined 228.615 nm and 228.802 nm, respectively. Table 1 shows the proportion of metal derived from ICP. It can be seen that the percentage of the metal atoms in the coordination polymer is consistent with the sample name.

Table 1. ICP detection of different metal coordination.

Sample name	Atomic number ratio Co/Cd
Co1Cd3TCPP	0.48
Co1Cd1TCPP	0.83
Co3Cd1TCPP	1.04

3.4. CV testing of sodium nitrite

Figure 3 shows the CV detection of sodium nitrite for the metal porphyrin coordination polymer in PBS (pH=7.0). The peak at -0.79V is labeled as peak 1 and belongs to the oxidation peak of Cd(II)/Cd, the peak at 0.20V is labeled as peak 2 and belongs to the oxidation peak of MWCNTs, the peak at 0.80V is labeled as peak 3 and belongs to the oxidation peak of Co(III)/Co(II). The response of metal coordination polymer/MWCNTs/GCE to sodium nitrite was also studied. As it can be seen from the figure, when adding 0.5 mM sodium nitrite, the CV chart of the metal porphyrin coordination polymer/MWCNTs/GCE shows a significant increase in current in the range of -1~1V. It is indicating that the metal porphyrin coordination polymer/MWCNTs/GCE has an electrocatalytic activity for sodium nitrite. The response potential is the same as that of the metal porphyrin coordination polymer/GCE which does not add to the MWCNTs. However, the CV curve of MWCNTs/GCE in PBS with or without sodium nitrite does not significantly change, indicating that MWCNTs/GCE has no obvious catalytic effect on sodium nitrite.

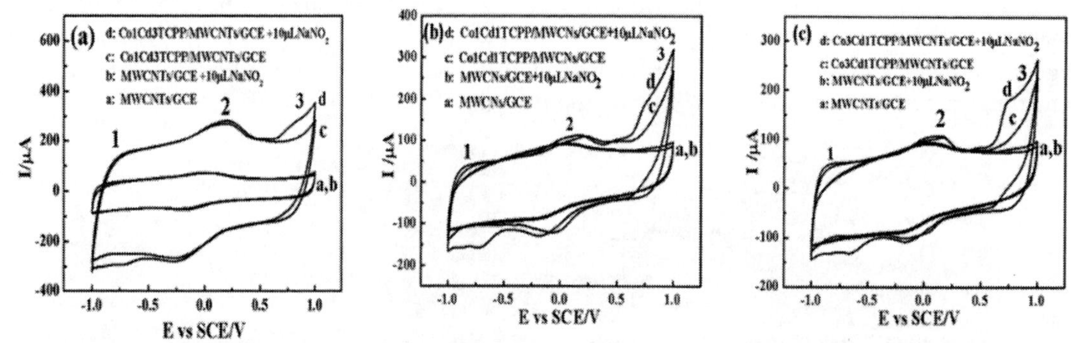

Figure 3. Comparison of the CV curves of 5×10^{-4} M $NaNO_2$ in PBS buffer solution (pH=7.0) as an electrochemical detection of the supporting electrolyte in metal porphyrin coordination polymer/MWCNTs/GCE and MWCNTs/GCE. The metal porphyrin coordination polymer represent that a is Co1Cd3TCPP, b is Co1Cd1TCPP, c is Co3Cd1TCPP.

3.5. I-t testing of sodium nitrite

The chronoamperometry (I-t) is more sensitive than cyclic voltammetry (CV). The CV showed that 0.85 V is the best potential, so sodium nitrite detection was carried out at this potential. Figure 4 to make the curve, the Co1Cd1TCPP has bigger current than Co1Cd3TCPP and Co3Cd1TCPP. Hence, the Co1Cd1TCPP/MWCNTs/GCE had the best performance for sodium nitrite detection. By the I-t detection at the optimum potential, with the concentration of sodium nitrite is increasing and the current value is also increasing. It is shown that the electrochemical sensor is responsive to sodium nitrite and that the response value of the current is linear relation to the increase in the concentration of sodium nitrite. With the increase of sodium nitrite concentration, the response current of Co1Cd1TCPP / MWCNTs / GCE is also increasing, and the effect is obvious. The electrochemical sensor sensitivity is 350.95 mA M^{-1} cm^{-2} and the slope of the calibration curve is 0.9997.

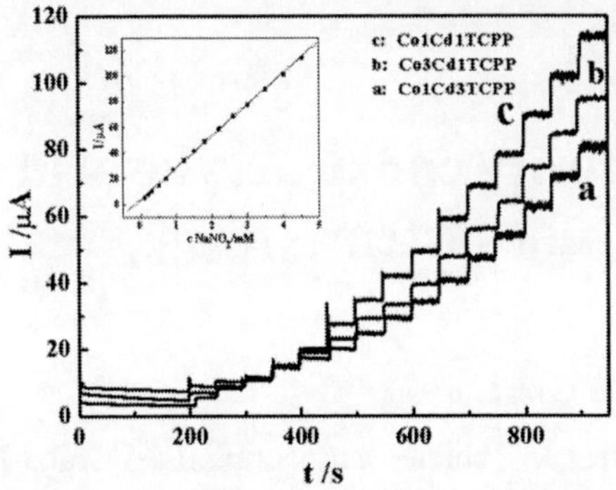

Figure 4. Comparison of the electrocatalytic oxidation sensitivity of three metal porphyrin coordination polymers to sodium nitrite at optimum potential.

4. Conclusion

In this paper, a series of metal porphyrin coordination polymers was synthesized by solvent thermal method. The electrochemical properties and the detection properties of sodium nitrite were studied by cyclic voltammetry (CV) and chronoamperometry (I-t). The results show that the Co1Cd1TCPP/MWCNTs/GCE have better catalytic effect on sodium nitrite than the Co1Cd3TCPP/MWCNTs/GCE and the Co3Cd1TCPP/MWCNTs/GCE. And the metal porphyrin coordination polymer/MWCNTs/GCE have satisfactory stability. At the same time, the metal porphyrin coordination polymers also enrich the application of electrochemical sensors.

Acknowledgements

The authors thank National Natural Science Foundation of China (Grant No. 21175068 and 21335004) for financial support. We also thank the Priority Academic Program Development of Jiangsu Higher Education Institutions.

References

[1] Yi H, et al 2016 Metal Organic Frameworks Combining $CoFe_2O_4$ Magnetic Nanoparticles as Highly Efficient SERS Sensing Platform for Ultrasensitive Detection of N-Terminal Pro-Brain Natriuretic Peptide *Acs Applied Materials & Interfaces* **8** 12

[2] Eddaoudi M, et al 2002 Systematic design of pore size and functionality in isoreticular MOFs and their application in methane storage *Science* **295** 5554

[3] Li J R, R J Kuppler, and H C Zhou 2009 Selective gas adsorption and separation in metal-organic frameworks *Chemical Society Reviews* **38** 5

[4] Su S, et al 2012 An active-site-accessible porous metal-organic framework composed of triangular building units: preparation, catalytic activity and magnetic property *Chemical Communications* **48** 90

[5] Morozan A, and F Jaouen 2012 Metal organic frameworks for electrochemical applications *Energy & Environmental Science* **5** 11

[6] Suslick K S, et al 2005 Microporous Porphyrin Solids *Cheminform* **38** 4

[7] Sharma C V K, et al 2012 Design Strategies for Solid-State Supramolecular Arrays Containing Both Mixed-Metalated and Freebase Porphyrins *Journal of the American Chemical Society* **121** 6

[8] Smythe N C, et al 2012 A heterobimetallic metal-organic framework with tunable reactive metal sites: synthesis, characterization, and reactivity *Dalton Transactions* **41** 26

IOP Conf. Series: Materials Science and Engineering **292** (2017) 012108 doi:10.1088/1757-899X/292/1/012108

Preparation of low viscosity epoxy acrylic acid photopolymer prepolymer in light curing system

P Li[1], J Y Huang and G Z Liu

School of Mechatronics & Vehicle Engineering, East China Jiaotong University, Nanchang, China

[1] E-mail: ecjtulipeng@126.com

Abstract. With the integration and development of materials engineering, applied mechanics, automatic control and bionics, light cured composite has become one of the most favourite research topics in the field of materials and engineering at home and abroad. In the UV curing system, the prepolymer and the reactive diluent form the backbone of the cured material together. And they account for more than 90% of the total mass. The basic properties of the cured product are mainly determined by the prepolymer. A low viscosity epoxy acrylate photosensitive prepolymer with a viscosity of 6800 mPa • s (25 ℃) was obtained by esterification of 5 hours with bisphenol A epoxy resin with high epoxy value and low viscosity.

1. Introduction

The photocurable material is prepared on the basis of prepolymer, adding specific reactive diluents, photoinitiator and various additives. Under the irradiation of the curing light source, the photoinitiator in the liquid light-curing material is excited to become free radical or cation, which causes the material containing unsaturated double bond in the material to form a solidification reaction.

In the UV curing system, the prepolymer and the reactive diluent form the backbone of the cured material together. They account for more than 90% of the total mass [1]. The basic properties of the cured product are mainly determined by the prepolymer. According to the mechanism of light curing, the suitable prepolymer is also different [2].

(1) Free radical polymerization mechanism

The suitable prepolymers are resins with C=C unsaturated double bonds, such as propylene oxide (CH2=CH-COO-), methacrylic oxide (CH2=C (CH3) -COO-), vinyl (C=C) and allyl (CH2=CH-CH2-) and so on.

Sort by the polymerization rate: CH2=CH-COO- > CH2=C (CH3)-COO- > C=C > CH2=CH-CH2-.

(2) Cationic polymerization mechanism

The applicable prepolymer mainly includes epoxy resin, epoxy functionalized polysiloxane resin, and resin containing vinyl ether functional group.

Modern industrialized acrylated prepolymers are predominantly present: acrylated epoxy resins, acrylated urethanes, acrylated polyesters, and acrylated polyacrylates. And the former two are the most important. Table 1 lists the performance of common prepolymers [3].

Table 1. The performance of common prepolymers.

Type	Curing rate	Tensile strength	Flexibility	Strength	Chemical resistance	Anti-yellowing
Epoxy acrylate	Fast	High	Not good	High	Excellent	Middle to bad
Polyurethane acrylate	Fast	Adjustable	Good	Adjustable	Good	Adjustable
Polyester Acrylate	Adjustable	Medium	Adjustable	Medium	Good	Not good
Polyether acrylate	Adjustable	Low	Good	Low	Not good	Good
Acrylate resin	Fast	Low	Good	Low	Not good	Excellent
Unsaturated polyester resin	Slow	High	Not good	High	Not good	Not good

The selection of prepolymer in light curing system mainly takes the following factors into consideration: low viscosity, light curing speed, good physical and mechanical properties, glass transition temperature, curing shrinkage, low toxicity and low irritation [4].

Acrylic series resin is the second generation resin of light curing system. It is introduced into the functional group of the crosslinking reaction at the end of some prepolymer by introducing double bond into the acrylate [5]. Common polyester acrylate, polyether acrylate, epoxy acrylate, polyurethane acrylate, melamine acrylate. Among them, epoxy acrylate (EA) attracted much attention because of easy synthesis, strong adhesion, chemical resistance, high strength characteristics [6].

Epoxy acrylate can be divided into bisphenol A epoxy acrylate resin, novolac epoxy acrylate, epoxy acrylate and epoxy acrylate oil according to the structure. Among them, the bisphenol A type epoxy acrylate is the most stable [7]. Bisphenol A epoxy acrylate is obtained by reacting bisphenol A epoxy resin with (meth) acrylic acid [8]. Bisphenol A epoxy acrylate was prepared from resin and acrylic acid as prepolymer of light repairing agent [9].

2. Experimental approach

2.1. Reaction mechanism

Epoxy acrylates are formed by the reaction of epoxy resin and acrylic acid,and the preparation conditions vary depending on the type of EA. Catalysts that promote the reaction of epoxy groups and hydroxyl groups include strong bases and organic amines. The reaction between the epoxy group and the hydroxyl group is an exothermic reaction. It is very important to select a suitable polymerization inhibitor to prevent radical polymerization of unsaturated groups on acrylic acid during the preparation of epoxy acrylates. The common inhibitors are phenols (such as hydroquinone, etc.). The synthesis of epoxy acrylate is as follows [10].

2.1.1. Main reaction.

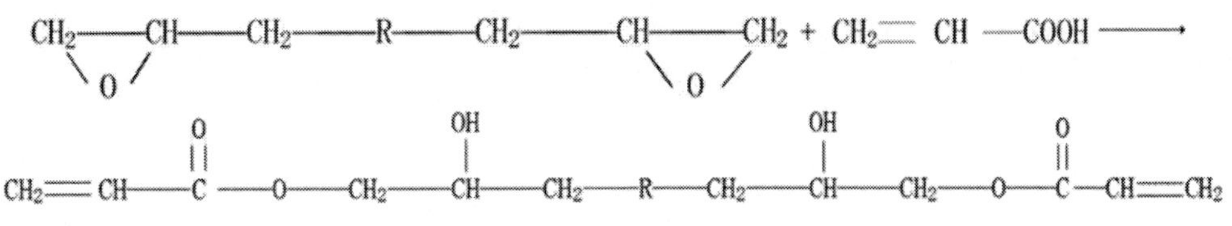

Where R is:

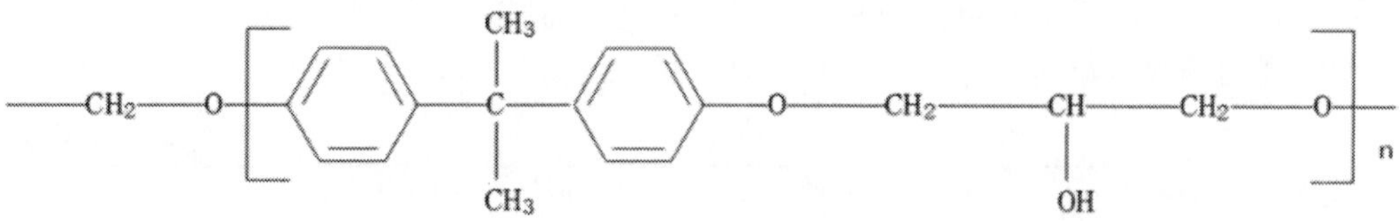

2.1.2. Side reaction. The side reactions mainly include: The self - polymerization reaction occurs after epoxy ring – opening and reaction of epoxy groups with active hydrogen to form hydroxyl groups.

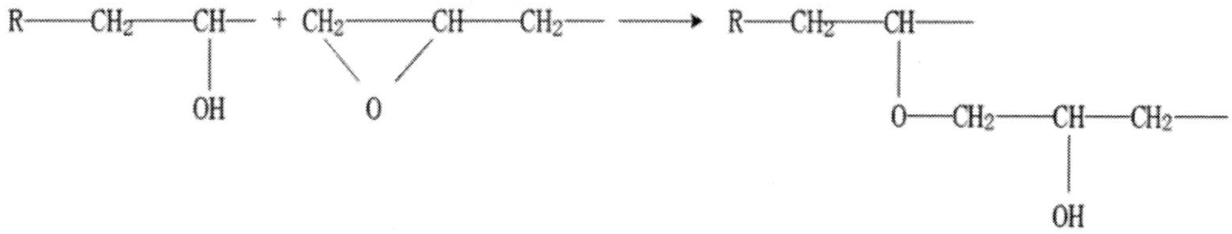

2.2. Evaluation of prepolymer quality

Measures to measure the quality of epoxy acrylate mainly consist of acidity value and viscosity [11]. In the reaction, the acidity value indicates the degree to which the reaction (esterification) proceeds (the conversion of the esterification reaction). The viscosity indicates the molecular size of the resin. In the experiment, a reasonable preparation process was selected by comparing the acidity value and viscosity of different reaction times [12].

2.2.1. Determination of acidity value. Weigh the quantitative samples dissolved in alcohol, adding phenolphthalein, titration with standard solution of potassium hydroxide; calculate the acidity value of the epoxy acrylic resin according to the following formula.

$$AV = \frac{C \times V \times 56.11}{m} \tag{1}$$

AV——acidity value V——the volume of KOH solution is consumed
C——concentration of KOH solution m——sample weight

The acidity value before reaction can be calculated according to formula (2)

$$AV_0 = \frac{K \times 56 \times 1000}{G} \tag{2}$$

AV_0—initial acidity value G—the total mass of the reactants

K—molar number of acids

The reaction conversion P:

$$P = \left(1 - \frac{AV}{AV_0}\right) \times 100\% \tag{3}$$

2.2.2. Determination of viscosity. The reaction product (prepolymer) of epoxy resin with acrylic acid is Newton fluid. There are many methods for measuring viscosity of Newton fluid, such as capillary method, rotation method, etc. This experiment uses falling-ball method [13].

2.3. Experimental equipment

Experimental equipment is shown in figure 1. It includes: thermostat water bath, mixer, three-necked flask, thermometer, dropper, and so on.

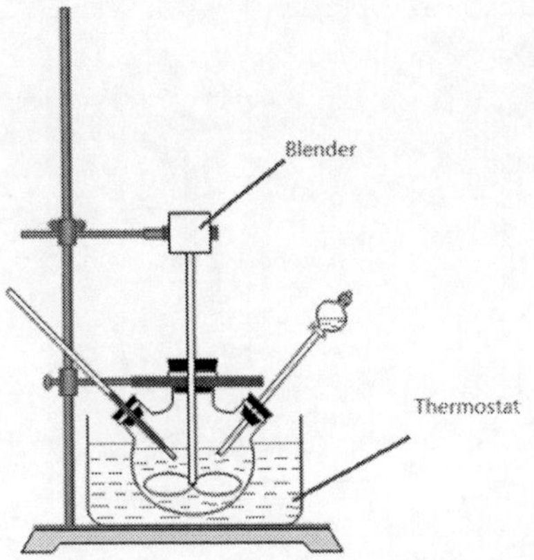

Figure 1. Experimental equipment.

2.4. Experimental material
Bisphenol A epoxy resin: E-51, acrylic acid (analytically pure), 95% ethanol (analytically pure), KOH (analytically pure), phenolphthalein (analytically pure)

2.5. Experimental method
(1) Preparation of phenolphthalein indicator and concentration of 0.1mol / L KOH-ethanol solution.

(2) Heat the thermostat water bath at a temperature of 90℃. Take proper amount of E-51, put it into three-necked flask, and heat three-necked flask in the thermostat water bath.

(3) Add acrylic acid to the three-necked flask slowly and stir constantly during the drip process. After completion of the drop in the corresponding time, heat the thermostat water bath into 100℃ and continue stirring.

(4) Measure the acidity value of the reactants in the three-necked flask at intervals of a certain time.

(5) After two hours of reaction, take a certain amount of reactants at intervals of one hour and measure the viscosity of the reactants using falling-ball method.

3. Experimental results

3.1. Acidity value and conversion rate in the reaction process
(1) According to formula (2), the initial acidity value before reaction is 155.6mg/g.

(2) In the process of reaction, the amount of acid was measured at regular time, and the acidity value and conversion rate were calculated according to formula (1) and formula (3). Table 2 shows the change of acidity value during the reaction. Figure 2 and figure 3 show the acidity value and conversion rate of reactants in different reaction stages.

Table 2. The acidity value changes during the reaction.

Reaction time	0.5h	1h	2h	3h	4h	5h	6h
Acidity value (mg/g)	100.8	72.8	44.8	28	16.8	5.6	5.6
Conversion rate (%)	35.2	53.2	71.2	82	89.2	96.4	96.4

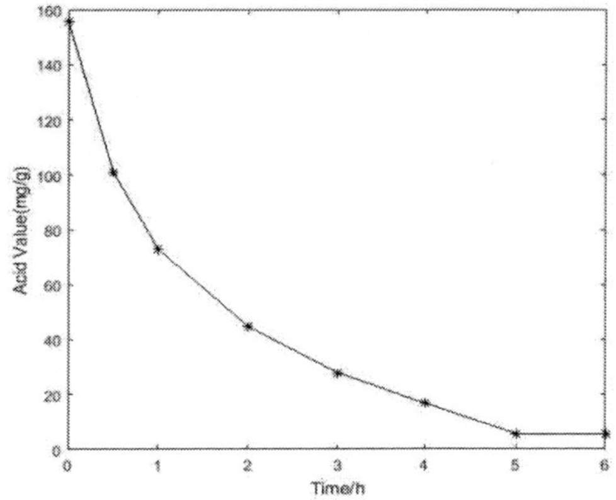

Figure 2. Acidity value-Reaction time.

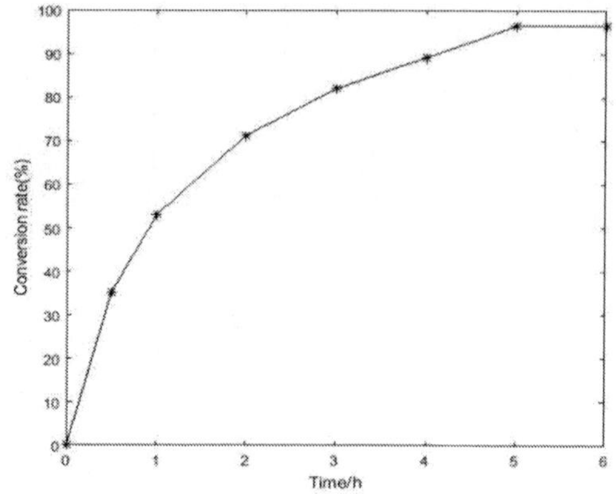

Figure 3. Conversion rate-Reaction time.

3.2. The change of viscosity during the reaction

In the reaction process, take proper amount of reactant and measure its viscosity at regular intervals by falling-ball method [14]. Then determine the change of viscosity under different temperature and reaction time. Table 3 shows the viscosity of prepolymer at different reaction time and environment temperature. Figure 4 and 5 show the viscosity and temperature-viscosity curves at different stages of esterification.

Table 3. The viscosity of prepolymer at different reaction time and environment temperature.

reaction time	viscosity of prepolymer (mPa·s)					
	15℃	25℃	35℃	45℃	55℃	65℃
2h	2752	1250	617	225	182	124
3h	5692	3086	1067	592	363	142
4h	13783	4313	2009	825	526	238
5h	34727	6800	3236	914	613	247
6h	41000	14278	5086	2022	793	278

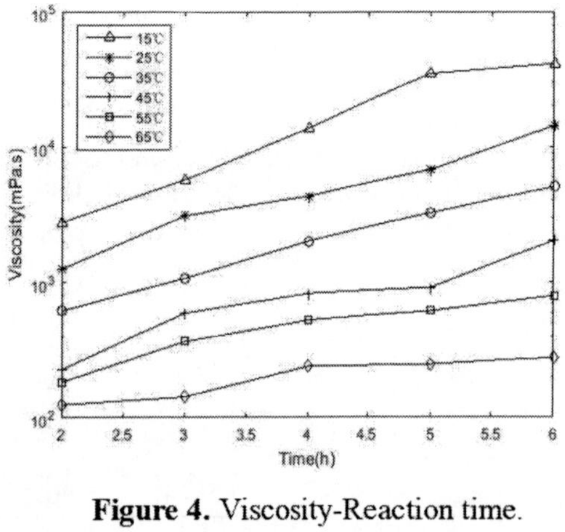

Figure 4. Viscosity-Reaction time.

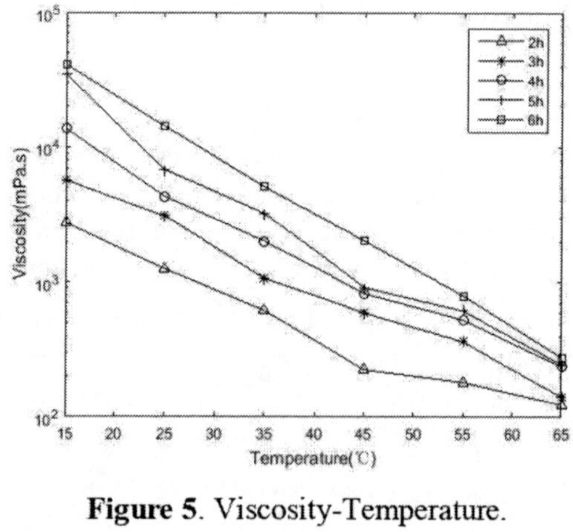

Figure 5. Viscosity-Temperature.

3.3. Experimental results

Experimental results show that:

(1) The initial esterification speed of epoxy resin and acrylic acid is faster. After 1 hour esterification, the conversion rate can reach 53.2%, and the corresponding acidity value is reduced to 72.8 mg/g.

(2) With the prolonging of esterification time, the conversion rate and acidity value gradually tend to be stable. When the reaction proceeds to 5 hours, the acidity value reaches a steady value of 5.6 mg/g. And the conversion rate of esterification is over 95%. As the reaction time increases, the conversion and the change of acidity value is no longer obvious. It indicates that the esterification reaction is basically over.

(3) As shown in figure 4, the viscosity of the reaction product of epoxy resin and acrylic acid increase with the progress of the reaction. The increase in viscosity is due to the fact that the macromolecule product is produced with the esterification reaction.

(4) As shown in figure 5, the viscosity of the reactants decreases with increasing temperature and the difference in viscosity of the reactants in the different esterification stages decreases as the temperature increases. When the temperature is 25 ℃ ~ 35 ℃, the viscosity of each reactant is relatively stable. Thus, the use of a viscosity of 25 ℃ to characterize the reactant viscosity is comparable.

(5) When the esterification reaction was carried out for 5 hours, the acidity value and the conversion rate of the reactants had reached a stable value. At this point, the viscosity of the reactant is 6800m Pa•s(25℃).

4. Conclusion

(1) The reaction temperature was divided into two stages (low temperature dropping and constant temperature reaction). It can effectively prevent the self-polymerization of acrylic acid and gel of hydroxyl groups of epoxy resin and acrylic acid. It is beneficial to reduce the viscosity of prepolymer and guarantee its quality by dividing the reaction temperature into two stages (low temperature dropping and constant temperature reaction) and strictly controlling the reaction temperature and reaction time.

(2) A low viscosity epoxy acrylate photosensitive prepolymer with a viscosity of 6800 mPa•s(25℃) was obtained by esterification of 5 hours with bisphenol A epoxy resin with high epoxy value and low viscosity. Compared with the viscosity value of partially low viscosity epoxy acrylic acid, the viscosity of epoxy acrylic acid prepared in this study is greatly reduced, which can meet the requirement of viscosity of light-repair. Table 4 is the viscosity comparison of the different prepolymers [1].

Table 4. Viscosity comparison of the different prepolymers.

Company	Self-made	Sartomer	UCB s.a.	Cognis	Cognis
Product code	——	CN121	EB6040	3215	3660
Chemical Name	Low viscosity EA	Low viscosity EA	Low viscosity EA	Low viscosity EA	Low viscosity EA
Viscosity(mPa•s, 25℃)	6800	57500	25000	15000	7000

Funding

This work is supported by the National Natural Science Foundation of China (Grant No. 51365012).

Reference

[1] Zhou X Y, Ma J J and Jiang L 2003 Research and composition of UV curable coatings *N chem.* **31** 20-23

[2] Miguelalmeida M E, Azevedo M L and Rachedjúnior F A 2012 Effect of light-activation with different light-curing units and time intervals on resin cement bond strength to intraradicular dentin *Braz. Dent. J.* **23** 362

IOP Conf. Series: Materials Science and Engineering **292** (2017) 012108 doi:10.1088/1757-899X/292/1/012108

[3] Pourreau D B and Smyth S 2005 New low-viscosity acrylic-urethane prepolymers and their acrylated oligomers for moisture and UV-curable coatings *Jct Coatingstech.* **2** 42-48
[4] Visvanathan A, Ilie N and Hickel R 2007 The influence of curing times and light curing methods on the polymerization shrinkage stress of a shrinkage-optimized composite with hybrid-type prepolymer fillers *Dent. Mater.* **23** 777-784
[5] Tillet G, Boutevin B and Ameduri B 2011 Chemical reactions of polymer crosslinking and post-crosslinking at room and medium temperature *Prog.Polym.Sci.* **36** 191-217
[6] Su Y C, Cheng L P and Cheng K C 2012 Synthesis and characterization of UV- and thermo-curable difunctional epoxyacrylates *Mater. Chem. Phys.* **132** 540-549
[7] Gao M, Deng G and Ren B 2009 Synthesis and Application of Photosensitive Novolac Epoxy Acrylate Resins *PCI* **39** 35-38
[8] Takizawa T and Ohmori N 2011 Photocurable liquid resin and photocurable liquid resin produced by such method
[9] Taib E R J, Abdullah L C and Min M A 2017 Physico-chemical characterisation of epoxy acrylate resin from jatropha seed oil *Pigm Resin Technol* **3**
[10] Jakob M and Nogai S 2011 Cross-linkable monomers and polymers and the use thereof.US EP 2181129 B1
[11] Su Y C, Cheng L P and Cheng K C 2012 Synthesis and characterization of UV- and thermo-curable difunctional epoxyacrylates *Mater.Chem.Phys* **132** 540-549
[12] Aminian A 2017 Predicting the effective viscosity of nanofluids for the augmentation of heat transfer in the process industries *J. Mol. Liq.* **229** 300-308
[13] Zhu K and Su Y 2017 Measuring viscosity coefficient of liquid by video image method *Exp. Tech. Mgmt.***34** 30-32
[14] Kuznetsov I A and Kuznetsov A V 2017 Using Resampling Residuals for Estimating Confidence Intervals of the Effective Viscosity and Forchheimer Coefficient *Transport Porous Med.*1-9

Synthesis and properties of amphiphilic hyperbranched polyethers as pigment dispersant

Q Xu[1], Y J Zhou[2, 3], S J Long[2], Y G Liu[2] and J H Li[2]

[1] Institute of Biology, Guizhou Academy of Sciences, Guiyang 550009, China
[2] Institute of Analysis and Testing, Guizhou Academy of Sciences, Guiyang 550002, China

[3] E-mail: zhouyuanjing-8006@163.com

Abstract. Hyperbranched polymers possess prominent properties such as low viscosity, good solubility, high rheological property, environmental non-toxic,and so on, which have potential applications in coatings. In this study, the amphiphilic hyperbranched polyethers (AHPs) consisting of hydrophobic hyperbranched polyethers core and hydrophilic poly (ethylene glycol) arms with different degree of branching (DB) under various reaction temperatures was prepared by the cation ring-opening polymerization. Their structures were characterized by IR, ^{13}CNMR and GPC. Their dispersion properties for pigment particles were investigated. The AHP_{47} with 0.47 DB was found to have good dispersion properties for Yellow HGR. This work would provide experimental data and theoretical foundation for the application of hyperbranched polyethers in environmental protection coating.

1. Introduction

Coating pigment particles are prone to form agaggregates because of their high surface energy, which affected seriously the quality of the coating. To make use of high-quality dispersant in pigment dispersing process is the best strategy. Dispersants play a very important role in the production of coatings. The stability of the dispersion system can avoid many coating problems and coating film defects. If the formula is reasonable, adding dispersant properly can effectively reduce the cost and improve the coating performance. Good pigment dispersant can have the following functions: (1) enhancing the luster and increasing the effect of levelling; (2) preventing floating; (3) improving the tinting strength; (4) reducing viscosity and increasing pigment loading; (5) reducing the flocculation; (6) preventing back coarse and increasing the storage stability; (7) increasing the colour saturation; (8) increasing transparency (organic pigment) or covering power (inorganic pigment); (9) improving the grinding efficiency and reducing production cost; (10) prevent sedimentation [1, 2].

Hyperbranched polymers with a special topological structure possess many unique and useful properties, such as high degree of branching (DB), lack of chain entanglement, high solubility, low melting point and low solution viscosity [3-5]. They have received considerable attention in recent years, and have potential applications in coatings [6, 7]. In this paper, a series of hyperbranched polyethers with different DB were synthesized under various reaction temperatures, and the amphiphilic hyperbranched polyether with a poly(3-ethyl-3-oxetanemethanol) core and multiple linear poly-(ethylene glycol) arms (AHP) was obtained as the pigment dispersant. Their structures and dispersion properties for pigment particles were investigated. It would be helpful to promote the hyperbranched polyethers for applying in environmental protection coating.

2. Experimental

2.1. Reagents and methods

Boron trifluoride etherate $BF_3O(C_2H_5)_2$ was purchased from Fluka. All other chemical reagents used in this experiment were obtained from commercial sources and were of the highest purity available. CH_2Cl_2 was dried by distillation in presence of CaH_2 under reduced pressure. Fourier transform infrared spectroscopy (FTIR) tests were conducted on a Spectrum 100 FTIR spectrometer (PerkinElmer, US) using KBr discs. ^{13}C NMR spectra were recorded on a Varian MERCURYplus 400 NMR spectrometer operating at 100 MHz. Molecular weight distributions were determined by using gel permeation chromatography (GPC) on Waters 2690 separations module using a waters 2410 refractive index detector with THF as an eluentand polystyrene as calibration standards. The viscosity was measured by Brookfield digital viscometer. The sagging test was carried out the sagging coating machine with 50-270 μm film thickness.

2.2. Synthesis and characterization

The synthetic route of amphiphilic hyperbranched polyether (AHP) was shown in figure 1. According to the reaction formula, the detailed synthesis process was described as below.

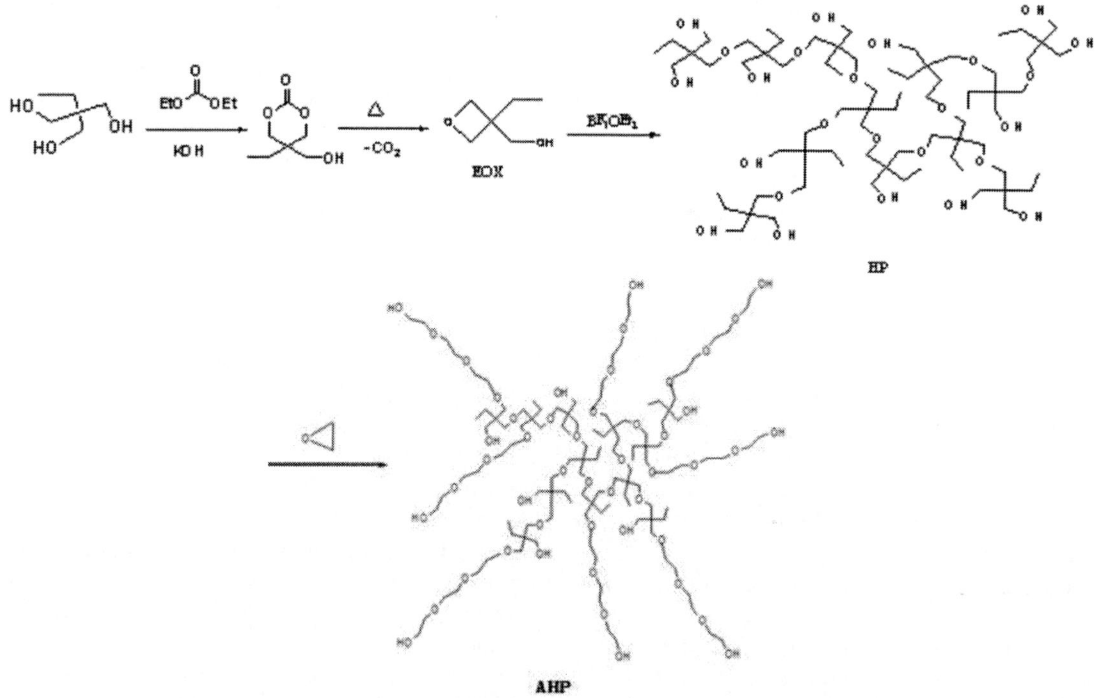

Figure 1. Synthetic routes of amphiphilic hyperbranched polyether (AHP)

2.2.1. Synthesis of 3-ethyl-3-oxetanemethanol (EOX). 3-Ethyl-3-oxetanemethanol (EOX) was prepared from trimethylolpropane according to the previous report with minor modifications [8]. Briefly, trimethylolpropane (26.8 g), diethyl carbonate (23.6 g) and potassiumhydroxide (10 mg) dissolved in absolute alcohol (0.5 mL) was refluxed at 105°C for 4h, and the mixture was heated above 180°C for 2h. The crude product was subjected to reduced pressure distillation, and the pure product (EOX) was obtained as a fraction boiling at 84–84.5 °C at 2.8 mmHg.FT-IR (cm^{-1}): 3413 (O-H), 2958, 2875 (C-H), 1050 (C-O), 976 (C-O-C).

2.2.2. Synthesis of hyperbranched polyethers (HPs). HPs were synthesized by cation ring-opening polymerization according to the previous report with minor modifications [9, 10]. Into a four-necked round-bottomed flask equipped with a PTFE stirrer, a funnel and a thermometer, nitrogen was flowed for 30 min to expel the oxygen from the system, then 80 mL of solvent CH_2Cl_2 and 6.4 mL of catalyst $BF_3O(C_2H_5)_2$ (0.05 mol) were added via syringe, espectively. The monomer EOX (0.1 mol, 11.6 mL) was introduced through the funnel within 10 min, and the reaction temperature was kept constant throughout the polymerization process. The polymerization was quenched with ethanol after 48 h. Then the mixture was precipitated into distilled water and immersed overnight. Finally, the resulting solid was obtained by filtration and then dried at 80°C under high vacuum. The same reaction was repeated at different temperatures. The samples prepared at 0°C, 10°C, 30°C and 50°C were named as HP_{35}, HP_{41}, HP_{47} and HP_{51}, respectively, according to the DB value. FT-IR (KBr, cm^{-1}): 3450, 2956, 2927, 2877, 1471, 1383, 1105, 1048, 752. ^{13}C NMR (100 MHz): δ (ppm) 7.9 ($-CH_3$), 22.8($-CH_2-$), 43.5(–C), 63.0 ($-CH_2OH$), 67.0– 72.0 ($-CH_2-O-CH_2-$).

2.2.3. Synthesis of amphiphilic hyperbranched polyethers (AHPs). HP and ethylene oxide were stirred for 24 h, and the hyperbranched polyether with a poly(3-ethyl-3-oxetanemethanol) core and multiple linear poly-(ethylene glycol) arms was obtained. FT-IR (KBr, cm^{-1}): 3347, 2958, 2930, 2881, 1480, 1385, 1111, 1050, 758. ^{13}C NMR (100 MHz): δ (ppm) 7.9 ($-CH_3$), 23.5 ($-CH_2-$), 43.7(–C), 62.0 ($-CH_2OH$), 67.0– 73.0 ($-CH_2-O-CH_2-$).

3. Result and discussion

3.1. DB analysis

The DB of a hyperbranched polymer is a measurement on the content of branches in the molecular structure and is considered to be a main structural feature affecting the properties. DB is expressed below in equation (1) [11]. L, D and T represent linear unit, dendritic unit and terminal unit, respectively.

$$DB = \frac{D+T}{D+T+L} \quad (1)$$

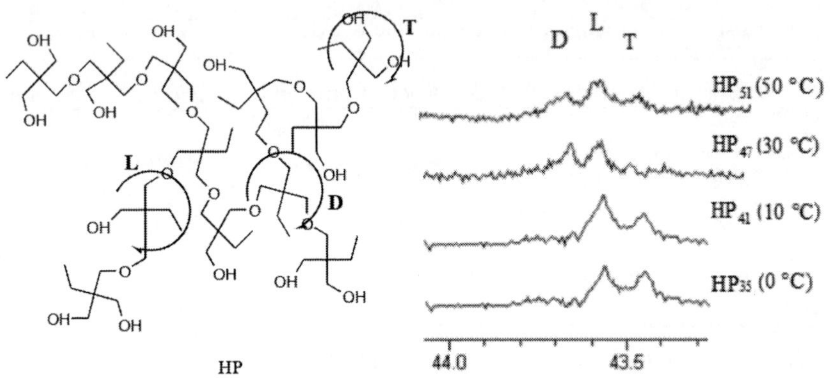

Figure 2. Structure of HP and ^{13}C NMR spectra of HP for the variation of the three peaks near 43.5 ppm in detail.

From figure 2, it showed in more detail the magnified signal region near 43.5 ppm in ^{13}C NMR spectra, which was attributed to quaternary carbon atoms of the D, L and T. [12] The DB value can be calculated by integrating the signals of different quaternary carbon atoms. The DB values for the samples prepared at 0°C, 10°C, 30°C and 50°C were 0.35, 0.41, 0.47, and 0.51, and the sample was labelled respectively HP_{35}, HP_{41}, HP_{47} and HP_{51}, which showed that the DB increased with increasing reaction temperature. That is, the structure of the resulting polyether depends on the reaction temperature: at low temperature, the linear addition is preferred, which leads to mostly linear products; at higher temperatures, the addition reaction becomes a random process resulting in hyperbranched polymers, which are consistent with previous reports [9, 10].

3.2. GPC analysis

The molecular weights of amphiphilic hyperbranched polyethers (AHPs) were determined by gel permeation chromatography (GPC), and the result was listed in table 1. As seen from the table, the weight average molecular weight (Mw) of the AHP varied from 2300 to 3400, and showed unregularity.

Table 1. The synthesis condition and characterization data of AHP.

Sample	T (°C)	Yield (%)	DB	Mw
HP_{35}	0	83	0.35	3037
HP_{41}	10	85	0.41	3379
HP_{47}	30	90	0.47	2385
HP_{51}	50	88	0.51	2655

3.3. Dispersion performance analysis

Pigments are usually insoluble in the used medium, and mostly in the form of aggregates. In order to obtain good coloring power, transparency, chromaticity, and so on, the aggregate of pigment must be prevented. If the pigment does not have good dispersion, many defects are likely to occur, such as flocculation, loss of light, color offset, floating, sedimentation, etc. In this paper, the viscosity, color and sagging performance were discussed. Table 2 showed the dispersion performance of AHPs for Yellow HGR in xylene and propylene glycol methyl ether acetate. It showed that AHP with different DB had different dispersion performance for Yellow HGR. The AHP_{47} with 0.47 DB owned the lowest viscosity and good color, and the sagging began from 150 μm film thickness, which was superior to other AHPs.

Table 2. The dispersion performance of AHPs on the Yellow HGR.

Sample	Viscosity (5rpm, cps)	Viscosity (50rpm, cps)	Color	Sagging (μm)
AHP_{35}	1668	1078	too light	75
AHP_{41}	1179	803	consistent	88
AHP_{47}	896	457	good	150
AHP_{51}	995	769	light	100

As previously reported that the structure of dispersing agent had important influence on the dispersion of pigment [13-16], the AHP_{47} with medium DB and lowest Mw among the products of AHP (figure 2) might interact with pigment particles better, and improve the dispersion performance. As seen from the table 2, the AHP_{47} could obviously reduce the viscosity of the color paste obtained, thus the loading capacity of the pigment could be increased and the production efficiency could be improved.

4. Conclusion

The amphiphilic hyperbranched polyethers (AHPs) with a poly(3-ethyl-3-oxetanemethanol) core and multiple linear poly-(ethylene glycol) arms were synthesized by simple and easy chemical reaction method. The AHPs owned good dispersion performance for coating pigment particles. Among the products of AHP, AHP_{47} with 0.47 DB had good potentiality in coating applications. The further study is underway.

Acknowledgments
This work was financially supported by the Science and Technology Project of Guizhou Province, China ([2016]1139, JZ [2015]2006, Z [2015]6013) and Youth Fund Project of Guizhou Academy of Sciences, Guizhou province, China (J [2016]20).

References
[1] Pan Y 2017 *Paint & Coatingd Industry* **2** 58
[2] Jin Z 2016 *Modern Paint and Finishing* **1** 39
[3] Qi M, Huang W, Xiao G, Zhu X, Gao C and Zhou Y 2017 *ActaPolym.Sin.* **2** 214
[4] Gurunathan T, MohantyS and Nayak, SK 2016 *Polym-Plast Technol.* **55** 92
[5] Ning Z, Wang X and Chen X. 2016 *Modern Chemical Industry* **10** 16
[6] Jana, T 2016 *Paintindia* **66** 59
[7] Zheng Y, Li S, Weng Z and Gao C 2015 *Chem. Soc. Rev* .**44** 4091
[8] Pattison D B 1957 *J. Am. Chem. Soc.***79** 3455
[9] Mai Y, Zhou Y, Yan D and Lu H 2003 *Macromolecules* **36** 9667
[10] Zhu X, Chen L, Yan D, Chen Q, Yao Y, Xiao Y, Hou J and Li J 2004 *Langmuir* **20** 484
[11] Hawker C J, Lee R and Frechet J M 1991 *J. Am. Chem. Soc.* **113** 4583
[12] Chen Y, Bednarek M, Kubisa P and Penczek S 2002 *J. Polym. Sci. Part A: Polym. Chem.* **40** 1991
[13] Zhu N, Ma D, Wang Y, Yang Q and Zhu J 2016 *Chem.Reagent* **12** 10
[14] Dong S, Liu G, Yang G and Bao J 2015 *Dyestuffs and Coloration* **6** 35
[15] Ren Q, Wang L, Li J, Deng J, Fang J, Wang C and Chen J 2014 *Journal of Chemical Industry and Engineering* **6** 2378
[16] Chen W, Yang C, Lu D, Liu A, Chai H and Song W 2013 *Textile Auxiliaries* **10** 20

The application of epoxy resin coating in grounding grid

Q Hu[1], Z R Chen[2], L J Xi[2], X Y Wang[2, 4] and H F Wang[3]

[1] Guizhou Electric Power Grid Co., Ltd. Electric Power Research Institute, Guiyang, 550002, China
[2] Northeast electric power university, Jilin, 132000, China
[3] Beijing Energy group Shenzhen Yu Lake Electric Power Co., Ltd, Shenzhen, 518000, China

[4] E-mail: tjkddflmg@163.com

Abstract. Epoxy resin anticorrosion coating is widely used in grounding grid corrosion protection because of its wide range of materials, good antiseptic effect and convenient processing. Based on the latest research progress, four kinds of epoxy anticorrosive coatings are introduced, which are structural modified epoxy coating, inorganic modified epoxy coating, organic modified epoxy coating and polyaniline / epoxy resin composite coating. In this paper, the current research progress of epoxy base coating is analyzed, and prospected the possible development direction of the anti-corrosion coating in the grounding grid, which provides a reference for coating corrosion prevention of grounding materials.

1. Introduction

Grounding grid is used as the grounding and lightning protection of substation ac and dc equipment, which plays an important part in the safe operation of power system transmission and distribution. The corrosion rate of grounding materials in soil is affected by soil pH, water content, salt content, organic matter content, microbial activity and other factors. The investigation results of many substations in China show that the effective lifetime of grounding materials in grounding grid is not more than 10 years, and in extreme cases, 3-4 years is seriously corroded and cannot be used continuously [1]. As an anticorrosive emerging technology, coating technology has been widely used in the field of anticorrosion. Epoxy resin - based anticorrosion coating is widely used because of its advantages such as good hydrolysis resistance, high strength, strong adhesion and good air tightness. Epoxy resin also has many shortcomings, such as brittleness, shock resistance, flexibility, low temperature resistance, UV resistance, linear difference. Therefore, it is necessary in order to modify the epoxy resin to improve its mechanical and corrosion resistance. At present, the modification of epoxy resin mainly focuses on the modification of epoxy resin structure, modification of inorganic filler and organic filler.

2. Structure modified epoxy resin

The introduction of specific functional groups in epoxy resin can improve the mechanical properties and anticorrosive properties of coatings. For example, the molecular structure of the naphthalene ring has low thermal expansion coefficient, excellent heat resistance and water resistance, but due to its rigid structure, the toughness of epoxy resin solidification is poor. Therefore, on the basis of the basic molecular skeleton of naphthalene, a good flexible fat chain is introduced, which makes the epoxy resin have a better flexibility [2]. Feng hao et al. create a modified epoxy resin that used polyetherketone-cadro(PEK-C) as a toughening agent and aromatic amine as curing agent, when

Wt.%(PEK-C) =45%~50%, the glass transition temperature (Tg) of modified epoxy resin is 179.8℃, and the roller peel strength is 123.60 N·mm/mm, which showed great thermal resistance [3]. According to Wang's study, side chain fluoridated epoxy resin was synthesized by laurylfluo-enanthol and bisphenol a epoxy resin. The results show that the fluorine content of the fluoridated epoxy resin is enriched on the surface, and the fracture elongation and impact strength are increased by 41% and 26.8% respectively compared with the bisphenol a epoxy resin, and the AC impedance is increased by more than 2 orders of magnitude. The toughness and electrical properties of fluoridated epoxy resin is improved remarkably [4].

3. Inorganic filler modified epoxy resin

Inorganic fillers can reduce the cost and fluidity of coatings, and improve the heat resistance, wear resistance, thermal conductivity and other mechanical properties of coatings.

Scale particle modified epoxy resin, such as glass flake, mica scale, aluminum powder graphite, etc, the most commonly used is glass scale packing. Glass flake is about 5 microns thickness for two-dimensional shards of glass. After joining resin are arranged parallel overlapping, the corrosive medium in speed is greatly delayed. Li bi-yuan et al. fill the curing epoxy resin with carbon black, anhydrous magnesium sulfate, aluminum oxide, found that the resistance to acid and alkali resistance of the epoxy resin modified have improved significantly. And compared with anhydrous Magnesium Sulfate and alumina, the carbon black showed the most noticeable improvement on the acid and alkali resistance of epoxy coatings. When the mass ratio of carbon black and epoxy resin was 2:20 and 1:20 respectively, the alkali resistance and acid resistance of the coatings increased by 10 times and 5 times respectively [5].

Fiber packing modified epoxy coatings conclude short glass fiber, carbon fiber, basalt fiber etc. Ye guo-rui et al. made a kind of modification graphene oxide by coupling agent and introduced it into the basalt fiber, and the modified basalt fiber epoxy resin was made. The results showed that the breaking strength of basalt fiber was enhanced by 30.8%, and the interlaminar shear strength of the composite fiber reinforced with graphene/basalt fiber/epoxy resin was increased by 10.6% [6]. Wang shu-juan et al. adding carbon fiber to epoxy resin, and was found that when composite modified carbon fiber content was 1.0%, the tensile strength of epoxy resin composites increased by 59.7%, the impact strength increased by 53.3% and the fracture elongation increased by 50.0%. In the meanwhile, the conductivity of epoxy resin was obviously improved [7].

Nanoparticles modified epoxy coatings conclude mica, silica, titanium dioxide, etc. Liu gang et al. added nano alumina to epoxy resin, and found that the anti-crack and expansion ability of the resin was improved obviously, while the other mechanical properties were basically unchanged [8]. Sun hong-mei et al. have added nano-silica to epoxy resin and experimented with electrical properties, and was found that the breakdown strength and dielectric loss of modified resin were improved, and the influence of resistivity and dielectric constant was not significant. Therefore, the addition of nano-silica can inhibit the electrical treeing of the resin and reduce the aging speed of the equipment [9].

4. Organic filler modified epoxy resin

Using rubber as modifier to modify epoxy resin can reduce the molecular internal stress, thus achieving the effect of enhancing the toughness and strength of paint. The nano-rubber modified epoxy resin prepared by Lin yan-yan et al., when the nano-rubber content was 6phr, the shear strength of the resin increased by 16.1%, the fracture elongation increased by 78.7% and the fracture energy increased by 3.65 times [10]. Xu li et al. modified the epoxy resin with the end hydroxyl butadiene rubber, and was found that the hydroxyl group reacted with the epoxide group in epoxy resin, were obtained good comprehensive mechanical properties and impact toughness, meanwhile the insulation resistance and dielectric strength increased, the dielectric constant and loss factor decreased, and the insulation capacity was improved [11].

5. Polyaniline /epoxy composited resin

Polyaniline is a kind of polymer material monomer with aniline, synthesized by chemical or electrochemical oxidation reaction. It has good environmental stability, simple synthesis method, low toxicity, easy material and high conductivity, also have protective properties to various materials such as carbon steel, Zn, Ti, Al and Cu. It has unique anti-pitting and anti-scratch performance, therefore, it is considered as a promising anticorrosion coating.

But there are many deficiencies in polyaniline, especially the insolubility, which greatly hinders the application of it. Polyaniline showed low solubility in a variety of organic solvents, such as carbon tetrachloride, trichloromethane and toluene. Although in N-methyl pyrrolidone (NMP), dimethyl sulfoxide (DMSO), dimethyl formamide (DMF) the solubility is higher, but the cost and environmental protection of the solvent limits on the application of polyaniline. It's low dispersion and adhesion in resin coatings also greatly reduce the anti-corrosion ability of polyaniline composite coatings. Therefore, the polyaniline material needs to be modified, composite or doped to synthesize composite materials with good anticorrosion properties.

Acid doped polyaniline coatings mainly have organic acid doping and inorganic acid doping, however, due to the large size of organic acids, the diffusion rate is slow and the doping effect is not good. Therefore, most of them are doped with inorganic acid. In 2007, Liu yi-jun et al. synthesized a conductive polyaniline by doped with hydrochloric acid, and then compounded it with epoxy resin to obtain a conductive coating, the electrical conductivity reached 10-8~10-5S/m [12]. In the same year, Fan hai-jun et al. found that the conductivity of polyaniline was strongly related to the strength of acid when it was made with inorganic acid as dopant. So they used perchloric acid as a dopant, as the result, the polyaniline films doped with perchloric acid have stronger electrical conductivity than the hydrochloric acid one [13].

Due to the sacrificial anode protection of metal, the dispersion and blocking action of metal oxides, the combination of metal or metal oxide with polyaniline can strengthen the anticorrosion of polyaniline. Wang hua et al. synthesized the polyaniline (PANI)/TiO2 composites by chemical oxidation method, dissolve and paint it on the surface of 304 stainless steel, then observe the corrosion behavior of the sample in 3.5% NaCl solution. The result showed that PANI-TiO2 composite coating has better protection to 304 stainless steel than polyaniline coating. When the content of TiO2 is 5%, the corrosion resistance of the composite coating is the best, and the corrosion potential is 381mV higher than that of the bare stainless steel, and the protection efficiency is 95.44% [14].

Huang de-yong et al. made the composite materials of polyaniline and glass scales by chemical oxidation. When the quality ratio of polyaniline and glass fiber was 2:1, the crystal performance of it was not much affected, but it's thermal stability was greatly improved. As the result, the epoxy resin polymerization coating prepared by this composite material greatly enhanced the anticorrosion effect and impact resistance [15].

6. Conclusion

In this paper, variety of modification methods for epoxy resin and polyaniline anticorrosive coating are enumerated. However, in the grounding grid, the corrosion resistance of the coating is not the only need to be considered, but also the coating has certain requirements for electrical conductivity. If the resistance of the coating is too large, the grounding performance of the grounding grid will be reduced. When the power grid system fails, the short circuit current cannot be discharge through the grounding network in time, which will cause serious accidents. In the future, the modification methods of epoxy resin can be studied in the following aspects.

Select more kinds of film-forming resins for comparison.

Intrinsic type conductive anticorrosive coating is a coating made of conductive polymer as matrix, such as polypyrrole, polyaniline and polythiophene. Owing to its inherent characteristics, the conductive ability is limited. Therefore, different kinds of film forming resins can be considered to improve their conductivity in the case of corrosion protection of grounding grids with superior electrical conductivity.

Using different dispersants to improve the dispersibility of nano materials.

At present, there are many researches on carbon nano materials in coatings, but few of them are put into use, because nanoparticles have small particle size, large specific surface area and high intermolecular force, so it is easy to agglomerate and disperse in coatings, which leads to the decrease of the overall performance of the coatings. It is necessary to change the surface structure of nano material by dispersing agent and show the affinity with organic material [16].

References

[1] Yan A J, Chen Y and Feng Y J 2010 *Corro. Sci. Prot. Technology* **197** 199
[2] Xu Y J 2008 *Dal. Univer. Technology* **16** 20
[3] Feng H, Qu C Y and Wang D Z 2015 *Chin. Adhe. Agent* **27** 30
[4] Wang C Z and Zuo Y 2009 *Thermo. Resin* **10** 13
[5] Li B Y, Ma L and Chen P 2017 *J. Minji. Coll.* **85** 89
[6] Ye G R, Yan Y W and Cao H L 2014 *J. Comp. Materials* **1402** 1408
[7] Wang S J. 2013 *J. Uni.* **20** 31
[8] Liu G, Zhang D J and Zhang H 2010 *Materi. Engineering* **47** 53
[9] Sun H M, Zhang K and Liu J N 2016 *Insul. Material* **12** 16
[10] Lin Y Y, Wang X L and Zhou Y Q 2016 *Plas. Industry* **32** 35
[11] Xu L, Zhou W Y and Gong Y 2016 *Chin. Plastics* **35** 39
[12] Liu Y J, Du Y C and Wang H Y 2007 *Coat. Industry* **16** 18
[13] Fan HJ, Gong C K and Zhang R G 2007 *Plasti. Technology* **40** 45
[14] Wang H and Wang H. 2015 *Elec. Coating* **94** 99
[15] Huang D 2015 *J. Uni.* **20** 22
[16] Liao G F, Zhang L and Zhang S L 2016 *Materi. Herald* **32** 38

IOP Conf. Series: Materials Science and Engineering **292** (2017) 012111 doi:10.1088/1757-899X/292/1/012111

The synthesis and characterization of p(nBA-DEAEMA) polymer

X M Jiang[1], J Zhu and Y Y Duan

Key Lab of Biorheological Science and Technology, Ministry of Education, Bioengineering College, Chongqing University, Chongqing, CHINA

[1] E-mail: jiangxuemei@cqu.edu.cn

Abstract. In this paper, the polymer of p (nBA-DEAEMA) was polymerized and characterized. The structure of copolymer was characterized by FT-IR and ^{1}HNMR, and the analysis of particle size, appearance and rheology was determined too. The results showed that, the p (nBA-DEAEMA) was synthesised successful, and it could become a surface modification for the inner face of artificial trachea, after further optimization.

1. Introduction

Trachea stenosis that is caused by the tracheal diseases and wounds is a common clinical disease with high mortality rates, and for these diseases, the unique curative treatment at present is the reconstruction of tracheal defect after extensive resection [1-2]. However, there are so many complications such as the airway restenosis and serious infection in the reconstructed trachea, and this is because the inner surface of the artificial trachea is not covered by cilium integrally so the wastes could not be cleared with cilium but store in the airway for a long time to become the hotbeds for bacteria to cause the serious infection, at the same time, the smooth inside wall leads to the running of the neighbouring tissue inwards to produce the proliferation with granulation tissue even restenosis [3]. That's the reason for the failure of reconstruction. In view of this situation, a new idea is provided to solve this problem based on the development of biomedical engineering and polymer science, which is coated a layer of polymer brush with cilia-like film cover on the inner surface of the reconstructed trachea to simulate the airway cilia layer. Combining the preparation and molding technology of polymer materials with the artificial cilia research, the clinical application of artificial trachea would have a new prospect. At present, the research of artificial cilia has shown a great charm with benefitting from its effectiveness to fluid treatment, but they still cannot meet the mechanical and biocompatible requirements as the cilia-like material. So, in addition to the function of removing the airway waste, we need to have a good biocompatibility and the ability to restrain the tissue from the surrounding tissue to the inner wall of the artificial trachea. Through the synthesis and study of the biological compatibility and mechanical properties of artificial cilia, to solve the problem of waste removal of artificial trachea, has very important theoretical significance and clinical application value.

The application of acrylate and its derivatives in the field of biomedical material had a long history, and they owned so many good properties as water resistance, strong adhesion, flexibility and elasticity [4]. As a fine material for surface modification, polymer brushes, not only in improving the biocompatibility [5] of the matrix but also in obtainment of functional material [6-9], had intensive studies yet.

In this paper, we used two acrylate monomers to synthesize a polymer and combined the molding process of Velcro to help achieving a topology of cilia-like surface. n-butyl acrylate (nBA) and 2-(diethylamino) ethyl methacrylate (DEAEMA) monomers were selected to copolymerize for preparing the polymer brush and molding a cilia-like film to lead the cilia-like behavior. Some properties of copolymer and cilia-like film were characterized, and then its function to clear away the airway wastes was explored with a preliminary research. The results showed a potential of this material to be a surface modification for the inner face of the artificial trachea.

2. Materials and methods

2.1. Chemicals

DEAEMA, nBA, cethltrimethylammonium chloride solution (CTAC, 25 wt.% in H_2O), tetrahydrofuran (THF), acetone and 0.1N volumetric standard solutions of sodium hydroxide (NaOH) and hydrochloride acid (HCl) were purchased from Sigma-Aldrich Chemical Co. 2,2'-Azobis (2-Methylpropionitrile) (AIBN, >98%+), potassium bromide (KBr, spectral purity, >99%), and deuterium dimethyl sulfoxide (DMSO-d6, spectral purity, >99%) was purchased from Adamas Reagent Co. Ltd.

2.2. Synthesis of copolymer

p(nBA-DEAEMA) copolymer was synthesized using the semi continuous emulsion polymerization process that was adapted for a small-scale polymerization. At room temperature, 25mL of double dionized (DDI) water, 0.62g of CTAC, 5.6g of monomers and 0.1g of oil-soluble initiator AIBN were prepared to form pre-emulsion at 300 rpm. Then, the other reaction flask was immersed in a water bath preheated to 72℃ and purged continuously with N_2 gas, which was first charged with 27mL of DDI water. At this point, the pre-emulsion was fed at 110μL/min into the vessel over a period of 3h. After completion of pre-emulsion feeding, the reaction was continued for additional 3h. The resulting colloidal dispersion was frozen drying after cooling to ambient temperature to obtain the copolymer. p(DEAEMA) and p(nBA) homopolymers were prepared using the same method.

2.3. Characterization of copolymer

The Molecular weight was determined using GPC (PL-120) equipped with a 1515 HPLC pump and a 214 nm model refractive index detector. Each sample was precipitated in THF and eluted through a HT2 and HT6E column. Elution time was referenced against polystyrene standard, and the molecular weight and polydispersity of the copolymer were 1803289 Daltons and 1.809559. Generally speaking [10], higher the molecular weight of polymer was, better the mechanical and thermal properties were but weaker the process ability was. So, high molecular weight and wide polydispersity had been obtained at the same time to ensure a smooth molding and a suitable mechanical performance for the cilia-like film in this part. FT-IR spectra were collected using Lambda 900 UV/NIR/MIR spectrometer (Perkin Elmer, USA) and Proton NMR (^1H NMR) spectra were acquired using the Nuclear Magnetic Resonance Spectrometer (500MHz, BRUKER OPTICS, Swiss). For FT-IR, a certain amount of freeze-drying copolymer colloids into acetone until dissolved completely, and drop a little sample solution onto the KBr blank flake with pipette, then test sample until the acetone evaporated completely. As anticipated, characteristic broadening and a shift of the carbonyl ester band at 1732 cm^{-1} with the band due to aminoethyl groups at 1568 cm^{-1} confirm copolymerization. For ^1H NMR measurement, 2.78% (w/v) of p (nBA/DEAEMA) copolymer was dissolved in DMSO-d6, and typical resonances at 0.85 ($-CH_2-CH_3$) with shoulder peaks and 3.9-4.1 ppm ($-O-CH_2$) arise from the copolymerization of nBA units, whereas the peaks at 1.2 ($-C-CH_3$), 3.0($-CH_2-N$), 1.0($-N-(CH_2CH_3)_2$), and 3.9-4.1 ppm ($-O-CH_2$) are due to DEAEMA units. Particle size analysis was performed using an analyzer (zetasizernano ZS90, Malvern, UK). Auto calibration against standard buffer solutions was completed before the titration. Standard HCl and NaOH were utilized to adjust pH values of copolymer dispersions. We got the results that the particle size increased from 44nm to 53nm as the pH values increased from 2 to 3 firstly and then decreased to 39nm as the pH values increased to 12 at

25℃. This behavior was attributes to the protonation of the tertiary amine functional groups that caused the colloidal particles to swell as the copolymer chains repel each other due to electrostatic interactions. Full protonation and swelling were achieved under acidic conditions, which was responsible for significantly larger particle sizes detected at pH value of 3. Also morphology analysis showed the consistent size as the particle size analysis that was 40-50nm. Most studies believed that the particle size of colloids and size distribution of emulsion played a main role in the viscosity of high-solid content emulsion. The particles obtained in our paper were nanoparticles, and not only could the micro size enable to get more dense film to improve its mechanical properties, but also could get enlarger surface area of the film to make contact with the mucus fully, so more effective mechanical response behavior was produced and there was little influence on the properties of material due to the pH value. The resulting copolymer colloids were dissolved in glycerin to be prepared some equidifferent emulsions with 10-80mg/mL, and then tested under 25℃ and 37℃. The rheological behavior was measured by rheometer (Gemini HRnano, Marvin, UK).

3. Results and Discussion

Figure 1 illustrated the FT-IR spectra and of p(nBA) and p(DEAEMA) homopolymers and p(nBA/DEAEMA) copolymer, respectively. As seen in trace A, it showed the FT-IR spectrum of p(nBA/DEAEMA) copolymer. As anticipated, selected bands in the 2850-3000 cm^{-1} and 1350-1500 cm^{-1} regions due to C-H stretching and CH_2/CH_3 deformation modes overlap with the p(nBA) and p(DEAEMA) spectra. Characteristic broadening and a shift of the carbonyl ester band at 1732 cm^{-1} with the band due to aminoethyl groups at 1568 cm^{-1}confirm copolymerization.

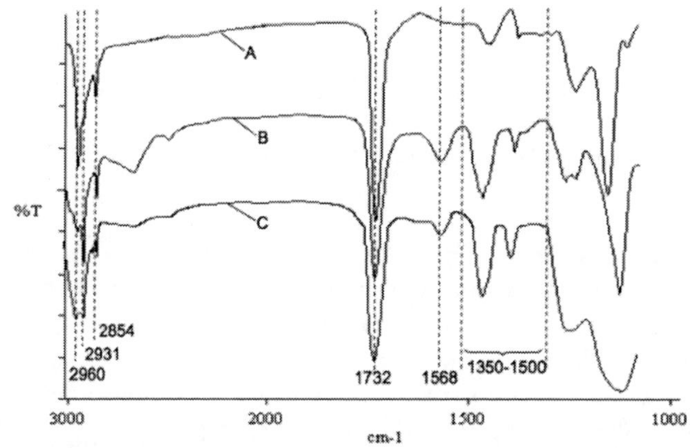

Figure 1. FT-IR spectrums of homopolymers and copolymer.

Further evidence of copolymerization was illustrated in the 1H NMR spectrum of p(nBA/DEAEMA) recorded at 25℃ in DMSO. As shown in figure 2, typical resonances at 0.85 (-CH_2-CH_3) with shoulder peaks and 3.9-4.1 ppm (-O-CH_2) arise from the copolymerization of nBA units, whereas the peaks at 1.2 (-C-CH_3), 3.0(-CH_2-N), 1.0(-N-$(CH_2CH_3)_2$), and 3.9-4.1 ppm (-O-CH_2) are due to DEAEMA units. The resonance at 2.5 ppm was due to DMSO solvent and 3.4 ppm was due to the water in DMSO. These spectroscopic data demonstrate that nBA and DEAEMA monomers were copolymerized for the feed ratio of 1:1 during the copolymerization.

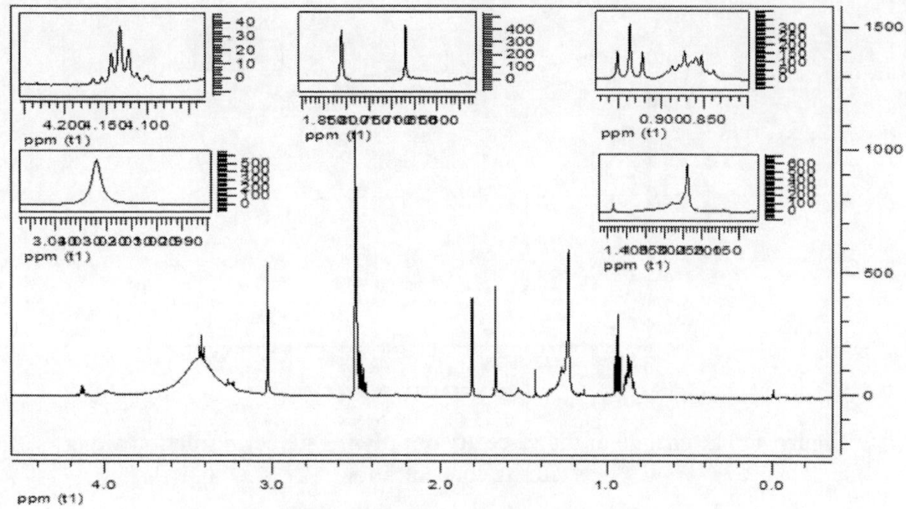

Figure 2. NMRspectrum of copolymer.

The particle size of this colloidal suspension was analyzed as a function of pH 2-12 at the 25℃ temperature. Figure 3illustrated the size distribution by volume of p(nBA/DEAEMA) copolymer colloidal dispersions at pH 7 and its pH responses. As seen, the particle size increased from 44nm to 53nm as the pH values increased from 2 to 3 firstly and then decreased from 53nm to 39nm as the pH values increased from 3 to 12 at 25℃. This behaviour was attributes to the protonation of the tertiary amine functional groups that caused the p(nBA/DEAEMA) colloidal particles to swell as the copolymer chains repel each other due to electrostatic interactions. Full protonation and swelling were achieved under acidic conditions, which was responsible for significantly larger particle sizes detected at pH value of 3.

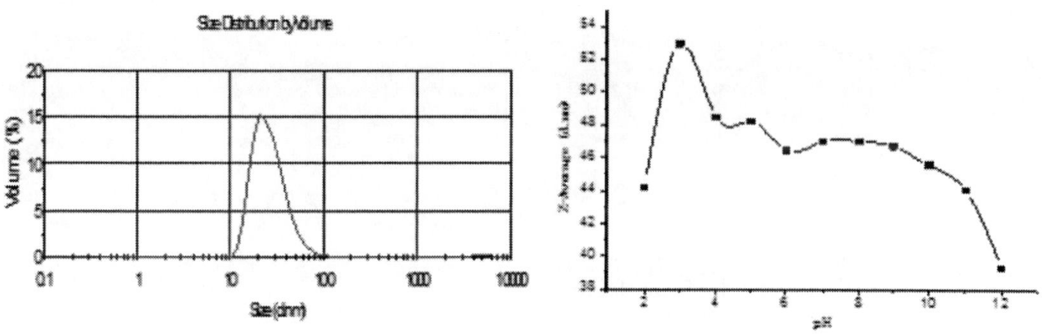

Figure 3. The particle size of copolymer at pH 7 and its pH responses.

The follows showed the rheological property of p(nBA/DEAEMA) colloidal particles in an solution state.

Generally speaking, the relationship should to be a linear one between the viscosity of polymer solution and its concentration, but in this article, as figure 4 shown, it's a non-linear relationship. Whether 25℃ or 37℃, there was a deflexed phenomenon within the concentration range of 60-80mg/ml.While the difference was that, when the concentration of polymer solution reached70mg/ml, the viscosity decreased more at 25℃ comparing with 37℃.

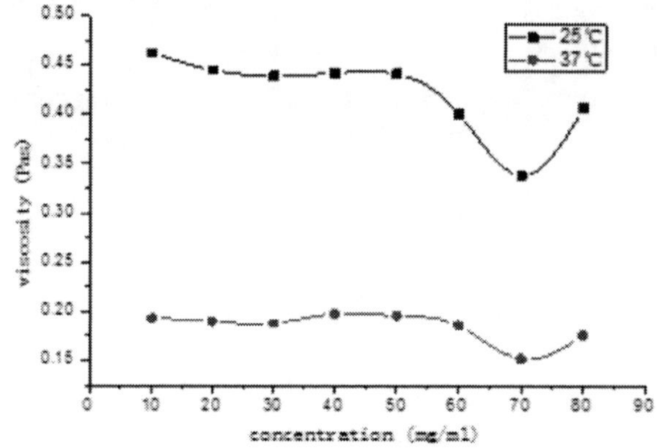

Figure 4. The change of the viscosity of polymer glycerin solution along with its concentration

Figure 5 illustrated the modulus changes of p(nBA/DEAEMA) glycerin solution with different concentration under 25℃ and 37℃, respectively. Overall, the elastic modulus of polymer solution both 25℃ and 37℃ had no obvious transformation and maintained at the same level and its complex modulus had the same changed trends with its viscous modulus, but its viscous modulus(complex modulus) reduced by 55% approximately as a whole at 37℃, compared with 25℃. We were pleasantly surprised to see that different variation of viscous modulus was taken place at different temperature. At 25℃, the viscous modulus of polymer glycerin solution showed decreases first and then increase, and reached a minimum at the concentration of 70mg/ml. In contrast, at 37℃, its viscous modulus showed increase first and then decrease, and reached a maximum at the concentration of 40mg/ml.

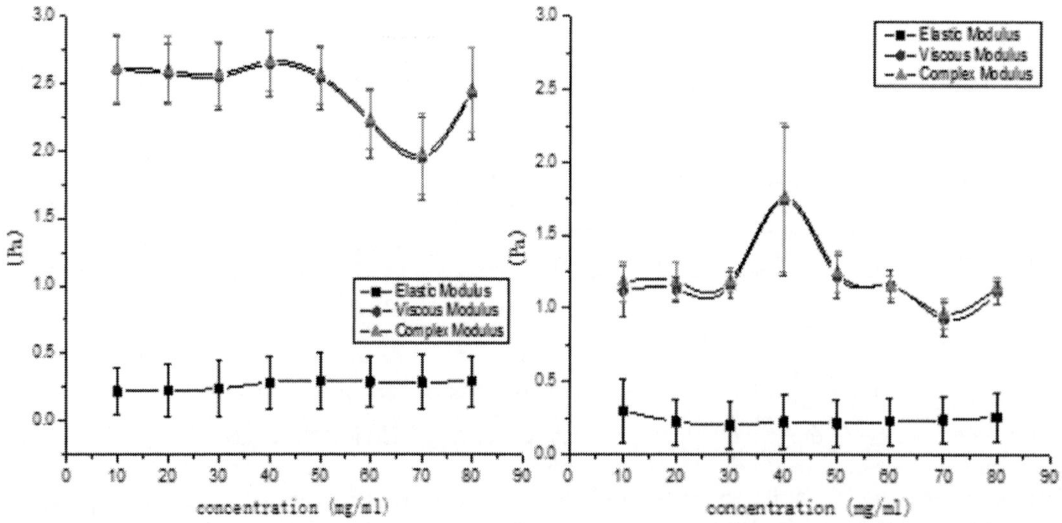

Figure 5. The modulus changes of polymer p(nBA/DEAEMA) glycerin solutionwith different concentration under 25℃(left) and 37℃(right).

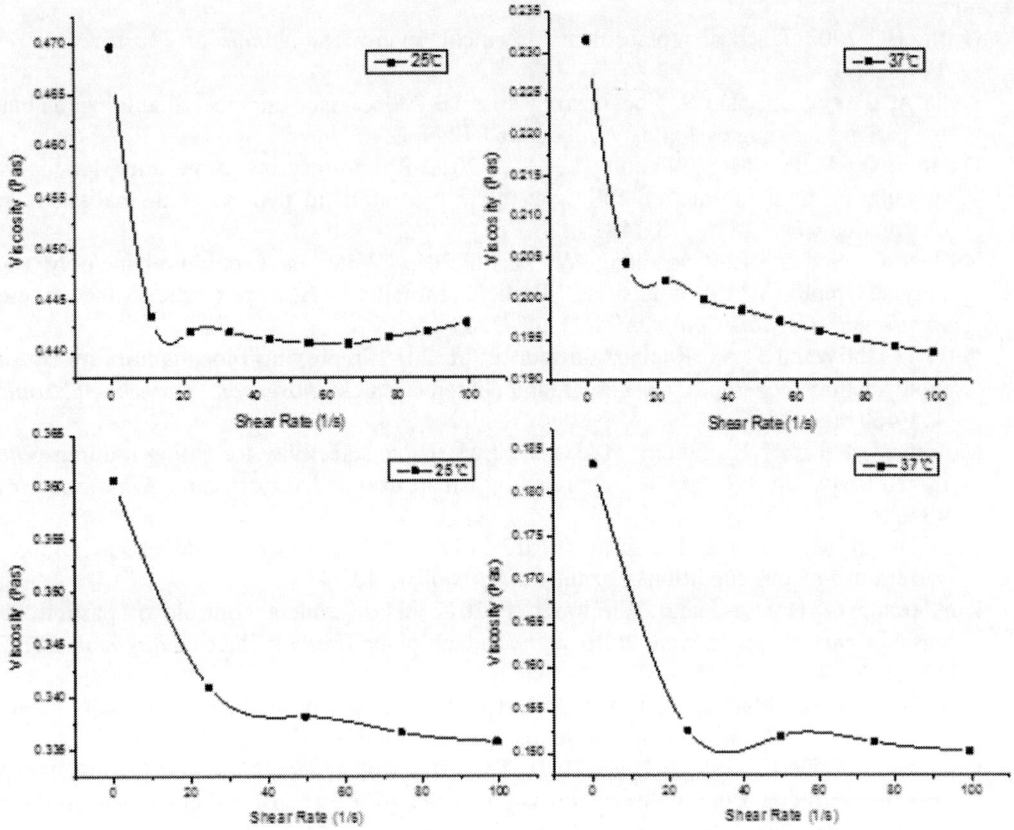

Figure 6. The flow curve (fixed-frequency: 1Hz) of polymer p(nBA/DEAEMA) glycerin solutionatthe concentration of 40mg/ml(up) and 70mg/ml(down).

We can see from the figure 6 that, the viscosity of polymer glycerin solution decreased along with the increase of the shear rate whether different temperature or different concentration, and it's a pseudo plastic fluid. However, there was a rapid decrease in the range of 0-20 s^{-1} but a slow reduction to steady gradually in the range of 20-100s^{-1} at the concentration of 40mg/ml under 25℃, and the situation under 37℃ was a continuous decline in the whole range of the experimental shear rate at the same concentration. In the same experiment, it showed a reverse trend at the concentration of 70mg/ml, that was, the sustaining downward of viscosity appeared at 25℃ while it stabilized from 20s^{-1} at 37℃.

4. Conclusion

We have prepared a cilia-like film with the two monomers to mimic the natural cilium successfully. The cilia-like layer could exist stably on the matrix and keep certain rigidity away from falling under the action of shear force, and it could product some effective mechanical behaviors to propel the wastes that attached on the airway mucus. This research could provide a new idea to solve the problem that the inner surface of the artificial trachea was not covered by airway cilium integrally and it has a very important value of theory and clinical application.

Funding

This work was supported by the National Natural Science Foundation of China (31400805) and Fundamental Research Funds for the Central Universities (No.106112017CDJQJ238815).

References

[1] Grillo HC 2002 Tracheal replacement: a critical review *The Annals of Thoracic Surgery* **73** 1995-2004

[2] Atala A, Bauer SB, Soker S, Yoo JJ and Retik AB 2006 Tissue engineered autologous bladders for patients needing cystoplasty *Lancet* **367** 1241-1246

[3] David E O, Archan M S, Xiudong L, et al. 2012 Respiratory infections increase the risk of granulation tissue formation following airway stenting in patients with malignant airway obstruction *Chest* **141** 1473-1481

[4] Yu-hua G, Shu-cai L, Gao-sheng W, et al. 2012 Waterborne polyurethane/poly (n-butyl acrylatstyrene) hybrid emulsions: Particle formation, film properties, and application *Progress in Organic Coatings* **74(1)** 248-256

[5] Peter L, ChiiWann L and Raafat Mansour, et al. 2013 Improving biocompatibility by surface modification techniques on implantable bioelectronics *Biosensors and Bioelectronics* **47** 451-460

[6] Motornov M, Tam T K., Pita M, et al. 2009 Switchable selectivity for gating ion transport with mixed polyelectrolyte brushes: approaching smart drug delivery systems *Nanotechnology* **20** 434006

[7] Motoyasu K, Masami T and Atsushi T 2012 Polyelectrolyte brushes: a novel stable lubrication system in aqueous conditions *Faraday discussions* **7** 403-412

[8] Dunderdale G, Howse J and Fairclough P 2012 pH-dependent controle of particle motion through surface interactions with patterned polymer brush surfaces *Langmuir* **28** 12955-12961

[9] Tam Tsz Kin, Pita Marcos, Motornov Mikhail, et al. 2010 Modified electrodes with switchable selectivity for cationic and anionic redox species *Electroanalysis* **22** 35-40

[10] K Magniez, C De Lavigne, B L Fox 2010 The effects of molecular weight and polymorphism on the fracture and thermo-mechanical properties of a carbon-fibre composite modified by electrospun poly(vinylidene fluoride) membranes *Plymer* **51** 2585-2596

IOP Conf. Series: Materials Science and Engineering **292** (2017) 012112 doi:10.1088/1757-899X/292/1/012112

Experimental and numerical analysis of the constitutive equation of rubber composites reinforced with random ceramic particle

D M Luo[1], Y Xie, X R Su and Y L Zhou

Department of Civil Engineering, Foshan University, Foshan, 528000

[1] E-mail: dongmei_luo@126.com

Abstract. Based on the four classical models of Mooney-Rivlin (M-R), Yeoh, Ogden and Neo-Hookean (N-H) model, a strain energy constitutive equation with large deformation for rubber composites reinforced with random ceramic particles is proposed from the angle of continuum mechanics theory in this paper. By decoupling the interaction between matrix and random particles, the strain energy of each phase is obtained to derive the explicit constitutive equation for rubber composites. The tests results of uni-axial tensile, pure shear and equal bi-axial tensile are simulated by the non-linear finite element method on the ANSYS platform. The results from finite element method are compared with those from experiment, and the material parameters are determined by fitting the results from different test conditions, and the influence of radius of random ceramic particles on the effective mechanical properties are analyzed.

1. Introduction

Some classical theoretical models, such as the Vogit upper-limit theory [1], Reuss lower-limit theory [2], Eshelby equivalent inclusion theory [3] and Mori-Tanaka method [4] and so on are satisfactory to analyze the mechanical property of particles composites with small deformation. However, this methods are not so precise for the analysis of rubber composites with large deformation. There are two kinds of methods to establish the constitutive equation for different materials, the first is the crystal slip model based on the meso-scopic plastic theory, which is limited for the complexity of slip direction. The second is the material constitutive relationship based on phenomenal theory, which is obtained by combining the mathematical method with experimental results. Based on the Neo-Hooken model, Guo Zaoyang [5] proposed a strain-energy function theory to analyze the mechanical properties of fiber composites with finite deformation, and the accuracy of the results were verified by numerical simulation. Huang Xiaoshuang [6] proposed an an-isotropic visco-super elastic constitutive model, and the parameters in the model are determined by uni-axial tension experiment, and the results are verified by the finite element method. By comparing the results from Matlab numerical simulation, Huang Jianlong [7] thought that the Yeoh strain-energy model is prior to the Mooney-Rivlin model to analyze the large deformation of rubber composites. In order to analyze the influence of reinforcement on effective mechanical properties of composites, Li Qing [8-9] added the carbon black with different volume fraction into the rubber materials to perform the mechanical experiment, the results show that the stiffness and initial modulus increases with the increase of volume fraction of carbon black. Dong Jinping [10] Put forward a kind of transverse isotropic hyperelastic constitutive model to describe the rubber composites reinforced with short fiber, combining with some test results, its application in numerical simulation is studied, and the result is satisfactory. According to the large strain hardening phenomenon of the rubber, Liu Feng [11] proposed a strain energy function to describe the constitutive relationship of rubber material, and the least square method is applied to fit the constitutive parameters of rubber. The good consistence with the experimental results establishes a foundation for the study of the large deformation of rubber. In the above researches, the strain invariant is always taken as variant to establish the strain energy function, which is reliable to describe the rubber with isotropic and in-compressible, but it is inappropriate to analyze non-linear rubber composites. In this paper, a new strain energy function which considering the interaction between random particles and rubber matrix is proposed, and the parameters are determined by fitting the curves from the uni-axial tension, equal bi-axial tension and pure shear experiments, and the stress-strain relationship is simulated by finite element method on the ANSYS platform, and the results are good consistent with those from the experiment.

2. Classical models of strain energy functions for rubber materials

A plenty of hyper-elastic constitutive models for rubber materials were proposed in the past years, and most of them were based on the concept of strain energy function, and the second Piola-Kirchhoff theory is utilized to establish the relationship between the hyper-elastic constitutive equation and the strain energy function [8].

$$\sigma = 2\frac{\partial W}{\partial C} \tag{1}$$

Where C is the strain tensor, and W is the strain energy function, which is determined by different supposed models.

Table 1. Classical models of strain energy functions for rubber materials.

Model	Common expression	Parameter meaning
Mooney-Rivlin (M-R)	$W = C_{10}(I_1 - 3) + C_{01}(I_2 - 3) + \frac{1}{d}(J-1)^2$	where C_{10}, C_{01} and d are the Mooney constants, which are determined by experimental method, and I_1, I_2 is the first and the second invariant, respectively. For in-compressible materials, $J=0$.
Yeoh	$W = \sum_{i=1}^{N} C_{i0}(I_1 - 3)^i + \sum_{k=1}^{N}\frac{1}{d_k}(J-1)^{2k}$	where N, C_{i0} and d_k are material constants, which are also determined by experimental method.
Ogden	$W = \sum_{i=1}^{3}\frac{\mu_i}{\alpha_i}(\lambda_1^{\alpha_1} - 1)$	Where, $\alpha_1 = 1.3, \alpha_2 = 5.0, \alpha_3 = -2.0$, $\mu_2 = 0.003\mu_0, \mu_3 = -0.0237\mu_0$, $\mu_1 = 1.491\mu_0$, are determined by Treloar test [12], $\mu 0$ is the initial poisson's ratio of rubber
Neo-Hookean (N-H)	$W = \frac{\mu}{2}(I_1 - 3)$	Where μ is the shear modulus.

The typical models for strain energy function showed in table 1 includes the polynomial model (M-R model), the first-order and third-order reduced polynomial model (N-H and Yeoh model), and the high-order polynomial model (Ogden model), and the difference among the four models were determined by different parameters in the equations, and the typical models and their parameters are shown in table 1. The purpose in this paper is to construct a nonlinear constitutive equation for rubber composites on the basis of the typical models, and the interaction between ceramic particles and rubber matrix is considered in the proposed equation, and the corresponding parameters in the equation are determined by fitting the results from numerical simulation on the platform of ANSYS to those from the experiments.

3. Constitutive equation for rubber composites with random ceramic particles

The proposed constitutive model in this paper is

$$W = W^M(I_1, I_2, I_3) + W^F(I_0) + W^{FM}(I_{10}) \tag{2}$$

where $w^M(I_1,I_2,I_3)$ is one of any classical model with a coefficient of $A=(1-v_P)\mu_m+v_P\mu_p$, where v_p is the volume fraction of ceramic particles, $\mu m, \mu p$ is the poisson's ratio of matrix and particles, respectively. $W^F(I_0)$ is the strain energy of ceramic particle. The elastic modulus of random particles are much greater than that of rubber matrix, and the particles can be treated as rigid body, so $W^F(I_0)=0$, $W^{FM}(I_{10})$ is the strain energy for the interaction between rubber and random ceramic particles.

In this paper, the random particles are considered as the infinitely small **and infinite density** orthogonality fabric, **so** the strain energy produced by the interaction between particles and matrix is the change of the angle between the orthogonal fibers for the particles (as shown in figure 1 and figure 2).

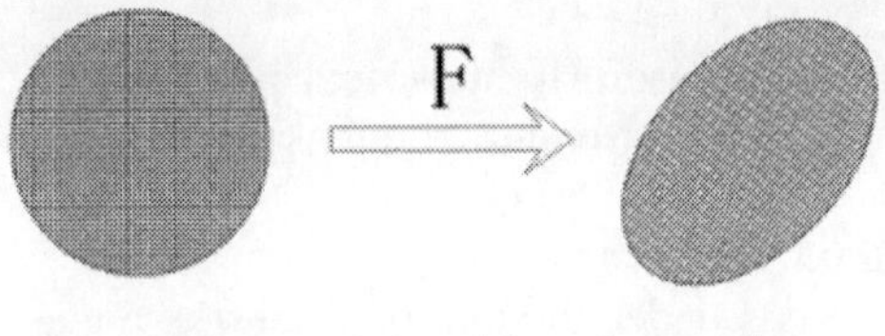

Figure 1. The particle with orthogonal woven.

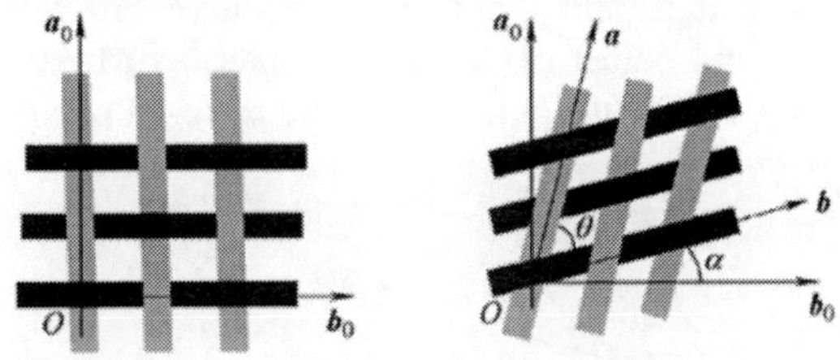

Figure 2. The change of the angle between fabrics Supposing.

$$I_{10} = \Delta\theta(I_4, I_6, I_8) \tag{3}$$

For the orthogonal fabric, since the angle has nothing to do with I_6, Equation (6) becomes

$$I_{10} = \Delta\theta \approx (I_4 I_8)^{-0.5} a_0 \cdot C \cdot b_0 - a_0 \cdot b_0 \tag{4}$$

and the strain energy for the interaction between rubber and random particles can be written as

$$W^{FM}(I_{10}) = W^{FM}(\Delta\theta) \approx (I_4 I_8)^{-0.5} a_0 C b_0 - a_0 b_0 = \sum_2^4 s_i (I_{10})^i \tag{5}$$

In which, C is the shear strain tensor, and

$$I_4 = a_0 \cdot C \cdot a_0 = \lambda_a^{\ 2}, \ I_8 = b_0 \cdot C \cdot b_0 = \lambda_b^{\ 2} \tag{6}$$

Where, a_0 and b_0 is the unit vector along the fiber direction for the orthogonal fabric as shown in figure 2, and λ_a, λ_b are the tensile rate of the orthogonal fibers, I_4, I_6 and I_8 all are the invariant. It is important to determine the shear parameters by experimental methods.

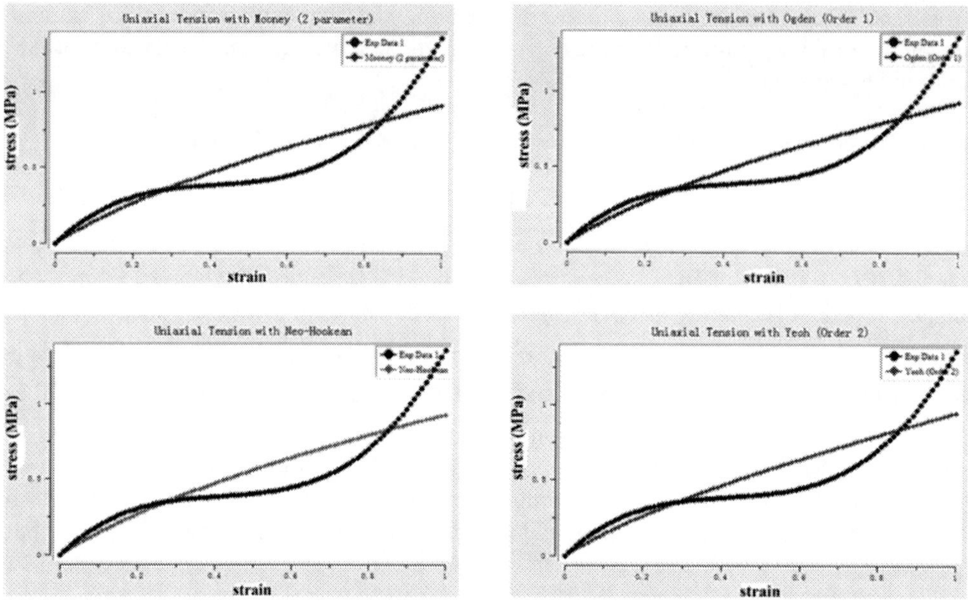

Figure 3. Fitting the test results to numerical simulation for different models with uni-axial tension.

4. Determination of material parameters

The test results for rubber materials under the uni-axial tensile, equal bi-axial tensile and pure shear are selected from reference [9]. By substituting the constitutive models into the ANSYS platform, the finite element results are simulated, and the non-linear curve are selected to fit the stress-strain relationship for both from test and numerical simulation as shown from figure 3 to figure 5, and the average curves for each test are shown in figure 6, then the material parameters are obtained in table 2.

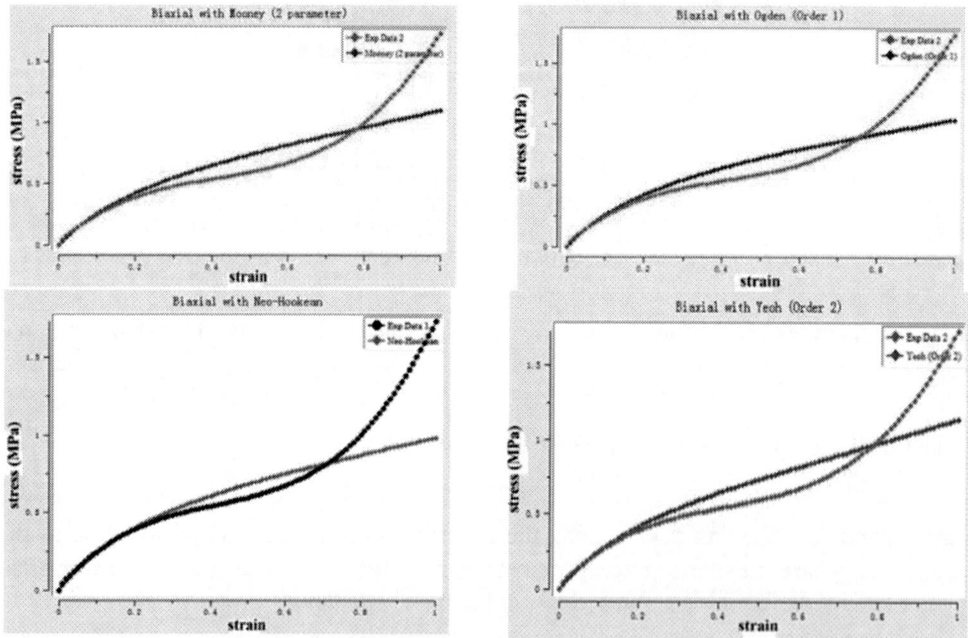

Figure 4. Fitting the test results to numerical simulation for different models with Equal bi-axial tension.

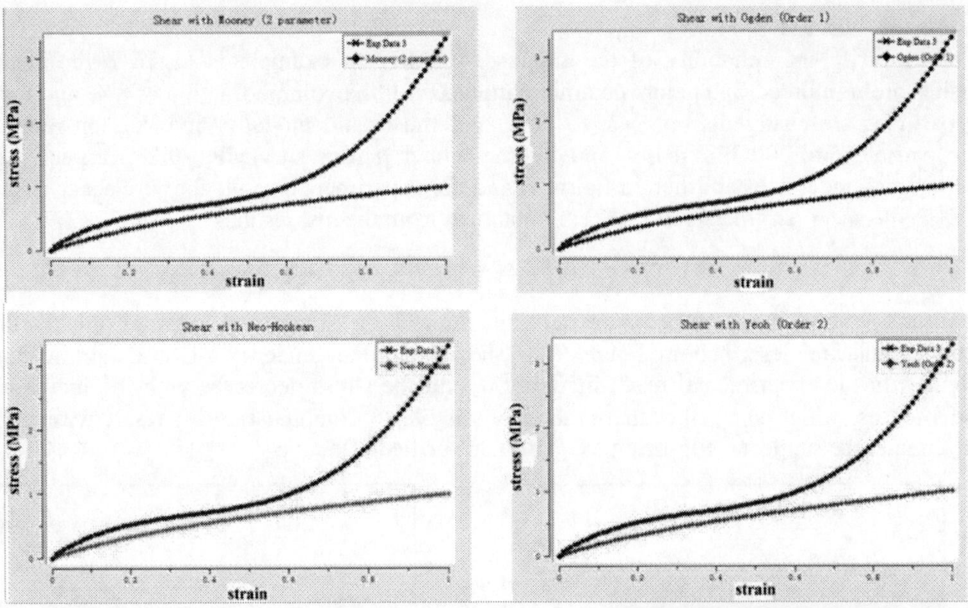

Figure 5. Fitting the test results to numerical simulation for different models with pure shear.

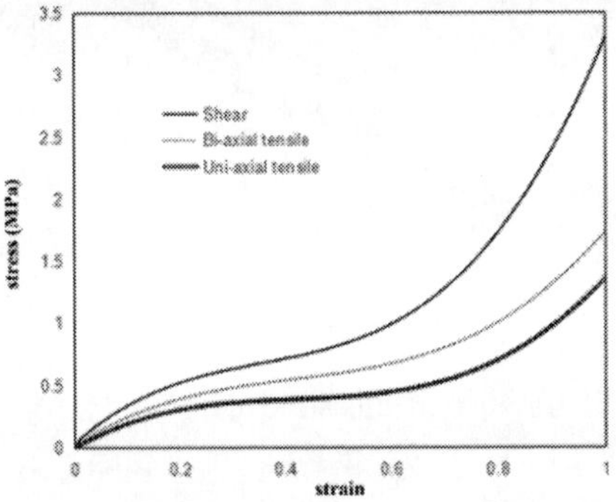

Figure 6. Stress-strain relationship for different test conditions (shear,uni- and bi-axial tensile).

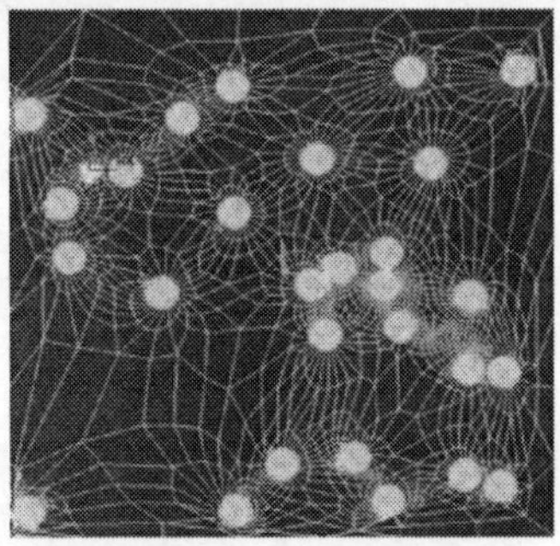

Figure 7. Finite element model.

Table 2. Material parameters for different models with different test methods.

	Uni-axial tensile	Pure shear	Bi-axial tensile	Three test methods
M-R	$C_{10}=0.227$	——	$C_{10}=0.183$	$C_{10}=0.258$
	$C_{01}=0.057$	——	$C_{01}=0.038$	$C_{01}=0.0055$
Yeoh	$C_{10}=0.232$	$C_{10}=0.343$	$C_{10}=0.225$	$C_{10}=0.259$
	$C_{20}=0.013$	$C_{20}=0.069$	$C_{20}=0.0116$	$C_{20}=0.0031$
Ogden	$\alpha_1=0.41$	$\alpha_1=0.338$	$\alpha_1=0.291$	$\alpha_1=0.537$
	$\mu_1=2.281$	$\mu_1=3.889$	$\mu_1=3.243$	$\mu_1=1.98$
N-H	$\mu=0.525$	$\mu=0.98$	$\mu=0.554$	$\mu=0.531$

IOP Conf. Series: Materials Science and Engineering **292** (2017) 012112 doi:10.1088/1757-899X/292/1/012112

5. Results and discuss

In order to verify the reliability of the supposed model, an example is tested here. The rubber composites are reinforced by random ceramic particles with the volume fraction of 8%, and the radius of the particles are changed from 2nm to 6nm, and the elastic modulus and Poisson ratio for the ceramic particles are 700GPa, 0.499, and for the rubber matrix are 4MPa, 0.23, respectively. The Finite Element model is established in figure 7, and the experimental result for pure shear is shown in figure 8, so the shear parameters in Eq.(7) are obtained from the test results:

$$s_2 = -10.752, s_3 = 11.386, s_4 = -1.675 \quad (7)$$

Substituting the obtained shear parameters into Eq.(7), the stress-strain relationship for different particle's radius are described in figure 9. It shows that the tendency of stress-strain change is consistent with the experimental result in figure 8, and the stress decreases with the increase of the particle's radius, which show that the method is reliable to combine the test result with numerical simulation, and the supposed constitutive equation is verified effectively.

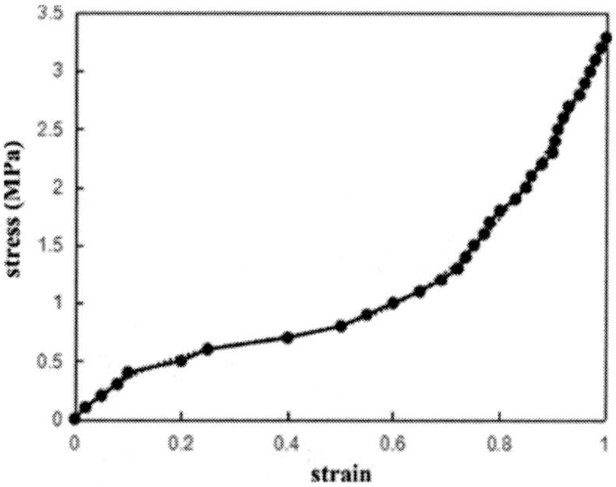

Figure 8. Stress-strain relationship for shear deformation.

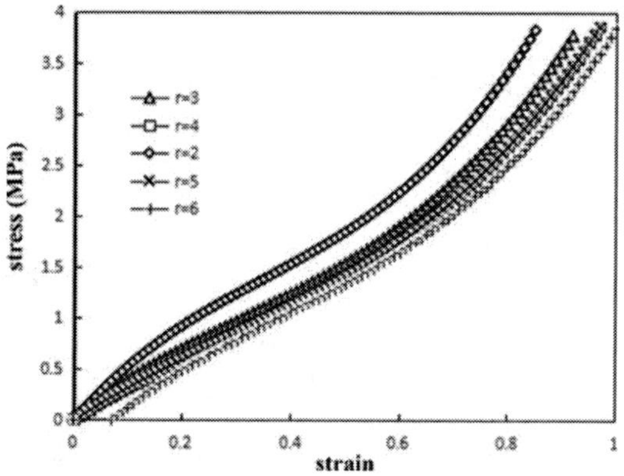

Figure 9. Stress-strain relationship for different particle's radius.

6. Conclusions

In this paper, a strain energy constitutive equation with large deformation for rubber composites reinforced with random ceramic particles is proposed from the angle of continuum mechanics theory. The results show that it is reliable to fit the material parameters for three kinds of typical experimental conditions (uni-axial tension, equal bi-axial tension and pure shear) with finite element method, and the proposed constitutive relationship is satisfactory to describe the non-linear stress-strain curves for rubber composites reinforced with random ceramic particles. The radius of random particles has a significant effect on the mechanical properties for rubber composites.

Acknowledgment

This work is supported by the NSFC (10772047/A020206, 11172066/A020305), Guangdong Natural Science Foundation of the P. R. China (S2011010004874), Foshan Science and Innovate Platform for University and Hospital (2016AG100341). Work partially supported by Foshan University Free Exploration Fund (2016LGZ06) and the Training Plan to Innovation and Enterprise of National University Students, Foshan University.

References

[1] Voigt W 1889 Ueber die Beziehung zwischen den beiden Elasticitäts constanten isotroper Körper *Annalen der Physik* **274(38)** p 573–587
[2] Reuss A 1929 Berechnung der Fließgrenze von Mischkristallen aufGrund der Plastizitäts

IOP Conf. Series: Materials Science and Engineering **292** (2017) 012112 doi:10.1088/1757-899X/292/1/012112

bedingung für Einkristalle *Zeitschrift für Angewandte Mathematik und Mechanik* **9** p 49–58

[3] Eshelby J D 1957 The determination of the elastic field of an ellipsoidal incclusion, and related problems *Proceedings of the royal society of London series a-Mathematical and physical Sciences* **241(1226)** p 376-396

[4] Mori T and K Tanaka 1973 Average stress in matrix and average elastic energy of materials with misfitting inclusions *Acta Metallurgica* **21(5)** p 571-574

[5] Guo Z Y, et al. 2014 Mechanical modeling of incompressible particle-reinforced neo-Hookean composites based on numerical homogenization *Mechanics of Materials* **70(3)** p 1-17

[6] Huang X S, Peng X Q, Zhang B C 2016 An Anisotropic Visco-hyperelastic Constitutive Model for Cordrubber Composites *Chinese Journal of Theoretical and Applied Mechanics* **48** p 140-145

[7] Huang J L, Xie G J, Liu Z W 2008 Finite element analysis of super-elastic rubber materials based on the Mooney-Rivlin and Yeoh model *China Rubber/Plastics Technology and Equipment* **34** p 22-26

[8] Li Q, Yang X X 2013 Study on Macroscopic and Microscopic Mechanical Behavior of Carbon Black Filled Rubber Composite *Journal of Mechanical Engineering* **49(18)** p 132-139

[9] Li Q, Yang X X 2014 N330 carbon black enhances the mechanical properties of natural rubber materials *Experimental mechanics* **29(1)** p 42-50

[10] Dong J P, Zhang Z Q 2016 The establishment and application of the short fiber reinforced rubber constitutive model of isotropic hyperelastic theory based on transverse view *Journal of computational mechanics* **2** p 231-237

[11] L F, Zhu Y F, Li L. J, Huang X Q 2006 The constitutive relation of large strain hardening of rubber material *Journal of south China university of technology* **34** p 9-12

[12] Yan S, Wang W 2014 Determination of material parameters for rubber-like hyperelastic constitutive models *Rubber Introduction* **61** p 453-457

IOP Conf. Series: Materials Science and Engineering **292** (2017) 012113 doi:10.1088/1757-899X/292/1/012113

Experimental study on emission of VOCs from tanker using hollow fiber membrane absorption method with different absorbents

J Y Zhou, B F Wang[1], L H Nie, J X Lu, Y J Hao and R R Xu

School of Petrochemical and Energetic Engineering, Zhejiang Ocean University, Zhoushan 316000, China

[1] E-mail: 646522786@qq.com

Abstract. China's oil dependence is getting higher and higher, 90% of oil import is transported by sea. Tankers will produce a lot of VOCs during loading and unloading, so the prevention of such pollution has become increasingly urgent. The hollow fiber membrane absorption method combined the characteristics of the absorption method for the treatment of high concentration and large flow of VOCs and the advantage of low energy consumption of membrane method .At present, the research on the recovery of oil and gas is relatively few. In this paper, the effect of membrane absorption on the recovery of oil and gas was investigated. The different absorbent affected the oil vapor recovery, the experimental results showed that the performance of absorbent of AbsFOV-97 was better than that of heat conductive oil.

1. Introduction

In 2015, China's apparent consumption of oil was estimated at 5.43 tons, up slightly by 25 million tons from a year earlier [1]. China's net imports of foreign crude oil reached 328 million tons in 2015, up 6.4 percent from the previous year, the growth rate was up 0.6 percentage points from 2014. China's oil consumption continued to grow at a moderate and low speed, and its foreign dependence exceeded 60% and make for the first time in history, reaching 60.6% [2]. 90% of China's imported oil is completed by sea shipping. Oil is a complex mixture of thousand of hydrocarbon compounds. Some volatile organic vapors(VOCs) will evaporate from the oil during the loading and unloading of tankers, the lower the boiling point of the oil, the more the times of loading and unloading, the greater the changes of temperature and pressure ,the more the loss of oil evaporation [3]. At present, the various VOCs program both at home and abroad includes resource non-recyclable and recyclable, the former such as direct combustion, catalytic combustion, photocatalysis, nano-materials self-catalyzed, the latter such as absorption, adsorption, condensation and membrane recovery technology, etc [4].Conventional absorption methods had the advantages of being able to treat the VOCs of high concentration and high flux ,while the membrane separation method had the characteristics of low energy consumption ,simple process, high recovery rate, high operational flexibility, high degree of automation, no need for maintenance, no secondary pollution, environmental protection , energy saving, safe and reliable [5], the contrast of several traditional techniques is shown in table 1. Therefore, the combination of this two methods, membrane-based absorption coupling technology, was of great concern because of its characteristics of large gas-liquid contact area, large mass transfer rate, high operational flexibility, no flooding, no fog entrainment and simple equipment etc [6]. The selection of absorbent was the key point of membrane absorption technique, because the absorbent was acted as a primary selective action on membrane absorption [7] .That is to say, the adsorbent has a direct effect on the selectivity and absorption rate of hydrocarbon components. Hollow fiber membrane absorption technology has been widely studied for the removal of CO_2 and SO_2 in flue gas [8-11].But the case of recycling VOCs with hollow fiber membranes has not yet appeared, this paper studied the recovery process and effect of the emission of VOCs during loading and unloading in tankers in the application of hollow fiber membranes with .different absorbents. In the mid-1980s, scientists developed a new composite membrane, nanofiltration (UF) membrane, which has a higher permeation flux [12] but we chose to use PVDF membrane was because that its characteristic of heat resistance, corrosion resistance, radiation resistance, chemical stability, mechanical high strength and good toughness was more suitable for the production of hollow fiber membrane.

Table 1. The comparison of several traditional technologies.

Recovery method	Advantages	Disadvantages
absorption	The method is simple and can handle mixed gases.	Low efficiency; high performance and equipment selection conditions
adsorption	Its operation is simple and the efficiency is high.	The adsorption of activated carbon will release heat, so there are security risks; the activated carbon need to replaced regularly; high operating costs.
condensation	The quality of recycled products is pure.	Large energy consumption; it cannot deal with low concentration of VOCs; high technical requirements.
membrane separation	Easy to use; service life long; small occupied area.	Large investment; high technical requirements.

2. Experimental

2.1. *Experimental apparatus and materials*

GC9720 gas chromatograph was supplied by Zhejiang Fuli Analytical Instruments Co. Ltd. PVDF hollow fiber membrane module was made in laboratory, its specifications was L50cm × D10cm, table 2 lists the solution composition, AbsFOV-97 absorbent was supplied by Changzhou university.

Table 2. Solution composition of PVDF hollow fiber membrane.

Spinning dope	Content (%)
PVDF	17%
DMAC	78%
PVP	5%

2.2. *Experimental methods*

The experimental process is shown in figure 1, the VOCs that needed for the experiment was produced by the way of nitrogen bubbling in the gasoline, the VOCs flowed from the inside of the hollow fiber membrane, and the outside was filled with absorbent. Then, the absorbent was evacuated and regenerated by a vacuum pump in the side of the shell. As a result, the raw gas was obtained from 1 # sampling port, the purified gas was obtained from 2 # sampling port, the condensed gas after absorbed was obtained from 3 # sampling port, and then these gas was sampled by gas sampling bag for gas chromatography analysis. Set the flow rate to 500 mL / min.

3. Results and discussions

3.1. Gas chromatographic analysis of raw VOCs before absorption

When the absorber was heat conductive oil, the results of gas chromatography analysis of 1# raw VOCs before absorption is shown in figure 2. From figure 2, it could be seen that the content of n-butane was 1.6099%, the content of propylene was 0.0010%, the content of isobutene was 0.3057%, the content of cis -2- butene was 0.2403%, the content of 1,3- butadiene was 1.9331%, and the content of 2- methyl -2- butene was 0.0006%,thus, it could be calculated that the concentration of the raw gas VOCs before absorbed was 4.0906%.

When the absorber was AbsFOV-97, the results of gas chromatography analysis of 1# raw VOCs before absorption is shown in figure 3. From figure 3, it could be seen that the content of propane was 0.3257%,the content of propylene was 0.2599%,the content of n-butane was 0.4988%, the content of allene was 0.0009%, the content of anti -2- butene was 0.0002%, the content of cis-2-butene was 0.2277%, the content of 1,3-butadiene was 0.8923%, the content of 2- methyl -2- butene was 0.8773%,thus, it could be calculated that the concentration of the raw gas VOCs before absorbed was 3.0828%.

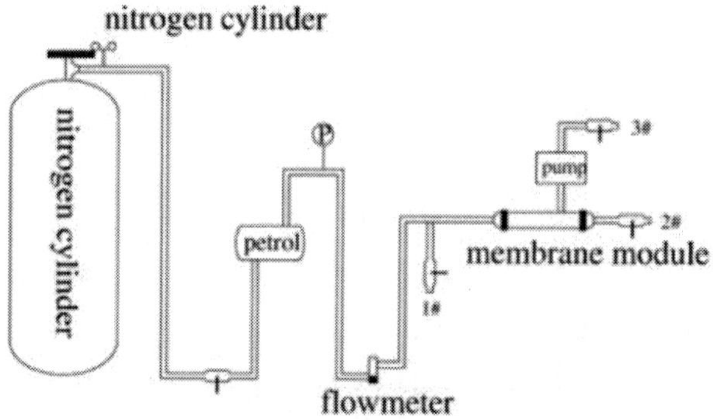

Figure 1. Experimental process.

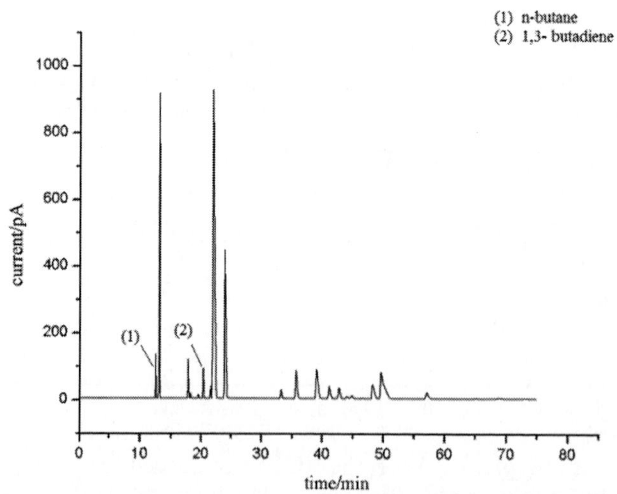

Figure 2. Gas chromatography of 1# raw VOCs before absorption (absorbent for heat conduction oil).

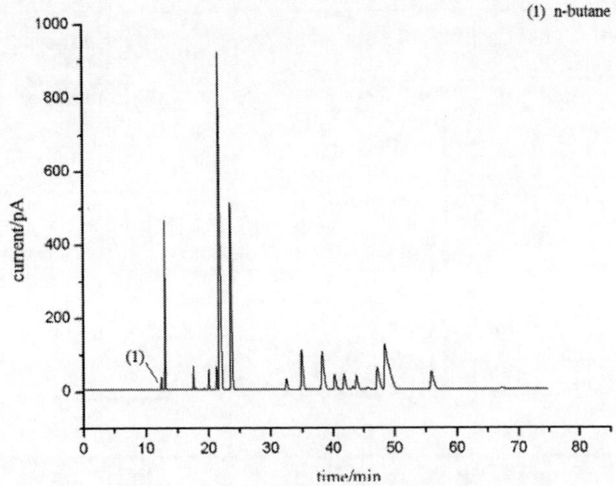

Figure 3. Gas chromatography of 1# raw VOCs before absorption (absorbent for AbsFOV-97).

3.2. Gas chromatographic analysis of purified VOCs after absorption

When the absorber was heat conductive oil, the results of gas chromatography analysis of 2# purified VOCs after absorption is shown in figure 4. From figure 4, it could be seen that the content of methane was 0.2297%, the content of ethane was 0.1724%, the content of ethylene was 0.0185%, the content of propylene was 0.0171%, the content of n-butane was 0.3341%, the content of trans-2-butene was 1.4695% The content of isobutene was 0.0274%, the content of cis-2-butene was 0.1056%, the content of 1,3-butadiene was 1.0798%, and the content of 2-methyl-2-butene was 2.1188%,thus, it could be calculated that the concentration of purified VOCs after absorption was 5.5729%.

When the absorber was AbsFOV-97, the results of gas chromatography analysis of 2# purified VOCs after absorption is shown in figure 5. From figure 5, it could be seen that the content of ethane was 0.0323%, the content of ethylene was 0.2426%, the content of propylene was 0.5685%, the content of n-butane was 0.0776%, the content of butene-1 was 1.2046%, the content of cis-2-butene was 0.0956%, the content of 1,3-butadiene was 1.0198%, thus, it could be calculated that the concentration of purified VOCs after absorption was 3.241%.

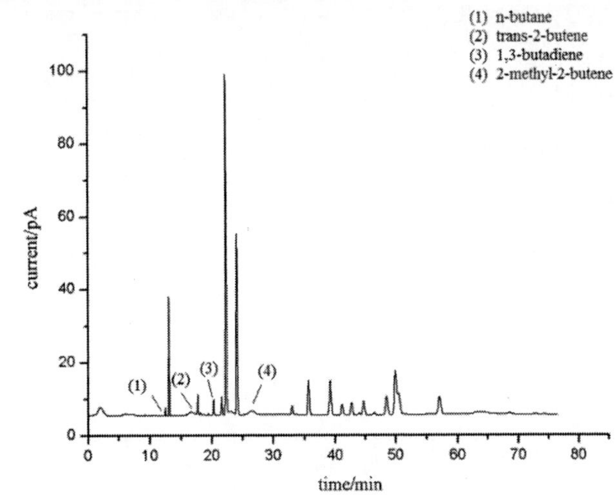

Figure 4. Gas chromatography of 2# purified VOCs after absorption (absorbent for heat conduction oil).

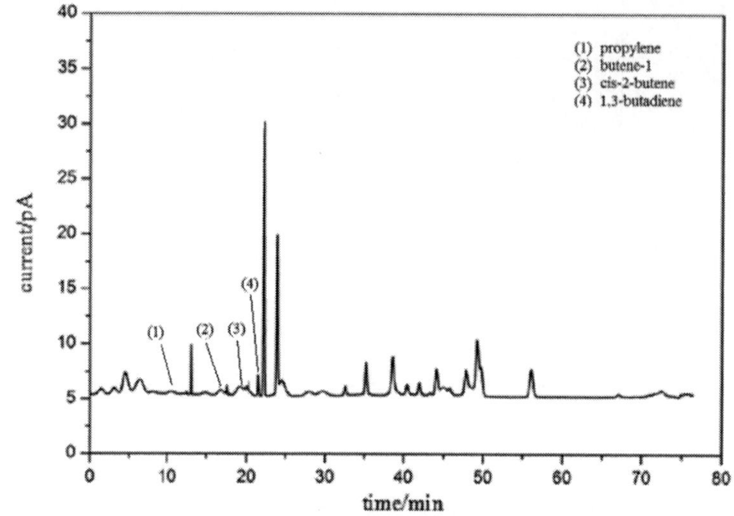

Figure 5. Gas chromatography of 2# purified VOCs after absorption(absorbent for AbsFOV-97).

3.3. Gas chromatographic analysis of concentrated gas VOCs after absorption

When the absorber was heat conductive oil, the results of gas chromatography analysis of 3# concentrated VOCs after absorption is shown in figure 6. From figure 6, it could be seen that the content of propylene was 0.0471%, the content of n-butane was 0.0005%, the content of propylene was 0.0022%, the content of isobutene was 0.2771%, the content of cis-2-butene was 0.2207%,the content of 1,3-butadiene was 0.0021% and the content of 2-methyl-2-butene was 0.0024%,thus, it could be calculated that the concentration of concentrated VOCs after absorption was 0.5521% .

When the absorber was AbsFOV-97, the results of gas chromatography analysis of 3# concentrated VOCs after absorption is shown in figure 7. From figure 7, it could be seen that the content of propane was 0.0002%, the content of n-butane was 0.2930%, the content of propadiene was 0.0005%, the content of cis-2-butene was 0.0010%, and the content of 1,3-butadiene was 0.0001%,thus, it could be calculated that the concentration of concentrated VOCs after absorption was 0.2948% .

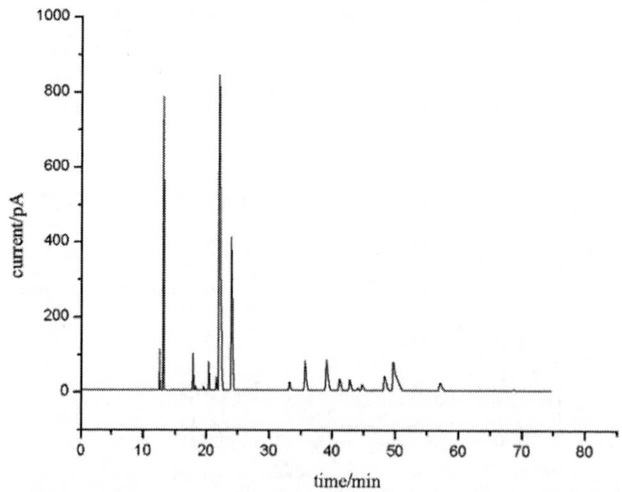

Figure 6. Gas chromatography of 3# concentrated VOCs after absorption (absorbent for heat conduction oil).

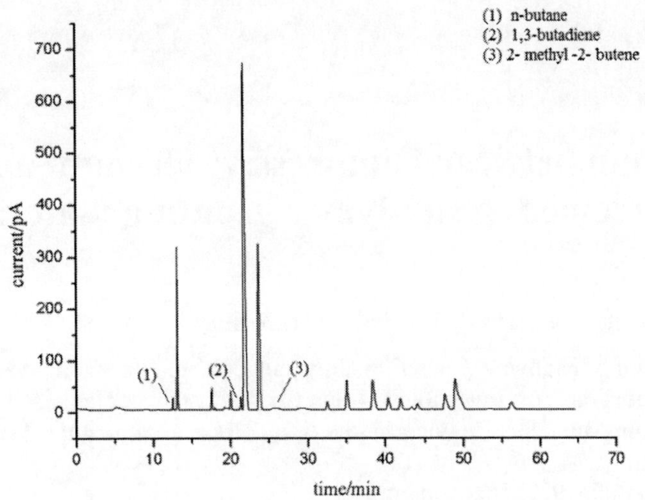

Figure 7. Gas chromatography of 3# concentrated VOCs after absorption(absorbent for AbsFOV-97).

4. Conclusion

When we used heat conductive oil as absortant, we can respectively obtained that the content of VOCs of raw gas, purified and concentrated gas was 4.0906%, 5.5729% and 0.5521% from figure 2, figure 4 and figure 6. When we used AbsFOV-97 as absortant, we can respectively obtained that the content of VOCs of raw gas, purified and concentrated gas was 3.0828%, 3.241% and 0.2948% from figure 3, figure 5and figure 7. It can be seen from the above experimental results that the removal rate with the absorbent of heat conductive oil was 86.50%, the removal rate with the absorbent of AbsFOV-97 was 90.44%. The performance of AbsFOV-97 was better than that of heat conductive oil, and the smaller the gas flow rate, the better the oil and gas recovery effect of the membrane absorption device, then when the gas flow rate was below 50mL/min, the removal rate of hydrocarbons should be about 99.50%.

References

[1] Chai J, Wang S B, Wang S Y and Guo J 2012 *J. .Energies* **5** *577-598*
[2] CNPC economic & technology research institute 2016 *Development report on domestic and foreign oil and gas industry in 2015* (Beijing: CNPC economic&technology research institute)
[3] Huang W Q 2011 *Basic theory and application of oil and gas recovery* (Beijing:China Petrochemical Press) pp 7-10
[4] Huang W Q, Wang D L,Li F and Wang H N 2012 *J. Oil and Gas Storage and Transportation* **31** 641-646
[5] Strathmann H 2001 *J. AIChE* **47** 1077-87
[6] Gabelman A and Huang S T 1999 *J. Journal of Membrane Scicence* **159** 61-106
[7] Shelekhina A B and Beckmanb I N 1992 *J. Journal of Membrane Science* **73** 73-85
[8] S Boributh, R Jiraratananon and K Li 2013 *J. Journal of Membrane Science* **429** 459-472
[9] Lai C F, Yang B, Zhang G L, Wen C S, Cao R and Wu R R 2012 *J. Journal of Chemical Industry and Engineering(China)* **63** 500-507
[10] Luis P, Garea A and Irabien A 2012 *J. Chemical Engineering and Processing: Process Intensification* **52** 151-154
[11] Cui Z D, Hou C N, Chen Y and Guan Y P 2012 *J. Membrane Science and Technology* **31** 84-88
[12] Miao J, Zhang L C and Lin H C 2013 *J. Chemical Engineering Science* **87** 152-159

IOP Conf. Series: Materials Science and Engineering **292** (2017) 012114 doi:10.1088/1757-899X/292/1/012114

The relationship between compressive strength and flexural strength of pavement geopolymer grouting material

L Zhang[1, 2], X X Han[1], J Ge[3] and C H Wang[1, 4]

[1] School of Highway, Chang'an University, Xi'an, 710064, Shaanxi, China
[2] Henan Transport Investment Group Co., Ltd., 450052, Henan, China
[3] Tianjin Municipal Engineering Design & Research Institute, Tianjin, 300051, China

[4] E-mail: 2539526581@qq.com

Abstract. To determine the relationship between compressive strength and flexural strength of pavement geopolymer grouting material, 20 groups of geopolymer grouting materials were prepared, the compressive strength and flexural strength were determined by mechanical properties test. On the basis of excluding the abnormal values through boxplot, the results show that, the compressive strength test results were normal, but there were two mild outliers in 7days flexural strength test. The compressive strength and flexural strength were linearly fitted by SPSS, six regression models were obtained by linear fitting of compressive strength and flexural strength. The linear relationship between compressive strength and flexural strength can be better expressed by the cubic curve model, and the correlation coefficient was 0.842.

1. Introduction

With the rapid development of the road traffic industry, the total mileage of highways in China has increased rapidly, by the end of 2016, the total mileage of the highway has reached 4.6963 million kilometers [1]. But the road design, the construction of the shortcomings and traffic flow, the rapid increase in axle load made the use of the road after a period of time often subsidence, cracks, slurry pump and other diseases. It is very important to take effective measures to reinforce and strengthen the pavement structure. Grouting technology has some characteristics such as fast reinforcement, small impact for traffic and other advantages of the road reinforcement. As an important part of grouting technology, grouting material directly affects the effect of grouting reinforcement. There are three main types of grouting material. The existing cement grouting material is poor in stability and bleeding water easily [2]. Chemical grouting material is costly and most of them are poisonous [3]. Polymer grouting material can only fill the void of the structural layer, cannot bond loose material between layers, the effect of structural reinforcement is not good [4]. Therefore, the future direction of building materials is still green, excellent performance and low price [5]. Geopolymer grouting material is a kind of green material. It has the advantages of early strength, corrosion resistance, good durability, low energy consumption, low price and so on. Geopolymer grouting material is an ideal reinforcing material for pavement structure, it is widely used in the field of grouting technology. Researchers in the field of road engineering carried out a large number of experimental studies on geopolymer grouting materials. Zhang Haiyan et al studied found that either at ambient temperature or after exposure to high temperature the strength of geopolymer paste, mortar and concrete prepared in this study can almost reach or exceed that of cement paste, mortar and concrete [6]. Yuan Bo analyzed the road performance of geopolymer grouting material through experiments. Found that after the grouting down the deflection on the surface enhance the obvious, can effectively extend the road life [7]. Wang Jian et al studied the influence rule of volume fraction of sodium silicate and powder proportion on initial/final setting time, uniaxial compressive strength, erosion resistance and dry shrinkage, the optimum mix amount of sodium silicate was determined to be 20% [8]. Yan C et al research on durability performance of novel double solution grouting material with metakaolin and the microstructure of this grouting material was analyzed [9]. Krivenko et al research a kind of shrinkage-compensating alkali activated slag cement mortars, this material can crack repair and joint grouting in concretes, the grouting material has good mechanical properties, durability and low-cost [10]. Cristelo N et al assessing using fly ash-based geopolymer material can significantly reduce CO_2 emissions [11]. Idrissi et al studied the effect of water content on microstructure and mechanical properties of geopolymer grouting material [12]. However, the existing research focuses on optimizing the preparation of geopolymer grouting material with multiple function and analyzing its microstructure, and the influence of different amount of materials on its performance was studied. But there is little research on the relationship between compressive strength and flexural strength.

Therefore, twenty groups of geopolymer grouting materials were prepared, the compressive strength and flexural strength were determined by mechanical properties test, on the basis of excluding the abnormal values through boxplot, the compressive strength and flexural strength were linearly fitted by SPSS. In order to determine the correlation between the compressive strength and the flexural strength of geopolymer grouting materials, this lays the foundation for the further study of the geopolymer grouting materials.

2. Test raw materials

The main raw materials used in this paper are 300 mesh blast furnace slag, water glass solution, industrial grade sodium hydroxide (sodium hydroxide content⩾99%) water reducing agent and water. The main components and basic index of blast furnace slag, water glass and water reducing agent are shown in table 1, table 2, table 3.

Table 1. Chemical composition of blast furnace slag.

Technical index	SiO_2	Al_2O_3	CaO	MgO	Fe_2O_3	MnO	S
Blast furnace slag	35.06	14.38	41.12	7.93	0.58	0.17	0.76

Table 2. Basic index of water glass.

Water glass	Na_2O (wt%)	SiO_2 (wt%)	Density (g/cm^3)	Water insoluble (wt%)	Modulus
Requirements	⩾8.2	⩾26.0	1.368~1.394	⩽40	3.1~3.4
measured	9	30	1.39	≤40	3.3

Table 3. Technical index of water reducing agent.

Technical index	Colour	Bulk density (kg/m^3)	Loss on drying (%)	PH value
Water reducing agent	white	430~740	<4.0	9~11.4

3. Test method

The sodium hydroxide solution was prepared by mixing sodium hydroxide with 80% water in proportions. After cooling the solution to room temperature and blended it with water glass solution to prepare alkaline activator, mixed the water reducing agent with the remaining 20% of the water. According to the "Test methods for water requirement of normal consistency, setting time and soundness of the portland cement"(GBT 1346-2011 of China)cleaned the mixing pan and mixing blade with a damp cloth, mixed the blast furnace slag powder and alkaline activator in the slurry mixer. After 120s low speed stirring, stopped stir and added residual water reducing agent solution, and stirred the mixture at high speed for 120s in the stirrer.

The polymer grouting material was prepared by refering to the method above, 40mm × 40 mm × 160mm specimens were molded by artificial pound and eliminate the air. To avoid the molded specimen water evaporation and water precipitation covered with fresh-keeping film, put specimens at temperature 20℃ ± 5℃, relative humidity greater than 50% of the environment for 20 to 24 hours. Then start mould stripped. Scraped off the surface of specimen, then put the specimen up in 20±1℃ water for curing. The curing specimen was removed from the water 15 minutes before the test age (7days). The compressive strength and flexural strength were tested referring to the "method of testing cements-determination of strength" (GB17671-1999 of China).

In order to facilitate the actual application process to calculate the composition of the geopolymer grouting material, based on blast furnace slag powder. The mass ratio of water glass solution to blast furnace slag powder was SS / SL, the mass ratio of sodium hydroxide to blast furnace slag was SH / SL. The mass ratio of water to blast furnace slag was W/SL. The mix design of pavement geopolymer grouting material is shown in the table 4.

Table 4. Proportioning design of road geopolymers grouting material.

Numbering	SL:SS:SH:W	Numbering	SL:SS:SH:W	Numbering	SL:SS:SH:W	Numbering	SL:SS:SH:W
1	1:0.09:0.07:0.37	2	1:0.09:0.09:0.47	3	1:0.09:0.11:0.42	4	1:0.11:0.07:0.47
5	1:0.11:0.09:0.42	6	1:0.09:0.07:0.42	7	1:0.09:0.07:0.45	8	1:0.09:0.08:0.42
9	1:0.09:0.08:0.45	10	1:0.09:0.08:0.47	11	1:0.10:0.07:0.42	12	1:0.10:0.07:0.45
13	1:0.10:0.07:0.47	14	1:0.10:0.11:0.45	15	1:0.11:0.08:0.45	16	1:0.11:0.08:0.42
17	1:0.07:0.07:0.42	18	1:0.07:0.09:0.37	19	1:0.07:0.11:0.47	20	1:0.11:0.11:0.37

4. The relationship between the flexural strength and the compressive strength

Using the boxplot in SPSS（Statistical Product and Service Solutions）, the abnormal values of compressive strength and flexural strength of 20 groups of geopolymer grouting materials were determined and eliminated. Draw the boxplot shown in figure 1, the internal and external limits are not shown in figure 1.

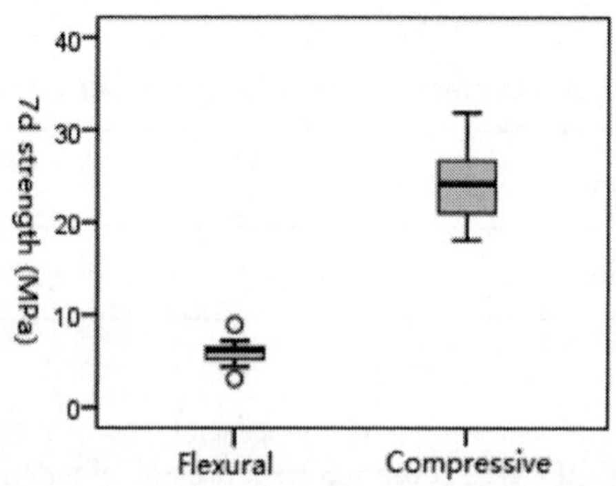

Figure 1. Compressive strength and flexural strength of 7days with different geopolymer grouting material.

Through the SPSS boxplot to verify, different geopolymer grouting materials have two mild outliers in the flexural strength test data in 7days, Flexural strength of 8.9MPa at SL:SS:SH:W=1:0.09:0.07:0.37, Flexural strength of 3.1MPa at SL:SS:SH:W=1:0.10:0.11:0.45, taking into account the composition of the two kinds of geopolymer grouting materials, there is a possibility that the flexural strength is higher or lower. Therefore, these two moderate anomalies are included in the analysis of the relationship between compressive strength and flexural strength of the geopolymer grouting material 7days.

The compressive strength and flexural strength of 7days grouting material are analyzed by linearity, logarithm, quadratic curve, cubic curve, power and six indexes. The result is shown in figure 2.

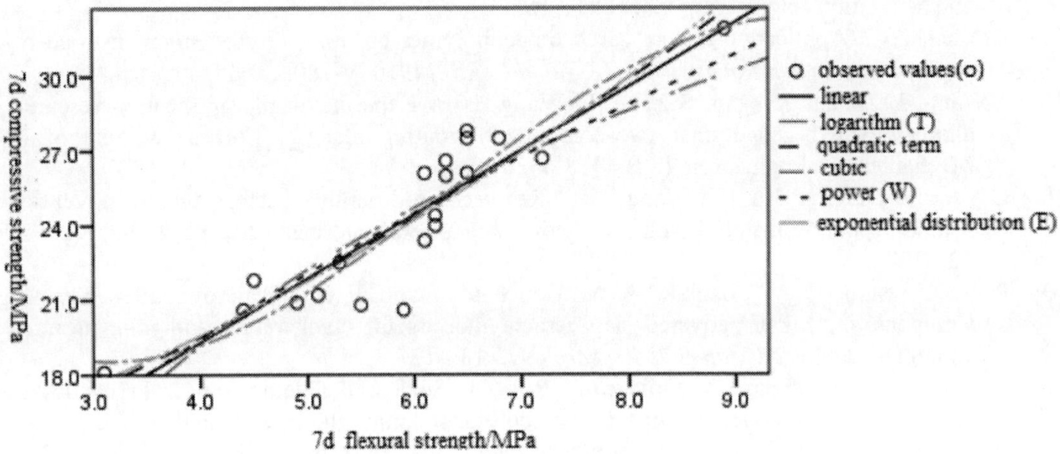

Figure 2. Regression analysis of flexural strength and compressive strength of geopolymer grouting material in 7days.

As you can see from figure 2, the cubic curve has the best fitting effect and the R^2 value is the largest, the relationship between the flexural strength and the compressive strength of the 7days geopolymer material can be expressed by $y = 32.076 - 9.627x + 2.027x^2 - 0.106x^3$, $R^2 = 0.842$, the fitted equation can better reflect the relationship between them.

5. Conclusions

In this paper, 20 groups of geopolymer grouting material were prepared, the compressive strength and flexural strength were determined by mechanical properties test, on the basis of excluding the abnormal values through boxplot, the compressive strength and flexural strength were linearly fitted by SPSS, the results show that, the compressive strength test results were normal, but there were two mild outliers in 7days flexural strength test, the conversion relationship between compressive strength and flexural strength of six ground geopolymer grouting material in 7days is established, the cubic curve can better reflect the relationship between them. The correlation coefficient is 0.842. In the follow-up study, the actual effect of 20 groups of different geopolymer grouting materials will be tested. This will provide a theoretical basis for further study and engineering application of geopolymer grouting material.

References

[1] Ministry of Transport 2017 Interpretation of the national statistical bulletin on toll roads in 2016, *China Communications Bulletin* 06-**29**(003)

[2] J F Wang, H B Zhang and X M Guan 2011 The study on micro-fine cement grouting matial applying in coal wall reinforcement *Journal of Henan university of technology (natural science edition)* **(02)** 145-148

[3] X Y Yu, N Zhang, H W Zhang, C Y Chen and D Y Qian 2010 The study on treating large roof

fall accident by polymer chemical grouting *Safety in CoalMines* **41**(6) 10-13

[4] S M Tong, Y Zhang, L Q Yang and P N Feng 2015 Experimental study on compressive strength of road polymer grouting materals *Western China Communications Science & Technology* **(11)** 17-19

[5] N P Sharifi, A A N Shaikh and A R Sakulich 2017 Application of phase change materials in gypsum boards to meet building energy conservation goals. *Energy & Buildings* **138** 455-467

[6] H Y Zhang, Q I Shu-Liang, L Cao 2015 Mechanical performance comparison of geopolymer paste, mortar and concrete after exposure to high temperature Journal of Disaster Prevention and Mitigation Engineering v.35(1) 11-16

[7] B Yuan 2013 Application and research of geopolymer grouting reinforcement in road repair engineering, *Urban Roads Bridges&Flood Control* **(9)** 179-180.

[8] J Wang, L Zhang, X Feng, S Zhao, H Wang, 2015 Experiment and application research on alkali-activated geopolymer two-component grouting material Chinese Journal of Rock Mechanics and Engineering (S2) 4418-4425

[9] C Yan, QJ Ding, JP Xu, HX Wang, 2011 Research on durability performance of novel double solution grouting material with metakaolin Advanced Materials Research 250-253(250-253) 722-727

[10] P V Krivenko, G V Vozniuk, A M Goncharova and E S Kavalerova 2014 Shrinkage-Compensating alkali activated slag cement mortars for crack repair and joint grouting in concretes. *Advanced Materials Research* **923** 134-137

[11] N Cristelo, T Miranda, DV Oliveira, I Rosa, E Soares, P Coelho and L Fernandes 2015 Assessing the production of jet mix columns using alkali activated waste based on mechanical and financial performance and CO_2, (eq) emissions *Journal of Cleaner Production* **102**(2) 447-460

[12] A C E Idrissi, E Roziere, A Loukili and S Darson 2016 Design of geopolymer grouts: the effects of water content and mineral precursor *European Journal of Environmental and Civil Engineering*

Kinetics of waterborne fluoropolymers prepared by one-step semi-continuous emulsion polymerization of chlorotrifluoroethylene, vinyl acetate, butyl acrylate andVeova 10

H Z Liu[1], M H Wang[1], Z F Wang[2] and J M Bian[1, 3]

[1] Key Laboratory of Materials Modification by Laser, Ion and Electron Beams(Ministry of Education), School of Physics, Dalian University of Technology, Dalian 116024, China

[2] Wenzhou Institute of Biomaterials and Engineering, CNITECH, Chinese Academy of Sciences, Wenzhou, 325001, China

[3] E-mail: jmbian@dlut.edu.cn

Abstract. Due to using gaseous fluorine monomer with toxicity, waterborne fluoropolymers are synthesized by semi-continuous high-pressure emulsion polymerization method which differs from free-pressure emulsion polymerization. To dates, the research on preparing process and kinetics for high-pressure emulsion polymerization is reported relatively less, which hinders researchers from understanding of mechanisms for monomer-fluorinated emulsion polymerization. The paper also provides a new method by element auxiliary analysis to calculate kinetics parameters of high-pressure emulsion polymerization. Based on aforementioned consideration, waterborne fluoropolymers were prepared by copolymerization of chlorotrifluoroethylene (CTFE), vinyl acetate (VAc), butyl acrylate (BA) and vinyl ester of versatic acid (Veova 10) using potassium persulfate as initiator and mixed surfactants. The kinetics of emulsion polymerization of waterborne fluoropolymers was then investigated. Effects of emulsifier concentration, initiator concentration, and polymerization temperature on polymerization rate (Rp) were evaluated, and relationship was described as $Rp \propto [I]^{0.10}$ and $Rp \propto [E]^{0.12}$. The apparent activation energy was determined to be 33.61 $kJ \cdot mol^{-1}$. Moreover, the relative conversion rate of CTFE with the other monomers was observed, and results indicated that CTFE monomer more uniformly copolymerized with the other monomers. The resulting emulsion properties and pressure change in an autoclave were evaluated at different stirring rates. The initial reaction time, defined as the beginning time of dropwise addition, was determined by the change in solid content and particle size of emulsion.

1. Introduction

Fluoropolymers have been among the most important materials used in various fields because of their chemical structure and unique properties [1]. Recently, studies on environment-friendly waterborne fluoropolymers have gained increasing interest [2, 3]. Numerous studies have been conducted on different types of waterborne fluoropolymers [4-8], including the preparation [9-15] and performance evaluation of fluoropolymers [4, 16-19].

In the past decades, studies on the kinetics and mechanism of CTFE (chlorotrifluoroethylene)-based copolymer have also been undertaken. These studies include topics on radical copolymerization of CTFE with 1-fluoromethyl vinyl benzene [20], emulsion copolymerization of CTFE with VAc (vinyl acetate) and vinyl propionate [21], thermal degradation of homopolymers and copolymers of CTFE with MMA (methyl methyl acrylate) [22] and thermal degradation of copolymers for VDF (vinyldenefluoride)-CTFE [23, 24]. Thomas *et al.* [25] provided a survey on the kinetics of polymerization of CTFE with several monomers such as VAc in bulk and in solution. However, the kinetics of emulsion polymerization was not further investigated. Synthesis and characterization of waterborne fluoropolymer by semi-continuous emulsion polymerization of CTFE with VAc, BA (butyl acrylate), and Veova 10(a vinyl ester of a highly branched decanoic acid) was completed in our previous study [26]. To the best of our knowledge, no studies on the kinetics of semi-continuous emulsion polymerization of CTFE with VAc, BA, and Veova10 have been reported. This inadequacy hinders the understanding of the mechanism of emulsion polymerization, which affects production control and polymerization engineering design.

In this paper, we will report the kinetics of the quadric-copolymerization system of CTFE-VAc-Veova10-BA implemented by semi-continuous emulsion polymerization. Thus, CTFE was directly added into the autoclave, and the other monomers were dropwise-added into the autoclave within 3 h according to 173ml/h, 138ml/h and 115ml/h, sequentially. The relationships of the initiator concentration, emulsifier concentration, and reaction temperature with the polymerization rate were evaluated. The activation free energy (Ea) of the polymerization system was calculated. The copolymerization rates between the CTFE monomer and the other monomers were investigated. In addition, the effects of stirring rate on the properties of waterborne fluoropolymers were evaluated.

2. Experimental

2.1. Materials

CTFE, VAc, BA and Veova 10 were supplied from Dalian Zhenbang Fluorocarbon Paint Stock Co., Ltd. (Dalian, China) as polymerization grade monomers and were used as received. NP-10 (octylphenolpolyoxyethylene ether), 600#A (styrylphenolpolyoxyethylene ether), SDS (sodium dodecyl sulfate) were obtained from Liaoning OxiranChemstry Co., Ltd (Liaoyang, China). Potassium persulfate (KPS) and sodium bicarbonate ($NaHCO_3$) were purchased and used as initiator and battering agent, respectively. The deionized water was obtained by ion exchange and used in polymerization. AMP-95(2-amino-2-methyl-1-propanol) was purchased from Dow Chemicals (Midmland, MI, USA) and used as pH adjusting agent.

2.2. Emulsion polymerization

A specific amount of deionized water, which was 20% of the amount of the emulsifier and sodium bicarbonate, was added directly into the autoclave and dissolved evenly by stirring. After the autoclave were vacuumed and filled with nitrogen three times, all of the CTFE and a certain amount of the mixed monomers were added into the reactor and emulsified for 30 min. Under required stirring, the temperature of the reactant reached 60°C. Feeding with 20% of potassium persulphate solution, polymerization began, causing the temperature of the reactor to gradually increase because of exothermic reaction. After maintaining the polymerization for 30 min, the remaining emulsified monomers and the initiator solution were added dropwise into the reactor within 3 hours. Polymerization was maintained for another 3 hours. The experimental formulation was presented in table 1. 2-5 grams of the samples were removed from the reactor every 30 min for detection.

Table 1. Experimental formulation of the waterborne fluoropolymer.

	CTFE	VAc	Veova10	BA	Emulsifier[a]	KPS	$NaHCO_3$	H_2O
Charge(g)	80.00	8.74	2.91	2.91	3.34-6.23	0.35-1.05	0.90	212.19
Feed(g)	0	141.24	47.09	47.09	9.34-17.46	1.15-4.52	0	181.97
Total(g)	80.00	149.98	50.00	50.00	12.68-23.69	1.86-5.57	0.90	394.16

[a]Emulsifier mixture includes NP-10, 600#A, and SDS (weight ratio of 5:4:1).

2.3. Nonvolatile and fluoride content

About 0.5000-1.000 g of emulsion samples was placed in a tin foil box and then dried for 24 hours at room temperature. Nonvolatile content (*N.V.%*) was calculated by formula 1:

$$N.V.\%=\frac{W_3-W_1}{W_2}\times 100\% \tag{1}$$

Where W_1 is the weight of the tin foil box, W_2 is the weight of emulsion, W_3 is the total weight of the dried emulsion and the tin foil box.

Fluorine content was detected by the lanthanum nitrate method[27]. Sample weighing 10-25 mg was wrapped in a filter paper flag and then clipped to a platinum wire. As much as 8 ml of water was added to 250 ml of Molotov cocktails. Oxygen was passed for about 20 s, light filtered, and inserted rapidly for the sample to burn and decompose completely. Molotov cocktails were shaken to facilitate the complete absorption of smoke into the solution. The stopper was rinsed using platinum with 10 ml of acetone. A drip of perchlorate solution ($HClO_4$ = 2 mol/L) was dropped into the water bath at 70 °C for 5 min. Subsequently, 1 g of hexamethylenetetramine, 20 ml of acetone, and four drops of hematoxylin indicator (2.5 g/L) were added into the Molotov cocktails. Under shaking condition, the titrate lanthanum nitrate standard solution ($C_{[1/3\ la\ (NO3)3]}$ =0.1mol/L) and when the color of the solution in the Molotov cocktails changes from yellow to pale blue, the consumption volume *V* was recorded and blank experiment (V_0) were conducted. The fluorine content of the samples was calculated by formula 2.

$$F\% = \frac{C_L \times (V - V_0) \times 19.00 \times 100}{m \times 1000} \tag{2}$$

Where *F%* is the fluorine content of the samples; C_L is the $1/3La(NO_3)_3$ molar concentration of the standard solution, mol/L; *m* is the sample quality, g; 19.00 is the molar quality of fluorine atoms, g/mol.

2.4. Conversion

The emulsion was removed from the autoclave (2-5 g). The nonvolatile and fluorine contents of each phase were evaluated as described in section 2.3.

First, the conversion of CTFE was determined by formula 3.

$$W_{CTFE*} = \frac{W_L}{\frac{48.92\%}{N.V.\% \times F\%} - 1} \tag{3}$$

Where W_L denotes the instant liquid materials of the system in the autoclave, which is equivalent to the sum of the polymerization reactor base and the dropwise-added material; W_{CTFE*} is the quality of CTFE that has been converted into a fluoropolymer; *48.92%* is the fluoride content of CTFE.

Conversion of monomers other than CTFE was determined by formula 4.

$$W_{m*} = (W_L + W_{CTFE*}) \times N.V.\% - W_{CTFE*} \tag{4}$$

Where W_{m*} is the quality of the other monomers that have been converted to polymers.

Finally, the conversion rate (*C*) of the system at each phase was calculated by formula 5.

$$C=\frac{W_{CTFE*}+W_{m*}}{W_{CTFE}+W_m}\times 100\% \tag{5}$$

Where W_{CTFE} is the total quality of CTFE, W_m is the total quality of the monomers except for CTFE.

2.5. Calculation of the polymerization rate (Rp)

Emulsion polymerization is generally divided into three stages: acceleration stage, constant speed stage, and slowdown stage [28]. The constant speed stage begins when micelles disappear, the number

IOP Conf. Series: Materials Science and Engineering **292** (2017) 012115 doi:10.1088/1757-899X/292/1/012115

of latex grains becomes constant, and the polymerization rate remains constant. In accordance with this theory and by applying linear regression, the curve slope (*dc/dt*) at different reaction stages (*t*) and monomer conversion (*C*) during polymerization were calculated. The polymerization reaction rate Rp (mol/L·s) was calculated by formula 6:

$$Rp=\frac{dc}{dt}\times[M_0] \tag{6}$$

Where, $[M_0]$ is the amount of monomer (mol)/water volume in the system (L).

3. Results and Discussion

3.1. Effects of the initiator dosage on polymerization rate

Figure 1 shows the relationship of the monomer conversion with reaction time at different initiator concentrations. As the initiator dosage is increased, the polymerization reaction rate also increases. This effect is attributed to the increased rate of free radical generation in the polymerization system. The nucleation rate simultaneously increases, and so polymerization is accelerated. Figure 2 presents the relation curve between the pressure of CTFE steam and reaction time at different initiator concentrations inside the autoclave. As the initiator concentration increases, the autoclave pressure decreases faster, this further demonstrates markedly faster polymerization with increasing initiator dosage.

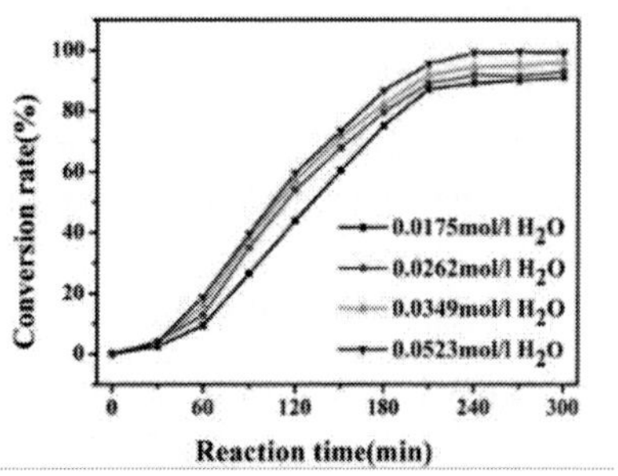

Figure 1. Influence of initiator dosage on polymerization rate.

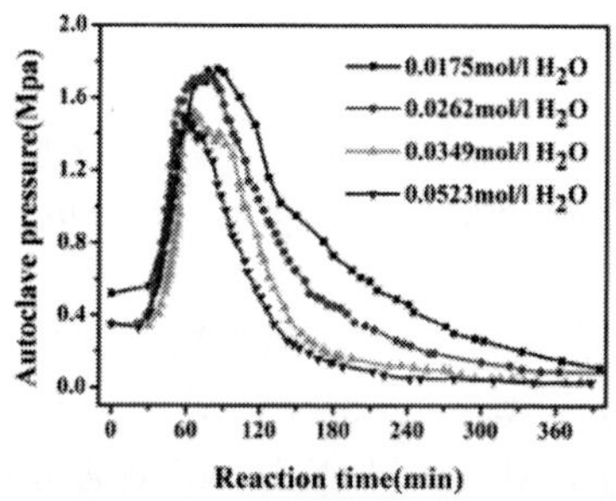

Figure 2. The change of autoclave pressure with reaction time.

By using linear regression of the slash part in figure 1, the relationship between -lnRp and -ln[I] is determined. As shown in figure 3, the slope of the straight line is 0.10, so $Rp\propto[I]^{0.10}$ is obtained, which differs from the classic emulsion polymerization index ($Rp\propto[I]^{0.4}$). The experimental result indicates that the polymerization rate is affected relatively less by the initiator concentration. As we know, the initiator concentration is controlled by dropwise addition technology, and is relatively low in reaction system, and thereby affect and reduce the reactive center.

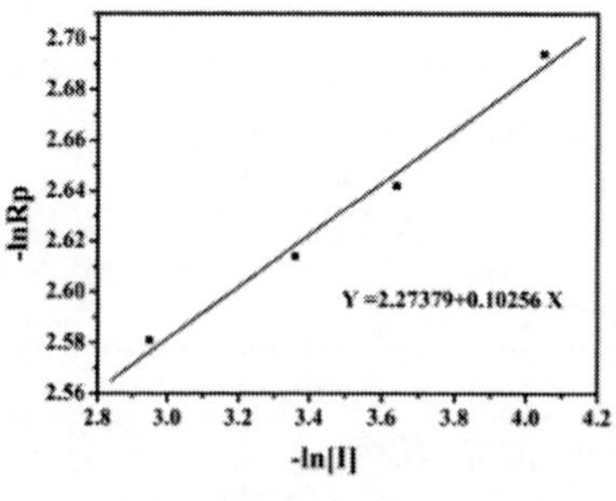

Figure 3. Relationship between -lnRp and -ln[I].

3.2. *Effects of emulsifier concentration on polymerization rate*

The effects of emulsifier dosage on polymerization rate are shown in figure 4. A higher emulsifier concentration results in faster polymerization. Increasing the emulsifier concentration leads to more micelles, thereby generates more emulsion grains. This occurrence indicates the presence of more reaction centers, and so the polymerization rate increases. This effect conforms to the basic principle of emulsion polymerization.

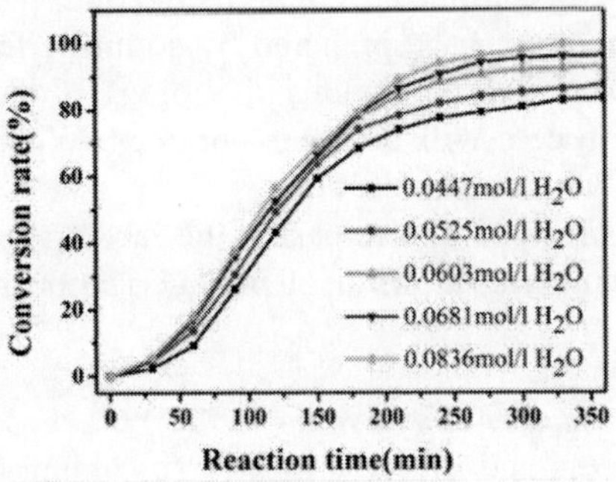

Figure 4. Influence of emulsifier dosage on reaction rate.

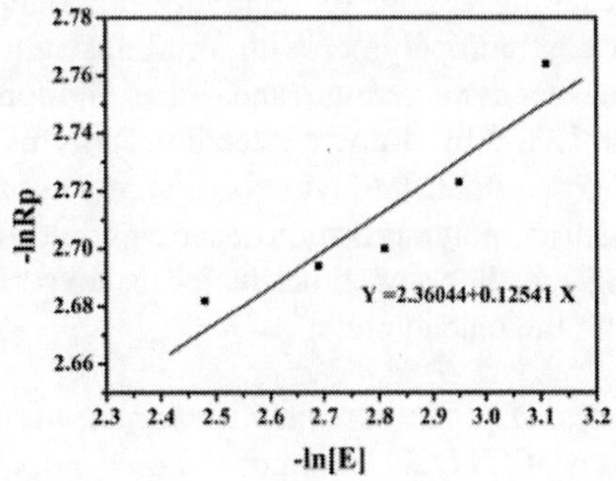

Figure 5. Relationship of -lnRp and -ln[E].

By using linear regression of the slash part in figure 4, the relationship between -lnRp and ln[E] was obtained, as shown in figure 5. The linear slope is 0.12, and the result is $Rp \propto [E]^{0.12}$, which differs from the classic emulsion polymerization index($Rp \propto [E]^{0.6}$). This outcome indicates that the emulsifier concentration only slightly influences polymerization. This result also relates the polymerization process. The emulsifier is fed into the autoclave through two parts: one part is entirely added into the reactor to produce micelles and an active reactive site, which significantly affects the polymerization rate. The other part is dropwise added to the reactor, whose key function is to maintain the emulsion particles stably.

3.3. *Relationship between the conversion rates of CTFE and other monomers with reaction time*

In this study, CTFE is added into the polymerization autoclave by disposable feeding way, whereas other monomers are fed dropwise. At different stages of reaction, the sample is removed from the autoclave, and the conversion rates of the CTFE and other monomers are calculated. The results are shown in figure 6.

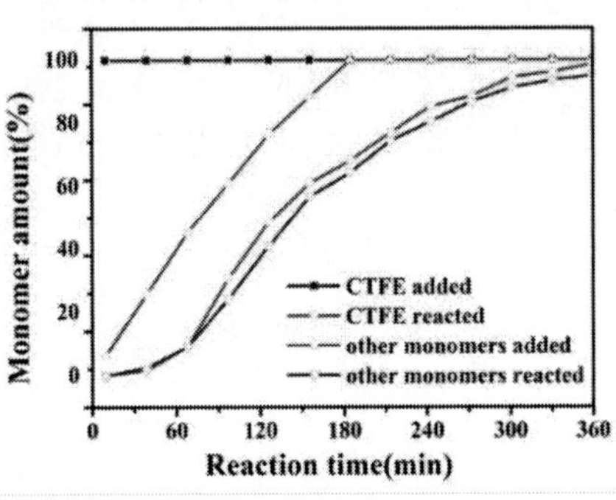

Figure 6. Change in the curve of charge and reaction of monomer with time.

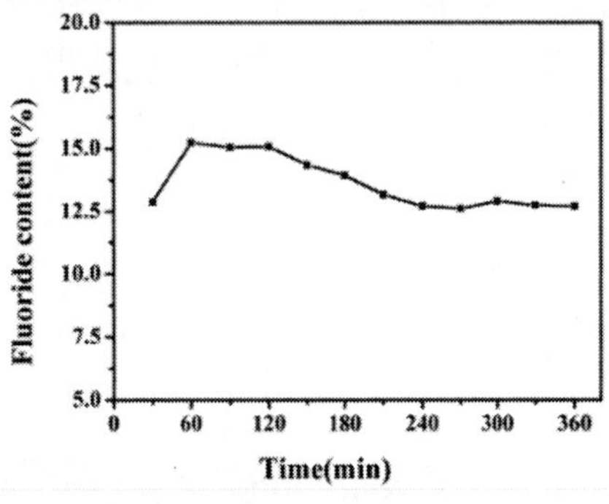

Figure 7. Change in fluoride content at different reaction stages with time.

As shown in figure 6, although CTFE involves one-step loading, it tends to copolymerize with other monomers and not self-polymerize. The dropwise addition of the monomer is faster than the speed of polymerization, CTFE and the other monomers are filled excessively in the system, which shows the macro-polymerization rate is controlled by the polymerization reaction rate. Figure 6 shows that the conversion of CTFE is relatively faster and in accordance with the copolymerization reaction rate of the other monomers. Thus, water-based fluoropolymers exhibit a relatively uniform copolymer structure. Fluorine contents at different stages of reaction are shown in figure 7. Fluorine content at the dropwise stage is relatively high. With a longer reaction time, the fluoride change becomes gradually stable, reaching a state of relatively homogeneous and synchronized copolymerization. Therefore, water-based fluoropolymer with an almost uniform structure can be prepared by adding CTFE via a disposable feeding syringe and other monomer mixtures by continuous dropping. If the CTFE monomer is fed by dropwise feeding, polymerization process will becomes complex. Whereas this technique can simplify polymerization process and facilitate industrialization.

In addition, polymerization reaction could remain inadequate by decreasing the feeding speed and increasing the dropping time, which is favorable for removing the heat of polymerization heat and controlling the reaction rate.

3.4. Relationship between the initial reaction time and monomer conversion

About 80 g of CTFE, 14.56 g of the other mixed monomers and 216.78 g of base (including initiator, emulsifier, pH buffer, and H_2O) are added into the autoclave (see table 1). The reactants are heated to 60 °C and time is started. A sample of about 2-5 g is removed every 8 min to detect nonvolatile and fluorine content. The conversion rate of monomers (including CTFE and the other monomers) are calculated. The results are listed in figure 8. As shown in the figure, monomer conversion is faster at 8-16 min and almost reaches 18% at 25-30 min. After this point, the monomer conversion slows down, suggesting that the base reaction of seed emulsion has been completed. It is defined as the initial reaction time, which is generally determined for about 25 min. After this period, the rest emulsified monomer can be added dropwise.

The particle size of the sample is evaluated at different stages, and the average particle size of the sample by weighted average is calculated. The results are shown in figure 9. When the reaction time is 8 min, the average particle size of the emulsion is 36 nm; when reaction time is 24 min, the average particle size of the emulsion is 73 nm. The average particle size changes slowly as the reaction time is extended. This result is in agreement with the conversion experimental data.

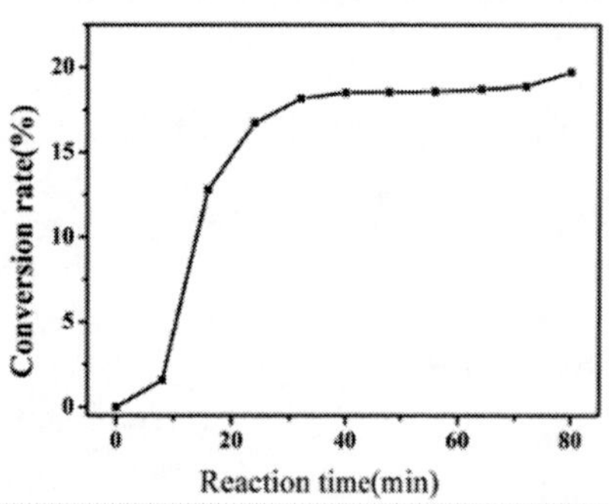

Figure 8. Change in the conversion rate of initial monomer with reaction time.

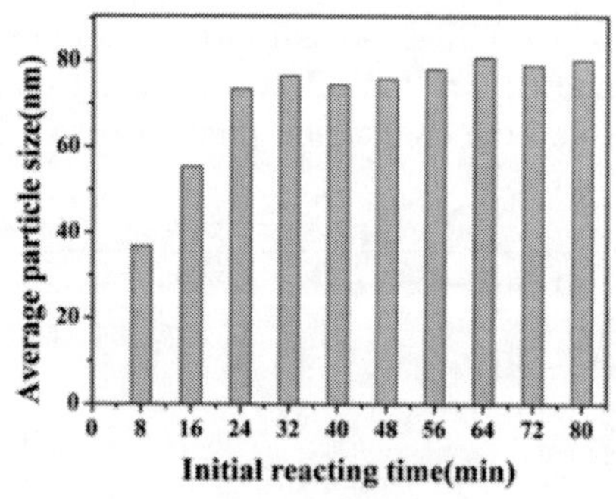

Figure 9. Change in the particle size of the emulsion at the initial reaction stage.

3.5. Effect of reaction temperature on polymerization rate

The influence of temperature on the polymerization rate is shown in figure 10. The results show that the polymerization rate increases with reaction temperature increasing, and no induction period is clearly observed during the initial stage of the reaction. It is because the initiator decomposition constant and chain growth rate constant increases with temperature increasing, which resulted in the increasing of monomer diffusion velocity to micelles, thereby increasing the polymerization rate.

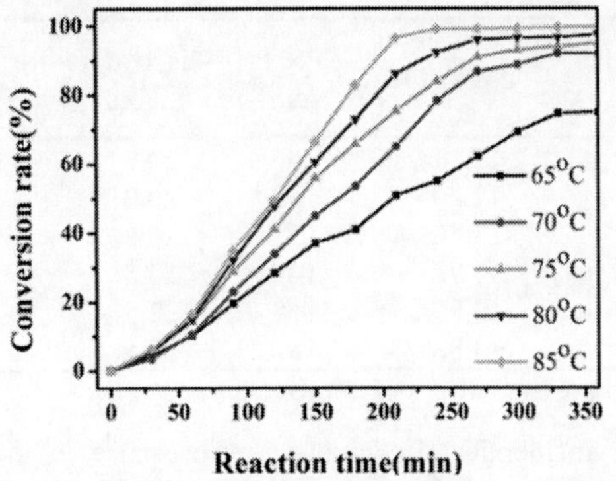

Figure 10. Effect of temperature on polymerization rate.

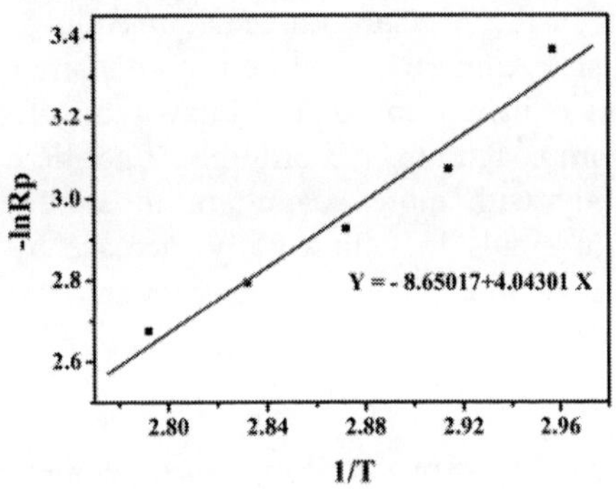

Figure 11. Relationship of -lnRp and 1/T.

By using linear regression of slash part in figure 10, the relationship between -lnRp and 1/T is determined, as shown in figure 11, the linear slope is obtained, and the apparent Ea of the polymerization system is calculated by formula 7 (Arrhenius equation formula), the result is 33.61 kJ/mol.

$$\ln k = -\frac{Ea}{RT} + \ln A \tag{7}$$

3.6. Effect of stirring speed on polymerization rate and characteristics of emulsion

Stirring speed is an important process condition for emulsion polymerization. Six different stirring speeds are used for the CTFE-VAc-BA-Veova10 quardri-copolymerization system. The reactor pressure is monitored by the gauge of autoclave. The particle size is evaluated by laser particle size distribution (MALVERN DTS 5101). The sample is diluted 1:2000 with deionized water and dispersed for 3 min in an ultrasonic bath. Viscosity is evaluated using 2# rotor of rotational viscometer (NDJ-1 style, Shanghai Jingke, China) at 30 rpm. After polymerization is completed, the coagulum is collected from the residue of the resulting emulsion by filtering. The coagulum is calculated by formula 8. The data are presented in table 2.

$$\text{coagulum\%} = \frac{\text{coagulum dried}}{\text{discharge amount}} \times 100\% \tag{8}$$

Table 2. Characteristic of water-based fluoropolymer.

Stirring rate /rpm	Nonvolatile /%	Fluoride content /%	PSD /nm	Viscosity /mPa.s	Reactor pressure[a] /Mpa	Coagulum /%
100	45.05	12.11	145	20.0	0.28	0.035
150	46.78	12.20	153	20.0	0.14	0.002
200	46.82	13.50	157	26.5	0.22	0.062
250	45.58	11.72	149	17.5	0.06	0.093
300	45.05	13.87	168	22.5	0.16	0.101
400	44.79	13.88	181	17.5	0.09	a lot

[a] Reactor pressure is the residual pressure after polymerization.

The stirring speed strongly influences the change of pressure in autoclave and the amount of coagulum, as shown in table 2. These results relate to CTFE as gas monomer which can polymerize with other monomers only by entering into liquid reactants. With stirring speed increasing, gaseous CTFE is dissolved much more in the liquid phase, so polymerization reaction speeds up. This phenomenon can be observed from the pressure on reaction ending. Whereas, the stability of the reactant in autoclave declines when keeping the condition of emulsifier unchanged, and the condition of high temperature and faster stirring speed is unfavorable for the stability of emulsion polymerization. So it leads to more coagulum. Accordingly, fluoride content also increases slightly, while nonvolatile decrease slightly with more coagulum. In addition, the viscosity of emulsion becomes smaller, and the PSD of emulsion will slightly increase by micellar's decrease used for nucleation at constant monomer amount. Thus, an appropriate stirring rate of 200-300 rpm is recommended.

4. Conclusions

Waterborne fluoropolymer emulsion based on four kinds of monomers of CTFE-VAc-BA-Veova 10 was synthesized by the semi-continuous seed emulsion polymerization with the dropwise addition technique. CTFE was fed directly in the reactor, and the other monomers were dropwise added during polymerization. The results of the fluorine content at different stage and conversion of CTFE and other monomers demonstrated that polymerization process was relatively homogeneous. The polymerization rates were accelerated when increasing concentration of initiator and emulsifier, and raising reaction temperature. The kinetics relationship was described as $Rp \propto [I]^{0.10}$, $Rp \propto [E]^{0.12}$, and Ea of the polymerization system was 33.61 kJ/mol. The faster stirring speed led to more coagulum amount and bigger particle diameter, and made nonvolatile and viscosity of emulsion lower.

Acknowledgements

This work is supported by the Fundamental Research Funds for the Central Universities (DUT16LAB11).

References

[1] H Teng 2012 *Appl. Sci.***2** 496-512
[2] N Sumi, I Kimura, M Ataku, T Maekawa and R Parker 2009 *JCT Coatings Tech.* **6** 24-28
[3] F Levine, J L Scala and W Kosik 2010 *Prog.Org.Coat.***69** 63-72
[4] N Sumi, I Kimura, M Ataku and T Maekawa 2008 *THE WATERBORNE SYMPOSIUM Advances in Sustainable Coating Technologies* Jan 30-Feb 1 New Orleans LA USA
[5] Y Nakatani and K Ichikawa 2008 *US Pat.*7468403 **B2**
[6] R A Iezzi, S Gaboury and K Wood 2000 *Prog.Org.Coat.***40** 55-60
[7] K I Suresh, T Pakula and E Bartsch 2007 *Macromol. React. Eng.***1** 253-263
[8] J Gao, X Wang, Y Wei and W Yang 2006 *J. Fluorine Chem.* **127** 282-286
[9] K Satoru, M kazuhiko and T Kentaro 1999 *JP* 99-116633
[10] S Kobayashi, K Maeda, Y Hirashima, S Hasegawa and K Sutsumi 1999 *US Pat.*5859123

[11] T E McCarthy, Y Chen and P Petersen 1999 *US Pat.*5880204
[12] T Takayanagi and M Yamabe 2000 *Prog.Org.Coat.***40** 185-190
[13] S Cheng, Y Chen and Z Chen 2002 *J. Appl. Polym. Sci.***85** 1147-1153
[14] C C Granel, L Hedhli, R Amin-sanayei and K A Wood 2004 *US Pat.*6833414 **B20**
[15] Z Chen, X Cui, W Jiang, X Cheng and H Wang 2006 *J.Appl.Polym. Sci.***99** 558-562
[16] M Yamauchi, T Hirono, S Kodama and M Mastuo 1996 *JOCCA-Surf.Coat.Int.***79** 312
[17] A Asakawa, M Unoki, T Hirono and T Takayanagi 2000 *J. Fluorine Chem.* **104** 47-51
[18] K A Wood 2001 *Prog.Org.Coat.* **43** 207-213
[19] L-P Sunga, S Vicinib, D L Ho, L Hedhlic, C Olmsteadc and K A Wood 2004 *Polymer***45** 6639-6646
[20] G Kostov, M Tredwell, V Gouverneur and B Ameduri 2007 *J.Poly.Sci.:Part A:Polym. Chem.* **45** 3843-3850
[21] D L Murray, H J Harwood, S M M Shendy and l Piirma 1995 *Polymer***36** 3841-3848
[22] S Zulfiqar, M Rizvi, A Munir, A Ghaffar and I C McNeill 1996 *Polym.Degrad.Stabil.***52** 341-348
[23] A Singh, P K Soni, M Singh and A Srivastava 2012 *Thermochim.Acta.***548** 88-92
[24] A Singh, P K Soni, T Shekharam, A Srivastava 2013 *J. Appl. Polym. Sci.***127** 1751-1757
[25] W M Thomas, M T Oshaughnessy 1953 *J.Polym.Sci.* **XI** 455-470
[26] H Liu, J Bian, Z Wang and C-J hou 2017 *Molecules***22** 184
[27] C Wang, C He 1980 *Gaofenzi Tongxun***5** 309-312
[28] C S Chern 2006 *Prog. Polym. Sci.* **31** 443-486

IOP Conf. Series: Materials Science and Engineering **292** (2017) 012116 doi:10.1088/1757-899X/292/1/012116

Quantum chemical calculation of imidization mechanism of polyamic acid

H Y Gu, L X Song, J L Zhang, L Wang, Z Y Qi and T Zhang[1]

Key Laboratory of Inorganic Coating Materials CAS, Shanghai Institute of Ceramics, Chinese Academy of Sciences, Shanghai 200050, China

[1] E-mail: tzhang@mail.sic.ac.cn

Abstract. In this paper, the imidization mechanism of polyamic acid to prepare polyimide has been studied by the density functional theory (DFT) method. Our results have shown that the imidization would proceed via two different reaction paths. The activation barrier via hydrogen transfer to closer carboxylate oxygen in the COOH is 220.35 kJ/mol while the activation barrier via hydrogen transfer to hydroxyl oxygen in the COOH is 220.70 kJ/mol. Both reaction paths would run in the actual thermal imidization process with closer high activation barriers. The high energy barrier interpreted the fact that additional energy or catalyst is usually needed during imidization. Moreover, the intrinsic imidization mechanism indicated is critical to further improve the imidization process.

1. Introduction

Polyimides (PIs), a kind of succesFsfully commercialized polymer, have been used extensively in a wide variety of applications such as in aerospace and microelectronics due to their unique physical and chemical properties such as excellent thermal stability, high strength, good dielectric and chemical resistance [1-3]. Since high molecular weight polyimide were firstly synthesized by a two-stage polycondensation of pyromellitic dianhydride with diamines in 1955, the interest in this class of polymers has grown steadily [4-6].

The typical synthesis to prepare polyimides consists of a two-step scheme in which the first step involves polymerization of a processable polyamic acid (PAA) intermediate, followed by a second cyclization or imidization step of polyamic acid to yield the final polyimide. This preparative pathway is representative of most of the early aromatic polyimide work and remains the most practical and widely utilized method of polyimide preparation to date. Imidization is commonly completed through conventional thermal treatment or chemical imidization. Imidization process is critical to the production and performance of polyimide. At present, extensive experimental data on different aspects of the thermal and chemical imidization of polyamide acid have been accumulated [7]. However, the experimental data alone are insufficient to lay down the unambiguous mechanism of the imidization. There are different assumptions about the intrinsic mechanism of the direct imide ring formation. Quantum chemical calculation was confirmed to be useful to interpret the experimental data by semi-empirical quantum chemical methods (MNDO) [7]. Quantum chemical calculation should become more fruitful on investigating the reaction mechanism with the increase of calculation precision and velocity [8-9].

Herein, the reaction mechanism of imidization in PAA has been studied by the usage of density functional theory (DFT). Reaction kinetics is explored within the framework of transition state theory.

In detail, we calculate the minimum energy reaction paths (MERPs) during imidization in polyamic acid using the B3LYP exchange and corre-lation functionals.

2. Theory

In transition state theory, a reaction is a movement from the valley or minimum representing the reactants to the valley of the products on the potential energy surface. The path across the landscape, including any intermediates and transition states, describing the reaction mechanism. The highest point along the lowest energy reaction path connecting reactants and products is the transition states, while the transition structure refers to the geometric structure of the first-order saddle point.

2.1. Computational methods

To determine the minimum energy reaction paths and the transition states (TSs), we used the GEDIIS method as implemented in Gaussian 09 version. The geometry optimization and vibrational frequency calculations were done using the B3LYP exchange and correlation functionals with the B3LYP/6-31G(d,p) basis set. Vibrational frequencies are computed by determining the second derivatives of the energy with respect to the Cartesian nuclear coordinates and then transforming to mass-weighted coordinates. Frequency values represent the property of the stable point. Local minima have all positive frequencies while transition states have one imaginary frequency. Zero-point energy corrections on the potential energy surfaces is also obtained during the frequency calculation. Dynamic reaction path (DRP) based on an intrinsic reaction coordinate (IRC) or minimum energy path is followed in one or both directions from the transition state at the same level.

2.2. Quantum chemical model

Reactant configuration was simplified as the computational model considering both the calculation precision and calculation time. The available evidence shows that it is sufficient to consider model polyamic acid as the type shown in figure 1 [7].

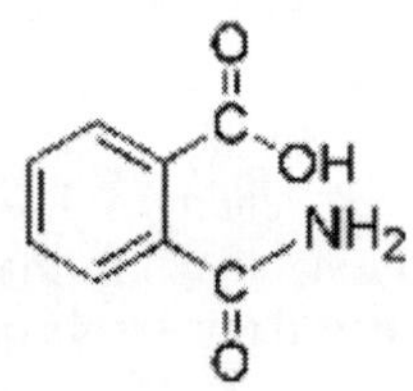

Figure 1. The computational model of polyamic acid.

3. Results and discussion

3.1. Molecular structures of polyamic acid

We have done an energy minimization for all reactants and products. Reaction process is based on the characteristics of the reactants. Thus, structural optimization of the polyamic acid was analyzed first.

The different optimized geometries of polyamic acid are shown in figure 2. The mirror conformation is ignored on account of the same energy. Based on the position of the carbonyl, hydroxyl and amino groups, there are four different conformations. Namely, the carbonyl groups are both inside the loop (figure 2a), the carbonyl groups are both outside of the ring (figure 2b), the carbonyl group attached to the $-NH_2$ group is inside the ring while the carbonyl group attached to OH is outside of the ring (figure 2c) and the opposite (figure 2d).

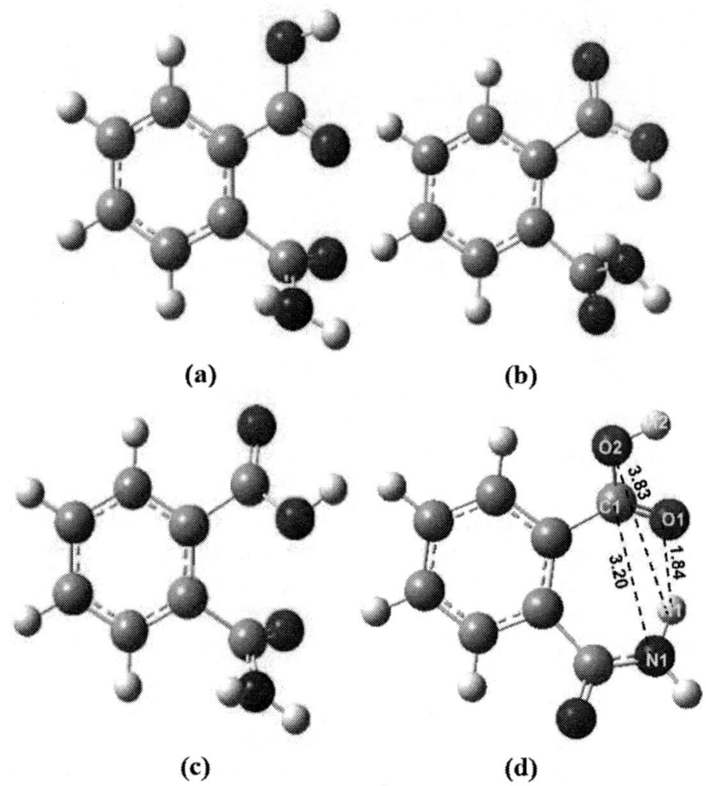

Figure 2. Typical conformations of polyamic acid (unit: Angstrom).

The absolute energy value of the different conformations from (a) to (d) in figure 2 are -1547518.7 kJ/mol, -1547494.5 kJ/mol, -1547517.8 kJ/mol and -1547524.1 kJ/mol, respectively. The conformation in (d) possesses the lowest energy, 5.4-29.6 kJ/mol lower than several other conformations. Hence, it is the most likely stable conformation of the polyamic acid. The minimum energy reaction paths would be calculated based this stable quantification model.

3.2. Minimum energy reaction paths

The cyclization of neutral polyamic acid should be expected to run via the bonding of C-N and the transfer of hydrogen in the NH2 to oxygens in the COOH. The hydrogen could transfer to carboxylate oxygen or hydroxyl oxygen in the COOH, which results in two different reaction paths. The carboxylate oxygen is closer to the hydrogen. Figure 3 is the calculated structures of transition state and intermediate product via the transfer of hydrogen to the closer carboxylate oxygen (O1). The imidization proceeds via the tetrahedral intermediate (INT) and four-membered transition states TS1 and TS2 in the stepwise mechanism. Relative energy of stationary points on the potential energy surface in the reaction of PAA→PI+H_2O without any catalysis are shown in figure 4. The imidization energy barrier (TS1) via this path is 220.35 kJ/mol. The activation barrier in the second step is 147.32 kJ/mol, which is 73.02 kJ/mol lower than that in the first step.

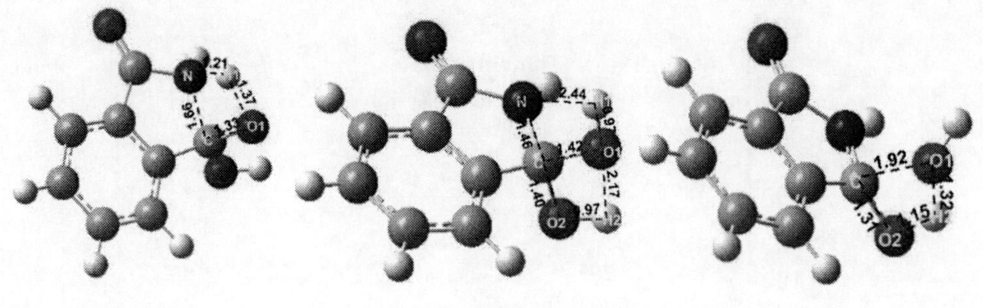

(a) (b) (c)

Figure 3. Molecular structures of the transition states and the intermediate via the transfer of H to O1 (unit: Angstrom); (a) Transition states TS1; (b) Intermediate; (c) Transition states TS2.

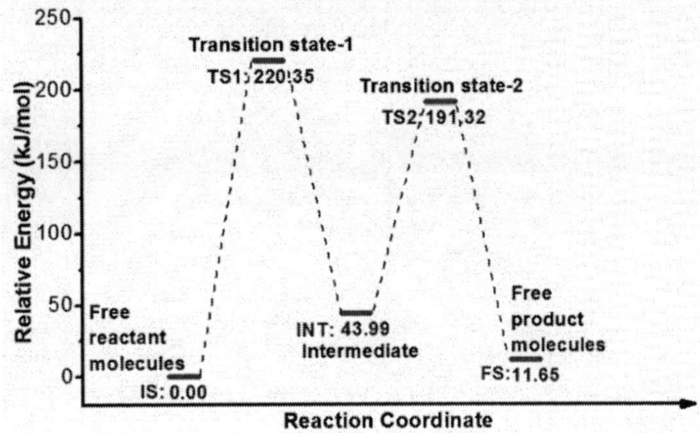

Figure 4. Calculated relative energy of stationary points on the potential energy surface in the reaction of PAA→PI+H_2O via the transfer of H to O1 reaction coordinate.

Figure 5 shows the calculated structure of the transition state via the transfer of hydrogen to the hydroxyl oxygen (O2). The imidization proceeds via the transition states TS3 in the S_N2 mechanism. Relative energy of stationary points on the potential energy surface in the reaction of PAA→PI+H_2O without any catalysis are shown in figure 6. The imidization energy barrier via this path is 220.70 kJ/mol.

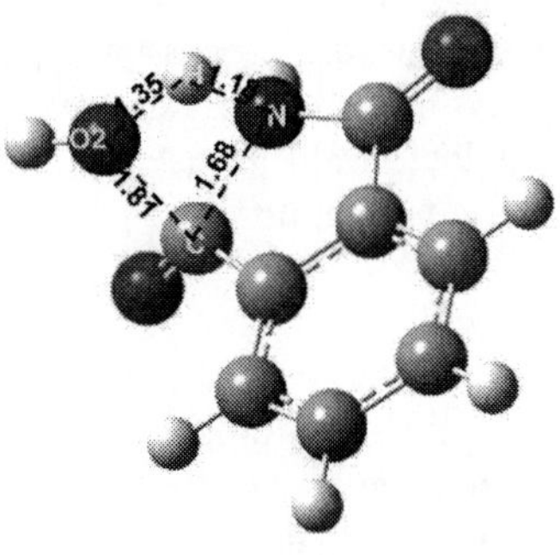

Figure 5. Molecular structure of the transition states TS3 via the transfer of H to O2 (unit: Angstrom).

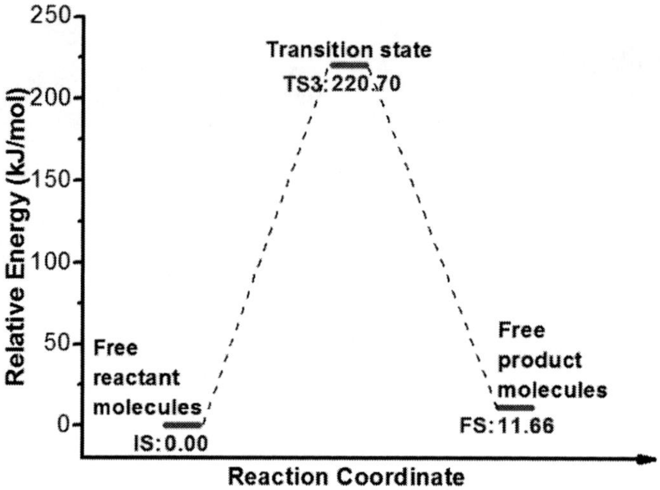

Figure 6. Calculated relative energy of stationary points on the potential energy surface in the reaction of PAA→PI+H_2O via the transfer of H to O2 reaction coordinate.

After passing the transition states as shown in figure 3 and figure 5, the product polyimide and water (H_2O) was obtained by the formation of a more stable five-membered ring. We have confirmed that reactants (polyamic acid) and products (polyimide and water) can both be obtained from the transition states based on an intrinsic reaction coordinate (IRC) or minimum energy path.

As a result of B3LYP calculations of transition states, the energy barrier according to the S_N2 mechanism is a little higher. During the actual imidization process, both reaction paths above would run based on very close activation barrier. This is different from the results calculated by MNDO method [7], which indicated that the stepwise mechanism was more favorable than the concerted mechanism.

Generally, the imidization is characterized by high activation barriers. This is in agreement with the intense heating of polyamic acid to obtain imidization. The energy barrier of TS1 calculated by MNDO method is 283 kJ/mol [7]. It should be noted that experimental activation energies of the thermal imidization of polyamic acid are in the range of 80-130 kJ/mol [9], which is much lower than the calculated reaction barrier. Besides, the result of B3LYP calculations is closer to the experimental data than the MNDO calculations. The high values of the calculated energy barriers are accounted for not only by the shortcomings of the calculation method, but also by the absence of consideration of such medium effects as catalysis of the cyclization by the carboxyl groups of polyamic acid, the difference of electric susceptibility from unity, etc. [7]. The possibility of catalysis by carboxyl groups suggests that one of the reasons that the thermal process does not yield a completely cyclized final polyimide may be a depletion of catalytically active carboxyl groups in imidization. Based on the above results, additional energy or catalyst could be used to change the paths of C-N bonding and hydrogen transfer to lower the imidization temperature.

4. Conclusion

The minimum energy reaction paths during the imidization of polyamic acid have been studied in the current work. As a result of the calculations, the imidization would proceed via two different reaction paths. The activation barrier via hydrogen transfer to the closer carboxylate oxygen in the COOH is 220.35 kJ/mol while the activation barrier via hydrogen transfer to hydroxyl oxygen in the COOH is 220.70 kJ/mol. Both reaction paths would run in the actual imidization process based on very close activation barrier. In general, the imidization is characterized by high activation barriers. This is in agreement with the intense heating of polyamic acid to obtain imidization. Based on the above conclusions, we could use the catalyst or additional energy to change the paths of C-N bonding and hydrogen transfer to lower the imidization temperature.

Acknowledgments
The current work was supported by the Laboratory foundation of Chinese Academy of Sciences (Grant No. 16S085) and the innovation fund of Computational Materials Center from SICCAS (Grant No. Y75ZC2120G).

References
[1] Kurosawa T, Higashihara T and Ueda M Polyimide memory: a pithy guideline for future applications 2013 *Polym. Chem-Uk.* **4** 16-30
[2] Liaw D J, Wang K L, Huang Y C, Lee K R, Lai J Y and Ha C S Advanced polyimide materials: Syntheses, physical properties and applications 2012 *Prog. Polym. Sci.* **37** 907-974
[3] Vanherck K, Koeckelberghs G and Vankelecom I F J 2013 Crosslinking polyimides for membrane applications: A review *Prog. Polym. Sci.* **38** 874-896
[4] Ghosh M K and Mittal K L 1996 *Polyimides: fundamentals and applications* (New York: Mercel Dekker)
[5] Mittal K L 2007 *Polyimides and other high temperature polymers: synthesis, characterization and applications* (New York: CRC Press)
[6] Abadie M J M 2012 *High Performance Polymers-Polyimides Based-From Chemistry to Application* (Croatia: InTech)
[7] Bessonov M I and Zubkov V A 1993 *Polyamic acids and polyimides: synthesis, transformations, and structure* (New York: CRC Press)
[8] Böhm O, Leitsmann R, Plänitz P, Radehausa C, Schallerc M and Schreiberb M 2012 Silylation of silicon bonded hydroxyl groups by silazanes and siloxanes containing an acetoxy group. N-trimethylsilylimidazole vs. dimethyldiacetoxysilane *Comput. Theor.* **991** 44-47
[9] Bessonov M I 1987 *Polyimides--thermally stable polymers* (New York: Plenum Press)

IOP Conf. Series: Materials Science and Engineering **292** (2017) 012117 doi:10.1088/1757-899X/292/1/012117

Hydrogenolysis of α-methylbenzyl alcohol to ethylbenzene over Pd/C catalyst

J Feng[1], Y H Zhong and S H Dai

Chongqing Key Laboratory of Industrial Fermentation Microorganism, College of Chemistry and Chemical Engineering, Chongqing University of Science and Technology, Chongqing 401331, China

[1] E-mail: fengjianscu@163.com

Abstract. The hydrogenolysis of α-methylbenzyl alcohol (MBA) to ethylbenzene (EB) over Pd/C catalyst was studied. The XRD and TEM results show that Pd nanoparticles are well dispersed on the carbon support with good crystallinity. There is no 1-cyclohexylethanol or ethylcyclohexane in the products, indicating that Pd/C is excellent for inhibiting the hydrogenation of the aromatic ring. Alcohol solvents are beneficial to increase the catalytic activity because of their strong polarity and good solubility.

1. Introduction

Catalytic hydrogenolysis is an important technique for the reduction of C–OH bond to the corresponding C–H bond. This transformation is widely used in organic synthesis [1-3] and biomass conversion [4-6]. In particular, the hydrogenolysis of C–OH bond in benzylic alcohols is highly desired because of the rapid development of producing value-added chemicals from lignin [7]. α-Methylbenzyl alcohol (MBA) is a typical benzylic alcohol, which is often used as a model compound to study the hydrogenolysis of C–OH bond [8]. MBA is also a byproduct in the production of propylene oxide employing the ethylbenzene hydroperoxide oxidation method [9]. If MBA can be hydrogenolyzed back to ethylbenzene (EB) with high selectivity (Scheme 1), the production of propylene oxide will achieve in a recycling process, thereby improving the economic benefit of the manufacturers.

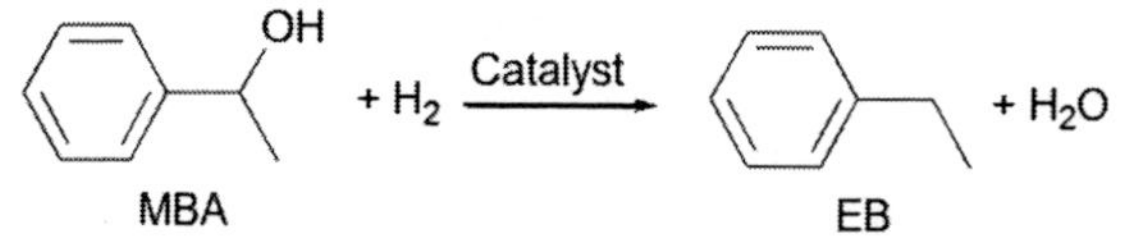

Scheme 1. Hydrogenolysis of α-Methylbenzyl alcohol (MBA) to ethylbenzene (EB).

The hydrogenolysis of MBA has been studied over several heterogeneous catalysts [8-12], mainly including Pd, Pt, Ru and Rh. It is found that Pt, Ru or Rh based catalysts usually lead to benzene ring hydrogenation, exhibiting poor selectivity to EB [9-11]. In contrast, Pd is a superior hydrogenolysis catalyst, which can greatly lower the risk of hydrogenation of the aromatic ring [9, 11]. For example, Pd/Co_3O_4 was reported as a satisfactory catalyst for the hydrogenolysis of MBA by Musolino and co-workers [8]. Their results showed that a different kinetic behavior can be observed on Pd/TiO_2, Pd/SiO_2 and Pd/Co_3O_4. Lin et al. [12] explored a green Pd catalytic system by using CO_2-expanded methanol and compressed CO_2/water as reaction media. CO_2 could decrease the mass transfer resistance, and increase the hydrogen solubility; however, the operating procedure was complicated and unmanageable. In addition, Pd based catalysts are also active for the transfer hydrogenolysis of MBA using hydrogen donors (formic acid and its salts) instead of molecular hydrogen [13-15]. Although the transfer hydrogenolysis approach seems safer, the cost of this process is more expensive. Furthermore, under the transfer hydrogenolysis condition, MBA often dehydrogenates to the undesired product acetophenone in a large amount because of the competing disproportionation reaction [15, 16].

In our previous work [13], we have found that Pd/C was superior to other supported Pd catalysts (Pd/Al_2O_3, Pd/NaY, Pd/SiO_2 and Pd/TiO_2) in the transfer hydrogenolysis reaction due to its moderate acid density feature. However, comparatively little research has focused on the systematically study of the traditional hydrogenolysis of MBA over Pd/C using molecular hydrogen. As a matter of fact, it is not a problem to utilize hydrogen for production in the modern industry. This inspires us to investigate the catalytic performance of Pd/C in the traditional hydrogenolysis process. The Pd/C catalyst was characterized by several techniques and evaluated by changing relevant parameters. Pd/C was found to be an efficient catalyst for the hydrogenolysis of MBA to EB under optimized conditions.

2. Experimental

The Pd/C (5%) catalyst was prepared by impregnation technique [13]. MBA (98%) was purchased from Aladdin. Other reagents were commercially obtained as analytical reagent samples and used as received. BET surface area of Pd/C was measured on an auto adsorption/desorption analyzer (ZXF-6, Northwest Chemical Industry). Powder X-ray diffraction (XRD) measurements were performed on a diffractometer (Philips) equipped with a Cu Kα radiation source (λ = 0.15406 nm). Transmission electron microscopy (TEM) images were obtained by JEM-1200EX. The hydrogenolysis of MBA was carried out in a 20 mL autoclave batch reactor. The catalyst was pre-activated in the same autoclave before each reaction test. A typical run was performed as follows. Pd/C (0.1 g), MBA (1 g) and ethanol (5 mL) were added to the autoclave. The reactor was sealed and purged repeatedly with hydrogen to eliminate air, and then pressurized to the necessary hydrogen pressure. The reaction mixture was stirred when it was heated to the desired temperature. After the reaction, the products were analyzed by a gas chromatograph (Agilent 6890N, Agilent Technologies).

3. Results and discussion

The BET surface area of the Pd/C catalyst was determined to be 585 m^2/g, a little lower than the carbon support (620 m^2/g, provided by the supplier). Figure 1 shows the XRD pattern of the Pd/C catalyst. The major phase presented in the diffractogram is amorphous carbon, which is accompanied by some graphite phase at $2\theta = 26.4°$ and $50.4°$ (JCPDS Card #41-1487). The diffraction peaks at 2θ = 40.1°, 46.7°, 68.1°, 82.1° and 86.6° are attributed to metallic Pd (JCPDS Card #05-0681). These sharp peaks indicate that the crystallinity of the Pd nanoparticles is good. According to the Scherrer equation, the mean Pd particle size was 13.0 nm. A representative TEM image of the Pd/C catalyst is shown in figure 2. The Pd nanoparticles are dispersed evenly over the carbon hosts, with a mean size of 11.3 nm. However, the size distribution is a bit wide, as shown by the histograms (± 1.9 nm). The mean Pd particle size determined by TEM is lower than that of XRD, possibly because some small Pd particles can't be detected by XRD.

As was mentioned above, an undesired side reaction in the hydrogenolysis of MBA is the hydrogenation of the aromatic ring [9-11]. In this work, EB is always the main product when the reaction is performed on Pd/C. A small amount of styrene and acetophenone are formed as the major byproducts. Styrene is produced by the dehydration of MBA on the acidic sites of the support [9, 17]. The formation of acetophenone is attributed to the dehydrogenation of MBA on the Pd active sites [13, 16]. No 1-cyclohexylethanol or ethylcyclohexane is detected in the products, suggesting that Pd/C is excellent for suppressing the hydrogenation of the benzene ring.

IOP Conf. Series: Materials Science and Engineering **292** (2017) 012117 doi:10.1088/1757-899X/292/1/012117

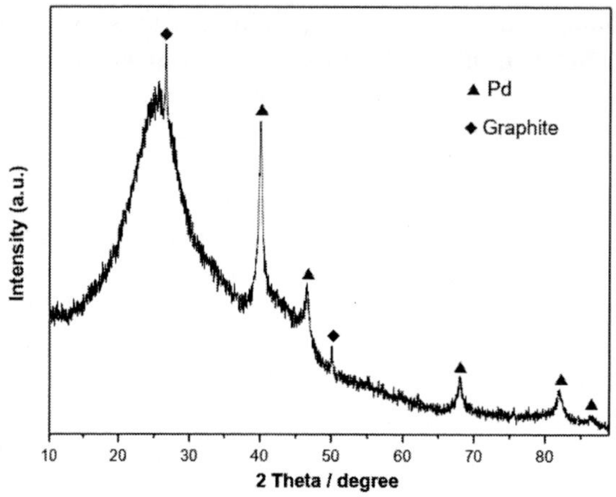

Figure 1. XRD pattern of Pd/C catalyst.

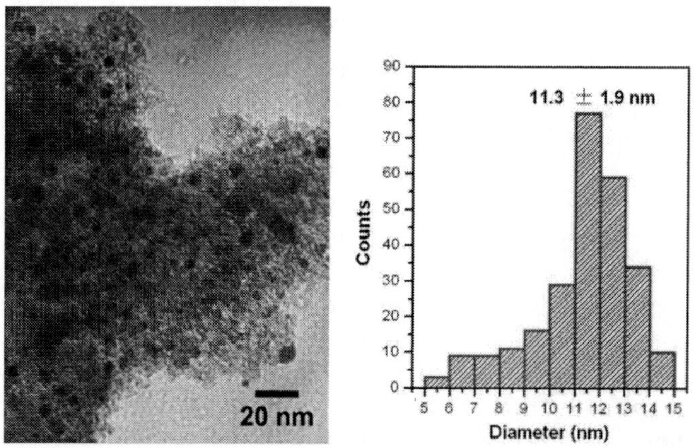

Figure 2. TEM image and particle size distribution of Pd/C catalyst.

Figure 3 presents the effect of reaction temperature on the MBA conversion and the product selectivity. As the temperature increased from 30 °C to 70 °C, there was a uniform increase in the MBA conversion from 66.8% to 99.9%. The reaction can proceed nearly completed at 70 °C. This temperature is even lower than that used in the transfer hydrogenolysis of MBA over Pd/C [13,15]. The selectivity to EB reached up to 97.1% at 50 °C, and it changed slightly even if the reaction temperature continued to increase. The high selectivity to EB is of great importance because it will enable the recycling process of the propylene oxide production to become really feasible and quite simple [13].

Figure 4 shows the effect of hydrogen pressure on the catalytic performance of Pd/C. The hydrogen pressure had a significant impact on the MBA conversion when it was below 1 MPa, after which the conversion changed slightly. This is consistent with the results obtained by the Pd/HMOR90 catalyst [9]. It is known that the solubility of hydrogen is proportional to the pressure [18]. Therefore, the increasing MBA conversion may be owing to the enhanced hydrogen concentration in the solution with increasing pressure. However, the concentration of hydrogen probably reached a certain limit and did not increase obviously when the pressure was higher than 1 MPa. Interestingly, the product selectivity showed little changes as the hydrogen pressure increased, but lower pressures (<1 MPa) can ensure an excellent selectivity to EB.

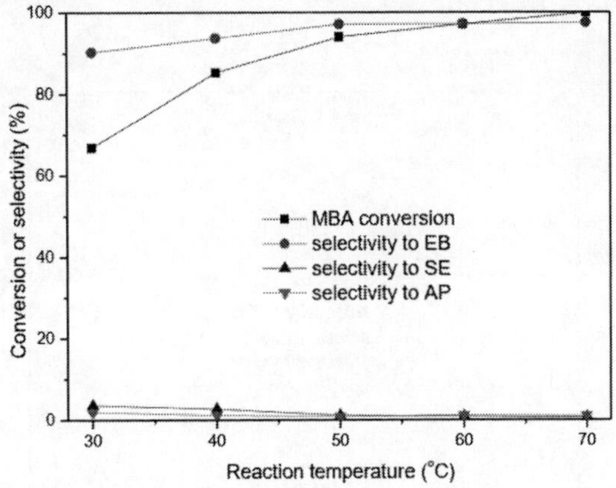

Figure 3. Effect of reaction temperature on the hydrogenolysis of MBA over Pd/C. EB: ethylbenzene; SE: styrene; AP: acetophenone.

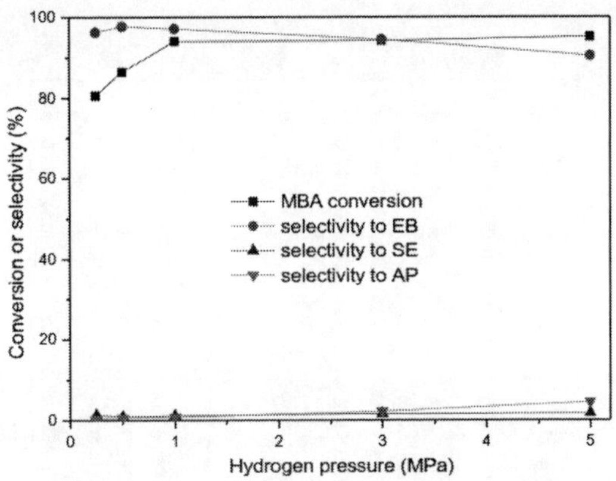

Figure 4. Effect of hydrogen pressure on the hydrogenolysis of MBA over Pd/C.

Shown in figure 5 is the reaction profile for the hydrogenolysis of MBA over Pd/C. It appears that as the conversion of MBA neared completion, the hydrogenation of styrene became difficult, which results in the decrease of EB selectivity. It has been revealed that styrene is an intermediate product in the hydrogenolysis process, and it can be hydrogenated to EB on the metal sites [9]. As the reaction progress, the carbon deposition may occur on the catalyst, leading to the blocking of Pd sites for hydrogenation. This is a reasonable explanation for the variation tendency of the product selectivity in figure 5.

Solvent choice is known to be important for the hydrogenolysis of lignin [19, 20]. Considering that benzylic alcohols are abundant functional group in lignin [7], we believe that solvent has a significant effect on the hydrogenolysis of MBA. Therefore, a series of commonly used solvents, including methanol, ethanol, n-propanol, n-hexane, and cyclohexane were evaluated. As illustrated in figure 6, all the tested solvents showed good selectivity to EB (> 90%), and the best results were achieved with methanol and ethanol (both were about 97%). As for the MBA conversion, the reactions conducted in alcohol solvents were obviously better than that in n-hexane or cyclohexane. Generally, the polarity and solubility are two main factors that can influence the solvent effect on the hydrogenolysis reaction [20]. Methanol, ethanol and propanol are excellent polar solvents. The interaction between alcohol solvents and the hydroxyl group of MBA can increase the substrate solubility, thereby facilitating its contact with the catalyst. From the viewpoint of its low toxicity and cost, ethanol is the best solvent for the hydrogenolysis of MBA to EB.

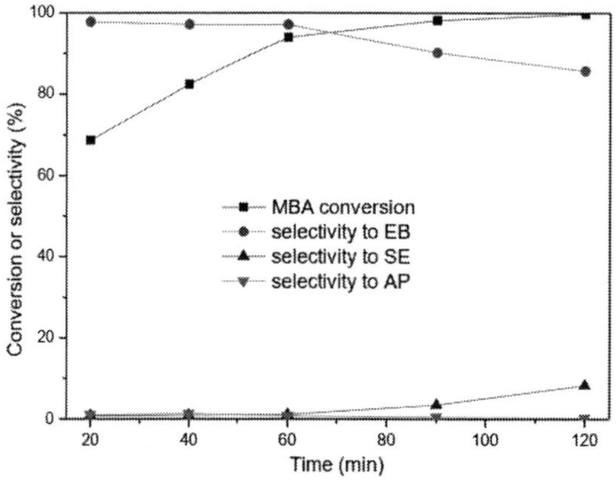

Figure 5. Reaction profile for the hydrogenolysis of MBA over Pd/C.

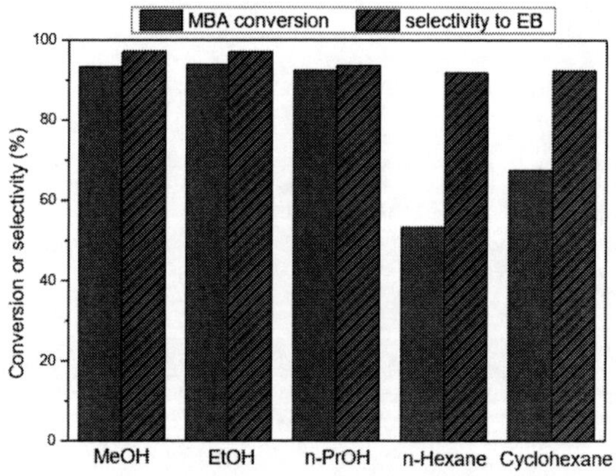

Figure 6. Hydrogenolysis of MBA in different solvents over Pd/C.

4. Conclusions

The efficient hydrogenolysis of MBA to EB was achieved over Pd/C catalyst. The reaction temperature, hydrogen pressure, reaction time and solvent can significantly influence the catalytic performance of Pd/C. Under optimized conditions (5 mL ethanol, 50 °C, 1 MPa, 1 h), the conversion of MBA and the selectivity to EB can be high up to 94.0% and 97.1%, respectively. The successful transformation of MBA to EB makes the production of propylene oxide can be realized in a recycling process.

Acknowledgments

This work was supported by the Scientific and Technological Research Program of Chongqing Municipal Education Commission (No. KJ1601309), and the Open Fund of Chongqing Key Laboratory of Industrial Fermentation Microorganism, Chongqing University of Science and Technology (No. LIFM201708).

References

[1] Ranade V S and Prins R 2000 *Chem. Eur. J.* **6** 313
[2] Muzart J 2005 *Tetrahedron* **61** 9423
[3] Thakar N, Polder N F, Djanashvili K, Bekkum H, Kapteijn F and Moulijn J A 2007 *J. Catal.* **246** 344
[4] Schlaf M 2006 *Dalton Trans.* **39** 4645
[5] Corma A, Iborra S and Velty A 2007 *Chem. Rev.* **107** 2411
[6] Feng J and Xu B 2014 *Prog. React. Kinet. Mech.* **39** 1
[7] Zakzeski J, Bruijnincx P C A, Jongerius A L and Weckhuysen B M 2010 *Chem. Rev.* **110** 3552
[8] Musolino M G, Mauriello F, Busacca C and Pietropaolo R 2015 *Top. Catal.* **58** 1077
[9] Kwak B S, Kim T J and Lee S I 2003 *Appl. Catal. A* **238** 141
[10] Sanders D K, Lin S D and Vannice M A 1994 *J. Catal.* **147** 375
[11] Sato T, Sato O, Arai K, Mine E, Hiyoshi N, Rode C V and Shirai M 2006 *J. Supercrit. Fluids* **37** 87
[12] Lin H W, Yen C H, Hsu H and Tan C S 2013 *RSC Adv.* **3** 17222
[13] Feng J, Yang C F, Zhang D L, Wang J B, Fu H Y, Chen H and Li X J 2009 *Appl. Catal. A* **354** 38
[14] Mao H, Liao X P and Shi B 2011 *Catal. Commun.* **12** 1177
[15] Sawadjoon S, Lundstedt A and Samec J S M 2013 *ACS Catal.* **3** 635
[16] Keresszegi C, Mallat T and Baiker A 2001 *New J. Chem.* **25** 1163
[17] Kwak B S and Kim T J 1999 *Appl. Catal. A* **188** 99
[18] Feng J, Zhang Y Q, Xiong W, Ding H and He B 2016 *Catalysts* **6** 51
[19] Wang X Y and Rinaldi R 2012 *ChemSusChem* **5** 1455
[20] Shu R Y, Zhang Q, Ma L L, Xu Y, Chen P R, Wang C G and Wang T J 2016 *Bioresour. Technol.* **221** 568

IOP Conf. Series: Materials Science and Engineering **292** (2017) 012118 doi:10.1088/1757-899X/292/1/012118

H_2O-Polyaluminium chloride-TBAB as synergistic catalysts for the synthesis of cyclic carbonate

X J Liu[1], P Yan and Y Han

Zaozhuang University and Lunan Coal Institute, Zaozhuang, 277100, Shandong province, China

[1] E-mail: lxj3786749@126.com

Abstract. An efficient catalytic system consisting of H_2O, Polyaluminium chloride (PAC) and Tetrabutylammonium bromide (TBAB) was applied to the cycloaddition of carbon dioxide (CO_2) to epoxides under mild conditions. Their catalytic cycloaddition activities were found to be well correlated with H_2O and polyaluminium chloride, which had a synergetic effect with the halide anion of TBAB. The presence of H_2O and PAC could remarkably improve the yield of propylene carbonate (PC) by which the reaction yield is about 4-5 times higher than TBAB. alone.The catalytic system also exhibited excellent cycloaddition activities for various epoxide substrates.

1. Introduction

Catalytic conversion of CO_2 has been attracting more and more attention due to the growing concern about the greenhouse effect and as an inexpensive, nontoxic, nonflammable, renewable, and highly abundant single carbon-atom (C1 resource) building block for organic synthesis [1-5]. The synthesis of the five-membered cyclic carbonates from epoxides and CO_2 is one of the most promising methods [6-9]. Not only is the reaction green for 100% atom efficiency but also the product cyclic carbonates have found wide applications as polar aprotic solvents [10-11], electrolytes [12], precursors of polymeric materials [13], and fine chemical intermediates [14]. In recent decades, a number of catalysts have been reported for this method of synthesizing cyclic carbonate, including organic bases [15], ionic liquids [16], transition metal complexes [17], metal oxides [18], and functional organic polymers [19]. Among these, Ionic liquids (ILs) have attracted much attention as a result of their unique properties, such as nonvolatility, nonflammability, and recyclability. However, the preparation of ionic liquids is high cost and time consuming and most of the ionic liquids are water or air sensitive. Therefore, how to improve the catalytic activity of simple ionic liquid has become an interesting topic. One strategy has focused on introducing a certain amount of cocatalysts (e.g., Brønsted acids [20], Lewis acids [21], water [22], alcohols [23]) into the traditional IL catalysts. Another efficient alternative strategy to improve the reactivity of ILs is to incorporate functional moieties (e.g., hydroxyl groups [24], carboxyl group [25], and even surface silanol groups [26] or metal complexes [27]) at the chain ends of ILs. Among these studies, it is accepted that there is an excellent synergistic effect between the hydrogen bond donor (the activation of epoxides) and the halide anion of ILs. Owing to functionalized IL catalysts being difficult to synthesize, numerous IL/co-catalyst binary catalytic systems have been reported up to now [28]. However, the exploration of cheap, nontoxic, highly active, easy post-processing co-catalysts is still highly desirable. In some catalyst systems, the cyclo-addition yields of CO_2 and epoxides can be significantly increased with the unsatisfactory reaction selectivity when a small.amount of water is introduced into the reaction system [24].Though aluminum-containing complexes have high selectivity as active catalysts [29], catalyst synthesis was a complex process. Based on the above research reports, we are going to find a simpler way to improve the yield of the cyclo-addition CO_2 with various epoxides by using the catalytic system consisting of water, inorganic aluminium salt, and TBAB under relatively mild conditions. It was revealed that hydroxy group present on the water and bromide anion within TBAB exhibited excellent synergistic effect in promoting the reaction. More interestingly, the yield and selectivity from the cycloaddition of carbon dioxide to epoxides were dramatically increased by addition of PAC. Moreover, Polyaluminium chloride (PAC) as heterogeneous catalyst is a cheap, stable, readily available, water-tolerant, and easily recyclable cocatalyst compared to other types of additives in the coupling reactions. To the best of our knowledge, we first report PAC as catalyst for the synthesis of cyclic carbonate. Thus, the catalytic system consisting of water, PAC, and TBAB can act as a mild and highly effective catalyst system for the synthesis of cyclic carbonate.

2. Experimental:

2.1. Materials and instruments

Carbon dioxide with purity of 99.99% was purchased from Zaozhuang Meixin Industrial gases Co. Ltd. Epoxides, TBAB, Al_2O_3, $Al(OH)_3$ were purchased from Aladdin Chemical Co. Ltd. PAC ($Al_n(OH)_mCl_{(3n-m)}$, pale yellow powder) was purchased from Adamas Reagent Co. Ltd and dried in vacuum at 100 ℃ for 3 hours before using. All other chemicals were purchased from Sinopharm Chemical Reagent Co. Ltd.

^{1}HNMR spectra were recorded on a picoSpin 80 in CD_3Cl at ambient temperature. GC analyses were performed on a Varian CP-3800 equipped with a capillary column (Varian CP-Wax 52 CB), using a flame ionization detector. MS was obtained using Agilent 7000 B.

2.2. Cycloaddition of epoxides with CO_2

All the cycloaddition reactions were conducted in a 100 mL stainless steel reactor equipped with a magnetic stirrer and self-acting temperature control system. In the typical procedure, desired amounts of catalyst (TBAB/H_2O/PAC) and propylene oxide (PO) were added into the reactor. Then, CO_2 was charged in the reactor and the pressure was adjusted to 1.5 MPa, and the stirring was started. The reactor was heated and maintained at 120 °C for 1 h, and the pressure was kept constant by means of a CO_2 cylinder connected to the reactor during the reaction. After the reaction was completed, the reactor was cooled to 0°C in an ice-water bath, and the remaining CO_2 was released. The catalyst (PAC) was separated from the system by centrifugalization, then washed with ethyl acetate (2×5mL), dried under vacuum and reused for another run. The product yield was obtained by silica gel column chromatography (100-200 mesh silica gel, 12:1 petroleum ether(60-90℃): EtOAc) afforded target product.

3. Results and discussion

3.1. Screening of catalysts

In the investigations of synergistic effects between acidic and basic functional groups [30-31], researchers pointed out that an epoxide was activated by acid functional groups through the O atom of epoxide, The coordination of both acidic and basic functional groups promoted the reaction. Water molecules can be activated to produce acidic functional groups by inorganic aluminum salts under certain conditions [32-33], and therefore three types of aluminum compounds are used for the reaction of propylene oxide (PO) and CO_2 to produce propylene carbonate (PC). The results are summarized in table 1. Almost no product was detected when H_2O and PAC were used alone (entries 1, 10). TBAB alone could catalyze the cycloaddition, but the yield of PC was low (entry 2). Although the addition of water obviously promoted the reaction, the yield is still low (entry 3). Due to the formation of propylene glycol, the reaction selectivity was reduced with increasing amounts of water, even if the reaction yield was improved (entries 3, 4). The addition of the aluminum compounds increased the yield of the reaction with a high selectivity (entries 5-7), and there are no polycarbonates are formed. The order of the activity of aluminum compounds was found to be PAC > Al_2O_3 > $Al(OH)_3$. Obviously, the reaction yield decreased when only using PAC and TBAB as catalysts without adding water (entries 8, 9). Therefore, the key to improve the yield of the reaction with a high selectivity was that H_2O and PAC were used together in the presence of TBAB (entries 1, 7, 8, and 10).

Table 1. Synthesis of PC catalyzed by different catalysts[a]

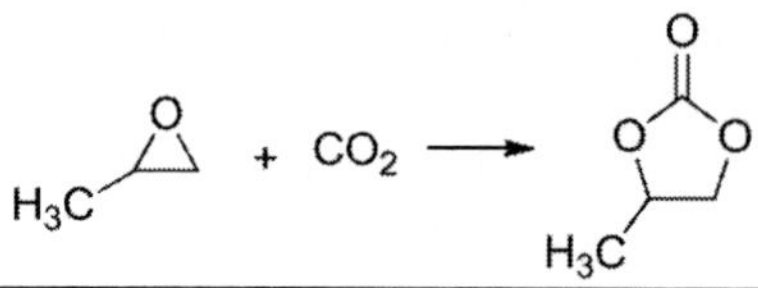

Entry	Catalyst	co-catalyst	Yield (%)[b]	Selectivity (%)[c]
1	H_2O	None	0	-
2	None	TBAB	23	99
3	H_2O	TBAB	52	99
4[d]	H_2O	TBAB	78	95
5	$H_2O + Al_2O_3$	TBAB	93	96
6	$H_2O + Al(OH)_3$	TBAB	59	98
7	H_2O + PAC	TBAB	99	99
8	PAC	TBAB	37	99
9[e]	PAC	TBAB	42	99
10	PAC	None	0	-

[a] Reaction conditions: PO (100 mmol), H_2O (2 mmol),TBAB (0.74 mmol), Co-catalyst (0.39g) 120 °C, 1.5 MPa, 1 h. [b] Isolated yield. [c] Determined by GC. [d] H_2O (0.42g, 23 mmol). [e] PAC (0.78g)

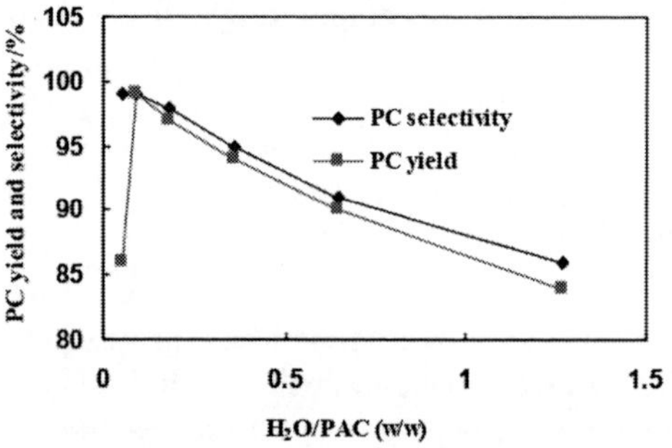

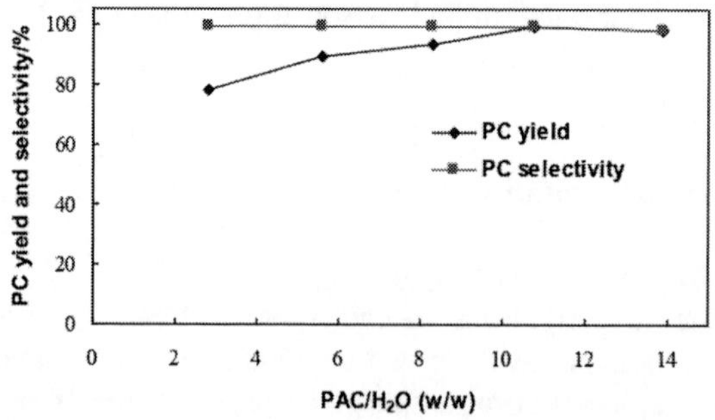

Figure 1. Effect of different proportions of H_2O/PAC. **Figure 2.** Effect of amount of PAC.

3.2. Effect of different proportion of H_2O and PAC

Since H_2O-PAC-TBAB as synergistic catalysts showed good catalytic activity, it was used to examine the effects of reaction parameters (catalyst amount, temperature, and CO_2 pressure) on the cycloaddition reaction of PO with CO_2. In order to further investigate the effect of the amount of water in the synthesis rate of PC, experiments were firstly carried out in which the ratio of H_2O/PAC was increased from 0.05 to 1.27(w/w). As shown in figure 1, only 86% of PC yield was observed at the ratio of 0.05. And the maximal PC yield (99%) obtained at the ratio of 0.09, while further increasing the ratio of H_2O/PAC from 0.09 to 1.27 caused an obvious decrease in the yield of PC though PO conversion kept at 100%. PC selectivity also decreases quickly from 99% to 86% correspondingly. It can be explained by the side reaction between water and PO to produce corresponding chemical, 1,3-propylene glycol (PG). Under the same amount of H_2O (2mmol H_2O/100mmol PO), the ratio of PAC/H_2O also could remarkably affect the synthesis rate of cyclic carbonate (figure 2). When the ratio of PAC/ H_2O was 10.8, PC yield could reach 99% and no change even if increasing the ratio of PAC/ H_2O.

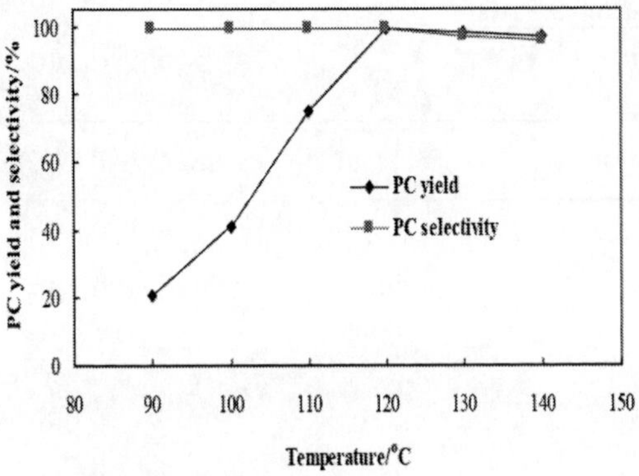

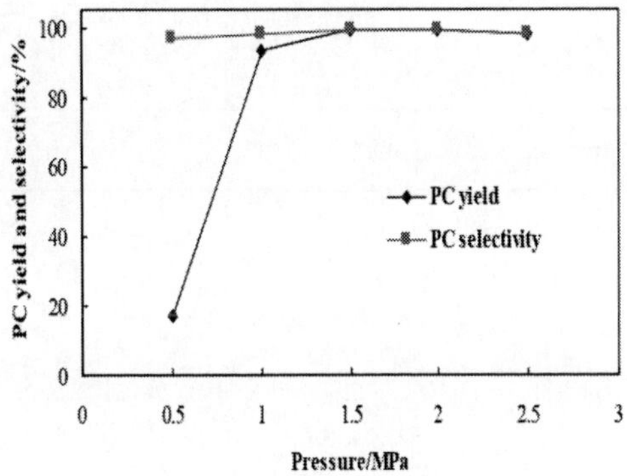

Figure 3. Effect of different reaction temperature. **Figure 4.** Effect of CO_2 pressure.

3.3 Effect of reaction temperature

Figure 3 clearly shows that reaction temperature has a strong effect on PC yield. If catalytic reactions carried out at 90°C, 100°C, 110 °C, and 120°C, PC yield were 21%, 41%, 75%, and 99%, respectively. The relationship between temperature and isolated yield was attributed to higher reactivity at higher temperatures. Furthermore, at reaction temperatures of 130°C and 140°C, yield variability did not change significantly.

3.4 Effect of CO_2 pressure

As seen in figure 4, the PC yield increased from 17% to 99% when the CO_2 pressure increased from 0.5 MPa to 1.5 MPa. Upon a further increase in the pressure (2.5 MPa), the PC yield also did not change significantly. During these catalytic reactions, the optimal CO_2 pressure can effectively increase the solubility of CO_2 in epoxides, enabling the reaction equilibrium to shift towards carbonate formation.

3.5 Reusability of the catalyst system

The stability of PAC was examined. There is no observable loss of catalytic activity over the PAC catalyst that even was reused for 6 times. The used PAC can be easily recovered by centrifugation, and washed with ethyl acetate. After drying, the PAC was used for the next run. The PC yield and selectivity were always above 98% across the cycles.

3.6 Cycloaddition of CO_2 with various epoxides

In order to survey the applicability of the catalyst system, we conducted cycloaddition reactions using epoxides of different kinds (table 2). It is apparent that the catalyst system exhibited high efficiency for all of the monosubstituted terminal epoxides in 1h (entries 1–4, table2). Disubstituted epoxides (entries 5-6) showed lower activity than mono-substituted terminal epoxides (entries 1-4), and required a long time (24h) to get a high yield, probably due to the high steric hindrance offered by the steric substituent.

3.7 Reaction mechanism

The catalytic activity of H_2O/TBAB was obviously improved by adding PAC for the cycloaddition of CO_2 to epoxides. However, the catalytic system PAC/TBAB showed the low catalytic activity even if the dosage of PAC was increased (table 1. entries 3, 7-9). According to the literature [34], PAC and H_2O can form a multimer according to a certain percentage. Therefore, water molecules can be activated to produce acidic functional groups by inorganic aluminum salts under certain conditions [32-33].

Table 2 Cycloaddition of CO_2 to various epoxides[a].

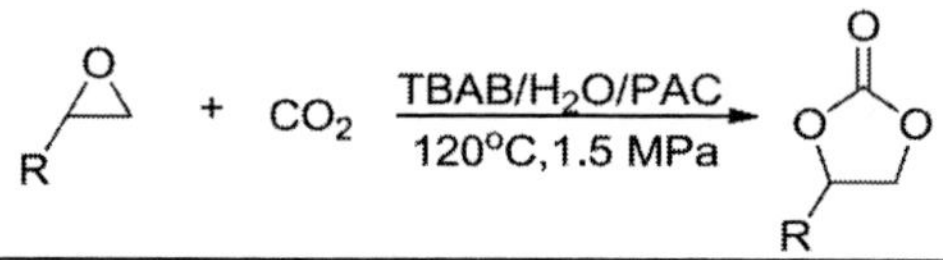

Entry	Substrate	Product	Yield (%)[b]	Selectivity (%)[c]
1	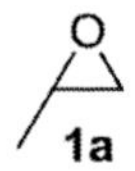 1a	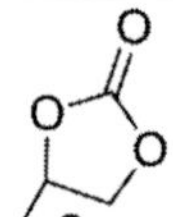 2a	99	99
2	 1b	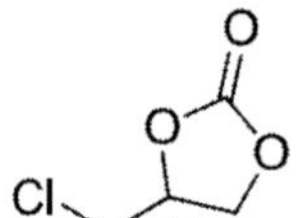 2b	98	99
3	 3a	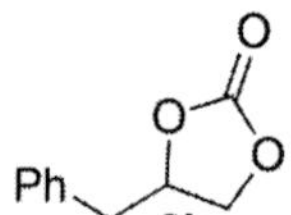 3b	99	99
4	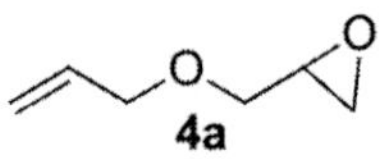 4a	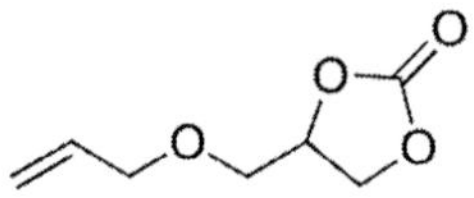 4b	98	99
5[d]	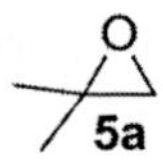 5a	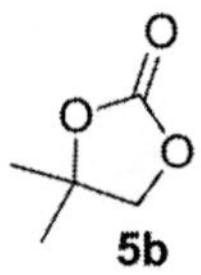 5b	94	97
6[d]	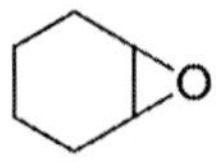 6a	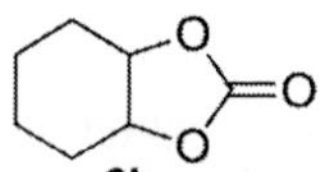 6b	99	98

[a]Reaction conditions: Epoxide (100 mmol), TBAB(0.74 mmol), H_2O(2 mmol), PAC(0.39g),120°C, CO_2(1.5MPa), and 1h.
[b]Isolated yield.
[c]Determined by GC.
[d]24h

The coordination of both acidic and basic functional groups (-OH/Br^-) promoted the reaction. Based on the above result, a possible mechanism for the synergetic effect of catalyst system was proposed (Scheme 1). The multimer formed by water and PAC enhanced the coordination effect of the H atom of water molecule with the O atom of epoxide through a hydrogen bond which resulted in the polarization of C–O bonds. At the same time, the halide anion made the nucleophilic attack on the less sterically hindered β-carbon atom of the epoxide. As a result, the ring of the epoxide was opened easily (Scheme 1, step 1). Then, the interaction occurred between the oxygen anion and CO_2, forming a carbonate anion (Scheme 1, step 2) which would be transformed into a cyclic carbonate by the intramolecular substitution of the halide in the next step (Scheme 1, step 3). Next, the catalytic system got back into its original state for the next catalytic cycle.

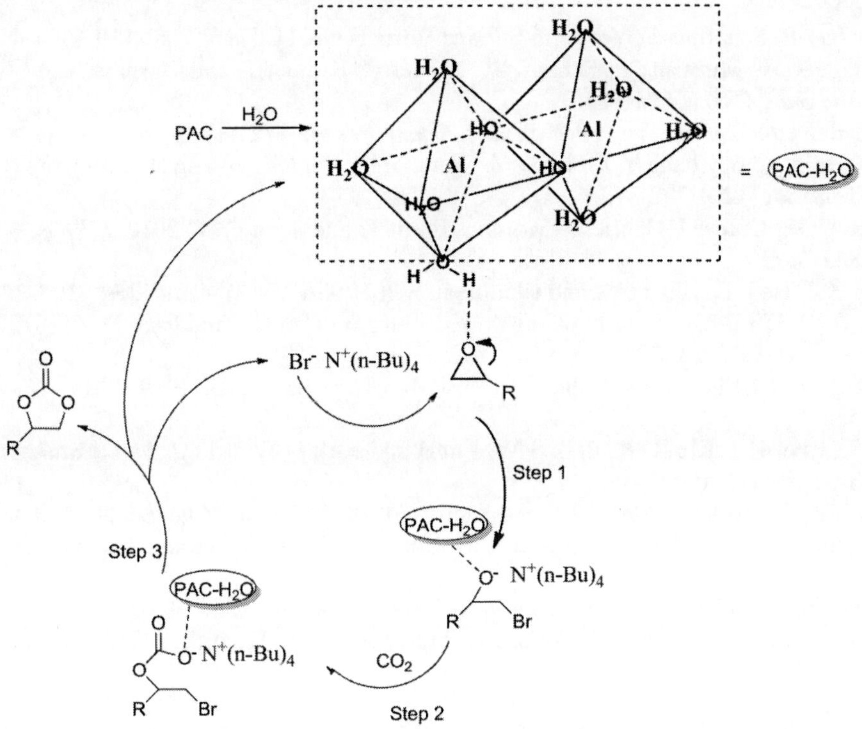

Scheme 1. Proposed mechanism for the cycloaddition reaction of CO_2 to epoxides catalyzed by TBAB/H_2O/PAC.

4. Conclusion

In conclusion, H_2O-PAC-TBAB as synergistic catalysts catalyzed the conversion of different organic epoxides into a 5-membered cyclic carbonate at rather mild reaction conditions (T=120°C, p (CO_2) =1.5MPa, 1h). Also PAC could play a significant role in determining the selectivity in addition to the activity of the catalyst. Being cheap, recyclable, water-tolerant, and environment-benign, PAC is a promising industrial catalyst for the cycloaddition of carbon dioxide (CO_2) to epoxides under mild conditions.

Acknowledgement
This research was supported by Natural Science Foundation of Shandong Province, China (ZR2010BL005) and Doctoral Startup Foundation of Zaozhuang University (1020704).

References

[1] Martin C, Fiorani G and Kleij A W 2015 *ACS Catal.* **5** 1353-1370
[2] Alessandro D M D, Smit B and Long J R 2010 *Angew. Chem. Int. Ed.* **49** 6058-6082
[3] Darensbourg D J 2010 *Inorg. Chem.* **49** 10765-10780
[4] Parvulescu V I, Hardacre C 2007 *Chem. Rev.* **107** 2615-2665
[5] Patil R M., Han H., Chai C L L, Ghanwat A A, Ganugapati S and Gnaneshwar R 2015 *Polym Bull.* **72** 2489-2501

[6] Shaikh A A G, Sivaram S 1996 *Chem Rev.* **96** 951-976
[7] Sakakura T, Kohno K 2009 *Chem Commun.* 1312-1330
[8] Yan P, Jing H W 2009 *Adv. Synth. Catal.* **351** 1325-1332
[9] North M, Pasquale R and Young C 2010 *Green Chem.* **12** 1514-1539.
[10] Postupna O O, Kolesnik Y V, Kalugin O N and PrezhdoO V 2011 *J Phys Chem B.* **115** 14563-14571
[11] Schäffner B, Schäffner F, Verevkin S P and Börner A 2011 *Chem. Rev.* **110** 4554-4581
[12] Schmitz R W, Murmann P, Schmitz R, Müller R, Krämer L and Kasnatscheew J 2014 *Prog. Solid State Chem.* **42** 65-84
[13] Lu X B, Ren W M and Wu G P 2012 *Acc. Chem. Res.* **45** 1721-1735
[14] Lu X B, Liang B, Zhang Y J, Tian Y Z, Wang Y M, Bai C X, Wang H and Zhang R 2004 *J. Am. Chem. Soc.* **126** 3732-3733
[15] Yu K M K., Curcic I, Gabriel J, Morganstewart H and Tsang S C 2010 *J. Phys. Chem. A.* **114** 3863-3872
[16] Yang Z Z, He L N, Miao C X and Chanfreau S 2010 *Adv. Synth. Catal.* **352** 2233-2240
[17] Tian D W, Liu B Y, Zhang L, Wang X Y, Zhang W, Han LN and Park D W 2012 *J. Ind. Eng. Chem.* **18** 1332-1338
[18] Yamaguchi K, Ebitani K, Yoshida T, Yoshida H and Kaneda K 1999 *J. Am. Chem. Soc.* **121** 4526-4527
[19] Han L, Choi H J, Kim D K, Park S W, Liu B and Park D W 2011 *J. Mol. Catal. A: Chem.* **338** 58-64
[20] Whiteoak C J, Nova A, Maseras F and Kleij A W 2012 *ChemSusChem.* **5** 2032-2038
[21] Wilhelm M E, Anthofer M H, Reich R M, Elia V D, Basset J M and Cokoja M 2014 *Catal Sci Technol.* **4** 1638-1643
[22] Sun J, Ren J Y, Zhang S J and Cheng W G 2009 *Tetrahedron Lett.* **50** 423-426
[23] Wilhelm M E, Anthofer M H, Cokoja M, Markovits I I E, Herrmann W A and Kuhn F E 2014 *ChemSusChem.* **7** 1357-1360
[24] Sun J, Zhang S J, Cheng W G and Ren J Y 2008 *Tetrahedron Lett.* **49** 3588-3591
[25] Roshan K R, Jose T, Kim D, Cherian K A and Park D W 2014 *Catal. Sci. Technol.* **4** 963-970
[26] Sakai T, Tsutsumi Y and Ema T 2008 *Green Chem.* **10** 337-341
[27] Chang T, Jin L L and Jing H W 2009 *ChemCatChem.* **1** 379-383
[28] Aoyagi N, Furusho Y and Endo T 2013 *J. Polym. Sci. Pol. Chem.* **51** 1230-1242
[29] Ren W M, Liu Y and Lu X B 2014 *J. Org. Chem.* **79** 9771-9777
[30] Sun J, Han L J, Cheng W G, Wang J Q, Zhang X P and Zhang S J 2011 *ChemSusChem.* **4** 502-507
[31] Udayakumar S, Lee M K, Shim H L and Park D W 2009 *Appl. Catal. A.* **365** 88-95
[32] Miura K, Hashimoto K, Fukui H, Yamada E and Shimokawa S. 1985 *J Phys Chem* **89** 5098-5101
[33] Maekawa K, Takahara K, Kajiwara T and Watanabe *M 2011 Transactions of the Society of Instrument & Control Engineer* **47** 150-156
[34] Sun Z, Zhao H D, Tong H G, Wang R F and Zhu F Z 2006 *Chin. J. Struct. Chem.* **25** 1217-1227

IOP Conf. Series: Materials Science and Engineering **292** (2017) 012119 doi:10.1088/1757-899X/292/1/012119

Hydrothermal synthesis and photocatalytic degradation ability of nickel phosphide micro/nano materials

J S Yan[1], H B Liu, Y S E and F W He

Liaoning Institute of Science and Technology, Benxi 117004, PR China

[1] E-mail: yanjs910@lnist.edu.cn

Abstract. $Ni_{12}P_5$ and Ni_2P catalysts were successfully synthetized by using water as solvent at a low temperature by hydrothermal method. The influence of molar ratio of P/Ni and hydrothermal reaction temperature was studied. The results showed that with initial P/Ni mole increment, pure tetragonal nickel phosphide ($Ni_{12}P_5$) could be transformed into pure hexagonal nickel phosphide (Ni_2P), while with temperature increment, Ni_2P could be gradually transformed into Ni_{12}P5. X-ray diffraction (XRD) and transmission electron microscopy (TEM) were used to characterize the phase and morphology of the as-prepared catalysts. The catalysts exhibited better photocatalytic degradation abilities for methylene blue under 365nm ultraviolet irradiation.

1. Introduction

In the past a few years, transitional metal phosphide, in particularly, nickel phosphides have been considered as a new type catalyst in the field of hydroprocessing, hydrogenation of nitro-compounds, adsorption of heavy metals and photocatalytic degradation reaction of organic dyes [1-3]. Besides, Ni_2P or CoP exihibited high catalytic activity and stability toward the hydrogen or oxygen evolution reaction [4-7]. Moreover, Ni_2P has been an attractive LIB anode material because of its high capacities and low polarization [8].In numerous of preparation methods of nickel phosphides, hydrothermal synthesis method has many advantages of low cost, mild preparation conditions, non-pollution, morphology and size controlled [2]. Also, stoichiometric ratio of nickel phosphide is very rich, P^{3-} ion is not stable in water system, so by changing the reaction parameters such as hydrothermal temperature, time, ratio of precursors, solvent or surfactant type, different crystalline phases and morphology of nickel phosphide products can be obtained such as dendric Ni_2P [9], hollow spherical Ni_2P, $Ni_{12}P_5$ [10, 11], unchin-like Ni_2P, Ni-P [3, 12], honeycomb $Ni_{12}P_5$ [13]. In this paper, $Ni_{12}P_5$ and Ni_2P catalysts were successfully synthetized by controlling molar ratio of P/Ni and hydrothermal reaction temperature at a low temperature by hydrothermal method. Furthermore, the photocatalytic degration of as-prepared catalysts for methylene blue was investigated.

2. Experimental section

2.1. *Materials and instruments*

$Ni(NO_3)_2 \cdot 6H_2O$, $(NH_4)H_2PO_4$, red phosphorus, ethanol and methylene blue were purchased from Sinopharm Chemical Reagent Co. Ltd. Photocatalytic device with 250w high pressure mercury lamp was home-made, main emission wavelength of which is 365nm.

2.2. *Hydrothermal synthesis of nickel phosphide*

Taking molar ratio 4/1 of phosphorus to nickel for example. Take 0.0375 mol of nickel nitrate and 50 ml of distilled water into a 100 ml beaker, fully stirring until the solution is clarified, then add 0.15 mol grinded red phosphorus, put the beaker into a water bath with electromagnetic stirrer for Ostwald ripening 2 h, put the suspension into a 100mL of high-pressure reaction kettle with teflon as interior material, sealed, followed by drying at 120 °C. After it, the reaction kettle was cooled to room temperature. The obtained black solid material in the kettle was collected and centrifuged to remove impurities with hot distilled water and ethanol in turn, finally vacuum drying for 12 hours under 60 ℃. Operation process for different molar ratio of phosphorus to nickel (1//1, 2/1, 6/1, 8/1) is the same as the above.

2.3. *Photocatalytic degradation experiment of methylene blue*

Add 0.05 g $Ni_{12}P_5$ and Ni_2P nanoparticles to 50 ml of 10 mg/L of methylene blue solution, respectively. Adopt magnetic stir for 15 min to make nickel phosphide evenly dispersed in the solution, then put the solution in photocatalytic device under UV illumination, keeping the distance from the lamp to the solution about 20 cm. During photocatalytic degradation, take a sample every 30 minutes followed high speed centrifuge and take out supernatant fluid, then measure absorbance of the solution at 665 nm for different reaction time by T6 ultraviolet-visible spectrophotometer. The procedure of blank test is the same as the above, but only differenc is no nickel phosphide catalyst. The degradation rate of methylene blue solution was caclulated by $(A_0-A_1-\Delta A)/A0$, where A_0 and A_1 is the absorbance of solution before and after photoccatalytic experiment, respectively, ΔA is the absorbance changes of methylene blue solution in blank test.

2.4. *Catalysts characterization*

The XRD patterns were measured on a Rigaku D/Max 2500X diffractometer using nickel filtered Cu Kα radiation at 40 kV and 80 mA. Transmission electron microscope (TEM) images of catalysts were obtained on a JEM2100 model. The sample was placed on a 200-mesh copper grid coated with carbon.

3. Results and Discussion

3.1. *The influence of initial molar ratio of P/Ni*

The XRD patterns of as-prepared products (figure 1) shows the effect of the different initial molar ratio of P/Ni on the crystal phase of as-prepared products. The peaks at 2θ of 8.1°, 14.0°, 19.9°, 24.4°, 28.2 °, 32.8 ° can be attributed to the characteristic peaks of $Ni_{11}(HPO_3)_8(OH)_6$(PDF44-1327)when initial n(P)/n(Ni) is 1/1or 2/1. When initial n(P)/n(Ni) is 4/1, the characteristic peaks at 2θof 32.7°, 35.8°, 38.4°, 41.7°, 44.4°, 47.0°, 49.0° can be attributed to $Ni_{12}P_5$(PDF22-1190). When initial molar ratio of P/Ni was increased to 6/1 or 8/1, the characteristic peaks at 2θ of 40.8°, 44.6°, 47.3°, 54.4^0 can be attributed to Ni_2P (PDF03-0953). The results shows the as-prepared products tend to transform into Ni_2P from $Ni_{12}P_5$ with initial molar ratio of p/Ni increased. The smallest molar ratio of p/Ni for synthesis of $Ni_{12}P_5$ is 4/1, while for synthesis of Ni_2P is 6/1 or 8/1.

3.2. *The influence of hydrothermal temperature*

The XRD patterns of as-prepared products (figure 2) shows the effect of hydrothermal temperature ranging from 90°C to 180°C on the crystal phase of as-prepared products. The molar ratio of p/Ni and hydrothermal temperature was fixed at 6 and 12 h. When the hydrothermal temperature is at 90°C, the strong peaks at 2θ of 40.8°, 44.6°, 47.3°, 54.4^0 occurred, which can be attributed to Ni_2P (PDF03-0953), but accompanied with the weak characteristic peaks of $Ni_{12}P_5$ at 2θ of 38.4°, 49.0°, suggesting the main product is Ni_2P while a small amount of $Ni_{12}P_5$ is also present. When the hydrothermal temperature is at 120°C, the characteristic peaks of Ni_2P are sharper while no other peaks can be observed, indicating high purity and crystallinity of as-prepared Ni_2P catalysts. When the hydrothermal temperature is at 150°C, the characteristic peaks of $Ni_{12}P_5$ occurred while there is a weak characteristic peak of Ni_2P at

2θ of 40.8°, indicating the main product is $Ni_{12}P_5$ while there is trace of Ni_2P. When the hydrothermal temperature is at 180°C, the position of peaks are basically the same as that at 150°C, however, the characteristic peaks of Ni_2P disappear, indicating the high purity of as-prepared $Ni_{12}P_5$. Therefore, the effect of hydrothermal temperature on the crystal phase of as-prepared products is great, the increment of hydrothermal temperature is in favor of the transformation of $Ni_{12}P_5$ from Ni_2P phase.

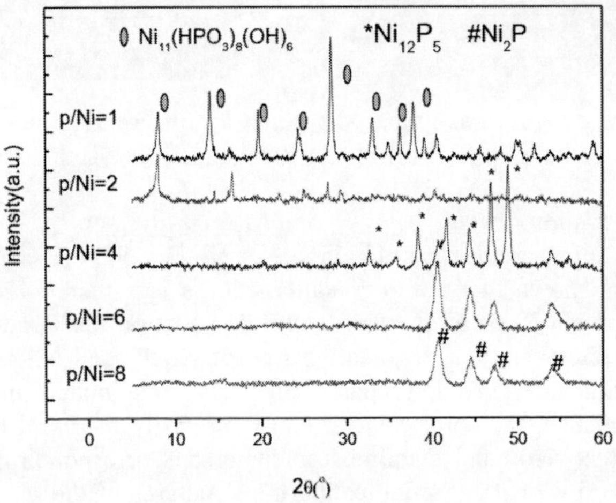

Figure 1. XRD patterns of samples prepared for different n(P)/n(Ni).

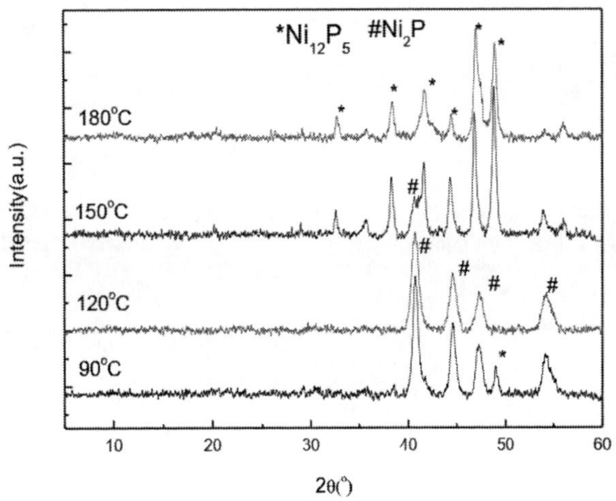

Figure 2. XRD patterns of samples prepared at different temperature for P/Ni=8.

3.3. *Analysis of TEM images and textural properties*

Figure 3 shows the TEM images of $Ni_{12}P_5$ and Ni_2P catalysts. As is shown in figure 3(a) that the morphology of as-prepared $Ni_{12}P_5$ presents clustered aggregates composed of a large number of spherical particles, the size of which is almost micron level. As is shown in figure 3(b) that the morphology of as-prepared Ni_2P particle spread out more evenly, the size of which is about 20 nm. Besides, the BET specific surface area result of $Ni_{12}P_5$ and Ni_2P is 2.70m^2/g and 4.21m^2/g, respectively. Generally, the larger specific surface area is, the higher contact area is, and the higher absorption ability and photocatlaytic degration ability of catalysts will be owing to quantum size effect.

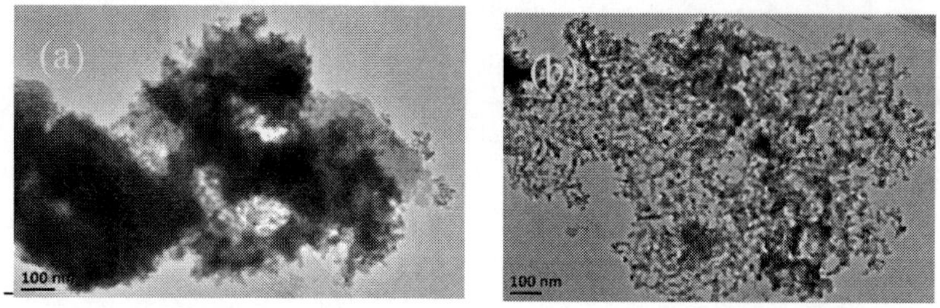

Figure 3. TEM images of as-prepared product for different molar ratio of P to Ni (a)$Ni_{12}P_5$(n(P)/n(Ni)=4/1)(b) Ni_2P(n(P)/n(Ni)=8/1).

3.4. *Photocatalytic degradation activity of $Ni_{12}P_5$ and Ni_2P catalysts*

Figure 4 shows that the photocatalytic degradation rate of $Ni_{12}P_5$ and Ni_2P catalysts for methylene blue with reaction time. As is shown that the degradation rate is less than 7.12% when the catalysts is absent. In the presence of $Ni_{12}P_5$ or Ni_2P catalyst, the degradation rate gradually increase and reach constant at 210min, the photocatalytic degradation rate of $Ni_{12}P_5$ and Ni_2P catalysts is 79.74% and 85.90%, respectively, indicating the as-prepared products have better photocatalytic degradation ability. After reaction finished, the solid catalyst could be easily recoverd by simple filtration. The catalytic activity of the recoverd catalysts almost unchanged. Correspondingly, the XRD patterns of that were the same as the fresh catalysts, indicating the structures of the recoverd catalysts were still retained.

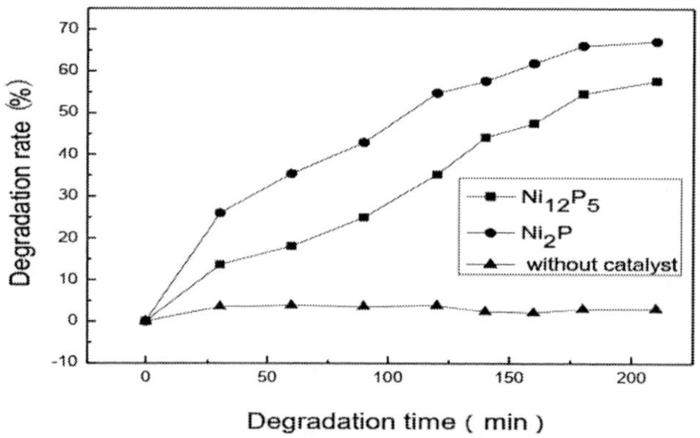

Figure 4. Degradation rate of $Ni_{12}P_5$ 和 Ni_2P catalysts for methyl orange.

4. Conclusions

Ni_2P and $Ni_{12}P_5$ can be controllably synthesised achieved by changing the molar ratio of p/Ni and reaction temperature. The increment of molar ratio of P/Ni is advantageous to transform the tetragonal $Ni_{12}P_5$ phase into hexagonal Ni_2P phase, while the increment of thermal temperature is advantageous to transform the hexagonal Ni_2P phase into the tetragonal $Ni_{12}P_5$ phase. Compared with $Ni_{12}P_5$, Ni_2P shows the better photocatalytic degradation ability for methylene blue, which of $Ni_{12}P_5$ and Ni_2P is 79.74% and 85.90 %, respectively.

Acknowledgements

The study was financially supported by the projects of Natural Science Foundation of Liaoning Province (No. 20170540475 & No. 201602404), Doctoral Scientific Research of Liaoning Province(20170520417) and Scientific Research Funds of Liaoning Institute of Science &Technology (RXYJ2015001)

References

[1] Oyama S. T., Gott T., Zhao H. 2009, *Catal. Today*, **143**(1-2) 94

[2] Ni Y O, Jin L, Hong J 2011 *Nanoscale* **3**(1) 196

[3] Ni Y. H., Mi K., Cheng C. 2011 *Chem. Commun.* **47**(20) 5891

[4] Jiang P., Liu Q, Ge Ch. J., Cui W., Pu Z.H., Abdullah M. A., Sun X. P. 2014 *J. Mater. Chem. A,* **2** 14634

[5] Wang J.,Ma X., Qu F., Abdullah M. A.,and Sun X. P. 2017*Inorg. Chem.* **56** 1041

[6] Zhang R., Tang Ch., Kong R., Du G., Abdullah M. A., Chen L. ,Sun X.P. 2017 *Nanoscale* **9** 4793

[7] Cui Liang, Qu Fengli, Liu Jingquan, Du Gu, Abdullah M. A., Sun X.P.2017 *ChemSusChem*, **10** 1370

[8] Liu Sh. , Feng J. , Bian X. , Liu J. , Xu H., An Y. 2017 *Energy & Environmental Science* **10** 1222

[9] Liu S L, Liu X Z, Xu L Q 2007 *J. Cryst. Growth* **304**(2) 430.

[10] Wang X., Wan F., Gao Y. 2008 *J. Cryst. Growth* **310**(10) 2569.

[11] Li J., Ni Y. H., Liao K. M. 2009 *J. Colloid Interface* Sci. **332** 231

[12] Yonghong, Ni, Jieding Wei. 2013 *China Patent*(In Chinese). 10594337.5

[13] Ni Y. H., Liao K. M., Li J. 2010 *Cryst. Eng. Comm.* **12**(5) 1568

IOP Conf. Series: Materials Science and Engineering **292** (2017) 012120 doi:10.1088/1757-899X/292/1/012120

Research status and development of application fields in enzyme technology

Y B Ji[1, 2], S W Wang[1], M Yu[1, 2, 3], X Ru[1], C Wei[1], H J Zhu[1], Z Y Li [1], H Zhao[1], A N Qiao[1], S Z Guo[1] and L Lu[1]

[1] Research Center on Life Science and Environmental Science, Harbin University of Commerce, Harbin 150076, China
[2] Engineering Research Center of Natural Anticancer Drugs, Ministry of Education, Harbin 150076, China

[3] E-mail: yumiao913@163.com

Abstract. Biological enzymes are catalyzed by living cells, most of which are proteins, and very few are RNA. Biological engineering as a new high-tech has been rapid development, Enzyme manufacturing and application areas are gradually expanding, In this paper, the status and progress of the application of enzyme technology are reviewed by reviewing the literature. and aims to provide reference for the application of enzyme technology and provide scientific basis for its future research and development in new field.

1. Introduction

Enzyme technology refers to the use of enzymes in a certain bioreactor catalytic conversion of material technology. The research of enzyme technology plays an important role in the application of enzymes in industrial production. Enzyme technology contains a lot of content, such as directional enzyme catalysis, increased enzyme selectivity and in large-scale synthesis reaction conditions to make the enzyme more stable, enzyme technology to improve the enzyme engineering. It is mainly about the nature of the molecular level, So far, the enzyme technology research is more abundant, in recent years, enzyme technology has also been rapid development, this paper summarizes the field of application of enzyme technology, its application has been throughout the industrial, pharmaceutical, agricultural, chemical analysis, environmental protection, Energy development and life science theory research and other aspects. At the same time, the types of enzymes, technical applications, key development and sources were classified, and the problems and application prospects of enzyme technology were discussed.

2. Introduction to enzymes

Enzymes are macromolecules with biocatalytic functions. The enzymes are highly specific and have a wide range. There are currently more than 4000 enzymes found, and enzymes in organisms are far greater than this number. There are many ways to classify enzymes. This paper lists several types of enzymes commonly used in various fields. See table 1 for details.

Table 1. Enzyme overview table.

Enzymes	source	Application areas	Key development
Soil enzyme	Microbiological, animal and plant secretion andRemains of the remains	agriculture	Improve the soil, improve crop yields and so on
Amylase	Animals, plants, microbes	Food production	Juice processing Vegetable processing, syrup manufacturing, and glucose processing and so on.
Curing enzyme	Immobilization method	Environmental protection	Water purification, environmental testing, treatment of white pollution and so on.
Protease	Animals, plants, microbes	Medicine field	Production of drugs, treatment of diseases and so on
Microbial enzymes	Animals, plants, microbes	Biochemical production	Fermentation, material synthesis and so on
Biological enzymes	Living cells	Textile industry, energy development	Textile industry energy development, the development of new substances, deinking treatment, cotton fabric degumming and so on

From the table 1 can be analyzed for a wide range of applications, including industrial, pharmaceutical, agricultural, chemical analysis, environmental protection, energy development,life sciences research and other aspects, so far, the industrial use of enzyme preparations are basically divided into Two categories: The first category is hydrolytic enzymes, including amylase, cellulase, protease, lipase, pectinase, lactase, etc., accounting for more than 75% of market sales. At present, about 60% of the enzyme preparation has been produced with genetically modified strains. The second category is non-hydrolytic enzymes, accounting for about 10% of market sales, and there is a tendency to increase year by year, mainly for the analysis of reagents and pharmaceutical industry. The food industry and the feed industry account for the largest share of the application, accounting for 45% of the total sales, 32% of the detergent, 11% for the textile industry, 7% for the paper industry and 4% for the chemical industry.

3. Application of enzyme technology

3.1. Application of enzyme technology in agriculture

Soil enzyme is a kind of biologically active substance with catalytic ability.The results showed that the degree of resource abundance and complexity of the community in the moderate degradation succession stage increased the stability of the grassland community, which was beneficial to the increase of the community productivity, resulting in the increase of plant litterfall to the soil and the increase of soil fertility [1].The results showed that long-term organic and inorganic fertilizers could significantly increase soil bacterial diversity and change the community structure of soil bacteria and fungi, improve soil enzyme activity, improve farmland ecosystem productivity and improve ecosystem productivity Health has a better effect [2].The results show that acetylcholinesterase can be used for

IOP Conf. Series: Materials Science and Engineering **292** (2017) 012120 doi:10.1088/1757-899X/292/1/012120

rapid qualitative screening of organophosphorus and carbamate pesticide residues in agricultural products [3].

3.2. Application of enzyme technology in food industry

Enzyme technology can be used in the food industry to increase production processes, increase production and reduce waste of resources. Sources of enzymes for the food industry include animals, plants and microbes [4]. Zhou Changchun [5] and others believe that enzyme technology can be applied to the fruit and vegetable juice processing industry, the literature shows that immobilized enzyme method can remove the bitterness of raw materials, and elaborated immobilized enzyme method, the enzyme can also be applied to carrots Juice production and clarification of litchi juice. Liu Yi [6].Through the analysis of the application of enzyme technology in grain processing, fruit and vegetable processing and meat processing, the main application scope of enzyme technology in food processing is expounded, which shows that enzyme technology can also be applied to food safety testing , The main method for the enzyme-linked immunoassay and enzyme biosensor method. Wu Jie [7]and others believe that the enzyme can change the baking food structure, improve food flavor and color, and can produce food grade oligosaccharides, increase the use of by-products, such as: in the small lobster processing 70 % of the material, and the protein mass fraction of about 35%.

3.3. Application of enzyme technology in environmental protection

With the rapid development of modern biotechnology and the gradual improvement of human environmental protection awareness, biotechnology has been widely used in environmental protection work. Modern biotechnology with high efficiency, low consumption, good safety and other advantages for environmental protection work provides a lot of convenience [8]. Song Jia [9] and others reported that enzyme technology can be applied to environmental monitoring, waste gas treatment, sewage purification, municipal solid waste treatment, etc, Yao Bu-xuan [10] suggested that the combination of biosynthesis can effectively play the synergistic effect of the biological enzyme and coenzyme function in the AAO wastewater treatment process, so that the COD can be reduced from 1500mg / L to 100mg / L, and the secondary Pollution, with significant economic and social benefits. Wang Wei-ping [11] believe that the application of biological enzyme technology is simple, rapid and sensitive advantages in pesticide residue detection. It has a broad application prospect, and has far-reaching significance for human health and environmental protection. Environmental protection work is inseparable from the application of biotechnology, biotechnology has played an irreplaceable important role in environmental protection work.

3.4. Application of enzyme technology in pharmaceutical industry

Biotechnology has been introduced into the pharmaceutical industry, making the bio-pharmaceutical industry become one of the most active and fastest-growing industries. At present, the enzyme technology has been widely used in the pharmaceutical industry, biosensors, monoclonal antibody diagnosis and so on [12]. In addition, the enzyme technology can also be used for the extraction of active ingredients of traditional Chinese medicine and pharmaceutical production and so on. The results showed [13] that under the optimum conditions: pectinase 0.3%, hydrolysis temperature 50, hydrolysis time 3h, enzyme PH3, jujube juice extraction rate can reach 51.89%, 35.93% of total sugar, soluble solids of 10.57%, the jujube juice, jujube rich natural color and good stability. Wang Hongyun [14] and others have shown that enzyme technology has a unique advantage in the extraction of traditional Chinese medicine, the enzyme can be more moderate to the decomposition of medicinal materials, promote the release of active ingredients, improve the extraction rate of effective ingredients. Ji Xue-hui [15] and others believe that enzyme technology can improve the extraction of polysaccharides, flavonoids, oils, alkaloids and saponins.

3.5. Application of enzyme technology in biochemical production

Enzyme technology can also be used in the production of biochemistry,Perfumes can be produced by biochemical methods in the spice industry,biochemical means can produce spices compound in the spice industrys, Ju Jing-qiang [16] believe that aspergillus fruitzae KM-1 strain can be used to quantitatively produce 2-heptanone, 2-nonanone and 2-undecanone using octanoic acid, decanoic acid and dodecanoic acid contained in coconut oil, and γ -decalactone of 10 carbon atoms was obtained from castor oil, The production method of γ -decalactone of 12 carbon atoms was obtained.D-arabinose attracting great attention from the community because of its many drug activity, microbial enzyme can produce D-aluo sugar [17], Zhang Xiaoling [18]believe that the fermentation process of enzymatic preparation of vinegar has the characteristics of slow, medium, fast, stable, short fermentation time, fast synthesis of acetic acid and good flavor in the process of alcohol fermentation.

3.6. Other applications of enzyme technology

Zheng Jie [19]believe that compared with the chemical deinking method, waste paper enzyme deinking time is short, less energy consumption, deodorization easy to bleach, pulp whiteness is high, especially in the increasingly prominent environmental problems in today's world , More environmentally friendly enzymatic deinking has a very good application prospects. The technology developed by Buckman [20] can handle different types of waste paper raw materials, bringing more potential benefits for different paper mills: improving slurry filtration performance; reducing steam consumption by 4% to 8%; Improve paper machine speed 5% to 10%; improve the ratio of waste paper; to improve paper uniformity and strength; white water system is more clean. Enzyme technology can also be used for the development of new energy sources, such as the development of new functional sugars. In addition, bio-enzyme technology has been in some edible fungus deep processing has been rapid development because of its high efficiency, mild conditions, low energy consumption, low pollution, simplify the process and other advantages [21].

4. Conclusion

Enzyme technology has the advantages of low investment, low energy consumption, good effect and so on, and the enzyme has a high degree of specificity, the company's production efficiency and product quality has been greatly improved, increasing the economic income of enterprises, the extensive application of enzyme technology significantly improve the quality of life of the people. However, the environment required for the enzyme is harsh, and it is difficult to ensure the enzyme activity in industrial production, and there is a problem that the use efficiency is low, resulting in a problem that the use cost is increased. Therefore, it is of great potential to improve the efficiency of enzyme usage and improve the activity of enzymes in harsh environment. It is also the direction for future researchers to work hard.

Acknowledgments

This work is supported by Postdoctoral foundation of China (2015M581467); Natural Science Foundation of Heilongjiang Province (D201138); Key Science and Technology Research Project of Ministry of Education (211045); Doctoral Foundation of Ministry of Education (20112332110003); Heilongjiang Postdoctoral special Foundation (LBH-TZ1613); Heilongjiang Postdoctoral Grant (LBH-Z15107); The Scientific Research Team Program of Harbin University of Commerce (2016TD002); Harbin Special Foundation for Young Technological Innovative Talented Person (2013RFQXJ150).

References

[1] Hu L, Wang C L, Wang C T, et al 2014 Changes of soil enzyme activities and microbial community structure in alpine meadow under different degradation succession stages in Sanjiangyuan area *Journal of Grassland Science* **23(03)** 8-19

[2] Lu H F, Zheng J W, Yu X C, et al 2015 Effects of Long - term Inorganic Organic Fertilizer on Microbial Community Diversity and Enzyme Activity in Paddy Soil Derived*Journal of Plant*

Nutrition and Fertilizer **21(03)** 632-643

[3] Zhu S M, Zhou C N, He J S, et al 2014 Rapid detection of pesticide residues based on enzymeinhibition method *Journal of Agricultural Engineering* **30(06)** 242-248

[4] Zhang H Y, Zheng X D, Xu P, et al 2002 Enzyme technology and its application in food industry *Grain and oil processing and food machinery* **(06)** 31-33

[5] Zhou C C 2004 Application of Enzyme Technology in Fruit and Vegetable Juice Processing Industry *China Food Additives* **(05)** 59-62

[6] Liu Y 2015 Application of Enzyme Technology in Food Processing and Detection *Food engineering* **(03)** 12-14

[7] Wu J , Gao Z C 2016 Application of Enzyme Technology in Food Processing *Shandong Industrial Technology* **(22)** 232-233

[8] Tang Q, Li Z S 2002 Application and Prospect of Biotechnology in Environmental Protection *Environmental Pollution Control Technology and Equipment* **(10)** 28-35

[9] Song J 2014 Application and Prospect of Biotechnology in Environmental Protection *Chemical Management* **(15)** 214

[10] Yao B X 2009 Application of Biological Enzyme Technology in Coking Wastewater Treatment *Journal of Fuel and Chemical Engineering* **40(06)** 40-41 + 44

[11] Wang W P, Wu Y, He L J, et al 2008 Application of Biological Enzyme Technology in Rapid Detection of Pesticide Residues *Analysis and Testing Technology and Instruments* **(02)** 72-78

[12] Yang J, Zhang Y B, Wu W T 2013 Immobilized Enzyme Technology and Its Application in Medicine *Journal of Pharmaceutical Biotechnology* **20(06)** 553-556

[13] Peng Y T, Sun Y E, Wang W D 2012 Application of Enzyme Technology in Extraction of Jujube Juice *Food Industry* **33(01)** 31-33

[14] Wang H Y, Li M 2014 Application of Enzyme Technology in Polysaccharide Extraction of Dendrobium *Journal of Yunnan Traditional Chinese Medicine* **35(06)** 81-83

[15] Ji X H, Zhang H 2011 Application of Enzyme Technology in Extraction and Separation of Traditional Chinese Medicine **29(10)** 2365-2367

[16] Ju J Q, Chen F L 2010 Production of Spice Compounds by Biochemical Methods *Chinese Journal of Food Additives* **(05)** 139-143

[17] Wang M L 2015 Productionof D-Alanosine by Microbial Enzyme *Shandong Food Fermentation* **(01)** 34-36

[18] Zhang X L 1999 Discussion on Improving the Utilization Rate of Starch by Reforming Acetic Acid Fermentation Process *Journal of Food & Beverage* **(01)** 20-21

[19] Zheng J, Sun G W, Zhang S J, et al 2013 Application of Biological Enzyme Technology in Deinking of Waste Paper *Heilongjiang Papermaking* **41(04)** 10-15

[20] 2014 Application of Barkman Filterase Technology in Packaging Paper*International Paper* **33(06)** 70-72

[21] Li P W, Wang S, Liang R R, et al 2014 Application of Microorganism and Enzyme Technology in the Development of New Functional Sugar *Amino Acids and Biological Resources* **36(04)** 1-7

IOP Conf. Series: Materials Science and Engineering **292** (2017) 012121 doi:10.1088/1757-899X/292/1/012121

The preparation and activity of Cu-Fe-Zr-Ce based catalysts for water gas shift

H D Wu[1], T S Liu and H Z Liu

College of Biological and Chemical Engineering, Guangxi University of science and technology, Liuzhou, 545006, P.R.China

[1] E-mail: whongda@126.com

Abstract. CeO_2-ZrO_2 composite oxide was synthesized with precipitation method as support and $Cu_aFe_b(ZrCe_4)_8O_x$ catalysts were prepared by impregnation; X-ray diffraction, H_2 temperature program reduction,and scanning electron microscope techniques were jointly used to characterize the crystal phases and reduction properties of catalysts. Then the activity of catalysts in water gas shift was studied, thus investigated how catalyst composition impacted the water gas shift. Conclusions drew from the results can be briefly stated. $Cu_aFe_b(ZrCe_4)_8O_x$ was provided with stable cubic crystalline framework and Cu and Fe, as the active components,was highly dispersed on the surface of supports in the form of CuO and Fe_2O_3 respectively. The strong interactions between copper and iron component enhanced the reducing capacity of CuO and Fe_2O_3. $Cu_aFe_b(ZrCe_4)_8O_x$ catalysts exhibited high catalytic activity and selectivity while the main active components were Cu and Fe_3O_4. The CO conversion rate reached 96% when $Cu_7Fe_3(ZrCe_4)_8O_x$ catalysts was used in water gas shift at 623K and the only products were H_2 and CO_2. The activity was still desirable even the catalysts was applied at 723K.

1. Introduction

Water Gas Shift (WGS) is widely used in industrial manufacturing processes such as the synthesis of ammonia, purification of automotive exhaust, exploitation of hydrogen resource. From a thermodynamic point of view, this shift ($CO+H_2O \rightarrow H_2+CO_2$) is a reversible exothermic reaction, and a lower reaction temperature is more favorable to the positive occurrence of the reaction which means catalysts with high activity at low temperature are required [1]. Cu-Zn catalyst is well accepted in WGS, however sintering deactivation makes it hard to dedicate in long-term reaction which leads to the development of other WGS catalysts including noble metals and transition metals as active components such as Pt, Ru and Au [2, 3]. Even though high activity and stability at low temperature are exhibited in the WGS, the high cost and scarcity of resources have limited the application of precious metal based catalysts. Due to the reasons above, low-cost catalysts with high activity and stability at low temperature have become the focus of the research.

It was reported that the CeO_2 crystal structure contains rich oxygen vacancies which is ideal for oxygen storage and oxygen evolution, thus makes it possible to promote a fast redox conversion [4, 5]. Also the high probability of strong interaction with other metal can improve the activity and selectivity of catalysts thus makes ceria a popular support of catalysts. Unfortunately, the stability of pure CeO_2 needs to be improved since it would be sintered at high temperature [6]. Recent researches shown that catalytic activity and thermal stability of catalysts could be dramatically improved by mixing stable

IOP Conf. Series: Materials Science and Engineering **292** (2017) 012121 doi:10.1088/1757-899X/292/1/012121

cation such as Zr^{4+} into CeO_2 [7]. Other studies suggested relatively approving catalytic activity of copper-based catalyst and iron-based catalyst when they were applied in WGS [8, 9], yet less reports about Cu-Fe-Ce-Zr based catalysts in WGS had been found. Based on this, preparation method, characterization results and catalytic performance of catalysts in WGS will be introduced in this article.

2. Experiment

Catalyst support was prepared by coprecipitation method and the active component is loaded by impregnation method. The catalyst activity was evaluated in a fixed bed microreactor. Then X-ray Powder Diffraction Method (XRD) was used to determined the crystal structure of catalyst. And at last the reduction characteristics of catalysts were studied by temperature-programmed reduction (TPR).

2.1. Catalyst preparation

The CeO_2-ZrO_2 composite oxide supports, molar ratio of Zr:Ce were 1:4, were prepared by co-precipitation method with starting material of cerous nitrate - zirconium nitrate aqueous solution and precipitant of ammonium carbonate. A certain amount of cetyltrimethylammonium bromide (CTAB) was added separately into the prepared ammonium carbonate aqueous solution with a concentration of $0.5 mol\cdot L^{-1}$ and mixed solution of $Zr(NO_3)_4$ and $Ce(NO_3)_3$ with a Ce^{3+} + Zr^{4+} ion concentration of $0.5 mol\cdot L^{-1}$. Added dropwise the mixed solution of $Zr(NO_3)_4$ and $Ce(NO_3)_3$ into aqueous ammonium carbonate solution with vigorous stirring, and kept the stirring for a time. After filtration, the filtrate cakes were washed by deionized water and anhydrous ethanol separately for several times, and then dried at 80℃ for 8h. Catalyst supports were obtained after calcination at 400℃and 600℃ each for 4h. Copper nitrate and iron nitrate were loaded by impregnation method and after a desiccation and calcinations at 600℃ for 3h Cu_aFe_b $(ZrCe_4)_8O_x$ catalyst was obtained with the Cu:Fe molar ratios of 9:1, 8:2, 7:3, 6:4, 5:5, 4:6, 2:8.

2.2. Evaluation of Catalytic activity

Water Gas Shift reaction was carried out in a down flow fixed bed quartz tubular reactor（inner diameter 10 mm） under atmospheric pressure and 500 mg catalyst (particle diameter 0.4～0.8 mm) was loaded. Before the reaction, the catalyst would be reduced with H_2 ($30\ mL\cdot min^{-1}$) at 400℃ for 1h. Then the reactor with the sample was cooled down to room temperature swept with CO ($10\ mL\cdot min^{-1}$). Using a micro-pump at the rate of $0.036\ mL\cdot min^{-1}$, highly purified water was infused in to a vaporizer heated to 200℃ and then introduced to the reactor. The obtained gas mixture was cooled at 0℃ using a condenser and the dry gas mixture was analyzed by a SP-6800A equipped with TCD detector. Afterward, the calibrated gas mixture of known composition was used for quantification.

2.3. Catalyst characterization

The examined sample was pretreated in situ in Ar stream (600℃, $20\ mL\cdot min^{-1}$, 40min) before the TPR test. Then the reactor with the sample was cooled down to room temperature, after which the stream was replaced with H_2 (5%)-Ar (95%) mixtures with the rate of $40\ mL\cdot min^{-1}$. Generally, the TPR measurements were carried out at $10℃\cdot min^{-1}$. X-Ray Diffraction (XRD) patterns were collected on a 8D ADVANCE powder diffractometer using Cu Kα radiation (Germany, 60KV, 300mA).

3. Results and discussion

It was characterized by XRD measurements that $ZrCe_4O_x$ composite oxide, as the support of catalyst, retained the lattice characteristics of CeO_2 which was the cubic crystalline phase. The lattice structure of this support was stable as catalyst was subjected to a hydrogen reduction process and applied in the WGS reaction process. The active component of the catalyst, such as copper and iron, was highly dispersed on the surface of the support. The copper component mainly existed as Cu. In the meanwhile, no diffraction peaks of iron component were detected which indicated that the iron component was highly dispersed on the surface of the support, as stated above. The results of TPR measurements exhibited strong interactions between these two components which might lead to partial reduction of

CuO and Fe_2O_3 by hydrogen at lower temperature. Thereinto CuO could be reduced to Cu_2O in the above-mentioned reduction at temperature lower than 100℃, while Fe_2O_3 could be reduced to Fe_3O_4 at around 225℃. Besides, as $Cu_aFe_b(ZrCe_4)_8O_x$ catalysts had better performance in catalytic activity of WGS when the temperature was between 400℃~500℃, it seemed that the Cu and Fe_3O_4 were the active component in WGS reaction. The loading of active component and the ratio of Cu to Fe, moreover, indicated their impacts on catalytic activity to some degree.

3.1. XRD analysis

Figure 1-L1 shows XRD patterns of $ZrCe_4O_x$ support. In the case of $ZrCe_4O_x$, the diffraction patterns are in good agreement with those of single ceria typical cubic structure(2θ=28.56°, 33.08°, 47.48°, 56.33°; 59.09°; 69.40°; 76.70°; 79.07°; 88.41°; JCPDS:34-0394). Moreover, no characteristic diffraction peaks of the doped zirconium or zirconia appear, which means Zr^{4+} ion has permeated the lattice of CeO_2. Figure 2 and figure 3 illustrating the XRD patterns of cubic-phase CeO_2 have no significant change in all $Cu_aFe_b(ZrCe_4)_8O_x$ catalysts which indicates a stable crystal structure of cubic-phase CeO_2 in support after high temperature reduction.

Figure 2 demonstrates XRD patterns of several fresh catalysts of $Cu_aFe_b(ZrCe_4)_8O_x$. No characteristic diffraction peak of CuO is observed when copper content is relatively low (see figure 2-a1), which suggests its high dispersion over the supports. However with the increase of copper content, the characteristic diffraction patterns of monoclinic structure CuO (2θ=35.5°; 38.73°; JCPDS: 45-0937) appear clearly and their intensity grows with copper content. This phenomenon implies that CuO nanocrystallite grows up progressively with the amount of copper. Figure 3 shows XRD patterns of catalysts with different molar ratios of Cu:Fe after the reaction under temperature of 400℃. No diffraction peak of CuO can be observed in this picture which means CuO has been reduced to Cu. However, there is no diffraction peak of "Cu" phase recorded by XRD when copper content is lower (see figure 3-a2), because copper components are highly dispersed over the supports. But diffraction peak of cubic Cu appears with increasing percentage in composition of Cu (2θ=43.4°, 50.8°; JCPDS: 65-9743). The experimental results shown that XRD peaks were strengthened with the increasing percentage of Cu in composition. The used catalyst was analyzed by XRD so that the existence form of copper components in water gas shift reaction could be investigated. Two phases, cubic Cu and monoclinic CuO(small-signal), were observed by XRD, as presented in figure 1-L2. The figure demonstrated the transition of Cu to CuO after the reaction with water ($Cu + H_2O \rightarrow CuO + H_2$).

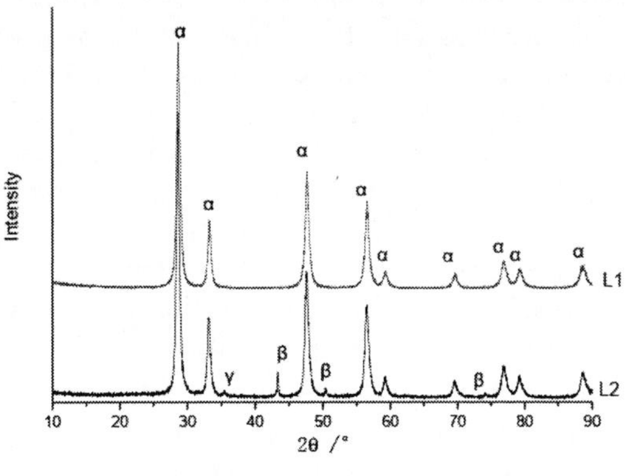

Figure 1. XRD patterns of catalyst used and fresh support.
L1: $(ZrCe_4)O_x$ L2: $(Cu_7Fe_3)(ZrCe_4)_8O_x$
α: CeO_2, β: Cu, γ: CuO

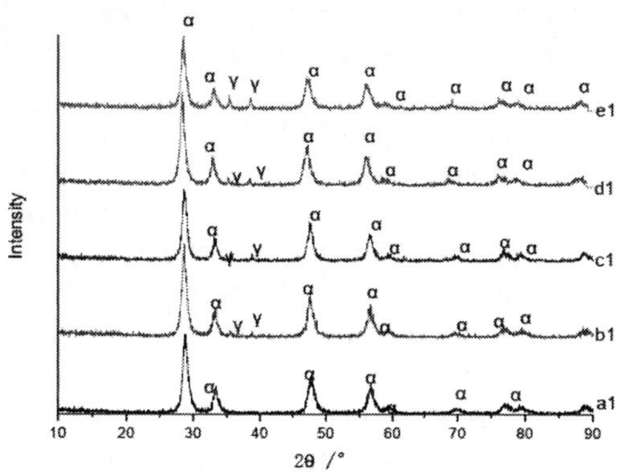

Figure 2. XRD patterns of fresh catalyst.
a1: $(Cu_5Fe_5)(ZrCe_4)_8O_x$ b1: $(Cu_6Fe_4)(ZrCe_4)_8O_x$
c1: $(Cu_7Fe_3)(ZrCe_4)_8O_x$ d1: $(Cu_8Fe_2)(ZrCe_4)_8O_x$
e1: $(Cu_9Fe_1)(ZrCe_4)_8O_x$ α: CeO_2, β:Cu, γ:CuO

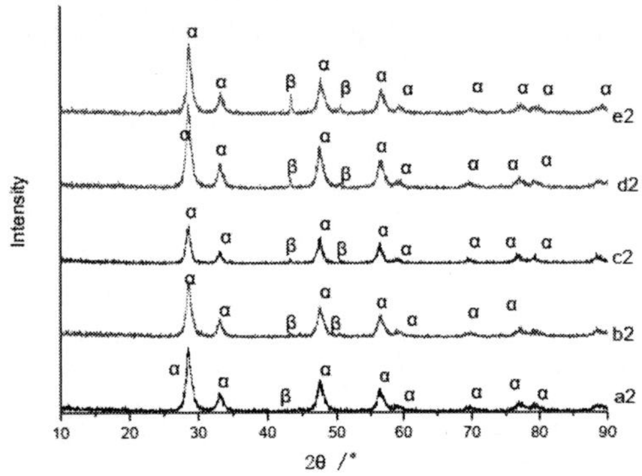

Figure 3. XRD patterns of after 400 ℃ reduction.
a2: $(Cu_5Fe_5)(ZrCe_4)_8O_x$ b2: $(Cu_6Fe_4)(ZrCe_4)_8O_x$
c2: $(Cu_7Fe_3)(ZrCe_4)_8O_x$ d2: $(Cu_8Fe_2)(ZrCe_4)_8O_x$
e2: $(Cu_9Fe_1)(ZrCe_4)_8O_x$ α:CeO_2, β:Cu, γ:CuO

Figure 2 and 3 do not imply any characteristic diffraction peaks of the doped iron or ferric oxide, which might due to the high dispersion of Fe_2O_3 on the supports promoted by the coexistent copper species and the overlapping effect of copper species on iron species. In fact, the 120℃ decomposition temperature of ferric nitrate is lower than copper nitrate's 170℃ which means during the calcination, $Fe(NO_3)_3$ will be decomposed prior to $Cu(NO_3)_2$ and the products will exist as Fe_2O_3 in crystalline form on the surface of supports. With the increase of calcination temperature, CuO, as the decomposition product of $Cu(NO_3)_2$, will be mixed with Fe_2O_3 and adhere to the surface of Fe_2O_3 crystalline grain and become a disincentive of grain growth [10, 11]. As the grain size of Fe_2O_3 is too small to be detected, there is no record during the XRD analysis process of iron oxide phase. On the contrary, CuO which exists on the surface of complex oxide, has much lower growth activation energy and higher growth rate. These features make it easy to shape larger size crystals so that produce a diffraction peaks during XRD analysis.

3.2. TPR analysis

Figure 4 indicates the TPR profiles of $Cu_aFe_b(ZrCe_4)_8O_x$ catalysts. In the figure demonstrating H_2-TPR process under the condition of 10℃·min^{-1}, catalysts with higher level of iron components content （see curve b, c, d, e in figure 4） present three reduction peaks of copper components under the stage of low temperature (marked as t1, t2, t3 in figure 4), and the peak temperatures are 108℃, 150℃, 200℃ respectively. However, with the decrease of iron content, t1 become lower and lower and it even disappear in the TPR of the catalysts with the lowest iron content such as $(Cu_9Fe_1)(ZrCe_4)_8O_x$ (see curve f in figure 4). Concerning the fact that no diffraction peaks of CuO in XRD result of $(Cu_5Fe_5)(ZrCe_4)_8O_x$ (figure 2-a1), its t1 peak can be attributed to reduction peak of CuO which has the characters of high dispersion and strong interaction with Fe_2O_3 ($2CuO + H_2 \rightarrow Cu_2O + H_2O$). While t2 can be cataloged to reduction peak of CuO with larger crystalline grain ($2CuO + H_2 \rightarrow Cu_2O + H_2O$) and t3 goes to overlapping reduction peaks of Cu_2O ($Cu_2O + H_2 \rightarrow 2Cu + H_2O$) and the remaining CuO.

The attribution of t3 peak can be obtained by analyzing the sum of three reduction peak areas of copper components (figure 4). In figure 4, it is obvious that the sum of t1 and t2 peak areas is smaller than that of t3, which indicates the incomplete reduction of $CuO \rightarrow Cu_2O$ under 150℃, while the reaction of $Cu_2O \rightarrow Cu$ has already begun at the temperature lower than 200℃. To have a better understand of to what extent the reduction of $CuO \rightarrow Cu_2O$ takes place under 150℃, TPR experiment is implemented at the heating rate of 1℃·min^{-1} (see figure 5). It is observed that the temperature of three reduction peaks of copper component all move toward a low temperature zone and the peak temperature of t1, t2, t3 are 65℃, 110℃, 150℃ respectively. The same phenomenon that sum of t1 and t2 peak area is lower than t3 peak area appears at this stage. Afterward, these two figures with controlled catalysts in reduction reactions for 2h under 150℃ and 200℃ was reexamined respectively, and the reduction peak area of the former (150℃) was still smaller than the latter one (200℃). The results above verify the fact that the reduction of $CuO \rightarrow Cu_2O$ cannot transform completely under 150℃, however the process of $Cu_2O \rightarrow Cu$ has taken place before 200℃ (the temperature would be 150℃ with a heating rate of 1℃·min^{-1}), that means there is a cross phenomenon between the two reduction. The above-mentioned can be explained by the overlapping effect of the outside Cu_2O on the inside CuO. As Cu_2O is reduced by H_2 and transform into Cu under a relatively high temperature, without the coverage of Cu_2O, the left CuO can be reduced by H_2. Previous study [12] asserts that during the reduction process, CuO transform into Cu^+ which is easy to absorb H_2, thus stimulates the electron transfer in the catalyst system and reduction of the inside CuO, and makes t3 peak far more bigger than t2 peak.

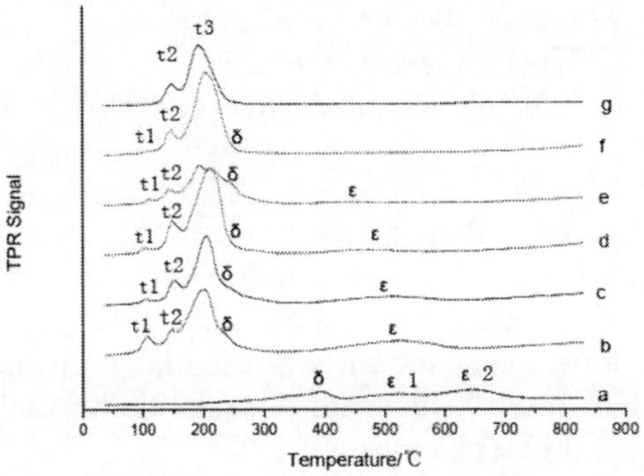

Figure 4. TPR profiles of $(Cu_aFe_b)(ZrCe_4)_8O_x$.
a: $(Fe_{10})(ZrCe_4)_8O_x$ b: $(Cu_5Fe_5)(ZrCe_4)_8O_x$
c: $(Cu_6Fe_4)(ZrCe_4)_8O_x$ d: $(Cu_7Fe_3)(ZrCe_4)_8O_x$
e: $(Cu_8Fe_2)(ZrCe_4)_8O_x$ f: $(Cu_9Fe_1)(ZrCe_4)_8O_x$
g: $(Cu_{10})(ZrCe_4)_8O_x$

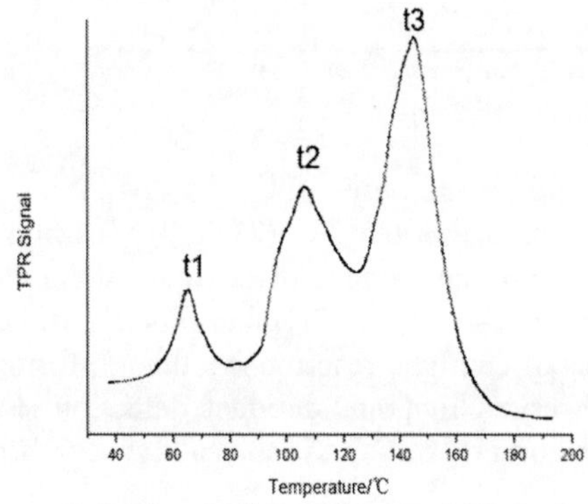

Figure 5. TPR profiles of $(Cu_7Fe_3)(ZrCe_4)_8O_x$. (The TPR profiles of measurements were carried out at 1℃·min^{-1})

There are two reduction peaks at the high temperature zone in the TPR profile of $Cu_aFe_b(ZrCe_4)_8O_x$, marked as δ and ε respectively (see figure 4(b, c, d, e, f)) and they can be attributed to reduction hydrogen consumption peak of iron component in catalysts. δ, as the shoulder peak, its peak temperature is around 225℃ and is regarded as the reduction hydrogen consumption peak of $Fe_2O_3 \rightarrow Fe_3O_4$ reduction step. The other broad peak (ε) with peak temperature of 520℃ is attributed to the reduction hydrogen consumption peak of the transition $Fe_3O_4 \rightarrow (FeO) \rightarrow Fe$. It is also observed that there are three-step mechanism in TPR of pure Fe_2O_3 [13], as the reduction peaks at 400℃, 500℃ and 655℃ could be assigned to the step reductions of $Fe_2O_3 \rightarrow Fe_3O_4 \rightarrow FeO \rightarrow Fe$. Apparently, there are two significant differences between our TPR results of $Cu_aFe_b(ZrCe_4)_8O_x$ and early research. One is that the reduction process of $Fe_2O_3 \rightarrow FeO$ appears in a relatively low temperature zone of $Cu_aFe_b(ZrCe_4)_8O_x$'s TPR curve. The other is that there is only one broad peak of $Fe_3O_4 \rightarrow (FeO) \rightarrow Fe$ which is a the two-step reaction on $Cu_aFe_b(ZrCe_4)_8O_x$'s TPR curve. A TPR text of (Fe_{10}) $(ZrCe_4)_8O_x$ is implemented in order to find out what leads to these discrepancies (see figure 4-a). Three reduction peaks of (Fe_{10}) $(ZrCe_4)_8O_x$ appear at 400℃, 520℃ and 630℃, and the XRD results of these three peaks (figure 6) indicate three steps reduction of $Fe_2O_3 \rightarrow Fe_3O_4 \rightarrow FeO \rightarrow Fe$. It is clear that during the whole TPR test of $Cu_aFe_b(ZrCe_4)_8O_x$, δ peak appears at a much lower temperature (225℃) than the first reduction peak temperature (400℃) of (Fe_{10}) $(ZrCe_4)_8O_x$, because copper has a dispersion effect on Fe_2O_3 and there are interactions between them which strengthen Fe_2O_3's ability of being reduced, thus lower the required temperature of reduction. Again, copper component plays the vital role in the $Fe_3O_4 \rightarrow (FeO) \rightarrow Fe$ process, which make ε a broad peak in $Cu_aFe_b(ZrCe_4)_8O_x$'s TPR test and is similar to what has been reported [14].

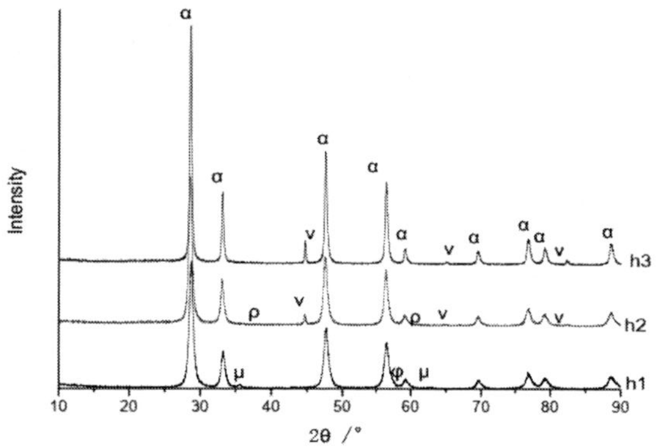

Figure 6. XRD spectras of $(Fe_{10})(ZrCe_4)_8O_x$.
h1:catalyst after 400℃ reduction
h2:catalyst after 520℃ reduction
h3:catalyst after 630℃reduction
α:CeO_2, μ:Fe_2O_3, ρ:FeO, ν:Fe, φ:Fe_3O_4

3.3. *Catalytic activity*

It was found that $(Cu_aFe_b)(ZrCe_4)_8O_x$ catalyst showed good catalytic activity in the reaction of water gas shift and no obvious inactivation was detected which suggests a good stability. Gas phase product only contains CO, H_2, CO_2, but no CH_4, indicating that the methanation reaction did not happen in the process of catalytic reaction. Methanol, formaldehyde, formic acid and carbon-containing organic was not observed in liquid product detection. Figure 7 and figure 8 illustrate the relationship between conversion rate of CO and the catalyst composition under different temperature.

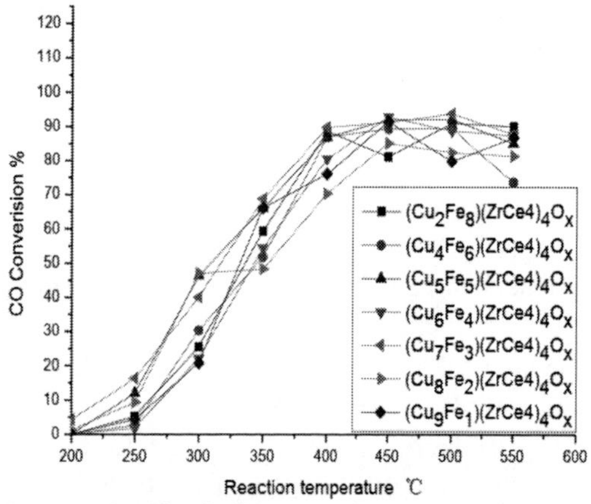

Figure 7. The influence of Cu:Fe molar ratio on catalytic activity.

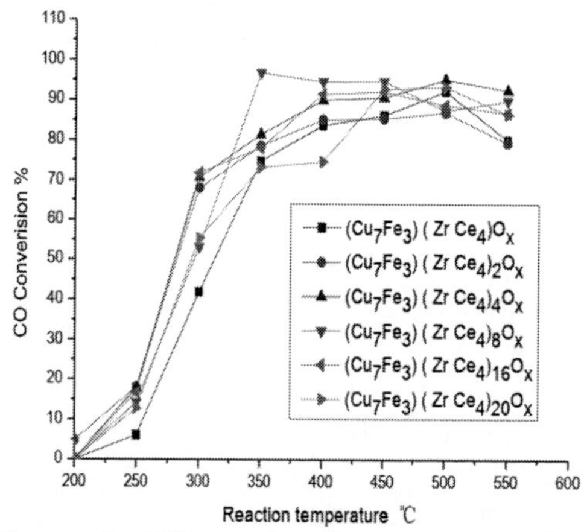

Figure 8. Effects of active component load amount.

Figure 7 displays that loaded Cu:Fe ratio has certain influences on the catalytic activity, and the catalytic activity reaches the top when the Cu:Fe ratio is 7:3, while the conversion rate of CO reached to 89% under 400℃. However, the catalytic activity will decrease no matter the Cu:Fe>7:3 or Cu:Fe<7:3. This result indicates that iron is a vital active component and there is a synergistic effect between iron and copper species during the water gas shift reaction. Such effects would change with the Cu/Fe ratio which ensured a proper proportion of active component that effectively increased catalytic activity [15].

Figure 8 presents the activity of catalysts in water gas shift with different loading amount of active component. It is clear that with the increase of active component's loading amount, the activity of the catalyst enhanced gradually, but excess active component can also decrease catalytic activity. $(Cu_7Fe_3)(ZrCe_4)_8O_x$ is the one with the highest catalytic activity. At 350℃, CO conversion rate reaches 96%, which can maintain a high catalytic activity even in the period of high temperature up to 450℃. The effects of loading amount of active component on the catalytic activity has also been reported in literature [16, 17]. Research on oxidation mechanism of CO indicated that the activated

CO adsorbed on the surface of active components (such as Cu, Fe, Ni) would be transformed into CO_2 when reacted with active oxygen species or active hydroxyl groups on the surface of the supports (such as CeO_2) [15, 18]. Thus there would be a satisfactory conversion rate from CO to CO_2 when the activated CO concentration of the catalyst surface matched the concentration of active oxygen or active hydroxyl groups. For this reason, when the amount of active component on the carrier surface was too large, the surface of the carrier would be overcovered and the concentration of active oxygen species or active hydroxyl groups on the surface of catalysts would decrease, leading to the decrease of the catalyst activity. The recently released report [19] further proved that the oxygen vacancy of the doped ceria carrier was often larger than that of the pure ceria, therefore there was a higher concentration of the surface active oxygen or active hydroxyl oxygen of the doped carrier.

4. Conclusion

The ZrO_2-CeO_2 composite oxide prepared by coprecipitation method was provided with stable cubic crystalline skeleton. Copper and iron, as the active components, having been loaded by impregnation method to obtain catalysts, was highly dispersed on the surface of supports in the form of CuO and Fe_2O_3 respectively. The strong interactions between copper and iron component enhanced the reducing capacity of CuO and Fe_2O_3. The elemental copper(Cu^0) and Fe_3O_4 were the main active components in WGS and the activity of catalyst would be affected by active component loading amount and Cu/Fe ratio.The prepared $Cu_aFe_b(ZrCe_4)_8O_x$ catalysts exhibited high catalytic activity and stability in WGS when temperature was between 400℃～500℃. The only two products, H_2 and CO_2, indicated favorable selectivity of this catalyst series.

Acknowledgments
This work was supported by Guangxi Natural Science Foundation (Guangxi 0542018)

References

[1] Ratnasamy C, Wagner J P 2009 *Catalysis Reviews-Science and Engineering* **51** 325.
[2] Kalamaras C M, Americanou S and Efstathiou A M 2011 *Journal of Catalysis* **279** 287
[3] Yu Q Q, Chen W, Li Y, Jin M S and Sou Z H 2010 *Catalysis Today* **158** 324
[4] Li L, Zhan Y Y, Zheng Q, Zheng Y H, Chen C Q, She Y S, Lin X Y and Wei K M 2009 *Catalysis Letters* **130** 532
[5] Campbell C T, Peden C H F 2005 *Science* **309** 713
[6] Song X L, Qiu G Z and Qu P 2004 *Journal of Hunan University*(Natural Sciences) **31** 13
[7] SatoT, Dosaka K and Ishitsuka H 1993 *Journal of Alloys and Compounds* **193** 274
[8] Cai L N, Hu Z H, Branton P and Li W C 2014 *Chinese Journal of Catalysis* **35** 159
[9] Khan A, Chen P, Boolchand P and Smirniotis P G 2008 Journal of. Catalysis **253** 91
[10] Shi J, Wu S H and Wang S R 2006 *Chinese Journal of Applied Chemistry* **23** 1396
[11] Liu H F, Peng T J and Sun H J 2009 *Journal of Instrumental Analysis* **28** 7
[12] Liu Y, Sun H L and Liu Q S 2001 *Chinese Journal of Catalysis* **22** 523
[13] Zielinski J, Zglinicka I, Znak L and Kaszkur Z 2010 *Applied Catalysis A: General* **381** 191
[14] Iheva L I, Andreeva D H and Andreev A A 1997 *Thermochimica Acta* **292** 169
[15] Saw E T, Oemar U, Ang M L, Kusa H and Kawi S 2016 *Catalysis Science & Technology* **6** 5336
[16] Gunawardana P V D S, Lee H C and Kim D H 2009 *International Journal of Hydrogen Energy* **34** 1336
[17] Li L, Song L, Chen C Q, Zhang Y J, g Zhan Y Y, Lin X Y, Zheng Q, Wang H D, Ma H X, Ding L H and Zhu W 2014 *International Journal of Hydrogen Energy* **39** 19570
[18] Benedetto A D, Landi G and Lisi L 2017 *International Journal of Hydrogen Energy* **42** 12262
[19] Ma D W, Lu Z S, Tang Y A, Li T X, Tang Z J and Yang Z X 2014 *Physics Letters A* **378** 2570

IOP Conf. Series: Materials Science and Engineering **292** (2017) 012122 doi:10.1088/1757-899X/292/1/012122

A new sandwich type transition metal complex based on tungstobismuthates and 1-methylimidazole ligands exhibiting photocatalytic properties

Q Wu[1, 2], D Q Liang[1, 2], H D Ju[1, 2], K Liu[1, 2], Y Cui[1, 2], W L Li[1, 2], B L Wang[1, 2], F P Ye[3], Y Zheng[3], Y M Wu[4], J Y Li[1], Y Lei[1], Y Xu[1], T Liu[1] and Y Wang[1, 5]

[1] Department of Chemistry Science & Technology, Kunming University, Kunming 650214, P. R. China

[2] Center for Disease Control and Prevention, Chengdu Military Region, Kunming, 650118, 11 People' s Republic of China

[3] Yunnan Engineering Technology Research Center for Plastic Films, Kunming University, Kunming, China

[4] School of International Cultures and Education, Yunnan University of Finance and Economics, Yunnan, Kunming 650221, P. R. China

[5] E-mail: sweetwy2011@163.com

Abstract. A new sandwich-type transition metal complex based on tungstobismuthates and 1-methylimidazole ligands $Na_7(mim)_2[\{Na(H_2O)_2\}_3\{Co^{II}(mim)\}_3(Bi^{III}W^{VI}{}_9O_{33})_2]\cdot 9H_2O$ (**1**) (mim = 1-methylimidazole) has been successfully isolated in aqueous solution. The composition and structure were established by the single crystal X-ray diffraction analysis, elemental analysis, IR spectroscopy. Single-crystal X-ray diffraction and TG analyses indicate that compound **1** consists of two tri-vacant $[\alpha\text{-B-}Bi^{III}W^{VI}{}_9O_{33}]^{9-}$ units linked through three cobalt ions. All the Co^{II} ions were sandwiched in the middle and bonded with terminal oxo atoms from $[\alpha\text{-B-}Bi^{III}W^{VI}{}_9O_{33}]^{9-}$ units as well as 1-methylimidazoles. Photocatalytic experiment indicates that compound **1** exhibit good catalytic activities for photodegradation of RhB under UV irradiation.

1. Introduction

Polyoxometalates (POMs), as a unique class of versatile metal–oxygen cluster complexes, have been attracting extensive interest due to not only their aesthetically topological properties but also their potential applications in catalysis, photochemistry, electro-chromism and magnetism [1-5]. Although this class of inorganic compounds has been known for more than a hundred years, a great number of novel polyoxoanions with unexpected shapes, sizes and properties are still being discovered [6-18].

In this aspect, transition-metal-substituted polyoxometalate (TMSP) is one of the most important species owing to their variety of building blocks, adjustable connection modes and tunable transition-metal ions that being accommodated. Among the class of TMSPs, one of the greatest subclasses is the sandwich-type polyoxoanions which are obtained by accommodate paramagnetic transition-metal cations between two lacunary polyoxoanions [19-26]. However, just as summarize by Xu and coworkers [27], although numerous sandwich-type polyoxoanions have been synthesized, almost all known structurally characterized sandwich-type POMs are based on trivacant tungstophosphate, -silicate, and –arsenate fragments, and little research was done on species that containing subvalent main group atom Bi^{III} as heteroatom up to present, such as, $[Bi_2W_{22}O_{74}(OH)_2]_{12}$ as well as its derivatives $[Bi_2W_{20}M_2O_{70}(H_2O)_6]^{(12-2n)-}$ (M^{n+}=Fe^{III}, Co^{II}, Zn^{II}, Ni^{II}, Mn^{II}), $[Sn_3\{WO_2(OH)\}(WO_2)_4(BiW_9O_{33})_4]^{21-}$ [28-31], $[Ln_3(H_2O)_{18}Bi_2W_{22}O_{76}]^{5-}$ (Ln^{3+}=Ce^{III}, Eu^{III}) [32, 33], constructure from trivacant $[b\text{-}B\text{-}(BiW_9O_{33})]^{9}$ units have been synthesized. On the other hand, tungstobismuthate anions containing $[a\text{-}B\text{-}BiW_9O_{33}]^{9-}$ moieties are also contains $[H_3BiW_{18}O_{60}]^{6-}$ [34], $[(BiW_9O_{33})_3Bi_6(OH)_3(H_2O)_3(V_4O_{10})]^{12-}$ [35], $[M_3(H_2O)_3(a\text{-}B\text{-}BiW_9O_{33})_2]^{(18-3n)}$ (M^{n+}= Fe^{III},Cr^{III},Mn^{II},Co^{II},Ni^{II},Cu^{II}) [36], and $[(VO)_3(a\text{-}B\text{-}BiW_9O_{33})_2]^{12-}$ [37]. Moreover, until now the reports made on the combination of organic ligand and tungstobismuthates are still pretty much unexplored.

In order to enrich the family of organic ligand decorated tungstobismuthate, in this paper, 1-methylimidazole was introduced as the N-containing ligand to modify tungstobismuthate and a novel sandwich-type compound $Na_7(mim)_2[\{Na(H_2O)_2\}_3\{Co^{II}(mim)\}_3(Bi^{III}W^{VI}{}_9O_{33})_2]\cdot 9H_2O$ (**1**) (mim = 1-methylimidazole) was successfully obtained. As far as we know, compound 1 represent for the first time that organic ligands simultaneously decorated to the tungstobismuthate frame work and serve as template molecules to organize the anions three-dimensional arrangement.

2. Experimental section

2.1. Materials and methods

All chemicals were commercially purchased and used without further purification. Elemental analyses Bi, W, Mn, and Na were analyzed on a PLASMA-SPEC (I) ICP atomic emission spectrometer. IR spectrum was recorded in the range 400-4000 cm^{-1} on an Alpha Centaurt FT/IR Spectrophotometer using KBr pellets. A 752 PC UV-vis spectrophotometer was used to obtain UV-vis absorption spectrum, Perkin–Elmer TGA7 instrument was used for TG analyses under flowing N_2 with a heating rate of 10 °C min^{-1}.

Synthesis of $Na_7(mim)_2[\{Na(H_2O)_2\}_3\{Co^{II}(mim)\}_3(Bi^{III}W^{VI}{}_9O_{33})_2]\cdot 9H_2O$ 1

$Na_2WO_4\cdot H_2O$ (0.66 g, 2.0 mmol) was dissolved in 20 mL distilled water with vigorous stirring for 10 min. Then, a little amount of insoluble substance was removed by filtration. To above solution, 10 mL of HCl (0.1M) aqueous solution containing the $Bi(NO_3)_3\cdot H_2O$ (0.05g, 0.1mmol) was added dropwise within 10 min. The mixture was kept on stirring for 1 hour at room temperature until the solution turned clear. During this time, $Co(NO_3)_2\cdot 6H_2O$ (0.06 g, 0.2 mmol (dissolved in 3 mL 1 mol/L NaOH solution) and 0.042g 1-methylimidazole was slowly added to the clear solution. The pH value of the resulting reaction mixture was adjusted to 8 at room temperature and further stirred for another 30 min at 40 °C and then cooled to the room temperature. After filtration, the purple filtrate was kept for slow evaporation at room temperature. After one week, purple-red block-like crystalline products of **1** were obtained. (yield 36.7 % based on W). Anal. Calcd for $C_{20}H_{30}Bi_2Co_3N_{10}Na_6O_{96}W_{18}$ **1**: Bi, 6.97; W, 55.26; Co, 2.95; Na, 2.30 %; Found: P, 7.02; W, 54.23; Co, 2.79; Na, 2.34 %.

2.2. X-ray crystallography

The crystallographic data were collected at 150 K on the Rigaku R-axis Rapid IP diffractometer using graphite monochromatic Mo-Kα radiation (λ = 0.71073 Å) and IP technique. Multi-scan absorption correction was applied. The crystal data of **1** was solved by the direct method and refined by the full-matrix least-squares method on F^2 using the SHELXTL-97 crystallographic software package [38]. During the refinement, The restraint command 'isor' was employed to restrain N3, N6, C7, C9, N4, C20, C13, C14 atoms so as to avoid the ADP and NPD problems on them. This command led to a restraint number 48. The polyoxoanion, 1-methylimidazole ligands and most of the water molecules were anisotropically refined, while solvent water molecules were just refined isotropically due to their unusual anisotropic thermal parameters and obvious disorder problems. H atoms on lattice water molecules cannot be found from the residual peaks and were directly included in the final molecular formula. The detailed crystal data and structure refinement for **1** were given in table 1.

Table 1. Crystal data and structure refinement for **1**.

	1
Empirical formula	$C_{20}H_{60}Bi_2Co_3N_{10}Na_6O_{96}W_{18}$
M	6018.52
λ/Å	0.71073
T/K	150(2) K
Crystal dimensions/mm	0.20× 0.18 × 0.16
Crystal system	Triclinic
Space group	P-1
Z	2
D_c/Mg m^{-3}	3.545
μ/mm^{-1}	22.055
θ Range/°	3.04 – 25.00
Reflections collected / unique	38030 / 18591
R_{int}	0.1661
Data/restraints/parameters	18591 / 48 / 926
$R_1(I > 2\sigma(I))$ [a]	0.06204
wR_2 (all data) [b]	0.1375
Goodness-of-fit on F^2	1.037
$\Delta\rho_{max, min}$/ e Å^{-3}	2.922, -4.018

$^aR_1 = \sum||F_O|-|F_C||/\sum|F_O|$; $^bwR_2 = \sum[w(F_O^2-F_C^2)^2]/\sum[w(F_O^2)^2]^{1/2}$

2.3. Crystal Description

The single crystal X-ray diffraction result shows that the polyoxoanion of **1** is mainly composed of two $[\boldsymbol{\alpha}\text{-B-Bi}^{III}W^{VI}{}_9O_{33}]^{9-}$ moieties which are linked together by three 1-methylimidazole coordinated cobalt units (see figure 1). In the polyoxoanion, all the lacunary Keggin fragments $[Bi^{III}W^{VI}{}_9O_{33}]^{9-}$ (*abbr.* {BiW$_9$}) exhibit the typical trivacant B-α-Keggin-type structure. Such a structural feature can be viewed as a mother α-Keggin unit that lost three edge-sharing {WO_6} octahedra. In these fragments, all Bi atoms exhibit a tri-coordination environment and all W centers display the octahedral coordination geometry. The bond lengths of Bi-O and the bond angles of O-Bi-O are in the range of 2.088 (16) ~ 2.138 (17) Å and 86.6(6) ~ 89.4(6) °, respectively. The bond lengths of W–O and bond angles of O–W–O vary from 1.696(16) ~ 2.310(16) Å and 72.1(6) ~ 171.6(6) °, respectively. 1-methylimidazole groups associated with the $[\boldsymbol{\alpha}\text{-B-Bi}^{III}W^{VI}{}_9O_{33}]^{9-}$ fragments stabilize the metal centers. In the equatorial plane, each cobalt ion coordinates to four oxygen atoms (μ_3-O) that derive from two different $[\alpha\text{-B-Bi}^{III}W^{VI}{}_9O_{33}]^{9-}$ fragments, while the apical position is occupied by a mim molecule to form a distorted square-pyramidal {MO4N} group. The bond lengths of Co–O(N) are in the range 2.000(17) 2.083(17) Å, and the bond angles of O(N)–Co–O vary from 83.8(6) to 97.4(7)°.

Interestingly, different from most of previous organic ligands decorated tungstobismuthates all the mim molecules in **1** are slope at an angle to the equatorial plane. Three dihedral angles are 35.688(1011), 22.060(786) and 13.313(1082), respectively. Meanwhile, three sodium ions between transition metal ions in the equatorial plane are hexacoordinated with four oxygen atoms that derive from $[Bi^{III}W^{VI}{}_9O_{33}]^{9-}$ units and two apically coordinated water molecules.

Figure 1. Polyhedral and ball-and-stick representation of **1**, lattice water molecules and free sodium ions are omitted for clarity.

Another interesting feature of compound **1** is mim molecules act as the template molecules and interact with adjacent anionic $[\{Na(H_2O)_2\}_3\{Co^{II}(mim)\}_3(Bi^{III}W^{VI}{}_9O_{33})_2]^{7-}$ units by van der Waals' force (see figure 2). Ascribing to the co-existence of mim molecules, the packing arrangement of **1** exhibits different three-dimensional structure. Meanwhile, to our knowledge, this is the first time that organic ligands simultaneously decorated to the tungstobismuthate framework and serve as template molecules to organize the anions three-dimensional arrangement.

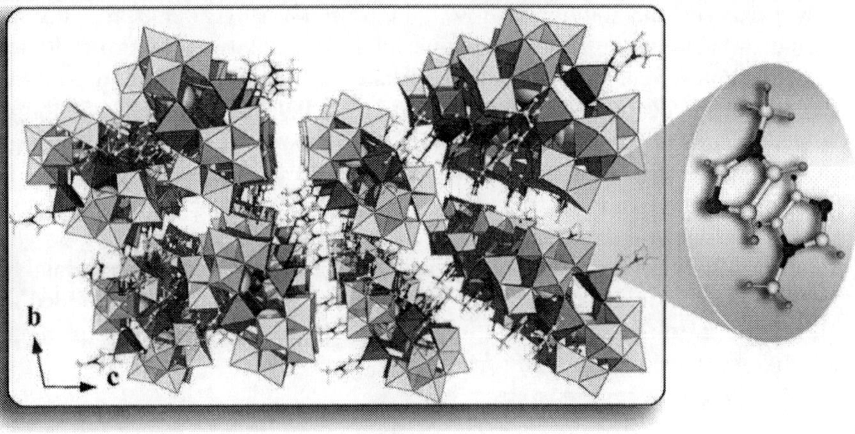

Figure 2. Ball-and-stick view of the 3-D supramolecular framework of **1** along the C-axis, water molecules omitted for clarity.

2.4. IR, UV-vis, and TG analysis

The IR spectrum of **1** shows a broad peak at 3520 cm^{-1} and strong peak at 1620 cm^{-1} attributed to the lattice water molecules. The characteristic peaks at 1091, 939, 852, 706 and 429cm^{-1} correspond to ν(Bi-O), ν(W=O_d), ν(W-O_b) and ν(W-O_c) vibrations, respectively. The UV-vis spectrum of **1** displays two main peaks at 210 nm and 260 nm, which are attributable to the O→W LMCT bands. The broad bands in the visible region of 320 ~ 380 nm are mainly ascribed to the absorption of Co^{2+} ions [39]. Thermogravimetric (TG) analysis of **1** (see figure 3) was performed under N_2 at a rate of 10°C min^{-1} in the range 20–600°C on compound exhibits two steps of weight loss. The first one of ca.5.2% in the temperature range of 35~ 380 °C is attributed to the loss of all lattice water molecules. This value is consistent with the calculated value of 4.9% (~ 15 H_2O). The second weight loss of ca. 6.92 % between 400 and 460 °C is probably the result of the loss of imi molecules.

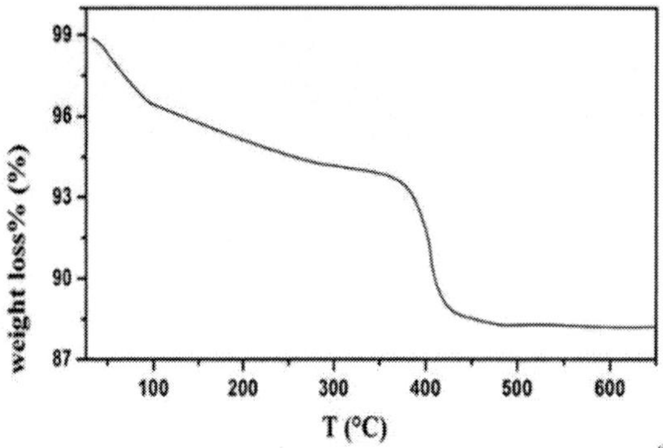

Figure 3. View of thermogravimetric analysis curve of **1**.

2.5. Photocatalytic properties

A wide range of POM possess photocatalytic ability of the degradations of recalcitrant organic pollutants under UV irradiations by oxidization of organic materials, however, the research made on amino-acid is still unexplored. We also investigated the photocatalytic performances of compounds **1** for the photocatalytic activity of RhB with UV light through a typical process: compounds **1** was dissolved in the solutions (2.40×10^{-5} mol L^{-1}) at pH=3.5 including Rhodamine-B (RhB) which were adjusted with either NaOH or $HClO_4$ to dilute aqueous solutions of, then vigorously stirred for 10min in the dark. Then exposed the solution to UV light from 125W Hg lamp at a distance of 4-5cm between the liquid surface and the lamp. Continuously stir the solution during irradiations. Every 60 minutes, 3.0ml of samples were taken out of the beaker for analysis. As shown in figure 4(a), the absorption peak of of RhB decreased obviously from 1.4 to 0.4 and 300min (λ=554nm). In addition, the concentrations of RhB*(K)* versus reaction time *(t)* of compound 1 were plotted in figure 4(b). The conversions of RhB*(K)* can be expressed as $K=(C_0-C_t)/C_0$, where C_0 represents the UV-vis intensity of RhB at the original reaction time (t=0), while I_t is the UV-vis absorption intensity at a certain irradiation time t. The calculation results show that the conversion of RhB is 75.67% by using **1** as the photocatalyst after 240min irradiation. Additionally, after three runs of photocatalytic tests, the photoactivity of compound **1** did not display significant loss when RhB was re-added in the system, see figure 4b. These results suggest that compound **1** may be potential photocatalysts with photocatalytic activities in the reduction of some organic dye.

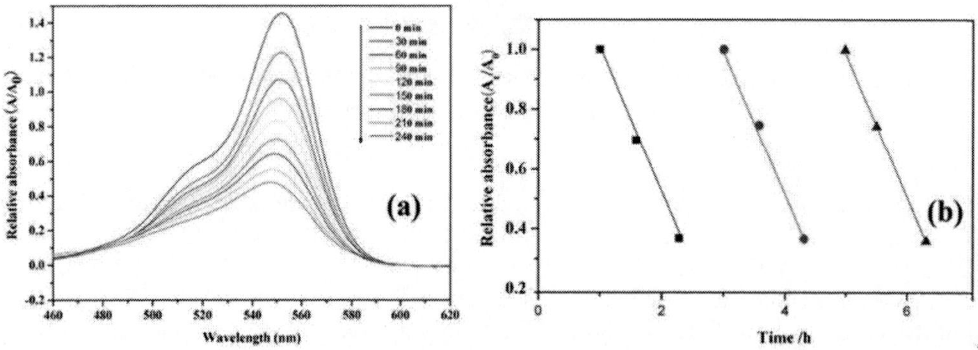

Figure 4. (a) Temporal UV-visible absorption spectral changes observed for the RhB solution as a function of irradiation time in the presence of 2×10^{-5} M compound **1**. (b) Three time of RhB degradation test by compound **1**.

3. Conclusions

In this work, we obtained a novel 1-methylimidazole decorated sandwich-type tungstobismuthate. Compared to previous organic ligand substituted tungstobismuthate anions, compound **1** shows for the first time that organic ligands simultaneously decorated to the tungstobismuthate frame work and serve as template molecules to organize the anions three-dimensional arrangement. Photocatalytic experiment indicated that compound **1** exhibits high catalytic activity for photodegradation of RhB with UV irradiation. Further research will focus on the reactions between various other lacunary tungstobismuthate anions and different N-containing heterocycle ligands in order to obtain new type of organic ligands decorated tungstobismuthate with novel structures and interesting properties.

Acknowledgment

This work was supported by Agricultural plastic film and products research and development of Plateau area (No. DH1606), the National Natural Science Foundation of China (No. 21702083; 51562019; 31760257; 21201090;). Applied Basic Research Programs of Science and Technology Commission Foundation of Yunnan Province (2012FD045)

References

[1] S Omwoma, C T Gore, Y Ji, C Hu and Y F Song 2015 *Coordin Chem Rev* **286** 17-29.
[2] H Liu, C Y Song, R W Huang, Y Zhang, H Xu, M J Li, S Q Zang and G G Gao 2016 *Angew Chem Int Edit* **55** 3699-3703
[3] M P Santoni, G S Hanan, B Hasenknopf 2014 *Coordin Chem Rev* **281** 64-85
[4] W J Luo, J Hu, H L Diao, B Schwarz, C Streb and Y F Song 2017 *Angew Chem Int Edit* **56** 4941-4944
[5] E Coronado, C Giménez-Saiz and C J Gómez-García 2005 *Coord. Chem. Rev.* **249** 1776
[6] B Li, J W Zhao, S T Zheng and G Y Yang 2009 *Inorg. Chem. Commun.* **12** 69
[7] G Xue, X Liu, H Xu, H Hu, F Fu and J Wang 2008 *Inorg. Chem.* **47** 2011
[8] C P Pradeep, D L Long, C Streb, L Cronin 2008 *J. Am. Chem. Soc.* **130** 14946
[9] H. Pang, C. Zhang, D. Shi and Y. Chen 2008 *Cryst. Growth Des.* **8** 4476
[10] H N Miras, D L Long, P Kögerler, L Cronin 2008 *Dalton Trans.* 214
[11] S S Mal, M H Dickman and U Kortz 2008 *Chem. Eur. J.* **14** 9851
[12] D L Long, C Streb, Y F Song, S Mitchell, L Cronin 2008 *J. Am. Chem. Soc.* **130** 1830
[13] U Lee, H C Joo, K M Park, S S Mal, U Kortz, B KeitaL Nadjo 2008 *Angew. Chem., Int. Ed.* **47**, 793
[14] S Konaka, Y Ozawa and A Yagasaki 2008 *Inorg. Chem. Commun.* **11** 1267
[15] X K Fang, P Kögerler 2008 *Chem. Commun.* 3396
[16] X Fang, P Kögerler 2008 *Angew. Chem., Int. Ed.* **47** 8123
[17] L H Bi, B Li, L X Wu 2008 *Inorg. Chem. Commun.* **11** 1184
[18] Ginger E Sigmon, Daniel K Unruh, J Ling, B Weaver, M Ward, L Pressprich, A Simonetti and Peter C Burns 2009 *Angew. Chem. Int. Ed.* **48** 2737
[19] L Wang, K Yu, J Zhu, B B Zhou, J R Liu and G Y Yang 2017 Dalton T **46** 2874-2883
[20] R Sato, K Suzuki, T Minato, K Yamaguchi and N Mizuno 2016 *Inorg Chem* **55** 2023-2029
[21] W H Fang, W D Wang, G Y. Yang 2015 *Dalton T* **44** 12546-12549
[22] W L Chen, Y Li, Y Wang, E B Wang and Z Su 2007 *Dalton Trans.* 4293
[23] D Aravena, D Venegas-Yazigi, E Ruiz 2016 *Inorg Chem* **55** 6405-6413
[24] W L Chen, Y Li, Y Wang, E B Wang and Z M Zhang 2008 *Dalton Trans.* **8** 865
[25] D Aravena, D Venegas-Yazigi, E Ruiz 2016 *Inorg Chem* **55** 6405-6413
[26] Q Wu, Y G Li, Y H Wang, E B Wang, Z M Zhang, R Clérac 2009 *Inorg. Chem.* **48** 1606
[27] H Liu, L Xu, G Gao, F Li, and N Jiang 2008 *J. Mol. Struct.* **878**(1-3) 124
[28] I Loose, E Droste, M Bossing, H Pohlmann, M H Dickman, C Rosu, M T Pope and B Krebs 1999 *Inorg. Chem.* **38** 688
[29] D Rodewald, Y Jeannin 1998 *C.R. Acad. Sci. Paris Ser. IIc.* **1** 175
[30] M Bossing, A Noh, I Loose and B Krebs 1998 *J. Am. Chem.Soc.* **120** 7252
[31] K Bernt, D Elisabeth, P Markus and V Guido 2000 *C.R. Acad. Sci. Paris Ser. IIc.* **3** 205
[32] B Krebs, I Loose, M Bösing, A Nöh, E Droste 1998 *C.R. Acad. Sci. Paris Ser.* IIc. **1** 351

[33] Y J Wang, H H Zhang, C C Huang, X H Yu, Q Z Sun and R Q Sun 2004 *Chinese J. Struct. Chem.* **23** 902
[34] Y Ozawa, Y Sasaki 1987 *Chem. Lett.* **16** 923
[35] B Botar, T Yamase, E Ishikawa 2000 *Inorg. Chem. Commun.* **3** 579
[36] R Rusu, A L Barra, L David, M Rusu, C Rosu, O Cozar and G Marcu 2001 *J. Chem. Soc. Dalton Trans.* 2879
[37] B Botar, T Yamase and E Ishikawa 2001 *Inorg. Chem. Commun.* 551
[38] G M Sheldrick 1997 SHELXL-97 Program for Crystal Structure Refinement *University of Göttingen, Göttingen, Germany* (b) G. M Sheldrick 1997 SHELXS97 Program for Crystal Structure Solution *University of Göttingen, Göttingen, Germany*
[39] M T Pope 1983 Heteropoly and Isopoly Oxometalates *Springer: Berlin*

Preparation and characterization of antimony barium composite oxide photocatalysts

X P Han[1, 2], B H Yao[1, 3], Q H Pan[2], C Pen[1] and C L Zhang[2]

[1] Institute of Water Conservancy and Hydro-Electric Power, Xi'an University of Technology, Xian 710054, China

[2] College of Material and Chemical Engineering, Hainan University, Haikou 570228, China

[3] E-mail: bhyao@xaut.edu.cn

Abstract. In this paper, two kinds of antimony barium composite oxide photocatalysts have been prepared by two methods and characterized by XRD and SEM. The photocatalytic activity was evaluated by a photocatalytic reactor and an ultraviolet spectrophotometer. The results showed that-$BaSb_2O_5$•$4H_2O$, $BaSb_2O_6$ two kinds of antimony barium composite oxide photocatalysts were successfully prepared in this experiment and they showed good photocatalytic properties. In addition, $BaSb_2O_6$ morphology showed more regular microstructure and better catalytic performance.

1. Introduction

Photocatalytic material has been broadly applied in environmental pollution controlling fields. Developments of new type of photocatalyst are important direction of photocatalystic research. The photocatalytic activity of p zone composite oxides whose d layer fulling of electron is widely attracted interests by the scientists [1], such as indium acid salt (In^{3+}), stannate (Sn^{4+}), antimonate (Sb^{5+}), germanate (Ge^{4+}), gallate (Ga^{3+}). Antimony resources are particularly rich in China. The research, development and utilization of antimony compounds are of great significance.

There are many reports on the study of antimony compounds as photocatalysts. Monteiroa O C etc. [2] prepared antimony complex oxide by self-combustion method and studied its photocatalytic properties; Zuo G L etc. [3] successfully prepared the $Sb_2O_3/\alpha\text{-}Fe_2O_3$ nano heterojunction composite photocatalyst by hydrothermal method and impregnation method; Lamba R etc. [4] synthesized Sb_2O_3-ZnO spindle shape nanocomposites and could degrade methylene blue effectively. Li D Z etc. [5] studied $CaSb_2O_5(OH)_2$ and $Cd_2Sb_2O_{6.8}$, and found their degradation efficiency on benzene contamination were higher than $P_{25}(TiO_2)$; Lin X P etc. [6] prepared MSb_2O_6 (M = Ca, Sr and Ba) by solid-phase method, and observed that photocatalytic effect of $BaSb_2O_6$ was the best through comparation with three antimonites. Chen J etc. [7, 8] synthesized micro-flowers $BaSb_2O_6$ by one-step method and studied their photocatalytic activity.

In this study, antimony barium composite oxides were prepared by hydrothermal method with different antimony sources. The characterization and catalytic properties of the composite were also studied.

2. Experimental section

2.1. Reagents and instruments

All chemicals (Sb_2O_5, $KSb(OH)_6$, $Ba(CH_3COO)_2$, $Ba(NO_3)_2$, NaOH, HCl, methylene blue) were analytical reagent grade and used without further purification; Water used in all experiments was deionized water. Main instruments: KSL-1100X muffle furnace; BL-GHX-V photochemical reaction apparatus; 721 ultraviolet visible spectrophotometer; D8advance X ray diffractometer; Hitachi S-3000N scanning electron microscope (SEM).

2.2. Prepare Sb/Ba oxide composites by using Sb_2O_5 as antimony source

0.005mol of antimony pentoxide was firstly added to some water to form Sb_2O_5 solution, and then potassium hydroxide was added to adjust the pH of above solution to neutral. Then another amount of 0.01mol potassium hydroxide was added, followed by stirring for 10~15min. After that, 0.005mol barium nitrate was added into the system with stirring to make it fully dissolved. Transfer the system into a high pressure reactor at 200 ℃ to carry out the hydrothermal reaction for 4h. Then cooling, and wash the white precipitate from the bottom of the reactor with deionized water for 2-3 times and anhydrous ethanol thoroughly. Finally, the product was dried in an oven at 80℃ for 4h, cooled and collected the outcomes.

2.3. Prepare antimony barium oxide composites with $KSb(OH)_6$ as antimony source

0.32g barium acetate was firstly weighed and dissolved in 10mL deionized water. Then 25 mL, 0.1 mol/L potassium pyroantimonate solution was added dropwise into the system and adjust the pH to 3 by dilute hydrochloric acid. After stirring continuously at room temperature for 12 h, the system was put in a high pressure autoclave and heated in a 200 ° C incubator to hydrothermal react for 4.5 h. Then cooling, and wash the white precipitate from the bottom of the reactor with deionized water for 2-3 times and anhydrous ethanol for 1 time. At last, dry the product in a oven at 60℃ for 4h, then cool, grind and collect the outcomes.

2.4. Photocatalysis experiment

The photocatalytic activity of the samples was evaluated by the methylene blue (MB) decolorization degradation system. Degradation of methylene blue by photocatalyst was carried out in BL-GHX-V photochemical reaction apparatus.

A mixture of methylene blue at a concentration of 14mg/L and 0.05g barium antimonate photocatalyst were added to a quartz tube of the photoreactor. The light source was high pressure UV mercury lamp (500W). Before turning on the light, air was introduced in to start the dark adsorption reaction for 20min, so that the process of the photocatalyst attaching on the methylene blue can reach an adsorption-desorption balance. Then extract a 0min sample as a reference, open the UV lamp, the experiment process should be in a environment with cooling water and air through the whole time, collect samples every 20min, remove the photocatalyst by the microporous membrane, measured the absorbance value of the solution in the 721 UV-visible spectrophotometer with 664nm wavelength.

The activity of the catalyst was characterized by the decolorization rate D% of the MB solution. The calculation method is as follows:

$$D\% = \frac{A_0 - A_t}{A_0} \times 100\% \tag{1}$$

A_0: absorbance value, which indicates that the dye solution in the reaction system absorbed in the dark after 20min;

A_t: absorbance value of the dye solution at reaction time t.

3. Results and analysis

3.1. XRD analysis of antimony/barium oxide composites

The samples prepared by 1.2 and 1.3 were subjected to XRD analysis with CuKα radiation, the current and voltage was 40 mA and 40 kV, respectively. The scanning speed was 10 °/min from 10 to 70 °. The results are shown in figure 1 and 2.

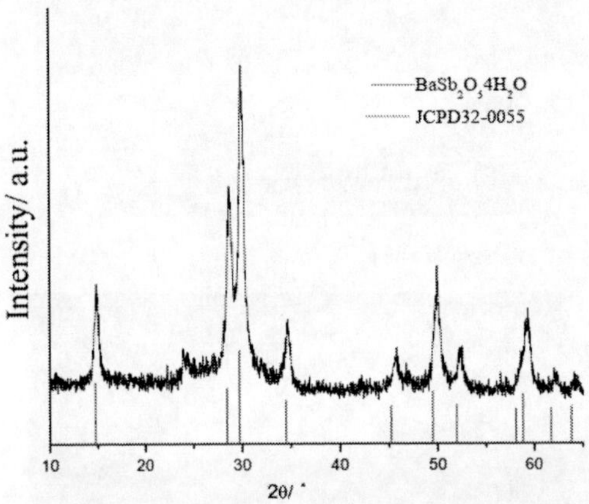

Figure. 1 XRD pattern of $BaSb_2O_5•4H_2O$.

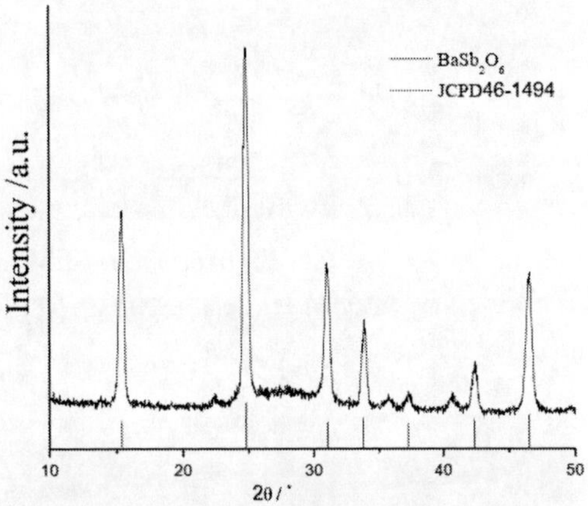

Figure. 2 XRD pattern of $BaSb_2O_6$.

It can be seen from figure 1 that the main diffraction peaks of the samples appear at 14.73°, 28.41°, 29.71°, 49.50° and 58.80°, respectively. All the characteristic peaks are the same of the international XRD diffraction data standard card of JCPDS No. 32-0055, indicating that the prepared sample is $BaSb_2O_5•4H_2O$. The product obtained by the preparation method of Sb_2O_5 as antimony source is $BaSb_2O_5•4H_2O$, the reaction principle is as follows:

$$\begin{gathered} 2Sb_2O_5 + 2Ba(NO_3)_2 + 4KOH + 6H_2O = \\ 2BaSb_2O_5 \cdot 4H_2O + 4KNO_3 + O_2 \end{gathered} \tag{2}$$

It can be seen from figure 2 that the main diffraction peaks of the samples appear at the 2θ angles of 15.37°, 24.77°, 31.03°, 33.77°, 42.34° and 46.51°, respectively. All the characteristic peaks are the same of the international XRD diffraction data standard card of JCPDS No. 46-1494, indicating that the prepared sample is $BaSb_2O_6$ and the product obtained by the preparation of $KSb(OH)_6$ as antimony source is $BaSb_2O_6$. The reaction principle is as follows:

$$Ba(CH_3COO)_2 + 2KSb(OH)_6 = BaSb_2O_6 + 2CH_3COOK + 6H_2O \tag{3}$$

3.2. SEM analysis of the antimony/barium oxide composites

S-3000N scanning electron microscopy was used to observe the morphology of the prepared samples. Before the test, a small amount of fully dried and polished samples were fixed on the conductive tape, sprayed with gold and observed under vacuum condition. The SEM results of $BaSb_2O_5•4H_2O$ and $BaSb_2O_6$ samples are shown in figure 3a.b and 4a.b.c.d, respectively.

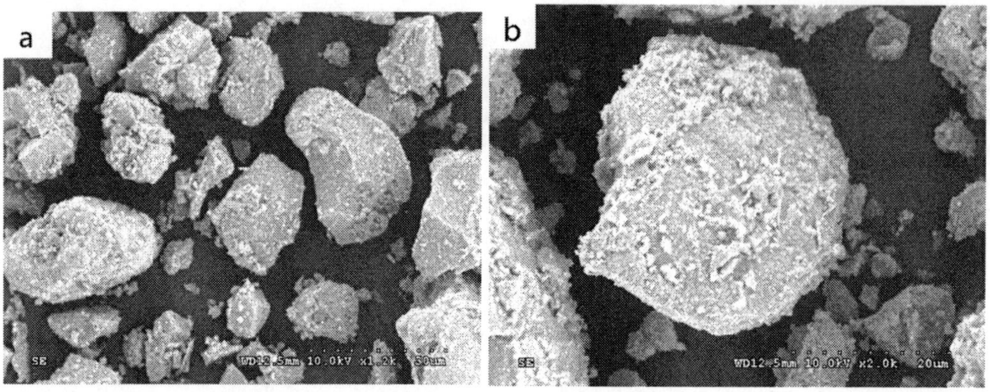

Figure. 3 a.b. SEM images of $BaSb_2O_5{\bullet}4H_2O$.

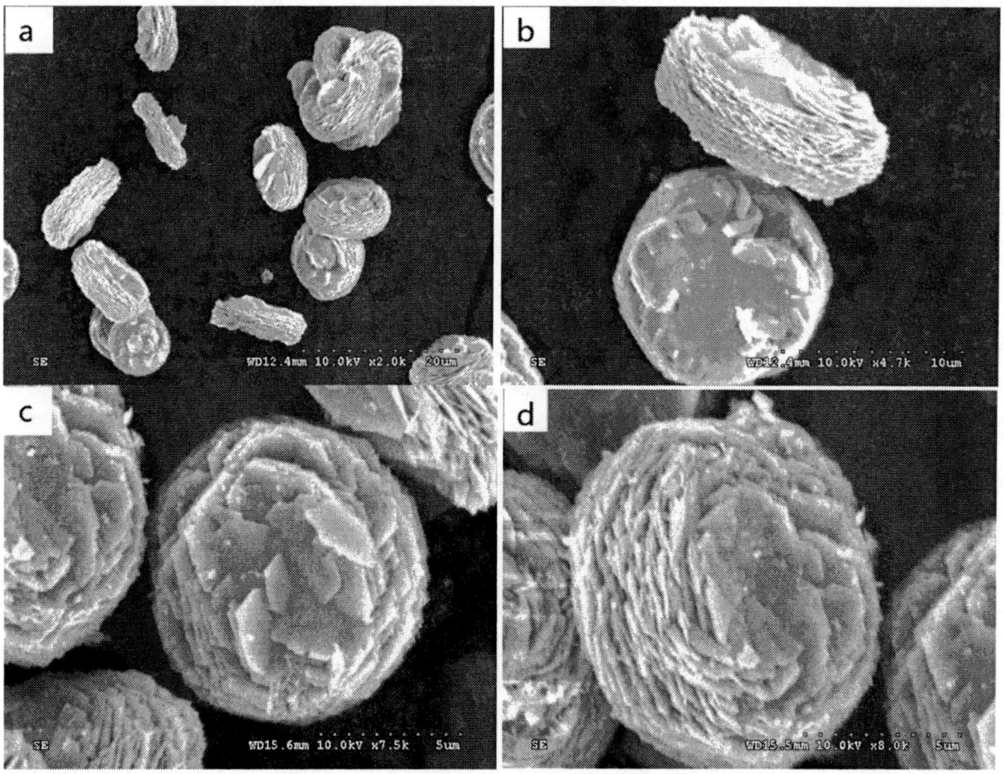

Figure. 4 a.b.c.d. SEM images of $BaSb_2O_6$.

It can be seen from figure 3a.b that the $BaSb_2O_5{\bullet}4H_2O$ is an irregular rocky structure with uneven size. The rough surface and porous structure were readily observed, and there's no obvious agglomeration between particles.

From figure 4a.b.c.d, it can be seen that the prepared $BaSb_2O_6$ particles are lamellar and cake-like structure, round and thick. The center and side of the grains are clear and compact, and look like a "natural open flowers", the size of the grains are almost of the same and they are dispersed uniformly. The diameter of one single particle was about 10μm.

3.3. Photocatalytic property of antimony/barium oxide composites

The prepared $BaSb_2O_5{\cdot}4H_2O$ and $BaSb_2O_6$ were subjected to photocatalytic reaction by the method depicted in section 1.4. The results are shown in figure 5.

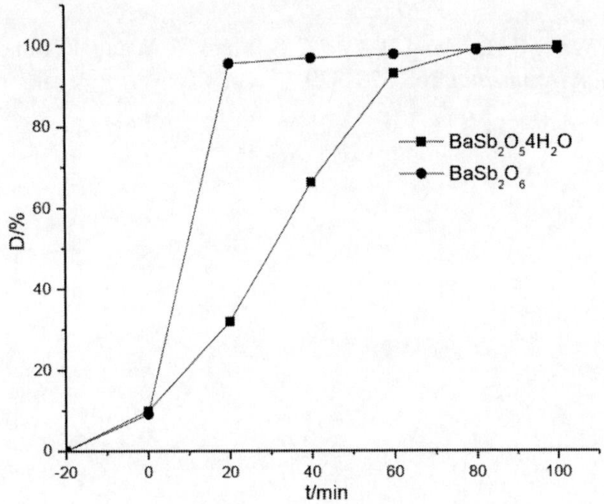

Figure. 5 The degradation curve of MB with $BaSb_2O_5{\bullet}4H_2O$ and $BaSb_2O_6$.

It can be seen from figure 5 that the photocatalytic efficiency of $BaSb_2O_5{\bullet}4H_2O$ and $BaSb_2O_6$ are good. After 100 minutes' lighting, the degradation rate of methylene blue reached higher than 99%. The degradation efficiency of $BaSb_2O_6$ to methylene blue was higher than that of $BaSb_2O_5{\bullet}4H_2O$ at the same lighting time period, which may be attributed to the lamellar microstructure of $BaSb_2O_6$.

4. Conclusions

In this study, the preparation of antimony/barium oxide composite photocatalyst with two kinds of substances as antimony sources was discussed. The results showed that the product obtained by hydrothermal reaction between Sb_2O_5 and $Ba(NO_3)_2$ under alkaline condition is $BaSb_2O_5{\bullet}4H_2O$; the product prepared by hydrothermal reaction between barium pyroantimonate and $Ba(CH_3CH_2COO)_2$ under acidic condition is $BaSb_2O_6$ powder, which has a regular microstructure, look like a "natural open flowers" and the size is uniform, non-agglomerate. Both $BaSb_2O_5{\bullet}4H_2O$ and $BaSb_2O_6$ have good photocatalytic efficiency on methylene blue under ultraviolet light, but $BaSb_2O_6$ was better, which was attributed to its own layered structure.

The results of this study will provide reference for the study of antimony composite oxides as photocatalysts.

Acknowledgements

This work was financially supported by Scientific research project of Hainan higher education institutions(Hnky2015-2), Hainan Natural Science Foundation(20162018), Hainan Natural Science Foundation (217055), Hainan Natural Science Foundation (217018).

References

[1] Tian M K, Shang guan W F, Ou yang Z Y and Wang S J 2005 *Journal of Functional Materials* **36** 1489
[2] Monteiroa O C, Marquesb R and Carvalhob M D 2013 *Materials Chemistry and Physics* **119** 418
[3] Zuo G L, Ye H Y and Li R L S 2015 *Journal of Nanyang Institute of Technology* **04** 103-107
[4] Lamba R, Umar A, Mehta SK and Kansal S K 2015 Orcid Ceramics International **4** 5429-38
[5] Li D Z and Fu X Z 2005 *Scientia Sinica Chimica* **42** 415
[6] Lin X P, Wu J J, Lu X J, Shan Z C, Wang W D and Huang F Q 2009 *Physical Chemistry Chemical Physics* **11** 10047
[7] Chen J, Li D Z, Hu J H, Chen W, Wang J X, Hu Y, Fu X Z and Shao Y 2012 *CrystEngComm*

14 8382
[8] Chen J, Li D Z, Wang J B, Wang P, Cao C S, Shao Y, Wang J X and Xian J J 2015 *Applied Catalysis B: Environmental* **163** 323-329